Lecture Notes in Computer Science 4109

Dit-Yan Yeung James T. Kwok
Ana Fred Fabio Roli
Dick de Ridder (Eds.)

Structural, Syntactic, and Statistical Pattern Recognition

Joint IAPR International Workshops
SSPR 2006 and SPR 2006
Hong Kong, China, August 17-19, 2006
Proceedings

Springer

Volume Editors

Dit-Yan Yeung
James T. Kwok
Hong Kong University of Science and Technology
Department of Computer Science and Engineering
Clear Water Bay, Kowloon, Hong Kong, China
E-mail: {dyyeung, jamesk}@cse.ust.hk

Ana Fred
Technical University of Lisbon, Telecommunications Institute
Department of Electrical and Computer Engineering, Lisbon, Portugal
E-mail: afred@lx.it.pt

Fabio Roli
University of Cagliari, Department of Electrical and Electronic Engineering
Cagliari, Italy
E-mail: roli@diee.unica.it

Dick de Ridder
Delft University of Technology
Faculty of Electrical Engineering, Mathematics and Computer Science
Information and Communication Theory Group, Delft, The Netherlands
E-mail: d.deridder@tudelft.nl

Library of Congress Control Number: 2006930416

CR Subject Classification (1998): I.5, I.4, I.2.10, I.3.5, G.2-3

LNCS Sublibrary: SL 6 – Image Processing, Computer Vision, Pattern Recognition, and Graphics

ISSN 0302-9743
ISBN-10 3-540-37236-9 Springer Berlin Heidelberg New York
ISBN-13 978-3-540-37236-3 Springer Berlin Heidelberg New York

Springer is a part of Springer Science+Business Media

springer.com

Typesetting: Camera-ready by author, data conversion by Scientific Publishing Services, Chennai, India
Printed on acid-free paper SPIN: 11815921 06/3142 5 4 3 2 1 0

Preface

This volume in the Springer *Lecture Notes in Computer Science* (LNCS) series contains 103 papers presented at S+SSPR 2006, which was the fifth time that the SPR and SSPR workshops organized by Technical Committees TC1 and TC2 of the International Association for Pattern Recognition (IAPR) were held together as joint workshops. It was also the first time that the joint workshops were held in the Far East, at the beautiful campus of the Hong Kong University of Science and Technology (HKUST), on August 17–19, 2006, right before the 18th International Conference on Pattern Recognition (ICPR 2006), also held in Hong Kong.

SPR 2006 and SSPR 2006 together received 217 paper submissions from 33 countries. This volume contains 99 accepted papers, with 38 for oral presentation and 61 for poster presentation. In addition to parallel oral sessions for SPR and SSPR, there were also some joint oral sessions with papers of interest to both the SPR and SSPR communities. A recent trend that has emerged in the pattern recognition and machine learning research communities is the study of graph-based methods that integrate statistical and structural approaches. For this reason, a special joint session on graph-based methods was co-organized by Technical Committee TC15 to explore new research issues in this topic. Moreover, invited talks were presented by four prominent speakers: Robert P.W. Duin from Delft University of Technology, The Netherlands, winner of the 2006 Pierre Devijver Award; Tin Kam Ho from Bell Laboratories of Lucent Technologies, USA; Thorsten Joachims from Cornell University, USA; and B. John Oommen from Carleton University, Canada.

We would like to take this opportunity to thank all members of the SPR and SSPR Program Committees and the additional reviewers for their professional support in reviewing the submitted papers. We thank all the Advisory Committee members, Shun-ichi Amari, Terry Caelli, Robert P.W. Duin, Anil K. Jain, Erkki Oja, Harry Shum, and Tieniu Tan, for their invaluable advice on organizing this event. We also thank all the authors, the invited speakers, the Organizing Committee members, the sponsors, and the editorial staff of Springer for helping to make this event a success and helping to produce this high-quality publication in the LNCS series.

August 2006

Dit-Yan Yeung
James T. Kwok
Ana Fred
Fabio Roli
Dick de Ridder

S+SSPR 2006

General Chair

Dit-Yan Yeung

Dept. of Computer Science and Engineering
Hong Kong University of Science and Technology
Clear Water Bay, Kowloon
Hong Kong, China
dyyeung@cse.ust.hk

Local Chair

James T. Kwok

Dept. of Computer Science and Engineering
Hong Kong University of Science and Technology
Clear Water Bay, Kowloon
Hong Kong, China
jamesk@cse.ust.hk

Webmaster

Dick de Ridder

Information & Communication Theory Group
Faculty of Electrical Engineering, Mathematics & Computer Science
Delft University of Technology
Delft, The Netherlands
d.deridder@tudelft.nl

Supported by

International Association for Pattern Recognition (IAPR)
Hong Kong University of Science and Technology (HKUST)

SSPR Committee

Co-chairmen

Ana Fred

Telecommunications Institute
Dept. of Electrical and Computer Engineering
Instituto Superior Técnico
Technical University of Lisbon
Lisbon, Portugal
afred@lx.it.pt

James T. Kwok

Dept. of Computer Science and Engineering
Hong Kong University of Science and Technology
Clear Water Bay, Kowloon
Hong Kong, China
jamesk@cse.ust.hk

Program Committee

SPR Committee

Co-chairmen

Fabio Roli

Dept. of Electrical and
Electronic Engineering
University of Cagliari
Cagliari, Italy
roli@diee.unica.it

Dit-Yan Yeung

Dept. of Computer Science and
Engineering
Hong Kong University of
Science and Technology
Clear Water Bay, Kowloon
Hong Kong, China
dyyeung@cse.ust.hk

Program Committee

Reviewers

The Program Committees for both SPR and SSPR were kindly assisted by:

Ghada Badr (Canada)
Hong Chang (China)
Donatello Conte (Italy)
Saverio De Vito (Italy)
Jiří Filip (Czech Republic)
Jiří Grim (Czech Republic)
Xuming He (Canada)
Zoe Hoare (UK)
Francesco Isgro (Italy)
Martin Kampel (Austria)
Alessandro Limongiello (Italy)
James Maclean (Canada)
Claudio Marrocco (Italy)
Luisa Mico (Spain)
Francisco Moreno-Seco (Spain)
Andrea Passerini (Italy)
Gennaro Percannella (Italy)
Faisal Qureshi (Canada)
Julien Ros (France)
Petr Somol (Czech Republic)
Domenico Sorrentino (Italy)
David Tax (The Netherlands)
Michael Villamizar (Spain)
Jun Wang (The Netherlands)
Changhua Wu (USA)
Kai Zhang (China)
Dragos Calitoiu (Canada)
James Charles (UK)
Guang Dai (China)
Florent Dupont (France)
Pasquale Foggia (Italy)
Yll Haxhimusa (Austria)
Emile A. Hendriks (The Netherlands)
Adrian Ion (Austria)
Mike Jamieson (Canada)
Jeroen Lichtenauer (The Netherlands)
Marco Loog (Denmark)
Diego Macrini (Canada)
Claudio Mazzariello (Italy)
Mario Molinara (Italy)
Jose Oncina (Spain)
Elżbieta Pękalska (UK)
Petra Perner (Germany)
Michael Reiter (Austria)
Paolo Simeone (Italy)
Sébastien Sorlin (France)
Nathan Srebro (Canada)
Ivor Tsang (China)
Gang Wang (China)
Christian Wolf (France)
De-Chuan Zhan (China)

Table of Contents

Invited Talks

SSPR

Image Analysis

Vision

Character Recognition

Bayesian Networks

Graph-Based Methods

Similarity and Feature Extraction

Poster Papers

Image and Video

Vision

Kernel-Based Methods

Recognition and Classification

Similarity and Feature Extraction

Document Analysis

Graph-Based Methods

SPR

Recognition and Classification I

Image Analysis

Facial Image Analysis I

Recognition and Classification II

Representation

Feature Selection

Clustering

Unsupervised Learning

Dimensionality

Representation

Biometrics

Applications

Structured Output Prediction with Support Vector Machines

Thorsten Joachims

Cornell University, Ithaca, NY, USA
tj@cs.cornell.edu
http://www.joachims.org

Abstract. This abstract accompanying a presentation at S+SSPR 2006 explores the use of Support Vector Machines (SVMs) for predicting structured objects like trees, equivalence relations, or alignments. It is shown that SVMs can be extended to these problems in a well-founded way, still leading to a convex quadratic training problem and maintaining the ability to use kernels. While the training problem has exponential size, there is a simple algorithm that allows training in polynomial time. The algorithm is implemented in the SVM-Struct software, and it is discussed how the approach can be applied to problems ranging from natural language parsing to supervised clustering.

1 Introduction

Over the last decade, much of the research on discriminative learning has focused on problems like classification and regression, where the prediction is a single univariate variable. But what if we need to predict complex objects like trees, orderings, or alignments? Such problems arise, for example, when a natural language parser needs to predict the correct parse tree for a given sentence, when one needs to optimize a multivariate performance measure like the F1-score, or when predicting the alignment between two proteins.

This abstract accompanies the presentation at S+SSPR 2006, discussing a support vector approach and algorithm for predicting such complex objects. It summarizes our recent work [10,20,21,11,12,9] on generalizing conventional classification SVMs to a large range of structured outputs and multivariate loss functions, and connects these results to related work [4,3,5,1,16,19,13,23]. While the generalized SVM training problems have exponential size, we show that there is a simple algorithm that allows training in polynomial time. The algorithm is implemented in the SVM-Struct software[1], and it is discussed how the approach can be applied to problems ranging from natural language parsing to supervised clustering.

2 Problems That Require Structured Outputs

While many prediction problems can easily be broken into multiple binary classification problems, other problems require an inherently structured prediction.

[1] Available at svmlight.joachims.org

D.-Y. Yeung et al. (Eds.): SSPR&SPR 2006, LNCS 4109, pp. 1–7, 2006.

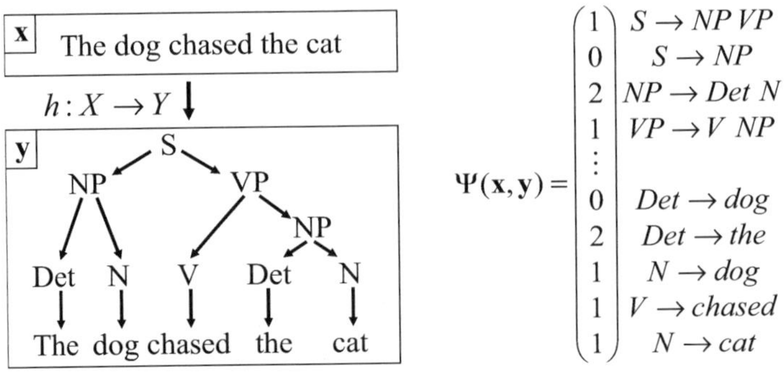

Fig. 1. Illustration of the NLP parsing problem

Consider, for example, the problem of natural language parsing. For a given sentence $\mathbf{x}$, the goal is to predict the correct parse tree $\mathbf{y}$ that reflects the phrase structure of the sentence. This is illustrated on the left-hand side of Figure 1. Training data of sentences that are labeled with the correct tree is available (e.g. from the Penn Tree Bank), making this prediction problem accessible for supervised learning.

Compared to binary classification, the problem of predicting complex and structured outputs differs mainly by the choice of the outputs $\mathbf{y}$. What are common structures that we might want to predict?

Trees: We have already discussed the problem of natural language parsing (see e.g. [14]), where a prediction $\mathbf{y} \in \mathcal{Y}$ is a tree.

Sequences: A problem related to parsing is that of part-of-speech tagging (see e.g. [14]). Given a sentence $\mathbf{x}$ represented as a sequence of words, the task is to predict the correct part-of-speech tag (e.g. "noun" or "determiner") for each word. While this problem could be phrased as a multi-class classification task, it is widely acknowledged that predicting the sequence of tags as a whole allows exploiting dependencies between tags (e.g. it is unlikely to see a verb after a determiner). Similar arguments also apply to tagging protein or gene sequences.

Alignments: For comparative protein structure modelling, it is necessary to predict how the sequence of a new protein with unknown structure aligns against another sequence with know structure (see e.g. [8]). Given the correct alignment, it is possible to predict the structure of the new protein. Therefore, one would like to predict the sequence alignment operations that "best" aligns two sequences according to some cost model.

Equivalence Relation: Noun-phrase co-reference resolution (see e.g. [15]) is the problem of clustering the noun phrases in one documents by whether they refer to the same entity. This can be thought of as predicting an equivalence relation, where training examples are the correct partitionings for some documents. More generally, this problem can be thought of as supervised clustering [9] — training a clustering algorithm to produce the desired kinds of clusters.

While these application problems appear quite different, we will show that they all can be approached in a similar way. In particular, the SVM algorithm we describe in the following is able to address each of these problems.

3 An SVM Algorithm for Structured Outputs

Formally, we consider the problem of learning a function

$$h : \mathcal{X} \longrightarrow \mathcal{Y}$$

where $\mathcal{X}$ is the space of inputs, and $\mathcal{Y}$ is the space of (multivariate and structured) outputs. In the parsing examples, $\mathcal{X}$ is the space of sentences, and $\mathcal{Y}$ is the space of trees over a given set of non-terminal grammar symbols. To learn h, we assume that a training sample of input-output pairs

$$S = ((\mathbf{x}_1, \mathbf{y}_1), \dots, (\mathbf{x}_n, \mathbf{y}_n)) \in (\mathcal{X} \times \mathcal{Y})^n$$

is available and drawn i.i.d. from a distribution $P(X, Y)$. The goal is to find a function h from some hypothesis space $\mathcal{H}$ that has low prediction error, or, more generally, low risk

$$\mathcal{R}_P^{\triangle}(h) = \int_{\mathcal{X}\times\mathcal{Y}} \triangle(\mathbf{y}, h(\mathbf{x}))\, dP(\mathbf{x}, \mathbf{y}) \,.$$

$\triangle(\mathbf{y}, \hat{\mathbf{y}})$ is a loss function that quantifies the loss associated with predicting $\hat{\mathbf{y}}$ when $\mathbf{y}$ is the correct output value. Furthermore, we assume that $\triangle(\mathbf{y}, \mathbf{y}) = 0$ and $\triangle(\mathbf{y}, \mathbf{y}') \geq 0$ for $\mathbf{y} \neq \mathbf{y}'$. We follow the Empirical Risk Minimization Principle [22] to infer a function h from the training sample S. The learner evaluates the quality of a function $h \in \mathcal{H}$ using the empirical risk $\mathcal{R}_S^{\triangle}(h)$ on the training sample S.

$$\mathcal{R}_S^{\triangle}(h) = \frac{1}{n} \sum_{i=1}^{n} \triangle(\mathbf{y}_i, h(\mathbf{x}_i))$$

Support Vector Machines select an $h \in \mathcal{H}$ that minimizes a regularized Empirical Risk on S. For conventional binary classification where $\mathcal{Y} = \{-1, +1\}$, SVM training is typically formulated as the following convex quadratic optimization problem [6,22].

OP 1 (Classification SVM (primal))

$$\min_{\mathbf{w}, b, \xi_i \geq 0} \frac{1}{2}\mathbf{w}^T\mathbf{w} + \frac{C}{n} \sum_{i=1}^{n} \xi_i$$

$$s.t.\ \ \forall i \in \{1, ..., n\}:\ \mathbf{y}_i(\mathbf{w}^T\mathbf{x}_i + b) \geq 1 - \xi_i$$

To generalize SVM training to structured outputs, we formulate an optimization problem that is similar to multi-class SVMs [7] and generalizes the Perceptron approach described in [4]. The idea is to learn a discriminant function $f : \mathcal{X} \times \mathcal{Y} \to \mathbb{R}$ over input/output pairs from which we can derive a prediction by maximizing f over all $\mathbf{y} \in \mathcal{Y}$ for a specific given input $\mathbf{x}$.

$$h_{\mathbf{w}}(\mathbf{x}) = \operatorname*{argmax}_{\mathbf{y}\in\mathcal{Y}} f_w(\mathbf{x}, \mathbf{y})$$

We assume that $f_{\mathbf{w}}(\mathbf{x}, \mathbf{y})$ takes the form of a linear function

$$f_w(\mathbf{x}, \mathbf{y}) = \mathbf{w}^T \Psi(\mathbf{x}, \mathbf{y})$$

where $\mathbf{w} \in \Re^N$ is a parameter vector and $\Psi(\mathbf{x}, \mathbf{y})$ is a feature fector describing the match between input $\mathbf{x}$ and output $\mathbf{y}$. Intuitively, one can think of $f_w(\mathbf{x}, \mathbf{y})$ as a compatibility function that measures how well the output $\mathbf{y}$ matches the given input $\mathbf{x}$.

The specific form of Ψ depends on the nature of the problem and special cases will be discussed subsequently. Using natural language parsing as an illustrative example, f_w can be chosen to be isomorphic to a Probabilistic Context Free Grammar (PCFG) (see e.g. [14]). Each node in a parse tree $\mathbf{y}$ for a sentence $\mathbf{x}$ corresponds to grammar rule g_j, which in turn has a score w_j. All valid parse trees $\mathbf{y}$ (i.e. trees with a designated start symbol S as the root and the words in the sentence $\mathbf{x}$ as the leaves) for a sentence $\mathbf{x}$ are scored by the sum of the w_j of their nodes. This score can thus be written in the form of Eq. 1, where $\Psi(\mathbf{x}, \mathbf{y})$ denotes the histogram vector of how often each grammar rule g_j occurs in the tree $\mathbf{y}$. This is illustrated on the right-hand side of Figure 1. $h_{\mathbf{w}}(\mathbf{x})$ can be efficiently computed by finding the structure $\mathbf{y} \in \mathcal{Y}$ that maximizes $f_{\mathbf{w}}(\mathbf{x}, \mathbf{y})$ via the CKY algorithm (see e.g. [14]).

For training the weights $\mathbf{w}$ of the linear discriminant function, we generalize the standard SVM optimization problem as follows [1,10,20,21]. A similar formulation was independently proposed in [16].

OP 2 (Structural SVM (primal))

$$\min_{\mathbf{w},\xi} \; \frac{1}{2}\mathbf{w}^T\mathbf{w} + \frac{C}{n}\sum_{i=1}^{n}\xi_i$$

$$s.t. \;\; \forall \mathbf{y} \in \mathcal{Y} : \mathbf{w}^T(\Psi(\mathbf{x}_i, \mathbf{y}_i) - \Psi(\mathbf{x}_i, \mathbf{y})) \geq \Delta(\mathbf{y}_i, \mathbf{y}) - \xi_i$$

The objective is the conventional regularized risk used in SVMs. The constraints state that for each training example $(\mathbf{x}_i, \mathbf{y}_i)$ the score $\mathbf{w}^T\Psi(\mathbf{x}_i, \mathbf{y}_i)$ of the correct $\mathbf{y}_i$ must be greater than the score $\mathbf{w}^T\Psi(\mathbf{x}_i, \mathbf{y})$ of all incorrect $\mathbf{y}$ by a difference of $\Delta(\mathbf{y}_i, \mathbf{y})$. Δ is an application dependent loss function that measures how different the two structures $\mathbf{y}_i$ and $\mathbf{y}$ are. Intuitively, the larger the loss, the further should the score be away from that of the correct training label $\mathbf{y}_i$. ξ_i is a slack variable shared among constraints from the same example, since in general the problem is often not separable. Note that $\sum \xi_i$ is an upper bound on the training loss $\mathcal{R}_S^{\Delta}(h)$.

Input: $S = ((\mathbf{x}_1, \mathbf{y}_1), \ldots, (\mathbf{x}_n, \mathbf{y}_n))$, $C > 0$, $\epsilon > 0$.
$K = \emptyset$, $\mathbf{w} = 0$, $\xi = 0$
repeat
- $K_{org} = K$
- for i from 1 to n
  - $\mathbf{y} = \text{argmax}_{\mathbf{y} \in \mathcal{Y}} \left[\Delta(\mathbf{y}_i, \mathbf{y}) + \mathbf{w}^T \Psi(\mathbf{x}_i, \mathbf{y})\right]$ # find most violated constraint
  - if $\mathbf{w}^T(\Psi(\mathbf{x}_i, \mathbf{y}_i) - \Psi(\mathbf{x}_i, \mathbf{y})) < \Delta(\mathbf{y}_i, \mathbf{y}) - \xi_i - \epsilon$ # violated more than ϵ?
    - $K = K \cup \left\{\mathbf{w}^T(\Psi(\mathbf{x}_i, \mathbf{y}_i) - \Psi(\mathbf{x}_i, \mathbf{y})) \geq \Delta(\mathbf{y}_i, \mathbf{y}) - \xi_i - \epsilon\right\}$
    - $(\mathbf{w}, \xi) = argmin_{\mathbf{w}, \xi} \frac{1}{2}\mathbf{w}^T\mathbf{w} + \frac{C}{n}\sum_{i=1}^{n} \xi_i$ subject to K.

until $(K = K_{org})$
Output: $\mathbf{w}$

Fig. 2. Cutting plane algorithm for training Structural SVMs

While the training problem is obviously still convex and quadratic, it typically has exponentially many constraints. For most choices of $\mathcal{Y}$ (e.g. sequences and trees), the cardinality of $\mathcal{Y}$ is exponential in the maximum size of $\mathbf{x}$ — and so is the number of constraints in OP2. This makes solving OP2 intractable using off-the-shelf techniques. However, it has been shown that the cutting plane algorithm in Figure 2 can be used to efficiently approximate the optimal solution of this type of optimization problem [21,12]. The algorithm starts with an empty set of constraints, adds the most violated constraint among the exponentially many during each iteration, and repeats until the desired precision $\epsilon > 0$ is reached. It can be proved that only a polynomial number of constraints will be added before convergence [21,12]. One crucial aspect of the algorithm, however, is the use of an oracle that can find the most violated constraint among the exponentially many possible constraints in polynomial time. That is, we need to compute

$$\underset{y \in \mathcal{Y}}{\text{argmax}}[\Delta(\mathbf{y}_i, \mathbf{y}) + \mathbf{w}^T \Psi(\mathbf{x}_i, \mathbf{y})]. \tag{1}$$

For many $\mathcal{Y}$, feature mappings Ψ, and the loss functions Δ, this problem can be solved via dynamic programming. For trees, for example, the argmax in Eq. (1) can be computed using the CKY algorithm, if Ψ follows from a context-free grammar and Δ is any loss function that can be computed from the contingency table [21]. The running time of the overall learning algorithm is then polynomial in the number of training examples, the length of the sequences, and ϵ [21,12].

An alternative to the cutting plane algorithm is the algorithm proposed in [19]. It applies when the loss function Δ decomposes linearly and the argmax in Eq. (1) can be computed using a linear program that is guaranteed to have an integer solution.

4 Application Examples and Related Work

It has been shown for a range of application problems and structures $\mathcal{Y}$ that SVM training is feasible and benficial. The work in [20,21] shows how structural SVMs can be applied to natural language parsing, sequence alignment, taxonomic classification, and named-entity recognition. More work on highly expressive models for parsing is given in [17], and the use of structural SVMs for protein threading is described in [10,12]. An alternative approach to alignment is [18]. Work on sequence tagging for natural language problems and OCR is given in [16,1]. Image segmentation is addressed in [2]. While traditional generative training can and has been used for many structural prediction problems in the past, the studies mentioned above have repeatedly shown that discriminative training gives superior prediction performance.

Conditional Random Fields (CRFs) [13] are the most popular alternative discriminative training methods for structured prediction problems. Like large-margin approaches, they also have shown excellent performance on a variety of problems. Instead of optimizing a regularized empirical risk for a user-defined loss function like in the SVM approach, CRFs optimize a regularized conditional likelihood. While they can be applied to many of the problems mentioned above, there is little direct comparison between SVM and CRF training yet.

Other training approaches for structured models include the perceptron algorithm and reranking approaches [4,3,5]. The structural SVM approach extends these. A very different approach to structured prediction is proposed in [23], implementing the structured prediction as a multivariate regression problem after mapping the structures into Euclidian space.

5 Summary

This paper provides a short summary of methods for Support Vector Machine training with structured outputs. In particular, it shows how a cutting-plane method can be used to solve the training problem efficiently despite an exponential number of constraints. Pointers towards applications and further reading provide a starting point for further exploration of this area of research.

References

1. Y. Altun, I. Tsochantaridis, and T. Hofmann. Hidden Markov support vector machines. In *International Conference on Machine Learning (ICML)*, 2003.
2. D. Anguelov, B. Taskar, V. Chatalbashev, D. Koller, D. Gupta, and G. Heitz. Discriminative learning of markov random fields for segmentation of 3d scan data. In *International Conference on Computer Vision and Pattern Recognition (CVPR)*, 2005.
3. M. Collins. Discriminative reranking for natural language parsing. In *International Conference on Machine Learning (ICML)*, 2000.
4. M. Collins. Discriminative training methods for Hidden Markov Models: Theory and experiments with perceptron algorithms. In *Proceedings of the Conference on Empirical Methods in Natural Language Processing*, 2002.
5. M. Collins and N. Duffy. New ranking algorithms for parsing and tagging: Kernels over discrete structures, and the voted perceptron. In *Conference of the Association for Computational Linguistics (ACL)*, 2002.

6. Corinna Cortes and Vladimir N. Vapnik. Support–vector networks. *Machine Learning Journal*, 20:273–297, 1995.
7. K. Crammer and Y. Singer. On the algorithmic implementation of multiclass kernel-based vector machines. *Journal of Machine Learning Research (JMLR)*, 2:265–292, 2001.
8. R. Durbin, S. Eddy, A. Krogh, and G. Mitchison. *Biological Sequence Analysis*. Cambridge University Press, 1998.
9. T. Finley and T. Joachims. Supervised clustering with support vector machines. In *International Conference on Machine Learning (ICML)*, 2005.
10. T. Joachims. Learning to align sequences: A maximum-margin approach. online manuscript, August 2003.
11. T. Joachims. A support vector method for multivariate performance measures. In *International Conference on Machine Learning (ICML)*, 2005.
12. T. Joachims, T. Galor, and R. Elber. Learning to align sequences: A maximum-margin approach. In B. Leimkuhler et al., editor, *New Algorithms for Macromolecular Simulation*, volume 49 of *LNCS*, pages 57–68. Springer, 2005.
13. J. Lafferty, A. McCallum, and F. Pereira. Conditional random fields: Probabilistic models for segmenting and labeling sequence data. In *Proceedings of the Eighteenth International Conference on Machine Learning*, pages 282–289, San Francisco, 2001. Morgan Kaufmann.
14. C. Manning and H. Schütze. *Foundations of Statistical Natural Language Processing*. MIT Press, Cambridge, MA, 1999.
15. V. Ng and C. Cardie. Improving machine learning approaches to coreference resolution. In *Annual Meeting of the Assoc. for Comp. Linguistics (ACL)*, 2002.
16. B. Taskar, C. Guestrin, and D. Koller. Maximum-margin markov networks. In *Neural Information Processing Systems (NIPS)*, 2003.
17. B. Taskar, D. Klein, M. Collins, D. Koller, and C. Manning. Max-margin parsing. In *Empirical Methods in Natural Language Processing (EMNLP)*, 2004.
18. B. Taskar, S. Lacoste-Julien, and D. Klein. A discriminative matching approach to word alignment. In *Empirical Methods in Natural Language Processing (EMNLP)*, 2005.
19. Ben Taskar. *Learning Structured Prediction Models: A Large Margin Approach*. PhD thesis, Stanford University, 2004.
20. I. Tsochantaridis, T. Hofmann, T. Joachims, and Y. Altun. Support vector machine learning for interdependent and structured output spaces. In *International Conference on Machine Learning (ICML)*, 2004.
21. I. Tsochantaridis, T. Joachims, T. Hofmann, and Y. Altun. Large margin methods for structured and interdependent output variables. *Journal of Machine Learning Research (JMLR)*, 6:1453 – 1484, September 2005.
22. V. Vapnik. *Statistical Learning Theory*. Wiley, Chichester, GB, 1998.
23. J. Weston, O. Chapelle, A. Elisseeff, B. Schölkopf, and V. Vapnik. Kernel dependency estimation. In *Advances in Neural Information Processing Systems 15*, Cambridge, MA, 2003. MIT Press.

On the Theory and Applications of Sequence Based Estimation of Independent Binomial Random Variables*

B. John Oommen[1,**], Sang-Woon Kim[2], and Geir Horn[3]

[1] Professor and Fellow of the IEEE, School of Computer Science, Carleton University, Ottawa, Canada : K1S 5B6
oommen@scs.carleton.ca

[2] Senior Member, IEEE, Dept. of Computer Science and Engineering, Myongji University, Yongin, 449-728 Korea
kimsw@mju.ac.kr

[3] SIMULA Research Laboratory, Martin Linges Vei 15-25, Fornebu, Norway
geirho@simula.no

Abstract. We re-visit the age-old problem of estimating the parameters of a distribution from its observations. Traditionally, scientists and statisticians have attempted to obtain strong estimates by "extracting" the information contained in the observations taken *as a set*. However, generally speaking, the information contained in the *sequence* in which the observations have appeared, has been ignored - i.e., except to consider dependence information as in the case of Markov models and n-gram statistics. In this paper, we present results which, to the best of our knowledge, are the first reported results, which consider how estimation can be enhanced by utilizing both the information in the observations *and in their sequence of appearance*. The strategy, known as Sequence Based Estimation (SBE) works as follows. We first quickly allude to the results pertaining to computing the Maximum Likelihood Estimates (MLE) of the data when the samples are taken individually. We then derive the corresponding MLE results when the samples are taken two-at-a-time, and then extend these for the cases when they are processed three-at-a-time, four-at-a-time etc. In each case, we also experimentally demonstrate the convergence of the corresponding estimates. We then suggest various avenues for future research, including those by which these estimates can be fused to yield a superior overall cumulative estimate of the parameter

* The work of the first author was done while visiting at Myongji University, Yongin, Korea. The first author was partially supported by NSERC, the Natural Sciences and Engineering Research Council of Canada, a grant from the Korean Research Foundation, and a grant from the SIMULA Research Laboratory in Norway. This work was generously supported by the Korea Research Foundation Grant funded by the Korea Government(MOEHRD-KRF-2005-D00004).

** The first author dedicates this paper to the memory of Mr. Sigurd Bratlie and Mrs. Rakel Bratlie, whose lives were instrumental in changing his life - in every sense of the word. "Thanks, Brother and Sister Bratlie. *What do I have that I did not receive*"?

D.-Y. Yeung et al. (Eds.): SSPR&SPR 2006, LNCS 4109, pp. 8–21, 2006.

of the distribution. We believe that our new estimates have great potential for practitioners, especially when the cardinality of the observation set is *small*.

1 Introduction

Estimation is a fundamental issue that concerns every statistical problem. Typically, the practitioner is given a set of observations involving the random variable, and his task is to estimate the parameters which govern the generation of these observations. Since, by definition, the problem involves random variables, decisions or predictions related to the problem are in some way dependent on the practitioner obtaining reliable estimates on the parameters that characterize the underlying random variable. Thus, if a problem can be modelled using a random variable which is binomially (or multinomially) distributed, the underlying statistical problem involves estimating the binomial (or multinomial) parameter(s) of the underlying distribution.

The theory of estimation has been studied for hundreds of years [1,3,8]. It is also easy to see that the learning (training) phase of a statistical pattern recognition system is, indeed, based on estimation theory [4,7,18]. Estimation methods generally fall into various categories, including the Maximum Likelihood Estimates (MLE) and the Bayesian family of estimates [1,3,4] which are well-known for having good computational and statistical properties.

To explain the contribution of this paper, we consider the strategy used for developing the MLE of the parameter of a distribution, $f_X(\theta)$, whose parameter to be estimated is θ. The input to the estimation process is the set of points $\mathcal{X} = \{x_1, x_2, \ldots, x_N\}$, which are assumed to be generated independently and identically as per the distribution, $f_X(\theta)$. The process involves deriving the likelihood function, i.e., the likelihood of the distribution, $f_X(\theta)$, generating the sample points $\mathcal{X}$ given θ, which is then maximized (by traditional optimization or calculus methods) to yield the estimate, $\hat{\theta}$. The general characteristic sought for is that the estimate $\widehat{\theta}_{MLE}$ converges to the true (unknown) θ with probability one, or in a mean square sense. Bayesian and MLE estimates generally possess this desirable phenomenon.

Suppose now that the user received $\mathcal{X}$ as a sequence of data points as in a typical real-life (or real-time) application such as those obtained in a data-mining application involving sequences, or in data involving radio or television news items[1]. The question which we have asked ourselves is the following: "Is there any information in the fact that in $\mathcal{X}$, x_i specifically precedes x_{i+1} ?". Or in a more general case, "Is there any information in the fact that in $\mathcal{X}$, the **sequence** $x_i x_{i+1} \ldots x_{i+j}$ occurs $n_{i,i+1,\ldots i+j}$ times?". Our position is that even though $\mathcal{X}$ is generated by an i.i.d. process, there is information in these pieces of sequential information, and we propose here a method by which these pieces

[1] We are currently investigating how we can utilize SBEs to yield a superior classification scheme for a real-life problem involving news files from the CBC.

of information can be "maximally" utilized. The estimates that we propose are referred to as the Sequence Based Estimators (SBE).

As far as we know, there are no available results which utilize sequential information in obtaining such estimates[2]. Indeed, even the results we have here are only for the case when the distribution is Binomial. Although some preliminary results for the multinomial case are available, these are merely alluded to. The paper also leads to a host of open problems.

Once we have obtained the SBE estimates based on the occurrence and sequential information, the next question is that of combining all these estimates together to yield a single meaningful estimate. We propose to achieve this by using techniques from the theory of fusion - excellent studies of which are found in [9] and [10]. The paper also includes some specific applications of SBEs.

The paper is organized as follows. Section 2 lists the SBE estimation results obtained when sequential information is used to estimate the parameter of the Binomial distribution, and the sequences are processed two-at-a-time. This is followed in Section 3 by cases when the data is analyzed in sequences three-at-a-time, four-at-a-time respectively. Section 4 discusses the open problems that are currently unsolved, namely those involving the *fusing* of the individual SBEs and those which we have encountered in classifying artificial and real-life data. Section 5 concludes the paper.

Contributions of the Paper: The contributions of this paper are:

1. This paper lists the first reported results for obtaining the maximum likelihood estimates (called the Sequence Based Estimates (SBEs)) of the parameter of a binomial distribution when the data is processed both as a *set* of observations and as a *sequence* by which the samples occur in the set.
2. The paper contains the formal results[3] and verification for the cases when the sequence is processed in pairs, and in subsequences of length 3 and 4.
3. The paper lists a few potential strategies by which SBE estimators can be fused to yield a superior estimate utilizing the MLE and the SBEs.
4. The paper lists a few potential schemes by which the MLE and SBE estimators can be used in pattern classification and other applications.

To the best of our knowledge, all of these are novel to the field of estimation, learning and classification.

Throughout this paper we assume that we are estimating the parameters of a binomial distribution. The binomial distribution is characterized by two parameters, namely, the *number* of Bernoulli trials, and the parameter characterizing

[2] The question of utilizing and estimating sequential information is not entirely new. It has long been used in syntactic pattern recognition, in estimating the bigram and n-gram probabilities of streams of data and grammatical inference, and in the learning problem associated with modelling channels using Hidden Markov Models [2,4]. But all of these methods further emphasize the dependence between the occurrences. We show that such information can be gleaned even if the occurrences are *independent*.

[3] The paper lists at least 17 results. But as the proofs of many of the theorems are quite similar, the details of the proofs are merely alluded to in the interest of brevity.

each Bernoulli trial. In this regard, we assume that the number of observations is the number of trials. Thus, all we have to do is to estimate the *Bernoulli* parameter for each trial. Thus, in terms of notation, if X is a binomially distributed random variable, which takes on the value of either '1' or '2'[4], we assume that X obeys the distribution S, where $S = [s_1, s_2]^T$, where, $s_1 + s_2 = 1$, and

X = '1' with probability s_1

= '2' with probability s_2 ,

Then, the following elementary result about the MLE of S is given below.

Result 1. Let X be a binomially distributed random variable, obeying the distribution S, where $S = [s_1, s_2]^T$. If $\mathcal{X} = \{x_1, x_2, \ldots, x_N\}$ is a realization of a sequence of occurrences of X, where each x_i is either '1' or '2', the MLE of s_i is $\widehat{s}_i = \frac{n_i}{N}$, where n_i is the number of occurrences of 'i' in $\mathcal{X}$. □

Notation 1: To be consistent, we introduce the following notation.

- X is a binomially distributed random variable, obeying the distribution S.
- $\mathcal{X} = \{x_1, x_2, \ldots, x_N\}$ is a realization of a sequence of occurrences of X, where each x_i is either '1' or '2'.
- Let $\langle j_1 j_2 \ldots, j_k \rangle$ be the sequence examined in the set $\mathcal{X}$, where each j_m, $(1 \leq m \leq k)$, is either a '1' or '2'. Then, the SBE for s_i obtained by examining the sequence $\langle j_1 j_2 \ldots, j_k \rangle$ will be given by $\widehat{s}_i\,|_{<j_1 j_2 \ldots, j_k>}$. □

Example of Notation 1: The SBE of s_1 obtained by examining all occurrences of the sequence '$\langle 121 \rangle$' will be be given by $\widehat{s}_1\,|_{<121>}$, and the SBE of s_2 obtained by examining all occurrences of the sequence '$\langle 2122 \rangle$' will be be given by $\widehat{s}_2\,|_{<2122>}$. Observe, trivially, that

$$\widehat{s}_2\,|_{<j_1 j_2 \ldots, j_k>} = 1 - \widehat{s}_1\,|_{<j_1 j_2 \ldots, j_k>}.$$

We shall now derive the explicit form of $\widehat{s}_i\,|_{<j_1 j_2 \ldots, j_k>}$ for various instantiations of sequences $\langle j_1 j_2 \ldots, j_k \rangle$. It is well known that the MLE converges with probability 1 and in the mean square sense to the true underlying parameter. Thus, all the estimates given in the following Sections/subsections converge (w. p. 1, and in the mean square sense) to the true underlying value of the parameter as the number of samples increases.

2 SBEs Using Pair-Wise Sequential Information

In this Section, we consider (analytically and experimentally) the estimation of the binomial parameter when we analyze the sequence of information by processing it in pairs. All the proofs of the results in this paper are either merely sketched or omitted in the interest of brevity, but can be found in [11].

Theorem 1. *Using Notation 1,* $\widehat{s}_1\,|_{<11>}$*, the SBE of* s_1 *obtained by examining the occurrences of '*$\langle 11 \rangle$*' is:*

[4] We depart from the traditional notation of the random variable taking values of '0' and '1', so that the notation is consistent when we deal with vectors.

$$\widehat{s_1}\,|_{<11>} = \sqrt{\frac{n_{11}}{N-1}} \qquad (1)$$

where n_{11} *is the number of occurrences of* '$< 11 >$' *in* $\mathcal{X}$.

Proof. The number of sequences of length two[5] in $\mathcal{X}$ is $N-1$. Of these, we observe n_{11} which have the value '$\langle 11\rangle$'. Consider now a random variable ξ_{11} which yields the outcome of either obtaining two consecutive 1's or not. ξ_{11} is a Bernoulli random variable whose distribution is :

$\xi_{11} =$ '11' with probability s_1^2

$\neq$ '11' with probability $1 - s_1^2$.

The MLE of the Bernoulli parameter of ξ_{11} is $\frac{n_{11}}{N-1}$, and thus,

$\widehat{s_1^2} = \frac{n_{11}}{N-1}$

whence $\widehat{s_1}\,|_{<11>} = \sqrt{\frac{n_{11}}{N-1}}$ and the result follows. □

Theorem 2. *Using Notation 1,* $\widehat{s_1}\,|_{<22>}$ *, the SBE of* s_1 *obtained by examining the occurrences of* '$\langle 22\rangle$' *is:*

$$\widehat{s_1}\,|_{<22>} = 1 - \sqrt{\frac{n_{22}}{N-1}} \qquad (2)$$

where n_{22} *is the number of occurrences of* '$< 22 >$' *in* $\mathcal{X}$.

Proof. The proof is similar to the proof of Theorem 1, except that we first solve for $\widehat{s_2}\,|_{<22>}$ and then obtain $\widehat{s_1}\,|_{<22>}$. The details are omitted. □

Theorem 3. *Using Notation 1,* $\widehat{s_1}\,|_{<12>}$ *and* $\widehat{s_1}\,|_{<21>}$ *, the SBEs of* s_1 *obtained by examining the occurrences of* '$\langle 12\rangle$' *and* '$\langle 21\rangle$'*, respectively, can be obtained if and only if the roots of the quadratic equation given below are :*

1. $\widehat{s_1}\,|_{<12>}$ *is the real root of* $\lambda^2 - \lambda + \frac{n_{12}}{N-1} = 0$ *whose value is closest to* $\widehat{s_1}$.
2. $\widehat{s_1}\,|_{<21>}$ *is the real root of* $\lambda^2 - \lambda + \frac{n_{21}}{N-1} = 0$ *whose value is closest to* $\widehat{s_1}$.

Proof. The proof of the result is found in [11]. □

A simple study of the patterns[6] that can occur will demonstrate that n_{21} differs from n_{12} by at most unity. Thus, the corresponding estimates $\widehat{s_1}\,|_{<21>}$ and $\widehat{s_1}\,|_{<12>}$ are almost the same. The ensemble estimates are, however, different.

Example: Let us suppose that $\mathcal{X}$ is the set $\{1, 2, 1, 2, 1, 1, 1, 2, 2, 1, 1\}$ where the elements occur in the specified order. Then : $n_1 = 7, n_{11} = 3, n_{12} = 3, n_{21} = 3$, and $n_{22} = 1$. Thus, the SBEs are: $\widehat{s_1}\,|_{<1>} = \frac{7}{11} = 0.6364$.

[5] The number of *distinct* sequences of length two is $\frac{N}{2}$. But since the elements of $\mathcal{X}$ are drawn independently and identically, there are $N-1$ consecutive pairs ("drawn with replacement") to be considered. More details of this are found in [11].

[6] The proverb "What goes up must come down !" is applicable here.

$\widehat{s}_1 |_{<11>} = \sqrt{\frac{3}{10}} = 0.5477$.

$\widehat{s}_1 |_{<12>} = Root(\lambda^2 - \lambda + \frac{3}{10} = 0)$. In this case, the roots of the quadratic are complex. Hence the quantity n_{12} can provide us no information about s_1. Similarly, the quantity n_{12} also leads to complex roots and so can provide us no information about s_1.

Finally, $\widehat{s}_2 |_{<22>} = \sqrt{\frac{1}{10}} = 0.3162$, and hence, $\widehat{s}_1 |_{<22>} = 0.6838$. □

Experimental Results: We present the results of our simulations[7] on synthetic data for the case when the sequence is processed in pairs. The SBE process for the estimation of the parameters for binomial random variables was extensively tested for numerous distributions, but in the interest of brevity, we merely cite one specific example. Also, to make the comparison meaningful, we have followed the "traditional" MLE computation (i.e., the one which does not utilize the sequential information) using the identical data stream. In each case, the estimation algorithms were presented with random occurrences of the variables for $N = 1,953,125$ (i.e, 5^9) time instances.

In the case of the SBE, the true underlying value of s_1 was computed using each of the estimates, $\widehat{s}_1 |_{<11>}$, $\widehat{s}_1 |_{<12>}$, $\widehat{s}_1 |_{<21>}$ and $\widehat{s}_1 |_{<22>}$, and the results are tabulated in Table 1. This table reports the values of the estimates as time progresses. However, to demonstrate the true convergence properties of the estimates, we have also reported the values of the ensemble averages of the estimates in Table 1, taken over an ensemble of 100 experiments, which are given in the second line of each row. The convergence of every single estimate is remarkable.

The reader should observe that the MLE and SBE taken for a *single* experiment are much more sporadic. This can be observed from Table 1. It is here that we believe that the SBE will find its niche, namely to enhance the MLE estimate using the information gleaned from the various SBEs.

3 SBEs Using Subsequences of Length Three and Four

We first consider the case when subsequences of length 3 are processed. The following results, whose proofs are found in [11], are true.

Theorem 4. *Using Notation 1, $\widehat{s}_1 |_{<111>}$, the SBE of s_1 obtained by examining the occurrences of '$\langle 111 \rangle$' is:*

$$\widehat{s}_1 |_{<111>} = \sqrt[3]{\frac{n_{111}}{N-2}} \tag{3}$$

where n_{111} is the number of occurrences of '$< 111 >$' in $\mathcal{X}$. □

Theorem 5. *Using Notation 1, $\widehat{s}_1 |_{<222>}$, the SBE of s_1 obtained by examining the occurrences of '$\langle 222 \rangle$' is:*

[7] In the tables, values of *unity* or *zero* represent the cases when the roots are complex or when the number of occurrences of the event concerned are zero.

Table 1. A table of the value of the MLE, $\widehat{s_1}$, and the SBEs $\widehat{s_1}\,|_{<11>}$, $\widehat{s_1}\,|_{<22>}$, $\widehat{s_1}\,|_{<12>}$, and $\widehat{s_1}\,|_{<21>}$, at time 'N', where the latter SBEs were estimated by using the results of Theorems 1, 2, and 3 respectively. The values of the second line of each row are the ensemble averages of the corresponding estimates, taken over an ensemble of 100 experiments.

N	$\widehat{s_1}$	$\widehat{s_1}\,\|_{<11>}$	$\widehat{s_1}\,\|_{<22>}$	$\widehat{s_1}\,\|_{<12>}$	$\widehat{s_1}\,\|_{<21>}$
5^1 (5)	0.8000	0.8660	1.0000	1.0000	0.5000
	0.7300	0.6686	0.8539	0.6000	0.6150
5^2 (25)	0.8000	0.8165	0.7959	0.8536	0.7887
	0.7212	0.7240	0.7267	0.6744	0.6816
5^3 (125)	0.7920	0.8032	0.7800	0.8111	0.8111
	0.7210	0.7215	0.7213	0.7121	0.7132
5^4 (625)	0.7456	0.7489	0.7375	0.7563	0.7532
	0.7248	0.7237	0.7282	0.7205	0.7206
5^5 (3,125)	0.7200	0.7226	0.7143	0.7277	0.7277
	0.7244	0.7240	0.7254	0.7231	0.7231
5^6 (15,625)	0.7199	0.7210	0.7171	0.7234	0.7233
	0.7246	0.7245	0.7249	0.7243	0.7243
5^7 (78,125)	0.7245	0.7244	0.7248	0.7242	0.7241
	0.7249	0.7248	0.7249	0.7248	0.7248
5^8 (390,625)	0.7252	0.7253	0.7250	0.7255	0.7255
	0.7250	0.7250	0.7249	0.7250	0.7250
5^9 (1,953,125)	0.7244	0.7243	0.7245	0.7242	0.7242
	0.7250	0.7250	0.7250	0.7250	0.7250

$$\widehat{s_1}\,|_{<222>} = 1 - \sqrt[3]{\frac{n_{222}}{N-2}} \tag{4}$$

where n_{222} is the number of occurrences of '$< 222 >$' in $\mathcal{X}$. □

Theorem 6. *Using Notation 1, the SBEs of s_1 obtained by examining the occurrences of subsequences which contain a* single *'2' such as '$\langle 211\rangle$', '$\langle 121\rangle$', and '$\langle 112\rangle$', can be obtained as the real root of the cubic equation given below :*

1. *$\widehat{s_1}\,|_{<211>}$ is the real root of $\lambda^3 - \lambda^2 + \frac{n_{211}}{N-2} = 0$ whose value is closest to $\widehat{s_1}$.*
2. *$\widehat{s_1}\,|_{<121>}$ is the real root of $\lambda^3 - \lambda^2 + \frac{n_{121}}{N-2} = 0$ whose value is closest to $\widehat{s_1}$.*
3. *$\widehat{s_1}\,|_{<112>}$ is the real root of $\lambda^3 - \lambda^2 + \frac{n_{112}}{N-2} = 0$ whose value is closest to $\widehat{s_1}$.*

□

Theorem 7. *Using Notation 1, the SBEs of s_1 obtained by examining the occurrences of subsequences which contain* two *'2's such as '$\langle 122\rangle$', '$\langle 212\rangle$', and '$\langle 221\rangle$', can be obtained as the real root of the cubic equation given below :*

1. *$\widehat{s_2}\,|_{<122>}$ is the real root of $\lambda^3 - \lambda^2 + \frac{n_{122}}{N-2} = 0$ whose value is closest to $\widehat{s_2}$, whence the estimate $\widehat{s_1}\,|_{<122>}$ can be obtained as $1 - \widehat{s_2}\,|_{<122>}$.*

2. $\widehat{s_2}\,|_{<212>}$ *is the real root of* $\lambda^3 - \lambda^2 + \frac{n_{212}}{N-2} = 0$ *whose value is closest to* $\widehat{s_2}$, *whence the estimate* $\widehat{s_1}\,|_{<212>}$ *can be obtained as* $1 - \widehat{s_2}\,|_{<212>}$.
3. $\widehat{s_2}\,|_{<221>}$ *is the real root of* $\lambda^3 - \lambda^2 + \frac{n_{221}}{N-2} = 0$ *whose value is closest to* $\widehat{s_2}$, *whence the estimate* $\widehat{s_1}\,|_{<221>}$ *can be obtained as* $1 - \widehat{s_2}\,|_{<221>}$. □

Experimental Results: We now present the results of our simulations on synthetic data for the cases studied in the previous sub-section, namely for the case when the sequence is processed in subsequences of length three. To make the comparison (with the pairwise computation) meaningful, we again report the result when the true value of s_1 is 0.725.

In the case of the SBE, the true underlying value of s_1 was computed using each of the estimates, SBEs $\widehat{s_1}\,|_{<111>}$, $\widehat{s_1}\,|_{<222>}$, $\widehat{s_1}\,|_{<211>}$, $\widehat{s_1}\,|_{<121>}$, $\widehat{s_1}\,|_{<112>}$, $\widehat{s_1}\,|_{<122>}$, $\widehat{s_1}\,|_{<212>}$, and $\widehat{s_1}\,|_{<221>}$, and the results are tabulated in Table 2 as a function of the number of samples processed.

Again, the reader should observe that the MLE and SBE taken for a *single* experiment are not as smooth - especially when the number of samples processed is small. This can be observed from Table 2. In practice, this is augmented by the fact that the SBEs sometimes lead to complex solutions or to unrealistic solutions when the number of samples processed is small. But fortunately, things "average" out as time proceeds.

Table 2. A table of the value of the MLE, $\widehat{s_1}$, and the SBEs $\widehat{s_1}\,|_{<111>}$, $\widehat{s_1}\,|_{<222>}$, $\widehat{s_1}\,|_{<211>}$, $\widehat{s_1}\,|_{<121>}$, $\widehat{s_1}\,|_{<112>}$, $\widehat{s_1}\,|_{<122>}$, $\widehat{s_1}\,|_{<212>}$, and $\widehat{s_1}\,|_{<221>}$, at time 'N', where the latter SBEs were estimated by using the results of Theorems 4, 5, 6, and 7, respectively. The values of the second line on each row mean the ensemble averages of the corresponding estimates, taken over an ensemble of 100 experiments.

N	$\widehat{s_1}$	$\widehat{s_1}\,\|_{<111>}$	$\widehat{s_1}\,\|_{<222>}$	$\widehat{s_1}\,\|_{<211>}$	$\widehat{s_1}\,\|_{<121>}$	$\widehat{s_1}\,\|_{<112>}$	$\widehat{s_1}\,\|_{<122>}$	$\widehat{s_1}\,\|_{<212>}$	$\widehat{s_1}\,\|_{<221>}$
5^1	0.8000 0.7300	0.8736 0.5154	1.0000 0.9497	0 0.4700	1.0000 0.5300	1.0000 0.4900	1.0000 0.9100	1.0000 0.9100	1.0000 0.9500
5^2	0.8000 0.7212	0.8050 0.7173	1.0000 0.8359	0 0.4769	0.8903 0.5724	0.7921 0.4629	0.7610 0.7237	1.0000 0.8299	0.7610 0.7189
5^3	0.7920 0.7210	0.7958 0.7199	0.7989 0.7530	0.7083 0.4534	0.8556 0.4906	0.7083 0.4351	0.7703 0.7161	0.9052 0.7305	0.7703 0.7158
5^4	0.7456 0.7248	0.7521 0.7226	0.7178 0.7316	0.7697 0.4653	0.7627 0.4474	0.7697 0.4723	0.7510 0.7273	0.7459 0.7253	0.7510 0.7271
5^5	0.7200 0.7244	0.7229 0.7236	0.7102 0.7251	0.7260 0.5607	0.7503 0.4780	0.7260 0.5607	0.7173 0.7258	0.7282 0.7247	0.7173 0.7258
5^6	0.7199 0.7246	0.7231 0.7244	0.7133 0.7247	0.7443 0.7076	0.7318 0.6717	0.7447 0.7076	0.7200 0.7251	0.7137 0.7250	0.7200 0.7251
5^7	0.7245 0.7249	0.7253 0.7248	0.7232 0.7245	0.7341 0.7246	0.7192 0.7234	0.7342 0.7246	0.7260 0.7251	0.7199 0.7248	0.7260 0.7251
5^8	0.7252 0.7250	0.7256 0.7250	0.7254 0.7249	0.7296 0.7251	0.7274 0.7250	0.7296 0.7251	0.7247 0.7250	0.7238 0.7249	0.7247 0.7250
5^9	0.7244 0.7250	0.7243 0.7250	0.7245 0.7249	0.7239 0.7252	0.7233 0.7249	0.7239 0.7252	0.7245 0.7250	0.7243 0.7249	0.7245 0.7250

We now extend the previous cases to consider the scenario when the sequential information is processed in subsequences of length four.

Theorem 8. *Using Notation 1,* $\widehat{s}_1 |_{<1111>}$ *, the SBE of* s_1 *obtained by examining the occurrences of '*$\langle 1111 \rangle$*' is:*

$$\widehat{s}_1 |_{<1111>} = \sqrt[4]{\frac{n_{1111}}{N-3}} \tag{5}$$

where n_{1111} *is the number of occurrences of '*$< 1111 >$*' in* $\mathcal{X}$*.* □

Theorem 9. *Using Notation 1,* $\widehat{s}_1 |_{<2222>}$ *, the SBE of* s_1 *obtained by examining the occurrences of '*$\langle 2222 \rangle$*' is:*

$$\widehat{s}_1 |_{<2222>} = 1 - \sqrt[4]{\frac{n_{2222}}{N-2}} \tag{6}$$

where n_{2222} *is the number of occurrences of '*$< 2222 >$*' in* $\mathcal{X}$*.* □

To simplify matters, we deal with the rest of the cases that involve four-at-a-time subsequences, by sub-dividing them into the cases when the subsequences contain *one* '2', *two* '2's, and *three* '2's, respectively. In each case, we shall deal with all the corresponding subsequence patterns in a single theorem.

Theorem 10. *Using Notation 1, the SBEs of* s_1 *obtained by examining the occurrences of subsequences which contain a* single *'2', can be obtained by the real roots (if any) of the quartic equations given below :*

1. $\widehat{s}_1 |_{<2111>}$ *is the real root of* $\lambda^4 - \lambda^3 + \frac{n_{2111}}{N-3} = 0$ *whose value is closest to* $\widehat{s}_1$*.*
2. $\widehat{s}_1 |_{<1211>}$ *is the real root of* $\lambda^4 - \lambda^3 + \frac{n_{1211}}{N-3} = 0$ *whose value is closest to* $\widehat{s}_1$*.*
3. $\widehat{s}_1 |_{<1121>}$ *is the real root of* $\lambda^4 - \lambda^3 + \frac{n_{1121}}{N-3} = 0$ *whose value is closest to* $\widehat{s}_1$*.*
4. $\widehat{s}_1 |_{<1112>}$ *is the real root of* $\lambda^4 - \lambda^3 + \frac{n_{1112}}{N-3} = 0$ *whose value is closest to* $\widehat{s}_1$*.*

□

Theorem 11. *Using Notation 1, the SBEs of* s_1 *obtained by examining the occurrences of subsequences which contain* two *'2's, can be obtained by the real roots (if any) of the quadratic* (not quartic !!!) *equations given below :*

1. $\widehat{s}_1 |_{<1122>}$ *is the real root of* $\lambda^2 - \lambda + \sqrt{\frac{n_{1122}}{N-3}} = 0$ *with value closest to* $\widehat{s}_1$*.*
2. $\widehat{s}_1 |_{<1212>}$ *is the real root of* $\lambda^2 - \lambda + \sqrt{\frac{n_{1212}}{N-3}} = 0$ *with value closest to* $\widehat{s}_1$*.*
3. $\widehat{s}_1 |_{<1221>}$ *is the real root of* $\lambda^2 - \lambda + \sqrt{\frac{n_{1221}}{N-3}} = 0$ *with value closest to* $\widehat{s}_1$*.*
4. $\widehat{s}_1 |_{<2112>}$ *is the real root of* $\lambda^2 - \lambda + \sqrt{\frac{n_{2112}}{N-3}} = 0$ *with value closest to* $\widehat{s}_1$*.*
5. $\widehat{s}_1 |_{<2121>}$ *is the real root of* $\lambda^2 - \lambda + \sqrt{\frac{n_{2121}}{N-3}} = 0$ *with value closest to* $\widehat{s}_1$*.*
6. $\widehat{s}_1 |_{<2211>}$ *is the real root of* $\lambda^2 - \lambda + \sqrt{\frac{n_{2211}}{N-3}} = 0$ *with value closest to* $\widehat{s}_1$*.* □

Theorem 12. *Using Notation 1, the SBEs of* s_1 *obtained by examining the occurrences of subsequences which contain* three *'2's, can be obtained by determining the real roots (if any) of the quartic equations given below and then subtracting their value from unity as below:*

1. $\widehat{s_1}\,|_{<1222>}$ *is the quantity* $[1$ - $Root(\lambda^4 - \lambda^3 + \frac{n_{1222}}{N-3} = 0)]$
2. $\widehat{s_1}\,|_{<2122>}$ *is the quantity* $[1$ - $Root(\lambda^4 - \lambda^3 + \frac{n_{2122}}{N-3} = 0)]$
3. $\widehat{s_1}\,|_{<2212>}$ *is the quantity* $[1$ - $Root(\lambda^4 - \lambda^3 + \frac{n_{2212}}{N-3} = 0)]$
4. $\widehat{s_1}\,|_{<2221>}$ *is the quantity* $[1$ - $Root(\lambda^4 - \lambda^3 + \frac{n_{2221}}{N-3} = 0)]$

where each of the above estimates is the value closest to $\widehat{s_1}$. □

Experimental Results: The simulation results for the the case when the sequence is processed in subsequences of length four is presented below. The experimental settings are identical to the ones used in the case of processing it in pairs and in subsequences of length three, namely, when s_1 is 0.725, and $N = 1,953,125$ (i.e, 5^9) time instances.

Table 3 lists the values of the SBEs, computed using each of the estimates, $\widehat{s_1}\,|_{<1111>}$, $\widehat{s_1}\,|_{<2222>}$, $\widehat{s_1}\,|_{<2111>}$, $\widehat{s_1}\,|_{<1122>}$, and $\widehat{s_1}\,|_{<1222>}$, and their ensemble averages. The other cases when the subsequences with *one* '2', *two* '2's, and *three* '2's (the other cases listed in Theorems 10, 11 and 12) are identical to the ones reported and so omit them here for ease of readability.

Again, we observe that the convergence of every single estimate is remarkable. For example, the traditional MLE, $\widehat{s_1}$, had the ensemble average of 0.7210 when only $N = 125$ symbols were processed. This value became 0.7248 when $N = 625$ symbols were processed, which converged to 0.7250 when $N = 5^9$. By way of comparison, for the same case, the SBE, $\widehat{s_1}\,|_{<1222>}$, had the ensemble average of 0.7444 when only $N = 125$ symbols were processed. It had the value 0.7319 after $N = 625$ symbols were processed, and as in the case of $\widehat{s_1}$ became increasingly closer to the true value as N increased. In this case, when $N = 5^9$, the value of $\widehat{s_1}\,|_{<1222>}$, was also exactly 0.7250. This was also true for the other SBEs.

In this case, the solutions to the equations were often complex initially (i.e., for small values of 'N'). But as time proceeded, the number of occurrences of the outcomes was more reasonable, and the solution obtained converged as expected.

4 Open Issues and Potential Applications of SBEs

As mentioned earlier, we believe that there are a host of open problems which concern the family of SBEs. We shall highlight them in the following subsections.

Higher Order SBEs: Till now, we have considered how we can obtain effective SBEs by considering subsequences of lengths 2, 3 and 4 respectively. There is no reason why we cannot consider subsequences of even longer length. Without much ado, we list (without proof) the form the SBEs would take for a few simple cases when subsequences of length 5 are analyzed. Indeed, using Notation 1, we can state that:

1. $\widehat{s_1}\,|_{<11111>}$, the SBE of s_1 obtained by examining the occurrences of '$\langle 11111 \rangle$' is : $\widehat{s_1}\,|_{<11111>} = \sqrt[5]{\frac{n_{11111}}{N-4}}$. □

Table 3. A table of the values of the MLE, $\widehat{s_1}$, and the SBEs $\widehat{s_1}|_{<1111>}$, $\widehat{s_1}|_{<2222>}$, $\widehat{s_1}|_{<2111>}$, $\widehat{s_1}|_{<1122>}$, and $\widehat{s_1}|_{<1222>}$, at time 'N', where the latter SBEs were estimated by using the results of Theorems 8, 9, 10, 11 and 12 respectively. The other cases when the subsequences with *one* '2', *two* '2's, and *three* '2's (the other cases listed in Theorems 10, 11 and 12) are identical.

N	$\widehat{s_1}$	$\widehat{s_1}\|_{<1111>}$	$\widehat{s_1}\|_{<2222>}$	$\widehat{s_1}\|_{<1222>}$	$\widehat{s_1}\|_{<1122>}$	$\widehat{s_1}\|_{<2111>}$
5^1 (5)	0.8000 0.7300	0.8409 0.3645	1.0000 0.9916	1.0000 0.8800	1.0000 0.8100	1.0000 0.6600
5^2 (25)	0.8000 0.7212	0.7765 0.6867	1.0000 0.9428	1.0000 0.8354	0.6918 0.5944	0.6918 0.4880
5^3 (125)	0.7920 0.7210	0.7920 0.7178	1.0000 0.8832	0.7811 0.7444	0.7625 0.6577	0.7625 0.3152
5^4 (625)	0.7456 0.7248	0.7492 0.7223	0.7619 0.7468	0.6976 0.7319	0.7402 0.7255	0.7659 0.3755
5^5 (3,125)	0.7200 0.7244	0.7247 0.7236	0.7412 0.7249	0.6942 0.7261	0.7069 0.7250	0.7143 0.4050
5^6 (15,625)	0.7199 0.7246	0.7232 0.7244	0.7090 0.7244	0.7161 0.7250	0.7221 0.7246	0.7196 0.4922
5^7 (78,125)	0.7245 0.7249	0.7253 0.7249	0.7205 0.7245	0.7248 0.7245	0.7285 0.7251	0.7277 0.6649
5^8 (390,625)	0.7252 0.7250	0.7256 0.7250	0.7257 0.7247	0.7252 0.7250	0.7250 0.7250	0.7257 0.7249
5^9 (1,953,125)	0.7244 0.7250	0.7241 0.7250	0.7244 0.7248	0.7246 0.7250	0.7250 0.7251	0.7247 0.7249

2. $\widehat{s_1}|_{<22222>}$, the SBE of s_1 obtained by examining the occurrences of '$\langle 22222 \rangle$' is : $\widehat{s_1}|_{<22222>} = 1 - \sqrt[5]{\frac{n_{22222}}{N-4}}$. □

We believe that deriving the expressions for other higher order SBEs is not expedient. Obtaining them would involve explicitly solving algebraic equations which are higher than of a *quintic* order, and it is well known that this is intractable (other than by resorting to numerical methods).

We conclude this section by stating that the question of how we can effectively *compute* the SBEs for orders of 5 and higher is still effectively open.

Fusing the MLE and the SBEs to Yield a Superior Estimate: One of the most interesting problems that still remains open involves the question of how the MLE and the SBEs can be *fused* to yield a superior estimate. Rather than discuss "specifics", let us assume that we have obtained a set of estimates $\Phi = [\phi_0, \phi_1, \phi_2, \ldots \phi_D]^T$, where, for simplicity, we use the notation that ϕ_0 is the traditional MLE, and the other ϕ_i's are the SBEs. Thus, for example, an instantiation of Φ could be the 7-component vector:

$\Phi = [\widehat{s_1}, \widehat{s_1}|_{<11>}, \widehat{s_1}|_{<12>}, \widehat{s_1}|_{<111>}, \widehat{s_1}|_{<222>}, \widehat{s_1}|_{<1111>}, \widehat{s_1}|_{<1222>}]^T$.

The aim of the fusing exercise is to combine the information in the components of Φ to obtain an even more superior estimate $\widehat{\widehat{s_1}}$.

The first question that needs to be answered is the following: If the traditional MLE and all the SBEs converge to the *same true value*, what is the advantage of such a fusing process? The answer lies simply in the fact that although the traditional MLE and all the SBEs converge *asymptotically* to the same value, they all have *completely different values*[8] when the number of samples examined is "small". Thus, for example, when the number of samples examined is only 125 and the true value of s_1 is 0.7250, the value of Φ is:

$\Phi = [0.7920, 0.8032, 0.7800, 0.8111, 0.7958, 0.7989, 0.7920, 0.7811]^T$.

Observe that the traditional MLE is 0.7920 (quite distant from the true value of 0.7250), while other SBEs are closer to the true value. Thus, it would be advantageous to seek a scheme which uses these different "descriptors" of s_1 to lead to a more accurate estimate of s_1. In all these estimates, we consciously discard elements of Φ which are *unity* or *zero*, as these represent the cases when the solution of the underlying equation don't lead to a realistic estimate.

We have designed four different fusion methods that use the D components of Φ. The details of the methods are omitted here in the interest of brevity, but can be found in [11], but this entire avenue is open to further research.

Classification Using the MLE and the SBE: Another major possibility for further research involves combining the classifier decisions obtained by the MLEs and the various SBEs. In the study of pattern recognition, classifier combination has received considerable attention because of its potential to improve the performance of individual classification systems. The basic idea is to solve each classification problem by designing a *specific* classifier, and then to combine the classifiers in some way to achieve reduced classification error rates. Indeed, the choice of an appropriate fusion method can further improve on the performance of the combination. Various Classifier Fusion Schemes (CFS) have been proposed in the literature - excellent studies are found in [9,?]. We have designed a few different fusion classification methods for SBEs (omitted here in the interest of brevity, but included in [11]), but here too the ground is fertile, and we believe that a lot of research remains to be accomplished.

Non Pattern Recognition Potential Applications of SBEs: Apart from the obvious applications in Pattern Recognition (PR) (mentioned above), there are numerous situations where data arrives in a sequence, and where it is possible to assign a binary indicator variable to the arriving data. This section introduces two such example applications that are representative for wide classes of use. We note, in passing, that thse methods can be used for even broader application areas when we seek methods to obtain the sequence estimates so as to improve the quality and convergence of lower order estimators. In particular, this may allow the use of estimation in real-time control applications where the convergence of MLEs are too slow compared to the time constant of the system being controlled.

Network Transmission Quality: The Internet is omnipresent today, and the Internet Protocol (IP) is the dominating protocol for long haul data communication. However, the IP itself does not provide any transmission guarantees, and

[8] This is reminiscent of the fairy tale of the seven blind men who each described an elephant with completely different descriptions. While each of the descriptors was, in itself, inaccurate, the composite picture was accurate!

packets may freely be delayed, delivered out of order, or even discarded in the case of network congestion. To remedy this situation most Internet applications requiring reliable and sequential delivery use the Transmission Control Protocol (TCP) on top of the IP, namely, the TCP/IP [14]. For applications which cannot accept neither long delays nor jitter (delay variation) nor out of order delivery the TCP is not the solution, and various other protocols and mechanisms to achieve the desired transmission quality in the Internet have been proposed [17]. Providing strict transmission quality guarantees per flow can be done, but this requires intelligence installed in every router along the transmission path [15]. Further, if a flow crosses multiple ISPs it will mix with other traffic between the ISPs, and the end-user's SLA with its ISP will not extend to the ISPs further down-stream. Currently, no end-to-end transmission guarantees for a flow in the Internet can be given beyond that of the TCP; this is an active research topic [5].

By categorizing a packet within the delay bound as a *Success* (say, with value '1') and a dropped packet as a *Failure* (with value '2'), the situation fits the SBE framework presented in this paper. The delay sensitive flow can be admitted if the failure rate, $\hat{s}_2$, estimated from the probing is below a certain limit. A side effect of using the SBE instead of the MLE would be the advantage of being able to gather statistics on the probability of a *sequence* of subsequent failures. More details of this proposition are found in [11].

Arithmetic Coding: Arithmetic coding [6,12] has the ability to serve two purposes: Lossless data compression [16] and the assignment of unique signatures for sets and databases. The fundamental idea is to encode a sequence of symbols taken from an *alphabet* as a sequence of bits, so that more frequently occurring symbols are assigned a lesser number of bits, where the assignment can be achieved either static or adaptive.

Given the encoded bit string the SBE is directly applicable. Using the SBE on the bit stream the decoder might be able to predict the incoming symbol before it has completely arrived. Such a look-ahead would be a result of combining the relative frequencies of '0' and '1' in the stream with the relative occurrence of the *sub-sequences*. The added benefit will be that the encoder can use the information of the symbol it is set to decode immediately, i.e. update the frequency table of the model *before* encoding the symbol. The details of this opportunity, including rendering it less vulnerable to transmission errors (see [11] needs to be further investigated, and is a topic for further research.

5 Conclusions

In this paper, we considered the age-old problem of estimating the parameters of a distribution from its observations. Unlike the method that is customarily employed, (which processes the information contained in the observations taken *as a set*), we demonstrate how the estimation can be enhanced by utilizing both the information in the observations *and in their sequence of appearance*. In this regard, we have derived the corresponding MLE results when the samples are taken two-at-a-time, three-at-a-time, four-at-a-time etc. In each case, we also experimentally demonstrated the convergence of the corresponding estimates. We have visited the various strategies by which these estimates could be fused to yield superior overall cumulative estimates. Our results demonstrate that the strategy is very promising, and that it has potential applications in fused PR systems, and in the domains of enhancing Internet protocols and data encoding.

References

1. P. Bickel and K. Doksum. *Mathematical Statistics: Basic Ideas and Selected Topics*, volume I. Prentice Hall, Second Edition, 2000.
2. H. Bunke. Structural and Syntactic Pattern Recognition. *Handbook of Pattern Recognition and Computer Vision*, World Scientific-25, 1993.
3. G. Casella and R. Berger. *Statistical Inference.* Brooks/Cole Pub. Co., Second Edition, 2001.
4. R. Duda, P. Hart, and D. Stork. *Pattern Classification.* John Wiley and Sons, Inc., New York, NY, Second Edition, 2000.
5. W.-C. Feng, D. D. Kandlur, D. Saha, and K. G. Shin, "Adaptive packet marking for maintaining end-to-end throughput in a differentiated-services internet," *IEEE/ACM Transactions on Networking*, vol. 7, pp. 685–697, Oct. 1999.
6. J. Glen and G. Langdon, "Arithmetic coding," *IBM Journal of Research and Development*, vol. 28, pp. 135–149, Mar. 1984.
7. R. Herbrich. *Learning Kernel Classifiers: Theory and Algorithms.* MIT Press, Cambridge, Massachusetts, 2001.
8. B. Jones, P. Garthwaite, and Ian Jolliffe. *Statistical Inference.* Oxford University Press, Second Edition, 2002.
9. J. Kittler, M. Hatef, R. P. W. Duin and J. Matas, "On combining classifiers", *IEEE Trans. Pattern Anal. and Machine Intell.*, vol. 20, No. 3, pp. 226 - 239, Mar. 1998.
10. L. I. Kuncheva, "A theoretical study on six classifier fusion strategies", *IEEE Trans. Pattern Anal. and Machine Intell.*, vol. 24, no. 2, pp. 281 - 286, Feb. 2002.
11. B. J. Oommen, S.-W. Kim, and G. Horn, "On the Estimation of Binomial Random Variables Using Occurrence and Sequential Information". *Unabridged version of this paper.*
12. J. Rissanen and J. Glen G. Langdon, "Arithmetic coding," *IBM Journal of Research and Development*, vol. 23, pp. 149–162, Mar. 1979.
13. A. S. Tanenbaum, *Computer Networks.* New Jersey: Prentice Hall, 4th ed., 2003.
14. U. o. S. C. Information Science Institute, "Transmission control protocol (TCP)." http://www.ietf.org/rfc/rfc0793.txt, sep 1981. Clarifications, corrections, extensions and a textbook introduction can be found in [13].
15. P. P. White, "RSVP and integrated services in the internet: A tutorial," *IEEE Communications Magazine*, pp. 100–106, May 1997.
16. I. H. Witten, R. M. Neal, and J. G. Cleary, "Arithmetic coding for data compression," *Communications of the ACM*, vol. 30, pp. 520–540, Jun. 1987.
17. X. Xiao and L. M. Ni, "Internet QoS: A big picture," *IEEE Network*, vol. 13, pp. 8–18, Mar. 1999.
18. A. Webb. *Statistical Pattern Recognition.* John Wiley & Sons, N.York, Second Edition, 2002.

Symmetries from Uniform Space Covering in Stochastic Discrimination

Tin Kam Ho

Bell Labs, Lucent Technologies
tkh@research.bell-labs.com

Abstract. Studies on ensemble methods for classification suffer from the difficulty of modeling the complementary strengths of the components. Kleinberg's theory of stochastic discrimination (SD) addresses this rigorously via mathematical notions of enrichment, uniformity, and projectability of a model ensemble. We explain these concepts via a very simple numerical example that captures the basic principles of the SD theory and method. We focus on a fundamental symmetry in point set covering that is the key observation leading to the foundation of the theory. We believe a better understanding of the SD method will lead to developments of better tools for analyzing other ensemble methods.

1 Introduction

Methods for classifier combination, or ensemble learning, can be divided into two categories: 1) *decision optimization* methods that try to obtain *consensus* among a *given* set of classifiers to make the best decision; 2) *coverage optimization* methods that try to *create* a set of classifiers that can do well for all possible cases under a *fixed* decision combination function.

Decision optimization methods rely on the assumption that the given set of classifiers, typically of a small size, contain sufficient expert knowledge about the application domain, and each of them excels in a subset of all possible input. A decision combination function is chosen or trained to exploit the individual strengths while avoiding their weaknesses. Popular combination functions include majority/plurality votes[19], sum/product rules[14], rank/confidence score combination[12], and probabilistic methods[13]. These methods are known to be useful in many applications where reasonably good component classifiers can be developed. However, the joint capability of the classifiers sets an intrinsic limitation that a decision combination function cannot overcome. A challenge in this approach is to find out the "blind spots" of the ensemble and to obtain an additional classifier that covers them.

Coverage optimization methods use an automatic and systematic mechanism to generate new classifiers with the hope of covering all possible cases. A fixed function, typically simple in form, is used for decision combination. This can be training set subsampling, such as stacking[22], bagging[2], and boosting[5], feature subspace projection[10], superclass/subclass decomposition[4], or other

D.-Y. Yeung et al. (Eds.): SSPR&SPR 2006, LNCS 4109, pp. 22–40, 2006.

methods for randomly perturbing the classifier training procedures[6]. Open questions in these methods are 1) how many classifiers are enough? 2) what kind of differences among the component classifiers yields the best combined accuracy? 3) how much limitation is set by the form of the component classifiers?

Apparently both categories of ensemble methods run into some dilemma. Should the component classifiers be weakened in order to achieve a stronger whole? Should some accuracy be sacrificed for the known samples to obtain better generalization for the unseen cases? Do we seek agreement, or differences among the component classifiers?

A central difficulty in studying the performance of these ensembles is how to model the complementary strengths among the classifiers. Many proofs rely on an assumption of statistical independence of component classifiers' decisions. But rarely is there any attempt to match this assumption with observations of the decisions. Often, global estimates of the component classifiers' accuracies are used in their selection, while in an ensemble what matter more are the local estimates, plus the relationship between the local accuracy estimates on samples that are close neighbors in the feature space.[1]

Deeper investigation of these issues leads back to three major concerns in choosing classifiers: discriminative power, use of complementary information, and generalization power. A complete theory on ensembles must address these three issues simultaneously. Many current theories rely, either explicitly or implicitly, on ideal assumptions on one or two of these issues, or have them omitted entirely, and are therefore incomplete.

Kleinberg's theory and method of stochastic discrimination (SD)[15][16] is the first attempt to explicitly address these issues simultaneously from a mathematical point of view. In this theory, rigorous notions are made for discriminative power, complementary information, and generalization power of an ensemble. A fundamental symmetry is observed between the probability of a fixed model covering a point in a given set and the probability of a fixed point being covered by a model in a given ensemble. The theory establishes that, these three conditions are sufficient for an ensemble to converge, with increases in its size, to the most accurate classifier for the application.

Kleinberg's analysis uses a set-theoretic abstraction to remove from consideration algorithmic details of classifiers, feature extraction processes, and training procedures. It considers only the classifiers' decision regions in the form of point sets, called *weak models*, in the feature space. A collection of classifiers is thus just a sample from the power set of the feature space. If the sample satisfies a uniformity condition, i.e., if its coverage is unbiased for any local region of the feature space, then a symmetry is observed between two probabilities (w.r.t. the feature space and w.r.t. the power set, respectively) of the same event that a point of a particular class is covered by a component of the sample. Discrimination between classes is achieved by requiring some minimum difference in each component's inclusion of points of different classes, which is trivial to satisfy. By

[1] There is more discussion on these difficulties in a recent review[8].

way of this symmetry, it is shown that if the sample of weak models is large, the discriminant function, defined on the coverage of the models on a single point and the class-specific differences within each model, converges to poles distinct by class with diminishing variance.

We believe that this symmetry is the key to the discussions on classifier combination. However, since the theory was developed from a fresh, original, and independent perspective on the problem of learning, there have not been many direct links made to the existing theories. As the concepts are new, the claims are high, the published algorithms appear simple, and the details of more sophisticated implementations are not known, the method has been poorly understood and is sometimes referred to as mysterious.

It is the goal of this lecture to illustrate the basic concepts in this theory and remove the apparent mystery. We present the principles of stochastic discrimination with a very simple numerical example. The example is so chosen that all computations can be easily traced step-by-step by hand or with very simple programs. We use Kleinberg's notation wherever possible to make it easier for the interested readers to follow up on the full theory in the original papers. Our emphasis is on explaining the concepts of uniformity and enrichment, and the behavior of the discriminant when both conditions are fulfilled. For the details of the mathematical theory and outlines of practical algorithms, please refer to Kleinberg's original publications[15][16][17][18].

2 Symmetry of Probabilities Induced by Uniform Space Covering

The SD method is based on a fundamental symmetry in point set covering. To illustrate this symmetry, we begin with a simple observation. Consider a set $S = \{a, b, c\}$ and all the subsets with two elements $s_1 = \{a, b\}$, $s_2 = \{a, c\}$, and $s_3 = \{b, c\}$. By our choice, each of these subsets has captured $2/3$ of the elements of S. We call this ratio r. Let us now look at each member of S, and check how many of these three subsets have included that member. For example, a is in two of them, so we say a is captured by $2/3$ of these subsets. We will obtain the same value $2/3$ for all elements of S. This value is the same as r. This is a consequence of the fact that we have used all such 2-member subsets and we have not biased this collection towards any element of S. With this observation, we begin a larger example.

Consider a set of 10 points in a one-dimensional feature space F. Let this set be called A. Assume that F contains only points in A and nothing else. Let each point in A be identified as q_0, q_1, ..., q_9 as follows.

$$\begin{matrix} \cdot & \cdot & \cdot & \cdot & \cdot & \cdot & \cdot & \cdot & \cdot & \cdot \\ q_0 & q_1 & q_2 & q_3 & q_4 & q_5 & q_6 & q_7 & q_8 & q_9 \end{matrix}$$

Now consider the subsets of F. Let the collection of all such subsets be $\mathcal{M}$, which is the power set of F. We call each member m of $\mathcal{M}$ a *model*, and we restrict our consideration to only those models that contain 5 points in A, therefore each

Table 1. Models m_t in $M_{0.5,A}$ in the order of $M = m_1, m_2, ..., m_{252}$. Each model is shown with its elements denoted by the indices i of q_i in A. For example, $m_1 = \{q_3, q_5, q_6, q_8, q_9\}$.

m_t	elements	m_t	elements	m_t	elements	m_t	elements	m_t	elements	m_t	elements
m_1	35689	m_{43}	12689	m_{85}	24578	m_{127}	01469	m_{169}	02468	m_{211}	02458
m_2	01268	m_{44}	04569	m_{86}	23568	m_{128}	03679	m_{170}	35678	m_{212}	13457
m_3	04789	m_{45}	01245	m_{87}	01267	m_{129}	04579	m_{171}	03589	m_{213}	24689
m_4	25689	m_{46}	01458	m_{88}	01257	m_{130}	01237	m_{172}	34679	m_{214}	03478
m_5	02679	m_{47}	15679	m_{89}	05679	m_{131}	24789	m_{173}	12346	m_{215}	23589
m_6	34578	m_{48}	12457	m_{90}	24589	m_{132}	45689	m_{174}	12458	m_{216}	24679
m_7	13459	m_{49}	02379	m_{91}	04589	m_{133}	16789	m_{175}	35789	m_{217}	02456
m_8	01238	m_{50}	02568	m_{92}	12467	m_{134}	13479	m_{176}	02358	m_{218}	05689
m_9	12347	m_{51}	12357	m_{93}	13578	m_{135}	02349	m_{177}	35679	m_{219}	12789
m_{10}	01579	m_{52}	14678	m_{94}	02369	m_{136}	13469	m_{178}	13458	m_{220}	02346
m_{11}	34589	m_{53}	12678	m_{95}	12469	m_{137}	03678	m_{179}	01459	m_{221}	23489
m_{12}	03459	m_{54}	23567	m_{96}	04567	m_{138}	23679	m_{180}	03479	m_{222}	23467
m_{13}	23459	m_{55}	02789	m_{97}	14679	m_{139}	46789	m_{181}	14789	m_{223}	12489
m_{14}	02457	m_{56}	24567	m_{98}	13467	m_{140}	01468	m_{182}	23678	m_{224}	14589
m_{15}	02368	m_{57}	13569	m_{99}	45678	m_{141}	03689	m_{183}	03456	m_{225}	25678
m_{16}	02689	m_{58}	01259	m_{100}	03469	m_{142}	02478	m_{184}	13456	m_{226}	12579
m_{17}	01368	m_{59}	23479	m_{101}	34789	m_{143}	23457	m_{185}	01568	m_{227}	03458
m_{18}	13589	m_{60}	03579	m_{102}	45679	m_{144}	02347	m_{186}	01578	m_{228}	01569
m_{19}	14579	m_{61}	12368	m_{103}	01358	m_{145}	01289	m_{187}	01678	m_{229}	45789
m_{20}	23468	m_{62}	23578	m_{104}	01379	m_{146}	01369	m_{188}	12367	m_{230}	12358
m_{21}	26789	m_{63}	02345	m_{105}	01236	m_{147}	01356	m_{189}	12345	m_{231}	02579
m_{22}	15678	m_{64}	01479	m_{106}	01679	m_{148}	12379	m_{190}	25679	m_{232}	01457
m_{23}	04578	m_{65}	03569	m_{107}	13689	m_{149}	02569	m_{191}	02367	m_{233}	05789
m_{24}	04679	m_{66}	01346	m_{108}	12479	m_{150}	34678	m_{192}	01256	m_{234}	01247
m_{25}	02459	m_{67}	24568	m_{109}	14568	m_{151}	24569	m_{193}	13679	m_{235}	03467
m_{26}	12569	m_{68}	01359	m_{110}	15689	m_{152}	03578	m_{194}	04689	m_{236}	12359
m_{27}	01269	m_{69}	12459	m_{111}	01258	m_{153}	02359	m_{195}	04568	m_{237}	02567
m_{28}	06789	m_{70}	01239	m_{112}	12389	m_{154}	01234	m_{196}	12578	m_{238}	12356
m_{29}	01689	m_{71}	24678	m_{113}	03568	m_{155}	01345	m_{197}	12468	m_{239}	02469
m_{30}	01248	m_{72}	01347	m_{114}	23689	m_{156}	02348	m_{198}	03468	m_{240}	13468
m_{31}	12456	m_{73}	01467	m_{115}	23478	m_{157}	03457	m_{199}	34569	m_{241}	02479
m_{32}	13579	m_{74}	04678	m_{116}	34568	m_{158}	02357	m_{200}	12369	m_{242}	36789
m_{33}	34689	m_{75}	12589	m_{117}	23569	m_{159}	01235	m_{201}	13489	m_{243}	13568
m_{34}	12679	m_{76}	01348	m_{118}	14689	m_{160}	01378	m_{202}	12567	m_{244}	02467
m_{35}	12568	m_{77}	14569	m_{119}	23789	m_{161}	14567	m_{203}	02489	m_{245}	01589
m_{36}	34579	m_{78}	01789	m_{120}	01246	m_{162}	23458	m_{204}	02678	m_{246}	01478
m_{37}	01389	m_{79}	01367	m_{121}	23579	m_{163}	56789	m_{205}	13567	m_{247}	15789
m_{38}	23469	m_{80}	12478	m_{122}	01456	m_{164}	34567	m_{206}	01357	m_{248}	01349
m_{39}	24579	m_{81}	25789	m_{123}	23456	m_{165}	01249	m_{207}	01278	m_{249}	02356
m_{40}	02589	m_{82}	01489	m_{124}	03789	m_{166}	03489	m_{208}	02578	m_{250}	14578
m_{41}	01567	m_{83}	03567	m_{125}	05678	m_{167}	02389	m_{209}	12348	m_{251}	13789
m_{42}	13478	m_{84}	12349	m_{126}	13678	m_{168}	12378	m_{210}	01279	m_{252}	02378

model has a size that is 0.5 of the size of A. Let this set of models be called $M_{0.5,A}$. Some members of $M_{0.5,A}$ are as follows.

$$\{q_0, q_1, q_2, q_3, q_4\}$$
$$\{q_0, q_1, q_2, q_3, q_5\}$$
$$\{q_0, q_1, q_2, q_3, q_6\}$$
$$\ldots$$

There are $C(10, 5) = 252$ members in $M_{0.5,A}$. Let M be a pseudo-random permutation of members in $M_{0.5,A}$ as listed in Table 1. We identify models in this sequence by a single subscript such that $M = m_1, m_2, ..., m_{252}$. We expand a collection M_t by including more and more members of $M_{0.5,A}$ in the order of the sequence M as follows. $M_1 = \{m_1\}$, $M_2 = \{m_1, m_2\}$, ..., $M_t = \{m_1, m_2, ...m_t\}$.

Since each model covers some points in A, for each member q in A, we can count the number of models in M_t that include q, call this count $N(q, M_t)$, and calculate the ratio of this count over the size of M_t, call it $Y(q, M_t)$. That is, $Y(q, M_t) = Prob_{\mathcal{M}}(q \in m | m \in M_t)$. As M_t expands, this ratio changes and we show these changes for each q in Table 2. The values of $Y(q, M_t)$ are plotted in Figure 1. As is clearly visible in the Figure, the values of $Y(q, M_t)$ converge to

Table 2. Ratio of coverage of each point q by members of M_t as M_t expands

	$N(M_t, q)$										$Y(M_t, q)$									
M_t	q_0	q_1	q_2	q_3	q_4	q_5	q_6	q_7	q_8	q_9	q_0	q_1	q_2	q_3	q_4	q_5	q_6	q_7	q_8	q_9
M_1	0	0	0	1	0	1	1	0	1	1	0.00	0.00	0.00	1.00	0.00	1.00	1.00	0.00	1.00	1.00
M_2	1	1	1	1	0	1	2	0	2	1	0.50	0.50	0.50	0.50	0.00	0.50	1.00	0.00	1.00	0.50
M_3	2	1	1	1	1	1	2	1	3	2	0.67	0.33	0.33	0.33	0.33	0.33	0.67	0.33	1.00	0.67
M_4	2	1	2	1	1	2	3	1	4	3	0.50	0.25	0.50	0.25	0.25	0.50	0.75	0.25	1.00	0.75
M_5	3	1	3	1	1	2	4	2	4	4	0.60	0.20	0.60	0.20	0.20	0.40	0.80	0.40	0.80	0.80
M_6	3	1	3	2	2	3	4	3	5	4	0.50	0.17	0.50	0.33	0.33	0.50	0.67	0.50	0.83	0.67
M_7	3	2	3	3	3	4	4	3	5	5	0.43	0.29	0.43	0.43	0.43	0.57	0.57	0.43	0.71	0.71
M_8	4	3	4	4	3	4	4	3	6	5	0.50	0.38	0.50	0.50	0.38	0.50	0.50	0.38	0.75	0.62
M_9	4	4	5	5	4	4	4	4	6	5	0.44	0.44	0.56	0.56	0.44	0.44	0.44	0.44	0.67	0.56
M_{10}	5	5	5	5	4	5	4	5	6	6	0.50	0.50	0.50	0.50	0.40	0.50	0.40	0.50	0.60	0.60
...						...										...				
M_{159}	81	80	79	79	79	77	82	78	74	86	0.51	0.50	0.50	0.50	0.50	0.48	0.52	0.49	0.47	0.54
M_{160}	82	81	79	80	79	77	82	79	75	86	0.51	0.51	0.49	0.50	0.49	0.48	0.51	0.49	0.47	0.54
M_{161}	82	82	79	80	80	78	83	80	75	86	0.51	0.51	0.49	0.50	0.50	0.48	0.52	0.50	0.47	0.53
M_{162}	82	82	80	81	81	79	83	80	76	86	0.51	0.51	0.49	0.50	0.50	0.49	0.51	0.49	0.47	0.53
M_{163}	82	82	80	81	81	80	84	81	77	87	0.50	0.50	0.49	0.50	0.50	0.49	0.52	0.50	0.47	0.53
M_{164}	82	82	80	82	82	81	85	82	77	87	0.50	0.50	0.49	0.50	0.50	0.49	0.52	0.50	0.47	0.53
M_{165}	83	83	81	82	83	81	85	82	77	88	0.50	0.50	0.49	0.50	0.50	0.49	0.52	0.50	0.47	0.53
M_{166}	84	83	81	83	84	81	85	82	78	89	0.51	0.50	0.49	0.50	0.51	0.49	0.51	0.49	0.47	0.54
M_{167}	85	83	82	84	84	81	85	82	79	90	0.51	0.50	0.49	0.50	0.50	0.49	0.51	0.49	0.47	0.54
M_{168}	85	84	83	85	84	81	85	83	80	90	0.51	0.50	0.49	0.51	0.50	0.48	0.51	0.49	0.48	0.54
...						...										...				
M_{243}	120	120	123	122	122	122	124	120	120	122	0.49	0.49	0.51	0.50	0.50	0.50	0.51	0.49	0.49	0.50
M_{244}	121	120	124	122	123	122	125	121	120	122	0.50	0.49	0.51	0.50	0.50	0.50	0.51	0.50	0.49	0.50
M_{245}	122	121	124	122	123	123	125	121	121	123	0.50	0.49	0.51	0.50	0.50	0.50	0.51	0.49	0.49	0.50
M_{246}	123	122	124	122	124	123	125	122	122	123	0.50	0.50	0.50	0.50	0.50	0.50	0.51	0.50	0.50	0.50
M_{247}	123	123	124	122	124	124	125	123	123	124	0.50	0.50	0.50	0.49	0.50	0.50	0.51	0.50	0.50	0.50
M_{248}	124	124	124	123	125	124	125	123	123	125	0.50	0.50	0.50	0.50	0.50	0.50	0.50	0.50	0.50	0.50
M_{249}	125	124	125	124	125	125	126	123	123	125	0.50	0.50	0.50	0.50	0.50	0.50	0.51	0.49	0.49	0.50
M_{250}	125	125	125	124	126	126	126	124	124	125	0.50	0.50	0.50	0.50	0.50	0.50	0.50	0.50	0.50	0.50
M_{251}	125	126	125	125	126	126	126	125	125	126	0.50	0.50	0.50	0.50	0.50	0.50	0.50	0.50	0.50	0.50
M_{252}	126	126	126	126	126	126	126	126	126	126	0.50	0.50	0.50	0.50	0.50	0.50	0.50	0.50	0.50	0.50

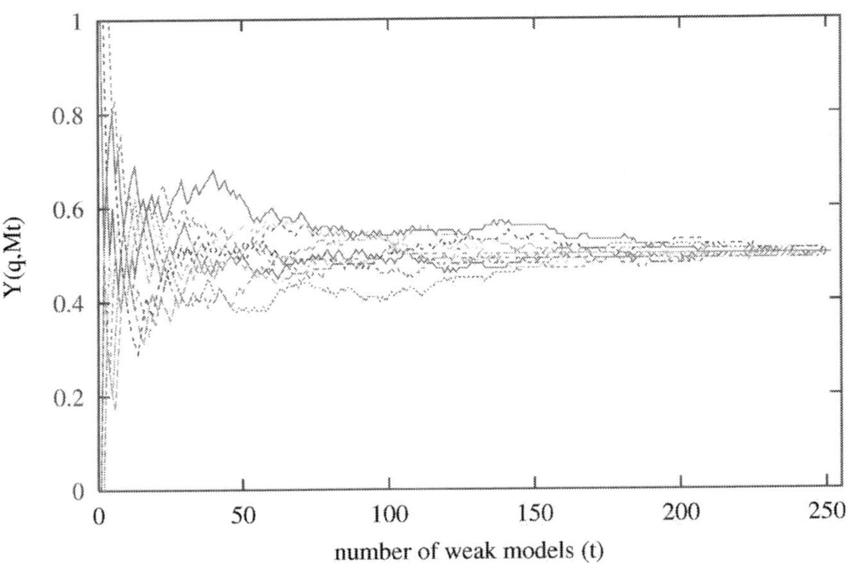

Fig. 1. Plot of $Y(q, M_t)$ versus t. Each line represents the trace of $Y(q, M_t)$ for a particular q as M_t expands.

0.5 for each q. Also notice that because of the randomization, we have expanded M_t in a way that M_t is not biased towards any particular q, therefore the values of $Y(q, M_t)$ are similar after M_t has acquired a certain size (say, when $t = 80$). When M_t=$M_{0.5,A}$, every point q is covered by the same number of models in

M_t, and their values of $Y(q, M_t)$ are identical and is equal to 0.5, which is the ratio of the size of each m relative to A (recall that we always include 5 points from A in each m).

Formally, when $t = 252$, $M_t = M_{0.5,A}$, from the perspective of a fixed q, the probability of it being contained in a model m from M_t is

$$Prob_{\mathcal{M}}(q \in m | m \in M_{0.5,A}) = 0.5.$$

We emphasize that this probability is a measure in the space $\mathcal{M}$ by writing the probability as $Prob_{\mathcal{M}}$. On the other hand, by the way each m is constructed, we know that from the perspective of a fixed m,

$$Prob_F(q \in m | q \in A) \;=\; 0.5.$$

Note that this probability is a measure in the space F. We have shown that these two probabilities, w.r.t. two different spaces, have identical values. In other words, let the membership function of m be $C_m(q)$, i.e., $C_m(q) = 1$ iff $q \in m$, the random variables $\lambda q C_m(q)$ and $\lambda m C_m(q)$ have the same probability distribution, when q is restricted to A and m is restricted to $M_{0.5,A}$. This is because both variables can have values that are either 1 or 0, and they have the value 1 with the same probability (0.5 in this case). This symmetry arises from the fact that the collection of models $M_{0.5,A}$ covers the set A uniformly, i.e., since we have used all members of $M_{0.5,A}$, each point q have the same chance to be included in one of these models. If any two points in a set S have the same chance to be included in a collection of models, we say that this collection is S-uniform. It can be shown, by a simple counting argument, that uniformity leads to the symmetry of $Prob_{\mathcal{M}}(q \in m | m \in M_{0.5,A})$ and $Prob_F(q \in m | q \in A)$, and hence distributions of $\lambda q C_m(q)$ and $\lambda m C_m(q)$.

The observation and utilization of this duality are central to the theory of stochastic discrimination. A critical point of the SD method is to enforce such a uniform cover on a set of points. That is, to construct a collection of models in a balanced way so that the uniformity (hence the duality) is achieved without exhausting all possible models from the space.

3 Two-Class Discrimination

Let us now label each point q in A by one of two classes c_1 (marked by "x") and c_2 (marked by "o") as follows.

x	x	x	o	o	o	o	x	x	o
q_0	q_1	q_2	q_3	q_4	q_5	q_6	q_7	q_8	q_9

This gives a training set TR_i for each class c_i. In particular,

$$TR_1 = \{q_0, q_1, q_2, q_7, q_8\},$$

and

$$TR_2 = \{q_3, q_4, q_5, q_6, q_9\}.$$

How can we build a classifier for c_1 and c_2 using models from $M_{0.5,A}$? First, we evaluate each model m by how well it has captured the members of each class. Define ratings r_i $(i = 1, 2)$ for each m as

$$r_i(m) = Prob_F(q \in m | q \in TR_i).$$

For example, consider model $m_1 = \{q_3, q_5, q_6, q_8, q_9\}$, where q_8 is in TR_1 and the rest are in TR_2. TR_1 has 5 members and 1 is in m_1, therefore $r_1(m_1) = 1/5 = 0.2$. TR_2 has (incidentally, also) 5 members and 4 of them are in m_1, therefore $r_2(m_1) = 4/5 = 0.8$. Thus these ratings represent the quality of the models as a description of each class. A model with a rating 1.0 for a class is a perfect model for that class. We call the difference between r_1 and r_2 the *degree of enrichment* of m with respect to classes $(1, 2)$, i.e., $d_{12} = r_1 - r_2$. A model m is *enriched* if $d_{12} \neq 0$. Now we define, for all enriched models m,

$$X_{12}(q, m) = \frac{C_m(q) - r_2(m)}{r_1(m) - r_2(m)},$$

and let $X_{12}(q, m)$ be 0 if $d_{12}(m) = 0$. For a given m, r_1 and r_2 are fixed, and the value of $X(q, m)$ for each q in A can have one of two values depending on whether q is in m. For example, for m_1, $r_1 = 0.2$ and $r_2 = 0.8$, so $X(q, m) = -1/3$ for points q_3, q_5, q_6, q_8, q_9, and $X(q, m) = 4/3$ for points q_0, q_1, q_2, q_4, q_7. Next, for each set $M_t = \{m_1, m_2, ..., m_t\}$, we define a discriminant

$$Y_{12}(q, M_t) = \frac{1}{t} \sum_{k=1}^{t} X_{12}(q, m_k).$$

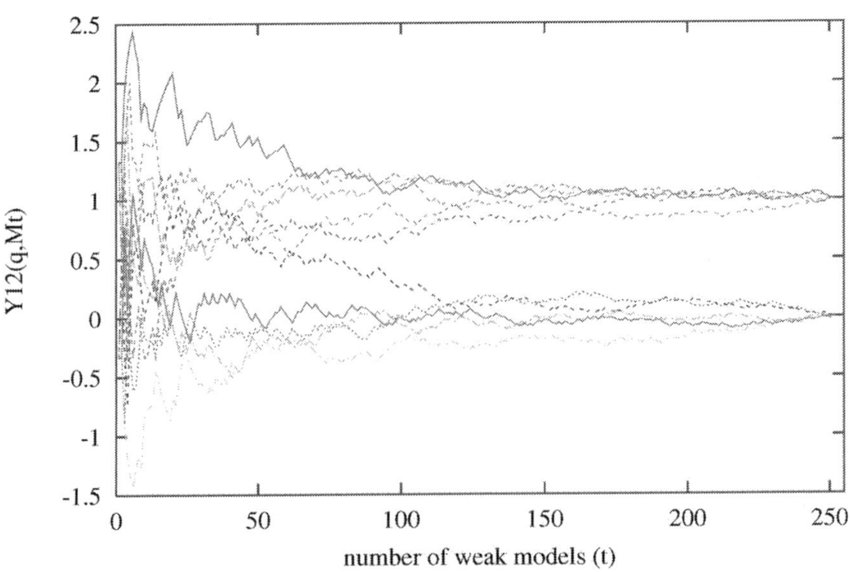

Fig. 2. Plot of $Y_{12}(q, M_t)$ versus t. Each line represents the trace of $Y_{12}(q, M_t)$ for a particular q as M_t expands.

Table 3. Changes of $Y_{12}(q, M_t)$ as M_t expands. For each t, we show the ratings for each new member of M_t, the values X_{12} for this new member, and Y_{12} for the collection M_t up to the inclusion of this new member.

M_t	m_t	r_1	r_2	$r_1 - r_2$	$X_{12}(q, m_t)$ if $q \in m_t$	$X_{12}(q, m_t)$ if $q \notin m_t$	$Y_{12}(q, M_t)$ q_0	q_1	q_2	q_3	q_4	q_5	q_6	q_7	q_8	q_9
M_1	m_1	0.20	0.80	-0.60	-0.33	1.33	1.33	1.33	1.33	-0.33	1.33	-0.33	-0.33	1.33	-0.33	-0.33
M_2	m_2	0.80	0.20	0.60	1.33	-0.33	1.33	1.33	1.33	-0.33	0.50	-0.33	0.50	0.50	0.50	-0.33
M_3	m_3	0.60	0.40	0.20	3.00	-2.00	1.89	0.22	0.22	-0.89	1.33	-0.89	-0.33	1.33	1.33	0.78
M_4	m_4	0.40	0.60	-0.20	-2.00	3.00	2.17	0.92	-0.33	0.08	1.75	-1.17	-0.75	1.75	0.50	0.08
M_5	m_5	0.60	0.40	0.20	3.00	-2.00	2.33	0.33	0.33	-0.33	1.00	-1.33	0.00	2.00	0.00	0.67
M_6	m_6	0.40	0.60	-0.20	-2.00	3.00	2.44	0.78	0.78	-0.61	0.50	-1.44	0.50	1.33	-0.33	1.06
M_7	m_7	0.20	0.80	-0.60	-0.33	1.33	2.28	0.62	0.86	-0.57	0.38	-1.28	0.62	1.33	-0.10	0.86
M_8	m_8	0.80	0.20	0.60	1.33	-0.33	2.17	0.71	0.92	-0.33	0.29	-1.17	0.50	1.12	0.08	0.71
M_9	m_9	0.60	0.40	0.20	3.00	-2.00	1.70	0.96	1.15	0.04	0.59	-1.26	0.22	1.33	-0.15	0.41
M_{10}	m_{10}	0.60	0.40	0.20	3.00	-2.00	1.83	1.17	0.83	-0.17	0.33	-0.83	0.00	1.50	-0.33	0.67
...																
M_{159}	m_{159}	0.60	0.40	0.20	3.00	-2.00	1.02	1.05	0.89	0.18	-0.01	-0.18	0.07	0.95	1.09	-0.06
M_{160}	m_{160}	0.80	0.20	0.60	1.33	-0.33	1.02	1.05	0.89	0.19	-0.01	-0.18	0.06	0.95	1.09	-0.06
M_{161}	m_{161}	0.40	0.60	-0.20	-2.00	3.00	1.03	1.03	0.90	0.21	-0.02	-0.19	0.05	0.93	1.11	-0.04
M_{162}	m_{162}	0.40	0.60	-0.20	-2.00	3.00	1.04	1.04	0.88	0.19	-0.03	-0.20	0.07	0.94	1.09	-0.02
M_{163}	m_{163}	0.40	0.60	-0.20	-2.00	3.00	1.06	1.06	0.89	0.21	-0.02	-0.21	0.06	0.92	1.07	-0.04
M_{164}	m_{164}	0.20	0.80	-0.60	-0.33	1.33	1.06	1.06	0.90	0.21	-0.02	-0.21	0.05	0.92	1.07	-0.03
M_{165}	m_{165}	0.60	0.40	0.20	3.00	-2.00	1.07	1.07	0.91	0.19	0.00	-0.22	0.04	0.90	1.05	-0.01
M_{166}	m_{166}	0.40	0.60	-0.20	-2.00	3.00	1.05	1.08	0.92	0.18	-0.01	-0.20	0.06	0.91	1.03	-0.02
M_{167}	m_{167}	0.60	0.40	0.20	3.00	-2.00	1.06	1.06	0.93	0.20	-0.02	-0.21	0.05	0.89	1.04	0.00
M_{168}	m_{168}	0.80	0.20	0.60	1.33	-0.33	1.06	1.07	0.94	0.20	-0.03	-0.21	0.04	0.90	1.05	-0.01
...																
M_{243}	m_{243}	0.40	0.60	-0.20	-2.00	3.00	1.04	0.99	1.05	0.03	-0.01	-0.02	0.02	0.96	0.96	-0.02
M_{244}	m_{244}	0.60	0.40	0.20	3.00	-2.00	1.05	0.98	1.06	0.03	0.00	-0.03	0.03	0.97	0.95	-0.03
M_{245}	m_{245}	0.60	0.40	0.20	3.00	-2.00	1.06	0.98	1.04	0.02	0.00	-0.02	0.02	0.96	0.96	-0.02
M_{246}	m_{246}	0.80	0.20	0.60	1.33	-0.33	1.06	0.98	1.04	0.02	0.00	-0.02	0.02	0.96	0.96	-0.02
M_{247}	m_{247}	0.60	0.40	0.20	3.00	-2.00	1.05	0.99	1.03	0.01	-0.01	-0.01	0.01	0.97	0.97	-0.01
M_{248}	m_{248}	0.40	0.60	-0.20	-2.00	3.00	1.03	0.98	1.03	0.00	-0.01	0.01	0.03	0.97	0.97	-0.01
M_{249}	m_{249}	0.40	0.60	-0.20	-2.00	3.00	1.02	0.99	1.02	-0.01	0.00	0.00	0.02	0.98	0.98	0.00
M_{250}	m_{250}	0.60	0.40	0.20	3.00	-2.00	1.01	1.00	1.01	-0.02	0.01	0.01	0.01	0.99	0.99	-0.01
M_{251}	m_{251}	0.60	0.40	0.20	3.00	-2.00	1.00	1.00	1.00	0.00	0.00	0.00	0.00	1.00	1.00	0.00
M_{252}	m_{252}	0.80	0.20	0.60	1.33	-0.33	1.00	1.00	1.00	0.00	0.00	0.00	0.00	1.00	1.00	0.00

As the set M_t expands, the value of Y_{12} changes for each q. We show, in Table 3, the values of Y_{12} for each M_t and each q, and for each new member m_t of M_t, r_1, r_2, and the two values of X_{12}. The values of Y_{12} for each q are plotted in Figure 2.

In Figure 2 we see two separate trends. All those points that belong to class c_1 have their Y_{12} values converging to 1.0, and all those in c_2 converging to 0.0. Thus Y_{12} can be used with a threshold to classify an arbitrary point q. We can assign q to class c_1 if $Y_{12}(q, M_t) > 0.5$, and to class c_2 if $Y_{12}(q, M_t) < 0.5$, and remain undecided when $Y_{12}(q, M_t) = 0.5$. Observe that this classifier is fairly accurate far before M_t has expanded to the full set $M_{0.5,A}$. We can also change the two poles of Y_{12} to 1.0 and -1.0 respectively by simply rescaling and shifting X_{12}:

$$X_{12}(q, m) = 2\left(\frac{C_m(q) - r_2(m)}{r_1(m) - r_2(m)}\right) - 1.$$

How did this separation of trends happen? Let us now take a closer look at the models in each M_t and see how many of them cover each point q. For a given M_t, among its members, there can be different values of r_1 and r_2. But because of our choices of the sizes of TR_1, TR_2, and m, we have only a small set of distinct values that r_1 and r_2 can have. Namely, since each model has 5 points, there are only six possibilities as follows.

no. of points from TR_1	0	1	2	3	4	5
no. of points from TR_2	5	4	3	2	1	0
r_1	0.0	0.2	0.4	0.6	0.8	1.0
r_2	1.0	0.8	0.6	0.4	0.2	0.0

Note that in a general setting r_1 and r_2 do not have to sum up to 1. If we included models of a larger size, say, one with 10 points, we can have both r_1 and r_2 equal to 1.0. We have simplified matters by using models of a fixed size and training sets of the same size. According to the values of r_1 and r_2, in this case we have only 6 different kinds of models.

Now we take a detailed look at the coverage of each point q by each kind of models, i.e., models of a particular rating (quality) for each class. Let us count how many of the models of each value of r_1 and r_2 cover each point q, and call this $N_{M_t,r_1,TR_1}(q)$ and $N_{M_t,r_2,TR_2}(q)$ respectively. We can normalize this count by the number of models having each value of r_1 or r_2, and obtain a ratio $f_{M_t,r_1,TR_1}(q)$ and $f_{M_t,r_2,TR_2}(q)$ respectively. Thus, for each point q, we have "a profile of coverage" by models of each value of ratings r_1 and r_2 that is described by these ratios. For example, point q_0 at $t = 10$ is only covered by 5 models $(m_2, m_3, m_5, m_8, m_{10})$ in M_{10}, and from Table 3 we know that M_{10} has various numbers of models in each rating as summarized in the following table.

r_1	0.0	0.2	0.4	0.6	0.8	1.0
no. of models in M_{10} with r_1	0	2	2	4	2	0
$N_{M_{10},r_1,TR_1}(q_0)$	0	0	0	3	2	0
$f_{M_{10},r_1,TR_1}(q_0)$	0	0	0	0.75	1.0	0

r_2	0.0	0.2	0.4	0.6	0.8	1.0
no. of models in M_{10} with r_2	0	2	4	2	2	0
$N_{M_{10},r_2,TR_2}(q_0)$	0	2	3	0	0	0
$f_{M_{10},r_2,TR_2}(q_0)$	0	1.0	0.75	0	0	0

We show such profiles for each point q and each set M_t in Figure 3 (as a function of r_1) and Figure 4 (as a function of r_2) respectively.

Observe that as t increases, the profiles of coverage for each point q converge to two distinct patterns. In Figure 3, the profiles for points in TR_1 converge to a diagonal $f_{M_t,r_1,TR_1} = r_1$, and in Figure 4, those for points in TR_2 also converge to a diagonal $f_{M_t,r_2,TR_2} = r_2$. That is, when $M_t = M_{0.5,A}$, we have for all q in TR_1 and for all r_1, $Prob_{\mathcal{M}}(q \in m | m \in M_{r_1,TR_1}) = r_1$, and for all q in TR_2 and for all r_2, $Prob_{\mathcal{M}}(q \in m | m \in M_{r_2,TR_2}) = r_2$. Thus we have the symmetry in place for both TR_1 and TR_2. This is a consequence of M_t being both TR_1-uniform and TR_2-uniform.

The discriminant $Y_{12}(q, M_t)$ is a summation over all models m in M_t, which can be decomposed into the sums of terms corresponding to different ratings r_i for either $i = 1$ or $i = 2$. To understand what happens with the points in TR_1, we can decompose their Y_{12} by values of r_1. Assume that there are t_x models in M_t that have $r_1 = x$. Since we have only 6 distinct values for x, M_t is a union of 6 disjoint sets, and Y_{12} can be decomposed as

$$\begin{aligned} Y_{12}(q, M_t) = {} & \tfrac{t_{0.0}}{t}\left[\tfrac{1}{t_{0.0}}\textstyle\sum_{k_{0.0}=1}^{t_{0.0}} X_{12}(q, m_{k_{0.0}})\right] + \tfrac{t_{0.2}}{t}\left[\tfrac{1}{t_{0.2}}\textstyle\sum_{k_{0.2}=1}^{t_{0.2}} X_{12}(q, m_{k_{0.2}})\right] + \\ & \tfrac{t_{0.4}}{t}\left[\tfrac{1}{t_{0.4}}\textstyle\sum_{k_{0.4}=1}^{t_{0.4}} X_{12}(q, m_{k_{0.4}})\right] + \tfrac{t_{0.6}}{t}\left[\tfrac{1}{t_{0.6}}\textstyle\sum_{k_{0.6}=1}^{t_{0.6}} X_{12}(q, m_{k_{0.6}})\right] + \\ & \tfrac{t_{0.8}}{t}\left[\tfrac{1}{t_{0.8}}\textstyle\sum_{k_{0.8}=1}^{t_{0.8}} X_{12}(q, m_{k_{0.8}})\right] + \tfrac{t_{1.0}}{t}\left[\tfrac{1}{t_{1.0}}\textstyle\sum_{k_{1.0}=1}^{t_{1.0}} X_{12}(q, m_{k_{1.0}})\right]. \end{aligned}$$

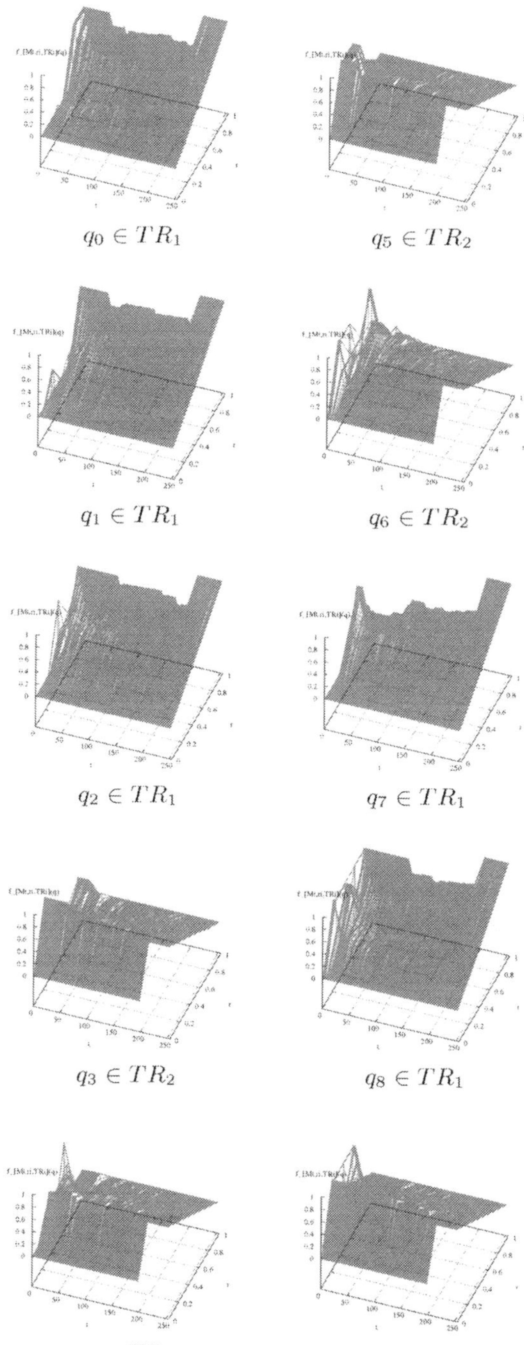

Fig. 3. $f_{M_t,r_1,TR_1}(q)$ for each point q and set M_t. In each plot, the x axis is t that ranges from 0 to 252, the y axis is r that ranges from 0 to 1, and the z axis is f_{M_t,r_1,TR_1}.

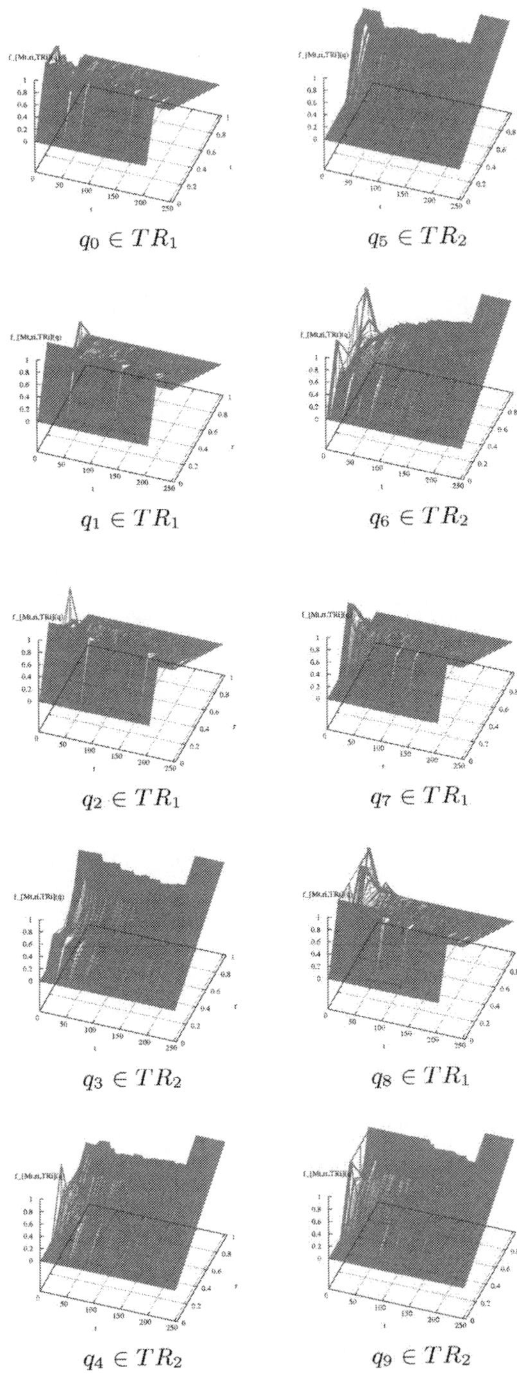

Fig. 4. $f_{M_t,r_2,TR_2}(q)$ for each point q and set M_t. In each plot, the x axis is t that ranges from 0 to 252, the y axis is r that ranges from 0 to 1, and the z axis is f_{M_t,r_2,TR_2}.

The factor in the square bracket of each term is the expectation of values of X_{12} corresponding to that particular rating $r_1 = x$. Since r_1 is the same for all m contributing to that term, by our choice of sizes of TR_1, TR_2, and the models, r_2 is also the same for all those m relevant to that term. Let that value of r_2 be y, we have, for each (fixed) q, each value of x and the associated value y,

$$E(X_{12}(q, m_x)) = E(\frac{C_{m_x}(q) - y}{x - y}) = \frac{E(C_{m_x}(q)) - y}{x - y} = \frac{x - y}{x - y} = 1.$$

The second to the last equality is a consequence of the uniformity of M_t: because the collection M_t (when $t = 252$) covers TR_1 uniformly, we have for each value x, $Prob_{\mathcal{M}}(q \in m | m \in M_{x,TR_1}) = x$, and since $C_{m_x}(q)$ has only two values (0 or 1), and $C_{m_x}(q) = 1$ iff $q \in m$, we have the expected value of $C_{m_x}(q)$ equal to x. Therefore

$$Y_{12}(q, M_t) = \frac{t_{0.0} + t_{0.2} + t_{0.4} + t_{0.6} + t_{0.8} + t_{1.0}}{t} = 1.$$

In a more general case, the values of r_2 are not necessarily equal for all models with the same value for r_1, so we cannot take y and $x - y$ out as constants. But then we can further split the term by the values of r_2, and proceed with the same argument.

A similar decomposition of Y_{12} into terms corresponding to different values of r_2 will show that $Y_{12}(q, M_t) = 0$ for those points in TR_2.

4 Projectability of Models

We have built a classifier and shown that it works for TR_1 and TR_2. How can this classifier work for an arbitrary point that is not in TR_1 or TR_2? Suppose that the feature space F contains other points p (marked by ","), and that each p is close to some training point q (marked by ".") as follows.

., ., ., ., ., ., ., ., ., .,
q_0, p_0 q_1, p_1 q_2, p_2 q_3, p_3 q_4, p_4 q_5, p_5 q_6, p_6 q_7, p_7 q_8, p_8 q_9, p_9

We can take the models m as regions in the space that cover the points q in the same manner as before. Say, if each point q_i has a particular value of the feature v (in our one-dimensional feature space) that is $v(q_i)$. We can define a model by ranges of values for this feature, e.g., in our example m_1 covers q_3, q_5, q_6, q_8, q_9, so we take

$$\begin{aligned} m_1 = &\{q | \tfrac{v(q_2)+v(q_3)}{2} < v(q) < \tfrac{v(q_3)+v(q_4)}{2}\} \cup \\ &\{q | \tfrac{v(q_4)+v(q_5)}{2} < v(q) < \tfrac{v(q_6)+v(q_7)}{2}\} \cup \\ &\{q | \tfrac{v(q_7)+v(q_8)}{2} < v(q)\}. \end{aligned}$$

Thus we can tell if an arbitrary point p with value $v(p)$ for this feature is inside or outside this model.

We can calculate the model's ratings in exactly the same way as before, using only the points q. But now the same classifier works for the new points p, since we can use the new definitions of models to determine if p is inside or outside each model. Given the proximity relationship as above, those points will be assigned to the same class as their closest neighboring q. If these are indeed the true classes for the points p, the classifier is perfect for this new set. In the SD terminology, if we call the two subsets of points p that should be labeled as two different classes TE_1 and TE_2, i.e., $TE_1 = \{p_0, p_1, p_2, p_7, p_8\}$, $TE_2 = \{p_3, p_4, p_5, p_6, p_9\}$, we say that TR_1 and TE_1 are M_t-indiscernible, and similarly TR_2 and TE_2 are also M_t-indiscernible. This is to say, from the perspective of M_t, there is no difference between TR_1 and TE_1, or TR_2 and TE_2, therefore all the properties of M_t that are observed using TR_1 and TR_2 can be projected to TE_1 and TE_2. The central challenge of an SD method is to maintain projectability, uniformity, and enrichment of the collection of models at the same time.

5 Developments of SD Theory and Algorithms

5.1 Algorithmic Implementations

The method of stochastic discrimination constructs a classifier by combining a large number of simple discriminators that are called *weak models*. A weak model is simply a subset of the feature space. In summary, the classifier is constructed by a three-step process: (1) weak model generation, (2) weak model evaluation, and (3) weak model combination. The generator enumerates weak models in an arbitrary order and passes them on to the evaluator. The evaluator has access to the training set. It rates and filters the weak models according to their capability in capturing points of each class, and their contribution to satisfying the uniformity condition. The combiner then produces a discriminant function that depends on a point's membership in each model, and the models' ratings. At classification, a point is assigned to the class for which this discriminant has the highest value. Informally, the method captures the intuition of gaining wisdom from random guesses with feedback.

Weak model generation. Two guidelines should be observed in generating the weak models:

(1) *projectability*: A weak model should be able to capture enough points both inside and outside the training set so that the solution can be projectable to points not included in the training set. Geometrically, this means that a useful model must be of certain minimum size, and it should be able to capture points that are considered *neighbors* of one another. To guarantee similar accuracies of the classifier (based on similar ratings of the weak models) on both training and testing data, one also needs an assumption that the training data are *representative*. Data representativeness and model projectability are two sides of the same question. More discussions of this can be found in [1]. A weak model defines a *neighborhood* in the space, and we need a training sample in a neighborhood of every unseen sample. Otherwise, since our only knowledge of the class

boundaries is derived from the given training set, there is no basis for inference concerning regions of the feature space where no training samples are given.

(2) *simplicity of representation*: A weak model should have a simple representation. That means, the membership of an arbitrary point with respect to a model must be cheaply computable. To illustrate this, consider representing a model as a listing of all the points it contains. This is practically useless since the resultant solution could be as expensive as an exhaustive template matching using all the points in the feature space. An example of a model with a simple representation is a half-plane in a two-dimensional feature space.

Conditions (1) and (2) restrict the type of weak models yet by no means reduce the number of candidates to any tangible limit. To obtain an unbiased collection of the candidates with minimum effort, random sampling with replacement is useful. The training of the method thus relies on a stochastic process which, at each iteration, generates a weak model that satisfies the above conditions.

A convenient way to generate weak models randomly is to use a type of models that can be described by a small number of parameters. Then a stream of models can be created by pseudo-random choices on the values of the parameters. Some example types of models that can be generated this way include (1) half-spaces bounded by a threshold on a randomly selected feature dimension; (2) half-spaces bounded by a hyperplane of equi-distance to two randomly selected points; (3) regions bounded by two parallel hyperplanes perpendicular to a randomly selected axis; (4) hypercubes centered at randomly selected points with edges of varying lengths; and (5) balls (based on the city-block metric, Euclidean distance, or other dissimilarity measures) centered at randomly selected points with randomly selected radii. A model can also be a union or intersection of several regions of these types. An implementation of SD using hyper-rectangular boxes as weak models is described in [9].

A number of heuristics may be used in creating these models. These heuristics specify the way random points are chosen from the space, or set limits on the maximum and minimum sizes of the models. By this we mean restricting the choices of random points to, for instance, points in the space whose coordinates fall inside the range of those of the training samples, or restricting the radii of the balls to, for instance, a fraction of the range of values in a particular feature dimension. The purpose of these heuristics is to speed up the search for acceptable models by confining the search within the most interesting regions, or to guarantee a minimum model size.

Enrichment enforcement. The enrichment condition is relatively easy to enforce, as models biased towards one class are most common. But since the strength of the biases ($|d_{ij}(m)|$) determines the rate at which accuracy increases, we tend to prefer to use models with an enrichment degree further away from zero.

One way to implement this is to use a threshold on the enrichment degree to select weak models from the random stream so that they are of some minimum quality. In this way, one will be able to use a smaller collection of models to yield a classifier of the same level of accuracy. However, there are tradeoffs involved in doing this. For one thing, models of higher rating are less likely to appear in the stream, therefore more random models have to be explored in order to find a sufficient number of higher quality weak models. And once the type of model is fixed and the value of the threshold is set, there is a risk that such models may never be found.

Alternatively, one can use the most enriched model found in a pre-determined number of trials. This also makes the time needed for training more predictable, and it permits a tradeoff between training time and quality of the weak models.

In enriching the model stream, it is important to remember that if the quality of weak models selected is allowed to get too high, there is a risk that they will become training set specific, that is, less likely to be projectable to unseen samples. This could present a problem since the projectability of the final classifier depends on the projectability of its component weak models.

Uniformity promotion. The uniformity condition is much more difficult to satisfy. Strict uniformity requires that every point be covered by the same number of weak models of every combination of per-class ratings. This is rather infeasible for continuous and unconstrained ratings.

One useful strategy is to use only weak models of a particular rating. In such cases, the ratings $r_i(m)$ and $r_j(m)$ are the same for all models m enriched for the discrimination between classes i and j, so we need only to make sure that each point is included in the same number of models. To enforce this, models can be created in groups such that each group partitions the entire space into a set of non-overlapping regions. An example is to use the leaves of a fully-split decision tree, where each leaf is perfectly enriched for one class, and each point is covered by exactly one leaf of each tree. For any pairwise discrimination between classes i and j, we can use only those leaves of the trees that contain only points of class i. In other words, $r_i(m)$ is always 1 and $r_j(m)$ is always 0. Constraints are put in the tree-construction process to guarantee some minimum projectability.

With other types of models, a first step to promote uniformity is to use models that are unions of small regions with simple boundaries. The component regions may be scattered throughout the space. These models have simple representations but can describe complicated class boundaries. They can have some minimum size and hence good projectability. At the same time, the scattered locations of component regions do not tend to cover large areas repeatedly.

A more sophisticated way to promote uniformity involves defining a measure of the lack of uniformity and an algorithm to minimize such a measure. The goal is to create or retain more models located in areas where the coverage is thinner. An example of such a measure is the count of those points that are covered by a less-than-average number of previously retained models. For each point x in the class c_0 to be positively enriched, we calculate, out of all previous models used for that class, how many of them have covered x. If the coverage is less than the average for class c_0, we call x a weak point. When a new model is created, we check how many such weak points are covered by the new model. The ratio of the set of covered weak points to the set of all the weak points is used as a merit score of how well this model improves uniformity. We can accept only those models with a score over a pre-set threshold, or take the model with the best score found in a pre-set number of trials. One can go further to introduce a bias to the model generator so that models covering the weak points are more likely to be created. The later turns out to be a very effective strategy that led to good results in our experiments.

5.2 Alternative Discriminants and Approximate Uniformity

The method outlined above allows for rich possibilities of variations in SD algorithms. The variations may be in the design of the weak model generator, or in ways to enforce the enrichment and uniformity conditions. It is also possible to change the definition of the discriminant, or to use different kinds of ratings.

A variant of the discriminating function is studied in detail in [1]. In this variant, the ratings are defined as

$$r'_i(m) = \frac{|m \cap TR_i|}{|m \cap TR|},$$

for all i. It is an estimate of the posterior probability that a point belongs to class i given the condition that it is included in model m. The discriminant for class i is defined to be:

$$W_i(q) = \frac{\sum_{k=1,...,p_i} C_m(q) r'_i(m)}{\sum_{k=1,...,p_i} C_m(q)}.$$

where p_i is the number of models accumulated for class i.

It turns out that, with this discriminant, the classifier also approaches perfection asymptotically provided that an additional *symmetry* condition is satisfied. The symmetry condition requires that the ensemble includes the same number of models for all permutations of $(r'_1, r'_2, ..., r'_n)$. It prevents biases created by using more (i, j)-enriched models than (j, i)-enriched models for all pairs (i, j)[1]. Again, this condition may be enforced by using only certain particular permutations of the r' ratings, which is the basis of the *random decision forest* method[7][10]. This alternative discriminant is convenient for multi-class discrimination problems.

The SD theory establishes the mathematical concepts of enrichment, uniformity, and projectability of a weak model ensemble. Bounds on classification accuracy are developed based on strict requirements on these conditions, which is a mathematical idealization. In practice, there are often difficult tradeoffs among the three conditions. Thus it is important to understand how much of the classification performance is affected when these conditions are weakened. This is the subject of study in [3], where notions of near uniformity and weak indiscernibility are introduced and their implications are studied.

5.3 Structured Collections of Weak Models

As a constructive procedure, the method of stochastic discrimination depends on a detailed control of the uniformity of model coverage, which is outlined but not fully published in the literature[17]. The method of random subspaces followed these ideas but attempted a different approach. Instead of obtaining weak discrimination and projectability through simplicity of the model form, and forcing uniformity by sophisticated algorithms, the method uses complete, locally pure partitions given by fully split decision trees[7][10] or nearest neighbor classifiers[11] to achieve strong discrimination and uniformity, and then explicitly forces different generalization patterns on the component classifiers. This is done by training large capacity component classifiers such as nearest neighbors and decision trees to fully fit the data, but restricting the training of each classifier to a coordinate subspace of the feature space where all the data points are projected, so that classifications remain invariant in the complement subspace. If there is no ambiguity in the subspaces, the individual classifiers maintain maximum accuracy on the training data, with no cases deliberately chosen to be sacrificed, and thus the method does not run into the paradox of sacrificing some training points in the hope for better generalization accuracy. This is to create a collection of weak models in a structured way.

However the tension among the three factors persists. There is another difficult tradeoff in how much discriminating power to retain for the component classifiers. Can every one use only a single feature dimension so as to maximize invariance in the complement dimensions? Also, projection to coordinate subspaces sets parts of the decision boundaries parallel to the coordinate axes. Augmenting the raw features by simple transformations[10] introduces more flexibility, but it may still be insufficient for an arbitrary problem. Optimization of generalization performance will continue to depend on a detailed control of the projections to suit a particular problem.

6 Conclusions

The theory of stochastic discrimination identifies three and only three sufficient conditions for a classifier to achieve maximum accuracy for a problem. These are just the three elements long believed to be important in pattern recognition: discrimination power, complementary information, and generalization ability. It sets a foundation for theories of ensemble learning. Many current questions on classifier combination can have an answer in the arguments of the SD theory: What is good about building the classifier on weak models instead of strong models? Because weak models are easier to obtain, and their smaller capacity renders them less sensitive to sampling errors in small training sets[20][21], thus they are more likely to have similar coverage on the unseen points from the same problem. Why are many models needed? Because the method relies on the law of large numbers to reduce the variance of the discriminant on each single point. How should these models complement each other? The uniformity condition specifies exactly what kind of correlation is needed among the individual models.

Finally, we emphasize that the accuracy of SD methods is not achieved by intentionally limiting the VC dimension[20] of the complete system; the combination of many weak models can have a very large VC dimension. It is a consequence of the symmetry relating probabilities in the two spaces, and the law of large numbers. It is a structural property of the topological space given by the points and their combinations. The observation of this symmetry and its relationship to ensemble learning is a deep insight of Kleinberg's that we believe can lead to a better understanding of other ensemble methods.

Acknowledgements

The author thanks Eugene Kleinberg for many discussions over the past 15 years on the theory of stochastic discrimination, its comparison to other approaches, and perspectives on the fundamental issues in pattern recognition.

References

1. R. Berlind, *An Alternative Method of Stochastic Discrimination with Applications to Pattern Recognition*, Doctoral Dissertation, Department of Mathematics, State University of New York at Buffalo, 1994.
2. L. Breiman, "Bagging predictors," *Machine Learning*, **24**, 1996, 123-140.
3. D. Chen, *Estimates of Classification Accuracies for Kleinberg's Method of Stochastic Discrimination in Pattern Recognition*, Doctoral Dissertation, Department of Mathematics, State University of New York at Buffalo, 1998.
4. T.G. Dietterich, G. Bakiri, "Solving multiclass learning problems via error-correcting output codes," *Journal of Artificial Intelligence Research*, **2**, 1995, 263-286.
5. Y. Freund, R.E. Schapire, "Experiments with a New Boosting Algorithm," *Proceedings of the Thirteenth International Conference on Machine Learning*, Bari, Italy, July 3-6, 1996, 148-156.
6. L.K. Hansen, P. Salamon, "Neural network ensembles," *IEEE Transactions on Pattern Analysis and Machine Intelligence*, **PAMI-12**, 10, October 1990, 993-1001.
7. T.K. Ho, "Random Decision Forests," *Proceedings of the 3rd International Conference on Document Analysis and Recognition*, Montreal, Canada, August 14-18, 1995, 278-282.
8. T.K. Ho, "Multiple classifier combination: Lessons and next steps," in A. Kandel, H. Bunke, (eds.), *Hybrid Methods in Pattern Recognition*, World Scientific, 2002.
9. T.K. Ho, E.M. Kleinberg, "Building Projectable Classifiers of Arbitrary Complexity," *Proceedings of the 13th International Conference on Pattern Recognition*, Vienna, Austria, August 25-30, 1996, 880-885.
10. T.K. Ho, "The random subspace method for constructing decision forests," *IEEE Transactions on Pattern Analysis and Machine Intelligence*, **20**, 8, August 1998, 832-844.
11. T.K. Ho, "Nearest Neighbors in Random Subspaces," *Proceedings of the Second International Workshop on Statistical Techniques in Pattern Recognition*, Sydney, Australia, August 11-13, 1998, 640-648.
12. T.K. Ho, J. J. Hull, S.N. Srihari, "Decision Combination in Multiple Classifier Systems," *IEEE Transactions on Pattern Analysis and Machine Intelligence*, **PAMI-16**, 1, January 1994, 66-75.

13. Y.S. Huang, C.Y. Suen, "A method of combining multiple experts for the recognition of unconstrained handwritten numerals," *IEEE Transactions on Pattern Analysis and Machine Intelligence*, **PAMI-17**, 1, January 1995, 90-94.
14. J. Kittler, M. Hatef, R.P.W. Duin, J. Matas, "On combining classifiers," *IEEE Transactions on Pattern Analysis and Machine Intelligence*, **PAMI-20**, 3, March 1998, 226-239.
15. E.M. Kleinberg, "Stochastic Discrimination," *Annals of Mathematics and Artificial Intelligence*, **1**, 1990, 207-239.
16. E.M. Kleinberg, "An overtraining-resistant stochastic modeling method for pattern recognition," *Annals of Statistics*, **4**, 6, December 1996, 2319-2349.
17. E.M. Kleinberg, "On the algorithmic implementation of stochastic discrimination," *IEEE Transactions on Pattern Analysis and Machine Intelligence*, **PAMI-22**, 5, May 2000, 473-490.
18. E.M. Kleinberg, "A mathematically rigorous foundation for supervised learning," in J. Kittler, F. Roli, (eds.), *Multiple Classifier Systems*, Lecture Notes in Computer Science 1857, Springer, 2000, 67-76.
19. L. Lam, C.Y. Suen, "Application of majority voting to pattern recognition," *IEEE Transactions on Systems, Man, and Cybernetics*, **SMC-27**, 5, September/October 1997, 553-568.
20. V. Vapnik, *Estimation of Dependences Based on Empirical Data*, Springer-Verlag, 1982.
21. V. Vapnik, *Statistical Learning Theory*, John Wiley & Sons, 1998.
22. D.H. Wolpert, "Stacked generalization," *Neural Networks*, **5**, 1992, 241-259.

Structural Inference of Sensor-Based Measurements

Robert P.W. Duin[1] and Elżbieta Pękalska[1,2]

[1] ICT group, Faculty of Electr. Eng., Mathematics and Computer Science
Delft University of Technology, The Netherlands
r.duin@ieee.org, e.m.pekalska@tudelft.nl
[2] School of Computer Science, University of Manchester, United Kingdom

Abstract. Statistical inference of sensor-based measurements is intensively studied in pattern recognition. It is usually based on feature representations of the objects to be recognized. Such representations, however, neglect the object structure. Structural pattern recognition, on the contrary, focusses on encoding the object structure. As general procedures are still weakly developed, such object descriptions are often application dependent. This hampers the usage of a general learning approach.

This paper aims to summarize the problems and possibilities of general structural inference approaches for the family of sensor-based measurements: images, spectra and time signals, assuming a continuity between measurement samples. In particular it will be discussed when probabilistic assumptions are needed, leading to a statistically-based inference of the structure, and when a pure, non-probabilistic structural inference scheme may be possible.

1 Introduction

Our ability to recognize patterns is based on the capacity to generalize. We are able to judge new, yet unseen observations given our experience with the previous ones that are similar in one way or another. Automatic pattern recognition studies the ways which make this ability explicit. We thereby learn more about it, which is of pure scientific interest, and we construct systems that may partially take over our pattern recognition tasks in real life: reading documents, judging microscope images for medical diagnosis, identifying people or inspecting industrial production.

In this paper we will reconsider the basic principles of generalization, especially in relation with sensor measurements like images (e.g. taken from some video or CCD camera), time signals (e.g. sound registered by a microphone), and spectra and histograms (e.g. the infra-red spectrum of a point on earth measured from a satellite). These classes of measurements are of particular interest since they can very often replace the real object in case of human recognition: we can read a document, identify a person, recognize an object presented on a monitor screen as well as by a direct observation. So we deal here with registered signals which contain sufficient information to enable human recognition in an almost natural way. This is an entirely different approach to study the weather patterns

D.-Y. Yeung et al. (Eds.): SSPR&SPR 2006, LNCS 4109, pp. 41–55, 2006.

from a set of temperature and air pressure measurements than taken by a farmer who observes the clouds and the birds.

The interesting, common aspect of the above defined set of sensor measurements is that they have an observable structure, emerging from a relation between neighboring pixels or samples. In fact we do not perceive the pixel intensity values themselves, but we directly see a more global, meaningful structure. This structure, the unsampled continuous observation in space and/or time constitutes the basis of our recognition. Generalization is based on a direct observation of the similarity between the new and the previously observed structures.

There is an essential difference between human and automatic pattern recognition, which will be neglected here, as almost everywhere else. If a human observes a structure, he may directly relate this to a meaning (function or a concept). By assigning a word to it, the perceived structure is named, hence recognized. The word may be different in different languages. The meaning may be the same, but is richer than just the name as it makes a relation to the context (or other frame of reference) or the usage of the observed object. On the contrary, in automatic recognition it is often attempted to map the observations directly to class labels without recognizing the function or usage.

If we want to simulate or imitate the human ability of pattern recognition it should be based on object structures and the generalization based on similarities. This is entirely different from the most successful, mainline research in pattern recognition, which heavily relies on a feature-based description of objects instead of their structural representations. Moreover, generalization is also heavily based on statistics instead of similarities.

We will elaborate on this paradoxical situation and discuss fundamentally the possibilities of the structural approach to pattern recognition. This discussion is certainly not the first on this topic. In general, the science of pattern recognition has already been discussed for a long time, e.g. in a philosophical context by Sayre [1] or by Watanabe on several occasions, most extensively in his book on human and mechanical recognition [2]. The possibilities of a more structural approach to pattern recognition was one of the main concerns of Fu [3], but it was also clear that, thereby, the powerful tools of statistical approaches [4,5,6,7] should not be forgotten; see [8,9,10].

Learning from structural observations is the key question of the challenging and seminal research programme of Goldfarb [10,11,12]. He starts, however, from a given structural measurement, the result of a 'structural sensor' [13] and uses this to construct a very general, hierarchial and abstract structural description of objects and object classes in terms of primitives, the Evolving Transformation System (ETS) [11]. Goldfarb emphasizes that a good structural representation should be able to generate proper structures. We recognize that as a desirable, but very ambitious direction. Learning structures from examples in the ETS framework appears still to be very difficult, in spite of various attempts [14].

We think that starting from such a structural representation denies the quantitative character of the lowest level of senses and sensors. Thereby, we will again face the question how to come to structure, how to learn it from examples given the numeric outcomes of a physical measurement process, that by its organization in time and space respects this structure. This question will not be solved here, as it is one of the most basic issues in science. However, we hope that a contribution is made towards the solution by our a summary of problems and possibilities in this area, presented from a specific point of view.

Our viewpoint, which will be explained in the next sections, is that the feature vector representation directly reduces the object representation. This causes a class overlap that can only be solved by a statistical approach. An indirectly reducing approach based on similarities between objects and proximities of their representations, may avoid, or at least postpone such a reduction. As a consequence, classes do not overlap intrinsically, by which a statistical class description can be avoided. A topological- or domain-based description of classes may become possible, in which the structural aspects of objects and object classes might be preserved. This discussion partially summarizes our previous work on the dissimilarity approach [15], proximities [16], open issues [17] and the science of pattern recognition [18].

2 Generalization Principles

The goal of pattern recognition may be phrased as the derivation of a general truth (e.g. the existence of a specified pattern) from a limited, not exhaustive set of examples. We may say that we thereby generalize from this set of examples, as the establishment of a general truth gives the possibility to derive non-observed properties of objects, similar to those of observed examples.

Another way to phrase the meaning of generalization is to state that the truth is *inferred* from the observations. Several types of inference can be distinguished:

Logical inference. This is the original meaning of inference: a truth is derived from some facts, by logical reasoning, e.g.
1. Socrates is a man.
2. All man are mortal.
3. Consequently, Socrates is mortal.

It is essential that the conclusion was derived before the death of Socrates. It was already known without having observed it.

Grammatical inference. This refers to the grammar of an artificial language of symbols, which describes the "sentences" that are permitted from a set of observed sequences of such symbols. Such grammars may be inferred from a set of examples.

Statistical inference. Like above, there are observations and a general, accepted or assumed, rule of a statistical (probabilistic) nature. When such a rule is applied to the observations, more becomes known than just the directly collected facts.

Structural inference. This is frequently used in the sense that structure is derived from observations and some general law. E.g. in some economical publications, "structural inference" deals with finding the structure of a statistical model (such as the set of dependencies) by statistical means [19]. On the contrary, "structural inference" can also be understood as using structural (instead of statistical) properties to infer unobserved object properties.

Empirical inference. This term is frequently used by Vapnik, e.g. in his recent revised edition of the book on structural risk minimization [20]. It means that unnecessary statistical models are avoided if some value, parameter, or class membership has to be inferred from observational data. It is, however, still based on a statistical approach, in the sense that probabilities and expectations play a role. The specific approach of empirical inference avoids the estimation of statistical functions and models where possible: do not estimate an entire probability density function if just a decision is needed.

It should be noted that in logical, statistical and empirical inferences object properties are inferred by logical, statistical and empirical means, respectively. In the terms of "grammatical inference" and "structural inference", the adjective does not refer to the means but to the goal: finding a grammar or a structure. The means are in these cases usually either logical or statistical. Consequently, the basic tools for inference are primarily logic and statistics. They correspond to knowledge and observations. As logic cannot directly be applied to sensor data, statistical inference is the main way for generalization in this case.

We will discuss whether in addition to logic and statistics, also structure can be used as a basic means for inference. This would imply that given the structure of a set of objects and, for instance, the corresponding class labels, the class label of an unlabeled object can be inferred. As we want to learn from sensor data, this structure should not be defined by an expert, but should directly be given from the measurements, e.g. the chain code of an observed contour.

Consider the following example. A professor in archeology wants to teach a group of students the differences in the styles of A and B of some classical vases. He presents 20 examples for each style and asks the students to determine a rule. The first student observes that the vases in group A have either ears or are red, while those of group B may also have ears, but only if they are blue (a color that never occurs for A). Moreover, there is a single red vase in group B without ears, but with a sharp spout. In group A only some vases with ears have a spout. The rule he presents is: `if (ears` $\wedge$ `not_blue)` $\vee$ `(red` $\wedge$ `no_ears` $\wedge$ `no_spout) then A else B`. The second student measures the sizes (weight and height) of all vases, plots them on a 2D scatter plot and finds a straight line that separates the vases with just two errors. The third student manually inspects the vases from all sides and concludes that the lower part is ball-shaped in group A and egg-shaped in group B. His rule is thereby: `if ball-shaped then A, if egg-shaped then B`.

The professor asked the first student why he did not use characteristic paintings on the vases for their discrimination. The student answered that they were not needed as the groups could have perfectly been identified by the given properties. They may, however, be needed if more vases appear. So, this rule works for the given set of examples, but does it generalize?

The second solution did not seem attractive to the professor as some measurement equipment is needed and, moreover, two errors are made! The student responded that these two errors showed in fact that his statistical approach was likely better than the logical approach of the first student, as it was more general (less overtrained). This remark was not appreciated by the professor: very strange to prove the quality of a solution by the fact that errors are made!

The third student seemed to have a suitable solution. Moreover, the shape property was in line with other characteristics of the related cultures. Although it was clear what was meant by the ball-ness and the egg-ness of the vase shapes, the question remained whether this could be decided by an arbitrary assistant. The student had a perfect answer. He drew the shapes of two vases, one from each group, on a glass window in front of the table with vases. To classify a given vase, he asked the professor to look through each of the two images to this vase and to walk to and from the window to adjust the size until a match occurs.

We hope that this example makes clear that logic, statistics and structure can be used to infer a property like a class label. Much more has to be explained about how to derive the above decision rules by automatic means. In this paper, we will skip the logical approach as it has little to do with the sensory data we are interested in.

3 Feature Representation

We will first shortly summarize the feature representation and some of its advantages and drawbacks. In particular, it will be argued how this representation necessarily demands a statistical approach. Hence, this has far reaching consequences concerning how learning data should be collected. Features are object properties that are suitable for their recognition. They are either directly measured or derived from the raw sensor data. The feature representation represents objects as vectors in a (Euclidean) feature space. Usually, but not always, the feature representation is based on a significant reduction. Real world objects cannot usually be reconstructed from their features. Some examples are:

- Pieces of fruit represented by their color, maximum length and weight.
- Handwritten digits represented by a small set of moments.
- Handwritten digits represented by the pixels (in fact, their intensities) in images showing the digits.

This last example is special. Using pixel values as features leads to pixel representations of the original digits that are reductions: minor digit details may not be captured by the given pixel resolution. If we treat, however, the digital picture of a digit as an object, the pixel representation is complete: it represents the object in its entirety. This is not strange as in handling mail and money transfers, data typists often have to recognize text presented on monitor screens. So the human recognition is based on the same data as used for the feature (pixels) representation.

Note that different objects may have identical representations, if they are mapped on the same vector in the feature space. This is possible if the feature representation reduces the information on objects, which is the main cause for class overlap, in which objects belonging to different classes are identically represented.

The most common and most natural way to solve the problem of class overlap is by using probability density functions. Objects in the overlap area are assigned to the class that is the most probable (or likely) for the observed feature vector. This not only leads to the fully Bayesian approaches, based on the estimation of class densities and using or estimating class prior probabilities, but also to procedures like decision trees, neural networks and support vector machines that use geometrical means to determine a decision boundary between classes such that some error criterion is minimized.

In order to find a probability density function in the feature space, or in order to estimate the expected classification performance for any decision function that is considered in the process of classifier design, a set of objects has to be available that is representative for the distribution of the future objects to be classified later by the final classifier. This last demand is very heavy. It requires that the designer of a pattern recognition system knows exactly the circumstances under which it will be applied. Moreover, he has to have the possibility to sample the objects to be classified. There are, however, many applications in which it is difficult or impossible. Even in the simple problem of handwritten digit recognition it may happen that writing habits change over time or are location dependent. In an application like the classification of geological data for mining purposes, one likes to learn from existing mining sites how to detect new ones. Class distributions, however, change heavily over the earth.

Another problem related to class overlap is that densities are difficult to estimate for more complete and, thereby, in some sense better representations, as they tend to use more features. Consequently, they have to be determined in high-dimensional vector spaces. Also the geometrical procedures suffer from this, as the geometrical variability in such spaces is larger. This results in the paradox of the feature representation: more complete feature representations need larger training sets or will deteriorate in performance [21].

There is a fundamental question of how to handle the statistical problem of overlapping classes in case no prior information is available about the possible class distributions. If there is no preference, the No-Free-Lunch-Theorem [22] states that all classifiers perform similarly to a random class assignment if we look over a set of problems on average. It is necessary to restrict the set of problems significantly, e.g. to compact problems in which similar objects have similar representations. It is, however, still an open issue how to do this [23]. As long as the set of pattern recognition problems is based on an unrealistic set, studies on the expected performance of pattern recognition systems will yield unrealistic results. An example is the Vapnik-Chervonenkis error bound based on the structural risk minimization [20]. Although a beautiful theoretical result is obtained, the prescribed training set sizes for obtaining a desired (test) performance are far from being realistic. The support vector machine (SVM), which is based on structural risk minimization, is a powerful classifier for relatively small training sets and classes that have a small overlap. As a general solution for overlapping classes, as they arise in the feature space, it is not suitable. We will point this out below.

We will now introduce the idea of domain-based classifiers [24]. They construct decision boundaries between classes that are described just by the domains they cover in the feature space (or in any representation space) and do not depend on (the estimates of) probability distributions. They are, thereby, insensitive to ill-sampled training sets, which may even be selectively sampled by an expert. Such classifiers may be beneficial for non-overlapping, or slightly overlapping classes and are optimized for distances instead of densities. Consequently, they are sensitive to outliers. Therefore, outliers should be removed in the firs step. This is possible as the training set can be sampled in a selective way. Domain-based classification may be characterized as taking care of the structure of the classes in the feature space instead of their probability density functions.

If Vapnik's concept of structural risk minimization [20] is used for optimizing a separation function between two sets of vectors in a vector space, the resulting classifier is the maximum margin classifier. In case no linear classifier exists to make a perfect separation, a kernel approach may be used to construct a non-linear separation function. Thanks to the reproducing property of kernels, the SVM becomes then a maximum margin hyperplane in a Hilbert space induced by the specified kernel [25]. The margin is only determined by support vectors. These are the boundary objects, i.e. the objects closest to the decision boundary $f(\mathbf{x};\boldsymbol{\theta})$ [26,25]. As such, the SVM is independent of class density models. Multiple copies of the same object added to the training set do not contribute to the construction of the SVM as they do for classifiers based on some probabilistic model. Moreover, the SVM is also not affected by adding or removing objects of the same class that lie further away from the decision boundary. This decision function is, thereby, a truly domain-based classifier, as it optimizes the separation of class domains and class density functions.

For nonlinear classifiers defined on nonlinear kernels, the SVM has, however, a similar drawback as the nonlinear neural network. The distances to the decision boundary are computed in the output Hilbert space defined by the kernel and not in the input space. A second problem is that the soft-margin formulation [26], the traditional solution to overlapping classes, is not domain-based. Consider a two-class problem with the labels $y \in \{-1, +1\}$, where $y(\mathbf{x})$ denotes the true label of $\mathbf{x}$. Assume a training set $X = \{\mathbf{x}_i, y(\mathbf{x}_i)\}_{i=1}^n$. The optimization problem for a linear classifier $f(\mathbf{x}) = \mathbf{w}^\mathsf{T}\mathbf{x} + w_0$ is rewritten into:

$$\begin{array}{ll} \min_{\mathbf{w}} & ||\mathbf{w}||^2 + C\sum_{\mathbf{x}_i \in X} \xi(\mathbf{x}_i), \\ s.t. & y(\mathbf{x}_i) f(\mathbf{x}_i) \geq 1 - \xi(\mathbf{x}_i), \\ & \xi(\mathbf{x}_i) \geq 0, \end{array} \quad (1)$$

where $\xi(\mathbf{x}_i)$ are slack variables accounting for possible errors and C is a trade-off parameter. $\sum_{\mathbf{x}_i \in X} \xi(\mathbf{x}_i)$ is an upper bound of the misclassification error on the training set, hence it is responsible for minimizing *a sum of error contributions.* Adding a copy of an erroneously assigned object will affect this sum and, thereby, will influence the sought optimum $\mathbf{w}$. The result is, thereby, based on a mixture of approaches. It is dependent on the distribution of objects (hence statistics) as well as on their domains (hence geometry).

A proper domain-based solution should minimize the class overlap in terms of distances and not in terms of probability densities. Hence, a suitable version of the SVM should be derived for the case of overlapping domains, resulting in the *negative margin SVM* [24]. This means that the distance of the furthest away misclassified object should be minimized. As the signed distance is negative, the negative margin is obtained. In the probabilistic approach, this classifier is unpopular as it will be sensitive to outliers. As explained above, outliers are neglected in domain-based classification, as they have to be removed beforehand.

Our conclusion is that the use of features yields a reduced representation. This leads to class overlap for which a probabilistic approach is needed. It relies on a heavy assumption that data are drawn independently from a fixed (but unknown) probability distribution. As a result, one demands training sets that are representative for the probability density functions. An approach based on distances and class structures may be formulated, but conflicts with the use of densities if classes overlap.

4 Proximity Representation

Similarity or dissimilarity measures can be used to represent objects by their proximities to other examples instead of representing them by a preselected set of features. If such measurements are derived from original objects, or from raw sensor data describing the objects fully (e.g. images, time signals and spectra that are as good as the real objects for the human observer), then the reduction in representation, which causes class overlap in the case of features, is circumvented. For example, we may demand that the dissimilarity of an object to itself is zero and that it can only be zero if it is related to an identical object. If it can be assumed that identical objects belong to the same class, classes do not overlap. (This is not always the case, e.g. a handwritten '7' may be identical to a handwritten '1').

In principle, such proximity representations may avoid class overlap. Hence, they may offer a possibility to use the structure of the classes in the representation, i.e. their domains, for building classifiers. This needs a special, not yet well studied variant of the proximity representation. Before a further explanation, we will first summarize two variants that have been worked out well. This summary is an adapted version of what has been published as [16]. See also [15].

Assume we are given a representation set R, i.e. a set of real-world objects that can be used for building the representation. $R = \{p_1, p_2, \ldots, p_n\}$ is, thereby, a set of prototype examples. We also consider a proximity measure d, which should incorporate the necessary invariance (such as scale or rotation invariance) for the given problem. Without loss of generality, let d denote dissimilarity. An object x is then represented as a vector of dissimilarities computed between x and the prototypes from R, i.e. $d(x, R) = [d(x, p_1), d(x, p_2), \ldots, d(x, p_n)]^{\mathsf{T}}$. If we are also given an additional labeled training set $T = \{t_1, t_2, \ldots, t_N\}$ of N real-world objects, our proximity representation becomes an $N \times n$ dissimilarity matrix $D(T, R)$, where $D(t_i, R)$ is now a row vector. Usually R is selected out of T (by various prototype selection procedures) in a way to guarantee a good tradeoff between the recognition accuracy and the computational complexity. R and T may also be different sets.

The k-NN rule can directly be applied to such proximity data. Although it has good asymptotic properties for metric distances, its performance deteriorates for small training (here: representation) sets. Alternative learning strategies represent proximity information in suitable representation vector spaces, in which traditional statistical algorithms can be defined. So, they become more beneficial. Such vector spaces are usually determined by some local or global embedding procedures. Two approaches to be discussed here rely on a linear isometric embedding in a pseudo-Euclidean space (where necessarily $R \subseteq T$) and the use of proximity spaces; see [16,15].

Pseudo-Euclidean linear embedding. Given a symmetric dissimilarity matrix $D(R, R)$, a vectorial representation X can be found such that the distances are preserved. It is usually not possible to determine such an isometric embedding into a Euclidean space, but it is possible into a pseudo-Euclidean space $\mathcal{E} = \mathbb{R}^{(p,q)}$. It is a $(p+q)$-dimensional non-degenerate indefinite inner product space such that the inner product $\langle\cdot,\cdot\rangle_{\mathcal{E}}$ is positive definite on $\mathbb{R}^p$ and negative definite on $\mathbb{R}^q$ [10]. Then, $\langle\mathbf{x},\mathbf{y}\rangle_{\mathcal{E}} = \mathbf{x}^{\mathsf{T}}\mathcal{J}_{pq}\mathbf{y}$, where $\mathcal{J}_{pq} = \mathrm{diag}\,(I_{p\times p}; -I_{q\times q})$ and I is the identity matrix. Consequently, $d^2_{\mathcal{E}}(\mathbf{x},\mathbf{y}) = \langle\mathbf{x}-\mathbf{y},\mathbf{x}-\mathbf{y}\rangle_{\mathcal{E}} = d^2_{\mathbb{R}^p}(\mathbf{x},\mathbf{y}) - d^2_{\mathbb{R}^q}(\mathbf{x},\mathbf{y})$, hence $d^2_{\mathcal{E}}$ is a difference of square Euclidean distances found in the two subspaces, $\mathbb{R}^p$ and $\mathbb{R}^q$. Since $\mathcal{E}$ is a linear space, many properties related to inner products can be extended from the Euclidean case [10,15].

The (indefinite) Gram matrix G of X can be expressed by the square distances $D^{\star 2} = (d^2_{ij})$ as $G = -\frac{1}{2}JD^{\star 2}J$, where $J = I - \frac{1}{n}\mathbf{1}\mathbf{1}^{\mathsf{T}}$ [10,27,15]. Hence, X can be determined by the eigendecomposion of G, such that $G = Q\Lambda Q^{\mathsf{T}} = Q|\Lambda|^{1/2}\mathrm{diag}(\mathcal{J}_{p'q'};0)\,|\Lambda|^{1/2}Q^{\mathsf{T}}$. $|\Lambda|$ is a diagonal matrix of first decreasing p' positive eigenvalues, then decreasing magnitudes of q' negative eigenvalues, followed by zeros. Q is a matrix of the corresponding eigenvectors. X is uncorrelated and represented in $\mathbb{R}^k$, $k = p'+q'$, as $X = Q_k|\Lambda_k|^{1/2}$ [10,27]. Since only some eigenvalues are significant (in magnitude), the remaining ones can be disregarded as non-informative. The reduced representation $X_r = Q_m\,|\Lambda_m|^{1/2}$, $m = p+q < k$, is determined by the largest p positive and the smallest q negative eigenvalues. New objects $D(T_{test}, R)$ are orthogonally projected onto $\mathbb{R}^m$; see [10,27,15]. Classifiers based on inner products can appropriately be defined in $\mathcal{E}$. A linear classifier $f(\mathbf{x}) = \mathbf{v}^{\mathsf{T}}\mathcal{J}_{pq}\mathbf{x} + v_0$ is e.g. constructed by addressing it as $f(\mathbf{x}) = \mathbf{w}^{\mathsf{T}}\mathbf{x} + v_0$, where $\mathbf{w} = \mathcal{J}_{pq}\mathbf{v}$ in the associated Euclidean space $\mathbb{R}^{(p+q)}$ [10,27,15].

Proximity spaces. Here, the dissimilarity matrix $D(X, R)$ is interpreted as a data-dependent mapping $D(\cdot, R)\colon X \to \mathbb{R}^n$ from some initial representation X to a vector space defined by the set R. This is the *dissimilarity space* (or a similarity space, if similarities are used), in which each dimension $D(\cdot, p_i)$ corresponds to a dissimilarity to a prototype $p_i \in R$. The property that dissimilarities should be small for similar objects (belonging to the same class) and large for distinct objects, gives them a discriminative power. Hence, the vectors $D(\cdot, p_i)$

can be interpreted as 'features' and traditional statistical classifiers can be defined [28,15]. Although the classifiers are trained on $D(\cdot, R)$, the weights are still optimized on the complete set T. Thereby, they can outperform the k-NN rule as they become more global in their decisions.

Normal density-based classifiers perform well in dissimilarity spaces [27,28,15]. This especially holds for summation-based dissimilarity measures, summing over a number of components with similar variances. Such dissimilarities are approximately normally distributed thanks to the central limit theorem (or they approximate the χ^2 distribution if some variances are dominant) [15]. For instance, for a two-class problem, a quadratic normal density based classifier is given by $f(D(x,R)) = \sum_{i=1}^{2} \frac{(-1)^i}{2} (D(x,R) - \mathbf{m}_i)^\mathsf{T} S_i^{-1} (D(x,R) - \mathbf{m}_i) + \log \frac{p_1}{p_2} + \frac{1}{2} \log \frac{|S_1|}{|S_2|}$, where $\mathbf{m}_i$ are the mean vectors and S_i are the class covariance matrices, all estimated in the dissimilarity space $D(\cdot, R)$. p_i are the class prior probabilities. By replacing S_1 and S_2 by the average covariance matrix, a linear classifier is obtained.

The two learning frameworks of pseudo-Euclidean embedding and dissimilarity spaces appear to be successful in many problems with various kinds of dissimilarity measures. They can be more accurate and more efficient than the nearest neighbor rule, traditionally applied to dissimilarity data. Thereby, they provide beneficial approaches to learning from structural object descriptions for which it is more easy to define dissimilarity measures between objects than to find a good set of features. As long as these approaches are based on a fixed representation set, however, class overlap may still arise as two different objects may have the same set of distances to the representation set. Moreover, most classifiers used in the representation spaces are determined based on the traditional principle of minimizing the overlap. They do not make a specific use of principles related to object distances or class domains. So, what is still lacking are procedures that use class distances to construct a structural description of classes. The domain-based classifiers, introduced in Section 3, may offer that in future provided that the representation set is so large that the class overlap is (almost) avoided. A more fundamental approach is described below.

Topological spaces. The topological foundation of proximity representations is discussed in [15]. It is argued that if the dissimilarity measure itself is unknown, but the dissimilarity values are given, the topology cannot, as usual, be based on the traditional idempotent closures. An attempt has been made to use neighborhoods instead. This has not resulted yet in a useful generalization over finite training sets.

Topological approaches will aim to describe the class structures from local neighborhood relations between objects. The inherent difficulty is that many of the dissimilarity measures used in structural pattern recognition, like the normalized edit distance, are non-Euclidean, and even sometimes non-metric. It has been shown in a number of studies that straightforward Euclidean corrections are counter productive in some applications. This suggests that the non-Euclidean aspects may be informative. Consequently, a non-Euclidean topology would be needed. This area is still underdeveloped.

A better approach may rely on two additional sources of information that are additionally available. These are the definition of the dissimilarity measure and the assumption of class compactness. They may together tell us what is really local or how to handle the non-Euclidean phenomena of the data. This should result in a topological specification of the class structure as learned from the training set.

5 Structural Representation

In the previous section we arrived at *a structure of a class* (or a concept), i.e. the structural or topological relation of the set of all objects belonging to a particular class. This structure is influenced by the chosen representation, but is in fact determined by the class of objects. It reflects, for instance, the set of continuous transformations of the handwritten digits '7' that generate exclusively all other forms that can be considered as variants of a handwritten '7'. This basically reflects the concept used by experts to assign the class label. Note, however, that this rather abstract structure of the concept should be clearly distinguished from the structure of individual objects that are the manifestations of that concept.

The *structure of objects*, as presented somewhere in sensory data of images, such as time signals and spectra, is directly related to shape. The shape is a one- or multi-dimensional set of connected boundary points that may be locally characterized by curvature and described more globally by morphology and topology. Note that the object structure is related to an outside border of objects, the place where the object ends. If the object is a black blob in a two-dimensional image (e.g. a handwritten digit) then the structure is expressed by the contour, a one-dimensional closed line. If the grey-value pixel intensities inside the blob are relevant, then we deal with a three-dimensional blob on a two-dimensional surface. (As caves cannot exist in this structure it is sometimes referred to as a 2.5-dimensional object).

It is important to realize that the sensor measurements are characterized by a sampling structure (units), such as pixels or time samples. This sampling structure, however, has nothing to do with the object structure. In fact, it disturbs it. In principle, objects (patterns describing real objects) can lie anywhere in an image or in a time frame. They can also be rotated in an image and appear in various scales. Additionally, we may also vary the sampling frequency. If we analyze the object structure for a given sampling, then the object is "nailed" to some grid. Similar objects may be nailed in an entirely different way to this grid. How to construct structural descriptions of objects that are independent of the sampling grid on which the objects are originally presented is an important topic of structural pattern recognition.

The problem of structural inference, however, is not the issue of representation itself. It is the question how we can establish the membership of an object to a given set of examples based on their structure. Why is it more likely that a new object X belongs to a set A than a set B? A few possible answers are presented below.

1. X is an example of A, because the object in A ∪ B that is most similar to X belongs to A. This decision may depend on the accidental availability of particular objects. Moreover, similarity should appropriately be defined.
2. X is an example of A, because the object from A ∪ B that is most easily transformed to X belongs to A. In this case similarity relies on the effort of transformation. This may be more appropriate if structures need to be compared. The decision, however, still depends on a single object. The entire sets or classes simply store examples that may be used when other objects have to be classified.
3. X is an example of A, because it can more easily be generated by transforming the prototype of set A than by transforming the prototype of set B. The *prototype* of a set may be defined as the (hypothetical) object that can most easily be transformed into any of the objects of the set. In this assignment rule (as well as in the rule above) the definition of transformation is universal, i.e. independent of the considered class.
4. X is an example of A, because it can more easily be transformed from a (hypothetical) prototype object by the transformations T_A that are used to generate the set A than by the transformations T_B that are used to generate the set B. Note that we now allow that the sets are generated from possibly the same prototype, but by using different transformations. These are derived (learnt) from the sets of examples. The transformations T_A and T_B may be learnt from a training set.

There is a strong resemblance with the statistical class descriptions: classes may differ by their means as well as by the shape of their distributions. A very important difference, however, between structural and statistical inference is that for an additional example that is identical to a previous one changes the class distribution, but not the (minimal) set of necessary transformations.

This set of assignment rules can easily be modified or enlarged. We like to emphasize, however, that the natural way of comparing objects, i.e. by accounting for their similarity, may be defined as the effort of transforming one structure into another. Moreover, the set of possible transformations may differ from class to class. In addition, classes may have the same or different prototypes. E.g. a sphere can be considered as a basic prototype both for apples as well as for pears. In general, classes may differ by their prototypes and/or by their set of transformations.

What has been called easiness in transformation can be captured by a measurable cost, which is an example of a similarity measure. It is, thereby, related to the proximity approaches, described above. Proximity representations are naturally suitable for structural inference. What is different, however, is the use of statistical classifiers in embedded and proximity spaces. In particular, the embedding approach has to be redefined for structural inference as it makes use of averages and the minimization of an expected error, both statistical concepts. Also the use of statistical classifiers in these spaces conflicts with structural inference. In fact, they should be replaced by domain-based classifiers. The discussed topological approach, on the other hand, fits to the concept of structural inference.

The idea that transformations may be class-dependent has not been worked out by us in the proximity-based approach. There is, however, not a fundamental objection against the possibility to attribute set of objects, or even individual objects in the training set with their own proximity measure. This will very likely lead to non-Euclidean data, but we have shown ways how to handle them. What is not studied is how to optimize proximity measures (structure transformations) over the training data. A possibility might be to normalize for differences in class structure by adapting the proximity measures that determined these structures.

There is, however, an important aspect of learning from structures that cannot currently be covered by domain-based classifiers built for a proximity representation. Structures can be considered as assemblies of more primitive structures, similarly as a house is built from bricks. These primitives may have a finite size, or may also be infinitesimally small. The corresponding transformations from one structure into another become thereby continuous. In particular, we are interested in such transformations as they may constitute the compactness of classes on which a realistic set of pattern recognition problems can be defined. It may be economical to allow for locally-defined functions in order to derive (or learn) transformations between objects. For instance, while comparing dogs and wolves, or while describing these groups separately, other transformations may be of interest for the description of ears then for the tails. Such a decomposition of transformations is not possible in the current proximity framework, as it starts with relations between entire objects. A further research is needed.

The automatic detection of parts of objects where different transformations may be useful for the discrimination (or a continuous varying transformation over the object) seems very challenging, as the characteristics inside an object are ill-defined as long as classes are not fully established during training. Some attempts in this direction have been made by Paclík [29,30] when he tries to learn the proximity measure from a training set.

In summary, we see three ways to link structural object descriptions to the proximity representation:

- Finding or generating prototypical objects that can easily be transformed into the given training set. They will be used in the representation set.
- Determining specific proximity measures for individual objects or for groups of objects.
- Learning locally dependent (inside the object) proximity measures.

6 Discussion and Conclusions

In this paper, we presented a discussion of the possibilities of structural inference as opposed to statistical inference. By using the structural properties of objects and classes of a given set of examples, knowledge such as class labels is inferred for new objects. Structural and statistical inference are based on different assumptions with respect to the set of examples needed for training and for the object representation. In a statistical approach, the training set has to be representative for the class distributions as the classifiers have to assign objects to the most probable class. In a structural approach, classes may be assumed to be separable. As a consequence, domain-based classifiers may be used [18,24]. Such classifiers, which are mainly still under development, do not need training sets that are representative for the class distributions, but which are representative for the class domains. This is greatly advantageous as these domains are usually stable with respect to changes in the context of application. Training sets may thereby be collected by a selective, instead of unselective sampling.

The below table summarizes the main differences between representations based on features (F), proximities (P) and structures (S) for the statistical and structural inference.

	Statistical inference	Structural inference
F	Features reduce; statistical inference is almost obligatory.	The structural information is lost by representing the aspects of objects by vectors and/or due to the reduction.
P	Proximity representations can be derived by comparing pairs of objects (e.g. initially described by features or structures). Statistical classifiers are built in proximity spaces or in (pseudo-Euclidean) embedded spaces.	Transformations between the structures of objects may be used to build proximity representations. Classes of objects should be separated by domain-based classifiers.
S	Statistical learning is only possible if a representation vector space is built (by features or proximities), in which density functions can be defined.	Transformations might be learnt by using a domain-based approach that transforms one object into another in an economical way.

This paper summarizes the possibilities of structural inference. In particular, the possibilities of the proximity representation are emphasized, provided that domain-based learning procedures follow. More advanced approaches, making a better usage of the structure of individual objects have to be studied further. They may be based on the generation of prototypes or on trained, possibly local transformations, which will separate object classes better. Such transformations can be used to define proximity measures, which will be further used to construct a proximity representation. Representations may have to be directly built on the topology derived from object neighborhoods. These neighborhoods are constructed by relating transformations to proximities. The corresponding dissimilarity measures will be non-Euclidean, in general. Consequently, non-Euclidean topology has to be studied to proceed in this direction fundamentally.

Acknowledgments. This work is supported by the Dutch Organization for Scientific Research (NWO).

References

1. Sayre, K.: Recognition, a study in the philosophy of artificial intelligence. University of Notre Dame Press (1965)
2. Watanabe, S.: Pattern Recogn. Human and Mechanical. Academic Press (1974)
3. Fu, K.: Syntactic Pattern Recognition and Applications. Pretice-Hall (1982)
4. Fukunaga, K.: Introduction to Statistical Pattern Recogn. Academic Press (1990)
5. Duda, R., Hart, P., Stork, D.: Pattern Classification. John Wiley & Sons, Inc. (2001)
6. Webb, A.: Statistical Pattern Recognition. John Wiley & Sons, Ltd. (2002)

7. Jain, A., Duin, R.P.W., Mao, J.: Statistical pattern recognition: A review. IEEE Transactions on Pattern Analysis and Machine Intelligence **22**(1) (2000) 4–37
8. Bunke, H., Günter, S., Jiang, X.: Towards bridging the gap between statistical and structural pattern recognition: Two new concepts in graph matching. In: International Conference on Advances in Pattern Recognition. (2001) 1–11
9. Fu, K.: A step towards unification of syntactic and statistical pattern recognition. IEEE Transactions on Pattern Analysis and Machine Intelligence **8** (1986)
10. Goldfarb, L.: A new approach to pattern recognition. In Kanal, L., Rosenfeld, A., eds.: Progress in Pattern Recognition. Volume 2. Elsevier Science Publishers BV (1985) 241–402
11. Goldfarb, L., Gay, D.: What is a structural representation? Fifth variation. Technical Report TR05-175, University of New Brunswick, Fredericton, Canada (2005)
12. Goldfarb, L.: What is distance and why do we need the metric model for pattern learning? Pattern Recognition **25**(4) (1992) 431–438
13. Goldfarb, L., Golubitsky, O.: What is a structural measurement process? Technical Report TR01-147, University of New Brunswick, Fredericton, Canada (2001)
14. Gutkin, A., Gya, D., Goldfarb, L., Webster, M.: On the articulatory representation of speech within the evolving transformation system formalism. In Goldfarb, L., ed.: Pattern representation and the future of pattern recognition, ICPR 2004 Workshop Proceedings, Cambridge, United Kingdom (2004) 57–76
15. Pękalska, E., Duin, R.P.W.: The Dissimilarity Representation for Pattern Recognition. Foundations and Applications. World Scientific, Singapore (2005)
16. Pękalska, E., Duin, R.P.W.: Learning with general proximity measures. In: Pattern Recognition in Information Systems. Volume 6. (2006)
17. Duin, R.P.W., Pękalska, E.: Open issues in pattern recognition. In: Computer Recognition Systems. Springer, Berlin (2005) 27–42
18. Duin, R.P.W., Pękalska, E.: The science of pattern recognition. Achievements and perspectives. (2006, submitted)
19. Dawid, A., Stone, M., Zidek, J.: Marginalization paradoxes in Bayesian and structural inference. J. Royal Stat. Soc., B **35** (1973) 180–223
20. Vapnik, V.: Estimation of Dependences based on Empirical Data, 2nd ed. Springer Verlag (2006)
21. Jain, A.K., Chandrasekaran, B.: Dimensionality and sample size considerations in pattern recognition practice. In Krishnaiah, P.R., Kanal, L.N., eds.: Handbook of Statistics. Volume 2. North-Holland, Amsterdam (1987) 835–855
22. Wolpert, D.: The Mathematics of Generalization. Addison-Wesley (1995)
23. Duin, R.P.W., Roli, F., de Ridder, D.: A note on core research issues for statistical pattern recognition. Pattern Recognition Letters **23**(4) (2002) 493–499
24. Duin, R.P.W., Pękalska, E.: Domain-based classification. Technical report, TU Delft (2005) `http://ict.ewi.tudelft.nl/~{}duin/papers/Domain_class_05.pdf`.
25. Vapnik, V.: Statistical Learning Theory. John Wiley & Sons, Inc. (1998)
26. Cristianini, N., Shawe-Taylor, J.: Support Vector Machines and other kernel-based learning methods. Cambridge University Press, UK (2000)
27. Pękalska, E., Paclík, P., Duin, R.P.W.: A Generalized Kernel Approach to Dissim. Based Classification. Journal of Machine Learning Research **2**(2) (2002) 175–211
28. Pękalska, E., Duin, R.P.W., Paclík, P.: Prototype selection for dissimilarity-based classifiers. Pattern Recognition **39**(2) (2006) 189–208
29. Paclík, P., Novovicova, J., Duin, R.P.W.: Building road sign classifiers using a trainable similarity measure. Journal of Intelligent Transportations Systems (2006)
30. Paclík, P., Novovicova, J., Duin, R.P.W.: A trainable similarity measure for image classification. In: 17th Int. Conf. on Pattern Recognition. (2006)

A Multiphase Level Set Evolution Scheme for Aerial Image Segmentation Using Multi-scale Image Geometric Analysis

Wang Wei, Yang Xin, and Cao Guo

Institute of Image Processing and Mode Recognition, Shanghai Jiaotong University, Shanghai, 200240, P.R. China
{wangwei2002, yangxin, gcao_jn}@sjtu.edu.cn

Abstract. This paper describes a new aerial images segmentation algorithm. The algorithm is based upon the knowledge of image multi-scale geometric analysis which can capture the image's intrinsic geometrical structure efficiently. The Contourlet transform is selected to represent the maximum information of the image and obtain the rotation invariant features of the image. A modified Mumford-Shah model is built to segment the aerial image by a necessary level set evolution. To avoid possible local minima in the level set evolution, we control the value of weight numbers of features in different evolution periods in this algorithm, instead of using the classical technique which evolve in a multi-scale fashion.

1 Introduction

Nowadays, with the development of sensor technology, the resolution of remotely sensed images has become higher, with more information being contained than before. Consequently, many remotely sensed image processing algorithms have appeared. Most of them are focused on the segmentation or classification of man-made objects.

The two main methods in the study of man-made object segmentation are: model-based algorithms and feature-based algorithms.

Model-based algorithms include the works of Jia Li[1] , A.L.Reno[2] , J.L.Solka[3] etc. These algorithms can segment man-made objects precisely. However, it is very difficult to build a precise estimation model due to the complexity of remotely sensed images. Moreover, the computation of the estimated parameters of the model is inevitably complex and time-consuming.

Feature-based algorithms include the works of Mark.J Carlotto[4] , Stephen Levitt[5] etc. These initial studies consider the low level features of the image. Recent studies integrate high level analysis of the features of color, texture, height and so on.

The remotely sensed images segmentation methodology proposed in this paper is based on the knowledge of image multi-scale geometric analysis, which can extract the features of the image efficiently. How to obtain the rotation invariant features is described in the paper. In order to classify the remotely sensed images, a modified Mumford-Shah model is introduced to integrate the rotation invariant features, while the level set method is responsible for the image evolution.

D.-Y. Yeung et al. (Eds.): SSPR&SPR 2006, LNCS 4109, pp. 56 – 64, 2006.

This paper is organized as follows: Section 2 introduces the Contourlet transform and the feature extraction method which is based on the knowledge of image multi-scale geometric analysis. Section 3 introduces the modified Mumford-Shah model. Section 4 elaborates on the new aerial images segmentation algorithm. The outputs of experiments are presented and illuminated in Section 5 and the conclusions of the paper are listed in Section 6.

2 Feature Extraction Based on Image Multiscale Geometric Analysis

2.1 Image Multi-scale Geometric Analysis and Contourlet Transform

The wavelet transform is widely used in many fields, but it still has some limitations. E.J.Candès [6] indicates that wavelets provide a very sparse representation for piecewise smooth 1-D signals but fail to do so for multi-dimensioned signals. Minh N. Do[7] compared 2-D separable wavelet transform with multi-scale geometric analysis. As we find in the Fig.1: Multi-scale geometric analysis is more efficient than wavelet transform because of those elongated shapes and multiple directions along the contour.

In 2003, Minh N.Do introduced the contourlet transform[8] which can be regarded as a discrete version of the curvelet transform. It solved most of the problems that the curvelet transform had met with, but it still has a redundancy ratio of about 33%. Although the crisp-contourlets[9] were later generated to reduce the redundancy ratio, DFB[10] applications still exists in the low-frequency component. Truong T. Nguyen and Soontorn Oraintara[11] developed the theory of multi-resolution DFB which can be uniformly and maximally decimated. They introduced the uniform DFB(uDFB) and the non-uniform DFB(nuDFB) in their paper .

In this paper, the contourlet transform is efficient enough to extract the features of the remotely sensed image since the features of aerial Images are mainly concentrated in the middle and high frequency component. The contourlet transform is briefly described as Fig.2.

The contourlet transform consists of the Laplacian pyramid and the DFB. The union can be described as pyramidal DFB(PDFB). In each scale decomposition, the

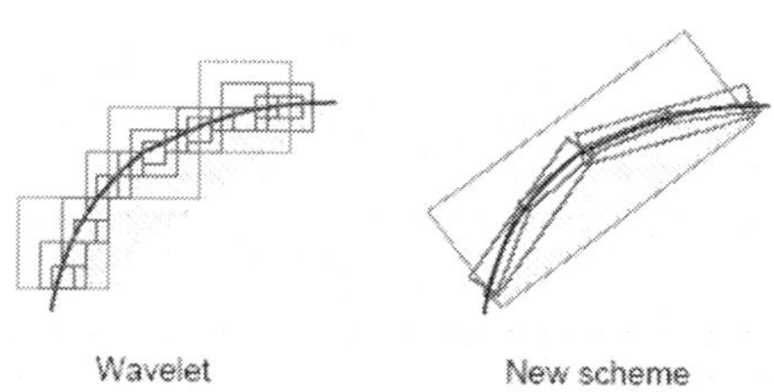

Fig. 1. Wavelet transform versus the new scheme[7]

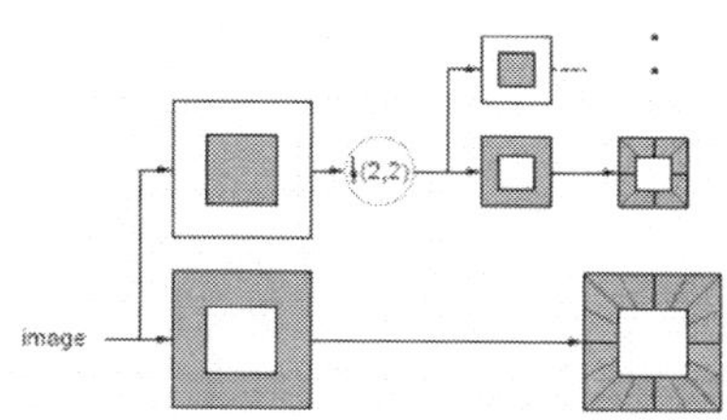

Fig. 2. The contoulet filter bank[7]

Laplacian pyramid separates the low-frequency component from the rest of the components, and then the DFB is applied to the rest.

2.2 The Extraction of Rotation Invariant Features

The features of aerial Images are mainly concentrated in the middle and high frequency components, while the low-frequency components usually contain the gray scale information. So we only need extract the features of the middle and high frequency components which we are interested in. The contourlet transform can meet our needs and avoid the complexities of the DFB brought upon by the uDFB and nuDFB[11].

Manesh Kokare[12] proposed a new rotationally invariant feature extraction method, in which the images are decomposed into different sub-bands by DT-CWT and DT-RCWF, then the final rotation invariant wavelet features are obtained from those sub-bands. Referring to Manesh Kokare's method, the rotation invariant contourlet features can be extracted as follows:

To calculate the features of a certain point in a remotely sensed image, we select a block with a size of 16×16 or 32×32, with a certain point in the center of the block. Then, we decompose this block into three levels by the contourlet transform. As to the first two levels of contourlet decompositions, we use a three levels DFB decomposition to get an eight-directional frequency partitioning for each level; as to the final contourlet decomposition, the wavelet transform is used to obtain 4 different sub-bands. In the end, the targeted block is decomposed into 20 sub-bands, just as Fig 3 shows.

Then the rotation invariant contourlet features in each level can be calculated, supposing that j denotes jth level, the size of sub-band w_j^i is $m\times n$, the feature of w_j^i is calculated as follows:

$$E_j^i = \frac{1}{NM}\sum_{l=1}^{N}\sum_{k=1}^{M}\left|x_j^i(l,k)\right| \tag{1}$$

$$\mu_j^i = \frac{1}{NM}\sum_{l=1}^{N}\sum_{k=1}^{M} x_j^i(l,k) \tag{2}$$

$$\sigma_j^i = \left[\frac{1}{N\times M}\sum_{l=1}^{N}\sum_{K=1}^{M}\left(x_j^i(l,k)-\mu_j^i\right)^2\right]^{\frac{1}{2}} \tag{3}$$

Where E_i^j is the energy of w_j^i, σ_i^j is the standard deviation of w_j^i, μ_i^j is the mean of w_j^i, $x_i^j(l,k)$ is the coefficients of w_j^i located in (l,k).

While the level j equal 1 or 2, the rotation invariant features is given by (4), while the level j equal 3, the rotation invariant features is given by (5).

$$E_j = \frac{1}{8}\sum_{i=0}^{7} E_j^i \qquad \sigma_j = \frac{1}{8}\sum_{i=0}^{7} \sigma_j^i \tag{4}$$

$$E_3 = \frac{1}{2}\left(E_3^1 + E_3^2\right) \qquad \sigma_3 = \frac{1}{2}\left(\sigma_3^1 + \sigma_3^2\right) \tag{5}$$

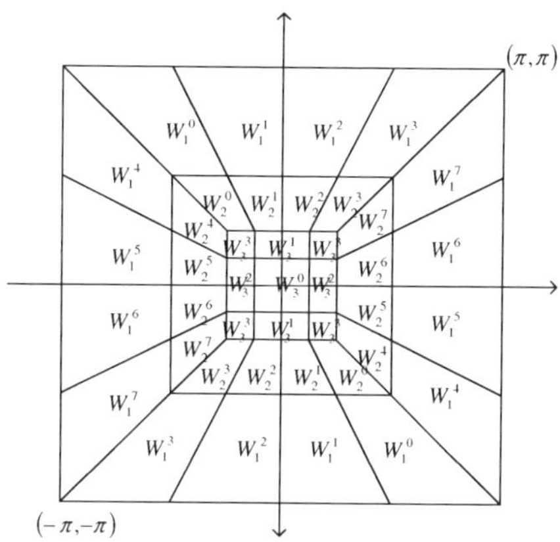

Fig. 3. Frequency partition of the three levels decompositions

The final six dimension rotation invariant features is given by

$$feature = \begin{bmatrix} E_3 & E_2 & E_1 & \sigma_3 & \sigma_2 & \sigma_1 \end{bmatrix} \bullet \begin{bmatrix} K_1 & 0 & 0 & 0 & 0 & 0 \\ 0 & K_2 & 0 & 0 & 0 & 0 \\ 0 & 0 & K_3 & 0 & 0 & 0 \\ 0 & 0 & 0 & K_4 & 0 & 0 \\ 0 & 0 & 0 & 0 & K_5 & 0 \\ 0 & 0 & 0 & 0 & 0 & K_6 \end{bmatrix} \tag{6}$$

Where $K_1 \sim K_6$ are weight numbers.

3 Mumford-Shah Model and the Modification

The Mumford-Shah model[13] is a commonly used model in image segmentation, based on this Chan and Vese proposed a multi-phase level set framework[14] for image segmentation. In the piecewise constant case, n phases can be represented by m level set functions, where m= $\log_2 n$. In this framework, there exist interactions between each level set function, which will reduce the speed of evolution.

In order to speed up the aerial images segmentation algorithm, we still use n instead of $\log_2 n$ level set functions to represent n phases. This method avoids the interaction between different level set functions. However, it has brought about new problems, such as vacuum or overlapping points, which can be regarded as payment for pursuing fast segment algorithm. After finishing the evolution, we need to classify these points by a strategy that will be discussed in later chapters.

The active contour evolving method can combine other features besides the grey level features. Jean-Francois Aujol, Gilles Aubert, and Laure Blanc-Féraud[15] presented a supervised classification model based upon a variational approach. The wavelet features are taken into consideration in this model. Cao Guo[16] proposed a simplified Mumford-Shah model in which the features of fractal error metric and the DCT coefficients of texture edges are considered.

In the situation of supervised classification that the mean values of each region are pre-known, we can obtain the jth energy function as follows:

$$F_j(C_j, c_o, c_{bj}) = u \cdot \text{Length}(C_j) + \lambda_1 \cdot \int_{inside(C)} \left| feature - \overline{feature_{oj}} \right|^2 dxdy + \lambda_2 \cdot \int_{outside(C)} (\left| feature - \overline{feature_{bj}} \right|^2) dxdy \tag{7}$$

Where *feature* are the six dimension features of point (x, y), $\overline{feature_{oj}}$ denotes the mean feature of the jth region, $\overline{feature_{bj}}$ is a changing value decided by the position (x, y), $\overline{feature_{bj}}$ is selected from one of the n pre-known mean values except $\overline{feature_{oj}}$.

Function(7) can be represented in another form as follows:

$$F_j(\phi_j, c_o, c_{bj}) = u \cdot \int_\Omega \delta(\phi_j) |\nabla \phi_j| dxdy + \lambda_1 \cdot \int_\Omega \left| feature - \overline{feature_{oj}} \right|^2 H(\phi_j) dxdy + \lambda_2 \cdot \int_\Omega (\left| feature - \overline{feature_{bj}} \right|^2) [1 - H(\phi_j)] dxdy \tag{8}$$

Where ϕ_j is the jth level set function.

The associated Euler-Lagrange equations to (8) give the following expression:

$$\begin{cases} \overline{feature_{oj}} = \dfrac{\int_\Omega feature(x, y) H(\phi_j) dxdy}{\int_\Omega H(\phi_j) dxdy}, \\ \left| feature(x, y) - \overline{feature_{bj}} \right|^2 = \min\left(\left| feature(x, y) - \overline{feature_{oi}} \right|^2 \right), \quad 1 \le i \le n, i \ne j; \\ \dfrac{\partial \phi_j}{\partial t} = \delta(\phi_j) \left[u \nabla \cdot \dfrac{\nabla \phi_j}{|\nabla \phi_j|} - \lambda_1 \cdot \left[feature(x, y) - \overline{feature_{oj}} \right]^2 + \lambda_2 \cdot \left[feature(x, y) - \overline{feature_{bj}} \right]^2 \right]; \\ \phi_j(0, x, y) = \phi_0(x, y); \end{cases} \tag{9}$$

where $H(Z) = \begin{cases} 1 & Z > 0 \\ 0 & Z < 0 \end{cases}$, $\delta(x)$ is the Dirac function.

To avoid possible local minima in the level set evolution, one classical technique is to evolve in a multi-scale fashion[17]. The evolution result from the lower resolution is selected to be the initial contour of the next evolution in the higher resolution. Instead of using the classical technique referred to above, we control the value of $K_1 \sim K_6$ in different evolution periods in this algorithm. In the beginning stages of the resolution, the features of lower resolution are applied with bigger weight. When the level set evolves into the more constant stages, the value of $K_1 \sim K_6$ changes to ensure the features of higher resolution are applied with bigger weight. The changing weighting numbers will lead the geodesic flow to the correct position not only in the lower resolution but also in the higher resolution.

4 Description of the Aerial Images Segment algorithm

The aerial image segmentation algorithm proposed in this paper is a supervised method. The procedure of segmentation can be described as follows:

Step 1: First of all, select the representative sections of different classes from the aerial image and save these sections into the list.

Step 2: Set the weighting numbers $K_1 \sim K_6$ to 1, then calculate the norm feature $\overline{feature_{oi}}$ of each section in the list. Calculate the feature $feature(x, y)$ of every point in the aerial image. Save $\overline{feature_{oi}}$ as $\overline{feature_{oi}}\_sav$ and save $feature$ as $feature\_sav$.

Step 3: Referring to Chan and Vese [14], initial closed curves in the aerial image are given in this algorithm, just as Fig 4(b) shows.

Step 4: The parameters are initially set as: $K_1 = K_4 = 1.2$, $K_2 = K_5 = 1.2$, $K_3 = K_6 = 0.6$. Refresh the values of $\overline{feature_{oi}}$ and $feature$ according to $\overline{feature_{oi}}\_sav$ and $feature\_sav$. The curve begins to evolve as described in the equation (9).

Step 5: When the difference between the two evolving steps is smaller than a pre-defined threshold as T_1, set the parameters as $K_1 = K_4 = 0.6$, $K_2 = K_5 = 1.2$, $K_3 = K_6 = 1.2$. Refresh the values of $\overline{feature_{oi}}$ and $feature$ again. Keep on the evolution.

Step 6: When the difference between two evolving steps is smaller than a pre-defined threshold as T_2, set the parameters as $K_1 = K_4 = 0.6$, $K_2 = K_5 = 1$, $K_3 = K_6 = 1.4$. Refresh the values of $\overline{feature_{oi}}$ and $feature$ again. Keep on the evolution.

Step 7: Update and evolve the level set function ϕ_j and check whether the criterion of termination is met or not. If the criterion of termination is met, the area inside the closed curves is the area of the jth class object. Start the ϕ_{j+1} evolution, repeat the steps of 4~7.

Step 8: After all the level set functions evolution have finished, check the whole image to find the vacuum or overlapped points. Calculate the mean features of these points and their neighboring points. Classify these points to their nearest class in terms of the calculated mean feature $\overline{feature}_{oi}$.

Step 9: According to the result of evolution, differentiate each region using different colors.

5 Experiment Results and Discussion

In these experiments, the criterion of termination is met when the difference between two evolving steps is smaller than a pre-defined threshold as 0.015 or the evolution reach 20 times. Set the parameters as T_1=0.45, T_2=0.15, $\lambda_1 = \lambda_2$=0.2.

The original aerial image with a size of 495×385 is shown in Fig 4(a), while the initial closed curves in the aerial image are shown in Fig 4(b).

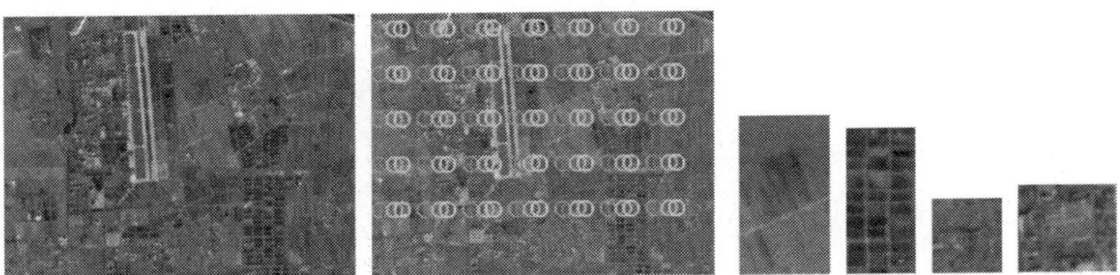

(a) (b) (c) (d) (e) (f)

Fig. 4. The aerial image to be classified. (a) The aerial image. (b) Initial conditions. (c) ~(f) Different represent regions, the positions of these regions are {(253,71),(307,167)}, {(353,252),(397,344)}, {(39,119),(83,167)} and {(111,44),(170,96)}, respectively, on rectangular coordinates of the aerial image.

The supervised method is taken to segment this aerial image into four kinds of regions, for which the selected representative regions are shown in Fig 4(c)~(f). The segmentation results are shown in Fig 5.

(a) (b)

Fig. 5. Segmentation results. (a) Result of evolution. (b) Differentiate each region using different colors.

The experiment results of segment aerial images with two and three classes are illustrated in Fig. 6 and Fig. 7, respectively. Satisfying experiment results are achieved by using the algorithm which is proposed in this paper.

More experiment results are shown as below:

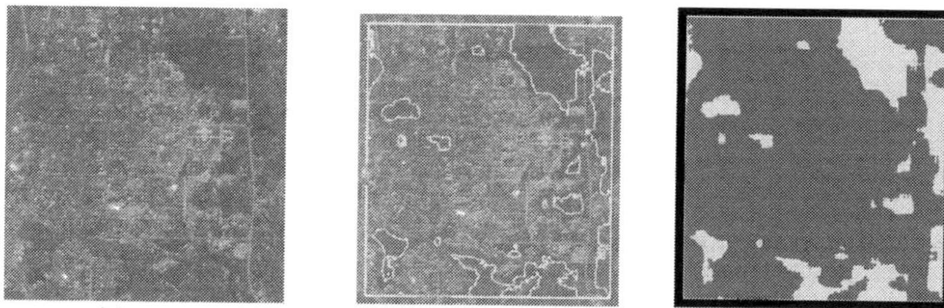

(a) (b) (c)

Fig. 6. Aerial image to be classified. The aerial image. (b) Result of evolution. (c) Differentiate each region by different colors.

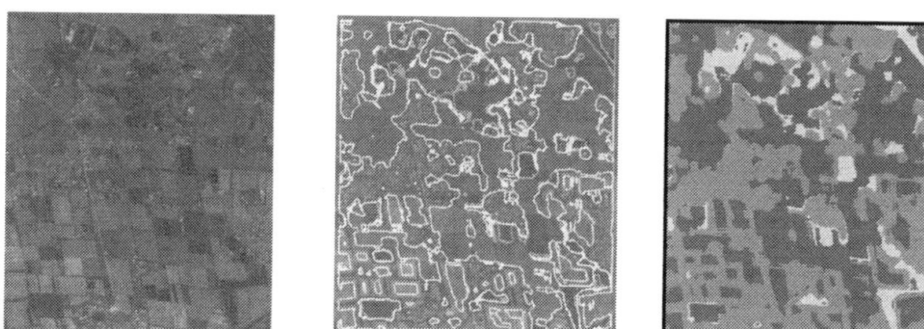

(a) (b) (c)

Fig. 7. Aerial image to be classified. (a) The aerial image. (b) Result of evolution. (c) Differentiate each region by different colors.

6 Conclusion

In this paper, a new supervised aerial images segmentation algorithm is presented. It is built on the basis of the multi-phase Mumford-Shah model. The rotation invariant contourlet features are obtained upon the knowledge of image multi-scale geometric analysis. In order to achieve a fast aerial images segmentation speed, several level set formulations are used to minimize the Mumford-Shah energy functions with contourlet features constraints. The proposed method is proven to be effective by the results of experiments.

References

1. Jia Li, Amir Najmi, Robert M.Gray. Image Classification by a Two-Dimensional Hidden Markov Model. IEEE Transactions on Signal Processing, 2000, 48(2), pp.517-533.
2. Reno, A.L., Booth, D.M. Using models to recognise man-made objects, Visual Surveillance.1999. Second IEEE Workshop on, 26 June 1999 pp.33 – 40.

3. J.L.Solka, D.J.Marchette, B.C.Wallet. Identification of Man-Made Regions in Unmanned Aerial Vehicle Imagery and Videos. IEEE Transactions on PAMI, 1998, 20(8), pp.852-857.
4. Carlotto,M.J. Detecting Man-Made Features in SAR Imagery. Geoscience and Remote Sensing Symposium, 1996. IGARSS '96. 'Remote Sensing for a Sustainable Future. International , Volume: 1 , 27-31 May 1996.
5. Stephen Lebitt, Farzin Aghdasi. Texture Measures for Building Recognition in Aerial Photographs. Communications and Signal Processing, 1997. COMSIG '97., Proceedings of the 1997 South African Symposium on , 9-10 Sept, 1997.
6. E J Candès. Ridgelets :Theory and Applications. USA:Department of Statistics ,Stanford University ,1998.
7. Minh N. Do. Contourlets: A new directional multiresolution image representation. Conference Record of the Asilomar Conference on Signals, Systems and Computers, v 1, 2002, pp. 497-501.
8. Minh N. Do. Contourlets and Sparse Image Expansions. Proceedings of SPIE - The International Society for Optical Engineering, v 5207, n 2, 2003, pp. 560-570.
9. Y. Lu and M. N. Do, Crisp-contourlets: A critically sampled directional multiresolution image representation, in Proc. SPIE Conf. Wavelet Applications Signal Image Process. , San Diego, CA, Aug. 2003.
10. R. H. Bamberger and M. J. T. Smith, A filterbank for the directional decomposition of images: Theory and design, IEEE Trans. Signal Process., vol. 40, no. 7, pp. 882–893, Apr. 1992.
11. Truong T. Nguyen and Soontorn Oraintara. Multiresolution Direction Filterbanks: Theory, Design, and Applications. IEEE Transactions on Signal Processing, Vol.53, No. 10, October 2005, pp.3895-3905.
12. Manesh Kokare, P.K. Biswas and B.N. Chatterji. Rotation Invariant Texture Features Using Rotated Complex Wavelet For Content Based Image Retrieval. 2004 International Conference on Image Processing(ICIP), pp.393-396.
13. Mumford D, Shah J. Optimal approximation by piece wise smooth functions and associated variational problems. Communications on Pure and Applied Mathematics, 1989,42(5), pp.577-685.
14. Luminita A.Vese and Tony F.Chan, A Multiphase Level Set Framework for Image Segmentation Using the Mumford and Shah Model, International Journal of Computer Vision, 2002, 50(3), pp.271-293.
15. Jean-Francois Aujol, Gilles Aubert, and Laure Blanc-Féraud. Wavelet-Based Level Set Evolution for Classification of Textured Images. IEEE TRANSACTIONS ON IMAGE PROCESSING, Vol. 12, No. 12, DECEMBER 2003, pp. 1634-1641.
16. Cao Guo,Yang xin and Mao, Zhihong. A two-stage level set evolution scheme for man-made objects detection in aerial images Proceedings of the IEEE Computer Society Conference on Computer Vision and Pattern Recognition, v 1, Proceedings - 2005 IEEE Computer Society Conference on Computer Vision and Pattern Recognition, CVPR 2005, 2005, p 474-479.
17. Geiger, D., Gupta, A., Costa, L.A., and Vlontzos, J. 1995. Dynamic programming for detecting, tracking, and matching deformable contours. *IEEE-PAMI*, 17(3).

Experiments on Robust Image Registration Using a Markov-Gibbs Appearance Model

Ayman El-Baz[1], Aly Farag[1], and Georgy Gimel'farb[2]

[1] Computer Vision and Image Processing Laboratory
University of Louisville, Louisville, KY 40292
{elbaz, farag}@cvip.louisville.edu
http://www.cvip.louisville.edu

[2] Department of Computer Science, Tamaki Campus
University of Auckland, Auckland, New Zealand
g.gimelfarb@auckland.ac.nz

Abstract. A new approach to align an image of a textured object with a given prototype under its monotone photometric and affine geometric transformations is experimentally compared to more conventional registration algorithms. The approach is based on measuring similarity between the image and prototype by Gibbs energy of characteristic pairwise co-occurrences of the equalized image signals. After an initial alignment, the affine transformation maximizing the energy is found by gradient search. Experiments confirm that our approach results in more robust registration than the search for the maximal mutual information or similarity of scale-invariant local features.

1 Introduction

The goal of image registration is to co-align two or more images of the same or similar objects acquired by different cameras, at different times, and from different viewpoints. Thus the images have to be photometrically and geometrically transformed in order to make them closely similar. Co-aligned images provide more complete information about the object and allow for building adequate object models.

Registration is a must in many applications, e.g. medical imaging, automated navigation, change detection in remote sensing, multichannel image restoration, cartography, automatic quality control in industrial vision, and so on [1]. Feature based registration relies on easily detectable local areal, linear, and point structures in the images, e.g. water reservoirs and lakes [2], buildings [3], forests [4], urban areas [5], straight lines [6], specific contours [7], coast lines [8], rivers, or roads [9], road crossings [10], centroids of water areas, or oil and gas pads [11]. In particular, the scale invariant feature transform (SIFT) [12] can reliably determine a collection of point-wise correspondences between two images under the affine geometric transformation and local contrast/offset photometric transformations. But these methods work only with distinctive and non-repetitive local features.

Alternative area-based registration, e.g. the least square correlation obviates the need for feature extraction due to direct matching of all image signals [13].

D.-Y. Yeung et al. (Eds.): SSPR&SPR 2006, LNCS 4109, pp. 65–73, 2006.

However, the correlation assumes spatially uniform contrast/offset transformations and a central-symmetric pixel-wise noise with zero mean. As a result, it frequently fails under non-uniform and spatially interdependent photometric transformations caused by different sensors and varying illumination. Phase correlation and spectral-domain (Fourier-Mellin transform based) methods [14] are less sensitive to the correlated and frequency dependent noise and non-uniform time-variant illumination but allow for only very limited geometric transformations.

Recent image registration by maximizing mutual information (MI) [15] presumes a most general type of photometric transformations, namely, any monotone transformation of the corresponding signals in one of the images. The similarity between two images is measured by the Kullback-Leibler divergence of a joint empirical distribution of the corresponding signals from the joint distribution of the independent signals. This approach performs the best with multi-modal images [15] and thus is widely used in medical imaging. The joint distribution is usually estimated using Parzen windows [16] or discrete histograms [17]. But the MI is invariant also to some non-monotone photometric transformations that change the images too much. The unduly extensive invariance of the MI hinders the registration accuracy.

This paper considers one further area-based registration method assuming that a textured object and its prototype have similar but not necessarily identical visual appearance under affine geometric and monotone photometric transformations of the corresponding signals. The latter transformations are suppressed by equializing both the prototype and the image area matched to it. The equalized prototype is described with a characteristic set of Gibbs potentials estimated from statistics of pairwise signal co-occurrences. The description implicitly considers each image as a spatially homogeneous texture with the same statistics. In contrast to more conventional area-based registration techniques, the similarities between the statistics rather than pixel-to-pixel correspondences are involved.

2 MGRF Based Image Registration

Basic notation. Let $\mathcal{Q} = \{0, \ldots, Q-1\}$; $\mathbf{R} = [(x,y) : x = 0, \ldots, X-1; y = 0, \ldots, Y-1]$ be a finite set of scalar image signals (e.g. gray levels) and a rectangular arithmetic lattice, respectively. The latter supports digital images $g : \mathbf{R} \rightarrow \mathcal{Q}$, and its arbitrary-shaped part $\mathbf{R}_\mathsf{p} \subset \mathbf{R}$ supports a certain prototype of an object of interest.

Let a finite set $\mathcal{N} = \{(\xi_1, \eta_1), \ldots, (\xi_n, \eta_n)\}$ of (x,y)-coordinate offsets define neighbors $\{((x+\xi, y+\eta), (x-\xi, y-\eta)) : (\xi, \eta) \in \mathcal{N}\} \wedge \mathbf{R}_\mathsf{p}$ interacting with each pixel $(x,y) \in \mathbf{R}_\mathsf{p}$. The set $\mathcal{N}$ produces a neighborhood graph on $\mathbf{R}_\mathsf{p}$ specifying translation invariant pairwise interactions. The latter are restricted to n families $C_{\xi,\eta}$ of second order cliques $c_{\xi,\eta}(x,y) = ((x,y), (x+\xi, y+\eta))$ of the graph. Interaction strength in each family is specified with the Gibbs potential function $\mathbf{V}^\mathsf{T}_{\xi,\eta} = \left[V_{\xi,\eta}(q,q') : (q,q') \in \mathcal{Q}^2\right]$ of the signal co-occurrences in the clique. The total interaction strength is given by the potential vector $\mathbf{V}^\mathsf{T} = \left[\mathbf{V}^\mathsf{T}_{\xi,\eta} : (\xi,\eta) \in \mathcal{N}\right]$ where T indicates the transposition.

MGRF based appearance model. The monotone (order-preserving) transformations of the image signals may occur due to different illumination or sensor characteristics. To make the registration (almost) insensitive to these transformations, both the prototype and conforming to it part of each image are equalized using cumulative empirical probability distributions of their signals on $\mathbf{R}_{\mathsf{p}}$. In line with a generic MGRF model with multiple pairwise interaction [18], the probability $P(g) \propto \exp(E(g))$ of an object g aligned with the prototype g° on $\mathbf{R}_{\mathsf{p}}$ is proportional to the Gibbs energy $E(g) = |\mathbf{R}_{\mathsf{p}}|\mathbf{V}^{\mathsf{T}}\mathbf{F}(g)$ where $\mathbf{F}^{\mathsf{T}}(g) = [\rho_{\xi,\eta}\mathbf{F}^{\mathsf{T}}_{\xi,\eta}(g) : (\xi,\eta) \in \mathcal{N}]$ is the vector of the scaled empirical probability distributions of signal co-occurrences over each clique family; $\rho_{\xi,\eta} = \frac{|\mathcal{C}_{\xi,\eta}|}{|\mathbf{R}_{\mathsf{p}}|}$ is the relative size of the family; $\mathbf{F}_{\xi,\eta}(g) = [f_{\xi,\eta}(q,q'|g) : (q,q') \in \mathcal{Q}^2]^{\mathsf{T}}$ with $f_{\xi,\eta}(q,q'|g) = \frac{|\mathcal{C}_{\xi,\eta;q,q'}(g)|}{|\mathcal{C}_{\xi,\eta}|}$ are the empirical probabilities of signal co-occurrences, and $\mathcal{C}_{\xi,\eta;q,q'}(g) \subseteq \mathcal{C}_{\xi,\eta}$ is a subfamily of the cliques $c_{\xi,\eta}(x,y)$ supporting the same co-occurrences ($g_{x,y} = q$, $g_{x+\xi,y+\eta} = q'$) in g.

The co-occurrence distributions and the Gibbs energy for the object are determined over $\mathbf{R}_{\mathsf{p}}$, i.e. within the prototype boundary after an object is geometrically transformed to be aligned with the prototype. To account for the transformation, the initial image is resampled to the back-projected $\mathbf{R}_{\mathsf{p}}$ by interpolation.

The appearance model consists of the neighborhood $\mathcal{N}$ and the potential $\mathbf{V}$ to be learned from the prototype. The approximate MLE of $\mathbf{V}$ is proportional to the scaled centered empirical co-occurrence distributions for the prototype [18]:

$$\mathbf{V}_{\xi,\eta} = \lambda\rho_{\xi,\eta}\left(\mathbf{F}_{\xi,\eta}(g^\circ) - \frac{1}{Q^2}\mathbf{U}\right); \; (\xi,\eta) \in \mathcal{N}$$

where $\mathbf{U}$ is the vector with unit components. The common scaling factor λ is also computed analytically; it is approximately equal to Q^2 if $Q \gg 1$ and $\rho_{\xi,\eta} \approx 1$ for all $(\xi,\eta) \in \mathcal{N}$. In our case it can be set to $\lambda = 1$ because the registration needs only relative potential values and energies.

Learning the characteristic neighbors. To find the characteristic neighborhood set $\mathcal{N}$, the top relative energies $E_{\xi,\eta}(g^\circ) = \rho_{\xi,\eta}\mathbf{V}^{\mathsf{T}}_{\xi,eta}\mathbf{F}_{\xi,\eta}(g^\circ)$ for the clique families, i.e. the scaled variances of the corresponding empirical co-occurrence distributions, have to be separated for a large number of low-energy candidates.

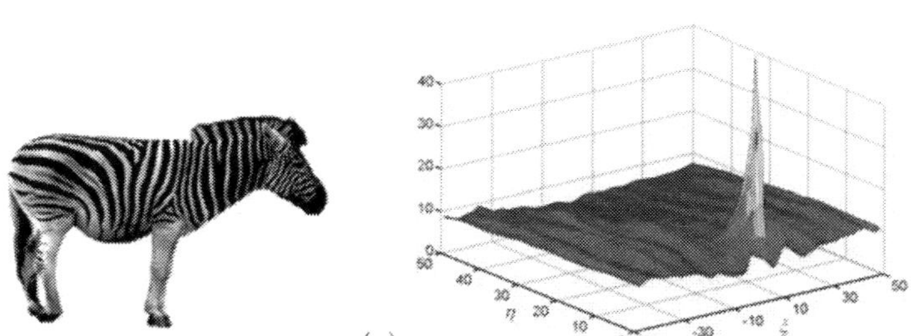

(a) (b)

Fig. 1. Zebra prototype (a) and the relative interaction energies (b) for the clique families in function of the offsets (ξ,η)

Figure 1 shows a zebra prototype and its Gibbs energies $E_{\xi,\eta}(g^\circ)$ for the 5,100 clique families with the inter-pixel offsets $|\xi| \leq 50$; $0 \leq \eta \leq 50$.

To automatically select the characteristic neighbors, let us consider an empirical probability distribution of the energies as a mixture of a large "non-characteristic" low-energy component and a considerably smaller characteristic high-energy component: $P(E) = \pi P_{\mathrm{lo}}(E) + (1-\pi)P_{\mathrm{hi}}(E)$. Because both the components $P_{\mathrm{lo}}(E)$, $P_{\mathrm{hi}}(E)$ can be of arbitrary shapes, we closely approximate them with linear combinations of positive and negative Gaussians. For both the approximation and the estimation of π, we use the efficient EM-based algorithms introduced in [19].

The intersection of the approximated low- and high-energy distributions gives an energy threshold θ for selecting the characteristic neighborhood $\mathcal{N} = \{(\xi,\eta) : E_{\xi,\eta}(g^\circ) \geq \theta\}$, that is, the threshold solves the equation $P_{\mathrm{hi}}(\theta) = P_{\mathrm{lo}}(\theta)\pi/(1-\pi)$. The above example results in the threshold $\theta = 28$ producing the 168 characteristic neighbors shown in Fig. 2 together with the corresponding relative pixel-wise energies $e_{x,y}(g^\circ)$ over the prototype:

$$e_{x,y}(g^\circ) = \sum_{(\xi,\eta)\in\mathcal{N}} V_{\xi,\eta}(g^\circ_{x,y}, g^\circ_{x+\xi,y+\eta})$$

Appearance-based registration. Let $g_{\mathbf{a}}$ denote a part of the object image reduced to $\mathbf{R}_{\mathrm{p}}$ by the affine transformation $\mathbf{a} = [a_{11}, \ldots, a_{23}]$: $x' = a_{11}x + a_{12}y + a_{13}$; $y' = a_{21}x + a_{22}y + a_{23}$. To align with the prototype, the object g should be

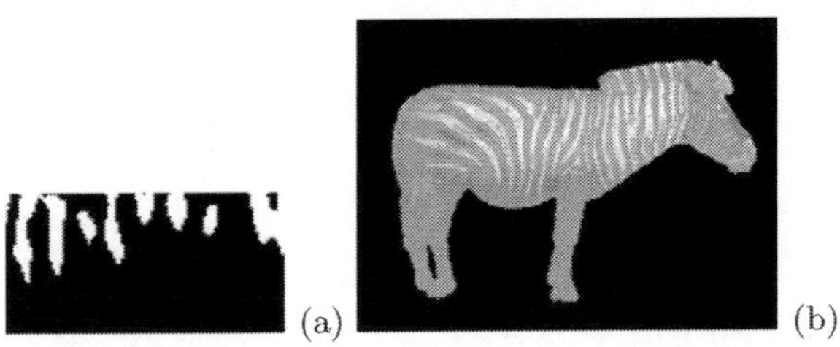

Fig. 2. Characteristic 168 neighbors among the 5100 candidates (a; in white) and the gray-coded relative pixel-wise Gibbs energies (b) for the prototype under the estimated neighborhood

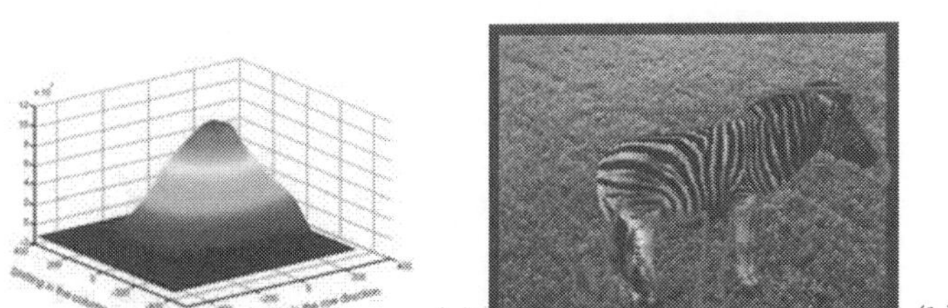

Fig. 3. Gibbs energies for the object's translations (a) with respect to the prototype and the resulting initial relative position of the object

affinely transformed to (locally) maximize its relative energy $E(g_{\mathbf{a}}) = \mathbf{V}^{\mathsf{T}}\mathbf{F}(g_{\mathbf{a}})$ under the learned appearance model $[\mathcal{N}, \mathbf{V}]$.

The initial transformation is a pure translation with $a_{11} = a_{22} = 1$; $a_{12} = a_{21} = 0$, ensuring the most "energetic" overlap between the object and prototype. The energy for the different translations (a_{13}, a_{23}) of the object relative to the prototype and the chosen initial position (a^*_{13}, a^*_{23}) maximizes this energy are shown in Fig. 3.

Then the gradient search for the local energy maximum closest to the initial point in the affine parameter space selects the six parameters $\mathbf{a}$. Figure 4 (a) illustrates the final alignment by back-projecting the prototype's contour to the object.

3 Experimental Results and Conclusions

Experiments have been conducted with several types of images. Below we discuss results obtained for zebra images available on the Internet (they include both artificial collages and natural photos) and for natural medical images such as dynamic contrast enhanced magnetic resonance imaging (DCE-MRI) of human kidneys and low dose computed tomography (LDCT) images of human lungs. These image types are commonly perceived as difficult for both the area- and feature-based registration. The like results have been obtained for other images of complex textured objects, e.g. starfish images available on the Internet and MRI of human brain. In total, we used in these experiments 24 zebra, 40 starfish, 200 kidney, 200 lungs, and 150 brain images.

We compared our approach to three popular conventional techniques, namely, to the area-based registration using the MI [15] or the normalized MI [17] and to the feature-based registration by establishing inter-image correspondences with the SIFT [12]. Results for the above zebra image are shown in Fig. 4. The SIFT-based alignment fails because the SIFT could not establish accurate correspondences between the similar zebra stripes (see Fig. 5).

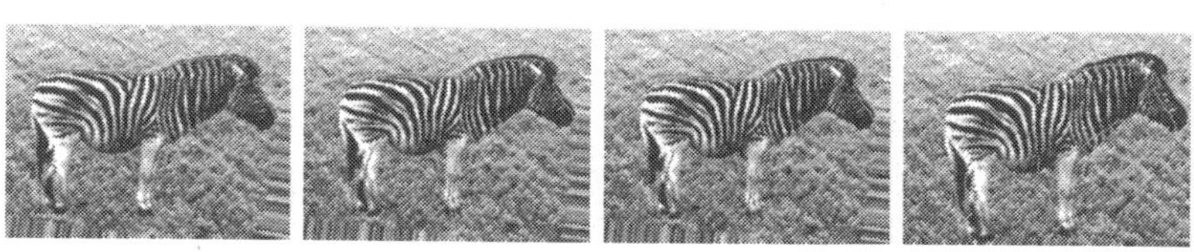

Fig. 4. From left to right: our, MI-, NMI (normalized MI)-, and SIFT-based registration

Fig. 5. Corresponding points by SIFT

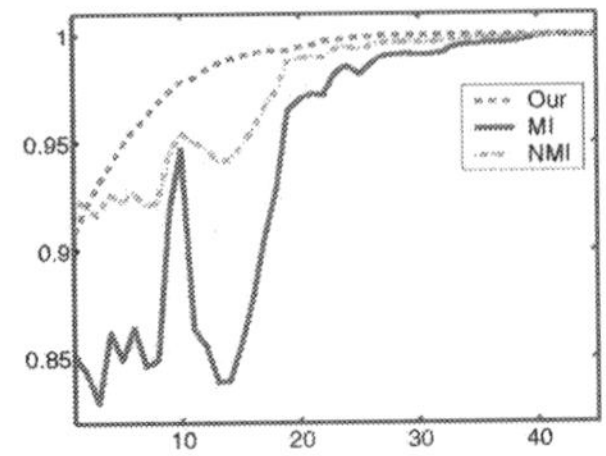

Fig. 6. Gibbs energy, MI, and NMI values at the successive steps of the gradient search

The lower accuracy of the MI- and NMI-based alignment comparing to our approach can stem from a notably different behavior of the MI / NMI and the Gibbs energy values in the space of the affine parameters. Figure 6 presents these values for the affine parameters that appear at successive steps of the gradient search for the maximum energy. Both the MI and NMI have many local maxima that potentially hinder the search, whereas the energy is close to unimodal in this case.

In the above example the object aligned with the prototype differed mainly by its orientation and scale. Figure 7 shows more diverse zebra objects and results of their Markov-Gibbs appearance-based and MI-based alignment with the prototype in Fig. 1(a). The results are illustrated by the back-projection of the prototype contour onto the objects. Visually, these results suggest that our approach has the better performance. To quantitatively evaluate the registration accuracy, the manually segmented masks of the co-aligned objects are averaged in Fig. 8. The common matching area for our approach (91.6%) is considerably larger than for the MI-based registration (70.3%).

(a) (b) (c)

Fig. 7. Original zebras (a) aligned with our (b) and the MI-based (c) approach

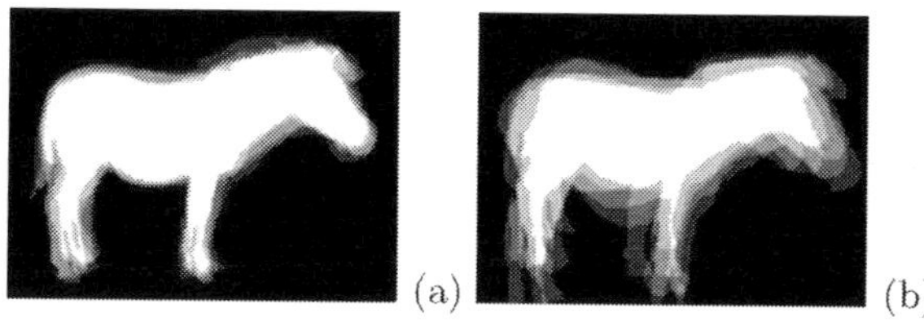

Fig. 8. Overlap between the object masks aligned with our (a; 91.6%) and the MI-based approaches (b; 70.3%)

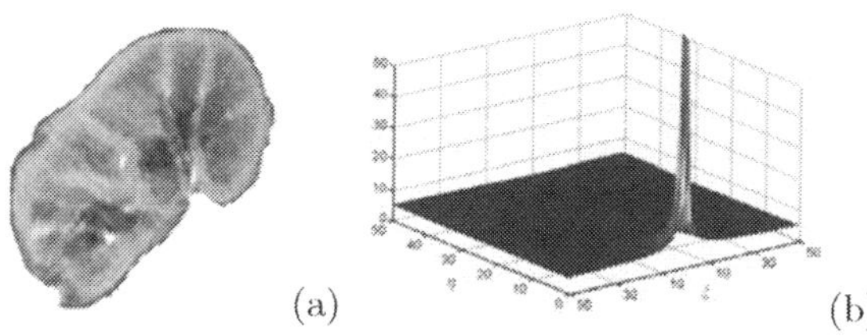

Fig. 9. Kidney image (a) and relative interaction energies (b) for the clique families in function of the offsets (η, ξ)

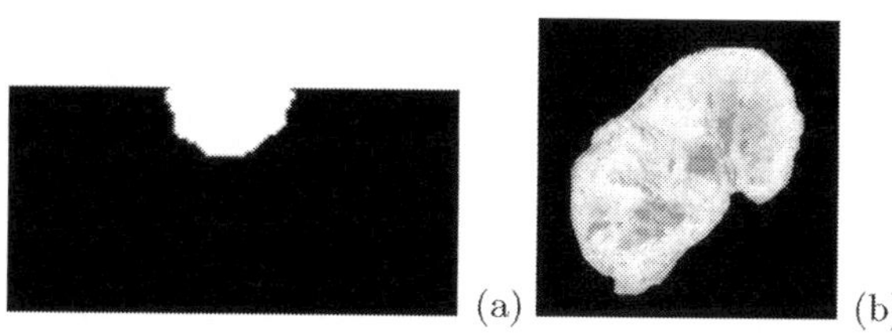

Fig. 10. (a) Most characteristic 76 neighbors among the 5,100 candidates (a; in white) and the pixel-wise Gibbs energies (b) for the prototype under the estimated neighborhood

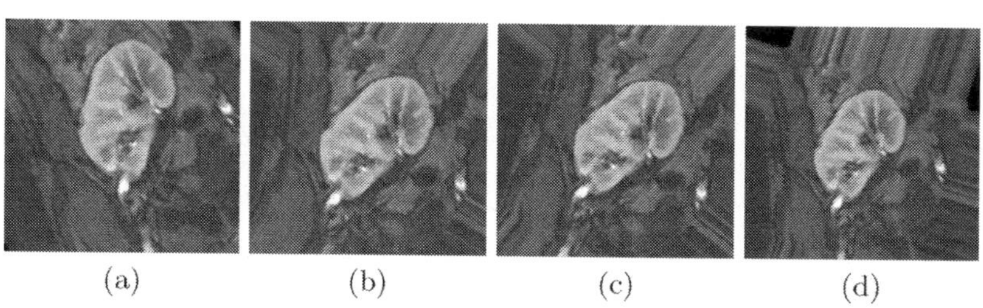

Fig. 11. Initialization (a) and our (b), the MI- (c), and the SIFT-based (d) registration

Similar results obtained for the kidney images are shown in Figs. 9–13: the common matching area 90.2% is for our approach vs. 62.6% for the MI-based one. Therefore, image registration based on our Markov-Gibbs appearance model is more robust and accurate than popular conventional algorithms. Due to reduced variations between the co-aligned objects, it results in more accurate average shape models that are useful, e.g. in image segmentation based on shape priors.

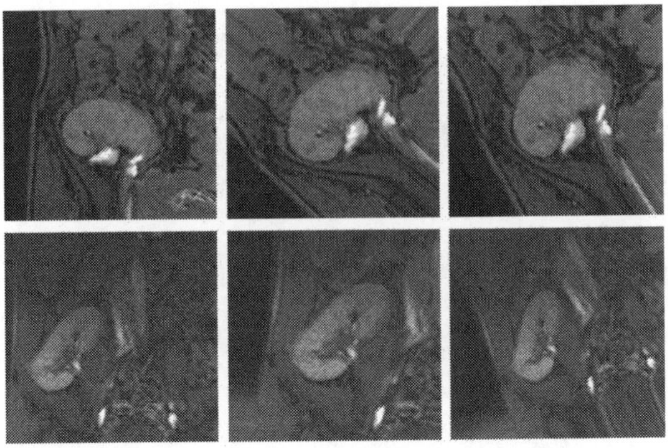

(a) (b) (c)

Fig. 12. Original kidneys (a) aligned with our (b) and the MI-based (c) approach

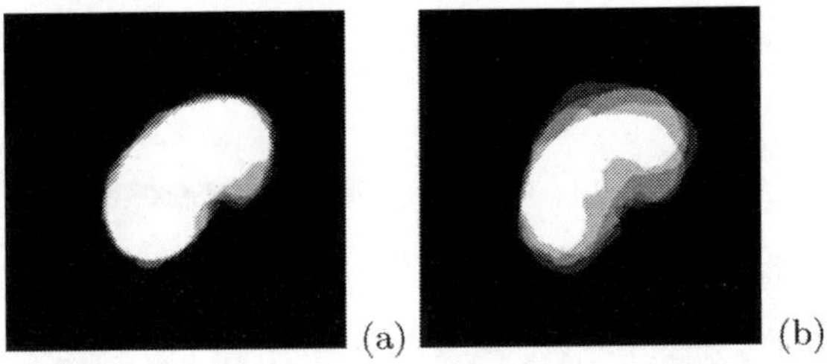

Fig. 13. Overlap between the object masks aligned with our (a; 90.2%) and the MI-based (b; 62.6%) approach

References

1. B. Zitova and J. Flusser, "Image registration methods: a survey," *Image and Vision Computing*, vol. 21, pp. 977–1000, 2003.
2. M. Holm, "Towards automatic rectification of satellite images using feature based matching," *Proc. Int. Geoscience and Remote Sensing Symp. IGARSS'91, Espoo, Finland*, 1991, pp. 2439–2442, 1991.
3. J. W. Hsieh, H. Y. M. Liao, K. C. Fan, M. T. Ko, and Y. P. Hung, "Image registration using a new edge-based approach," *Computer Vision and Image Understanding*, vol. 67, pp. 112–130, 1997.
4. M. Sester, H. Hild, and D. Fritsch, "Definition of ground control features for image registration using GIS data," *Proc. Symp. on Object Recognition and Scene Classification from Multispectral and Multisensor Pixels*, CD-ROM, Columbus, Ohio, 1998.
5. M. Roux, "Automatic registration of SPOT images and digitized maps," *Proc. IEEE Int. Conf. on Image Processing ICIP'96*, Lausanne, Switzerland, 1996, pp. 625–628.
6. Y. C. Hsieh, D. M. McKeown, and F. P. Perlant, "Performance evaluation of scene registration and stereo matching for cartographic feature extraction," *IEEE Trans. Pattern Analysis and Machine Intelligence*, vol. 14, pp. 214–237, 1992.

7. X. Dai and S. Khorram, "Development of a feature-based approach to automated image registration for multitemporal and multisensor remotely sensed imagery," *Proc. Int. Geoscience and Remote Sensing Symp. IGARSS'97, Singapore, 1997*, pp. 243–245, 1997.
8. D. Shin, J. K. Pollard, and J. P. Muller, "Accurate geometric correction of ATSR images," *IEEE Trans. Geoscience and Remote Sensing*, vol. 35, pp. 997–1006, 1997.
9. E. H. Mendoza, J. R. Santos, A. N. C. S. Rosa, and N. C. Silva, "Land Use/land Cover Mapping in Brazilian Amazon Using Neural Network with Aster/terra Data," *Proc. Geo-Imagery Bridging Continents, Istanbul, Turkey, 2004*, pp. 123–126, 2004.
10. S. Growe and R. Tonjes, "A knowledge based approach to automatic image registration," *Proc. IEEE Int. Conf. on Image Processing ICIP'97, Santa Barbara, California, 1997*, pp. 228–231, 1997.
11. J. Ton and A. K. Jain, "Registering landsat images by point matching," *IEEE Trans. Geoscience and Remote Sensing*, vol. 27, pp. 642–651, 1989.
12. D. G. Lowe, "Distinctive Image Features from Scale-Invariant Keypoints," *Int. J. of Computer Vision*, vol. 60, pp. 91–110, 2004.
13. Pope and J. Theiler, "Automated Image Registration (AIR) of MTI Imagery," *Proc. SPIE 5093*, vol. 27, pp. 294–300, 2003.
14. H. Foroosh, J. B. Zerubia, and M. Berthod, "Extension of phase correlation to subpixel registration," *IEEE Trans. Image Processing*, vol. 11, pp. 188–200, 2002.
15. P. Viola, "Alignment by Maximization of Mutual Information," *Ph.D. dissertation, MIT*, Cambridge, MA, 1995.
16. P. Thevenaz and M. Unser, "Alignment An efficient mutual information optimizer for multiresolution image registration," *Proc. IEEE Int. Conf. on Image Processing ICIP'98, Chicago, USA, 1998*, pp. 833–837, 1998.
17. C. Studholme, D. L. G. Hill, and D. J. Hawkes, "An overlap invariant entropy measure of 3D medical image alignment," *Pattern Recognition*, vol. 32, pp. 71–86, 1999.
18. G. Gimelfarb and A. A. Farag, "Texture Analysis by accurate identification of simple Markov models," *Cybernetics and Systems Analysis*, vol. 41, no. 1, pp. 37–49, 2005.
19. G. Gimelfarb, A.A. Farag and A. El-Baz, "Expectation-Maximization for a linear combination of Gaussians," *Proc. of 18th IAPR Int. Conf. on Pattern Recognition (ICPR-2004), Cambridge, UK, August 2004*, pp. 422–425, 2004.

Fully Automatic Segmentation of Coronary Vessel Structures in Poor Quality X-Ray Angiogram Images

Cemal Köse

Department of Computer Engineering, Faculty of Engineering,
Karadeniz Technical University, 61080 Trabzon, Turkey
ckose@ktu.edu.tr

Abstract. In this paper a fully automatic method is presented for extracting blood vessel structures in poor quality coronary angiograms. The method extracts blood vessels by exploiting the spatial coherence in the image. Accurate sampling of a blood vessel requires a background elimination technique. A circular sampling technique is employed to exploit the coherence. This circular sampling technique is also applied to determine the distribution of intersection lengths between the circles and blood vessels at various threshold depths. After this sampling process, disconnected parts to the centered object are eliminated, and then the distribution of the intersection length is examined to make the decision about whether the point is on the blood vessel. To produce the final segmented image, mis-segmented noisy parts and discontinuous parts are eliminated by using angle couples and circular filtering techniques. The performance of the method is examined on various poor quality X-ray angiogram images.

1 Introduction

To exploit blood vessels of human body, several medical imaging techniques such as X-ray, Computed Tomography (CT), and Magnetic Resonance (MR) are used. Extraction of blood vessels in a medical image with lack of contrast pose, drift in image intensity and noisy signal is a significant challenge in medical imaging. Automated systems and high processing throughput are needed in computationally intensive tasks including visualization of coronary blood flow and three-dimensional reconstruction of vascular structure from biplane medical images [1], [2], [3], [4], [5]. Previously developed methods for blood vessel segmentation in medical images are limited by at least one of the following drawbacks. Firstly, these methods may be applicable for limited morphologies. Secondly, user involvement is needed to select the region of interest. Thirdly, lack of adaptive capabilities may result in poor quality of segmentation under varying image condition. Lastly, blood vessel segmentation process requires a large computational effort [6], [7], [8], [9], [10]. These blood vessel segmentation techniques may be classified under following titles; pattern recognition, model based, tracking and propagation, neural network, and artificial intelligent based techniques [11], [12], [13], [14], [15].

In this paper, a method is presented to segment coronary artiograms in a medical image. This proposed method generates a complete segmentation of vessels in a medical image without user intervention. The method can handle complex structures such

D.-Y. Yeung et al. (Eds.): SSPR&SPR 2006, LNCS 4109, pp. 74 – 82, 2006.

as sharp curved, branched vessels, and vessels with varying length on a noisy and changing background. The method firstly fitters, and then extracts the background image of a medical image. Secondly, intersections between sampling circles and sampled blood vessel are determined to calculate the intersection distribution. The dominant intersections are checked to segment the vessel structure in the medical image. Finally, a circular filtering technique is used to remove small noisy fragments on the image. In Section 2, the proposed segmentation method is described. The performance of the method is examined on real images in various qualities. The results are given in Section 3 and finally the conclusions and future work are discussed in Section 4.

2 Description of the Segmentation Method

The proposed segmentation method exploits the spatial coherence existing in a medical image by considering neighboring pixels around the current one being processed. Therefore, the effect of local discontinuities and disorder are tolerated, and recognition of normal and distorted blood vessels in a noisy image is improved on fully automatic segmentation. The basic steps in automatic coronary segmentation are (1) filtering and extracting whole background image, (2) eliminating the pixel under the background threshold depth, applying the circular sampling to the pixels that are not eliminated in the previous step and applying proper Bezier spline to make more smoother samples along the scan-line in the circular sample, (3) eliminating the noisy and non-vessel parts by using angle couples at several levels over the threshold depth, (4) separating the disconnected parts from the sample, (5) determination of the blood vessel and circle intersections at several levels over the threshold depth, (6) calculating the intersection distributions and dominant intersection lengths, and then segmenting the image, and (7) finally circular filtering of whole image.

2.1 Eliminating the Background Effect

The background in a medical image affects the segmentation of the blood vessel in the image negatively. If the background is not eliminated correctly, the circular sampling technique will mis-sample the object. Therefore, a technique is needed to prevent this background effect. Here, the sampling circle gets larger so does the scan-lines then, mis-sampling occurs as illustrated in Fig. 1.a and b. Elimination of this effect is very important to produce a correctly segmented image. The elimination process is shown in Fig. 1.c and d. The background effect elimination approach described here uses an averaging technique that calculates the average intensity within the region of interests with a dimension of [(2N+1)x(2N+1)]. The centre point of this region is at current pixel point (m,n). The average $\overline{X}$ is given by Equation (1). The standard deviation of pixels in the region is given by Equation (2). Finally, threshold depth is calculated by using equation (3).

$$\overline{X}(m,n) = [\sum_{k=-N}^{N} \sum_{\ell=-N}^{N} I(i+k, j+\ell)]/(2N+1)^2 . \qquad (1)$$

$$\sigma(m,n)=\sqrt{\{\,[\sum_{k=-N}^{N}\sum_{\ell=-N}^{N}\left(\,I(m+k,n+\ell)-\overline{X}(m,n)\right)^{2}]/(2N+1)^{2}\,\}}\ . \qquad (2)$$

$$T = I_{Pxl(I,J)} - \{\overline{X}_{Pxl(I,J)} - \upsilon[\sigma(m-1,n)\alpha + (1.0-\alpha)\sigma(m,n)]\}\ . \qquad (3)$$

Where T is the depth of background threshold and I(m,n) is the intensity value of the current pixel. The parameters υ and α in Equation (3) are experimentally determined as 0.25 and 0.75, respectively.

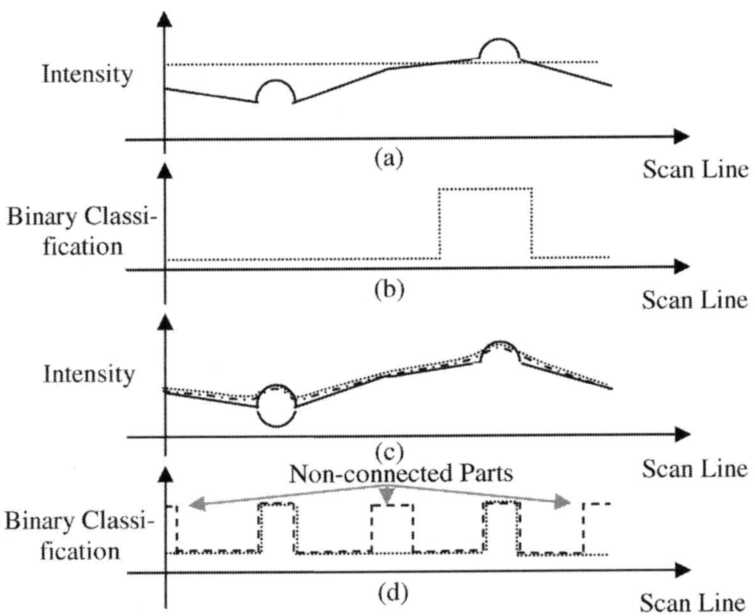

Fig. 1. Sampling along the circular scan-line (a) without (c) with background effect elimination, and corresponding binary classifications (b) and (d)

Accurate choice of the length of the averaging area is important. A large averaging area flattens the background whereas small averaging area does not cover enough background information. To eliminate most of the non-vessel-like structures, the background threshold depths are calculated at each pixel by Equation (3). This threshold depth is not used to produce final segmented image. It is used to make a pre-classification to eliminate the pixels, which are not a part of a blood vessel.

2.2 Circular Sampling Technique and Eliminating Non-vessel and Nosily Areas

The circular sampling technique samples a structure around a sampling point by extending the sampling circles spatially. Thus, it enables the segmentation process to

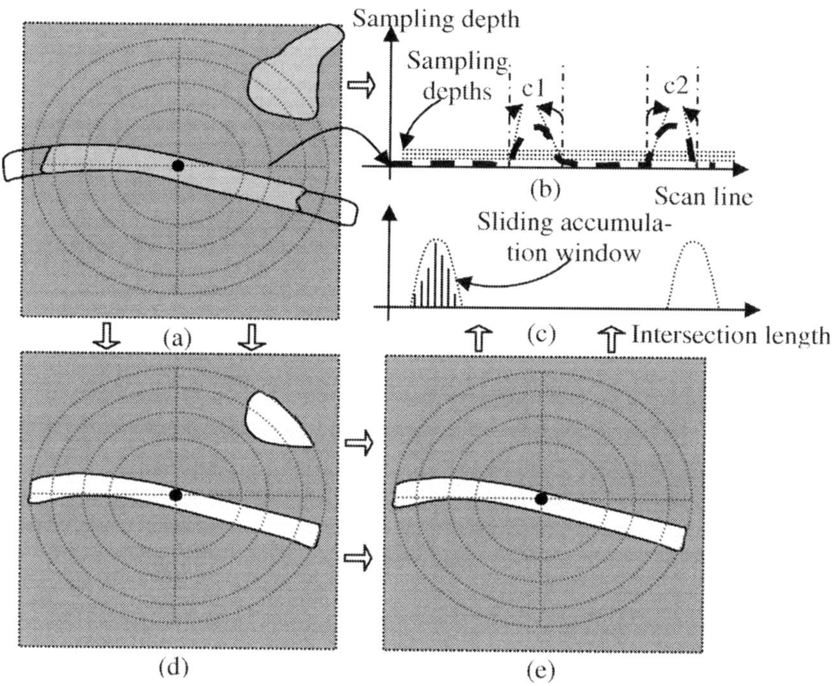

Fig. 2. (a) Circular sampling, (b) the intensity along a scan line of a circle, (c) distribution of intersection lengths of a circle and blood vessel, (d) circular sampling at a certain background threshold depth on a flat background and (e) removing the disconnected parts

exploit the spatial coherence that exists in the vessel structures on an angiogram. Here, the points around the sampling point on the image are sampled at a certain threshold depth related to the background level by using the circular sampler in Fig. 2. After the circular sampling, some noisy structures can be mis-segmented because random distribution of intersections could be concentrated at some lengths. Here, the Bezier spline is used to filter the samples along each scan line as illustrated in the Fig. 2.b. In order to reduce the number of the mis-segmentations, a technique illustrated in the figure, is employed. Here, these noisy structures are removed by counting the angle couples. The angle couples along the sampling circle's scan-line are calculated at the current pixel by accounting the several neighboring pixel intensities in the scan-line. Slope of each line producing the angle couples must be over a threshold degree (for example 45^0). These angle couples are calculated on each circular scan lines for each pixel after the agiogram is sampled, as illustrated in Fig. 2.a. A typical circular scan line with the angle couples is illustrated in Fig. 2.b. The number of angle couples is used to eliminate the noisy pixels. A pixel may be considered as non-vessel when the number of the angle couples for the current pixel is less than 4. If the number of the angle couples is far more than expected (such as 50), the pixel is classified as noisy.

2.3 Separating the Disconnected Parts from the Sample of Interest and Calculating the Distribution of Intersection Lengths

After a blood vessel on the image is sampled at various background depths, as illustrated in Fig. 2.d, the disconnected parts in the sampling scope are removed as shown in Fig. 2.e. Separating noisy or disconnected part from the sampled vessel slice is important because the parts may cause the generation of wrong segmentation results. The centre of each sampling point should be positioned on blood vessel so that enlarging sampling circles from the centre are used to eliminate the parts that are not a part of a blood vessel. Firstly each sample point along the scan-lines of the circles is pre-classified by using a binary classifier. If a sampling point along a scan-line of a circle is pre-classified as a part of a vessel that has no connection to the centre, the sample should be signed as background. This enables us to determine the circle and blood vessel intersection length distribution correctly. Therefore, only the considered blood vessel is accounted and other misclassified parts in the focused area are eliminated.

After the separation of the disconnected parts the circle blood vessel intersection lengths are accumulated according to the lengths. When a pixel is tested to determine whether it is on a blood vessel or not, it is expected that the accumulation distribution should be concentrated around a certain length. A sliding accumulation function is used to calculate this accumulation value. Width and weight parameters of the function vary according to the length of the blood vessels. If the blood vessel is narrow, the width of the function is narrow. When blood vessel's length gets larger, the circle-vessel intersection length distribution spreads as shown in Fig. 2.c. To produce the distribution of intersection lengths, the intersection lengths are accumulated by Equation (4).

$$\{D(l) = D(l) + 1\}_{I(m,n)}\,. \tag{4}$$

were, $D(l)$ represents the circles and blood vessel intersection accumulation distribution array, and $\{\}_{I(m,n)}$ represents the current depth and pixel. This accumulation process is done at several background threshold values depending on the deviation of the intensity around the current pixel.

This distribution function has also to be normalized according to the length of the intersection because more intersection occurs for narrow blood vessels than wide blood vessels. Finally, the measured length accumulation density value is over a certain threshold, the pixel is considered as blood vessel.

2.4 Decision Criteria and Threshold

The distribution of the intersection lengths, the peak values of the dominant intersections, and the relative values (to all intersections) of these dominant intersections are very important to determine a correct decision threshold value. Equation (5) is used to calculate the dominant or the maximum circle blood vessel intersections, where $M_k\{D(\ell)_{I(m,n)}\}$ represents the strength of the dominant intersection length along the intersection accumulation array, $a(\ell)$ is the weighting array used to calculate the dominant intersections. Equation (6) is used to make decision about whether the current pixel is on a blood vessel or not. Then, the equation is used to segment the image and $S_{I(m,n)}$ represents the segmentation result. If the density is less than the bottom threshold T_b, the pixel is considered as background. If the density is above the upper threshold T_u, the pixel signed as artery. If the density is in between these two thresholds, then second decision rule is applied for the correct segmentation. We experimentally found that choosing T_b and T_u as 8.5 and 12, respectfully, yields satisfactory results.

$$M_k\{D(\ell)_{I(m,n)}\} = \{\sum_{\ell=-M}^{M} a(\ell)A(k+\ell)_{I(m,n)}\}, \quad \mathrm{M} \le \mathrm{k} < \mathrm{Max\_Length}. \tag{5}$$

$$S_{I(m,n)} = \begin{cases} \text{Background} & \text{if} \quad M_k\{D(\ell)_{I(m,n)}\} < \mathrm{T_b} \\ \text{Undecided} & \text{if} \quad \mathrm{T_b} \le M_k\{D(\ell)_{I(m,n)}\} \le \mathrm{T_u} \\ \text{Artery} & \text{if} \quad M_k\{D(\ell)_{I(m,n)}\} > \mathrm{T_u} \end{cases} . \tag{6}$$

The second decision rule checks the normalized second and third peak intersection in the distribution to make a more precise decision. If these peak intersections (relative to their intersection length) on the same branch of the vessel from the current centre are not evident and if the density is larger than a threshold value, the pixel signed as artery.

2.4 Circular Filtering

Generally, the vessels in a medical image are continuous and long structures. On the other hand, sometimes background and noisy structures could be detected as vessel structures even though they are more often discontinuous and short vessel like structures rather than long and continuous vessel structures. These mis-segmented parts are removed from the final image by using the circular filtering technique. Here, all pixels signed as blood at the previous stage are taken for further examination. The center of the circle is positioned at the current pixel to be examined, and then, the radius of the circle is increased to test whether the segmented structure is a small discontinuous part or not. If there was no pixel signed as a vessel along the enlarging circles' scan line, the pixel is considered as a non-blood vessel and set to background.

2.5 Fast Segmentation

Full resolution segmentation produces a better quality segmented images but it is more expensive than the half, quarter or fast segmentation. To accelerate the segmentation process, fewer pixels than full resolution calculation can be visited during the segmentation of images. Three approaches can be applied to speed-up the segmentation process. The first way (half segmentation) of doing this is that the image can be processed at every other pixel (or more) on vertical and horizontal lines. The second approach (quarter segmentation) processes every fourth pixel along the horizontal and vertical scan lines. Then, neighboring pixels are signed by using a simple decision rule such as background threshold depth decision rule. Although this approach is quite fast, it produces poor quality segmentation results, especially for the thin blood vessels. The third way (fast segmentation) of speeding-up the segmentation process is to apply the whole decision processes to the neighboring pixels of a pixel segmented as blood vessel by using half segmentation approach.

3 Results

The performance of the method is tested on several real images with several difficulties. In the first experiment, the accuracy of the method on poor quality contrast angiogram image was evaluated. The Fig. 3.a show a low contrast image and

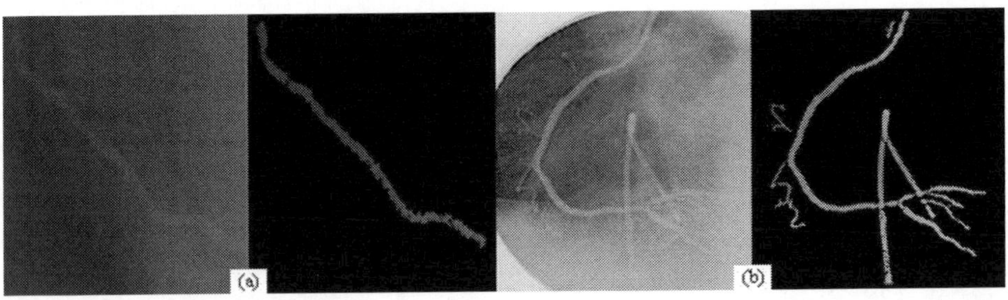

Fig. 3. (a) A poor quality angiogram and its segmented image, and (b) an angiogram image with many branches and its segmented image

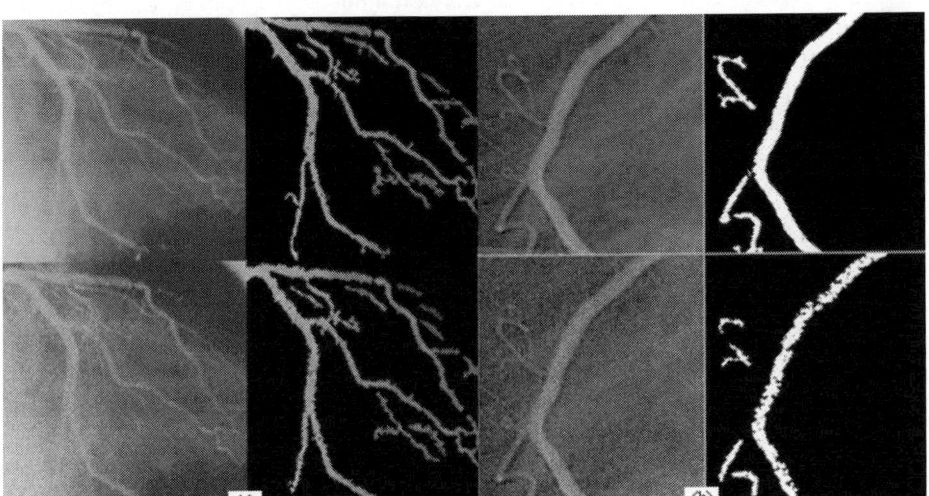

Fig. 4. (a) Noise added complex angiogram images and their segmented images and (b) Two other noise added complex angiogram images and their segmented images

corresponding segmented image, respectively. The performance of this method at the side of the blood vessel is slightly low but whole blood vessel is successfully extracted and tracked.

In the second experiment, the ability of the method was tested to extract branching arteries in a complex angiogram. Fig. 3.b shows an artery with many branches on varying background. The corresponding segmented images are also shown in Fig. 3.b. As seen from the result, the method successfully follows the branched arteries. The aim of the final experiment is to test accuracy of the method for several noise levels. For this purpose, two different images were selected. Fig. 4.a and Fig. 4.b (first and third images on the first and second lines) show noise added real images. Here, selected noise levels are 5, 20, 5, and 30 grey levels, respectively. Fig. 4.a and Fig. 4.b (second and fourth images on the first and second lines) show the corresponding segmented images. The method results in slightly noisy vessel edges but the whole blood vessel is successfully extracted and tracked. The results indicate that the proposed method yields accurate results even in complex and too noisy images.

4 Conclusion and Discussion

In this work a blood vessel segmentation technique is applied to extract the structure of the blood vessels in two-dimensional medical images. The technique exploits the spatial coherency that exists in two-dimensional medical image and works on each pixel on the image for extracting the structure of blood vessel. This fully automatic technique is robust to noise, low contrast and varying background, and able to extract vascular structures without human intervention. To eliminate small noisy parts and fragments at the final image, a circular filtering technique is used and quality of segmentation is improved. An elliptical filter may be considered as a future work to get further improvement.

When these segmentation results are compared to the results of the other methods such as the model based approach, this proposed method is quite successful in exploiting the whole vessel structure in a medical image except the branching areas and some long vessel like structures. Thus, the proposed method may not be very successful around the branching vessel area where the intersection length distribution gets more complicated. On the other hand, long vessel like structure in medical image may easily be excluded in other user intervened methods but the proposed method may not be that successful.

The segmentation method was run on P4-3.2 GHz PC. The segmentation durations for the poor quality image (300x220 pixels) is given in Fig. 3.a are about 73.2, 18.9, 10.1, and 34.5 seconds for full, half, quarter, fast resolution segmentation respectively. Typical durations for an image with dimension of 600x700 pixels (Fig. 3.b) is about 169.6, 43.4, 22.1 and 78.7 seconds for full, half, quarter and fast resolution segmentation, respectively. The durations were computed for noise added real images. The segmentation durations of noisy images (550x330 pixels) shown in Fig. 4.b are162.0 and 51.6 seconds, respectively, whereas those of the corresponding original images are 148.3 and 45.3 seconds.

References

1. Coste E., Vasseur C., Rousseau J.: 3D reconstruction of the cerebral arterial network from stereotactic DSA. Medical Physics, Vol. 26. (1999) 1783-1793
2. Kitmura K., J. Tobis M., Sklansky J.: Estimating the 3D skeleton and transverse areas of coronary arteries from biplane angiograms. IEEE Trans. Med. Imaging, Vol. 17. (1988) 173-187
3. Eichel P., Delph E. J., Koral K., Buda A. J.: A method for fully automatic definition of coronary areterial edges from cineangiograms. IEEE Trans. Med. Imaging, Vol. 7. (1988) 315-320
4. Kayikcioglu T., Gangal A., Turhal M.: Reconstructing coronary arterial segments from three projection boundaries. Pattern Recognition Letters, Vol. 21. (2001) 611-624
5. Kayikcioglu T., Gangal A., Turhal M., Köse, C.: A surface based method for detection of corronary boundaries in poor quality X-ray angiogram images. Pattern Elsevier Recognition Letters, Vol. 23. (2002) 783-802
6. Klein A. K., Amin A.: Quantities coronary angiography with deformable spline models. IEEE Trans. Med. Imag., Vol. 16. (1997) 468-482
7. Dhawan A. P., Aratal L.: Segmentation of medical images through competitive learning. Computer Methods and Programs in Biomedicine, Vol. 40. (1993) 203-215
8. Kirbas C., Francis K., Queck H.: A review of vessel extraction Techniques and Algorithms. Vision Interfaces and Systems Laboratory, Department of Computer Science and Engineering, Wright State University, Dayton Ohio (2002).
9. Suri J. S., Liu K.C., Reden L., Laxminarayan S.: A review on MR vascular image processing: Skeleton versus nonskeleton approaches. IEEE Transaction on Information Technology in Biemedicine, Vol. 6. (2002)
10. Kottke D. P., Sun Y.: Adaptive segmentation of coronary angiograms. In Proc. 14th Northeast Bioeng. Conf. (1988) 278-290
11. Kottke D. P., Sun Y.: Segmentation of coronary arteriograms by iterative ternary classification. IEEE Transaction on Biomedical Engineering, Vol. 37. (1990) 778-785
12. Francis K., Quek H., Kirbas C.: Vessel extraction in medical images by wave-propagation and trace-back. IEEE Transaction on Medical Imaging, Vol. 20. (2001)
13. Gudmundsson M., El-Kwae E.A., Kabuka M.R.: Edge Detection in medical images using a genetic algorithm. IEEE Transaction on Medical Imaging, Vol. 17. (1998) 469-474
14. Osher S., Sethian J. A.: Fronts propagating with curvature dependent speed: Algorithms based on Hamilton Jacobi formulation. Journal of Computational Physics, Vol. 79. (1988)
15. Yanagihara Y., Sugahara T., Sugimoto N.: Extraction of vessel in brain using fast x-ray CT images. Systems and Computers in Japan, Vol. 25. (1994)78-85

Smoothing Tensor-Valued Images Using Anisotropic Geodesic Diffusion

Fan Zhang and Edwin R. Hancock

Department of Computer Science, University of York,
York, YO10 5DD, UK
{zfan, erh}@cs.york.ac.uk

Abstract. This paper considers the feature space of DT-MRI as a differential manifold with an affine-invariant metric. We generalise Di Zenzo's structure tensor to tensor-valued images for edge detection. To improve the quality of the edges, we develop a generalised Perona-Malik method for smoothing tensor images. We demonstrate our algorithm on both synthetic and real DT-MRI data.

1 Introduction

Diffusion tensor magnetic resonance imaging (DT-MRI) [1] endows each voxel a 3×3 symmetric positive-definite matrix, which measures the anisotropic behaviour of water diffusion in the white matter of the brain. The feature space of DT-MRI data is no longer a linear space, but a curved convex half-cone in R^{n^2}. Thus edge or interface [2] detection, which is important for segmentation and registration, is more complicated than in scalar-valued or vector-valued images. In an attempt to overcome these difficulties Feddern et al [3] generalise Di Zenzo's [4] concept of structure tensors to tensor-valued images for level-set motions. Using the same structure tensor, O'Donnell, et al [2] introduced a more sophisticated gradient estimation method for DT-MRI edge detection. The structure tensor used simply considers diffusion tensors as vectors in R^{n^2}. However, it neglects the constraints between components induced by symmetry and positive-definiteness of the tensor.

In this paper we consider the space of diffusion tensors as a differential manifold with an affine invariant metric. In this way we generalise the Di Zenzo's structure tensor to tensor-valued images. In order to reduce the influence of noise and obtain a high quality edge detector, we show how to extend the Perona and Malik [5] anisotropic diffusion method to tensor-valued images. To do this we make use of the exponential map of the tensor data and use geodesic marching. The idea of using a manifold of diffusion tensors, has been recently used to analyse the principle geodesic modes [6] of tensor data and the segmentation of DT-MRI [7]. Pennec et al [8] has developed a framework for the analysis of statistical data residing on manifolds, and has generalised the operations of interpolation, isotropic and anisotropic regularisation for DT-MRI.

D.-Y. Yeung et al. (Eds.): SSPR&SPR 2006, LNCS 4109, pp. 83–91, 2006.

2 Space of Diffusion Tensors

Let $\Sigma(r)$ be the set of $r \times r$ real matrices and $GL(r)$ be its subset of non-singular matrices which is a Lie group. Recall that in $\Sigma(r)$ the Euclidean inner product, which is known as the Frobenius inner product, is defined as $\langle A, B\rangle_F = \mathrm{tr}(A^T B)$, where $\mathrm{tr}(\cdot)$ denotes the trace and superscript T denotes the transpose.

For a matrix A whose eigen-decomposition is $A = UDU^T$, the exponential of A is given by convergent series

$$\exp\ A = \sum_{k=0}^{\infty} \frac{1}{k!} A^k = U\exp(D)U^T, \tag{1}$$

and the inverse logarithm of A is given by

$$\log\ A = -\sum_{k=1}^{\infty} \frac{(I - A)^k}{k} = U\log(D)U^T, \tag{2}$$

where I is the identity matrix.

Let $S(r)$ be the space of $r \times r$ symmetric matrices and $S^+(r)$ be the space of symmetric positive-definite matrices. Thus, the feature space M of DT-MRI is identified with $S^+(3)$. Through the identity mapping

$$\psi : P \in S^+(r) \to (\sigma_{11}, ..., \sigma_{ij}), \quad i \leq j, \quad i, j = 1, ..., r, \tag{3}$$

$S^+(r)$ is isomorphic with an open subset U of the R^m where $m = \frac{1}{2}r(r+1)$. Thus we could consider $S^+(r)$ as a m-dimensional differential manifold with (U, ψ) as the coordinate system. At each point $P \in S^+(r)$ the tangent space $T_P S^+(r)$ is equal to $S(r)$. So a basis of $T_P S^+(r)$ can be defined as

$$\frac{\partial}{\partial \sigma_{ij}} \leftrightarrow E_{ij} \in S(r), \quad i \leq j, \quad i, j = 1, ..., r, \tag{4}$$

and

$$E_{ij} = \begin{cases} 1_{ii} & \text{if} \ \ i = j \\ 1_{ij} + 1_{ji} & \text{if} \ \ i \neq j, \end{cases} \tag{5}$$

where 1_{ij} means the $r \times r$ matrix with a 1 at element (i, j) and 0 elsewhere.

We can turn $S^+(r)$ into a Riemannian manifold by introducing a Riemannian metric g at P

$$g(\frac{\partial}{\partial \sigma_{ij}}, \frac{\partial}{\partial \sigma_{kl}}) = g(E_{ij}, E_{kl}) = \mathrm{tr}(P^{-1} E_{ij} P^{-1} E_{kl}). \tag{6}$$

This is the same as the positive-definite inner product used by [6,9,8] , i.e., $\langle A, B\rangle_P = \mathrm{tr}(P^{-1} A P^{-1} B)$, $A, B \in T_P S^+(r)$, which is invariant under group actions of $GL(r)$.

Thus, for a smooth curve $C : [a, b] \to S^+(r)$ in $S^+(r)$, the length of $C(t)$ can be computed via the invariant metric

$$\ell(C) = \int_a^b \|C'(t)\|_{C(t)} = \int_a^b \sqrt{\mathrm{tr}(C(t)^{-1}C'(t))^2}, \tag{7}$$

which is also invariant under $GL(r)$, i.e., $C(t) \mapsto GC(t)G^T, \quad G \in GL(r)$. The distance between two points $A, B \in S^+(r)$ is the infinum of lengths of curves connecting them, i.e.,

$$d(x, y) := \underset{C}{\mathrm{argmin}} \; \{\ell(C) \mid C(a) = A, C(b) = B\}\,. \tag{8}$$

The curve satisfying this infinum condition is a geodesic. In $S^+(r)$ the geodesic with initial point at I and tangent vector $W \in T_I S^+(r)$ given by $\exp(tW)$. Using invariance under group action $GL(r)$, an arbitrary geodesic $\Gamma(t)$ such that $\Gamma(0) = P$ and $\Gamma'(0) = W$ is given by

$$\Gamma_{(P,W)}(t) = P^{\frac{1}{2}} \exp(tP^{-\frac{1}{2}} W P^{-\frac{1}{2}}) P^{\frac{1}{2}}\,. \tag{9}$$

Thus, the geodesic distance between two points A and B in $S^+(r)$ is

$$d(A, B) = \left\|\log(A^{-1}B)\right\|_F = \sqrt{\sum_{i=1}^{n} (\log\lambda_i)^2}, \tag{10}$$

where λ_i are the eigenvalues of $A^{-1}B$.

We can relate an open subset of the tangent space $T_P S^+(r)$ to a local neighbourhood of P in $S^+(r)$ using the exponential map $\mathrm{Exp} : \Omega \subset T_P S^+(r) \to S^+(r)$, which is defined as $\mathrm{Exp}_P(W) = \gamma_{(P,W)}(1)$. Geometrically, $\mathrm{Exp}_P(W)$ is a point of $S^+(r)$ obtained by marking out a length equal to $|W|$ commencing from P, along a geodesic which passes through P with velocity equal to $\frac{W}{|W|}$. From Equation 9, it follows that

$$\exp_P(W) = P^{\frac{1}{2}} \exp(P^{-\frac{1}{2}} W P^{-\frac{1}{2}}) P^{\frac{1}{2}}\,. \tag{11}$$

Since $\exp_P$ is a local diffeomorphism, it has an inverse map, the so-called logarithmic map $\mathrm{Log}_P : S^+(r) \to B_\epsilon(0) \subset T_P S^+(r)$ where $\mathrm{Log}_P(\gamma_{(P,W)}(t)) = tW$. Thus, for a point A near P it also follows

$$\log_P(A) = P^{\frac{1}{2}} \log(P^{-\frac{1}{2}} A P^{-\frac{1}{2}}) P^{\frac{1}{2}}\,. \tag{12}$$

3 Generalised Structure Tensor

For tensor-valued images, the image features live on a m-dimensional manifold $M = S^+(r)$, $m = \frac{1}{2}r(r+1)$ (r=3 for DT-MRI), which we call the feature space or feature manifold. An image is a map from a domain Ω to M, i.e., $f : \Omega \in R^n \to M$, where $n = 2$ for planar images and $n = 3$ for volume images.

Drawing on ideas from Di Zenzo's pioneering work [4], we can generalise the structure tensor to tensor-valued images. At each point $x = (x_1, ..., x_n) \in \Omega$ we wish to find the direction with maximal variations in the image. For two points $x = (x_1, ..., x_n)$ and $x' = (x'_1, ..., x'_n)$, the difference of vector image values is the $f(x') - f(x)$. As the distance between the two points $\|x' - x\|$ becomes infinitesimal, the local variation df of the image values is given by

$$df = \sum_{i=1}^{n} \frac{\partial f}{\partial x_i} dx_i. \tag{13}$$

Then the square vector norm is

$$df^2 = \sum_{i=1}^{n} \sum_{j=1}^{n} (\frac{\partial f}{\partial x_i} \cdot \frac{\partial f}{\partial x_j}) dx_i dx_j, \tag{14}$$

where $\frac{\partial f}{\partial x_i}$ is the directional derivative of f along x_i. If we define h as $h_{ij} := \frac{\partial f}{\partial x_i} \cdot \frac{\partial f}{\partial x_j}, i, j = 1, ..., n$, we have the following quadratic form

$$df^2 = dx^T h dx, \quad \text{where} \quad dx = (dx_1, ..., dx_n)^T. \tag{15}$$

When $m \geq n$, we can consider an image f as a n-dimensional manifold H embedded in M, i.e., $\Phi : H \to M$. Then $\frac{\partial f}{\partial x_i}$, $i = 1, ..., n$ is a basis of the tangent space $T_x H$ at x. Thus, the quadratic form in Equation 15 is the first fundamental form of the manifold H [10,4]. It also follows that h is the metric tensor of H, which is symmetric and semi-positive definite. The quantity h is sometimes called the structure tensor in image processing. We note that a similar idea of considering an image as a surface embedded in a space-feature space has been used in [11] for scalar and color image smoothing.

For tensor-valued images, since the feature space is not Euclidean R^m but a curved manifold M with Riemannian metric g, we can not calculate the metric h of H directly. Since we have already investigated the space of M in Section 2 and introduced the metric g of M, we can overcome this problem by inducing the metric tensor h of H from the embedding $\Phi : H \to M$. Let $x_1, ..., x_n$ be the local coordinates of H, than the embedding map is

$$\begin{aligned} &(x_1, ..., x_n) \to \{\Phi_1 = \sigma_{11}(x_1, ..., x_n), ..., \Phi_k = \\ &\sigma_{ij}(x_1, ..., x_n), ..., \Phi_m = \sigma_{rr}(x_1, ..., x_n)\}, \end{aligned} \tag{16}$$

where $i \leq j$. Since the metric tensor h measures the element length of arc ds_H in H as

$$ds_H^2 = \sum_{k=1}^{n} \sum_{l=1}^{n} h_{kl} dx_k dx_l. \tag{17}$$

Similarly, for the metric tensor g on the manifold M we have

$$ds_M^2 = \sum_{i=1}^{n} \sum_{j=1}^{n} g_{ij} d\Phi_i d\Phi_j. \tag{18}$$

The embedding Φ is isometric, which means that the element length appearing in Equation 17 and 18 are equal. Using the rule of change of coordinates $d\Phi_i = \frac{\partial \Phi_i}{\partial x_k} dx_k$, the induced Riemannian metric tensor h on H is

$$h_{kl} = \sum_{i=1}^{n} \sum_{j=1}^{n} g_{ij} \frac{\partial \Phi_i}{\partial x_k} \frac{\partial \Phi_j}{\partial x_l}. \tag{19}$$

The metric tensor (or structure tensor) h of H characterises the local geometry of images. The maximum (or minimum) change of f is in the direction $v = (dx_1,...,dx_2)$, $\|v\| = 1$ that maximizes or minimizes the quadratic form df^2 in Equation 14. The maximum λ_+ and minimum λ_- eigenvalues of the structure tensor h give the maximum and mininum rate of changs of f at a given point. Their corresponding eigenvectors θ_+, θ_- are the directions of maximum and minimum changes.

For planar images where $n = 2$, $\lambda_\pm, \theta_\pm$ are given by

$$\begin{aligned} \lambda_\pm &= \frac{h_{11}+h_{22}\pm\sqrt{(h_{11}-h_{22})^2+4h_{12}^2}}{2} \\ \theta_\pm &= (2h_{12},\ h_{22} - h_{11} \pm \sqrt{(h_{11} - h_{22})^2 + 4h_{12}^2})^T. \end{aligned} \tag{20}$$

When the image is scalar-valued, $\theta_+ = \frac{\nabla I}{\|\nabla I\|}$, $\theta_- = \frac{\nabla I^T}{\|\nabla I\|}$, $\lambda_+ = \|\nabla I\|^2$, $\lambda_- = 0$. Thus, the gradient is always perpendicular to the edges for scalar images because $\lambda_- = 0$. However, for multi-valued images, such as color images and tensor-valued images, we also need to consider the minimum rate of change λ_-. It is the values of λ_+ together with λ_- that discriminate different local geometries. If $\lambda_+ \approx \lambda_- \approx 0$, the image changes at an equal rate in all directions, so the image surface is almost flat at this point. Thus there are no edges or corners here. If $\lambda_+ \approx \lambda_- >> 0$, there is a saddle point of the image surface, and the corresponding point is a corner. If $\lambda_+ >> \lambda_-$, there is step and the corresponding point is an edge. Let N be the gradient norm used to detect edges and corners in images. Three different combinations exist in the literature [12], i.e., $N_1 = \lambda_+$ [10], $N_2 = \sqrt{\lambda_+ - \lambda_-}$ [13], and $N_3 = \sqrt{\lambda_+ + \lambda_-} = \sqrt{\mathrm{tr}(h)}$ [14,15,16]. The combination N_1 neglects λ_-, and thus is the case in gray-scale images. The combination N_2 can not detect corners where $\lambda_+ \approx \lambda_- >> 0$. The combination N_3 can be used to detect both edges and corners. For volume images where $n = 3$, we could use either N_1 or N_3 for edge detection.

4 Anisotropic Diffusion

In order to obtain high quality edges, it is necessary to smooth noise before performing edge detection. In [5], Perona and Malik reported an edge preserving smoothing method using anisotropic diffusion. They use anisotropic diffusion equation to evolve gray-scale images $f(x, y) : \Omega \subset R^2 \to R$

$$\frac{\partial f}{\partial t} = \mathrm{div}(\rho(\|\nabla f\|)\nabla f), \tag{21}$$

where $\rho = e^{-\frac{\|\nabla f\|^2}{k}}$ or $\rho = \frac{1}{1+\frac{\|\nabla f\|^2}{k}}$. Their idea is to halt the heat-flow process at object boundaries. To do this they control the diffusivity using the magnitude of the image gradient. When the gradient is large, which indicates the existence of a likely edge, the value of diffusivity is small. When the gradient is small, on the other hand, the value of diffusivity is large. This method has been improved by using more sophisticated diffusion flows [17]. Here, we generalise the Perona-Malik method to tensor-valued images.

The Perona-Malik method discretises Equation 21 on a square lattice and uses the numerical scheme

$$\begin{aligned} f_{i,j}^{t+1} = f_{i,j}^{t} + \lambda\,[\,&\rho_{x_+}(f_{i-1,j} - f_{i,j}) + \rho_{x_-}(f_{i+1,j} - f_{i,j}) \\ &+\rho_{y_+}(f_{i,j-1} - f_{i,j}) + \rho_{y_-}(f_{i,j+1} - f_{i,j})\,]_{i,j}^{t} \,. \end{aligned} \tag{22}$$

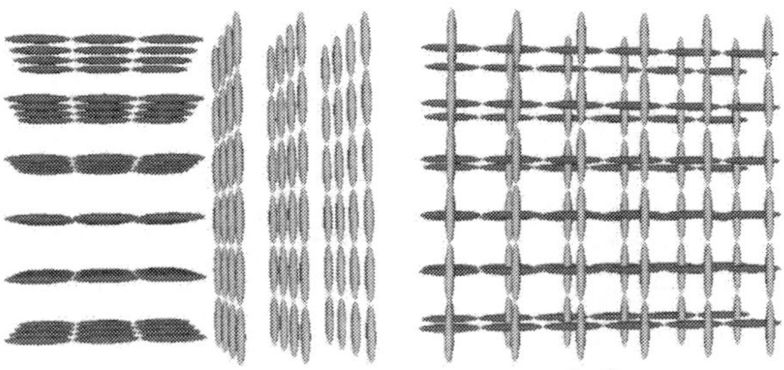

(a1) (a2)

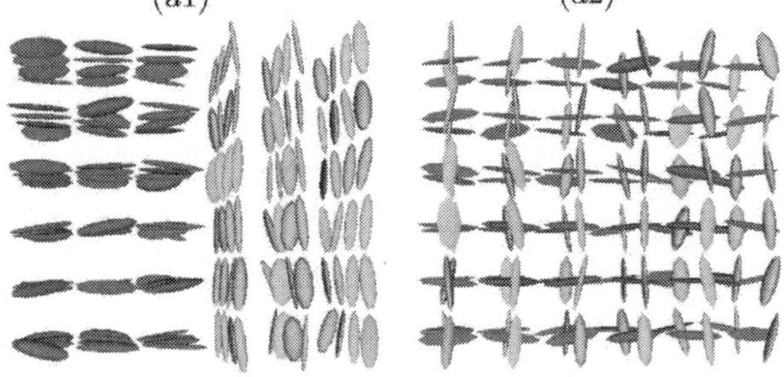

(b1) (b2)

(c1) (c2)

Fig. 1. (a1) and (a2): Synthetic tensor fields. (b1) and (b2): Corresponding noisy tensor fields. (c1) and (c2): Filtered results using generalised Perona-Malik anisotropic diffusion with $\lambda = 0.25, k = 1, 5$ iterations.

For tensor-valued images, since the feature space M is curved, we should use the intrinsic subtraction $\oplus$ and addition $\ominus$ operators on M [8] for the purposes of numerical implementation. That is, we let the image values $f(x, y)$ at the location (x, y) diffuse by marching along the geodesics emanating from this location. For two points A, B on M, we define $A \oplus B = \mathrm{Exp}_A(B)$ and $A \ominus B = \mathrm{Log}_A(B)$. For tensor-valued images, we have the following numerical scheme

$$\begin{aligned} f_{i,j}^{t+1} = \mathrm{Exp}_{f_{i,j}^t} \{ \lambda [& \rho_{x_+} \mathrm{Log}_{f_{i,j}}(f_{i-1,j}) \\ & + \rho_{x_-} \mathrm{Log}_{f_{i,j}}(f_{i+1,j}) + \rho_{y_+} \mathrm{Log}_{f_{i,j}}(f_{i,j-1}) \\ & + \rho_{y_-} \mathrm{Log}_{f_{i,j}}(f_{i,j+1})]_{i,j}^t \} , \end{aligned} \tag{23}$$

where $\rho_{x_+} = \exp(- \left\| \mathrm{Log}_{f_{i,j}}(f_{i-1,j}) \right\|^2 / k)$ and similar definition for others.

5 Experiments

We have applied our Riemannian edge detector and the generalised Perona-Malik anisotropic diffusion to synthetic and real-world tensor-valued images.

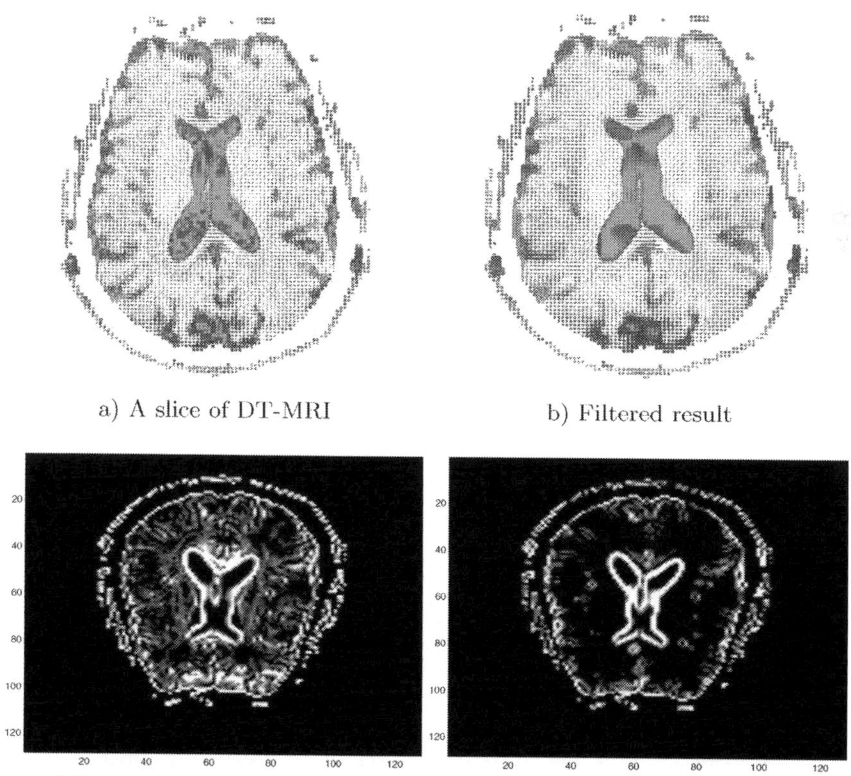

a) A slice of DT-MRI

b) Filtered result

c) Riemannian structure tensor

d) Euclidean structure tensor

Fig. 2. Real DT-MRI example

Fig. 1 shows the results of the generalised anisotropic diffusion on two synthetic noisy tenor fields. We first generate two noise-free tensor fields with different complexity. Field (a1) is fairly simple, while (a2) contains crossing fibers. We corrupt the tensor fields by adding the same quantity of independent and identically distributed (IID) additive noise to eigenvectors and eigenvalues of the tensors respectively. We then apply our algorithm to regularise the noisy tensor fields (b1) and (b2), and the results are shown in (c1) and (c2). The resulted fields show that the generalised anisotropic diffusion well preserves the interfaces between regions and recovers the fine details of the structures, whilst smoothing out the noise.

We have also tested our method on a real-world DT-MRI volume. Fig.2 shows the results of a sample slice. Subfigure (a) is the tensor image visualised using ellipsoids. (b) is the filtered result after applying the anisotropic diffusion. (c) and (d) are the trace of our Riemannian structure tensor and the Euclidean structure tensor [3,2] of the filtered slice respectively. The results shows that the Riemannian structure tensor is more sensitive for edge detection and can detect the fibres inside the image.

6 Conclusions

In this paper we have introduced the structure tensor and the anisotropic diffusion to tensor-valued images. We consider images as surfaces embedded in the space of tensors, which is a differential manifold with an affine-invariant metric. The structure tensor is then the same as the metric tensor of the image surface. Anisotropic diffusion is generalised for tensor-valued images using the exponential map and geodesic marching. Experiments shows that the generalised anisotropic diffusion is efficient to eliminate noise, and our Riemannian structure tensor is more sensitive for edge detection than the Euclidean one.

References

1. P.J. Baser, J. Mattiello and D. LeBihan : Mr diffusion tensor spectroscopy and imaging. Biophysical Journal **66** (1994) 259–267
2. L. O'Donnell, W. Grimson, C.-F. Westin : Interface detection in diffusion tensor mri. In: Proceedings of MICCAI 2004. (2004) 360–367
3. C. Feddern, J. Weickert, B. Burgeth : Level-set methods for tensor-valued images. (In: Proc. IEEE Workshop on Variational, Geometric and Level Set Methods in Computer Vision) 65–72
4. S. Di Zenzo : A note on the gradient of a multi-image. Computer Vision, Graphics, and Image Processing **33** (1986) 116–125
5. Perona, P., Malik, J.: Scale-space and edge dectection using anisotropic diffusion. IEEE Trans. Pattern Anal. and Machine Intell. **12**(7) (1990) 629–639
6. P. Fletcher and S. Joshi : Principal geodesic analysis on symmetric spaces: Statistics of diffusion tensors. (In: Proc. Computer Vision Approaches to Medical Image Analysis, ECCV Workshop (2004))

7. C. Lenglet, M. Rousson, R. Deriche and O. Faugeras : A riemannian approach to diffusion tensor images segmentation. In: Proceedings of IPMI 2005. (2005) 591–602
8. X. Pennec, P. Fillard, N. Ayache : A riemannian framework for tensor computing. International Journal of Computer Vision **66**(1) (2006) 41–66
9. M. Moakher : A differential geometric approach to the geometric mean of symmetric positive-definite matrices. SIAM Journal on Matrix Analysis and Applications **26**(3) (2005) 735–747
10. A. Cumani : Edge detection in multispectral images. CVGIP: Graphical Models and Image Processing **53**(1) (1991) 40–51
11. N. Sochen, R. Kimmel and R. Malladi: A general framework for low level vision. IEEE Trans. on Image Processing **7**(3) (1998) 310–318
12. Tschumperle, D.: PDE's Based Regularization of Multivalued Images and Applications. PHD Thesis, University of Nice-Sophia Antipolis (2002)
13. G. Sapiro and D.L. Ringach : Anisotropic diffusion of multivalued images with application to color filtering. IEEE Trans. on Image Processing **5**(11) (1996) 1582–1586
14. P. Blomgren and T.F. Chan, : Color tv: Total variation methods for restoration of vector-valued images. IEEE Trans. on Image Processing **7**(3) (1998) 304–309
15. B. Tang, G. Sapiro AND V. Caselles : Diffusion of general data on non-flat manifolds via harmonic maps theory: The direction diffusion case. International Journal of Computer Vision **36**(2) (2000) 149–161
16. A. Pardo and G. Sapiro : Vector probability diffusion. IEEE Signal Processing Letters (**8**(4))
17. F. Catte, P. L. Lions, J. M. Morel and T. Coll : Image selective smoothing and edge detection by nonlinear diffusion. SIAM J. Num. Anal. **29** (1992) 182–193

Diffusion of Geometric Affinity for Surface Integration

Roberto Fraile and Edwin Hancock

Department of Computer Science
University of York, YO10 5DD UK

Abstract. A combinatorial method is used to reconstruct a surface by integrating a field of surface normals. An affinity function is defined over pairs of adjacent locations. This function is based on the surface's principal curvature directions, which are intrinsic and can be estimated from the surface normals. The values of this locally supported function are propagated over the field of surface normals using a diffusion process. The surface normals are then regularised, by computing the weighted sum of the affinity evolved over time. Finally, the surface is reconstructed by integrating along integration paths that maximise the total affinity. Preliminary experimental results are shown for different degrees of evolution under the presence of noise.

1 Introduction

Directional information about surfaces, in the form of surface normals or gradients, is involved in several computer vision problems such as Shape-from-Shading and Photogrammetric Stereo, or, more recently, diffusion tensor magnetic resonance (DT-MRI). Integration of a field of surface normals can be exact if the vector field is integrable, that is, if the measured curl is zero. Since this is not the case for most applications, due to measurement errors, it is necessary to develop methods to estimate the most likely surface from which the surface normals have been obtained. Figure 1 illustrates surface integration over facial data.

Most surface integration methods use a variational approach [1,2,3,4], which consist of defining a suitable functional

$$J(S) = \int\int E(S, \nabla S, \mathbf{n})$$

where S is the surface to be estimated, and $\mathbf{n}$ is the surface normal information provided. Frankot-Chellappa [1] project the gradient field on integrable Fourier basis functions, and variations of this method use other families of integrable basis functions [4].

Alternatively, considering a discrete field of surface normals as a labelled grid graph, surface height estimates can be obtained by integrating along paths of the graph. Local information can be used to construct space-filling integration paths, for example, an affinity function can be defined for graph edges corresponding to affinity between the vertexes' surface normals.

D.-Y. Yeung et al. (Eds.): SSPR&SPR 2006, LNCS 4109, pp. 92–99, 2006.

The path integration approach is optimal under the assumption that at least there is one path joining every pair of surface locations, over which the height increments can be estimated. Graph-spectral methods are used to find those paths within the set of all possible paths.

Robles-Kelly [5] and Klette [6] describe path integration methods. In [5] a path is constructed using a graph-spectral analysis over the affinity matrix, defined using estimates of the sectional curvature. Several non-overlapping paths are used to cover the entire graph.

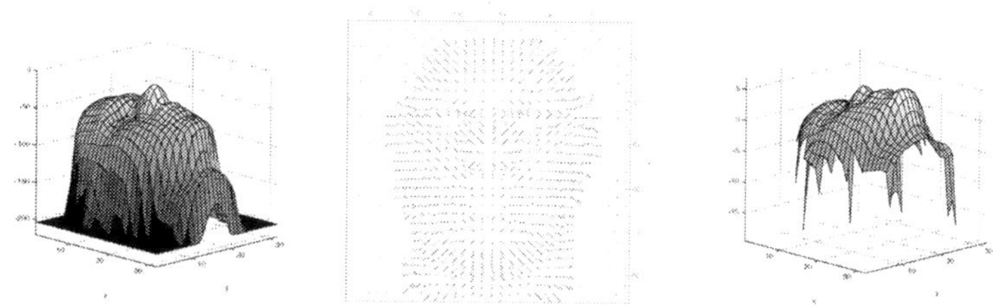

(a) Ground truth (b) Surface normals (c) Reconstruction

Fig. 1. An example of surface integration from directional data

The algorithm presented in this paper uses an intrinsic geometric property as affinity. This affinity function is used for regularisation of the surface normals. and for making a choice of integration paths. The affinity function is based on the surface's principal curvature directions, which can be estimated from the surface normals and are representation invariant. First, the affinity is propagated according to a diffusion process; second, the surface normals are modified according to those local reliability estimates; and third, the surface is integrated, from the modified surface normals. The paths overlap, thus avoiding the need for segmentation.

The values of this locally supported function are propagated over the field of surface normals following a diffusion procedure described in [7,8]. A discrete Laplacian operator is used to construct the heat kernel, which can then be evaluated at different times.

The surface normals are then regularised, by computing the weighted sum of the affinity evolved over time. Integration paths that span the entire surface and minimise the total affinity are computed using a Minimum Spanning Tree algorithm. The surface is reconstructed applying the trapezium rule piecewise along each integration path. In this approach the surface S is modelled by a spanning tree T, and the following cost function is minimised:

$$J(T) = \sum_{(i,j)\in T} \frac{1}{A_{ij}}$$

where A_{ij} is the affinity of edge (i, j).

We now proceed by describing a combinatorial method for surface integration given a local affinity function. In the following sections, the geometric affinity function is motivated, and the affinity is propagated to non-neighbouring surface locations. The field of surface normals is updated according to this extended affinity, and the combinatorial surface integration method is applied to the modified surface normals.

2 Surface Integration

In this section a combinatorial method to integrate a surface from a field of surface normals is described. This method only requires a definition of affinity between ajacent locations, and assumes that, for every two locations, the path that minimises the affinity function also minimises the integration error.

Given two adjacent locations i and j, and their surface normals $\mathbf{n}_i$ and $\mathbf{n}_j$, it is possible to estimate the height difference by applying the trapezium rule. An affinity function assigns a number between 0 and 1 to each pair of adjacent surface normals i and j, estimating the reliability of integration along the edge joining i and j. Therefore an affinity matrix is defined $A = \{a_{ij}\}_{i,j \in V}$.

Consider the graph $G = (E, V)$ whose vertexes V are the surface normal locations, and whose edges E are all the pairs of adjacent locations (typically the 4-neighbours). Let W be the array whose entries are inverse of the entries in A. In this way we have assigned weights to the edges of the graph.

A first requirement for a path integration method is that every two locations are connected by a path, so that the height can be estimated for all locations. A second requirement is that there is only one path between every two points, so that surface height estimates are unique. Therefore the paths will form a spanning tree T over G. A spanning tree that minimises the total weights given by W, or, equivalently, maximises the total affinity, can be obtained by using a Minimum Spanning Tree algorithm [9].

The surface can then be reconstructed by integration along the edges of T, applying the trapezium rule to the surface normals. This optimisation procedure is independent of the affinity function, and the problem of obtaining adecuate height estimates is reduced to defining an appropriate affinity function. Figure 2 illustrates the notion of integration tree.

For this path-based integration method, it requires that minimising the affinity function is equivalent to finding a path over which integration is valid. Note

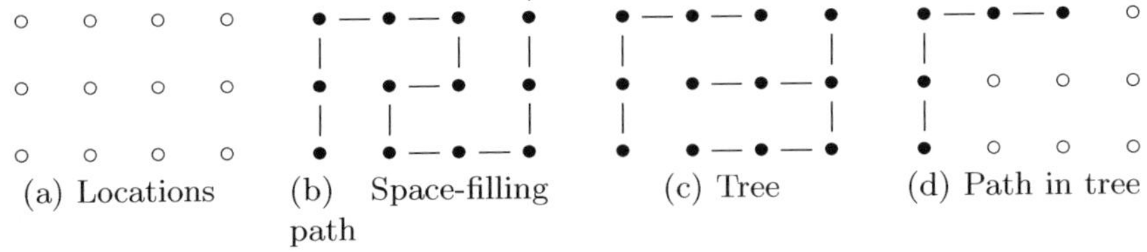

Fig. 2. Paths defined over the set of locations. Each location is labeled with a 3D surface normal vector. The surface is integrated over a path using the trapezium rule.

that this method does not require the field of surface normals to have zero curl, which would imply that *every* path joining every two locations provides correct height estimates.

The practical use of such a path-based integration method is limited by the extent to which the affinity function can capture global structure of the graph, and also by the quality of the surface normals. The field of surface normals contains redundant information because the trapezium rule only uses one of the two components of the gradient. It seems therefore reasonable to integrate over a modified field of surface normals in which the redundant information has been propagated along the graph G. The proposed way of doing this involves a local geometric affinity function, and a diffusion process to take into account the global structure of the graph G.

3 Geometric Affinity: Principal Curvature Directions

The affinity function needs to be defined for each pair of adjacent locations. In order to assign greater affinity to lower risk of integration errors, it should be monotonic with any distance defined over the surface normals as Euclidean vectors. It is also desirable that the affinity function does not depend on the way the surface normals have been sampled from the surface.

The field of surface normals is sampled from a surface, it is therefore possible to estimate intrinsic properties of the surface from the surface normals. One such intrinsic property is the principal curvature and its directions. A measure of geometric affinity between two surface locations is given by how close the direction linking them is to the minor principal curvature direction. (In regions where the principal curvature directions are not defined, the affinity can be considered the same in all directions). Figure 3 illustrates the principal curvature directions for a simple surface.

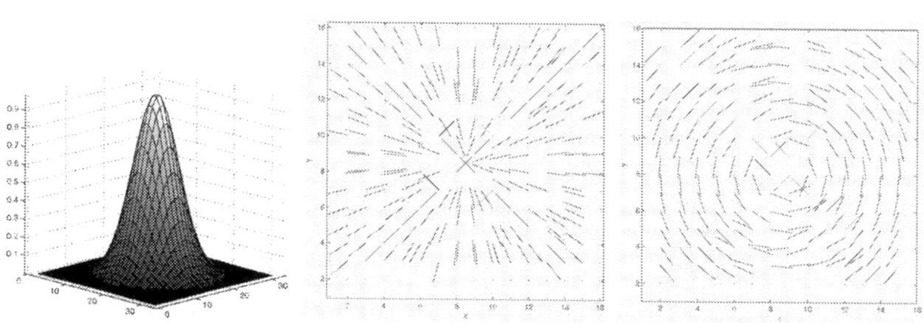

(a) Surface (b) Major directions (c) Minor directions

Fig. 3. Principal curvature directions estimated from a field of surface normals

To estimate the principal curvature directions from the surface normals, consider S, the surface function. The Hessian matrix, which can be used to calculate the principal curvature directions, is constructed using the second derivatives at each point.

The Hessian matrix at each location,

$$H = \begin{pmatrix} \partial_{xx} & \partial_{xy} \\ \partial_{yx} & \partial_{yy} \end{pmatrix} S$$

can be obtained by estimating the second partial derivatives of S at each point using finite differences over the first partial derivatives.

The first derivatives $\partial_x S$ and $\partial_y S$ are obtained from the normal vector (n_x, n_y, n_z)

$$\begin{pmatrix} n_x \\ n_y \\ n_z \end{pmatrix} = -n_z \begin{pmatrix} \partial_x S \\ \partial_y S \\ -1 \end{pmatrix}$$

The partial derivatives $\partial_x S$ and $\partial_y S$ are defined whenever n_z is nonzero. This is the case, for example, when data is available in shape-from-shading.

The second derivatives can be approximated using a finite difference operator. Let $\mathbf{v}_i$ be the coordinates of location i, and $\mathbf{e}$ a unit 2D vector.

$$\partial_{\mathbf{e}} f(\mathbf{v}_i) \simeq \Delta_{\mathbf{e}} f(\mathbf{v}_i) = \frac{f(\mathbf{v}_i + \mathbf{e}) - f(\mathbf{v_i} - \mathbf{e})}{2}$$

The eigensystem of H consists of the principal curvatures and their directions. For most surfaces the eigenvectors $\mathbf{h}_{\min}$ and $\mathbf{h}_{\max}$ of H are not linearly dependent, and form a basis of the 2D space.

The edge joining two adjacent locations l_1 and l_2, corresponds to a direction vector $\mathbf{e}$, which can be represented in the eigensystem of H:

$$\mathbf{e} = \alpha \mathbf{h}_{\min} + \beta \mathbf{h}_{\max}$$

Therefore an geometric measure of affinity α to each pair of adjacent locations l_1 and l_2. Greater geometric affinity corresponds to directions closer to the minor principal curvature direction.

We have obtained an affinity function that can be used to calculate an affinity matrix A, which is symmetric with size $|V| \times |V|$. Without loss of generality, we can assume that A is normalised so that each row adds up to 1. This measure of affinity is only defined locally, we now proceed to propagate affinity across the field of surface normals.

4 Propagating Affinity Using the Heat Kernel

The affinity matrix A embodies local information, in the form of geometric affinity between adjacent locations. The matrix A can be considered as a transition probability between locations in the graph of locations. A path joining two locations i and j is a subset or E and therefore has an probability induced by A.

We would like to calculate the affinity between every two locations, not only those which are neighbouring in the graph. To do so, let us consider a random walk over G whose transition probability is given by A, and calculate the probability of the random walk joining locations i and j in t steps. This corresponds to a diffusion process in which the conductivity is given by A [7,8].

Let $\hat{L}$ be the normalised Laplacian associated to A, $\hat{L} = D^{-\frac{1}{2}} L D^{\frac{1}{2}}$, where $L = D - A$ and D is the diagonal matrix such that $D_{ii} = \sum_j A_{ij}$. The matrix $\hat{L}$ can be seen as a discrete approximation of the continuous Laplacian operator in the following diffusion equation [7]

$$\partial_t H = -\hat{L} H \tag{1}$$

The solution $H(t)$ can be calculated by matrix exponentiation

$$H(t) = e^{-t\hat{L}}$$

Each entry (i, j) of the matrix $H(t)$ can be interpreted as the probability of a random walk joining locations i and j, after t steps, given the transition probability A. Note that the size of both A and $H(t)$ is $|V| \times |V|$.

The presence of a path in G of high affinity between two nodes i and j increases the probability $H(t)_{ij}$. The matrix $H(t)$ can now be used to modify the surface normals.

5 Updating the Surface Normals

While the field of surface normals contains redundant information, integration along paths, which effectively are subsets of E, discards all surface normal information not in the direction of the edges that form the path. In order to make use of redundant information before the integration step, let us modify the surface normals using the transition probability $H(t)$ obtained in the previous section. This is a generalisation of a simple average of neighbouring surface normals, and is similar to subjecting the surface normals to a process of anisotropic diffusion [10].

The updated normals corresponding to a random walk of length t, $N(t) = \{\mathbf{n}(t)_i\}_{i \in V}$, are defined as a weighted sum of the surface normals:

$$\mathbf{n}(t)_i = \sum_j H(t)_{ij} \mathbf{n}_j \tag{2}$$

where the indexes i and j visit all locations.

As a result, the field of surface normals $N(t)$ is the weighted sum over all locations, with weights given by the probability of a random walk of length t joining those two locations. The transition probability for the random walks was given by the geometric affinity matrix A. We will use the modified surface normals $N(t)$ to perform path integration.

6 Experiments

Experiments have been performed with surface normal data corresponding to a human face. A field of surface normals was the input, and the output was the

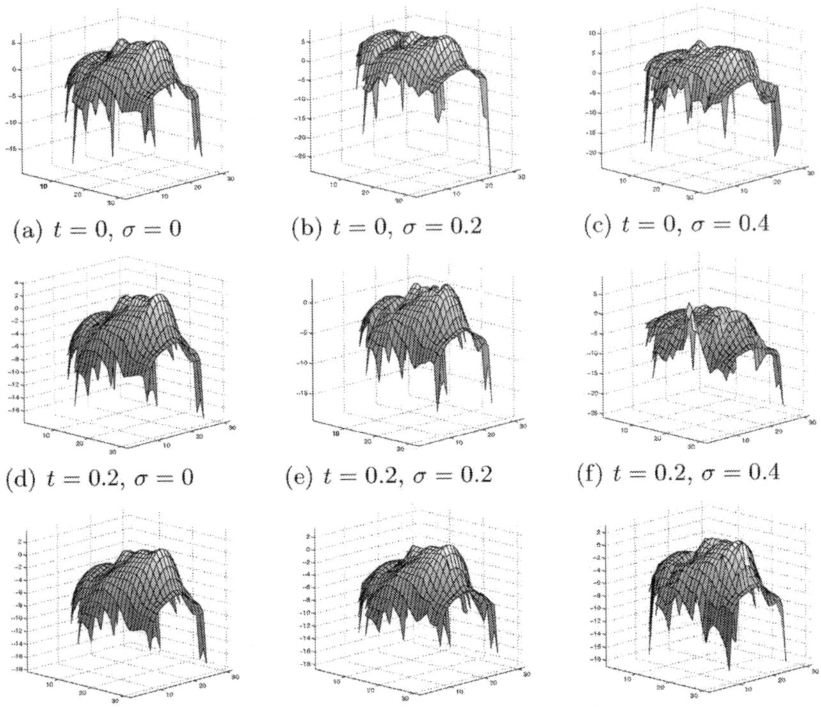

(a) $t = 0$, $\sigma = 0$ (b) $t = 0$, $\sigma = 0.2$ (c) $t = 0$, $\sigma = 0.4$

(d) $t = 0.2$, $\sigma = 0$ (e) $t = 0.2$, $\sigma = 0.2$ (f) $t = 0.2$, $\sigma = 0.4$

(g) $t = 0.4$, $\sigma = 0$ (h) $t = 0.4$, $\sigma = 0.2$ (i) $t = 0.4$, $\sigma = 0.4$

Fig. 4. Surface reconstruction for varying levels of Gaussian noise σ (columns) and time t (rows)

reconstructed surfaces. The performance was assessed with Gaussian noise added to the field of surface normals, in order to simulate a source of measurement noise. The experiment parameters are therefore the standard deviation of the Gaussian noise, and the time parameter t used to evaluate the heat kernel.

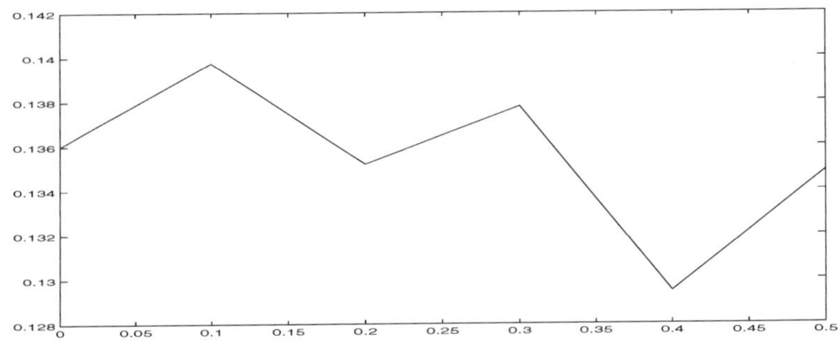

Fig. 5. Reconstruction error for $t = 0.4$, measured in RMS as a function of σ. The minimum corresponds to the bottom-right reconstruction in Figure 4.

The diffusion parameter t produced a reasonable smoothing of the field of surface normals for values below 1. For example, using diffusion parameter $t = 0.4$, the reconstruction is not affected by noise of parameter $\sigma < 0.5$. Figure 4 illustrates the reconstruction from a field of size 32×32, for varying noise and time. (A raised chin usually corresponds to a lower reconstruction error, even when detail is not recovered). Figure 5 shows the RMS error between the ground truth and the reconstructed surface, for this value of t.

7 Conclusion

We have presented a method for path-based surface integration whose only parameters are a local affinity function and a time parameter t. We have also presented an affinity function based on an intrinsic geometric property of the surface being reconstructed, namely the principal curvature directions. The affinity is propagated over the graph of surface locations using a diffusion process. The result is used to re-weight the field of surface normals in order to make use of its spatial redundancy.

A future direction for this work is to interpret the diffusion process in the anisotropic diffusion framework presented by [4], and to state explicitly the relationship between the affinity function and the error model for the field of surface normals in a probabilistic setting.

References

1. Frankot, R.T., Chellappa, R.: A method for enforcing integrability in shape from shading algorithms. IEEE Trans. Pattern Analysis and Machine Intelligence **10** (1988) 439–451
2. Zhang, R., Tsai, P.S., Cryer, J., Shah, M.: Shape-from-shading: a survey. IEEE Transactions on Pattern Analysis and Machine Intelligence **21** (1999) 690–706
3. Agrawal, A., Chellappa, R.: An algebraic approach to surface reconstruction from gradient fields. In: Proceedings of ICCV 2005. (2005) 23–114
4. Agrawal, A., Chellappa, R.: What is the range of surface reconstructions from a gradient field. To appear in the proceedings of ECCV 2006 (2006)
5. Robles-Kelly, A., Hancock, E.: Steady state random walks for path estimation. In: Structural, Syntactic, and Statistical Pattern Recognition. Volume 3138 of LNCS., Springer (2004) 143–152
6. Klette, R., Schlüns, K.: Height data from gradient fields. In: Proceedings of SPIE. Number 2908 (1996) 204–215
7. Kondor, R.I., Lafferty, J.: Diffusion kernels on graphs and other discrete structures. ICML 2002 (2002)
8. Zhang, F., Qiu, H., Hancock, E.: Evolving spanning trees using the heat equation. In: Int. Conf. on Computer Analysis of Images and Patterns. (2005)
9. Gibbons, A.: Algorithmic Graph Theory. Cambridge University Press (1985)
10. Perona, P., Malik, J.: Scale space and edge detection using anisotropic diffusion. IEEE Transactions on Pattern Analysis and Machine Intelligence **12** (1990) 629–639

Comparative Study of People Detection in Surveillance Scenes

A. Negre, H. Tran, N. Gourier, D. Hall, A. Lux, and J.L. Crowley

Institut National Polytechnique de Grenoble, Laboratory GRAVIR, INRIA Rhone-Alpes, France

Abstract. We address the problem of determining if a given image region contains people or not, when environmental conditions such as viewpoint, illumination and distance of people from the camera are changing. We develop three generic approaches to discriminate between visual classes: ridge-based structural models, ridge-normalized gradient histograms, and linear auto-associative memories. We then compare the performance of these approaches on the problem of people detection for 26 video sequences taken from the CAVIAR database.

1 Introduction

Many video-surveillance systems require the ability to determine if an image region contains people. This problem can be considered as a specific case of object classification in which there are only two object classes: person and non-person. Object classification in general is difficult because it has to face different kinds of imaging conditions. People detection is even harder due to the high variation of human appearance, gait, as well as the small size of human region which prevents face or hand recognition. Numerous efficient appearance-based approaches exist for object recognition [8,2]. However, such techniques tend to be computationally expensive. Video-surveillance systems must run at video-rate and thus require a trade-off between precision and computing time.

To speed up the classification, simpler methods have been proposed. In [4], the authors only use compactness measure computed on the region of interest to classify car, animal or person. This measure is simple but sensitive to scale and affine transformations. Moreover, this method highly depends on segmentation, which remains a primitive problem. In [1] and [12], the contour is used to modelize deformable shapes of a person. However, the person must be represented by a closed contour. These methods strongly depend on contour detection or segmentation techniques.

This paper presents three methods for determining the presence of people in an imagette. Two methods use ridges as structural features to model people: the structural method uses a set of main human components like legs, torso, and the statistical method describes humans by modified SIFT based descriptor. The third method uses global appearance information of the detected region to discriminate between person and non-person. This method inherits strong points of

D.-Y. Yeung et al. (Eds.): SSPR&SPR 2006, LNCS 4109, pp. 100–108, 2006.

appearance based vision: simplicity and independance from the detection technique. In the following, we expose each method and compare their performance. Our objective is to show the advantages as well as drawbacks of appearance-based object classification approaches and structural feature based approaches, experimented in case of people. This comparative study motivates the use of a multi-layer object classifier to improve the detection rate.

2 Local Feature Extraction in Scale-Space

Everyday objects typically exhibit significant features at several different scales. To describe such structures of different sizes, images must be analysed in scale space. The scale-space representation of an image is a continuous space $L(x, y, \sigma)$ obtained by convolution of the image $I(x, y)$, with a Gaussian $G(x, y; \sigma)$: $L(x, y, \sigma) = G(x, y; \sigma) * I(x, y)$ where $G(x, y; \sigma) = \frac{1}{2\pi\sigma^2} e^{-(x^2+y^2)/2\sigma^2}$.

Natural interest points are local extrema in Laplacian scale space. Such points correspond to the center of blob-like structures and are widely used as key-points for scale invariant indexing and matching. Such a description provides a reliable method for object detection and description. However, natural interest points are well suited for compact objects, but tend to become unstable in the presence of elongated objects.

We extend natural interest points to describe elongated objects with natural interest lines. In addition of providing a more reliable scale normalization, natural interest lines also provide local orientation information and affine normalization. As with natural interest points, the value of σ for the maximal scale corresponds to the half-width of the object. At this scale, the amplitude of the Laplacian exhibits a ridge. The mathematical definition of a ridge point on a surface is as follows: given a scale space $L(x, y, \sigma)$, a ridge point at scale σ is a point at which the signal $L(x, y, \sigma)$ has a local extremum in the direction of the largest surface curvature. The ridge detection method used in this paper is described in full detail in [9].

3 Human Recognition Based on Structural Model

To represent a person in a structural manner, some authors use silhouettes [1,4], or skeletons [5] and study changes of the model (like head, hand, legs, ...) in the time to analyse person movement. This representation strongly depends on the segmentation algorithm which is a primitive problem in computer vision. Ridges represent centerlines of an oblong structure. At an appropriate scale, it represents a skeleton of the object. Ridges at several scales capture more information about the object.

Figure 1 shows imagettes of a person extracted from a walking sequence of the CAVIAR[1] database. On these imagettes, we overlay ridges and blobs (extrema of Laplacian in 3 dimensions) detected in the region of interest. It is interesting

[1] http://homepages.inf.ed.ac.uk/rbf/CAVIAR/caviar.htm

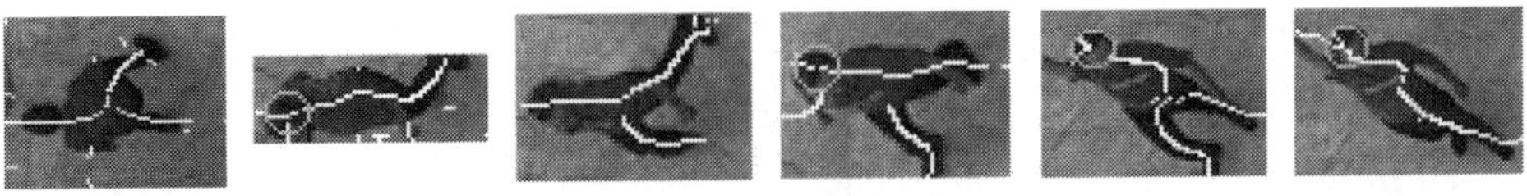

Fig. 1. Different configurations of a person represented by ridges (lines) and blobs (circles) at scale $\sigma = 4\sqrt{2}$

to see that ridges not only represent torso, legs and other significant parts of a person, but also changes in configuration of the person. We propose to model a person by using ridges representing person parts, more precisely torso and legs.

3.1 Extracting Ridges in Region of Interest

Given a region of interest, we want to know at which scale ridges should be detected. If the region perfectly fits the person, the scale to detect ridges corresponding to torso is exactly equal to the half of the region width and the scale to detect ridge corresponding to legs is quarter width. This is straightforward for a rectangle. If the region is defined by a contour, the width and the height of a region are deduced from its second moments.

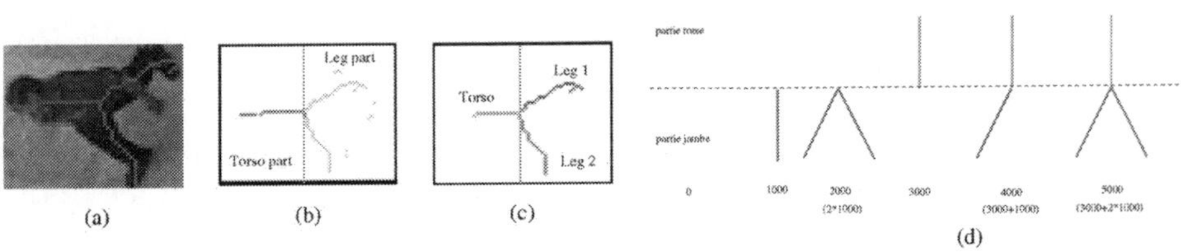

Fig. 2. (a) Ridges detected at scale related to the width of region. (b-c) Seleted ridges corresponding to torso and legs of person. (d) 5 configuration possibilities for each person.

Experimentation on ridge detection shows that with the use of the Laplacian, some ridges representing the same structures of objects are repeated at several scales. This also happens with persons: ridges detected at torso scale in the leg part represent well the legs (as we see in figure 1). Therefore, we propose to begin with ridges detected only at torso scale. In this manner, we only work at the scale corresponding to the size of the person.

3.2 Determining Major Ridges Corresponding to Torsos and Legs

Knowing the orientation of a person, we cut the region into two parts by the smaller main axis (figure 2b) and take for torso part the longest ridge, the second longest for leg part (figure 2c). The detected ridges have to be significant in energy and length. Only ridges having length and average Laplacian bigger than a threshold are considered. There may be no ridge satisfying the above condition in torso part or there is only zero/one ridge in leg part. This is the case of a person wearing a T-shirt or a trouser of same colour as the background or a partially hidden person. It is not important because it makes the model robust to partial occlusion. Using ridges, a person can be in one of the configurations presented in figure 2d.

3.3 Constructing Descriptors

We represent a configuration of a person by a vector of 10 components determined from 3 ridges detected previously: (N, θ_1, len_1, dis_1, θ_2, len_2, dis_2, θ_3, len_3, dis_3). The first component is the number n of ridges we take from torso part and leg part of the region of interest. n can be 0, 1, 2, 3. As n = 1 (torso ridge or leg ridge) and n = 2 (torso ridge + leg ridge or 2 leg ridges) do not represent an unique configuration. We assign a weight to each ridge in the model in function of its importance (for example 1 for leg ridge and 3 for torso ridge). n is now converted into a sum of weighted ridge number. This means $\{0, 1, 2, 3, 4, 5\}$.

The nine following components are 3 triplets (angle between ridge and main axis, ridge length normalized to scale, distance from ridge center to region center normalized to scale). Among the ten components in the descriptor, the first component is the most significant because it represents the configuration of a person. For this reason, we give a strong weight to the first component (1000 in our experimentation), and normalize all other components by their maximal values. These values are learnt from the groundtruth: $\theta_{max} = 2\pi$, $len_{max} = 35, dis_{max} = 17$.

4 Ridge Normalized Gradient Histograms

Based on observation that human silhouette can be represented by a long ridge, we propose an another approach that describes human region by a SIFT based descriptor. More precisely, we extract the main ridge to obtain a local reference invariant to orientation and scale. A gradient histogram is computed in this reference system.

4.1 Computing Ridge Properties

The first step consists in detecting and separating each ridge structure in scale space. We begin to compute ridges at each scale level as seen in the previous section. In order to obtain video-rate performance, a pyramidal algorithm described in [2] is used to compute the Laplacian scale space. Ridge structures are obtained by connected component analysis in this scale space.

We then obtain a set of ridge points $X_{n=1..N} = (x_n y_n s_n)^T$ where x_n and y_n represent the position in the image and s_n represent the scale. In order to obtain a local reference of the ridge, we compute the first and second moments of these feature points. For more robustness, each point is weighted by its Laplacian value. As we work in a down-sampled pyramid, we weight each point by 2^{k_n} where k_n represents the stage in the pyramid. The result of ridge description is a set of ridge lines, characterized by the position of the center of gravity of the ridge points, as well as the orientation of the ridge (x, y, σ, θ). In the following section, we will see how to use such a representation to describe and to recognize objects.

4.2 Statistical Description of Ridges

We experiment a statistical description of ridges inspired by the SIFT descriptor [6] and Gaussian Receptive Field Histograms [7]. The descriptor is based on an array of gradient histograms. Our original contribution is to normalize each gradient measure using the intrinsic scale and the orientation of the most contrasted ridge in the imagette (cf. fig.3). After building a local reference from ridge parameters, the gradient (L_x, L_y) is computed for each pixel in the imagette at a scale $\sigma_c = \alpha\sigma_i$ where σ_i is the average scale of the ridge and α is a constant. A typical value of α is 0.5. This scale is found empirically and corresponds to the boundary information of the blob described by the ridge.

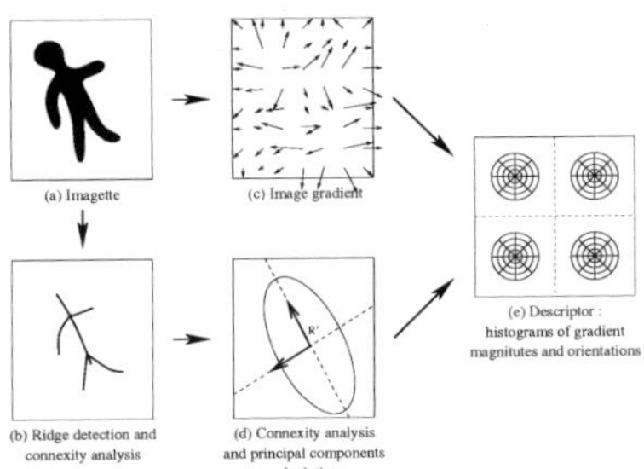

Fig. 3. Calculation of the ridge descriptor : ridge extraction and connectivity analysis are computed to obtain a set of ridge objects (b). The main ridge is selected and the first and second order moments are computed to obtain a local reference (d). The descriptors are then created by computing the image gradient (c), rotated by the principal direction of the ridge. The gradient orientation and magnitude are then accumulated into histograms (e).

Gradient magnitude is normalized by the average amplitude of the Laplacian of the ridge in order to correct for variations in illumination. The gradient orientation is rotated relatively to the orientation of $\mathcal{R}'$. This normalized gradient field of the imagette is divided into four regions, and the statistics of the gradient magnitudes and orientations for each region is collected in a histogram (fig.3(e)). A Gaussian weighting function γ is used to assign more importance to centered points. The γ function is defined by the ridge properties :

$$\gamma(x, y) = e^{-\frac{x_{\mathcal{R}'}^2}{2\sigma_1^2} - \frac{y_{\mathcal{R}'}^2}{2\sigma_2^2}}$$

Where $(x_{\mathcal{R}'}, y_{\mathcal{R}'})$ are the position of the considering point in the reference $\mathcal{R}'$ and λ_1 is the greatest eigenvalue of the ridge covariance matrix. When the histogram is computed, a four-point linear interpolation is used to distribute the value of the gradient in adjacent cells, in order to minimize boundary effects. Moreover, to make comparisons, the gradient histogram is normalized into each region.

5 Recognizing People Using Linear Auto-associative Memories

As a global approach, auto-associative memories use the entire appearance of the region of interest. The main advantage of this kind of approach is that no landmarks or model has to be computed, only the objects has to be detected. Global approaches can also handle very low resolutions. A popular method for template matching is PCA [10], but this tends to be sensitive to alignment, and the number of dimensions has to be specified. Neural nets also have been used. However, the number of cells in hidden layers is chosen arbitrarily.

We adapt auto-associative memory neural networks by using the Widrow-Hoff learning rule [11]. As in ridge extraction, the tracker detects bounding boxes and main orientation for each object in the scene. We use these informations to create grey value imagettes normalized in size and orientation as in [3]. This normalization step provides robustness to size, chrominance, alignment and orientation.

5.1 Linear Auto-associative Memories

Linear auto-associative memories are a special case of one-layer linear neural networks where input patterns are associated with each other. Each cell corresponds to an input pattern [11]. Auto-associative memories aim to associate each image with its respective class, and to recognize learned images when input images are degraded or partially occluded. We describe a grey-level input image by a normalized vector $x = \frac{x'}{\|x'\|}$. m images of n pixels of the same class are stored into a n x m matrix $X = \big(x_1, \ldots, x_m\big)$. The linear auto-associative memory of the class k is represented by the connexion matrix W_k. The reconstructed image y_k is obtained by computing the product between the source image x and the connexion weighted matrix $W_k : y_k = W_k \cdot x$. We measure the similarity between the source image and a class k of images by taking the cosine between x and y_k : $cos(x, y) = x \cdot y^T$. A score of 1 corresponds to a perfect match. The connexion matrix W_k^0 is initialized with the standard Hebbian learning rule $W_k^0 = X_k \cdot X_k^T$. Reconstructed images with Hebbian learning are equal to the first eigenface of image class. To improve recognition abilities of the neural network, we learn W_k with the Widrow-Hoff rule.

5.2 Widrow-Hoff Correction Rule

The Widrow-Hoff correction rule is a classical local supervised learning rule. It aims to minimize the difference between desired and given responses for each ell of the memory. At each presentation of an image, each cell modifies its weights from the others. Images X of the same class are presented iteratively with an adaptation step η until all are classified correctly. This corresponds to a PCA with equalized eigenvalues. As a result, the connexion matrix W_k becomes spherically normalized. The Widrow-Hoff learning rule can be described by:

$$W_k^{t+1} = W_k^t + \eta \cdot (x - W_k^t \cdot x) \cdot x^T$$

In-class images are little degraded by multiplying with the connexion matrix. In opposite, extra-class images are strongly degraded. Imagettes of the same class are used for training an auto-associative memory using the Widrow-Hoff correction rule. Prototypes of image classes can be recovered by exploring the memory. In opposite, prototypes can not be recovered with non-linear memories. Auto-associative classification of different class is obtained by comparing input and reconstructed images. The class which obtains the highest score is selected. We train two auto-associative memories for classes 0 and $n \geq 1$ persons.

6 Comparative Performance Evaluation

We evaluate the three techniques in the context of video-surveillance by determining if an image region contains people or not. Our training database consists of 12 video sequences which contain about 20000 people whose regions of interest are labelled in CAVIAR database. The two ridge-based methods compute human descriptors from imagettes in the training sequences and learn the descriptors by using KMeans algorithm. 34 human descriptors have been learnt in the first method and 30 in the second. The third method based on associative memories needs to learn people examples as well as non-people examples. For this, we created two sequences of the background and taken random imagettes from these sequences. Two matrices have been learnt and they are considered as people model and non-people model. For test, we use 14 sequences including 12 other sequences in CAVIAR database and 2 background sequences. These sequences contain 9452 people and 4990 non-people regions. Ridge-based methods measure the similarity as the euclidian distance between two vectors of descriptors in the first method and the χ^2 distance in the second method. The third method computes directly the cosine between the imagette with the recontructed imagettes. The three similarity measures are normalized and thresholded to dertemine the presence of people.

Table 1. Comparaison of recognition methods

Method	People		Others	
	Recall	Precision	Recall	Precision
Ridge based Structual Model	0.80	0.90	0.80	0.70
Ridge based Normalized Histogram	0.90	0.93	0.80	0.73
Linear Auto-associatives Memories	0.99	0.96	0.70	0.90
Modified SIFT	0.77	0.90	0.75	0.51

Table 1 shows the perfomance of 4 human classification techniques: three techniques presented in the previous sections and one technique using SIFT descriptor computed at the most significative interest point detected in the imagette. This method uses the same technique for learning and testing than the second method. We can observe that the technique based on associative memories performs best. The reason is that this method has learnt person examples as well as non-person examples as the two first methods based on ridge learnt only person examples. If we do not train a non-people class, it gives the worst result because this method used only one model to represent all variations in the human classe. So it can not discriminate non-peolple from people. This method is good for people identification and can help for split-merge detection.

The statistical descriptor computed on ridge region gives better results than the structural descriptor. This is explained by the fact that the first method considers also one ridge as human model. Consequently, all regions containing one ridges are classified as people regions. This method requires more parameters and human knowledge than ridge histograms, but can recover people configuration. The second method gives good result in general case but presents some drawbacks when human is partially occulted or affected by light or shadow. In these cases, the detected ridge does not correspond to the global shape of the human. Therefore, the descriptor is built on nearby region but not centered on human region. Modified SIFT performs worst, because interest points are less stable than ridges for representing elongated structure like human shape. Linear auto-associative memories are disrupted when people walk through shadow areas, but can recognize configurations which do not exhibit ridges, such as people crouching down.

7 Conclusion

We proposed 3 different approaches for entity recognition in video sequences. Two approaches are based on local features: the ridge configuration model and the ridge normalized gradient histograms. The third one, linear auto-associative memories, is based on global appearance. Ridge normalized gradient histograms are robust to illumination changes, whereas auto-associative memories are sensitive to it. Ridge configuration models are robust to global illumination changes, but are disrupted in case of local changes. Ridge normalized gradient histograms also provide an estimation of the size and orientation of the object. As a global approach, auto-associative memories do not need to compute a model for persons and run at video-rate, but have to learn a 0 person class to be efficient. Ridge-based approaches can be disrupted by neighboorhoods of pixels, whereas auto-associoative memories are robust to partial changes in the imagette.

We believe all three approaches can be extended to other cognitive vision problems. Ridge configuration models can be useful for gait and number of people estimation. However, this method requires specific adaptation to other object categories. Ridge normalized gradient histograms are well-suited to the discrimination of other objects, provided that these objects exhibit a main ridge. We can improve the recognition process by combining all three methods: Ridge-based methods localize objects and detect their size and main orientation using their main ridge. The image region can be normalized into a fixed size imagette to be compared to appearance prototypes constructed by linear auto-associative memories or ridge normalized gradient histograms. People configuration and gait can be described by ridge structural model.

References

1. A. M. Baumberg and D. C. Hogg. Learning flexible models from image sequences. Technical report, University of Leeds, October 1993.
2. J. L. Crowley D. Hall and V. Colin de Verdière. View invariant object recognition using coloured receptive fields. *Machine GRAPHICS and VISION*, 9(2):341–352, 2000.
3. N. Gourier, D. Hall, and J.L. Crowley. Estimating face orientation from robust detection of salient facial features. In *Proceedings of Pointing 2004, ICPR International Workshop on Visual Observation of Deictic Gestures*, pages 17–25, August 2004.
4. I. Haritaoglu, D. Harwood, and L. S. David. Hydra: Multiple people detection and tracking using silhouettes. In *Second IEEE Workshop on Visual Surveillance*, Fort Collins, Colorado, 26 June 1996.
5. M. K. Leung and Y. H. Yang. First sight: A human body outline labeling system. *IEEE Transaction on Pattern Analysis and Machine Intelligence*, 17(4):359–377, April 1995.
6. D. G. Lowe. Distinctive image features from scale-invariant keypoints. In *IJCV*, volume 60, pages 91–110, 2004.
7. B. Schiele and J.L. Crowley. Recognition without correspondence using multidimensional receptive field histograms. 36(1):31–50, January 2000.
8. C. Schmid. *Appariement d'images par invariants locaux de niveaux de gris.* PhD thesis, Institut National Polytechnique de Grenoble, 1996.
9. H. Tran and A. Lux. A method for ridge detection. In *Asean Conference on Computer Vision*, pages 960–966, Jeju, Korea, January 2004.
10. M. Turk and A. Pentland. Eigenfaces for recognition. *Cognitive Neuroscience*, 3(1):71–96, 1991.
11. D. Valentin, H. Abdi, and A. O'Toole. Categorization and identification of human face images by neural networks: A review of linear auto-associator and principal component approaches. *Journal of Biological Systems*, 2:413–429, 1994.
12. L. Zhao. *Dressed Human Modeling, Detection, and Part Localization.* PhD thesis, The Robotics Institute Carnegie Mellon University, 2001.

A Class of Generalized Median Contour Problem with Exact Solution

Pakaket Wattuya and Xiaoyi Jiang

Department of Mathematics and Computer Science
University of Münster, Germany
{wattuya, xjiang}@math.uni-muenster.de

Abstract. The ability to find the average of a set of contours has several applications in computer vision including prototype formation and computational atlases. While contour averaging can be handled in an informal manner, the formal formulation within the framework of generalized median as an optimization problem is attractive. In this work we will follow this line. A special class of contours is considered, which start from the top, pass each image row exactly once, and end in the last row of an image. Despite of the simplicity they frequently occur in many applications of image analysis. We propose a dynamic programming approach to exactly compute the generalized median contour in this domain. Experimental results will be reported on two scenarios to demonstrate the usefulness of the concept of generalized median contours. In the first case we postulate a general approach to implicitly explore the parameter space of a (segmentation) algorithm. It is shown that using the generalized median contour, we are able to achieve contour detection results comparable to those from explicitly training the parameters based on known ground truth. As another application we apply the exact median contour to verify the tightness of a lower bound for generalized median problems in metric space.

1 Introduction

The ability to find the average of a set of contours has several applications in computer vision including prototype formation and computational atlases. While contour averaging can be handled in an informal manner as done in [1,11], the formal formulation within the framework of generalized median as an optimization problem is attractive. In this work we will follow this line.

Given a set of n patterns $C_1, C_2, \ldots, C_n$ in an arbitrary representation space U, we assume a distance function $d(p,q)$ to measure the dissimilarity between any two patterns $p, q \in U$. Then, the generalized median $\overline{C}$ is defined by:

$$\overline{C} = \arg\min_{C \in U} \sum_{i=1}^{n} d(C, C_i) \qquad (1)$$

This concept has been successfully applied to strings [7,9] and graphs [5] in structured pattern recognition.

D.-Y. Yeung et al. (Eds.): SSPR&SPR 2006, LNCS 4109, pp. 109–117, 2006.

If a contour is coded by a string, then the same procedure can be adapted to averaging contours [7]. However, this general approach suffers from high computational complexity. It is proved in [4] that computing the generalized median string is NP-hard. Sim and Park [12] proved that the problem is NP-hard for finite alphabet and for a metric distance matrix. Another result comes from computational biology. The optimal evolutionary tree problem there turns out to be equivalent to the problem of computing generalized median strings if the tree structure is a star (a tree with $n+1$ nodes, n of them being leaves). In [13] it is proved that in this particular case the optimal evolutionary tree problem is NP-hard. The distance function used is problem dependent and does not even satisfy the triangle inequality. All these theoretical results indicate the inherent difficulty in finding generalized median strings, or equivalently the generalized median contours. Not surprisingly, researchers make use of domain-specific knowledge to reduce the complexity [9] or resort to approximate approaches [7].

In this work we consider a special class of contours for which the generalized median can be found by an efficient algorithm based on dynamic programming. We first motivate our work by giving some background information about this class of contours. Then, the algorithm for finding the exact solution is described in Section 3. In Section 4 we describe two applications of generalized median computation: exploring the parameter space of a contour detection algorithm and tightness evaluation of a lower bound of generalized median problems in metric space. Finally, some discussion conclude the paper.

2 Class of Contours

The class of contours considered in this work is defined as follows:

Definition 1. *For a given $M \times N$ image a contour $C = p_1 p_2, \ldots, p_M$ is a sequence of points drawn from the top to the bottom, where p_i, $i = 1, \ldots, M$, is a point in the i-th row. The points p_i and p_{i+1}, $i = 1, \ldots, M-1$, of two successive rows are continuous.*

These contours start from the top, pass each image row exactly once, and end in the last row.

At the first glance the question may arise why such simple contours are of use in practice. Some thoughts, however, reveal that there do exist several situations, where we are directly or indirectly faced with this class of contours. In medical imaging it is typical for the user to specify some region of interest (ROI) and then to find some contours within the ROI. As an example, Figure 1 shows

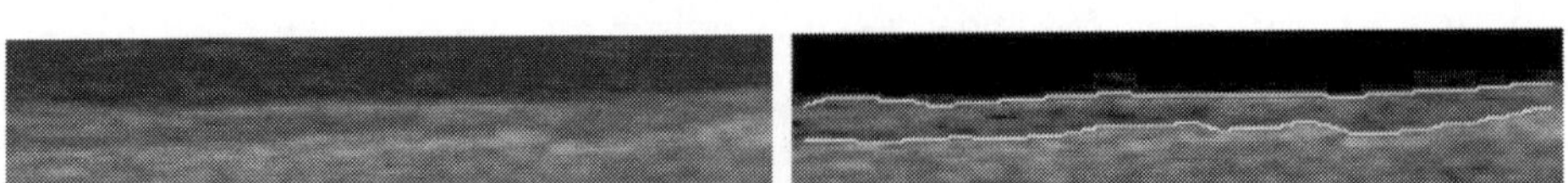

Fig. 1. ROI in a CCA B-mode sonographic image (left) and detected layer of intima and adventitia (right)

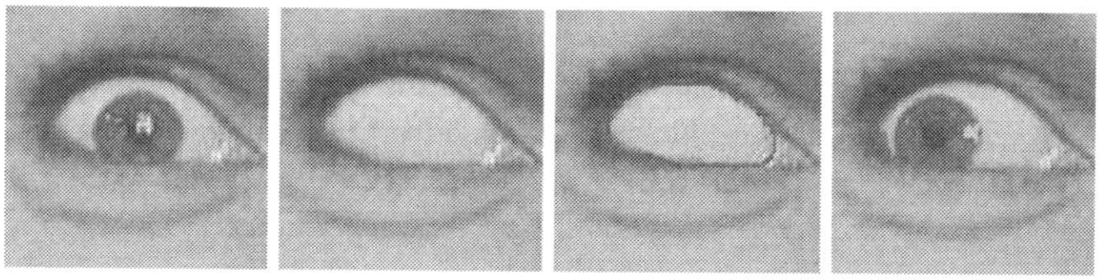

(a) (b) (c) (d)

Fig. 2. Detection of closed contour: (a) input image; (b) removal of iris; (c) detection of eye contour; (d) strabismus simulation

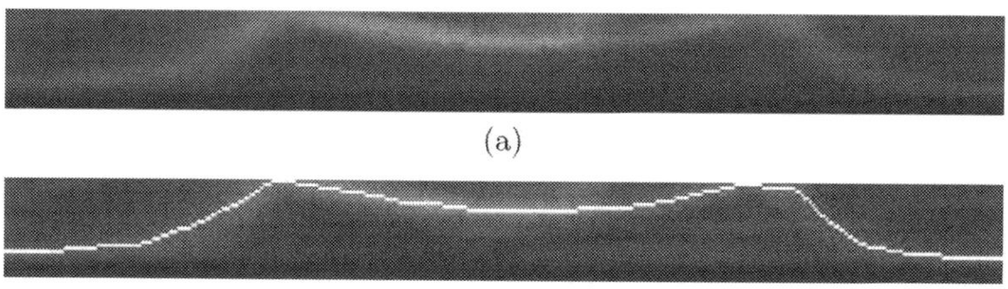

(a)

(b)

Fig. 3. Polar space for contour detection: (a) polar space; (b) optimal path.

a ROI in a CCA (Common Carotid Artery) B-mode sonographic image. The task is to detect the layer of intima and adventitia for computing the intima-media thickness which is an important index in modern medicine. Details of this application and an algorithm for automatic layer detection can be found in [2]. Essential to the current work is the fact that both the intimal layer and the adventitial layer are examples of the contour class defined above (although we have to rotate the image by 90 degrees). This application reflects a typical situation in medical image analysis.

The same fundamental principle can be extended to deal with closed contours. For this purpose we need a point p in the interior of the contour. Then, a polar transformation with p being the central point brings the original image into a matrix, in which a closed contour becomes a contour from top to bottom afterwards. Note that this technique works well for all star-shaped contours including convex contours as a special case. As an example, Figure 2 shows a problem of eye contour detection taken from [8]. In the image after removal of iris, the eye contour is detected as a closed contour based on the interior reflection point. The polar space representation related to Figure 2(b) can be seen in Figure 3(a) where the intensity is replaced by a measure of edge magnitude. In this space we are faced with the same contour detection problem as in Figure 1. The result is shown in Figure 3(b) and Figure 2(c) after projecting back into the image space. The task in this application is then to simulate strabismus by replacing the iris. The eye contour serves to restrict the region, within which the newly positioned iris lies. For (almost) convex contours the selection of the origin of polar space is not critical. In the general case of star-shaped contours, however, it must be chosen within the area, in which the complete contour can be seen.

The two situations above and others appear in a variety of applications. They indicate the broad applicability of the class of contours considered in this paper and thus justify to investigate them in their own right.

The concept of generalized median in Eqn. (1) can be easily adapted to our domain by specifying a distance function between two contours. Since each point p_i of a contour $P = p_1 p_2, \ldots, p_M$ has a constant y-coordinate i, we use p_i to represent its x-coordinate only in the following in order to simplify the notation. Given this convention, the distance between two contours P and Q can be defined by the k-th power of the Minkowski distance:

$$d(P,Q) = \sum_{i=1}^{M} (p_i - q_i)^k \tag{2}$$

In this case the representation space U contains all *continuous* contours from top to bottom of an input $M \times N$ image.

3 Computation of Generalized Median Contours

Given n contours $C_1, C_2, \ldots, C_n$, the task is to determine a contour $\overline{C}$ such that the sum of distances between $\overline{C}$ and all input contours is minimized. It is important to notice that we cannot solve this problem of generalized median contours by computing the optimal value for each of the M rows *independently*, which could be done, for instance, by enumerating all possibilities between the leftmost and rightmost point in the row. Doing it this way, we encounter the trouble of generating a discontinuous resultant contour.

Our proposed method is formulated as a problem of finding an optimal path in a graph based on dynamic programming. We first generate a two-dimensional $M \times N$ cost matrix of the same size as the image, in which every element corresponds to an image point. Each element is assigned a *Local_Goodness* value, which measures its suitability of being a candidate point on the generalized median contour we are looking for. According to the distance given in Eqn. (2) the *Local_Goodness* value is simply:

$$Local\_Goodness(i,j) = \sum_{l=1}^{n} (x_{li} - j)^k, \quad 1 \le i \le M,\ 1 \le j \le N$$

where x_{li} represents the x-coordinate of the l-th contour C_l in i-th row. Generally, small *Local_Goodness* values indicate better candidates. As a matter of fact, the optimality of a candidate for $\overline{C}$ is measured by the sum of its *Local_Goodness* values over all image rows.

Dynamic programming is applied to search for an optimal path in a cumulative cost matrix CC. The cumulative cost of a node (i,j) is computed as:

$$CC(i,j) = \min_{l=-1,0,1} \{CC(i-1, j+l)\} + Local\_Goodness(i,j) \tag{3}$$

for $2 \leq i \leq M,\ 1 \leq j \leq N$. This means that a contour point (i, j) has three potential predecessors $(i-1, j-1)$, $(i-1, j)$, $(i-1, j+1)$ in the previous row. In addition, the choice of a transition from a point in i-th row to a predecessor in the $(i-1)$-th row is made based on the lowest cumulative cost of the predecessors. The computation of CC starts by initializing the first row by:

$$CC(1, j) = Local\_Goodness(1, j), \quad 1 \leq j \leq N$$

Then, the cumulative cost matrix CC is filled row by row from left to right by using Eqn. (3).

The node in the last row of matrix CC with the lowest value gives us the last point of the optimum path. To determine this path, a matrix of pointers is created at the time of computing the matrix CC. The optimum path, which corresponds to the generalized median contour, is determined by starting at the last point and following the pointers back to the first row. Using this dynamic programming technique, we are able to compute the generalized median contour exactly.

The computational complexity of the algorithm amounts to $O(MNn)$ while $O(MN)$ space is required. Note that the search space of dynamic programming can be substantially reduced. For each row we only need to consider the range bounded by the leftmost and rightmost point from all input contours in that row. The size of this reduced search space depends on the variation of input data. The less variation of the input data, the more the reduction effect. Most likely, this reduction results in a computational complexity of $O(Mn)$ only. The proposed algorithm was implemented in Matlab on a Pentium IV 2.1 GHz PC. As an example, the computation time for 250 input contours of 105 points each with 0.00 standard deviation in the input data is 10 milliseconds. At an increased level of data variation of 81.74 standard deviation, 90 milliseconds were recorded. We can conclude that the dynamic programming approach delivers an efficient way of exactly computing the generalized median of contours.

4 Experimental Results

We have conducted a series of experiments using both synthetic and real data. In the following we report some results to illustrate two applications of the concept of generalized median contours.

4.1 Test Images and Contour Data

Both studies are based on CCA B-mode sonographic images [2]. An image dataset was established which consists of 23 such images of 105 columns each. They are actually ROI cut out of larger images. Each image contains two contours of interest: intima (y_1) and adventitia (y_2). Both contours run from left to right of an image. If we turn the images by 90 degrees, then we are faced with the problem of optimally masking the two contours of length 105 each from top to bottom.

Table 1. Performance measures of parameter training and generalized median (GM) approaches on 5 test sets

	y_1 (intima)		y_2 (adventitia)	
Test set	Parameter training	GM	Parameter training	GM
1	48.98	49.77	60.59	50.18
2	48.68	49.37	53.56	52.82
3	51.09	51.16	51.79	51.26
4	49.90	50.66	46.83	47.08
5	46.53	46.53	50.03	48.07
average	49.04	49.50	52.56	49.88

Each image has its ground truth contours manually specified by an experienced physician. This information is used for an objective, quantitative comparison with automatic detection results. The similarity measure is simply the distance function in Eqn. (2). In all our tests we have fixed k of the distance function to $k = 1$.

4.2 Exploring Parameter Space Without Ground Truth

Segmentation algorithms mostly have some parameters and their optimal setting is not a trivial task. In recent years automatic parameter training has become popular. Typically, a training image set with (manual) ground truth segmentation is assumed to be available. Then, a subspace of the parameter space is explored to find out the best parameter setting. For each parameter setting candidate a performance measure is computed in the following way:

- Segment each image of the training set based on the parameter setting;
- Compute a performance measure by comparing the segmentation result and the corresponding ground truth;
- Compute the average performance measure over all images of the training set.

The optimal parameter setting is given by the one with the largest average performance measure. Since fully exploring the subspace can be very costly, space subsampling [10] or genetic search [3] has been proposed.

While this approach is reasonable and has been successfully practiced in several applications, its fundamental disadvantage is the assumption of ground truth segmentation. The manual generation of ground truth is always painful and thus a main barrier of wide use in many situations.

We propose to apply the concept of generalized median for implicitly exploring the parameter space without the need of ground truth segmentation. It is assumed that we know a reasonable subspace of the parameter space (i.e. a lower and upper bound for each parameter), which is sampled into a finite number $\mathcal{M}$ of parameter settings. Then, we run the segmentation procedure for all the $\mathcal{M}$ parameter settings and compute the generalized median of the $\mathcal{M}$ segmentation results. The rationale behind our approach is that the median segmentation tends to be a good one within the explored parameter subspace.

This idea has been verified on the database described above within the contour detection algorithm [2]. It has two parameters and a reasonable parameter subspace is divided into 250 samples. The database is partitioned into a training set of 10 images and a test set of 13 images. The training set is then used to find the optimal parameter setting among the 250 candidates, which is applied to the test set. The average performance measure over the 13 test images is listed in Table 1. Note that the testing procedure is repeated 5 times for different partitions of the 23 images into training and test set. On the other hand, the generalized median approach has no knowledge of the ground truth segmentation. It simply detects 250 contours and computes their generalized median. The average performance measure of the 13 generalized median contours in the test set as shown in Table 1 indicates that basically no real performance differences exist between these two approaches. Without using any ground truth information, the generalized median technique is able to produce contours of essentially identical quality as the training approach.

5 Verification of Optimal Lower Bound for Generalized Median Problems in Metric Space

The computation of generalized median patterns is typically an NP-complete task. Therefore, research efforts are focused on approximate approaches. One essential aspect in this context is the assessment of the quality of the computed approximate solutions. Since the true optimum is unknown, the quality assessment is not trivial in general. A recent work [6] presented the lower bound for this purpose.

Referring to the notation in Eqn. (1), an approximate computation method gives us a solution $\tilde{C}$ such that

$$\mathrm{SOD}(\tilde{C}) \;=\; \sum_{i=1}^{n} d(\tilde{C}, C_i) \;\geq\; \sum_{i=1}^{n} d(\overline{C}, C_i) \;=\; \mathrm{SOD}(\overline{C})$$

where SOD stands for sum of distances and $\overline{C}$ represents the (unknown) true generalized median. The quality of $\tilde{C}$ can be measured by the difference $\mathrm{SOD}(\tilde{C}) - \mathrm{SOD}(\overline{C})$. Since $\overline{C}$ and thus $\mathrm{SOD}(\overline{C})$ are unknown in general, we resort to a lower bound $\Gamma \leq \mathrm{SOD}(\overline{C})$ and measure the quality of $\tilde{C}$ by $\mathrm{SOD}(\tilde{C}) - \Gamma$. Note that the relationship

$$0 \;\leq\; \Gamma \;\leq\; SOD(\overline{C}) \;\leq\; SOD(\tilde{C})$$

holds. Obviously, $\Gamma = 0$ is a trivial, and also useless, lower bound. We require Γ to be as close to $\mathrm{SOD}(\overline{C})$ as possible. This tightness can be quantified by $\mathrm{SOD}(\overline{C}) - \Gamma$ with a value zero for the ideal case.

In [6] the tightness of the lower bound has been tested in the domain of strings and graphs. Since the computation of generalized strings and graphs is exponential, only approximate solutions have been considered there.

Ideally, the tightness should be investigated in domains where we know the true generalized median. The current work provides us a means of validating

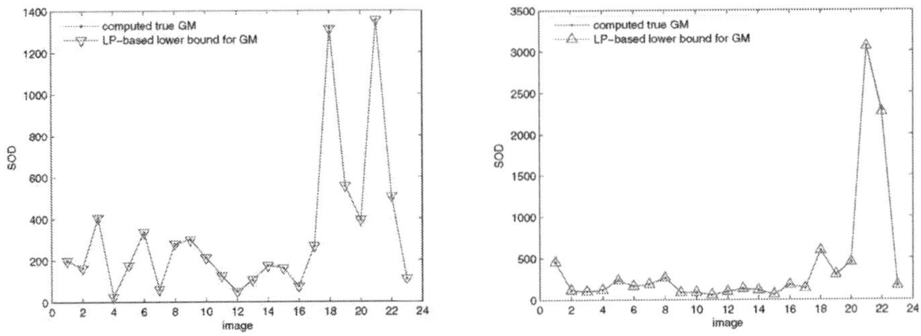

Fig. 4. Tightness of lower bound Γ for 50 y_1 contours (intima, left) and 50 y_2 contours (adventitia, right) contours for all 23 images

the tightness under ideal conditions. For this purpose we sampled 50 parameter settings of the parameter subspace[1]. For each image, we thus compute 50 contours and afterwards their exact generalized median $\overline{C}$ by the dynamic programming technique proposed in this paper. In Figure 4 both the lower bound Γ and SOD($\overline{C}$) for all 23 images are plotted. Obviously, these two values are so similar that no difference is visible. This is clearly a sign of good tightness of the lower bound Γ. Although this statement is made for the particular case of contours, it builds a piece of the mosaic of validating the tightness in many problem spaces.

6 Conclusions

In this paper we have considered a special class of contours which start from the top, pass each image row exactly once, and end in the last row of an image. Despite of the simplicity they frequently occur in many applications of image analysis. We have proposed a dynamic programming approach to exactly compute the generalized median contour in this domain.

Experimental results have been reported on two scenarios, in which the concept of generalized median plays a very different role. In the first case we have postulated a general approach to implicitly explore the parameter space of a (segmentation) algorithm. It was shown that using the generalized median contour, we are able to achieve contour detection results comparable to those from explicitly training the parameters using a training set with known ground truth. This performance is remarkable and should be further investigated in other contexts.

Having a generalized median problem with exact solution is interesting in its own right for the specific problem domain. From a more general point of view,

[1] The reason for selecting only 50 instead of 250 as in other experiments lies in the high computation time and space requirement of the lower bound computation which is based on linear programming.

the exact solution gives us a means to verify the tightness of the lower bound for generalized median computation under ideal conditions. We have performed the verification which shows the high tightness. As part of our efforts in verifying the tightness of the lower bound using a variety of generalized median problems with exact solution, the current work represents a valuable contribution.

References

1. V. Chalana and Y. Kim. A methodology for evaluation of boundary detection algorithms on medical images. IEEE Trans. on Medical Imaging, 16(5): 642–652, 1997.
2. D. Cheng, X. Jiang, A. Schmidt-Trucksäss and K. Cheng. Automatic intima-media thickness measurement of carotid artery wall in B-mode sonographic images. Proc. of IEEE International Symposium on Biomedical Imaging: From Nano to Macro, 912–915, 2006.
3. L. Cingue, R. Cucciara, S. Levialdi, S. Martinez, and G. Pignalberi. Optimal range segmentation parameters through genetic algorithms. Proc. of 15th Int. Conf. on Pattern Recognition, Vol. 1, 474–477, Barcelona, 2000.
4. C. de la Higuera and F. Casacuberta. Topology of strings: Median string is NP-complete. Theoretical Computer Science, 230(1/2): 39–48, 2000.
5. X. Jiang, A. Münger, and H. Bunke. On median graphs: Properties, algorithms, and applications. IEEE Trans. on Pattern Analysis and Machine Intelligence, 23(10): 1144–1151, 2001.
6. X. Jiang and H. Bunke. Optimal lower bound for generalized median problems in metric space. In: Structural, Syntactic, and Statistical Pattern Recognition (T.Caelli, A.Amin, R.P.W.Duin, M.Kamel, and D.de Ridder, Eds.), Springer-Verlag, 143–151, 2002.
7. X. Jiang, K. Abegglen, H. Bunke, and J. Csirik. Dynamic computation of generalized median strings. Pattern Analysis and Applications, 6(3): 185–193, 2003.
8. X.Jiang, S.Rothaus, K.Rothaus, and D.Mojon. Synthesizing face images by iris replacement: Strabismus simulation. Proc. of Int. Conf. on Computer Vision Theory and Applications, 41–47, Setuba, Portugal, 2006.
9. D. Lopresti and J. Zhou. Using consensus sequence voting to correct OCR errors. Computer Vision and Image Understanding, 67(1): 39-47, 1997.
10. J. Min, M. Powell, and K.W. Bowyer. Automated performance evaluation of range image segmentation algorithms. IEEE Trans. on SMC – Part B, 34(1): 263-271, 2004.
11. T.B. Sebastian, P.N. Klein, and B.B. Kimia. On aligning curves. IEEE Trans. on Pattern Analysis and Machine Intelligence, 25(1): 116–125, 2003.
12. J.S. Sim and K. Park. The consensus string problem for a metric is NP-complete. Journal of Discrete Algorithms, 2(1), 2001.
13. L. Wang and T. Jiang. On the complexity of multiple sequence alignment. Journal of Computational Biology, 1(4): 337–348, 1994.

Structural and Syntactic Techniques for Recognition of Ethiopic Characters

Yaregal Assabie and Josef Bigun

School of Information Science, Computer and Electrical Engineering
Halmstad University, SE-301 18 Halmstad, Sweden
{Yaregal.Assabie, Josef.Bigun}@ide.hh.se

Abstract. OCR technology of Latin scripts is well advanced in comparison to other scripts. However, the available results from Latin are not always sufficient to directly adopt them for other scripts such as the Ethiopic script. In this paper, we propose a novel approach that uses structural and syntactic techniques for recognition of Ethiopic characters. We reveal that primitive structures and their spatial relationships form a unique set of patterns for each character. The relationships of primitives are represented by a special tree structure, which is also used to generate a pattern. A knowledge base of the alphabet that stores possibly occurring patterns for each character is built. Recognition is then achieved by matching the generated pattern against each pattern in the knowledge base. Structural features are extracted using direction field tensor. Experimental results are reported, and the recognition system is insensitive to variations on font types, sizes and styles.

1 Introduction

Ethiopia is among the few countries in the world which have a unique alphabet of its several languages. The Ethiopic alphabet has been in use since the 5th century B.C.[5] and the present form of the alphabet is obtained after passing through many improvements. At present, the alphabet is widely used by Amharic which is the official language of Ethiopia, and a total of over 80 million people inside as well as outside Ethiopia are using this alphabet for writing.

Research on Ethiopic OCR is a recent phenomenon and there are only few papers presented in conferences [3],[6]. Moreover, it has been difficult to develop a good recognition system for Ethiopic characters due to the complex composition of their basic graphical units. In this paper, we present a novel approach to recognize Ethiopic characters by employing structural and syntactic techniques in which each character is represented by a pattern of less complex structural features called primitives [2] and their spatial relationships. Each character forms a unique set of patterns which are generated from the relationships of primitives. The structural features and their relationships are extracted by using direction field tensor. The characters expressed in terms of primitives and their relationships remain similar under variations on the size, type and style of characters. Accordingly, the present results are novel contributions towards the development of a general Ethiopic OCR system that works independent of the appearance and characteristics of the text.

D.-Y. Yeung et al. (Eds.): SSPR&SPR 2006, LNCS 4109, pp. 118 – 126, 2006.

2 Ethiopic Alphabet

The most common Ethiopic alphabet used by Amharic language has 34 basic characters and other six orders derived from the basic forms making a total of 238 characters. The alphabet is conveniently written in tabular format of seven columns as shown in Table 1, where the first column represents the base character and the other columns represent modifications of the base character.

Table 1. Part of the Ethiopic Alphabet

1st(ä) order	2nd(u) order	3rd(i) order	4th(a) order	5th(e) order	6th(ə) order	7th(o) order
ሀ hä	ሁ hu	ሂ hi	ሃ ha	ሄ he	ህ hə	ሆ ho
ለ lä	ሉ lu	ሊ li	ላ la	ሌ le	ል lə	ሎ lo
ሐ hä	ሑ hu	ሒ hi	ሓ ha	ሔ he	ሕ hə	ሖ ho
⋮	⋮	⋮	⋮	⋮	⋮	⋮

2.1 Structural Analysis

Ethiopic characters are considered to have the most attractive appearance when written with thick appendages, vertical and diagonal strokes, and thin horizontal lines. Most of the horizontal strokes in Ethiopic characters are only a few pixels wide and sometimes they do not exist, especially in degraded documents. Thus, prominent structural features in the alphabet are appendages, vertical and diagonal lines. These prominent features form a set of primitive structures. In this research, we reveal 7 primitive structures which are interconnected in different ways to form a character. Primitives differ from one another in their structure type, relative length, orientation, and spatial position. The classes of primitives are given below.

- **Long Vertical Line (LVL).** A vertical line that runs from the top to bottom level of the character. The primitive is found in characters like ሀ, በ, and ዘ.
- **Medium Vertical Line (MVL).** A vertical line that touches either the top or the bottom level (but not both) of a character. ሰ, ኸ, and ና are some of the characters that have these primitives.
- **Short Vertical Line (SVL).** A vertical line that touches neither the top nor the bottom level of the character. It exists in characters like ሐ, ቀ, and ሰ.
- **Long Forward Slash (LFS).** A forward slash primitive that runs from the top to the bottom level of a character. It is found in few characters like ሄ, ሥ, and ጎ.
- **Medium Forward Slash (MFS).** A forward slash primitive that touches either the top or the bottom level (but not both) of a character. This primitive is also found in few characters like አ, ፉ, and ነ.
- **Backslash.** A line that deviates from the vertical line position to the right when followed from top to bottom. The characters ል, ሰ, and ሸ have such primitives.
- **Appendages.** Structures which have almost the same width and height. These primitives are found in many characters. Examples are ሄ, ት, and ቹ.

2.2 Spatial Relationships of Primitives

The unique structure of characters is determined by primitives and their interconnection. The interconnection between primitives describes their spatial relationship. A primitive structure can be connected to another at one or more of the following regions of the structure: *top* (t), *middle* (m), and *bottom* (b). The spatial relationship between two primitives α and β connected only once is represented by the pattern αzβ, where z is an ordered pair (x,y) of the connection regions *t*, *m*, or *b*. In this pattern, α is connected to β at region x of α, and β is connected to α at region y of β. Moreover, the primitive α is also said to be spatially located to the left of β. Thus, the spatial relationship is described by *spatial position* (left and right) and *connection region* (t, m, and b).

There may also be two or three connections between two primitives. The first connection detected as one goes from top to bottom is considered as the *principal connection*. Other additional connections, if there exist, are considered as *supplementary connections*. The principal connection between two primitives is an ordered pair formed by the possible combinations of the three connection regions. This will lead to nine principal connection types as represented by the set: {(t,t),(t,m),(t,b),(m,t),(m,m),(m,b),(b,t),(b,m),(b,b)}.

The principal connection (t,t), i.e., two primitives both connected at the top, has five types of supplementary connections: {(m,b),(b,m),(b,b),(m,m)+(b,m),(m,m)+(b,b)}. The principal connection (t,m) has only one supplementary connection: {(b,m)}. The principal connection (m,t) has three types of supplementary connections: {(m,b),(b,m),(b,b)}. The rest principal connections do not have any supplementary connection. This makes up the possibility of two primitives to be connected in 18 different ways: 9 principal connections alone and 9 principal with supplementary connections. This is shown in Table 2 with example characters in brackets.

Table 2. Connection types between two primitive structures

Principal Connection	Supplementary Connections					
	None	(m,b)	(b,m)	(b,b)	(m,m)+(b,m)	(m,m)+(b,b)
(t,t)	(በ)	(ሆ)	(ሞ)	(ዐ)	(ዓ)	(ፀ)
(t,m)	(ሐ)		(ቀ)			
(t,b)	(ሰ)					
(m,t)	(ሐ)	(ቀ)	(ፍ)	(ቢ)		
(m,m)	(ዘ)					
(m,b)	(ሠ)					
(b,t)	(ነ)					
(b,m)	(ሣ)					
(b,b)	(ሀ)					

2.3 Representation

Primitives are connected to the left and right of another primitive at one of its three connection regions. To the right of a primitive, two different primitives can also be connected at the middle as in the case of ቱ. Therefore, a maximum of three primitives can be connected to the left of another primitive and up to four primitives can be connected to the right. To represent this relationship, a special tree structure having three left nodes and four right nodes is proposed as shown in Fig. 1. Each node in the tree stores data about the type of the primitive itself, the type of connections with its parent primitive, and the spatial positions of primitives connected to it.

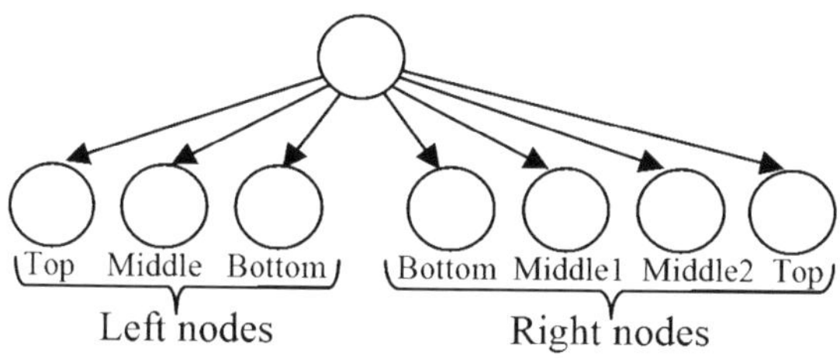

Fig. 1. General tree structure of characters

A primitive is appended to its parent primitive at one of the seven child nodes in the tree based on the principal connection that exists between them. For implementation, primitives are represented by a two digit numerical code as shown in Table 3. The first digit represents their relative length, spatial position and/or structure, and the second digit represents their orientation.

Table 3. Numerical codes assigned to primitive structures

Primitive Types	Numerical Codes
LVL	98
MVL	88
SVL	78
LFS	99
MFS	89
Backslash	87
Appendages	68

Each connection between two primitives is also assigned a two digit number which represents the *left* and *right* spatial positions. The three connection regions, i.e., top, middle, and bottom are represented by the numbers 1, 2, and 3 respectively. For example, using this approach, the connection *(m,b)* is represented by *23*. Connection types with two or more connections between primitives are assigned a numerical code formed by the concatenation of the numerical codes of the respective connections. For example, the numerical code of the connection type *(t,t)+(m,m)+(b,m)* is *112232*. When there is a primitive without being connected to any other primitive in the character, a connection type of *none* (with code number *44*) is used. In this case, the primitive is appended to one of other primitives based on the closeness in their spatial position. The connection type of the root primitive, which has no any parent, is also *none*.

2.4 Building Primitive Tree and Pattern Generation

The first step in building a primitive tree is identifying the root primitive. Variation in setting root primitive results in a different tree structure which will adversely affect the pattern generated for a character. To build a consistent primitive tree structure for each character, a primitive which is spatially located at the left top position of the character is used as the root primitive. The following recursive algorithm is developed to build primitive tree of characters. The function is initially invoked by passing the root primitive as a parameter.

```
BuildPrimitiveTree (Primitive)
{
    BuildPrimitiveTree(LeftTopPrimitive)
    BuildPrimitiveTree(LeftMidPrimitive)
    BuildPrimitiveTree(LeftBotPrimitive)
    BuildPrimitiveTree(RightBotPrimitive)
    BuildPrimitiveTree(RightMid1Primitive)
    BuildPrimitiveTree(RightMid2Primitive)
    BuildPrimitiveTree(RightTopPrimitive)
}
```

Examples of primitive trees built by the above algorithm are shown in Fig. 2. After building the primitive tree, a string pattern is generated by using in-order traversal of the tree (*left*{top, mid, bottom}, *parent*, *right*{bottom, middle1, middle2, top}). By starting on the root primitive, in-order traversal of the tree generates a unique set of patterns for each character. The algorithm is implemented using a recursive function in a similar way as building the primitive tree.

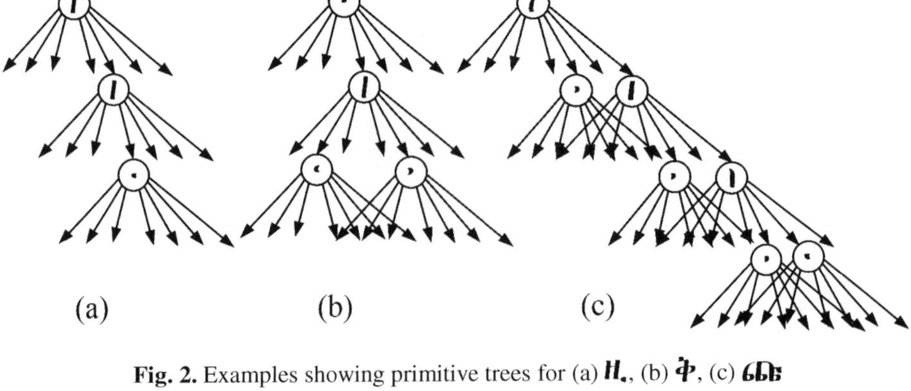

Fig. 2. Examples showing primitive trees for (a) ዚ, (b) ቕ, (c) ፏ

2.5 Alphabet Knowledge Base

The geometric structures of primitives and their spatial relationships remain the same under variations on fonts and their sizes. In Fig. 3*a*, all the different font types and sizes of the character በ are described as two Long Vertical Lines both connected at the top. This is represented by the pattern {44,98,11,98}. As there is no structural difference between a Long Vertical Line and its bold version, Fig. 3*b* is also represented by the same pattern as in the case of Fig. 3*a*. In Fig. 3*c*, the character is described as two Long Forward Slashes both connected at the top and it is represented by the pattern {44,99,11,99}. Therefore, any form of the character በ is represented as a set of patterns {{44,98,11,98},{44,99,11,99}}. Accordingly, the knowledge base of the alphabet consists of a set of possibly occurring patterns of primitives and their relationships for each character. This makes the proposed recognition technique tolerant of variations in the parameters of fonts.

(a) (b) (c)

Fig. 3. (a) The Ethiopic character በ with different font types and sizes of 12 and 18, (b) bold style of ***a***, (c) italic style of ***a***

3 Extraction of Structural Features Using Direction Field Tensor

A local neighborhood in an image where the gray value changes only in one direction, and remains constant in the orthogonal direction, is said to have Linear Symmetry (LS) property [1]. The LS property of an image can be estimated by analyzing the direction field tensor. The direction tensor, also called the structure tensor [4],[7], is a 3D field tensor representing the local direction of pixels. For a local neighborhood *f(x,y)* of an image *f*, the direction tensor S is computed as a 2x2 symmetric matrix using Gaussian derivative operators D_x and D_y.

$$S = \begin{pmatrix} \iint (D_x f)^2 dxdy & \iint (D_x f)(D_y f) dxdy \\ \iint (D_x f)(D_y f) dxdy & \iint (D_y f)^2 dxdy \end{pmatrix} \tag{1}$$

Linear symmetry exists at edges where there are gray level changes and it can be estimated by eigenvalue analysis of the direction tensor using complex moments of order two which are defined as follows.

$$I_{20} = \iint ((D_x + iD_y) f)^2 \, dxdy \tag{2}$$

$$I_{11} = \iint \left| (D_x + iD_y) f \right|^2 dxdy \tag{3}$$

The complex partial derivative operator $D_x + iD_y$ is defined as:

$$D_x + iD_y = \frac{\partial}{\partial x} + i\frac{\partial}{\partial y} \tag{4}$$

The value of I_{20} is a complex number where the argument is the local direction of pixels in double angle representation and the magnitude is a measure of the local LS strength. The scalar I_{11} measures the amount of gray value changes in a local neighborhood of pixels. Direction field tensor, which is a 3D tensor field, can also be conveniently represented by the 2D complex I_{20} and 1D scalar I_{11}. The complex image I_{20} can be displayed in color as shown in Fig. 4 where the hue represents direction of pixels in double angle representation.

Due to the Gaussian filtering used in the computation of direction tensor, the LS strength (magnitude) at the orthogonal cross-section of edges in the image forms a Gaussian of the same window size. Therefore, the cross-section of lines in the I_{20} image can be reduced to a skeletal form (one pixel size) by taking the point closest to the mean of the Gaussian formed by the LS strength in the orthogonal direction. The skeletal form of I_{20} image is then used for extraction of structural features.

Fig. 4. (a) Scanned document, (b) I_{20} of ***a*** where hue represents direction, (c) skeletal form of ***b*** without direction information

Before extracting structural features, characters are segmented into individual components. In the skeletal form of the I_{20} image, horizontal spaces that lack LS (LS strength <0.05 after normalization) are used to segment text lines, and vertical spaces that lack LS are used to segment characters in the text lines. Rectangular boxes in Fig. 5 show segmented characters of Ethiopic text. Since the direction of pixels is represented by double angle, the angle θ obtained from the argument of I_{20} is in the range of 0 to 180 degree. The direction of pixels at the edges of primitives is close to 0 and 180 degrees and can be converted to the range of 0 to 90 degrees by $\varepsilon = abs(90-\theta)$ so that ε for primitives is consistently close to 90 degree. In this study, pixels with $\varepsilon >= 30^{o}$ and having strong LS property (LS strength >=0.05) are considered as parts of primitives, and those with $\varepsilon < 30^{o}$ and having strong LS property are considered as parts of connectors. A primitive in the grayscale image will have

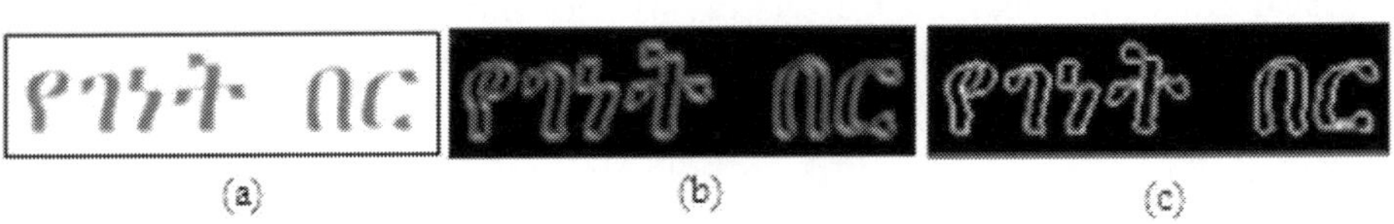

Fig. 5. (a) Character segmentation mapped to the original image, (b) skeletal form of the I_{20} image where the red and purple colors show the left and right edges of primitives respectively, and the green color shows connectors, (c) extracted primitives shaded with blue color

two lines (left and right edges) in the skeletal form of the I_{20} image. Primitive structures are then constructed by the two matching lines. The group information about direction and spatial position of pixels in a primitive are finally used to classify the primitives.

4 The Recognition Process

A general recognition system of Ethiopic characters is proposed as shown in Fig. 6. Characters are segmented by making use of direction field tensor. Structural features are then extracted and a pattern of their spatial relationships is generated for each segmented character. A character is said to be recognized if the string pattern generated from primitive tree has a matching pattern in the knowledge base. Pattern matching is done by comparing the string pattern generated from the image against with each string pattern stored in the knowledge base. The similarity of each comparison is computed and the most similar pattern is considered to decide whether the string pattern is recognized or not. This is done by setting a threshold of similarity.

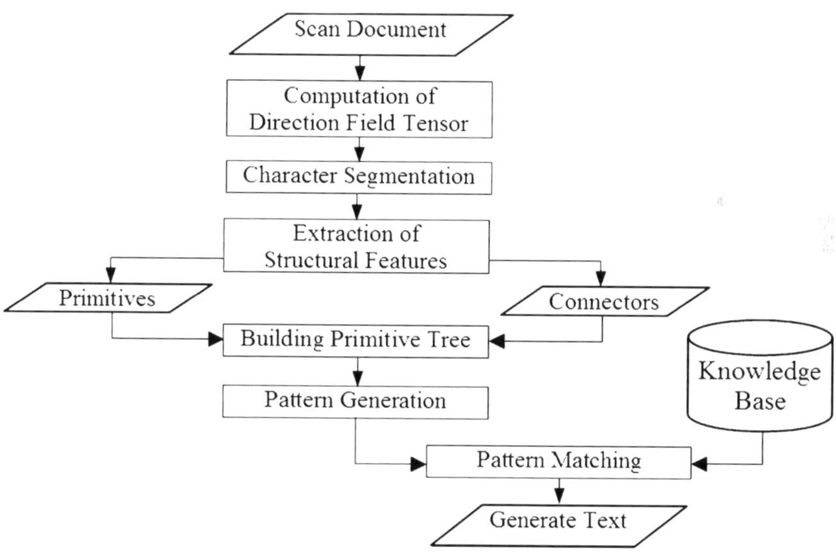

Fig. 6. Flowchart of the recognition process

5 Experiment

The size of the Gaussian window used for filtering operations is determined by the size of fonts in the document. For example, a Gaussian window of pixel-size 3x3 was efficient for texts with font size of 12, and a window size of 5x5 was found to be better for font sizes of 18.

There is no standard image database of Ethiopic text developed for testing character recognition systems. Thus, the experiment was done on images of about thirty pages scanned from newspapers, books and clean printouts that contain characters of different fonts and sizes varying from 12 to 18. Images taken from clean printouts show better recognition due to their relatively better quality. A recognition rate of 92% was achieved for clean printouts and the recognition accuracy for newspaper and books was 86%. However, there was no difference in recognition accuracy due to variation in fonts, and larger font sizes tend to be recognized slightly better than their smaller versions.

The structural and syntactic method used for recognition of Ethiopic characters is efficient to uniquely identify the characters. Recognition errors mainly come from poor quality of documents. Character segmentation errors also affect the overall character recognition accuracy. The other process that hampers the recognition process is extraction of connectors. This is due to the fact that horizontal lines in Ethiopic characters are very thin and sometimes absent especially in degraded and low quality documents. The algorithm used to extract primitives works well even in noisy documents.

6 Conclusion and Future Work

In this paper, a novel approach is proposed for Ethiopic character recognition. Structural and syntactic techniques are effectively used to uniquely represent the complex structure of characters by the relationships of less complex primitive structures. To this end, direction field tensor is used as a tool for extraction of structural features. The use of Gaussian separable filters to compute direction field tensor made the computation time minimal. The recognition accuracy can still be improved to a higher level by working more on character segmentation, extraction of structural features and pattern matching algorithms. Extraction of structural features can be further improved by applying statistical techniques. The process of pattern matching and classification is expected to perform better by using neural networks. In general, the recognition system is insensitive to variations on the size, type and other parameters of characters and therefore, the overall research activity will lead to the development of efficient OCR software for Ethiopic script.

References

1. J. Bigun, *Vision with Direction: A Systematic Introduction to Image Processing and Vision*, Springer, Heidelberg, 2006.
2. H. Bunke and A. Sanfeliu, *Syntactic and Structural Pattern Recognition: Theory and Applications,* World Scientific, Singapore, 1990.
3. J. Cowell and F. Hussain, "Amharic character recognition using a fast signature based algorithm," *Proc. Fourth Int'l Conf. Information Visualization*, 2003, pp. 384-389.
4. W. Forstner, "A framework for low level feature extraction", in J. Eklundh (ed.), *Lecture Notes in Computer Science*, vol. 801, Springer-Verlag, Berlin, 1994, pp. 383-394.
5. A. S. Gerard, *African Language Literatures: An introduction to the literary history of sub-Saharan Africa*, Three Continents Press, Washington, 1981.
6. L. Premaratne, Y. Assabie and J. Bigun, "Recognition of modification-based scripts using direction tensors," *ICVGIP'04*, Kolkata, 2004, pp. 587-592.
7. J. Weickert, "Coherence-enhancing shock filters," in B. Michaelis and G. Krell (eds.), *Lecture Notes in Computer Science*, vol. 2781, Springer-Verlag , Berlin, 2003, pp 1-8.

Context Driven Chinese String Segmentation and Recognition

Yan Jiang[1], Xiaoqing Ding[1], Qiang Fu[1], and Zheng Ren[2]

[1] Department of Electronic Engineering, Tsinghua University, Beijing, China, 100084
{jyan, dxq, fuq}@ocrserv.ee.tsinghua.edu.cn
[2] Siemens AG, D-78467 Konstanz, Germany
zheng.ren@siemens.com

Abstract. This paper presents a context driven segmentation and recognition method for handwritten Chinese characters. We follow a split-merge technique in character segmentation. In this process, a Chinese text line is first pre-segmented into a sequence of radicals, which are then merged according to a cost function combining both recognition confidence and contextual cost. Two strategies are also proposed for implementation: bi-gram based merging and lexicon driven merging. In the former one, we generate a set of merging paths which are then evaluated by Viterbi algorithm. The radicals' best merging method is given by the path with the highest score. In the latter strategy, a lexicon is preset and compared with the radicals to determine both radicals' merging and candidate character selection. Experiments show that contextual information plays a crucial role in Chinese character segmentation and could obviously improve the segmentation and recognition results.

1 Introduction

Single character recognition has achieved impressive progress both in accuracy and speed in the past 40 years. However, it could not remarkably benefit a document reading system directly because some practical difficulties. For example, it is hard to extract text lines from a complex layout document containing both graphs and characters in different fonts and size. Though a text line is perfectly extracted, character segmentation is another ineluctable and decisive step since a general classifier could only recognized a single-character image at a time.

Recently, there many papers considering character segmentation of digits, western and eastern languages. According to Casey ([1]), these methods are categorized into three basic strategies: structural analysis, recognition based and holistic tactic. In this paper, we suggest these methods are concluded into two levels according to the information sources they are rested on (Fig.1). Low level methods make use of information directly from image and high level methods utilize contextual and grammatical constraints originated from prior knowledge.

Character segmentation is still an obstacle in Chinese OCR especially for off-hand case, it should deal with diverse writing styles, a large character set and complex character structures. Moreover, characters are written with touching

D.-Y. Yeung et al. (Eds.): SSPR&SPR 2006, LNCS 4109, pp. 127–135, 2006.

and overlapping in scripts. According to the recent work, low-level methods are effective to remove touching and overlapping by accurately locating the segmentation points for touching and overlapping strokes. But it may come with another problem that a character is wrongly separated into different parts.

In conventional methods, character segmentation contains two steps. The first step is pre-segmentation, which decomposes a text line into a series of radicals. Secondly, these radicals are reunited into characters. Previously, only low level information is considered in the second step, which is shown to be unreliable in practice. A Chinese character may be composed of some parts, each of them is indeed a character itself. Low level methods perform inadequately and always segment a character into more than one parts. Recent papers are considering to involve contextual relationship in this process. In western languages, these context methods are always based on a word dictionary ([5]). However, the methods for oriental languages are quite different from that for western languages. Liu ([3]) proposes a lexicon driven way for Japanese address reading. Takahiro([7]) introduces bi-gram in his likelihood function for on-line Japanese handwriting recognition. But related work has seldom been done for Chinese up to now.

In this paper, we introduce contextual restrictions by incorporating bi-gram and lexicon-dictionary in character segmentation. According to the experiments, we can see that both segmentation rate and recognition rate could get improved.

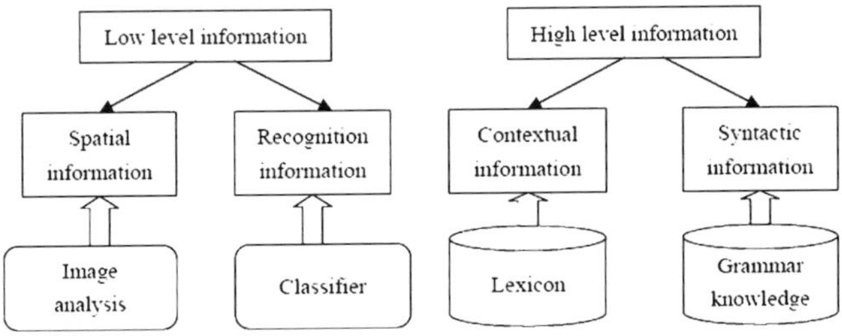

Fig. 1. Character segmentation strategies

2 Pre-segmentation

A Chinese character could be regarded as a composition of some primitive components. In pre-segmentation, we cite Xue's work([6]) to extract these components (Fig.2(a)). He extracts connected components from a text line, which are then merged into strokes (Fig.2(c)). These strokes are assembled to form radicals. Each radical should be warranted as just one part of a character (Fig.2(d)). A segmentation graph is accordingly established (Fig.2(e)), which is directed and acyclic. An arc corresponds to a certain radical combination and a path from the first node to the last represents a merging way for radicals. We assign each arc a cost to evaluate the likeness for a merged image of being a real character([6]).

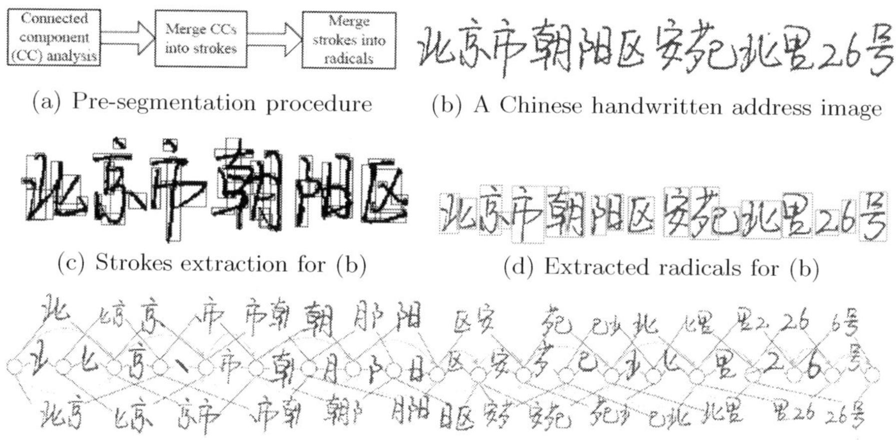

(a) Pre-segmentation procedure (b) A Chinese handwritten address image

(c) Strokes extraction for (b) (d) Extracted radicals for (b)

(e) Segmentation graph for (d)

Fig. 2. Pre-segmentation for Chinese handwritten texts

3 Bi-gram Based Segmentation

3.1 Introduction of Bi-gram Model

Let $X = x_1 x_2 \ldots x_T$ be a sequence of character images. A classifier commonly gives some hypotheses for each image, for example, $c_{t,1} c_{t,2} \ldots c_{t,M}$ denote the candidates for x_t. Post-processing selects characters from each candidate set and composes the most likely string $c_{1,k_1} c_{2,k_2} \cdots c_{T,k_T}$, where $1 \le k_t \le M, 1 \le t \le T$.

$$c_{1,k_1^*}, c_{2,k_2^*}, \ldots, c_{T,k_T^*} = \arg\max_{1 \le k_t \le M, 1 \le t \le T} P(c_{1,k_1}, c_{2,k_2}, \ldots, c_{T,k_T} | x_1, x_2, \ldots, x_T) \quad (1)$$

By Bayesian formula, we have

$$\begin{aligned} & P(c_{1,k_1}, c_{2,k_2}, \ldots, c_{T,k_T} | x_1, x_2, \ldots, x_T) \\ & = \frac{P(x_1, x_2, \ldots, x_T | c_{1,k_1}, c_{2,k_2}, \ldots, c_{T,k_T}) P(c_{1,k_1}, c_{2,k_2}, \ldots, c_{T,k_T})}{P(x_1, x_2, \ldots, x_T)} \end{aligned} \quad (2)$$

Assuming that the classifier's decision for the current image is independent of the previous image ([9]), we have

$$P(x_1, x_2, \ldots, x_T | c_{1,k_1}, c_{2,k_2}, \ldots, c_{T,k_T}) = \prod_{i=1}^{T} P(x_i) \times \prod_{i=1}^{T} \frac{P(c_{i,k_i} | x_i)}{P(c_{i,k_i})} \quad (3)$$

In natural language processing (NLP), N-gram model assumes that only the n adjacent characters before the given character make sense. In bi-gram ($n = 2$), $P(c_{1,k_1} c_{2,k_2} \cdots c_{T,k_T})$ is simplified to $P(c_{1,k_1}) \prod_{i=2}^{T} P(c_{i,k_i} | c_{i-1,k_{i-1}})$. Instead of Eq.(1), we turn to maximize $P(c_{1,k_1}) \prod_{i=2}^{T} P(c_{i,k_i} | c_{i-1,k_{i-1}}) \prod_{i=1}^{T} P(c_{i,k_i} | x_i)$ to form the most likely string. The maximum of the above criteria could be regarded as a hybrid of context and recognition cost. In our method, given a merging path in the segmentation graph, the maximum of H (Eq.(4)) is applied in evaluation.

$$H = \frac{1}{T}[\log P(\mathrm{c}_{1,k_1}) + \sum_{i=2}^{T} \log P(\mathrm{c}_{i,k_i}|\mathrm{c}_{i-1,k_{i-1}}) + \sum_{i=1}^{T} \log P(\mathrm{c}_{i,k_i}|\mathrm{x}_i)] \quad (4)$$

3.2 Confidence Estimation

A general character classifier outputs a set of sorted characters with ascending distances. However, distance measure is not discriminating in judging whether an input image is a real character or not. On the other hand, classifiers based on different learning algorithm would output different types of distances, which makes it inconceivable for further discussion. As shown above, distance measure is required to be transformed into probability measure. We will briefly review some basic transformation techniques. Suppose we have M candidate hypotheses $\mathrm{c}_1, \mathrm{c}_2, \ldots, \mathrm{c}_M$ for image x with corresponding ascending distances $d_1 \leq d_2 \leq \ldots \leq d_M$. (5) gives a set of experimental transformations for posterior probability estimation ([8]). In [4], Liu proposes his method based on Gauss distribution assumption (6), where variance parameter θ is estimated from training samples.

$$P(\mathrm{c}_i|\mathrm{x}) = \begin{cases} 1/d_i/\sum_{j=1}^{j=M}(1/d_j) & (5.1) \\ 1/d_i^2/\sum_{j=1}^{j=M}(1/d_j^2) & (5.2) \\ 1/(d_i - d_1 + 1)/\sum_{j=1}^{j=M}[1/(d_j - d_1 + 1)] & (5.3) \end{cases} \quad (5)$$

$$P(\mathrm{c}_i|\mathrm{x}) = \frac{\exp((d_i - d_1)/\theta)}{\sum_{j=1}^{j=M} \exp((d_j - d_1)/\theta)} \quad (6)$$

3.3 Implementations

In post-processing, character images are fixed prior to candidate selection. However, in radical merging step, we don't know how radicals will be organized. The number of possible ways of merging increases exponentially with respect to the number of radicals. It is necessary to discuss some more applicable methods. In this section, two implementations are provided for bi-gram driven way.

Beam search is an optimization of the best first search algorithm where only a predetermined number of paths are kept as candidates. If more paths than a threshold are generated, the worst paths are discarded.

Bi-gram Driven Beam Search (BDBS)
$\mathrm{s}_1, \mathrm{s}_2, \ldots, \mathrm{s}_N$—the pre-segmented radicals
$\mathrm{S}_{i,j}$—radical combination of radical $\mathrm{s}_i\mathrm{s}_{i+1}\ldots\mathrm{s}_j$
$\mathrm{c}_h(\mathrm{S}_{i,j})$—the h-th candidate character for radical combination $\mathrm{S}_{i,j}$, $1 \leq h \leq M$

Initialization step. For a predefined integer L, we test first L radical combinations $\mathrm{S}_{1,1}$, $\mathrm{S}_{1,2}$,... $\mathrm{S}_{1,L}$ and recognize them. If $\log P(\mathrm{c}_h(\mathrm{S}_{1,j})|\mathrm{S}_{1,j}) + \log P(\mathrm{c}_h(\mathrm{S}_{1,j}))$

is more than C, we add $< 1, j, c_h(S_{1,j}), \log P(c_h(S_{1,j})|x_{1,j}) + \log P(c_h(S_{1,j})), 1 >$ to the node list as a valid expansion, in which, the fifth element records the number of characters up to current merging.

Expanding step. For each node $< i, j, c_h(S_{i,j}), Q, n >$ in the list, we expand the node list as follows. We recognize $S_{j+1,j+1}$, $S_{j+1,j+2}, \ldots$ $S_{j+1,j+L}$, if there exist p, q that satisfy $j + 1 \le p \le j + L$, $1 \le q \le M$ and $\log P(c_q(S_{j+1,p})|S_{j+1,p}) + \log P(c_q(S_{j+1,p})|c_h(S_{i,j})) > C$, we update this node to $< j+1, p, c_q(S_{j+1,p}), Q + \log P(c_q(S_{j+1,p})|S_{j+1,p}) + \log P(c_q(S_{j+1,p})|c_h(S_{i,j})), n+1 >$. Meanwhile, if there are multiple choices, all the valid expansions must be added too.

Pruning step. In beam search, only B_0 nodes are allowed to be kept. If B, the number of nodes, exceeds B_0, we prune the redundant nodes for the sake of efficiency. All the nodes in the list are reordered according to the descending average scores. That is, node $< i, j, c, Q, n >$ is ranked according to the average score $\frac{Q}{n}$. The nodes with the smallest $B - B_0$ average scores are removed.

In another implementation, We first apply K-shortest algorithm ([2]) to the segmentation graph in order to generate a set of path hypotheses for evaluation.

If the character images, merged according to a certain path, are denoted by $x_1, x_2, \ldots, x_T$ and the recognized characters of the t-th image are $c_{t,1}, c_{t,2}, \cdots,$ $c_{t,M}$, the maximum of H(see Eq.(4)) of this path is computed by Viterbi algorithm. In following procedure, $Q(p,q)$ denotes the accumulative total of the logarithm of the probability value for the most likely string from the start character to $c_{p,q}$.

Bi-gram Driven Viterbi Evaluation (BDVE)
Step 1. For $1 \le q \le M$, we set $Q(1,q) = \log P(c_{1,q}) + \log P(c_{1,q}|x_1)$.
Step 2. For $2 \le p \le n_k$ and $1 \le q \le M$, we calculate $Q(p,q)$ as follows: $Q(p,q) = \max_{1 \le j \le M} \{Q(p-1,j) + \log P(c_{p,q}|c_{p-1,j})\} + \log P(c_{p,q}|x_p)$.
Step 3. We output $\frac{\max_{1 \le q \le M} Q(T,q)}{T}$ as the maximum of H.

The maximum number of radicals in a character is usually less than 6. For the worst case, the complexity of REA is at most $O(6N + KN \log 6)$, the complexity of Viterbi algorithm is $O(M^2T)$. Accordingly, generating K-shortest paths is very fast, however, the size of the candidate set M controls the total time.

In Fig.(3), we make a comparison between REA+BDVE method and the minimal spatial cost method. The proposed method achieves a segmentation rate of 100%, much better than the rate, 77%, given by the latter method. The merging paths recommended by the two methods are also illustrated in Fig.2(e).

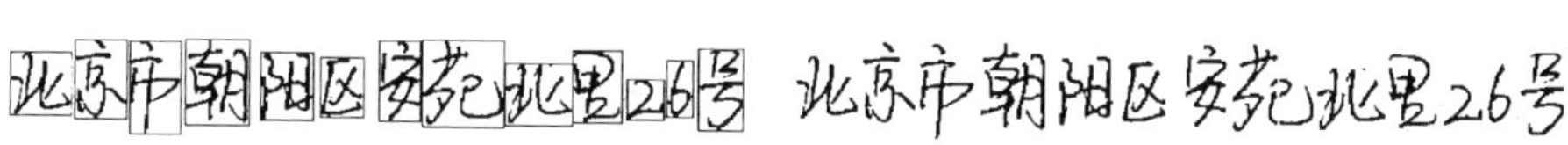

(a) Result of REA+BDVE (b) Result of minimal spatial cost path

Fig. 3. Segmentation results comparison

4 Lexicon Based String Segmentation and Recognition

In the above, we try to promote the string recognition rate by improving segmentation and recognition of characters. However, perfect character segmentation may not be necessary, fractional characters in a string could also help in unique identification against a lexicon dictionary. The differences between western languages and oriental ones result in distinct strategies in holistic string recognition. In English, a string image is divided into word images first and then recognized by dictionary matching. However, there are no gaps between characters in Chinese, that means a Chinese string must be dealt with at the same time. Liu([3]) uses a trie structure to organize address lexicons for Japanese address reading. He also adopts a split-merge strategy, determining merging path for radicals, identifying characters for images and matching the text line with a certain lexicon. This process is implemented by beam search from the left-most radical to the end. More than 110,000 items are considered in his method and each lexicon is limited within ten characters. Unlike the sequential characteristic of Liu's method, we propose a novel algorithm that could start from all the substrings. By this adaption, we could hurdle the problem that one segmentation error or one character mis-classification may potentially break down the whole process.

Optimal Substring Alignment Algorithm (OSAA)
$\mathrm{w}_1\mathrm{w}_2\ldots\mathrm{w}_m$—a given Chinese string
$R[p][q]$—the set of all the substrings that start with w_q and contain p characters

Step 1. For a predefined integer L, we recognize $\mathrm{S}_{i,i+j}$, where $0 \leq j \leq L-1$ and $1 \leq i \leq i+j \leq n$. If the candidate set of $\mathrm{S}_{i,i+j}$ contains w_k, we add $< i, j, k, k, c >$ to $R[1][k]$, where c is the confidence score.
Step 2. After initializing $R[1][k]$ for $1 \leq k \leq m$, for $1 \leq p \leq m, 1 \leq q \leq m-p+1$, we compute $R[p][q]$ as follows: if there exists $< i_1, j_1, k_{1,1}, k_{1,2}, c_1 >$ in $R[p-1][q]$ and $< i_2, j_2, k_{2,1}, k_{2,2}, c_2 >$ in $R[1][p+q-1]$ that satisfy $i_2 = j_1+1$, we then add a new element $< i_1, j_2, k_{1,1}, k_{2,2}, \frac{c_1 \times (p-1)+c_2}{p} >$ to $R[p][q]$.
Step 3.1. For a given threshold D, we select all the substrings $< i_l, j_l, k_{l,1}, k_{l,2}, c_l >, 1 \leq l \leq B$ from $R[p][q], p \geq D$ satisfying $1 \leq i_1 \leq j_1 < i_2 \leq j_2 < \ldots < i_B \leq j_B \leq n$ and $1 \leq k_{1,1} \leq k_{1,2} < k_{2,1} \leq k_{2,2} < \ldots < k_{B,1} \leq k_{B,2} \leq m$.
Step 3.2. For the selected substrings, we traverse all the characters and their possible corresponding radical combinations which are not covered by the substrings. By this means, each character in the given string could find its corresponding radicals. Then we calculate the value of a certain cost function which is designed to evaluate both dissimilarities between images and characters and differences between characters and the given lexicon. In step 3.1 and 3.2, we simply apply depth-first search to traverse all the possible cases.

Comparing with the bi-gram driven way, lexicon driven method is seen as a more strict rule to control both radical merging and candidate selection process.

5 Experiments and Discussions

We collect 1141 handwritten Chinese address lines including 14,970 characters written by different people. In bi-gram based case, we test both BDBS and REA+BDVE implementations. We found that BDBS may be attacked by a intervened error and thus result in poorer performance comparing with REA+BDVE (in Table 1, we adopt Eq.(6) for recognition confidence estimation, $K = 15N, M = 10$). In the following, we will mainly discuss REA+BDVE.

Table 2 compares confidence estimation methods (Eq.(5)&Eq.(6))and selection strategies of K. Different ways of estimation don't result in obvious distinctions. We are inclined to select K according to the number of the radicals. For a text line with fewer radicals, this adaption will save time, on the other hand, more paths will be examined if a text line has more radicals. In the following experiments, we use $K = 15N$ and Eq.(4) without specification.

Table 1. Comparison of BDBS and REA+BDVE

	BDBS ($B_0 = 40$)	REA+BDVE ($K = 15N, M = 10$)
Segmentation rate (%)	82.75	92.53
Time (s)	24.7	0.5

We then compare different factors in segmentation. In the columns of Table 3, we test the minimal spatial cost path, the maximal recognition confidence path, the minimal recognition distance path (i.e., we assign each arc the confidence/distance of the best candidate given for the image associated with the arc) and the best contextual ranked path (i.e., recognition confidence is omitted) respectively. Noticing the results from different criteria, contextual relation is most important.

Generally, we select 10 candidates for each character image in recognition (i.e. $M = 10$), since the size greatly affects the computation time of Viterbi. As shown in Table 4, extending the candidate set size will not improve the results obviously, however, brings about a rapid increase in time consumption.

Table 2. REA+BDVE segmentation results

	(5.1)	(5.2)	(5.3)	(6)
$K = 200, M = 10$	91.34	91.73	92.62	92.20
$K = 15N, M = 10$	91.50	91.99	92.89	92.53

Table 3. Analysis of different factors in segmentation

	Spatial cost segmentation	Recognition confidence segmentation	Recognition distance segmentation	Contextual relationship segmentation
Correct rate (%)	81.94	53.46	3.13	90.61

Table 4. Candidate set size for segmentation

	$M = 5$	$M = 10$	$M = 50$
Average time for Viterbi per text line (ms)	37	137	3368
Character segmentation rate (%)	92.02	92.53	92.91

Table 5. Right path distribution in the K-shortest paths

$K = 200$	$K = 400$	$K = 600$	$K = 800$	$K = 1000$
81.69	86.25	88.17	89.48	90.27

Table 6. Correct rate for different numbers of candidate paths

	$K = 200$	$K = 500$	$K = 1000$
Segmentation rate (%)	92.20	92.59	92.63
Total time (ms)	410	622	861

Table 7. Bi-gram driven segmentation results for a general document

Segmentation rate (%)	Recognition rate (%)
92.9	84.9
88.2	79.0

We cannot assure that the right answer must be in the candidate set, though K is very large. In Table 5, we give the rate of the correct merging path in the first K paths. However, sub-optimal paths are always included in the K-shortest paths, which are applicable to achieve a acceptable segmentation rate, so enlarging K will not benefit the segmentation rate remarkably (Table 6).

In the proposed method, we use an averaged score considering different number of characters merged, otherwise, if we don't take the difference into consideration as in [7], we may encounter a little drop in the segmentation rate. For example, if H is replaced by $\tilde{H} = \log P(\mathrm{c}_{1,k_1}) + \sum_{i=2}^{T} \log P(\mathrm{c}_{i,k_i}|\mathrm{c}_{i-1,k_{i-1}}) + \sum_{i=1}^{T} \log P(\mathrm{c}_{i,k_i}|\mathrm{x}_i)$, the segmentation rate drops from 92.5% to 90.0%.

The above experiments utilize the bi-gram training on the address lexicons of Beijing. If we turn to a more general bi-gram, trained on "People's Daily" of 2000 (covering politics, economics, science and etc.), we get a segmentation rate of 84.59%. Comparing with the rate of 92.53%, the general bi-gram weakens the contextual relationship of specific documents and degrades the performance, however, the result exceeds the performance of minimal spatial cost path.

Furthermore, we extend our idea to a more general case. We collect some text lines containing 238 characters from a technical document, by using the bi-gram of "People's daily", both segmentation and recognition rates get improved (Table 7).

OSAA is designed for lexicon driven holistic string recognition. We use 500 handwritten address lines and a database containing more than 370,000 lexicons in our experiments and achieve a string recognition rate of 86.8%.

6 Conclusions and Discussions

This paper presents a context driven way for unconstrained off-line handwritten Chinese characters segmentation. Unlike the previous techniques based on low level information, we pay more attention to the application of contextual knowledge in this process, which may be more useful as revealed by the experiments.

Contextual information could effectively determine how to merge the radicals, however, low level information is essential to get these radicals. Noticing that there have been many papers considering various techniques based on low level pre-segmentation to remove touching and overlapping for Chinese scripts, our method could be easily extended to the merging step of those methods.

Acknowledgements. This work has been funded by Siemens AG under contract number 20030829 - 24022SI202. The author would also thank to the anonymous reviewers for their kindly and helpful advice.

References

1. Richard G. Casey, Eric Lecolinet: A Survey of Methods and Strategies in Character Segmentation. IEEE Trans. PAMI **18**(7), (1996) 690–706
2. Víctor M. Jimenez and Andrés Marzal: Computing the K shortest paths: A new algorithm and an experimental comparison. *Proc. 3rd WAE*, 1999, 15–29. LNCS vol. 1668. Springer
3. Chenglin Liu, Masashi Koga, Hiromichi Fujisawa: Lexicon-driven Segmentation and Recognition of Handwritten Character Strings for Japanese Address Reading.IEEE Trans. PAMI **24**(11), (2002) 1425–1437
4. Chenglin Liu, Masaki Nakagawa: Precise Candidate Selection for Large Character Set Recognition by Confidence Evaluation.IEEE Trans. PAMI **22**(6),(2000) 636–642
5. Stefano Messelodi, Carla Maria Modena: Context Driven Text Segmentation and Recognition. Pattern Recognition Letters **17**(1), (1996) 47–56
6. Junliang Xue, Xiaoqing Ding, et al: Location and Interpretation of Destination Addresses on Handwritten Chinese Envelopes. Pattern Recognition Letters **22**(6),(2001) 639–656
7. Takahiro Fukushima, Masaki Nakagawa, On-line Writing-box-free Recognition of Handwritten Japanese Text Considering Character Size Variations, *Proc. 15th ICPR*, 359–363.
8. Xiaofan Lin, 1998. Theory and Application of Confidence Analysis and Multiple Classifier Combination in Character Recognition. Ph.d. dissertation, Tsinghua University.
9. Yuanxiang Li, 2001. The Research on Chinese Character Recognition Using Contextual Information. Ph.d. dissertation, Tsinghua University.

Support Vector Machines for Mathematical Symbol Recognition

Christopher Malon[1], Seiichi Uchida[2], and Masakazu Suzuki[1]

[1] Engineering Division, Faculty of Mathematics, Kyushu University
6–10–1 Hakozaki, Higashi-ku, Fukuoka, 812–8581 Japan
[2] Faculty of Information Science and Electrical Engineering, Kyushu University
6–10–1 Hakozaki, Higashi-ku, Fukuoka, 812–8581 Japan

Abstract. Mathematical formulas challenge an OCR system with a range of similar-looking characters whose bold, calligraphic, and italic varieties must be recognized distinctly, though the fonts to be used in an article are not known in advance. We describe the use of support vector machines (SVM) to learn and predict about 300 classes of styled characters and symbols.

1 Introduction

Optical character recognition problems were considered very early in the development of support vector machines, with promising results [1]. However, the problem of OCR for mathematical documents is substantially more difficult than standard OCR problems for three principal reasons:

1. Although a variety of fonts is used in mathematical literature, when reading any single paper, it is important to keep appearances of italic, bold, roman, calligraphic, typewriter, and blackboard bold letters distinguished.
2. A rich set of symbols is used, and distinctions between letters may be more subtle than within the character set of a typical human language.
3. The symbols are not arranged in a simple one–dimensional pattern. Subscripts, superscripts, fractional relationships, and accents occur, and may be nested [2].

The *Infty Project[7]* in the Suzuki Laboratory at Kyushu University is developing *Infty Reader* software [5] to perform OCR of scanned mathematical journal articles, and produce output in languages that allow symbol relationships to be properly encoded, including TeX and MathML. Although Infty Reader nominally achieved 99 percent accuracy of single–letter recognition before this investigation (October 2005), its failure to distinguish certain common symbols would be bothersome to any serious user.

The Infty Project defined entities for about 600 characters and symbols used in mathematical research, and created a ground truth database identifying their appearances in page–by–page scans of hundreds of journal articles. Many character pairs could be distinguished in different styles by simple clustering techniques

D.-Y. Yeung et al. (Eds.): SSPR&SPR 2006, LNCS 4109, pp. 136–144, 2006.

applied to directional features measured in a mesh grid. Runtime accuracy exceeded 99% , but hundreds of letter pairs remained consistently problematic. We aim to improve the accuracy of single–character recognition through the use of support vector machines.

2 Test Data

The Infty character set comprises 1,571 Roman and Greek letters, numbers, and mathematical symbols, divided into 46 categories according to purpose and style. Further details of these characters appear in [8].

The Infty Project has selected journal articles representing diverse fields of higher mathematics, taken from a thirty year period. These articles were scanned page by page at 600 dpi to produce bitmap image files. The Infty OCR engine extracted the symbols from each page and recognized each symbol as a character

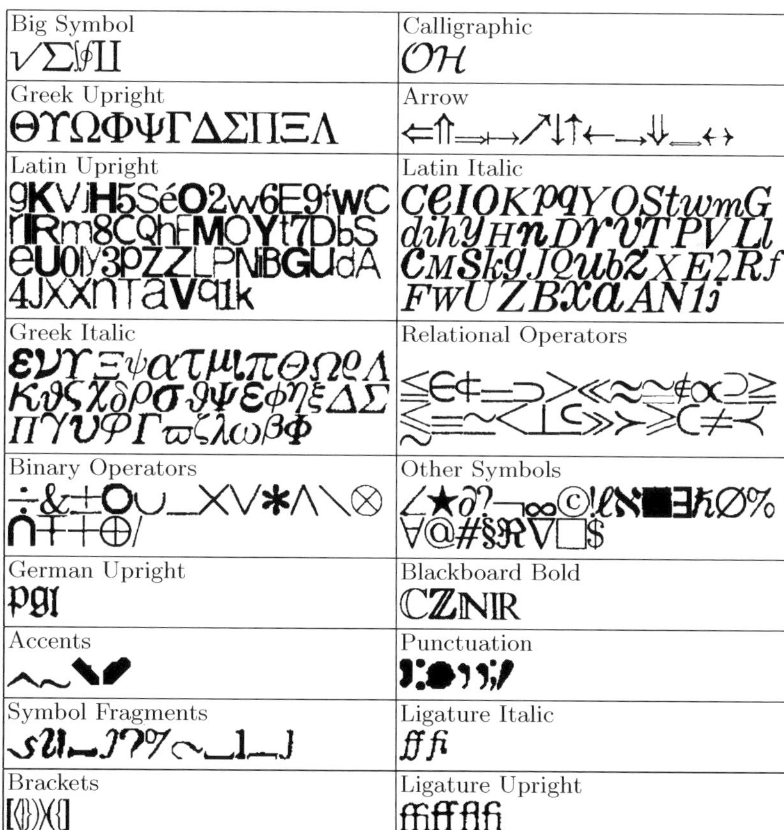

Fig. 1. Symbols with 10 training, selection, and testing samples

from the Infty character set. College and graduate mathematics students manually inspected and corrected the results.

The results of this process appear in a "ground truth" database. Namely, for each character on a scanned page, a bitmap framed by the bounding box of that character is taken from the page. This bitmap is tagged with the correct identification of the symbol. [1] "Link information" describing the character's relationship to others on the page (subscripts, superscripts, limits, portions of a fraction, *etc.*) is also available in the database, but it is not utilized in the present study.

In fact, the Infty project has produced two databases of this kind. One, called InftyCDB-1, is freely available for research purposes upon request, and is summarized in [8]. The other is used internally by the Infty Reader OCR engine. We use the latter database in this experiment, because it has more data, and because it makes it easier to compare our results with those of the actual Infty Reader. Our data sample consists of 284,739 character symbols extracted from 363 journal articles. There are 608 different characters represented.

At random, we divide the 363 articles into three parts consisting of 121 articles each. The data from the corresponding articles is marked as "training", "selection", or "testing" accordingly. To make sure we had enough data to train and evaluate our classifiers, we examined only the characters with at least ten samples in training, selection, and testing portions of the database. This left 297 characters, pictured in Figure 1.

3 Directional Features

Given an instance of a symbol, let w be its width and h be its height. Our feature vectors consist of the aspect ratio ($\frac{h}{w}$), followed by 160 floating–point coordinates of mesh directional feature data.

This mesh data is divided into "tall", "square", and "short" blocks of 48, 64, and 48 coordinates respectively. When the aspect ratio of a character exceeds 1.3, the tall block contains directional feature data computed from a 3×4 mesh; otherwise it contains zero–valued entries. When the aspect ratio of a character is between $\frac{1}{1.7}$ and 1.7, the square block contains directional feature data from a 4×4 mesh; otherwise it contains zero–valued entries. When the aspect ratio of a character is less than $\frac{1}{1.3}$, the short block contains directional features computed from a 4×3 mesh; otherwise it contains zero–valued entries. Thus, for any symbol, one or two (but never three) of the blocks are assigned nonzero entries.

We describe roughly the algorithm for associating directional feature data to an $m \times n$ mesh block. Divide the original bitmap horizontally into m equally sized lengths, and vertically into n equally sized lengths. Assign a "chunk" of four coordinates of the block to each of the $m \times n$ grid positions; initially, their values

[1] Some bitmaps would not be identifiable solely on the basis of this bitmap. For example, a hyphen could not be distinguished from an underscore, without knowing its relationship to the baseline on the page when it was scanned. The original position on the page is part of the database, but this information was discarded prior to our experiment.

are zero. These four coordinates represent the horizontal and vertical directions, and two diagonals.

The contribution of part of the outline's direction to the mesh features is determined from its position in the bitmap, using a partition of unity. Given a positive integer r, our r–fold partition of unity consists of functions $p_i^r : [0,1] \to [0,1]$, $i = 0, \ldots, r-1$, with the property that p_i^r is supported on $[\frac{i-1}{r}, \frac{i+2}{r}]$.

Discard every isolated black pixel from the original bitmap. In the remaining bitmap, trace every outline between white and black pixels, following its chain code description. When visiting the pixel in location (x, y) during this trace, identify the direction (horizontal, vertical, diagonal one, or diagonal two) where the next pixel in the outline will be. For every i, $0 \leq i < m$, and every j, $0 \leq j < n$, add $p_i^m(\frac{x}{w}) \cdot p_j^n(\frac{y}{h})$ to the coordinate of the (i, j) chunk representing that direction.

After completing the trace of each outline component, divide all the values by the perimeter of the bounding box. This result gives the values to be entered in the corresponding block of the feature vector.

4 Naive Classifier

Typically, a support vector machine learns a binary classification. There are various techniques for putting SVM's together to distinguish multiple classes; a comparison of some popular methods (1–vs–1, 1–vs–all, and the Directed Acyclic Graph) may be found in [4]. Except for the 1–vs–all method, these methods require the construction of $O(n^2)$ classifiers to solve an n–class classification problem. Because the Infty character set includes more than 1,500 entities, this seemed unnecessarily burdensome. Therefore, we try to extract an easier part of the classification problem that can be solved without SVM.

Taking the data assigned to the "training" portion of the database, we compute the mean feature vectors for the instances of each symbol. We create a naive classifier that assigns an input to the class whose mean feature vector is nearest, by Euclidean distance.

We run this naive classifier on the "selection" portion of the database, to produce a confusion matrix. The (i, j) entry of this matrix counts the number of samples in which a character truly belonging to class i was assigned to class j by this rule. The 297 by 297 confusion matrix we produced had 947 nonzero off-diagonal entries, an average of 3.2 misrecognitions per character.

We consider some of the misrecognitions to be too difficult for any classifier to resolve on the basis of our mesh of directional features. Particularly, we do not expect bold and non–bold variants of the same character to be distinguishable. Also, we do not expect upper and lower case variants of the letters C, O, P, S, V, W, X, and Z to be distinguishable in the same style, or in styles that are identical except for boldness. Disregarding misrecognitions of these two kinds, 896 other nonzero off–diagonal entries remain in the confusion matrix.

For 62 of the 297 characters with ten training, selection, and testing samples, the naive classifier recognized less than half of the selection samples correctly.

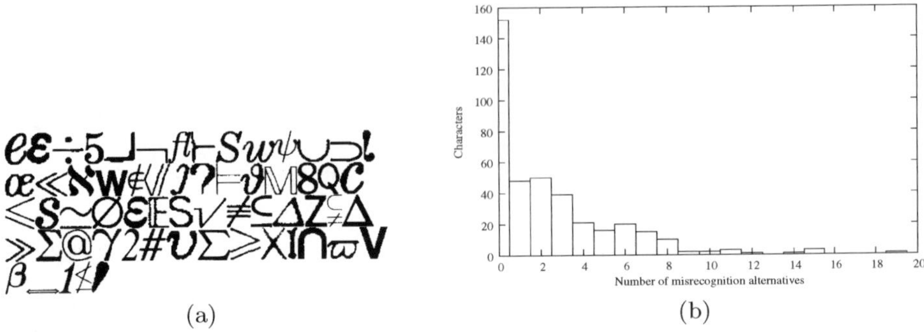

Fig. 2. The naive classifier: (a) Characters the naive classifier usually fails to recognize. (b) Histogram of distinct misrecognitions of an input character by the naive classifier.

These characters are displayed in Figure 2 (a). In comparison, ninety percent accuracy is achieved for 197 of the 297 symbols, 95 percent accuracy for 163 symbols, and 99 percent accuracy for 123 symbols.

Although the confusion matrix is relatively sparse, certain troublesome characters have many misrecognition results, as can be seen in Figure 2 (b). For 95 of the 297 characters, at least four distinct characters occur as misrecognition results. Eleven letters (plain '1', '4', 'E', 'I', 'l', 'r', 's', 't', ' " ', and italic 'γ' and 'ψ') had ten or more distinct characters appear as misrecognition results.

At runtime, the naive classifier will be used to assign each letter to a cluster of possible candidates, consisting of the recognition result and the other candidates most likely to have produced that recognition result (as determined by our confusion matrix). The harder problem of distinguishing between the letters in each of these clusters will be assigned to support vector machines.

5 Linear SVM

Within each cluster, we will use the 1–to–1 approach to multiclass classification. This requires first creating a binary SVM for each pair of classes in the cluster.

Because they are simple and can be computed quickly, we begin our experiment with SVM's that use the linear kernel:

$$K(\boldsymbol{x}, \boldsymbol{y}) = \boldsymbol{x} \cdot \boldsymbol{y}. \tag{1}$$

The naive classifier, when restricted to two classes, can be thought of as the linear classifier determined by the hyperplane equidistant from the two cluster centers. The support vector method enables us to search for hyperplanes in the original feature space that perform better on the training data.

There are no kernel parameter choices needed to create a linear SVM, but it is necessary to choose a value for the soft margin (C) in advance. Then, given training data with feature vectors $\boldsymbol{x_i}$ assigned to class $y_i \in \{-1, 1\}$ for $i = 1, \ldots, l$, the support vector machines solve

$$\min_{\boldsymbol{w},b,\boldsymbol{\xi}} \quad \tfrac{1}{2}K(\boldsymbol{w},\boldsymbol{w}) + C\textstyle\sum_{i=1}^{l}\xi_i \tag{2}$$
$$\text{subject to } y_i(K(\boldsymbol{w},\boldsymbol{x_i}) + b) \geq 1 - \xi_i$$
$$\xi_i \geq 0$$

where $\boldsymbol{\xi}$ is an l–dimensional vector, and $\boldsymbol{w}$ is a vector in the same feature space as the $\boldsymbol{x_i}$ (see, *e.g.*, [3]). The values $\boldsymbol{w}$ and b determine a hyperplane in the original feature space, giving a linear classifier. *A priori*, one does not know which value of soft margin will yield the classifier with the best generalization ability. We optimize this choice for best performance on the selection portion of our data, as follows.

Our basic parameter search method, here and in the following sections, is a grid search method that generalizes to any number of dimensions. For each candidate parameter assignment, we train an SVM with those parameters on the training portion of our data. Then we measure its performance on the instances of two classes that appear in the selection data. The score of the parameter is the minimum of the accuracy on the first class's input and the accuracy on the second class's input. *Hereafter, "accuracy" by itself, in the context of a binary classification problem, will refer to this score.*

Often, grid searches require a search interval to be specified for each dimension. Our approach requires only an initial parameter choice, and then grows the search region outward, until performance stops improving. Initial choices may matter under this grid search algorithm, if the algorithm terminates before reaching a selection of parameters that produces globally optimal results. This possibility seems unlikely as long as the resulting SVM performs better than random guessing in each case. The linear SVM problem has only the soft margin C as a parameter, and we set it initially to be 1024.

Table 1 displays the accuracy achieved by the linear SVM selected, on the testing data for pairs of symbols that the naive classifier sometimes confused.

We compared the chosen linear SVM classifier's performance on the letters where the naive classifier did not reach 100% accuracy, to the performance of the naive classifier. The 896 misrecognitions of the naive classifier comprise 795 unordered pairs of symbols. For nine of these pairs, both the naive classifier and the linear SVM always misassigned one of the two characters. Figure 3 (a) compares the performance of the two methods on the remaining 786 pairs. Of

Table 1. Linear SVM performance

Accuracy >	Number of pairs	Accuracy >	Number of pairs
Total	795	.9	750
0	783	.95	742
.5	774	.97	720
.6	770	.99	684
.7	767	.995	650
.8	759	.999	609

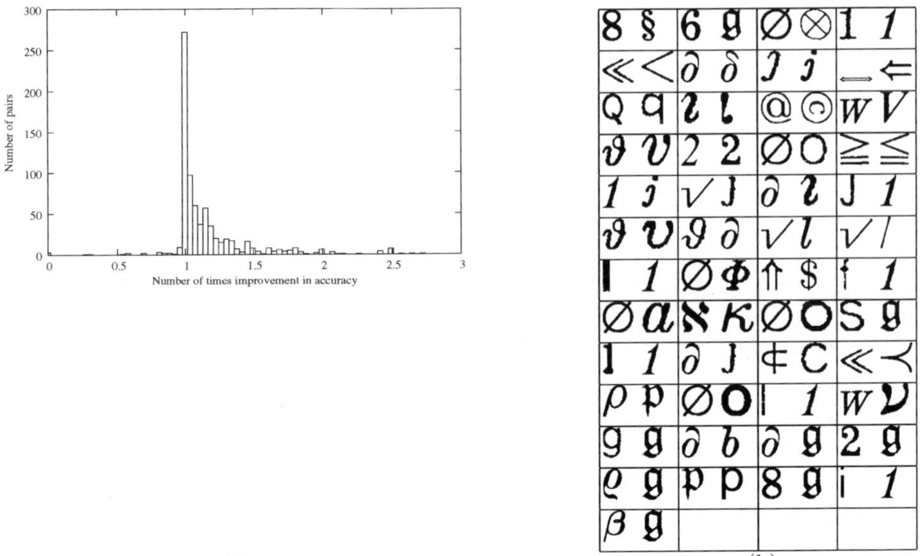

(a) (b)

Fig. 3. Linear SVM improvement compared to naive classifier: (a) Histogram. (b) Pairs with two–fold improvement.

the 786 pairs, 34 did not perform as well under the linear SVM as with the naive classifier. The exact same performance was achieved on 95 pairs, and improvement occurred on 657 pairs. The histogram does not report the 24 symbols with more than a three–fold improvement in accuracy. Thirteen of these symbols received zero accuracy from the naive classifier, for an infinite improvement in performance.

Figure 3 (b) illustrates the cases where the linear SVM achieved double the accuracy of the naive classifier.

6 Gaussian SVM

Just by using linear kernel support vector machines, our symbol recognition rates dramatically improved, but the use of a linear kernel severely limits the potential benefit of a support vector machine. The use of a Gaussian (radial) kernel

$$K(\boldsymbol{x}, \boldsymbol{y}) = e^{-\gamma \|\boldsymbol{x}-\boldsymbol{y}\|^2} \tag{3}$$

in the SVM problem (2) effectively transforms the input feature space into an infinite–dimensional one, where the search for an optimal separating hyperplane is carried out. Classifiers of this form may perform better on classes whose feature data is not linearly separable in the original feature space. However, the addition of the parameter γ in the kernel definition makes the parameter search two–dimensional, adding computational expense to the selection of a classifier.

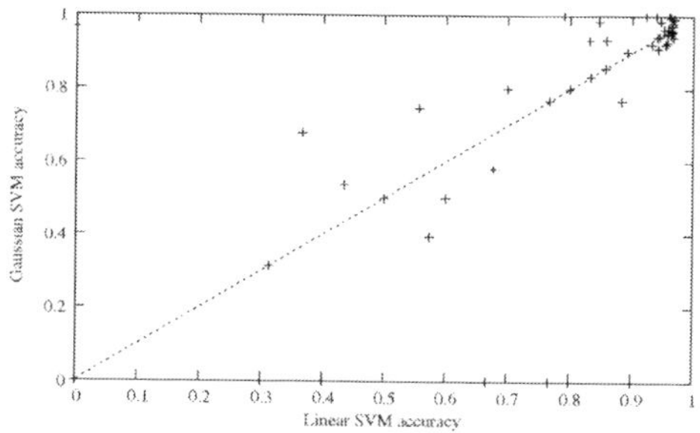

(a)

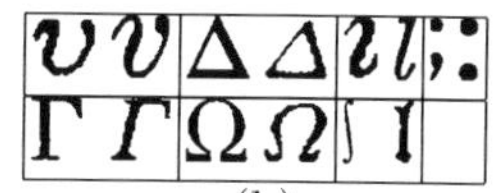

(b)

Fig. 4. Comparison of linear and Gaussian SVM: (a) Accuracies. (b) Pairs with 10% improvement from linear to Gaussian kernel.

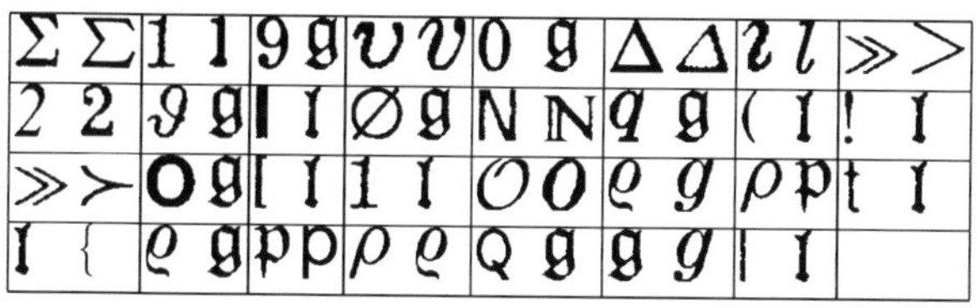

Fig. 5. Pairs with under 80% accuracy by Gaussian SVM

According to a result of Keerthi and Lin [6], given a soft margin C, the sequence of Gaussian SVM classifiers with kernel parameter γ and soft margin $\frac{C}{2\gamma}$ converges pointwise, as $\gamma \to 0$, to the linear SVM classifier with soft margin C. Thus, if our parameter search is wide enough, we should achieve higher accuracy with the Gaussian kernel than with the linear one.

We constructed Gaussian–kernel SVM classifiers for the 75 pairs of letters that the linear kernel failed to distinguish with 97% accuracy. A comparison of the performance of the chosen classifiers for each kernel type is given in Figure 4 (a). In Figure 4 (b), we display the eight pairs on which the Gaussian SVM performed with at least 10% higher accuracy than the linear SVM. The 31 pairs where Gaussian SVM accuracy falls below 80% are shown in Figure 5.

7 Conclusion

Even with the simplest kernel, the support vector method is strong enough to achieve good generalization accuracy on an optical character recognition problem that causes difficulty for simpler classification methods. We believe that our SVM results may be the best classification possible on the basis of the mesh of directional features we are using.

To distinguish the characters that confuse our SVM classifier, we plan to add new features. For example, by counting the number of connected components in a symbol, we could distinguish many variants of the greater–than sign (>). We also plan to record the convexity or concavity of a symbol as traced along its outline, to distinguish various nearly vertical characters. These features will be the topic for a future paper.

To our surprise, the SVM's we constructed with the Gaussian kernel did not show significantly stronger performance on the testing data. We attribute this phenomenon to the simple nature of our mesh of directional features. We plan to repeat this comparison after attaching a greater variety of features to our data.

References

[1] Cortes, C., and Vapnik, V. Support-vector networks. *Mach. Learn. 20*, 3 (1995), 273–297.

[2] Eto, Y., and Suzuki, M. Mathematical formula recognition using virtual link network. In *Sixth International Conference on Document Analysis and Recognition (ICDAR)* (2001), IEEE Computer Society Press, pp. 430–437.

[3] Hsu, C.-W., Chang, C.-C., and Lin, C.-J. A practical guide to support vector classification. `http://www.csie.ntu.edu.tw/ %7Ecjlin/papers/guide/guide.pdf`, July 2003.

[4] Hsu, C.-W., and Lin, C.-J. A comparison of methods for multi-class support vector machines. *IEEE Transactions on Neural Networks 13* (2002), 415–425.

[5] The infty project. `http://infty.math.kyushu-u.ac.jp`.

[6] Keerthi, S. S., and Lin, C.-J. Asymptotic behaviors of support vector machines with gaussian kernel. *Neural Comput. 15*, 7 (2003), 1667–1689.

[7] Suzuki, M., Tamari, F., Fukuda, R., Uchida, S., and Kanahori, T. Infty: an integrated ocr system for mathematical documents. In *DocEng '03: Proceedings of the 2003 ACM symposium on Document engineering* (New York, NY, USA, 2003), ACM Press, pp. 95–104.

[8] Suzuki, M., Uchida, S., and Nomura, A. A ground-truthed mathematical character and symbol image database. In *ICDAR '05: Proceedings of the Eighth International Conference on Document Analysis and Recognition (ICDAR'05)* (Washington, DC, USA, 2005), IEEE Computer Society, pp. 675–679.

Bayesian Class-Matched Multinet Classifier

Yaniv Gurwicz and Boaz Lerner

Pattern Analysis and Machine Learning Lab
Department of Electrical & Computer Engineering
Ben-Gurion University, Beer-Sheva 84105, Israel
{yanivg, boaz}@ee.bgu.ac.il

Abstract. A Bayesian multinet classifier allows a different set of independence assertions among variables in each of a set of local Bayesian networks composing the multinet. The structure of the local network is usually learned using a joint-probability-based score that is less specific to classification, i.e., classifiers based on structures providing high scores are not necessarily accurate. Moreover, this score is less discriminative for learning multinet classifiers because generally it is computed using only the class patterns and avoiding patterns of the other classes. We propose the Bayesian class-matched multinet (BCM^2) classifier to tackle both issues. The BCM^2 learns each local network using a detection-rejection measure, i.e., the accuracy in simultaneously detecting class patterns while rejecting patterns of the other classes. This classifier demonstrates superior accuracy to other state-of-the-art Bayesian network and multinet classifiers on 32 real-world databases.

1 Introduction

Bayesian networks (BNs) excel in knowledge representation and reasoning under uncertainty [1]. Classification using a BN is accomplished by computing the posterior probability of the class variable conditioned on the non-class variables. One approach is using Bayesian multinets. Representation by a multinet explicitly encodes asymmetric independence assertions that cannot be represented in the topology of a single BN using a several local networks that each represents a set of assertions for a different state of the class variable [2]. Utilizing these different independence assertions, the multinet simplifies graphical representation and alleviates probabilistic inference in comparison to the BN [2]-[4]. However, although found accurate at least as other BNs [3], [4], the Bayesian multinet has two flaws when applied to classification. The first flaw is the usual construction of a local network using a joint-probability-based score [4], [5] which is less specific to classification, i.e., classifiers based on structures providing high scores are not necessarily accurate in classification [4], [6]. The second flaw is that learning a local network is based on patterns of only the corresponding class. Although this may approximate the class data well, information discriminating between the class and other classes may be discarded, thus undermining the selection of the structure that is most appropriate for classification.

We propose the Bayesian class-matched multinet (BCM^2) classifier that tackles both flaws of the Bayesian multinet classifier (BMC) by learning each local network

D.-Y. Yeung et al. (Eds.): SSPR&SPR 2006, LNCS 4109, pp. 145 – 153, 2006.

using a detection-rejection score, which is the accuracy in simultaneously detecting and rejecting patterns of the corresponding class and other classes, respectively. We also introduce the tBCM^2 which learns a structure based on a tree-augmented naïve Bayes (TAN) [4] using the SuperParent algorithm [7]. The contribution of the paper is three fold. First is the suggested discrimination-driven score for learning BMC local networks. Second is the use of the entire data, rather than only the class patterns for training each of the local networks. Third is the incorporation of these two notions into an efficient and accurate BMC (i.e., the tBCM^2) that is found superior to other state-of-the-art Bayesian network classifiers (BNCs) and BMCs on 32 real-world databases.

Section 2 of the paper describes BNs and BMCs. Section 3 presents the detection-rejection score and BCM^2 classifier, while Section 4 details experiments to compare the BCM^2 to other BNCs and BMCs and their results. Section 5 concludes the work.

2 Bayesian Networks and Multinet Classifiers

A BN model B for a set of n variables $\boldsymbol{X}=\{X_1,\ldots,X_n\}$, having each a finite set of mutually exclusive states, consists of two main components, $B=(G,\Theta)$. The first component G is the model structure that is a directed acyclic graph (DAG) since it contains no directed cycles. The second component is a set of parameters $\boldsymbol{\Theta}$ that specify all of the conditional probability distributions (or densities) that quantify graph edges. The probability distribution of each $X_i \in \boldsymbol{X}$ conditioned on its parents in the graph $\boldsymbol{Pa}_i \subseteq \boldsymbol{X}$ is $P(X_i=x_i| \boldsymbol{Pa}_i) \in \boldsymbol{\Theta}$ when we use X_i and $\boldsymbol{Pa}_i$ to denote the ith variable and its parents, respectively, as well as the corresponding nodes.

The joint probability distribution over $\boldsymbol{X}$ given a structure G that is assumed to encode this probability distribution is given by [1]

$$P(\boldsymbol{X}=\boldsymbol{x} \mid G)=\prod_{i=1}^{n} P(X_i=x_i \mid \boldsymbol{Pa}_i, G) \tag{1}$$

where $\boldsymbol{x}$ is the assignment of states (values) to the variables in $\boldsymbol{X}$, x_i is the value taken by X_i, and the terms in the product compose the required set of local conditional probability distributions $\boldsymbol{\Theta}$ quantifying the dependence relations. The computation of the joint probability distribution (as well as related probabilities such as the posterior) is conditioned on the graph. A common approach is to learn a structure from the data and then estimate its parameters based on the data frequency count. In this study, we are interested in structure learning for the local networks of a BMC.

A BN entails that the relations among the domain variables be the same for all values of the class variable. In contrast, a Bayesian multinet allows different relations, i.e., (in)dependences for one value of the class variable are not necessarily those for other values. A BMC [2]-[5], [8], [9] is composed of a set of local BNs, $\{B_1,\ldots,B_{|C|}\}$, each corresponds to a value of the $|C|$ values of the class node C. The BMC can be viewed as generalization of any type of BNC when all local networks of the BMC have the same structure of the BNC [4]. Although a local network must be searched for each class, the BMC is generally less complex and more accurate than a BNC. This is because usually each local network has a lower number of nodes than the

BNC, as it is required to model a simpler problem. The computational complexity of the BMC is usually smaller and its accuracy higher than those of the BNC since both the complexity of structure learning and number of probabilities to estimate increase exponentially with the number of nodes in the structure [2].

A BMC is learned by partitioning the training set into sub-sets according to the values of the class variable and constructing a local network B_k for $\boldsymbol{X}$ for each class value $C=C_k$ using the kth sub-set. This network models the kth local joint probability distribution $P_{B_k}(\boldsymbol{X})$. A multinet is the set of local BNs $\{B_1,\ldots,B_{|C|}\}$ that together with the prior $P(C)$ on C classify a pattern $\boldsymbol{x}=\{x_1,\ldots,x_n\}$ by choosing the class $C_K \;\forall K \in [1,|C|]$ maximizing the posterior probability

$$C_K = \arg\max_{k\in[1,|C|]} \left\{ P(C=C_k \mid \boldsymbol{X}=\boldsymbol{x}) \right\}, \tag{2}$$

where

$$P(C=C_k \mid \boldsymbol{X}=\boldsymbol{x}) = \frac{P(C=C_k, \boldsymbol{X}=\boldsymbol{x})}{P(\boldsymbol{X}=\boldsymbol{x})} = \frac{P(C=C_k)P_{B_k}(\boldsymbol{X}=\boldsymbol{x})}{\sum_{i=1}^{|C|} P(C=C_i)P_{B_i}(\boldsymbol{X}=\boldsymbol{x})}. \tag{3}$$

In the Chow-Liu multinet (CL multinet) [4], the local network B_k is learned using the kth sub-set and based on the Chow-Liu (CL) tree [10]. This maximizes the log-likelihood [4], which is identical to minimizing the KL divergence between the estimated joint probability distribution based on the network P_{B_k} and the empirical probability distribution for the sub-set $\hat{P}_k$ [5],

$$\mathrm{KL}(\hat{P}_k, P_{B_k}) = \sum_{\boldsymbol{x}} \hat{P}_k(\boldsymbol{X}=\boldsymbol{x}) \cdot \log\left[\frac{\hat{P}_k(\boldsymbol{X}=\boldsymbol{x})}{P_{B_k}(\boldsymbol{X}=\boldsymbol{x})}\right]. \tag{4}$$

Thus, the CL multinet induces a CL tree to model each local joint probability distribution and employs (2) to perform classification. Further elaborations to the construction of the CL tree may be found in [3]. Also we note that the CL multinet was found superior in accuracy to the naïve Bayes classifier (NBC) and comparable to the TAN [4]. Other common BMCs are the mixture of trees model [9], the recursive Bayesian multinet (RBMN) [8] and the discriminative CL tree (DCLT) BMC [5].

3 The Bayesian Class-Matched Multinet Classifier

We suggest the Bayesian class-matched multinet (BCM^2) that learns each local network using the search-and-score approach. The method searches for the structure maximizing a discrimination-driven score that is computed using training patterns of all classes. Learning a local network in a turn rather than both networks simultaneously has computational benefit regarding the number of structures that need to be considered. First we present the discrimination-driven score and then the tBCM^2 that is a classifier based on the TAN [4] and searched using the SuperParent algorithm [7].

The BCM² Score. We first make two definitions: (a) a pattern $\boldsymbol{x}$ is *native* to class C_k if $\boldsymbol{x} \in C_k$ and (b) a pattern $\boldsymbol{x}$ is *foreign* to class C_k if $\boldsymbol{x} \in C_j$ where $j \in [1,|C|]$ and $j \neq k$. We partition the dataset D into test (D_{ts}) and training (D_{tr}) sets, the latter is further divided into internal training set T used to learn candidate structures and a validation set V used to evaluate these structures. Each training pattern in D_{tr} is labeled for each local network B_k as either native or foreign to class C_k depending on whether it belongs to C_k or not, respectively. In each iteration of the search for the most accurate structure, the parameters of each candidate structure are learned using T in order to construct a classifier that can be evaluated using a discrimination-driven score on the validation set. After selecting a structure, we update its parameters using the entire training set (D_{tr}) and repeat the procedure for all other local networks. The derived BCM² can be then tested using (2).

The suggested score evaluates a structure using the ability of a classifier based on this structure in detecting native patterns and rejecting foreign patterns. The score S_x for a pattern $\boldsymbol{x}$ is determined based on the maximum a posteriori probability, i.e.,

$$S_x = \begin{cases} 1, \textit{if } \{P(C=C_k \mid X=x_n^k) \geq P(C \neq C_k \mid X=x_n^k)\} \textit{ or } \{P(C \neq C_k \mid X=x_f^k) > P(C=C_k \mid X=x_f^k)\} \\ 0, \textit{if } \{P(C=C_k \mid X=x_n^k) < P(C \neq C_k \mid X=x_n^k)\} \textit{ or } \{P(C \neq C_k \mid X=x_f^k) \leq P(C=C_k \mid X=x_f^k)\} \end{cases}, \quad (5)$$

where $\boldsymbol{x}_n^k$ and $\boldsymbol{x}_f^k$ are native and foreign patterns to C_k, respectively. The first line in (5) represents correct detection (classification of a native pattern to C_k) or correct rejection (classification of a foreign pattern to a class other than C_k), whereas the second line represents incorrect detection of a native pattern or incorrect rejection of a foreign pattern. By identifying *TP* (true positive) as the number of correct detections and *TN* (true negative) as the number of correct rejections made by a classifier on all the $|V|$ validation patterns in V, we define the detection-rejection measure (*DRM*)

$$DRM = \frac{\sum_{x \in V} S_x}{|V|} = \frac{(TP + TN)}{|V|}, \qquad DRM \in [0,1]. \quad (6)$$

That is, for each local network and each search iteration, we select the structure that the trained classifier based on this structure simultaneously detects native patterns and rejects foreign patterns most accurately. Both correct detection and correct rejection contribute equally to the score although any other alternative is possible.

TAN-Based BCM². We propose a TAN-based BCM² (*t*BCM²) that utilizes the *DRM* and SuperParent algorithm searching the TAN space. The SuperParent (SP) algorithm has reduced computational cost compared to hill-climbing search (HCS) and it expedites the search [7]. In each iteration, we determine the best edge to add to a structure by finding a good parent and then the best child for this parent.

Following [7] we define: (a) an *Orphan* is a node without a parent other than the class node, (b) a *SuperParent* (SP) is a node extending edges to all orphans simultaneously (as long as no cycles are formed) and (c) a *FavoriteChild* (FC) of an SP is the orphan amongst all orphans that when connected to the SP provides a structure having the highest value of the *DRM*. We initialize the search for each local network using the NBC structure and employ the value of *DRM* it provides as the current *DRM* value. Each iteration of the search comprises of two parts. First, we make each node an SP in turn and choose the SP that if added to the structure would provide the highest value of the *DRM*. Second, we find the FC for this SP and add the edge between them to the structure if this edge increases the current value of the *DRM*. We update the current value of the *DRM* and continue the search as long as the *DRM* value increases and more than one orphan remains unconnected to an SP. Since in each iteration we connect one variable at the most, the maximum number of iterations and edges that can be added to the initial structure is n-1 (yielding the TAN structure). We repeat this procedure for all $|C|$ local networks terminating with the tBCM^2, as is exemplified in the following pseudo code:

```
1. For k=1:|C|    // index of the local network B_k
1.1 Start with the NBC structure as the current structure of the kth local network. In
     all stages, use T to learn the structure and V to calculate the structure DRM.
1.2 For g=1:n-1     // index of iteration
1.2.1 Find the SP yielding the structure having the highest DRM.
1.2.2 Find the FC for this SP.
1.2.3 If the edge SP → FC improves the DRM value of the current structure, update
         the structure with this edge and employ the structure as the current structure.
        Else: Return the current structure as the kth local network and go to 1.
1.3 Return the current structure as the kth local network and go to 1.
2. Calculate the parameters of each local network using D_tr and return the tBCM^2.
```

Although both the CL multinet and tBCM^2 learn a multinet based on the TAN, the two algorithms differ in a several main issues. First, the CL multinet is learned using a constraint-based approach [11] based on the CL tree algorithm [10] or an extended version of this algorithm [3], while the tBCM^2 is learned by employing the search-and-score approach [11]. Second, the former algorithm establishes for each class a CL tree that maximizes a joint-probability-based measure, whereas the latter algorithm employs a discrimination-driven score for structure learning. Third, the CL multinet utilizes only the class patterns for learning each local network, whereas the tBCM^2 utilizes all patterns. Fourth, the CL multinet always adds n-1 edges even when some variables are completely independent, while the tBCM^2 stops adding edges when there is no improvement in the score of a local network.

Finally we note that the worst case computational complexity of the tBCM^2 (excluding the cost of parameter learning) is $O(3\cdot|C|\cdot|V|\cdot n^3/2)$, which incurs if the algorithm does not end before finding the maximum possible number of SPs [12]. As an example, Figure 1 demonstrates the four local networks learned by the tBCM^2 for the UCI repository Car database [13] along with the corresponding *DRM* values.

4 Experimental Results

Between the *DRM* and Classification Accuracy. Since the *DRM* is measured for each local network separately and using the validation set and the classification accuracy is measured for the *t*BCM^2 and the test set, we studied the relation between the two scores. We started the search for each local network with the NBC structure and identified an iteration by the addition of an edge between an SP and its FC. Whenever all the local networks had completed an iteration, we computed the values of *DRM* they achieve, the average *DRM* value and the test accuracy of the *t*BCM^2 that used these networks. We repeated this procedure until all local networks completed learning (i.e., all final structures were found). Networks that completed learning before their counterpart networks, contributed their final *DRM* values to the calculation of the average *DRM* in each following iteration. Figure 2a presents the relation between the average *DRM* value of the local networks and the classification accuracy of the *t*BCM^2 for increasing numbers of iterations of the SP algorithm and the UCI repository Nursery database [13]. This database is large (i.e., providing reliable results) and has relatively many variables that introduce numerous possible edge additions in each search iteration, thereby the database enables testing structure

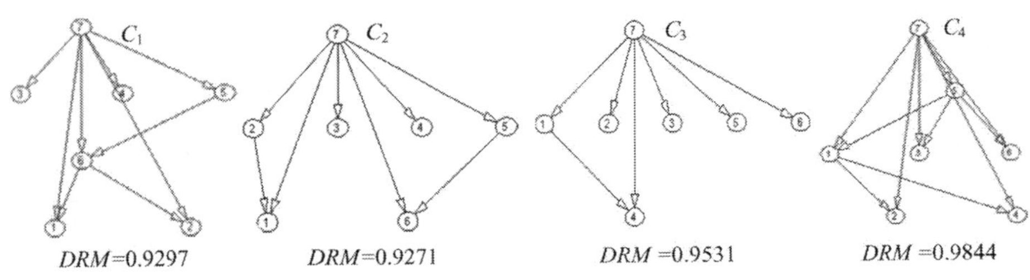

Fig. 1. The four local networks and associated *DRM* values of the *t*BCM^2 for the Car database

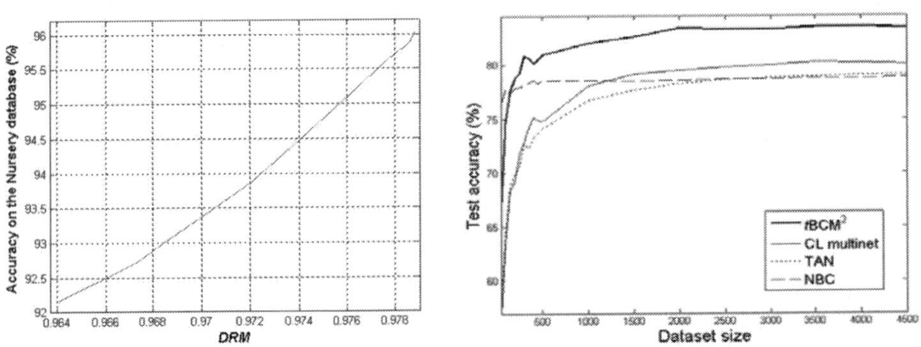

Fig. 2. (a) The relation between the average *DRM* and the *t*BCM^2 classification accuracy for increasing numbers of iterations of the SP algorithm and the UCI Nursery database. (b) Learning curves for the *t*BCM^2, CL multinet, TAN and NBC for the Waveform-21 database.

learning extensively. The figure shows that the classification accuracy increases monotonically with the average *DRM* value.

Learning Curves. Figure 2b presents learning curves for the $t\text{BCM}^2$, CL multinet, TAN and NBC for the large UCI repository Waveform-21 database [13]. Each of ten random replications of the database was partitioned into ten sets. One set was reserved for the test, and the other nine sets were added incrementally to the training set. Each classifier was trained using the increased-size training set and tested on the same test set following each increase. The accuracy was repeatedly measured for all data replications and averaged. Figure 2b demonstrates that the NBC and CL multinet have, respectively, the smallest and largest sensitivity to the sample size. The former classifier has lesser sensitivity since it needs to estimate only few parameters so even a small sample size provides the classifier its asymptotic accuracy. The $t\text{BCM}^2$ is less sensitive than the CL multinet for two reasons. First, the $t\text{BCM}^2$ may have fewer edges for each of its local networks than the CL multinet (Section 3) and therefore it needs to estimate less parameters. Second, the $t\text{BCM}^2$ utilizes all the data whereas the CL multinet employs only the class data. In addition we note that except for a very small sample size, the $t\text{BCM}^2$ is superior to all other classifiers for this database. Similar conclusions are drawn for most of the other databases.

Classification Accuracy. Table 1 demonstrates the superior classification accuracy of the $t\text{BCM}^2$ in comparison to the NBC, TAN, CL multinet and RBMN for 32 databases of the UCI repository. Out of the databases, the $t\text{BCM}^2$ accomplishes higher accuracy than the CL multinet on 24 databases, identical accuracy on 3 databases and inferior accuracy on 5 databases. It achieves higher accuracy than the TAN on 28 databases and inferior accuracy on 4 databases. The $t\text{BCM}^2$ also outperforms the NBC on 90% of the databases. Twenty-two databases are tested using CV10 and the remaining (large) databases using holdout. On the former databases, the $t\text{BCM}^2$ reaches higher accuracy than the CL multinet classifier on 16 of the databases with statistical significance of 95% (t-test with α=0.05) on 12 of the databases and the CL multinet classifier achieves higher accuracy than the $t\text{BCM}^2$ on 4 of the databases without statistical significance for none of them. Also for these 22 databases, the $t\text{BCM}^2$ accomplishes higher accuracy than the TAN on 18 of the databases with statistical significance of 95% (α=0.05) for 13 of them and the TAN achieves higher accuracy than the $t\text{BCM}^2$ on 4 of the databases with statistical significance of 95% (α=0.05) for 1 of the databases.

In addition, Table 1 exemplifies the $t\text{BCM}^2$ superiority to the RBMN [8] for those databases for which results are provided. Also, we compare the $t\text{BCM}^2$ to the DCLT algorithm [5] for the only two databases for which results are given in [5]. We find for the Hepatitis database accuracies of 89.25% and 90.4% and for the Voting database accuracies of 92.18% and 93.97% for the DCLT and $t\text{BCM}^2$ classifiers, respectively. Finally, Table 1 presents also the average classification accuracies of the inspected methods over all 32 databases. The table shows that the $t\text{BCM}^2$ (89.64%) is superior on average to the NBC (85.74%), TAN (87.41%) and CL multinet (87.45%).

Table 1. Classification accuracies of the *t*BCM^2 and other classifiers on 32 databases from [13]. Bold font represents the highest accuracy for a database.

Database	NBC	TAN	CL multinet	RBMN	*t*BCM^2
Adult	83.61	85.83	85.11	NA	**87.33**
Australian	85.36 (±2.14)	84.15 (±2.17)	85.22 (±2.09)	85.21	**88.38 (±2.32)**
Balance	**91.85 (±2.54)**	85.44 (±2.07)	84.63 (±1.81)	NA	90.88 (±1.03)
Breast	97.51 (±0.94)	96.12 (±1.99)	96.34 (±1.00)	95.75	**98.24 (±1.14)**
Car	85.71 (±2.33)	89.81 (±1.89)	**94.10 (±0.98)**	93.06	93.92 (±0.81)
Cmc	51.66 (±2.97)	52.00 (±1.10)	50.85 (±1.82)	NA	**52.85 (±1.65)**
Corral	85.06 (±4.59)	96.06 (±2.44)	99.23 (±2.93)	NA	**100.0 (±0.00)**
Crx	85.98 (±1.85)	85.67 (±2.72)	86.14 (±2.79)	**90.05**	88.89 (±2.59)
Cytogenetic	77.94	81.14	80.30	NA	**82.87**
Flare	79.82 (±1.66)	82.54 (±1.17)	82.55 (±0.94)	**86.87**	83.35 (±1.21)
Hayes	**81.88 (±4.25)**	75.00 (±3.27)	63.13 (±4.86)	NA	80.63 (±3.13)
Hepatitis	85.23 (±1.27)	86.01 (±1.78)	86.54 (±2.00)	NA	**90.40 (±1.52)**
Ionosphere	91.16 (±2.34)	91.44 (±2.57)	**93.92 (±2.01)**	NA	93.03 (±2.69)
Iris	93.67 (±2.99)	93.33 (±2.16)	93.33 (±2.16)	NA	**95.83 (±2.21)**
Krkp (Chess)	87.32	92.31	93.02	94.18	**95.03**
Led-7	74.41	73.76	73.10	NA	**75.89**
Lymphography	83.19 (±3.93)	84.57 (±5.47)	79.81 (±5.05)	NA	**85.57 (±5.16)**
Mofn-3-7-10	85.05 (±1.80)	91.06 (±2.01)	90.63 (±2.46)	90.53	**94.43 (±2.30)**
Monks	96.39 (±1.68)	98.73 (±1.41)	**98.92 (±1.09)**	NA	**98.92 (±1.09)**
Mushroom	97.40	99.47	99.47	NA	**100**
Nursery	89.17	91.09	93.89	91.06	**96.03**
Pendigit	85.72	94.32	**96.62**	NA	96.04
Segment	91.34 (±0.83)	94.09 (±1.04)	94.42 (±1.23)	89.35	**96.13 (±1.23)**
Shuttle	98.45	99.61	**99.92**	97.21	**99.92**
Splice (DNA)	96.33	89.65	96.74	87.52	**97.98**
Tic Tac Toe	69.62 (±1.96)	**75.07 (±2.64)**	73.07 (±2.41)	NA	72.65 (±1.45)
Tokyo	91.45 (±1.82)	92.01 (±2.19)	92.39 (±1.55)	NA	**93.94 (±1.98)**
Vehicle	62.42 (±2.67)	70.82 (±2.51)	69.93 (±2.91)	**73.64**	68.54 (±2.65)
Voting	90.96 (±2.62)	93.99 (±2.16)	93.97 (±2.46)	**96.55**	93.97 (±2.46)
Waveform-21	78.60	78.94	79.69	77.79	**83.82**
Wine	98.27 (±1.65)	98.03 (±1.55)	98.27 (±1.65)	NA	**98.98 (±1.21)**
Zoo	92.00 (±4.66)	**95.08 (±4.25)**	93.09 (±5.03)	NA	94.08 (±4.17)
Average	85.74	87.41	87.45	---	**89.64**

5 Summary and Concluding Remarks

We propose the *t*BCM^2 which is a multinet classifier that learns each local network based on a detection-rejection measure, i.e., the accuracy in simultaneously detecting and rejecting, respectively, the corresponding class and other class patterns. The *t*BCM^2 uses the SuperParent algorithm to learn for each local network a TAN having only augmented edges that increase the classifier accuracy. Evaluated on 32 real-world databases, the *t*BCM^2 demonstrates on average superiority to the NBC, TAN, CL multinet and RBMN classifiers. The advantage of the *t*BCM^2 to the TAN is related to the facts that the former classifier is a multinet that is learned using a discrimination-driven score, and the advantage of the *t*BCM^2 to the CL multinet is attributed to the score of the former and the facts that it usually learns a smaller number of parameters and use the whole data for training.

In further work, we will make parameter learning discriminative rather than generative and apply the BCM^2 to less restricted structure spaces, such as augmented naïve and general Bayesian networks.

Acknowledgment. This works was supported in part by the Paul Ivanier Center for Robotics and Production Management, Ben-Gurion University, Beer-Sheva, Israel.

References

1. Pearl, J.: Probabilistic Reasoning in Intelligent Systems: Networks of Plausible Inference. Morgan-Kaufmann, San-Francisco (1988)
2. Geiger, D., Heckerman, D.: Knowledge representation and inference in similarity networks and Bayesian multinets. Artificial Intelligence 82 (1996) 45-74
3. Cheng, J., Greiner, R.: Learning Bayesian belief network classifiers: Algorithms and system. In Proc. 14th Canadian Conf. on Artificial Intelligence (2001) 141-151
4. Friedman, N., Geiger, D., Goldszmidt, M.: Bayesian network classifiers. Machine Learning 29 (1997) 131-163
5. Huang, K., King, I., Lyu, M. R.: Discriminative training of Bayesian Chow-Liu multinet classifier. In Proc. Int. Joint Conf. Neural Networks (2003) 484-488
6. Kontkanen, P., Myllymaki, P., Sliander, T., Tirri, H.: On supervised selection of Bayesian networks. In Proc. 15th Conf. on Uncertainty in Artificial Intelligence (1999) 334-342
7. Keogh, E.J., Pazzani, M.J.: Learning the structure of augmented Bayesian classifiers. Int. J. on Artificial Intelligence Tools 11 (2002) 587-601
8. Pena, J.M., Lozano, J.A., Larranaga, P.: Learning recursive Bayesian multinets for data clustering by means of constructive induction. Machine Learning 47 (2002) 63-89
9. Meila, M., Jordan, M.I.: Learning with mixtures of trees. J. of Machine Learning Research 1 (2000) 1-48
10. Chow, C.K., Liu, C.N.: Approximating discrete probability distributions with dependence trees. IEEE Trans. Info. Theory 14 (1968) 462-467
11. Spirtes, P., Glymour, C., Scheines, R.: Causation, Prediction and Search. 2nd edn. MIT Press, Cambridge MA (2000)
12. Gurwicz, Y.: Classification using Bayesian multinets. M.Sc. Thesis. Ben-Gurion University, Israel (2004)
13. Blake, C.L., Merz, C.J.: UCI Repository of machine learning databases. http://www.ics.uci.edu/~mlearn/MLRepository.html, 1998

Bayesian Network Structure Learning by Recursive Autonomy Identification

Raanan Yehezkel and Boaz Lerner

Pattern Analysis and Machine Learning Lab
Department of Electrical & Computer Engineering
Ben-Gurion University, Beer-Sheva 84105, Israel
{raanany, boaz}@ee.bgu.ac.il

Abstract. We propose the recursive autonomy identification (RAI) algorithm for constraint-based Bayesian network structure learning. The RAI algorithm learns the structure by sequential application of conditional independence (CI) tests, edge direction and structure decomposition into autonomous sub-structures. The sequence of operations is performed recursively for each autonomous sub-structure while simultaneously increasing the order of the CI test. In comparison to other constraint-based algorithms d-separating structures and then directing the resulted undirected graph, the RAI algorithm combines the two processes from the outset and along the procedure. Thereby, learning a structure using the RAI algorithm requires a smaller number of high order CI tests. This reduces the complexity and run-time as well as increases structural and prediction accuracies as demonstrated in extensive experimentation.

1 Introduction

A Bayesian network (BN) is a graphical model that efficiently encodes the joint probability distribution for a set of variables [1]. Learning the model structure from data by considering all possible structures exhaustively is infeasible as the number of possible structures grows exponentially with the number of nodes [2]. Hence, structure learning requires either sub-optimal heuristic search algorithms or algorithms which are efficient under certain assumptions. In the constraint-based (CB) approach [3], [4], a structure edge is learned if meeting a constraint, called conditional independence (CI) test, derived from comparing the value of a statistical or information-theory-based test of conditional independence to a threshold. Meeting such constraints enables the formation of an undirected graph that is further directed based on causality inference rules [5]. Once the structure is learned, the model parameters are usually computed from the relative frequencies of variable states as represented in the data.

Most of the CB algorithms, such as IC [5], PC [4] and TPDA [3], construct a BN in two stages. The first stage is learning associations between variables for constructing an undirected structure. This requires an exponentially growing number of CI tests with the number of nodes. The PC and TPDA algorithms reduce the number of CI tests using some assumptions to restrict the space of possible structures. The second

D.-Y. Yeung et al. (Eds.): SSPR & SPR 2006, LNCS 4109, pp. 154–162, 2006.

stage in most algorithms is directing edges using inferred causality performed in two steps: finding and directing V-structures and inductively directin additional edges [5]. Edge direction is unstable, i.e., small errors in the input to the stage yield large errors in its output [4]. To eliminate the instability, the algorithms separate the two stages trying to minimize in the first stage erroneous decisions about independence caused by large condition sets that are more likely to be incorrect and also lead to poorer estimation of dependences due to the curse-of-dimensionality.

We propose the recursive autonomy identification (RAI) algorithm, which is a CB model that learns the structure of a BN by sequential application of CI tests, edge direction and structure decomposition into autonomous sub-structures. This sequence of operations is performed recursively for each autonomous sub-structure. In each recursive call, the order of the CI test is increased similarly to the PC algorithm [4]. By performing CI tests of low order (i.e., tests employing small conditions sets) before those of high order, the RAI algorithm performs more reliable tests that save performing less reliable tests. By considering directed rather than undirected edges, the RAI algorithm avoids unnecessary CI tests and performs tests using smaller condition sets. Smaller condition sets are represented more precisely than larger sets in the dataset, i.e., the RAI algorithm diminishes the curse-of-dimensionality while testing using smaller condition sets and thereby it improves the prediction accuracy. Repeated recursively for autonomies decomposed from the graph, both mechanisms reduce computational and time complexities, database queries and structural errors.

Section 2 provides preliminaries and Section 3 introduces the RAI algorithm. Section 4 presents experimental evaluation of the RAI algorithm regarding structural correctness, complexity and prediction accuracy. Section 5 summarizes the study.

2 Preliminaries

A BN B(G,Θ) is a model for representing the joint probability distribution for a set of variables $\boldsymbol{X}=\{X_1,\ldots,X_n\}$. The structure G(**V**,**E**) is a directed acyclic graph (DAG) composed of **V**, a set of nodes representing the variables ***X***, and **E**, a set of directed edges connecting the nodes. An edge manifests dependence between the nodes connected by the edge while the absence of an edge demonstrates independence between the nodes. A directed edge $X_i \rightarrow X_j$ connects a child node X_j to its parent X_i. ***Pa***(*X*,G) is the set of *X's* parents in G. The set of parameters Θ holds local conditional probabilities over ***X***, $P(X_i \mid \boldsymbol{Pa}(X_i,\mathrm{G}))\ \forall i$ that quantify the edges. The joint probability distribution for ***X*** represented by a BN is $P(X_1,\ldots X_n)=\prod_{i=1}^{n} P(X_i \mid \boldsymbol{Pa}(X_i,G))$ [1]. We also use the term partially directed graph (PDG), i.e., a graph which may have both directed and undirected edges and has at most one edge between any pair of nodes. We use this term in learning a graph starting from a complete undirected graph and eliminating and directing edges until uncovering a graph representing a family of Markov equivalent structures (pattern) of the true BN (i.e., the graphs have the same sets of adjacencies and V-structures) [4], [5]. $\boldsymbol{Pa}_p(X,\mathrm{G})$, ***Adj***(*X*,G) and ***Ch***(*X*,G) are respectively the sets of potential parents, adjacent nodes (two nodes connected by an edge) and children of *X* in a

PDG, $\boldsymbol{Pa_p}(X,\mathrm{G})=\boldsymbol{Adj}(X,\mathrm{G})\backslash\boldsymbol{Ch}(X,\mathrm{G})$. We use $X \perp\!\!\!\perp Y | \boldsymbol{S}$ to indicate that X and Y are independent given a set of nodes $\boldsymbol{S}$ and employ also the notion of d-separation [5]. Next, we define d-separation resolution evaluating d-separation for different sizes of condition sets, an exogenous cause to a graph and an autonomous sub-structure.

Definition 1 – ***d-separation resolution:*** The resolution of a d-separation relation between a pair of non-adjacent nodes in a graph is the size of the smallest condition set that d-separates the two nodes (see Figure 1 for an example).

Definition 2 – ***d-separation resolution of a graph****:* The d-separation resolution of a graph is the highest d-separation resolution in the graph (see Figure 2).

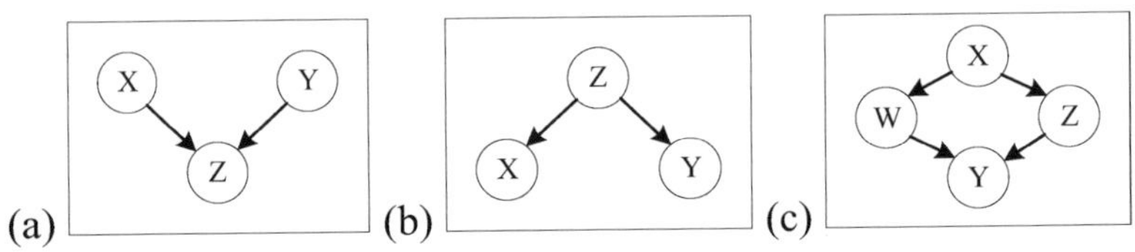

Fig. 1. Examples of d-separation resolutions of (a) 0, (b) 1 and (c) 2 between nodes X and Y

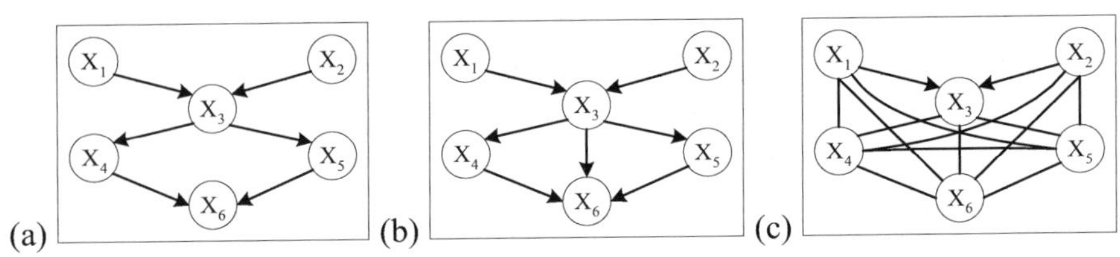

Fig. 2. Examples of graph d-separation resolutions of (a) 2, (b) 1 and (c) 0

Definition 3 – ***exogenous cause****:* Y is an exogenous cause to $\mathrm{G}(\mathbf{V},\mathbf{E})$ if $Y \notin \mathbf{V}$ and $\forall X \in \mathbf{V}$, $Y \in \boldsymbol{Pa}(X)$ or $Y \notin \boldsymbol{Adj}(X)$ [5].

Definition 4 – ***autonomous sub-structure****:* A sub-structure $\mathrm{G}^{\mathrm{A}}(\mathbf{V}^{\mathbf{A}},\mathbf{E}^{\mathbf{A}})$ in $\mathrm{G}(\mathbf{V},\mathbf{E})$ s.t $\mathbf{V}^{\mathrm{A}} \subset \mathbf{V}$, $\mathbf{E}^{\mathrm{A}} \subset \mathbf{E}$ is autonomous given a set $\mathbf{V}_{\mathbf{ex}}$ of exogenous nodes to G^{A} if $\forall X \in \mathbf{V}^{\mathrm{A}}$, $\boldsymbol{Pa}(X,\mathrm{G}) \subset \{\mathbf{V}^{\mathrm{A}} \cup \mathbf{V}_{\mathbf{ex}}\}$. If $\mathbf{V}_{\mathbf{ex}}$ is empty, we say that the sub-structure is autonomous.

An autonomous sub-structure G^{A} holds the Markov property, i.e., two non-adjacent nodes in G^{A} are d-separated given nodes in G^{A} or exogenous causes to G^{A} (Figure 3).

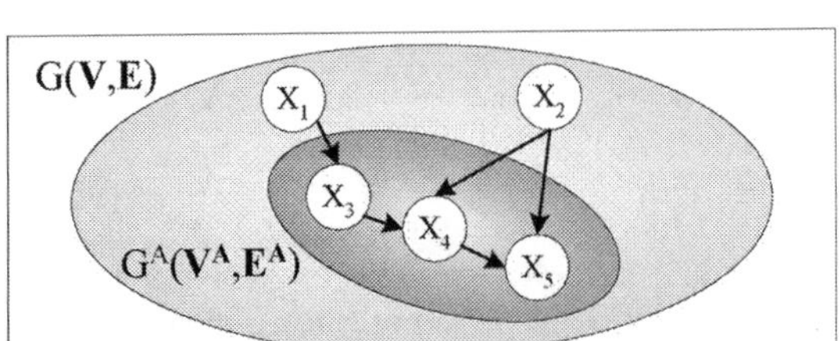

Fig. 3. An autonomous sub-structure G^{A} in G having exogenous nodes X_1 and X_2

3 Recursive Autonomy Identification

Starting from a complete undirected graph and proceeding from low to high graph d-separation resolution, the RAI algorithm uncovers the correct pattern of a structure by recursively performing the sequence: (1) test of CI between nodes followed by the removal of edges related to independences, (2) edge direction according to inferred causality rules, and (3) graph decomposition into autonomous sub-structures.

CI testing of order *n* between nodes *X* and *Y* is performed by thresholding the value of a criterion measuring the dependence between the nodes conditioned on a set of *n* nodes (the condition set) from the parents of *X* or *Y*. The set is determined by the Markov property [5], i.e., if *X* is directed into *Y* then only *Y*'s parents are included in the set. Commonly, this criterion is the χ^2 goodness of fit test [4] or conditional mutual information [3]. Directing edges is conducted according to causality rules [5]. Given an undirected graph and a set of independences, the following two steps are performed consecutively. First, V-structures are identified and the corresponding edges are directed. A V-structure $X \rightarrow Z \leftarrow Y$ is defined if 1) *X* and *Y* are unconnected neighbors of *Z*, 2) *X* and *Y* are marginally independent and 3) *X* and *Y* are dependent conditioned on *Z*. In the second step, also called the inductive stage, an edge *Y*—*Z* is directed as $Y \rightarrow Z$ if: 1) there exists an edge $X \rightarrow Y$ where *X* and *Z* are not adjacent and there is no arrowhead at *Y*, 2) there is a chain $Y \rightarrow X \rightarrow Z$, or 3) two chains $Y — X \rightarrow Z$ and $Y — W \rightarrow Z$ exist. This step is continued until no more edges can be directed complying with two restrictions: a) no V-structures in addition to those of the first step and b) no directed cycles are created. Finally, decomposition into autonomous sub-structures reveals the structure hierarchy. It also allows the performance of fewer CI tests that are conditioned on a large number of potential parents, and hence reduces complexity. The RAI algorithm identifies ancestor and descendant sub-structures; the formers are autonomous and the latters are autonomous given nodes of the formers.

An iteration of the RAI algorithm starts with knowledge produced in the previous iteration and the current d-separation resolution, *n*. Previous knowledge includes G_{start}, a structure having d-separation resolution of *n*-1, and $\mathbf{G}_{ex}$, a set of structures having each possible exogenous causes to G_{start}. In the first iteration, $n = 0$, $G_{start}(\mathbf{V},\mathbf{E})$ is the complete undirected graph and $\mathbf{G}_{ex}=\varnothing$. Given a structure G_{start} having d-separation resolution *n*-1, the RAI algorithm seeks independences between adjacent nodes conditioned on sets of size *n*, resulting in a structure having d-separation resolution of *n*. After applying causality rules in order to direct the edges, a partial topological ordering is obtained in which parent nodes precede their descendants. This ordering is partial as not all the edges can be directed, so nodes connected by undirected edges have equal topological order. Using this partial topological ordering, the algorithm decomposes the structure into ancestor and descendent autonomous sub-structures in order to reduce the complexity of the successive stages. A descendant sub-structure is established by identifying the lowest topological order nodes (either a single node or a several nodes having the same lowest order). We will refer to a single descendent sub-structure although it may consist of a several non-connected sub-structures. This structure is autonomous given nodes of higher topological order composing ancestor sub-structures. The algorithm first learns ancestor sub-structures and only then the descendant sub-structure in order to consider for each pair of nodes of the descendent sub-structure condition sets that (possibly) have smaller numbers of parents. Each ancestor or descendent sub-structure is further learned by recursive calls to the algorithm. Figures 4 and 5 show respectively the RAI algorithm and a manifesting example.

The RAI algorithm is composed of four stages (A to D in Figure 4) and an exit condition checked before the execution of each stage. The purpose of Stage A is to shrink the link between $\mathbf{G}_{ex}$ and G_{start}, the latter is having d-separation resolution of n-1, and direct the edges of G_{start}. This is achieved by CI testing using condition sets of size n between nodes in $\mathbf{G}_{ex}$ and nodes in G_{start}, removing edges corresponding to independences and directing those remaining edges that can be directed. Stage B performs the same for edges in G_{start}. In both stages, the condition set includes nodes of $\mathbf{G}_{ex}$ and G_{start}. Stage B also identifies the partial topological ordering of nodes and decomposes the current graph into ancestor and descendant sub-structures. Stage C is a recursive call to the RAI algorithm for learning each ancestor sub-structure with order n+1. Similarly, Stage D of the algorithm is a recursive call to the RAI for learning the descendant sub-structure with order n+1 while assuming that the ancestor sub-structures have been fully learned (having the maximal d-separation resolution).

Figure 5 sketches stages in learning an example graph. Figure 5a shows the true structure we wish to uncover. Initially, G_{start} is the complete undirected graph, n=0 and $\mathbf{G}_{ex}$ is empty so Stage A is skipped. In Stage B1, pairs of nodes are CI tested given empty condition sets i.e., marginal independence, which yields the removal of the edges between node X_1 and nodes X_3, X_4 and X_5 (Figure 5b). The causal relations inferred in Stage B2 are shown in Figure 5c. The nodes having the lowest topological order (X_2, X_6, X_7) are grouped into a descendant sub-structure G_D (Stage B3) while the remaining nodes form two unconnected ancestor sub- structures, G_{A_1} and G_{A_2} (Stage B4) (Figure 5d). In Stage C, the algorithm is called recursively for each of the ancestor sub-structures with n=1, $G_{start}=G_{A_i}$ and $\mathbf{G}_{ex}=\varnothing$. Since sub-structure G_{A_1} contains a single node, the exit condition for the structure is satisfied. While calling $G_{start}=G_{A_2}$, Stage A is skipped and Stage B1 identifies that $X_4 \perp\!\!\!\perp X_5 | X_3$ thus removes X_4—X_5. No causal relations are identified so the nodes have equal topological order and they are grouped to from a descendant sub-structure. The recursive call for this sub-structure with n=2 is returned immediately since the exit condition is satisfied (Figure 5e). In Stage D, the RAI is called with n=1, $G_{start}=G_D$ and $\mathbf{G}_{ex}=\{G_{A_1},G_{A_2}\}$. In Stage A1 relations $X_1 \perp\!\!\!\perp \{X_6,X_7\}|X_2$, $X_4 \perp\!\!\!\perp \{X_6,X_7\}|X_2$ and $\{X_3,X_5\} \perp\!\!\!\perp \{X_2,X_6,X_7\}|X_4$ are identified and the corresponding edges are removed (Figure 5f). In Stage A2, X_2 is identified as a parent of X_6 and X_7 (Figure 5g). Stage B1 identifies that $X_2 \perp\!\!\!\perp X_7|X_6$ and Stage B2 identifies X_6 as a parent of X_7 leading, respectively, to the removal of $X_2 \rightarrow X_7$ and direction of $X_6 \rightarrow X_7$ (Figure 5h). Then, in Stages B3 and B4, X_7 and $\{X_2, X_6\}$ are identified as a descendant and an ancestor sub-structures, respectively. Further recursive calls are returned immediately and the resulting PDG (Figure 5h) represents a family of Markov equivalent structures of the true structure (Figure 5a).

Main function $G_{out} = RAI(n, G_{start}(\mathbf{V_{start}}, \mathbf{E_{start}}), G_{ex}(\mathbf{V_{ex}}, \mathbf{E_{ex}}))$

Exit condition

If all nodes in G_{start} have less than n+1 potential parents exit.

A. Thinning the link between G_{ex} and G_{start} and directing G_{start}

1. $\forall Y$ in G_{start} and its parent X in $\mathbf{G_{ex}}$, if $\exists \mathbf{S} \subset \{\boldsymbol{Pa}(Y, \mathbf{G_{ex}}) \backslash X \cup \boldsymbol{Pa}_p(Y, G_{start})\}$ and $|\mathbf{S}| = n$ s.t $X \perp\!\!\!\perp Y | \mathbf{S}$, then remove the edge between X and Y.
2. Direct the edges using causality inference rules.

B. Thinning, directing and decomposing G_{start}

1. $\forall Y$ and its potential parent X both in G_{start}, if $\exists \mathbf{S} \subset \{\boldsymbol{Pa}(Y, \mathbf{G_{ex}}) \cup \boldsymbol{Pa}_p(Y, G_{start}) \backslash X\}$ and $|\mathbf{S}| = n$ s.t $X \perp\!\!\!\perp Y | \mathbf{S}$, then remove the edge between X and Y.
2. Direct the edges using causality inference rules.
3. Group nodes having the lowest topological order into a descendant sub-structure G_D.
4. Remove G_D from G_{start} temporarily and define the resulting unconnected structures as ancestor sub-structures $G_{A_1}, \ldots, G_{A_k}$.

C. Ancestor sub-structure decomposition

for $i = 1$ to k, call $RAI(n+1, G_{A_i}, \mathbf{G}_{ex})$

D. Descendant sub-structure decomposition

1. Define $\mathbf{G}_{D\_ex} = \{G_{A_1}, \ldots, G_{A_k}, \mathbf{G}_{ex}\}$ as the exogenous structure to G_D.
2. Call $RAI(n+1, G_D, \mathbf{G}_{D\_ex})$

Fig. 4. The RAI algorithm

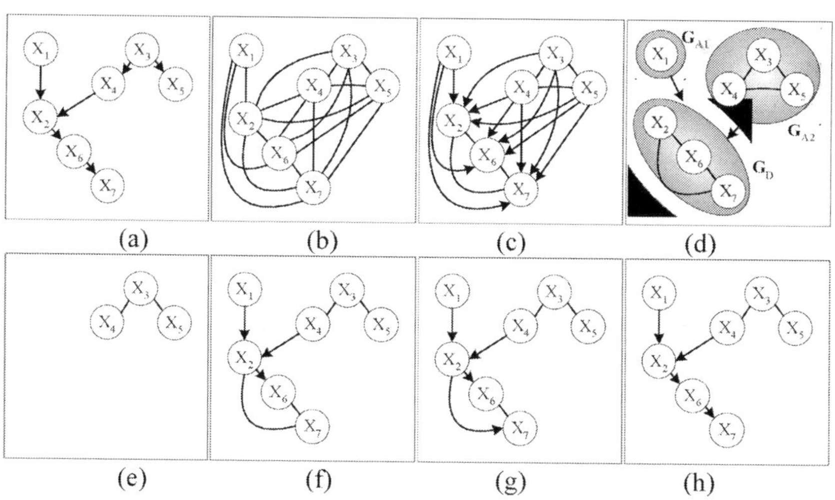

Fig. 5. Learning an example structure. a) The true structure and structures learned in Stages (see Figure 4) b) B1, c) B2, d) B3 and B4, e) C, f) D and A1, g) D and A2 and h) D, B1 and B2.

4 Experiments and Results

Synthetic data. The complexity of the RAI algorithm was compared to that of the PC algorithm by the number of CI tests required to learn synthetically generated structures. We learned graphs of sizes (numbers of nodes) between 6 and 15. We used 3,000 randomly generated graphs restricted by a maximal fan-in value of 3, i.e., every node has at most 3 parents and at least one node has 3 parents. The implementation was aided by the Bayes net toolbox (BNT) [6] and BNT structure learning package [7]. Figure 6a shows the percentage of CI tests saved using the RAI algorithm compared to the PC algorithm as a function of the condition set size for different graph sizes. The figure shows that the percentage of CI tests saved using the RAI algorithm increases with both graph and condition set sizes. For example, the save in CI tests for a graph of size 15 and condition sets of size 4 is more than 70%.

ALARM network. Correctness of the learned structure was evaluated using the ALARM network [8], which is a widely accepted structure learning benchmark since the true graph is known. The RAI algorithm was compared to the PC, TPDA and K2 [2] algorithms using 10 databases of 10,000 randomly generated cases each. The node ordering required for the K2 algorithm was determined by learning a maximum weighted spanning tree (MWST) and selecting a root node [7]. We also evaluated the K2 algorithm with the true ordering, which is inferred from the known network. We identify these two versions as K2 (MWST) and K2 (true), respectively. Since the TPDA algorithm uses for CI testing the conditional mutual information (CMI) [3], we employ this test also for the RAI and PC algorithms and selected thresholds of $3 \cdot 10^{-3}$, $3 \cdot 10^{-3}$ and $2 \cdot 10^{-3}$ for the RAI, TPDA and PC algorithms, respectively. Structural correctness was evaluated by measuring the root mean square of extra and missing edges errors in the learned structure compared with the true structure (i.e., the total error). The smallest total error of 1.3% was achieved by the RAI algorithm compared to errors of 6.32%, 2.94%, 6.76% and 4.68% of the TPDA, PC, K2 (MWST) and K2 (true) algorithms, respectively. This superiority of the RAI algorithm was validated using a *t*-test with 1% significance level. Complexity was measured by the total number of log operations (logarithms, multiplications and divisions) required for calculating CMI in CI testing. As Figure 6b shows, the PC and TPDA algorithms require respectively, 521% and 394% more log operations than the RAI algorithm.

Real-world data. The RAI prediction accuracy was evaluated using databases of the UCI Repository [9]. Continuous variables were discretized and instances with missing values were removed. All databases were analyzed using a CV5 experiment except the large "chess", "mofn 3-7-10", "nursery" and "shuttle" databases which were analyzed using the holdout method. CI tests were carried out using the χ^2 test, which is recommended for the PC algorithm [4], with thresholds chosen for each algorithm and database in order to maximize the prediction accuracy on a validation set. Parameter learning was performed assuming Dirichlet prior distribution having zero hyper-parameters leading to the maximum likelihood solution [1]. Prediction accuracies of the RAI and PC algorithms as well as the run-time saved by using the RAI are summarized in Table 1. The accuracy is also compared to those of the TPDA algorithm [10] and three classifiers reported in [11], namely, the naïve Bayesian classifier (NBC), tree augmented naïve (TAN) Bayes and a BN learned using the minimum description length (MDL) score. Databases in Table 1 for which accuracies are not reported in [10] or [11] are represented by empty entries. On thirteen of the nineteen databases, the RAI algorithm improves accuracy on the PC algorithm, on five it keeps accuracy intact and on the remaining "iris" database it deteriorates accuracy. Since

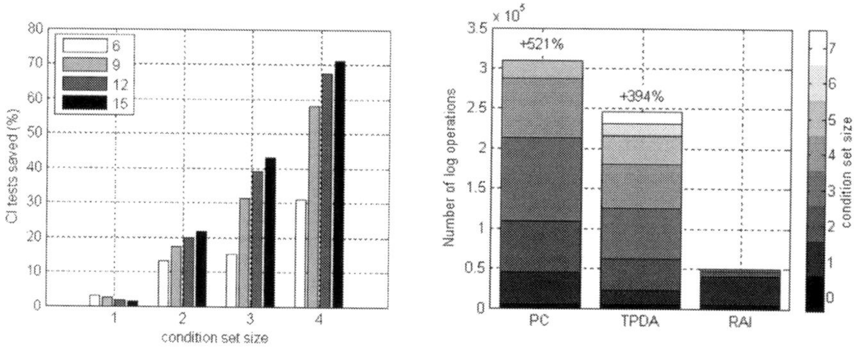

(a) (b)

Fig. 6. (a) Percentage of CI tests saved by using the RAI algorithm compared to the PC algorithm as a function of the condition set size and number of nodes in the graph – 6, 9, 12 or 15 (gray shades). (b) The numbers of log operations required by the PC, TPDA, and RAI algorithms for learning the ALARM network. Gray shades represent different condition set sizes for the CI tests. Percentages on the tops of the bars are with reference to the RAI algorithm.

Table 1. Mean prediction accuracy of classifiers based on the RAI, PC, TPDA, NBC, MDL and TAN. Bold font emphasizes the highest accuracy for a database. Also shown is the run-time cut due to the RAI compared to the PC algorithm. Standard deviationa are ommited.

Database	RAI (%)	PC (%)	TPDA (%)	NBC (%)	MDL (%)	TAN (%)	Run-time cut (%)
australian	85.51	85.51		**86.23**	**86.23**	81.3	**6.05**
breast	96.49	95.46		**97.36**	96.92	95.75	**71.87**
car	**92.94**	85.07	86.11				**91.10**
chess	93.53	93.15	94.65	87.15	**95.59**	92.40	**80.65**
cleve	81.41	76.67		**82.76**	81.39	79.06	**39.60**
cmc	**51.12**	50.92					**14.22**
corral	**98.52**	84.53		85.88	97.60	95.32	**87.94**
crx	**86.38**	**86.38**		86.22	85.60	83.77	**25.25**
flare C	**84.30**	**84.30**	82.27	79.46	82.74	82.74	**20.38**
iris	93.33	**96.00**		93.33	94.00	93.33	**19.10**
led7	**73.59**	73.31					**91.74**
mofn 3-7-10	**93.16**	81.45		86.43	85.94	91.70	**67.70**
nursery	**93.06**	**93.06**	89.72				**89.70**
shuttle (s)	**99.22**	98.40		98.34	99.17	98.86	**38.94**
tic-tac-toe	**75.57**	74.74					**36.52**
vehicle	**70.22**	63.93		58.28	61.00	67.86	**13.15**
vote	**95.87**	95.64	95.17	90.34	94.94	89.20	**46.06**
wine	**87.07**	85.44					**29.11**
zoo	88.95	88.95					**13.63**

the accuracies for the TPDA, NBC, MDL-based and TAN classifiers are borrowed from the original papers, no examination of statistical significance of the results could have been performed. Therefore, any advantage of a classifier over another classifier for a specific database is not necessarily statistically significant. However, averaging the prediction accuracy over the twelve databases for which we have results for all but the TPDA algorithm shows that the RAI, PC, NBC, MDL-based and TAN classifiers achieve average accuracies of 89.8, 86.8, 86.0, 88.4 and 87.6%, respectively.

5 Summary

We demonstrate that the RAI algorithm requires less CI tests of high order than the PC algorithm, and the percentage of tests saved by the RAI algorithm increases with the sizes of the network and condition set for the CI test. The RAI algorithm reconstructs the ALARM network with significantly less errors than the PC, TPDA and K2 algorithms and has a considerably smaller computational complexity. In addition, the structure learned by the RAI algorithm yields a classifier that is mostly more accurate than those learned using the PC, TPDA, NBC, TAN and MDL-based algorithms.

Acknowledgment. This works was supported in part by the Paul Ivanier Center for Robotics and Production Management, Ben-Gurion University, Beer-Sheva, Israel.

References

1. Heckerman, D. A tutorial on learning with Bayesian networks. MS TR-95-06, March 1995.
2. Cooper, G. F., and Herskovits, E. A Bayesian method for the induction of probabilistic networks from data. Machine Learning, Vol. 9, 309-347, 1992.
3. Cheng, J., Bell, D. and Liu, W. Learning Bayesian networks from data: an efficient approach based on information theory. Sixth ACM Int. Conf. on Information and Knowledge Management, pages 325-331, 1997.
4. Spirtes, P., Glymour, C., and Scheines, R. Causation, Prediction and Search, 2nd edition, MIT Press, 2000.
5. Pearl, J. Causality: Models, Reasoning, and Inference. Cambridge. 2000.
6. Murphy, K. Bayes net toolbox for Matlab. Computing Science and Statistics, Vol. 33, 2001.
7. Leray, P., and Francois, O. BNT structure learning package: documentation and experiments. PSI TR, 2004.
8. Beinlich, I. A., Suermondt, H. J., Chavez, R. M., and Cooper, G. F. The ALARM monitoring system: A case study with two probabilistic inference techniques for belief networks. Second European Conf. on Artificial Intelligence in Medicine, pages 246-256, 1989.
9. Newman, D. J., Hettich, S., Blake, C. L., and Merz, C. J. UCI Repository of machine learning databases. U. of California, Irvine, Dept. of Information and Computer Science 1998.
10. Cheng, J., and Greiner, R. Comparing Bayesian network classifiers. Fifteenth Conf. on Uncertainty in Artificial Intelligence, pages 101-107, 1999.
11. Friedman, N., Geiger, D. and Goldszmidt, M. Bayesian network classifiers. Machine Learning, Vol. 29, 131-161, 1997.

Fast Suboptimal Algorithms for the Computation of Graph Edit Distance

Michel Neuhaus*, Kaspar Riesen, and Horst Bunke

Institute of Computer Science and Applied Mathematics, University of Bern
Neubrückstrasse 10, CH-3012 Bern, Switzerland
{mneuhaus, riesen, bunke}@iam.unibe.ch

Abstract. Graph edit distance is one of the most flexible mechanisms for error-tolerant graph matching. Its key advantage is that edit distance is applicable to unconstrained attributed graphs and can be tailored to a wide variety of applications by means of specific edit cost functions. Its computational complexity, however, is exponential in the number of vertices, which means that edit distance is feasible for small graphs only. In this paper, we propose two simple, but effective modifications of a standard edit distance algorithm that allow us to suboptimally compute edit distance in a faster way. In experiments on real data, we demonstrate the resulting speedup and show that classification accuracy is mostly not affected. The suboptimality of our methods mainly results in larger inter-class distances, while intra-class distances remain low, which makes the proposed methods very well applicable to distance-based graph classification.

1 Introduction

Graph matching refers to the process of evaluating the structural similarity of graphs. The main advantage of a description of patterns by graphs instead of vectors is that graphs allow for a more powerful representation of structural relations. In the most general case, vertices and edges are labeled with arbitrary attributes. One of the most flexible error-tolerant graph matching methods applicable to unconstrained graphs is based on graph edit distance [1]. However, the error-tolerant nature of edit distance — unlike exact graph matching methods such as subgraph isomorphism or maximum common subgraph — potentially allows every vertex of a graph to be mapped to every vertex of another graph. The time and space complexity of edit distance computation is therefore very high. Consequently, the edit distance can be computed for graphs of a rather small size only.

In recent years, a number of methods addressing the high computational complexity of graph edit distance computation have been proposed. A common way

* Supported by the Swiss National Science Foundation NCCR program *Interactive Multimodal Information Management (IM)2* in the Individual Project *Multimedia Information Access and Content Protection*.

D.-Y. Yeung et al. (Eds.): SSPR&SPR 2006, LNCS 4109, pp. 163–172, 2006.

to make graph matching more efficient is to restrict considerations to special classes of graphs. Examples include the classes of planar graphs [2], bounded-valence graphs [3], trees [4], and graphs with unique vertex labels [5]. A number of graph matching methods based on genetic algorithms have been proposed [6]. Genetic algorithms offer an efficient way to cope with large search spaces, but are non-deterministic and suboptimal. If the structural matching problem is formulated as a vertex labeling problem, relaxation labeling techniques can be used for graph matching [7]. While in some cases such graph matching methods may perform efficiently, it seems to be rather difficult to apply them to strongly distorted data. Recently, a suboptimal edit distance algorithm has been proposed [8] that requires the vertices of graphs to be planarly embedded, which is satisfied in many, but not all computer vision applications of graph matching. In [9], the authors propose an edit distance method based on bipartite matching. The main drawback of their method is that no edge information is used in the bipartite matching step of the algorithm.

In this paper, we address the issue of efficient edit distance computation in a different way. We exploit the fact that exact edit distance algorithms typically explore large areas of the search space that are not relevant for certain classification tasks. We propose simple variants of a standard edit distance algorithm that make the computation substantially faster, but keep the resulting suboptimal distances sufficiently accurate.

2 Graph Edit Distance

The key idea of graph edit distance is to define the dissimilarity of two graphs by the minimal amount of distortion that is needed to transform one graph into the other. The distortion model is defined by a number of underlying vertex and edge edit operations. The most common set of graph edit operations consists of an insertion, a deletion, and a substitution operation on vertices and edges. Given a source and a target graph, the idea is to remove some vertices and edges from the source graph, relabel some of the remaining vertices and edges, and possibly insert some vertices and edges such that eventually the target graph is obtained. A sequence of edit operations that transform the source graph into the target graph is called an *edit path* between source and target graph. Moreover, cost functions are introduced measuring the strength of the distortion caused by each edit operation. These cost functions are used to decide whether an edit path represents weak modifications only or a significant amount of structural distortion. If there exists an inexpensive edit path between two graphs, these graphs are considered structurally similar in terms of the underlying edit operation model and edit cost functions; if no such edit path exists, the graphs are considered dissimilar. Consequently, the *edit distance* of two graphs is defined by the minimum cost edit path between the two graphs [1]. In the following, we denote a graph by $g = (V, E, \mu, \nu)$, where V denotes a finite set of vertices, $E \subseteq V \times V$ a set of directed edges, $\mu : V \to L$ a vertex labeling function assigning each vertex an attribute from L, and $\nu : E \to L$ an edge labeling function. The substitution of a vertex u by a vertex v is denoted by $u \to v$, the insertion of u by $\varepsilon \to u$, and the deletion of u by $u \to \varepsilon$.

The computation of edit distance is usually carried out by means of a tree search algorithm. Provided that a few weak conditions are satisfied in the definition of edit costs, it is sufficient to consider only a finite number of edit paths to find one with minimum costs. The most widely used method for edit distance computation is based on the A* algorithm [10]. The A* algorithm is a best-first algorithm that attempts to retrieve an optimal path from a search tree based on heuristic information. The idea is to use a search tree to represent the considered optimization problem in a tree data structure, such that the root node represents the starting point, inner nodes correspond to partial solutions, and leaf nodes to complete solutions. A search tree is dynamically constructed at runtime by iteratively creating successor nodes linked by edges to the currently considered node. The A* search algorithm is characterized by a heuristic function that estimates the expected costs of the best route from the root through the current node to a leaf node. At each step during tree traversal, the most promising node — the one with the lowest heuristic cost value — from the set of nodes to be processed is chosen. Formally, for a node of the search tree p, we use $g(p)$ to denote the costs of the optimal path from the root node to the current node p found by A* so far and $h(p)$ to denote the estimated costs from p to a leaf node. The sum $g(p) + h(p)$ gives the heuristic assessment of node p. If the estimated costs $h(p)$ are always lower than, or equal to, the real costs, the algorithm is known to be admissible, that is, an optimal path from the root node to a leaf node is guaranteed to be found by this procedure [10].

In graph edit distance, unlike exact graph matching algorithms, vertices of the source graph can potentially be mapped to any vertex of the target graph. Given two graphs, the A* search tree for edit distance is constructed by considering vertices of the first graph one after the other. An A* algorithm for the computation of graph edit distance is given in Alg. 1. Let us assume that the vertices of the first graph are processed in the order $(u_1, u_2, \ldots)$. All possible edit operations are constructed simultaneously for each vertex, that is, the removal of the vertex (line 12) or the substitution of the vertex by any unprocessed vertex of the second graph (line 11), which produces a number of successor nodes in the search tree. Note that edit operations on edges are implied by edit operations on their adjacent vertices. If all vertices of the first graph have been processed, the remaining vertices of the second graph can be inserted into the graph in a single step (line 14). The set of partial edit paths OPEN consists of the search tree nodes to be considered in the next step. The currently most promising node p of the search tree, or partial edit path, is the one minimizing the A* search costs $g(p)+h(p)$ (line 5). When a complete edit path is obtained in this way, it is guaranteed to be an optimal one and is returned as the solution (line 7). In cases where the edit distance computation takes longer than a predefined threshold, the corresponding distance is set to infinity.

The function $g(p)$ measuring the costs from the root node to the current node p is simply set equal to the cost of the partial edit path accumulated so far. In

Algorithm 1. Computation of graph edit distance by A* algorithm

Input: Non-empty graphs $g_1 = (V_1, E_1, \mu_1, \nu_1)$ and $g_2 = (V_2, E_2, \mu_2, \nu_2)$, where $V_1 = \{u_1, \ldots, u_{|V_1|}\}$ and $V_2 = \{v_1, \ldots, v_{|V_2|}\}$

Output: A minimum-cost edit path from g_1 to g_2 e.g. $p_{min} = \{u_1 \to v_3, u_2 \to \varepsilon, \ldots, \varepsilon \to v_6\}$

Initialize OPEN to the empty set
For each vertex $w \in V_2$, insert the substitution $\{u_1 \to w\}$ into OPEN
Insert the deletion $\{u_1 \to \varepsilon\}$ into OPEN
loop
  Remove $p_{min} = \arg\min_{p \in \text{OPEN}} \{g(p) + h(p)\}$ from OPEN
  if p_{min} is a complete edit path **then**
    Return p_{min} as the solution
  else
    Let $p_{min} = \{u_1 \to v_{i_1}, \ldots, u_k \to v_{i_k}\}$
    if $k < |V_1|$ **then**
      For each $w \in V_2 \setminus \{v_{i_1}, \ldots, v_{i_k}\}$, insert $p_{min} \cup \{u_{k+1} \to w\}$ into OPEN
      Insert $p_{min} \cup \{u_{k+1} \to \varepsilon\}$ into OPEN
    else
      Insert $p_{min} \cup \bigcup_{w \in V_2 \setminus \{v_{i_1}, \ldots, v_{i_k}\}} \{\varepsilon \to w\}$ into OPEN
    end if
  end if
end loop

the simplest scenario, the estimated lower bound $h(p)$ of the costs from p to a leaf node is set to zero for all p. This means that no heuristic information of the potentially best search direction is used at all, and one actually performs a breadth-first search. In the remainder of this paper, this method will be referred to as PLAIN-A*. The other extreme would be to compute for a partial edit path the actual optimal path to a leaf node, that is, perform a complete edit distance computation for each node of the search tree. In this case, the function $h(p)$ is not a lower bound, but the exact value of the optimal costs. Of course, the computation of such a perfect heuristic is both unreasonable and untractable.

Somewhere in between the two extremes, one can define a function $h(p)$ evaluating how many edit operations have to be performed in a complete edit path at the least [11]. The method we use in this paper is very intuitive and can be computed efficiently. In the following, assume that a partial edit path at a position in the search tree is given, and let the number of unprocessed vertices of the first graph g_1 and second graph g_2 be n_1 and n_2, respectively. For an efficient estimation of the optimal remaining edit operations, we first attempt to perform as many vertex substitutions as possible, since a substitution is often less expensive than a deletion followed by an insertion. To this end, we potentially substitute each of the n_1 vertices from g_1 with any of the n_2 vertices from g_2. To obtain a lower bound of the exact edit costs, we accumulate the costs of the $\min\{n_1, n_2\}$ least expensive of these vertex substitutions and the costs of $\max\{0, n_1 - n_2\}$ vertex deletions or $\max\{0, n_2 - n_1\}$ vertex insertions. Any of the selected substitutions that is more expensive than a deletion followed by an insertion operation is replaced by the latter. This procedure only considers the most optimistic way to edit the remaining part of g_1 into the remaining part of g_2, and the estimated costs therefore constitute a lower bound of the exact cost. In the following, we refer to this method as HEURISTIC-A*.

3 Fast Suboptimal Edit Distance Algorithms

The methods described in the previous section find an optimal edit path between two graphs. Unfortunately, the computational complexity of the edit distance algorithm, whether or not heuristics are used to govern the tree traversal process, is exponential in the number of vertices of involved graphs. This means that the running time and space complexity may be huge even for reasonably small graphs. In practice we are able to compute the edit distance of graphs typically containing 12 vertices at most. In this paper, we therefore propose two edit distance variants that are conceptually very simple, but lead to a significant speedup of the computation. These methods do not generally return the optimal edit path, but only a suboptimal one.

3.1 A*-Beamsearch

The first method is based on beam search. Instead of expanding all sucessor nodes in the search tree, only a fixed number s of nodes to be processed are kept in the OPEN set at all times. Whenever a new partial edit path is added to the OPEN set in Alg. 1, only the s partial edit paths p with the lowest costs $g(p) + h(p)$ are kept, and the remaining partial edit paths in OPEN are removed. This means that not the full search space is explored, but only those nodes are expanded that belong to the most promising partial matches. For similar graphs, it is clear that edit operations of an optimal path have low costs. Therefore if only the partial edit paths with lowest costs are considered, we will obtain an edit path that is nearly optimal, which will result in a suboptimal distance close to the exact distance. For dissimilar graphs, the suboptimal distance will remain large. In the following, this method with parameter s is referred to as PLAIN-A*-BEAMSEARCH(s) or HEURISTIC-A*-BEAMSEARCH(s), respectively, depending on whether or not heuristic information is used in the tree search procedure.

3.2 A*-Pathlength

In the second variant, we exploit an observation from edit distance systems in practice. If graphs with a rather large number of vertices are given, it may very well be that a considerable part of an optimal edit path is constructed in the first few steps of the tree traversal, because most substitutions between similar graphs have small costs. Whenever the first significantly more expensive edit operation occurs (in the optimal edit path), this node will prevent the tree search algorithm from quickly reaching a leaf node and unnecessarily make it expand a large part of the search tree. We therefore propose an additional weighting factor favoring long partial edit paths over shorter ones. Formally, instead of evaluating $g(p) + h(p)$ in Alg. 1 (line 5), we use

$$\frac{g(p) + h(p)}{t^{|p|}} ,$$

with parameter $t > 1$. The term $|p|$ denotes the number of edit operations in partial edit path p. We refer to this method as PLAIN-A*-PATHLENGTH(t) or HEURISTIC-A*-PATHLENGTH(t), respectively.

4 Experimental Results

The methods we propose for speeding up the computation of graph edit distance are suboptimal in the sense that only an approximate edit distance value is obtained. In fact, from the description above it is clear that the approximate distance value will be equal to, or larger than, the exact distance value, since the suboptimal methods find an optimal solution in a subspace of the complete search space. In this section, we measure the speedup of the suboptimal methods and analyze the accuracy of the suboptimal distance.

To address the classification problems considered in this paper, we apply k-nearest-neighbor classifiers in conjunction with edit distance. Given a labeled set of training graphs, an unknown graph is assigned to the class that occurs most frequently among the k closest graphs (in terms of edit distance) from the training set. Hence, we assume that graphs belonging to the same class should be similar. In the experiments, insertion and deletion costs are set to constant values, and substitution costs are set proportional to the Euclidean distance of involved labels. To optimize these edit cost parameters, we first determine a set of parameters that is optimal on a validation set. The validated parameters are then applied to the independent test set. Note that the parameters are optimized once for the exact distance and then used throughout all optimal and suboptimal computations.

We first evaluate the distances on a graph database representing distorted letter drawings. In this experiment, we consider the 15 capital letters that consist of straight lines only (A, E, F, ...). For each class, a prototype line drawing is manually constructed. We then apply distortion operators to the prototype line drawings, resulting in randomly shifted, removed, or added lines. Using this procedure, we are able to generate arbitrarily large sample sets of drawings with arbitrarily strong distortions. These drawings are then converted into graphs by representing ending points of lines by vertices and lines by edges. Each vertex is labeled with a two-dimensional attribute giving its position. The graph database used in our experiments consists of a training set, a validation set, and a test set, each of size 150. The letter graphs consist of 4.6 nodes and 4.4 edges on the average.

To obtain a visual representation of the accuracy of the suboptimal methods, we plot for each pair of test and training pattern its exact (horizontal axis)

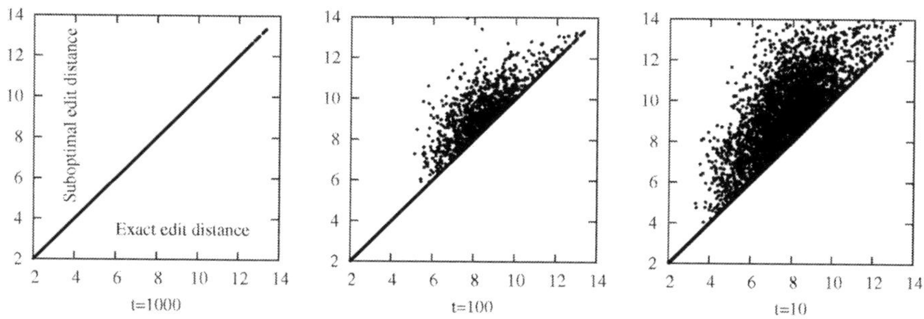

Fig. 1. Distance accuracy of PLAIN-A*-BEAMSEARCH(t) for $t = 1000, 100, 10$

and suboptimal (vertical axis) distance value. The respective illustrations are shown in Fig. 1. For $t = 1000$, we find that the suboptimal method does not differ considerably from the exact method in terms of distance. If the suboptimal method is constrained to $t = 100$ or $t = 10$ items in the OPEN list, however, it often results in larger distances. Additionally evaluating the running time of the edit distance computation, we observe that the suboptimal methods (for $t = 10, 100$) are faster than the exact method by several orders of magnitude.

The crucial question is whether the larger distances belong to graphs of the same class or graphs from different classes. In the latter case, the increased inter-class distance will not negatively affect the classification accuracy. In Table 1 we give the classification accuracy of three nearest-neighbor classifiers and the average time it takes to compute a single edit distance. The traditional edit distance algorithms are denoted by PLAIN-A* and HEURISTIC-A*, and the suboptimal methods proposed in this paper are referred to as PLAIN-A*-BEAMSEARCH, PLAIN-A*-PATHLENGTH, HEURISTIC-A*-BEAMSEARCH, and HEURISTIC-A*-PATHLENGTH. It turns out that the speedup of the suboptimal methods is significant, while the accuracy remains high for most configurations. The speedup of PLAIN-A*-BEAMSEARCH for decreasing parameter is clearly visible. Concerning the accuracy, suboptimal methods can even be observed to outperform the two exact methods in some cases. This means that a suboptimal algorithm may be able to correct misclassifications by assigning higher costs to pairs of graphs from different classes than the exact algorithm. The suboptimal method PLAIN-A*-PATHLENGTH(1.05) achieves the best classification accuracy of 86.7% among all methods and is more than 3 times, or 16 times, respectively, faster than the exact methods. Note that the performance of the two exact methods need not be identical, since in some cases the running time of the faster HEURISTIC-A* may be below and that of the slower PLAIN-A* above the predefined timeout threshold.

For a more thorough evaluation of the classification accuracy, we apply the proposed methods to the problem of image classification. Images are converted into attributed graphs by segmenting them into regions, eliminating regions that are irrelevant for classification, and representing the remaining regions by ver-

Table 1. Letter Database: Classification accuracy and average running time

Method	1-NN	3-NN	5-NN	Time (ms)
PLAIN-A*	82.0	80.7	81.3	2200
PLAIN-A*-BEAMSEARCH(1000)	82.0	80.7	82.7	620
PLAIN-A*-BEAMSEARCH(100)	81.3	79.3	81.3	40
PLAIN-A*-BEAMSEARCH(10)	76.7 ○	74.7 ○	72.0 ○	13
PLAIN-A*-PATHLENGTH(1.05)	79.3	80.0	86.7	132
PLAIN-A*-PATHLENGTH(1.1)	77.3	79.3	82.7	2
HEURISTIC-A*	82.0	80.7	82.7	468
HEURISTIC-A*-BEAMSEARCH(100)	82.0	80.7	82.0	18
HEURISTIC-A*-PATHLENGTH(1.1)	79.3	82.7	84.0	8

○ Statistically significantly worse than PLAIN-A* and HEURISTIC-A* ($\alpha = 0.05$)

Table 2. Image Database: Classification accuracy and average running time

Method	1-NN	3-NN	5-NN	Time (ms)
PLAIN-A*	46.3	48.2	44.4	10
PLAIN-A*-BEAMSEARCH(10)	46.3	48.2	48.2	8
PLAIN-A*-BEAMSEARCH(5)	48.2	50.0	44.4	6
PLAIN-A*-PATHLENGTH(1.1)	48.2	50.0	46.3	5
HEURISTIC-A*	46.3	48.2	44.4	20
HEURISTIC-A*-BEAMSEARCH(10)	46.3	44.4	48.2	16
HEURISTIC-A*-PATHLENGTH(1.1)	50.0	48.2	51.9	15

Table 3. Fingerprint Database: Classification accuracy and average running time

Method	1-NN	3-NN	5-NN	Time (ms)
Approximate method [13]	82.6	83.8	84.4	11
PLAIN-A*	—	—	—	— [1]
PLAIN-A*-BEAMSEARCH(50)	87.4 •	87.8 •	87.6 •	167
PLAIN-A*-BEAMSEARCH(40)	85.6 •	88.2 •	88.0 •	74
PLAIN-A*-BEAMSEARCH(10)	72.0 ○	72.8 ○	72.4 ○	9
PLAIN-A*-PATHLENGTH(...)	—	—	—	— [1]
HEURISTIC-A*	—	—	—	— [1]
HEURISTIC-A*-BEAMSEARCH(50)	87.4 •	87.8 •	87.6 •	218
HEURISTIC-A*-PATHLENGTH(...)	—	—	—	— [1]

○ Statistically significantly worse than reference method [13] ($\alpha = 0.05$)
• Statistically significantly better than reference method [13] ($\alpha = 0.05$)
[1] Empty entries indicate computation failure due to lack of memory

tices and the adjacency of regions by edges [12]. Our image database consists of 5 classes (*city, countryside, people, snowy, streets*) and is split into a training set, a validation set, and a test set of size 54. On the average, the graphs consist of 2.8 nodes and 2.5 edges. The nearest-neighbor classification performance and the running time of the edit distance computation using the exact algorithms and the proposed suboptimal algorithms are given in Table 2. Note that in this application HEURISTIC-A* takes significantly longer for the edit distance computation than PLAIN-A*. This means that the computational overhead of the heuristic evaluation of future costs in the search tree cannot be compensated for by a faster tree traversal, mostly because the graphs under consideration, and hence also the constructed search tree, are rather small. Generally, the decrease of the running time is not massive, but the accuracy of the suboptimal methods is at least as high as that of the exact methods. Particularly PLAIN-A*-PATHLENGTH(1.1) outperforms the exact methods PLAIN-A* and HEURISTIC-A* and is at least twice as fast.

Finally, we apply the proposed methods to the difficult problem of fingerprint classification. To this end, we construct graphs from fingerprint images of the NIST-4 database by extracting characteristic regions in fingerprints and converting the result into attributed graphs [13]. We use a validation set of size 300 and a training set and test set both of size 500. On the average, the fingerprint graphs consist of 5.2 nodes and 8.6 edges. In our experiment, we address the 4-class problem (classes *arch, left loop, right loop, whorl*). In Table 3, in addition to the systems described in this paper, we also give the results of another method [13]. Note that for this dataset, the exact edit distance PLAIN-A* and HEURISTIC-A* cannot be computed because the search tree grows too large. The results clearly demonstrate that the classification accuracy of the suboptimal methods, for moderate running times, is very high.

Summarizing we conclude that although the edit distance computed by the proposed suboptimal methods is not always close to the exact edit distance, this problem mainly pertains to pairs of graphs from different classes and therefore does not negatively affect the classification performance. The suboptimal methods offer more flexibility in terms of tradeoff between speed and accuracy than the exact edit distance.

5 Conclusions

One of the main problems of graph edit distance is its exponential computational complexity, which makes its application feasible for small graphs only. In this paper, we propose two simple variants of a standard tree search algorithm for edit distance. The idea is to explore not the full search space, but only a subspace of promising candidates. The two proposed methods are related to beam search and to a re-weighting of edit operation costs. With these simple modifications, it turns out that a significant speedup of the edit distance computation can be achieved. At the same time, the classification accuracy of the suboptimal methods remains high on all datasets — and is sometimes even higher than the one of the exact method. This means that the suboptimality mainly leads to an increase of inter-class distances, while intra-class distances, which are highly relevant for classification, are not strongly affected. We provide an experimental evaluation and demonstrate the usefulness of our methods on semi-artificial line drawings, on scenery images, and fingerprints.

References

1. Sanfeliu, A., Fu, K.: A distance measure between attributed relational graphs for pattern recognition. IEEE Transactions on Systems, Man, and Cybernetics (Part B) **13** (1983) 353–363
2. Hopcroft, J., Wong, J.: Linear time algorithm for isomorphism of planar graphs. In: Proc. 6th Annual ACM Symposium on Theory of Computing. (1974) 172–184
3. Luks, E.: Isomorphism of graphs of bounded valence can be tested in ploynomial time. Journal of Computer and Systems Sciences **25** (1982) 42–65

4. Torsello, A., Hidovic-Rowe, D., Pelillo, M.: Polynomial-time metrics for attributed trees. IEEE Transactions on Pattern Analysis and Machine Intelligence **27** (2005) 1087–1099
5. Dickinson, P., Bunke, H., Dadej, A., Kraetzl, M.: On graphs with unique node labels. In: Proc. 4th Int. Workshop on Graph Based Representations in Pattern Recognition. LNCS 2726, Springer (2003) 13–23
6. Cross, A., Wilson, R., Hancock, E.: Inexact graph matching using genetic search. Pattern Recognition **30** (1997) 953–970
7. Christmas, W., Kittler, J., Petrou, M.: Structural matching in computer vision using probabilistic relaxation. IEEE Transactions on Pattern Analysis and Machine Intelligence **17** (1995) 749–764
8. Neuhaus, M., Bunke, H.: An error-tolerant approximate matching algorithm for attributed planar graphs and its application to fingerprint classification. In: Proc. 10th Int. Workshop on Structural and Syntactic Pattern Recognition. LNCS 3138, Springer (2004) 180–189
9. Hlaoui, A., Wang, S.: A node-mapping-based algorithm for graph matching (2006) To appear.
10. Hart, P., Nilsson, N., Raphael, B.: A formal basis for the heuristic determination of minimum cost paths. IEEE Transactions of Systems, Science, and Cybernetics **4** (1968) 100–107
11. Bunke, H., Allermann, G.: Inexact graph matching for structural pattern recognition. Pattern Recognition Letters **1** (1983) 245–253
12. Le Saux, B., Bunke, H.: Feature selection for graph-based image classifiers. In: Proc. 2nd Iberian Conf. on Pattern Recognition and Image Analysis. LNCS 3523, Springer (2005) 147–154
13. Neuhaus, M., Bunke, H.: A graph matching based approach to fingerprint classification using directional variance. In: Proc. 5th Int. Conf. on Audio- and Video-Based Biometric Person Authentication. LNCS 3546, Springer (2005) 191–200

A Spectral Generative Model for Graph Structure

Bai Xiao and Edwin R Hancock

Department of Computer Science,
University of York, York Y01 5DD, UK

Abstract. This paper shows how to construct a generative model for graph structure. We commence from a sample of graphs where the correspondences between nodes are unknown *ab initio*. We also work with graphs where there may be structural differences present, i.e. variations in the number of nodes in each graph and the edge-structure. The idea underpinning the method is to embed the nodes of the graphs into a vector space by performing kernel PCA on the heat kernel. The coordinates of the nodes are determined by the eigenvalues and eigenvectors of the Laplacian matrix, together with a time parameter which can be used to scale the embedding. Node correspondences are located by applying Scott and Longuet-Higgins algorithm to the embedded nodes. We capture variations in graph structure using the covariance matrix for corresponding embedded point-positions. We construct a point distribution model for the embedded node positions using the eigenvalues and eigenvectors of the covariance matrix. We show how to use this model to both project individual graphs into the eigenspace of the point-position covariance matrix and how to fit the model to potentially noisy graphs to reconstruct the Laplacian matrix. We illustrate the utility of the resulting method for shape-analysis using data from the COIL database.

1 Introduction

The literature describes a number of attempts aimed at developing probabilistic models for variations in graph-structure. Some of the earliest work was that of Wong, Constant and You [4], who capture the variation in graph-structure using a discretely defined probability distribution. Bagdanov and Worring [3] have overcome some of the computational difficulties associated with this method by using continuous Gaussian distributions. For problems of graph matching Christmas, Kittler and Petrou [1], and Wilson and Hancock [2] have used simple probability distributions to measure the similarity of graphs. There is a considerable body of related literature in the graphical models community concerned with learning the structure of Bayesian networks from data [5].

Recently there has been some research aimed at applying central clustering techniques to cluster graphs. However, rather than characterising them in a statistical manner, a structural characterisation is adopted. For instance, both Lozano and Escolano [7], and Bunke et al. [8] summarize the data using a supergraph. Each sample can be obtained from the super-graph using edit operations. However, the way in which the super-graph is learned or estimated is not statistical in nature. Jain and Wysotzki, adopt a geometric approach which aims to embed graphs in a high dimensional space by means of the Schur-Hadamard inner product [9]. Central clustering methods are then

D.-Y. Yeung et al. (Eds.): SSPR&SPR 2006, LNCS 4109, pp. 173–181, 2006.

deployed to learn the class structure of the graphs. The embedding offers the advantage that it is guaranteed to preserve structural information present. Unfortunately, the algorithm does not provide a means of statistically characterising the modes of structural variation encountered.

Hence, the methods described in the literature fall well short of constructing genuine generative models from which explicit graph structures can be sampled. The aim in this paper use ideas from the spectral analysis of graphs to construct a simple and explicit generative model for graph-structure. To this end, we use the heat-kernel embedding to construct a generative model for graph-structure. We use the heat-kernel to map the nodes of a graph to positions in a vector space. Our aim is to construct a statistical model that can account for the distribution of embedded point-positions for corresponding nodes in a sample of graphs. A reference graph is selected, and the correspondences between the nodes of each sample graph and the reference graph are established using the point-matching method of Scott and Longuet-Higgins [6]. We capture variations in graph structure using the covariance matrix for the corresponding embedded point-positions. We construct a point distribution model for the embedded node positions using the eigenvalues and eigenvectors of the covariance matrix. We show how to use this model to both project individual graphs into the eigenspace of the point-position covariance matrix and to fit the model to potentially noisy graphs to reconstruct the Laplacian matrix. We illustrate the utility of the resulting method for shape-analysis. Here we perform experiments on the COIL data-base, and show that the model can be used to both construct pattern spaces for sets of graphs and to cluster graphs.

2 Heat Kernel Embedding

We are interested in using the heat-kernel to embed the nodes of a graph in a vector space. To commence, suppose that the graph under study is denoted by $G = (V, E)$ where V is the set of nodes and $E \subseteq V \times V$ is the set of edges. Since we wish to adopt a graph-spectral approach we introduce the adjacency matrix A for the graph where the elements are

$$A(u,v) = \begin{cases} 1 \text{ if} u,v \in E \\ 0 \text{ otherwise} \end{cases} \tag{1}$$

We also construct the diagonal degree matrix D,whose elements are given by $D(u,u) = \sum_{v \in V} A(u,v)$. From the degree matrix and the adjacency matrix we construct the Laplacian matrix $L = D - A$, i.e. the degree matrix minus the adjacency matrix. The normalised Laplacian is given by $\hat{L} = D^{-\frac{1}{2}} L D^{-\frac{1}{2}}$. The spectral decomposition of the normalised Laplacian matrix is $\hat{L} = \Phi \Lambda \Phi^T$, where $\Lambda = diag(\lambda_1, \lambda_2, ..., \lambda_{|V|})$ is the diagonal matrix with the ordered eigenvalues as elements and $\Phi = (\phi_1|\phi_2|....|\phi_{|V|})$ is the matrix with the ordered eigenvectors as columns. Since $\hat{L}$ is symmetric and positive semi-definite, the eigenvalues of the normalised Laplacian are all positive. The eigenvector associated with the smallest non-zero eigenvector is referred to as the Fiedler-vector. We are interested in the heat equation associated with the Laplacian, i.e. $\frac{\partial h_t}{\partial t} = -\hat{L} h_t$, where h_t is the heat kernel and t is time. The heat kernel can hence be viewed as describing the flow of information across the edges of the graph with time. The rate of flow is determined by the Laplacian of the graph. The solution to the heat equation is found by exponentiating the Laplacian eigen-spectrum, i.e. $h_t = \Phi \exp[-t\Lambda]\Phi^T$.

We use the heat kernel to map the nodes of the graph into a vector-space. Let Y be the $|V| \times |V|$ matrix with the vectors of co-ordinates as columns. The vector of co-ordinates for the node indexed u is hence the uth column of Y. The co-ordinate matrix is found by performing the Young-Householder decomposition $h_t = Y^T Y$ on the heat-kernel. Since $h_t = \Phi \exp[-\Lambda t]\Phi^T$, $Y = \exp[-\frac{1}{2}\Lambda t]\Phi^T$. Hence, the co-ordinate vector for the node indexed u is

$$y_u = (\exp[-\frac{1}{2}\lambda_1 t]\phi_1(u), \exp[-\frac{1}{2}\lambda_2 t]\phi_2(u), ..., \exp[-\frac{1}{2}\lambda_{|V|} t]\phi_{|V|}(u))^T$$

The kernel mapping $\mathcal{M} : V \to \mathcal{R}^{|V|}$, embeds each node on the graph in a vector space $\mathcal{R}^{|V|}$. The heat kernel $h_t = Y^T Y$ can also be viewed as a Gram matrix, i.e. its elements are scalar products of the embedding co-ordinates. Consequently, the kernel mapping of the nodes of the graph is an isometry. The squared Euclidean distance between the nodes u and v is given by

$$d_E(u, v)^2 = (y_u - y_v)^T (y_u - y_v) = \sum_{i=1}^{|V|} \exp[-\lambda_i t](\phi_i(u) - \phi_i(v))^2 \tag{2}$$

3 Generative Model

Our aim is to construct a generative model that can be used to represent the statistical variations in a sample of graphs. Let the sample be $T = \{G_1, G_2, ..., G_k, ...G_K\}$ where the kth graph $G_k = (V_k, E_k)$ has node-set V_k and edge-set E_k. The result of performing heat-kernel embedding of the nodes of the kth graph is a matrix of co-ordinates Y_k.

Our aim in this paper is to construct a generative model that can be used to describe the distribution of embedded node co-ordinates for the sample of graphs. Since the graphs contain different numbers of nodes, we truncate the co-ordinate matrices to remove the spatial dimensions corresponding to insignificant eigen-modes of the kernel matrix. Hence, we retain just the first N rows of each co-ordinate matrix. For the graph G_k the truncated node co-ordinate matrix is denoted by $\hat{Y}_k$.

3.1 Node Correspondences

To construct the generative model, we require correspondences between the nodes of each sample graph and the nodes of a reference structure. Here we take the reference graph to be the graph in the sample with the largest number of nodes. This graph has index $k^* = \arg\max_{G_k \in T} |V_k|$.

To locate the correspondences between the nodes of each sample graph and those of the reference graph, we use the Scott and Longuet-Higgins algorithm. The algorithm uses the distances between the reference graph nodes and the nodes of the sample graph

k to compute an affinity matrix. Let $\hat{y}_k^i$ is the ith column vector of the truncated co-ordinate matrix $\hat{Y}_k$, i.e. the co-ordinates of the node $i \in V_k$. For the node i of the sample graph G_k and the node j the affinity matrix element is

$$R_{k,k^*}(i,j) = \exp[-\frac{1}{\sigma^2}(\hat{y}_k^i - \hat{y}_{k^*}^j)^T(\hat{y}_k^i - \hat{y}_{k^*}^j)]$$

where σ is a scaling parameter.

According to Scott and Longuet-Higgins [10] if R_{k,k^*} is a positive definite $|V_k| \times |V_{k^*}|$ matrix, then the $|V_k| \times |V_{k^*}|$ orthogonal matrix R_{k,k^*}^* that maximises the quantity $Tr[R_{k,k^*}(R_{k,k^*}^*)^T]$ may be found by performing singular value decomposition. To do this they perform the matrix factorisation $R_{k,k^*} = V\Delta U^T$, where V is a $|V_D| \times |V_D|$ orthogonal matrix, U is a $|V_{k^*}| \times |V_{k^*}|$ orthogonal matrix and Δ is a $|V_k| \times |V_{k^*}|$ matrix whose off-diagonal elements $\Delta_{i,j} = 0$ if $i \neq j$ and whose "diagonal" elements $\Delta_{i,i}$ are non-zero. Suppose that E is the matrix obtained from Δ by making the diagonal elements $\Delta_{i,i}$ unity. The matrix R_{k,k^*}^* which maximises $Tr[R_{k,k^*}(R_{k,k^*}^*)^T]$ is $R_{k,k^*}^* = VEU^T$. The element $R_{k,k^*}^*(i,j)$ indicates the strength of association between the node $i \in V_k$ in the graph G_k and the node $j \in V_{k^*}$ in the reference graph. The rows of R_{k,k^*}^*, index the nodes in the graph G_k, and the columns index the nodes of the reference graph G_{k^*}. If $R_{k,k^*}^*(i,j)$ is both the largest element in row i and the largest element in column j then we regard these nodes as being in one-to-one correspondence with one-another. We record the state of correspondence using the matrix C_{k,k^*}. If the pair of nodes (i,j) satisfies the row and column correspondence condition, then we set $C_{k,k^*}(i,j) = 1$, otherwise $C_{k,k^*}(i,j) = 0$.

3.2 Embedded Point Distribution Model

Once we have correspondences to hand, then we can construct the generative model for the set of graphs. To do this we model variations in the positions of the embedded points using a point distribution model. We commence by computing the mean point positions. The matrix of mean-position co-ordinates and the associated covariance matrix are

$$\hat{X} = \frac{1}{T}\sum_{k \in T} C_{k,k^*}^T \hat{Y}_k$$

$$\Sigma = \frac{1}{T}\sum_{k \in T}(C_{k,k^*}^T \hat{Y}_k - \hat{X})(C_{k,k^*}^T \hat{Y}_k - \hat{X})^T$$

To construct the point-distribution model, we perform the eigendecomposition $\Sigma = \Psi\Gamma\Psi^T$ where $\Gamma = diag(\gamma_1, \gamma_2, ..., \gamma_K)$ is the diagonal matrix of ordered eigenvectors and $\Psi = (\psi_1|.....|\psi_K)$ is the matrix with the correspondingly ordered eigenvectors as columns.

We deform the mean-embedded node positions in the directions of the leading eigenvectors of the point-position covariance matrix Σ. Let $\tilde{\Psi}$ be the result of truncating the matrix Ψ after S columns and let b be a parameter-vector of length S. We convert the mean point position matrix with a long vector form. Let $Col_i(\hat{X})$ be the

ith column of the mean-point position matrix $\hat{X}$. The long vector is given by $\hat{Z} = (Col_1^T(\hat{X}), Col_2^T(\hat{X}), ...)$. The long vector corresponding to deformed point set position is $\tilde{Z} = \hat{Z} + \tilde{\Psi} b$. The matrix with deformed point position as column is $\tilde{X}$.

An observed configuration of embedded nodes $\tilde{Y}$ may be fitted to the model. To do this the best fit parameters estimated using the least squares procedure

$$b^* = \arg\min_b (\tilde{Y} - \hat{X} - \tilde{\Psi} b)^T (\tilde{Y} - \hat{X} - \tilde{\Psi} b)$$

The best-fit parameter vector is $b^* = \tilde{\Psi}^T(\tilde{Y} - \hat{X})$ and the reconstructed set of embedded point positions is $\tilde{Y}^* = \hat{X} + \tilde{\Psi}\tilde{\Psi}^T(\tilde{Y} - \hat{X})$. From the reconstructed point-positions we can recover the Laplacian matrix for the corresponding graph. The heat-kernel for the reconstructed embedded graph is $h_t^* = (\tilde{Y}^*)^T(\tilde{Y}^*) = \exp[-\hat{L}^* t]$ and the Laplacian is hence $\hat{L}^* = -\frac{1}{t}\ln\{(\tilde{Y}^*)^T(\tilde{Y}^*)\}$. From the reconstructed Laplacian we can compute the corresponding adjacency matrix

$$A^* = D - D^{\frac{1}{2}} \hat{L}^* D^{\frac{1}{2}} = D + \frac{1}{t} D^{\frac{1}{2}} \ln\{(\tilde{Y}^*)^T(\tilde{Y}^*)\} D^{\frac{1}{2}}.$$

Finally, the similarity of a pair of graphs can be measured using the difference in their best-fit parameter vectors. Since the parameter-vector is just the projection of the corresponding graph into the eigenspace of the model, the difference is parameter vectors is related to the distance between graphs in the eigenspace. Suppose that the graphs G_{k_1} and G_{k_2} have best fit parameter vectors $b_{k_1}^*$ and $b_{k_2}^*$ respectively. The Euclidean distance between the parameter vectors is

$$d^2(k_1, k_2) = (b_{k_1}^* - b_{k_2}^*)^T (b_{k_1}^* - b_{k_2}^*) = (\hat{Y}_{k_1} - \hat{Y}_{k_2})^T \tilde{\Psi}\tilde{\Psi}^T (\hat{Y}_{k_1} - \hat{Y}_{k_2})$$

4 Experiments

In this section we provide some experimental evaluation of our generative model for real-world data. We use the COIL data-base. The data-set contains multiple views of the same object in different poses with respect to the camera. Example images from the data-set are shown in Figures 1. We extract the feature points using the methods of [11]. We have extracted graphs from the images by computing the Voronoi tessellations of the feature-points, and constructing the region adjacency graph, i.e. the Delaunay triangulation, of the Voronoi regions.

Fig. 1. Three objects from the COIL data-base

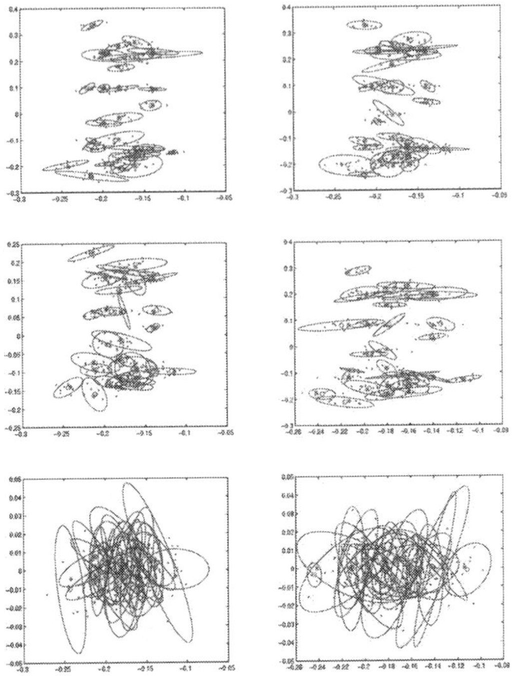

Fig. 2. Embedded point positions and fitted covariance ellipsoids with varying t(from left to right, top to bottom $t = 0.001, 0.01, 0.1, 1, 10, 100$) for the heat kernel

In Figure 2 we show the result of projecting the nodes into the space spanned by the leading two eigenvectors of the heat-kernel. The different panels in the figure are for different values of t ,from left to right and top to bottom the t are $0.001, 0.01, 0.1, 1,$ $10, 100$. For this experiment we have taken 15 images from the duck sequence. Each

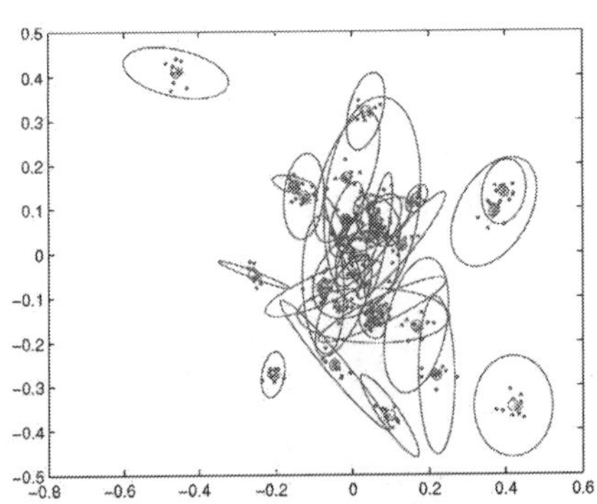

Fig. 3. Embedded point positions and fitted covariance ellipsoids for Laplacian matrix

blue point in the embedding corresponds to a single node of one of the 15 sample graphs. Superimposed on the node-positions as red-points are the locations of the mean node positions. Around each mean node position we have drawn an ellipse. The major and minor axes of the ellipse are in the principal directions of the eigenvectors of the node-position covariance matrix and the lengths of the semi-major axes are the corresponding eigenvalues. There are a number of features to note from this figure. First, for small values of t the embedded points form relatively compact clusters. Second, there is a significant variation in the size and directions of the ellipses. The compactness of the clusters supports the feasibility of our embedding approach and the variation in the ellipses underpins the need for a relatively complex statistical model to describe the distribution of embedded point positions. As the value of t increases then so the overlap of the ellipses increases. For comparison Figure 3 shows the result of repeating the embedding by using the Laplacian spectrum. The node-clusters are more overlapped than those obtained with the heat kernel for small values of t.

To investigate the role of the number of Laplacian eigenmodes in the reconstruction of the graph-structure we have examined the value of the Froebenius norm $F = ||A - A^*||$ between the original graph adjacency matrix A and the reconstructed adjacency matrix A^* computed by fitting the generative model. In Figure 4 we show the value of F as a function of the number of eigenmodes used. The different curves in the

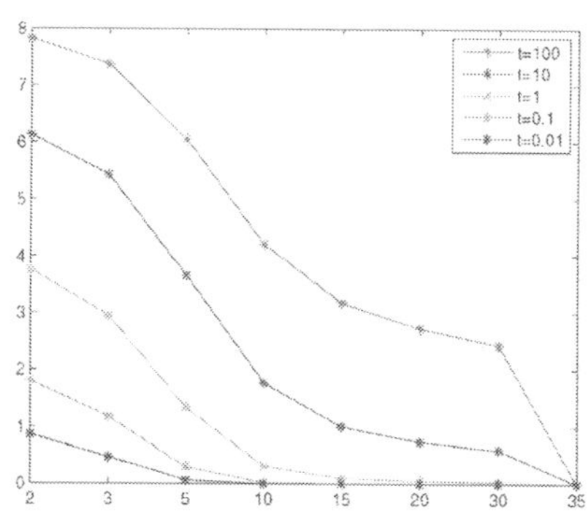

Fig. 4. Froebenius norm as a function of number of eigenmodes

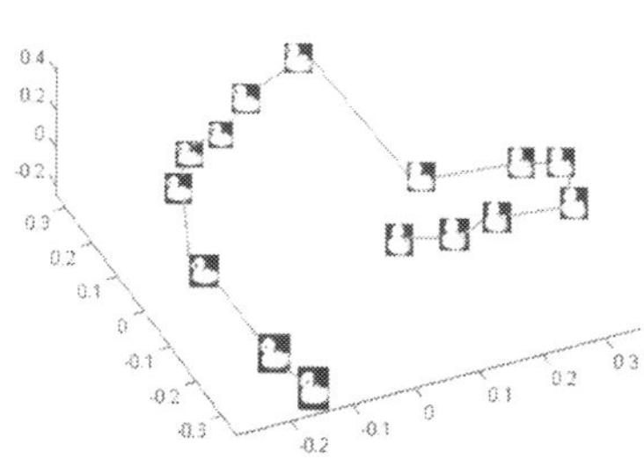

Fig. 5. Eigen-projection of graphs from 15 images in duck sequence

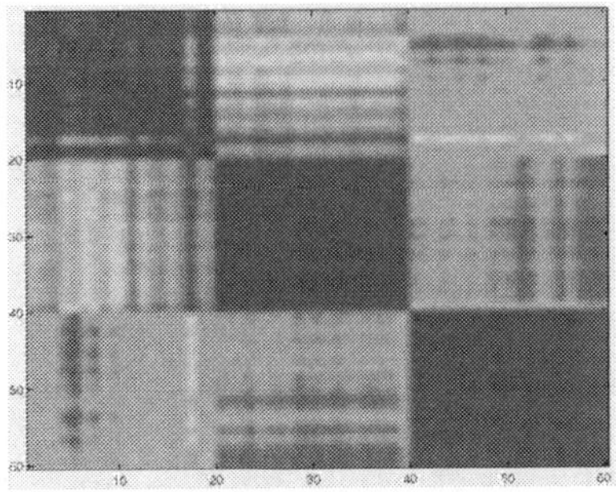

Fig. 6. Distance matrix for fitted parameter vectors

plot are for different values of t. The best reconstructions are obtained with small values of t and an increasing number of eigenmodes.

In Figures 5 we show the result of projecting the embedded node vectors for the graphs extracted from the duck sequence in the COIL data-base onto the eigenvectors of the embedded node position covariance matrix Σ. We have placed a thumbnail image at the location specified by the first three components of the parameter-vector b. The line connecting the thumbnails corresponds to the sequence order of the original images. The main feature to note is that neighboring images in the sequence are close together in the eigenspace.

We have also experimented with the generative model as a means of clustering graphs. In Figure 6 we show the matrix distances between the best fit parameter vectors. The main feature to note is that there is a clear block structure emerges corresponding to the different objects.

5 Conclusions

In this paper we have used the heat-kernel embedding of graphs to construct a generative model for graph structure. The mapping allows nodes of the graphs under study to be embedded as points in a vector-space. The idea underpinning the generative model is to construct a point-distribution model for the positions of the embedded nodes. The required correspondences needed to construct this model are recovered using the Scott and Longuet-Higgins algorithm. The method proves to be effective for computing distances between graphs and also for clustering graphs.

Our future plans revolve around the use of a mixture model to describe the positions of the embedded nodes, and to assess uncertainty in the computation of correspondence.

References

1. W.J. Christmas, J. Kittler, and M. Petrou. Structural matching in computer vision using probabilistic relaxation. *IEEE Transactions on Pattern Analysis and Machine Intelligence*, 17(8):749–764, 1995.
2. R.C. Wilson and E.R. Hancock. Structural matching by discrete relaxation. *IEEE Transactions on Pattern Analysis and Machine Intelligence*, 19(6):634–648, June 1997.

3. A.D. Bagdanov and M. Worring, First Order Gaussian Graphs for Efficient Structure Classification *Pattern Recognition*, **36**, pp. 1311-1324, 2003.
4. A.K.C Wong, J. Constant and M.L. You, Random Graphs *Syntactic and Structural Pattern Recognition*, World Scientific, 1990.
5. D. Heckerman, D. Geiger and D.M. Chickering, Learning Bayesian Networks: The combination of knowledge and statistical data *Machine Learning*, **20**, pp. 197-243, 1995.
6. G.L.Scott,H.C.Longuett-Higgins An algorithm for associating the features of two images *Proceedings of the Royal Society of London*, **244**, pp. 21-26, 1991.
7. M. A. Lozano and F. Escolano, ACM Attributed Graph Clustering for Learning Classes of Images In *Graph Based Representations in Pattern Recognition*, LNCS 2726, pp.247-258, 2003.
8. H.Bunke et al., Graph Clustering Using the Weighted Minimum Common Supergraph. In *Graph Based Representations in Pattern Recognition*, LNCS 2726, pp.235-246, 2003.
9. B. J. Jain and F. Wysotzki, Central Clustering of Attributed Graphs. *Machine Learning*, Vol. 56, pp. 169-207, 2004.
10. G.L.Scott and H.C.Longuet-Higgins, An Algorithm for Associating the Features of two Images. *Proceedings of the Royal Society of London*, Vol. 244, pp. 21-26, 1991.
11. C.G.Harris, and M.J.Stephens, "A Combined Corner and Edge Detector", *Fourth Alvey Vision Conference*, pp. 147–151, 1994.

Considerations Regarding the Minimum Spanning Tree Pyramid Segmentation Method*
(Why Does it Always Find the Lady?)

Adrian Ion, Walter G. Kropatsch, and Yll Haxhimusa

Vienna University of Technology, Pattern Recognition and Image Processing Group, Favoritenstr. 9/1832, A-1040 Vienna, Austria
{ion, krw, yll}@prip.tuwien.ac.at

Abstract. The minimum spanning tree pyramid is a hierarchical image segmentation method. We study it's properties and the regions it produces. We show the similarity with the watershed transform and present the method in a domain in which this is easy to understand. For this, a short overview of both methods is given. Catchment basins are contracted before their neighbouring local maximas. Smooth regions surrounded by borders with maximal local variation are selected. The maximum respectively minimum variation on the border of a region is larger than the maximum respectively minimum variation inside the region.

1 Introduction

Image segmentation is the process of partitioning the image into salient parts, i.e. partitioning the image into regions, such that each region is homogeneous with respect to some criteria such as greyvalue, colour, or texture. A segmentation method should have the folowing [1,2]: create a hierarchy, capture perceptually important groupings, and run in linear time.

The presented work is motivated by the desire to further understand and improve the results of one such segmentation method, the minimum spanning tree pyramid (MST Pyramid) [3], and better fit it to the necessities of higher level processing [4]. During the past years, we have had the chance to use and test different implementations of the MST Pyramid method and even though random selection mechanisms are used [5] and in most of the cases the neighbourhood graph of an image does not have a unique minimum spanning tree, the most important entities, like the lady in Fig. 1, could always be found in the produced results.

After benchmarking the method using human made segmentations [6] the necessity for a more analytical approach has risen, for which the results are presented here. While looking in detail at the properties of the method, a certain similarity with the watershed transfrom [7] has also been observed and is included in this discussion.

* This paper was supported by the Austrian Science Fund under grants FSP-S9103-N04 and P18716-N13.

D.-Y. Yeung et al. (Eds.): SSPR&SPR 2006, LNCS 4109, pp. 182–190, 2006.

This paper is organised as follows: Sections 2 and 3 contain a short description of the MST Pyramid and the Watershed transfrom, Section 4 presents the results of our study, Section 5 contains the outlook, and we end with the conclusions in Section 6.

2 The MST Pyramid Segmentation

Initially developed in the dual graph contraction and dual irregular pyramid framework and recently adapted to 2D combinatorial maps and combinatorial map pyramids [8], the MST Pyramid method [3] takes as input a weighted neighbourhood graph (NG) and produces a hierarchy of partitions by using the minimum spanning tree (MST) algorithm by Borůvka [9] and region internal/external contrast concepts [1].

Algorithm 1. MST Pyramid segmentation

Input: Attributed neighbourhood graph G_0.

$k = 0$
repeat
  $ME_k = \bigcup$ smallest edge around each vertex of G_k
  $CE_k =$ edges from ME_k connecting two regions having larger internal contrast than the external contrast between them {contrast test step}
  $G_{k+1} = (G_k$ with the edges from CE_k contracted)
  $k = k + 1$
until $G_k = G_{k-1}$

Output: An attributed neighbourhood graph at each level of the pyramid $(G_0, G_1, ..., G_k)$.

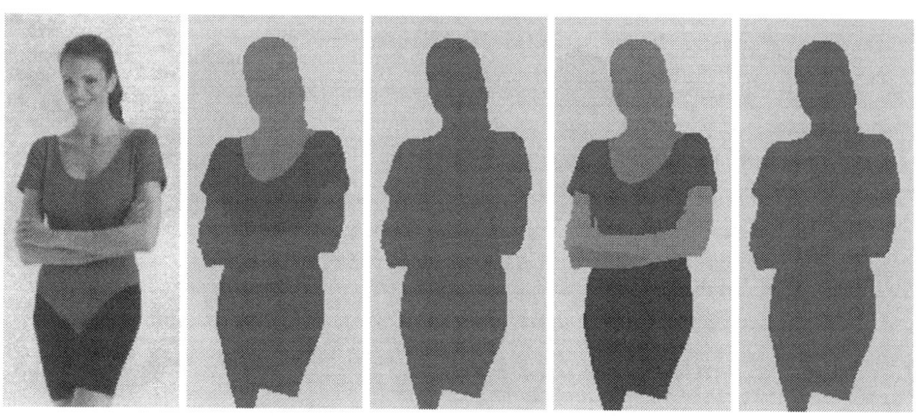

a) $|V_0| = 30\,276$ b) $|V_{40}|=12$ c) $|V_{42}|=3$ d) $|V_{37}|=11$ e) $|V_{40}|=2$

Legend: number of components in the specified level of the pyramid

Fig. 1. Levels of the MST Pyramid segmentation of the image of a Woman: with (b,c) and without (d,e) the contrast test step

To apply this method for image segmentation, the input NG is obtained by associating a vertex to each pixel and connecting two neighbouring vertices by an edge weighted with the distance of the two pixel values in some featurespace (we have experimented with difference in greyscale and RGB colour). Internal contrast of a region is defined as the biggest weight of the edges of it's MST. External contrast between two neighbouring regions is defined as the smallest weight of the edges connecting vertices from the two regions. Algorithm 1. shows a description of the MST Pyramid method, and Fig. 1 shows some results. Step 4. of the Algorithm is called the contrast step. More details can be found in [10].

3 The Watershed Segmentation

A well known method used for segmentation but not only, the watershed transform has it's origins in mathematical morphology. An intuitive way to view it is that of a landscape (topographic surface) being flooded by water (rain), and the watersheds being the lines which separate the different domains of attraction of rain over the relief [11]. Another way to imagine it, is to think of the landscape with holes made in the local minima, being immersed in water. Starting at these holes (local minima), catchment basins fill with water and the watersheds are the dams build in the places where two such catchment basins would meet to stop them from merging.

Fig. 2. Watershed segmentation of the image of a Woman's head

As mentioned above, the method can be applied to any topographic surface, and in the case of segmentation, it is most often applied to the gradient image of the image to process. The resulting catchment basins define the segments of the image. For a survey of existing methods that can be used to obtain the watershed transfrom and a detailed description see [7]. Fig. 2 shows an example result.

4 Understanding Global Properties of the MST Pyramid

Local decisions taken when merging regions make the MST Pyramid method well suited for parallel processing. On the other side, having global information makes estimating, characterising, and influencing the results much easier.

After doing experiments, we have noticed that the majority of edges filtered by step 4 (contrast test) of Algorithm 1. pass the test, and that removing the filter and just contracting all the proposed edges does not significantly change the results in most of the levels of the pyramid (a discussion of this, follows at the end of Section 4.2). Because of this, we have simplified the modell for the current study and removed the contrast test (step 4 in Algorithm 1.) from it.

4.1 Case Study - A 1D Image

Let I be a 1D image, defined as $I(p), p = 1, \ldots, m$. For a certain p, $I(p)$ identifies the pixel at position p in the image, $I(p_1)$ and $I(p_2)$ are neighbours if $|p_1 - p_2| = 1$. The neighbourhood graph NG=(V, E) of such an image is a chain of vertices $v \in V$ (one for each pixel in the original image), with the vertices associated to each two neighbouring pixels joined by an edge $e \in E$, and it's minimum spanning tree is the graph itself (See Fig. 3a,b).

The edge graph EG=(VE, EE) of a graph is a graph where each vertex $ve \in VE$ represents an edge in the original graph (in our case NG), and two vertices are joined by an edge $ee \in EE$ if their corresponding edges in the original graph share a common vertex. (The EG will be used to show the similarity with the watershed segmentation).

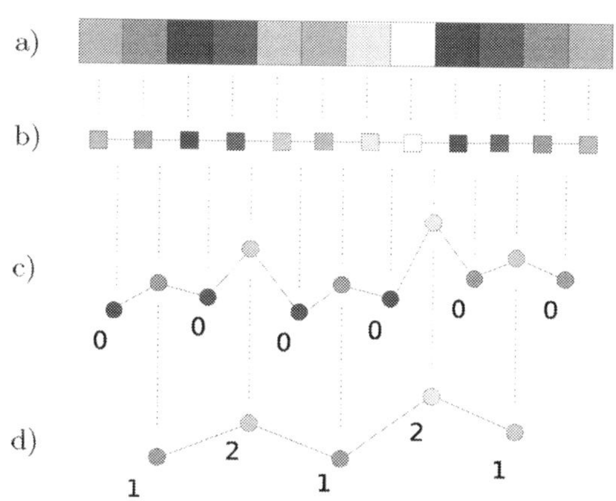

Fig. 3. MST based contraction of a 1D image: a) Image; b) associated NG; c) associated EG with survival levels specified (higher vertex position means larger weight); d) associated EG of second pyramid level, with survival levels specified

In the rest of the section, the numbering of vertices and edges in both the NG and the EG is done depending on the position of the associated element in the image, i.e. in the NG, v_i is associated to $p(i)$ and e_i is the edge connecting v_i with v_{i+1}, and in the EG, ve_i is associated to e_i respectively to the edge connecting v_i with v_{i+1} (See Fig. 3b,c).

A Step in the MST Pyramid. We recall that edges in the NG are attributed with the difference in some featurespace of the two neighbouring pixels' values. The same value is used to attribute their associated vertices in the EG.

When searching for edges to be contracted, the MST Pyramid method selects the smallest edge connecting one vertex in the NG with it's neighbours. In the case of unequal values this results in a unique solution.

Because in our case, in one selection step any edge $e_i, i = 1, \ldots, m-1$ connecting two vertices v_i and v_{i+1} is part of two such tests, we conclude that e_i is selected if $e_i < e_{i+1}$ or $e_i < e_{i-1}$ or one of its bounding vertices is a leaf. Which, in it's associated EG, is equivalent to ve_i is not a local maximum or ve_i is a leaf i.e. $\neg(ve_i > \max(ve_{i+1}, ve_{i-1})) \vee (i \in \{1, m-1\})$. This means, that in one such step only local maxima survive (See Fig. 3b,c).

What Happens Further in the MST Pyramid? The selected edges are contracted i.e. the new NG contains only the surviving (non-selected) edges and each group of vertices connected by the selected (non-surviving) edges are merged into one single vertex. In the EG this is equivalent with removing all the selected vertices and connecting each two surviving vertices if they were connected by a path of non-surviving vertices. (See Fig. 3c,d). The whole process of selection-contraction is repeated until no more contraction is possible.

Characterising the Regions. The initial aim of the present study was to try to characterise the regions produced by the method i.e. given an image and a connected region in it (a cut), to be able to say what properties (internal/external) must the edges inside, outside, and on the region-border have, such that the region is produced as one segment in one of the levels of the hierarchy. In the case of our 1D image, this is reduced to: given the image and 2 edges, how can we best characterise the region between the two edges?

Recall that in one MST Pyramid step, from the level below only local maxima survive, which is equivalent to applying the watershed transform, on the gradient image of our 1D image (See Table 1).

Table 1. Domain similarity of the MST Pyramid and the Watershed segmentation

	Borůvka MST Pyramid	**Watershed segmentation**
Domain	edge graph / derivative along edge in the NG	gradient image / derivative in each pixel
Method	local maxima survive	

Each local maximum that survives to a certain level k, defines in each level $l_i, i = 1, \ldots, k$, on each side, an attraction area. (See Fig.4) These attraction areas contain only values smaller than that of the local maximum and depend on it and its neighbours (up to the local maxima that survived to level l_i). The higher we go in the pyramid, the larger these attraction areas become, and two such neighbouring regions, defined by 2 neighbouring local maxima, define a catchment basin which will be merged in the next step. (See Fig.4)

Let $ve_i(k)$ and $ve_j(k)$ be the two neighbouring local maxima that define the two attraction areas $aa_1 = \{ve_{i+1}(k), ve_{i+2}(k), ..., ve_q(k)\}$ and $aa_2 = \{ve_{q+1}(k),$

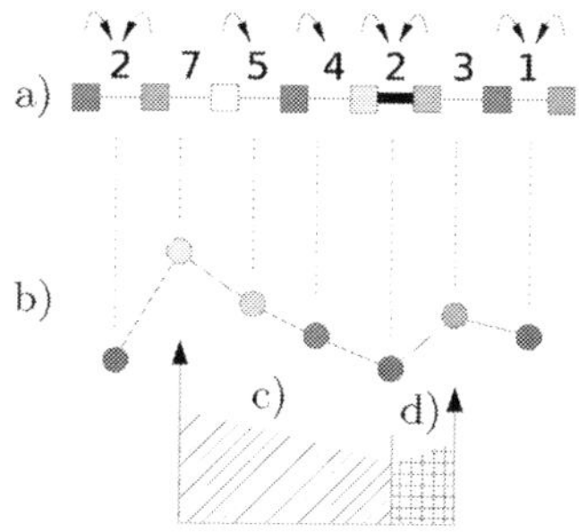

Fig. 4. Attraction areas in a catchment basin: a) NG, each vertex chooses it's smallest edge; b) associated EG; c) attraction area of edge 7 from the NG; d) attraction area of edge 2 from the NG

$ve_{q+2}(k), ..., ve_{j-1}(k)\}$, with $i < q < j$ denoting vertex indices in our chain-like edge graph and k denoting a level in our pyramid. From the above, we get:

$$\max(ve_i(k), ve_j(k)) > \max(ve_{i+1}(k), ..., ve_{j-1}(k))$$

$$\min(ve_i(k), ve_j(k)) > \min(ve_{i+1}(k), ..., ve_{j-1}(k))$$

for any level k with $ve_i(k)$ and $ve_j(k)$ being local maxima in level k (they survive to level $k+1$) and $ve_{i+1}(k), ..., ve_{j-1}(k)$ not being local maxima. If we recursively follow the previous we get that:

$$\max(ve_i(k_1), ve_j(k_1)) > \max(ve_{i+1}(k_2), ..., ve_{j-1}(k_2))$$

$$\min(ve_i(k_1), ve_j(k_1)) > \min(ve_{i+1}(k_2), ..., ve_{j-1}(k_2))$$

for $k_1 > k_2$, i.e. the biggest of the values surrounding a certain region in level k_1 is larger then biggest of all the values from any level $k_2 < k_1$ below. The previous holds for the smallest also. So, the maximum edge weight on the border of any region in any level, is larger than the maximum edge weight inside, i.e. maximum variation on the border of a region is larger then maximum variation inside the region, and the same holds for the minimum.

4.2 The 2D Case

To continue in the same line of ideas, we present the MST Pyramid edge selection mechanism for a 2D image, in a domain in which the presented similarities with the watershed method remain valid. For this, we do not focus on finding the minimum spanning tree (MST) itself, but on the way the MST Pyramid selects edges for contraction and thus constructs the MST by creating increasingly bigger parts from smaller ones.

For a given 2D Image and its associated NG. We determine the edge graph (EG) of the MST of the NG (MST_NG). Each vertex from the EG is attributed with the weight of its associated edge from the MST_NG. (See Fig. 5a,b)

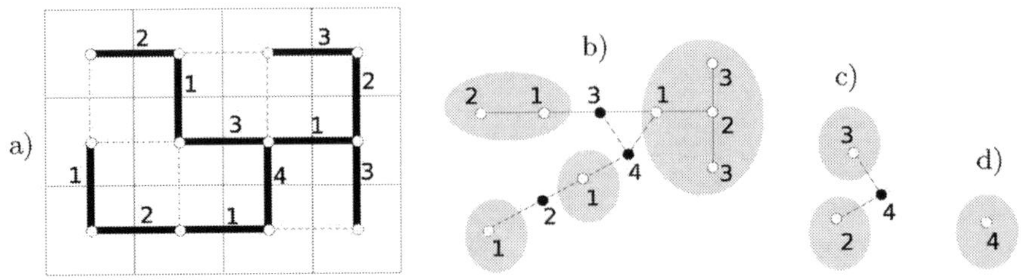

Fig. 5. MST based contraction, 2D case: a) image (thin continuous line) with associated NG (dashed line), it's MST (thick line) with its edge weights; b) EG of the MST_NG (vertices of the same component are white and in the same grey ellipse, local maxima i.e. surviving vertices are black); c) EG - second level; d) EG - third level

According to Algorithm 1., in one MST Pyramid selection step each vertex from the NG selects the smallest edge around it (which is guaranteed to be on the MST of the NG). In the context of the EG of the MST_NG of the image, this can be described in a watershed like manner as follows:

1. **Initial configuration:** all the vertices in the EG have no labels and their attribute is the weight of the their corresponding edges in the NG;
2. From minimum to maximum **progressively threshold** the values in the vertices and each unlabelled vertex with a value below the threshold:
   - gets a unique numeric label, if no neighbours are numerically labelled yet (we found a new catchment basin/local minimum),
   - gets the unique numeric label of its labelled neighbours, if no 2 neighbours have different numeric labels (belong to different watersheds) and at least one is numerically labelled (watershed increases),
   - is labelled as "don't contract/survive", if none of the previous apply

At the end of each such step, the vertices labelled with the same numeric value are joined and they define connected regions (their corresponding edges in the MST_NG are contracted). A new EG is obtained by keeping the vertices with no numeric labels, the edges connecting them, and additionally connecting any 2 such vertices if in the labelled graph from the current step, they could be connected by paths made only of numerically labelled vertices. (See Fig. 5b,c,d).

As in the case of the 1D image, in each step local maxima from the previous level survive, and the properties observed in the 1D image case study remain valid for the MST_NG of a 2D image. Let r be a connected region in a 2D image. Let NG=(V, E) be its associated neighbourhood graph. Let $E_c \subset E$ be the cut-edges connecting the vertices $V_r \subset V$, associated to the pixels of r, to the rest of NR $(V \setminus V_r)$. Also let $E_r = \{(v_i, v_j) \in E \mid v_i, v_j \in V_r\}$, MST_NG=$(V_{mst}, E_{mst})$ the MST of NG, $E_{cmst} = E_c \bigcap E_{mst}$, and $E_{rmst} = E_r \bigcap E_{mst}$. If r is a region produced by the MST Pyramid (without the contrast step) then:

$$\max(E_{rmst}) < \max(E_{cmst}),$$
$$\min(E_{rmst}) < \min(E_{cmst}),$$
$$R_{mst} = (V_r, E_{rmst}) \text{ is a connected graph.}$$

The above also explains why most of the edges pass the contrast test (step 4 in Algorithm 1.), and why this step does not significantly change the results. The purpose of the contrast step is to ensure that the algorithm produces regions with small variation surrounded by borders with large variation, but this is allready achieved in most of the cases by the edge selection mechanism in step 3. Where the results differ significantly is that without using the contrast step, the pyramid always reaches an apex. The results when using the contrast step are better if we are looking for a segmentation that spans just one pyramid level and we have small regions. Here the additional condition stops these small regions to be merged with the surrounding while the rest of the graph is contracted.

Because of the way edges from the NG are attributed (distance in some featurespace), having the MST also gives us an upper bound on the weights of all the other edges. The difference between the values of two neighbouring pixels is less or equal to the sum of the weights of the edges along the path connecting their associated vertices in the MST of the NG.

5 Outlook

The previous study should help in improving the method and using it as a basis for reaching higher level abstraction. We plan to add the slope when calculating the edge weights to prevent "leakage". Knowing the properties of the method allows us to easily controll it and insert a priory information from e.g. a successfull previous segmentation, or a high level process. Knowing the properties of the regions produced allows us to select a "best segmentation" that spans multiple levels and which can be used by higher level processes that need only one segmentation, or as a start seed for the ones that are able to use hierarchies but use a single segmentation at some instance of time (e.g. object recognition).

6 Conclusion

We have presented a set of properties of the regions produced by the MST Pyramid segmentation method and showed its similarity with the watershed transform of an image. Attraction regions are contracted before their neighbouring local maxima. Smooth parts of the image surrounded by borders with maximal local variation are selected. Maximum and respectively minimum variation on the border of a region is bigger then the maximum and respectively minimum variation inside the region. Internal/external contrast conditions do not affect too much the lower levels of the pyramid.

References

1. Felzenszwalb, P.F., Huttenlocher, D.P.: Efficient graph-based image segmentation. International Journal of Computer Vision **59** (2004) 167–181
2. Shi, J., Malik, J.: Normalized cuts and image segmentation. IEEE Transactions on Pattern Analysis and Machine Intelligence **22** (2000) 888–905

3. Haxhimusa, Y., Kropatsch, W.G.: Segmentation graph hierarchies. In: Proceedings of Joint Workshops on Structural, Syntactic, and Statistical Pattern Recognition S+SSPR. Volume 3138 of Lecture Notes in Computer Science., Lisbon, Portugal (2004)
4. Keselman, Y., Dickinson, S.J.: Generic model abstraction from examples. IEEE Trans. Pattern Anal. Mach. Intell. **27** (2005) 1141–1156
5. Kropatsch, W.G., Haxhimusa, Y., Pizlo, Z., Langs, G.: Vision pyramids that do not grow too high. Pattern Recognition Letters **26** (2005) 319–337
6. Haxhimusa, Y., Ion, A., Kropatsch, W.G.: Evaluating graph-based segmentation algorithms. In: Proceedings of the 18th Internation Conference on Pattern Recognition, Hong Kong (2006)
7. Roerdink, J.B.T.M., Meijster, A.: The watershed transform: Definitions, algorithms and parallelization strategies. Fundamenta Informaticae **41** (2000) 187–228
8. Brun, L., Kropatsch, W.G.: Irregular Pyramids with Combinatorial Maps. In: Proceedings of Joint Workshops on Structural, Syntactic, and Statistical Pattern Recognition S+SSPR. Volume 1876 of Lecture Notes in Computer Science., Alicante, Spain (2000) 256–265
9. Neštřil, J., Miklovà, E., Neštřilova, H.: Otakar Borůvka on minimal spanning tree problem translation of both the 1926 papers, comments, history. Discrete Mathematics **233** (2001) 3–36
10. Haxhimusa, Y.: Structurally Optimal Dual Graph Pyramid and its Application in Image Partitioning. PhD thesis, Vienna University of Technology, Faculty of Informatics, Institute of Computer Aided Automation, Pattern Recognition and Image Processing Group (2006)
11. Meyer, F.: Graph based morphological segmentation. In Kropatsch, W.G., Jolion, J.M., eds.: 2nd IAPR-TC15 Workshop on Graph-based Representations in Pattern Recognition. Volume 126., Vienna, Austria, OCG (1999) 51–60

A Random Walk Kernel Derived from Graph Edit Distance

Michel Neuhaus* and Horst Bunke

Institute of Computer Science and Applied Mathematics, University of Bern
Neubrückstrasse 10, CH-3012 Bern, Switzerland
{mneuhaus, bunke}@iam.unibe.ch

Abstract. Random walk kernels in conjunction with Support Vector Machines are powerful methods for error-tolerant graph matching. Because of their local definition, however, the applicability of random walk kernels strongly depends on the characteristics of the underlying graph representation. In this paper, we describe a simple extension to the standard random walk kernel based on graph edit distance. The idea is to include global matching information in the local similarity evaluation of random walks in graphs. The proposed extension allows us to improve the performance of the random walk kernel significantly. We present an experimental evaluation of our method on three difficult graph datasets.

1 Introduction

For more than thirty years, a huge variety of methods have been developed addressing the problem of graph matching [1]. In recent years, a novel class of algorithms based on kernel machines has gained a significant amount of interest in the pattern recognition community. The basic idea of kernel machines is to map the classification problem from the pattern domain to a vector space implicitly defined in terms of a kernel function [2]. In the context of graph matching, kernel machines allow us to apply vector space operations to graphs by embedding the space of graphs in a vector space. Provided that the definition of a kernel function that is suitable for the pattern matching problem under consideration is given, a large number of algorithms for pattern analysis and recognition can readily be applied, including principal component analysis, Fisher discrimination analysis, and Support Vector Machines [2].

Various kernel functions have been proposed to solve the graph matching problem as well as the related string matching problem. In a common approach, the similarity of patterns is defined in terms of similar substructures they contain [3,4]. Another approach employs the definition of a Schur-Hadamard inner product on graphs [5]. Based on the notion of random walks in graphs, several kernels have been developed [6,7]. While kernel methods provide a powerful way

* Supported by the Swiss National Science Foundation NCCR program *Interactive Multimodal Information Management (IM)2* in the Individual Project *Multimedia Information Access and Content Protection*.

D.-Y. Yeung et al. (Eds.): SSPR&SPR 2006, LNCS 4109, pp. 191–199, 2006.

to analyse and classify graphs, they are, in some cases, limited in terms of the flexibility of their structural matching process.

In this paper we aim at enhancing a standard random walk kernel by information derived from the well-established error-tolerant graph edit distance measure [8,9]. In the remainder of this paper, we will briefly introduce graph edit distance and the random walk kernel, describe the extension we propose, and demonstrate the usefulness of our method in classification experiments.

2 Graph Edit Distance

Graph edit distance is one of the most universal graph matching methods in the sense that edit distance is not restricted to special classes of graphs, such as planar graphs, bounded valence graphs, or graphs labeled with discrete attributes. The key idea is to measure the structural dissimilarity of two graphs by the minimal amount of distortion that is needed to transform one graph into the other [8,9]. The only requirement for graph edit distance to be applicable is that an underlying distortion model must be given such that the strength of distortions can be measured. Hence, graph edit distance can be computed for graphs with arbitrary node and edge relations and any kind of node and edge labels.

More formally, let $g = (V, E, \mu, \nu)$ denote a graph g consisting of a finite set of nodes V, a set of directed edges $E \subseteq V \times V$, a node labeling function $\mu : V \rightarrow L$ assigning an attribute from L to each node, and an edge labeling function $\nu : E \rightarrow L$. The label alphabet L is often defined as a finite set of labels, $L = \{\alpha, \beta, \gamma, \ldots\}$, or a Euclidean vector space, $L = \mathbb{R}^n$. We then define a number of distortion, or edit, operations on graphs. A standard set of graph edit operations consists of an insertion, a deletion, and a substitution operation of nodes and edges. An edge deletion is equivalent to the removal of an edge from a graph, and a node substitution results in the replacement of a node label by another one. Further required is a cost function assigning each edit operation a penalty cost value, such that weak edit operations have low costs and strong edit operations have high costs. For instance, slightly changing a label should, in most cases, result in lower costs than strongly changing the same label. The key idea of graph edit distance is that for two structurally similar graphs only a few weak edit operations are needed to convert one graph into the other. By contrast, for two quite different graphs, a larger number of edit operations of greater strength are needed to make the two graphs identical to each other. Consequently, the edit distance of two graphs g and g' is defined by the minimum cost sequence of edit operations transforming g into g',

$$d(g, g') = \min_{(e_1,\ldots,e_k) \in E(g,g')} \sum_{i=1}^{k} c(e_i) \ . \tag{1}$$

A sequence of edit operations transforming one graph into the other is also called an edit path. Note that $E(g, g')$ denotes the set of edit paths from g to g', and c is a function assigning costs to edit operations.

The edit distance of graphs is usually computed by means of a tree search procedure [9]. As every node can potentially be substituted by any other node, it can be shown that the computational complexity of edit distance is exponential in terms of space and time. In practice, it turns out that the computation of exact edit distance is limited to graphs with up to 12 nodes, typically. In this paper, we therefore resort to an approximate edit distance algorithm [10] in those cases where the exact distance cannot be computed.

The edit distance of graphs is normally used in conjunction with a k-nearest-neighbor classifier. For an unknown input graph, we compute the edit distance to a number of prototype graphs and assign the input graph to the most frequent class among the k closest prototypes.

3 Random Walk Kernels

The objective in this section is to define error-tolerant graph similarity measures, or kernel functions, that can be used in conjunction with kernel machines [2]. The main advantage of kernel based classifiers for structured data is that the classification problem can be formulated in a vector space related to the original pattern space solely by definition of a kernel function. Given a valid kernel function, it can be proven that there exists a vector space with its inner product being equal to the kernel function. This allows us to run a number of algorithms for classification and pattern analysis in the implicitly existing vector space without explicitly mapping the graphs to the elements of the vector space. In our experiments, we apply the kernel functions in conjunction with one of the most prominent and best performing kernel based classifiers, the Support Vector Machine (SVM) [2].

We proceed by first describing a well-known random walk kernel for discretely labeled graphs [6] and its extension to continuously labeled graphs [7]. Then we suggest modifications to make the random walk kernel more robust.

3.1 Discretely Labeled Graphs

The original random walk kernel is defined by means of the direct product graph [6]. The direct product of two graphs $g = (V, E, \mu, \nu)$ and $g' = (V', E', \mu', \nu')$ is the graph $(g \times g') = (V_\times, E_\times, \mu_\times, \nu_\times)$ given by

$$
\begin{aligned}
V_\times &= \{(v, v') \in V \times V' : \mu(v) = \mu'(v')\} \text{ and} \\
E_\times &= \{((u, u'), (v, v')) \in V_\times^2 : (u, v) \in E \wedge (u', v') \in E' \wedge \nu(u, v) = \nu'(u', v')\} \ .
\end{aligned}
\tag{2}
$$

The labeling functions of the product graph are defined by $\mu_\times(v, v') = \mu(v) = \mu'(v')$ and $\nu_\times((u, u'), (v, v')) = \nu(u, v) = \nu'(u', v')$. In other words, in the direct product graph $(g \times g')$, we simply identify pairs of nodes of both graphs with identical labels and pairs of edges with identical labels, constituting the nodes and edges of the product graph. The adjacency matrix $A_\times$ of $(g \times g')$ is then defined as

$$
[A_\times]_{(u,u'),(v,v')} = \begin{cases} 1 & \text{if } ((u, u'), (v, v')) \in E_\times \ , \\ 0 & \text{otherwise} \ . \end{cases}
\tag{3}
$$

Note that the adjacency matrix is a $|V_\times| \cdot |V_\times|$-matrix containing at position (i, j) value 1 if node i is connected to node j by an edge in $(g \times g')$, and value 0 otherwise. From the adjacency matrix $A_\times$ of the direct product, one can then derive the graph kernel with weighting parameter $\lambda \geq 0$ according to the formula [6]

$$k_\times(g, g') = \sum_{i,j=1}^{|V_\times|} \left[\sum_{n=0}^{\infty} \lambda^n A_\times^n \right]_{ij} . \tag{4}$$

If $\lambda < 1$, it is sufficiently accurate to evaluate infinite sums by their first few dominant addends only.

The kernel can be interpreted as a measure of the number of matching labeled random walks in both graphs. That is, if the sequence of node and edge labels encountered on a random walk in g matches the sequence of node and edge labels of a random walk in g', this contributes a certain amount to the overall similarity $k_\times(g, g')$. The graph kernel reflects the intuitive understanding that two graphs are similar if there are a large number of identical random walks in both graphs.

3.2 Continuously Labeled Graphs

The main limitation of the kernel defined above is that it is only applicable to graphs with discretely labeled nodes and edges. If a random walk in g differs from a random walk in g' only in a single node label, the two walks are considered completely different and are therefore not taken into account. Unfortunately, most graphs extracted from real-world data contain a significant amount of noise, and attributes with continuous values are mostly used to describe non-discrete data. For these reasons, an extension of the original random walk kernel has been proposed [7]. The idea is not to evaluate if two walks are identical, but rather if they are similar. This modified kernel is applicable to graphs with continously labeled nodes and edges.

To obtain the modified kernel, we leave out the label equality conditions in Eq. 2, resulting in a modified direct product $(g \times g')$, and define the adjacency matrix of $(g \times g')$ by

$$[A_\times]_{(u,u'),(v,v')} = \begin{cases} k((u,u'),(v,v')) & \text{if } ((u,u'),(v,v')) \in E_\times , \\ 0 & \text{otherwise} , \end{cases} \tag{5}$$

where the kernel function k measuring the similarity of pairs of nodes (u, u') and (v, v') is given by

$$k((u,u'),(v,v')) = k_{node}(u,u') \cdot k_{edge}((u,v),(u',v')) \cdot k_{node}(v,v') . \tag{6}$$

This function is defined with respect to underlying kernels k_{node} evaluating the similarity of two node labels and k_{edge} evaluating the similarity of two edge labels. In our experiments, we use standard RBF kernels for this purpose. Note that the adjacency matrix defined in Eq. 5 can be interpreted as a fuzzy adjacency matrix, where the adjacency value of two nodes of the product graph is high if the corresponding pairs of nodes and pairs of edges have similar labels, and low otherwise.

Plugging the adjacency matrix from Eq. 5 into the kernel function in Eq. 4, we obtain the modified kernel that can be applied to continuously labeled graphs [7].

3.3 Edit Distance Enhancement

As will be shown in the next section, the random walk kernel defined above is very powerful on certain datasets, but may perform poorly on other data compared to a standard edit distance based nearest-neighbor classifier. In the case of the random walk kernel, the similarity of graphs is defined by accumulating the similarity of local parts of the graphs. For certain graph representations, however, there are global matching constraints that need to be taken into account. In such a case it may be more appropriate to apply other graph matching methods, such as the one based on edit distance. Experiments confirm that random walk kernels and edit distance methods address the graph matching problem in complementary ways, and one approach usually performs significantly worse or better than the other one.

The main objective in this paper is to bring together the best from both worlds: The flexibility of graph edit distance and the power of random walk kernels. The basic idea is to enhance the random walk kernel with an edit distance matching at the global level. This allows us to integrate global information into the local random walk matching process. To this end, let us assume that an optimal edit path from g to g' has been computed, and let $S = \{v_1 \rightarrow v'_1, v_2 \rightarrow v'_2, \ldots\}$ denote the set of node substitutions present in the optimal path. We then proceed by defining the adjacency matrix of the direct product graph $(g \times g')$ by

$$[A_\times]_{(u,u'),(v,v')} = \begin{cases} k((u,u'),(v,v')) & \text{if } ((u,u'),(v,v')) \in E_\times \text{ and} \\ & u \rightarrow u' \in S \text{ and } v \rightarrow v' \in S \;, \\ 0 & \text{otherwise} \;. \end{cases} \quad (7)$$

In other words, we restrict the random walks to nodes that satisy the optimal node-to-node correspondences identified by the edit distance computation. This adjacency matrix is then used with the kernel function given in Eq. 4.

4 Experimental Results

In this section, we offer an evaluation of the proposed enhanced random walk kernel in comparison to two baseline systems. In the first baseline system, the edit distance of graphs is computed (see Sec. 2), and test graphs are classified according to the k most similar graphs from a labeled training set. In the second baseline system, the similarity of graphs is evaluated by means of the traditional random walk kernel (see Sec. 3.2), and an SVM is used for classification. The third system, our proposed method, is based on the enhanced random walk kernel defined in Sec. 3.3.

The first database consists of line drawings representing capital letters. To obtain a noisy sample set of letters, we iteratively apply distortions to clean letter prototypes. The distorted line drawings are then converted into graphs by

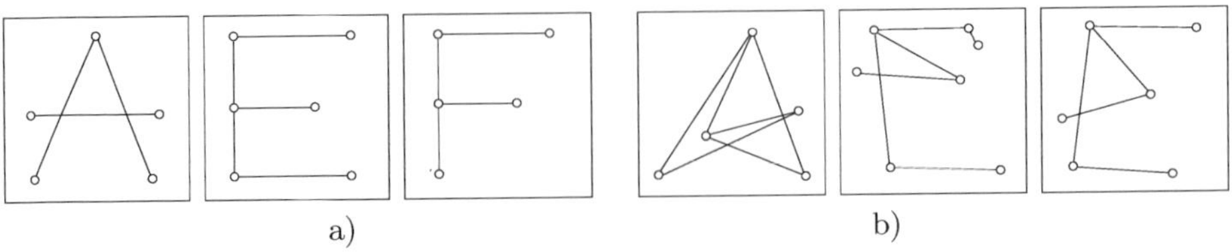

Fig. 1. Illustration of a) three clean letters and b) three distorted letters A, E, F

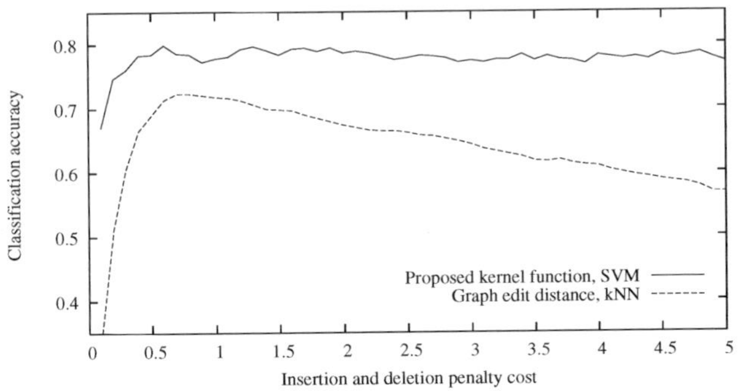

Fig. 2. Influence of insertion and deletion penalty cost on classification accuracy

representing end points of lines by nodes and lines by edges. Nodes are labeled with the two-dimensional position of the corresponding end point. Following this procedure, we construct a training set and validation set of size 150 each, and a test set of size 750. The database consists of 15 classes of letters (A, E, F, H, I, K, L, M, N, T, V, W, X, Y, Z). An illustration is provided in Fig. 1.

In a first experiment, we focus on the influence of edit costs on the classification accuracy. For this purpose we only consider the k-nearest-neighbor edit distance based classifier and the SVM with the edit distance enhanced kernel. The edit costs of node (or edge) insertions and deletions essentially determine how likely a node (edge) is to be substituted by another node (edge). If insertion and deletion costs are low, only a few inexpensive substitutions will occur in an optimal edit path. Conversely, if insertion and deletion costs are high, the edit distance algorithm will tend to substitute as many nodes (edges) as possible. For an edit distance based nearest-neighbor classifier, the resulting cost of an optimal edit path is crucial for the performance. It is therefore important to carefully adjust insertion and deletion costs, as well as any other edit cost parameter. In the case of the random walk kernel proposed in this paper, on the other hand, we are interested in promising node-to-node correspondences, rather than a particular distance value. This means that in the case of the proposed method there is no need for an extensive optimization of the edit cost parameters. This issue can very well be observed in Fig. 2, where the classification accuracy of an edit distance based nearest-neighbor classifier and an SVM based on the proposed kernel function is shown for various insertion and deletion penalty costs. As

a) b)

Fig. 3. Example images from the Lesaux database, a) *city* and b) *countryside*

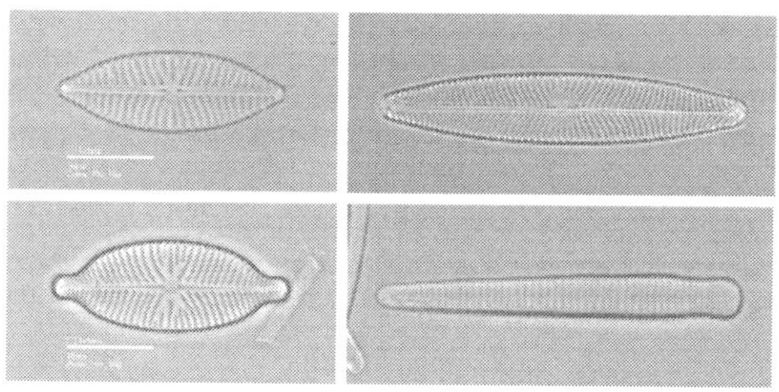

Fig. 4. Example images from the Diatom database (four different classes)

expected, the accuracy of the traditional edit distance classifier strongly depends on the actual edit costs, while the proposed method exhibits a roughly constant behavior for penalty costs above a certain threshold. It should also be noted that the proposed method clearly outperforms the nearest-neighbor classifier.

We next compare the classification accuracy of the two baseline classifiers with the proposed method. To this end, we classify graphs from the Letter database described above, from the Lesaux database, and the Diatom database. The Lesaux database [11] consists of graphs representing images from five classes (*city, countryside, people, snowy, streets*). Graphs are extracted from images by running a region segmentation process and removing those segments that are deemed irrelevant for classification. The remaining regions are then turned into a region adjacency graph with labels describing the dominant colors of the region. We use a training set, validation set, and test set of size 54 each. For two example images, see Fig. 3. The Diatom database [12] consists of 110 microscopic images of diatoms, evenly split into training set, validation set, and test set. The recognition task is to classify diatoms from the test set according to 22 classes. The images are represented by attributed region adjacency graphs. Four example diatom images from different classes are shown in Fig. 4. The various parameters of the classifiers (such as edit cost parameters and weighting factor λ) are first optimized on the validation set and then applied to the independent test set.

The classification accuracy of the three methods under consideration determined on the independent test set is given in Table 1. There are two entries in

Table 1. Comparison of classification accuracy

	Letter database	Lesaux database	Diatom database
Edit distance, kNN	69.3	48.2*	63.9*
Random walk, SVM	75.7*	33.3	44.4
Proposed, SVM	74.7*	51.9*	58.3*

* Marked classification rates do not differ significantly. Unmarked classification rates are significantly lower than marked ones ($\alpha = 0.05$).

each column of this table marked with an asterisk. These two entries are, in each column, not significantly different from each other (on a statistical significance level of $\alpha = 0.05$). However, the unmarked entry in each column is significantly smaller than the two marked ones. It can clearly be observed that the two traditional methods — the edit distance based k-nearest-neighbor classifier and the standard random walk kernel — perform quite different on all datasets. One of the two methods is always significantly better than the other one. The proposed random walk kernel enhanced by edit distance information, on the other hand, performs as good as the better method throughout our experiments. That is, while the traditional edit distance method and random walk kernel method emphasize a certain aspect of the graph matching problem, the proposed kernel function combines the information in an advantageous manner. By applying the method proposed in this paper, we obtain a robust classifier that succeeds well on all tested datasets without recourse to the characteristics of the underlying graphs. Our method can be regarded as an extension to the standard random walk kernel that leads to a statistically significant improvement of the graph matching performance on the Lesaux database and the Diatom database.

5 Conclusions

In this paper, we propose an extension of a standard random walk kernel for graphs. It can be observed, on graphs extracted from real-world data, that random walk kernels offer an interesting alternative to traditional edit distance based graph classifiers in the sense that they address the graph matching problem in a different way. One some datasets, the edit distance measure is the most suitable method for graph matching; on other datasets, edit distance is outperformed by random walk kernels and Support Vector Machines. The method we propose is based on the idea that it is advantageous to include graph matching information from the global level in the random walk kernel defined locally based on the similarity of walks in graphs. By constraining the random walk kernel to pairs of nodes that satisfy the global node-to-node correspondence, instead of any pairs of nodes, we obtain a system that combines the flexibility of graph edit distance with the classification power of the random walk kernel. The proposed kernel offers a classification accuracy that is at least as good as the better one of the two baseline methods — graph edit distance and standard random walk kernels — and significantly better than the other one. The performance is evaluated on a semi-artificial line drawing dataset and two real-world image datasets.

References

1. Conte, D., Foggia, P., Sansone, C., Vento, M.: Thirty years of graph matching in pattern recognition. Int. Journal of Pattern Recognition and Artificial Intelligence **18** (2004) 265–298
2. Shawe-Taylor, J., Cristianini, N.: Kernel Methods for Pattern Analysis. Cambridge University Press (2004)
3. Watkins, C.: Dynamic alignment kernels. In Smola, A., Bartlett, P., Schölkopf, B., Schuurmans, D., eds.: Advances in Large Margin Classifiers. MIT Press (2000) 39–50
4. Haussler, D.: Convolution kernels on discrete structures. Technical Report UCSC-CRL-99-10, University of California, Santa Cruz (1999)
5. Jain, B., Geibel, P., Wysotzki, F.: SVM learning with the Schur-Hadamard inner product for graphs. Neurocomputing **64** (2005) 93–105
6. Gärtner, T., Flach, P., Wrobel, S.: On graph kernels: Hardness results and efficient alternatives. In Schölkopf, B., Warmuth, M., eds.: Proc. of the 16th Annual Conf. on Learning Theory. (2003) 129–143
7. Borgwardt, K., Ong, C., Schönauer, S., Vishwanathan, S., Smola, A., Kriegel, H.P.: Protein function prediction via graph kernels. Bioinformatics **21** (2005) 47–56
8. Sanfeliu, A., Fu, K.: A distance measure between attributed relational graphs for pattern recognition. IEEE Transactions on Systems, Man, and Cybernetics (Part B) **13** (1983) 353–363
9. Bunke, H., Allermann, G.: Inexact graph matching for structural pattern recognition. Pattern Recognition Letters **1** (1983) 245–253
10. Neuhaus, M., Bunke, H.: An error-tolerant approximate matching algorithm for attributed planar graphs and its application to fingerprint classification. In: Proc. 10th Int. Workshop on Structural and Syntactic Pattern Recognition. LNCS 3138, Springer (2004) 180–189
11. Le Saux, B., Bunke, H.: Feature selection for graph-based image classifiers. In: Proc. 2nd Iberian Conf. on Pattern Recognition and Image Analysis. LNCS 3523, Springer (2005) 147–154
12. Ambauen, R., Fischer, S., Bunke, H.: Graph edit distance with node splitting and merging and its application to diatom identification. In Hancock, E., Vento, M., eds.: Proc. 4th Int. Workshop on Graph Based Representations in Pattern Recognition. LNCS 2726, Springer (2003) 95–106

Edit Distance for Ordered Vector Sets: A Case of Study

Juan Ramón Rico-Juan and José M. Iñesta*

Departamento de Lenguajes y Sistemas Informáticos
Universidad de Alicante, E-03071 Alicante, Spain
{juanra, inesta}@dlsi.ua.es

Abstract. Digital contours in a binary image can be described as an ordered vector set. In this paper an extension of the string edit distance is defined for its computation between a pair of ordered sets of vectors. This way, the differences between shapes can be computed in terms of editing costs. In order to achieve efficency a dominant point detection algorithm should be applied, removing redundant data before coding shapes into vectors. This edit distance can be used in nearest neighbour classification tasks. The advantages of this method applied to isolated handwritten character classification are shown, compared to similar methods based on string or tree representations of the binary image.

Topics: Dominant Points, Pattern Recognition, Structural Pattern Recognition.

1 Introduction

The description of an object contour in a binary image as a string [1] using Freeman codes [2] or using a tree representation structure [3,1] is widely used in pattern recognition. For using these structures in a recognition task, the edit distance is often used as a measure of the differences between two instances. Both, string edit distances [4] and tree edit distances [5] are used, depending on the data structures utilised for representing the problem data. In this paper, in order to obtain a representation of the object contour from a binary image, an ordered vector set is extracted, and an edit distance measure is defined between pairs of instances of this representation. This measure is an extension of the string edit distance, adding two new rules and changing vectors by symbols.

Freeman chain codes keep very fine details of the shapes since they code the relations between every pair of adjacent pixels of the contours. To avoid computation time and in order to remove irrelevant details, a dominant point detection algorithm is needed. The goal is to reduce the features that represent a binary image in order to remove redundant data to compute the distance faster, keeping the final classification time low and good error rates.

* Work supported by the Spanish CICYT under project TIC2003-08496-CO4 and Generalitat Valenciana I+D+i under project GV06/166.

D.-Y. Yeung et al. (Eds.): SSPR&SPR 2006, LNCS 4109, pp. 200–207, 2006.

The remainder of this paper consists of four sections. In section 2, two different representations of the same binary image are extracted. In section 3, a new distance based in ordered vector set is defined. In section 4, the results of experiments in a classification task, applying string and ordered vector set edit distances are presented. Finally in section 5, the conclusions and future word are presented.

2 Feature Extraction from a Binary Image

The goal of the ordered vector set is to describe the contour of an object using the least possible number of elements. The classical representation of a contour

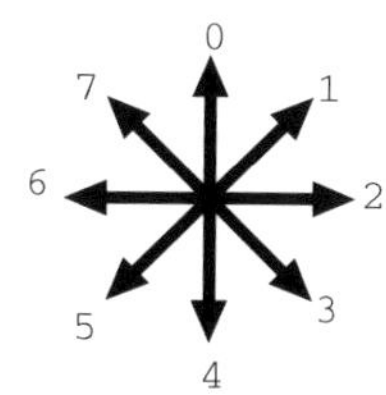

Fig. 1. Freeman 2D code

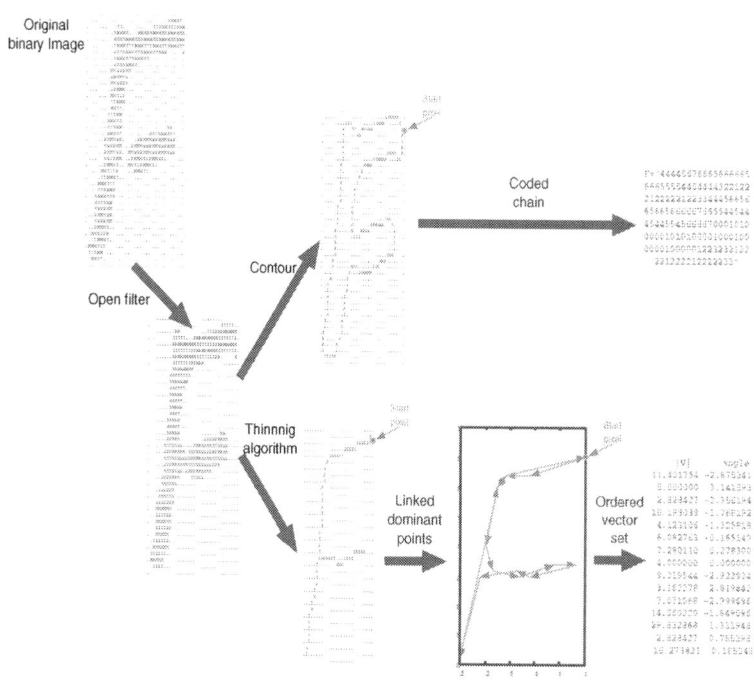

Fig. 2. General scheme. From the binary image, morphological filters are applied to correct gaps and spurious points. Thus, contour and skeleton are obtained. From the first, the chain code is obtained and from the second, the ordered vector set is extracted using a dominant point selection algorithm.

in a binary image links the contour pixels with their neighbors using 0 to 7 (see Fig. 1) codes which represent a discrete number of 2D directions. This way, a string that represents the contour is obtained (Fig. 2 top-right).

This kind of feature extraction assumes that all linked pixels are of equal importance. If we select the most representative points of the contour and link all these points, a compact representation of 2D figures is obtained, with less features than using Freeman codes.

The idea is to select a set of dominant points in a contour [6,7], link those points following the contour of the figure using 2D vectors, and then use these ordered vector set to represent the image (Fig. 2 bottom-right).

In a particular application of handwritten character recognition, it is recommended to apply some filter operations to original image before extracting and coding the contours [8] including an opening filter [9] and a thinning algorithm [10] in order to remove noise and redundant information.

3 Ordered Vector Set Edit Distance

The string edit distance definition [4] is based on three edit operations: insertion, deletion, and substitution. Let Σ the alphabet, $A, B \in \Sigma^*$ two finite strings of characters, and Λ is a null character. $A\langle i\rangle$ is the ith character of the string A; $A\langle i:j\rangle$ is the substring form the ith to jth characters of A, both inclusive.

An edit operation is a pair $(a,b) \in (\Sigma \cup \{\Lambda\})^2 : (a,b) \neq (\Lambda,\Lambda)$. So, the basic edit operations are substitution $a \rightarrow b$, insertion $\Lambda \rightarrow b$ and deletion $a \rightarrow \Lambda$. If a generic cost function is associated to each operation $\gamma_s(a \rightarrow b)$, the cost of the sequence of edit operations that transforms a finite string A in B is defined as

$$d_s(A,B) = \min \begin{cases} \gamma_s(\Lambda \rightarrow B\langle 1\rangle) + d_s(A, B\langle 2:|B|\rangle) & |B| \geq 1 \\ \gamma_s(A\langle 1\rangle \rightarrow \Lambda) + d_s(A\langle 2:|A|\rangle, B) & |A| \geq 1 \\ \gamma_s(A\langle 1\rangle \rightarrow B\langle 1\rangle) + d_s(A\langle 2:|A|\rangle, B\langle 2:|B|\rangle) & |A| \geq 1 \wedge |B| \geq 1 \\ 0 & |A| = 0 \wedge |B| = 0 \end{cases}$$

The similar idea of an ordered string is extended to an ordered vector set. Let $V, W \in (\mathbb{R}\times[0,2\pi])^*$ a finite set of vectors and Λ is a null vector. $V\langle i\rangle$ is the vector ith in the set V, $V_N\langle i\rangle$ is the norm and $V_\alpha\langle i\rangle$ is the angle of the ith vector; $V\langle i:j\rangle$ is the subset from ith to jth component vectors of V, both included.

Now, an edit operation is a pair $(v,w) \in (\mathbb{R}\times[0,2\pi])$, $(v,w) \neq (\Lambda,\Lambda)$: $(v,w^*) \cup (v^*,w)$. So, the basic edit operations are substitution (1 to 1) $v \rightarrow w$, substitution (1 to N) called fragmentation $v \rightarrow w^+$, substitution (N to 1) called consolidation $v^+ \rightarrow w$, insertion $\Lambda \rightarrow w$ and deletion $v \rightarrow \Lambda$. Here, we have considered the case that one vector could be replaced by N, or vice versa.

When using dominant points, it is usual that a small change in the contour generates a new dominant point, so when comparing two prototypes 1 vector in the first prototype can be similar to N continuous vectors from the second prototype.

The cost of sequence of edit operations that transforms a finite ordered vector set V into W, if we establish a cost function $\gamma_v(v^*, w^*)$, is defined as

$$d_v(V, W) = \min \begin{cases} \gamma_v(\Lambda \to W\langle 1\rangle) + d_v(V, W\langle 2:|W|\rangle) & |W| \geq 1 \\ \gamma_v(V\langle 1\rangle \to \Lambda) + d_v(V\langle 2:|V|\rangle, W) & |V| \geq 1 \\ \gamma_v(V\langle 1\rangle \to W\langle 1\rangle) + d_v(V\langle 2:|A|\rangle, W\langle 2:|B|\rangle) & |V| \geq 1 \wedge |W| \geq 1 \\ \min_{j\in[2,|W|]} \gamma_v(V\langle 1\rangle \to W\langle 1:j\rangle) + d_v(V\langle 2:|V|\rangle, B\langle j+1:|W|\rangle) & |W| > 2 \\ \min_{i\in[2,|V|]} \gamma_v(V\langle 1:i\rangle \to W\langle 1\rangle) + d_v(V\langle j+1:|V|\rangle, B\langle 2:|W|\rangle) & |V| > 2 \\ 0 & |V| = 0 \wedge |W| = 0 \end{cases}$$

In a similar way to the efficient (dynamic programming technique) algorithm proposed in [4] for computing the string edit distance, it can be extended to compute the ordered vector set edit distance in the following way:

```
Function vectorEditDistance(V,W)
   D[0,0] := 0;
   for i := 1 to |V| do D[i,0] := D[i-1,0] + γ_v (V⟨i⟩ → Λ);
   for j := 1 to |W| do D[0,j] := D[0,j-1] + γ_v (Λ → W⟨j⟩);
   for i := 1 to |V| do
     for j := 1 to |W| do
       m_1 := D[i-1,j-1] + γ_v (V⟨i⟩ → W⟨j⟩);
       m_2 := D[i-1,j] + γ_v (V⟨i⟩ → Λ);
       m_3 := D[i,j-1] + γ_v (Λ → W⟨j⟩);
       m := ∞;
       for k := 1 to |V| do
         if (i-k) ≥ 0 then
             m := min{m, D[i-k,j-1] + γ_v (V⟨i-k : i⟩ → W⟨j⟩)};
       endfor
       for k := 1 to |W| do
         if (j-k) ≥ 0 then
             m := min{m, D[i-1,j-k] + γ_v (V⟨i⟩ → W⟨j-k : j⟩)};
       endfor
       D[i,j] := min(m, m_1, m_2, m_3);
     endfor
   endfor
return D[i,j]
```

The complexity of the string edit distance algorithm is proportional to the length of both strings, $\mathcal{O}(|A|\,|B|)$. In the case of the *vectorEditDistance*, it has three nested loops and the complexity is $\mathcal{O}(|V|\,|W| \max\{|V|\,|W|\}\,\mathcal{O}(\gamma_v))$, but if we consider that a vector can be replaced by a fixed constant number of vectors and the function γ_v defined bellow, the complexity is reduced to $\mathcal{O}(|V|\,|W|)$. Thus, the cost is similar to that of the string edit distance.

To compute the difference between one vector and a set of N vectors, used in *vectorEditDistance*, the following function is utilised:

```
Function γ_v (V⟨k⟩ → W⟨i : j⟩)
   float auxN := 0, aunAng := 0, r := 0, rSubs := 0, rLeft := 0
   auxN := V_N⟨k⟩  //Norm single vector
   auxAng := V_α⟨k⟩  //Angle single vector
   for l := i to j do
      if auxN ≥ 0 then //Left norm single vector
         rSubs := rSubs + auxN * closest(auxAng, W_α⟨l⟩)
         auxAng := W_α⟨l⟩
      endif
      auxN := auxN − W_N⟨l⟩
   endfor
   if auxN ≥ 0 then //Left norm single vector
      rLeft := auxN * kInsertion
   else //Norms W vectors > V
      rLeft := −auxN * kDeletion
   endif
return rSubs + rLeft
```

where $\text{closest}(angle1, angle2)$ returns the smallest angle between both parameters, resulting a value in $[0, \pi]$. The $\text{kInsertion} = \text{kDeletion} = \pi/2$ is the maximum possible difference between two angles.

The functions $\gamma_v (V\langle i.j\rangle \to W\langle k\rangle)$ and $\gamma_v (V\langle i\rangle \to W\langle j\rangle)$ are similar. In the first case, the parameters change the order and in the second case, both parameters are unitary vectors.

The insertion and deletion functions are defined as $\gamma_v (\Lambda \to W\langle j\rangle) = |W\langle j\rangle| * \text{kInsertion}$ and $\gamma_v (V\langle i\rangle \to \Lambda) = |V\langle j\rangle| * \text{kDeletion}$.

4 Experiments

Three algorithms have been compared based on different contour descriptions:

1. Classical Freeman chain code extracted from the object contour in the binary image. Any point reduction method is applied.
2. The ordered vector set extracted from the dominant points computed by the algorithm described in [7], that will be referred as non collinear dominant points (NCDP).
3. The new structure based in the ordered vector set extracted from dominant points described in [6]. In this article, $1 -$ curvature and $k -$ curvature algorithms are defined in order to select dominant points using these measures. The authors showed that the obtained dominant points were similar for both curvature measures, so we utilised the faster one: $1 -$ curvature.

In the preliminary trials tested, the algorithm $1-$curvature obtained lower error rates than NCDP. Thus, the k parameter in the *vectorEditDistance* function was tuned when applied to $1-$curvature. The k parameter is the maximum number of continuous vectors that was set to $k = 1$.

A classification task using the NIST SPECIAL DATABASE 3 of the National Institute of Standards and Technology was performed using the different contour descriptions enumerated above to represent the characters. Only the 26 uppercase handwritten characters were used. The increasing-size training samples for the experiments were built by taking 500 writers and selecting the samples randomly. The nearest neighbour (NN) technique was used for perform classification.

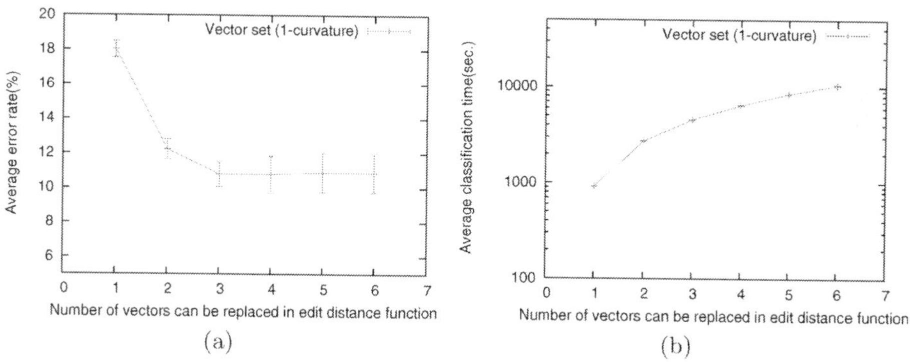

Fig. 3. Results for NN classification of characters obtained with ordered vector set ($1-$curvature), different training set (200 examples per class) and test set (50 samples per class and 26 character classes) as a function of different number of vectors that can be replaced in a substitution operation in the vector edit distance: (a) average error rate $\pm$ standard deviation; (b) average classification time

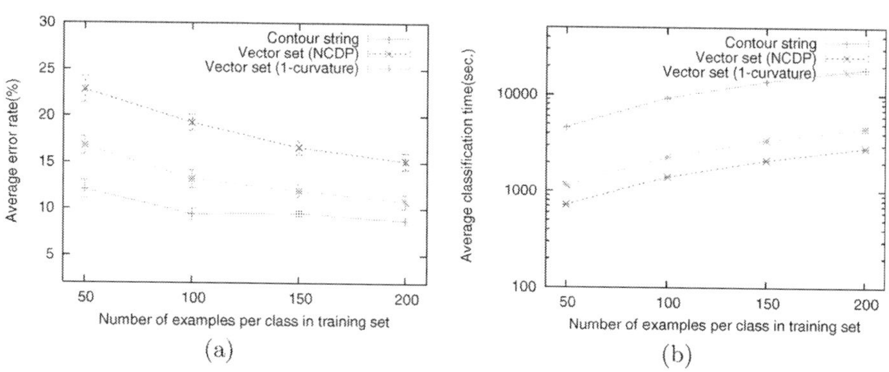

Fig. 4. Results for NN classification of characters obtained with different contour representations as a function of different training example sizes: (a) average error rate $\pm$ standard deviation; (b) average classification time

Figure 3 shows the comparison between the error rate in the vector classification task evaluated for different sizes, k (*vectorEditDistance*). This experiment shows that the error rate decreases linearly when the k grows to a limit. If k grows the number of computations increases as well the classification time. In this case, we found the lowest error rate with the lowest k, so the optimal parameter value was $k = 3$.

The figure 4 shows the classification error rate and the time used in the classification of 50 examples per class as a function of different training set.

In all cases the use of Freeman chain codes generates a lower error rate (less than 9%) in recognition than using ordered vector sets, although the classification time is much higher. Thus, the ordered vector set description based on dominant points $1-$curvature [6] is a good trade-off choice. It obtains also a low error rate (less than 11%) and it is 10 times faster than using the Freeman chain codes.

5 Conclusions and Future Work

The computation of the edit distance between ordered vector sets that represent the contour of an object in a binary image (based on dominant point computation using 1-curvature) is one order of magnitude faster than using Freeman chain codes, and it has just a slightly higher error rate when using it for recognition. The edit distance defined in this paper to compare ordered vector sets has similar complexity than that of string edit distance. Since the size of the ordered vector set is significatively lower than that of strings for representing the same object, the time needed for computing the distance needed for classification is much lower.

As it can be seen in the results section the error rate using ordered vector set based on dominant points is similar to that of using the Freeman chain code.

As future work we planned to use some special labels for each vector to describe the curved shape of the original image in order to obtain a better description of the binary image contour and decrease the error rate in this classification task. Another possible line of future work is to apply algorithms such as [11] in order to optimise the cost functions for the ordered vector set edit distance.

References

1. Rico-Juan, J.R., Micó, L.: Comparison of AESA and LAESA search algorithms using string and tree edit distances. Pattern Recognition Letters **24(9)** (2003) 1427–1436
2. Freeman, H.: On the encoding of arbitrary geometric configurations. IRE Transactions on Electronic Computer **10** (1961) 260–268
3. Rico-Juan, J.R., Micó, L.: Some results about the use of tree/string edit distances in a nearest neighbour classification task. In Goos, G., Hartmanis, J., van Leeuwen, J., eds.: Pattern Recognition and Image Analysis. Number 2652 in Lecture Notes in Computer Science, Puerto Andratx, Mallorca, Spain, Springer (2003) 821–828
4. Wagner, R.A., Fischer, M.J.: The string-to-string correction problem. J. ACM **21** (1974) 168–173

5. Zhang, K., Shasha, D.: Simple fast algorithms for the editing distance between trees and related problems. SIAM Journal of Computing **18** (1989) 1245–1262
6. Teh, C.H., Chin, R.T.: On the detection of dominant points on digital curves. IEEE Trans. Pattern Anal. Mach. Intell. **11** (1989) 859–872
7. Iñesta, J.M., Buendía, M., Sarti, M.A.: Reliable polygonal approximations of imaged read objects though dominant point detection. Pattern Recognition **31** (1998) 685–697
8. Rico-Juan, J.R., Calera-Rubio, J.: Evaluation of handwritten character recognizers using tree-edit-distance and fast nearest neighbour search. In Iñesta, J.M., Micó, L., eds.: Pattern Recognition in Information Systems, Alicante (Spain), ICEIS PRESS (2002) 326–335
9. Serra, J.: Image Analysis and mathematical morphology. Academic Press (1982)
10. Carrasco, R.C., Forcada, M.L.: A note on the Nagendraprasad-Wang-Gupta thinning algorithm. Pattern Recognition Letters **16** (1995) 539–541
11. Ristad, E., Yianilos, P.: Learning string-edit distance. IEEE Transactions on Pattern Analysis and Machine Intelligence **20** (1998) 522–532

Shape Retrieval Using Normalized Fourier Descriptors Based Signatures and Cyclic Dynamic Time Warping

Andrés Marzal, Vicente Palazón, and Guillermo Peris*

Dept. Llenguatges i Sistemes Informàtics,
Universitat Jaume I de Castelló, Spain
{amarzal, palazon, peris}@lsi.uji.es

Abstract. The WARP system defines a dissimilarity measure between shapes described by their contours which is based on Dynamic Time Warping of Fourier Descriptors based signatures. These signatures are invariant to translation, scaling, rotation, and selection of the starting point. However, identical shapes present ambiguous signatures and similar shapes may yield significantly different signatures. Differences affect rotation and starting-point of the signatures, which may lead to poor performance in classification and shape retrieval tasks. We propose a different signature method to provide true rotation invariance and a Cyclic Dynamic Time Warping dissimilarity measure to achieve true starting-point invariance in shape comparisons.

1 Introduction

Content-based image retrieval is being increasingly demanded in many applications: digital libraries, broadcast media selection, multimedia editing, etc. [7]. In order to be effective in classification an retrieval tasks, shape descriptions, combined with (dis)similarity measures, must be robust to noise and invariant to transformations such as translation, scaling, and rotation.

Recently, Bartolini *et al.* have proposed a new Discrete Fourier Transform based approach to represent and compare shapes: the WARP System [1]. The normalized, low-frequency Fourier Descriptors (FDs) (including phase information) are used to reconstruct the original shape. We will refer to the reconstructed shape with the term *signature*. The signature is a good approximation to the original shape and contains a small number of points. Moreover, it is a sequence of complex values with a canonical starting point, which makes it amenable to be compared to other signatures by means of standard sequence comparison methods. The WARP system uses Dynamic Time Warping (DTW) in order to compare sequences [6]. In [1], some experiments on the SQUID Demo and MPEG-7 CE-Shape-1 databases show that the WARP system outperforms other indexable curvature-based shape descriptors and FDs-based signatures that do not take into account phase information.

* This work has been supported by the Spanish *Ministerio de Ciencia y Tecnología* and FEDER under grant TIC2002-02684.

D.-Y. Yeung et al. (Eds.): SSPR&SPR 2006, LNCS 4109, pp. 208–216, 2006.

The WARP system presents two drawbacks: (1) reconstructing the shape contour from normalized FDs produces signatures with an ambiguity modulo a rotation of π radians [2] (which also affects the starting-point selection); and (2) perceptually similar shapes may have significantly different signatures (in orientation and starting point selection), which leads to poor performance of DTW-based comparisons. In order to solve these problems, we propose (1) a different encoding of the shape contour (which is based on the derivative of the reconstructed contour), and (2) to compare derivative-based signatures by means of a Cyclic Dynamic Time Warping dissimilarity measure.

The paper is organized as follows: In Sect. 2, some notation is introduced. In Sect. 3, the WARP system is reviewed and the observed drawbacks are pointed out. A simple improvement to the WARP system which provides better rotation invariance and a Cyclic Dynamic Time Warping procedure that provides starting point invariance when comparing signatures are presented in Sect. 4. In Sect. 5, experimental results on image retrievals tasks for the SQUID Demo and MPEG-7 CE-Shape-1 databases compare the different methods. Finally, some conclusions are presented in Sect. 6.

2 Notation

Shapes can be coded as a cyclic sequence of points along the contour. A cyclic sequence can be viewed as the set of sequences obtained by cyclically shifting a representative sequence (i.e., by choosing different starting points).

Let $\mathbb{C}^*$ be the closure of $\mathbb{C}$, the field of complex numbers, under a concatenation operator and let $a = a_0 a_1 \dots a_{m-1} \in \mathbb{C}^*$ be a sequence of m points (complex values) describing a (counter-clockwise) contour[1]. A *cyclic shift* σ of a is a mapping $\sigma : \mathbb{C}^* \to \mathbb{C}^*$ defined as $\sigma(a_0 a_1 \dots a_{m-1}) = a_1 a_2 \dots a_{m-1} a_0$. Let σ^k denote the composition of k cyclic shifts and let σ^0 denote the identity. Two sequences a and $\hat{a}$ are cyclically equivalent if $a = \sigma^k(\hat{a})$ for some integer k. A cyclic sequence is an equivalence class $[a] = \{\sigma^k(a) : 0 \leq k < m\}$. Any of its members is a representative (non-cyclic) sequence.

3 The WARP System

Dynamic Time Warping (DTW) of sequences of 2D points describing shapes is sensitive to changes in position, scale, orientation of contours and to selection of their starting points. Therefore, DTW does not lead to good dissimilarity measures when the original, cyclic sequences describing shapes are used. The WARP images retrieval system [1] is based on the DTW-based comparison of compact, normalized signatures of shapes. These signatures are obtained by applying the Inverse Discrete Fourier Transform (IDFT) to the shape's Fourier Descriptors (FDs) after a normalization procedure.

[1] Note that $a_0 a_1 \dots a_{m-1}$ does not denote the product of m complex numbers, but their concatenation to form a sequence.

The Discrete Fourier Transform (DFT) of a sequence $a = a_0a_1 \ldots a_{m-1}$ is an ordered set of complex values $A = (A_{-m/2}, \ldots, A_{-1}, A_0, A_1, \ldots, A_{m/2-1})$ where $A_i = \sum_{0 \leq k < m} a_k e^{-j2\pi ki/m}$ and $j = \sqrt{-1}$. These coefficients are the FDs and model the contour of a shape as a composition of ellipses revolving at different frequencies [2]. The main ellipse is centered at the contour centroid, A_0, and translation of the contour only affects this descriptor. Scaling by a factor α scales the FDs by α. Rotating the shape by an angle θ yields a phase shift of θ in the FDs. The cyclic shift $\sigma^k(a)$ produces a linear phase shift of $2\pi ki/m$ to A_i.

The A_0 descriptor can be set to 0 in order to provide invariance to translation. Let us consider the polar representation of the descriptors: $A_i = r_i e^{j\theta_i}$. The value of A_1 is the length of the main axis of the basic (low frequency) ellipse; therefore, dividing all the descriptors by r_1 provides invariance to scale. Invariance to rotation can be obtained by substracting $(\theta_{-1} + \theta_1)/2$ (the orientation of the basic ellipse) to each θ_i. Invariance with respect to the starting point can be achieved by adding $i(\theta_{-1} - \theta_1)/2$ to each θ_i. In principle, the shape can be reconstructed to a canonical form (invariant to translation, scaling, rotation, and starting point) by computing the IDFT. Noise in the contour can be reduced by taking only $M \ll m$ low frequency components before computing the IDFT. The WARP system only uses the $M = 32$ lower frequency FDs before computing the IDFT. The resulting shape is a more compact, canonical representation of the original one: a *signature*.

Let $a = a_0a_1 \ldots a_{m-1}$ and $b = b_0b_1 \ldots, b_{n-1}$ be two sequences. An *alignment* between a and b is a sequence of pairs (i_0, j_0), (i_1, j_1), $\ldots$, (i_{k-1}, j_{k-1}) such that (a) $0 \leq i_\ell < m$ and $0 \leq j_\ell < n$ for $0 \leq \ell < k$; (b) $0 \leq i_{\ell+1} - i_\ell \leq 1$ and $0 \leq j_{\ell+1} - j_\ell \leq 1$ for $0 \leq \ell < k-1$; and (c) $(i_\ell, j_\ell) \neq (i_{\ell+1}, j_{\ell+1})$ for $0 \leq \ell < k-1$. The pair (i_ℓ, j_ℓ) is said to *align* a_{i_ℓ} with b_{j_ℓ}. The weight of an alignment is defined as $\sum_{0 \leq \ell < k} \delta(a_{i_\ell}, b_{j_\ell})$, where δ is a "local dissimilarity" function that the WARP system defines as $\delta(a_i, b_j) = |a_i - b_j|^2$. An optimal alignment is an alignment of minimum weight.

The DTW dissimilarity measure $D(a, b)$ is defined as $\sqrt{d(m-1, n-1)}$, where $d(m-1, n-1)$ is the weight of an optimal alignment and is defined as[2]

$$d(i,j) = \begin{cases} \delta(a_0, b_0), & \text{if } i = j = 0; \\ d(i-1, j) + \delta(a_i, b_0), & \text{if } i > 0 \text{ and } j = 0; \\ d(i, j-1) + \delta(a_0, b_j), & \text{if } i = 0 \text{ and } j > 0; \\ \min \left\{ \begin{array}{l} d(i-1, j-1), \\ d(i-1, j), \\ d(i, j-1) \end{array} \right\} + \delta(a_i, b_j), & \text{if } i > 0 \text{ and } j > 0. \end{cases} \tag{1}$$

This equation can be solved by Dynamic Programming in $O(mn)$ time: the problem is reduced to the computation of an optimal path in the *warping graph*, a weighted, acyclic graph with $O(mn)$ arcs. Fig. 1 depicts the complete WARP

[2] The recursive equation in [1, page 144] contains a typo: the square root should be applied only to $d(m-1, n-1)$, and not to all $d(i, j)$.

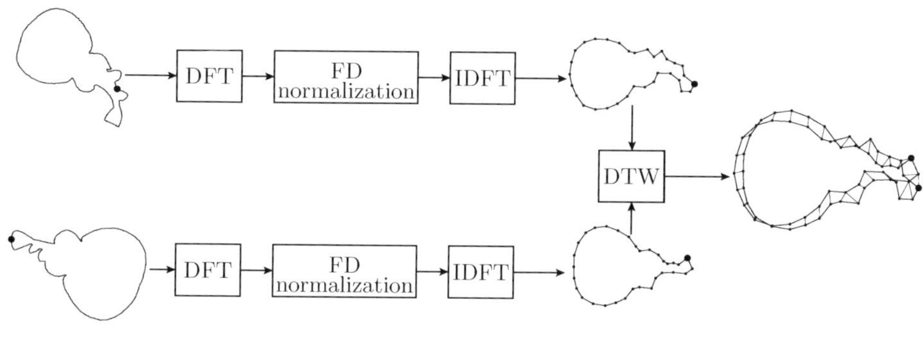

Fig. 1. The WARP system: shapes are compared by means of DTW on the IDFT of normalized FDs

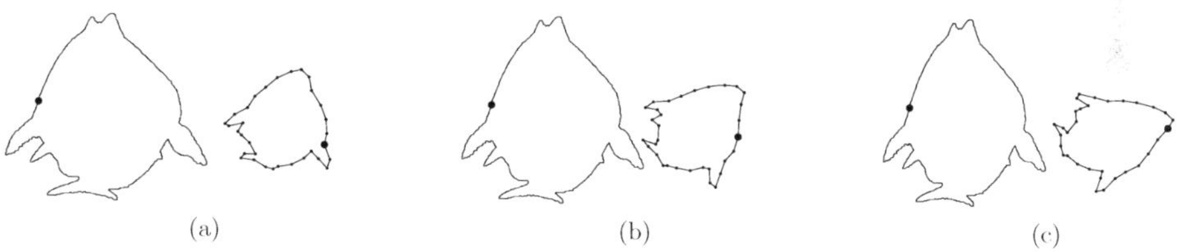

Fig. 2. (a) Original shape and its normalized version. (b) The same shape compressed in the X axis and its normalized version, which has a different rotation and starting point. (c) A bit more compressed shape and its normalized version, which is also different.

comparison procedure. The DTW computation in the WARP system is $O(M^2)$, where $M \ll m, n$, since comparisons are performed on signatures.

3.1 Drawbacks of the WARP System

It should be noted that subtracting $(\theta_{-1} + \theta_1)/2$ to the orientation of all FDs only provides rotation invariance modulo π radians [2]. The WARP system does not consider this ambiguity. Anyway, let us consider that the rotation ambiguity is not present. The basic idea of the WARP system is that, after normalization, all shapes have a canonical version with a "standard" centroid, scale, rotation, and starting point and thus, can be compared by means of the DTW dissimilarity measure. But this is a flawed reasoning: invariance is only achieved for different translations, scalings, rotations, and starting points *of the same shape.* Different shapes (even similar ones) may differ substantially in their normalized orientation and starting point. Fig. 2 shows three perceptually similar figures (in fact, the second and third ones have been obtained from the first one by slightly compressing the horizontal axis) whose normalized version are significantly different in terms of *orientation and starting point.* This problem appears frequently in shapes whose basic ellipse is almost a circle. Invariance to rotation and starting point election should be provided by a different method.

In the next section, we present an alternative signature which provides better rotation invariance for similar shapes and a dissimilarity measure which is not affected by the starting point of the signature.

4 Cyclic Dynamic Time Warping: A Rotation and Starting-Point Invariance Comparison

We have seen that the signature of similar shapes may present different orientations (Fig. 2). True rotation invariance can be obtained by taking the derivative of the normalized shape, i.e., replacing a'_i by $a'_i - a'_{(i-1) \bmod M}$. We need this derivative signature to use the dissimilarity measure that is detailed next.

When two signatures have "equivalent" starting points, DTW provides a good dissimilarity measure. However, we have seen that similar shapes can present very different starting points. It is useful to consider the problem under the framework of cyclic alignments, i.e., alignments between cyclic sequences.

Let $[a] = [a_0 a_1 \ldots a_{m-1}]$ and $[b] = [b_0 b_1 \ldots b_{n-1}]$ be two cyclic sequences. A *cyclic alignment* between $[a]$ and $[b]$ is a sequence of pairs (i_0, j_0), (i_1, j_1), $\ldots$, (i_{k-1}, j_{k-1}) such that, for $0 \leq \ell < k$, (a) $0 \leq i_\ell < m$ and $0 \leq j_\ell < n$; (b) $0 \leq i_{(\ell+1) \bmod m} - i_\ell \leq 1$ and $0 \leq j_{(\ell+1) \bmod n} - j_\ell \leq 1$; and (c) $(i_\ell, j_\ell) \neq (i_{(\ell+1) \bmod m}, j_{(\ell+1) \bmod n})$. The *weight of a cyclic alignment* (i_0, j_0), (i_1, j_1), $\ldots$, (i_{k-1}, j_{k-1}) is defined as $\sum_{0 \leq \ell < k} \delta(a_{i_\ell}, b_{j_\ell})$, where δ is the local dissimilarity measure. An optimal cyclic alignment is a cyclic alignment of minimum weight.

The Cyclic Dynamic Time Warping (CDTW) dissimilarity measure $\hat{D}([a], [b])$ is defined as the square root of the weight of the optimal cyclic alignment between a and b. First, we are going to show that the optimal cyclic alignment can be defined in terms of alignments between non-cyclic sequences, i.e., in terms of $D(\cdot, \cdot)$; then, we will present an efficient procedure to compute it.

Lemma 1. *If $m, n > 1$ and (i_0, j_0), (i_1, j_1), $\ldots$, (i_{k-1}, j_{k-1}) is an optimal alignment between two sequences $a_0 a_1 \ldots a_{m-1}$ and $b_0 b_1 \ldots b_{n-1}$, there is at least one ℓ such that $i_\ell \neq i_{(\ell+1) \bmod m}$ and $j_\ell \neq j_{(\ell+1) \bmod n}$.*

Proof: Any alignment including (i_ℓ, j_ℓ), $(i_\ell + 1, j_\ell)$, and $(i_\ell + 1, j_\ell + 1)$ can be "improved" by removing $(i_\ell + 1, j_\ell)$, since $\delta(a_{i_\ell+1}, b_{j_\ell}) \geq 0$. Analogously, any alignment including (i_ℓ, j_ℓ), $(i_\ell, j_\ell + 1)$, and $(i_\ell + 1, j_\ell + 1)$ can be "improved" by removing $(i_\ell, j_\ell + 1)$. □

Lemma 2. *The CDTW dissimilarity between $[a] = [a_0 a_1 \ldots a_{m-1}]$ and $[b] = [b_0 b_1 \ldots b_{n-1}]$, $\hat{D}([a], [b])$, can be computed as $\min_{0 \leq k < m} \min_{0 \leq \ell < n} D(\sigma^k(a), \sigma^\ell(b))$.*

Proof: Trivial when $m = 1$ or $n = 1$. Let us consider that $m, n > 1$ and let (i_0, j_0), (i_1, j_1), $\ldots$, (i_{k-1}, j_{k-1}) be an optimal alignment. Let ℓ be an index such that $i_\ell \neq i_{(\ell+1) \bmod m}$ and $j_\ell \neq j_{(\ell+1) \bmod n}$ (by Lemma 1). The weight of this cyclic alignment is $D(\sigma^{(i_\ell+1) \bmod m}(a), \sigma^{(j_\ell+1) \bmod n}(b))$, which is considered by the double minimization. □

According to Lemma 2, the value of $\hat{D}([a],[b])$ can be trivially computed in $O(m^2n^2)$ time by solving mn recurrences like equation (1). Maes showed in [4] that the Cyclic Edit Distance (CED), a related dissimilarity measure, can be computed in $O(m^2n)$ time by performing cyclic shifts only on one of the sequences. This observation finally led to a $O(mn \lg m)$ time algorithm. Is it possible to perform cyclic shifts on only one of the sequences when computing the CDTW? The answer is no: in general, $\hat{D}([a],[b])$ is neither $\min_{0 \le k < m} D(\sigma^k(a), b)$ nor $\min_{0 \le k < n} D(a, \sigma^k(b))$, as the following counter-example shows: let z and w be two complex numbers such that $\delta(z,w) = 1$; the value of $\hat{D}([zwz],[wzw])$ is 0, since $D(zzw, zww) = 0$, but $D(zwz, wzw) = 3$ and $D(wzz, wzw) = D(zzw, wzw) = D(zwz, zww) = D(zwz, wwz) = 1$. *Therefore, an equivalent of Maes' algorithm for the CED computation cannot be directly applied to CDTW dissimilarity computation.*

Theorem 1. *The CDTW dissimilarity between cyclic sequences $[a]$ and $[b]$ can be computed as* $\hat{D}([a],[b]) = \min_{0 \le k < m} \left(\min(D(\sigma^k(a), b), D(\sigma^k(a)a_k, b))\right)$.

Proof: Each alignment induces a segmentation on a and a segmentation on b. All the values in a segment are aligned with the same value of the other cyclic sequence (Lemma 1). There is a problem when $b_{n-p-1}, b_{n-p}, \ldots b_{n-1}$ and $b_0, b_1, \ldots b_q$, for some p, $q \ge 0$, should belong to the same segment of b. In that case, the optimal path cannot be obtained by simply shifting a, since b_{n-1} must be aligned with the last value of $\sigma^k(a)$ and b_0 must be aligned with its first value, i.e., they cannot belong to the same segment. The sequence $\sigma^k(a)a_k$ allows to align $b_{n-p}b_{n-p+1} \ldots b_n$ and $b_0 b_1 \ldots b_q$ with the first value of $\sigma^k(a)$, since a_k also appears at the end of $\sigma^k(a)a_k$. □

The value of $D(\sigma^k(a), b)$ and $D(\sigma^k(a)a_k, b)$, for each k, can be obtained by computing shortest paths in an *extended warping graph* similar to the extended

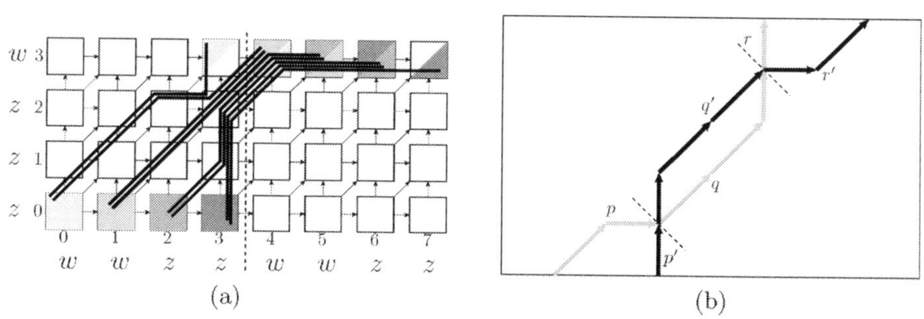

Fig. 3. (a) Extended warping graph for $a = wwzz$ and $b = zzzw$, where z and w are complex numbers such that $\delta(z,w) = 1$. Arcs ending at node (i,j) are weighted $\delta(a_i, b_j)$. The optimal alignment for $[a]$ and $[b]$ is the minimum weight path starting from any colored node in the lower row and ending at a node containing the same color in the upper row (all path candidates are shown with thick lines). (b) Optimal crossing paths can be avoided: if the weight of the subpath q is greater than the weight of the subpath q', the black path can be improved by traversing q' instead of q.

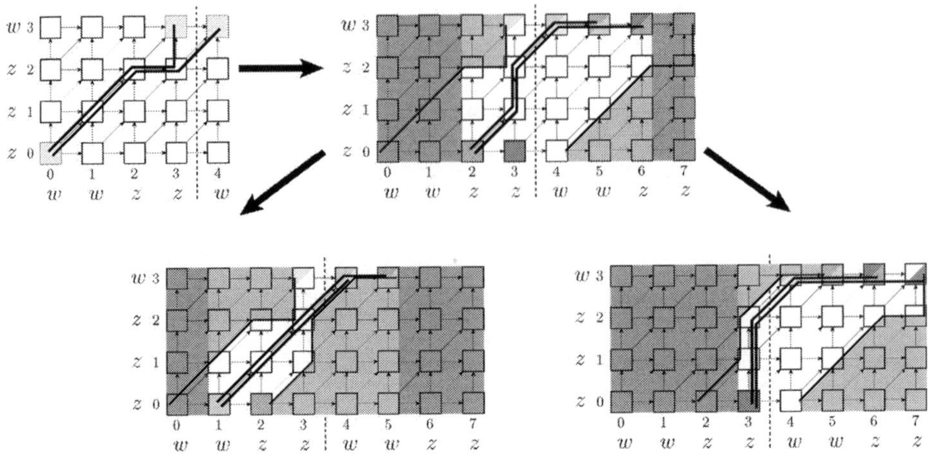

Fig. 4. Divide-and-Conquer procedure to compute the CDTW dissimilarity between the sequences of Fig. 3. First, the optimal alignment (path) between a and b and between $\sigma^0(a)a_0$ and b is computed. The first optimal path is used as a left and right frontier in the extended graph: only the white region must be explored to compute the optimal alignment between $\sigma^2(a)$ and b and between $\sigma^2(a)a_2$ and b. This idea is applied recursively to the computation of the other optimal alignments, but using also the optimal alignment between $\sigma^2(a)$ and b as a new left or right frontier.

edit graph defined by Maes [4] (see Fig. 3 (a)). Since the non-crossing property of edit paths also holds for alignment paths (see Fig. 3 (b)), the Divide-and-Conquer approach proposed by Maes can be applied to CDTW. The reader is addressed to [4] to obtain a complete description of the Divide-and-Conquer procedure, which is depicted in Fig. 4. It should be taken into account that, unlike in Maes' algorithm, the optimal path starting at $(k, 0)$ can finish either at node $(k+m-1, n-1)$ or $(k+m, n-1)$.

When applied to signatures, the running time of the algorithm is $O(M^2 \log M)$: each recursive step divides the search space in two halves and all recursive operations at the same recursion level require total $O(M^2)$ time.

5 Experiments

In [1], the WARP system was tested on a labeled version of the SQUID Demo database and the MPEG-7 Core Experiment CE-Shape-1 (part B). We have performed comparative experiments with the same test sets.

The SQUID Demo database consists of 1100 contours of marine species and is used as a demonstration application of the Shape Queries Using Image Databases system [5]. The original database does not divide the contours into classes. Bartolini *et al.* manually classified 252 images into 10 semantic categories[3]. They

[3] Seahorses (5 images), seamoths (6), sharks (58), soles (52), tonguefishes (19), crustaceans (4), eels (26), u-eels (25), pipefishes (16), and rays (41).

conducted some precision (P) versus recall (R) shape retrieval experiments with 30 query images from the 10 semantic categories. For each query, images in the same category were considered relevant and all the others were considered irrelevant. Since we do not know which query images were used, we have run queries on the 252 labeled shapes.

Fig. 5 shows the precision/recall graph for 3 retrieval procedures: (i) WARP: the standard WARP system; (ii) Derivative: derivative of the reconstructed contour as a shape signature and DTW-based comparison; (iii) CDTW: derivative of the reconstructed contour and comparison by means of the Cyclic Dynamic Time Warping dissimilarity. It can be seen that the two methods proposed in this work improve the WARP results. The signature based on the derivative provides a better precision/recall curve, thus confirming that the WARP system is sensitive to variation of orientation in the signatures of similar (but not identical) shapes. There is also a significant difference between CDTW-comparison and the other methods.

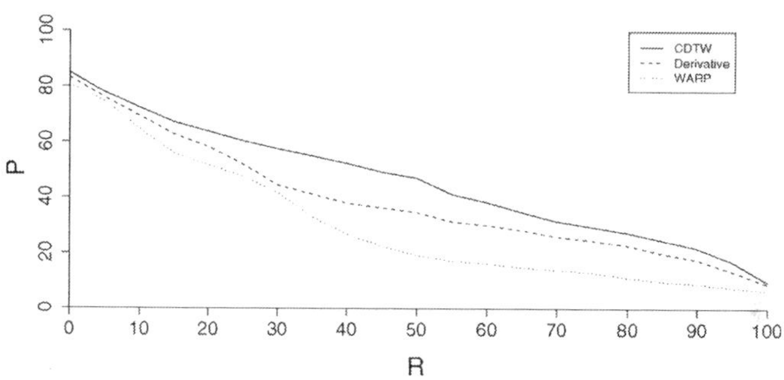

Fig. 5. Precision/Recall results on the SQUID Demo database

In [1], the WARP system was also compared to a Curvature Scale Space (CSS) based image retrieval on the same MPEG-7 experiment presented in [3]. A CSS-based query system obtained an average precision of 37.72% (the maximum precision attainable in that experiment is 50%) and the WARP system obtained a 29.25% average precision. Bartolini *et al.* explain in [1] that the CSS system is an approximate query processing algorithm that can easily lead to false dismissals (filtering out best-matching images) by discarding shapes with an aspect ratio greater than a user threshold. Other techniques with similar or slightly better results are not suitable for efficient indexing and, thus, can only be used in small-size databases. Using the derivative-based signature, the average precision is 31.29%. The precision raises to 34.17% when the CDTW is used.

6 Conclusions

In this work, we have critically studied the WARP system, detected some drawbacks, and presented several ways to improve its precision/recall behavior on shape-based image retrieval tasks: (a) using the original signatures derivative, (b) using the signature derivative with a CDTW comparison. Proposal (a) provides better results than the WARP system and proposal (b) offers the best precision/recall.

The CDTW dissimilarity has been defined and an algorithm to compute it in $O(M^2 \log M)$ for two signatures of length M has been presented. We have shown that the Cyclic Edit Distance algorithm presented by Maes cannot be directly extended to CDTW: two conventional DTW dissimilarities must be computed for each cyclic shift of one sequence. Fortunately, one of these dissimilarities can be obtained as a subproduct of the computation of the other.

Acknowledgments

The authors wish to thank S. Abbasi, F. Mokhtarian, and J. Kittler for making the SQUID database publicly available and to I. Bartolini, P. Ciaccia, and M. Patella for providing their labeled version of the SQUID database.

References

1. I. Bartolini, P. Ciaccia, and M. Patella. WARP: Accurate Retrieval of Shapes Using Phase of Fourier Descriptors and Time Warping Distance. *IEEE Transactions on Pattern Analysis and Machine Intelligence*, 27(1):142–147, 2005.
2. A. Folkers and H. Samet. Content-based Image Retrieval Using Fourier Descriptors on a Logo Database. In *Proc of the 16th Int Conf on Pattern Recognition*, pages 521–524, 2002.
3. J. Latecki, R. Lakämper, and U. Eckhardt. Shape descriptors for non-rigid shapes with a single closed contour. In *Proc. of the IEEE Conf. on Computer Vision and Pattern Recognition*, pages 424–429, 2000.
4. M. Maes. On a Cyclic String-to-String Correction Problem. *Information Processing Letters*, 35:73–78, 1990.
5. F. Mokhtarian, J. Kittler, and S. Abbasi. Shape queries using image databases. `http://www.ee.surrey.ac.uk/Research/VSSP/imagedb/demo.html`.
6. D. Sankoff and J. Kruskal, editors. *Time warps, string edits, and macromolecules: the theory and practice of sequence comparison.* Addison-Wesley, Reading, MA, 1983.
7. T. Sikora. The mpeg-7 visual standard for content description – an overview. *IEEE Transactions on Circuits and Systems for Video Technology*, 11(6):696–702, 2001.

Watermarking for 3D CAD Drawings Based on Three Components

Ki-Ryong Kwon[1], Suk-Hwan Lee[2,*], Eung-Joo Lee[3], and Seong-Geun Kwon[4]

[1] Division of Electronic, Computer & Telecommunication Eng., Pukyong National Univ.
krkwon@pknu.ac.kr
[2] Dept. of Information Security, TongMyong University
skylee@tit.ac.kr
[3] Dept. of Information Communication Eng., TongMyong University
ejlee@tit.ac.kr
[4] Mobile Communication Division, SAMSUNG electronics co
seonggeunkwon@hanmail.net

Abstract. Currently there has been much interested in developing the watermarking for 3D graphic data of mesh model or NURBS. However, the watermarking technique based on 3D CAD drawing leaves something to be desired. This paper proposes a watermarking technique for 3D design drawing using the components of Line, 3DFACE and ARC based on vertex that prevent the infringement of copyright from unlawfulness reproductions and distribution. By experimental result, we confirmed the invisibility of embedded watermarks as well as the robustness in geometrical attacks and file conversion to DWG, DXF, DWT and DWS.

Keywords: 3D CAD Drawing, Watermarking.

1 Introduction

With the rapid increase of the multimedia information in information-communication technology, the intellectual property and copyright protection has been made at issue. Generally there are two technologies for the intellectual property and copyright protection; cryptography and watermarking. The cryptography technology cut off the access of the unauthorized person after the multimedia information is encrypted. However, it cannot prevent the unlawful action of an authorized person and cannot solve the problem that some copyright owners assert their ownerships for one content. To solve the problems of the cryptography, there have been much researched in watermarking technology, which is the end-step in information security and protects the copyright of owner by embedding the watermark into the multimedia information.

A lot of research has been carried out to protect the copyright protection of image, video, and audio [1],[2]. Recently, 3D polygonal models, such as VRML, MPEG-4, have become very popular leading the development of 3D watermarking algorithms to protect the copyright of 3D graphic models with the extending technique of the image watermarking. 3D polygonal model are usually represented by a mesh defined by

* Corresponding author.

D.-Y. Yeung et al. (Eds.): SSPR&SPR 2006, LNCS 4109, pp. 217–225, 2006.

the coordinates and connectivity of vertices in a 3D Cartesian coordinate system. Ohbuchi et al. presented the watermarking for 3D polygonal model through geometric and topological modification [3]. Mao et al. presented the watermarking for 3D geometric model through the triangle subdivision [4]. Beneden et al. also presented an algorithm that adds a watermark by modifying the normal distribution of the model geometry [5]. Kanai et al. presented the watermarking for 3D polygons using the multiresolution wavelet decomposition [6].

Many design drawings in the industrial filed have been designed by 3D CAD tools. However, because of the unlawful reproduction of the architectural drawings, the construction industry has been financially damaged. Many researches have not been carried out to protect the copyright of 3D CAD compared with 3D polygonal model. Unlike 3D polygonal model, 3D CAD drawings can be designed by the basic components; LINE, ARC, CIRCLE, and 3DFACE. 3DFACE is similar as polygon of 3D polygon model. We presented 2D CAD watermarking that the watermark is embedded into the position of vertex in two components, LINE and Arc [7]. This algorithm needs to know the original vertices for watermark extracting.

This paper proposed the watermarking for 3D CAD drawing based on 3D vector data, which is a public watermarking using Line, Arc, and 3DFACE components in 3D CAD drawing. The watermark is embedded into the length of Line component in the embedding primitives, the circular radius of Arc component, and the length ratio of two sides in 3DFACE component. Thus, according to the distribution of three components in 3D CAD drawing, the embedding component can be determined. The results of experiment verify that the proposed algorithm is imperceptible and robust against file format conversion, move, scaling, rotation, component cropping, and layer cropping.

2 The Proposed 3D CAD Watermarking

2.1 3D CAD Drawing

The basic system of 3D CAD drawings consists of HEADER, TABLE, BLOCK, ENTITY, EOF sections. The shape of drawings is designed with the basic component of Line, Arc, Circle, and 3DFACE based on vector data in ENTITY section as shown in Fig. 1. The structure of each component is explained clearly in the following sections. 3D CAD drawings can be attacked intentionally or non-intentionally by CAD tools. The general attacks in CAD tools are followed as;

1) File Format Conversion: CAD drawings are easily converted to the format in AutoCAD such as DXF, DWG, DWT, and DWS.
2) Geometrical attack: There are translation, scaling, rotation, and cropping.
3) Layer cutting: The drawings in AutoCAD are composed of several layers. Users are able to cut some layers illegally.

The watermark in CAD drawings must be robust against the above attacks.

2.2 Watermark Embedding

The proposed watermarking embeds the watermark into the components of Line, Arc, and 3DFACE respectively according to the structure of 3D CAD drawings, as shown in Fig. 2. Three components are obtained from ENTITY section of CAD data. The

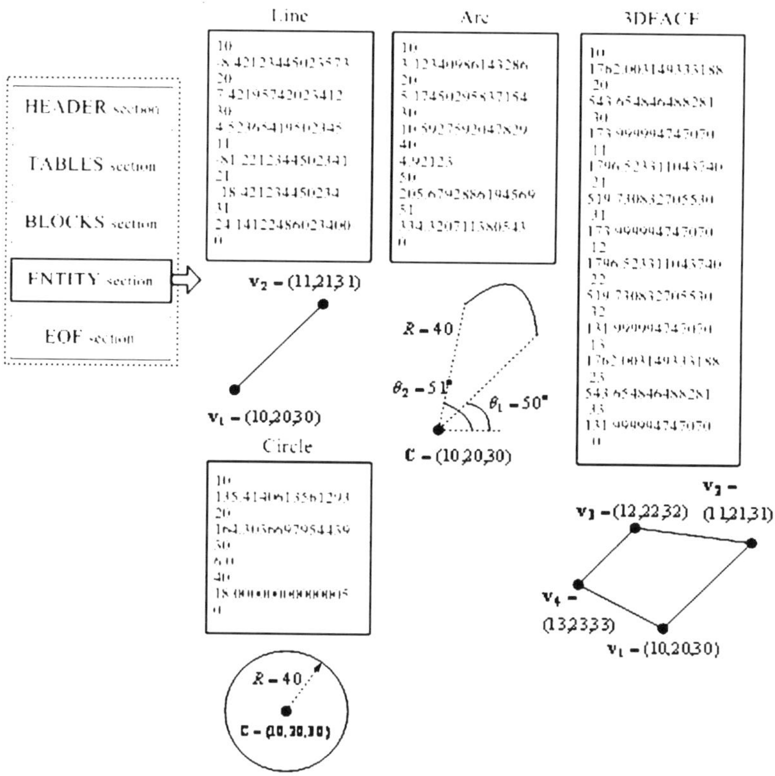

Fig. 1. The Structure of 3D CAD data

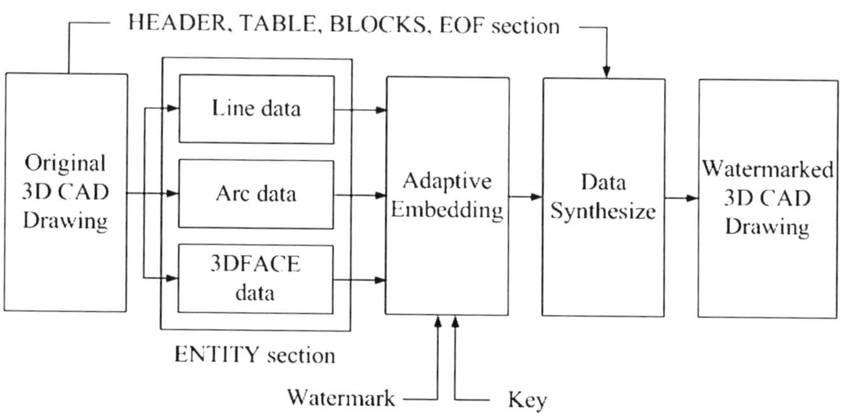

Fig. 2. Proposed 3D CAD watermark embedding system

embedding primitives are the length in Line, the radius of curvature in Arc, and the length ratio of two sides in 3DFACE. They are stored to extract the watermark as the key.

2.1.1 Line Component

A Line component consists of the start point $\mathbf{v}_s = (x_s, y_s, z_s)$ and the end point $\mathbf{v}_e = (x_e, y_e, z_e)$ as shown in Fig. 1. Each point may be connected to the neighborhood points. If a point is varied by the watermark, the neighborhood points will be varied together. The set of embedding primitive that consists of an arbitrary point and the neighborhood points connected to a point is gathered to embed the watermark. A bit of binary watermark is embedded into one point of a center line in an embedding primitive. Thus, a n th watermark bit w_n is embedded into the length l_n of center line as follows; if $w_n = 1$, then $l_n \geq \bar{l}_n$. Otherwise, $l_n < \bar{l}_n$. $\bar{l}_n$ is an average length of the connected Lines. To change the length of line according to the watermark, the coordinate of the point must be changed imperceptibly considering the neighborhood points. Fig. 3 (a) shows a center line $L_n = \{\mathbf{v}_1\mathbf{v}_2\}$ and 3 neighborhood lines in an embedding primitive. $\mathbf{v}_0$ is connected to $\mathbf{v}_1$ and $\mathbf{v}_3, \mathbf{v}_4$ are connected to $\mathbf{v}_2$. The search regions of $\mathbf{v}_1, \mathbf{v}_2$ which can be changed invisibly are determined respectively to be below the coordinate values of the connected points. Thus, the search region of $\mathbf{v} = (x, y, z)$ is $[\mathbf{v} - \Delta\mathbf{v}, \mathbf{v} + \Delta\mathbf{v}]$, $\Delta\mathbf{v} = 0.5 \times \min | t - \mathbf{v}_k^t |_{\mathbf{v}_t \in C(\mathbf{v}), t \in \{x,y,z\}}$ where $C(\mathbf{v})$ represents the points that are connected to $\mathbf{v}$. Two points $\mathbf{v}_1, \mathbf{v}_2$ are changed to $\mathbf{v}_1' = \mathbf{v}_1 \pm \alpha$, $\mathbf{v}_2' = \mathbf{v}_2 \pm \alpha$ alternatively within the search regions until satisfies the condition as follows; $\min_{\theta_{1k} \in \Theta_1} | \theta_{1k} - \theta_{1k}' | < \varepsilon$, $\min_{\theta_{2k} \in \Theta_2} | \theta_{2k} - \theta_{2k}' | < \varepsilon$ where $\theta_{1k} = \cos^{-1}\left(\frac{\overrightarrow{\mathbf{v}_1 \cdot \mathbf{v}_2} \cdot \overrightarrow{\mathbf{v}_1 \cdot \mathbf{v}_{1k}}}{\| \overrightarrow{\mathbf{v}_1 \cdot \mathbf{v}_2} \| \| \overrightarrow{\mathbf{v}_1 \cdot \mathbf{v}_{1k}} \|}\right)$, $\theta_{2k} = \cos^{-1}\left(\frac{\overrightarrow{\mathbf{v}_1 \cdot \mathbf{v}_2} \cdot \overrightarrow{\mathbf{v}_2 \cdot \mathbf{v}_{2k}}}{\| \overrightarrow{\mathbf{v}_1 \cdot \mathbf{v}_2} \| \| \overrightarrow{\mathbf{v}_2 \cdot \mathbf{v}_{2k}} \|}\right)$ and $\mathbf{v}_{1k} \in C(\mathbf{v}_1)$, $\mathbf{v}_{2k} \in C(\mathbf{v}_2)$.

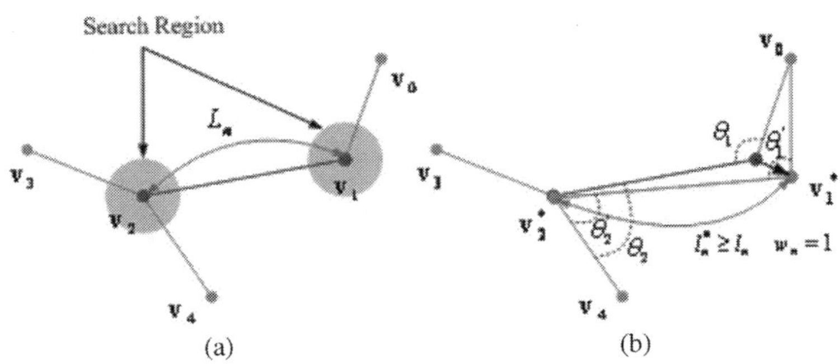

(a) (b)

Fig. 3. (a) An embedding primitive in Line components and (b) embedding the watermark into a embedding primitive

2.1.2 Arc Component

An Arc component consists of two points $\mathbf{v}_0, \mathbf{v}_1$, a circle center point C, a circle radius R, a standard point of angle P, two angles θ_0, θ_1 between points and P as shown in

Fig. 4 (a). The watermark bit is embedded into a circle radius R in the randomly selected Arc component. If a watermark bit w_n is 0, C is moved forward to $\mathbf{m} = (\mathbf{v}_0 + \mathbf{v}_1)/2$. Otherwise, C is moved backward to $\mathbf{m}$. C must be moved within the limit region that the differential ratio of the curvature $\Delta\kappa = \kappa - \kappa' = 1/R - 1/R'$ is below ε. R' is the circle radius of Arc with C' which is moved according to the watermark bit.

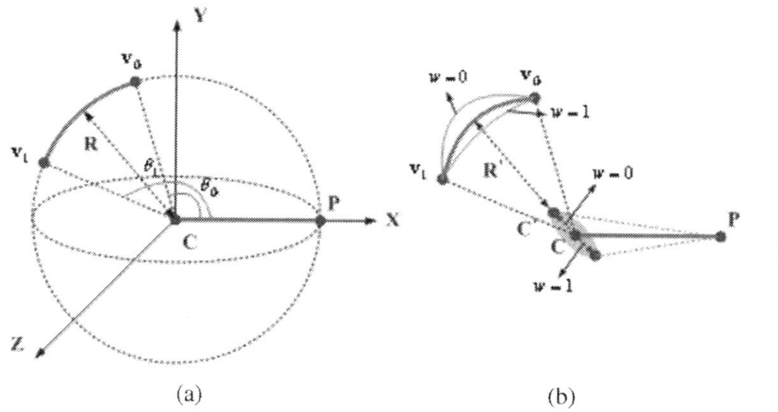

Fig. 4. (a) The structure of an Arc component and (b) the watermark embedding according to the center point **C**

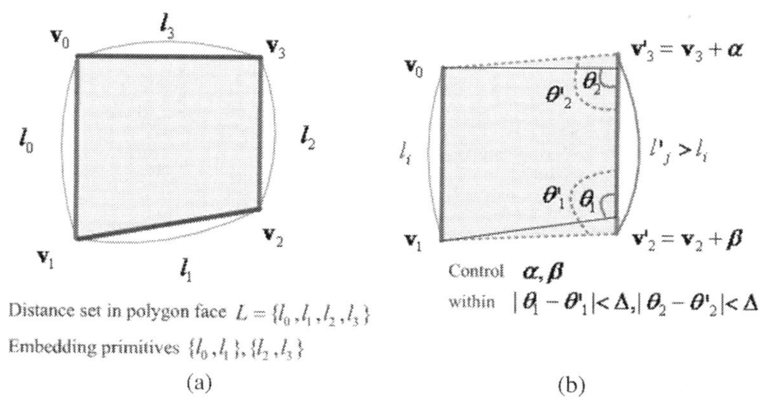

Fig. 5. (a) The structure of 3DFACE component with 4 faces and (b) the embedding using the ratio of two distances when $w = 0$

2.1.3 3DFACE Component

3DFACE represents 3D polygon surface that can be in general use for the surface modeling in 3D CAD drawing. In this paper, 3DFACE components over quadrilateral are used for the watermark embedding. In the randomly selected a 3DFACE component, the set of the distance between two points is obtained; $L = \{l_0, l_1, \cdots l_n\}$,

$l_i = |\mathbf{v}_i\mathbf{v}_{i+1}|$. An arbitrary distance pair $\{l_i, l_j\}$, $l_i, l_j \in L$, $i < j$, $|l_i - l_j| < \Delta$ without the shared point is randomly selected where $l_i = |\mathbf{v}_i\mathbf{v}_{i+1}|$ is a reference distance, $l_j = |\mathbf{v}_j\mathbf{v}_{j+1}|$ is the changeable distance.

The watermark bit w is embedded into the ratio $\alpha = l_i / l_j$ of distance pair; $\alpha \geq 1$ if $w = 1$ and $\alpha < 1$ otherwise. To change the distance ratio according to the watermark bit, two points $\mathbf{v}_j, \mathbf{v}_{j+1}$ in the changeable distance l_j must be changed to be parallel the reference distance l_i within the invisible range as shown in Fig. 4. The invisible ranges of two points are $|\theta_j - \theta_j^{'}| < \Delta$ and $|\theta_{j+1} - \theta_{j+1}^{'}| < \Delta$ respectively. θ_{j+k} and $\theta_{j+k}^{'}$ are the angles of two lines with the original intersection point, $\mathbf{v}_{j+k}$, or the changed point $\mathbf{v}_{j+k}^{'}$, $k = 0,1$; $\theta_{j+k} = \cos^{-1}\left(\frac{\overrightarrow{\mathbf{v}_{j+k}\mathbf{v}_{j+k+1}} \cdot \overrightarrow{\mathbf{v}_{j+k}\mathbf{v}_{j+k-1}}}{\|\overrightarrow{\mathbf{v}_{j+k}\mathbf{v}_{j+k-1}}\| \|\overrightarrow{\mathbf{v}_{j+k}\mathbf{v}_{j+k-1}}\|}\right)$.

2.3 Watermark Extracting

The watermark is extracted from the watermarked drawing using two points of the embedded Line component, a circle radius of the embedded Arc component, and the embedding primitive of the embedded 3DFACE component on the same as the embedding algorithm. But to extract the watermark in the watermarked drawing scaled to an arbitrary factor, the re-scaling process is performed by using $\bar{l}^*$, $\bar{R}^*$, $\bar{A}^*$, which are an average length, circle radius, area of all the embedded Line, Arc, 3DFACE. All components are re-scaled to dilated or shrinked until all ratios $\bar{l}^* / \bar{l}'$, $\bar{R}^* / \bar{R}'$, $\bar{A}^* / \bar{A}'$ are 1. $\bar{l}'$, $\bar{R}'$, $\bar{A}'$ are an average length, circle radius, area of Line, Arc, 3DFACE in the attacked drawing. It takes long time to re-scale closely to original scale factor.

3 Experimental Results

To evaluate the performance of the proposed watermarking, we used 3D CAD drawings designed by AutoCAD 2002 software; Campus, Watch, and Office drawings as shown in Fig. 6. The watermark was used as bit stream generated by a Gaussian random sequence. The length of watermark can be determined by the component distribution of the 3D CAD drawing. There are a number of Line components in Campus1 drawing, Arc components in Watch drawing, 3DFACE components in Campus2, and Line, 3DFACE components in Office as shown in Fig. 6. Among these components in each drawing, we selected 500 components for the embedding primitives. The 3D CAD drawings that are watermarked by the proposed algorithm are shown in Fig. 7. In this figure, a subjective evaluation was used to verify that the watermark was invisible. For the objective evaluation for visibility, we used SNR of each embedding components, which are SNR of points in Line, 3DFACE and radius in Arc. The SNR is defined as

$$SNR = 10\log_{10}\frac{\mathrm{var}(\|\mathbf{v} - \mathbf{M}\|)}{\mathrm{var}(\|\mathbf{v} - \mathbf{v}'\|)} \qquad (1)$$

where var(**v**) is a variance of random variable **v** and **M** is the center points in each components. Table 1 shows that SNRs of the watermarked drawings are about 39.89-42.50dB according to the number of components. These values verify the good quality.

To evaluate the robustness, the watermarked drawings were attacked by file format conversion, RST (translation, scaling, rotation), cropping, and layer cutting, compare with Jang's algorithm [7]. Since Jang's algorithm can apply to 2D CAD using Line

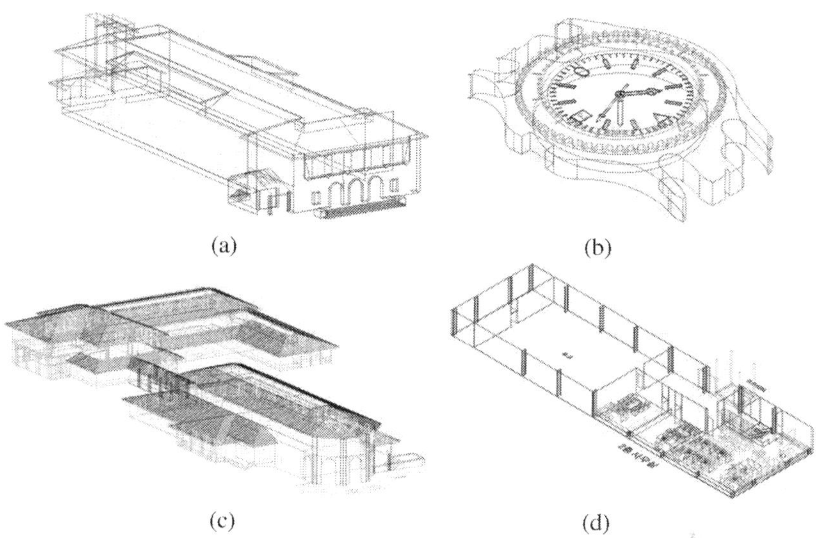

Fig. 6. 3D CAD drawings in AutoCAD; (a) Campus1, (b) Watch, (c) Campus2, and (d) Office

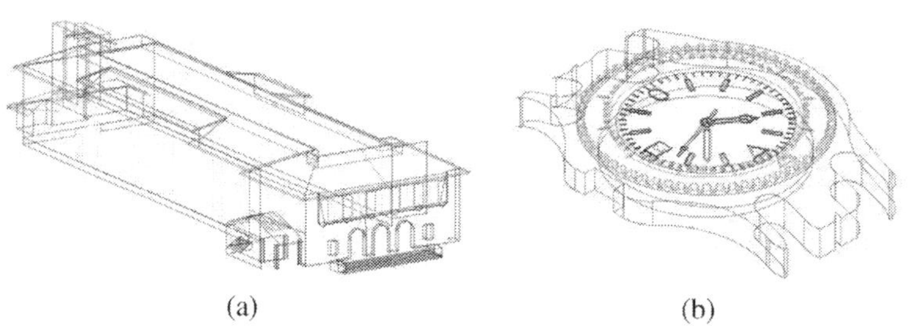

Fig. 7. The watermarked (a) Campus1 and (b) Watch drawings

Table 1. SNR of the watermarked 3D CAD drawings

Test drawing	Campus	Watch	Campus	Floor
Embedding component (Number)	Line (1,130)	Arc (1,770)	3DFACE (551)	Line(8,738) 3DFACE (3,927)
SNR	40.12dB	41.05dB	39.89dB	42.50dB, 41.33dB

and Arc components, our experiment extends to 3D component and embeds the binary watermark for according to the condition of the proposed algorithm. In file format conversion, the watermarked drawings were converted to DXF, DWG, DWT, and DWS by using AutoCAD. But, the watermark can be extracted without bit error in any file format. For geometrical attacks, the watermarked drawings were translated to arbitrary point, dilated or shrinked to arbitrary scaling factor, rotated to arbitrary

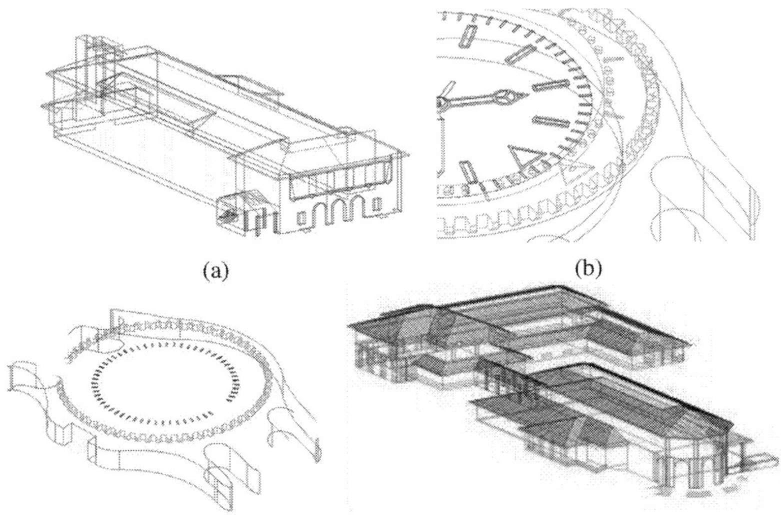

(a) (b) (c) (d)

Fig. 8. (a) A cropped Campus1 to 30% of Line components, (b) a scaled Watch to 3 times, and (c) Watch, (d) Campus2 with layer cutting

Table 2. BER of the extracted watermark

Test drawing	Attack	Jang	Proposed
Campus1 (Line)	Format Conversion (DXF,DWG,DWT,DWS)	-	-
	RST	0.10	-
	30% cropping	0.37	0.13
	Layer cutting	0.30	0.24
Watch (Arc)	Format Conversion (DXF,DWG,DWT,DWS)	-	-
	RST	-	-
	30% cropping	0.29	0.08
	Layer cutting	0.23	0.15
Campus2 (3DFACE)	Format Conversion (DXF,DWG,DWT,DWS)	x	-
	RST	x	-
	30% cropping	x	0.22
	Layer cutting	x	0.28
Floor (Line, 3DFACE)	Format Conversion (DXF,DWG,DWT,DWS)	-	-
	RST	0.08	-
	30% cropping	0.27	0.03
	Layer cutting	0.21	0.07

angle, and cropped to 30% of the embedding components. Since the proposed algorithm embeds the watermark into the length in Line, the radius of curvature in Arc, and the length ratio of two sides in 3DFACE, it has not effect on translation and rotation. In scaled drawing, the watermark has to be extracted after performing re-scaling process. But the conventional algorithm has about 10% bit error in Line component. Table 2 verifies that all watermark bits can be extracted without bit error. But in cropping, BER (bit error rate) is less about 0.18 than the conventional algorithm according to the cropping percentage of components. Furthermore, we cut an arbitrary layer in the watermarked drawings that similar as the cropping of all components in a layer. In this case, the bit error occurred less about 0.06-0.14 than the conventional algorithm, which effects on the number of the embedding components in a layer. Since Floor drawings were embedded the watermark into two components, BER is an average of BERs in two components. The above results verified that the watermark still alive above 78% in any attacks. BER represents the bit error rate of the extracted watermark.

4 Conclusions

This paper presented a watermarking for 3D CAD drawings using Line, Arc, and 3DFACE components. The embedding components can be selected randomly or by the component distribution in drawing. The watermark is embedded into the length in Line component, the radius of curvature in Arc, and the length ratio of two sides in 3DFACE. Experimental results verified that the proposed watermarking has the robustness against Format conversion, RST, cropping, and layer cutting as well as the invisibility in a view of component SNR.

Acknowledgement

This work was supported by Korea Research Foundation Grant (KRF-2004-002-D00289).

References

1. I.J.Cox, J.Kilian, T.Leighton, T.Shamoon (1995) Secure Spread Spectrum Watermarking for Multimedia, NEC Research Institute Tech Rep.: 95-10
2. C. Podilchuk, W.Zeng (1998) Image Adaptive Watermarking Using Visual Models, IEEE Journal on Selected Areas in Communication: Vol. 16, No. 4. 525-539
3. R. Ohbuchi, H. Masuda, M. Aono (1998) Watermarking Three-Dimensional Polygonal Models Through Geometric and Topological Modification, IEEE JSAC: 551-560
4. X. Mao, M. Shiba, A. Imamiya (2001) Watermarking 3D Geometric Models Through Triangle Subdivision, Proc. of Security and Watermarking of Multimedia Contents III,IS&T/SPIES's Electronic Imaging : 253-260
5. O. Benedens (1999) Geometry-Based Watermarking of 3D Models, IEEE CG&A: 46-55
6. S. Kanai, H. Date, T. Kishinami (1998) Digital Watermarking for 3D Polygons using Multiresolution Wavelet Decomposition, Proc. Sixth IFIP WG 5.2 GEO-6: 296-307
7. Bong-Ju Jang, Ki-Ryong Kwon, Kwang-Seok Moon, Young Huh (2003) A New Digital Watermarking for Architectural Design Drawing Using LINEs and ARCs Based on Vertex, IWDW2003: 565-579

Hierarchical Video Summarization Based on Video Structure and Highlight

Yuliang Geng, De Xu, and Songhe Feng

Institute of Computer Science and Technology,
Beijing Jiaotong University, Beijing, 100044, China
gengyuliang@hotmail.com

Abstract. Video summarization is a significant scheme to organize massive video data, and implement a meaningful rapid navigation of video. In this paper, we propose a hierarchical video summarization approach based on video structure and highlight. We extract video structure unit, and measure the unit (frame, shot and scene) importance rank based on visual and audio attention models. According to the unit importance rank, the skim ratio and key frame ratio are assigned to the different video units. Thus we achieve a hierarchical video summary. Experimental results show the excellent performance of the approach.

1 Introduction

With recent advance in digital video technologies, the amount of video data has grown enormously, so quick browsing a video and getting its main content becomes a crucial problem. Video summarization is a significant scheme to organize massive video data, and implement a meaningful rapid navigation of video. Video summarization technique has attracted attention of many researchers in recent years. There are two fundamentally different approaches for video summarization: static summary and dynamic skimming. Static summary is a collection of key frames selected from video sequence, many approaches are proposed to extract and organize key frames, such as video table of contents [1], storyboard [2], and pictorial video summary [3]. Dynamic skimming consists of a collection of video clips selected from video sequence. There are two main approaches for video skimming extraction: one is the predefined event-based approach in which the events are detected and ranked to create video skimming. For example, in sport video [4,5], goal, foul, and touchdown are detected as important events and composite video skimming. The other is a bottom-up approach, which employs special features to analyze the video content [7,8,9]. In [7], authors use audio and video tempo to simulate human's emotion feeling and extract meaningful skim. Literature [8] constructs a user attention curve based on visual, audio attention model to abstract video skimming. In [9], each scene is modeled as a graph, and its optimal skimming is created with graph dynamic programming.

As mentioned above, the static summary based on key frames covers the total video content, but it cannot reflect video semantic content effectively because it loses audio and temporal attributes. The dynamic skimming emphasizes video

D.-Y. Yeung et al. (Eds.): SSPR&SPR 2006, LNCS 4109, pp. 226–234, 2006.

highlight and preserves audio and temporal attributes, but it sacrifices the content integrity. In this paper, we integrate the advantages of static summary and dynamic skimming, and propose an effective approach for multilevel video summarization based on video structure and highlight. First we extract the video structure and measure the unit (frame, shot and scene) importance rank based on visual and audio attention models. According to the unit importance rank, the skim ratio and key frame number of video summary are assigned to different video units. Thus a hierarchical video summary is generated. The block diagram of the video summarization approach is shown in Fig. 1, which gives a 3-level video summary that consists of scene level summary, shot level summary and sub-shot level summary from bottom to up. The hierarchical video summarization approach can provide viewers a multilevel summary with different granularity. In the scene level summary, the viewers can obtain an overview of a video, and can grasp the highlight plots rapidly. In the next level summary, the viewers can further obtain more concise video highlight scenes. In general, our approach not only maintains the content integrity but also emphasizes highlight scenes that may attract viewers' attention.

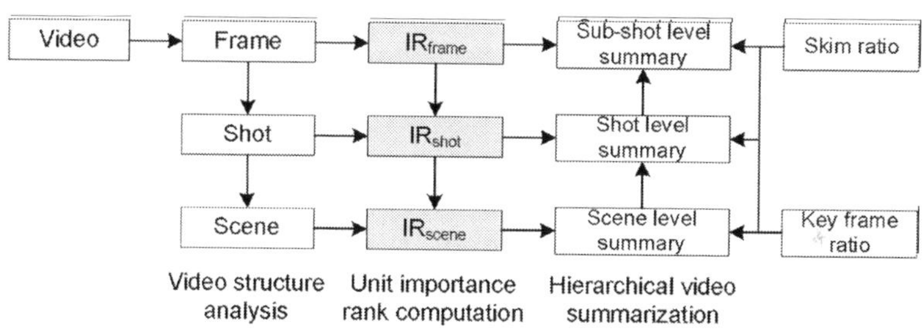

Fig. 1. Block diagram of the hierarchical video summarization approach

The organization of the paper is as follows. Section 2 gives an overview of video structure analysis. We present, in Section 3, the unit importance rank computation based on attention models in detail. Then a hierarchical video summarization approach is proposed in Section 4. Section 5 and 6 give the experimental results and draw the conclusions.

2 Video Structure Analysis

Shot and scene are usually two basic temporal units in video structure analysis. A shot is defined as a single continuous recording made by a camera. A scene consists of a series of related shots (in time, space, etc.), which is a higher-level semantic unit and reflects the narrative structure of a film. We employ singularity detection with wavelet to detect shot boundary [10]. Then we exploit the cinematic rules as a guideline to identify the video scenes [11]. In this step, three main scene categories are identified: dialogue scene, action scene and dialogue with action scene. Thus, we achieve the hierarchical structure of video data.

3 Unit Importance Rank Computation Based on Attention Model

In this section, we compute video unit (frame, shot and scene) importance rank based on visual and audio attention models. And the unit importance rank is regarded as an effective measurement for the highlight of video unit.

3.1 Audio Attention Model

As loudness is a fundamental component of film sound, it plays an important role in defining the overall sonic texture of film. A film usually startles the viewers by exploiting abrupt and extreme shifts in loudness, which is called changes in dynamics [12]. A rough analysis of the loudness can be gained by the square of signal amplitude. In order to stabilize the signal, a Gauss filtering of loudness amplitudes is performed. The loudness amplitude is normalized to archive comparability. Then we utilize the difference between loudness peak $E^{\mathrm{A}}_{\mathrm{peak}}$ and loudness mean $E^{\mathrm{A}}_{\mathrm{mean}}$ to measure loudness dynamic change. Meanwhile, the loudness mean is another important factor in loudness attention measurement. So we define the loudness attention as

$$M_{\mathrm{loud}} = E^{\mathrm{A}}_{\mathrm{mean}} \cdot (E^{\mathrm{A}}_{\mathrm{peak}} - E^{\mathrm{A}}_{\mathrm{mean}}) \tag{1}$$

This metric is similar to the audio saliency attention model proposed by Ma [8], but we more emphasize the dynamic change of loudness. In experiment, the audio signal is sampled at 22.05KHz and each audio frame contains 512 samples shifted by 128 samples from the previous audio frame. The audio feature extraction is based on the audio frame. Here one-second sliding window is used to compute loudness attention M_{loud}.

From the viewpoint of human aural perception, various sounds usually play different roles in attracting the audience attention. So we first classify the audio stream into four classes of semantic segments: silence, speech, music, and environment sound [13]. We assign a weight for each audio semantic segment according to its semantic class.

Obviously, speech usually gives audience more meaningful narrative content, but a long speech scene with low loudness may not attract viewers' attention. While an excellent action scene often accompanies the environment sound with high loudness. Here we unify the sound events, such as explosion, whistle and collision, into the environment sounds, and don't identify them respectively. There are two music effects: harmonic sound and inharmonic sound. Harmonic sounds are perceived as more comfortable, and often are accompanied with mild scene content. While inharmonic sounds often implicate that an unpredictable event may happen, or a worrying event is happening, and can more arouse audience's attention. In the scene construction, the length of the harmonic sound is longer than that of the inharmonic sound.

With above analysis, we define the weights of various audio semantic segments. The weight of a speech segment at time t is defined as

$$w_s(t) = \begin{cases} -(t - t_{start})/t_{Th} + 2 & if \quad t - t_{start} < t_{Th} \\ 1 & else \end{cases} \tag{2}$$

where t_{Th} is a given threshold, and t_{start} is the start time of the speech segment.

The weight of a music segment at time t is defined as

$$w_m(t) = \exp(MinLM - L_{music}) + 1 \tag{3}$$

where $MinLM$ denotes the minimum length among all the music segments, and L_{music} is the length of the current music segment.

The weight of a silence segment $w_z(t)$ is set as 1, and the weight of an environment segment $w_e(t)$ is set as 1.5.

Thus, the audio attention value at the tth second is computed as

$$M_{audio}(t) = w(t) \cdot M_{loud}(t) \tag{4}$$

where $M_{loud}(t)$ is the loudness attention value at the tth second, and $w(t)$ is the weight of the corresponding semantic segment. For example, if the audio segment at the tth second is music, $w(t)$ is set as $w_m(t)$.

3.2 Visual Attention Model

As motion is an intrinsic nature of video and implicates some semantic cues in visual perception, we combine the camera motion and local motion to compute visual attention value.

First, we employ a qualitative method to estimate the camera motion category, which employs motion vectors mutual relationship to implement camera motion classification [14]. As the camera motion continuity, we utilize a sliding window to filter abnormal camera motion. Similar to camera attention weighted strategy [8], we assign different weight w_c for a given video frame according to its camera motion category.

Then, the visual attention value of the ith frame is represented as

$$M_{visual}(i) = w_c(i) \cdot E^M(i) \tag{5}$$

where $E^M(i)$ is the motion activity of the ith frame, which is defined as the standard deviation of motion vector magnitudes because it can measure local motion intensity effectively.

3.3 Unit Importance Rank Computation

Because the visual attention value is a metric based on video frame, and the audio attention value is a metric based on second, we first unify the measurement units to frame according to the video frame rate. Then the visual and audio attention values are normalized by using Gauss normalization formula, and are denoted as $\bar{M}_{visual}(i)$ and $\bar{M}_{audio}(i)$. The attention value at the ith frame is defined as a linear combination of the audio and visual attention values.

$$IR_{frame}(i) = \alpha \cdot \bar{M}_{visual}(i) + \beta \cdot \bar{M}_{audio}(i) \tag{6}$$

where α and β are the preassigned weights and used to be a tradeoff between the visual and audio attention values.

The shot importance rank of the shot j is defined as

$$IR_{\text{shot}}(j) = \sum_i IR_{\text{frame}}(i)/N_{\text{frame}}(j) \tag{7}$$

where $N_{\text{frame}}(j)$ is the video frame number of the shot j.

We employ three main components to determine the scene importance rank, namely, shot cut frequency, visual and audio attention values. In the film editing, filmmaker often uses a series of short shots to create tense or strong atmosphere. The shot cut frequency of shot j is defined as the inverse of shot length and is normalized as $SF(j)$. We define the scene importance rank of the scene k as

$$IR_{\text{scene}}(k) = a \cdot \sum_j (IR_{\text{shot}}(j) \cdot N_{\text{frame}}(j))/(\sum_j N_{\text{frame}}(j)) + b \cdot \sum_j SF(j)/N_{\text{shot}} \tag{8}$$

where N_{shot} is the shot number of the scene. a and b are the weight values.

4 Hierarchical Video Summarization

4.1 Scene Level Summary

Once the skim ratio SR and the key frame ratio KFR are given, we may assign them to each scene according to the scene importance rank. Before assigning the key frame number, we set the minimum of the key frame number of the various scene categories that are extracted in Section 2. For the dialogue scene, dialogist number, which can be archived in scene analysis, is used to determine the key frame number. The action scene should have three key frames at least to represent the attack, sustain and release of action scene. Here we use $MinKF(i)$ to represent the minimum of key frame number of the scene i. So the key frame number of the scene i is assigned as

$$KFN_{\text{scene}}(i) = \min(KFN_{\text{video}} \cdot IR_{\text{scene}}(i)/\sum_j IR_{\text{scene}}(j), MinKF(i)) \tag{9}$$

where $IR_{\text{scene}}(i)$ is the scene importance rank of scene i. KFN_{video} is the total number of key frames in the video sequence and is set as the nearest integer to $KFR \cdot L_{\text{video}}$. L_{video} is the total number of video frames in the video sequence.

For every scene, we utilize the C-mean clustering algorithm to locate the key frames according to its key frame number KFN_{scene}. Thus, we obtain the scene level summary that consists of a group of key frames.

Then, we select the first K scenes with the greatest scene importance ranks as skimming scenes according to the skim ratio. K is the maximum integer, which satisfies the inequality: $\sum_{k=1}^{K} L_{\text{scene}}(k)/L_{\text{video}} \leq SR, k \in \{\text{skimming scenes}\}$. $L_{\text{scene}}(k)$ is the total number of video frames in the scene k. The other scenes with low scene importance ranks are regarded as common scenes. Thus we obtain the scene level summary that consists of a group of skimming scenes.

4.2 Shot/Sub-shot Level Summary

The approach for shot level summarization is similar to the approach for scene level summarization. In this step, we need reset the minimum of the key frame number for each shot according to its camera motion category. Here the minimum of key frame number for still shot is set as 1, and other shot types are set as 2. We also need reassign the skim ratio for each scene, $SR_{\text{scene}}(i)$, according to its scene importance rank as Eq. (10) depicted. If $SR_{\text{scene}}(i)$ is less than a given threshold T_{SR}, we will discard this scene. Thus we obtain the shot level summary according to $SR_{\text{scene}}(i)$ and $KFR_{\text{scene}}(i)$. $KFR_{\text{scene}}(i)$ is the key frame ratio of the scene i and is set as $KFN_{\text{scene}}(i)/L_{\text{scene}}(i)$.

$$SR_{\text{scene}}(i) = \min(\frac{SR \cdot L_{\text{video}}}{L_{\text{scene}}(i)} \cdot \frac{IR_{\text{scene}}(i)}{\sum_j IR_{\text{scene}}(j)}, 1) \tag{10}$$

Next, we construct the sub-shot level summary. For a given shot, we reassign its key frame ratio and skim ratio as the same way. Then we extract its sub-shot around the maximum of attention value curve $IR_{\text{frame}}(i)$. The length of the sub-shot is determined by its skim ratio. The key frames are also extracted to represent the skimming shot according to its key frame ratio.

Thus we get a hierarchical and scalable video summary that is composed of static key frame sequence and dynamic skimming. As the video hierarchical structure is the basic element for filmmaker to construct story plots, the summary based on the video structure and unit important rank can provide a good tradeoff between the content integrity and content compactness. Additionally, users may adjust the summary by tuning the key frame ratio and skim ratio.

5 Experimental Results

The video summary is the logical layer of representation based on subjective semantics, and there are still no objective definition and evaluation criterion. So how to evaluate video summary is a difficult issue. In experiment, we invite test users including naive users and experienced users (engaged in video retrieval) to assess the performance of the proposed video summarization approach. We collect the test dataset from five various movie videos, namely, *Rain man*, *Ghost* are dramas; *Leon* and *The Shaolin Temple* are action movies; and *Shrek* is a cartoon movie. The total length of test dataset is about 75 minutes, which is composed of 878 shots and 53 scenes. All the video data is in MPEG-1 format with a frame rate of 30 fps, and the audio track was sampled at 22.05 KHz.

First, we carry out an experimental comparison to evaluate the performance of the key frame sequence of video summary between our approach (denoted as HVS) and storyboard technique (denoted as ST) [2]. Here we design two evaluation criteria, content compactness and content integrity, to evaluate the performance of these two approaches. For the content compactness, test users give an assessment of being too much, much, good, few and too few to key frame sequence, corresponding to quantitative scores: 0.1, 0.5, 1, 0.5 and 1. The content integrity means whether test users can capture the story plot from the key frame sequence by answering the questions, such as, "who", "where", "when", and "what". According to the accuracy of answers, the score of the content integrity is obtained. All the questions are selected from the user investigation report. For example, for the static summary, users pay more attention to whether they can get the information about the protagonists, location, and coarse events, which is the reason that these questions are provided in our evaluation scheme.

Fig. 2 gives the performance curves of the content integrity and the content compactness. As Fig. 2 illustrates, our approach can maintain the content integrity at different key frame ratio very well. When the key frame ratio is increasing, the content compactness is decreasing. Our approach (sub-shot level summary) got the best performance when the key frame ratio is set as 0.02. Our proposed method can provide a meaningful representation of video content because the key frames assignment and location are based on the semantic content of the video unit, while the storyboard based on the hierarchical clustering method cannot ensure the extracted key frames have semantic structure.

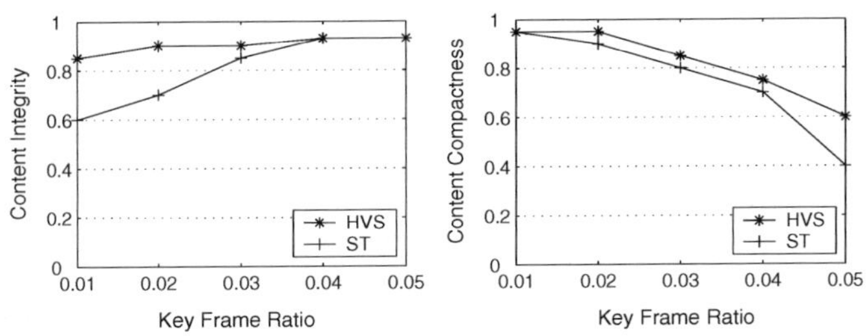

Fig. 2. Experimental comparison between our approach and storyboard method. Left: content integrity curve; right: content compactness curve.

Next, we evaluate the quality of the video skimming from two criteria: comprehensibility and highlight degree. For a good video skimming, like the trailer of a movie, the users more care whether its content is comprehended easily, and whether the summary is composed of the most excellent video clips. Because it is still a subjective problem to evaluate the video skimming, we only assess the video skimming by analyzing test users' answers to the test questions. Here we carry out an experimental comparison between our approach and the method (denoted as SAGO) proposed in [9]. Video skimming assessment is complex process. We first let the test users look through the video skimming from low to high skim ratio in turn. When the test users finished viewing the video skimming with a certain skim ratio, they need assess the video skimming according to the two criteria. Then the users continue their assessment with another skim ratio, and so on. After they finished all the video skimming, they may reassess these video skimming. The assessment is quantified to score from 0 to 1. Fig. 3 gives the experimental results.

As the comparison results shown, our proposed approach has a good performance. One important reason is that we extract video skimming under the

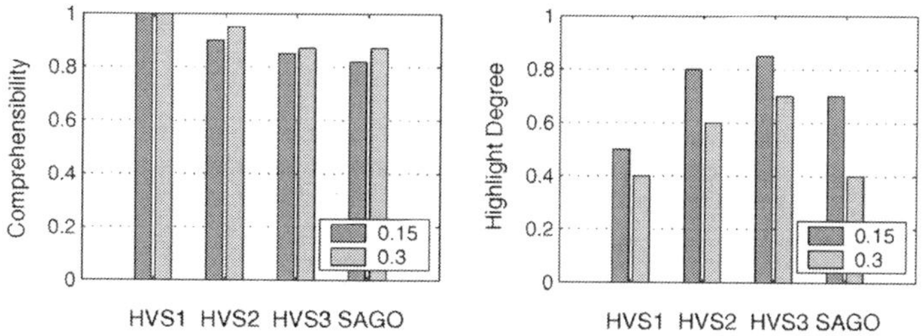

Fig. 3. Experimental comparison between our approach and the SAGO [9]. Left: comprehensibility assessment score; right: highlight degree assessment score. Notes: HVS1 denotes scene level summary, HVS2 denotes shot level summary, HVS3 denotes sub-shot level summary.

guideline of user attention value, which ensures the highlight degree of video summary. Another reason is that the hierarchical structure of video skimming keeps the integrity of semantic content. From Fig. 3 we can see, the video skimming with skim rate of 0.3 has higher comprehensibility score, and the video skimming with skim rate of 0.15 has higher highlight degree. In general, when the skim rate is set as 0.15, sub-shot level summary can archive the best experimental results.

6 Conclusions

We have addressed the main issues of the video summarization from the video structure analysis, unit importance rank computation to video summarization. As the video hierarchical structure is the basic element for filmmaker to construct story plots, and the unit importance rank is an effective measurement for the highlight of video unit, the approach for video summarization based on video structure and highlight can give us a better tradeoff between the content integrity, comprehensibility and the content compactness, highlight degree. Additionally, users can also adjust video summary by tuning the key frame ratio and skim ratio. In general, our proposed approach can provide us a multilevel and flexible video summary with different granularity. Experimental results have been reported in detail.

Acknowledgements

This research was supported by Science Foundation of Beijing Jiaotong University (Grant No. 2004SM013).

References

1. Rui, Y., Huang, T.S., Mehrotra, S.: Constructing Table-of-Content for Videos. ACM Multimedia Systems Journal, Special Issue Multimedia Systems on Video Libraries, Vol. 7, No. 5 (1999) 359-368

2. Hasebea, S., Mustafa M.S.: Constructing Storyboards Based on Hierarchical Clustering Analysis. In: Proceedings of Visual Communications and Image Processing, SPIE Vol. 5960 (2005) 437-445
3. Ma, Y.F., Zhang, H.J.: Video Snapshot: A Bird View of Video Sequence, In: Proceed-ings of the 11th International Multimedia Modelling Conference (2005) 94-101
4. Tjondronegoro, D.W., Chen, Y.P.P., Pham, B.: Classification of Self-Consumable Highlights for Soccer Video Summaries. In: Proceedings of IEEE ICME, Vol. 1 (2004) 579-582
5. Noboru, B., Yoshihiko, K., et al.: Personalized Abstraction of Broadcasted American Football Video by Highlight Selection, IEEE Transactions on Multimedia, Vol. 6, No. 4, (2004) 575-586
6. Rapantzikos, K., Tsapatsoulis, N., Avrithis, Y.: Spatiotemporal Visual Attention Architecture for Video Analysis. In: Proceedings of Multimedia Signal Processing (2004) 83-86
7. Lee S.H., Yeh, C.H., Kuo, C.C.J.: Video Skimming Based on Story Units via General Tempo Analysis. In: Proceedings of IEEE ICME, Vol. 2 (2004) 1099-1102
8. Ma, Y.F., Lu, L., Zhang, H.J., Li, M.J.: A User Attention Model for Video Summarization. In: Proceedings of ACM Multimedia, (2002) 533-542
9. Lu, S., King, I., Michael, R.L.: Video Summarization by Video Structure Analysis and Graph Optimization. In: Proceedings of IEEE ICME, Vol. 3 (2004) 1959-1962
10. Geng, Y.L., Xu, D.: A Unified Framework for Shot Boundary Detection. Journal of Image and Graphics (Chinese) Vol.10, No.5 (2005) 650-655
11. Geng, Y.L., Xu, D., Wu, A.M.: Effective Video Scene Detection Approach Based on Cinematic Rules. In: Proceedings of KES, LNCS, Vol. 3682 (2005) 1197-1204
12. Ohm, J.R, Multimedia Communication Technology Representation, Transmission and Identification of Multimedia Signals. Springer, Berlin Heidelberg (2004)
13. Lu, L., Jiang, H. and Zhang, H.J.: A Robust Audio Classification and Segmentation Method. In: Proceedings of ACM Multimedia, Vol. 9 (2001) 203-211
14. Zhu, X.Q., Xue, X.Y.: Qualitative Camera Motion Classification for Content-Based Video Indexing. In: Proceedings of IEEE PCM, LNCS, Vol. 2532, (2002) 1128 - 1136

Direct Curvature Scale Space in Corner Detection

Baojiang Zhong[1] and Wenhe Liao[2]

[1] Department of Mathematics, Nanjing university of Aeronautics & Astronautics, Nanjing, 210016, China
zhbj@nuaa.edu.cn
[2] College of Mechanical and Electrical Engineering, Nanjing university of Aeronautics & Astronautics, Nanjing, 210016, China
njwho@nuaa.edu.cn

Abstract. Curvature Scale Space (CSS) representation of planar curves is considered to be a modern tool in image processing and shape analysis. Direct Curvature Scale Space (DCSS) is defined as CSS that results from convolving the curvature of a curve with a Gaussian kernel directly. Recently a theory of DCSS in corner detection has been established. In the present paper the DCSS theory is considered to transform the DCSS image of a given curve into a tree organization, and then corners on the curve are detected and located in a multiscale sense. Experiments are conducted to show that the DCSS corner detector can work equally well as the CSS corner detector does on curves with multiple-size features, however, at much less computational cost.

1 Introduction

The scale space concept was introduced by Iijima [3] more than 40 years ago and became popular later on by the works of Witkin [13] and Koenderink [4]. Scale space analysis of a signal $f(x)$ is generally made by convolving it with a Gaussian kernel, treating the quadratic variance σ of Gaussian as a parameter of scale. Extrema in the first derivative, or, zero-crossings in the second derivative of the convolved signal are located at varying scales. The image on the (x, σ) plane showing the extrema or zero-crossings is called a scale space image.

Asada and Brady [1] extended the scale space concept to represent significant changes in curvature along a planar curve. The curve is expressed as a function $\varphi(s)$ of the orientation of the tangent φ against arc length s. Then $\varphi(s)$ is convolved with a Gaussian. Local positive maxima and negative minima in the first and second derivatives of the convolved function are located, resulting two scale space images on the (s, σ) plane. A set of curvature changes were selected as *primitives*. The scale space behavior of *Corner* and *Smooth join* was studied, and the behavior of other curvature primitives was illustrated.

Mokhtarian and Mackworth [5,6] developed Curvature Scale Space (CSS) by finding curvature zero-crossings of a curve at varying levels of detail. They

D.-Y. Yeung et al. (Eds.): SSPR&SPR 2006, LNCS 4109, pp. 235–242, 2006.

treated the curve as a 2-D signal. Considering a path length variable u, the curve is expressed in terms of two functions: $\{x(u), y(u)\}$. Then the two functions are convolved respectively. By encoding the curve with the curvature zero-crossings, a multiscale shape representation was formulated. The CSS representation has been selected as a shape contour descriptor for MPEG-7 [7].

Rattarangsi and Chin [10] employed CSS to detect corners on planar curves. A CSS image consisting of the maxima of absolute curvature is constructed. The scale space behavior of isolated single and double corner models was sketched[1], with which the CSS image is transformed into a tree organization and then corners are detected. Pei and Lin [9] also proposed a corner detector based on the scale space concept. Similar to Asada and Brady's representation, they treated planar curves as 1-D signals. Extreme curvature points are located by convolving the curvature of a curve directly. Pei and Lin studied a procedure of corner detection. However, no scale space analysis was provided.

During the last decade, scale space concept has attracted a wide interest in the field of shape representation, feature extraction, and object recognition, see, e.g. [2,8,11,12,14,15,16]. For distinguishing purpose, CSS resulting from direct curvature convolution is referred to as *Direct Curvature Scale Space* (DCSS). Note that since differentiation and convolution are commutative, Asada and Brady's representation relates to DCSS directly. Recently, the CSS theory in corner detection is re-established, and a DCSS theory is established [18]. The problem of how a planar curve shrinks in its scale space was also mathematically studied in [18].

Compared to CSS, DCSS is much cheaper in terms of computational cost. Based on the DCSS theory, we study a procedure of DCSS corner detection in a multiscale sense. In Section 2 the DCSS theory is reviewed. In Section 3 the procedure of DCSS corner detection is presented and experiments are conducted. Finally, in Section 4 the paper is concluded.

2 A Theory of Direct Curvature Scale Space

Corners of a planar curve correspond to points of high curvature. Let $\kappa(s)$ be the curvature function. The DCSS representation describes the curve at increasing levels of detail by convolving $\kappa(s)$ with a Gaussian $g(s, \sigma)$ directly. Denote by $\otimes$ the convolution operator. The convolved curvature function $\kappa(s, \sigma)$ is given by

$$\kappa(s, \sigma) = \kappa(s) \otimes g(s, \sigma).$$

It can be shown explicitly as

$$\kappa(s, \sigma) = \frac{1}{\sigma\sqrt{2\pi}} \int_{-\infty}^{+\infty} \kappa(u) e^{-\frac{(s-u)^2}{2\sigma^2}} du.$$

[1] Rattarangsi and Chin presented a theoretical study of the CSS representation in corner detection; however, due to a fundamental mistake in their work the CSS theory was not established correctly.

To determine corners at a given scale σ, we solve for all the locations that have maxima absolute curvature, $|\kappa(s,\sigma)|$, which are the positive maxima and negative minima of the curvature. A DCSS image of corners is then constructed by $\max_{s,\sigma} |\kappa(s,\sigma)|$. To investigate the properties of the DCSS image, the single corner Γ model and double corner END and STAIR models [1,10] are considered.

Figure 1 shows a Γ corner model and its DCSS image. The model properties are summarized in Property 1.

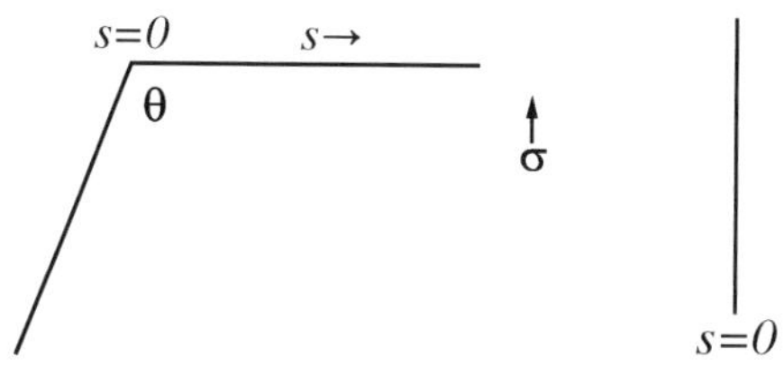

(a) Γ *model* (b) *DCSS image*

Fig. 1. Γ model produces a stationary and persistent scale space image

Property 1: A Γ corner model has a single stationary and persistent line pattern in direct curvature scale space independent of the corner angle θ and the scale parameter σ.

Consider an END corner model, which consists of two corner θ_l and θ_r with the same concavity separated by a width of $2w$, see Figure 2(a). Define

$$\lambda = \frac{\pi - \theta_r}{\pi - \theta_l}; s_l = \frac{\lambda - 1}{\lambda + 1} w.$$

Let s_r be implicitly given by

$$\ln(\lambda \frac{w - s_r}{w + s_r}) + \frac{2ws_r}{(w + s_r)(w - s_r)} = 0, s_r \in (0, w),$$

and define

$$\sigma_\lambda^2 = (w + s_r)(w - s_r).$$

Properties of the END corner model are summarized in Property 2-4 and its DCSS image is sketched in Figure 2(b).

Property 2: For an END model with $0 < \theta_l < \theta_r < \pi$, the DCSS line pattern of the strong corner θ_l is persistent and asymptotically stationary at $s = s_l$.

Property 3: For an END model with $0 < \theta_l < \theta_r < \pi$, the DCSS line pattern of the weak corner θ_r terminates at (s_r, σ_λ), and there it meets another kind of line pattern which consists of a set of minima of absolute curvature.

Property 4: An END model, with a corner separation $2w$ and $0 < \theta_l = \theta_r < \pi$, has a DCSS image symmetric with respect to $s = 0$. When $0 < \sigma < w$, the two absolute maxima move towards each other as σ increases. When $\sigma \geq w$, the two absolute maxima merge, forming a single stationary and persistent line pattern.

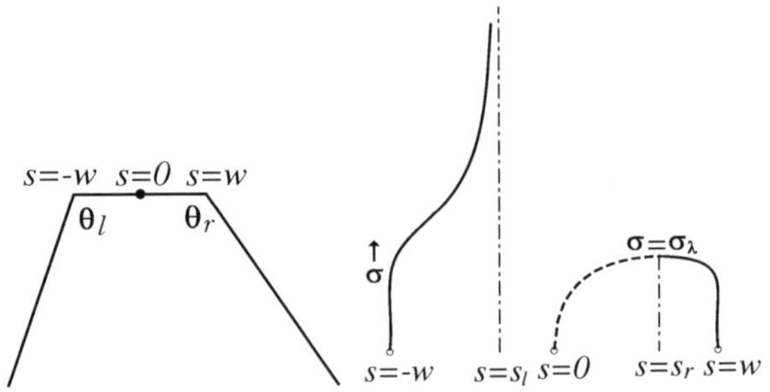

(a) *END model* (b) *DCSS image*

Fig. 2. END model and its DCSS image

Consider a STAIR corner model, which consists of two corner θ_l and θ_r of opposite concavity separated by a width of $2w$, see Figure 3(a). Its model properties are summarized in Property 5-7, and its DCSS image is sketched in Figure 3(b).
Property 5: For a STAIR model with $0 < \theta_l < 2\pi - \theta_r < \pi$, the DCSS line pattern of the strong corner θ_l is persistent and asymptotically stationary at $s = s_l$.
Property 6: For a STAIR model with $0 < \theta_l < 2\pi - \theta_r < \pi$, the DCSS line pattern of the weak corner θ_l is persistent, and it is bounded by $\sigma = \mu\sqrt{s}$ as σ increases, where μ is defined as $\mu = (-\frac{2w}{\ln(-\lambda)})^{1/2}$.
Property 7: A STAIR model with $0 < \theta_l = 2\pi - \theta_r < \pi$ produces a persistent DCSS image, which is symmetric with respect to $s = 0$. The two line patterns are bounded by $\sigma = -s$ and $\sigma = s$ respectively and they repel each other at a same rate as σ increases.

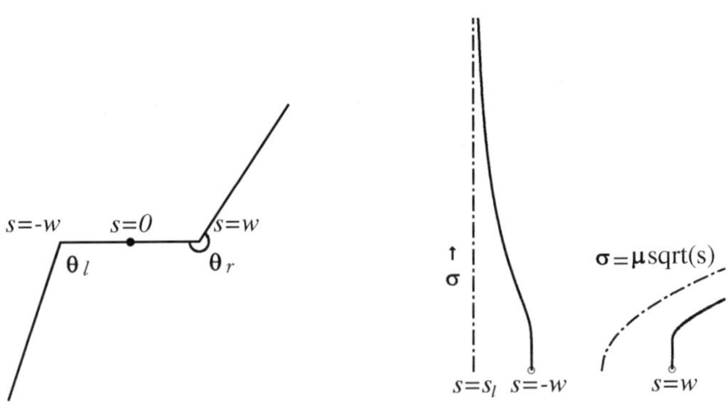

(a) *STAIR model* (b) *DCSS image*

Fig. 3. STAIR model and its DCSS image

To investigate the effect of corner separation of a double corner model on its DCSS image, we fix θ_l and θ_r, and vary w. The following property is established.
Property 8: The DCSS image of an END or STAIR model with corner separation $2w_i$ is linearly related to the DCSS image of another END or STAIR model with corner separation $2w_j$ by the ratio of the separations given by $\frac{w_i}{w_j}$.

3 DCSS Corner Detection

To detect corners on a planar curve, a tree organization similar to that in [13,10,11] is constructed from its DCSS image. To filter out quantization noise,

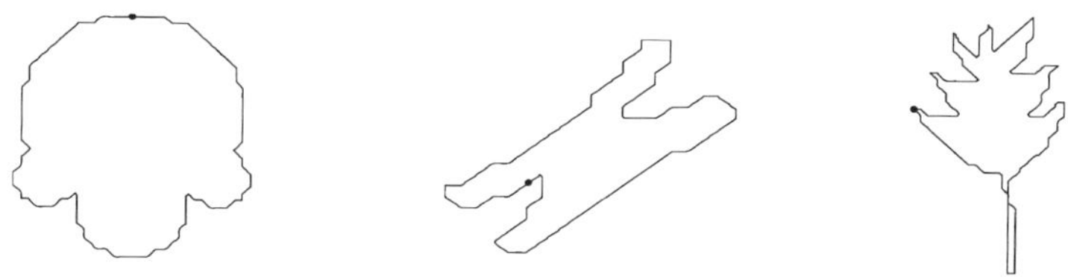

(a) *Test curves: in contour-clockwise; start point is indicated by a bullet*

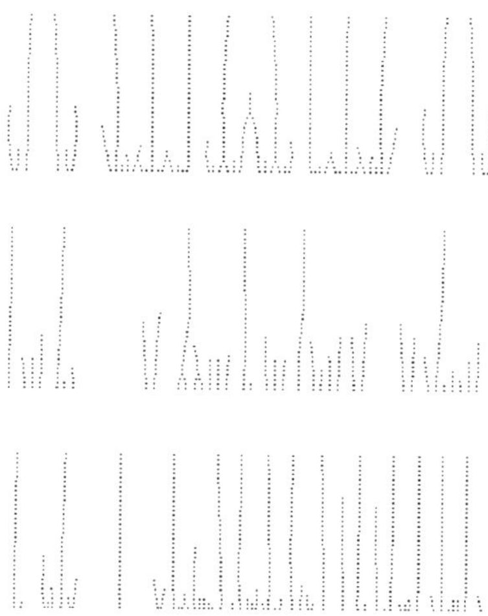

(b) *DCSS images*

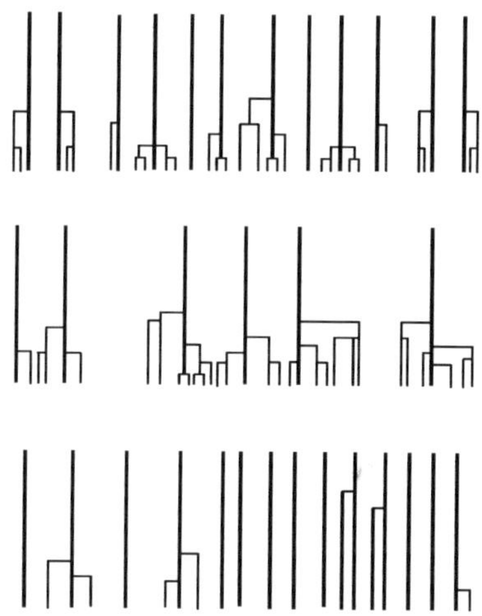

(c) *Tree organizations*

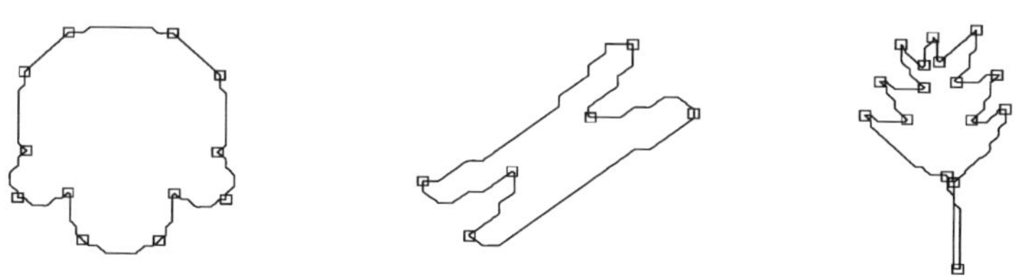

(d) *Corner detection results of DCSS*

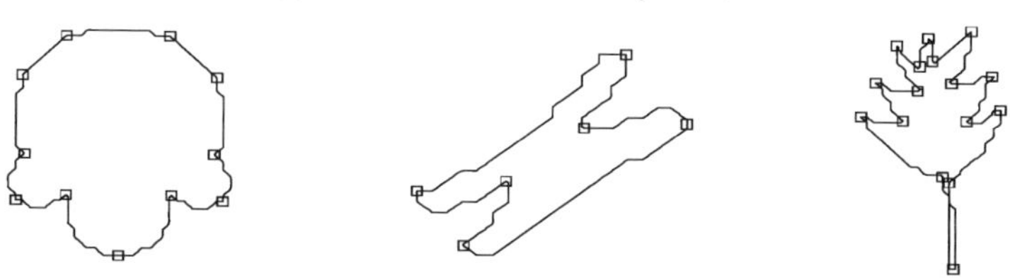

(e) *Corner detection results of CSS*

Fig. 4. DCSS corner detection and a comparison with CSS

line patterns caused by extremely low absolute maxima (say, below 0.01) are removed from the image at first. After this cleaning process, for each line pattern a vertical line located at the finest scale location of the line pattern is drawn. If two line patterns merge to become a single one, two vertical lines are drawn respectively, and a single vertical line located at the merging scale location is also drawn. The three vertical lines are joined by a horizontal line. For a non-survival line pattern, the vertical line is joined by a horizontal line to the nearest more persistent vertical line.

The model properties specified in Section 2 are then considered to parse the tree organization. A tree consisting of a single root corresponds to a single corner located at the finest scale location of the root. For a tree with offspring, the length of its root is compared with the height of its offspring. If the root length exceeds the height of the offspring, the root is declared stable, corresponding to a corner located at the orthogonal projection of the root on x-axis. Otherwise the search for corners proceeds to the offspring. If an offspring is a leaf whose length is longer than that of its parent, it is stable, corresponding to a corner located at the finest scale location of the leaf. If an offspring has its own offspring, the search for corners is applied to its family. Each tree of the organization is evaluated in the same manner.

A set of objects that have been commonly used in many previous studies are chosen as test curves. They are the Semicircles, the Chromosome, and the Leaf, see Figure 4(a) (Since DCSS is invariant under rotation, the start point of each curve is selected randomly). Figure 4(b) shows their DCSS images, and Figure 4(c) shows the corresponding tree organizations. The roots and leaves of the trees which correspond to corner points are indicated by bold lines. Corner detection results of DCSS are shown in Figure 4(d).

For comparison, Figure 4(e) shows the results of CSS corner detection. It can be seen that the DCSS corner detector compares favorably with CSS. For the last two curves, the two detectors have the same performance. For the first curve, we believe the performance of the DCSS detector is more reasonable: the semicircles with different radii have been distinguished by different corner patterns.

On the other hand, the DCSS corner detector outperforms the CSS corner detector very clearly with respect to computational cost. Table 1 summarizes a comparison between the CPU processing time of the two detectors spent on the three curves. The programs were implemented on a Pentium IV-2.80G PC with 512M memory. It can be seen that the DCSS detector requires only about 1/3 CPU time of the CSS detector. As a result, the efficiency of corner detection is improved significantly.

Table 1. Comparison of the CPU time of the CSS and DCSS detectors

	semicircle curve	chromosome curve	leaf curve
CSS	166 ms	190 ms	186 ms
DCSS	57 ms	63 ms	62 ms

4 Concluding Remarks

In this paper we have studied a procedure of applying DCSS to detect and locate corners on planar curves. Numerical results show that the DCSS corner detector can operate successfully on curves with multiple-size features, however, at much less computational cost compared to the CSS corner detector.

To compute a scale space image, DCSS is much cheaper than CSS. This can be appreciated by a glance at the computational complexity of the two representations.

CSS: Compute $\{x(s,\sigma), y(s,\sigma)\}$ and the curvature of the convolved curve at each scale;

DCSS: Compute the curvature at scale $\sigma = 0$, and compute $\kappa(s,\sigma)$ at each scale.

Since DCSS computes only one scale space instead of two, and computes the curvature of the curve (an expensive operation) only once, its computational cost is significantly less than that of CSS.

For noisy curves, a small amount of Gaussian smoothing is suggested to be a preprocessing step of DCSS [9]. In particular, during the smoothing process absolute maxima in curvature can be located as a by-product, and then a small part of the CSS image can be constructed. This brings to us a hybrid scheme to apply CSS and DCSS [17], by which corners can be located at the finest scale. The hybrid CSS/DCSS corner detector, including a switchover criterion between CSS and DCSS, has be discussed in detail in [18].

References

1. Asada, H., Brady, M.: The curvature primal sketch. IEEE Trans. Pattern Anal. Mach. Intell. 8 (1986) 2–14
2. Garrido, A., Blanca, N., Vente, M.: Boundary simplification using a mutiscale dominant-point detection algorithm. Pattern Recognition 31 (1998) 791–804
3. Iijima, T.: Basic theory on normalization of pattern (in case of typical one-dimensional pattern). Bulletin of the Electrotechnical Laboratory 26 (1962) 368–388
4. Koenderink, J.J.: The structure of images. Biol. Cybern. 50 (1984) 363–370
5. Mokhtarian, F., Mackworth, A.: Scale-based description and recognition of planar curves and two-dimensional shapes. IEEE Trans. Pattern Anal. Mach. Intell. 8 (1986) 34–43
6. Mokhtarian, F., Mackworth, A.: A theory of multi-scale, curvature-based shape representation for planar curves. IEEE Trans. Pattern Anal. Mach. Intell. 14 (1992) 789–805
7. Mokhtarian, F., Bober, M.: Curvature Scale Space Representation: Theory, Applications, and MPEG-7 Standardization. Kluwer Academic Publishers, Dordrecht, March 2003.
8. Mokhtarian, F., Abbasi, S.: Robust automatic selection of optimal views in multi-view free-form object recognition. Pattern Recognition 38 (2005) 1021–1031

9. Pei, S., Lin, C.: The detection of dominant points on digital curves by scale-space filtering. Pattern Recognition 25 (1992) 1307–1314
10. Rattarangsi, A., Chin, R.: Scale-based detection of corners of planar curves. IEEE Trans. Pattern Anal. Mach. Intell. 14 (1992) 430–449
11. Ray, B., Ray, K.: Corner detection using iterative Gaussian smoothing with constant windows size. Pattern Recognition 28 1995 1765–1781
12. Ray, B., Pandyan, R.: ACORD-an adaptive corner detector for planar curves. Pattern Recognition 36 (2003) 703–708
13. Witkin, A.P.: Scale-space filtering. In: Proc. Eighth Int. Joint Conf. on Artificial Intelligence, Karlsruhe, Germany (1983) 1019–1021
14. Xin, K., Lim, K., Hong, G.: A scale-space filtering approach for visual feature extraction. Pattern Recognition 28 (1995) 1145–1158
15. Zabulis, X., Sporring, J., Orphanoudakis, S.: Perceptually relevant and piecewise linear matching of silhouettes. Pattern recognition 38 (2005) 75–93
16. Zhang, D., Lu, G.: Review of shape representation and description techniques. Pattern Recognition 37 (2004) 1–19
17. Zhong, B.J., Liao W.H.: A hybrid method for fast computing the curvature scale space image. International Conference on Geometric Modeling and Processing, IEEE Computer Society Press (2004) 124–130
18. Zhong, B.J.: Research on Algorithms for Planar Contour Processing. PhD Dissertation, Nanjing university of Aeronautics & Astronautics (2006)

Aligning Concave and Convex Shapes

Silke Jänichen and Petra Perner

Institute of Computer Vision and applied Computer Sciences, IBaI,
Körnerstr. 10, 04107 Leipzig
pperner@ibai-institut.de, www.ibai-institut.de

Abstract. There are plenty of different algorithms for aligning pairs of 2D-shapes and point-sets. They mainly concern the establishment of correspon-dences and the detection of outliers. All of them assume that the aligned shapes are quite similar and belonging to the same class of shapes. But special problems arise if we have to align shapes that are very different, for example aligning concave shapes to convex ones. In such cases it is indispensable to take into account the order of the point-sets and to enforce legal sets of correspondences; otherwise the calculated distances are incorrect. We present our novel shape alignment algorithm which can handle such cases also. The algorithm establishes legal one-to-one point correspondences between arbitrary shapes, represented as ordered sets of 2D-points and returns a distance measure which runs between *0* and *1*.

Keywords: Shape Alignment, Correspondence Problem, Aligning Convex to Concave Shapes and vise-versa.

1 Introduction

The analysis of shapes and shape variation is of great importance in a wide variety of disciplines. It is especially interesting for biologists, since shape is one of the most concise features of an object class and may change over time due to growth or evolution. The problems of shape spaces and distances have been intensively studied by Kendall [1] and Bookstein [2] in a statistical theory of shape. In digital image processing the statistical analysis of shape is a fundamental task in object-recognition and classification [3] .

In all these applications, shapes of the same class are aligned and compared. The mapping of convex to concave pieces of the shapes rather indicates that wrong correspondences between elements have been established or that there are outliers [4]. However, there is a number of applications where we have to study the similarity between shapes of different classes. In that case we are faced with the problem to determine the similarity between convex and concave shapes.

We are describing our work on aligning arbitrary shape to each other and determining the similarity between them. It can happen that we have to compare convex to concave shapes. The natural shapes are acquired manually from real images [5]. The object shapes can appear with varying orientation, position, and scale in the

D.-Y. Yeung et al. (Eds.): SSPR & SPR 2006, LNCS 4109, pp. 243 – 251, 2006.

image. The shapes are arbitrary and there is nothing special about them. Our algorithm establishes symmetric and legal one-to-one point correspondences between arbitrary shapes, represented as ordered sets of 2D-points and returns a similarity value.

The paper is organized as follows. We describe the problem of shape alignment in Sect. 2. The algorithm for pair-wise alignment of the shapes and calculation of distances is proposed in Sect. 3 and evaluated in Sect. 4. Finally we give conclusions in Sect. 5.

2 The Problem of Alignment of 2-D Shapes

Consider two shape instances P and O defined by the point-sets $p_i \in R^2$, $i = 1, 2, \ldots, N_P$ and $o_k \in R^2$, $k = 1, 2, \ldots, N_O$ respectively. The basic task of aligning two shapes consists of transforming one of them (say P) so that it fits in some optimal way the other one (say O) (see Fig 1 left). Generally the shape instance $P = \{p_i\}$ is said to be aligned to the shape instance $O = \{o_k\}$ if a distance $d(P,O)$ between the two shapes cannot be decreased by applying a transformation ψ to P.

The problems of shape spaces and distances have been intensively studied [1], [2] in a statistical theory of shape. The well-known Procrustes distance [6], [7] between two point-sets P and O is defined as the sum of squared distances between corresponding points:

$$d(P,O) = \sum_{i=1}^{N_{PO}} \left\| \frac{(p_i - \mu_P)}{\sigma_P} - R(\theta)\frac{(o_i - \mu_O)}{\sigma_O} \right\|^2 , \qquad (1)$$

where $R(\theta)$ is the rotation matrix, μ_P and μ_O are the centroids of the object P and O respectively, σ_P and σ_O are the standard deviations of the distance of a point to the centroid of the shapes and N_{PO} is the number of point correspondences between the point-sets P and O. This example shows that the knowledge of correspondences is an important prerequisite for calculation of shape distances.

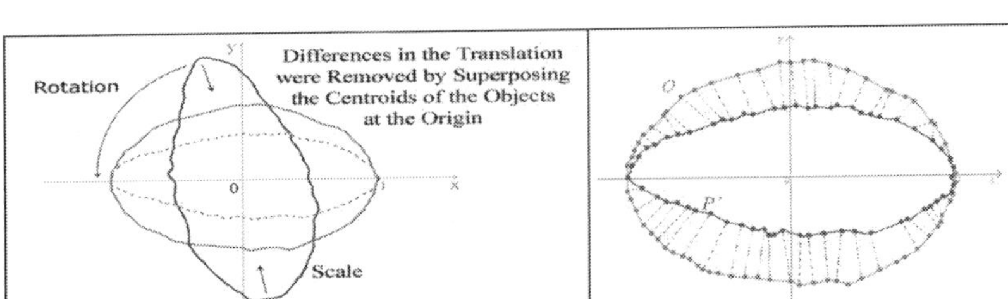

Fig. 1. Alignment of shape instances, superimposition, and calculation of correspondences

Various alignment approaches are known [8][9]. They mainly differ in the kind of mapping (i.e. similarity, rigid, affine) and the chosen distance measure. A survey of different distance measures used in the field of shape matching can be found in [10].

For calculating a distance between two shape instances the knowledge of corresponding points is required. If the shapes are defined by sets of landmarks [11], the knowledge of point correspondences is implicit. However, at the beginning of many applications this condition does not hold and often it is hard or even impossible to assign landmarks to the acquired shapes. Then it is necessary to automatically determine point correspondences between the points of two aligned shapes P and O, see (see Fig 1 right).

There has been done a lot of work concerning the problem of automatically finding point correspondences between two unknown shapes. An extension of the classical Procrustes alignment to point-sets of differing point counts is known as the *Softassign Procrustes Matching* algorithm [6]. It alternates between solutions for the correspondence, the spatial mapping, and the Procrustes rescaling.

Hill *et al.*[9] presented a greedy algorithm used as an iterative local optimization scheme to modify the correspondences, in order to minimize the distance between two polygon segments of shapes. Another popular approach to solving the correspondence problem is called *Iterative Closest Point* (*ICP*) developed by Besl and McKay [12]. In the original version of the *ICP* the complexity of finding for each point p_k in P the closest point in the point-set O is $O(N_P N_O)$ in the worst case. Marte *et al.*[13] improved this complexity by applying a spatial subdivision of the points in the set O. Fitzgibbon [14] replaced the closed-form inner loop of the *ICP* by the Levenberg-Marquardt algorithm, a non-linear optimization scheme. Another solution of the correspondence problem was presented by Belongie *et al.*[15]. He added to each point in the set a descriptor called shape context. In our work we solved the correspondence problem by a nearest-neigbor search algorithm [5].

One of the most essential demands on these approaches is symmetry. Symmetry means obtaining the same correspondences when mapping instance P to instance O and vise versa instance O to instance P. This requirement is often bound with the condition to establish one-to-one correspondences. This means a point o_k in shape instance O has exactly one corresponding point p_k in shape instance P. If we compare point sets with unequal point numbers under the condition of one-to-one mapping, it is clear that some points will not have a correspondence in the other point set. These points are called outliers.

Special problems arise if we have to align shapes that are very different, for example aligning concave to convex shapes. In these cases it is indispensable to take into account the order of the point-sets and to enforce legal sets of correspondences by not allowing inverse mapping of the points. To demonstrate this, see points o_2 and o_4 in Table 2(a). Suppose that a concave shape representing the letter C is compared with the shape of the letter O (see Table 1). If the pair-wise correspondences were established between nearest neighbored points by one-to-one mapping and by allowing inverse mapping, the resulting distance between both shape instances will be very small (Table 1 a).

Table 1. Illegal and legal sets of correspondences

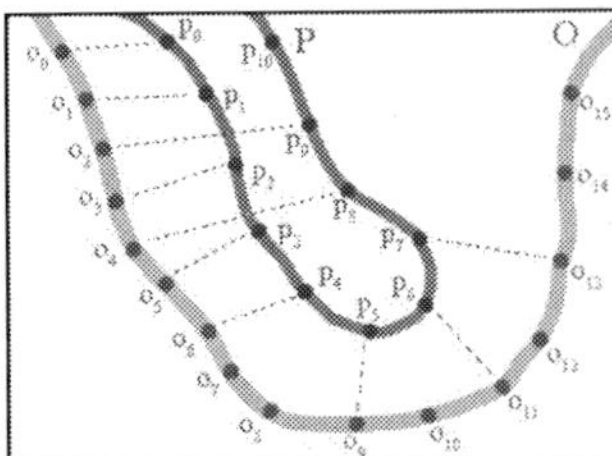	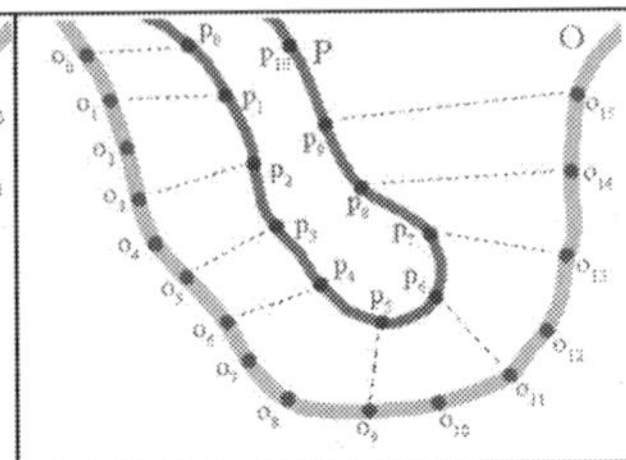
(a) Illegal correspondences inversions at o_2 and o_4	**(b)** Legal correspondences without any inversions

But intuitively we would say that these shapes are not very similar. Particularly in such cases it is necessary to regard the order of point correspondences and to remove correspondences if they produce inversions (see Table 1 b). Ultimately it can be seen that big distances are arising between corresponding points which leads to an increased distance measure.

Table 2. Establishing correspondences while mapping a concave and convex shape

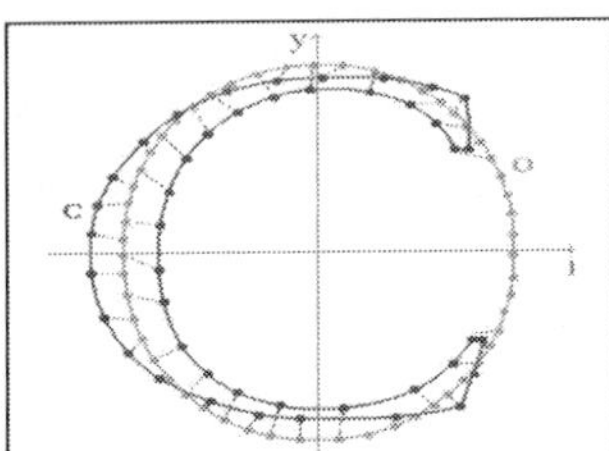	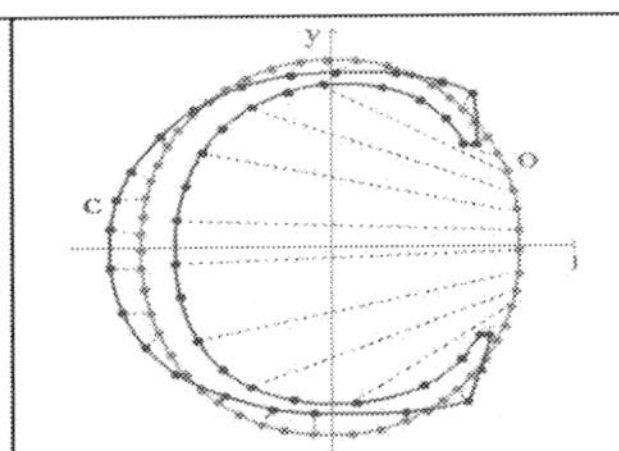	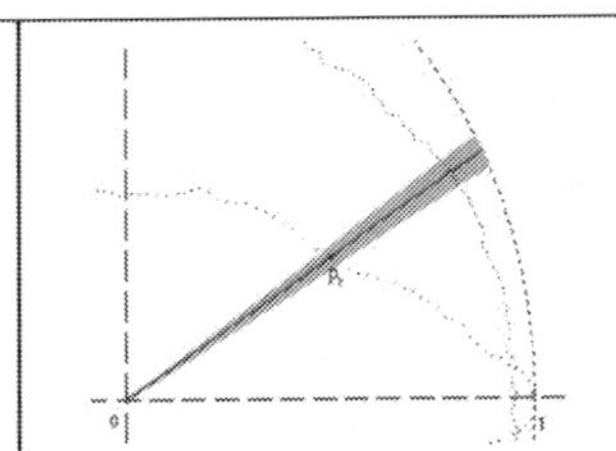
(a) Illegal correspondences with inversions	**(b)** Enforced legal correspondences	**(c)** Range for finding potential correspondents for p_k

3 Our Algorithm

The input into our algorithm (see table 3) is the rescaled shape P and O translated into its origin. This normalization ensures that the centroids are identical and that our similarity measure is running between *0* and *1*. The Euclidean distance between the two shapes P and O is calculated. We are also calculating the maximum distance and a score based on the sum between the maximum distances and the mean distance.

The algorithm is comprised of three main steps: (A) rotate shape, (B) calculate point correspondences, and (3) calculate the similarity score. The differences in rotation will be removed during our iterative alignment algorithm. In each iteration of this algorithm, the first shape is rotated stepwise by an angle $\nabla \psi$, while the second

Table 3. Outline of our shape alignment algorithm

```
Initialize ψ          /* stepwise rotation angle */
SET ψ_i = 0           /* actual rotation angle */

Input: Normalized Shape O and Shape P
Output: min{SCORE(P,O_i)}
REPEAT UNTIL ψ_i ≥ 2π or SCORE(P,O_i)=0
   (A) Rotate O with ψ_i = ψ_{i-1} + ψ
   (B) CalcCorrespondences( P , O_i )
   (C) CalcScore SCORE(P,O_i)
RETURN SCORE(P,O) = min{SCORE(P,O_i)}


SUB (B) CalcCorrespondences( P , O ) BEGIN
   Calculate γ_dev
   FOR EACH point p in P DO
      -Calculate orientation angle γ_p of p
      -Put into{CorrList(p)} all points o with angle γ_o
       where (γ_p − γ_dev) ≤ γ_o ≤ (γ_p + γ_dev)
      -IF {CorrList(p)} = EMPTY THEN
          Mark p as Outlier
      -ELSE
         -QuickSort{CorrList(p)} with ascending
          distances in relation to p
         -FOR EACH item k in CorrList(p)
            -IF k has no Correspondence on P THEN
                SET Correspondence between k and p
   If t + x < i < t THEN Remove p_i
END
```

shape is kept fixed. For every transformed point in the first shape we try to find a corresponding point on the second shape. For the establishment of point correspondences we demand the following facts: a. produce one-to-one point correspondences, remove illegal point-correspondences from the list of one-to-one point correspondences, c. determine points without a correspondence as outlier, and d. produce symmetric results, which is obtaining the same results when aligning instance P to instance O as when aligning instance O to P.

Based on the distance between these corresponding points the alignment score is calculated for this specific iteration step. When the first shape is rotated once around its centroid, finally that rotation is selected and applied where the minimum alignment score is calculated.

In this respect the algorithm is similar to our nearest neighbor-search algorithm proposed in [5]. The main difference is the way we calculate point correspondences. It was shown in Sect. 2 that the establishment of legal sets of correspondences is an

Table 4. Evaluation of symmetric property

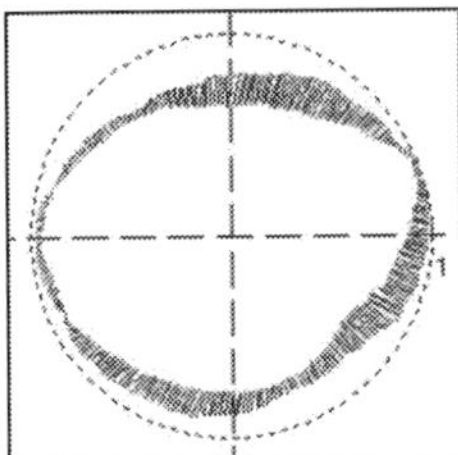	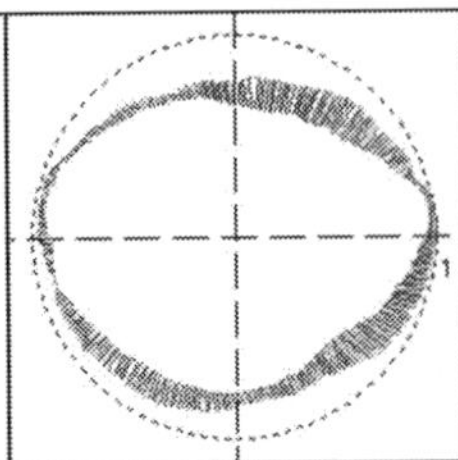	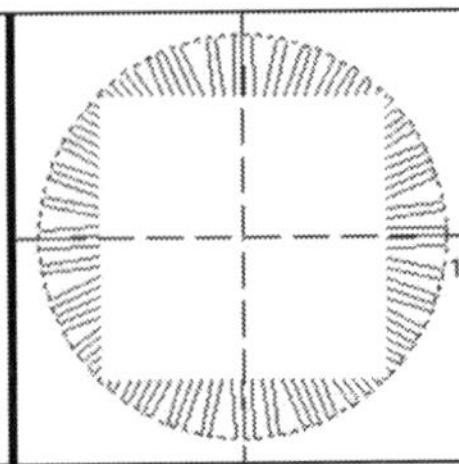	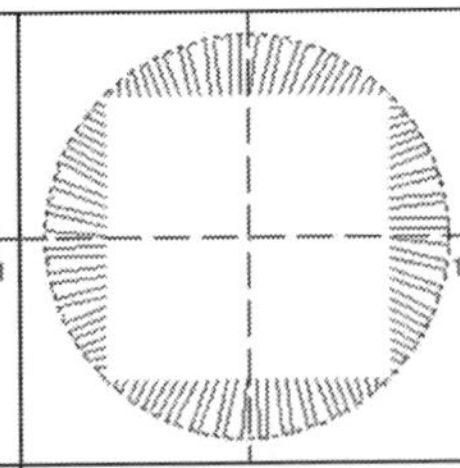
(a) *shape_12* (340 points) align to *shape_13* (340 points)	**(b)** *shape_13* (340 points) align to *shape_12* (340 points)	**(c)** *rect_mid* (116 points) *is* aligned to *circle* (144 points)	**(d)** *circle* (144 points) aligned to *rect_mid* (116 points)
$\psi = 0.2094$ $\bar{\varepsilon} = 0.0842$; $\varepsilon_{max} = 0.1835$ $Score = 0.1339$	$\psi = -0.2094$ $\bar{\varepsilon} = 0.0842$; $\varepsilon_{max} = 0.1835$ $Score = 0.1339$	$\bar{\varepsilon} = 0.1848$; $\varepsilon_{max} = 0.2921$ $Score = 0.2384$	$\bar{\varepsilon} = 0.1887$; $\varepsilon_{max} = 0.2904$ $Score = 0.2395$

important fact to distinguish between concave and convex shapes. The drawback of this requirement is that the set P of contour points p_i of the acquired shapes have to be an ordered set $(P, \leq)$.

Before the iterative algorithm starts we define a range where to search for potential correspondences. This range is defined by a maximum deviation of the orientation according to the centroid (see Table 2 c). This restriction will help us to produce legal sets of correspondences. The maximum permissible deviation of orientation γ_{dev} will be calculated in dependence of the amount of contour points n_O of the shape O, which is the instance that has more points than the other one. Our investigations showed that the following formula leads to a well-sized range

$$\gamma_{dev} = \pm \frac{4\pi}{n_O} \; . \qquad (2)$$

Let $t + x$ be an upper bound in the search area for a subset I of P if for every $i \in I$, we have $i \leq t + x$ and similarly, a lower bound in the search area for a subset I is an element t such that for every $i \in I, t \leq i$. Now, if we find more the one mapping between the point o and the points p_i within the search area, we remove the points $\{o, p_i\}$ having an ordering number i larger than the considered interval $\{t, t + x\}$ with $x = \frac{\gamma_{dev} \cdot n_o}{4\pi}$.

The complexity of the algorithm is $n\, N_O\, O(k \log_2 k)$. By introducing Bucket Sort instead of QuickSort we can reduce the complexity to linear complexity ($n\, N_O\, O(k)$).

4 Results

Table 5 shows some results of the alignment process. A point aligned to a circle gives the expected maximum dissimilarity value of *one* (Table 5 a), since *zero* means identity. If we align an ellipse to the circle and let this ellipse converge to a line, we get an increasing dissimilarity value which reaches the value 0.5 in case of a line (see Table 5 b- e). It can be seen that the dissimilarity value between the line and the circle is not exactly 0.5 (see Table 5 e). This is a small approximation error of the algorithm

Table 5. Exemplary results of our alignment process

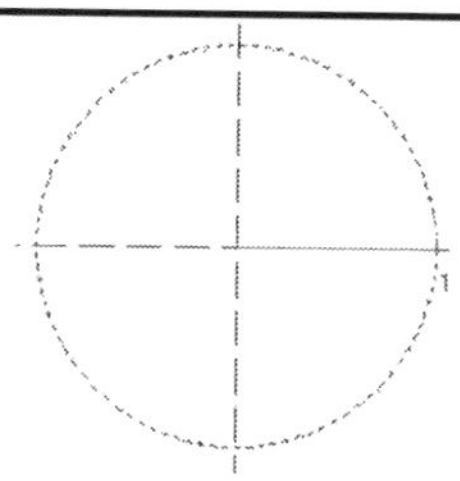 **(a)** *point* aligned to *circle* $\bar{\varepsilon} = 1$; $\varepsilon_{max} = 1$ $Score = 1$ Outlier included: *0*	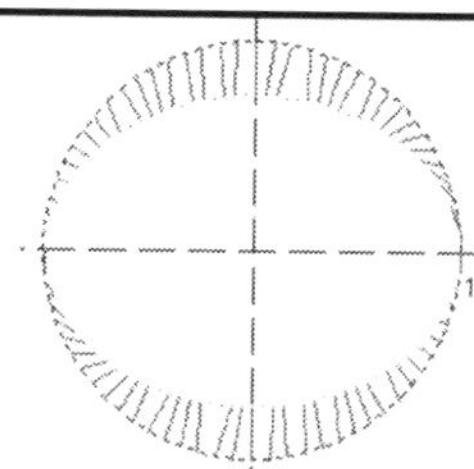 **(b)** *circle* aligned to *ellipse_1* $\bar{\varepsilon} = 0.1643$; $\varepsilon_{max} = 0.2532$ $Score = 0.2088$ Outlier included: *0*	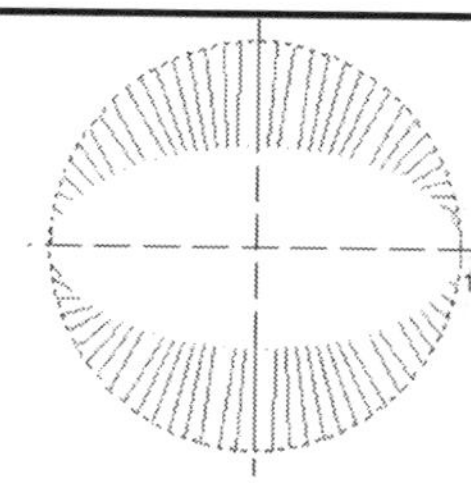 **(c)** *circle* aligned to *ellipse_2* $\bar{\varepsilon} = 0.3165$; $\varepsilon_{max} = 0.5016$ $Score = 0.4090$ Outlier included: 0
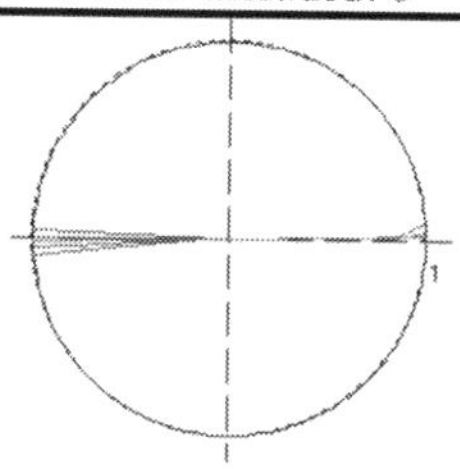 **(e)** *circle* aligned to *diameter* $\bar{\varepsilon} = 0.5112$; $\varepsilon_{max} = 1$ $Score = 0.7556$ Outlier included: *0*	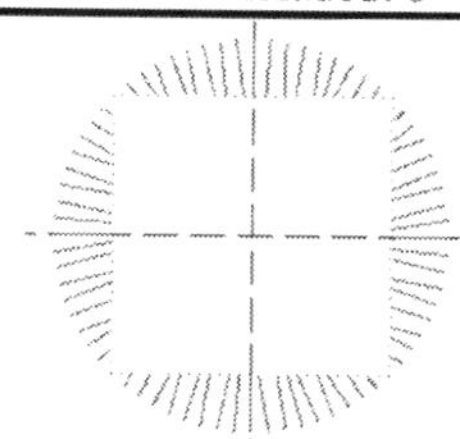 **(f)** *circle* aligned to *rect_mid* $\bar{\varepsilon} = 0.1924$; $\varepsilon_{max} = 0.2907$ $Score = 0.2415$ Outlier included: *0*	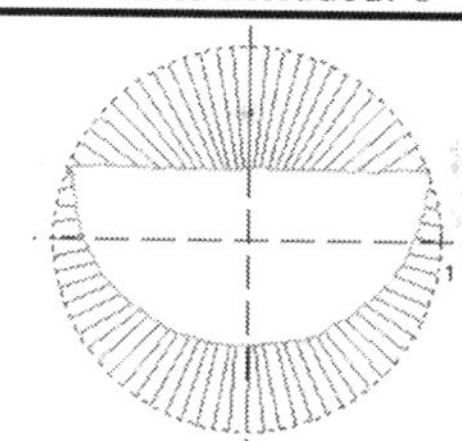 **(g)** *circle* al. to *semicircle* $\bar{\varepsilon} = 0.3642$; $\varepsilon_{max} = 0.6509$ $Score = 0.5076$ Outlier included: 0
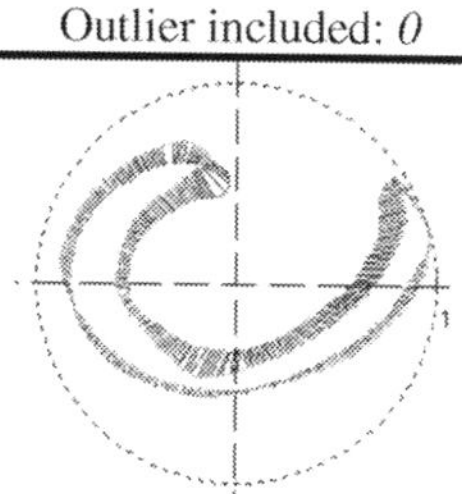 **(i)** *concave3* al. to *concave6* $\bar{\varepsilon} = 0.0777$; $\varepsilon_{max} = 0.1617$ $Score = 0.1197$ Outlier included: *0*	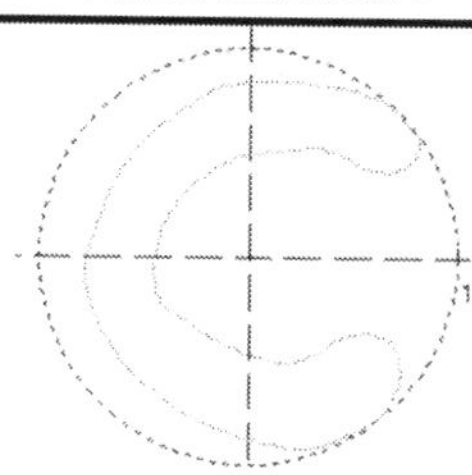 **(j)** concave1 al. to concave1 $\bar{\varepsilon} = 0$; $\varepsilon_{max} = 0$ $Score = 0$ Outlier included: *0*	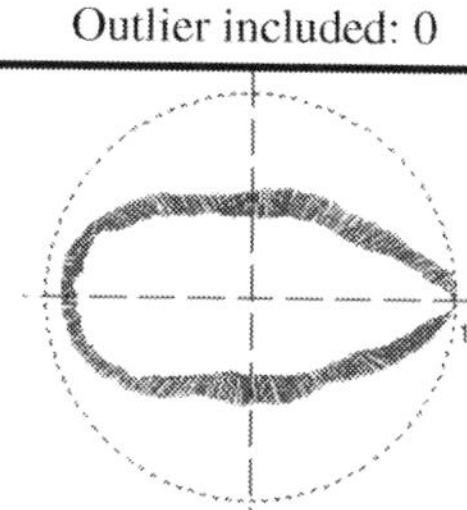 **(k)** *shape_2* al. to *shape_1* $\bar{\varepsilon} = 0.1015$; $\varepsilon_{max} = 0.1557$ $Score = 0.1286$ Outlier included: *0*

caused by the allowed search area for the correspondences. The alignment of other arbitrary shapes is shown in Table 5 f-p. The alignment of a concave object to the convex shape of a circle is shown in Table 5 h. The established correspondences are legal and a set of outliers was detected. Finally, this results in a high dissimilarity value.

In case both shapes have the same number of points, the symmetry of the similarity is given (see Table 4 a and Table 4 b). But the symmetry property does not exactly hold if a shape that consists of m points is aligned to a shape that consists of n points where $m > n$ (see Table 4 c and Table 4 d). The similarity value has a small deviation. This is because there are multiple choices to establish correspondences among the larger number of points of shape P to the smaller number of points of shape O. If the shape with the larger number of points has to be aligned to the shape with a lower number of points so that the symmetry criterion holds, some constraints are necessary that will be developed during further work.

In our study we are interested in determining the pair-wise similarity for clustering the set of acquired shapes into groups of similar shapes. The main goal is to learn for each of the established groups a generalized, representative shape. Finally, the set of generalized shapes is used for object recognition. From this point of view we do not need to enforce symmetric results ad hoc. The requirement was to result in a proper dissimilarity measure which holds under a wide variety of different shapes.

5 Conclusions

We have proposed a method for the acquisition of shape instances and our novel algorithm for aligning arbitrary 2D-shapes, represented by ordered point-sets of varying size. Our algorithm aligns two shapes under similarity transformation; differences in rotation, scale, and translation are removed. It establishes one-to-one correspondences between pairs of shapes and ensures that the found correspondences are symmetric and legal. The method detects outlier points and can handle a certain amount of noise. We have evaluated that the algorithm also works well if the aligned shapes are very different, like i.e. the alignment of concave and convex shapes. A distance measure which runs between *0* and *1* is returned as a result.

The methods are implemented in the program *CACM* (case acquisition and case mining)[16] *which runs on a* Windows *PC.*

References

1. D.G. Kendall, A Survey of the Statistical Theory of Shape, Statistical Science, Vol. 4, No. 2, pp. 87-120, 1989.
2. F.L. Bookstein, Size and Shape Spaces for Landmark Data in Two Dimensions, Statistical Science, Vol. 1, No. 2, pp. 181-242, 1986.
3. S. Belongie, J. Malik and J. Puzicha, Shape Matching and Object Recognition Using Shape Contexts, IEEE Transactions on Pattern Analysis and Machine Intelligence, Vol. 24, No. 24, pp. 509-522, 2002

4. N. Ueda and S. Suzuki, Learning Visual Models from Shape Contours Using Multiscale Convex/Concave Structure Matching, IEEE Trans. on Pattern Analysis and Machine Learning, vol. 15, No. 4, April 1993, p. 307-352.
5. P. Perner and S. Jänichen, Learning of Form Models from Exemplars, In: Ana Fred, Terry Caelli, Robert P. W. Duin, Aurelio Campilho, and Dick de Ridder (Eds.), Structural, Syntactic, and Statistical Pattern Recognition, Proceedings of the SSPR 2004, Lisbon/Portugal, Springer Verlag, lncs 3138, pp. 153-161, 2004.
6. A. Rangarajan, H. Chui and F.L. Bookstein, The Softassign Procrustes Matching Algorithm, Proc. Information Processing in Medical Imaging, pp. 29-42, 1997.
7. S. Sclaroff and A. Pentland, Modal Matching for Correspondence and Recognition, IEEE Trans. Pattern Analysis and Machine Intelligence, Vol. 17, No. 6, pp. 545-561, 1995.
8. J. Feldmar and N. Ayache, Rigid, Affine and Locally Affine Registration of Free-Form Surfaces, The International Journal of Computer Vision, Vol. 18, No. 3, pp. 99-119, 1996.
9. A. Hill, C.J. Taylor and A.D. Brett, A Framework for Automatic Landmark Identification Using a New Method of Nonrigid Correspondence, IEEE Transactions on Pattern Analysis and Machine Intelligence, Vol. 22, No. 3, pp. 241-251, 2000.
10. R.C. Veltkamp, Shape Matching: Similarity Measures and Algorithms, Shape Modelling International, pp. 188-197, 2001.
11. S.R. Lele and J.T. Richtsmeier, An Invariant Approach to Statistical Analysis of Shapes, Chapman & Hall / CRC 2001.
12. P Besl and N. McKay, A Method for Registration of 3-D Shapes, IEEE Trans. Pattern Analysis and Machine Intelligence, Vol. 14, No. 2, pp. 239-256, 1992.
13. O.-C. Marte and P. Marais, Model-Based Segmentation of CT Images, In South African Computer Journal, Vol. 28, pp. 54-59, 2002.
14. A.W. Fitzgibbon, Robust Registration of 2D and 3D Point Sets, In Proc. British Machine Vision Conference, Vol. II, pp. 411-420, Manchester, UK, 2001
15. S. Belongie, J. Malik and J. Puzicha, Shape Matching and Object Recognition Using Shape Contexts, IEEE Transactions on Pattern Analysis and Machine Intelligence, Vol. 24, No. 24, pp. 509-522, 2002.
16. P. Perner, S. Jähnichen, Case Acquisition and Case Mining for Case-Based Object Recognition, , In: Peter Funk, Pedro A. González Calero (Eds.), Advances in Case-Based Reasoning, Springer Verlag 2004, lnai 3155, pp. 616-629

Research on Iterative Closest Contour Point for Underwater Terrain-Aided Navigation

Wang Kedong[1], Yan Lei[2], Deng Wei[2], and Zhang Junhong[3]

[1] School of Astronautics, Beihang University, Beijing 100083, China
wangkd@buaa.edu.cn
[2] Institute of Remote Sensing and GIS, Peking University, Beijing 100871, China
[3] School of Automation Science and Electrical Engineering, Beihang University, Beijing 100083, China

Abstract. In order to provide underwater vehicle high-precision navigation information for long time, the coordinate properties of underwater terrain can be used to aid inertial navigation system (INS) by matching algorithm. Behzad and Behrooz (1999) introduce iterative closest contour point (ICCP) from image registration to underwater terrain matching and provide its exact form and prove its validity with an example. Bishop (2002) proves its validity systemically. However, their research considers that the matching origin is known exactly while it is seldom satisfied in practice. Simulation results show that ICCP is easy to diverge when the initial INS error is very large (such as 3km). To overcome the drawback, two enhancements are put forward. (1) The matching origin is added into matching process; (2) The whole matching process is divided into two phases: the coarse and the accurate. The coarse matching rules include mean absolute difference (MAD) and mean square difference (MSD) which is usually applied in terrain contour matching (TERCOM). The accurate matching is the ICCP optimization. Simulation results show that the updated ICCP matches application conditions very well and it is convergent with very high precision. Especially, when INS precision is not high, the updated ICCP matching process is more stable and its precision is higher than TERCOM's.

Keywords: ICCP, TERCOM, Pattern Recognition, Map Matching, Terrain-Aided Navigation.

1 Introduction

To provide underwater vehicle high-precision navigation information for long time, the coordinate property of underwater terrain can be used to aid inertial navigation system (INS) by matching algorithm. It is called terrain-aided navigation (TAN). A TAN system shown in Fig.1 is mainly composed of INS, sonar, terrain map, and matching algorithm. The core of the system is matching algorithm. The important part of matching algorithm is map pattern recognition and the recognition correctness decides the matching precision. Different recognition methods lead to different matching algorithms, such as terrain contour matching (TERCOM), Bayes, and etc. Since sonar is a single-point sensor, the matching process is one dimension. In order to improve matching precision, a series of measured points rather than a point are

D.-Y. Yeung et al. (Eds.): SSPR&SPR 2006, LNCS 4109, pp. 252–260, 2006.

accumulated to match with map. It is a kind of correlation matching which depends on terrain fluctuation and structure. With the appearance of underwater sonar array and multi-beam sonar, it is also possible to develop two-dimension matching algorithms. This article focuses on one-dimension matching algorithm. [1-8]

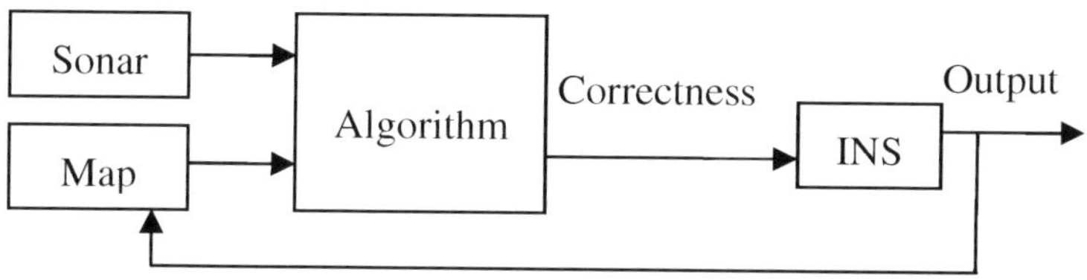

Fig. 1. The Scheme of Underwater Terrain-Aided Navigation System

Iterative closest contour point (ICCP) is a common image registration algorithm. Article [9] introduces ICCP to underwater TAN and provides its principle and procedure and proves its validity with an example. Article [10] proves its validity with different application conditions further. The other articles are also based on and similar with the research of the article [9, 10]. However, it is assumed in the articles that the matching origin is exactly known while it is seldom happened in practice except that INS is accurately aligned initially. In fact, due to the difference of underwater terrain fluctuation, some regions are suitable for matching while the others are unsuitable, i.e. there are matchable and unmatchable regions. In an unmatchable region, INS alone provides underwater vehicle navigation information and its error is accumulated with running time. When vehicle has passed an unmatchable region and then enters a matchable one, it is unsuitable to consider that the matching origin is exactly known any more due to the accumulated INS error. Therefore, it is necessary to update ICCP so that it can also be applied when the initial INS error is large, such as 2~5 nautical miles. (1 nautical mile is about 1.852 kilometer.)

As to the limitation of ICCP in principle when the initial INS error is very large, two enhancements are made firstly. Then, the updated ICCP is proved by simulation. In the end, the advantages of the updated ICCP are concluded by comparison with the existed ICCP and TERCOM.

2 Principle

2.1 The Existed ICCP (It Is Called ICCP-A in the Following)[9]

The principle of ICCP-A is shown in Fig.2. There is an actual path (the heavy line shown in Fig.2) which is called 'actual path' and composed of the points P_i' ($i = 1, 2, \cdots, M$) and M is the path length. INS provides a measured path (the fine line shown in Fig. 2) which is called 'INS path' and composed of the points P_i ($i = 1, 2, \cdots, M$). At the same time, sonar provides the corresponding measured water depth values c_i ($i = 1, 2, \cdots, M$) and each depth value corresponds to a contour. Due to INS error, there is difference between P_i and P_i' inevitably. The thought of ICCP is that P_i' should lie in c_i-contour and then P_i should be moved to the estimated points P_i'' ($i = 1, 2, \cdots, M$), which compose 'estimated path', and approached the contour with a certain rule. The approach process is realized by optimization.

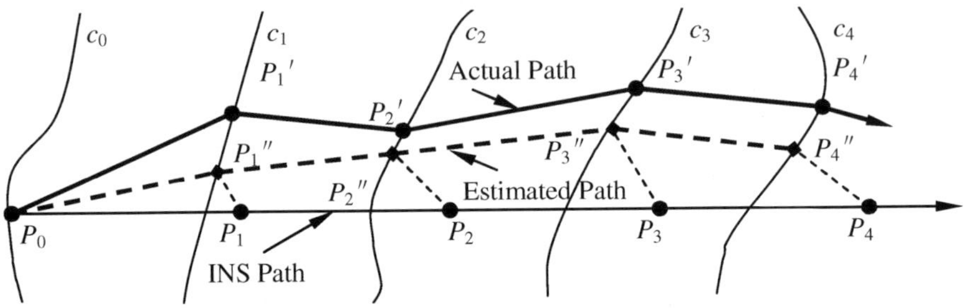

Fig. 2. The Sketch of the Current ICCP Algorithm

2.2 The Enhancement I (It Is Called ICCP-B in the Following)

The principle of ICCP-B is shown in Fig.3. The only difference between ICCP-B and ICCP-A is that the matching origin is also adjusted as a point of the matching path in ICCP-B while it is not adjusted in ICCP-A. Similar with ICCP-A, we can also establish the subject of optimization as follows:

$$E = d(\mathbf{x}_1, \ \mathbf{a}_1) + \sum_{i=2}^{M} d(\mathbf{x}_i - \mathbf{x}_{i-1}, \ \mathbf{a}_i - \mathbf{a}_{i-1}) + K \sum_{i=1}^{M} d(\mathbf{x}_i, \ \mathbf{y}_i), \quad (1)$$

in which E is the subject or the total constrain error, $d(\mathbf{p}, \mathbf{q})$ the Euclidean distance between $\mathbf{p}$ and $\mathbf{q}$, $\mathbf{x}_i$ the estimated position of P_i'', $\mathbf{y}_i$ the closest contour point or projection of $\mathbf{x}_i$ in c_i-contour, $\mathbf{a}_i$ the INS measured position, and K stiffness coefficient. As shown in Eq.(1), the first two items of the right side is to restrict the

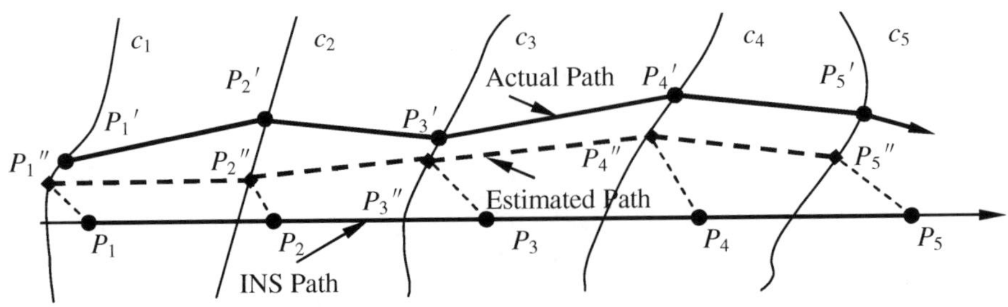

Fig. 3. The Sketch of ICCP-B

estimated path in the vicinity of the INS path and the third is to make the estimated path approach to the sonar measured value contour.

Shown in Fig.4 is the influence of vehicle speed and heading errors on one-step running. ρ_i and θ_i in Fig.4 are vehicle speed and heading errors respectively. $\mathbf{b}_{i+1}$ is the unit vector along ($\mathbf{a}_{i+1} - \mathbf{a}_i$) and $\mathbf{e}_{i+1}$ is the unit vector perpendicular to $\mathbf{b}_{i+1}$ (i.e. $\mathbf{b}_{i+1}$ rotated 90° counter-clockwise). The following equations can be expressed according to Fig.4.

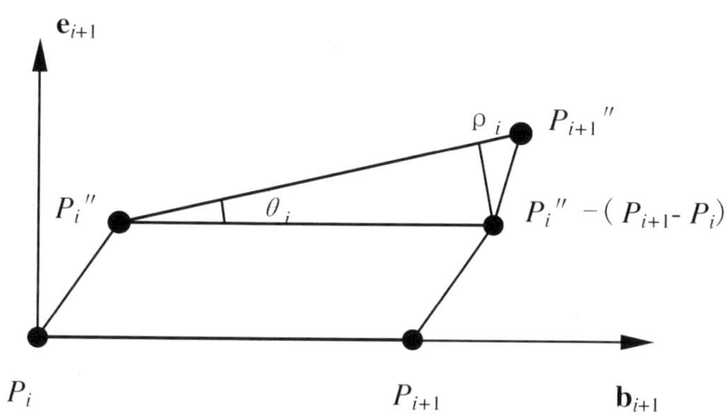

Fig. 4. The Influence of Vehicle Speed and Heading Errors on One-step Running

$$\begin{aligned} \mathbf{x}_i &= \mathbf{x}_{i-1} + \left(\|\mathbf{a}_i - \mathbf{a}_{i-1}\| + \rho_i\right)\left(\cos\theta_i \mathbf{b}_i + \sin\theta_i \mathbf{e}_i\right), \quad i > 1; \\ \mathbf{x}_1 &= \mathbf{a}_1 + \rho_1\left(\cos\theta_1 \mathbf{b}_1 + \sin\theta_1 \mathbf{e}_1\right), \quad i = 1. \end{aligned} \tag{2}$$

If ρ_i and θ_i are small and the second and higher order items are ignored, Eq.(2) can be approximated as follows.

$$\begin{aligned} \mathbf{x}_i &\approx \mathbf{x}_{i-1} + \mathbf{a}_i - \mathbf{a}_{i-1} + \rho_i \mathbf{b}_i + \zeta_i \mathbf{e}_i, \quad i > 1; \\ \mathbf{x}_1 &\approx \mathbf{a}_1 + \rho_1 \mathbf{b}_1 + \zeta_1 \mathbf{e}_1, \quad i = 1, \end{aligned} \tag{3}$$

in which $\zeta_i = \|\mathbf{a}_i - \mathbf{a}_{i-1}\|\theta_i$ and $\zeta_1 = \rho_1\theta_1$. Substitute Eq.(3) into (1) and obtain

$$E = \sum_{i=1}^{M}\left(\rho_i^2 + \zeta_i^2\right) + K\sum_{i=1}^{M}\|\mathbf{x}_i - \mathbf{y}_i\|^2. \tag{4}$$

If $\mathbf{a}_i$, ρ_i, and θ_i are known, $\mathbf{x}_i$ ($i = 1, 2, \cdots, M$) can be calculated by Eq.(3). $\mathbf{a}_i$ is the INS measured position and known, so $\mathbf{x}_i$ is the function of ρ_i and θ_i. Moreover, $\mathbf{y}_i$ can be decided by $\mathbf{x}_i$, so it is the implicit function of ρ_i and θ_i.

It is a nonlinear optimization problem in { ρ_i, θ_i }. The simplex method, which does not require the gradient, can be used to solve the minimization problem.

2.3 The Enhancement II (It Is Called ICCP-C in the Following)

It is known by comparing Fig.1 with 2 that ICCP-B is more adjacent to the practice than ICCP-A. However, it can be deduced that ICCP-B is still based on the following two hypotheses.

(1) The vehicle actual position is in the vicinity of the INS measured position.
(2) The vehicle actual position is in the vicinity of the sonar measured contour.

The hypothesis (2) is the basis of matching and can be satisfied if sonar's error is low enough. However, the hypothesis (1) can not be satisfied if the initial INS error is much large, which may lead to the divergence of ICCP-B. If a coarse matching is implemented before ICCP is applied, a coarse matched path, which should be closer to the actual path than the INS path, will be obtained. In the following accurate matching, the INS path is substituted with the coarse matched path and the following procedure is same as ICCP-B.

Mean square difference (MSD) and mean absolute difference (MAD), two of TERCOM matching rules, are applied here to construct the coarse matching. The two rules are expressed in Eq.(5) and (6).

$$J_{\mathrm{MAD}}(x, y) = \frac{1}{M}\sum_{i=1}^{M}\left|\left[h_t(i) - h_m(i) - \left(\bar{h}_t - \bar{h}_m\right)\right]\right|, \tag{5}$$

$$J_{\mathrm{MSD}}(x, y) = \frac{1}{M}\sum_{i=1}^{M}\left\{\left[h_t(i) - h_m(i)\right] - \left(\bar{h}_t - \bar{h}_m\right)\right\}^2, \tag{6}$$

in which $J_{\mathrm{MAD}}(x, y)$ and $J_{\mathrm{MSD}}(x, y)$ are the MAD and MSD values at the point (x, y), $h_t(i)$ the water depth of the i-th point in the estimated path in map, $h_m(i)$ the water value of the i-th point in the sonar measured path, $\bar{h}_t$ and $\bar{h}_m$ the mean value of the estimated and sonar measured paths respectively.

If the point at which the minimum MAD value is obtained is same as the point at which the minimum MSD value is obtained, the path at the point in map is the coarse estimated path. Otherwise, the absolute differences between the paths at the minimum MAD and MSD value points and the sonar measured path are calculated. The path with a smaller difference is decided as the coarse estimated path.

Up to now, the coarse matching is completed and the coarse estimated path, which substitutes for the INS measured path in the following accurate matching, is obtained. The accurate matching is same as ICCP-B. This is ICCP-C which is composed of the coarse and accurate matching phases.

3 Simulations

The main simulation conditions are listed in Table 1 and sonar error model is shown in Table 2. The 3-D image of a map is shown in Fig.5. The map is a 473×473 square grid and the grid distance is 58 meters. Its origin is (0°, 0°). The simulation origin is (0.03976°, 0.151599°) and vehicle sails along longitude from west to east. The path length M is 10 points. The compared object is matching error. In Table 1, g is gravity acceleration and 0.02° longitude (or latitude) error is about 3.149km. In Table 2, h is water depth.

Shown in Fig.6 are the simulation results of ICCP-A, ICCP-B, ICCP-C, and TERCOM.

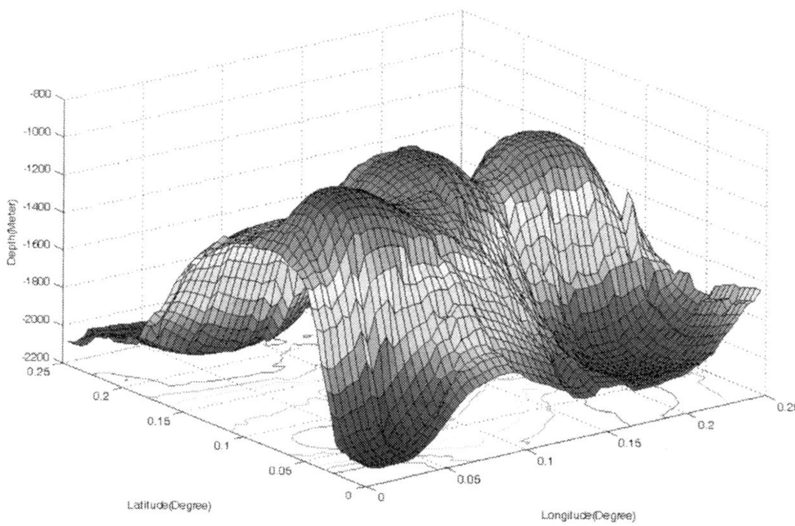

Fig. 5. The 3-D Image of the Map

Table 1. The Main Simulation Conditions

Vehicle	Initial Longitude and Latitude Errors (°)	0.02
	Sample Step (s)	14
	Speed (m/s)	4
Gyroscope	X-axis Bias (°/h)	0.001
	Y-axis Bias (°/h)	0.001
	Z-axis Bias (°/h)	0.001
	X-axis Random Walk (°/$h^{1/2}$)	0.001
	Y-axis Random Walk (°/$h^{1/2}$)	0.001
	Z-axis Random Walk (°/$h^{1/2}$)	0.001
Accelerator	X-axis Bias (g)	1×10^{-5}
	Y-axis Bias (g)	1×10^{-5}
	Z-axis Bias (g)	1×10^{-5}
	X-axis Random Walk (g·$s^{1/2}$)	1×10^{-5}
	Y-axis Random Walk (g·$s^{1/2}$)	1×10^{-5}
	Z-axis Random Walk (g·$s^{1/2}$)	1×10^{-5}

Table 2. Sonar Error Model

Depth (m)	Error
0~10	0.1m
10~100	0.4%× *h*
100~300	0.6%× *h*
300~1000	0.8%× *h*
1000~8000	1%× *h*
>8000	Not Workable

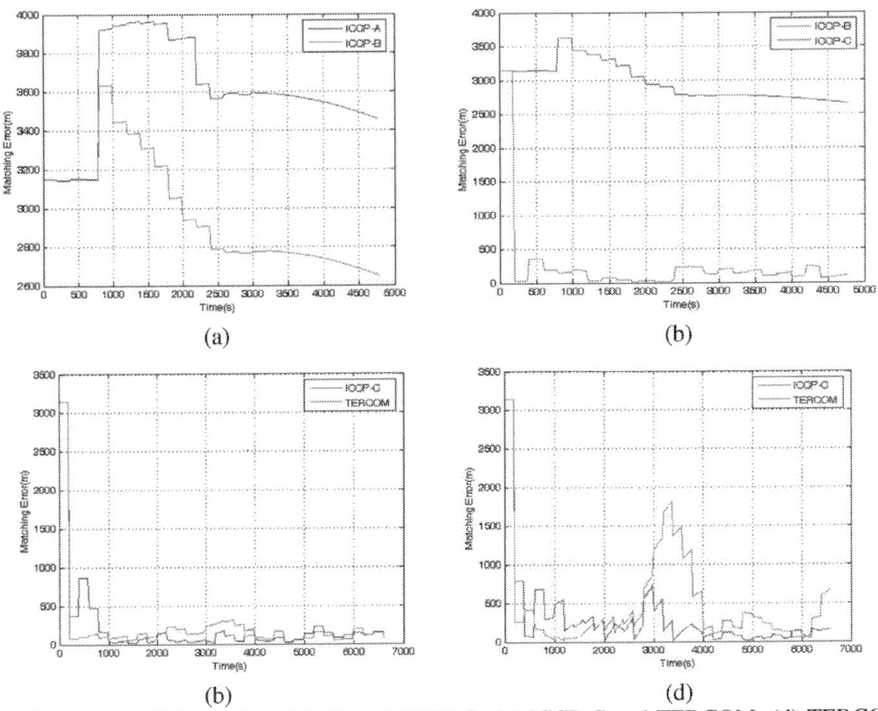

(a) ICCP-A and ICCP-B; (b) ICCP-B and ICCP-C; (c) ICCP-C and TERCOM; (d) TERCOM and ICCP-C with low-precision INS

Fig. 6. The Comparison of Simulation Results

It is known from Fig.6 (a) that the ICCP-B matching error is smaller than the ICCP-A's about 800m. The result proves that adjusting the matching origin with other points of the estimated path simultaneously in ICCP-B matches the condition that the initial INS error is very large (about 3.149km), but the optimization process is very difficult to converge. In programming, there is a threshold value for iterated times, i.e., when iterated times is larger than the threshold value, the optimization is terminated and the current path is put out. The path is just a local suboptimum usually. In fact, the latter part of the ICCP-B simulation shown in Fig.6 (a) is divergent and the

INS path is put out. The result shows that ICCP-B is also easy to diverge and its matching error is still too large to be used.

Shown in Fig. 6 (b) is that ICCP-C converges quickly and its matching error is reduced sharply and the convergent error is about as low as 200m. The result proves that the introduction of the coarse matching shrinks the searching window of the following accurate matching remarkably, which makes the accurate matching applied in the vicinity of the actual path and easy to converge. The enhancement is more effective.

Shown in Fig. 6 (c) is that the matching results of ICCP-C and TERCOM are similar except that ICCP-C's precision is higher than TERCOM's in some segments. The results show that the two algorithms' results are very good under the conditions.

Shown in Fig. 6 (d) are the matching results of ICCP-C and TERCOM under the other simulation conditions. The only differences between the conditions include all gyroscope bias and random walk values changing from 0.001 to 0.01 and all acceleration bias and random walk values changing from 1×10^{-5} to 1×10^{-4}. That is, the INS' precision here is much lower than the former. The results show that ICCP-C's matching precision is higher than TERCOM's. The main reason is that there is the optimization further in ICCP-C after the coarse matching and the estimated path is moved closer to the actual path by the optimization. There is no such adjusting procedure in TERCOM so that its matching result is easy to be influenced by the shape of the INS path. Especially, when INS' precision is low, there is much large difference between the INS path and the actual path, which will lead to much large miss-matching in TERCOM. The results prove that ICCP-C is more applicable than TERCOM due to the existence of optimization.

The results show that ICCP-C is the effective fusion of ICCP-B and TERCOM. Firstly, the MAD and MSD rules in the coarse matching can reduce the following accurate matching scope and decrease the chance of miss-matching sharply, which benefits improving matching precision and increasing the convergence speed; secondly, the optimization in the accurate matching can promote matching precision further and the advantage of the existence of the optimization is more obvious especially when INS' precision is low.

4 Conclusions

By analyzing the current ICCP, i.e. ICCP-A, and the application conditions, two enhancements are put forward.

(1) Add the matching origin into matching process and construct the updated algorithm ICCP-B.

(2) Construct the coarse matching by referring the MAD and MSD rules from TERCOM. The accurate matching is same as ICCP-B. The updated algorithm ICCP-C has two matching phases: the coarse and the accurate.

The simulation results prove the following points.

(1) The two enhancements are effective and ICCP-C's precision is highest under the same simulation conditions.

(2) The enhancement that adding the matching origin into matching process makes the updated ICCP, i.e. ICCP-B, is more coincident with the application conditions

than ICCP-A and is beneficial to improving matching precision, but ICCP-B is still easy to diverge due to the local suboptimal rather than the global optimal path is found when the initial INS error is large (such as 3km).

(3) The convergence speed and matching precision of ICCP-C are improved sharply due to the introduction of the coarse matching. Compared with TERCOM, ICCP-C's application scope is larger and its matching precision is higher especially when INS' precision is low. The main reason is that the existence of the coarse matching makes the optimization applied in the vicinity of the actual path, which matches the application conditions, and the optimization can improve the matching precision further. When the shape of the estimated path is much different from the shape of the actual path due to the low INS precision, the advantage of the existence of the optimization in ICCP-C is more prominent.

References

1. Yan, M. : Terrain Matching Algorithm Design and Simulation for Underwater TAN System. Beijing: Peking University thesis (2004)
2. Sistiaga, M., Opderbecke, J., Aldon, M.J., et al. : Map Based Underwater Navigation Using a Multibeam Echo Sounder. Oceans Conference Record (IEEE), 2 (1998) 747-751
3. Maurer, C., Aboutanos, G., Dawant, B., et al. : Registration of 3-D Images Using Weighted Geometrical Features. IEEE Transactions on Medical Imaging, 6 (1996) 836-849
4. Yamany, S.M., Ahmed, M.N., Hemayed, E.E., et al. : Novel Surface Registration Using the Grid Closest Point (GCP) Transform. IEEE International Conference on Image Processing, 3 (1998) 809-813
5. Miga, M.I., Sinha, T.K., Cash, D.M., et al. : Cortical Surface Registration for Image-guided Neurosurgery Using Laser-range Scanning. IEEE Transactions on medical Imaging, 8 (2003) 973-985
6. Yang, X., Sheng, Y., Guan, W., et al. : Adaptive Hill Climbing and Iterative Closest Point Algorithm for Multisensor Image Registration with Partial Hausdorff Distance. Proceedings of SPIE-The International Society for Optical Engineering, Vol.4051, (2000) 99-109
7. Kaneko, S., Kondo, T., Miyamoto, A. : Robust Matching of 3D Contours Using Iterative Closest Point Algorithm Improved by M-estimation. Pattern Recognition, 9 (2003) 2041-2047
8. Fountain, J. R. : Digital Terrain System. IEE Colloquium (Digest), 169 (1997) 4/1-4/6
9. Behzad, K.P., Behrooz, K.P. : Vehicle Localization on Gravity Maps. Proceedings of SPIE-The International Society for Optical Engineering, Vol.3693, (1999) 182-191
10. Bishop, G.C. : Gravitational Field Maps and Navigational Errors. IEEE Journal of Oceanic Engineering, 3 (2002) 726-737

Image-Based Absolute Positioning System for Mobile Robot Navigation

JaeMu Yun, EunTae Lyu, and JangMyung Lee[1]

[1]Department of Electronics Engineering, Pusan National University,
Busan 609-735, Korea
jmlee@pusan.ac.kr

Abstract. Position estimation is one of the most important functions for the mobile robot navigating in the unstructured environment. Most of previous localization schemes estimate current position and pose of a mobile robot by applying various localization algorithms with the information obtained from sensors which are set on the mobile robot, or by recognizing an artificial landmark attached on the wall or objects of the environment as natural landmarks. Several drawbacks about them have been brought up. To compensate the drawbacks, a new localization method that estimates the absolute position of the mobile robot by using a fixed camera on the ceiling in the corridor is proposed. And also, the proposed method can improve the success rate for position estimation, since it calculates the real size of an object. This is not a relative localization scheme which reduces the position error through algorithms with noisy sensor data, but a kind of absolute localization. The effectiveness of the proposed localization scheme is demonstrated through the experiments.

1 Introduction

By the end of the 21st century, robots may not be strangers to us anymore. Compared with in recent years, the useful range for robots has gradually spread to a wide variety of areas. Mobile robots are especially being used as a substitute for humans in unwelcoming environments or to do simple works those are either in or outside. In addition, they are used for investigating planets in space [1]. In such a mobile robot system, getting exact information on its current position is very important. The mobile robot mainly calculates its position with the data acquired from a rotary encoder which is connected to the wheel, and from a gyroscope sensor. However it couldn't perceive the correct position because of slippage, a rough surface, and sensor error such as gyroscope drift. Many solutions have been proposed to overcome these unavoidable errors. For example some researchers presented a method that estimates the current position by applying information obtained by a rotary encoder and an ultrasonic sensor by applying an EKF (extended Kalman filter) [2,3]. And a researcher updated the positioning of mobile robots by fusing data from multi-sensors such as magnetic compasses, gyroscopes, rotary encoders with the EKF [4]. These methods need much calculation for a mobile robot to

D.-Y. Yeung et al. (Eds.): SSPR&SPR 2006, LNCS 4109, pp. 261 – 269, 2006.

perform a task, which results in a sharp drop in the total system efficiency. Another disadvantage is a great localization uncertainty which is the result of the statistical error accumulated from sensors and control over long distances. Contrary to the methods mentioned above, which intended to reduce the position error with relative positioning sensors, the following method provides an absolute position regardless of the distance moved and working time of a mobile robot. And some researchers presented a method that estimates the position of a robot through geometric calculation, after it recognizes a landmark [5-6]. Even though a CCD camera set on a robot is used for avoiding obstacles and tracking objects and so on, in these methods, the camera system was consumed unnecessarily for a robot to search and recognize the exact landmark. Another robot equipped with a CCD camera, estimates its position by recognizing a characteristic topography or an object, and compares it with the model image saved in advance [8]. In general, some feature points are utilized such as a wall or a corner as landmark in the workspace. However it has low confidence in recognition and requires much calculation. Therefore, the processing speed of the system is low.

In this paper, to overcome these problems, a new localization method is proposed and illustrated in Fig. 1. A camera installed on the ceiling of the corridor is utilized for the localization of the mobile robot. The sequence of an absolute positioning system can be summarized as follows:

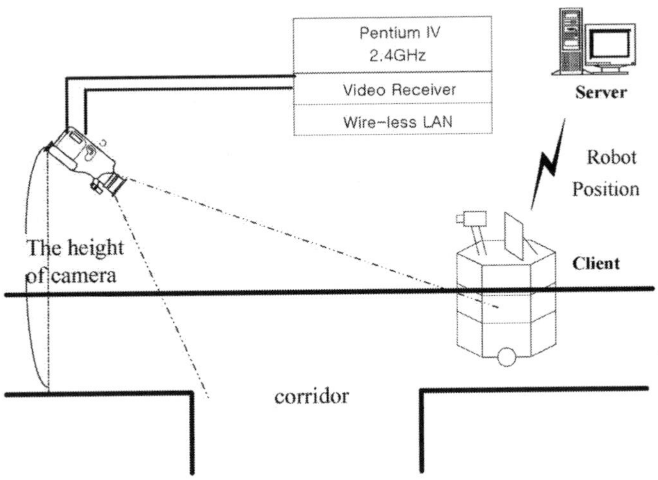

Fig. 1. Absolute positioning system

First, the system recognizes whether it is a moving robot or not, with a CCD camera. Secondly, if the object is a moving robot, the system obtains the position of the robot. Finally, the system transmits the position data to the robot for the localization.

2 Object Segmentation Through Image Process

Moving objects are extracted using the difference image which is obtained as the difference between an input image, which is being inputted consecutively, and a reference image, which is captured and stored in advance.

2.1 Image Pre-processing

A Gaussian mask is applied with the nine pixels for removing illumination dependent image noises, and modular four images which have 160 X 120 pixels for an image are used for image pre-processing.

2.2 Filtering and Labeling

A filtering method that has been used widely, a morphological filtering method is adopted. Through the labeling, objects are distinguished and their features are searched using labels.

2.3 Reference Image Modification

In order to extract a moving object in a dynamic environment correctly, the reference image needs to be updated dynamically instead of keeping the initial reference image. In Fig. 2, I_k^{rf} is an updated reference image that will be used for the next frames. Also, a mask image M_k^n is represented as follows:

$$M_k^n = \begin{Bmatrix} 1 & if\,(x,y) \in C_k^n, contour-set \\ 0 & otherwise. \end{Bmatrix} \quad (1)$$

This new reference image can be represented as (refer to Fig. 3.),

$$I_k^{rf} = \overline{M_k^n} I_k + M_k^n I_{k-1}^{rf} \,. \quad (2)$$

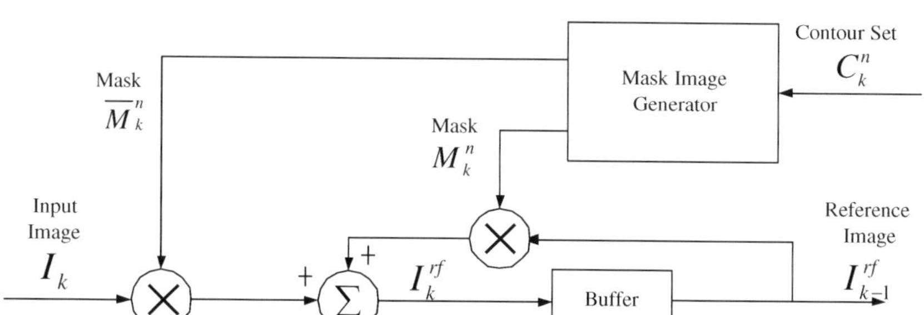

Fig. 2. Object segmentation model

(a) $\overline{M_k^n} I_k$ (b) $M_k^n I_k$ (c) $M_k^n I_{k-1}^{rf}$ (d) I_k^{rf}

Fig. 3. Update process of background image

3 Transformation from Image Coordinates to Real Coordinates

The distance between camera and object is obtained using a single camera so that, such distance can be represented as real coordinates [8]. As shown in Fig. 4, the solid square border in the center has a screen image for a mobile robot. This image is projection of the mobile robot on the corridor, which is in real three dimensions. Here, the image coordinates can be transformed to real coordinates to obtain the location of the robot.

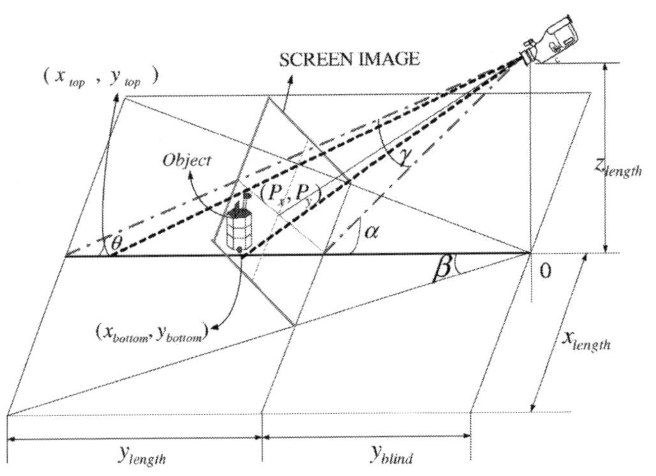

Fig. 4. Modeling for the correspondence between 2D image and 3D coordinates

$$\alpha = \tan^{-1}\left(\frac{z_{Length}}{y_{blind}}\right) \tag{3}$$

$$\beta = \tan^{-1}\left(\frac{x_{length}}{y_{blind} + y_{length}}\right) \tag{4}$$

$$\gamma = \alpha - \theta \tag{5}$$

where $\theta = \tan^{-1}\left(\frac{z_{length}}{y_{blind} + y_{length}}\right)$.

The screen image is described in detail as Fig. 5.

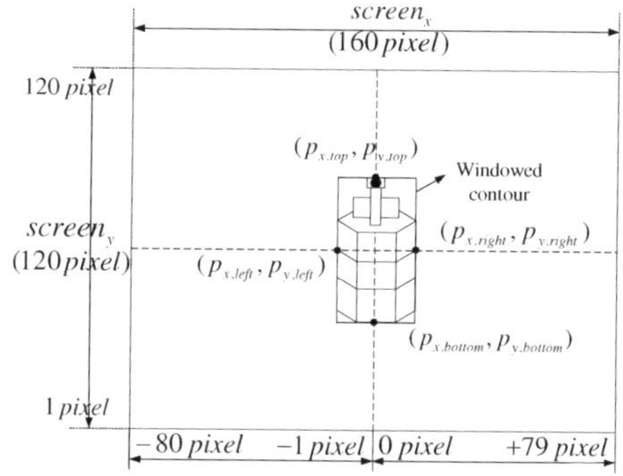

Fig. 5. Screen image

The real coordinates of the robot center on the floor, $(x_{robot\_position}, y_{robot\_position})$, can be calculated as follows:

$$y_{bottom} = z_{length} \times \tan[(90° - \alpha) + \gamma \times (\frac{p_{y,bottom}}{screen_y})] \tag{6}$$

The y-axis center of the robot can be obtained as,

$$y_{robot\_position} = y_{bottom} + (L/2) \tag{7}$$

where L is width of the robot. And,

$$x_{robot\_position} = y_{robot\_position} \times \tan \beta(\frac{2p_{x,bottom}}{screen_x}). \tag{8}$$

4 Feature Extraction

4.1 Height and Width of an Object

The height and width of the robot can be obtained using geometric analysis.

As shown in Fig. 6, the distance y_1 from the lowest coordinates of the object to the origin is calculated using y_{bottom} in Eq. (6) as,

$$y_1 = y_{bottom} - O \tag{9}$$

where O represents the origin.

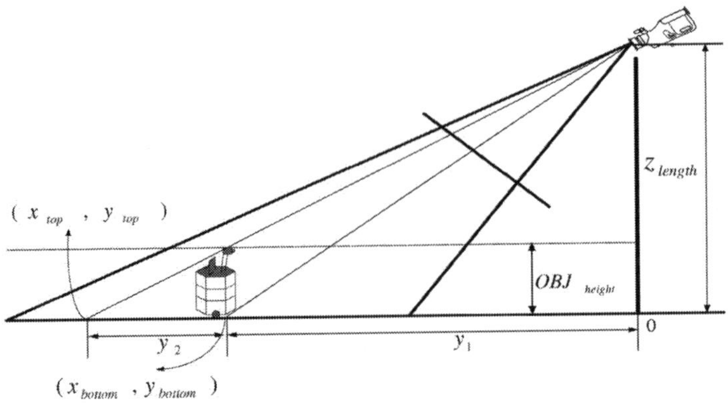

Fig. 6. Height measurement using a camera

In the same manner, y_{top} can be calculated from Eq. (6) by replacing y_{bottom} as y_{top} and $P_{y,bottom}$ as $P_{y,top}$. Therefore, the distance y_2 from the highest coordinates of the object to y_{bottom} is calculated as,

$$y_2 = y_{top} - y_{bottom} \,. \tag{10}$$

When the coordinates, y_1 and y_2 are obtained, the height of the robot, OBJ_{height} can be calculated as,

$$OBJ_{height} = \frac{z_{length} \times y_2}{(y_1 + y_2)} \tag{11}$$

from the similarity properties of triangles.

Following the same procedure, the width of the mobile robot can be obtained as follows:

The real length $length_{pixel}$ per pixel is calculated as follow:

$$length_{pixel} = OBJ_{height} / (P_{y,top} - P_{y,bottom}) \,. \tag{12}$$

Then, the width, OBJ_{width}, of the object is calculated as

$$OBJ_{width} = length_{pixel} \times (p_{x,right} - p_{x,left}) \,. \tag{13}$$

4.2 Extraction of Color Information

To recognize the mobile robot, the height, width and color information have been used for a neural network. Since most color cameras used for acquiring digital images utilize the RGB format, RGB values for the object image are obtained and represented as 8 bit data.

5 Experiments and Discussion

5.1 Mobile Robot for Experiments

Two mobile robots shown in Fig. 7 are used for experiments.

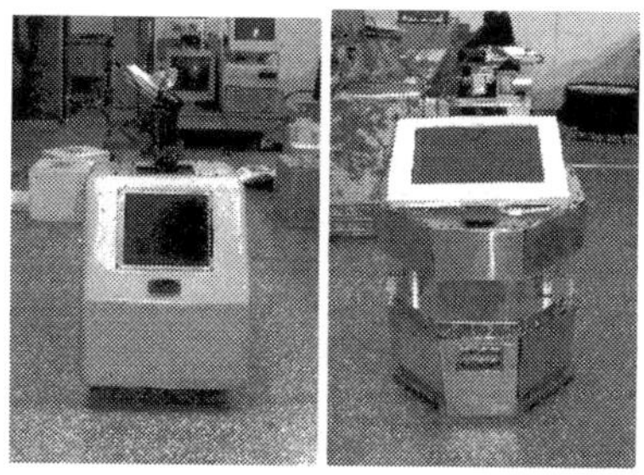

Fig. 7. Experimental mobile robots

5.2 Object Segmentation

The images of robots are extracted, which are navigating in corridor. The experimental results are shown in Fig. 8.

(a) Ref. Image (b) Input Image (c) Ext. Image

Fig. 8. Extraction of mobile robot images

5.3 Recognition of a Robot Through Neural Network

First of all, it is necessary to recognize an object to estimate the exact position of the robot. For this, a neural network is utilized to decide whether an extracted object in the image is a robot or not.

As shown in Table 1, with the size information, the success rate is improved a lot.

Table 1. Success rate of recognition

	recognition by using only color information		recognition by using color information and size information	
object	Number of trials	Number of success	Number of trials	Number of success
IRL-2002	40	30	40	35
Ziro3	40	32	40	38
people	20	12	20	18

5.4 Acquisition of a Robot Position and Results of Experiments

When a mobile robot is driven 10m forward, experimental results are shown in Fig. 9. Using only an encoder sensor and the kinematics of the mobile robot [7], there exists an approximately 40cm deflection along the x axis. However note that using the proposed method, the robot trajectory is kept close to the center line of a driven corridor.

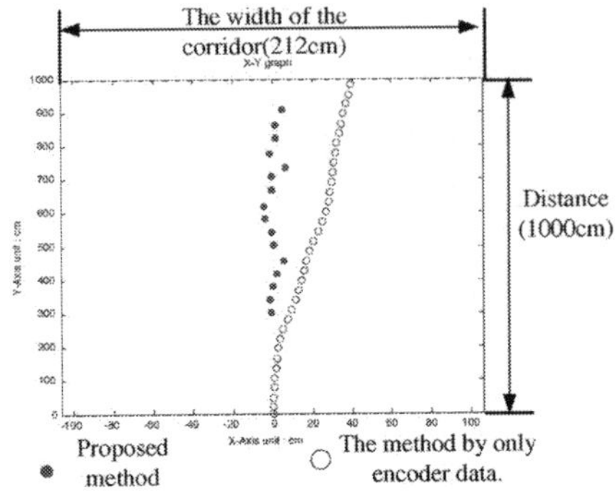

Fig. 9. Motion trajectory of a robot

Table 2. Position error by the proposed method

Distance from camera	Axis	Maximum error (cm)	Minimum error (cm)	Average error (cm)
3m	X axis error	0.573	0.201	0.420
	Y axis error	2.583	0.125	1.706
4m	X axis error	2.386	0.318	1.175
	Y axis error	4.833	0.364	3.073

As shown in Table 2, the further the robot moved from the camera, the greater the error became in real coordinates. The error in the x axis is influenced by both the distance and angle from the camera. Consequently, in the limited camera view area, the robot position is precisely recognized without missing the robot.

6 Conclusion

In this paper, a new localization method with a fixed camera is proposed, which utilizes the external monitoring camera information under the indoor environment. When a mobile robot is moving along the corridor, it helps the localization of robot by estimating the current position through the geometric analysis of the mobile robot image. The exact position of the mobile robot was obtained and demonstrated to be correct by the real experiments. And through the experiments, the advantages and efficiency of the proposed method are demonstrated illustratively.

For a future research topic, an efficient image processing scheme is necessary to improve and reduce the absolute error.

Acknowledgement

This work was supported by 'Research Center for Logistics Information Technology (LIT)' hosted by the Ministry of Education and Human Resources Development in Korea.

References

1. Clark F. Olson, "Probabilistic Self-Localization for Mobile Robots," IEEE Trans. on Robotics and Automation, vol. 16, no. 1, pp. 55 –66, Feb. (2000).
2. Leopoldo Jetto, Sauro Longhi and Giuseppe Venturini, "Development and Experimental Validation of an Adaptive Extended Kalman Filter for the Localizaiton of Mobile Robots," IEEE Trans. on Robotics and Automation, vol. 15, no. 2, pp. 219 –229, Apr. (1999).
3. A. Curran and K.J. Kyriakopoulos, "Sensor-Based Self-Localization for Wheeled Mobile Robots," Proc. of ICRA, vol. 1, pp. 8 –13, May (1993).
4. Ching-Chih Tsai, "A localization system of a mobile robot by fusing dead-reckoning and ultrasonic measurements," IEEE Trans. on Instrumentation and Measurement, Vol. 47, no. 5, pp.1399 –1404, Oct. (1998).
5. Hognbo Wang, Cheolung Kang, Shin-ichirou Tanaka and Takakazu Ishimatsu, "Computer Control of Wheel Chair by Using Landmarks," Proc. of KACC, Oct. (1995).
6. M. Mata, J.M. Armingol, A. de la Escalera and M.A. Salichs, "A visual landmark recognition system for topological navigation of mobile robots," Proc. of ICRA, Vol. 2, pp. 1124 – 1129, May (2001).
7. Il-Myung Kim, Wan-Cheol Kim, Kyung-Sik Yun and Jang-Myung Lee, "Navigation of a Mobile Robot Using Hand Gesture Recognition," Trans. on Control, Automation and Systems Engineering, vol. 8, no. 7, pp. 599-606, Jul. (2002).
8. Sung Yug Choi and Jang Myung Lee, "Applications of moving windows technique to autonomous vehicle navigation," Image and Vision Computing, pp. 120-130, Jan. (2006).

An Evaluation of Three Popular Computer Vision Approaches for 3-D Face Synthesis

Alexander Woodward, Da An, Yizhe Lin, Patrice Delmas,
Georgy Gimel'farb, and John Morris

Dept. of Computer Science, Tamaki Campus
The University of Auckland, Auckland, New Zealand
awoo016@ec.auckland.ac.nz,
{p.delmas, g.gimelfarb, j.morris}@auckland.ac.nz

Abstract. We have evaluated three computer approaches to 3-D reconstruction - passive computational binocular stereo and active structured lighting and photometric stereo - in regard to human face reconstruction for modelling virtual humans. An integrated experimental environment simultaneously acquired images for 3-D reconstruction and data from a 3-D scanner which provided an accurate ground truth. Our goal was to determine whether today's computer vision approaches are accurate and fast enough for practical 3-D facial reconstruction applications. We showed that the combination of structured lighting with symmetric dynamic programming stereo has good prospects with reasonable processing time and accuracy.

1 Introduction

Vision based 3-D facial reconstruction is appealing because it uses low-cost off-the-shelf hardware. Our main objective was to assess the usability of three of the most popular reconstruction techniques - computational binocular stereo, structured lighting and photometric stereo - for creating realistic virtual humans. Binocular stereo is of particular interest as it is a passive technique, whereas the other two actively project light onto the scene. Determining whether a passive approach can provide results competitive with active techniques is important.

Seeing and interacting with humans is commonplace in a person's everyday life. Indeed, most verbal and non-verbal communication uses part of the face. Facial modelling has therefore become a major issue for the successful design of human computer interfaces. The applications for facial modelling to create virtual humans are wide and varied, including surveillance, entertainment and medical visualisation [10]. Faces are highly emotive and consequently virtual humans are a powerful tool, often a necessary one, in a variety of multimedia applications.

Section 2 briefly surveys the state-of-the-art in face reconstruction techniques. Accuracy criteria relevant to face reconstruction and vision based 3-D reconstruction techniques are summarised in Section 3. The experimental setup is described in Section 4, Sections 5 and 6 discuss experimental results.

D.-Y. Yeung et al. (Eds.): SSPR&SPR 2006, LNCS 4109, pp. 270–278, 2006.

2 Previous Work

Facial reconstruction is a very specific task. Image based 3-D reconstructions appear most accurate when viewed under directions similar to those in which they were acquired. Rotations to novel views of the 3-D data often reveal the most prominent flaws in a reconstruction. However, performance analysis of vision based reconstruction has focused on a collection of arbitrarily chosen scenes [9]. We focused on human face reconstruction because of its identified importance. The techniques compared here have been described in detail [3,4,7,11,12].

Facial reconstruction from digital images reduces modelling time and allows for a personalised result. Almost all vision based techniques use a generic face model that is warped to the raw data.

Successful techniques [8] use data gathered from a 3-D scanner. Unfortunately the cost of 3-D scanning equipment makes this impractical for many situations.

3 Tested Reconstruction Algorithms

In contrast to previous work, we focus on more stringent error analysis and criteria for face reconstruction. The characteristic face feature areas - eyes, mouth, nose, etc - are especially important for reconstruction.

Accuracy of surface normal reconstruction, which is often neglected in existing analysis, is an important indicator of quality when a surface area exhibits an overall shift in depth but retains a low comparative depth variance measure. We included this measure to provide an extended reconstruction error analysis.

There are a large number of algorithms for 3-D reconstruction so we selected some of the most popular techniques in each of the chosen approaches.

Binocular Stereo. After comparing a set of implemented dense two-frame stereo algorithms, we chose the algorithms in Table 1 as they provide a cross-section of local and global techniques. Global algorithms incorporate an optimisation process over the entire domain and produce smoother results, but usually at the sacrifice of speed. Tthe algorithms used are described elsewhere [7,9].

Table 1. Tested Binocular Stereo Techniques

'Winner Takes All' Sum of Absolute Differences (SAD)[1]	- local algorithm
Dynamic Programming Method (DPM)[1]	- global algorithm
Symmetric Dynamic Programming Stereo (SDPS)[2]	- global algorithm
BVZ (Graph Cut based algorithm)[1]	- global algorithm
Belief–Propagation (BP)[3]	- global algorithm
Chen and Medioni (CM)[2]	- local algorithm

[1] Scharstein and Szeliski, http://cat.middlebury.edu/stereo/code.html
[2] Our own implementation
[3] Felzenszwalb and Huttenlocher, http://people.cs.uchicago.edu/~pff/bp/ [14]

Structured Lighting. Structured lighting techniques use active illumination to label visible 3-D surface points. Active illumination aims to simplify the surface reconstruction problem. Reconstruction time depends on a compromise between the number of images required (for complex coding strategies) and uniqueness of pixel label and thus ability to resolve ambiguities. The Gray code algorithm matches codes whereas both of the direct coding techniques project a light pattern that aids the correspondence process in a standard binocular stereo algorithm, cf. Table 2.

Table 2. Structured lighting techniques to test

Time-multiplexed structured lighting using Gray code
Direct Coding with a Colour Gradation Pattern
Direct Coding with a Colour Strip Pattern

We aim to determine whether a simpler single light projection coupled with a traditional stereo algorithm is competitive with a more complex coding scheme such as a Gray code constructed from multiple projections. An et al. give a more detailed description of the structured lighting techniques used [4].

Photometric Stereo. An albedo independent approach [5] with three light sources was used in this experiment. This technique assumes Lambertian scatterers, a parallel projection model and light sources situated at infinity. However this is a drastic simplification of reality. This paper focusses on assessing the gradient field integration component of photometric stereo. The algorithms were chosen to present both local and global techniques. Global algorithms incorporate an optimisation process over the entire field and produce smoother results. The presented gradient field integration techniques are described by Woodward and Delmas [12].

Table 3. Tested photometric stereo techniques

Frankot-Chellappa Variant (FCV)	- global algorithm
Four-Scan Method	- local algorithm
Shapelets (9 scales)	- local algorithm

4 Experimental Setup

A diagram of each sub-system is in Fig. 1. Images were taken automatically through specifically designed software and all data was processed in a batch manner. For each test subject, the facial region (about 800×700 pixels) was cut from the images for comparison.

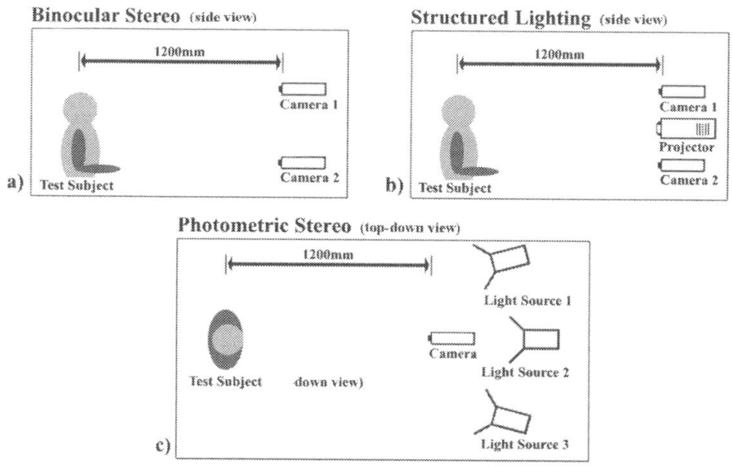

Fig. 1. System geometries for all techniques

A *Solutionix Rexcan 400* 3-D scanner (depth accuracy ~0.5 mm, planar resolution 0.23 mm) was used to obtain ground truth data for each test subject (see Figure 2).

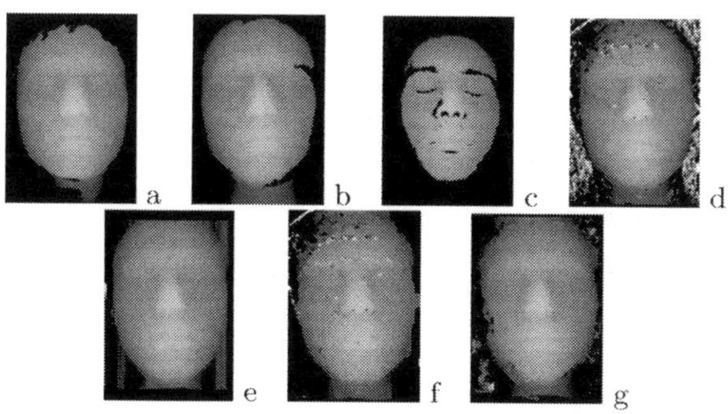

Fig. 2. Reconstruction examples: a) Ground truth, b) Gray code, c) FCV, d) SAD, e) SDPS, f) BVZ, g) CM

4.1 Binocular Stereo

A pair of *Canon EOS 10D* 6.3 Mpixel cameras was used for high resolution image acquisition. This allows for very dense disparity maps and accordingly a larger disparity range. Each camera lens has a measured focal length of 52 mm. The baseline separation between the two cameras was 175 mm. The cameras were aligned with their optical axes parallel, allowing for simplified reconstruction formulae. The test subject was placed approximately 1200 mm from the cameras.

4.2 Structured Lighting

This system used the same cameras as in binocular stereo. The main concern is the slow acquistion time that belies a potentially fast process when the appropriate hardware is available. With these cameras, it is in the order of tens of seconds.

An *Acer LCD Projector, model PL111*, was used to project light patterns into the scene. The device is capable of projecting an image of 800×600 pixels and has a focal length of $21.5 - 28$ mm.

4.3 Photometric Stereo

A system with three $150W$ light sources was used [5]. A *JVC KY-F55B* camera controlled automatically by a switching device connected to a computer captured the images. As shown in Figure 1c, the lights are positioned so as to be non-coplanar which is a requirement for the algorithm to work correctly.

4.4 System Calibration

A cubic calibration object with 63 circular calibration markings distributed evenly over two of its sides was used. Tsai's calibration technique was used [13].

A light calibration step must also be performed for the photometric stereo system. This determines the direction to the lights from an arbitrary scene origin. A calibration sphere was used for this process as directions can be determined analytically. The sphere was placed in the same location as the subject will be positioned during data acquisition.

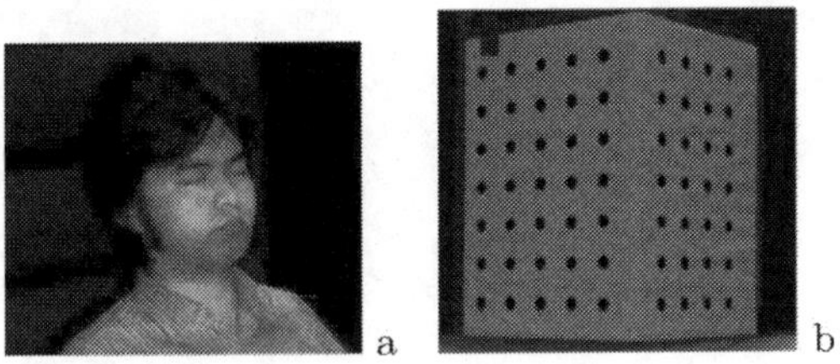

Fig. 3. (a) Test subject during acquisition with a projected colour pattern. (b)Calibration object for camera calibration.

4.5 Image Rectification

Stereo images were rectified by converting them to a standard epipolar stereo geometry. The rectification process transforms and resamples the images so that the resultant image pairs will meet the epipolar constraint.

To satisfy this requirement, an intuitive rectification is to rotate both cameras around their optical centres to a common orientation. The rotated cameras still comply with the pinhole camera model and the baseline remains intact. Using a calibration result (see Section 4.4), one can compute the baseline and a new common orientation from the pose of the two cameras. This method is similar to the method of Ayache and Hansen [1] which insists on neither the extrinsic nor the intrinsic parameters of a camera but the 3×4 perspective projection matrix. Our method utilises the extrinsic and intrinsic parameters, which is simpler and decouples the lens distortion from the aforementioned 3×4 matrix.

4.6 Data Processing

Data from the several experiments was aligned using a semi-automatic process involving 3-D object rigid transformations using the 3-D scanner software which allows for data manipulation and registration. After alignment all data was subsequently projected into disparity space and disparity maps were compared. Thus our primary accuracy metric was disparity (depth) deviations from the ground truth data. Throughout the experiment, it was found that 3-D data alignment is a difficult process and much care is needed. A small number of correspondences were entered manually to ensure correct registration.

5 Experimental Results

A Pentium 4 3.4 GHz machine with 2 Gbyte RAM computed the depth maps. The resultant face reconstructions and a ground truth of the test subject were compared. A set of 17 subjects were used.

The reconstruction accuracy metrics were: the percentage of pixels with absolute depth errors less than two disparity units ($P_{<2}$), the maximum (max) absolute pixel depth error , the mean ($e_{\rm mn}$) absolute pixel depth error, the standard deviation (σ_e) of errors, and the mean cosine error (MCE). Central differencing was used to estimate surface normals, and the MCE measures the quality of reconstruction of surface normals:

$$MCE = \left| \left(\frac{1}{M \times N} \sum_{i=1}^{M} \sum_{j=1}^{N} \mathbf{n}_{i,j} \bullet \mathbf{n}^*_{i,j} \right) - 1 \right| \tag{1}$$

where M, N are the image dimensions, $\mathbf{n}_{i,j}$ and $\mathbf{n}^*_{i,j}$ are the reconstructed surface and ground truth normals, respectively, and "$\bullet$" is the dot product operator. The MCE measures how close the reconstructed surface normals are to the ground truth, in particular, MCE $= 0$ if $\mathbf{n}_{i,j} = \mathbf{n}^*_{i,j}$, 1 if $\mathbf{n}_{i,j} \perp \mathbf{n}^*_{i,j}$, and 2 if $\mathbf{n}_{i,j}$ and $\mathbf{n}^*_{i,j}$ are collinear but with opposite directions.

The experimental results in Table 4 show that active reconstruction techniques consistently perform better than purely passive ones. Passive binocular stereo is greatly improved by supplementing the process with only a single light pattern (indicated as *Gradation* and *Strip* in Table 4).

Photometric stereo, although active in nature, is unable to recover true depth measurements due to the required gradient field integration step. None of the

Table 4. Average reconstruction accuracy and running time

Method	$P_{<2}$, %	max	e_{mn}	σ_e	MCE	Time, *sec*
Gray code	97	8	0.6	0.6	0.01	4.0
SDPS	89	13	1.0	0.9	0.09	6.0
SDPS + Gradation	90	13	1.0	1.0	0.11	.
SDPS + Strip	93	9	0.8	0.7	0.09	.
DPM	79	19	1.4	1.6	0.24	6.0
DPM + Gradation	84	13	1.2	1.2	0.25	.
DPM + Strip	92	13	0.8	0.8	0.14	.
BVZ	77	42	1.8	3.4	0.12	3517
BVZ + Gradation	83	31	1.3	1.5	0.09	.
BVZ + Strip	92	40	0.9	1.6	0.09	.
SAD	80	42	1.8	3.4	0.17	1.7
SAD + Gradation	85	32	1.2	1.7	0.16	.
SAD + Strip	93	35	0.8	1.3	0.09	.
BP	73	27	2.1	3.0	0.18	180
BP + Gradation	77	21	1.8	2.3	0.16	.
BP + Strip	89	18	1.0	1.2	0.16	.
CM	88	20	1.0	1.1	0.09	30.0
CM + Gradation	89	22	1.2	1.4	0.13	.
CM + Strip	92	21	0.9	1.1	0.10	.
PSM FCV	69	14	1.7	1.7	0.09	4.0
PSM Four-path	54	13	2.4	2.0	0.05	37.0
PSM Shapelet	71	12	1.7	1.7	0.04	153

Gradation and *Strip* refer to active projection of a Colour Gradation or Colour Strip pattern, respectively, on the object.

compared photometric stereo algorithms performed as well as the best offerings found in the other two approaches.

The performance of a pure Gray code approach is clearly superior to other techniques. It attains the lowest scores for all categories. Through effective formulation, it can handle coding errors that can happen in problem areas having low albedo or strong specularities, such as the eye regions [4] where PSM techniques usually fail.

The tuning of parameters is a difficult task. They are usually set with respect to the image size. It was found that global algorithms based on more complex optimisation techniques such as Belief Propagation (BP) [14] and the Graph Minimum Cut (BVZ) [2] did not perform as well as expected for human faces and relatively large disparity ranges. Thus our results differ from Scharstein and Szeliski's ranking of stereo algorithms [9] and the Middlebury Stereo Vision web page (www.middlebury.edu/stereo/). Our test has much higher resolution images and, in turn, much greater depth ranges. On facial images the accuracy of dynamic programming based algorithms was similar or even better than for

a)  b) 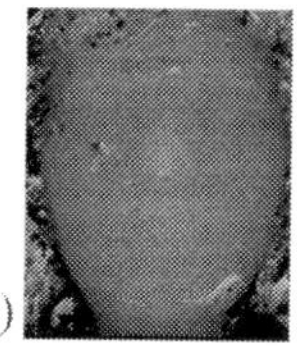 c) 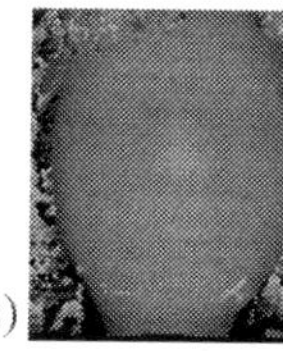

Fig. 4. Stereo (SAD) with and without a projected pattern. a) The colour strip pattern used, b) SAD without projected pattern, c) SAD with projected pattern.

these much more computationally complex (and supposedly better performing) BP and BVZ algorithms.

Colour projections that are similar to skin tones should be avoided in order to provide maximal contrast over the facial surface. The spatial frequency of projected patterns is important and needs to be high enough to provide uniqueness in matching. Thus a low frequency gradation pattern does not perform as well as a strip pattern.

6 Conclusion and Future Work

We introduced a framework and test bench for passive and active 3-D acquisition systems using three different approaches (binocular stereo, photometric stereo and structured lighting) and sixteen algorithms. We compared the data acquired to a benchmark with sub milli-metre depth accuracy using surface normal and depth map information.

All tested algorithms showed reconstruction errors that exceed the requirement for direct presentation of virtual humans and this is currently only remedied in postprocessing steps. Our experiments have shown that errors do not occur in specific areas of the face. Masking out specific regions that are highly textured, counter lowly textured, does not cause significant alterations in results.

Active methods such as structured lighting and photometric stereo have problems with specular, shadow and low albedo regions. Binocular stereo has problems dealing with texture-less regions of the face, the projection of a colour strip pattern saw a marked improvement in reconstruction accuracy. This can be easily seen in the example presented in Figure 4. The FCV algorithm performs at the forefront of the tested PSM algorithms when considering both accuracy and time complexity. Overall, the Gray code approach provides the expected best overall results. However, from these results it appears that the SDPS algorithm coupled with just a single strip pattern is a strong choice in terms of accuracy and time complexity.

We are currently assessing further algorithms, especially those for binocular stereo. The combination of active illumination and stereo vision (using the SDPS algorithms) shows the best potential for generating 3-D characters from a rig of video-cameras.

References

1. N. Ayache and C. Hansen. Rectification of Images for Binocular and Trinocular Stereovision. In *Proc. 9th Int. Conf. on Pattern Recognition, Rome, 1988.* IEEE CS Press: Los Alamitos, 1988.
2. Y. Boykov, O. Veksler, and R. Zabih. Fast approximate energy minimization via graph cuts. In *IEEE Trans. Pattern Analysis and Machine Intelligence*, vol. 23(11), pp. 1222–1239, 2001.
3. M. Chan, P. Delmas, G. Gimel'farb, and P. Leclercq. Comparative study of 3d face acquisition techniques. In A. Gagalowicz and W. Philips, eds., In *Proc. Int. Conf. Computer Analysis of Images and Patterns (CAIP'05), Versaille, France*, LNCS 3691, pp. 740–747, Sept. 2005.
4. D. An, A. Woodward, P. Delmas, and C. Chen. Comparison of Structured Lighting Techniques with a View for Facial Reconstruction. In *Proc. Image and Vision Computing New Zealand Conf.* , Dunedin, New Zealand, pp. 195–200, 2005.
5. R. Klette and K. Schluns. *Computer Vision - Three-dimensional Data from Images.* Springer: Berlin, 1998.
6. T. Kurihara and K. Arai. A transformation method for modeling and animation of the human face from photographs. In *Proc. Computer Animation'91 Conf., Tokyo*, pp.45–58, 1991.
7. P. Leclercq, J. Liu, M. Chan, A. Woodward, G. Gimel'farb, and P. Delmas. Comparative study of stereo algorithms for 3D face reconstruction. In *Proc. Int. Conf. on Advanced Concepts for Intelligent Vision Systems, Brussels, Belgium*, Sept. 2004.
8. Y. Lee, D. Terzopoulos, and K. Waters. Realistic modeling for facial animation. In *Proc. ACM SIGGRAPH'95 Conf.*, pp.55–62, 1995.
9. D. Scharstein and R. Szeliski. A taxonomy and evaluation of dense two-frame stereo correspondence algorithms. *Int. J. Computer Vision*, vol. 47(1), pp. 7–42, 2002.
10. Z. Wen and T.S. Huang. 3D Face Processing: Modeling, Analysis and Synthesis. *The International Series in Video Computing*, Vol. 8, Springer: Berlin, 2004.
11. A. Woodward and P. Delmas. Towards a low cost realistic human face modelling and animation framework. In *Proc. Image and Vision Computing New Zealand, Akaroa, Christchurch, New Zealand*, pp. 11–16, Nov. 2004.
12. A. M. Woodward and P. Delmas. Synthetic Ground Truth for Comparison of Gradient Field Integration Methods for Human Faces. In *Proc. Image and Vision Computing New Zealand, Dunedin, New Zealand*, pp. 155–160, Nov. 2005.
13. R.Y. Tsai. A Versatile Camera Calibration Technique for High Accuracy 3-D Machine Vision Metrology using Off the Shelf TV Cameras and Lenses. *Int. J. Robotics and Automation*, vol. 3(4), pp. 323–344, 1987.
14. P.F. Felzenszwalb, and D.P. Huttenlocher. Efficient Belief Propagation for Early Vision. In *Proc. IEEE Conf. on Computer Vision and Pattern Recognition*, IEEE CS Press: Los Alamitos, pp. 261–268, Jun. 2004.

Optical Flow Computation with Fourth Order Partial Differential Equations

Xiaoxin Guo, Zhiwen Xu, Yueping Feng, Yunxiao Wang, and Zhengxuan Wang

Key Laboratory of Symbol Computation and Knowledge Engineering of the Ministry of Education, College of Computer Science and Technology, Jilin University, Qianjin Street 2699#, Changchun, 130012, P.R. China
xiaoxin@mail.jl.cn, guoxx@jlu.edu.cn

Abstract. In this paper, we propose a new hybrid optical flow computation with fourth order partial differential equations (PDEs). The integration of local and global optical flow methods exploits fourth order PDEs rather than second order for the purpose of the improvement of smoothness and accuracy of the estimated optical flow field. Furthermore, we describe the implementation of the method in detail. The experiments show that the employment of fourth order PDEs benefits the improvement of the two aspects of the resulting optical flow field.

1 Introduction

Optical flow is the term used to indicate the distribution of velocity generated by the relative motion between an object and the camera, over the points of an image sequence, and carries important information which is valuable for analyzing dynamic scenes or motion in video. Optical flow is determined by the velocity vector of each pixel in each frame. One of the most appealing features of optical flow computation methods is perhaps their generality, which provides a basis for their application in a broad spectrum of computer vision applications. Several schemes have been devised for calculating optical flow based on two or more frames of a sequence. These schemes can be classified into two general categories: local methods, which may optimize some local energy-like expression, and global strategies, which attempt to minimize a global energy functional.

There exists a very large number of publications on optical flow computation [1][2]. Schnörr [3] sketched a framework for supplementing global energy functionals with multiple equations that provide local data constraints. He suggested to use the output of Gaussian filters shifted in frequency space [4] or local methods incorporating second-order derivatives [4][5], but did not consider methods of Lucas–Kanade type.

While the noise sensitivity of local differential methods has been studied intensively in recent years [6]-[11], the noise sensitivity of global differential methods has been analyzed to a significantly smaller extent. In this context, Galvin et al. [12] have compared a number of classical methods where small amounts of Gaussian noise had been added. Their conclusion was similar to the findings of Barron et al. [13]: the

D.-Y. Yeung et al. (Eds.): SSPR&SPR 2006, LNCS 4109, pp. 279 – 286, 2006.

global approach of Horn and Schunck is more sensitive to noise than the local Lucas–Kanade method.

This paper proposes a new hybrid optical flow computation, which incorporates global strategies into local methods. Different from other optical flow computations, our work uses fourth order partial differential equations (PDEs) instead of second order PDEs as a technique to avoid over-smooth effects while achieving good tradeoff between smoothness and faithfulness to the data. Because of the use of nonlinear penalty function, the optical flow computation can also achieves the goal of preserving discontinuity of optical flow field. In implementation, we adopt the method called successive overrelaxation (SOR) [17] to numerically approximate the optical flow equation with fourth order PDEs. The iterative method possesses good properties in both temporal and spatial computation complexity. The experimental results show the validity and applicability of the proposed method.

The paper is organized as follows. Section 2 presents optical flow computation including Lucas-Kanade and Horn-Schunck methods, and Section 3 proposes our method, involving basic ideas and model. We describe the implementation of the proposed method in Section 4. Experiments are presented in Section 5 and the paper is concluded in Section 6.

2 Optical Flow Computation

Consider an image sequence $g(x, y, t)$, where (x, y) denotes the location within a rectangular image domain Ω, and $t \in [0, T]$ denotes time. Many differential methods for optical flow are based on the assumption that the grey values of image objects in subsequent frames do not change over time:

$$g_x u + g_y v + g_t = 0, \tag{1}$$

where the displacement field (u, v) is called *optical flow*, and subscripts denote partial derivatives.

Evidently, this single equation is not sufficient to uniquely compute the two unknowns u and v (*aperture problem*): In order to cope with the aperture problem, Lucas et al. [14][15] proposed to assume that the unknown optical flow vector is constant within some neighborhood of size ρ. In this case it is possible to determine the two *constants* u and v at some location (x, y, t) from a weighted least square fit by minimizing the function

$$E_{LK}(u, v) = K_\rho * \left(\left(g_x u + g_y v + g_t \right)^2 \right), \tag{2}$$

where K_ρ is a Gaussian kernel with the standard deviation ρ, and $*$ is a convolution operator. The method is robust against noise. However, it constitutes the most severe drawback of local gradient methods: its flow fields are nondense.

In order to end up with dense flow estimates one may embed the optical flow constraint into a regularization framework. Horn et al. [16] have pioneered this class of global differential methods. They determine the unknown functions $u(x, y, t)$ and $v(x, y, t)$ as the minimizers of the global energy functional

$$E_{HS}(u,v) = \int_{\Omega} \left(\left(g_x u + g_y v + g_t \right)^2 + \alpha \left(|\nabla u|^2 + |\nabla v|^2 \right) \right) dxdy \, , \tag{3}$$

where the smoothness weight $\alpha > 0$ serves as a regularization parameter, and ∇ is a gradient operator. The use of the regularizer results in dense flow fields and makes subsequent interpolation steps obsolete. This is a clear advantage over local methods. Unfortunately, the method is sensitive to noise.

Since both local and global differential methods have complementary advantages and shortcomings, it would be interesting to construct a hybrid technique that constitutes the beneficial factors of two methods: It should combine the robustness of local methods with the density of global approaches.

An existing hybrid method employs the convolution kernel with standard deviation ρ for local methods and the optical flow constraint for global approaches. The estimated optical flow field is the solution of the minimization problem, given by the following functional

$$E(u,v) = \int_{\Omega} \left(K_{\rho} * \left(g_x u + g_y v + g_t \right)^2 + \alpha \left(\psi\left(|\nabla u| \right) + \psi\left(|\nabla v| \right) \right) \right) dxdy \, , \tag{4}$$

where $\psi(\cdot)$ is called a potential function. When $\rho = 0, \psi(\cdot) = (\cdot)^2$, the above equation will be reduced to Eq. (3).

3 Smoothness Constraints Using Fourth Order PDEs

Although these techniques using second order PDEs as smoothness constraints are able to achieve a good tradeoff between smoothness and optical flow constraints, they tend to cause over-smooth effect as a result of the fact that second order PDEs are strong constraints. This result is undesirable and is likely to cause a computer vision system to falsely recognize the motions of different object as ones belong to the same object.

This over-smooth effect is, to a large extent, inherent in the nature of second order PDEs. Since second order derivatives are zero only if the optical flow field is linear-monotonously changing, these PDEs for the two velocity components will evolve toward and settle down to an optical flow field with constant gradients if the field is infinite. For fields of limited support, however, symmetric boundary condition is usually employed in order to avoid motion distortion at the boundaries. Then these PDEs will evolve toward a constant field. Since these PDEs are usually designed such that optical flows in smooth areas evolve faster than those around rough areas in order to preserve discontinuity while removing noise, consistent flow areas will become flat faster than less consistent areas. Consequently, the optical flow field is likely to evolve at early stage into such a vector field that may be approximated by constant subfields. The boundaries of these subfields may coincide with true segmentations of the moving objects, but may result in incorrect motion estimations due to the over-smooth effect.

For this, we propose a novel smoothness constraints using fourth order PDEs for optical flow equations, forming a new hybrid method. First consider the following functional defined in the space of continuously varying vectors over a support of Ω:

$$E(u,v)=\int_{\Omega}\left(K_{\rho}*\left(g_x u+g_y v+g_t\right)^2+\alpha\left(\psi\left(\left|\Delta u\right|\right)+\psi\left(\left|\Delta v\right|\right)\right)\right)dxdy\,, \tag{5}$$

where Δ is a Laplacian operator. Different from Eq.(4), the smoothness term in Eq. (5) use Laplacian operators $\psi(|\Delta u|)+\psi(|\Delta v|)$ rather than gradient operators $\psi(|\nabla u|)+\psi(|\nabla v|)$. We require that the potential function $\psi(\cdot)\geq 0$ and is an increasing function:

$$\psi'(\cdot)>0\,, \tag{6}$$

so that the functional is an increasing function with respect to the smoothness of the field as measured by $|\Delta u|$ and $|\Delta v|$. Therefore, the minimization of the functional is equivalent to smoothing the optical flow field. The use of the potential function enables the construction of nonlinear equations. For example, $\psi(\cdot)$ can adopt a Huber function as a penalty function to control the discontinuity in the *a priori* model. The minimum of the functional may be found by solving the following Euler's equation for all $(x,y)\in\Omega$,

$$\Delta\left[\psi'\left(\left|\Delta u\right|\right)\Delta u/\left|\Delta u\right|\right]-K_{\rho}*\alpha^{-1}\left(g_x g_x u+g_x g_y v+g_x g_t\right)=0\,, \tag{7}$$

$$\Delta\left[\psi'\left(\left|\Delta v\right|\right)\Delta v/\left|\Delta v\right|\right]-K_{\rho}*\alpha^{-1}\left(g_y g_x u+g_y g_y v+g_y g_t\right)=0\,. \tag{8}$$

An optical flow field whose velocity components both satisfy a plane equation refers to a plane optical flow field. For its velocity components, their Laplacians are zero, so they satisfy Eq. (5). Therefore, a plane field is obviously a global minimum of the functional (5).

Let $\Omega_i, i=1,2,\ldots,n$ be a partition of Ω. For an approximated optical flow composed of plane subfields, we require that the plane subfields be such that the combined field is continuous. Therefore, the velocity components in any two adjacent subfields must be on different planes; otherwise, we can combine them as one. Let us denote $\partial\Omega_i$ as the boundary of portion Ω_i, then $\Omega_i-\partial\Omega_i$ is the interior of Ω_i. It is obvious that

$$\nabla u_i(x,y)=c_1,\ \nabla v_i(x,y)=c_2,(x,y)\in(\Omega_i-\partial\Omega_i)\,, \tag{9}$$

where c_1 and c_2 are both constant. So we have

$$\Delta u_i(x,y)=0,\Delta v_i(x,y)=0,(x,y)\in(\Omega_i-\partial\Omega_i)\,, \tag{10}$$

for $i = 1,2,\ldots, n$. Therefore,

$$\Delta u(x,y)=0,\Delta v(x,y)=0,(x,y)\in(\Omega-\partial\Omega)\,, \tag{11}$$

where $\partial\Omega=\cup_{i=1}^{n}\partial\Omega_i$. Since it is required that the velocity components in any two adjacent subfields be on different planes, we have

$$\nabla u_i\neq\nabla u_j,\text{or } \nabla v_i\neq\nabla v_j\,, \tag{12}$$

for any two adjacent portions Ω_i and Ω_j. This indicates that the gradient for the components is not continuous at the boundary $\partial\Omega$. So we have

$$\Delta u(x, y) = \infty, \text{or } \Delta v(x, y) = \infty . \tag{13}$$

If we require that

$$\psi'(\infty) = 0 , \tag{14}$$

we then have

$$\psi'(|\Delta u|)\Delta u/|\Delta u| = 0, \psi'(|\Delta v|)\Delta v/|\Delta v| = 0 , \tag{15}$$

for all $(x, y) \in \Omega$. Therefore, an optical flow field composed of plane subfields satisfies the Euler's equation.

4 Implementation

The differential equation (7) and (8) may be solved numerically using an iterative SOR method [17]. The SOR method is a good compromise between simplicity and efficiency. Assuming a space grid size of h, we discretize the space coordinates as follows:

$$x = ih, y = jh, i = 1, 2, \ldots, N, j = 1, 2, \ldots, N, \tag{16}$$

where $Nh \times Nh$ is the size of image support. We then employ a three-stage approach to calculate the constraint terms of Eq. (7) and (8). At the first stage, we calculate the Laplacians of the optical flow vector functions as

$$\Delta u^k_{i,j} = \left(u^k_{i+1,j} + u^k_{i-1,j} + u^k_{i,j+1} + u^k_{i,j-1} - 4u^k_{i,j}\right)/h^2 , \tag{17}$$

$$\Delta v^k_{i,j} = \left(v^k_{i+1,j} + v^k_{i-1,j} + v^k_{i,j+1} + v^k_{i,j-1} - 4v^k_{i,j}\right)/h^2 , \tag{18}$$

with symmetric boundary conditions:

$$\begin{aligned} u^k_{i,-1} &= u^k_{i,0}, u^k_{i,J+1} = u^k_{i,J}, v^k_{i,-1} = v^k_{i,0}, v^k_{i,J+1} = v^k_{i,J}, i = 1, 2, \ldots, N; \\ u^k_{-1,j} &= u^k_{0,j}, u^k_{I+1,j} = u^k_{I,j}, v^k_{-1,j} = v^k_{0,j}, v^k_{I+1,j} = v^k_{I,j}, j = 1, 2, \ldots, N. \end{aligned} \tag{19}$$

At the second stage, we calculate the value of the following functions

$$\varphi(\Delta u) = \psi'(|\Delta u|)\Delta u/|\Delta u| \text{ and } \varphi(\Delta v) = \psi'(|\Delta v|)\Delta v/|\Delta v| . \tag{20}$$

For convenience, the above equations can be discretized as

$$\varphi^k_{u(i,j)} = \varphi(\Delta u^k_{(i,j)}) \text{ and } \varphi^k_{v(i,j)} = \varphi(\Delta v^k_{(i,j)}) . \tag{21}$$

Finally, the numerical approximation to the differential equation (7) and (8) is given as

$$u_{(i,j)}^{k+1} = (1-\omega)\varphi_{u(i,j)}^{k} + \omega\Big[\sum\nolimits_{(p,q)\in\mathcal{N}^-(i,j)} \varphi_{u(p,q)}^{k+1} + \sum\nolimits_{(p,q)\in\mathcal{N}^+(i,j)} \varphi_{u(p,q)}^{k} - \tfrac{h^2}{\alpha}(g_x(i,j)g_y(i,j)v_{(i,j)}^{k} + g_x(i,j)g_t(i,j))\Big]\Big/\Big[|\mathcal{N}(i,j)| + \tfrac{h^2}{\alpha} g_x(i,j)g_x(i,j)\Big], \tag{22}$$

$$v_{(i,j)}^{k+1} = (1-\omega)\varphi_{v(i,j)}^{k} + \omega\Big[\sum\nolimits_{(p,q)\in\mathcal{N}^-(i,j)} \varphi_{v(p,q)}^{k+1} + \sum\nolimits_{(p,q)\in\mathcal{N}^+(i,j)} \varphi_{v(p,q)}^{k} - \tfrac{h^2}{\alpha}(g_y(i,j)g_x(i,j)u_{(i,j)}^{k+1} + g_y(i,j)g_t(i,j))\Big]\Big/\Big[|\mathcal{N}(i,j)| + \tfrac{h^2}{\alpha} g_y(i,j)g_y(i,j)\Big], \tag{23}$$

with symmetric boundary conditions

$$\begin{aligned}&\varphi_{u(i,-1)}^{k} = \varphi_{u(i,0)}^{k}, \varphi_{u(i,J+1)}^{k} = \varphi_{u(i,J)}^{k}, \varphi_{v(i,-1)}^{k} = \varphi_{v(i,0)}^{k}, \varphi_{v(i,J+1)}^{k} = \varphi_{v(i,J)}^{k}, i = 1,2,\ldots,N;\\ &\varphi_{u(-1,j)}^{k} = \varphi_{u(0,j)}^{k}, \varphi_{u(I+1,j)}^{k} = \varphi_{u(I,j)}^{k}, \varphi_{v(-1,j)}^{k} = \varphi_{v(0,j)}^{k}, \varphi_{v(I+1,j)}^{k} = \varphi_{v(I,j)}^{k}, j = 1,2,\ldots,N;\end{aligned} \tag{24}$$

where $|\mathcal{N}(i)|$ denotes the number of neighborhoods of pixel (i,j), and

$$\mathcal{N}^-(i,j) = \{(p,q)\in \mathcal{N}(i,j) \mid q < j \text{ or } q = j, p < i\}, \tag{25}$$

$$\mathcal{N}^+(i,j) = \{(p,q)\in \mathcal{N}(i,j) \mid q > j \text{ or } q = j, p > i\}. \tag{26}$$

5 Experiments

We now demonstrate the performance of the proposed optical flow computation using fourth order PDE. We use our scheme to computer optical flow and compare the results with those processed using second order PDE. For both PDEs we use the following function

$$\psi'(s) = s\big/\left(1+(s/\kappa)^2\right), \tag{27}$$

with $\kappa = 1$. Obviously, the above equation satisfies Eq. (6) and (14). Without its prototype function $\psi(s)$, we may directly use the iterative approach presented in Section 5. Besides, the experiment uses the following parameters α=950, ρ=4.55 and h=1.

We use two "Lena" images (Fig. 1 (a) and (b)) as original images between which the misalignment (2 degrees rotation) exists. Since the true optical flow field (Fig. 1 (e)) is known, it is convenient to conduct the experiments for quantitative comparison. Fig. 1 (c) and (d) are the two original images contaminated by Gaussian noise with the deviation $\sigma_n^2 = 0.006$. Two different optical flow computations, using second and fourth order PDE, respectively, are operated on the two degraded images. The results are shown in Fig. 1 (f) and (g). From the results, we see that the estimated flow fields are consistent with the true one.

In order to quantitatively compare the two schemes, we give for different noise levels the average angular errors, shown in Table 1, computed by

$$\arccos\left((u_c u_e + v_c v_e + 1)\big/\sqrt{(u_c^2 + v_c^2 + 1)(u_e^2 + v_e^2 + 1)}\right), \tag{28}$$

where (u_c, v_c) denotes the correct flow, and (u_e, v_e) is the estimated flow[13].

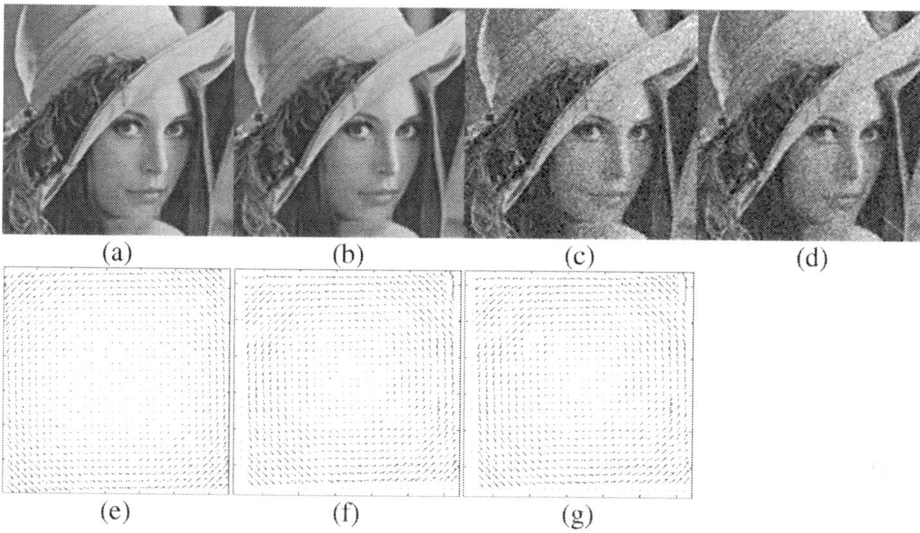

Fig. 1. Hybrid optical flow computations. (a) original image (frame 1); (b) original image (frame 2); (c) noisy image with Gaussian noise $\sigma_n^2 = 0.006$ (frame 1); (d) noisy image with Gaussian noise $\sigma_n^2 = 0.006$ (frame 2); (e) the true optical flow field; (f) the estimated optical flow field using 2nd order PDEs; (g) the estimated optical flow field using 4th order PDEs.

Table 1. Average angular errors computed with varying standard deviations σ_n^2 of Gaussian noise

σ_n^2	2nd order PDEs	4th order PDEs
0	2.632°	1.705°
0.0015	3.387°	3.100°
0.006	4.924°	4.609°

For the hybrid methods, the influence of constraints using second and fourth order PDEs, respectively, on the resulting flow field is different. For low noise levels, the accuracy for the latter is higher than that for the former. This indicates the usefulness of filling-in effect. For high noise levels, on the other hand, the latter can rival the former for robustness. Moreover, the hybrid method using fourth order PDEs doesn't reduce necessary smoothing.

6 Conclusions

In general, the optical flow computation with fourth order PDEs consider the two aspects: accuracy, which relies on the filling-in effect in flat areas, and robustness, which enhances the ability to resist noise. In addition, small angular errors show the ability to preserve the discontinuity of the flow field using fourth order PDEs.

Therefore, compared with the method with second order PDEs, the proposed method is superior to the former.

References

1. Mitiche, A., Bouthemy, P.: Computation and analysis of image motion: A synopsis of current problems and methods. International Journal of Computer Vision, (1996) 19(1):29–55
2. Stiller, C., Konrad, J.: Estimating motion in image sequences. IEEE Signal Processing Magazine, (1999) 16:70–91
3. Schnörr, C.: On functionals with greyvalue-controlled smoothness terms for determining optical flow. IEEE Transactions on Pattern Analysis and Machine Intelligence, (1993) 15:1074–1079
4. Fleet, D.J., Jepson, A.D.: Computation of component image velocity from local phase information. International Journal of Computer Vision, (1990) 5(1):77–104
5. Tretiak, Pastor, L.: Velocity estimation from image sequences with second order differential operators. In Proc. Seventh International Conference on Pattern Recognition, Montreal, Canada, (1984) 16–19
6. Uras, S., Girosi, F., Verri, A., V. Torre, A.: Computational approach to motion perception. Biological Cybernetics, (1988) 60:79–87
7. Bainbridge-Smith, Lane, R.G.: Determining optical flow using a differential method. Image and Vision Computing, (1997) 15(1):11–22
8. Fermüller, Shulman, D., Aloimonos, Y.: The statistics of optical flow. Computer Vision and Image Understanding, (2001) 82(1):1–32
9. Jähne: Digitale Bildverarbeitung. Springer: Berlin. (2001)
10. Ohta, N.: Uncertainty models of the gradient constraint for optical flow computation. IEICE Transactions on Information and Systems, (1996) E79-D(7):958–962
11. Simoncelli, E.P., Adelson, E.H., Heeger, D.J.: Probability distributions of optical flow. In Proc. 1991 IEEE Computer Society Conference on Computer Vision and Pattern Recognition, IEEE Computer Society Press: Maui, HI, (1991) 310–315
12. Galvin, B., McCane, B., Novins, K., Mason, D., Mills, S.: Recovering motion fields: An analysis of eight optical flow algorithms. In Proc. 1998 British Machine Vision Conference, Southampton, England. (1998)
13. Barron, J. L., Fleet, D. J., Beauchemin, S. S.: Performance of optical flowtechniques. International Journal of Computer Vision, (1994) 12(1):43–77
14. Lucas, B., Kanade, T.: An iterative image registration technique with an application to stereo vision. In: Proc. Seventh International Joint Conference on Artificial Intelligence, Vancouver, Canada, (1981) 674–679
15. Lucas, B. D.: Generalized image matching by the method of differences. PhD thesis, School of Computer Science, Carnegie–Mellon University, Pittsburgh, PA. (1984)
16. Horn, K. P., Schunk, B. G.: Determining optical flow. Aritificial Intelligence, (1981) 17:185–203
17. Young, D. M.: Iterative Solution of Large Linear Systems. Academic Press: New York. (1971)

Transforming Strings to Vector Spaces Using Prototype Selection

Barbara Spillmann[1], Michel Neuhaus[1], Horst Bunke[1], Elżbieta Pękalska[2], and Robert P.W. Duin[2]

[1] Institute of Computer Science and Applied Mathematics, University of Bern, Neubrückstrasse 10, CH-3012 Bern, Switzerland
[2] Faculty of Electrical Engineering, Mathematics and Computer Science, Mekelweg 4, 2628 CD Delft, Delft University of Technology, The Netherlands
{spillman, mneuhaus, bunke}@iam.unibe.ch, e.m.pekalska@tudelft.nl, r.duin@ieee.org

Abstract. A common way of expressing string similarity in structural pattern recognition is the edit distance. It allows one to apply the kNN rule in order to classify a set of strings. However, compared to the wide range of elaborated classifiers known from statistical pattern recognition, this is only a very basic method. In the present paper we propose a method for transforming strings into n-dimensional real vector spaces based on prototype selection. This allows us to subsequently classify the transformed strings with more sophisticated classifiers, such as support vector machine and other kernel based methods. In a number of experiments, we show that the recognition rate can be significantly improved by means of this procedure.

1 Introduction

Strings are one of the fundamental representation formalisms in structural pattern recognition [1]. Using a sequence of symbols rather than a vector of features often has some advantages. For example, the number of symbols in a string is variable and depends on the individual pattern under consideration, while in a feature vector we are forced to always use the same number of features, no matter how simple or complex a pattern is. In fact, strings have been successfully used in a number of applications, including digit recognition [2], shape classification [3,4], and bioinformatics [5].

In many tasks, one needs to measure distances between patterns. In case of string representations the standard distance function is the edit distance. This distance function is based on the minimum number of edit operations, such as insertion, deletion and substitution of symbols, required to transform one of two given strings into the other [6]. This distance can be computed in quadratic time with respect to the lengths of the two strings under consideration. Based on the edit distance one can easily implement classifiers of the nearest-neighbor type. However, more sophisticated classifiers, such as Bayes classifier, neural net, or

D.-Y. Yeung et al. (Eds.): SSPR&SPR 2006, LNCS 4109, pp. 287–296, 2006.

support vector machine, are not applicable in the domain of strings [7,8]. This is a serious drawback and restriction of string based pattern representation.

In the present paper we propose a transformation that maps elements from the domain of strings into real vector spaces. This transformation is intended to maintain the convenience and representational power of strings, but makes available, at the same time, the rich repository of classification tools developed in statistical pattern recognition. A transformation of graphs into vector spaces has been proposed recently [9]. In [10] general properties of embedding transformations have been discussed from various points of view. The method proposed in this paper is closely related to the dissimilarity based approach to pattern recognition proposed in [11,12]. However, while the main focus in [11,12] is on the transformation of feature vectors into dissimilarity spaces, and the possible gain in recognition accuracy obtained from this transformation, the main motivation of our approach is to build a bridge between structural and statistical pattern recognition by making the large spectrum of classifiers known from statistical pattern recognition available to string representations.

In the next section, we will introduce our terminology. Then, in Section 3, we will show how strings are transformed to n-dimensional real vector spaces, $\mathbb{R}^n$, based on various prototype selection procedures. Experimental results of the proposed method, applied to handwritten digit recognition using nearest-neighbor classifiers and support vector machines, are reported in Section 4. Finally, in Section 5, we present concluding remarks.

2 Basic Notation

Let A be a finite alphabet of symbols and A^* be the set of all strings over A. Furthermore, let ϵ denote the empty symbol. We can replace a symbol $a \in A \cup \{\epsilon\}$ by $b \in A \cup \{\epsilon\}$ and call this action an edit operation. More precisely, we refer to $a \rightarrow b$ as a substitution, $a \rightarrow \epsilon$ a deletion and $\epsilon \rightarrow a$ an insertion. In order to measure the dissimilarity of strings, a cost c is assigned to these edit operations: $c(a \rightarrow b)$, $c(a \rightarrow \epsilon)$ and $c(\epsilon \rightarrow a)$. Given a sequence $S = e_1, \ldots, e_n$ of edit operation, its cost is defined as $c(S) = \sum_{i=1}^{n} c(e_i)$. Considering two strings $x, y \in A^*$ and all sequences of edit operations that transform x into y, the edit distance, $d(x, y)$, of x and y is the sequence with minimum cost. The edit distance can be computed by dynamic programming in $O(nm)$ time and space, where n and m are the lengths of the two strings under consideration.

With the notation introduced above, the *set median string* and the *set marginal string* of a given set of strings can be defined as follows. If we denote a set of strings by $\mathcal{X}$, the set median string of $\mathcal{X}$, median ($\mathcal{X}$), is defined as the string $x_{mdn} \in \mathcal{X}$ that satisfies $x_{mdn} = \text{argmin}_{y \in \mathcal{X}} \sum_{x \in \mathcal{X}} d(x, y)$. It is a popular approximation of the generalized median string [13]. Similar to the set median we define the set marginal string, marginal ($\mathcal{X}$), of X as the string $x_{mrg} \in \mathcal{X}$ for which the sum of the edit distances to the remaining elements in $\mathcal{X}$ is maximal: $x_{mrg} = \text{argmax}_{y \in \mathcal{X}} \sum_{x \in \mathcal{X}} d(x, y)$. Obviously, set median and set marginal strings can be easily obtained by first computing all pairwise distances and then selecting the string with the minimum and maximum average distance, respectively.

3 Transforming Strings to Real Vector Spaces

The idea of our transformation approach is to select a number of prototypes out of a given set of strings. By characterizing an arbitrary string in terms of its edit distances to the prototypes, we obtain a vectorial description of the string. More precisely, a string can be transformed into a vector by calculating the edit distances to all the prototypes, where each distance represents one vector component. Formally, if we denote a set of strings (over an alphabet A) by $\mathcal{X} \subseteq A^*$ and a set of prototypes by $\mathcal{P} = \{p_1, \dots, p_n\} \subseteq \mathcal{X}$, the transformation $t_n^{\mathcal{P}} : \mathcal{X} \rightarrow \mathbb{R}^n$ is defined as a (not necessarily injective) function, where $t_n^{\mathcal{P}}(x) = (d(x,p_1), \dots, d(x,p_n))$ and $d(x,p_i)$ is the edit distance between the strings x and p_i. Obviously, the dimension of the vector space equals the number of prototypes.

3.1 Prototype Selection Methods

In the previous paragraph, the basic idea of our transformation from the string domain to a vector space has been described. However, no concrete prototype selection strategies have been considered. In the current subsection we will discuss possible algorithms for selecting prototypes from a given set of patterns.

Intuitively, a good selection strategy should satisfy the following three conditions. First, if some prototypes are similar—that is, if they are close in the space of strings—their distances to a sample string should vary only little. Hence, in this case, some of the respective vector components are redundant. Consequently, a selection algorithm should *avoid redundancies.* Secondly, to include as much structural information as possible in the prototypes, they should be *uniformly distributed* over the whole set of patterns. Thirdly, since outliers are likely to introduce noise and distortions, a selection algorithm should *disregard outliers.*

In this paper we will focus on four different class-independent selection algorithms, which we call *center prototype selector*, *border prototype selector*, *spanning prototype selector* and *k-medians prototype selector*. In the following, we will describe these selection algorithms and discuss them in terms of the above mentioned criteria.

Center Prototype Selector. As its name indicates, the *center prototype selector (c-ps)* selects prototypes situated in the center of a given set of strings. Considering the set median string to be the most central string, the set of i prototypes $\mathcal{P}_i \subseteq \mathcal{X}, i = 0, \dots, |\mathcal{X}|$, selected by the *c-ps*, is iteratively constructed as:

$$\mathcal{P}_i = \begin{cases} \emptyset & \text{if } i = 0, \\ \mathcal{P}_{i-1} \cup \{p_i\} & \text{if } 0 < i \leq |\mathcal{X}|, \end{cases} \qquad \text{where } p_i = \text{median}\,(\mathcal{X} \setminus \mathcal{P}_{i-1}).$$

For an intuitive illustration using points on the two-dimensional plane see Fig. 1a. Due to their central position all prototypes are structurally similar. Hence, many redundant prototypes occur. On the other hand, strings at the border are not considered, and thus, the set of prototypes is not negatively influenced by outliers. Obviously, the property of uniform distribution is not satisfied.

Border Prototype Selector. The *border prototype selector (b-ps)* acts just contrary to the *c-ps*. It selects prototypes from the border and is therefore based on marginal strings. The set of i prototypes $\mathcal{P}_i \subseteq \mathcal{X}, i = 0, \ldots, |\mathcal{X}|$, selected by the *border prototype selector* is defined as:

$$\mathcal{P}_i = \begin{cases} \emptyset & \text{if } i = 0, \\ \mathcal{P}_{i-1} \cup \{p_i\} & \text{if } 0 < i \leq |\mathcal{X}|, \end{cases} \quad \text{where } p_i = \text{marginal}\,(\mathcal{X} \setminus \mathcal{P}_{i-1}).$$

An illustration is given in Fig. 1b. Obviously, only few redundant prototypes are selected. However, there are no prototypes located in the center and the condition of uniform distribution is only partially fulfilled. Furthermore, it is to be expected that there are outliers among the prototypes selected by the *b-ps*.

Spanning Prototype Selector. A set of prototypes selected by the *spanning prototype selector (s-ps)* is given by the following iterative procedure. The first prototype is the set median string. Every further prototype is the string with the largest distance to the set of previously selected prototypes. Analog algorithms have been proposed for k-means initialization [14,15]. Formally, the set of i prototypes $\mathcal{P}_i \subseteq \mathcal{X}, i = 0, \ldots, |\mathcal{X}|$, selected by the *spanning prototype selector (s-ps)* is defined as

$$\mathcal{P}_i = \begin{cases} \emptyset & \text{if } i = 0, \\ \text{median}\,(\mathcal{X}) & \text{if } i = 1, \\ \mathcal{P}_{i-1} \cup \{p_i\} & \text{if } 1 < i \leq |\mathcal{X}|, \end{cases} \quad \text{where } p_i = \underset{x \in \mathcal{X} \setminus \mathcal{P}_{i-1}}{\text{argmax}} \; \min_{p \in \mathcal{P}_{i-1}} d(x,p).$$

Each additional prototype selected by the *s-ps* is the string located the furthest away from the already selected prototypes. Thus, the case that two prototypes are very close is avoided and hence also redundant prototypes are prevented. It is in the nature of this algorithm that new prototypes are selected from an area which hasn't been considered before. This leads to a good distribution of the prototypes. However, since outliers have a large distance to the other patterns, there is a certain chance for them to be selected. Fig. 1c illustrates the behavior of the *s-ps*.

K-Medians Prototype Selector. The *k-medians prototype selector (km-ps)* is based on the k-means clustering algorithm [16]. The idea is to find n clusters in the given set of data and to declare each cluster center, i.e. the set median of each cluster, to be a prototype. An illustration can be found in Fig. 1d.

The advantage of the prototypes selected by the *km-ps* is that they are evenly spread over the whole set of data. Similar strings are represented by the same

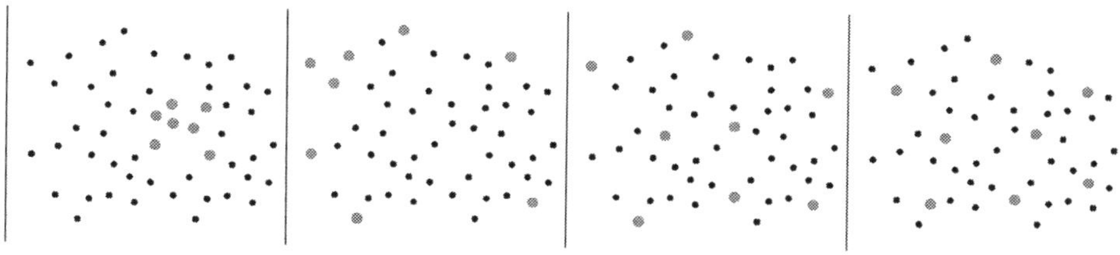

a) b) c) d)

Fig. 1. Illustration of the a) *c-ps*, b) *b-ps*, c) *s-ps* and d) *km-ps* algorithms. The larger dots represent the selected prototypes.

prototype. Hence, redundant prototypes are mostly avoided. Furthermore, outliers usually aren't positioned in the center of a k-medians cluster and thus the chance for them to be selected as a prototype is small.

4 Experimental Results

This section provides experimental classification results using the transformation schema introduced in Section 3. The prototype selection algorithms are tested on the *Pendigits* database described in [17] (original, unnormalized version). The original version contains 10,992 instances of handwritten digits 0 to 9, where 7,494 are used for training and 3,498 for testing (see Fig. 2). Each digit is originally given as a sequence of two-dimensional points. To obtain a suitable string representation, each digit curve is first approximated by a sequence $s = z_1, \ldots, z_n$ of vectors of constant length $|z_i| = l$, such that the start and end points of all z_i lie on the original curve.

A string can be generated by one of the following two methods. Either the sequence s of vectors is directly regarded as a string. Then the costs of the edit operations are defined as follows. A substitution has the costs $c(z_i \rightarrow z_j) = ||z_i - z_j||^{q_v}$, where q_v is a positive real value; for the costs of insertion and deletion we take the arithmetic mean of the extremal values (0 and $(2l)^{q_v}$) of the substitution costs, which is $2^{q_v-1}l^{q_v}$. This cost function is referred to as *vector cost function.* Another way of generating a string is to consider the sequence $\alpha_1, \ldots, \alpha_{n-1}$ of angles, where α_i is the angle between vectors z_i and z_{i+1}. In that case, the costs assigned to the edit operations are constantly set to $0 \leq q_a \leq \frac{\pi}{2}$ in case of angle insertions and deletions, and for substitutions the costs are given by the absolute difference of the two involved angles α_i and α_j, $c(\alpha_i \rightarrow \alpha_j) = |\alpha_i - \alpha_j|$. We call this cost function *angle cost function.*

To find a suitable string representation, the values of parameters l, q_v and q_a are optimized on a validation set, which consists of one fifth of the original training set. For this purpose we generate string representations for various combinations of parameter values and classify the validation set with a k-nearest-neighbor classifier, using the original training set minus the validation set as the set of labeled training items. Finally, we select the parameter combination of

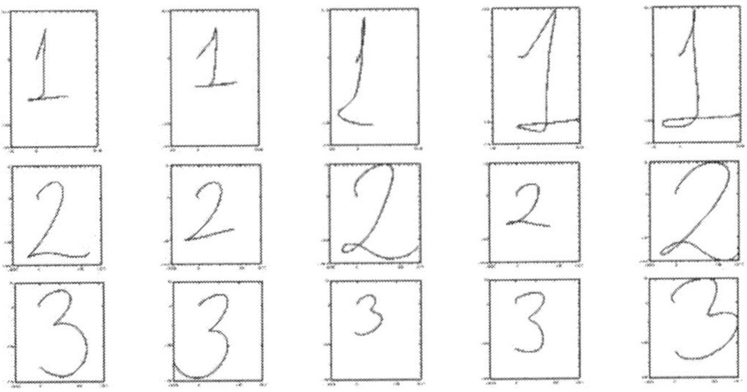

Fig. 2. Example patterns of the classes "1", "2" and "3"

l, q_v, q_a and k, that leads to the highest recognition rate. The value of parameter k is used at the same time to build a k-nearest-classifier in the string domain, which we use as a reference classifier for vector space classifiers.

Once the dataset is prepared, i.e. all the elements are represented as strings, the prototypes are selected from the training set that is made up of the remaining four fifths of the original training set (i.e. the part of the training set that is not used for validation). The prototypes are exclusively used for the purpose of mapping the data from the string to the vector space. Once the prototypes are selected, the complete dataset is mapped into the vector domain, without losing the partitioning into training, validation and test set. That is, each set still represents the same objects as in the string domain. After the dataset has been mapped to the vector domain, any classifier known from statistical pattern recognition can be trained by using the transformed validation and training sets, as described in the following.

The number of prototypes, i.e. the dimensionality of the vector space, and the prototype selection strategy as well as classifier parameters are determined on the validation set. That is, a number of possible vector space dimensions are considered for each selection strategy and one individual classifier is built for each combination of possible dimensionality and selection strategy. Then the validation set is classified with each of these classifiers. Finally, the parameter values for the dimensionality, the selection strategy, and the classifier leading to best performance on the validation set are selected. Then this classifier is taken to classify the test set. An overview of the classifiers we used in our experiments is given in the following.

First of all, we apply a kNN classifier not only in the string domain, but also in the vector space. The distance measure we use is the Minkowski metric $L_p(\boldsymbol{x}, \boldsymbol{y}) = (\sum_{i=1}^{n} |x_i - y_i|^p)^{\frac{1}{p}}$, where $\boldsymbol{x} = (x_1, \ldots, x_n) \in \mathbb{R}^n$ and $\boldsymbol{y} = (y_1, \ldots, y_n) \in \mathbb{R}^n$. In case of this classifier, both parameters k and p are optimized on the validation set and the training set (excluding the validation set) is used for finding the nearest neighbors.

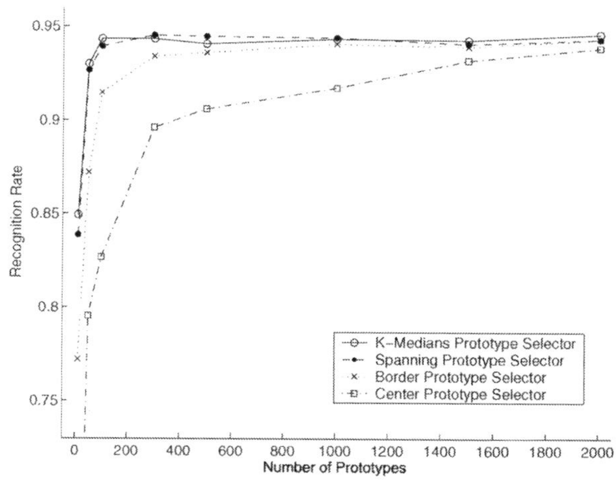

Fig. 3. Comparison of the four prototype selection strategies *c-ps*, *b-ps*, *s-ps* and *km-ps*: recognition rates of a 3NN classifier on the validation set depending on the number of prototypes

Another possibility is to apply the kNN classifier in a higher-dimensional feature space. This method uses kernel theory [18] which has become a popular subject in statistical pattern recognition. Instead of directly classifying the transformed strings, the patterns are mapped by a non-linear function $\Phi : \mathbb{R}^n \to \mathbb{R}^m (m > n)$ to a higher-dimensional real vector space $\mathbb{R}^m$, called feature space, in which the kNN classification is performed with the Euclidean distance L_2 as distance measure. In the feature space $\mathbb{R}^m$ an inner product $< \boldsymbol{x}, \boldsymbol{y} >$ exists and $\mathbb{R}^m$ is complete with respect to the norm $\|\boldsymbol{x}\| = \sqrt{< \boldsymbol{x}, \boldsymbol{x} >}$, defined by the inner product. This fact allows us to define kernel functions $k_\Phi(\boldsymbol{x}, \boldsymbol{y}) := \langle \Phi(\boldsymbol{x}), \Phi(\boldsymbol{y}) \rangle, (\boldsymbol{x}, \boldsymbol{y} \in \mathbb{R}^n)$. The kernel function $k_\Phi(\boldsymbol{x}, \boldsymbol{y})$ can then be regarded as a similarity measure in the vector space $\mathbb{R}^n$, and the Euclidean distance in the feature space $\mathbb{R}^m$ can be derived from it. With the use of this method, the explicit application of the mapping Φ can be avoided. In our experiments we used the following standard kernel functions: radial basis function, polynomial function, and sigmoid function [18].

Another classification method using kernel functions is the support vector machine (SVM) [8,19]. The key idea is to find a hyperplane that separates the data into two classes with a maximal margin. Such a hyperplane can only be found if the data are linearly separable. If linear separability is not fulfilled, a weaker definition of the margin, the soft margin, can be used. In this case, the optimal hyperplane is the one that minimizes the sum of errors and maximizes the margin. This optimization problem is usually solved by quadratic programming. In order to improve the classification of non-linearly separable data, an explicit mapping to a higher-dimensional feature space can be performed, or instead, a the above mentioned kernel function can be applied. For our experiments we use the LIBSVM library [20]. The kernel functions used in our experiments are the linear kernel and the radial basis function.

Fig. 3 illustrates the performance of the four prototype selection methods described in Section 3 with respect to the number of prototypes. It shows the recognition rate of a 3NN classifier on the validation set, where the digit curves are approximated by segments of length 20 and the angle cost function with $q_a = \frac{11}{36}\pi$ is used. (Note, the classifier parameter k is kept constant for this plot.) The number of prototypes n ranges from 10 to 2000. Generally, it can be observed that the recognition rate increases with an increasing number of prototypes. Once the recognition rate has reached a certain value, however, it roughly remains constant. For a small value of n, differences in the quality of each method can be detected, but the recognition rates become incrementally equal for larger n. That is, while the *c-ps* and the *b-ps* clearly perform worse than the *s-ps* and *km-ps* for small n, the difference at $n = 2000$ almost disappears. We observe that selection strategies which uniformly distribute the prototypes, *s-ps* and *km-ps*, have a higher performance for smaller n.

In Tab. 1 the recognition rates on the test set with the above mentioned classifiers are listed. The table shows the results for both angle (`pen ang`) and vector cost function (`pen vec`). Three different partitions of the dataset into a validation, training and test set have been used. The term `pen1` refers to the original partitioning into training and test set. The experiments `pen2` and `pen3` are further setups, where the size of each set is unchanged, but different partitions have been performed. In order to show the performance of the transformation, we use the recognition rate of the kNN classifier in the string space as a reference value. Recognition rates printed in bold face refer to statistically significant better results at a significance level of 0.95.

Table 1. Recognition rates on the Pendigits dataset

	kNN string domain	kNN Mink. metric	kNN RBF kernel	kNN poly. kernel	kNN sig. kernel	SVM RBF	SVM lin.
`pen1 ang`	88.56	**90.99**	**89.51**	**89.48**	**89.74**	**90.99**	**94.54**
`pen2 ang`	92.48	92.96	91.97	92.00	92.32	**96.16**	**95.25**
`pen3 ang`	92.71	**93.43**	92.36	92.36	91.83	**95.83**	**95.70**
`pen1 vec`	97.48	97.60	97.06	97.06	97.08	**98.34**	97.88
`pen2 vec`	99.33	99.31	99.28	99.28	99.20	**99.68**	**99.57**
`pen3 vec`	99.33	99.25	99.12	99.12	99.04	99.55	99.31

First we observe that the application of the standard kNN classifier in the vector space (column kNN Mink. metric) leads to an improvement of the recognition rate over the kNN classifier in the string domain in four out of six cases. For all other kernel based kNN classifiers (columns kNN RBF kernel, kNN poly. kernel and kNN sig. kernel), an improvement is obtained in only one out of six cases. However, both SVMs demonstrate superior performance. The SVM with radial basis function kernel leads in all six cases to an improvement, five of which are statistically significant, and even the linear kernel SVM shows a higher recognition performance than the classifier in the string domain in all cases but one.

In [21], classification based on two MLP approaches has been performed on the same data. The recognition rates on the test set achieved in [21] are 95.26% and 94.25%, respectively. We note that both recognition rates are already outperformed by our kNN classifier in the string domain using the vector cost function. Nevertheless, a further improvement can be achieved by means of the proposed embedding procedure in conjunction with both SVMs.

5 Conclusion

In this paper we study the representation and classification of strings in n-dimensional real vector spaces. The transformation is accomplished with a prototype selection procedure, where each vector component of a transformed string represents the edit distance to one prototype.

We evaluate the transformation on strings extracted from the Pendigits database. The recognition rates of several kNN methods and support vector machines for the transformed strings are compared to a kNN classifier in the original string domain. We show that by means of SVM the recognition rate for strings can significantly be improved. However, the improvement of the correct classification rate in the considered task is just one contribution of this paper. From the general point of view, the methodology proposed in the paper opens new ways of embedding symbolic data structures, i.e. strings, into vector spaces using edit distance and prototype selection. Based on such an embedding, a large number of methods from statistical pattern recognition become available to string representations. In our future work we will study additional classifiers, such as Bayes classifier and neural net, as well as data dimensionality reduction and clustering tasks.

Acknowledgements

This work has been partially supported by the Swiss National Science Foundation NCCR program *Interactive Multimodal Information Management (IM)2* in the Individual Project *Multimedia Information Access and Content Protection* as well as by the Dutch Organization for Scientific Research (NWO).

References

1. Bunke, H., Sanfeliu, A.: Syntactic and Structural Pattern Recognition – Theory and Applications. World Scientific Publ. Co. (1990)
2. Cha, S.H., Shin, Y.C., Srihari, S.N.: Approximate stroke sequence matching algorithm for character recognition and analysis. In: 5th International Conference on Document Analysis and Recognition. (1999) 53–56
3. Bunke, H., Bühler, U.: Applications of approximate string matching to 2D shape recognition. Pattern Recognition **26** (1993) 1797–1812
4. Chen, S.W., Tung, S.T., Fang, C.Y., Cheng, S., Jain, A.K.: Extended attributed string matching for shape recognition. Computer Vision and Image Understanding **70** (1998) 36–50
5. Durbin, R., Eddy, S.R., Krogh, A., Mitchison, G.: Biological sequence analysis. Cambridge University Press, Cambridge, UK (1998)
6. Wagner, R.A., Fischer, M.J.: The string-to-string correction problem. Journal of the ACM **21** (1974) 168–173
7. Duda, R., Hart, P., Stork, D.: Pattern Classification. 2nd edn. Wiley, New York (2001)
8. Vapnik, V.: The Nature of Statistical Learning Theory. 2nd edn. Springer-Verlag (2000) ISBN: 0-387-98780-0.
9. Wilson, R.C., Hancock, E.R., Luo, B.: Pattern vectors from algebraic graph theory. IEEE Transactions on Pattern Analysis and Machine Intelligence **27** (2005) 1112–1124
10. Hjaltason, G.R., Samet, H.: Properties of embedding methods for similarity searching in metric spaces. IEEE Transactions on Pattern Analysis and Machine Intelligence **25** (2003) 530–549
11. Pękalska, E.: Dissimilarity representations in pattern recognition. PhD thesis, Delft University of Technology (2005)
12. Pękalska, E., Duin, R.P., Paclík, P.: Prototype selection for dissimilarity-based classifiers. Pattern Recognition **39** (2006) 189–208
13. Kohonen, T.: Median strings. Pattern Recognition Letters **3** (1985) 309–313
14. Katsavounidis, I., Kuo, C.C.J., Zhang, Z.: A new initialization technique for generalized lloyd iteration. IEEE Signal processing letters **1** (1994) 144–146
15. Juan, A., Vidal, E.: Comparison of four initialization techniques for the k -medians clustering algorithm. In: Proceedings of the Joint IAPR International Workshops on Advances in Pattern Recognition, London, UK, Springer-Verlag (2000) 842–852
16. Jain, A.K., Dubes, R.C.: Algorithms for clustering data. Prentice-Hall, Inc., Upper Saddle River, NJ, USA (1988)
17. Alpaydin, E., Alimoglu, F.: Department of Computer Engineering, Bogaziçi University, 80815 Istanbul Turkey (1998) `ftp://ftp.ics.uci.edu/pub/mlearn/databases/pendigits`.
18. Shawe-Taylor, J., Cristianini, N.: Kernel Methods for Pattern Analysis. Cambridge University Press (2004)
19. Vapnik, V.: Statistical Learning Theory. Wiley-Interscience (1998) ISBN: 0-471-03003-1.
20. Chang, C.C., Lin, C.J.: LIBSVM: a library for support vector machines. (2001) `http://www.csie.ntu.edu.tw/~cjlin/libsvm`.
21. Alimoglu, F., Alpaydin, E.: Combining multiple representations for pen-based handwritten digit recognition. Turk J Elec Engin **9** (2001)

Shape Categorization Using String Kernels

Mohammad Reza Daliri[1], Elisabetta Delponte[2], Alessandro Verri[2], and Vincent Torre[1]

[1] SISSA, Via Beirut 2-4, 34014 Trieste, Italy
[2] DISI, Universita degli Studi di Genova, Via Dodecaneso 35, 16146 Genova, Italy

Abstract. In this paper, a novel algorithm for shape categorization is proposed. This method is based on the detection of perceptual landmarks, which are scale invariant. These landmarks and the parts between them are transformed into a symbolic representation. Shapes are mapped into symbol sequences and a database of shapes is mapped into a set of symbol sequences and therefore it is possible to use support vector machines for categorization. The method here proposed has been evaluated on silhouettes database and achieved the highest recognition result reported with a score of 97.85% for the MPEG-7 shape database.

1 Introduction

The final goal of computer vision is to make machines as capable as humans in terms of visual perception and understanding [23]. Object recognition and classification has been extensively studied and analyzed in recent years, but current techniques are far from needed. An even more difficult task for a machine is to determine the category to which the object belongs, rather than to find out whether or not that particular object has been seen before. There are several reasons that make this problem so difficult. The first reason is related to the uncertainty about the level of categorization in which recognition should be done. Based on the research made by cognitive scientists [9], there are several levels at which categorization is performed. Another reason is the natural variability within various classes. The generality of a class is directly proportional to the within-class variation. Moreover, the characterization should be invariant to rotation, scale, translation and to certain deformations. Objects have several properties that can be used for recognition, like shape, color, texture, brightness. Each of these cues can be used for classifying objects. Biederman [4] suggested that edge-based representations mediate real-time object recognition. In his view, surface characteristics such as color and texture can be used for defining edges and can provide cues for visual search, but they play only a secondary role in the real-time recognition. There are two major approaches for shape-based object recognition: 1) boundary-based, that uses contour information [5], [20], [16], [3], [1], and 2) holistic-based representation, requiring more general information about the shape [18], [17]. In this paper, a new representation for categorization based on the extraction of the perceptually relevant landmarks is

D.-Y. Yeung et al. (Eds.): SSPR&SPR 2006, LNCS 4109, pp. 297–305, 2006.

proposed. Each shape is transformed into a symbolic representation, where each shape is mapped in a string of symbols. The present manuscript is organized as follows: Localization and extraction of landmarks are investigated in Section 2. The symbolic representation is presented in Section 3. Section 4 describes the feature space composed by string kernels. In Section 5 geometrical invariants features are described. The results are presented in Section 6.

2 Extraction and Localization of Landmarks

A database of black shapes over a white background (Silhouettes) was used [21] (Figure 1). In this case the extraction of the contour is straightforward and it is represented by the edge chain (x(j),y(j)) j=1,...,N where N is the chain or contour length. The next step is finding the gradient of the contour at the optimal scale. As suggested by Lindeberg [13], the local scale can be estimated considering the normalized derivatives: $G_\lambda = t^{\lambda/2}\sqrt{L_x^2 + L_y^2}$.

Fig. 1. Some sample shapes from MPEG7 database and Kimia database used in our experiments

Where L_x and L_y are the x and y derivatives of the original image convolved with the Gaussian filter $exp(-(x^2+y^2)/2t)$ with $t = \sigma^2$. These normalized derivatives $G_\lambda(t)$ depend on the value of the parameter λ. As discussed by Lindeberg [13] and Majer [15], the most convenient choice for the Gaussian step edges is $\lambda = 1/2$. The best scale was extracted with the Lindeberg formula and the gradient at this scale was computed by a simple 2-D gaussian filtering in X and Y direction in the image plane, $\overrightarrow{G} = (G_x, G_y)$. As the tangent vector is orthogonal to the gradient vector we obtain at the best scale, $\overrightarrow{T} = (T_x, T_y) = (G_y, -G_x)$. The curvature κ of a planar curve at a point P on the curve is defined as the instantaneous rate of change of the tangent's slope angle at point P with respect to arc length s: $\kappa = \frac{\partial \vec{T}}{\partial s}$. In order to calculate the derivative of each component of the tangent vector we convolve them by the first derivative of one dimensional Gaussian function:

$$\frac{\partial T_x}{\partial s} = \frac{\partial [T_x \otimes g(s,\sigma)]}{\partial s} = T_x \otimes [\frac{\partial g(s,\sigma)}{\partial s}] \quad (1)$$

so that: $g(s,\sigma) = \frac{1}{\sqrt{2\pi}\sigma} \exp(-\frac{s^2}{2\sigma^2})$.

This will be done for the Y component of the tangent vector. Because different shapes in our database have different length, it would be better to select a sigma related to the length of a shape contour, to formulate this statement we choose our sigma to be: $\sigma_1 = \sigma_0 \frac{l}{l_0}$, where l is the length of contour shape and, based on our experiment, we select $l_0 = 200$ and $\sigma_0 = 3$. Now, the curvature value will be calculated as follows: $\|\kappa\| = \sqrt{(\frac{\partial T_x}{\partial s})^2 + (\frac{\partial T_y}{\partial s})^2}$.

For having the complete calculation of the curvature we need to attribute a sign to it. Direction of tangent vector is a good representation for calculating the sign of curvature but it must be smoothed. We applied convolution to each component of the tangent vector with a one dimensional Gaussian function with $\sigma = 3$ for the small smoothing of the tangent vector to remove the noise. Now, with the smoothed tangent vector we can calculate the sign of curvature as follows: $Sign(\kappa) = sign[(T_{x,sm}(s), T_{y,sm}(s), 0) \times (T_{x,sm}(s-1), T_{y,sm}(s-1), 0)]$.

The complete definition of our curvature will be obtained by multiplying the value of curvature with its sign. The obtained curvature is noisy and in order to reduce it a non-linear filtering was used. The aim of the non-linear filtering was to smooth regions of low curvature and to leave unaltered regions of high curvature. We first compute the local square curvature as:

$$\overline{\kappa^2(n)} = \frac{1}{2\sigma_1 + 1} \sum_{i=-\sigma_1}^{\sigma_1} \kappa^2(n+i) \quad (2)$$

Non-linear filtering is performed by convolving the curvature with a one-dimensional Gaussian function, where the scale of filter is:

$$\sigma_2(n) = \sigma_{min} + \frac{\hat{\kappa}}{\overline{\kappa^2(n)}} \quad (3)$$

In our experiment good results were obtained by using the values of $\sigma_{min} = 0$ and $\hat{\kappa} = 0.02$. In this way a robust and perceptually relevant representation for the curvature of the shapes was obtained. Now, the local maxima (negative and positive peaks) of the curvature are detected and identified as landmarks in the original 2-D contours(Figure 2).

3 Symbolic Representation

In this section we will transform each shape into symbolic representation to be used for categorization. Firstly angles close to 180 degrees are removed. In what follows a "dictionary" is presented, allowing the transformation of the curvature representation into a string of symbols.

3.1 Labeling of Angles

Features detected as corners are quantized so that angles have either 45 or 90 or 135 degrees. These angles can have either a positive or a negative value of curvature. A total of 6 different corners are obtained which can be labeled as A1, A2 , ... up to A6.

3.2 Labeling of Curves

Curve parts have the average curvature between straight lines and sharp angles that with setting a threshold can be found. Curves are labeled either as concave (C1) or convex (C2), according to the sign of their average curvature.

3.3 Labeling Links Between Angles (and Curves)

Pieces of the contour of silhouettes linking two corners (or curves) are labeled in three ways: L1 if it is a straight line, L2 if it is not a straight line (and it is not a curve) but has an average positive curvature and L3 if, on average, has a negative curvature.

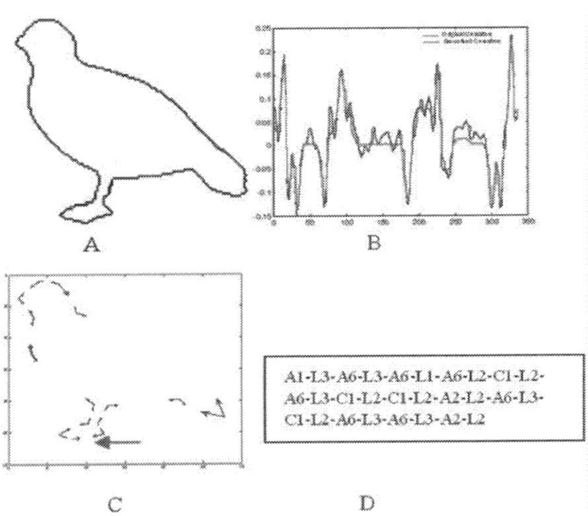

Fig. 2. A) A contour of shape from our database with associated numbers based on an arbitrary starting-point in the contour. B) Curvature profile and smoothed one in order to have perceptually relevant peaks as described in the text. C) Angle representation based on the maxima and peaks information of the curvature representation. D) Symbolic representation for the bird shape.

4 Creating the Feature Space

In our approach, shape categorization becomes similar to text categorization, where each string of symbols can be either a sentence or a separate document. A standard approach [11] to text categorization uses the classical text representation [19], which maps each document into a high-dimensional feature vector, where each entry of the vector represents the presence or the absence of a feature. Our approach makes use of specific kernels [14]. This specific kernel named string kernel, maps strings, i.e. the symbolic representation of the contour obtained in the previous section, into a feature space. In this high dimensional feature space all shapes have the same size. This transformation provides the desired rotational invariance and therefore the categorization system is also invariant to the initial symbol of the string describing the shape. The feature space in this case is composed by the set of all substrings of maximum length L of k-symbols. In agreement with a procedure used for text classification [14], the distance and therefore the similarity between two shapes is obtained by computing the inner product between their representations in the feature space. Their inner product is computed by making use of kernel functions [14], which compute the inner product by implicitly mapping shapes to the feature space. In essence, this inner product measures the common substrings of the symbolic representations of the two shapes: if their inner product is high the two shapes are similar. Substrings do not need to be contiguous, and the degree of contiguity of one substring determines its weight in the inner product. Each substring is weighted according to its frequency of appearance and on its degree of compactness, measured by a decay factor, λ, between (0,1) [14]. To create the feature space we need to search all possible substrings starting from each single-symbol to strings of length L composed by k symbols, which in our case are the 11 symbols introduced in Section 3. For each substring there is a weight in the feature space given by the sum of all occurrences of that sub-string considering the decay factor for non-contiguity. After creating the invariant feature space, we need to use a classifier to find the best hyper-planes between the different classes. Support Vector Machines (SVM) are a very successful class of statistical learning theory in high dimensional space [24]. For classification, SVMs operate by finding a hyper-surface in the space of possible inputs. In their simplest version they learn a separation hyper plane between two sets of points, the positive examples from the negative examples, in order to maximize the margin -distance between plane and closest point. Intuitively, this makes the classification correct for testing data that is near, but not identical to the training data. Further information can be found anywhere such as [6], [8].

5 Geometric Invariant Features

Beside the high dimensional feature space described in the previous section, a set of geometrical properties for each shape were measured. They consist of 16 different numbers that are normalized so to be invariant for rotation and size

Table 1. Some of the Geometric Invariant Features

Geometric Feature	Definition
Roughness	Perimeter/Convex Perimeter
Compactness or Circularity	$(Perimeter^2)/(4 * \pi *$Area of the shape)
Solidity	Number of pixels in the convex hull/the number of shape points
Rectangularity	Number of pixels in the bounding box/ the number of shape pixels
Normalized Major Axis Length	The length of the major axis of the ellipse that has the same second-moments as the shape
Normalized Minor Axis Length	The length of the minor axis of the ellipse that has the same second-moments as the shape
Elongation	Major Axis Length/Minor Axis Length
Normalized Equivalent Diameter	The diameter of a circle with the same area as the shape
Eccentricity	The ratio of the distance between the foci of the ellipse and its major axis length

transformation. Table 1 illustrates some of these geometrical features. For further information we refer readers to [7].

6 Experimental Results

In this section, some experiment results aiming at evaluating and comparing the proposed algorithm for shape classification will be presented. Firstly, a database extracted from Kimia's silhouette database [21] was used. Three different categories were considered composed by the category of birds consisting of 51 shapes, the category of mammals consisting of 178 shapes and the category of fish consisting of 79 shapes. Some shapes of the database were rotated and resized. We used LIBSVM [10] tools that support multi-class classification. To test the success of our classification the cross-validation leave-one-out method was used. In the first experiment the feature vector is created without inserting any information about the distance from the 2-D image of each shape. Different kernel functions with different parameters have been tested to reach the best result, but a simple linear kernel was the best. As discussed in section 4 it is possible to consider feature vectors with different maximum length of symbols and therefore we compared results obtained with categorization based on substrings with a maximum length of 3 and 4 symbols. As shown in Table 2 successful categorization

Table 2. Comparison of different maximum lengths for searching substring based on the classification rate($\lambda = 0.5$)

	Bird	Mammal	Fish
Substring with maximum length of 3	64.7%	87%	86%
Substring with maximum length of 4	66%	88.7%	84.8%

Table 3. Classification rate for the best parameter of $\lambda(0.3)$ after inserting the distance information between landmarks

Bird	Mammal	Fish
93.61%	92.69%	86.07%

Table 4. Classification rate for combined-features

Bird	Mammal	Fish
96.8%	97.75%	96.2%

increases with longer substring, but not so much to justify a significantly heavier computational load. As shown in Table 2, successful categorization for birds is worse than for mammals and fish, as there was a higher inter-class variation for birds with less number of samples. The database was enriched by creating new shapes by re-scaling (up to 1.5) and rotating, flipping and mirroring some of the bird shapes. After inserting new bird shapes, this category was consisting of 94 different bird images. The result was improved as shown in Table 3. In the second set of experiments we introduced also information on the distance between landmarks and different decaying factor (λ) similar to that used for text categorization [13] was tested. The best value for the parameter λ was equal to 0.3 and we set it to this value for further experiments (Table 3). Features obtained from the curvature do not catch important geometrical features of the shape to be categorized and therefore categorization based on mixed features was considered. Table 1 illustrates 17 different geometrical features which were computed for every shape and were added to the vector feature. These geometrical features consist of 17 different features such as roughness, elongation, compactness, rectangularity, convex area,..., that has been normalized so to have features invariant for size and rotation transformations. Table 4 illustrates results from cross-validation leave-one-out method combining geometric invariant and feature vector derived by string kernel. Finally the proposed categorization method was tested also on large shape database MPEG-7 CE-Shape-1 [12] consisting of 70 types of objects each having 20 different shapes. Geometrical invariant features listed in Table 1 were combined with feature vectors derived by string kernels. For the experiment one-against-one strategy, and cross-validation leave-one-out method (for each two different categories) was used. Table 5 reproduces a comparison of successful classification between the proposed methods and those available in the literature. As shown in Table 5 the combination of geometrical features (see Table 1) and landmarks extracted from the contour makes the proposed categorization rather successful and better than all previously proposed methods [22] and [2]. Some authors report retrieval accuracy over MPEG7 shape database, but as our method rely on learning module (SVM), it is useful for recognition and categorization not retrieval, so we can not report that accuracy here.

Table 5. Classification accuracy for different methods for the MPEG7 shape database

Method	Classification Accuracy
Chance probabilities [22]	97.1%
Normalized square distances [22]	96.9%
Racer [22]	96.8%
Polygonal representation and elastic matching [2]	97.79%
Proposed method in this paper	97.85%

7 Conclusion

In this paper an algorithm for object categorization based on shape information is proposed. In this model, landmarks from the shape contours are first extracted and then are transformed into a sequence of symbols. By using tools used for text categorization [14] and combining the information extracted from the contour with additional geometrical features a rather good categorization is achieved (see Table 5). The feature space representation makes our system completely invariant to affine transforms. The proposed method is expected to be robust for the partial occlusions, because it is based on the similarity of substrings, i.e. to local property of shapes.

References

1. K. Arbter, W.E. Snyder, H. Burkhardt, and G. Hirzinger. Application of affine-invariant fourier descriptors to recognition of 3-d objects. *IEEE PAMI*, 12(7):640–647, 1990.
2. E. Attalla and P. Siy. Robust shape similarity retrieval based on contour segmentation polygonal multiresolution and elastic matching. *Pattern Recognition*, 38(12):2229–2241, 2005.
3. S. Belongie, J. Malik, and J. Puzicha. Shape matching and object recognition using shape contexts. *IEEE PAMI*, 24:509–522, 2002.
4. I. Biederman and G. Ju. Surface versus edge-based determinants of visual recognitions. *Cognit. Psych.*, 20:38–64, 1988.
5. H. Blum. A transformation for extracting new descriptors of shape. In Weiant Wathen-Dunn, editor, *Models for the Perception of Speech and Visual Form*, pages 362–380. MIT Press, Cambridge, 1967.
6. C.J.C. Burges. A tutorial on support vector machines for pattern recognition. *Data Mining and Knowledge Discovery*, 2(2):121–167, 1998.
7. L.D.F. Costa and R.M.C. Junior. *Shape Analysis and Classification: Theory and Practice.* CRC Press, 2000.
8. N. Cristianini and J. Shawe-Taylor. *An Introduction to Support Vector Machines.* Cambridge University Press, 2000.
9. S. Edelman. *Representation and Recognition in Vision.* MIT Press, 1999.
10. R.E. Fan, P.H. Chen, and Lin C.J. Working set selection using the second order information for training svm. Technical report, Department of Computer Science, National Taiwan University, 2005.

11. T. Joachims. Text categorization with support vector machines: Learning with many relevant features. In Claire Nédellec and Céline Rouveirol, editors, *Proceedings of ECML-98, 10th European Conference on Machine Learning*, number 1398 in Lecture Notes in Computer Science, pages 137–142. Springer Verlag, Heidelberg, DE, 1998.
12. L.J. Latecki, R. Lakämper, and U. Eckhardt. Shape descriptors for non-rigid shapes with a single closed contour. In *IEEE Conf. on CVPR*, pages 424–429, 2000.
13. T. Lindeberg. Edge detection and ridge detection with automatic scale selection. *Int. J. of Comput. Vis.*, 30(2):77–116, 1998.
14. H. Lodhi, C. Saunders, J. Shawe-Taylor, N. Cristianini, and C.J.C.H. Watkins. Text classification using string kernels. *Journal of Mach. Learn. Res.*, 2:419–444, 2002.
15. P. Majer. The influence of the gamma-parameter on feature detection with automatic scale selection. In *Scale-Space '01: Proceedings of the Third International Conference on Scale-Space and Morphology in Computer Vision*, pages 245–254. Springer-Verlag, 2001.
16. F. Mokhtarian and A. Mackworth. Scale based description and recognition of planar curves and two-dimensional shapes. *IEEE PAMI*, 8(1):34–43, 1986.
17. H. Murase and S.K. Nayar. Visual learning and recognition of 3-d objects from appearance. *Int. J. of Comput. Vis.*, 14(1):5–24, 1995.
18. E. Rivlin and I. Weiss. Local invariants for recognition. *IEEE PAMI*, 17(3):226–238, 1995.
19. G. Salton, A. Wong, and C.S. Yang. A vector space model for automatic indexing. *Commun. ACM*, 18(11):613–620, 1975.
20. T. Sebastian, P.N. Klein, and B. Kimia. Recognition of shapes by editing their shock graphs. *IEEE PAMI*, 26(5):550–571, 2004.
21. D. Sharvit, J. Chan, H. Tek, and B.B. Kimia. Symmetry-based indexing of image databases. *J. of Visual Communication and Image Representation*, 9(4):366–380, 1998.
22. B.J. Super. Learning chance probability functions for shape retrieval or classification. In *IEEE Workshop on Learning in Computer Vision and Pattern Recognition at CVPR*, volume 6, page 93, 2004.
23. S. Ullman. *High-level Vision.* MIT Press, 1996.
24. V. Vapnik. *Statistical Learning Theory.* Wiley-Interscience, 1998.

Trace Formula Analysis of Graphs

Bai Xiao and Edwin R. Hancock

Department of Computer Science,
University of York, York Y01 5DD, UK

Abstract. In this paper, we explore how the trace of the heat kernel can be used to characterise graphs for the purposes of measuring similarity and clustering. The heat-kernel is the solution of the heat-equation and may be computed by exponentiating the Laplacian eigensystem with time. To characterise the shape of the heat-kernel trace we use the zeta-function, which is found by exponentiating and summing the reciprocals of the Laplacian eigenvalues. From the Mellin transform, it follows that the zeta-function is the moment generating function of the heat-kernel trace. We explore the use of the heat-kernel moments as a means of characterising graph structure for the purposes of clustering. Experiments with the COIL and Oxford-Caltech databases reveal the effectiveness of the representation.

1 Introduction

The Laplacian spectrum of a graph has found widespread use in computer vision for a number of applications including segmentation [2] and routing [3], graph clustering[10] and graph indexing[4]. For instance, the Fiedler vector [1] [12] [13], i.e. the eigenvector associated with the smallest non-zero eigenvalue, can be used to perform pairwise clustering of data. The Laplacian eigenvalues may be used to characterise graphs for the purposes of clustering. Several authors have explored the use of the Laplacian and related operators to map data to manifolds in a low dimensional space [9] [15] [16] [17] [5]. These methods share the feature of using the spectrum of the Laplacian matrix to map data specified in terms of a proximity matrix to a vector space. For instance in the Laplacian eigenmap [15], the mapping is determined by the raw Laplacian spectrum. The diffusion map [5] of Lafon and Coifman constructs the mapping by raising the Laplacian eigensystem to a negative integer power. This mapping is shown to preserve the distances between nodes under a random walk, or diffusion, on the graph. In the heat-kernel embedding of Lebanon and Lafferty [6], the embedding is based on the heat-kernel and this is found by exponentiating the Laplacian eigensystem.

The aim in this paper is to explore whether the trace of the heat-kernel [7] can be used for the purposes of characterising the properties of graphs. The trace of the heat kernel is found by summing a series of terms, each of which is the result of exponentiating a Laplacian eigenvalue with time. As a result the heat-kernel trace is a function whose parameters are the Laplacian eigenvalues and whose argument is time. Our aim in this paper is to explore whether the shape of this function can be used to characterise the corresponding graph. There are several ways in which this can be done. In spectral geometry, the heat kernel trace has been used to characterise the differential geometry

D.-Y. Yeung et al. (Eds.): SSPR&SPR 2006, LNCS 4109, pp. 306–313, 2006.

of manifolds [7] [8]. Here the spectrum of the Laplace-Beltrami operator is used to construct a trace-function. This function can be expanded as a polynomial series in time, and the co-efficients of the series can be related to the Ricci curvature tensor of the manifold. Unfortunately, the relationships between the elements of the Ricci curvature tensor and the co-efficients are difficult to determine, and are only tabulated up to third order [8]. For large graphs, the Laplacian can be viewed as a discrete approximation of the Laplace-Beltrami operator and this analysis can be carried over from manifolds to graphs [14].

However, in this paper we deal with rather small graphs and take a different approach. Our idea is to measure the shape of the heat-kernel trace by taking moments with respect to time. Using the Mellin transform it is straightforward to show that the moment generating function is related to the zeta function of the graph. The zeta function is a series found by exponentiating and summing the reciprocals of the non-zero eigenvalues of the Laplacian. We construct a feature-vector whose components are the values of the zeta-function with integer argument.

Experiments with real world data taken from the COIL and Caltech-Oxford databases reveal that the zeta-function provides useful features for clustering graphs and to outperform the Laplacian spectrum.

2 The Laplacian Eigensystem and the Heat-Kernel

To commence, suppose that the graph under study is denoted by $G = (V, E)$ where V is the set of nodes and $E \subseteq V \times V$ is the set of edges. Since we wish to adopt a graph-spectral approach we introduce the adjacency matrix A for the graph where the elements are

$$A(u, v) = \begin{cases} 1 \text{ if} u, v \in E \\ 0 \text{ otherwise} \end{cases} \tag{1}$$

We also construct the diagonal degree matrix D, whose elements are given by $D(u, u) = \sum_{v \in V} A(u, v)$. From the degree matrix and the adjacency matrix we construct the Laplacian matrix $L = D - A$, i.e. the degree matrix minus the adjacency matrix. The normalised Laplacian is given by $\hat{L} = D^{-\frac{1}{2}} L D^{-\frac{1}{2}}$. The spectral decomposition of the normalised Laplacian matrix is $\hat{L} = \Phi \Lambda \Phi^T$, where $\Lambda = diag(\lambda_1, \lambda_2, ..., \lambda_{|V|})(0 = \lambda_1 < \lambda_2 < ... < \lambda_{|V|})$is the diagonal matrix with the ordered eigenvalues as elements and $\Phi = (\phi_1|\phi_2|....|\phi_{|V|})$ is the matrix with the ordered eigenvectors as columns. Since $\hat{L}$ is symmetric and positive semi-definite, the eigenvalues of the normalised Laplacian are all positive. The eigenvector associated with the smallest non-zero eigenvector is referred to as the Fiedler-vector.

We are interested in the heat equation associated with the Laplacian, i.e.

$$\frac{\partial h_t}{\partial t} = -\hat{L} h_t \tag{2}$$

where h_t is the heat kernel and t is time. The heat kernel can hence be viewed as describing diffusion across the edges of the graph with time. The rate of flow is determined by the Laplacian of the graph. The solution to the heat equation is found by exponentiating the Laplacian eigenspectrum, i.e.

$$h_t = \sum_{i=1}^{|V|} \exp[-\lambda_i t]\phi_i \phi_i^T = \Phi \exp[-t\Lambda]\Phi^T \tag{3}$$

The heat kernel is a $|V| \times |V|$ matrix, and for the nodes u and v of the graph G the resulting element is

$$h_t(u,v) = \sum_{i=1}^{|V|} \exp[-\lambda_i t]\phi_i(u)\phi_i(v) \tag{4}$$

When t tends to zero, then $h_t \simeq I - \hat{L}t$, i.e. the kernel depends on the local connectivity structure or topology of the graph. If, on the other hand, t is large, then $h_t \simeq \exp[-t\lambda_2]\phi_2\phi_2^T$, where λ_2 is the smallest non-zero eigenvalue and ϕ_2 is the associated eigenvector, i.e. the Fiedler vector. Hence, the large time behavior is governed by the global structure of the graph.

The trace of the heat kernel is

$$Z(t) = Tr[h_t] = \sum_{i=1}^{|V|} \exp[-\lambda_i t] \tag{5}$$

To provide an illustration of the potential utility of the trace-formula, in Figure 1 we show four small graphs with rather different topologies. Figure 2 shows the trace of the heat kernel as a function of t for the different graphs. From the plot it is clear that the curves have a distinct shape and could form the basis of a useful representation to distinguish graphs. For instance, the more "dumbbell" shaped the graph the more strongly peaked the trace of the heat-kernel at the origin. This is due to the fact the spectral gap, i.e. the size of λ_2, determines the rate of decay of the trace with time, and this in turn is a measure of the degree of separation of the graph into strongly connected subgraphs or "clusters".

3 Zeta-Function and Heat-Kernel Trace Moments

The aim in this paper is to use the shape of the heat-kernel trace function as a means of characterising graph-structure. Our characterisation is found by taking moments of trace-function over time.

To commence our development, we consider the zeta function associated with the Laplacian eigenvalues. The zeta function is given by

$$\zeta(s) = \sum_{\lambda_i \neq 0} \lambda_i^{-s} \tag{6}$$

In other words, it is the result of exponentiating and summing the reciprocal of the non-zero Laplacian eigenvalues.

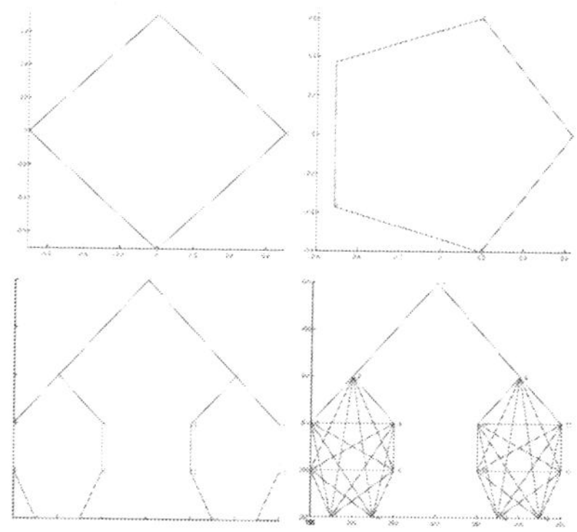

Fig. 1. Four graphs used for heat-kernel trace analysis

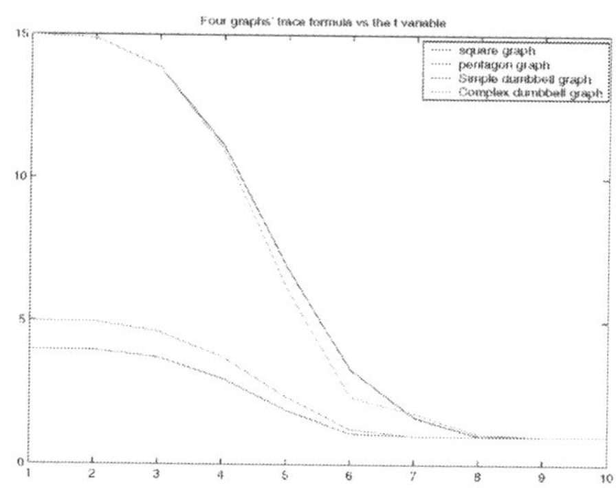

Fig. 2. Heat kernel trace as a function of t for four simple graphs

To establish the link between the zeta function and the trace of the heat-kernel we make use of the Mellin transform

$$\lambda_i^{-s} = \frac{1}{\Gamma(s)} \int_0^\infty t^{s-1} \exp[-\lambda_i t] dt \tag{7}$$

where

$$\Gamma(s) = \int_0^\infty t^{s-1} \exp[-t] dt \tag{8}$$

Hence, we can write the zeta function as a moment generating function

$$\zeta(s) = \frac{1}{\Gamma(s)} \int_0^\infty t^{s-1} \sum_{\lambda_i \neq 0} \exp[-\lambda_i t] dt \tag{9}$$

The sum of exponentials inside the integral is clearly linked to the trace of the heat-kernel. To show this we make use of the fact that

$$Tr[h_t] = C + \sum_{\lambda_i \neq 0} \exp[-\lambda_i t] \tag{10}$$

where C is the multiplicity of the zero eigenvalue of the Laplacian, or the number of connected components of the graph. Substituting this result back into the Mellin transform, we have

$$\zeta(s) = \frac{1}{\Gamma(s)} \int_0^{\infty} t^{s-1} \Big\{ Tr[h_t] - C \Big\} dt \tag{11}$$

As a result the zeta function is related to the moments of the heat-kernel trace. It is hence a way of characterising the shape of the heat kernel trace.

4 Experiments

We have experimented with the zeta-function characterisation of the heat-kernel trace. The data used for our study furnished by two data-bases used widely in the object recognition literature, namely the COIL data-base and Oxford-Caltech data-base. For the Coil data-base, we extract the feature points using the method of Harris and Stephens [18]. We have extracted graphs from the images by computing the Voronoi tessellations of the feature-points, and constructing the region adjacency graph, i.e. the Delaunay triangulation, of the Voronoi regions. Figure 3 shows some examples images with the extracted Delaunay graph overlayed for each of the four objects studied. For the Caltech-Oxford data-base, in Figure 4, we use Gestalt relation graphs between line-segments. For each image we extract line-segments using the Canny edge detector and contour polygonalisation. We treat each line-segment as a node in the relation graph. The weights between each pair of nodes are from the relative distance and relative angles attributes between the line-segments. The weighted links between the line segments capture the regular Gestalt-inspired relationships of proximity, parallelism, closure, and continuity [19]. The graphs used in our study are undirected and unattributed.

Both data-sets contain multiple images of either objects of the same class or views of the same object in different poses with respect to the camera. Example images from the data-sets are shown in Figures 3 and 4.

We commence by illustrating the behavior of the zeta-function for the images of objects from COIL data-base. From left-to-right and top-to-bottom in Figure 5 we show

Fig. 3. Example images of four objects from the COIL data-base with their Delaunay graphs overlayed

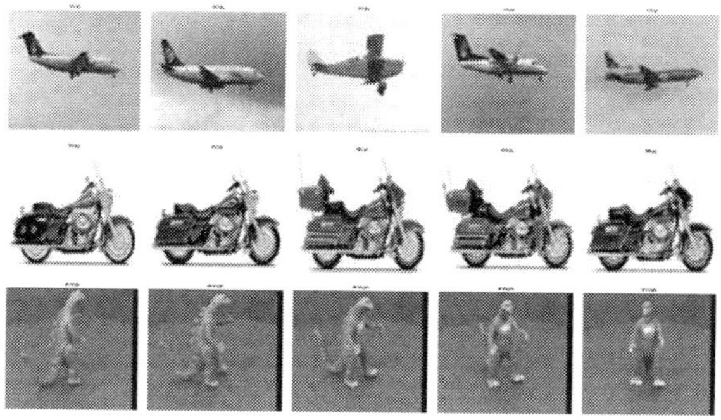

Fig. 4. Example images from the Oxford-Caltech Database

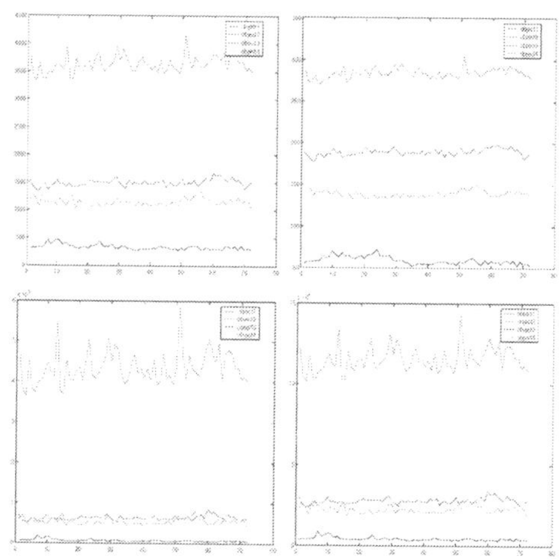

Fig. 5. Zeta function variation with view number

the values of $\zeta(1)$, $\zeta(2)$, $\zeta(3)$ and $\zeta(4)$ as a function of view-number for the four objects. The different curves in the four plots correspond to the different objects. The main feature to note is that the curves for the different objects are well separated, and that the individual values of the zeta-function do not vary significantly with view number. Moreover the fluctuations in the values are generally smaller than the differences between different objects. This feature is shown more clearly in Figure 6. Here we show the average value of the zeta-function moments as a function of the moment order. The different curves are for different objects. The error-bars show the standard deviation of the moment over the different views (instances) of the same object. The left-hand plot is for the COIL data and the right-hand plot for the Oxford-Caltech data. The moments do not overlap for the different objects, and the best separation is achieved for moments of intermediate order.

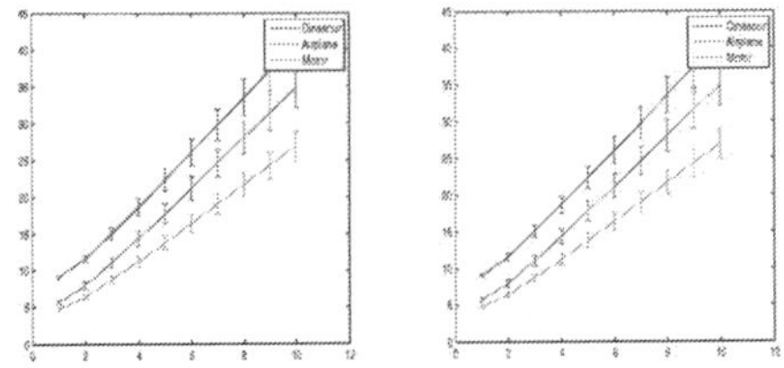

Fig. 6. Zeta function moments as a function or order for the different objects (COIL left and Oxford-Caltech right)

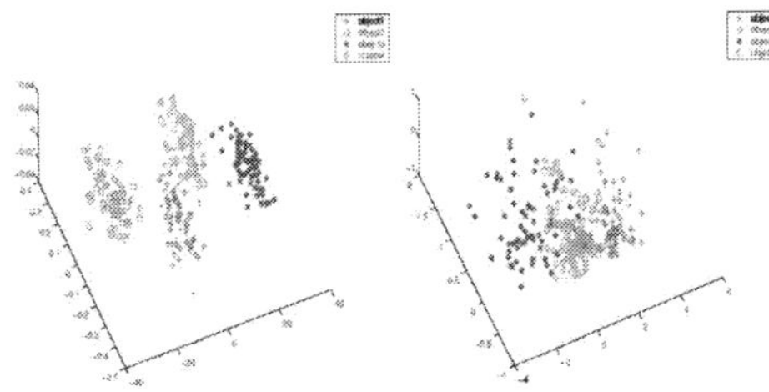

Fig. 7. Zeta-function and Spectral Clustering for the COIL database

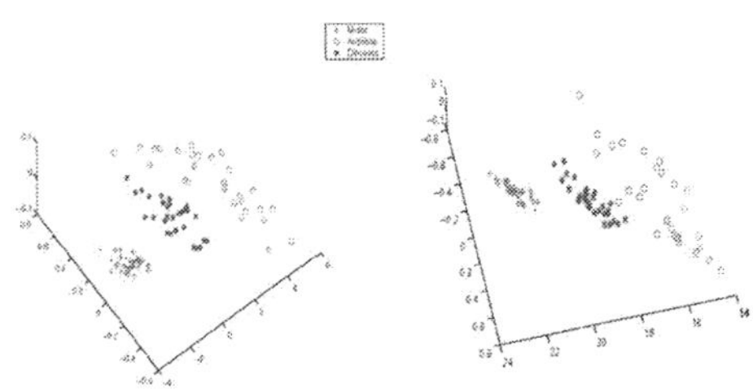

Fig. 8. Zeta-function and Spectral Clustering for the Oxford-Caltech database

In the left-hand panel of Figure 7 we show the result of performing principal components analysis on a feature-vector $f = (\zeta(1), \zeta(2)....., \zeta(10))^T$ which has as its components the zeta-function evaluated at the integers 1 to 10. Here we project the graphs onto the eigenspace spanned by the first three eigenvectors of the feature-vector covariance matrix. The different objects are denoted by points of a different color. The different objects are well separated in the eigenspace. For comparison the right-hand panel in Figure 7 we show the result of repeating this analysis on a vector of leading eigenvalues of the Laplacian matrix $f_\Lambda = (\lambda_1, \lambda_2...., \lambda_{10})^T$. In the eigenspace, the objects are severely overlapped, and the clustering is poorer. In Figure 8 we repeat the analysis of the zeta-function and Laplacian spectrum for the objects from the Caltech-Oxford database. Again, the best clusters are obtained using the zeta-function moments.

5 Conclusions

In this paper we have explored the use of the zeta-function as a means of characterising the shape of the heat-kernel trace for the purposes of graph-clustering. Using the Mellin transform, we have shown that the zeta-function is linked to the moment generating function of the heat-kernel trace. We have experimentally explored the use of the zeta-function as a means of characterising graphs for the purposes of clustering. The method works well on the COIL and Caltech-Oxford data-bases.

References

1. F.R.K.Chung. Spectral Graph Theory *American Mathematical Society*, 1997.
2. J.Shi and J.Malik. Normalized cuts and image segmentation. *IEEE PAMI*, 22:888–905, 2000.
3. J.E.Atkins, E.G.Bowman and B.Hendrickson. A spectral algorithm for seriation and the consecutive ones problem. *SIAM J. Comput.*, 28:297–310, 1998.
4. A.Shokoufandeh, S.Dickinson, K.Siddiqi, and S.Zucker. Indexing using a spectral encoding of topological structure. *IEEE Conf. on Computer Vision and Pattern Recognition*, 491–497, 1999.
5. R.R.Coifman and S.Lafon. Diffusion maps. *Applied and Computational Harmonic Analysis*, 2004.
6. J.Lafferty and G.Lebanon. Diffusion kernels on statistical manifolds. *Technical Reports CMU-CS-04-101*, 2004.
7. S.T.Yau and R.M.Schoen. Differential geometry. *Science Publication*, 1988.
8. P.B.Gilkey. Invariance theory, the heat equation, and the atiyah-singer index theorem. *Perish Inc.*, 1984.
9. Sam Roweis and Lawrence Saul. Nonlinear dimensionality reduction by locally linear embedding. *Science*, 290(5500):2323–2326 ,Dec.22, 2000.
10. Bin Luo, Richard C. Wilson and Edwin R. Hancock. Spectral embedding of graphs. Pattern Recognition, 36:2213–2230, 2003.
11. S.Rosenberg. The laplacian on a Riemannian manifold. *Cambridge University Press*, 2002.
12. B.Mohar Laplace eigenvalues of graphs - a Survey. *Discrete Math.*, 109:171–183, 1992.
13. L.Lovaz Random Walks on Graphs: A Survey. *Combinatorics, Paul Erds is eighty*, 2:353–397, 1996.
14. M.Hein, J.Y.Audibert and U.von Luxburg From graphs to manifolds – Weak and strong pointwise consistency of graph laplacian *18th Annual Conference on Learning Theory*, 470–485, 2005.
15. M.Belkin and P. Niyogi, "Laplacian Eigenmaps for Dimensionality Reduction and Data Representation", Neural Computation, **15**, pp. 1373–1396, 2003.
16. J.B. Tenenbaum, V.D. Silva and J.C.Langford, "A global geometric framework for non-linear dimensionality reduction", *Science*, **290**, pp. 586–591, 2000.
17. X.He and P. Niyogi, "Locality preserving projections", *NIPS03*.
18. C.G.Harris, and M.J.Stephens, "A Combined Corner and Edge Detector", *Fourth Alvey Vision Conference*, pp. 147–151, 1994.
19. S.Sarkar, and K.L.Boyer, "Quantitative measures of change based on feature organization: Eigenvalues and eigenvectors. Computer Vision and Image Understanding", *Computer Vision and Image Understanding*, pp. 110–136, 1998.

A Robust Realtime Surveillance System*

Byung-Joo Kim[1], Chang-Bum Lee[2], and Il-Kon Kim[3]

[1] Youngsan University Dept. of Information and Network Engineering, Korea
bjkim@ysu.ac.kr

[2] Youngsan University Dept. of Information and Network Engineering, Korea
cblee@ysu.ac.kr

[3] Kyungpook National University Dept. of Computer Science, Korea
ikkim@knu.ac.kr

Abstract. This paper describes a feature extraction method for realtime surveillance. Eigenspace models are a convenient way to represent set of images with widespread applications. In the traditional approach to calculate these eigenspace models, known as batch PCA method, model must capture all the images needed to build the internal representation. This approach has some drawbacks. Since the entire set of images is necessary, it is impossible to make the model build an internal representation while exploring a new person. Updating of the existing eigenspace is only possible when all the images must be kept in order to update the eigenspace, requiring a lot of storage capability. In this paper we propose a method that allows for incremental eigenspace update method by incremental kernel PCA for realtime surveillance. Experimental results indicate that accuracy of proposed method is comparable to batch KPCA and outperforms than APEX. Furthermore proposed method has efficiency in memory requirement compared to KPCA.

1 Introduction

Unsupervised surveillance gadgets aided by hi-tech visual information retrieval and indexing systems use computerized face recognition techniques that can recognizes faces from an image. There are two main approaches for face recognition[1]. The first approach is the feature based matching approach using the relationship between facial features[2]. The second approach is the template matching approach using the holistic features of the face images[2]. Template based techniques often follow the subspace method called eigenface originated by Turk and Pentland[3]. This technique is based on the Karhunen-Loeve transformation, which is also referred as PCA. It has gained great success and become a de facto standard and a common performance benchmark in face recognition. One of the attractive characteristics of PCA is that a high demension vector can be represented by a small number of orthogonal basis vectors. The conventional methods

* This study was supported by a grant of the Korea Health 21 R&D Project, Ministry of Health & Welfare, Republic of Korea (A05-0909-A80405-05N1-00000A).

D.-Y. Yeung et al. (Eds.): SSPR&SPR 2006, LNCS 4109, pp. 314–321, 2006.

of PCA such as singular value decomposion(SVD) and eigen-decomposition, perform in batch-mode with a computational complexity of $O(m^3)$ when m is the minimum value between the data dimension and the number of training examples. Undoubtedly these methods are computationally expensive when dealing with large scale problems where both the dimension and the number of training examples are large. To address this problem, many researchers have been working on incremental algorithms. Among them Chandrasekaran et al presented an incremental eigenspace update method using SVD[4]. Hall et al derived an eigen-decomposition based incremental algorithm and later extended their work to merge and split eigenspace models[5]. Another problem of PCA is that it only defines a linear projection of the data, the scope of its application is necessarily somewhat limited. It has been shown that most of the data in the real world are inherently non-symmetric and therefore contain higher-order correlation information that could be useful[6]. PCA is incapable of representing such data. For such cases, nonlinear transforms is necessary. Recently kernel trick has been applied to PCA and is based on a formulation of PCA in terms of the dot product matrix instead of the covariance matrix[7]. Kernel PCA(KPCA), however, requires storing and finding the eigenvectors of a $N \times N$ kernel matrix where N is a number of patterns. It is infeasible method for when N is large. This fact has motivated the development of incremental way of KPCA method which does not store the kernel matrix. In this paper we propose a method that allows for incremental eigenspace update method by incremental kernel PCA for vision learning and recognition. Paper is organized as follows. In Section 2 we will briefly explain the incremental PCA method. In Section 3 KPCA is introduced and to make KPCA incrementally, empirical kernel map method is explained. Experimental results to evaluate the performance of proposed method is shown in Section 4. Discussion of proposed method and future work is described in Section 5.

2 Incremental PCA

In this section, we will give a brief introduction to the method of incremental PCA algorithm which overcomes the computational complexity of standard PCA. Before continuing, a note on notation is in order. Vectors are columns, and the size of a vector, or matrix, where it is important, is denoted with subscripts. Particular column vectors within a matrix are denoted with a superscript, while a superscript on a vector denotes a particular observation from a set of observations, so we treat observations as column vectors of a matrix. As an example, A^i_{mn} is the ith column vector in an $m \times n$ matrix. We denote a column extension to a matrix using square brackets. Thus $[A_{mn}b]$ is an $(m \times (n+1))$ matrix, with vector b appended to A_{mn} as a last column.

To explain the incremental PCA, we assume that we have already built a set of eigenvectors $U = [u_j], j = 1, \cdots, k$ after having trained the input images $\mathbf{x}_i, i = 1, \cdots, N$. The corresponding eigenvalues are Λ and $\bar{\mathbf{x}}$ is the mean of input image. Incremental building of eigenspace requires to update these eigenspace to take into account of a new input image. Here we give a brief summarization of the method which is described in [5]. First, we update the mean:

$$\overline{x}' = \frac{1}{N+1}(N\overline{x} + x_{N+1}) \tag{1}$$

We then update the set of eigenvectors to reflect the new input image and to apply a rotational transformation to U. For doing this, it is necessary to compute the orthogonal residual vector $\hat{h} = (Ua_{N+1} + \overline{x}) - x_{N+1}$ where a_{N+1} is principal component and normalize it to obtain $h_{N+1} = \frac{h_{N+1}}{\|h_{N+1}\|_2}$ for $\| h_{N+1} \|_2 > 0$ and $h_{N+1} = 0$ otherwise. We obtain the new matrix of eigenvectors $U^{'}$ by appending h_{N+1} to the eigenvectors U and rotating them :

$$U' = [U, h_{N+1}]R \tag{2}$$

where $R \in \mathbf{R}_{(\mathbf{k+1})\times(\mathbf{k+1})}$ is a rotation matrix. R is the solution of the eigenspace of the following form:

$$DR = R\Lambda' \tag{3}$$

where Λ' is a diagonal matrix of new eigenvalues. We compose $D \in \mathbf{R}_{(\mathbf{k+1})\times(\mathbf{k+1})}$ as:

$$D = \frac{N}{N+1}\begin{bmatrix} \Lambda & 0 \\ 0^T & 0 \end{bmatrix} + \frac{N}{(N+1)^2}\begin{bmatrix} aa^T & \gamma a \\ \gamma a^T & \gamma^2 \end{bmatrix} \tag{4}$$

where $\gamma = h_{N+1}^T(x_{N+1} - \bar{x})$ and $a = U^T(x_{N+1} - \bar{x})$. Though there are other ways to construct matrix D[4][5], the only method ,however, described in [6] allows for the updating of mean.

2.1 Updating Image Representations

The incrementl PCA represents the input image with principal components $a_{i(N)}$ and it can be approximated as follows:

$$\widehat{x}_{i(N)} = Ua_{i(N)} + \bar{x} \tag{5}$$

To update the principal components $a_{i(N)}$ for a new image x_{N+1} , computing an auxiliary vector η is necessary. η is calculated as follows:

$$\eta = \left[U\widehat{h}_{N+1}\right]^T (\overline{x} - \overline{x}') \tag{6}$$

then the computation of all principal components is

$$a_{i(N+1)} = (R')^T \begin{bmatrix} a_{i(N)} \\ 0 \end{bmatrix} + \eta, \qquad i = 1, \cdots, N+1 \tag{7}$$

The transformations described above yield a model that represents the input images with the same accuracy as the previous one, therefore we can now discard the old subspace and the coefficients that represent the image in it. x_{N+1} is represented accurately as well, so we can safely discard it. The representation of all $N+1$ images is possible because the subspace is spanned by $k+1$ eigenvector. Due to the increase of the dimensionality by one, however, more storage is required to represent the data. If we try to keep a k-dimensional eigenspace, we lose a certain amount of information. In order to balance the storage requirements with the level of accuracy, it is needed for us to set the criterion on retaining the number of eigenvectors. There is no explicit guideline for retaining a number of eigenvectors.

In this paper we set our criterion on adding an eigenvector as $\lambda'_{k+1} > 0.7\bar{\lambda}$ where $\bar{\lambda}$ is a mean of the λ. Based on this rule, we decide whether adding u'_{k+1} or not.

3 Incremental KPCA

A prerequisite of the incremental eigenspace update method is that it has to be applied on the data set. Furthermore incremental PCA builds the subspace of eigenvectors incrementally, it is restricted to apply the linear data. But in the case of KPCA this data set $\Phi(x^N)$ is high dimensional and most of the time can not even be calculated explicitly. For the case of nonlinear data set, applying feature mapping function method to incremental PCA may be one of the solutions. This is performed by so-called *kernel-trick*, which means an implicit embedding to an infinite dimensional Hilbert space[9](i.e. feature space) F.

$$K(x,y) = \Phi(x) \cdot \Phi(y) \tag{8}$$

Where K is a given kernel function in an input space. When K is semi positive definite, the existence of Φ is proven[7]. Most of the case ,however, the mapping Φ is high-dimensional and cannot be obtained explicitly. The vector in the feature space is not observable and only the inner product between vectors can be observed via a kernel function. However, for a given data set, it is possible to approximate Φ by empirical kernel map proposed by Scholkopf[10] and Tsuda[11] which is defined as $\Psi_N : \mathbf{R}^d \to \mathbf{R}^N$

$$\begin{aligned} \Psi_N(x) &= [\Phi(x_1)\cdot\Phi(x), \cdots, \Phi(x_N)\cdot\Phi(x)]^T \\ &= [K(x_1,x), \cdots, K(x_N,x)]^T \end{aligned} \tag{9}$$

A performance evaluation of empirical kernel map was shown by Tsuda. He shows that support vector machine with an empirical kernel map is identical with the conventional kernel map[12]. The empirical kernel map $\Psi_N(x_N)$,however, do not form an orthonormal basis in $\mathbf{R}^N$, the dot product in this space is not the ordinary dot product. In the case of KPCA ,however, we can be ignored as the following argument. The idea is that we have to perform linear PCA on the $\Psi_N(x_N)$ from the empirical kernel map and thus diagonalize its covariance matrix. Let the $N \times N$ matrix $\Psi = [\Psi_N(x_1), \Psi_N(x_2), \ldots, \Psi_N(x_N)]$, then from equation (9) and definition of the kernel matrix we can construct $\Psi = NK$. The covariance matrix of the empirically mapped data is:

$$C_\Psi = \frac{1}{N}\Psi\Psi^T = NKK^T = NK^2 \tag{10}$$

In case of empirical kernel map, we diagonalize NK^2 instead of K as in KPCA. Mika shows that the two matrices have the same eigenvectors $\{u_k\}$[12]. The eigenvalues $\{\lambda_k\}$ of K are related to the eigenvalues $\{k_k\}$ of NK^2 by

$$\lambda_k = \sqrt{\frac{k_k}{N}} \tag{11}$$

and as before we can normalize the eigenvectors $\{v_k\}$ for the covariance matrix C_Ψ of the data by dividing each $\{u_k\}$ by $\sqrt{\lambda_k N}$. Instead of actually diagonalize the covariance matrix C_Ψ, the incremental KPCA is applied directly on the mapped data $\Psi = NK$. This makes it easy for us to adapt the incremental eigenspace update method to KPCA such that it is also correctly takes into account the centering of the mapped data in an incremental way. By this result, we only need to apply the empirical map to one data point at a time and do not need to store the $N \times N$ kernel matrix.

4 Experiment

To evaluate the performance of accuracy on eiegnspace update for incremental data we take nonlinear data. The disadvantage of incremental method is their accuracy compared to batch method even though it has the advantage of memory efficiency. So we shall apply proposed method to a simple toy data which will show the accuracy and memory efficiency of incremental KPCA compared to APEX model proposed by Kung[13] and batch KPCA. Next we will use images from the Columbia Object Image Library(COIL-20). The set is consisted of images of 20 objects rotated about their vertical axis, resulting in 72 images per objects. We used these images for testing the performance of incremental KPCA.

4.1 Toy Data

To evaluate the eigenspace update accuracy and memory efficiency of incremental KPCA compared to APEX and KPCA we take nonlinear data used by Scholkoff[8]. Totally 41 training data set is generated by:

$$y = x^2 + 0.2\varepsilon : \ \varepsilon \ from \ N(0,1), x = [-1,1] \tag{12}$$

First we compare feature extraction ability of incremental KPCA to APEX model. APEX model is famous principal component extractor based on Hebbian learning rule. Applying toy data to incremental KPCA we finally obtain 2 eigenvectors. To evaluate the performance of two methods on same condition, we set 2 output nodes to standard APEX model.

In table 1 we experimented APEX method on various conditions. Generally neural network based learning model has difficulty in determining the parameters; for example learning rate, initial weight value and optimal hidden layer node. This makes us to conduct experiments on various conditions. $\| w \|$ is norm of weight vector in APEX and $\| w \| = 1$ means that it converges stable minimum. $cos\theta$ is angle between eigenvector of KPCA and APEX, incremental

Table 1. Performance evaluation of incremental KPCA(IKPCA) and APEX

Method	Iteration	Learning Rate	$\lVert w_1 \rVert$	$\lVert w_2 \rVert$	$cos\theta_1$	$cos\theta_2$	MSE
APEX	50	0.01	0.6827	1.4346	0.9993	0.7084	14.8589
APEX	50	0.05				do not converge	
APEX	500	0.01	1.0068	1.0014	0.9995	0.9970	4.4403
APEX	500	0.05	1.0152	1.0470	0.9861	0.9432	4.6340
APEX	1000	0.01	1.0068	1.0014	0.9995	0.9970	4.4403
APEX	1000	0.05	1.0152	1.0470	0.9861	0.9432	4.6340
IKPCA	100		1	1	1	1	0.0223

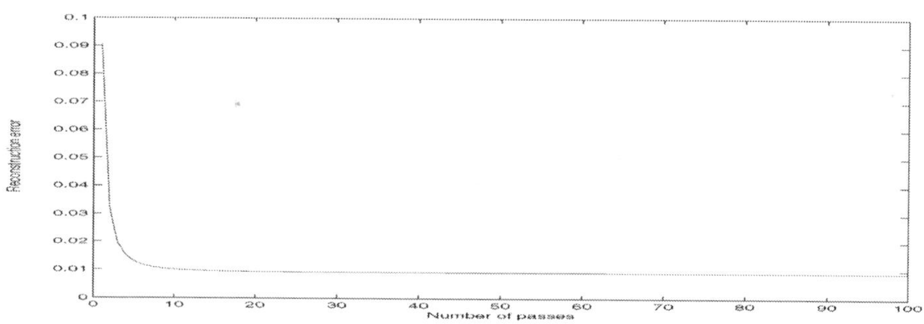

Fig. 1. Reconstruction error change by re-learning in incremental KPCA

KPCA respectively. $cos\theta$ of eigenvector can be a factor of evaluating accuracy how much incremental KPCA and APEX is close to accuracy of KPCA. Table 1 nicely shows the two advantages of incremental KPCA compared to APEX: first, performance of incremental KPCA is better than APEX; second, the performance of incremental KPCA is easily improved by re-learning. Another factor of evaluating accuracy is reconstruction error. Reconstruction error is defined as the squared distance between the Ψ image of x_N and reconstruction when projected onto the first i principal components.

$$\delta = |\Psi(x_N) - P_l\Psi(x_N)|^2 \tag{13}$$

In here P_l is the first i principal component. The MSE(Mean Square Error) value of reconstruction error in APEX is 4.4403 whereas incremental KPCA is 0.0223. This means that the accuracy of incremental KPCA is superior to standard APEX and similar to that of batch KPCA. Figure 1 shows the MSE value change for reconstruction error by re-learning in incremental KPCA. Relearning is similar meaning of epoch in neural network learning. We can see that the performance of incremental KPCA is easily improved by re-learning. Above results of simple toy problem indicate that incremental KPCA is comparable to the batch way KPCA and superior in terms of accuracy.

Table 2. Memory efficiency of incremental KPCA compared to KPCA on toy data

	KPCA	IKPCA
Kernel matrix	41 X 41	none
R matrix	none	3 X 3
D matrix	none	3 X 3
Efficiency ratio	93.3889	1

Next we will compare the memory efficiency of incremental KPCA compared to KPCA. In this experiments, incremental KPCA only needs D matrix and R matrix whereas KPCA needs kernel matrix. Table 2 shows the memory requirement of each method. Memory requirement of standard KPCA is 93 times more than incremental KPCA. We can see that incremental KPCA is more efficient in memory requirement than KPCA and has similar ability of eigenspace update accuracy. By this simple toy problem we can show that incremental KPCA has similar accuracy compare to KPCA and more efficient in memory requirement than KPCA.

4.2 Reconstruction Ability

To compare the reconstruction ability of incremental eigenspace update method proposed by Hall to APEX model we conducted experiment on face data. Applying this data to incremental eigenspace update method we finally obtain 30 Eigenvectors. As earlier experiment we set 30 output nodes to standard APEX

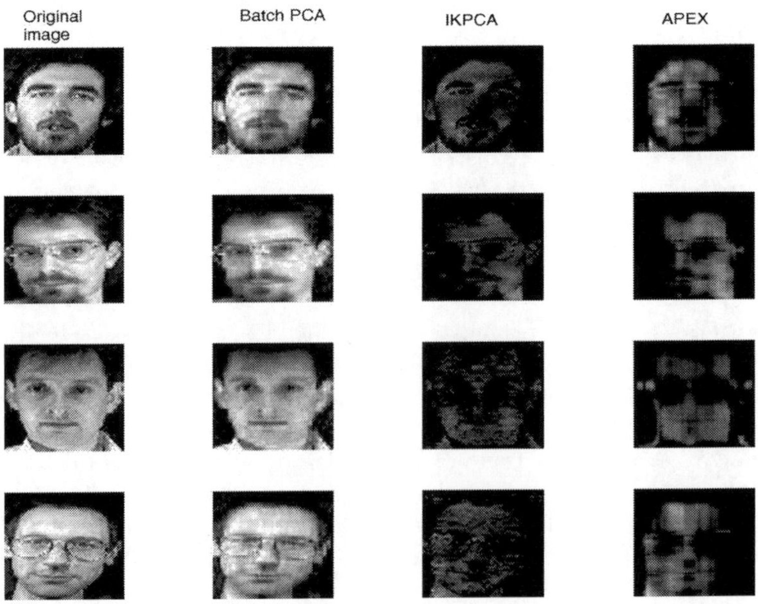

Fig. 2. Reconstructed image by incremental KPCA, APEX and batch PCA

method. Figure 2 shows the original data and their reconstructed images by incremental KPCA method, batch PCA and APEX respectively. The MSE(Mean Square Error) value of reconstruction error in APEX is 10.159 whereas incremental KPCA is 0.26941 and KPCA is 0.15762. This means that the accuracy of incremental KPCA is superior to standard APEX and similar to that of batch KPCA. We can see that reconstructed images by incremental KPCA update method is similar to original image and more clear compared to APEX method.

5 Conclusion and Remarks

A real time feature extraction for realtime surveillance system is proposed in this paper. We use incremental KPCA method in order to represent images in a low-dimensional subspace for realtime surveillance. Proposed method allows discarding the acquired images immediately after the update. By experimental results we can show that incremental KPCA has similar accuracy compare to KPCA and more efficient in memory requirement than KPCA. This makes proposed model is suitable for real time surveillance system. We will extend our research to realtime face recognition based on this research.

References

1. Chellappa, R. Wilson, C.L and Sirohey, S.: Human and machine recognition of faces:a survey Proc. of IEEE, vol.83, N0.5, May (1995) 705-740
2. Brunelli, R. and Poggio, T.:Face recognition:feature versus templates. IEEE Trans. PAMI, vol. 15, no. 10, (1993) 1042-1052
3. Turk, M. and Pentland, A.:Face recognition using eigenfaces. Proc. IEEE Conf. on CVPR, (1991) 586-591
4. Winkeler, J. Manjunath, B.S. and Chandrasekaran, S.:Subset selection for active object recognition. In CVPR, volume 2, IEEE Computer Society Press, June (1999) 511-516
5. Hall, P. Marshall, D. and Martin, R.: On-line eigenalysis for classification. In British Machine Vision Conference, volume 1, September (1998) 286-295
6. Softky, W.S and Kammen, D.M.: Correlation in high dimensional or asymmetric data set: Hebbian neuronal processing. Neural Networks vol. 4, Nov. (1991) 337-348
7. Gupta, H., Agrawal, A.K., Pruthi, T., Shekhar, C., and Chellappa., R.:An Experimental Evaluation of Linear and Kernel-Based Methods for Face Recognition. accessible at http://citeseer.nj.nec.com.
8. Murakami, H. Kumar.,B.V.K.V.:Efficient calculation of primary images from a set of images. IEEE PAMI, 4(5) (1982) 511-515
9. Vapnik, V. N.:Statistical learning theory. John Wiley & Sons, New York (1998)
10. Scholkopf, B. Smola, A. and Muller, K.R.:Nonlinear component analysis as a kernel eigenvalue problem. Neural Computation 10(5), (1998) 1299-1319
11. Tsuda, K.:Support vector classifier based on asymmetric kernel function. Proc. ESANN (1999)
12. Mika, S.:Kernel algorithms for nonlinear signal processing in feature spaces. Master's thesis, Technical University of Berlin, November (1998)
13. Diamantaras, K.I. and Kung, S.Y.:Principal Component Neural Networks: Theory and Applications. New York John Wiley & Sons, Inc (1996)

Ubiquitous Intelligent Sensing System for a Smart Home

Jonghwa Choi, Dongkyoo Shin, and Dongil Shin*

Department of Computer Science and Engineering, Sejong University, 98 Kunja-Dong Kwangjin-Gu, Seoul, Korea
jhchoi@gce.sejong.ac.kr, shindk@sejong.ac.kr, dshin@sejong.ac.kr

Abstract. We present the ubiquitous intelligent sensing system for a smart home in this paper. A smart home is intelligent space that studies patterns of home contexts that is acquired in a home, and provides automatic home services for the human. The ubiquitous intelligent sensing system acquires seven sensing contexts from four sensor devices. We utilize association rules of data mining and linear support machine to analyze context patterns of seven contexts. Also, we analyze stress rates of the human through the HRV pattern of the ECG. If the human is suffering from stress, the ubiquitous intelligent sensing system provides home service to reduce one's stress. In this paper, we present the architecture and algorithms of the ubiquitous intelligent sensing system. We present the management toolkit to control the ubiquitous intelligent sensing system, and show implementation results of the smart home using the ubiquitous intelligent sensing system.

1 Introduction

The computer industry has seen wonderful results in miniaturization and performance improvement of devices as a result of rapid technological development: central computing with mainframes (1950s through the 1980s), personal computers (1980s to present), and computer networks (1990s to present) [1]. A fourth era is now emerging as computers became ubiquitous, a technology more noticeable by its absence than its presence [2, 3]. This paper addresses the architecture of a ubiquitous intelligent sensing system and algorithms of components that compose the system. A future home is an intelligent space that realizes ubiquitous computing that offer the human a more comfortable life. Ubiquitous computer aims to "enhance" computer use by making many computers available throughout the physical environment, but making them effectively invisible to the user [4]. A smart home helps to provide a more comfortable home life to humans through intelligent sensing that offers a human automated service in a ubiquitous space. Studies of intelligent agents in a smart home have proceeded into many different a lot of directions. MS Easy Living provides a home service through location tracking of the human [5]. MavHome presented a prediction algorithm of home service in smart home, and applied data mining as its algorithm [6, 7].

We implemented four sensor devices (ECG sensor, home temperature sensor, network camera for human location and motion and facial expression sensor) that acquire home contexts from human and home. The sensor devices provide seven

* Corresponding author.

D.-Y. Yeung et al. (Eds.): SSPR&SPR 2006, LNCS 4109, pp. 322–330, 2006.

sensing contexts (ECG, pulse, body temperature, home temperature, human location, human motion and facial expression) to the ubiquitous intelligent sensing system. We apply LSVM (linear support vector machine) and association rules of data mining to analyze patterns of all contexts.

Section 2 gives related research work on an intelligent sensing system. Section 3 addresses architecture of the ubiquitous intelligent sensing system for the smart home. In section 4, we explain a detailed algorithm of the components that compose the intelligent sensing system. Section 5 presents implementation and experimental results. We conclude with section 6.

2 Related Studies

Perry and Dowdall documented the rationale and design of a multimodal interface to a pervasive/ubiquitous computing system that supports independent living by older people in their own homes [8]. The Smart-In-House project used a system of basic sensors to monitor a person's in-home activity; a prototype of the system is being tested within a subject's home [9]. They examined whether the system could be used to detect behavioral patterns and report the results. Alex and Brent discuss the use of computer vision in pervasive healthcare systems, specifically in the design of a sensing agent for an intelligent environment that assists older adults with dementia during the activity of daily living [10]. The 1:1 pro system constructed personal profiles based on the customer's transactional histories. The system used data mining techniques to discover a set of rules describing customer's behavior and supports human experts in validating the rules [11]. Kehagias and Petridis introduced the PREdictive MOdullar Fuzzy System (PREMOFS) to perform a time series classification. A PREMOFS consists of a bank of predictors and a fuzzy inference module. The PREMOFS is a fuzzy modular system that classifies the time series as one of a finite number of classes, using the full set of un-preprocessed past data to perform a recursive, adaptive and competitive computation of membership function, based on predictive power [12].

3 Architecture of Ubiquitous Intelligent Sensing System

Figure 1 shows the architecture of the ubiquitous intelligent sensing system for a smart home. Figure 1 presents ten sensing contexts, and we provide the intelligent sensing system with seven sensing contexts and time for pattern prediction of the home service. Eye tracker and voice sensing are processed by a rule-based algorithm.

Five home appliances have been connected to the wireless network in the laboratory. The ubiquitous intelligent sensing system predicts a home service pattern of the human, and offers automatic home service for the human. We use the SAPR (supervised algorithm for pattern recognition: a linear support vector machine) and the data miner (association rules) to analyze the home service for the human. The ubiquitous intelligent sensing system studies the human's home service pattern without providing a home service for the human during the learning phase. For example, when the human turns on the TV, the context extractor acquires the human's home service command (TV-On) and sensing contexts from four sensor devices, and transmits all data to the pattern recognition process. During the prediction phase of the home service, the ubiquitous intelligent sensing system acquires seven sensing contexts and the time from the sensor devices every three seconds, and it provides an automatic home service to the human through pattern analysis of previous home services. The context extractor acquires all contexts from sensor devices, and manages them. The context extractor processes contexts in two steps to analyze the pattern of eight contexts acquired from sensor devices. First, the context extractor normalizes eight contexts between 0.1 and 0.9 so that contexts are treated by input value (train input and test input) of the supervised algorithm for pattern recognition. Second, the context extractor stores all contexts in the database for creation of association rules. We applied the HHIML that is based on XML as the context's management structure. This helped to easily approach all contexts in other components, and presents an efficient interface to the web [13, 14]. The home service provider executes home service that is predicted from the supervised algorithm for pattern recognition and association rules of data mining.

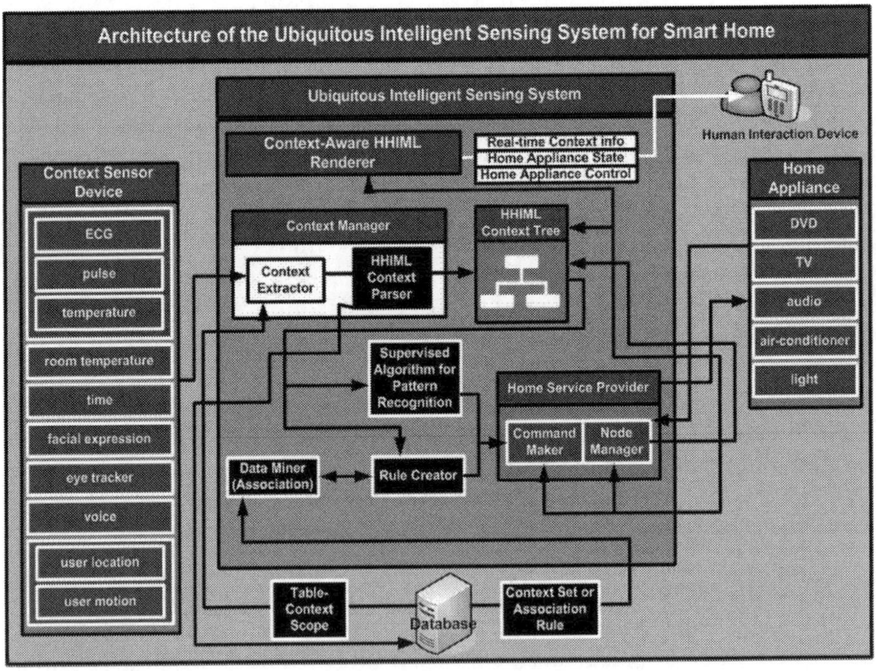

Fig. 1. Architecture of the ubiquitous intelligent sensing system for a smart home

4 Components of the Ubiquitous Intelligent Sensing System

4.1 Context Extraction and Processing

Figure 2 shows pictures of four sensor devices used to acquire seven contexts for the pattern recognition of home service.

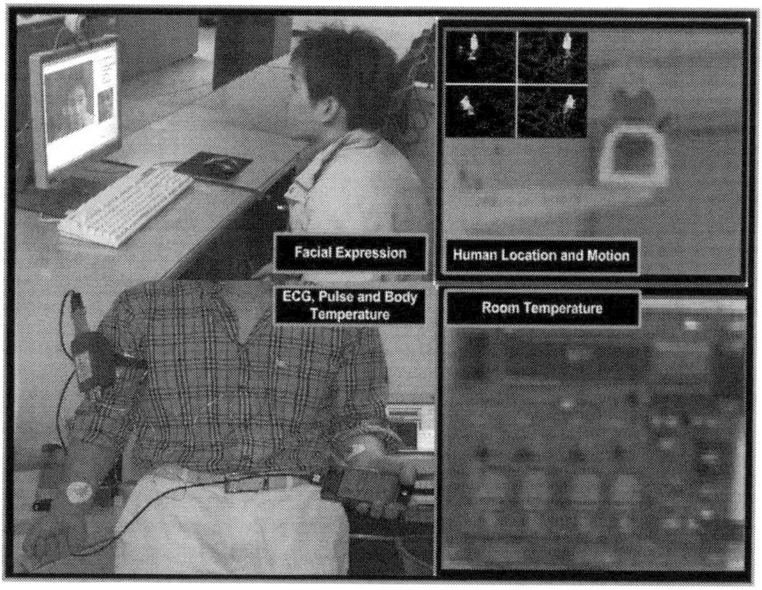

Fig. 2. Four sensor devices for extraction of seven contexts

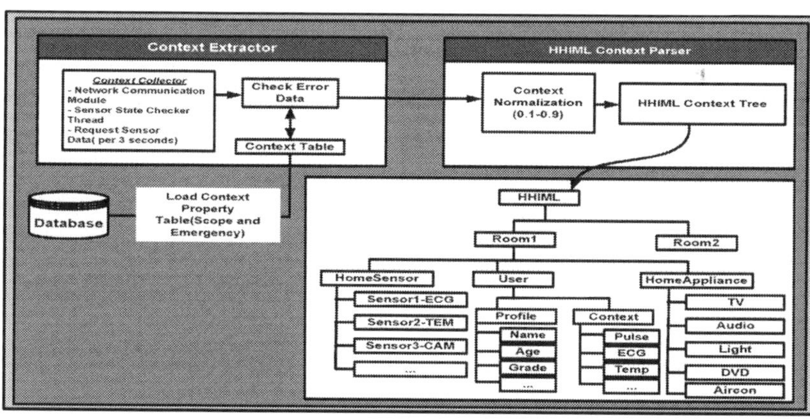

Fig. 3. Structure of the context extractor that takes charge of the sensing context's extraction and context processing

The four sensor devices use a WiFi for network communication with the ubiquitous intelligent sensing system. The facial expression sensor recognizes the human's expression by comparing it with seven standard expressions (blank, surprise, fear, sad, angry, disgust and happy). They are categorized as described in [15]. The human location is decided from absolute coordinates through analysis of raw images. The human's motion is recognize from six motions (lie, stand, sit, sit_gesture1, sit_gesture2 and sit_gesture3) using pattern recognition algorithms.

We acquire the ECG signal, pulse and temperature from the ECG sensor device. The pulse and the temperature are transmitted to the pattern recognition algorithm for pattern analysis of the home service: the ECG signal is used to predict human's stress. Figure 3 shows the structure of the context extractor that takes charge of sensing context's extraction and context processing.

The context extractor acquires contexts from the sensor devices that is presented in figure 2, and normalizes all contexts between 0.1 and 0.9. Then, it converts all contexts that are acquired in real-time into the HHIML's tree that is based on XML to manage all contexts efficiently.

4.2 Pattern Recognition of Human's Home Service

Figure 4 shows the structure of the pattern recognition algorithm that predicts the home service that the human wants.

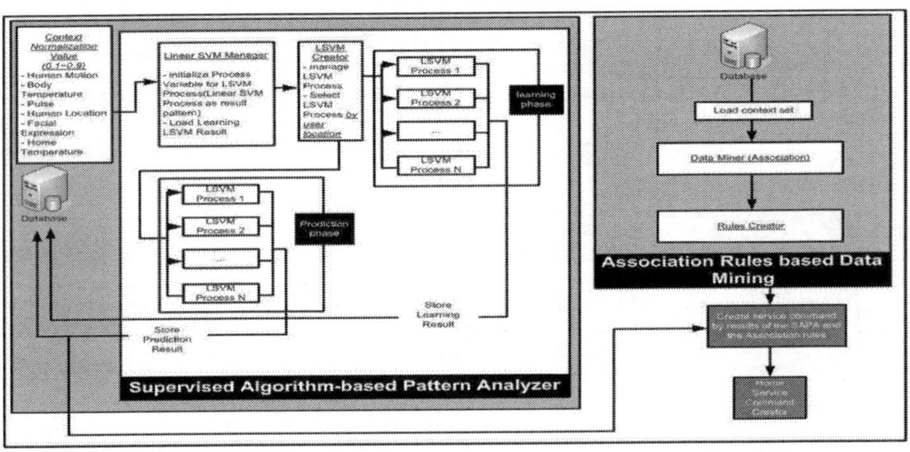

Fig. 4. Structure of the supervised algorithm-based pattern analyzer and the association rules based data mining

We applied LSVM by hierarchal structure for pattern recognition of three home services (TV, audio and DVD). The ubiquitous intelligent sensing system measures stress of a human from analysis of the ECG. If the human is suffering from stress, the ubiquitous intelligent sensing system provides a home service (soft music and a faint light) to comfort the human and reduce stress. The ECG (electrocardiogram) is an electric signal, which reflects the hearts pulse rate that is measurable on the body's surface. Figure 5 shows the ECG graph with P, Q, R, S and T values that were extracted from an ECG signal.

The ubiquitous intelligent sensing system provides a home service that reduces the stress of a human using the HRV (Heart Rate Variability) of the ECG signal [16, 17].

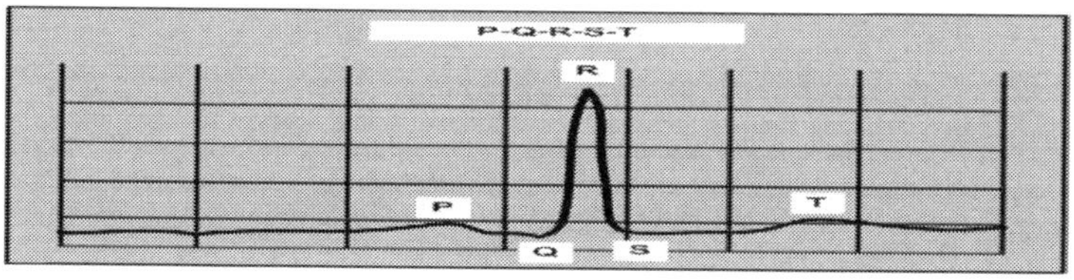

Fig. 5. Pulse rate of heart with P, Q, R, S, and T values extracted from the ECG

4.3 Sensing Context and Home Service Management

We handled all sensing contexts and control of home service using PDA and PC. Figure 6 shows management screen of all contexts and home service in the ubiquitous intelligent sensing system. (A) and (B) in figure 6 show the manager toolkit that controls the intelligent sensing system by PDA.

Fig. 6. The manager toolkit for management of the ubiquitous intelligent sensing system

(D) in figure 6 expresses the home's structure and sensing contexts in three dimensions according to the HHIML's structure, which was created from the ubiquitous intelligent sensing system.

5 Experiments and Evaluations

The ubiquitous intelligent sensing system provides an automatic home service that is predicted by the pattern recognition algorithm. All contexts were applied as LSVM's features and association rule's input data. The ubiquitous intelligent sensing system provides predicted home services if prediction results of two pattern recognition algorithms (the SAPR and the data miner) are the same. Figure 7 shows a distribution chart of sensing contexts that is presented by the data miner.

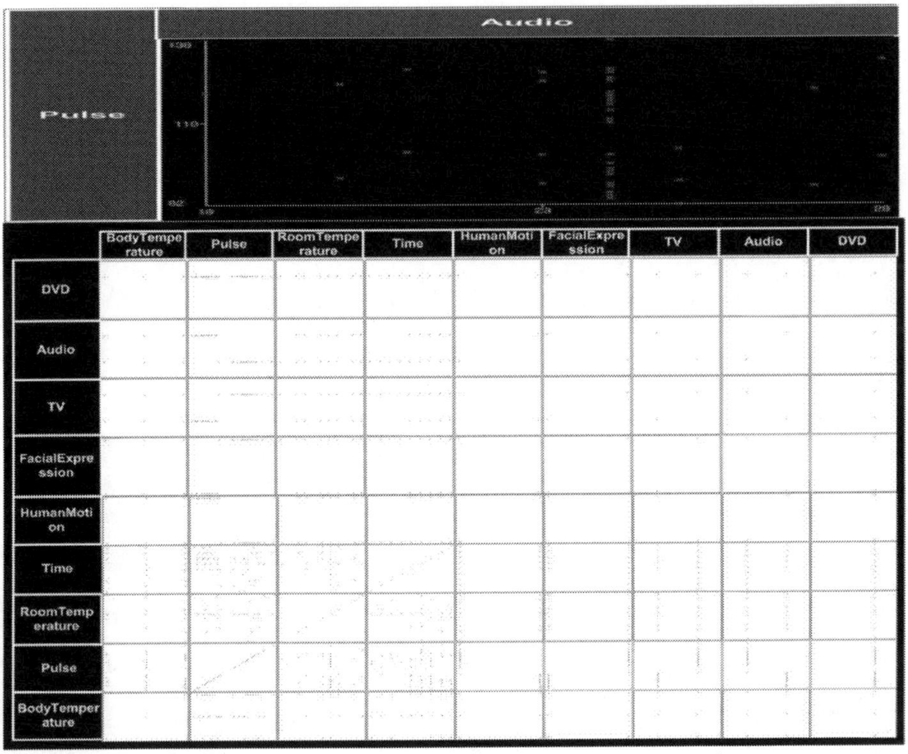

Fig. 7. Distribution chart of sensing contexts that is presented by the data miner

Table 1 shows the performance of the prediction component that is created by the LSVM and the association rules.

Table. 1. The performance of the prediction component that is created by the LSVM and the association rules

	Number of Service Prediction by the SAPA	Number of Service Prediction by the Association Rules	Number of Service Prediction by Integration Methods	Number of Rejected Service by Human	Precision on test set
TV	123	148	76	21	72.3%
DVD	156	111	90	24	73.3%
Audio	211	154	107	31	71.0%

Home service prediction of the ubiquitous intelligent sensing system showed an average 72.2%. We are progressing with experiments of additional sensing contexts to increase the system's performance. Also, we are testing how the performance of the system change by changing the importance of different feature sets.

The ECG is a sensing context that is applied to measure the human's stress. The ECG wave means a healthy state if it presents big oscillation within the standard scope. Otherwise, it means that the autonomic nervous system's ability to adapt to stress decreases. (a) shows a normal human's ECG wave, and (b) shows human's ECG wave with stress. If a human is suffering from stress, the ubiquitous intelligent sensing system provides home service (soft music and a faint light) that comforts to human and reduces stress.

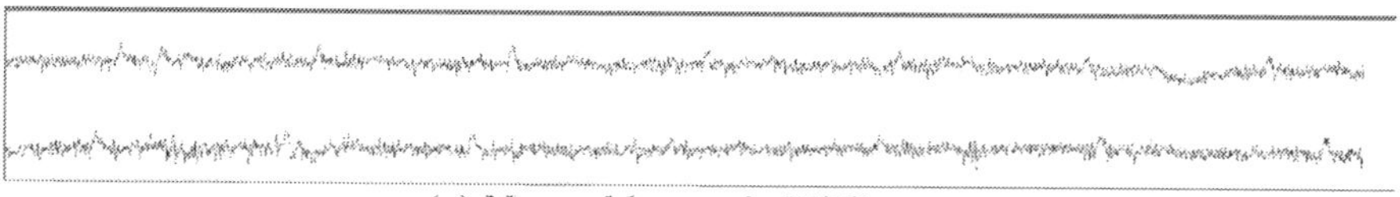

(a) Normal human's ECG wave

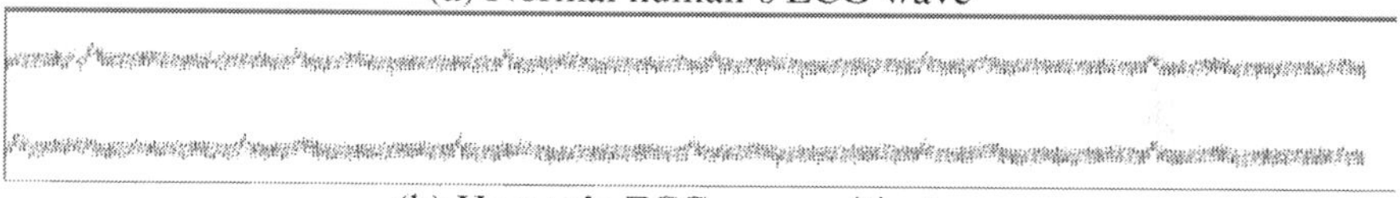

(b) Human's ECG wave with stress

Fig. 8. ECG signals that is extracted from the ECG sensor device

6 Conclusions

We presented the ubiquitous intelligent sensing system for a smart home in this paper. A smart home is intelligent space that studies pattern of home contexts that is acquired in a home, and provides an automatic home service for the human. We implemented a smart home using the ubiquitous intelligent sensing system. The ubiquitous intelligent sensing system acquires seven sensing contexts from four sensor devices. This paper uses association rules of data mining and linear support machine to analyze context patterns of seven contexts. Also, we analyze stress rate of human through the HRV pattern of the ECG. If human is suffering from stress, the ubiquitous intelligent sensing system provides home service that reduces human's stress. In this paper, we present the architecture of ubiquitous intelligent sensing system, explains algorithms of components that compose the ubiquitous intelligent sensing system. We present the management toolkit to control the ubiquitous intelligent sensing system, and shows pictures that is executed in the ubiquitous intelligent sensing system.

References

1. Petriu. E.M, Georganas. N.D, Petriu. D.C, Makrakis. D, Groza. V.Z.: Sensor-based information appliances. Instrumentation & Measurement Magazine. IEEE Volume 3. Issue 4. (2000) 31 - 35
2. J. Birnbaum.: Pervasive information systems. Communications of the ACM. vol 40. no 2. (1997) 40-41
3. M. L. Dertouzos.: The future of computing. Scientific American. vol 52. (1999) 52-55

4. M. Weiser.: Some computer science issues in ubiquitous computing. Commun ACM. vol 36. no 7. (1993) 75-84
5. Easy Living. http://research.microsoft.com/easyliving/
6. MavHome. http://mavhome.uta.edu/
7. D.J. Cook, M. Youngblood, E. Heierman, K. Gopalratnam, S. Rao, A. Litvin, and F. Khawaja.: Mavhome: An agent based smart home. Proceedings of the IEEE International Conference on Pervasive Computing and Communications. (2003) 521-524
8. Perry. M, Dowdall. A, Lines. L, Hone. K.: Multimodal and ubiquitous computing systems: supporting independent-living older users. Information Technology in Biomedicine. IEEE Transactions on. Volume 8. Issue 3. (2004) 258 - 270
9. Barger. T.S, Brown. D.E, Alwan. M.: Health-Status monitoring through analysis of behavioral patterns. Systems, Man and Cybernetics, Part A, IEEE Transactions on. Volume 35. Issue 1. (2005) 22 - 27
10. Mihailidis. A, Carmichael. B, Boger. J.: The use of computer vision in an intelligent environment to support aging-in-place, safety, and independence in the home. Information Technology in Biomedicine. IEEE Transactions on. Volume 8. Issue 3. (2004) 238 - 247
11. Adomavicius. G, Tuzhilin. A.: Using data mining methods to build customer profiles. Computer. Volume 34. Issue 2. (2001) 74 - 82
12. Kehagias. A, Petridis. V.: Predictive modular fuzzy systems for intelligent sensing. Systems, Man, and Cybernetics, IEEE International Conference on. Volume 4. (1996) 2816 – 2821
13. World Wide Web Consortium (W3C). Extensible Markup Language (XML) 1.0. Available: http://www.w3c.org/TR/REC-xml. (1998)
14. Gunhee Kim, Dongkyoo Shin, Dongil Shin.: Design of a middleware and HIML(Human Interaction Markup Language) for context aware services in ubiquitous computing environment. Lecture Notes in computer science. Vol 3207. (2004) 682-691
15. Charles, D.: The expression of the emotions in man and animals. Electronic Text Center. University of Virginia Library
16. Task Force of the European Society of Cardiology and the North American Society of Pacing and Electrophysiology.: Heart Rate Variability standards of measurement. physiological interpretation, and clinical use. Cir. 93. (1996) 1043-1065
17. Lombardi F and Sandrone G.: Heart rate variability and sympatho-vagal interaction after myocardial infarction. In: Malik M and Camm AJ, Heart Rate Variability. Armonk NY: Futura Publishing Company, Inc. (1995) 222-234

Object Recognition Using Multiresolution Trees

Monica Bianchini, Marco Maggini, and Lorenzo Sarti

DII - Università degli Studi di Siena
Via Roma, 56 - 53100 Siena - Italy
{monica, maggini, sarti}@dii.unisi.it

Abstract. This paper presents an object recognition method based on recursive neural networks (RNNs) and multiresolution trees (MRTs). MRTs are a novel hierarchical structure proposed to represent both the set of homogeneous regions in which images can be divided and the evolution of the segmentation process performed to determine such regions. Moreover, knowing the optimal number of regions that should be extracted from the images is not critical for the construction of MRTs, that are also invariant w.r.t. rotations and translations. A set of experiments was performed on a subset of the Caltech benchmark database, comparing the performances of the MRT and directed acyclic graph (DAG) representations. The results obtained by the proposed object detection technique are also very promising in comparison with other state-of-the-art approaches available in the literature.

1 Introduction

In graphical pattern recognition, data is represented as an arrangement of elements, that encodes both the properties of each element and the relations between them. Hence, patterns are modeled as labeled graphs where, in general, labels can be attached to nodes and edges.

In the last few years, a new connectionist model, that exploits the above definition of pattern, has been developed [1]. In fact, recursive neural networks (RNNs) have been devised to face one of the most challenging task in pattern recognition: realizing functions from graphs to vectors in an automatic and adaptive way. The original RNN model and its evolutions were recently applied to image processing tasks [2,3], obtaining interesting results. However, in order to exploit RNNs, a crucial role is played by the graphical representation of patterns, i.e. the way in which each image is represented by a graph. This choice affects the performances of the whole process.

In this paper we propose a new graphical representation of images based on multiresolution trees (MRTs), that are hierarchical data structures, somehow related to other representations used in the past to describe images. MRTs are generated during the segmentation of the images, like, for instance, quad–trees [4]. Other hierarchical structures, so as monotonic trees [5] or contour trees [6], can be exploited to describe the set of regions obtained at the end of the segmentation process, representing the inclusion relationship established among the

D.-Y. Yeung et al. (Eds.): SSPR&SPR 2006, LNCS 4109, pp. 331–339, 2006.

region boundaries. However, MRTs represent both the result of the segmentation, and the sequence of steps that produces the final set of regions. Moreover, the construction of MRTs does not depend on a priori knowledge on the number of regions needed to represent the images. Finally, MRTs combined with RNNs allow us to develop efficient object detection and object recognition systems.

This paper proposes an object recognition method and evaluates its performances on the Caltech benchmark dataset [7]. Two comparisons are presented: first, the images are represented by MRTs and directed acyclic graphs (DAGs) to assess which kind of structure allows to achieve better results; then the object recognition technique is compared against the state of the art methods based on vectorial representation and classical pattern recognition approaches [7,8,9,10].

The paper is organized as follows. In the next section, the RNN model is described, whereas in Section 3 the algorithm to extract MRTs is defined. Section 4 collects the experimental results and, finally, Section 5 draws some conclusions.

2 Recursive Neural Networks

Recursive neural networks were originally proposed to process directed positional acyclic graphs (DPAGs) [1,11]. More recently, an extended model, able to map rooted nonpositional graphs with labeled edges (DAGs-LE) into real vectors, was described [12]. This last RNN model is implemented based on a state transition function which has not a predefined number of arguments and which does not depend on the argument position. The different contribution of each child to the state of its parents depends on the label attached to the corresponding edges. At each node v, the total contribution $\overline{X}_v \in I\!R^n$ of the states of its children is computed as

$$\overline{X}_v = \sum_{i=1}^{od[v]} \Phi(X_{ch_i[v]}, L_{(v,ch_i[v])}, \theta_\Phi),$$

where $od[v]$ is the outdegree of the node v, i.e. the number of its children, $\Phi : I\!R^{(n+k)} \to I\!R^n$ is a function depending on a set of parameters θ_Φ, $X_{ch_i[v]} \in I\!R^n$ is the state of i-the child of node v, and $L_{(v,ch_i[v])} \in I\!R^k$ is the label attached to the edge $(v, ch_i[v])$. The state at the node v is then computed by another function $f : I\!R^{(n+m)} \to I\!R^n$ that combines the contribution of $\overline{X}_v$ with the node label $U_v \in I\!R^m$:

$$X_v = f(\overline{X}_v, U_v, \theta_f),$$

being f a parametric function depending on the parameters θ_f. Moreover, at the root node s, also an output function g is computed by another function g as

$$Y_s = g(X_s, \theta_g).$$

The functions Φ, f and g can be implemented by feedforward neural networks, in which the parameters θ_Φ, θ_f and θ_g are connection weights (see Figure 1(b)). As shown in Figure 1(c), the processing of an input graph is obtained by applying the recursive neural network (Figure 1(b)) recursively on the graph nodes, starting from the leaves. This processing scheme yields an *unfolding network* whose structure depends on the topology of the input graph. The state X_v at each node encodes a representation of the subgraph rooted at v.

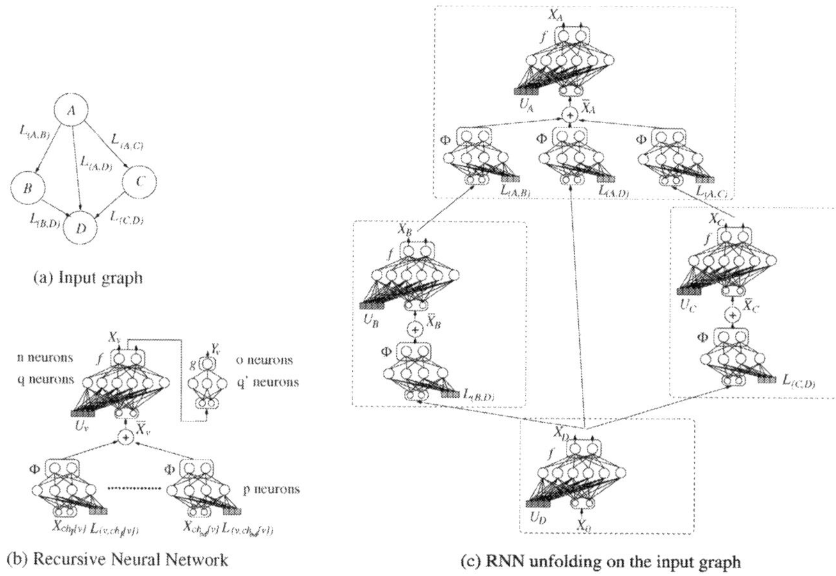

Fig. 1. The RNN processing scheme

RNNs can be trained to categorize images represented as graphs. Therefore an object recognition system can be developed based on a pool of RNNs where each network is specialized in recognizing a particular object class.

3 Multiresolution Trees

The neural network model proposed in Section 2 assumes to process structured data. Therefore a preprocessing phase that allows to represent images by graphs is needed, in order to exploit such model to perform any task on images (classification, localization or detection of objects, etc.). In the last few years, some image analysis systems based on RNNs and graph–based representation of images were proposed [12,13,14]. In these approaches, images are segmented to obtain a set of homogeneous regions that are subsequently represented by region adjacency graphs (RAGs). Then, since RNNs can process only directed structures, while RAGs are not, a further step is needed, to transform each RAG into a directed acyclic graph (DAG) [14] or into a set of trees [2]. The transformation from a RAG to a DAG can be obtained very efficiently, performing a breadth–first visit of the RAG, until each node is visited once. However, during this process each edge is transformed into a directed arc depending on arbitrary choices, in particular the starting node for the breadth–first visit. In fact, the assignment of a direction to the adjacency relation changes the semantics of this relation, that is naturally undirected. On the contrary, the mapping from a DAG to a set of trees allows us to preserve the structural information, but it is particularly time consuming. As a matter of fact, for each node belonging to the RAG, a breadth–first visit must be performed to obtain a tree.

In this section, the representation of images based on MRTs is proposed. This kind of structure presents two advantages: first MRTs can be processed directly by RNNs without the need of any transformation; second they somehow reduce the dependence of the representation from the choice of the number of regions to be extracted in the segmentation process.

Nowadays, there is no a universal theory on image segmentation, and all the existing methods are, by nature, ad hoc. Moreover, the determination of the exact number of regions that should be extracted from a given image is a very challenging task, and can affect the performances of the system that takes the computed representation as input. MRTs allow us to ignore this information, since they collect, at each level, a distinct segmentation of the input image, obtained during a region growing procedure.

Since an MRT is generated during the segmentation, we need to describe how a set of homogeneous regions is extracted from an input image. First, a K–means clustering of the pixels belonging to the input image is performed. The clustering algorithm minimizes the Euclidean distance (defined in the chosen color space) of each pixel from its centroid. The number of clusters computed during this step is determined considering the average texture of the input image, since such a parameter provides useful information about the complexity of the depicted scene. It is worth noting that the number of extracted regions is greater than the number of clusters, since each cluster typically corresponds to several connected components. At the end of the K–means, a region growing step is carried out to reduce the number of regions, and, at the same time, to generate the MRT. The region growing procedure is sketched in Algorithm 1.1. Actually, the proposed method assumes to merge together groups of homogeneous regions, with the aim of bounding the maximum outdegree of the MRT, and, consequently, its depth. The algorithm takes as parameters the set of regions ($Rset$) obtained at the end of the K–means, and $maxGroupSize$, the maximum number of regions that can belong to a group. The goal of the algorithm is to reduce the number of regions, and to compute the set of nodes V, and the set of edges E, that define the MRT.

Initially, the set V is updated exploiting the function *createNode* that creates a new node and computes the node label (a set of visual and geometric features). These nodes represent the leaves of the MRT. Then, for each region r belonging to $Rset$, a region group g is created and stored in $Gset$, using the function *createGroup*. The region r represents the seed of g. Moreover, each region adjacent to r is added to the group that has r as its seed. When the number of adjacent regions is greater than $maxGroupSize$, the color distance between r and its adjacent regions is computed, and only the $maxGroupSize$ nearest adjacent regions are added to g. Note that, after this step, $\bigcup_{i=1}^{|Gset|} g_i = I$, being I the whole image, but $\bigcap_{i=1}^{|Gset|} g_i \neq \emptyset$, and then $Gset$ must be rearranged with the aim of obtaining a partitioning of the image, such that $\bigcap_{i=1}^{|Gset|} g_i = \emptyset$.

Algorithm 1.1. `CreateMRT(`$Rset$`,`$maxGroupSize$`)`

```
{
  V←E←Gset←∅;
  for each r ∈ Rset
     V←V ⋃ {createNode(r)};
  while(|Rset|≥ maxGroupSize) {
     for each r ∈ Rset
        Gset←Gset ⋃ {createGroup(r)};
     Gset←cleanGroups(Gset,H());
     for each g ∈ Gset{
        newr←mergeGroup(g);
        Rset←Rset - members(g) ⋃ {newr};
        newn←createNode(newr);
        V←V ⋃ {newn};
        for each r ∈ members(g)
           E←E ⋃ {(newn,getNodeAssociatedWith(r))};
     }
     Gset←∅;
  }
  root←createNode(⋃_{i=1}^{|Rset|} r_i);
  V←V ⋃ {root};
  for each r ∈ Rset
     E←E ⋃ {(root,getNodeAssociatedWith(r))};
}
```

This phase is performed by the function *cleangroups*, that is described by Algorithm 1.2. This function takes as input $Gset$ and a homogeneity function H(), that is used to compute the degree of similarity of the regions that belong to a given group. The function H() is a parameter of the segmentation algorithm and must be chosen such that a high value of $H(g)$ corresponds to a high probability of merging g. First, the groups are sorted in descending order w.r.t. their homogeneity. As a matter of fact, if a region r belongs to a group having high homogeneity, the group that has r as its seed must be removed from $Gset$. Considering that *adjacent*(g) collects the set of regions that belong to g, except for its seed, and the function *getGroupBySeed*(r) returns the group that has r as its seed, the first iterative block of the algorithm performs the following steps. For each group g, the set *adjacent*(g) is analyzed. For each region r in *adjacent*(g), if r is the seed of a group that has lower homogeneity than g, then the group is removed from $Gset$, otherwise r is removed from *adjacent*(g). However, the goal of obtaining a partitioning of the original image is still not reached, because the algorithm removed only whole groups, but some regions can belong to more than one group. Then, the function *cleanGroups* scans again all the members of the groups looking for regions that belong to two or more groups, and, if they exist, removes them from the groups that have lower homogeneity.

Algorithm 1.2. `cleanGroups(`*Gset*`,H())`

```
{
  Gset←sort(Gset,H());
  for each g ∈ Gset
     for each r ∈ adjacent(g)
        if (H(g) ≥ H(getGroupBySeed(r)))
           Gset←Gset - getGroupBySeed(r);
        else
           adjacent(g)←adjacent(g)-r;
  for each g¹ ∈ Gset
     for each g² ∈ Gset and g¹ ≠ g²
        for each r¹ ∈ adjacent(g¹)
           for each r² ∈ adjacent(g²)
              if (r¹ = r²)
                 if (H(g¹) ≥ H(g²))
                       adjacent(g²)←adjacent(g²)-r²;
                 else
                       adjacent(g¹)←adjacent(g¹)-r¹;
}
```

At the end of Algorithm 1.2, *Gset* collects a partitioning of the whole image and Algorithm 1.1 merges together the regions that belong to the same groups. The set of regions *Rset* is updated considering the new regions, and a new level of the MRT is created. For each new region *newr*, a new node *newn* is created. Finally, *newn* is linked to the nodes that are associated with the regions merged to obtain *newr*.

The main loop of Algorithm 1.1 is repeated until the cardinality of *Rset* is greater than *maxgroupSize*. At the end of the main loop, the root node is added to the MRT and linked to all the nodes that correspond to the regions collected in *Rset*. All the nodes in the MRT have a label that describes visual and geometric properties of the associated regions, and also each edge has a label that collects some features regarding the merging process.

The proposed technique presents some advantages. First, segmentation methods based on region growing generally produce results that depend on the order in which the regions are selected during the merging process, and, as a side effect, the final set of regions is not invariant w.r.t. rotations, translations, and other transformations of the input image. Instead, the region growing method proposed in Algorithm 1.1 is independent from rotations and translations, since the regions are selected considering the order defined by the homogeneity function.

The main advantage of MRTs consists in being independent of the number of regions needed to represent the image. Actually, a distinct segmentation, with a different number of regions, is stored in each level of the tree. The key idea consists in exploiting the capabilities of adaptive models, like RNNs, to discover at which level the best segmentation is stored. Moreover, MRTs are invariant w.r.t. rotations and translations of the images and do not describe directly the topological arrangement of the regions, that however can be inferred considering both the geometric features associated to each node (for instance, the coordinates of the bounding box of each region can be stored in the node label) and the MRT structure. Finally, the region growing is performed merging together groups of regions instead of pairs, to avoid the generation of binary trees. As a matter of fact, the generation of binary trees implies the generation of deeper structures, and RNNs suffer in processing such structures, due to the "long-term dependency" phenomenon, that was originally investigated for recurrent neural networks [15].

4 Experimental Results

In order to evaluate the capability of MRTs to represent the contents of images, some experiments were carried out, addressing an object recognition problem. The experiments were performed using the Caltech Database[1], since it represents a popular benchmark for object class recognition. The Caltech database collects six classes of objects: motorbikes, airplanes, faces, cars (side view), cars (rear view), and spotted cats. Our experiments were focused on a subset of the dataset, that consists only of images from the motorbikes, airplanes, faces, and cars (rear view) classes.

Fig. 2. Samples of images from the Caltech database

For each class, three datasets were created: training, test, and cross–validation sets. The training and test sets collect 96 images each, while 48 images belong to the cross–validation set. For each set, half images correspond to positive examples, while the other images are examples of the negative class (i.e. images from the other classes). All the images were selected randomly from the Caltech database and segmented in order to obtain both MRTs and DAGs.

MRTs were obtained using Algorithm 1.1 and a maximum group dimension equal to 7. The homogeneity function, that affects directly the segmentation and the MRT generation, was chosen to be $H(g) = \frac{1}{\sigma^2}$, being σ^2 the color variance of the group g in the image color space. The node labels in the MRTs collect geometric and visual information, like area, perimeter, barycenter coordinates, momentum, etc., while color distance, barycenter distance, and the ratio obtained dividing the area of the child region by the area of the parent region, are

[1] The Caltech database is available at http://www.robots.ox.ac.uk/~vgg/data3.html

Table 1. Results obtained representing images by DAGs or MRTs. The second column shows the number of state neurons of the recursive network. The results are reported using the average ROC equal error rate, obtained performing ten learning runs.

	State neurons	Airplanes	Motorbikes	Faces	Cars(rear)
M	5	100	97.91	100	92.7
R	7	95.83	92.7	100	90.6
T	10	94.79	92.7	100	92.7
D	5	75	69.79	73.54	76.04
A	7	75	70.83	70.41	77
G	10	75	68.75	71.67	78.12

Table 2. Best results obtained using MRTs compared against results available in the literature. The results are reported using the average ROC equal error rate.

	RNNs and MRTs	Zhang [8]	Fergus [7]	Opelt [9]	Thureson [10]
Motorbikes	97.91	99	92.5	92.2	93.2
Airplanes	100	98.3	90.2	88.9	83.8
Faces	100	99.7	96.4	93.5	83.1

used as edge labels. With respect to the generation of DAGs, a modified version of Algorithm 1.1 was exploited. The instructions related to the MRT generation were removed, and the main loop was halted when the number of regions become smaller than the parameter that was used to determine the number of initial K–means clusters. Finally, at the end of the region growing phase, the DAG was generated following the steps described in [14]. The generated MRTs collect 400 nodes and are composed by 8 levels, on average, whereas DAGs contain about 70 nodes. For each class, several RNN classifiers were trained, using both MRTs and DAGs, in order to determine the best network architecture. The transition function f is realized by an MLP with $n+1$ hidden units (using the hyperbolic tangent as output function) and n linear outputs, being n the number of state neurons. The function Φ, that combines the state of each child with the corresponding edge label, is implemented by an MLP with a layer of $n+1$ sigmoid hidden units and n linear outputs. Finally, the output network g is an MLP with n inputs, $n-2$ sigmoid hidden units and one sigmoid output. The obtained results are reported in Table 1. Even if the main goal of the experiments is the comparison between MRTs and DAGs, Table 2 collects also a comparison between the presented object recognition system and other methods known in the literature, that were evaluated using the same benchmark database. The method based on MRTs definitely outperforms the DAG–based representation. Moreover, the comparison with the other methods reported in Table 2 shows very promising results, even if our experiments were performed considering only a subset of the Caltech database.

5 Conclusions

In this paper, we proposed a new hierarchical representation of images, based on multiresolution trees. An MRT represents, in a unique structure, the result of image segmentation, and the sequence of steps that produces the final set of regions. The performances of the proposed representation technique were evaluated addressing an object recognition task. A method based on RNNs and MRTs was proposed and evaluated on the Caltech benchmark dataset, showing promising results.

References

1. Frasconi, P., Gori, M., Sperduti, A.: A general framework for adaptive processing of data structures. IEEE Transactions on Neural Networks **9** (1998) 768–786
2. Bianchini, M., Maggini, M., Sarti, L., Scarselli, F.: Recursive neural networks learn to localize faces. Patter Recognition Letters (2005) 1885–1895
3. Bianchini, M., Maggini, M., Sarti, L., Scarselli, F.: Recursive neural networks for object detection. In: Proceedings of IEEE IJCNN. (2004) 1911–1915
4. Hunter, G.M., Steiglitz, K.: Operations on images using quadtrees. IEEE Transactions PAMI **1** (1979) 145–153
5. Song, Y., Zhang, A.: Monotonic tree. In: Proceedings of the 10th Intl. Conf.on Discrete Geometry for Computer Imagery, Bordeaux – France (2002)
6. Roubal, J., Peucker, T.: Automated contour labeling and the contour tree. In: Proceedings of AUTO-CARTO 7. (1985) 472–481
7. Fergus, R., Perona, P., Zisserman, A.: Object class recognition by unsupervised scale–invariant learning. In: Proceedings of IEEE CVPR. (2003) 264–271
8. Zhang, W., Yu, B., Zelinsky, G., Samaras, D.: Object class recognition using multiple layer boosting with heterogeneous features. In: Proceedings of CVPR. Volume 2. (2005) 323–330
9. Opelt, A., Fussenegger, M., Pinz, A., Auer, A.: Weak hypoteses and boosting for object detection and recognition. In: Proceedings of ECCV. Volume 2. (2004) 71–84
10. Thureson, J., Carlsson, S.: Appearance based qualitative image description for object class recognition. In: Proceedings of ECCV. Volume 2. (2004) 518–529
11. Küchler, A., Goller, C.: Inductive learning in symbolic domains using structure–driven recurrent neural networks. In Görz, G., Hölldobler, S., eds.: Advances in Artificial Intelligence. Springer, Berlin (1996) 183–197
12. Bianchini, M., Maggini, M., Sarti, L., Scarselli, F.: Recursive neural networks for processing graphs with labelled edges: Theory and applications. Neural Networks (2005) 1040–1050
13. de Mauro, C., Diligenti, M., Gori, M., Maggini, M.: Similarity learning for graph based image representation. Pattern Recognition Letters **24** (2003) 1115–1122
14. Gori, M., Maggini, M., Sarti, L.: A recursive neural network model for processing directed acyclic graphs with labeled edges. In: Proceedings of IEEE IJCNN. (2003) 1351–1355
15. Bengio, Y., Frasconi, P., Simard, P.: Learning long–term dependencies with gradient descent is difficult. IEEE Transactions on Neural Networks **5** (1994) 157–166

A Robust and Hierarchical Approach for Camera Motion Classification

Yuliang Geng, De Xu, Songhe Feng, and Jiazheng Yuan

Institute of Computer Science and Technology,
Beijing Jiaotong University, Beijing, 100044, China
gengyuliang@hotmail.com

Abstract. Camera motion classification is an important issue in content-based video retrieval. In this paper, a robust and hierarchical camera motion classification approach is proposed. As the Support Vector Machine (SVM) has a very good learning capacity with limited sample set and does not require any heuristic parameter, the SVM is first employed to classify camera motions into translation and non-translation motions in preliminary classification. In this step, four features are extracted as input of the SVM. Then, zoom and rotation motions are further classified by analyzing the motion vectors' distribution. And the directions of translation motions are also identified. The experimental results show that the proposed approach achieves a good performance.

1 Introduction

Camera motion classification is an important issue in content-based video retrieval. Taking no account of scene depth variation, there are four basic camera motion categories, namely, still, zoom (includes zoom in, zoom out), rotation and translation (includes panning right, tilting down, panning left and tilting up). Extracting camera motion will help understand higher-level semantic content, especially in some specific domains, such as sports video, movie video and surveillance video. Usually, zoom-in motion will give the details about the characters or objects do, or imply an important event may happen. Zoom-out motion gives a distant framing, which shows the spatial relations among the important figures, objects, and setting in a scene. Translation motion often indicates the dominant motion direction of a scene or gives an overview of mise en scene. So camera motion classification is essential to video structure analysis and higher semantic information extraction.

There are a number of methods proposed to detect camera motion in recent literatures [1,2,3,4]. Most of prior work is focused on parameter model estimation, such as affine model, perspective model, etc [1,2,3]. But high computational complexity and noise sensibility are still the main problems of parameter model estimation approach, especially in massive video analysis and retrieval. In [2] Huang *et al.* utilize feature points selection to improve performance of parameter estimation. In [3], Kumar *et al.* utilize parameter-decomposition estimation to reduce computational complexity. In fact, it is not necessary for video parsing

D.-Y. Yeung et al. (Eds.): SSPR&SPR 2006, LNCS 4109, pp. 340–348, 2006.

and understanding to extract accurate motion parameters. Qualitative camera motion classification helps improve computational performance and reduce noise influence. Zhu *et al.* [4] propose a qualitative method, which employs motion vectors mutual relationship to implement camera motion classification, and obtains a satisfying results.

In this paper, we propose an effective and efficient camera motion classification approach. First, cinematic rules are utilized to filter abnormal noise and foreground motion noise in preprocessing step. Then, the SVM is employed to classify camera motions into translation and non-translation motions in preliminary classification of camera motion. Finally, we refine the camera motion categories. In this step, the zoom and rotation motions are further distinguished, and the translation direction is also identified by analyzing the motion vectors' distribution. Experimental results validate the effectiveness of our proposed approach. The block diagram of our approach is shown in Fig. 1.

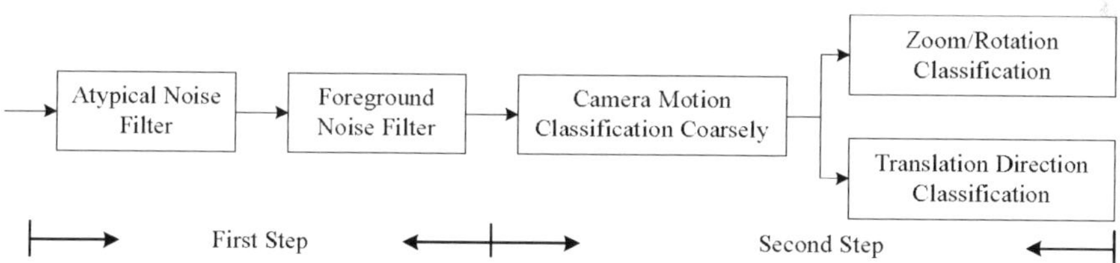

Fig. 1. Block diagram of the proposed approach

The organization of this paper is as follows. In Section 2, we represent the preprocessing step of camera motion classification, which is used to reduce abnormal noise and foreground motion noise. A robust and hierarchical approach for camera motion classification is proposed in Section 3. Section 4 and 5 give the experimental results and draw the conclusions.

2 The Preprocessing of Motion Vector Field (MVF)

Before estimating camera motion categories, we need filter motion noises in MVF because they might result in an error motion estimation. There are two different motion noises. One is abnormal motion noise that is generated from motion estimation by block-matching algorithm or optical flow equation. The other is foreground noise that is generated from the foreground object motion.

First, we filter the abnormal motion noise. In data analysis, Interquartile Range is an effective way to detect noise data in a given data set [5]. That is, any data that is less than $LQ - 1.5IQR$ or greater than $UQ + 1.5IQR$ is regarded as noise data, where LQ is the lower quartile, UQ is the upper quartile, and IQR is the interquartile range which is defined as $UQ - LQ$.

For a given MVF, we suppose that the magnitudes of motion vectors satisfy Gaussian distribution. We remove the abnormal motion vectors by computing the Interquartile Range, and denote the valid motion vector set as $\mathbf{V}_1$.

Then we further reduce the foreground noise. In mise en scene, filmmaker often places the foreground object on the center region of screen, also called attention region, to attract viewers' attention [6]. The attention region is determined by Golden Section spatial composition rule, which suggests dividing the screen into 3×3 regions in $3:5:3$ proportion in both directions. We denote the center region, that is attention region, as **C**, and the surrounding region as **B**. As the foreground objects and background have conspicuously different motion vectors, we compute the motion saliency map based on the valid motion vectors $\mathbf{V}_1$ as

$$S(i,j) = |E(i,j) - (\omega_1 \bar{E}_{\mathrm{B}} + \omega_2 \bar{E}_{\mathrm{C}})| \tag{1}$$

where $E(i,j)$ is the motion energy of block (i,j). ω_1, ω_2 are the preassigned weight values, and $\omega_1 \geq \omega_2$, $\omega_1 + \omega_2 = 1$. As the discussed above, the surrounding region plays more important role in camera motion classification, so we assign a greater value to ω_1 than ω_2. $\bar{E}_{\mathrm{B}}$, $\bar{E}_{\mathrm{C}}$ are the average motion energies of region B and C respectively.

Thus, we get the foreground motion region approximately by binarizing the motion saliency map. The binarization threshold is estimated in an adaptive method. We filter the foreground motion and achieve valid motion vector set, which is denoted as $\mathbf{V}_2$. The camera motion classification is based on $\mathbf{V}_2$.

3 Hierarchical Camera Motion Classification

The hierarchical approach for camera motion classification is composed of two steps as Section 3.1 and 3.2 depicted. Before camera motion classification, the still camera motion is detected. We regard the camera motion as still category if the average motion energy of the valid motion vectors is less than a given threshold TH_{still}. TH_{still} is an empirical value, and is set as 2.

3.1 Camera Motion Preliminary Classification Based on SVM

As the translation motions have similar motion vector fields, which are different from the ones for zoom and rotation motions. Namely, the MVF for translation is composed of parallel motion vectors with uniform magnitudes; and the MVF for zoom is composed of radial vectors whose magnitudes are proportional to their distance from the center of focus (COF). The motion vectors for zoom-in/zoom-out point inward to/outward from the COF. The vertical MVF for rotation has the same characteristic as the MVF for zoom.

As the discussed above, we first classify camera motions into two categories: translation and non-translation motion (includes rotation and zoom). Here, we extract four features to characterize the camera motions as follows.

1) Motion Direction Feature. The motion vector direction is classified into 12 categories: $(-15° + 30°i,\ 15° + 30°i)$, $i = 0, 1, \ldots, 11$. Let $H^{\mathrm{A}}(i)$ represent the percentage of motion vectors at the ith direction. Then the motion direction consistency is computed as

$$F^{\mathrm{AngEn}} = -\sum_{i=0}^{11} (H^{\mathrm{A}}(i) \log H^{\mathrm{A}}(i)) \tag{2}$$

2) Motion Direction Relationship. To characterize motion direction relationship, we first compute included angles among the valid motion vectors. Then the included angles are classified into 8 categories: $(22.5°i, 22.5°(i+1))$, $i = 0, 1, \ldots, 7$. Let $H^{\mathrm{I}}(i)$ represent the percentage of included angles at the ith direction. The motion direction relationship is characterized by the mean and entropy of angular histogram $H^{\mathrm{I}}(i)$.

$$F^{\mathrm{InAngMean}} = \sum_{i=0}^{7} (15 \times i \times H^{\mathrm{I}}(i))/(N_2(N_2 - 1)/2) \tag{3}$$

$$F^{\mathrm{InAngEn}} = -\sum_{i=0}^{7} (H^{\mathrm{I}}(i) \log H^{\mathrm{I}}(i)) \tag{4}$$

where N_2 is the size of the valid motion vector set $\mathbf{V}_2$.

3) Motion Energy Feature. We compute the motion energy histogram of valid motion vectors with 10 equally spaced bins, and denote it as $H^{\mathrm{M}}(i)$. Then the motion energy distribution is characterized as

$$F^{\mathrm{MagEn}} = -\sum_{i=0}^{9} (H^{\mathrm{M}}(i) \log H^{\mathrm{M}}(i)) \tag{5}$$

The four feature values can be taken as a feature vector, and each component is normalized by the Gauss normalization formula. Thus, we denote the feature vector as $\mathbf{F} = [\bar{F}^{\mathrm{AngEn}}, \bar{F}^{\mathrm{InAngMean}}, \bar{F}^{\mathrm{InAngEn}}, \bar{F}^{\mathrm{MagEn}}]$.

As the SVM has a very good learning capacity with limited sample set, and does not require any heuristic parameter [7], we select the SVM as the classifier in camera motion preliminary classification. We set $\mathbf{F}$ as the input vector of the SVM. There are three common kernel functions: Radial Basis Function $K(x, y) = \exp(-||\mathbf{x} - \mathbf{y}||^2/2\sigma^2)$, Polynomial Function $K(x,y) = (\mathbf{x} \cdot \mathbf{y} + b)^d$, and Sigmoid Kernel function $K(x,y) = \tanh[b(\mathbf{x} \cdot \mathbf{y}) - \theta)]$. So far, kernel function selection still relays on the experimental method. In Section 4, we'll discuss how to select kernel functions and determine its parameters in detail.

After the preliminary classification, we classify the camera motion into translation and non-translation motions.

3.2 Refine the Camera Motion Categories

Translation Motion Classification. For the translation motion, we identify its motion direction by computing the dominant motion direction histogram, $H^{\mathrm{ori}}(k)$, which represents the percentage of motion vectors at the kth direction, $(-45° + 90°k,\ 45° + 90°k)$.

$$H^{\mathrm{ori}}(k) = \sum_{j=-1}^{1} H^{\mathrm{A}}((3k + j) \bmod 12) \quad k = 0, 1, 2, 3 \tag{6}$$

We classify the translation motion into a specific direction whose bin value has the maximum.

Non-translation Motion Classification. For the non-translation motion, we further classify them into zoom and rotation motion. As discussed in Section 3.1, the motion vectors for zoom-in/zoom-out point inward to/outward from the COF, while the vertical motion vectors for rotation have same characteristic as zoom. So we can identify the zoom and rotation motion categories by detecting COF as follows.

Step 1. As the discussed in Section 2, the surrounding region **B** is composed of 8 subregions. We select one motion vector as key motion vector in each subregion respectively. The key motion vector is the one whose motion direction consists with the dominant motion direction of that subregion. If the number of the motion vectors whose magnitudes are equal to zero is greater than two thirds of the number of the total motion vectors in one subregion, we should not select key motion vector in this subregion. Thus, we get the key motion vector set $\{\mathbf{V}(x_i, y_i)\}$, where $\mathbf{V}(x_i, y_i)$ represents the motion vector of macro block (x_i, y_i).

Step 2. As discussed in Section 3.1, the straight line L_i through point (x_i, y_i) in direction of $\mathbf{V}(x_i, y_i)$ should pass through the COF in the MVF for zoom, so we compute the intersection points formed by pairwise intersection of straight lines (if they intersect) that are determined by the key motion vector set $\{\mathbf{V}(x_i, y_i)\}$.

Step 3. We calculate the centroid of the intersection points. A simply way is to compute the mean for the intersection points' position. We regard the centroid as the COF, and denote it as (x_0, y_0).

Step 4. We calculate the average distance, $dist((x_0, y_0), L_i)$, from (x_0, y_0) to straight line L_i that is determined by the key motion vector $\mathbf{V}(x_i, y_i)$.

$$\bar{Dist} = \frac{1}{N} \sum_{i=1}^{N} dist((x_0, y_0), L_i) \tag{7}$$

where N is the size of key motion vector set $\{\mathbf{V}(x_i, y_i)\}$. If the average distance $\bar{Dist}$ is less than TH_{zoom}, the camera motion is identified as zoom motion. TH_{zoom} is a given threshold and is set as one third of the MVF height.

Step 5. For each key motion vector $\mathbf{V}(x_i, y_i)$, we compute the inner-product between $\mathbf{V}(x_i, y_i)$ and $(x_i - x_0, y_i - y_0)$.

$$O_{\text{zoom}} = \sum_{i} \text{sgn}(dot(\mathbf{V}(x_i, y_i), (x_i - x_0, y_i - y_0))) \tag{8}$$

where sgn() is a sign function, which returns 1 if the element is greater than zero, 0 if it equals zero and -1 if it is less than zero. $dot(,)$ is a inner-product function. If $O_{\text{zoom}} > 0$, the camera motion is zoom in, otherwise is zoom out.

As the vertical MVF for rotation has the same characteristic with the MVF for zoom, we can identify the rotation motion as the same way.

4 Experimental Results

To evaluate the proposed approach for camera motion classification, we collect various video data from MPEG-4 test set and www.open-video.com. The video data set consists of *Apo13001*, *Bor10_009*, *Winn001002*, *Rotation*, and *Coastguard*. We analyze the camera motion category every ten frames because there is similar motion between consecutive frames. There are 2214 frames in total.

Fig. 2 gives several examples of motion noise reduction and feature vector extraction. The figures in the first column are the original video frames, and give the attention regions divided by Golden Section spatial composition rule. The figures in the second column are the corresponding MVFs. The figures in the third column give the experimental results of noise reduction, where the white regions indicate the detected motion noises. As the figures shown, the preprocessing step can filter most of abnormal and foreground motion noises effectively. The figures in the fourth column depict the motion feature vectors for various motion categories. The components of the motion feature vector for translation motion are often less than 0.5 and tend to 0, while the components of the motion feature vector for zoom or rotation motion, except for the second component that always changes between 0.4 and 0.5, are often greater than 0.5 and tend to 1.

So far, the experimental method is still a main way to select kernel function and its parameters. The optimal kernel function just corresponds to the specific application. In this section, we utilize k-fold Cross Validation method ($k = 5$)

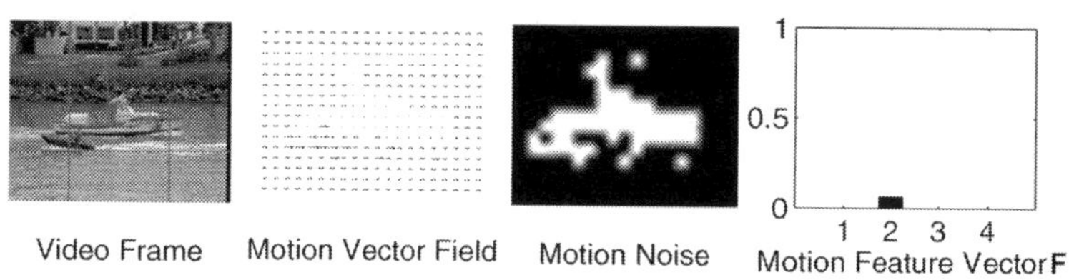

(a) Translation motion extracted from the 84th frame of *Coastguard*

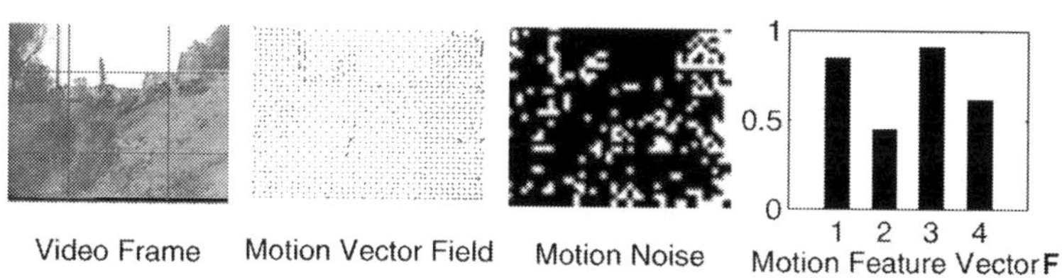

(b) Zoom motion extracted from the 689th frame of *Bor10_009*

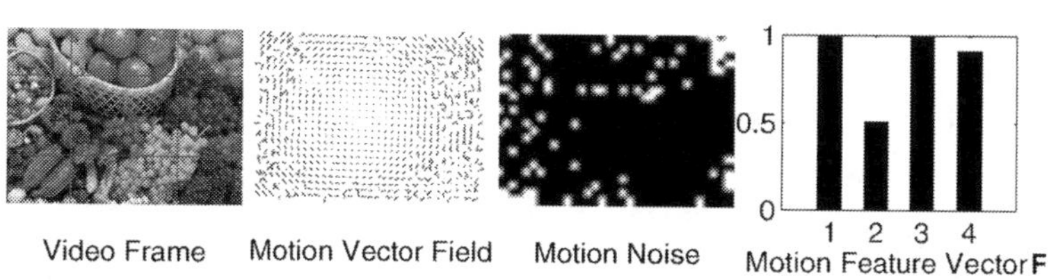

(c) Rotation motion extracted from the 12th frame of *Rotation*

Fig. 2. Examples of motion noise reduction and feature vector extraction for various camera motions to select kernel function, determine its parameters and restriction condition C. Fig. 3 gives the experimental results of parameter selection for various kernel function. In Fig. 3, each value of C corresponds with a curve, which represents the Cross-Validation error along the parameter of kernel function. Different plot symbols, namely, square, circle, star, triangle and diamond, respond with different values of C, 0.1, 0.5, 1, 10 and 100, respectively. We observe the classification performance does not improve obviously with changes in C, while training time increases obviously when the parameter value increases. For polynomial kernel function, parameters b and d have little effect on classification performance. For radial basis function, we achieve better performance when σ changes between 0.1 and 1. For sigmoid kernel function, we achieve the best experimental result when θ is set as -1 and b is set as 0.5. Taking account of the stability of the classifier and classification performance, we select polynomial kernel function ($b = 1$, $d = 2$ and $C = 1$). The experimental results verify the performance of the classifier based on SVM.

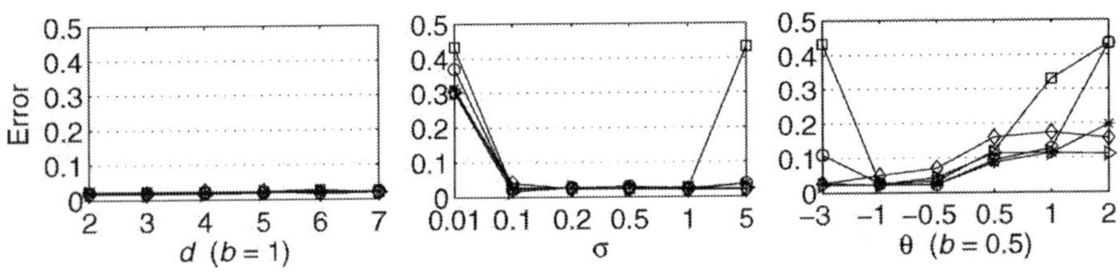

(a) Polynomial Kernel (b) Radial Basis Kernel (c) Sigmoid Kernel

Fig. 3. Kernel function and its parameters selection

Table 1 gives the experimental results. Here we only consider the number of the correct classification (CC) against the ground truth (GT) for various camera motion categories occurring in each video sequence. F. # is the abbreviation of the number of video frames. P. is the abbreviation of classification precision. The experimental results show that the proposed approach can deal with motion noise robustly and achieve satisfying performance. Although video sequence *Apo13001* and *Winn001002* have poor quality, the proposed approach still gets satisfying results. For video sequences *Bor10_009*, *Rotation* and *Coastguard*, our approach achieves higher precision because these video sequences have stable camera motion and high quality.

In experiment, we find that most of the false detections in the camera motion classification are due to that the video frames have very slight camera motion, and are falsely identified as still motion category. Smooth texture region detection is another problem because the smooth texture region often generates mass abnormal motion noises in the motion vector estimation. For example, video sequence *Winn001002* has a low precision just because some scenes are shoot in the sky. These are our further work.

Besides implementing the camera motion classification, the proposed approach can detect the COF for zoom or rotation motions accurately. Several examples

Table 1. Experimental results for camera motion classification (In the first row, camera motion categories: still, zoom in, zoom out, rotation, panning right, tilting down, panning left and tilting up are denoted as 1, 2, ..., 8.)

Video	F. #		Correct Classification #								P.(%)
			1	2	3	4	5	6	7	8	
Apo13001	962	GT	665	172	60	15	21	16	0	13	79.1
		CC	517	143	50	9	18	13	0	11	
Bor10_009	357	GT	83	48	97	0	129	0	0	0	88.2
		CC	77	39	81	0	118	0	0	0	
Winn001002	511	GT	309	72	50	0	56	18	6	0	80
		CC	259	57	36	0	38	14	5	0	
Rotation	236	GT	31	0	0	74	30	41	27	33	85.6
		CC	25	0	0	61	27	37	23	29	
Coastguard	148	GT	5	0	0	0	106	0	32	5	96
		CC	5	0	0	0	103	0	31	3	

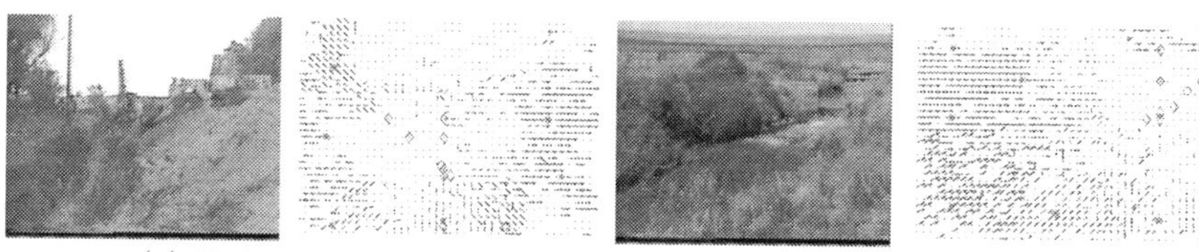

(a) Video Frame# 69 (b) Video Frame# 115

Fig. 4. Experimental results for COF detection

of original frames extracted from *Bor10_009* and their corresponding detection results are illustrated in Fig. 4. In figure, the star indicates the key motion vector in each subregion, the diamond indicates the intersection point determined by the key motion vectors, and the circle indicates the COF estimated by the key motion vectors. As the Fig. 4(a) shown, the object motion (the grab) and smooth texture region (the sky) are eliminated effectively by motion noise reduction, and the COF is correctly identified. When the COF is not at the center of the screen, as Fig. 4(b) shown, the proposed approach can also identify the motion category and detect the COF correctly.

5 Conclusions

We proposed a robust and hierarchical camera motion classification approach in the paper. First, the camera motions were classified into translation and non-translation motions based on the SVM. Then, the rotation and zoom motions were further distinguished, and the translation directions were also identified by analyzing the motion vectors' distribution. The experimental results shown that the proposed approach achieved a good performance. As camera motion can provide an important clue in content-based video parsing and understanding, our future work is to further improve the performance of camera motion classification, and to apply the camera motion classification into video semantic analysis.

Acknowledgements

This research was supported by Science Foundation of Beijing Jiaotong University (Grant No. 2004SM013).

References

1. Su, Y.P., Sun, M.T., Hsu, V.: Global Motion Estimation From Coarsely Sampled Motion Vector Field and the Applications. IEEE Transaction on Circuits and System Video Technology, Vol. 15, No. 2, (2005) 232 - 242
2. Huang, J.C., Hsieh, W.S.: Automatic Feature-Based Global Motion Estimation in Video Sequences. IEEE Transaction Consumer Electronics, Vol. 50, No. 3, (2004) 911 - 915
3. Kumar, S., Biswas, M., et al.: Global Motion Estimation in Frequency and Spatial Domain. In: Proceedings of IEEE ICASSP, (2004) 17 - 21
4. Zhu, X.Q., Xue, X.Y., et al.: Qualitative Camera Motion Classification for Content-Based Video Indexing. In: Proceedings of IEEE PCM, LNCS, Vol. 2532, (2002) 1128 - 1136
5. Fan, J.C., Mei, C.L.: Data Analysis (Chinese). Science Press, Beijing, China (2002)
6. Millerson, G.: The Technique of Television Production. 12th ed. Focal Publishers (1990)
7. Cristianini, N., Taylor, J.S.: An Introduction to Support Vector Machines. Cambridge University Press, Cambridge, UK (2000)

Time Series Analysis of Grey Forecasting Based on Wavelet Transform and Its Prediction Applications

Haiyan Cen, Yidan Bao, Min Huang, and Yong He

College of Biosystems Engineering and Food Science, Zhejiang University, 310029, Hangzhou, China
yhe@zju.edu.cn

Abstract. Grey forecasting based on GM (1,1) has become an important methodology in time series analysis. But due to the limitation of predicting non-stationary time series, an improved grey forecasting GM (1,1) model with wavelet transform was proposed. The time series data was first decomposed to different scales by wavelet transform with à trous algorithm previous of Mallat algorithm in the parallel movement of time series, and then the decomposed time series were forecasted by GM (1,1) model to obtain forecasting results of the original time series. Time series prediction capability of GM (1,1) combined with wavelet transform was compared with that of traditional GM (1,1) model and autoregressive integrated moving average (ARIMA) model to energy source consumption and production forecasting in China. To effectiveness of these methods, eighteen years of time series records (1985 to 2002) for energy source consumption and production were used. The forecasting result from GM (1,1) model with wavelet transform for the data from 2000 to 2002 presented highest precision of three models. It shows that the GM (1,1) model with wavelet transform is more accurate and performs better than traditional GM (1,1) and ARIMA model.

1 Introduction

Time series analysis was used to forecast the developing trend or changes in the future according to a data set arranged by time, and it has been applied widely in many different fields such as economics, sociology and science. Traditional statistical models including autoregressive (AR), moving average (MA), exponential smoothing, autoregressive moving average (ARMA) and autoregressive integrated moving average (ARIMA) [1] are the most popular time series methodologies, but their forecasting abilities are constrained by their assumption of a linear behavior and thus it is not very satisfactory. To improve forecasting non-linear time series events, alternative modeling approaches have been developed. Recently, non-statistical methods and techniques [2][3] have been applied to detect and predict the changes in the region of non-linear time series, like grey system, artificial neural network (ANN) and fuzzy logic systems that can find the characteristic of complex and random data and build accurate time series models, especially of grey system that is more effective for a data set with poor message [4].

Deng [5] first proposed the grey system to build a forecasting model according to real time series of controlling system. Grey system is used to study the object that

D.-Y. Yeung et al. (Eds.): SSPR&SPR 2006, LNCS 4109, pp. 349–357, 2006.

only presents a small part of information in the whole, and then deduce and obtain unknown or new information to develop the system. All that can be done is to find out some regular patterns from the data of time series which is called grey forecasting. The grey forecasting based on GM (1,1) model [6] can try describing those uncertain parameters which are important but lack measurable messages by grey parameters and grey series. The solution to grey differential equation of the GM (1,1) is an exponential function which is appreciate for the fitting of more stationary data but not fit for the fitting of data with seriously random fluctuation. The precision for prediction will be decreasing when it is used to handle the data with great fluctuation which results in many limitations in some fields. Thus, many researchers proposed new methods to improve the GM (1,1) model. Tien [7] did the research on the prediction of machining accuracy by the deterministic grey dynamic model DGDM (1,1,1). He [8] used grey-markov forecasting model for the electric power requirement in China. Liu [9] improved the stability of grey discrete-time systems.

Wavelet transform has been studied for many years by mathematicians and widely used in numerous applications. The wavelet transform is performed using translated and dilated versions of a single function, which is called a wavelet. It is a mathematical process that cut up data into different frequency components, and then study each component with a resolution matched to its scale. Because the signal becomes simpler in different frequency components, and is also smoothed by wavelet decomposition, the stationary of signals is better than that in non-decomposition. For the data with seriously random fluctuation, it is considered to use GM (1,1) model after it is processed by wavelet decomposition. Then the conventional grey forecasting can be used to predict these data series.

In this study, a new time series forecasting model technique the GM (1,1) with wavelet transform was proposed, and the application of this forecasting model was also presented. In addition, the other two models including traditional GM (1,1) and ARIMA were evaluated on the basis of their efficiency to provide accurate fits and operational forecasts on the history data of energy source consumption and production in China.

2 Research Methodologies

2.1 The ARIMA Model

Box and Jenkins [10] proposed an ARIMA (p,d,q) model which considers the last p-known values of the series as well as q of the past modeling errors as follows:

$$y_t = \sum_{i=1}^{p} \phi_i y_{t-i} + \sum_{j=1}^{q} \theta_j e_{t-j} + \varepsilon_t \tag{1}$$

First or second order differencing by d-times processes the problem of non-stationary mean, and logarithmic or power transformation of original data processes non-stationary variance.

To evaluate the appropriateness of the selected ARIMA model, the 'portmanteau' test Q, was also considered by calculating the statistical quantity Q as follows:

$$\hat{\rho}_k(e_t) = \frac{\sum_{t=1}^{N-k} e_t e_{t+k}}{\sum_{t=1}^{N} e_t^2} \quad (2)$$

$$Q = [N - D - \max(p,q)] \sum_{k=1}^{m} \hat{\rho}_k^2(e_t) \quad (3)$$

In the ARIMA (*p*,*d*,*q*) model, Q follows a $\chi^2\alpha(m-p-q)$ distribution. When $Q \le \chi^2\alpha(m-p-q)$, the ARIMA (*p*,*d*,*q*) model is appreciate and acceptable, otherwise the model is not valid and needs reformulation.

2.2 Wavelet Transform

Theoretical Background. The decomposition of signals is an important procession in Wavelet transforms, especially for the analysis of non-stationary time series. There are many algorithms of wavelet decomposition, such as Mallat [11] algorithm, which has been proposed to compute the discrete wavelet transform (DWT) coefficients. It needs two extractions from signals and leads to the reduction of signal spots. It is very disadvantage to be used in prediction. However, Wickerhauser [12] considered if the input discrete $f[n]$ has 2^L ($L \in N$) nonzero samples. There are L different scales, and each scale sequence and wavelet sequence's length are decimated by two with change of scale. Thus, this paper adopt a simple, quick algorithm called à trous algorithm without signal extraction [13]. By à trous algorithm, the sequence's length of decomposed time series won't change. What's more, the length of decomposed series in scales is equal to the length of original series. This algorithm overcomes the problem from Mallat algorithm. Besides, it is good for the reconstruction of decomposed series.

The à Trous Wavelet Transform. It is supposed that there is a time series $x(t)(t = 1, 2, ..., N)$ to be processed by wavelet decomposition, where N is the present time-point. It is arranged as $c_0(t) = x(t)$. The decomposition with à trous algorithm is as followed:

$$c_i(t) = \sum_{k=-\infty}^{+\infty} h(k) c_{i-1}(t + 2^i k) \ (i = 1, 2, ...) \quad (4)$$

$$w_i(t) = c_{i-1}(t) - c_i(t) \ (i = 1, 2, ...) \quad (5)$$

In the formula, $h(k)$ is a discrete low-pass filter. $c_i(t)$ and $w_i(t)$ $(i = 1,2,...,J)$ are the scaling coefficients and wavelet coefficients of scale i, and J is scale number. The number of different scales is under log(N) (N is the length of time series). $\{w_1, w_2, ..., w_J, c_J\}$ is called wavelet decomposition or wavelet transform series under scale J.

Then, the decomposed wavelet was reconstructed with à trous algorithm as follows:

$$c_0(t) = c_J(t) + \sum_{i=1}^{J} w_i(t) \tag{6}$$

The wavelet pass filter called Haar $h = (\frac{1}{2}, \frac{1}{2})$ was selected in à trous algorithm. Here is the derived formula for the decomposed series in corresponding space.

$$c_{i+1}(t) = \frac{1}{2}(c_i(t - 2^i) + c_J(t)) \tag{7}$$

$$w_{i+1}(t) = c_i(t) - c_{i+1}(t) \tag{8}$$

From formula (7) and (8), it is easy to find that the wavelet coefficient for any time spot can be calculated without the information after the time of t.

Besides, wavelet transform with à trous algorithm needs decomposed values outside signal boundaries. Although other strategies could be envisaged, we use a mirror approach $x(N\text{-}K)=x(N+K)$.This is tantamount to redefine the discrete filter associated with the scale function in the signal boundary region and to redefine the associated wavelet function in this region. It is hypothesized that future data is based on values in the immediate past. Not surprisingly there is discrepancy in fit in the succession of scales, which grows with scale as larger numbers of immediately past values are taken into account. The first values of our time series, which also constitute a boundary, can cause difficulty, but this is of no practical consequence.

2.3 GM (1,1) Grey Forecasting Model

It is supposed that $x^{(0)}$ is an original time series

$$x^{(0)} = \left\{x^{(0)}(1), x^{(0)}(2), \ldots, x^{(0)}(N)\right\} \tag{9}$$

and a new series is given by the accumulated generating operation (AGO)

$$x^{(1)} = \left\{x^{(1)}(1), x^{(1)}(2), \ldots, x^{(1)}(N)\right\} \tag{10}$$

where $x^{(1)}(t) = \sum_{i=1}^{t} x^{(0)}(i)$, $t = 1, 2, \ldots, N$.

According equation (10), the grey generated model, based on the series $x^{(1)} = \left\{x^{(1)}(1), x^{(1)}(2), \ldots, x^{(1)}(N)\right\}$ is given by the first-order differential equation

$$\frac{dx^{(1)}}{dt} + ax^{(1)} = u \tag{11}$$

where the coefficient a and grey control parameter u are the model parameters to be estimated. Then the least squares solution of $x^{(0)}$ is as followed:

$$\hat{x}^{(1)}(t+1)=(x^{(0)}(1)-\frac{u}{a})e^{-at}+\frac{u}{a}\quad(t=1,2,3,...) \tag{12}$$

where $\hat{b}=\begin{bmatrix}a\\u\end{bmatrix}=(B^TB)^{-1}B^TY$, $B=\begin{bmatrix}-1/2(x^{(1)}(1)+x^{(1)}(2)) & 1\\ -1/2(x^{(1)}(2)+x^{(1)}(3)) & 1\\ ... & ...\\ -1/2(x^{(1)}(N-1)+x^{(1)}(N)) & 1\end{bmatrix}$,

$Y=(x^{(0)}(2),x^{(0)}(3),...,x^{(0)}(N))$.

According to the equation (12), $x^{(1)}$ can be predicted. By inverse accumulated generated operation (IAGO) of $\hat{x}^{(1)}$, the forecasting value can be reversed.

$$\hat{x}^{(0)}(t+1)=\hat{x}^{(1)}(t+1)-\hat{x}^{(1)}(t)\ (t=1,2,3,...) \tag{13}$$

where $\hat{x}^{(0)}(t)(t=1,2,...,N)$ is the regression value of original data series $x^{(0)}(t)(t=1,2,...,N)$, and $\hat{x}^{(0)}(t)(t>N)$ is the predicting value of original data series.

2.4 GM (1,1) Grey Forecasting Model with Wavelet Transform

There is a time series $x(t)$ (1,2,…,N). It is processed by wavelet decomposition with à trous algorithm. And then the time series in every layer is reconstructed. It can be showed

$$x=w_1+w_2+...+w_J+c_J \tag{14}$$

In this equation, $w_i:\{w_i(1),w_i(2),...,w_i(N)\}$ is detail signals in the decomposed layer i. $c_J:\{c_J(1),c_J(2),...,c_J(N)\}$ is the approaching signals in the decomposed layer J. Thus, $x(t)=w_1(t)+w_2(t)+...+w_J(t)+c_J(t)$, where $x(t)$ is known at the time of $\{t|t<N\}$. The value next time can be obtained (equation (15)).

$$x(t+1)=w_1(t+1)+w_2(t+1)+...+w_J(t+1)+c_J(t+1) \tag{15}$$

The steps of prediction for $w_1(t+1),w_2(t+1),...,w_J(t+1),c_J(t+1)$ are as followed.

First, move the difference of information parallel in time series for every layer of w_i and c_J. Then, the series moved is used to evaluate the parameters in GM (1,1). After moving, $w_1(t+1),w_2(t+1),...,w_J(t+1),c_J(t+1)$ is predicted by GM (1,1) model. The value of forecasting is $\hat{w}_1(t+1), \hat{w}_2(t+1),..., \hat{w}_J(t+1), \hat{c}_J(t+1)$. Finally, the predicting value of original time series x is obtained,

$$\hat{x}(t+1) = \hat{w}_1(t+1) + \hat{w}_2(t+1) + ... + \hat{w}_J(t+1) + \hat{c}_J(t+1) \tag{16}$$

3 Examples

A new forecasting model GM (1,1) with wavelet transform, traditional GM (1, 1) model and ARIMA model proposed in this paper were applied to forecast energy source consumption and production in China. The duration considered in this study ranges from 1985 to 2002. These data shown in Table 1 were refereed from National Bureau of Statistic of China [14].

Energy source consumption and production are influenced by many factors, including the economy development, industry structure, weather, policy and so on. It was shown that the time series of energy source consumption and production in a country have serious random fluctuation. However, in order to harmonize the relationship between the high needs and reduction of energy source, it is more and more important to a country to make a prediction for energy source consumption and production in the future. From Table 1, it was found that the data were rising year by year, while fluctuating randomly.

Table 1. Energy source consumption and production of China from 1985 to 2002 (unit: ten million kilogram standard coal)

Year	Energy source consumption	Energy source production	Year	Energy source consumption	Energy source production
1985	76682	85546	1994	122737	118729
1986	80850	88124	1995	131176	129034
1987	86632	91266	1996	138948	132616
1988	92997	95801	1997	137798	132410
1989	96934	101639	1998	132214	124250
1990	98703	103922	1999	130119	109126
1991	103783	104844	2000	130297	106988
1992	109170	107256	2001	134914	120900
1993	115993	111059	2002	148222	138369

To eliminate the fluctuation of these two data sets, wavelet transform with à trous algorithm was used to data processing before GM (1,1). Fig. 1 shows the decomposed signals of low and high frequency in two layers. The non-stationary time series data was smoothed and parallel shifted by wavelet decomposition, and the fluctuation was considered noises decomposed into high frequency. The detail signals and approaching signals were used for wavelet reconstruction.

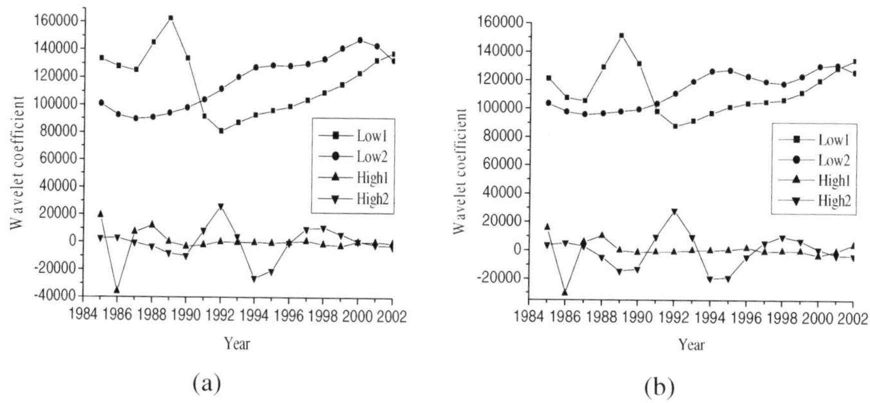

Fig. 1. Detail signals of (a) energy source consumption and (b) energy source production in low and high frequency of two layers with wavelet decomposition

In this study, wavelet transform was performed in software Matlab 7.0, and GM (1,1) and ARIMA model were achieved in DPS, a software of data processing system for Practical Statistics [1].

The forecasting results for energy source consumption and production from 2000 to 2002 by GM (1,1) with wavelet transform were shown in Table 2. Meanwhile, the forecasting values from traditional GM (1,1) and ARIMA were also obtained. In these three models, GM (1,1) with wavelet transform displayed the highest precision except one year forecasting on energy source production. The data without any processing used in GM (1,1) gave the lowest precision, which revealed the default of GM (1,1) model for non-stationary series only using poor information to forecast. Although ARIMA model seems better than GM (1,1), it needs more history data. For practical application, most of the forecasting problems are described under poor information in a data set, or even a small data. Thus, the improved GM (1,1) based on wavelet transform is promising for non-stationary or lack information time series forecasting.

Table 2. Forecasting results from three different models for the energy source consumption and production from 2000 to 2002 (unit: ten million kilogram standard coal)

Wavelet-GM (1,1)		GM (1,1)		ARIMA	
		Energy source consumption			
Prediction	Precision	Prediction	Precision	Prediction	Precision
136141	95.71%	149483	87.17%	140261	92.90%
139278	96.87%	155455	86.79%	128887	95.53%
143761	96.99%	161665	91.68%	135567	91.46%
		Energy source production			
114161	93.72%	133790	79.97%	103809	97.03%
120226	99.44%	137300	88.06%	112351	92.93%
126715	98.20%	126716	91.58%	113291	81.88%

4 Conclusions

Grey forecasting model is a good method for the time series changed smoothly. However, the predicting precision will reduce when it is used to forecast time series with serious random fluctuation. The non-stationary time series are influenced by many factors, which make the prediction more complexly. The wavelet transform is an effective tool for time history dynamic analysis of structures. The problem in GM (1,1) model can be solved by wavelet analysis, which can decompose the time series into different layers according to the different scales. As the case study shows that the accuracy of GM (1,1) model with wavelet transform in forecasting energy source consumption and production from 2000 to 2002 is higher than these in traditional GM (1,1) and ARIMA models. Thus, it is concluded that GM (1,1) model with wavelet transform is promising for time series analysis, especially for non-stationary time series.

Besides, there is a problem in wavelet decomposition. The selection of algorithm in wavelet decomposition is very important for the result of wavelet analysis. In this paper, a simple and quick algorithm called à trous algorithm without signal extraction was used in wavelet decomposition. By à trous algorithm, the length of decomposed series in scales is equal to the length of original series. This algorithm overcame the problem from Mallat algorithm. However, wavelet transform with à trous algorithm need decomposed values outside signal boundaries. Although other strategies could be envisaged, we used a mirror approach. It is necessary to study further whether this method is popular for any time series.

Acknowledgments

This study was supported by the Teaching and Research Award Program for Outstanding Young Teachers in Higher Education Institutions of MOE, P. R. C., Natural Science Foundation of China (Project No: 30270773), Specialized Research Fund for the Doctoral Program of Higher Education (Project No: 20040335034), and Science and Technology Department of Zhejiang Province (Project No. 2005C21094, 2005C12029).

References

1. Tang, Y.M., Feng, M.G.: Data Processing System for Practical Statistics. Science Publishing Company Press. Beijing (2002)
2. Wu, B., Chen, M.H.: Use of Fuzzy Statistical Technique in Change Periods Detection of Nonlinear Time Series. Applied Mathematics and Computation. Vol. 99. No. 2-3 (1999) 241-254
3. He, Y., Zhang, Y., Xiang, L.G.: Study of Application Model on BP Neural Network Optimized by Fuzzy Clustering. Lecture Notes in Artificial Intelligence. Vol. 3789 (2005) 712-720
4. Deng, J.L.: Control Problems of Grey System. Huazhong University of Science and Technology Press. Wuhan (1990) 1-2
5. Deng, J.L.: Grey Forecasting and Decision Making. Huazhong University of Science and Technology Press. Wuhan (1985)

6. Bao, Y.D., Wu, Y.P., He, Y.: A New Forecasting Model Based on the Combination of GM (1,1) Model and Linear Regression. Systems Engineering (Theory and Practice). Vol. 24. No. 3 (2004) 95-98
7. Tien, T.L.: A Research on the Prediction of Machining Accuracy by the Deterministic Grey Dynamic Model DGDM (1,1,1). Applied Matheatics and Computation. Vol. 161. No. 3 (2005) 923-945
8. He, Y., Huang, M.: A Grey_Markov Forecasting Model for the Electric Power Requirement in China. In:Gelbukh, A., Albornoz, A.D., Terashima, H.(eds.): Lecture Notes in Artificial Intelligence. Vol. 3789 (2005) 574-582
9. Liu, P.L., Shyr, W.J.: Another Sufficient Condition for the Stability of Grey Discrete-time Systems. Journal of the Franklin Institute. Vol. 342. No. 1 (2005) 15-23
10. Box, G.E.P., Jenkins, G.M.: Time Series Analysis. Forecasting and Control. Holden-Day. San Francisco (1976) 575
11. Li, X.B., Li, H.Q., Wang, F.Q., Ding, J.: A Remark on the Mallat Pyramidal Algorithm of Wavelet Analysis. Communications in Nonlinear Science and Numerical Simulation. Vol. 2. No. 4 (1997) 240-243
12. Wickerhauser, M.V.: Adapted Wavelet Analysis from Theory to Software. Wellesley. Massachusetts (1994) 213-235
13. Sun, Z.M., Jiang, X.W., Wang, X.F.: Application of à Trous Wavelet to Satellite Telemetry Data Recursive Prediction. Journal of Nanjing University of Science and Technology. Vol. 28. No. 6 (2004) 606-611
14. National Bureau of Statistic of China: China Statistical Year Book. China Statistics Press. Beijing (2004)

A Study of Touchless Fingerprint Recognition System

Chulhan Lee, Sanghoon Lee, and Jaihie Kim

Department of Electrical and Electronic Engineering, Yonsei University,
Biometrics Engineering Research Center (BERC),
Republic of Korea
devices@yonsei.ac.kr

Abstract. Fingerprint recognition systems are widely used in the field of biometrics. Many existing fingerprint sensors acquire fingerprint images as the user's fingerprint is contacted on a solid flat sensor. Because of this contact, input images from the same finger can be quite different and there are latent fingerprint issues that can lead to forgery and hygienic problems. For these reasons, a touchless fingerprint recognition system has been investigated, in which a fingerprint image can be captured without contact. While this system can solve the problems which arise through contact of the user's finger, other challenges emerge, for example, low ridge-valley contrast, and 3D to 2D image mapping. In this paper we discuss both the disadvantages and the advantages of touchless fingerprint systems and introduce the hardware and algorithm approach to solve the problems. We describe the structure of illuminator and the wavelength of light to acquire a high contrast fingerprint images. To solve the problem of 3D to 2D image mapping, we describe the method to remove the strong view difference fingerprint images. Experiments show that the touchless fingerprint system has better performance than the conventional touch based system.

1 Introduction

Many biometric features have been used to confirm the identity of a given human. Some of these recognition features have included iris, face, fingerprint, voice, hand geometry, and the retinal pattern of eyes. Among all these features, fingerprint recognition has been the most popular and reliable biometric feature for automatic personal identification. Various types of sensors (including optical, thermal, and capacitive sensors) have been developed in order to acquire good fingerprint images with appropriate characteristics. Also, a large variety of algorithms have been proposed in order to achieve better authentication performance. In spite of all these efforts to acquire good fingerprint images and enhance performance, there are a number of problems which occur when using conventional touch-based sensors.

1.1 Problems with Touch-Based Sensors

In order to acquire fingerprint images with conventional touch-based sensors, the user must place his finger on the flat window of the sensor. Because the skin

D.-Y. Yeung et al. (Eds.): SSPR&SPR 2006, LNCS 4109, pp. 358–365, 2006.

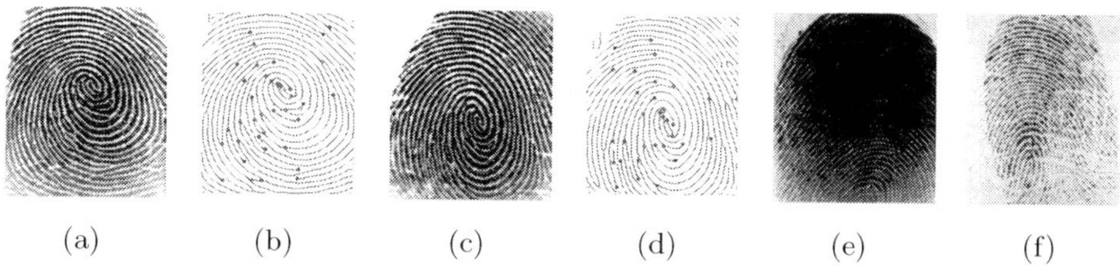

Fig. 1. Distorted images from a touch-based sensor. (a) and (c) are gray-value images, (b) and (c) are corresponding minutiae extracted images, and (e) and (f) show the effects of different strengths of impression.

of the finger is not flat, the user must apply enough pressure on the window to obtain sufficient size and achieve good image quality. However, this pressure produces unavoidable physical distortion in arbitrary directions, which is represented differently throughout every area of the same fingerprint image. Also, since the image varies with each impression, each fingerprint image from the same finger can appear quite different. Fig. 1 shows the images from touch-based sensor. Because of different pressure, the relative position and types of corresponding minutiae are different (Fig. 1(a)-(d)). Also, the sizes of the fingerprint areas and the quality of the fingerprint images are quite different (Fig. 1(e),(f)). These problems critically affect fingerprint recognition. There are also latent fingerprint problems. A latent fingerprint refers to the trail of the fingerprint on the window of the sensor. This can lead to hygienic problems as well as fraudulent use, such as the faking of fingerprints [1]. The issue of protecting biometric information and privacy is paramount in biometric technology. As mentioned above, previous touch-based sensor approaches can lead to several problems. To solve these problems, [2] proposed the elastic minutiae matching algorithm using the TPS (thin-plate spline) model. In this method, corresponding points were detected using local minutiae information, and global transformation was determined with the TPS model. Although the method produced higher matching scores when compared to the ridge-matching algorithm, it requires a good minutiae extraction algorithm that extracts well in even distorted images. However, if a fingerprint image is highly distorted, to extract usable minutiae is very difficult. [3] proposed detection of artificial fingers. This method showed that detection of a perspiration pattern in a time progression of fingerprint images can be used as an anti-spoofing measure. However, environmental factors (such as temperature and humidity) and user-specific factors (such as skin humidity and pressure) are not taken into account. In this paper, we investigate a touchless fingerprint system that fundamentally overcomes the problems involved in conventional touch-based sensors. This paper is organized as follows. In section 2, we explain the advantages and disadvantages of touchless fingerprint systems and introduce the hardware approach and the algorithm approach to address the disadvantages. In section 3, a comparison between touch-based sensors and touchless sensors is given in terms of recognition performance. Finally, conclusions and suggestions for future works are discussed in section 4.

2 Touchless Fingerprint System

In order to solve the innate problems of touch-based sensors, we studied a touchless fingerprint recognition system. [4] developed a touchless fingerprint sensor using a camera. They used a polarizer filter and a band-pass filter in order to acquire a good quality image. Unfortunately, there is no explanation between the illuminator and filters which affected image quality. We also used a camera to capture fingerprint images in our touchless fingerprint sensor. Advantages of using a camera include: i) the fingerprint image can be acquired without plastic distortion from contact pressure, ii) latent fingerprints do not appear on the sensor, iii) hygienic problems are reduced, and iv) a large image area can be captured quickly. The combination of distortion-free fingerprint images and large image areas is desirable in order to acquire many minutiae in the same relative location and direction at every instance. Therefore, this combination helps the authentication system to have low FAR and FRR. While there are strong advantages to this system, there are also new disadvantages. These can be classified into two areas: low contrast between the ridges and valleys, and 3D to 2D image mapping. Low ridge-valley contrast is caused by the motion blur of hand tremble, camera background noise, and a small Dof (Depth of field). Because the 3-dimensional object finger is projected onto a 2-dimensional image plane, the difference of the camera viewpoint caused by the 3-dimensional rotation of the finger can produce a small common area between the enrolled and input images. This leads to an increased false acceptance rate (FRR). Fig. 2 shows the sample images which represent the above-mentioned problems. In this paper, we propose the hardware approach in order to solve the low ridge-valley contrast and the algorithm approach in order to solve the difference between the camera viewpoints.

2.1 The Hardware Approach: Low Ridge-Valley Contrast Issue

Structure of Device. Motion blur generally stems from the long capture time relative to the slight motion of the finger. Although high-sensitivity sensors can handle fast shutter speed, images normally contain increased electrical background noise. Although a large aperture stop could be an alternative way to retain rapid shutter speed, this method is an unsuitable way to guarantee the required Dof, which is important to keep the variation range of the finger

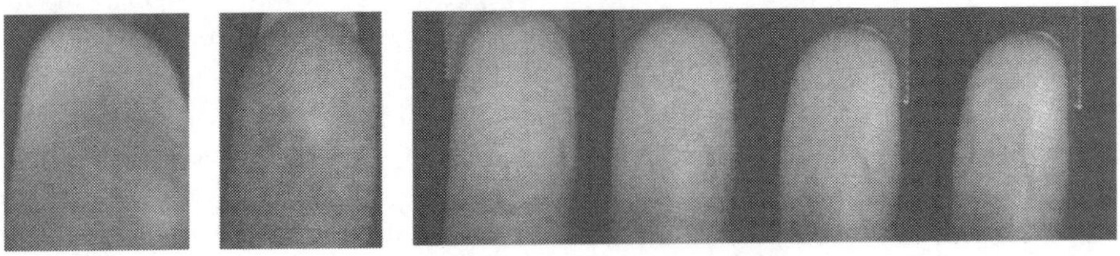

(a) (b) (c)

Fig. 2. (a) Motion-blurred image, (b) partially defocusing, (c) images from rolled fingers

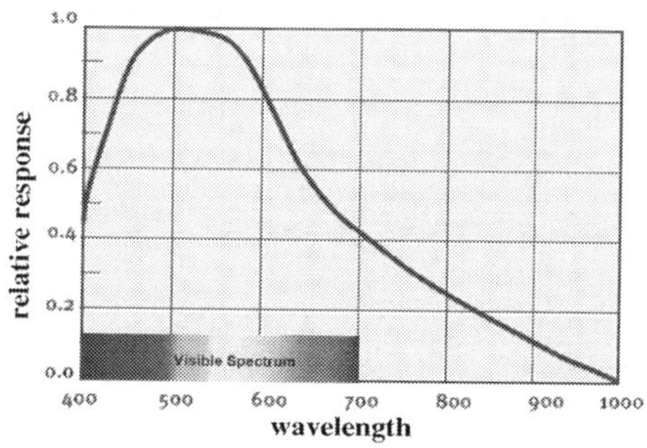

Fig. 3. The spectral sensitivities curve of the camera

position in focus. Hence, a small apparatus and appropriate lighting gadgetry are necessary. For this reason, we adopt a finger support to decrease motion blur and an appropriate illuminator to obtain uniform lighting and sufficient brightness. There are several types of illuminator structures which can be used to obtain good images in special applications [5]. For example, dark-field lights illuminate objects from their sides with specific patterns to emphasize shadows and enhance image quality. A ring-type illuminator, especially, is easy to build up and produces uniform illumination on the object, and on-axis lights illuminate the object relative to the camera axis. Among those types, the ring-type and their modified types simply and stably obtain uniform light conditions.

Wavelengths of Illuminators. To acquire good quality fingerprint images for fingerprint recognition, both the spectral sensitivities of the camera and the skin reflectance of the finger must be considered. The method proposed in [6] measured the light reflected from the skin using a high-resolution, high-accuracy spectrograph under precisely calibrated illumination conditions. This showed that the skin reflectance of humans is mainly influenced by both melanin and hemoglobin. However, the skin reflectance of the palm (including the fingers) is mainly influenced by oxygenated hemoglobin because of the scarcity of melanin in the palm. In this paper, the hemoglobin absorption spectrum was founded. The ratio of hemoglobin absorption is lower around 500nm and higher around 545nm and 575 nm. Because it is desirable to observe only the surface of the finger and remove the effect of hemoglobin in order to obtain a high-contrast fingerprint image, the wavelength at lower hemoglobin absorption must be chosen on the wavelength of the illuminator of the touchless fingerprint sensor. Fig. 3 shows the spectral sensitivities of the normal CCD. The sensitivities are high around 500nm. Considering both the skin reflectance of the finger and the spectral sensitivities of each camera, the wavelength of light on the touchless fingerprint sensor can be determined.

2.2 Software Approach: 3D to 2D Image Mapping Issue

Since touchless fingerprint sensors acquire 2-dimensional fingerprint images from 3-dimensional finger skins, a significant view difference is generated when rolling the finger in the image acquisition step. We divided the view difference image into a weak view difference image and a strong view difference image. The weak view difference image usually has a small influence on the fingerprint and is allowable through elastic matching by the tolerance parameters of the bounding box [7]. When the strong view difference image appears, ridges on the slanted skin are densified on the image (due to the 3D to 2D projections), they are apt to have type-changed, missed or even false minutiae. Moreover, the foreground becomes the near-boundary region and also becomes dark or out of focus. This decreases the number of usable minutiae as well as the good quality foreground, and also reduces the common area between the fingerprint impressions, which results in bad system performance. Therefore, it is desirable to reject images with strong view differences and instruct the user to retry the input. To determine the strong view difference image, we measure the distance between the core and the center axis of the finger. Fig. 4 shows the distance as determined by the rolled finger. The larger the distance between the core and the center axis, the more the image is rolled. Most fingerprints have at least one core (except fingerprints of the arch type). Though these fingerprints have no core point, we can detect one singular point that is invariant to rolling. We implemented the core point detection algorithm using [8]. This method is an accumulating mechanism similar to a Hough transform. Using a ridge orientation map, this algorithm follows a path that is perpendicular to the ridge orientation and accumulates a histogram. After finishing all searches, the location of the core point can be determined by

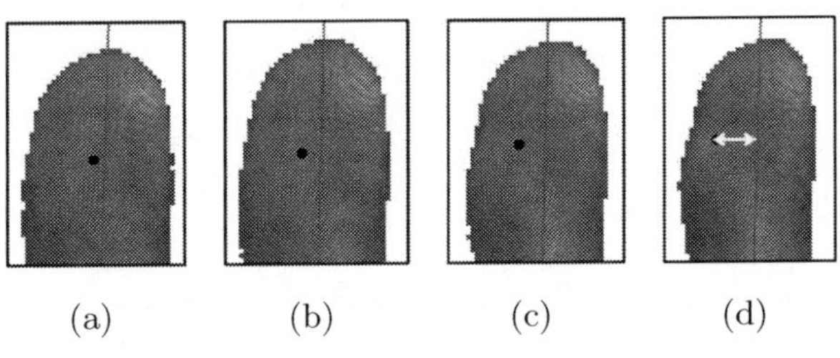

Fig. 4. The distance between the core and center axis of rolled finger. (a) 0 degree, (b) 10 degrees, (c) 20 degrees, (d) 40 degrees.

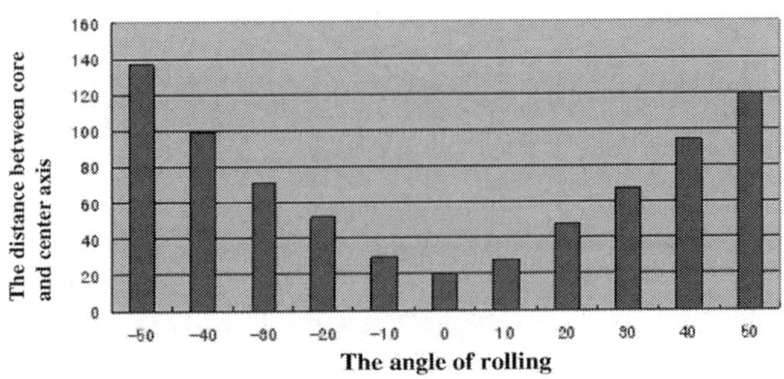

Fig. 5. The distance between the core and center axis of rolled finger

detecting the maximum position on the histogram. This method is robust to noise and can detect the maximum position for the specific arch type. The center axis of the finger is determined as the KL(Karhunen and Loeve) transform with the segmented image. The distance that determines the strong view difference image can be defined as follows:

$$\text{Distance} = \frac{|ax_c + by_c + c|}{\sqrt{a^2 + b^2}} \quad (1)$$

where (x_c, y_c) is the location of the core point and a, b and c are the coefficients of the center axis $ax + by + c = 0$. Fig. 5 shows the averaged distance of 50 fingers according to the sampled rolling angle. We can observe that the distance can discriminate between the strong view difference image and the weak view difference image.

3 Experimental Results

3.1 Comparison Between Touch-Based Sensors and Touchless Sensors

For this comparison, we collected 1630 fingerprint images from 163 fingers using a touchless-sensor and also a touch-based optical sensor (Ditgen FD1000 [9]). To avoid strong rolled fingerprint images, we mentioned this to the users

Table 1. The specification of sensors

	Touchless sensor	Touch-based sensor
Size of input window	24mm×34mm	17mm×19mm
Resolution	450 dpi	500dpi
Size of image	480×640 pixels (Fingerprint 320×450 Pixels)	280×320 pixels

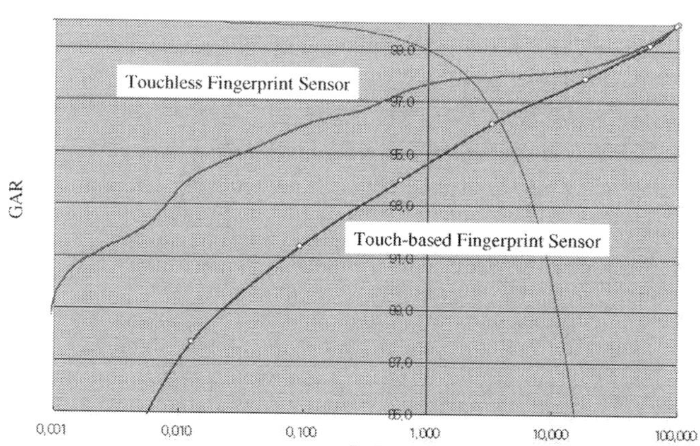

Fig. 6. The ROC of touch-based fingerprint recognition and touchless fingerprint recognition

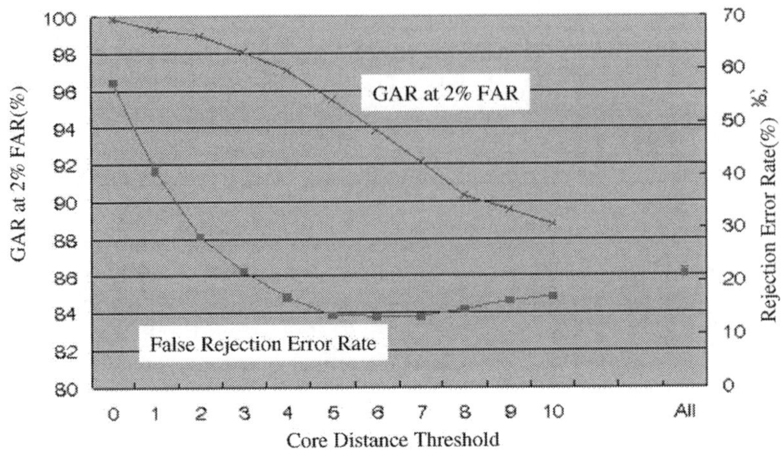

Fig. 7. Performance of the system (GAR at 2% FAR) and the false reject error rate

before acquiring the images. Table 1 shows the specifications of the two sensors. After acquiring the images, we adopted the same fingerprint image processing algorithm [10] and matching algorithm [11]. Fig. 6 shows the ROC of the two results. From these experimental results, we can observe that the performance of touchless fingerprint sensors is better than the performance of touch-based fingerprint sensors.

3.2 Rejecting the Strong View Difference Image: An Evaluation of the Method

1100 rolled finger images were collected from 50 fingers. Each finger was rolled from -50 degrees to +50 degrees with an interval of 10 degrees. Fig 7 shows the relationship between the performance of the system (GAR at 2% FAR) and the false reject error rate. The false reject error is composed of two types of errors. The type-I error is the number of reject images corresponding to the weak view difference image. The type-II error is the number of acceptance image corresponding to the strong view difference image. We defined the strong and weak view difference images by matching the algorithms with the matching threshold T. The matching threshold is defined by experiment 3.1.

4 Conclusions and Future Works

In fingerprint recognition, conventional touch-based sensors contain innate problems that are caused by the contact of the finger. Some examples of these problems are distortions in the fingerprint images, latent fingerprints and hygienic problems. To overcome these fundamental problems, we investigated a touchless fingerprint recognition system. While we obtained a large distortion-free fingerprint image, new problems needed to be dealt with, such as low contrast between ridges and valleys and 3D to 2D image mapping. To acquire good fingerprint images, we introduced a hardware approach that used the structure of the device and the wavelengths of light. Also, we proposed a strong view difference image rejection method using the distance between the core and the center axis of the finger in order to overcome the 3D to 2D image mapping problem. In the experiments, we compared the touchless fingerprint sensor with the touch-based fingerprint sensor in terms of performance and evaluated the proposed rejection method.

In future work, we will develop the matching algorithm that is invariant to 3D camera viewpoint change and compare fingerprint recognition system in images captured with touchless sensors and other touch-based sensors. In this comparison, we will compare not only verification performance but also image quality, the convenience of usage, the size of fingerprint image, and the number of true minutiae.

Acknowledgements

This work was supported by Korea Science and Engineering Foundation (KOSEF) through the Biometrics Engineering Research Center at Yonsei University.

References

1. T. Matsumoto, H. Matsumoto, K. Yamada, and S. Hoshino, "Impact of artificial "gummy" fingers on fingerprint systems", In Proceedings of SPIE Vol. Num.4677, Jan 2002.
2. Bazen, A.M. and Gerez, S.H., "Elastic minutiae matching by means of thin-plate spline models" Pattern Recognition, 2002. Proceedings. 16th International Conference on Vol. 2, pp. 985 - 988, Aug. 2002.
3. Derakhshani R., Schuckers S. A. C., Hornak L., and O'Gorman L. "Determination of vitality from a non-invasive biomedical measurement for use in fingerprint scanners." Pattern Recognition Journal, Vol.36, (2003) 383-396.
4. Hans J. Einighammer and Jens Elnighammer, "System for the Touchless Recognition of Hand and Finger Line", US6404904, 2002.
5. Pantazis Mouroulis, "Visual Instrumentation: Optical Design & Engineering Principles" McGraw-Hill Professional, 1999.
6. Angelopoulou Elli, "Understanding the Color of Human Skin. ", Proceedings of the 2001 SPIE conference on Human Vision and Electronic Imaging VI, SPIE Vol. 4299, pp. 243-251, May 2001.
7. Jain A., Lin H. and Bolle R., "On-line Fingerprint Verification," IEEE Trans. on Pattern Analysis and Machine Intelligence, Vol. 19, no. 4, pp.302-314, Apr. 1997.
8. Huvanandana, S., Changick Kim and Jenq-Neng Hwang, "Reliable and fast fingerprint identification for security applications", IEEE International Conference on Image Processing, vol. II, pp. 503-506, Vancouver, Canada, October 2000.
9. http://www.digent.co.kr
10. L. Hong, Y. Wan and A.K. Jain, "Fingerprint Image Enhancement: Algorithms and Performance Evaluation", IEEE Transactions on PAMI, Vol. 20, No. 8, pp.777-789, August 1998.
11. D. Lee, K. Choi and Jaihie Kim, "A Robust Fingerprint Matching Algorithm Using Local Alignment", International Conference on Pattern Recognition, Quebec, Canada, August 2002.

Development of a Block-Based Real-Time People Counting System

Hyun Hee Park[1], Hyung Gu Lee[1], Seung-In Noh[2], and Jaihie Kim[1]

[1] Department of Electrical and Electronic Engineering, Yonsei University, Biometrics Engineering Research Center(BERC), Republic of Korea
{inextg, lindakim, jhkim}@yonsei.ac.kr
[2] Samsung Electronics, 416, Maetan-3dong, Yeongtong-gu, Suwon-city, Gyeonggi-do, Republic of Korea

Abstract. In this paper, we propose a block-based real-time people counting system that can be used in various environments including shopping mall entrances, elevators and escalators. The main contributions of this paper are robust background subtraction, the block-based decision method and real-time processing. For robust background subtraction obtained from a number of image sequences, we used a mixture of K Gaussian. The block-based decision method was used to determine the size of the given objects (moving people) in each block. We divided the images into 72 blocks and trained the mean and variance values of the specific objects in each block. This was done in order to provide real-time processing for up to 4 channels. Finally, we analyzed various actions that can occur with moving people in real world environments.

1 Introduction

People counting systems can be used to count or track people, for example at the entrances of shopping malls and buildings. The information can then be used for surveillance purposes, to gather marketing data or to facilitate building management. The use of early automatic counting methods such as light beams, turnstiles and rotary bars led to various problems. These conventional methods could not count people accurately when many individuals passed through the sensors at the same time. To solve this problem, it is necessary for image processing-based approaches to be hance motivated. Thou-Ho et al.[1] presented a bi-directional counting rule, but this method failed in terms of measuring the fixed sizes of objects in the image regions. Terada and Yamaguchi[2] utilized a color camera to extract images of moving people, but the problem of direction-orientation remained intractable. Yoshida et al.[3] used stereo cameras to capture pairs of images, but this method still couldn't solve the problems of counting crowds and direction recognition. The above research[1]-[5] describes how image processing has been used to provide image data that is based on motion analysis, which assumes that the people are moving relative to a static background. In order to focus on dynamic backgrounds, Qi Zang et al.[6] proposed a method of robust background subtraction and maintenance.

D.-Y. Yeung et al. (Eds.): SSPR&SPR 2006, LNCS 4109, pp. 366–374, 2006.

In this paper, we apply the proposed method to practical and complicated environments such as building gates, escalators, and elevators with a large number of passing people. These practical environments present many problems when compared to limited and simplified environments. To solve these problems, we propose the following methods. First, for robust background subtraction from image sequences, we used a mixture of K Gaussian. Second, we divided images into 6×12 sub-blocks and calculated the mean and variance values of the extracted object size of each block. We then plotted these means and variances into a table. Third, we did not use complicated image processing techniques to recognize and track each person. We simply tracked masses of objects. We were able to improve processing time and counting accuracy by using this method. In this paper, we propose a people counting algorithm and present experimental results to verify the effectiveness of the proposed method.

2 Theoretical Approach

2.1 System Configuration

While Figure 1 shows the configuration of our system, Figure 2 shows the overall system block diagram. This is divided into two parts. The first part refers to moving object extraction in image sequences. The second part refers to tracking and counting decisions that are made using the extracted objects. The moving object extraction process consists of four parts. First, we used only the LL part of the Harr wavelet transformed images to remove noise from the input images and down-sample the images ($320 \times 240 \Longrightarrow 160 \times 120$) in order to improve processing time. Second, we produced a reference background model by using a mixture of K Gaussian distributions with N input images. Third, we extracted moving objects by calculating the background subtraction and frame differences between differing time ($t-2$, $t-1$, t) images. Last, we used a morphological mask to remove noise from the images and to fill in the large and small holes existing in the extracted objects. The object tracking and counting decision stage

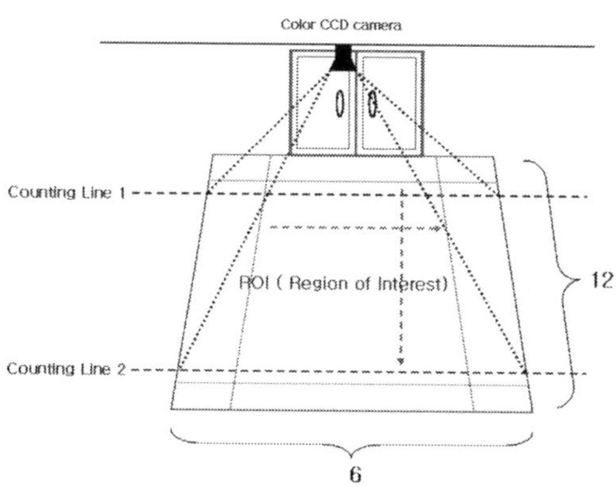

Fig. 1. System configuration

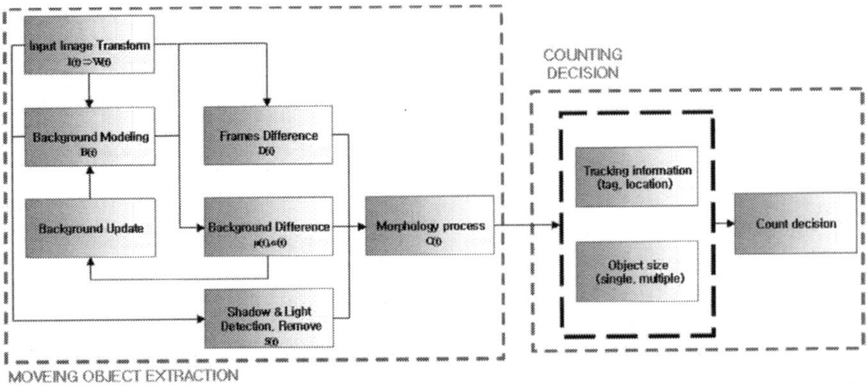

Fig. 2. System overall block diagram

consists of two parts. First, we analyzed the relationship between the extracted objects by using binary images that were obtained from the object extracting stage and compared them with previous ones in order to update correlations among objects and create information when new objects were extracted. Last, we decided to count when objects passed the ROI (region of interest).

2.2 Wavelet Transforms and Background Model

We used the Harr wavelet transform for two reasons; to remove noise from the images by using only a low frequency component, and to down-sample the images by using only the *LL* part. The first and most important step was to extract the moving objects from the background. Each background pixel was modeled by using a mixture of K Gaussian distributions. The K Gaussian distributions were evaluated by using a simple heuristic method to hypothesize which was most likely to be part of the background process. Each pixel was modeled by a mixture of K Gaussian distributions as stated in the formula, where W_t, $\mu_{i,t}$, K and $\sum_{i,t}$ are the input images, the mean value of the *ith* distribution, the number of distributions and the *ith* covariance matrix, respectively.

$$P(W_t) = \sum_{i=1}^{K} w_{i,t} * \eta(W_t, \mu_{i,t}, \textstyle\sum_{i,t}) \tag{1}$$

Previous studies used $K = 3$ for indoor scenes and $K = 5$ for outdoor scenes[6]. So, we made a reference background model by using a mixture of the K Gaussian distributions with the N wavelet transform images. We also made a reference background model with $K = 3$ empirically.

2.3 Shadow and Instant Change Detection

Shadow regions and instant changes of pixels intensity are the main reasons for undesired parts, which affect the final counting results. Therefore, it was necessary to detect and remove these undesired parts. We detected the shadow regions and instant change regions by using the formula below. Once we detected these two regions, we were easily able to remove them to obtain only the desired moving parts.

$$W_{out}(t) = \frac{R_w + G_w + B_w}{3}, B_{out}(t) = \frac{R_B + G_B + B_B}{3} \quad (2)$$

$$S(t) = \frac{W_{out}(t)}{B_{out}(t)} \quad (3)$$

Where $W_{out}(t)$ and $B_{out}(t)$ are the mean values of the R, G, and B components of the present input image and the reference background image, respectively. $S(t)$ is the ratio between $W_{out}(t)$ and $B_{out}(t)$. We were able to decide if a certain area was a shadow region or an instantly changing region by using $S(t)$. Generally, $0 < S(t) < 1$ is defined as a shadow region and $S(t) \geq 1$ is defined as an instantly changing region[7].

2.4 Moving Region Extraction Using Frame Differences

Many errors occur when we extract moving objects only by using background subtraction. To reduce these errors, we used the difference between the frames. The formula below is the procedure used to save $t-2$, $t-1$ and t images in memory and produce the difference images from them. First, we calculated $F(t)$ and $F(t-1)$ using two images. $D(t)$ was calculated from $F(t)$ and $F(t-1)$. $W(t)$ was the Harr wavelet transform image from the original image $I(t)$.

$$F(t-1) = W(t-1) - W(t-2), F(t) = W(t) - W(t-1) \quad (4)$$

$$D(t) = F(t) \vee F(t-1) \quad (5)$$

In this result, we used the extracted moving object combined with the background subtraction result.

2.5 Morphological Process

A morphological process was used to remove noise from the images and fill in small holes of the extracted objects. Practically, an extracted object usually contains large holes that cannot be filled by morphological process. For this problem, we used the masking method instead to fill both the large and small holes effectively and remove noise from the images as well[4]. One disadvantage of this method is the high computational expense. We proposed the modified masking method to reduce the processing time as follows. We produced a proper image to track by using a 5×5 main-mask, which can remove noise from images and fill in holes of the extracted objects. Figure 3 shows this process. This mask will change the pixel value from 1 to 0, if the pixel is determined as noise. Otherwise, it leaves the pixel value as it is. This process sets pixels to the majority pixel value of the mask by counting each pixel value. If a certain pixel value is

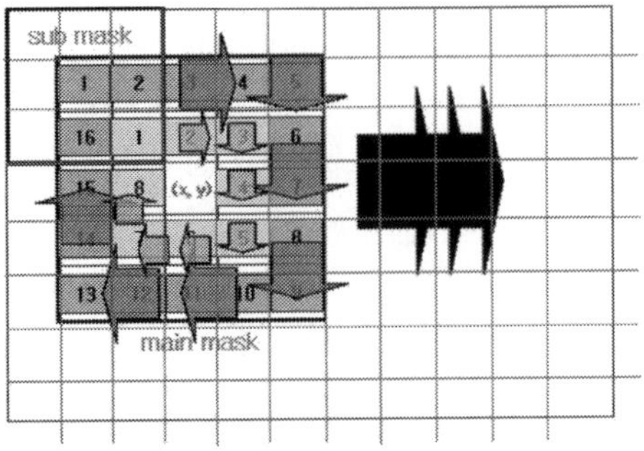

Fig. 3. 5×5 main-mask and 3×3 sub-mask used for the morphological process

determined to be 0, we move the mask to the next pixel whose value is 1. This process reduces the processing time. If a certain pixel value is determined to be 1, we use a 3×3 sub-mask for the surrounding 8 pixels of the center of the 5×5 main-mask and count each pixel value to change them $1 \rightarrow 0$, $1 \rightarrow 1$, $0 \rightarrow 1$, $0 \rightarrow 0$. After performing the above steps, we use the same 3×3 sub-mask for the other surrounding 16 pixels in the same way. Using this method, we fill in both the large and small holes of the extracted objects and remove the noise from the images at the same time. This would be impossible when using only the morphological process.

2.6 Block-Based Object Counting

The previous method used the same size for the extracted objects of any part of the image and counted objects passing through the interesting area by using this fixed size[1]. This is not an appropriate approach when working in real environments because the size of the extracted objects depends on the position in the image (with a camera at a height of $3.1 \sim 3.3m$). So we divided the images into blocks and calculated the mean and variance values of the size of each person for each block. This information was entered into a table. Figure 4 shows the calculating process. The mean and variance values were calculated from the trained images shown in Figure 5. People pass each block at least 10 times in the training images. We trained the images and made a mean and variance table in various environments, such as a $2.8mm$ lens with a camera at a height of $3.0m$, a $2.8mm$ lens with a camera at a height of $3.1m$, and so on. With this table, we were able to conveniently use this system in a wide range of different environments without training.

2.7 Counting Decision Rule

In tracking and counting procedures, it is necessary to analyze merging and splitting relations among people. For example, in light traffic hours, it is simple to count people because they often move independently of one another. However, in busier hours, people frequently merge and split. Therefore, during these busier

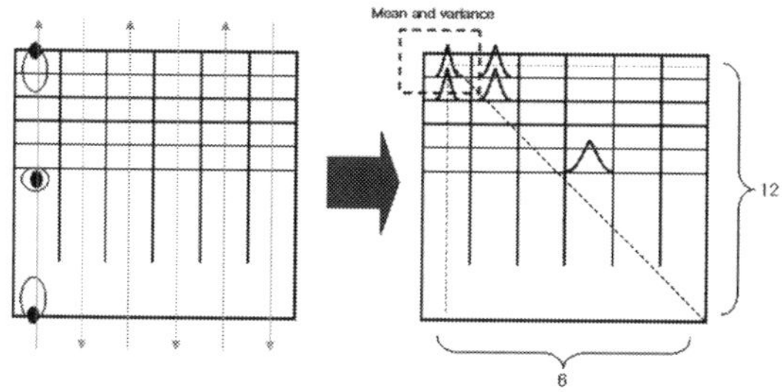

Fig. 4. Divided 6 × 12 blocks and calculation of mean and variance values

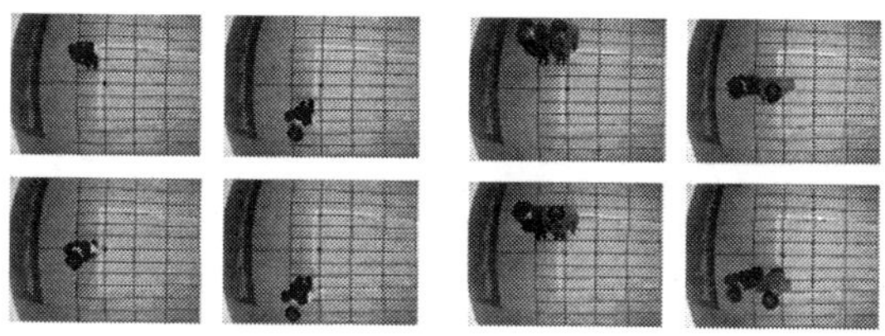

(a) (b)

Fig. 5. (a) Images for one-person training image, (b) Images for two-person training images

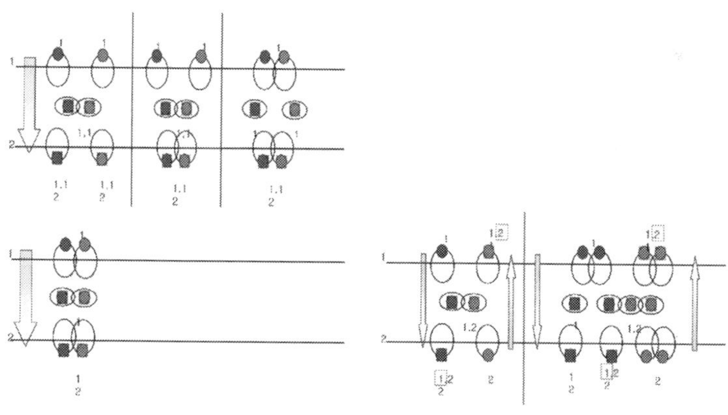

(a) (b)

Fig. 6. (a) Co-directional rule, (b) Bi-directional rule

times it is difficult to count the number of people passing through the ROI and estimate the direction of their paths. To solve this problem, researchers have used specific color information for each person. Although this method is helpful, it requires more processing time, because the system has to store and update all the color information for each person. Hence it is not suitable for real-time systems. It was therefore decided to assign 'tag' information to each moving person, in order to improve time efficiency. This tag information was maintained in the ROI and updated to track the direction of the paths. In this way, no additional image processing or information was needed. Figure 6 shows some examples of the proposed tagging rule in co-directional and bi-directional cases. In Figure 6-(a), when people stepped inside line 1 (the entrance counting line), they were given a label of 'Tag 1'. Similarly, when people stepped outside line 2 (exit counting line), they were given a label of 'Tag 2'. Then, the counting process could be easily performed according to changes of the value of the tag. Also, as shown in Figure 6-(b), the directions of the paths sometimes differ. But if we knew the entrance tag number of each person in advance, we could easily count the passing people by using the alteration of their tag numbers.

3 Experimental Result

We experimented with a camera on the ceiling at a height of H assuming that the height of an average person is h. H and h were measured at approximately 3.1~3.3m and 150~180cm, respectively. We used a general CCD camera and a 2.8mm lens. We performed the experiment under practical conditions using three different locations at three different times. Experiments were performed in a corridor, on an escalator, and in an entrance. These locations represented various environments such as light, shadow and highlights caused by strong light. These experimental environments are shown in Figure 7. We used about 20 000 sequential images for each environment. The first 100 frame images were pure background images, which we used to make a reference background model. We produced this model from the first 20 images in the experiments. Table 1 shows the people counting error rates that were recorded for each environment. The TPP (total passing people) parameter represents the counting of the number of people through the ROI (region of interest). The ACE (add counting error) parameter represents values that are higher than the correct count and the UCE (under counting error) parameter represents values that are lower than the correct count (since they exclude children and overweight people). We only trained people of average height and weight. The TCE (total counting error) parameter represents the sum of the ACE and the UCE parameters. Figure 8 shows the people extraction and tracking results in the various environments.

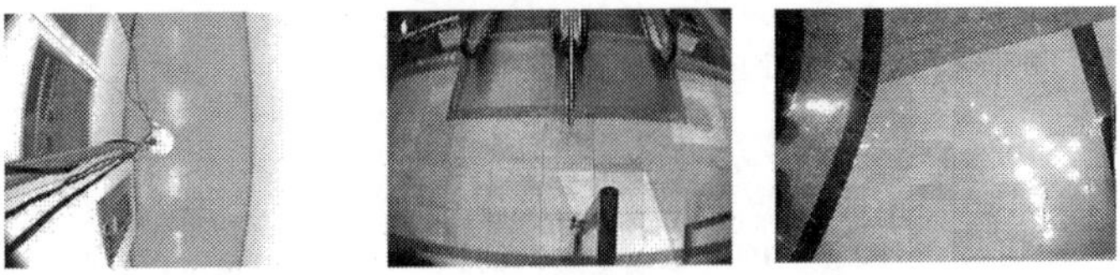

(a) (b) (c)

Fig. 7. Examples of the different environments: (a) Corridor, (b) Escalator, (c) Entrance

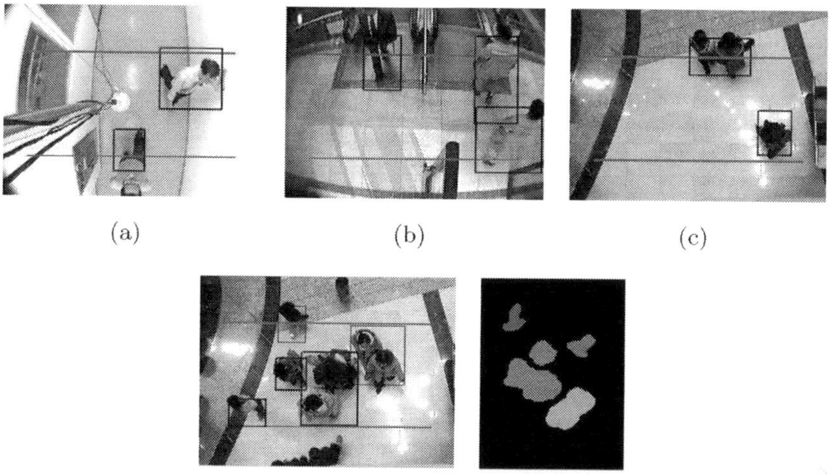

(a) (b) (c) (d) (e)

Fig. 8. Results obtained from the different environments: (a) Corridor, (b) Escalator, (c) Entrance, (d) Crowd image, (e) Segment of crowd image

Table 1. Error rates obtained in each environment

Environment	Entrance	Escalator	Corridor
ACE/TPP	10/192(4.16%)	2/207(0.96%)	0/64(0.000%)
UCE/TPP	8/192(5.21%)	6/207(2.89%)	2/64(3.125%)
TCE/TPP	18/192(9.37%)	8/207(3.85%)	2/64(3.125%)

4 Conclusion

In this paper, we proposed a people counting system that can be used to count and track people at entrances, elevators, or escalators where many people are moving. This system is useful for surveillance, building management, and marketing data. We proposed the block-based people counting system which divides an image into 6×12 blocks and trains the size of each person for each block. This proposed method does not detect each person but only tracks masses of objects and counts them by using the trained size of the person. This method improves both accuracy and processing time. We analyzed the time performance with a Pentium 4 3.2 GHz using a video at 320×240 frames (24 bits per pixel) in which our system obtained an average frame rate of 25 fps (performance obtained using the video shown in Figure 8). The counting accuracy was 100% when used with one or two moving people and about 90~94% when used with three or more moving people.

Acknowledgements

This work was supported by the Korea Science and Engineering Foundation (KOSEF) through the Biometrics Engineering Research Center (BERC) at Yonsei University.

References

1. Thou-Ho Chen, "An automatic bi-directional passing-people counting method based on color image processing", Security Technology, Proceedings. IEEE 37th Annual 2003 International Carnahan Conference on 14-16 Oct. 2003 PP. 200 - 207, 2003.
2. Kenji Terada and Jun'ichi Yamaguchi, "A System for Counting Passing People by Using the Color Camera", The Transactions of The Institute of Electrical Engineers of Japan.
3. K. Terada, D. Yoshida, S. Oe and J. Yamaguchi, "A Method of Counting the Passing People by Using the Stereo Images", Image Processing, 1999. ICIP 99. Proceedings. 1999 International Conference on Volume 2, 24-28, PP. 338 - 342, Oct. 1999.
4. Segen, J., "A camera-based system for tracking people in real time", Pattern Recognition, 1996., Proceedings of the 13th International Conference on Volume 3, 25-29, PP. 63 - 67, Aug. 1996.
5. Rossi, M., Bozzoli, A., "Tracking and counting moving people", Image Processing, 1994. Proceedings. ICIP-94., IEEE International Conference Volume 3, 13-16, PP. 212 - 216, Nov. 1994.
6. Qi Zang, Klette, R., "Robust background subtraction and maintenance", Pattern Recognition, 2004. ICPR 2004. Proceedings of the 17th International Conference on Volume 2, 23-26, PP. 90 - 93, Aug. 2004.
7. Hanzi Wang, Suter D., "A re-evaluation of mixture of Gaussian background modeling [video signal processing applications]",Acoustics, Speech, and Signal Processing, 2005. Proceedings. (ICASSP '05). IEEE International Conference on Volume 2, 18-23, PP. ii/1017 - ii/1020, March 2005.

A Classification Approach for the Heart Sound Signals Using Hidden Markov Models*

Yong-Joo Chung

Department of Electronics, Keimyung University,
Daegu, S. Korea
yjjung@kmu.ac.kr

Abstract. Stethoscopic auscultation is still one of the primary tools for the diagnosis of heart diseases due to its easy accessibility and relatively low cost. Recently, many research efforts have been done on the automatic classification of heart sound signals for supporting clinicians to make better heart sound diagnosis. Conventionally, automatic classification methods of the heart sound signals have been usually based on artificial neural networks (ANNs). But, in this paper, we propose to use hidden Markov models (HMMs) as the classification tool for the heart sound signal. In the experiments classifying 10 different kinds of heart sound signals, the proposed method has shown quite successful results compared with ANNs achieving average classification rate about 99%.

1 Introduction

Heart sound auscultation is still a very important method for the diagnosis of heart diseases with its easy accessibility and the relatively low cost. However, detecting symptoms of various heart diseases by auscultation requires a skill that takes years of experiences in the field. As the skill is not easy to acquire for junior clinicians, an automatic classification system of the heart sound signal would be very useful for assisting the clinicians to make better diagnosis of the heart disease [1][2].

The dynamic spectral characteristic and non-stationary nature of the heart sound signal makes the automatic classification difficult. As the heart sound signal comes from the human body, it may be highly variable from cycle to cycle and even according to the patient's conditions. To overcome these problems, a classifier which can take into account the variability should be used in the automatic diagnosis of the heart sound signal.

Artificial neural networks (ANNs) based approaches have been widely used in the classification of heart sound signals with some success [1],[2]. ANNs are known to be efficient in classifying complex patterns and have been traditionally used for problems in bio-medical and image pattern classification [3]. However, neural networks are not designed to be suitable for time sequential input patterns like heart sound signals. In fact, nearly all ANNs used for heart sound signals are just static pattern

* This work has been supported by The Advanced Medical Technology Cluster for Diagnosis and Prediction at KNU, which carries out one of the R&D Projects sponsored by the Korea Ministry Of Commerce, Industry and Energy.

D.-Y. Yeung et al. (Eds.): SSPR&SPR 2006, LNCS 4109, pp. 375–383, 2006.

classifiers. The heart sound signal should be segmented before use in training and the ANN can only recognize the segmented data. In real situation, it may not easy to segment the relevant parts of the heart sound signal because the automatic classification system should accept the continuous input signal. Even if we can successfully segment one cycle of the heart sound signal, the non-stationary characteristics within the cycle makes it inappropriate to consider the whole samples in the cycle as an input to the ANNs, because we may fail to focus on the time-varying characteristics within the cycle. For the efficient signal classification, methods which can analyze separately the stationary parts in a cycle and later combine them would be desirable. But, neural networks have difficulty in integrating the time-varying statistical characters of the heart sound signal.

In contrary, hidden Markov models (HMMs) have shown quite successful results in classifying time sequential patterns like speech signals [4]. HMMs with its Markov chain structure can inherently incorporate the time sequential character of the signal. By using the Gaussian mixture densities, the HMMs are also expected to faithfully represent the various spectral characteristics of the heart sound signal. The non-stationary nature of the heart sound signal may be well represented by the state transitions in HMMs. As will be shown later in this paper, we could find from the classification experiments that HMMs are very efficient for modeling the dynamic and non-stationary nature of the heart sound signal.

In the next section, we will explain in detail methods how to construct a classifier using HMMs and, in section 3, we show experimental results which demonstrate the feasibility of using HMMs in classifying the heart sound signal. And finally, we make conclusion in section 4.

2 Methods

2.1 Hidden Markov Models

The basic theory of HMMs has been published in a series of papers by Baum [5] in the late 1960s and early 1970s. The underlying assumption in the use of the HMM for signal modeling is that the signal can be well characterized as a parametric random process and the parameters of the random process can be estimated in a well-defined manner. The HMM has shown to provide a highly reliable way of recognizing time sequential patterns like speech. A simple three-state HMM is shown in Fig.1 [6].

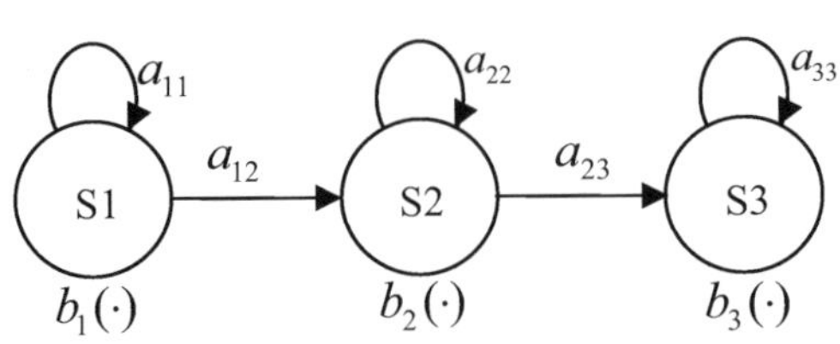

Fig. 1. A 3-state hidden Markov model

The transitions between states are controlled by the transition probability

$$a_{ij} = P(S_j | S_i). \tag{1}$$

And, the probability of generating an observation is determined by the output probability distributions in each state which are usually modeled by a mixture of multivariate Gaussian distributions. Given the observation $\mathbf{y}(t)$, the output probability distribution in state s_j is given by

$$b_j(y(t)) = \sum_{m=1}^{M} c_{jm} N(y(t); \mu_{jm}, \Sigma_{jm}). \tag{2}$$

where $N(y(t); \mu_{jm}, \Sigma_{jm})$ is a multi-variate Gaussian distribution, with mean vector μ_{jm} and covariance matrix Σ_{jm}, each mixture component having an associated weight c_{jm}.

The HMM, as an acoustic model is required to determine the probability of the time sequential observation data $Y = \{y(1), y(2), \cdots, y(T)\}$. This is done by computing the probability of the observation data $P(Y \mid M_i)$, $i = 1,2,\cdots,V$. Here M_i represents an HMM corresponding to a class in the classification problem and V is the total number of classes. The classification is performed by finding the class k which gives the best likelihood score.

$$k = \arg\max_{i=1,2,\cdots,V} P(Y \mid M_i) \tag{3}$$

In practice, Viterbi decoding is employed to find the class with the best likelihood score. Meanwhile, the estimation of the parameters ($\mu_{jm}, \Sigma_{jm}, c_{jm}$) of the HMM is an optimization problem based on some appropriate criterion. Maximum likelihood estimation (MLE) is the most popular one and it tries to match the HMM to the training data as closely as possible. For the efficient MLE training of the HMM, the Baum-Welch algorithm based on Expectation-Maximization (EM) technique is commonly used [7].

2.2 Classification of Heart Sound Signals Using HMMs

A cycle of heart sound signals consists of four elements. The first one called S1 is heard when the mitral and tricuspid valve is closed. The second one S2 is related with the closure of the aortic and pulmonary valve. The systolic and diastolic phase refers to the intervals between S1 and S2 during which any sound is hardly heard in the normal case. We used a four-state HMM to model a cycle of the heart sound signal as shown in Fig. 2. The number of states in the HMM may be determined based on the nature of the signal being modeled. We assumed that each state of the HMM corresponds to an element of the heart sound signal because the signal characteristics in each element are thought to be homogeneous. The spectral variability in each state is modeled using multiple mixture components. By trial and error, we determined the number of mixture components in each state to be 10.

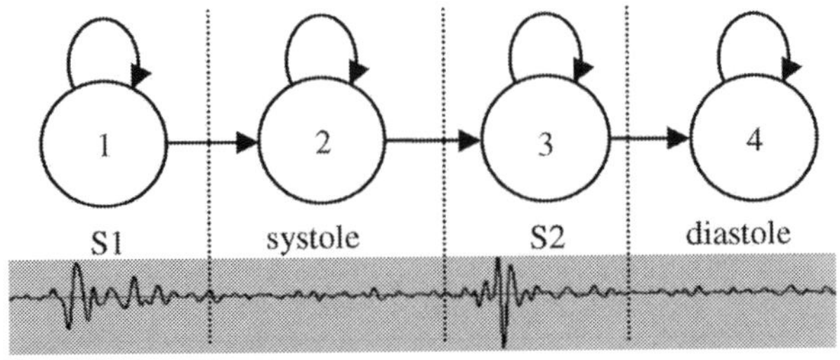

Fig. 2. An HMM for a cycle of the heart sound signal

In Fig. 3, we show the procedure of classifying the heart sound signal using the trained HMMs. The HMM parameters are estimated during the training procedure. For the initial parameter estimation, every cycle of the heart sound signal is manually segmented into 4 regions giving the statistical information corresponding to each element [7]. The feature vectors used are mel-frequency cepstral coefficients (MFCCs) and filter bank outputs both derived from the fast Fourier transform (FFT). MFCCs are popularly used for speech recognition [6] and the filter bank outputs have been usually employed for the spectral analysis of the heart sound signal.

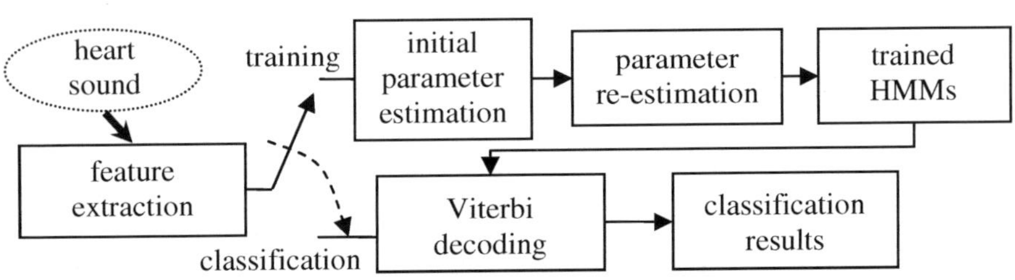

Fig. 3. The procedure of classifying the input heart sound signal using the trained HMMs

3 Results and Discussion

The heart sound data used for the experiments were obtained from the clinical training CDs for the physicians [9]. The original data were down sampled to 16 KHz and stored in a 16 bit resolution. The heart sound signal was already diagnosed and labeled as a specific heart condition. The classification experiments were done using 159 heart sound examples corresponding to 10 different heart conditions: normal sound, innocent murmur, AR (Aortic Regurgitation), AS (Aortic Stenosis), CA (Coarctation of the Aorta), MR (Mitral Regurgitation), MS (Mitral Stenosis), MVP (Mitral Valve Prolapse), TR (Tricuspid Regurgitation) and VSD (Ventricular Septal Defect).

An HMM was constructed for each type of heart condition using the corresponding data. To overcome the problem of small amount of data collected, the classification test was done by the Jack-Knifing method. In the process, the HMM is trained with all the available examples except the one which is used for the testing. This process is repeated so that all the examples can be used for the testing. The test results are then averaged to give the final classification rate.

The feasibility of modeling the heart sound signal using HMMs can be checked by the segmentation results which can be obtained in the Viterbi decoding. Using the training data, we could verify that each state of the HMM matches quite closely with the respective components of the heart sound signal as we expected. In Fig. 4, we show the matching relations between the HMM states and the heart sound signal waveforms as the number of mixture components varies. In the figure, the vertical lines represent the boundaries between the states.

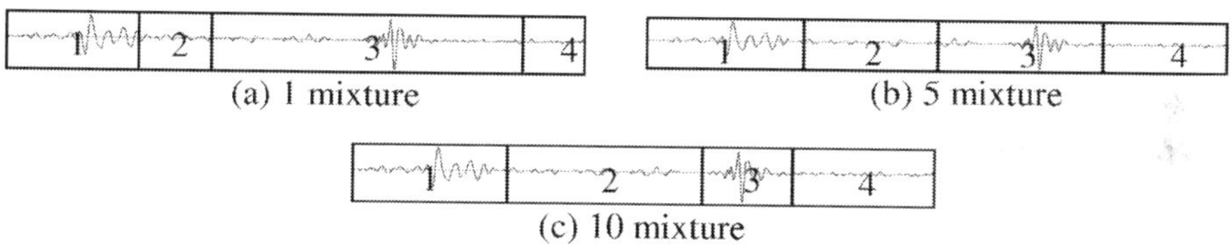

Fig. 4. The matching relations between the HMM states and the heart sound signal waveforms

Although the matching between them is poor when the number of mixtures is 1, it becomes quite accurate as we increase the number of mixture components to 10. With these segmentation results satisfying our expectation, we may conclude that the HMM is quite suitable for the stochastic modeling of the heart sound signal.

Conventional approaches for the heart sound signal classification have been usually based on ANNs [8]. ANNs are known to be able to discriminate complex patterns by generating nonlinear functions of the input. While they have proved useful in recognizing static patterns like spelled characters, they may not be tailored for time sequential input patterns. To compare the performance of the proposed HMM-based classifier with ANNs, we constructed an ANN and tested its performance in recognizing the heart sound signals. The ANN consists of 3 layers (input layer, hidden layer and output layer) connected in sequence. The nodes in each layer process the outputs from the previous layer or directly take in the input features of the heart sound signal. The number of nodes in the input layer is 210 equal to the dimension of the input feature vector and the number of nodes in the hidden and output layer is 20 and 10, respectively [8]. The node in the output layer corresponds to each class of the heart sound signals to be discriminated. In the training, we used 10 different classes of heart sound signals including the normal one.

We investigated the spectral characteristic of the heart sound signal by obtaining the normalized energy spectrum through the fast Fourier transform (FFT). In addition to the frame-level energy spectrum, the energy spectrum of the whole cycle of the heart sound signal was also obtained. The length of the cycle ranges from 500 ms to 1 sec and the frame length was fixed at 2.5 ms. The heart sound signal is processed on a frame basis in the HMM while the whole cycle is fed into the ANN as an input. So,

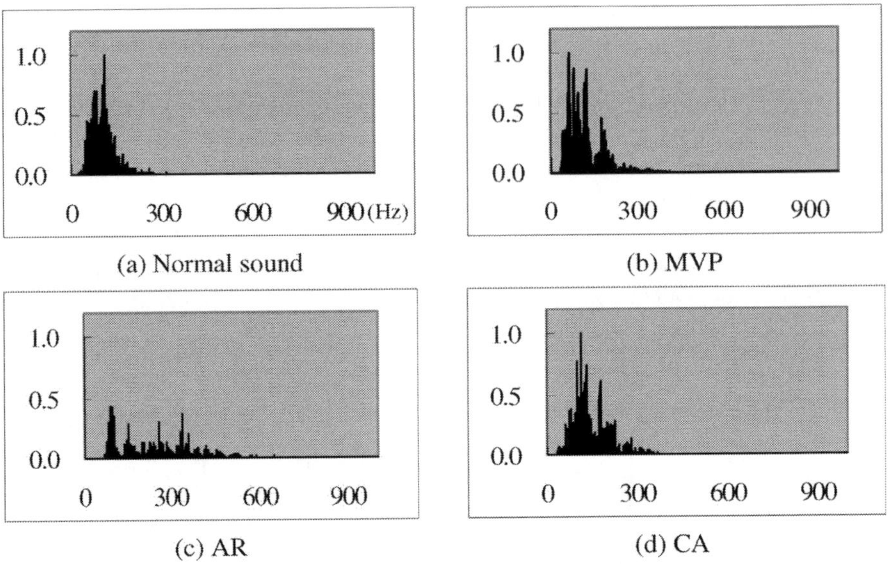

Fig. 5. Normalized energy spectrum of a cycle of the heart sound signal

we may expect that different input spectral characteristics will be modeled in the HMM and ANN, respectively. In Fig. 5, we show the normalized energy spectrum for the whole cycle of the heart sound signals. The heart sound signal seems to contain most of its energy between 0 and 300 Hz, although there are some energy for frequencies up to 600 Hz.

Contrary to the ANN, the input feature vector for the HMM is given on a frame basis. In Fig. 6, the frame-level energy spectrum is shown as time spans. In Fig. 6(a), we can find significant peaks in the energy spectrum at about 200 ms and 600 ms. They seem to correspond to S1 and S2, respectively and their frequency range is between 0 and 300Hz. Meanwhile, for the heart sound signals related with some diseases, there is considerable energy in the diastole and systole phase. In particular, in Fig. 6(c) and (d), we can see some energy peaks between S1 and S2 and their frequency range goes up to 600 Hz. For the MVP case in Fig. 6(b), although the signal waveform looks similar to the normal case, the S2 is usually split as can be seen in the spectrum.

In the experiments classifying heart sound signals, we used two different types of features. In Table 1, we show the classification results when using the MFCC and filterbank outputs. The frequency range for both types of features was relatively broad from 200 to 6400 Hz. The two kinds of features show similar results with marginal improvement obtained in the case of using filterbank outputs. Although the MFCC may be very adequate to speech signals, it was not quite successful to the heart sound signal. With these results, the filterbank outputs are used as the basic input features in the following experiments.

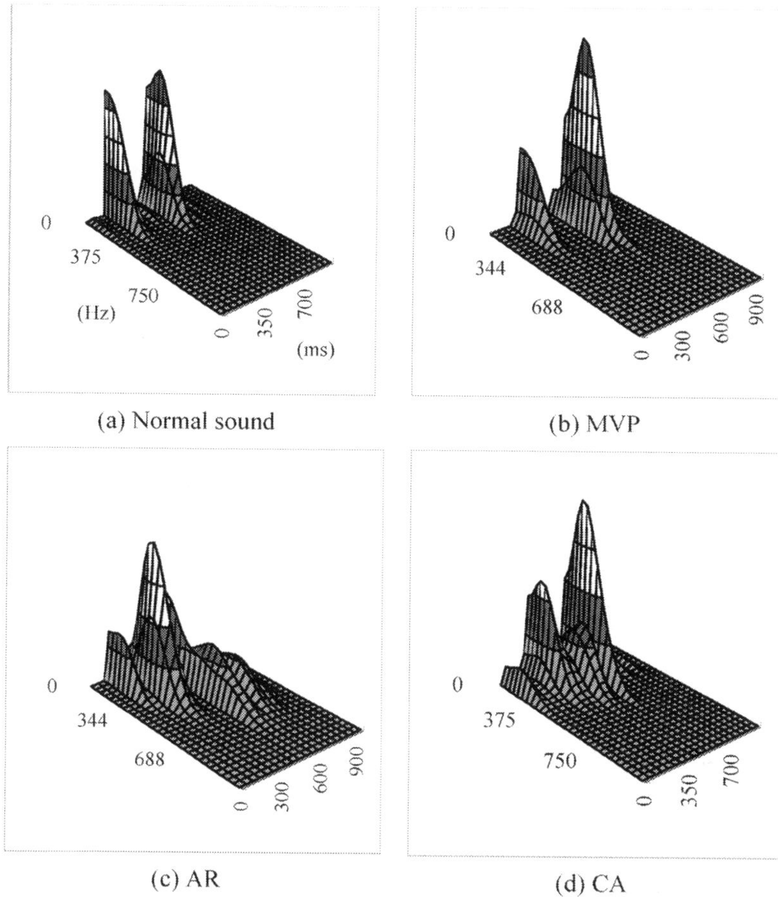

Fig. 6. Normalized frame-level energy spectrum variations with time

As mentioned earlier, the effective frequency range in the spectrum of the heart sound signal is usually well below 600 Hz. So, we experimented by varying the frequency ranges in the filterbank outputs. The frequency ranges considered are 0~900 Hz, 0~210 Hz, 200~900 Hz, 200~300 Hz and 30-900 Hz. The classification results are shown in Table 2. The 0~900 Hz and 200~900Hz ranges performed better than others although there were not significant differences in the classification results with the various frequency ranges. Also, the results in Table 2 were better than the previous results in Table1 indicating that it is important to consider only the relevant frequency ranges in obtaining the filterbank outputs.

To compare the performance of the proposed method with the conventional aproaches, an ANN was trained and tested in the classification experiments. Its struture is as described in the above and two different frequency ranges were considered in the filterbank outputs. From the results in Table 3, we can see that the performance of the ANN was quite inferior to the HMM. The HMM's ability to model efficiently the dynamic time sequential input patterns may be the reason for the superior performance over the ANN.

Table 1. The classification results depending on input features

	MFCC		Filterbank output	
	Accuracy(%)	Correct / Total	Accuracy(%)	Correct /Total
Normal sound	100	15/15	100	15/15
Innocent murmu r	92.86	13/14	92.86	13/14
AR	100	14/14	100	14/14
AS	100	18/18	100	18/18
CA	100	20/20	95	19/20
MR	100	21/21	100	21/21
MS	100	14/14	100	14/14
MVP	76.92	10/13	92.31	12/13
TR	100	20/20	100	20/20
VSD	100	10/10	100	10/10
Average	97.48	155/159	98.11	156/159

Table 2. The results with various frequency ranges in the filterbank outputs

Range(Hz)	Average(%)	Correct/Total
0~900	99.37	158/159
0~210	98.74	157/159
200~900	99.37	158/159
200~300	98.11	156/159
30~900	98.74	157/159

Table 3. The classification results of the ANN

Range(Hz)	Average(%)	Correct/Total
0~210	93.08	148/159
0~420	90.56	144/159

4 Conclusion

As a preliminary study of developing an automatic diagnosis system for heart diseases, we proposed a statistical classifier using HMMs. Although the number of diseases to be classified was fairly large compared with the previous works, it achieved a satisfactory classification rate about 99%. In particular, the proposed method showed quite superior performances compared with the ANN. This seems to come from the HMM's ability to cope with the dynamic time sequential input patterns. However, some of the heart signals were found to be difficult to discriminate. As the HMM is very flexible in modeling the signals, we may improve the discrimination between the models by careful investigation on the heart sound signal characteristics. For example, the number of states and mixture components can be varied depending on the signal types and the amount of the training data. Also features which can contribute more to discriminating between classes can be considered and various estimation criterions for the HMM parameters can be considered for the better modeling.

References

1. Cathers, I.: Neural Network Assisted Cardiac Asculation. Artif. Intell. Med. 7 (1995) 53–66
2. Bhatikar, S.R., DeGroff, C., Mahajan, R. L.: A Classifier Based on Artificial Neural Network Approach for Cardiac Auscultation in Pediatrics. Artif. Intell. Med. 33 (2005) 251–260
3. Lippmann, R. P.: An Introduction to Computing with Neural Nets, IEEE ASSP Magazine April (1987) 4-22
4. Rabiner, L. R.: A Tutorial on Hidden Markov Models and Selected Applications in Speech Recognition, Proceedings of the IEEE, 77, Feb. (1989)
5. Baum, L. E., Petrie, T., Soules, G., Weiss, N.:A Maximization Technique Occurring in the Statistical Analysis of Probabilistic Functions of Markov Chains, Annals of Mathematical Statistics 41 (1970) 164-171
6. Lee, K. F.,: Automatic Speech Recognition, Kluwer Academic Publishers (1989)
7. Rabiner, D. R.,: A Tutorial on Hidden Markov Models and Selected Applications in Speech Recognition IEEE Proceedings (1998)
8. DeGroff C, Bhatikar S, Hertzberg J, Shandas R, Valdes-Cruz L, Mahajan R, "Artificial neural network-based method of screening heart murmur in children." Circulation 103, 2711-6, 2001.
9. Mason D.,: Listening to the Heart, Hahnemann University (2000)

Structure Analysis Based Parking Slot Marking Recognition for Semi-automatic Parking System

Ho Gi Jung[1, 2], Dong Suk Kim[1], Pal Joo Yoon[1], and Jaihie Kim[2]

[1] MANDO Corporation Central R&D Center, Advanced Electronic System Team
413-5, Gomae-Dong, Giheung-Gu, Yongin-Si, Kyonggi-Do 446-901, Republic of Korea
{hgjung, greenhupa, pjyoon}@mando.com
http://www.mando.com/eng/main.sap
[2] Yonsei University, School of Electrical and Electronic Engineering
134, Sinchon-Dong, Seodaemun-Gu, Seoul 120-749, Republic of Korea
{hgjung, jhkim}@yonsei.ac.kr
http://cherup.yonsei.ac.kr

Abstract. Semi-automatic parking system is a driver convenience system automating steering control required during parking operation. This paper proposes novel monocular-vision based target parking-slot recognition by recognizing parking-slot markings when driver designates a seed-point inside the target parking-slot with touch screen. Proposed method compensates the distortion of fisheye lens and constructs a bird's eye view image using homography. Because adjacent vehicles are projected along the outward direction from camera in the bird's eye view image, if marking line-segment distinguishing parking-slots from roadway and front-ends of marking line-segments dividing parking-slots are observed, proposed method successfully recognizes the target parking-slot marking. Directional intensity gradient, utilizing the width of marking line-segment and the direction of seed-point with respect to camera position as a prior knowledge, can detect marking line-segments irrespective of noise and illumination variation. Making efficient use of the structure of parking-slot markings in the bird's eye view image, proposed method simply recognizes the target parking-slot marking. It is validated by experiments that proposed method can successfully recognize target parking-slot under various situations and illumination conditions.

1 Introduction

Semi-automatic parking system is a driver convenience system automating steering control required during parking operation. Because recently driver's interest about parking assist system increases drastically, car manufacturers and component providers are developing various kinds of parking assist systems [1][2]. Fig. 1 shows the configuration of semi-automatic parking system currently being developed. The system consists of six components: EPS (Electric Power Steering) for active steering, vision sensor acquiring rear-view image, ultra-sonic sensors measuring distances to nearby side/rear obstacles, touch screen based HMI (Human Machine Interface) pro viding information to driver and receiving command from driver, EPB (Electric

D.-Y. Yeung et al. (Eds.): SSPR&SPR 2006, LNCS 4109, pp. 384 – 393, 2006.

Parking Braking) automatically activating parking brake, and processing computer. Algorithms running on the processing computer consist of three components: target parking position designation, path planning generating trajectory from current position to target position, and path tracker which continuously estimates current position and controls steering system to achieve the planned path.

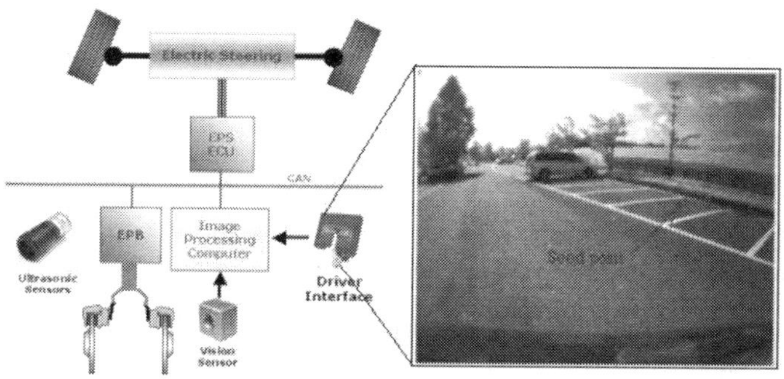

Fig. 1. System configuration of semi-automatic parking system

There are many kinds of methods for the target parking position designation: laser-scanner based method [3], SRR (Short Range Radar) network based method [1], computer vision based method, GPS/digital map based method [4], and driver's manual designation based method. Prius IPAS (Intelligent Parking Assist System), mass-produced by Toyota and AISIN SEIKI in 2003, is the example of manual designation method [5]. Computer vision based method, which can provides monitoring view of ongoing parking procedure, attracts more and more interests. Computer vision based method can be categorized into three kinds: method recognizing adjacent vehicles, method recognizing parking-slot markings, and method recognizing both adjacent vehicles and parking-slot markings. Nico Kaempchen developed a system localizing free parking space by recognizing adjacent vehicles with stereo vision based method [6]. Jin Xu developed monocular vision based parking-slot marking recognition using neural network [7]. In previous research, we developed stereo vision based parking-slot marking recognition considering adjacent vehicles [8]. AISIN SEIKI's next generation is expected to recognize adjacent vehicles three-dimensionally by motion stereo and provide rendered image from a virtual viewpoint suitable for the understanding of parking operation [9].

When driver designates a seed-point inside target parking-slot with touch screen as shown in Fig. 1, proposed method recognizes corresponding parking-slot marking as target parking position. Proposed method is designed not only to solve the discomfort of previous Prius's fully manual designation method, but also to eliminate the overweighed requirements of stereo vision based method, i.e. high-performance hardware and enormous computing power. After the compensation of fisheye lens distortion and the construction of bird's eye view image, marking line-segments crossed by the gaze from camera to seed-point are detected. Guideline, distinguishing parking-slots from roadway, can be easily detected by simply finding the nearest among the detected line-segments. Consecutively, separating line-segments are detected based on the detected guideline. Experimental results show that proposed method is simple and robust to noise and illumination change.

2 Bird's Eye View Construction

Proposed system compensates the fisheye distortion of input image and constructs bird's eye view image using homography. Installed rear view camera uses fisheye lens, or wide-angle lens, to cover wide FOV (Field Of View) during parking procedure. As shown in Fig. 2, input image through fisheye lens can capture wide range of rear scene but inevitably includes severe distortion. It is well known that the major factor of fisheye lens distortion is radial distortion, which is defined in terms of the distance from the image center [10]. Modeling the radial distortion in 5th order polynomial using Caltech calibration toolbox and approximating its inverse mapping by 5th order polynomial, proposed system acquires undistorted image as shown in Fig. 2 [11]. Homography, which defines one-to-one correspondence between coordinate in undistorted image and coordinate in bird's eye view image, can be calculated from the height and angle of camera with respect to the ground surface [8]. Bird's eye view is the virtual image taken from the sky assuming all objects are attached onto the ground surface. General pinhole camera model causes perspective distortion, by which the size of object image is changing according to the distance from camera. Contrarily, because bird's eye view image eliminates the perspective distortion of objects attached onto the ground surface, it is suitable for the recognition of objects painted on the ground surface. Final image of Fig. 2 is the bird's eye view image of the undistorted image. Hereafter, almost image processing is fulfilled in the bird's eye view image and characters in lower case bold face, e.g. **u**, represent coordinate or vector in the bird's eye view image.

Fig. 2. Construction procedure of bird's eye view image

3 Guideline Recognition

Parking-slot markings consist of one guideline and separating line-segments as shown in Fig. 3(a). To recognize the parking-slot markings, marking line-segment distinguishing parking-slots from roadway should be recognized at first. Because the line-segment is the reference of remaining recognition procedures, it is called guideline. Each parking slot is distinguished by two line-segments perpendicular to the guideline, which is called separating line-segment.

3.1 Marking Line-Segment Recognition by Directional Intensity Gradient

Proposed system recognizes marking line-segments using directional intensity-gradient on a line lying from seed-point to camera. As shown in Fig. 3(a), vector from seed-point to camera is represented by $\mathbf{v}_{\text{seed point-camera}}$ and its unit vector is represented by $\mathbf{u}_{\text{seed point-camera}}$. Fig. 3(b) shows the intensity profile of pixels on the line in the unit of pixel length s. If the start point $\mathbf{p}_s$ and unit vector $\mathbf{u}$ are fixed, the intensity of a pixel which is distant by s in the direction of $\mathbf{u}$ from $\mathbf{p}_s$, i.e. $I(\mathbf{p}_s+s\cdot \mathbf{u})$, is represented by simple notation I(s). Because the line crosses two line-segments, it can be observed that two intensity peaks with the width of ling-segment exist.

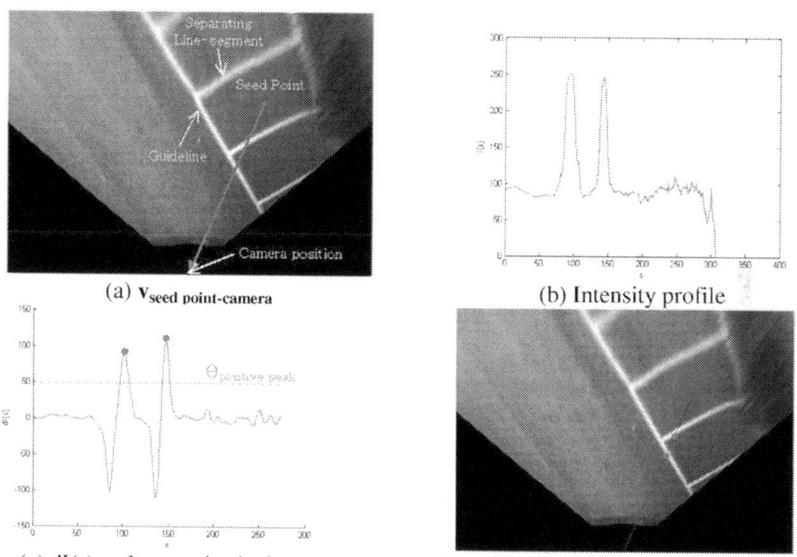

(a) $\mathbf{v}_{\text{seed point-camera}}$ (b) Intensity profile

(c) dI(s) and recognized edges (d) recognized marking line-segments

Fig. 3. Procedure of marking line-segment recognition

Equation (1) defines the directional intensity-gradient of a point $\mathbf{p}$ (x,y) with respect to vector $\mathbf{u}$, $dI(\mathbf{p},\mathbf{u})$. Because camera maintains a certain height and angle with respect to the ground surface, marking line-segment painted on the ground surface will appear with a fixed width W. Therefore, directional intensity-gradient using the average intensity of W/2 interval is robust to noise while detecting interesting edges.

$$dI(\mathbf{p},\mathbf{u})=\frac{1}{\frac{W}{2}}\sum_{i=1}^{\frac{W}{2}} I(\mathbf{p}-i\cdot\mathbf{u})-\frac{1}{\frac{W}{2}}\sum_{i=1}^{\frac{W}{2}} I(\mathbf{p}+i\cdot\mathbf{u}) \quad (1)$$

Fig. 3(c) shows the profile of directional intensity-gradient of the line with respect to $\mathbf{u}_{\text{seed point-camera}}$, i.e. $dI(\mathbf{p}_{\text{seed point}}+s\cdot\mathbf{u}_{\text{seed point-camera}}, \mathbf{u}_{\text{seed point-camera}})$, which is denoted by simple notation dI(s). Positive peaks correspond to the camera-side edges of marking line-segments and negative peaks correspond to the seed-point-side edges. Because camera-side edge is easy to follow, it is recognized as the position of marking line-segment. Threshold for positive peak detection, $\theta_{\text{positive peak}}$, is defined adaptively like equation (2). Fig. 3(c) shows established threshold and recognized positive peaks. Fig. 3 (d) shows the recognized marking line-segments in bird's eye view image.

$$\theta_{\text{positive peak}} = \frac{1}{3}\left(\max_s I(s) - \operatorname{avg}_s I(s)\right) \tag{2}$$

3.2 Recognition of Marking Line-Segment Direction

Proposed system detects the direction of marking line-segments using the directional intensity-gradient of local window and edge following based refinement. Edge following results can eliminate falsely detected marking line-segments.

The directional intensity-gradient of a point displaced by (dx,dy) from a center point $\mathbf{p}_c$ (x_c, y_c) can be calculated by $dI(\mathbf{p}_c+(dx,dy),\mathbf{u})$, which is denoted by simple notation dI(dx,dy) if $\mathbf{p}_c$ and $\mathbf{u}$ are fixed. Proposed system calculates the directional intensity-gradient, $dI(\mathbf{p}_{\text{cross}}+(dx,dy),\mathbf{u}_{\text{seed point-camera}})$, of (W+1)x(W+1) local window around the detected cross-points $\mathbf{p}_{\text{cross}}$. Here, dx and dy are in the range of –W/2~W/2. Fig. 4 shows the calculated dI(dx,dy) of local window around a cross-point. It can be observed that dI(dx,dy) array forms a ridge, of which direction is the same as the edge direction.

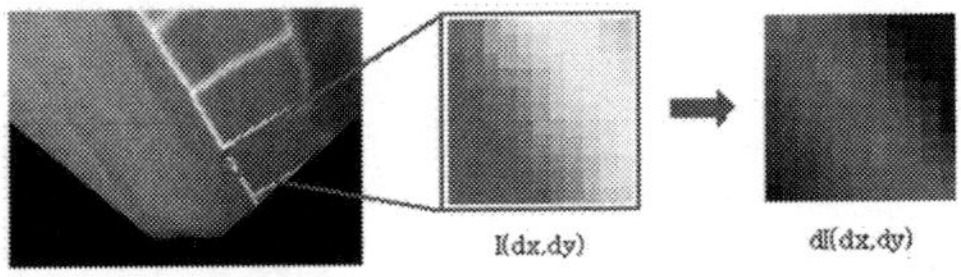

Fig. 4. Directional intensity-gradient of local window around a cross-point

To detect the direction of the ridge, proposed system introduces $\text{fitness}_{\text{ridge}}(\phi)$, which measures how well a line rotating by ϕ around the cross-point is similar to the ridge direction like equation (3). As shown in Fig. 5(a), $\text{fitness}_{\text{ridge}}(\phi)$ is the difference between two line-sums in dI(dx,dy). These lines are orthogonal to each other.

$$fitness_{ridge}(\phi)=\sum_{i=-\frac{W}{2}}^{\frac{W}{2}} dI(i\cdot\cos(\phi), i\cdot\sin(\phi)) - \sum_{i=-\frac{W}{2}}^{\frac{W}{2}} dI(i\cdot\cos(\phi+\tfrac{\pi}{2}), i\cdot\sin(\phi+\tfrac{\pi}{2})) \tag{3}$$

Fig. 5(b) shows calculated $\text{fitness}_{\text{ridge}}(\phi)$ in the range of 0~180° and it can be approximated by a cosine function whose frequency f_0 is 1/180° like equation (4).

To eliminate the effect of noise, estimated phase parameter is used to estimate the ridge direction like equation (5). In general, amplitude and phase parameter can be estimated by MLE (Maximum Likelihood Estimation) [12]. Estimated cosine function in Fig. 5(b) shows that the maximum value of $fitness_{ridge}(\phi)$ can be robustly detected.

$$fitness_{ridge}[n] = A \cdot \cos(2\pi f_0 n + \psi) + w[n]$$
$$\text{where, } n: \text{ integer index of } \phi,\ w[n]: \text{white gausian noise} \tag{4}$$

$$\phi_{ridge} = -\frac{\hat{\psi}}{2\pi f_0}, \text{ where } \hat{\psi} = \tan^{-1}\left(\frac{-\sum_{n=0}^{179} fitness_{ridge}[n] \cdot \sin(2\pi f_0 n)}{\sum_{n=0}^{179} fitness_{ridge}[n] \cdot \cos(2\pi f_0 n)} \right) \tag{5}$$

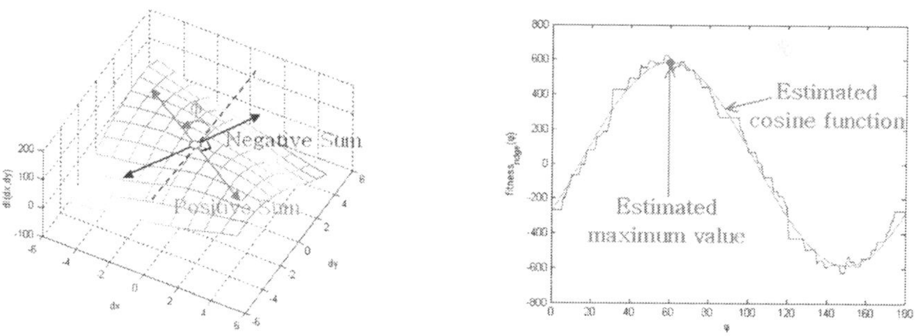

(a) dI(dx,dy) in 3D display

(b) Estimated maximum value of $fitness_{ridge}(\phi)$

Fig. 5. Estimation of line-segment direction by model based fitness estimation

Edge following, starting from the detected cross-point in the estimated edge direction, refines the edge direction and eliminates falsely detected cross-points. Edge position estimate of n+1 step can be calculated by cross-point $\mathbf{p}_{edge}[0]$ and edge direction of n step $\mathbf{u}_{edge}[n]$ like equation (6). Finding maximum of local directional intensity-gradient dI(t), defined like equation (7), updates the edge position of n+1 step like equation (8). $\mathbf{n}_{edge}[n]$ is the unit vector normal to $\mathbf{u}_{edge}[n]$ and $t_{max}[n]$ is the relative position maximizing dI(t) in $\mathbf{n}_{edge}[n]$ direction as shown in Fig. 6(a). Iterating edge following terminates if new edge strength $dI(t_{max}[n+1])$ is definitely smaller than the edge strength of cross point $dI(t_{max}[0])$, e.g. 70%. Proposed system rejects detected cross-points of which successful edge following iteration is smaller than a certain threshold $\theta_{\text{edge following}}$ to eliminate falsely detected marking line-segments. Consequently, refined edge direction $\mathbf{u}_{edge}[n+1]$ is set to a unit vector from $\mathbf{p}_{edge}[0]$ to $\mathbf{p}_{edge}[n+1]$. Fig. 6(b) shows the edge following results and refined direction.

$$\hat{\mathbf{p}}_{\mathbf{edge}}[n+1] = \mathbf{p}_{\mathbf{edge}}[0] + (n+1) \cdot ds \cdot \mathbf{u}_{\mathbf{edge}}[n] \tag{6}$$

$$dI(t) = dI(\hat{\mathbf{p}}_{\mathbf{edge}}[n+1] + t \cdot \mathbf{n}_{\mathbf{edge}}[n], \mathbf{n}_{\mathbf{edge}}[n]), \text{ where } t: -\tfrac{W}{2} \sim \tfrac{W}{2} \tag{7}$$

$$\mathbf{p}_{\mathbf{edge}}[n+1] = \hat{\mathbf{p}}_{\mathbf{edge}}[n+1] + t_{\max}[n+1] \cdot \mathbf{n}_{\mathbf{edge}}[n] \tag{8}$$

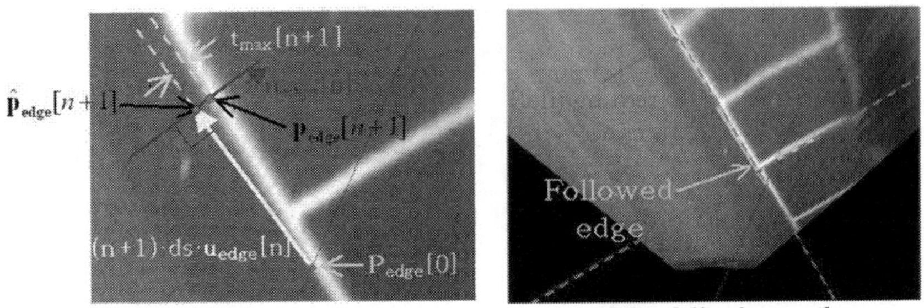

(a) Edge following method (b) Edge following results

Fig. 6. Edge following refines the direction of detected marking line-segments

3.3 Determination of Guideline

If the seed-point designated by driver is locating in a valid parking-slot, gaze-line from seed-point to camera should meet marking line-segments more than once making corresponding cross-points. Among marking line-segments validated by the edge following, guideline is the marking line-segment of which cross-point is nearest to the camera position. In other words, guideline has smallest distance between cross-point and camera, i.e. $|\mathbf{p}_{\mathbf{camera}}\text{-}\mathbf{p}_{\mathbf{edge}}[0]|$. Fig. 7 shows recognized guideline.

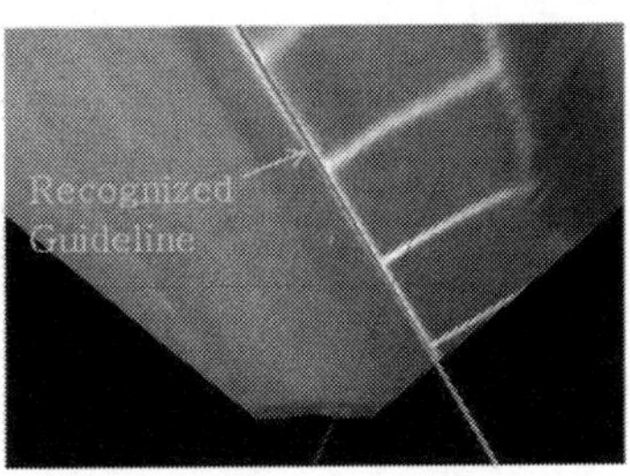

Fig. 7. Guideline recognized by structural relation between cross-points

4 Recognition of Target Parking-Slot

By searching separating line-segments bi-directionally from selection point $\mathbf{p}_{\mathbf{selection}}$ that is obtained by projecting the seed-point onto the guideline, proposed system recognizes the exact location of target parking-slot. $\mathbf{p}_{\mathbf{selection}}$ is calculated by cross-point $\mathbf{p}_{\mathbf{cross}}$ and guideline unit vector $\mathbf{u}_{\mathbf{guideline}}$ like equation (9).

$$\mathbf{p}_{\mathbf{selection}} = \mathbf{p}_{\mathbf{cross}} + \left(\mathbf{u}_{\mathbf{guideline}} \cdot (\mathbf{p}_{\mathbf{seed\ point}} - \mathbf{p}_{\mathbf{cross}})\right)\mathbf{u}_{\mathbf{guideline}} \tag{9}$$

$$I_{on}(s) = \frac{1}{\frac{W}{2}} \sum_{t=0}^{\frac{W}{2}} I(\mathbf{p}_{\mathbf{selection}} + s \cdot \mathbf{u}_{\mathbf{searching}} + t \cdot \mathbf{n}_{\mathbf{guideline}}) \tag{10}$$

$$I_{off}(s) = \frac{1}{\frac{W}{2}} \sum_{t=\frac{3}{2}W}^{2W} I(\mathbf{p}_{\mathbf{selection}} + s \cdot \mathbf{u}_{\mathbf{searching}} + t \cdot \mathbf{n}_{\mathbf{guideline}}) \tag{11}$$

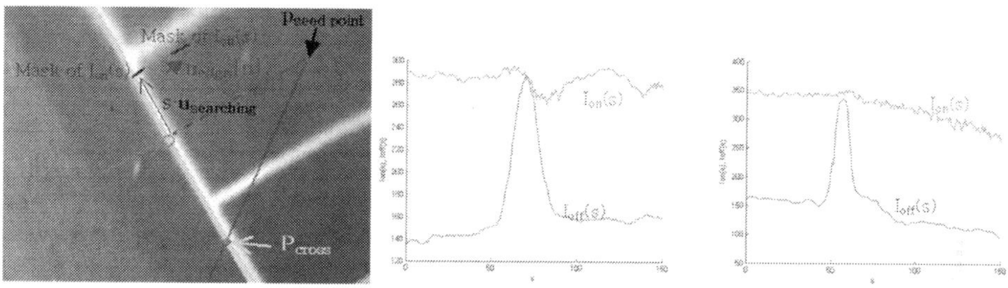

Fig. 8. Measuring $I_{on}(s)$ and $I_{off}(s)$ bi-directionally to find both-side 'T'-shape junctions

Searching 'T' -shape junction between guideline and separating line-segment can detect the position of the separating line-segment. Average intensity on the guideline marking, $I_{on}(s)$, is measured like equation (10) and the average intensity of neighboring region outward from camera, $I_{off}(s)$, is measured like equation (11). Here, $\mathbf{u}_{\mathbf{searching}}$ is either $\mathbf{u}_{\mathbf{guideline}}$ or $-\mathbf{u}_{\mathbf{guideline}}$ according to the search direction. Fig. 8(a) depicts the procedure of measuring $I_{on}(s)$ and $I_{off}(s)$. Fig. 8(b) and (c) shows the measured $I_{on}(s)$ and $I_{off}(s)$ in both directions. It can be observed that $I_{off}(s)$ is similar to $I_{on}(s)$ only around 'T'-shape junction. Therefore, the location of junction can be detected by thresholding the ratio of $I_{off}(s)$ to $I_{on}(s)$, named $L_{separating}(s)$. Fig. 9(a) and (b) shows detected junction and Fig. 9(c) shows recognized target parking-slot.

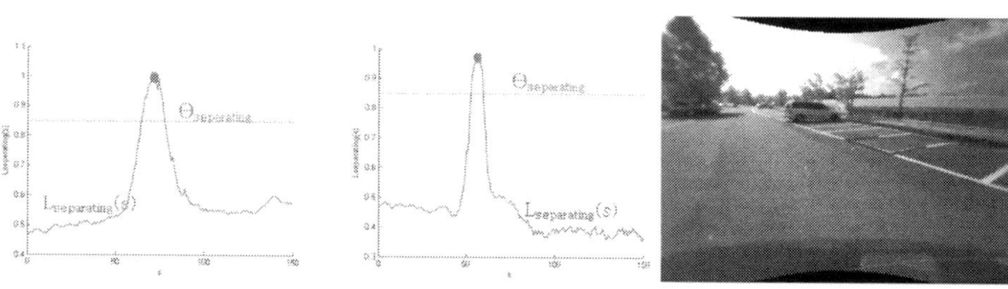

Fig. 9. $L_{separating}(s)$ can find separating line-segments irrespective of local intensity variation

5 Experimental Results and Conclusion

In bird's eye view image, objects above the ground surface are projected outward from camera. Therefore, only if guideline and the 'T'-shape junctions of target parking-slot are observed, proposed method can successfully detect cross-points,

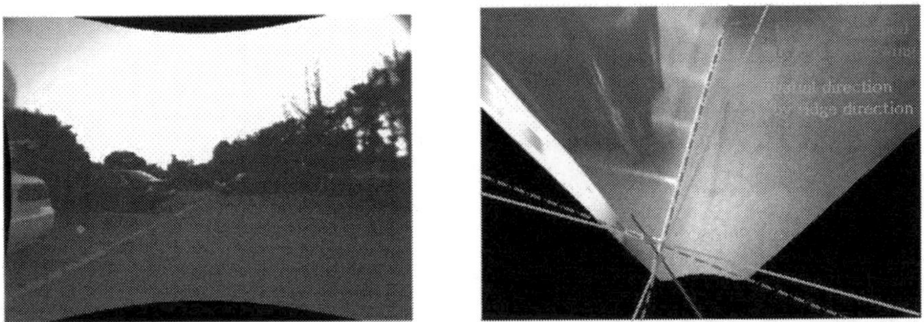

(a) Recognized target parking-slot (b) Direction refinement by edge following

Fig. 10. Case with adjacent vehicles and torn markings

related marking line-segments and separating line-segments as shown in Fig. 10. Fig 10(a) is captured against the light and markings are torn to be noisy. In Fig. 10(b), edge following based edge direction refinement overcomes the error of initial direction estimation.

Separating line-segment detection method considering locally changing illumination condition can successfully detect target parking-slot even if local intensities are different from each other. Fig. 11 shows that $L_{separating}(s)$ can compensate local intensity variation.

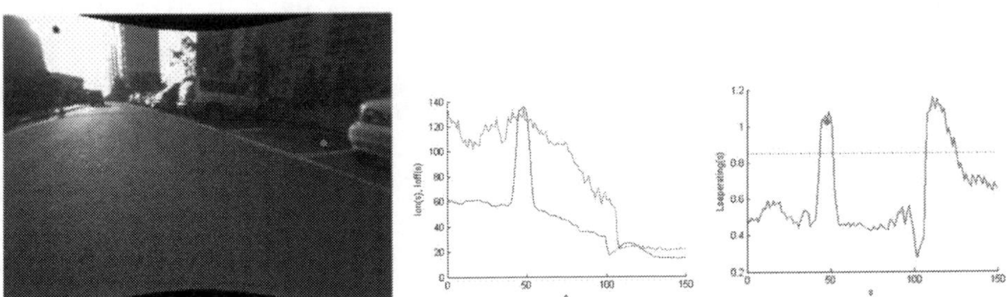

(a) Recognized target parking-slot (b) $I_{on}(s)$ and $I_{off}(s)$ (c) $L_{separating}(s)$

Fig. 11. Case with strong sunshine causing locally changing intensity

Major contribution of this paper is that because parked vehicles are projected outward from camera in bird's eye view image, if guideline and 'T'-shape junctions are observed, guideline can be detected simply by finding the cross-point nearest from camera. Because proposed method uses small portion of input image and fully utilizes structural characteristics of parking-slot markings in bird's eye view image, it can achieve concise implementation and deterministic performance.

References

1. Richard Bishop: Intelligent Vehicle Technology and Trends, Artech House Pub. (2005)
2. Randy Frank: Sensing in the Ultimately Safe Vehicle, SAE Paper No.: 2004-21-0055, Society of Automotive Engineers (2004)

3. Alexander Schanz, et al.: Autonomous Parking in Subterranean Garages – A Look at the Position Estimation, IEEE Intelligent Vehicle Symposium (2003), 253-258
4. Massaki Wada, et al.: Development of Advanced Parking Assistance System, IEEE Trans. Industrial Electronics, Vol. 50, No. 1 (2003), 4-17
5. Masayuki Furutani: Obstacle Detection Systems for Vehicle Safety, SAE Paper No.: 2004-21-0057, Society of Automotive Engineers (2004)
6. Nico Kaempchen, et al.: Stereo Vision Based Estimation of Parking Lots Using 3D Vehicle Models, IEEE Intelligent Vehicle Symposium (2002), 459-464 vol.2
7. Jin Xu, et al.: Vision-Guided Automatic Parking for Smart Car, IEEE Intelligent Vehicle Symposium (2000), 725-730
8. Ho Gi Jung, et al.: Stereo Vision Based Localization of Free Parking Site, CAIP 2005, LNCS 3691 (2005), 231-239
9. C. Vestri, et al.: Evaluation of a Vision-Based Parking Assistance System, IEEE Conf. Intelligent Transportation Systems (2005), 56-60
10. J. Salvi, et al.: A Comparative Review of Camera Calibrating Methods with Accuracy Evaluation, Pattern Recognition 35 (2002), 1617-1635
11. J. Y. Bouguet: Camera Calibration Toolbox for Matlab, http://www.vision.caltech.edu/bouguetj/calib_doc/index.html
12. Steven M. Key: Fundamentals of Statistical Signal Processing, Volume I: Estimation Theory, Prentice Hall Inc. (1993), 193-195

A Fast and Exact Modulo-Distance Between Histograms

Francesc Serratosa[1] and Alberto Sanfeliu[2]

[1] Universitat Rovira I Virgili, Dept. d'Enginyeria Informàtica i Matemàtiques, Spain
francesc.serratosa@.urv.net
[2] Universitat Politècnica de Catalunya, Institut de Robòtica i Informàtica Industrial, Spain
sanfeliu@iri.upc.es

Abstract. The aim of this paper is to present a new method to compare modulo histograms. In these histograms, the type of elements are cyclic, for instance, the hue in colour images. The main advantage is that there is an important time-complexity reduction respect the methods presented before. The distance between histograms that we present is defined on a structure called *signature*, which is a lossless representation of histograms.

We show that the computational cost of our distance is $O(z'^2)$, being z' the number of non-empty bins of the histograms. The computational cost of the algorithms presented in the literature depends on the number of bins of the histograms. In most of the applications, the obtained histograms are sparse, then considering only the non-empty bins makes the time consuming of the comparison drastically decrease.

The distance and algorithms presented in this paper are experimentally validated on the comparison of images obtained from public databases.

1 Introduction

A histogram of a set with respect to a measurement represents the frequency of quantified values of that measurement among the samples. Finding the distance or similarity between histograms is an important issue in pattern classification or clustering and image retrieval. For this reason, a number of measures of similarity between histograms have been proposed and used in computer vision and pattern recognition. Protein classification is one of the common histogram applications [9]. Moreover, if the ordering of the elements in the sample is unimportant, the histogram obtained from this set is a lossless representation of it and can be reconstructed from its histogram. Then, we can compute the distance between sets in an efficient way by computing the distance between their histograms.

The probabilistic approaches use histograms based on the fact that the histogram of a measurement provides the basis for an empirical estimate of the probability density function [1]. Computing the distance between probability density functions can be regarded as the same as computing the Bayes probability. This is equivalent to measuring the overlap between probability density functions as the distance. The *B-distance* [2], proposed by Kailath, measures the distance between populations. It is a value between 0 and 1 and provides bounds on the Bayes misclassification probability. An approach closely related to the *B-distance* was proposed by Matusita [3]. Finally, Kullback generalised the concept of probabilistic uncertainty or

D.-Y. Yeung et al. (Eds.): SSPR&SPR 2006, LNCS 4109, pp. 394–402, 2006.

"entropy" and introduced the *K-L-distance* measure [1,4] that is the minimum cross entropy.

Most of the distance measures presented in the literature (there is an interesting compilation in [5]) consider the overlap or intersection between two histograms as a function of the distance value but they do not take into account the similarity on the non-overlapping parts of the two histograms. For this reason, Rubner presented in [6] a new definition of the distance measure between histograms that overcomes this non-overlapping parts problem. It was called Earth Mover's Distance and it is defined as the minimum amount of work that must be performed to transform one histogram into the other one by moving distribution mass. They used the simplex algorithm [8] to compute the distance measure and the method presented in [7] to search a good initialisation.

We consider three types of measurements called nominal, ordinal and modulo. In a nominal measurement, each value of the measurement is a name an there is not any relation between them such as great than or lower than (e.g. the names of the students). In an ordinal measurement, the values are ordered (e.g. the age of the students). Finally, in the modulo measurement, measurement values are ordered but form a ring due to the arithmetic modulo operation (e.g. the angle in a circumference).

Cha presented in [5] three new algorithms to obtain the distance between one-dimensional histograms that use the Earth Mover's Distance. These algorithms computed the distance between histograms when the type of measurements where *nominal*, *ordinal* and *modulo* in $O(z)$, $O(z)$ and $O(z^2)$ respectively, being z the number of levels or bins.

Often, for specific set measurements, only a small fraction of the *bins* in a histogram contain significant information, that is, most of the *bins* are empty. This is more frequent when the dimensions of the element domain increase. In that cases, the methods that use histograms as fixed-sized structures obtain poor efficiency. For this reason, Rubner [6] presented the variable-size descriptions called *signatures*. In that representations, the empty bins where not explicitly considered.

Another method used to reduce the dimensionality of the data in the case that the statistical properties of the data are a priori known was shown in [10]. The similarity measures are improved by the smoothing projections that are applicable for reduction of the dimensionality of the data and also to represent sparse data in a more tight form in the projection subspace.

We presented in [12] the definition of the nominal, ordinal and modulo distances between histograms in which, only the non-empty bins where considered. In [11], the algorithms of these distances where shown, demonstrated and validated.

In this paper, we present the algorithm to compute the modulo distance between histograms that the computational cost depends only on the non-empty bins instead of the number of bins as it is in the algorithms presented in [5,6]. The time saving of our modulo-distance algorithm is higher than our nominal-distance or ordinal-distance algorithms due to the computational cost is quadratic instead of lineal.

The subsequent sections are constructed as follows. First, we define the histograms and signatures. Then in section 3 we define the modulo distance between signatures. In section 4, we depict the basic algorithm to compute the modulo distance between signatures. In section 5, we use our method to compare images obtained from databases. Finally, we conclude with emphasis of the advantage of using our distance between signatures and using the proposed algorithm.

2 Histograms and Signatures

In this section, we formally give a definition of histograms and signatures. The section finishes with a simple example to show the representations of the histograms and signatures given a set of measurements.

2.1 Histogram Definition

Let x be a measurement which can have one of T values contained in the set $X=\{x_1,...x_T\}$. Consider a set of n elements whose measurements of the value of x are $A=\{a_1,...a_n\}$ where $a_t \in X$.

The histogram of the set A along measurement x is $H(x,A)$ which is an ordered list consisting of the number of occurrences of the discrete values of x among the a_t. As we are interested only in comparing the histograms and sets of the same measurement x, $H(A)$ will be used instead of $H(x,A)$ without loss of generality. If $H_i(A)$, $1 \le i \le T$, denotes the number of elements of A that have value x_i, then $H(A)=[H_1(A), ...,H_T(A)]$ where

$$H_i(A) = \sum_{t=1}^{n} C_{i,t}^{A} \quad (1)$$

and the individual costs are defined as

$$C_{i,t}^{A} = \begin{cases} 1 & if\ a_t = x_i \\ 0 & otherwise \end{cases} \quad (2)$$

The elements $H_i(A)$ are usually called *bins* of the histogram.

2.2 Signature Definition

Let $H(A)=[H_1(A), ...,H_T(A)]$ and $S(A)=[S_1(A), ...,S_z(A)]$ be the histogram and the signature of the set A, respectively. Each $S_k(A)$, $1 \le k \le z \le T$ is composed by a pair of terms, $S_k(A)=\{w_k, m_k\}$. The first term, w_k, shows the relation between the signature $S(A)$ and the histogram $H(A)$. Thus, if the $w_k=i$ then the second term, m_k, is the number of elements of A that have value x_i, that is, $m_k=H_i(A)$ where $w_k<w_t \Leftrightarrow k<t$ and $m_k>0$.

The signature of a set is a lossless representation of its histogram in which the *bins* of the histogram that has value 0 are not expressed implicitly. From the signature definition, we obtain the following expression,

$$H_{w_k}(A) = m_k \quad where \quad 1 \le k \le z \quad (3)$$

2.3 Extended Signature

The **extended signature** is a signature in which the minimum number of empty bins has been added to assure that, given a pair of signatures to be compared, the number of bins is the same. Moreover, each bin in both signatures represents the same bin in the histograms.

2.4 Example

In this section we show a pair of sets with their histogram and signature representations. This example is used to explain the distance measures in the next sections. Figure 1 shows the sets *A* and *B* and their histogram representations. Both sets have 10 elements and values are contained from 1 to 8. Horizontal axis in the histograms represents the values of the elements and the vertical axis represents the number of elements that have this value, that is m_i. Empty bins are the ones that $m_i=0$.

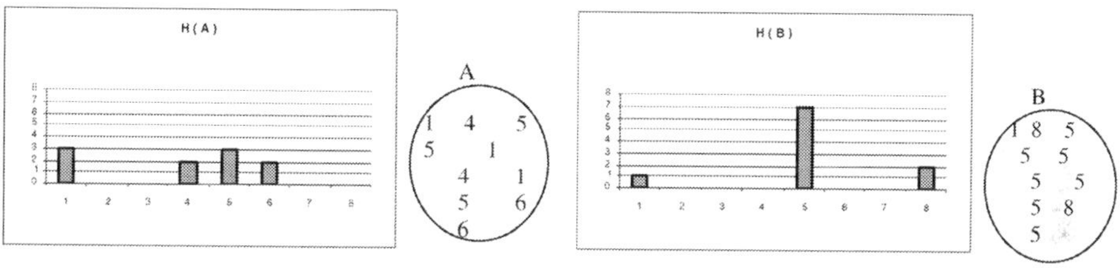

Fig. 1. Sets *A* and *B* and its histograms

Figure 2 shows the signature representation of the sets *A* and *B*. The length of the signatures is 4 and 3, respectively. The vertical axis represents the number of elements of each bin and the horizontal axis represents the bins of the signature. The set A has 2 elements with value 6 since this value is represented by the bin 4 ($W_4^A=6$) and the value of the vertical axis is 2 at bin 4. In the signature representation there is not any empty bin.

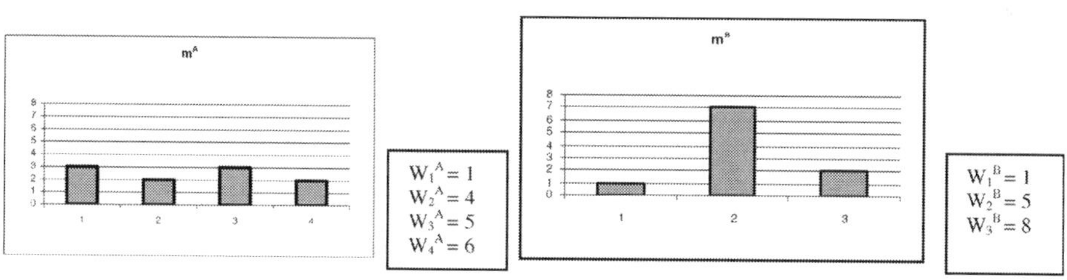

Fig. 2. Signature representation of the sets *A* and *B*

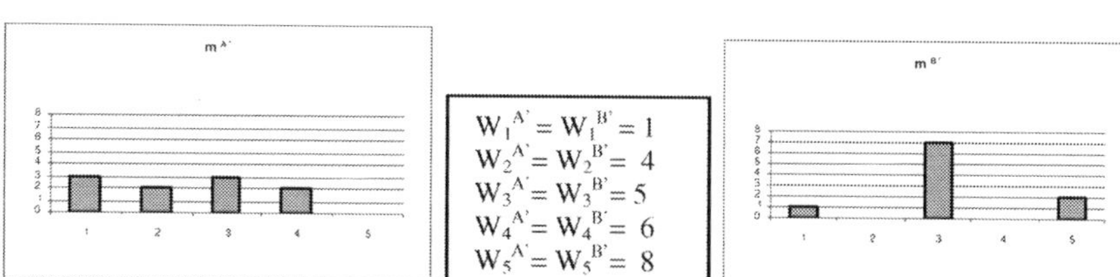

Fig. 3. Extended Signatures *A'* and *B'*. The number of elements m_i are represented graphically and the value of its elements is represented by w_i.

Figure 3 shows the extended signatures of the sets *A* and *B* with *5* bins. Note that the value that the extended signatures represent for each bin, w_i, is the same for both signatures. Moreover, in *A'* and *B'*, one and two empty bins have been added, respectively.

3 Modulo Distance Between Signatures

The aim of this section is to present the new distance between signatures. To do so, we first show de definition of the distance between histograms and then we move on the new definition of the distance between signatures. The algorithms used to obtain the extended signatures and the distances are described in the algorithms section.

For the following definition of the distance and also for the algorithms section, we assume that the extended signatures of *S(A)* and *S(B)* are *S(A')* and *S(B')*, respectively, where $S_i(A')=\{w_i^{A'},m_i^{A'}\}$ and $S_i(B')=\{w_i^{B'},m_i^{B'}\}$. The number of bins of *S(A)* and *S(B)* is z^A and z^B and the number of bins of both extended signatures is *z'*.

The distance value between two modulo measurement values is the interior difference of each element (see [5] for the proofs the metric property).

$$d_{\mathrm{mod}}(a,b)=\begin{cases}|a-b| & \text{if } |a-b|\le T/2\\ T-|a-b| & \text{otherwise}\end{cases} \tag{4}$$

The modulo distance between two histograms was presented in [6] as the minimum of work needed to transform one histogram to the other. Histogram *H(A)* can be transformed into histogram *H(B)* by moving elements to left or right and the total of all necessary minimum movements is the distance between them. There are two operations. Suppose an element *a* that belong to the bin *i*. One operation is *move left (a)*. This operation results that the element *a* belong to bin *i-1* and the cost to do so is 1. Another operation is *move right (a)*. Similarly, after the operation, *a* belongs to the bin *i+1* and the cost is 1. These operations are graphically represented by right-to-left arrows and left-to-right arrows.

In a modulo type histograms, the first bin and the last bin are considered to be adjacent to each other, and hence, it forms a closed circle, due to the nature of the data type. Transforming a modulo type histogram to another while computing their distance

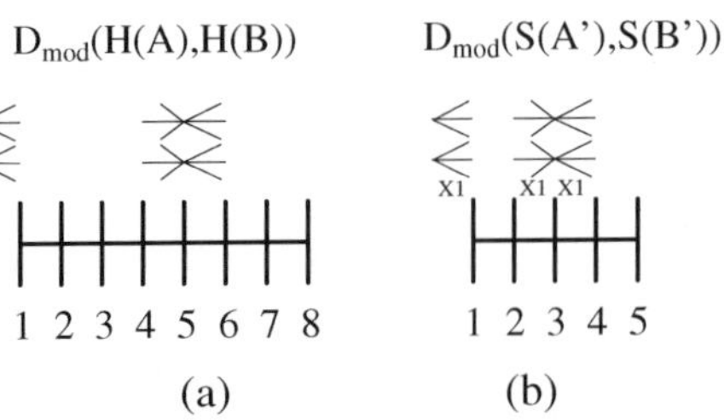

Fig. 4. Arrow representation of the modulo distance using (a) histograms and (b) signatures

should allow cells to move from the first bin to the last one or vice versa at a cost of a single movement. Thus, cells or blocks of earth can move from the fist bin to the last bin with the operation *move left (1)*. Similarly, blocks can move from the last bin to the first one with the operations *move right (T)*.

Figure 4 shows the arrows needed to transform (a) histogram *H(A)* to histogram *H(B)* and (b) the extended signature *S(A')* to *S(B')*.

The main difference between the histogram and signature case is that in the second one we have to take into consideration that the difference between bins is not constant. Our arrows have not a constant size (or constant cost) but they depend on the distance between bins. If element *a* belongs to the bin *i*, the operation *move left (a)* results that the element *a* belong to bin *i-1* and the cost to do so is $w_i - w_{i-1}$. Similarly, after the operation *move right(a)*, the element *a* belongs to the bin *i+1* and the cost is $w_{i+1} - w_i$.

The costs of the last and first movements are the addition of three terms. (a) The cost from the last bin of the signature, $w_{z'}$, to the last bin of the histogram, *T*. (b) The cost from the last bin of the histogram, *T*, to the first bin of the histogram, *1*. (c) The cost from the first bin of the histogram, *1*, to the first bin of the signature, w_1. Then, the costs are calculated as the length of these terms. The cost of (a) is $T-w_{z'}$, the cost of (b) is *1* (similarly to the cost between histograms) and the cost of (c) is w_1-1. Therefore, the final cost from the last bin to the first one or vice versa between signatures is $w_1-w_{z'}+T$.

Due to the modulo properties explained before, we can transform one signature or histogram into another one in several ways. Among these ways, there exists a minimum distance whose number of movements (or the cost of the arrows and the number of arrows) is the lowest. If there is a border line between bins that has both directional arrows, they are cancelled out. These movements are redundant and so the distance cannot be obtained through this configuration of arrows. To find the minimum configuration of arrows, we can add a complete chain in the histogram or signature of same directional arrows, then the opposite arrows on the same border between bins are cancelled out.

The modulo distance between signatures is defined as follows,

$$D_{\text{mod}}\left(S(A),S(B)\right)=\min_{c}\left\{\sum_{i=1}^{z'-1}\left[\left(w_{i+1}^{A'}-w_{i}^{A'}\right)\left|c+\sum_{j=1}^{i}\left(m_{j}^{A'}-m_{j}^{B'}\right)\right|\right]+\left(w_{1}^{A'}-w_{z'}^{A'}+T\right)\left|c\right|\right\} \quad (5)$$

The cost of the movement of blocks from the first bin to the last one or viceversa is $w_1-w_{z'}+T$ and the costs of the other movements is $w^{A'}_{i+1}-w^{A'}_{i}$. Moreover, *c* represents the chains of left arrows or right arrows added to the current arrow representation. The absolute value of *c* at the end of the expression is the number of chains added to the current representation. It is multiplied by the cost of the arrows from the last bin to the first one or vice versa.

Example. Figure 5 shows five different transformations of signature *S(A)* to signature *S(B)* and their related costs. The cost is the number of arrows multiplied by the length of the arrows (shown under the arrows). In the first transformation, one chain of right arrows are added (c=1). In the second one, no chains are added (c=0), thus the cost is the same than the ordinal distance. In the third to the last ones, 1, 2 and 3 chains of left arrows are added, respectively. We can see that the minimum cost is 6 and it is the case that c=-2, then the distance value is 6 for the modulo distance and 14 for the ordinal distance.

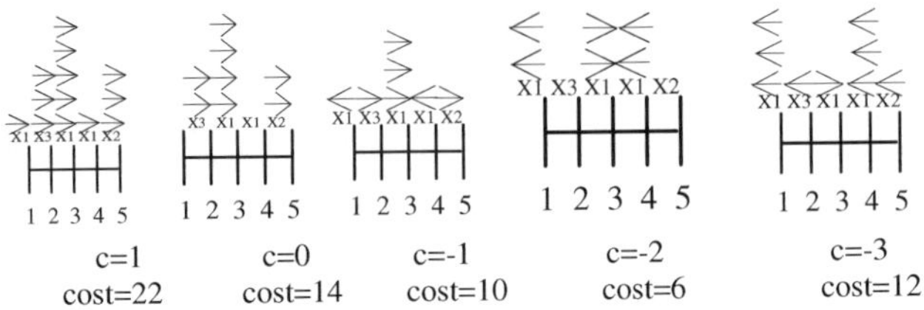

Fig. 5. Five different transformations of signature *S(A)* to the signature *S(B)* with their related *c* and the obtained cost

4 Algorithm

The process *Modulo_Distance* obtains the modulo distance of two signatures. Given two signatures, the process *Extended_Signature* obtains two minimum extended signatures in *O(z)* (the algorithm was presented in [11]).

```
Dmod = Modulo_Distance {S(A),S(B)}
{S(A'),S(B'),z'} = Extended_Signature {S(A),S(B)}
Dmod = 0 p[0] = m0^A' - m0^B'
for (i = 2 to z') p[i] = mi^A' - mi^B' + p[i-1]
for (i = 1 to z-1') Dmod += (wi+1^A' - wi^A') * abs(p[i])
do
    D2=0
    c = min positive {p[i] for 1≤i≤z'}
    Temp[i]=p[i]-c for 1≤i≤z'
    for (i = 1 to z'-1) D2 += (wi+1^A' - wi^A') * abs(Temp[i])
    if (Dmod > D2) Dmod = D2
    p[i]= Temp [i] for 1≤i≤z'
while(Dmod > D2)
do
    D2=0
    c = max negative {p[i] for 1≤i≤z'}
    Temp[i]=p[i]-c for 1≤i≤z'
    for (i = 1 to z'-1) D2 += (wi+1^A' - wi^A') * abs(Temp[i])
    if (Dmod > D2) Dmod = D2
    p[i]= Temp [i] for 1≤i≤z'
while(Dmod > D2)
```

Correctness of the Procedure

The arrow representation of minimum distance can be achieved from any arbitrary valid arrow representation by combination of two basic operations: Increasing the chains of right arrows (when the value of *c* is positive) or increasing the chains of left arrows (when the value of *c* is negative). The distance value can increase infinitely but there exists only one minimum among valid representations. In order to reach to the minima, first the algorithms tests for increasing positively c if whether it gives higher or lower distance value. If the distance reduces, keep applying the operations until no more reduction occurs. Then, the algorithms does the same operations but increasing negatively c. With these two actions, the algorithm guarantees that all possible combinations of correct representations of arrows are tested.

The procedure runs in $O(z'^2)$ time. The lines 1 to 3 obtain the ordinal distance. In the lines 4 – 11 chains of right arrows are added to the current arrow representation until there is no more reduction to the total number of arrows. This increment is considered in the algorithm by the variable c. Next, chains of left arrows are added in the similar manner (lines 12 – 19).

5 Validation of the Method and Algorithm

The method and algorithms presented in this paper are applied on histograms, independently on the kind of the original set from which they have been obtained, i.e. images [13], discretized probability-density functions [14],... The only condition to use our method is to know the type of elements of the original set: ordinal, nominal or modulo.

Table 1. Hue 2^8 bins. Modulo histogram

	Length	Increase Speed	Correct.	Decrease Correct.
Histo.	265	1	86%	1
Signa.	215	1.23	86%	1
Signa. 100	131	2.02	85%	0.98
Signa. 200	95	2.78	73%	0.84
Signa. 300	45	5.88	65%	0.75

Table 2. Hue 2^{16} bins. Modulo histogram

	Length	Increase Speed	Correct.	Decrease Correct.
Histo.	65,536	1	89%	1
Signa.	205	319.68	89%	1
Signa. 1	127	516.03	89%	1
Signa. 2	99	661.97	78%	0.87
Signa. 3	51	1285.01	69%	0.77

To show the validity of our new method, we have tested the modulo distance between histograms and between signatures. We used 1000 images (640 x 480 pixels) obtained from public databases. To validate the modulo distance, the histograms represent the hue coordinate with 2^8 levels (table 1) and with 2^{16} levels (table 2). Each of the tables below shows the results of 5 different tests. In the first and second files of the tables, the distance where computed between histograms and signatures, respectively. In the other three, the distance was computed between signatures but, with the aim of reducing the length of the signature (and so to increase the speed), the bins that had less elements than 100, 200 or 300 in table 1 and less elements than 1, 2 or 3 in table 2 where removed. The first column is the number of bins of the histogram (first cell) or signatures (the other four cells). The second column represents the increase of speed if we use signatures respect histograms. It is calculated as the ratio between the run time of the histogram method and the signature method. The third column is the average correctness. The last column represents the decrease of correctness due to using the signatures with filtered histograms. It is obtained as the ratio of the correctness of the histogram by the correctness of each filter.

Tables 1 and 2 show us that our method is much useful when the number of bins increases since the number of empty bins tends to increase. Note that in the case of the first filter (third experiment in the tables), there is no decrease in the correctness although there is much increase in the speed respect the signature method (second experiment in the tables).

6 Conclusions and Future Work

We have presented the modulo distance between signatures and the algorithm used to compute it. We have shown that signatures are a lossless representation of histograms and that computing the distance between signatures is the same than between histograms but with a lower computational time. We have validated this new algorithm with a huge amount of real images and we have realised that there is an important time saving do to most of the histograms are sparse. Moreover, if we apply filtering techniques on the histograms, the number of bins of the signatures reduces and so the run time of their comparison.

References

1. R.O. Duda, P.E. Hart & D.G. Stork, *Pattern Classification*, 2nd edition, Wiley, New York, 2000.
2. T. Kailath, "The divergence and bhattacharyya distance measures in signal selection", *IEEE Transactions Community Technol.* COM-15, 1, pp:52-60, 1967.
3. K. Matusita, "Decision rules, based on the distance, for problems of fit, two samples and estimation", *Annals Mathematic Statistics*, 26, pp: 631-640, 1955.
4. J.E. Shore & R.M. Gray, "Minimum cross-entropy pattern classification and cluster analysis", Transactions on Pattern Analysis and Machine Intelligence, 4 (1), pp: 11-17, 1982.
5. S.-H. Cha, S. N. Srihari, "On measuring the distance between histograms" *Pattern Recognition* 35, pp: 1355–1370, 2002.
6. Y. Rubner, C. Tomasi, and L. J. Guibas, "A Metric for Distributions with Applications to Image Databases" *International Journal of Computer Vision* 40 (2), pp: 99-121, 2000.
7. E. J. Russell. "Extension of Dantzig's algorithm to finding an initial near-optimal basis for the transportation problem", *Operations Research*,17,pp:187-191, 1969.
8. *Numerical Recipes in C: The Art of Scientific Computing*, ISBN 0-521-43108-5.
9. Y-P Nieh & K.Y.J. Zhang, "A two-dimensional histogram-matchinig method for protein phase refinement and extension", Biological Crystallography, 55, pp:1893-1900, 1999.
10. J.-K. Kamarainen, V. Kyrki, J. Llonen, H. Kälviäinen, "Improving similarity measures of histograms using smoothing projections", *Pattern Recognition Letters* 24, pp: 2009–2019, 2003.
11. F. Serratosa & A. Sanfeliu, "Signatures versus Histograms: Definitions, Distances and Algorithms", *Pattern Recognition* (39), Issue 5, pp. 921-934, 2006.
12. F. Serratosa & A. Sanfeliu, "A fast distance between histograms", *Lecture Notes and Computer Science 3773, pp: 1027 - 1035*, 2005
13. M. Pi, M.K. Mandal, A. Basu, "Image retrieval based on histogram of fractal parameters", *Multimedia, IEEE Transactions on*, Vol. 7 (4), 2005, pp. 597 – 605.
14. F. Serratosa, R. Alquézar y A. Sanfeliu, "Function-Described Graphs for modeling objects represented by attributed graphs", *Pattern Recognition*, 36 (3), pp. 781-798, 2003.

Using Learned Conditional Distributions as Edit Distance*

Jose Oncina[1] and Marc Sebban[2]

[1] Dep. de Lenguajes y Sistemas Informáticos, Universidad de Alicante (Spain)
oncina@dlsi.ua.es
[2] EURISE, Université de Saint-Etienne, (France)
marc.sebban@univ-st-etienne.fr

Abstract. In order to achieve pattern recognition tasks, we aim at learning an *unbiased* stochastic edit distance, in the form of a finite-state transducer, from a corpus of (*input,output*) pairs of strings. Contrary to the state of the art methods, we learn a transducer independently on the marginal probability distribution of the *input* strings. Such an unbiased way to proceed requires to optimize the parameters of a *conditional* transducer instead of a *joint* one. This transducer can be very useful in pattern recognition particularly in the presence of noisy data. Two types of experiments are carried out in this article. The first one aims at showing that our algorithm is able to correctly assess simulated theoretical target distributions. The second one shows its practical interest in a handwritten character recognition task, in comparison with a standard edit distance using *a priori* fixed edit costs.

1 Introduction

Many applications dealing with sequences require to compute the similarity of a pair (*input,output*) of strings. A widely-used similarity measure is the well known *edit distance*, which corresponds to the minimum number of operations, *i.e. insertions, deletions*, and *substitutions*, required to transform the *input* into the *output*. If this transformation is based on a random phenomenon and then on an underlying probability distribution, edit operations become random variables. We call then the resulting similarity measure, the *stochastic edit distance*.

An efficient way to model this distance consists in viewing it as a stochastic transduction between the input and output alphabets [1]. Stochastic finite-state transducers suffer from the lack of a training algorithm. To the best of our knowledge, the first published algorithm to automatically learn the parameters of a stochastic transducer has been proposed by Ristad and Yianilos [2,1]. They provide a stochastic model which allows us to learn a stochastic edit distance, in the form of a memoryless transducer (*i.e.* with only one state), from a corpus of similar examples, using the Expectation Maximization (EM) algorithm. During the last few years, the algorithm EM has also been used for learning other transducer-based models [3,4,5].

* This work was supported in part by the IST Programme of the European Community, under the PASCAL Network of Excellence, IST-2002-506778. This publication only reflects the authors' views.

D.-Y. Yeung et al. (Eds.): SSPR&SPR 2006, LNCS 4109, pp. 403–411, 2006.

Ristad and Yianilos define the stochastic edit distance between two strings x and y as (the minus logarithm of) the *joint* probability of the pair (x, y). In this paper, we claim that it would be much more relevant to express the stochastic edit distance from a *conditional* probability.

First, in order to correctly compute the edit distance, we think that the probabilities of edit operations over a symbol must be independent of those computed over another symbol. In other words, if the transformation of a string x into another one y does not require many edit operations, it is expected that the probability of the substitution of a symbol by itself should be high. But, as the sum of the probabilities of all edit operations is one, then the probability of the substitution of another symbol by itself can not obviously be too large. Thus, by using a joint distribution (summing to 1), one generates an awkward dependence between edit operations.

Moreover, we think that the primitive edit costs of the edit distance must be independent of the *a priori* distribution $p(x)$ of the input strings. However, $p(x)$ can be directly deduced from the joint distribution $p(x, y)$, as follows: $p(x) = \sum_{y \in Y^*} p(x, y)$, where Y^* is the set of all finite strings over the output alphabet Y. This means that this information is totally included in the joint distribution. By defining the stochastic edit distance as a function of the joint probability, as done in [1], the edit costs are then dependent of $p(x)$. However, if we use a conditional distribution, this dependence is removed, since it is impossible to obtain $p(x)$ from $p(y|x)$ alone.

Finally, although it is sensible and practical to model the stochastic edit distance by a memoryless transducer, it is possible that the *a priori* distribution $p(x)$ may not be modeled by such a very simple structure. Thus, by learning a transducer defining the joint distribution $p(x, y)$, its parameters can converge to compromise values and not to the true ones. This can have dramatic effects from an application standpoint. Actually, a widely-used solution to find an optimal output string y according to an input one x consists in first learning the joint distribution transducer and later deducing the conditional transducer dividing by $p(x)$ (more precisely by its estimates over the learning set). Such a strategy is then irrelevant for the reason we mentioned above.

In this paper we have developed a way to learn directly the conditional transducer. After some definitions and notations (Section 2), we introduce in Section 3 the learning principle of the stochastic edit distance proposed by Ristad and Yianilos [2,1]. Then, by simulating different theoretical joint distributions, we show that the *unique way*, using their algorithm, to find them consists in sampling a learning set of (x, y) pairs according to the marginal distribution (*i.e.* over the input strings) of the target joint distribution itself. Moreover, we show that for any other *a priori* distribution, the difference between the target and the learned model increases. To free the method from this bias, one must *directly* learn at each iteration of the algorithm EM the conditional distribution $p(y|x)$. Achieving this task requires to modify Ristad and Yianilos's framework. That is the goal of Section 4. Then, we carry out experiments that show that it is possible to correctly estimate a target distribution whatever the *a priori* distribution we use. Section 5 is devoted to compare both models (along with two versions of the classic edit distance) in a character recognition task.

2 Notation

An alphabet X is a finite nonempty set of symbols. X^* denotes the set of all finite strings over X. Let $x \in X^*$ be an arbitrary string of length $|x|$ over the alphabet X. In the following, unless stated otherwise, symbols are indicated by $a, b, \ldots$, strings by $u, v, \ldots, z$, and the empty string by λ. $\mathbb{R}^+$ is the set of non negative reals. Let $f(\cdot)$ be a function, from which $[f(x)]_{\pi(x,\ldots)}$ is equal to $f(x)$ if the predicate $\pi(x, \ldots)$ holds and 0 otherwise, where x is a (set of) dummy variable(s).

3 Stochastic Edit Distance and Memoryless Transducers

A joint memoryless transducer defines a joint probability distribution over the pairs of strings. It is denoted by a tuple (X, Y, c, γ) where X is the input alphabet, Y is the output alphabet, c is the *primitive* joint probability function, $c : E \to [0, 1]$ and γ is the probability of the termination symbol of a string. As $(\lambda, \lambda) \notin E$, in order to simplify the notations, we are going to use $c(\lambda, \lambda)$ and γ as synonyms.

Let us assume for the moment that we know the probability function c (in fact, we will learn it later). We are then able to compute the joint probability $p(x, y)$ of a pair of strings (x, y). Actually, the joint probability $p : X^* \times Y^* \to [0, 1]$ of the strings x, y can be recursively computed by means of an auxiliary function (forward) $\alpha : X^* \times Y^* \to \mathbb{R}^+$ or, symmetrically, by means of an auxiliary function (backward) $\beta : X^* \times Y^* \to \mathbb{R}^+$ as:

$$\begin{aligned} \alpha(x,y) &= [1]_{x=\lambda \wedge y=\lambda} \\ &+ [c(a,b) \cdot \alpha(x',y')]_{x=x'a \wedge y=y'b} \\ &+ [c(a,\lambda) \cdot \alpha(x',y)]_{x=x'a} \\ &+ [c(\lambda,b) \cdot \alpha(x,y')]_{y=y'b}. \end{aligned} \qquad \begin{aligned} \beta(x,y) &= [1]_{x=\lambda \wedge y=\lambda} \\ &+ [c(a,b) \cdot \beta(x',y')]_{x=ax' \wedge y=by'} \\ &+ [c(a,\lambda) \cdot \beta(x',y)]_{x=ax'} \\ &+ [c(\lambda,b) \cdot \beta(x,y')]_{y=by'}. \end{aligned}$$

And then, $p(x, y) = \alpha(x, y)\gamma$ or $p(x, y) = \beta(x, y)\gamma$.

Both functions (forward and backward) can be computed in $O(|x| \cdot |y|)$ time using a dynamic programming technique. This model defines a probability distribution over the pairs (x, y) of strings. More precisely,

$$\sum_{x \in X^*} \sum_{y \in Y^*} p(x, y) = 1,$$

that is achieved if the following conditions are fulfilled [1],

$$\gamma > 0, c(a,b), c(\lambda,b), c(a,\lambda) \geq 0 \quad \forall a \in X, b \in Y$$
$$\sum_{\substack{a \in X \cup \{\lambda\} \\ b \in Y \cup \{\lambda\}}} c(a,b) = 1$$

Given $p(x, y)$, we can then compute, as mentioned in [1], the stochastic edit distance between x and y. Actually, the stochastic edit distance $d_s(x, y)$ is defined as being

Table 1. Target joint distribution $c^*(a,b)$ and its corresponding marginal distribution $c^*(a)$

$c^*(a,b)$	λ	a	b	c	d	$c^*(a)$
λ	0.00	0.05	0.08	0.02	0.02	0.17
a	0.01	0.04	0.01	0.01	0.01	0.08
b	0.02	0.01	0.16	0.04	0.01	0.24
c	0.01	0.02	0.01	0.15	0.00	0.19
d	0.01	0.01	0.01	0.01	0.28	0.32

the negative logarithm of the probability of the string pair $p(x,y)$ according to the memoryless stochastic transducer.

$$d_s(x,y) = -\log p(x,y), \forall x \in X^*, \forall y \in Y^*$$

Let S be a finite set of (x,y) pairs of *similar* strings. Ristad and Yianilos [1] propose to use the expectation-maximization (EM) algorithm to find an optimal joint stochastic transducer. The EM algorithm consists in two steps (expectation and maximization) that are repeated until a convergence criterion is achieved.

Given an auxiliary $(|X|+1) \times (|Y|+1)$ matrix δ, the expectation step can be described as follows: $\forall a \in X, b \in Y$,

$$\delta(a,b) = \sum_{(xax',yby') \in S} \frac{\alpha(x,y)c(a,b)\beta(x',y')\gamma}{p(xax',yby')} \qquad \delta(\lambda,b) = \sum_{(xx',yby') \in S} \frac{\alpha(x,y)c(\lambda,b)\beta(x',y')\gamma}{p(xx',yby')}$$

$$\delta(a,\lambda) = \sum_{(xax',yy') \in S} \frac{\alpha(x,y)c(a,\lambda)\beta(x',y')\gamma}{p(xax',yy')} \qquad \delta(\lambda,\lambda) = \sum_{(x,y) \in S} \frac{\alpha(x,y)\gamma}{p(x,y)} = |S|,$$

and the maximization as:

$$c(a,b) = \frac{\delta(a,b)}{N} \quad \forall a \in X \cup \{\lambda\}, \forall b \in Y \cup \{\lambda\} \text{ where } N = \sum_{\substack{a \in X \cup \{\lambda\} \\ b \in Y \cup \{\lambda\}}} \delta(a,b).$$

To analyze the ability of Ristad and Yianilos's algorithm to correctly estimate the parameters of a target joint memoryless transducer, we carried out a series of experiments.

We simulated a target joint memoryless transducer from the alphabets $X = Y = \{\mathtt{a}, \mathtt{b}, \mathtt{c}, \mathtt{d}\}$, such as $\forall a \in X \cup \{\lambda\}, \forall b \in Y \cup \{\lambda\}$, the target model is able to return the primitive theoretical joint probability $c^*(a,b)$. The target joint distribution we used is described in Table 1[1]. The marginal distribution $c^*(a)$ can be deduced from this target such that: $c^*(a) = \sum_{b \in X \cup \{\lambda\}} c^*(a,b)$.

Then, we sampled an increasing set of learning input strings (from 0 to 4000 sequences) of variable length generated from a given probability distribution $p(a)$ over the input alphabet X. In order to simplify, we modeled this distribution in the form of an automaton with only one state[2] and $|X|$ output transitions with randomly chosen probabilities.

[1] Note that we carried out many series of experiments with various target joint distributions, and all the results we obtained follow the same behavior as the one presented in this section.

[2] Here also, we tested other configurations leading to the same results.

We used different settings for this automaton to analyze the impact of the input distribution $p(a)$ on the learned joint model. Then, given an input sequence x (generated from this automaton) and the target joint distribution $c^*(a,b)$, we sampled a corresponding output y. Finally, the set S of generated (x,y) pairs was used by Ristad and Yianilos's algorithm to learn an estimated primitive joint distribution $c(a,b)$.

We compared the target and the learned distributions to analyze the behavior of the algorithm to correctly assess the parameters of the target joint distribution. We computed an average difference between the both, defined as follows:

$$d(c,c^*) = \frac{\sum_{a\in X\cup\{\lambda\}}\sum_{b\in Y\cup\{\lambda\}}|c(a,b)-c^*(a,b)|}{2}$$

Normalized in this way, $d(c,c^*)$ is a value in the range $[0,1]$. Figure 1 shows the behavior of this difference according to various configurations of the automaton. We can note that the unique way to converge towards a difference near from 0 consists in using the marginal distribution $c^*(a)$ of the target for generating the input strings. For all the other ways, the difference becomes very large.

As we said at the beginning of this article, we can easily explain this behavior. By learning the primitive joint probability function $c(a,b)$, Ristad and Yianilos learn at the same time the marginal distribution $c(a)$. The learned edit costs (and the stochastic edit distance) are then dependent of the *a priori* distribution of the input strings, that is obviously awkward. To free of this statistical bias, we have to learn the primitive conditional probability function independently of the marginal distribution. That is the goal of the next section.

4 Unbiased Learning of a Conditional Memoryless Transducer

A conditional memoryless transducer is denoted by a tuple (X,Y,c,γ) where X is the input alphabet, Y is the output alphabet, c is the primitive conditional probability function $c: E\rightarrow[0,1]$ and γ is the probability of the termination symbol of a string. As in the joint case, since $(\lambda,\lambda)\notin E$, in order to simplify the notation we use γ and $c(\lambda|\lambda)$ as synonyms.

The probability $p: X^*\times Y^*\rightarrow[0,1]$ of the string y assuming the input one was a x (noted $p(y|x)$) can be recursively computed by means of an auxiliary function (forward) $\alpha: X^*\times Y^*\rightarrow\mathbb{R}^+$ or, in a symmetric way, by means of an auxiliary function (backward) $\beta: X^*\times Y^*\rightarrow\mathbb{R}^+$ as:

$$\begin{aligned}\alpha(y|x) &= [1]_{x=\lambda\wedge y=\lambda}\\ &+[c(b|a)\cdot\alpha(y'|x')]_{x=x'a\wedge y=y'b}\\ &+[c(\lambda|a)\cdot\alpha(y|x')]_{x=x'a}\\ &+[c(b|\lambda)\cdot\alpha(y'|x)]_{y=y'b}.\end{aligned}\qquad\begin{aligned}\beta(y|x) &= [1]_{x=\lambda\wedge y=\lambda}\\ &+[c(b|a)\cdot\beta(y'|x')]_{x=ax'\wedge y=by'}\\ &+[c(\lambda|a)\cdot\beta(y|x')]_{x=ax'}\\ &+[c(b|\lambda)\cdot\beta(y'|x)]_{y=by'}.\end{aligned}$$

And then, $p(y|x)=\alpha(y|x)\gamma$ and $p(y|x)=\beta(y|x)\gamma$.

As in the joint case, both functions can be computed in $O(|x|\cdot|y|)$ time using a dynamic programming technique. In this model a probability distribution is assigned conditionally to each input string. Then

$$\sum_{y \in Y^*} p(y|x) \in \{1, 0\} \quad \forall x \in X^*.$$

The 0 is in the case the input string x is not in the domain of the function[3]. It can be show that the normalization of each conditional distribution can be achieved if the following conditions over the function c and the parameter γ are fulfilled,

$$\gamma > 0, c(b|a), c(b|\lambda), c(\lambda|a) \geq 0 \quad \forall a \in X, b \in Y \tag{1}$$

$$\sum_{b \in Y} c(b|\lambda) + \sum_{b \in Y} c(b|a) + c(\lambda|a) = 1 \quad \forall a \in X \tag{2}$$

$$\sum_{b \in Y} c(b|\lambda) + \gamma = 1 \tag{3}$$

As in the joint case, the expectation-maximization algorithm can be used in order to find the optimal parameters. The expectation step deals with the computation of the matrix δ:

$$\delta(b|a) = \sum_{(xax',yby') \in S} \frac{\alpha(y|x)c(b|a)\beta(y'|x')\gamma}{p(yby'|xax')} \qquad \delta(b|\lambda) = \sum_{(xx',yby') \in S} \frac{\alpha(y|x)c(b|\lambda)\beta(y'|x')\gamma}{p(yby'|xx')}$$

$$\delta(\lambda|a) = \sum_{(xax',yy') \in S} \frac{\alpha(y|x)c(\lambda|a)\beta(y'|x')\gamma}{p(yy'|xax')} \qquad \delta(\lambda|\lambda) = \sum_{(x,y) \in S} \frac{\alpha(y|x)\gamma}{p(y|x)} = |S|.$$

In order to do the maximization step, we begin by normalizing the insertion cost because it appears in both normalization equations (eq. 2 and eq. 3). Then:

$$c(b|\lambda) = \frac{\delta(b|\lambda)}{N} \quad \text{where} \quad N = \sum_{\substack{a \in X \cup \{\lambda\} \\ b \in Y \cup \{\lambda\}}} \delta(b|a)$$

The value of γ is now fixed by eq. 3 as:

$$\gamma = \frac{N - N(\lambda)}{N} \quad \text{where} \quad N(\lambda) = \sum_{b \in Y} \delta(b|\lambda)$$

and $c(b|a)$ and $c(\lambda|a)$ are obtained working out the values in eq. 2 and distributing the probability proportionally to their respective expectations $\delta(b|a)$ and $\delta(\lambda|a)$. Then

$$c(b|a) = \frac{\delta(b|a)}{N(a)} \frac{N - N(\lambda)}{N} \qquad c(\lambda|a) = \frac{\delta(\lambda|a)}{N(a)} \frac{N - N(\lambda)}{N} \quad \text{where } N(a) = \sum_{b \in Y \cup \{\lambda\}} \delta(b|a).$$

We carried out experiments to assess the relevance of our new learning algorithm to correctly estimate the parameters of target transducers. We followed exactly the same

[3] If $p(x) = 0$ then $p(x, y) = 0$ and as $p(y|x) = \frac{p(x,y)}{p(x)}$ we have a $\frac{0}{0}$ indeterminism. We chose to solve it taking $\frac{0}{0} = 0$, in order to maintain $\sum_{y \in Y^*} p(y|x)$ finite.

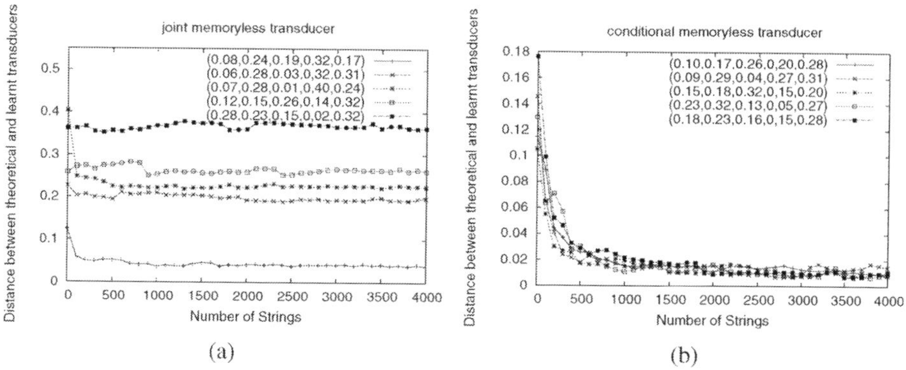

Fig. 1. Average difference between the target and the learned distributions according to various generations of the input strings using a joint (a) and a conditional (b) memoryless transducer. The tuples $(p_a, p_b, p_c, p_d, p_\#)$ represents the probabilities of the symbols a, b, c, d and the probability of ending in the stochastic automaton used to generate the input strings.

experimental setup as the one of the previous section, except to the definition of our difference $d(c, c^*)$. Actually, our new framework estimates $|X|$ conditional distributions. So $d(c, c^*)$ is defined as:

$$d(c, c^*) = \frac{(A + B\,|X|)}{2\,|X|}$$

where $A = \sum_{a \in X} \sum_{b \in Y \cup \{\lambda\}} |c(b|a) - c^*(b|a)|$ and $B = \sum_{b \in Y \cup \{\lambda\}} |c(b|\lambda) - c^*(b|\lambda)|$.

The results are shown in Figure 1. We can make the two following remarks. First, the different curves clearly show that the convergence toward the target distribution is independent of the distribution of the input strings. Using different parameter configurations of the automaton, the behavior of our algorithm remains the same, *i.e* the difference between the learned and the target conditional distributions tends to 0. Second, we can note that $d(c, c^*)$ rapidly decreases, *i.e.* the algorithm requires few learning examples to learn the target.

5 Application to the Handwritten Character Recognition

In order to assess the relevance of our model in a pattern recognition task, we applied it on the real world problem of handwritten digit classification. We used the NIST Special Database 3 of the National Institute of Standards and Technology, already used in several articles such as [6,7,8]. This database consists in 128 × 128 bitmap images of handwritten digits and letters. In this series of experiments, we only focus on digits written by 100 different writers. Each class of digit (from 0 to 9) has about 1,000 instances, then the whole database we used contains about 10,000 handwritten digits. Since our model handles strings, we coded each digit as contour chain following the feature extraction algorithm proposed in [6].

As presenting throughout this article, our method requires a set of (input,output) pairs of strings for learning the probabilistic transducer. While it is rather clear that pairs in the form of (noisy,unnoisy) strings constitute the most relevant way to learn an edit distance useful in a noise correction model, what must they represent in a pattern recognition task, with various classes, such as in handwritten digit classification? As already proposed in [1], a possible solution consists in building pairs of "similar" strings that describe the possible variations or distortions between instances of each class. In this series of experiments, we build pairs of (input,output) strings, where the input is a learning string, and the output is the corresponding nearest-neighbor in the learning set. The objective is then to learn a stochastic transducer that allows to optimize the conditional probabilities $p(output/input)$.

In the following series of experiments, we aim at comparing our approach (i) to the one of Ristad and Yianilos, and (ii) to the classic edit distance. Note that for the latter, we used two different matrices of edit costs. The first one is the most classic one, *i.e.* each edit operation has the same cost (here, 1). According to [7], a more relevant strategy would consist in taking costs proportionally to the relative angle between the directions used for describing a digit.

In order to assess each algorithm, the number of learning strings varied from 200 (20 for each class of digits) to 6,000 (600 for each class), with a step of 20 strings per class (resulting in 30 step iterations). The test accuracy was computed with a test set containing always 2,000 strings. For each learning size, we run 5 times each algorithm using 5 different randomly generated learning sets and we computed the average.

From Fig. 2, we can make the following remarks. First of all, learning an edit distance in the form of a conditional transducer is indisputably relevant to achieve a pattern recognition task. Whatever the size of the learning set, the test accuracy obtained using the stochastic edit distance is higher than the others. However, note that the difference decreases logically with the size of the learning set. Whatever the distance we choose, when the number of examples increases, the nearest-neighbor of an example x tends to be x itself. Interestingly, we can also note that for reaching approximately the same accuracy rate, the standard edit distance (using proportional costs) needs much more learning strings, and therefore requires a higher time complexity, than our approach.

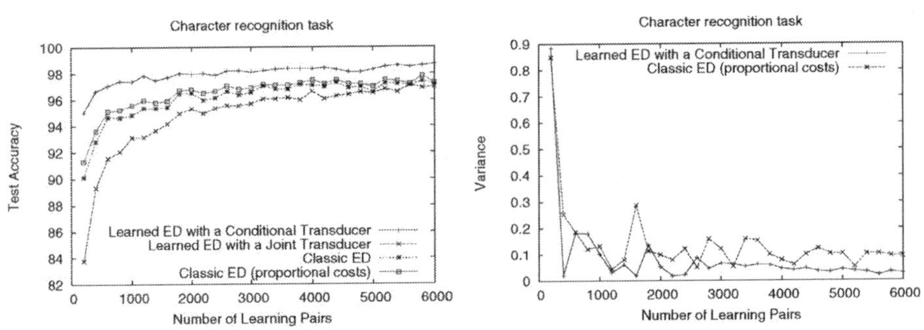

(a) (b)

Fig. 2. Evolution of the accuracy (a) and the variance (b) throughout the iterations in the character recognition task

Second, when the number of learning string pair is small, all the drawbacks with Ristad and Yianilos's method we already mentioned in the first part of this paper occur. Actually, while a nearest-neighbor is always a string belonging to the learning set, many learning strings are not present in the current (small) set of nearest-neighbors. Therefore, while all these strings (inputs and outputs) come from the same set of digits, the distribution over the outputs (the nearest-neighbors) is not the same as the distribution over the inputs (the learning strings). Of course, this bias decreases with the rise of the learning set size, but not sufficiently in this series of experiments for improving the performances of the classic edit distance.

To assess the level of stability of the approaches, we have computed a measure of dispersion on the results provided by the standard edit distance (with proportional costs) and our learned distance. Fig. 2 shows the behavior of the variance of the test accuracy throughout the iterations. Interestingly, we can note that in the large majority of the cases, our method gives a smaller variance.

6 Conclusion

In this paper, we proposed a relevant approach for learning the stochastic edit distance in the form of a memoryless transducer. While the standard techniques aim at learning a joint distribution over the edit operations, we showed that such a strategy induces a bias in the form of a statistical dependence on the input string distribution. We overcame this drawback by directly learning a conditional distribution of the primitive edit costs. The experimental results bring to the fore the interest of our approach. We think that our model is particularly suited for dealing with noisy data.

References

1. Ristad, E.S., Yianilos, P.N.: Learning string-edit distance. IEEE Trans. Pattern Anal. Mach. Intell. **20**(5) (1998) 522–532
2. Ristad, E.S., Yianilos, P.N.: Finite growth models. Technical Report CS-TR-533-96, Princeton University Computer Science Department (1996)
3. Casacuberta, F.: Probabilistic estimation of stochastic regular syntax-directed translation schemes. In: Proceedings of th VIth Spanish Symposium on Pattern Recognition and Image Analysis. (1995) 201–207
4. Clark, A.: Memory-based learning of morphology with stochastic transducers. In: Proceedings of the Annual meeting of the association for computational linguistic. (2002)
5. Eisner, J.: Parameter estimation for probabilistic finite-state transducers. In: Proceedings of the Annual meeting of the association for computational linguistic. (2002) 1–8
6. Gómez, E., Micó, L., Oncina, J.: Testing the linear approximating eliminating search algorithm in handwritten character recognition tasks. In: VI Symposium Nacional de reconocimiento de Formas y Análisis de Imágenes. (1995) 212–217
7. Micó, L., Oncina, J.: Comparison of fast nearest neighbour classifiers for handwritten character recognition. Pattern Recognition Letters **19** (1998) 351–356
8. Rico-Juan, J.R., Micó, L.: Comparison of aesa and laesa search algorithms using string and tree-edit-distances. Pattern Recognition Letters **24** (2003) 1417–1426

An Efficient Distance Between Multi-dimensional Histograms for Comparing Images

Francesc Serratosa and Gerard Sanromà

Universitat Rovira i Virgili, Dept. d'Enginyeria Informàtica i Matemàtiques, Spain
francesc.serratosa@.urv.cat, gerard.sanroma@urv.cat

Abstract. The aim of this paper is to present an efficient distance between n-dimensional histograms. Some image classification or image retrieval techniques use the distance between histograms as a first step of the classification process. For this reason, some algorithms that find the distance between histograms have been proposed in the literature. Nevertheless, most of this research has been applied on one-dimensional histograms due to the computation of a distance between multi-dimensional histograms is very expensive. In this paper, we present an efficient method to compare multi-dimensional histograms in O(2z), where z represents the number of bins. Results show a huge reduction of the time consuming with no recognition-ratio reduction.

1 Introduction

Finding the distance or similarity between histograms is an important issue in image classification or image retrieval since a histogram represents the frequency of the values of the pixels among the images. For this reason, a number of measures of similarity between histograms have been proposed and used in computer vision and pattern recognition. Moreover, if the position of the pixels is unimportant while considering the distance measure, we can compute the distance between images in an efficient way by computing the distance between their histograms.

Most of the distance measures presented in the literature (there is an interesting compilation in [1]) consider the overlap or intersection between two histograms as a function of the distance value but they do not take into account the similarity on the non-overlapping parts of the two histograms. For this reason, Rubner presented in [2] a new definition of the distance measure between n-dimensional histograms that overcomes this non-overlapping parts problem. It was called Earth Mover's Distance and it is defined as the minimum amount of work that must be performed to transform one histogram into the other one by moving distribution mass.

Often, for specific set measurements, only a small fraction of the *bins* in a histogram contain significant information, that is, most of the *bins* are empty. This is more frequent when the dimensions of the histograms increase. In that cases, the methods that use histograms as fixed-sized structures obtain poor efficiency. In the algorithm depicted by Rubner [2] to find the Earth Mover's Distance the empty-bins where not explicitly considered. They used the simplex algorithm [3] to compute the distance measure and the method presented in [4] to search a good initialisation. The computational cost of the simplex iteration is $O(z'^2)$, where z' is the number of

D.-Y. Yeung et al. (Eds.): SSPR&SPR 2006, LNCS 4109, pp. 412–421, 2006.

non-empty bins. The main drawback of this method is that the number of iterations is not bounded. Moreover, the initialisation cost is $O(2z')$.

To reduce the computational cost, Cha presented in [1] three algorithms to obtain the Earth Mover's Distance between one-dimensional histograms when the type of measurements where *nominal*, *ordinal* and *modulo* in $O(z)$, $O(z)$ and $O(z^2)$ respectively, being z the number of levels or bins.

Finally, Serratosa reduced more the computational cost in [5]. They presented three new algorithms to compute the Earth Mover's Distance between one-dimensional histograms when the type of measurements where *nominal*, *ordinal* and *modulo*. The computational cost were reduced to $O(z')$, $O(z')$ and $O(z'^2)$ respectively, being z' the number of non-empty bins.

It was presented in [6] an algorithm to compute the distance between histograms that the input was a built histogram (obtained from the target image) and another image. Then, it was not necessary to build the histogram of the image of the database to compute the distance between histograms.

Really few have been done to compare n-dimensional histograms except in [2]. The main drawback of the method presented in [2] is the computational cost. In this paper, we present an efficient algorithm to compute the distance between n-dimensional histograms with a computational cost of $O(2z)$. Our algorithm does not depend on the type of measurements (*nominal*, *ordinal* or *modulo*). In the next section, we define the histograms and types of values. In section 3, we give the definitions of distances between histograms and between sets and in section 4 we show the algorithm to compute the distance between histograms. In sections 5 and 6 we show the experimental validation of our algorithm and the conclusions.

2 Sets and Histograms

In this section, we formally give a definition of histograms. Moreover, we show a property obtained from the definition of the histograms that will be useful in the definitions of the distances given in the next section. Finally, we define the distance between pixel values.

2.1 Histogram Definition

Let x be a measurement which can have one of z values contained in the set $X=\{x_1,...x_z\}$. Each value can be represented in a T-dimensional vector as $x_i=(x_i^1, x_i^2,...,x_i^T)$. Consider a set of n elements whose measurements of the value of x are $A=\{a_1,...a_n\}$ where $a_t \in X$ being $a_t=(a_t^1, a_t^2,...,a_t^T)$.

The histogram of the set A along measurement x is $H(x,A)$ which is an ordered list consisting of the number of occurrences of the discrete values of x among the a_t. As we are interested only in comparing the histograms and sets of the same measurement x, $H(A)$ will be used instead of $H(x,A)$ without loss of generality. If $H_i(A)$, $1 \leq i \leq z$, denotes the number of elements of A that have value x_i, then $H(A)=[H_1(A), ...,H_z(A)]$ where

$$H_i(A)=\sum_{t=1}^{n} C_{it}^{A} \qquad (1)$$

and the individual costs are defined as

$$C_{i,t}^{A}=\begin{cases}1 & if\ a_t = x_i\\ 0 & otherwise\end{cases} \tag{2}$$

The elements $H_i(A)$ are usually called *bins* of the histogram. Note that z is the number of bins of the histogram. In a T-dimensional histogram with m values per each dimension, the number of bins is $z=m^T$.

2.2 Property of the Individual Costs

Given a value a_t, the addition of all the individual costs is 1.

$$\sum_{i=1}^{z} C_{i,t}^{A} = 1 \quad 1 \le t \le n \tag{3}$$

Proof
Given the t-element of the set A, this element has only one value a_t. Therefore, there is only one value of i such that $C_{i,t}^{A}=1$ (when $a_t = x_i$) and for all the other values of i, $C_{i,t}^{A}=0$ (that is, $a_t \neq x_i$). Then, the addition of all the values is one.

2.3 Type of Measurements and Distance Between Them

The distance between histograms presented in this paper is used as a fast method for comparing images and image retrieval. The most used colour representations are base on the R,G,B or H,S,I descriptors. The hue parameter (H) is a modulo-type measurement (measurement values are ordered but form a ring due to the arithmetic modulo operation) and the other parameters are ordinal-type measurements.

Corresponding to these types of measurements mentioned before, we define a measure of difference between two measurement levels $a=(a^1, a^2,\ldots,a^T) \in X$ and $b=(b^1, b^2,\ldots,b^T) \in X$ as follows:

$$d(a,b)=\sum_{j=1}^{T} S \text{ where } S=\begin{cases} m-\left|a^j-b^j\right| & if\ a^j-b^j \le m/2\ and\ a^j,b^j \in Modulo\ type\\ \left|a^j-b^j\right| & otherwise\end{cases} \tag{4}$$

This measure satisfy the following necessary properties of a metric. Since they are straightforward facts, we omit the proofs. The proof of the triangle inequality for the modulo distance is depicted in [1] for the one-dimensional case ($T=1$).

3 Distance Definitions

In this section we present the distance between sets $D(A,B)$ and the distance between their histograms $D(H(A),H(B))$. We proof that both satisfy the necessary properties of a metric and that the distance values are the same, $D(A,B) = D(H(A),H(B))$. To do so, we find a relation between the assignments between elements of the sets A and B while computing $D(A,B)$ and the assignments between *bins* while computing $D(H(A),H(B))$.

This is an important result since the computational cost of *D(A,B)* is exponential respect the number of the set elements, *n*, but the computational cost of *D(H(A),H(B))* is only quadratic respect the number of *bins* of the histogram *z*. Moreover, in most of the applications, *z* is much smaller than *n*. Another advantage is that the time consuming of the comparison is constant and does not depend on each set.

3.1 Distance Between Sets

Given two sets of *n* elements, *A* and *B*, the distance measure is considered as the problem of finding the minimum difference of pair assignments between both sets. That is, to determine the best one-to-one assignment *f* (bijective function) between the sets such that the sum of all the differences between two individual elements in a pair $a_i \in A$ and $b_{f(i)} \in B$ is minimised.

$$D(A,B) = \min_{\forall f A \to B} \left(\sum_{t=1}^{n} d\left(a_t, b_{f(t)}\right) \right) \tag{5}$$

We are interested only in the *D(A,B)* value rather than the assignment *f*. Nevertheless, we call f_{opt} as the assignment such that the distance is obtained, so we can redefine the distance as follows,

$$D(A,B) = \sum_{t=1}^{n} d\left(a_t, b_{f_{opt}(t)}\right) \tag{6}$$

3.2 Distance Between Histograms

The distance between histograms that we present here is a generalisation of the Earth Mover's Distance presented in [2]. Intuitively, given two T-dimensional histograms, one can be seen as a mass of earth properly spread in space, the other as a collection of holes in that same space. Then, the distance measure is the least amount of work needed to fill the holes with earth. Here, a unit of work corresponds to transporting a unit of earth by a unit of ground distance.

More formally, given two histograms *H(A)* and *H(B)*, where measurements can have one of *z* values contained in the set $X=\{x_1,...x_z\}$, the distance between the histograms *D(H(A),H(B))* is defined as follows,

$$D(H(A),H(B)) = \min_{\forall f A \to B} \left(\sum_{i,j=1}^{z} d\left(x_i, x_j\right) g_f(i,j) \right) \tag{7}$$

The flow between the *bins* of both histograms is represented by $g_f(i,j)$, that is, the mass of earth that is moved as one unit from the *bin i* to the *bin j*. The product $d(x_i,x_j)g_f(i,j)$ represents the work needed to transport this mass of earth. Similarly to equation (5), we can redefine the distance using the optimal assignment f_{opt},

$$D(H(A),H(B)) = \sum_{i,j=1}^{z} d\left(x_i, x_j\right) g_{f_{opt}}(i,j) \tag{8}$$

3.2.1 New Definition of the Flow Between Bins

In the definition of the distance between histograms presented in [2], the flow between histograms was shown to be a bi-dimensional matrix. The rows of the matrix represented the *bins* of one of the histograms and the columns represented the *bins* of the other histogram. Thus, each value of a matrix element was the flow between both *bins*. In that paper, there was no relation between the distance between the sets, *D(A,B)*, and the distance between the histograms of these sets, *D(H(A),H(B))*. For this reason, in the definition of the flow between *bins*, some constraints were needed to be imposed to match the distance definition to the transportation problem.

In our paper, we determine the flow between *bins* $g_f(i,j)$, as a function of the one-to-one assignment f between the sets A and B used to compute the distance $D(A,B)$ as follows,

$$g_f(i,j)=\sum_{t=1}^{n} C_{i,t}^{A} C_{j,f(t)}^{B} \qquad 1\le i,j\le z \tag{9}$$

were the costs C are given in (2).

With this new definition, we obtain two advantages; First, there is a relation between distances *D(A,B)* and *D(H(A),H(B))* through their definition. Second, the constraints arbitrarily imposed to the flow between *bins* in [2], were converted in deducted properties that make possible to naturally match the distance between histograms to the transportation problem.

3.2.2 Properties of the Flow $g_f(i,j)$

The flow between the *bin* i of the set A and the *bin* j of the set B through the assignment f fulfils the following three properties,

Property a) $g_f(i,j)\ge 0 \quad 1\le i,j\le z$

Property b) $\sum_{j=1}^{z} g_f(i,j)=H_i(A) \quad 1\le i\le z$

Property c) $\sum_{i=1}^{z} g_f(i,j)=H_j(B) \quad 1\le j\le z$

Proofs

Property (a) is a straightforward fact due to equations (2) and (9).

Property (b) Using equation (9), we obtain that $\sum_{j=1}^{z} g_f(i,j)=\sum_{j=1}^{z}\sum_{t=1}^{n} C_{i,t}^{A} C_{j,f(t)}^{B}$, and exchanging the sumatories, we obtain that $\sum_{j=1}^{z} g_f(i,j)=\sum_{t=1}^{n}\sum_{j=1}^{z} C_{i,t}^{A} C_{j,f(t)}^{B}$. Then, if we spawn the external sumatory, we have the following formula, $C_{i,1}^{A}\sum_{j=1}^{z} C_{j,f(1)}^{B}+C_{i,2}^{A}\sum_{j=1}^{z} C_{j,f(2)}^{B}+\ldots+C_{i,n}^{A}\sum_{j=1}^{z} C_{j,f(n)}^{B}$ that can be reduced to $C_{i,1}^{A}+C_{i,2}^{A}+C_{i,3}^{A}+\ldots+C_{i,n}^{A}$ do to equation (3) and considering that f is bijective. So, we arrive at the expression $\sum_{j=1}^{z} g_f(i,j)=\sum_{t=1}^{n} C_{it}^{A}=H_i(A)\cdot$

Property (c) Using equation (9), we obtain that $\sum_{i=1}^{z} g_f(i,j)=\sum_{i=1}^{z}\sum_{t=1}^{n} C_{i,t}^{A} C_{j,f(t)}^{B}$, and exchanging the sumatories and the order of the costs, we obtain that $\sum_{i=1}^{z} g_f(i,j)=\sum_{t=1}^{n}\sum_{i=1}^{z} C_{j,f(t)}^{B} C_{i,t}^{A}$. Then, if we spawn the external sumatory, we have the following formula, $C_{j,f(1)}^{B}\sum_{i=1}^{z} C_{i,1}^{A} + C_{j,f(2)}^{B}\sum_{i=1}^{z} C_{i,2}^{A} + \ldots + C_{j,f(n)}^{B}\sum_{i=1}^{z} C_{i,n}^{A}$. Finally, applying equation (3), this sumatory is reduced to $C_{j,f(1)}^{B} + C_{j,f(2)}^{B} + \ldots + C_{j,f(n)}^{B}$. And so, $\sum_{i=1}^{z} g_f(i,j)=\sum_{t=1}^{n} C_{j,f(t)}^{B} = H_j(B)$.

3.3 Properties of the Distances

We present in this section the metric properties of the distances between sets and histograms. Moreover, we show that the distance value of these distances is the same. To that aim, we first describe a lemma. We assume that there are two measurement sets *A* and *B* that have *n* elements contained in the set $X=\{x_1,\ldots x_z\}$.

Lemma
The distance between two elements of the sets *A* and *B* given an assignment *f*, can be obtained as the distance between *bins* as follows,

$$d\left(a_t, b_{f(t)}\right) = \sum_{i,j=1}^{z} C_{i,t}^{A} C_{j,f(t)}^{B} d\left(x_i, x_j\right) \quad 1 \le t \le n \quad f \text{ bijective} \tag{10}$$

Proof

By definition of the individual cost in equation (2), the only case that $C_{i,t}^{A} = 1$ and $C_{j,f(t)}^{B} = 1$ is when $a_t = x_i$ and $b_{f(t)} = x_j$ and so $d\left(a_t, b_{f(t)}\right) = d\left(x_i, x_j\right)$.

Properties
Property a) The distance measure *D(A,B)* between sets *A* and *B* satisfy the metric properties.

Property b) The distance value of distances between sets and histograms of these sets is the same, *D(A,B) = D(H(A),H(B))*.

Property c) The distance measure *D(H(A),H(B))* between histograms *H(A)* and *H(B)* satisfy the metric properties.

Proofs
Property (a): The proof of this property was depicted in [5]. Although in that paper, the histograms were defined one-dimensional, the proof was based on the distance between elements *d(a,b)* independently on the dimension of the elements *a* and *b*.

Property (b): If we apply equation (10) to substitute the distance between elements $d\left(a_t, b_{f_{opt}(t)}\right)$ in the definition of the distance between sets (6), we obtain the formula $D(A,B)=\sum_{t=1}^{n}\sum_{i,j=1}^{z} C_{i,t}^{A} C_{j,f_{opt}(t)}^{B} d\left(x_i, x_j\right)$. Then, rearranging the elements, we get $\sum_{i,j=1}^{z} d\left(x_i, x_j\right)\sum_{t=1}^{n} C_{i,t}^{A} C_{j,f_{opt}(t)}^{B}$.

Finally, if we substitute the equation of the flow (9) we obtain the final expression, $\sum_{i,j=1}^{z} d(x_i, x_j) g_{f_{opt}}(i, j) = D(H(A), H(B))$.

Property (c): The proof is simple since we have proved that the distance value is the same (property b) and that the distance measure between sets satisfy the metric property (property a).

4 Algorithm

In this section, we depict an efficient algorithm used to compute the distance between histograms based on a solution to the well-known transportation problem [3]. Suppose that several suppliers, each with a given amount of goods, are required to supply several consumers, each with a given limited capacity. For each pair of suppliers and consumers, the cost of transporting a single unit of goods is given. The transportation problem is then to find a least-expensive flow of goods from the suppliers to the consumers that satisfies the consumer's demand. Our distance between histograms can be naturally cast as a transportation problem by defining one histogram as the supplier and the other one as the consumer. The cost of transporting a single unit of goods is set to the distance between the *bin* of one histogram and the *bin* of the other one, $d(x_i,x_j)$. Intuitively, the solution of the transportation problem, $g_f(i,j)$, is then the minimum amount of "work" required to transform one histogram to the other one subjected to the constraints defined by the properties of the flow $g_f(i,j)$ (section 4.2.2).

The computational cost of the transportation problem is exponential, respect the number of suppliers and consumers, that is, the number of bins of the histograms, z. Fortunately, efficient algorithms are available. One of the most common solutions is the simplex algorithm (), which is an iterative method that the cost of one simplex iteration is $O(z^2)$. The main drawback is that the number of iterations is not bounded and that this method needs a good initial solution. The Russell method [4] is the most common method used to find the first solution with a computational cost of $O(2z-1)$.

In this paper, we present an efficient and not iterative algorithm (figure 1) with a computational cost of $O(2z-1)$.

Given a pair of bins from both histograms, i and j, our algorithm finds the amount of goods that can be transported, $g_f(i,j)$, and computes the cost of this transportation, $g_f(i,j)*d(x_i,x_j)$. The algorithm finishes when all the goods have been transported, that is, all the elements of the sets, n, have been considered. In each iteration, a pair of bins is selected by the function *next*, in a given order and considering that the bins are not empty. The order of the bins is set by the following energy function,

$$E(i, j) = Path\_Deviation_j(i) + Path\_Deviation_i(j) \tag{11}$$

The $Path\_Deviation_j(i)$ is the difference between the maximum cost from the bin i to any bin of the histogram and the real cost from this bin to the bin j,

$$Path\_Deviation_j(i) = \max\_dist(x_i) - d(x_i, x_j) \tag{12}$$

It represents the worst case that the good can be sent (supplier) or received (consumer) respect the best case.

```
Algorithm Histogram-Distance (H(A),H(B))
i,j = first()
while n > 0 // n: the number of elements of both sets
     gf(i,j) = min (Hi(A) , Hj(B))
     Hi(A) = Hi(A) - gf(i,j)
     Hj(B) = Hj(B) - gf(i,j)
     n = n - gf(i,j)
     D = D + gf(i,j) * d(xi,xj)
     i,j = next (i , j , H(A), H(B))
Return D //distance between histograms
```

Fig. 1. Algorithm that computes the distance between n-dimensional histograms

Theorem. The worst computational cost of the algorithm is $O(2z\text{-}1)$.

Proof. The pair of bins i,j generated by the function *next* forms a z X z matrix. In each iteration, one column or file (or both) of the matrix (depending if $H_i(A) = 0$ or $H_j(B) = 0$ is erased from the matrix (can not be used any more). Then, the worst case is the one that alternatively, one column is erased and after that one file is erased. Thus, the number of iterations is the number of columns plus the number of files less one.

5 Experimental Validation

We have used the coil image database [7] to validate our new algorithm and to show the usefulness of the histograms as the only information of the images. Only 20 objects were selected (figure 2). The test set was composed by 100 images (5 images

Fig. 2. Images taken at angle 5 of the 20 objects

of these 20 objects taken at the angles 5, 15, 25, 35 and 45). And the reference set was composed by other 100 images (5 images of the same objects taken at angles 0, 10, 20, 30, 40 and 50).

Table 1 (left) shows the number of correctly classified images (1-nearest neighbour) and (right) the average number of iterations of the inner loop of the algorithm in figure 1. The run time is proportional to the number of iterations. The first column is the number of bins (and bits) per each dimension. The number of colours is $bins^{nD}$. In the other columns, we show the results for 3 different 3D-histograms, 2 different 2D-histograms and 2 more 1D-histograms. The number of iterations underlined and in bold (right table) are the ones that all the images have been properly classified (99 or 100% in left table). If the recognition ratio is expected to be 99 or 100%, the best combination is HSV(2bits), CIELAB(3bits), HL(3bits) and HS(3bits).

Table 1. (left) Number of objects properly classified and (right) average number of iterations

Dimension	3D			2D		1D	
Bins(bits)	HSV	RGB	CIELAB	HS	HL	HUE	GREY
4 (2)	99	98	95	98	97	77	64
8 (3)	100	97	99	99	100	94	94
16 (4)	100	100	100	99	100	95	96
64 (6)	--	--	--	99	100	97	100
256 (8)	--	--	--	--	--	97	100

Dimension	3D			2D		1D	
Bins(bits)	HSV	RGB	CIELAB	HS	HL	HUE	GREY
4 (2)	53	32	19	20	20	6	6
8 (3)	**250**	120	55	70	**70**	14	13
16 (4)	**896**	**425**	**180**	219	**192**	29	26
64 (6)	--	--	--	1431	**1100**	95	**100**
256 (8)	--	--	--	--	--	229	**383**

Table 2 shows the worst number of iterations obtained from the theoretical cost. We realise that there is a huge difference between the real number of iterations (table 1 right) and the worst cases (table 2).

Table 2. Worst number of iterations obtained from the theoretical cost

	3D			2D		1D	
Bins (bits) X dimension	HSV	RGB	CIELAB	HS	HL	HUE	GREY
4 (2)	$2{*}4^3$-1= 127			$2{*}4^2$-1= 31		$2{*}4^1$-1= 7	
8 (3)	$2{*}8^3$-1= 1,023			$2{*}8^2$-1= 127		$2{*}8^1$-1= 15	
16 (4)	$2{*}16^3$-1= 8,191			$2{*}16^2$-1= 511		$2{*}16^1$-1= 31	
64 (6)	$2{*}64^3$-1= 524,287			$2{*}64^2$-1= 8,191		$2{*}64^1$-1= 127	
256 (8)	$2{*}256^3$-1= 33,554,431			$2{*}256^2$-1= 131,071		$2{*}256^1$-1= 511	

6 Conclusions and Future Work

We have presented a new distance between multi-dimensional histograms and an efficient algorithm to compute this distance. Our method is useful for comparing black&white or colour images and using H,S,I or R,G,B colour descriptors. The theoretical computational cost is $O(2z)$, being z the number of levels of the pixels. The experimental validation demonstrates that it is worth increasing the number of dimensions and reducing the number of bins per each dimension, i.e. HSV (2bits).

Moreover, the real number of iterations (or run time) is really lower than the theoretical one.

References

1. S.-H. Cha, S. N. Srihari, "On measuring the distance between histograms" *Pattern Recognition* 35, pp: 1355–1370, 2002.
2. Y. Rubner, C. Tomasi, and L. J. Guibas, "A Metric for Distributions with Applications to Image Databases" *International Journal of Computer Vision* 40 (2), pp: 99-121, 2000.
3. *Numerical Recipes in C: The Art of Scientific Computing*, ISBN 0-521-43108-5.
4. E. J. Russell. "Extension of Dantzig's algorithm to finding an initial near-optimal basis for the transportation problem", *Operations Research*, 17, pp: 187-191, 1969.
5. F. Serratosa & A. Sanfeliu, "Signatures versus Histograms: Definitions, Distances and Algorithms", *Pattern Recognition* (39), Issue 5, pp. 921-934, 2006.
6. F.-D. Jou, K.-Ch. Fan, Y.-L. Chang, "Efficient matching of large-size histograms", *Pattern Recognition Letters* 25, pp: 277–286, 2004.
7. http://www1.cs.columbia.edu/CAVE/research/softlib/coil-100.html

Finding Captions in PDF-Documents for Semantic Annotations of Images

Gerd Maderlechner, Jiri Panyr, and Peter Suda

Corporate Technology
Siemens AG,
D-81730 München, Germany

Abstract. The Portable Document Format (PDF) is widely-used in the Web and searchable by search engines, but only for the text content. The goal of this work is the extraction and annotation of images in PDF-documents, to make them searchable and to perform semantic image annotation. The first step is the extraction and conversion of the images into a standard format like jpeg, and the recognition of corresponding image captions using the layout structure and geometric relationships. The second step uses linguistic-semantic analysis of the image caption text in the context of the document domain. The result on a PDF-document collection with about 3300 pages with 6500 images has a precision of 95.5% and a recall of 88.8% for the correct image captions.

1 Introduction

This work is motivated by the following five facts: (1) in the world wide web nearly all searchable documents are in HTML-format. The next important format is the PDF-format (portable document format), with a proportion of about 3% of the searchable web (experimental result with Google and Yahoo). Since most PDF documents are larger than HTML-pages we estimate that about 10% of the searchable information is in PDF format [17]. The remaining document formats are below 1%. (2) PDF is a standard format for archiving all types of documents in libraries, government or companies. PDF has an open published specification. PDF/A is an ISO standard for archiving. (3) PDF is popular for electronic publishing because it is a page description format which preserves even complex layouts consisting of text, graphics and images on all output devices. (4). The existing image search engines like Google, Yahoo, Picsearch etc. do not consider images in PDF-documents. (5) Present text based image search engines use keywords and not semantic annotations of the images.

From this we conclude that it is worthwhile to consider the PDF format for image search. Furthermore the image search quality can be improved by semantic annotations of the image captions. This will allow image searching not only by keywords but using questions like "Show me the player who scored the goal 1:0 in the match Mexico-Costa Rica on August 17th, 2005". The semantic annotation can be applied also to image captions in HTML-pages. An example will be presented below (Fig. 5).

Semantic and index information for image understanding and searching may be obtained from the image content using image processing methods [1], or from some text describing the image [2]. Some approaches use a combination of either information [3].

D.-Y. Yeung et al. (Eds.): SSPR&SPR 2006, LNCS 4109, pp. 422–430, 2006.

This paper describes a new approach that does not use the image content analysis but relies on the recognition of existing image captions using layout analysis. We concentrate on the image caption recognition and do not go into details of the linguistic-semantic methods.

2 Related Work

2.1 PDF-Document Analysis

There exist many commercial and public domain programs for processing of PDF-documents [4, 5]. But the result of our investigation for the purpose of image capture recognition was disappointing. We found some papers [6, 7] on PDF-document analysis using the open source library xpdf and the programs pdftotext, pdfimages and pdf2html. These tools are helpful but we had to add considerable functions to the existing programs which is described in chapter 3.

Lovegrove et.al.[8] analyze PDF files using Adobe SDK with the goal to perform logic labeling of the layout objects, which also contains image captions with only few examples and no quantitative evaluation.

Chao and Lin [9] developed a proprietary system for PDF-layout analysis with a different purpose.

2.2 Image Caption Recognition

For *HTML* web pages there are many research and commercial systems available which use also image captions, e.g. Google image search: "Google analyzes the text on the page adjacent to the image, the image caption and dozens of other factors to determine the image content. Google also uses sophisticated algorithms to remove duplicates and ensures that the highest quality images are presented first in your results." [15].

Rohini Srihari [3] applied natural language processing to figure captions in newspapers in combination with face detection in the corresponding image. This work does not locate but takes image captions as granted. Her focus is on natural language processing and segmentation of faces in the images to associate them with person names in captions.

Rowe et.al. [10] stress the importance of captions for indexing of images. But they do not use layout (geometry) but only neighborhood in ASCII text representation. They determine statistically relevant presence or absence of particular keywords in the potential caption sentences. (MARIE-4 system).

Paek et.al.[11] classify photographs with corresponding captions into indoors and outdoors. They compare text based and image content based methods for classification. The text based method achieved 83% accuracy on a test set size of 1339.

3 Approach

The proposed approach consists of two steps: First Recognition of the image captions and second semantic annotation of the image based on the caption text (see Figure 1).

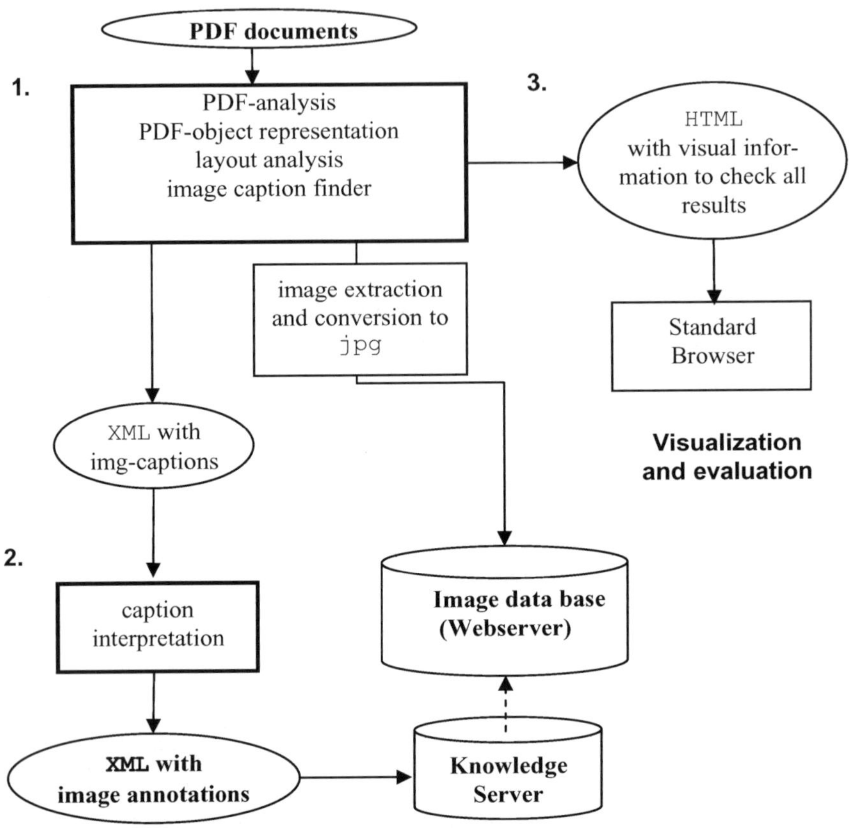

Fig. 1. Two step approach for image annotation in PDF documents: 1.Recognition of image captions and storage of extracted images in an image data base. 2. Interpretation of the captions and storage as semantic annotations of the extracted images in a knowledge server with references to the image data base. 3. The results of layout and caption analysis may be visualized for evaluation using a standard HTML-browser.

3.1 Recognition of the Image Captions

In contrast to the methods mentioned above (chapter 2) this approach tries to locate the existing *image captions* using the layout structure of the document, i.e. positions and sizes of the images and text objects and their geometric arrangement on each page in complete PDF-documents.

An *image caption* is defined as a *text block* which is intentionally placed (by the author resp. publisher) below or above the image to describe the semantics of the image. Left and right positions of captions are neglected currently because of their rare occurrence. A *text block* is defined as a visually separable unit of one or more text lines with homogeneous layout features.

The generation of text blocks applies a bottom-up process starting from the glyphs to build words, text lines, text blocks, and columns. This is similar to document image analysis methods [12, 13], but more accurate, because there are no distortions due to scanning noise or page skew. Instead of the pixel image we use the PDF objects and streams for text and image layout analysis. The images are located and saved in separate files after conversion to jpg format if necessary.

Fig. 2. Global resolution of Top-Bottom caption conflict using font attributes from the whole PDF-document. The bottom text block is recognized (cyan) as caption of image 1_11. The rectangles display the dimensions of the text lines and text blocks from the original PDF-page.

The main problem is to decide which text block is an image caption resp. which image has a caption. This problem is solved as a constraint satisfaction problem using generic layout rules. The rules are derived from the standard publishing rules for allocating text blocks and image captions using font, line, and block attributes.

The recognition process has three phases: (1) Neighborhood analysis starting from the images with local conflict resolution, (2) Neighborhood analysis starting from the text blocks and local conflict resolution, (3) Global resolution of the remaining ambiguities. The last step tries to determine dominant layout attributes in the whole document, like font type, font size and font style to discriminate the captions from other text blocks (see an example in Fig. 2).

3.2 Semantic Annotation of Caption Text

The semantic annotation of the caption text is performed in cooperation with the DFKI (German Center for Artificial Intelligence). It is based on linguistic-semantic analysis of the caption text and the whole document text using the SPROUT tool. A detailed description is given in [14].

4 Results and Discussion

The first test set is a document collection (corpus) of 290 PDF documents downloaded from the FIFA web site *http://fifaworldcup.yahoo.com/06/de/index.html* containing 3323 pages with a total of 6507 images. This corpus was chosen because this work is part of a larger project called SmartWeb [16], which has the goal to allow natural language questions to be answered automatically by semantic web technologies. A first use case is the soccer domain in the context of the FIFA WorldCup 2006 in Germany. The results are summarized in Table 1 and Table 2.

Table 1 shows the confusion matrix between the recognized image captions and the Ground Truth data of the test set. The diagonal entries show the correct results. For images without captions small images (below a size threshold) are separately shown, and all of them are correctly recognized (true negatives TN). From the remaining 2716 images without captions 3+8=11 images erroneously got an image caption (false positives FP). This proves the intended high specificity (TN/(TN+FP) = 99.83%) of our approach.

In total 76 image captions were not found (false negatives FN), from which 6 captions were left/right captions that are not yet considered in our approach.

The majority of captions are located below the images (caption type Bottom). There are 9 + 6 = 15 captions associated with the wrong images (Top/Bottom confusion), which we also count as false positives (FP) in Table 2.

Table 1. Confusion matrix between recognized types of caption and the ground truth (GT) for the test set of 290 PDF documents with a total number of 3323 pages and 6507 images

Ground-Truth:	**Recog-nized:**	Small images	Without captions	Top	Bot-tom	Left	Right	**Sum of GT images**
Small images (no caption)		3145	0	0	0	0	0	3145
Img without caption		0	2705	3	8	0	0	2716
Img with Top caption		0	49	135	9	0	0	193
Img with Bottom caption		0	21	6	420	0	0	447
Img with left caption		0	4	0	0	0	0	4
Img with right caption		0	2	0	0	0	0	2
Sum of images		3145	2781	144	437	0	0	6507

In Table 2 the results for top and bottom captions are summarized. In terms of the common quality measures of *precision* and *recall* the result is as follows:

Precision = TP / (TP + FP) = 555/(555+26) = 95.5% and Recall = TP / (TP + FN) = 555/(555+70) = 88.8%, whereas FP consists of 3+8=11 non-captions and 9 + 6 = 15 Top/Bottom confusion errors. The 5850 true negatives consist of 3145 small images and 2705 images recognized without caption.

The average processing time per document is 0.74 sec and per page about 0.06 sec on a standard PC with a 2.7 GHz Pentium 4 processor.

Table 2. Recognition results for image captions over the whole test set of 290 documents with 3323 pages and 6507 images. This results in a precision of 95.5% and recall of 88.8% for both top and bottom caption recognition.

Caption type	True positives	False positives	True negatives	False negatives
Top	135	9 (=6+3)		49
Bottom	420	17 (=9+8)		21
Both (sum)	*555*	*26*	*5850*	*70*

Fig. 3. Result of image caption recognition on a complex PDF page with several background images and images without captions. The only image caption (no. 4) of image 1_4 (cyan) with the text "Prof. Wahlster; Ministerpräsident Müller; Prof. Seibert, FORGIS" was correctly recognized.

In Figure 3 we show the result on a complex PDF-document with a lot of background images and many images without captions.

The second test was performed with a small set of PDF-documents which were converted from HTML to PDF using Adobes PDFprinter. The purpose of this test was to check the quality of the resulting HTML-files (Fig. 1, No. 3) by comparing it with the original HTML.

These PDF-documents may contain a large number of images per page consisting of small graphical objects in gif format because HTML does not support graphics format. The results were comparable to the first test set, but some new problems occurred: Some image captions belonged to the text of buttons, which was sometimes misleading. Figure 4 shows an image caption which was correctly located but does

Nachrichten

Statistik spricht für Spanien, Tschechien und die Türkei

11. November 2005
von FIFAworldcup.com

Foto vergrössern
Fotogalerie

Keine der drei europäischen Paarungen in der Relegation um die letzten drei verbleibenden Plätze für die FIFA Fussball-Weltmeisterschaft Deutschland 2006™ stellt eine Premiere dar. Auf der Grundlage früherer Begegnungen sprechen die Statistiken für Spanien, Tschechien und die Türkei. In der Partie zwischen Australien und Uruguay handelt es sich um die Neuauflage der Relegation von vor vier Jahren. Zudem spielen Trinidad und Tobago gegen Bahrain.

SCHWEIZ – TÜRKEI

Die Schweizer und Türken sind schon häufiger aufeinander getroffen. In der Qualifikation zum FIFA-Weltpokal™ Deutschland 1974 setzte sich die Türkei durch. In Basel erreichte sie am 9. Mai 1973 ein torloses Unentschieden und gewann im Rückspiel am 18. November 1973 mit 2:0 in Izmir.

Fig. 4. The image caption containing the text "Foto vergrössern, Fotogalerie" was correctly recognized, but does not describe the image content. This PDF file was generated from an HTML page. In the original HTML format the "image caption" is a button that has to be clicked by the user to display the image together with the actual caption from data base.

Jared Borgetti ist der neue Rekordtorschütze der mexikanischen National-mannschaft

Name:	Jared Borgetti
Team:	Mexico
Match:	Mexico - Costa Rica
Location:	Mexico City
Date:	2005-08-17
Scorer:	1:0
Minute:	62

Fig. 5. The Semantic Annotation of the figure caption (middle) recognized the name "Jared Borgetti" of the player. Using the semantic annotation of the whole document further data of the soccer event can be determined (right).

not describe the image content. We did not observe such image captions in original PDF documents. This weakness of the layout based caption recognition is obvious, but can be remedied by the following linguistic and semantic post processing.

The semantic annotation is not in the main focus of this paper (see [14]). An example of the results is given in Figure 5. Detailed results and discussion will be presented in a future paper.

5 Conclusion

This paper describes a new layout based approach to find image captions in PDF-documents. The application of layout rules to discriminate image caption text blocks from other text blocks is successful. This work complements existing systems for image indexing like Google image search, which do not support PDF-documents.

The method was tested on a test set of 290 PDF-documents containing about 3000 pages with about 6500 images. The precision and recall of correct image captions is 95.5% resp. 88.8%. Only 0.17% of images without captions are erroneously associated with a caption.

The subsequent semantic annotation of the image captions using the SPROUT tool is promising. This technique is applicable also to HTML-pages.

Applications of this work are not limited to searching in the web but also suitable for analysis of existing electronic archives of legacy documents in PDF format.

Acknowledgements

We would like to thank Paul Buitelaar and his colleagues from the German Research Center for Artificial Intelligence (DFKI) for providing the corpus of PDF documents and the linguistic-semantic annotation tools.

This work was supported in part by the German Federal Ministry of Education and Research under grant no. BMBF FKZ 01IMD01K.

References

1. Flickner, M., et. al.: Query by image and video content: the QBIC system. IEEE Computer 28 (9), 1995, 23--32
2. Sable, Carl L., Hatzivassiloglou, Vasileios: Text-based approaches for non-topical image categorization. Int. J. Digital Libraries (2000) 261–275
3. Srihari, Rohini K.: Automatic Indexing and Content-Based Retrieval of Captioned Images. IEEE Computer, September, (1995) 49-56
4. www.adobe.com
5. www.xpdf.com, www.foolabs.com
6. Kou, Zhenzhen, Cohen, W.W., Wang, R., Murphy, R.F.: Extracting information from text ands images for location proteomics. Proceedings of the 3rd ACM SIGKDD Int. Workshop on Data Mining in Bioinformatics, Washington DC, USA, Aug. (2003) 2-9
7. W.W. Cohen, R. Wang, R.F. Murphy: Understanding Captions in Biomedical Publications. Proceedings of the ninth ACM SIGKDD international conference on Knowledge discovery and data mining, Washington DC, USA, Aug. (2003) 499-504
8. William S. Lovegrove and Davis F. Brailsford, Document analysis of PDF files: methods, results and implications, Electronic Publishing, Vol. 8, 1995, 207 - 220
9. H. Chao and X. Lin, Capturing the Layout of Electronic Documents for the Reuse in Variable Data Printing, Proc. 7th International Conference on Document Analysis and Recognition, Seoul, Korea, August 2005, 940 - 944
10. Neil C. Rowe and Brian Frew, Automatic Caption Localization for Photographs on World Wide WebPages, Information Processing and Management, Volume 34, 1998, 95 - 107
11. S. Paek, C. L. Sable, and V. Hatzivassiloglou, Integration of Visual and Text Based Approaches for the Content Labeling and Classification of Photographs, ACM SIGIR Workshop on Multimedia Indexing and Retrieval, 1999
12. S. Mao, A. Rosenfeld, T. Kanungo, Document structure analysis algorithms: a literature survey Proc. SPIE Electronic Imaging, January 2003 SPIE Vol. 5010, 197-207

13. Gerd Maderlechner, Peter Suda, and Thomas Brückner, Classification of documents by form and content, Pattern Recognition Letters 18 (11-13), 1997, 1225-1231
14. M. Becker, W. Drozdzynski, H.-U. Krieger, J. Piskorski, U. Schäfer, F. Xu, SProUT, Shallow Processing with Unification and Typed Feature Structures, Proceedings of the International Conference on NLP (ICON 2002). December 18-21, Mumbai, India, 2002
15. www.google.com/help/faq_images.html (at 10.02.2006)
16. http://SmartWeb.dfki.de
17. Philipp Mayr, Das Dateiformat PDF im Web - eine statistische Erhebung, Informationswissenschaft & Praxis, Vol. 53, 2002, 475 - 481

Effective Handwritten Hangul Recognition Method Based on the Hierarchical Stroke Model Matching

Wontaek Seo and Beom-joon Cho*

Dept. of Computer Science, University of Maryland, College Park, MD 20742
Dept. of Computer Engineering, Chosun University, 375 Seosuk-dong, Dong-gu, Gwangju, Korea
Tel.: +82-62-230-7103; Fax: +82-62-230-7381
wtseo@cs.umd.edu, bjcho@chosun.ac.kr

Abstract. This study defines three models based on the stroke for handwritten Hangul recognition. Those are trainable and not sensitive to variation which is frequently founded in handwritten Hangul. The first is stroke model which consists of 32 stroke models. It is a stochastic model of stroke which is fundamental of character. The second is grapheme model that is a stochastic model using composition of stroke models and the last is character model that is a stochastic model using relative locations between the grapheme models. This study also suggests a new stroke extraction method from a grapheme. This method does not need to define location of stroke, but it is effective in terms of numbers and kinds of stroke models extracted from graphemes of similar shape. The suggested models can be adapted to hierarchical bottom-up matching, that is the matching from stroke model to character model. As a result of experiment, we obtain 88.7% recognition rate of accuracy that is better than those of existing studies.

1 Introduction

A character is composed by joint of several strokes. The union and location of each stroke become very important information in recognizing a character. Besides, the other existed information in a character can be aware as noises which are occurred through a user or an input device. As this view, recognizing a handwritten character by union and location of each stroke is very common process and this is considered as structural method. The structural method can be completed under the hypothesis which the position of each stroke becomes a most important information of recognizing an independent character [1, 2].

The most methods, which have used strokes in the past, have expressed strokes and their relation by heuristic. For relation between each stroke, they used slope between a strokes and surrounding strokes [5, 6]. There was an approach which used symbolic way between each stroke. The types of stroke are divided as horizontal, vertical, left

* Corresponding author.

D.-Y. Yeung et al. (Eds.): SSPR&SPR 2006, LNCS 4109, pp. 431 – 440, 2006.

diagonal, and right diagonal and the relationship of stroke is divided as L form, T joint, parallel and the others [7, 8].

Yet, this heuristic method is insufficient for practical uses because it is very sensitive with noise of input character. Furthermore, there is a limit which is difficult to be trained. In these days, statistic method has been introduced which uses graph modeling; a stroke is presented by probability of stroke slope and its length and the relationship of each stroke is presented by their relative location [3]. There is another method which uses a systematic relation between each stroke. In this method the relative information of each stroke are presented through statistic dependence [4].

In this research, a new stroke model will be presented and the composition method of grapheme models and character models will be introduced. And also, matching method of each model with statistic way and recognizing method of handwritten Hangul character by using characteristic of our own class composition will be introduced. In chapter 2, the characteristics of Hangul characters will be explained. Three new proposed models (stroke model, grapheme model, character model) and their composition and matching method will be explained in chapter 3, and experiment result will be shown in chapter 4, and the conclusion will be in chapter 5.

2 Characteristic of Hangul

2.1 Composition of Hangul

Hangul is a phonetic alphabet which one character has an independent sound, and it is constructed of consonant and vowel those are arranged on two dimensional spaces. Although there are only 24 vowels and consonants, the number of character which can be made through composing of them is 11,172. However, the practically being used characters are 2,350 and the commonly being used characters are only 520 of them.

Table 1. Shape and position of graphemes of Hangul

Groups	Grahpemes	Position
Basic consonant	ㄱ,ㄴ,ㄷ,ㄹ,ㅁ,ㅂ,ㅅ,ㅇ,ㅈ,ㅊ,ㅋ,ㅌ,ㅍ,ㅎ	FC, LC
Basic vowel	ㅏ, ㅑ, ㅓ, ㅕ, ㅣ	VV
	ㅗ,ㅛ,ㅜ,ㅠ,ㅡ	HV
Combined consonant	ㄲ, ㄸ, ㅃ, ㅆ, ㅉ	FC, LC
	ㄳ, ㄵ, ㄶ, ㄺ, ㄻ, ㄼ, ㄽ, ㄾ, ㄿ, ㅀ, ㅄ	LC
Combined vowel	ㅐ, ㅒ, ㅔ, ㅖ	VV
	ㅘ,ㅙ,ㅚ,ㅝ,ㅞ,ㅟ,ㅢ	HV + VV

Hangul has six important composition formats which consonant and vowel are arranged on 2 dimensional spaces. Every character is constructed by the six formats and there are rule which consonant and vowel are used for each format. Therefore, distinct recognition of formatting information will make us much easier to recognize a character.

There are 14 of basic consonant, 5 of double consonant which makes strong sounds, and 11 of repeated consonant which is composed of two different basic consonant. The First Consonant (FC) decides early stage of pronunciation, and the Last Consonant (LC) does later stage of pronunciation. Vowel settles the middle parts of pronunciation. There are 10 basic vowel and 11 combined vowel. There are two types of vowel according to the shape, Horizontal Vowel (HV) and Vertical Vowel (VV). As above, each pronunciation and the meaning are decided by the composition of 2~4 of vowel and consonant, which is shown in table 1.

2.2 Hierarchical Decomposition of Hangul

Hangul shows hierarchical joint structure: several strokes joint together to make a form of grapheme and the grapheme joint each other to make an independent character on the 2 dimensional spaces. Therefore, we can divide a character using opposite way of jointing: a character is divided into vowel and consonant depending on the format type and those are divided again into strokes. In here, a stroke means basic stroke such as '—', 'ㅣ' and composed stroke such as 'ㄱ', 'ㄴ' by Korean writing style.

3 Proposed Models

3.1 Stroke Model

Stoke model is suggested for effective modeling of stokes of lower parts in a point of top-down decomposition. The most common strokes in Hangul are horizontal, vertical, left diagonal, right diagonal and circle such as in 'ㅇ', 'ㅎ'. In this research, a model of jointed stroke rather than single one is proposed. Using four basic strokes, which are horizontal line, vertical line, left diagonal line and right diagonal line, we made 32 kinds of joint strokes which are shown in fig. 1. The circle stroke is excluded because those introduced strokes can make circle by jointing each other.

Each stroke models has parameters of edges and nodes. Each edge's probability distribution of directive angle is existed and the connect part node has probability distribution of connect angle of two edge.

A matching of stroke models is the first stage of matching process. After we set the ending point, connecting point and the bending point as a fixed node of the graph, as well as a attributed graph by extracting an edge between nodes, we find out a stroke model which is the best matching with the part of attributed graph. For this, production of sub graph which matches with stroke model is needed.

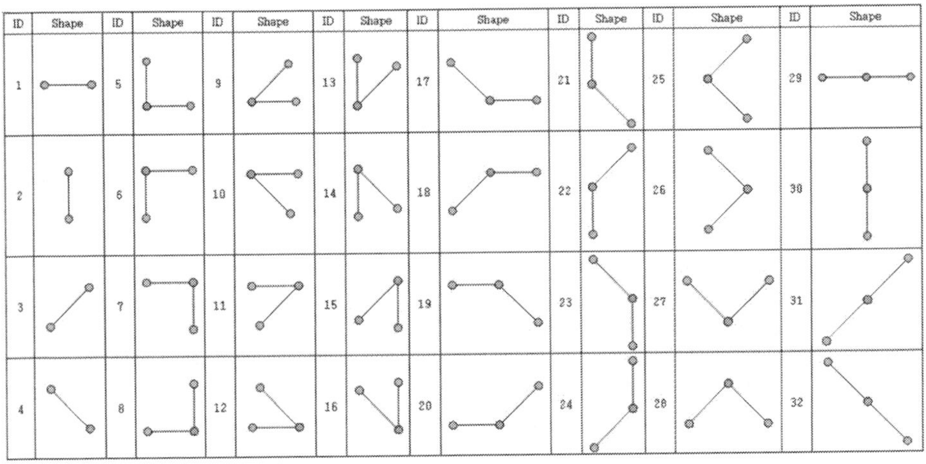

Fig. 1. Shape of proposed Stroke models

Fig. 2 shows an attributed graph decomposition method into sub graph for matching of stroke model with an example of grapheme 'ㅅ'. In the set of edges, if two edges share a node, sub graph which has two edges and a node is extracted to match with stroke models. This method is very successful way because there is no need to define the positional relation by heuristic, and there is only one stroke model combination with a grapheme. Strokes that are shown as a single stroke also have two strokes which are same directions.

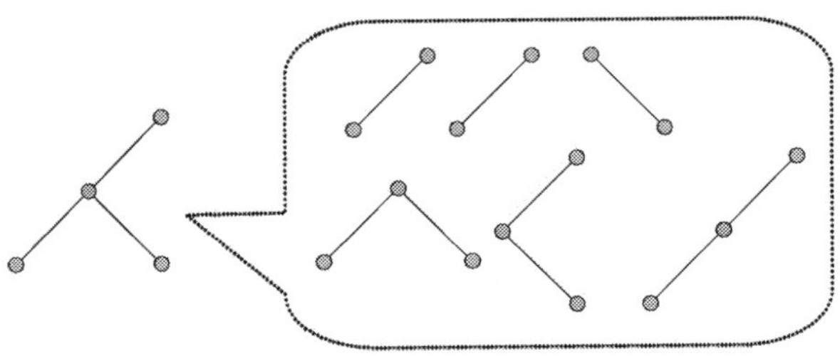

Fig. 2. Example of sub graph extraction from an attributed graph

The matching of output sub graph and the stroke models is calculated from multiplication of parameter of stroke models of direction d of sub graph's edge, and multiplication of parameter of stroke models of joint angle a of sub graph. The parameter of stroke models is probability distribution as explained above. Eq.(1) is the calculating the matching probability of stroke model.

$$P(M_S \mid X) = \prod_{x \in \forall E, k \in d, a} P(x_k) \tag{1}$$

X is a sub graph of attributed graph, and M_S is stroke model. As it is shown, matching probability of stroke model can be calculated from multiplication of probability of input.

3.2 Grapheme Model

Grapheme consists of combinations of several strokes, so it is presented as relationship of each stroke. As a result of using the extracting method, which is suggested previous, the kinds of stroke model and its frequency can be very good information. The result of extraction of stroke model from, 'ㅁ' and 'ㅂ' by using their matching method is shown at fig. 3. 'ㅂ' seems it is added only two more lines to 'ㅁ', but it's added 6 different strokes when we decompose to stroke model. Using this characteristic, grapheme model which has probability distribution of each stroke model frequency of occurrence is defined.

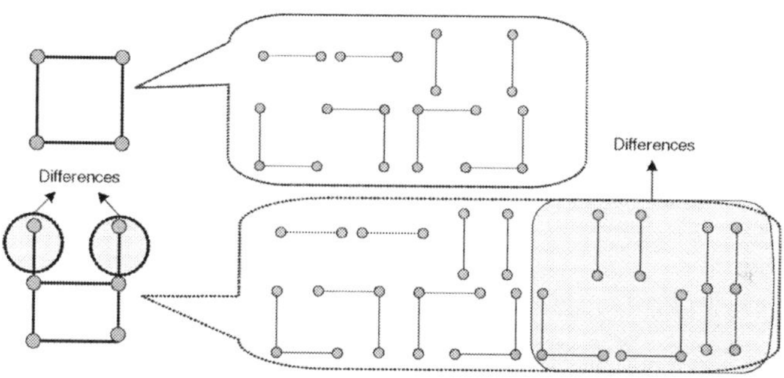

Fig. 3. Comparison of stroke models extracted from two graphemes of similar shape

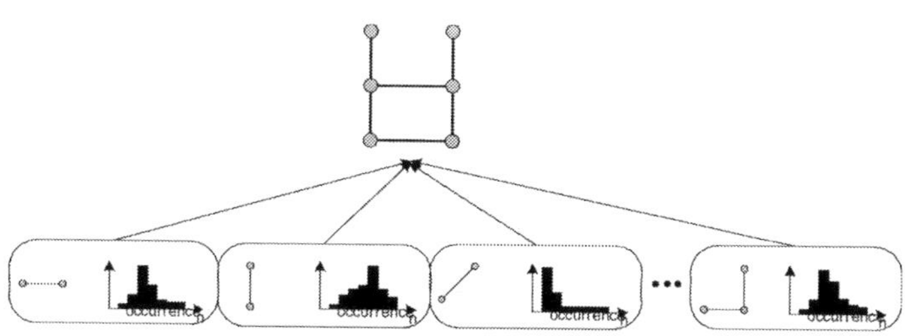

Fig. 4. Example of probability distribution of a grapheme model

There are 51 graphemes existed in Hangul as it is shown at table 1. Therefore 51 grapheme models are defined in this research.

Fig. 4 is an example of parameter of grapheme model of grapheme 'ㅂ'. Because it is composed of horizontal stroke and vertical stroke, their probability of high frequency is high, but the probability of high frequency of diagonal stroke's occurrence is low.

As shown in fig. 4, using probability distribution of occurrence frequency of 32 stroke models, 51 grapheme models are defined, and matching probability is defined by the average of occurrence frequency of 32 stroke models. The equation is shown at (2).

$$P(M_G \mid X) = \frac{1}{n} \times \sum_{i=1}^{n} P(O(S_n)) \qquad (2)$$

X is the set of stroke model, M_G is a grapheme model, n is the number of model, and S_n is the n th stroke model, and $O(S_n)$ is the number of occurrence of S_n, and $P(O(S_n))$ is the probability of S_n.

3.3 Character Model

Character is made through an arrangement of several graphemes on 2 dimensional spaces, and meaning and pronunciation are concluded by their location and variety. In this research, character model is defined by using this characteristic. Probability distribution of relative position of each grapheme is used to present the character model. The position of each grapheme is already defined according to the character type, but it is available only after we recognize a character, therefore the grapheme information of location cannot be used. Relative location of grapheme is defined as horizontal, vertical, right diagonal, and left diagonal relationship. The example of positional relationship is shown in Fig 5. The relation between 'ㅅ' and 'ㅏ' is horizontal relation, and the relation between 'ㅅ' and 'ㄹ' is left diagonal relation.

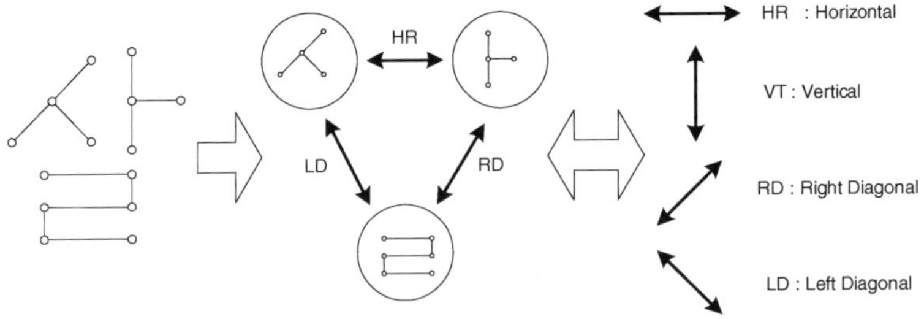

Fig. 5. Positional relationship between grapheme models

For probability matching with graphemes extracted from grapheme matching, character model has to constitute parameters. The parameters of character model are

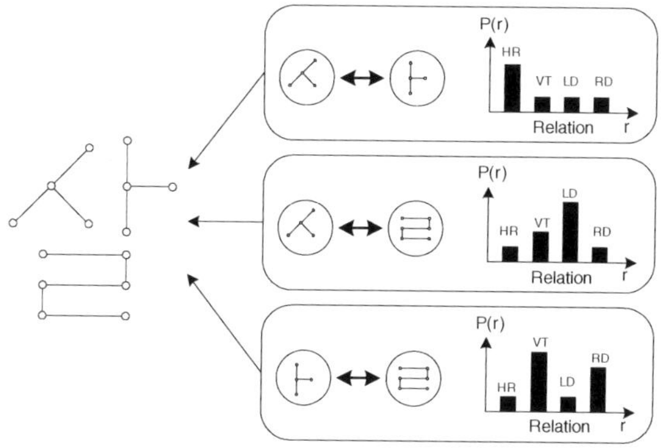

Fig. 6. Example of probability distribution of a character model

graphemes and probability distribution of their relative position. An example of the composition of character model is shown at Fig 6.

As graphemes and its positional relation which are existed in a character are defined by probability and compose an independent character model, the compositions work out for each of 2350 complete characters in Hangul.

By multiplication of parameters of every grapheme models extracted from grapheme matching stage in character model, matching probability of character is calculated. The formal equation is shown in Eq.(3).

$$P(M_C \mid X) = \prod_{x \in X} P_r(x \mid N(x)) \tag{3}$$

In above, X means outputs of matching stage of grapheme model, $N(x)$ means neighbor graph of x and $P_r(x \mid N(x))$ means the probability of relation of x and the neighbor x.

3.4 Recognition

Recognition in the statistic model means a process of deriving the maximum probability from a relation between data and a model. In this research, every graph model and its node and edge are models which are presented the distribution of probability, therefore calculating the maximum probability between a model and input is a matching method.

In this research, the method of calculating the recognition probability of character is shown in Eq.(4)..

$$P(M \mid X) = \prod_{i=1, x \subset X}^{n} P(M_{G_i} \mid x) \times P(M_C \mid X) \tag{4}$$

In the above, $P(M_G \mid x)$ means matching probability of model G_i to the set of input stroke x and $P(M_C \mid X)$ means matching probability of according character model. The character recognition probability is calculated when each grapheme multiplies $P(M_G \mid x)$ and $P(M_C \mid X)$. The character which produces the highest probability will be the matching character.

3.5 Hierarchical Bottom-Up Matching

In this research, the method of recognition through hierarchical bottom-up matching process using stroke, grapheme, and character model is proposed. First, stroke and its model need to be matched, and then, perform matching with grapheme's model using the extracted stroke model. In this process, for every condition of possible stroke model's set, we must perform the matching with grapheme's model. At last, the set which can compose the character at grapheme model's set has to be extracted and need to calculate the matching probability. The character model from the process and each grapheme that attends the matching are multiplied to get recognition probability and the character which has the highest probability can be the result of recognition.

4 Experimental Results

The experiment in this research used the database of common handwritten Hangul database PE92. PE92 consists of each 100 sets of character image to the total 2350 of Hangul character; 40 sets for training and 60 sets for test.

The probability of recognition is being compared while the numbers of characters are being limited, in this experiment, the 520 of practical character and 2350 of character can be found at table 2. As it shown in the table, we could get 90.5% of accuracy about 520 of characters and 88.7% for 2350 of characters. Also, as a result of considering the 5th recognition candidates, we could get 95.5% of accuracy for the 520 characters and 95.2% for the 2350 characters. In here, those are recognized as similar forms, but not often as different character.

Table 2. Recognition result of proposed method

Number of category (char.)		520	2350
Recognition rate(%)	1st	90.5	88.7
	2nd	92.3	90.3
	5th	95.5	95.2

When we compare the result of this research to the previous researches [3,4], we can claim that the method, which records the increasing 1% rate of accuracy with the standard of perfect characters, is much more excellent than other methods. [3] composes three kinds of models, primitive stroke model, grapheme model and character model using random graph modeling, but the definition of the models is quite different to those of models suggested in this paper. There are some problems which the grapheme modeling use stroke model extracted from the other stroke groups and it takes more time to match all different strokes. Oppositely, the method which is suggested in this paper uses only the stroke models which are extracted from identical stroke group of grapheme model matching by decomposing the stroke group. Therefore, this method brings the solution to the previous problems. In the study [4], the condition of limitation about grapheme production prevent the production of it in unusual spaces, yet the way of establishing the condition rely too much on heuristic. However, in our research, by using the relative probability in decomposing and composing of stroke group, we can present every process in a correct range of calculation of probability.

Table 3. Performance comparison to the previous works

method	hierarchical random graph[3]	stochastic relationship[4]	proposed
Rec. rate(%)	86.3	87.7	88.7

5 Conclusion

In this paper, we proposed the new stroke based models and matching methods for an effective Hangul recognition system. The stroke model is defined using 3 probability distributions which two are direction of edges and one is angle between two edges. 32 stroke models are defined based on the composition of 4 basic strokes. The grapheme model is defined based on the new stroke extraction method. The frequency of each stroke extracted from a grapheme is modeled as grapheme model. The character model is defined using relative position between each grapheme. Hierarchical bottom-up matching can be adapted to these three models, because the concept of model definition is started from the structural characteristic of Hangul. As a result of experiment, we could get the high performance of recognition of 88.7% compare to previous research as well as better result in the periodic problems.

The model introduced in this paper will be successful for not only for Hangul recognition, but also for the other characters such as Chinese which has also very complicated structure.

Acknowledgement

This study was supported in part by research funds from Chosun University, 2004.

References

1. A. K. C. Wong, D. E. Ghahraman, "Random Graphs: structural-contextual dichotomy", IEEE Trans. On Pattern Analysis and Machine Intelligence, Vol. 2, No. 4, pp. 341-348, 1980.
2. A. K. C. Wong and M. You, "Entropy and distance of random graphs with application to structural pattern recognition", IEEE Trans. On Pattern Analysis and Machine Intelligence, Vol. 7, No. 5, pp. 599-609, 1985.
3. Ho-Yon Kim, Jin H. Kim, "Hierarchical random graph representation of handwritten characters and its application to Hangul recognition", Pattern Recognition, 34, pp.187-201, 2001.
4. Kyung-won Kang, Jin H. Kim, "Utilization of Hierarchical, Stochastic Relationship Modeling for Hangul Character Recognition", IEEE PAMI. Vol. 26, No. 9, pp. 1185-1196, 2004.
5. F. H. Cheng, "Multi-stroke Relaxation Matching Method for handwritten Chinese Character Recognition", Pattern Recognition, Vol. 31, No. 4, pp. 401-410, 1998.
6. H. j. Lee, B. Chen, "Recognition of handwritten Chinese Characters via Short Line Segments", Pattern Recognition, Vol. 25, No. 5, pp. 543-552, 1992.
7. X. Zhang, Y. Xia, "The Automatic Recognition of Handprinted Chinese Characters – A Method of Extracting an Order Sequence of Strokes", Pattern Recognition Letters, Vol. 1, No. 4, pp. 259-265, 1983.
8. C. L. Liu, I. -J. Kim, J. H. Kim, "Model-based Stroke Extraction and matching for Handwritten Chinese Character Recognition", Pattern Recognition, Vol. 34, No. 12, pp. 2339-2352, 2001.

Graph Embedding Using Commute Time

Huaijun Qiu and Edwin R. Hancock

Department of Computer Science, University of York,
York, YO10 5DD, UK

Abstract. This paper explores the use of commute-time preserving embedding as means of data-clustering. Commute time is a measure of the time taken for a random walk to set-out and return between a pair of nodes on a graph. It may be computed from the spectrum of the Laplacian matrix. Since the commute time is averaged over all potential paths between a pair of nodes, it is potentially robust to variations in graph structure due to edge insertions or deletions. Here we demonstrate how nodes of a graph can be embedded in a vector space in a manner that preserves commute time. We present a number of important properties of the embedding. We experiment with the method for separating object motions in image sequences.

1 Introduction

The embedding of the nodes of a graph in a vector-space is an important step in developing structural pattern analysis algorithms. For instance, node embeddings are key for graph matching [17,5,4], graph-based clustering [18] and graph visualisation [9]. Although the embedding can be effected using a number of different techniques including those that are geometrically based [19] and those that are based on optimisation techniques [14], one of the simplest approaches is to adopt a graph-spectral approach [11,1]. This involves embedding the nodes of the graph under study using the eigenvectors or scaled eigenvectors of the Laplacian or adjacency matrix. For instance both Shapiro and Brady [17], and Kosinov and Caelli [4] use spectral methods to embed nodes of graphs in a vector space, and then use proximity to establish correspondences. Spectral embeddings have also been used to visualise complex graphs.

However, one of the problems of spectral embedding is stability under noise. From matrix perturbation it is well known that noise in an adjacency matrix causes the eigenvectors can rotate erratically under noise, and this means that the embedding coordinates are also unstable under noise. The aim in this paper is to explore the use of commute time as a means of stabilising the spectral embedding of graph nodes. The commute time between a pair of nodes on a graph is the expected time taken for a random walk to set-out and return. It is hence averaged over the set of all possible paths between each pair of nodes. In fact, commute time is a metric that can be computed using the Green's function or pseudo inverse of the graph Laplacian. In a recent series of papers Qiu and Hancock [13] have shown how commute time can give improved graph partitions and spectral clusterings.

The aim in this paper is to investigate whether the averaging of paths that is implicit to the computation of commute time can lead to improved embeddings. The embedding that preserves commute times is found the scaling the transpose of the Laplacian

D.-Y. Yeung et al. (Eds.): SSPR&SPR 2006, LNCS 4109, pp. 441–449, 2006.

eigenvector matrix by the pseudo-inverse of the Laplacian eigenvalues. We commence by performing a theoretical analysis that establishes the link between this embedding and the Laplacian eigenmap and the diffusion map. We then present some experiments that illustrate the practical utility of the embedding.

2 Commute Time and Commute Time Embedding

We denote a weighted graph by $\Gamma = (V, E)$ where V is the set of nodes and $E \subseteq V \times V$ is the set of edges. Let Ω be the weighted adjacency matrix satisfying

$$\Omega(u,v) = \begin{cases} w(u,v) & \text{if } (u,v) \in E \\ 0 & \text{otherwise} \end{cases}$$

Further let $T = diag(d_v; v \in V)$ be the diagonal weighted degree matrix with elements $d_u = \sum_{v=1}^{|V|} w(u,v)$. The *un-normalized* Laplacian matrix is given by $L = T - \Omega$ and the *normalized* Laplacian matrix is defined to be $\mathcal{L} = T^{-1/2} L T^{-1/2}$, and has elements

$$\mathcal{L}_\Gamma(u,v) = \begin{cases} 1 & \text{if } u = v \\ -\frac{w(u,v)}{\sqrt{d_u d_v}} & \text{if } u \neq v \text{ and } (u,v) \in E \\ 0 & \text{otherwise} \end{cases}$$

The spectral decomposition of the *normalized* Laplacian is $\mathcal{L} = \Phi \Lambda \Phi^T$, where $\Lambda = diag(\lambda_1, \lambda_2, ..., \lambda_{|V|})$ is the diagonal matrix with the ordered eigenvalues as elements satisfying: $0 = \lambda_1 \leq \lambda_2 \ldots \leq \lambda_{|V|}$ and $\Phi = (\phi_1 | \phi_2 | .... | \phi_{|V|})$ is the matrix with the ordered eigenvectors as columns.

Let G be the pseudo-inverse of the *normalized* Laplacian matrix satisfying $G\mathcal{L} = \mathcal{L}G = I - \phi_1 \phi_1^T$. Then we have

$$G(u,v) = \sum_{i=2}^{|V|} \frac{1}{\lambda_i} \phi_i(u) \phi_i(v) \tag{1}$$

From Equation 1, the eigenvalues of $\mathcal{L}$ and G have the same sign and $\mathcal{L}$ is positive semidefinite, and so G is also positive semidefinite. Since G is also symmetric (see [6] page 4), it follows that G is a kernel.

We note that the *commute time* $CT(u,v)$ is the expected time for the random walk to travel from node u to reach node v and then return. It can be computed using the Green's function G by

$$CT(u,v) = vol T^{-1/2} \left(G(u,u) + G(v,v) - 2G(u,v) \right) T^{-1/2} \tag{2}$$

As a result,

$$CT(u,v) = \sum_{i=2}^{|V|} \left(\sqrt{\frac{vol}{\lambda_i d_u}} \phi_i(u) - \sqrt{\frac{vol}{\lambda_i d_v}} \phi_i(v) \right)^2 \tag{3}$$

Hence, the embedding of the nodes of the graph into a vector space that preserves commute time has the co-ordinate matrix

$$\Theta = \sqrt{vol} \Lambda^{-1/2} \Phi^T T^{-1/2} \tag{4}$$

The columns of the matrix are vectors of embedding co-ordinates for the nodes of the graph. The term $T^{-1/2}$ arises from the normalisation of the Laplacian. The embedding is nonlinear in the eigenvalues of the Laplacian. This distinguishes it from principle components analysis (PCA) and locality preserving projection (LPP) [10] which are both linear. As we will demonstrate in the next section, the commute time embedding is just kernel PCA [16] on the Green's function. Moreover, it can be viewed as Laplacian eigenmap since they actually are minimizing the same objective function.

2.1 The Commute Time Embedding and the Laplacian Eigenmap

In the Laplacian eigenmap [3,2] the aim is to embed a set of points with co-ordinate matrix $\bar{\mathbf{X}} = \{\bar{\mathbf{x}}_1, \bar{\mathbf{x}}_2, ..., \bar{\mathbf{x}}_n\}$ from a R^n space into a lower dimensional subspace R^m with the co-ordinate matrix $\mathbf{Z} = \{\mathbf{z}_1, \mathbf{z}_2, ..., \mathbf{z}_m\}$. The original data-points have a proximity weight matrix Ω with elements $\Omega(j, j) = \exp[-||\bar{\mathbf{x}}_i - \bar{\mathbf{x}}_j||^2]$. The aim is to find the embedding that minimises the objective function

$$\epsilon = \sum_{i,j} \|\mathbf{z}_i - \mathbf{z}_j\|^2 \Omega(i, j) = tr(\mathbf{Z}^T L\mathbf{Z}) \tag{5}$$

where Ω is the edge weight matrix of the original data $\bar{\mathbf{X}}$.

To remove the arbitrary scaling factor and to avoid the embedding undergoing dimensionality collapse, the constraint $\mathbf{Z}^T T\mathbf{Z} = I$ is applied. The embedding problem becomes

$$\mathbf{Z} = \arg \min_{\mathbf{Z}^T T\mathbf{Z}=I} tr(\mathbf{Z}^T L\mathbf{Z}) \tag{6}$$

The solution is given by the lowest eigenvectors of the generalized eigen-problem

$$L\mathbf{Z} = \Lambda T\mathbf{Z} \tag{7}$$

and the value of the objective function corresponding to the solution is $\epsilon^* = tr(\Lambda)$.

For the commute-time embedding the objective function minimised is

$$\epsilon = \frac{\sum_{i,j} \|\mathbf{z}_i - \mathbf{z}_j\|^2 \Omega(i, j)}{\sum_i \mathbf{z}_i^2 d_i} = tr(\frac{\mathbf{Z}^T L\mathbf{Z}}{\mathbf{Z}^T T\mathbf{Z}}) \tag{8}$$

To show this, let $\mathbf{Z} = Y^T = (\sqrt{vol}\Lambda^{-1/2}\Phi^T T^{-1/2})^T$, we have

$$\begin{aligned}
\epsilon &= tr(\frac{\sqrt{vol}\Lambda^{-1/2}\Phi^T T^{-1/2} L T^{-1/2}\Phi\Lambda^{-1/2}\sqrt{vol}}{\sqrt{vol}\Lambda^{-1/2}\Phi^T T^{-1/2} T T^{-1/2}\Phi\Lambda^{-1/2}\sqrt{vol}}) \\
&= tr(\frac{\Lambda^{-1/2}\Phi^T \mathcal{L}\Phi\Lambda^{-1/2}}{\Lambda^{-1/2}\Phi^T\Phi\Lambda^{-1/2}}) \\
&= tr(\frac{\Lambda^{-1/2}\Lambda\Lambda^{-1/2}}{\Lambda^{-1}}) \\
&= tr(\Lambda) = \epsilon^*
\end{aligned} \tag{9}$$

Hence, the commute time embedding not only aims to maintain proximity relationships by minimizing $\sum_{u,v} \|\mathbf{z}_u - \mathbf{z}_v\|^2 \Omega_{u,v}$, but it also aims to assign large co-ordinate values to nodes (or points) with large degree (i.e. it maximizes $\sum_u \mathbf{z}_u^2 d_u$). Nodes with large degree are the most significant in a graph since they have the largest number or total weight of connecting edges. In the commute time embedding, these nodes are furthest away from the origin and are hence unlikely to be close to one-another.

Finally, we note that the objective function appearing in Equation (15) is identical to that used in the normalized cut. To show this let $\boldsymbol{\theta}$ be a dimensional indicator vector. The minimum value obtained by the normalized cut [18] is

$$\boldsymbol{\theta}_1 = \arg \min_{\boldsymbol{\theta}^T \mathbf{T} \mathbf{1}=0} \frac{\boldsymbol{\theta}^T (\mathbf{T} - \Omega)\boldsymbol{\theta}}{\boldsymbol{\theta}^T \mathbf{T} \boldsymbol{\theta}} \tag{10}$$

Hence comparing with Equation (8) it is clear that the objective function minimised by the commute time embedding is exactly the same as that minimized by the normalized cut, provided that the eigenvectors are scaled by the inverse of the corresponding non-zero eigenvalues. In the bipartition case, this does not make any difference since scaling will not change the distribution of the eigenvector components. However, in the multi-partition case, the scaling differentiates the importance of different eigenvectors. It is clear that the eigenvector corresponding to the smallest non-zero eigenvalue contributes the greatest amount to the commute time. Moreover, it is this eigenvector or Fiedler vector that is used in the normalized cut to bipartition the graphs recursively. In the case of the commute time embedding, the scaled eigenvectors are used as projection axes for the data. As a result if we project the data into the commute time embedding subspace, the normalized cut bipartition can be realized by simply dividing the projected data into two along the axis spanned by the Fiedler vector. Further partitions can be realized by projecting and dividing along the axes corresponding to the different scaled eigenvectors.

In Figure 2 we compare the result of embedding using commute time and the Laplacian eigenmap on a planar graph shown in Figure 1. The original graph is constructed by connecting two randomly generated planar graphs. The graph is un-weighted. We project the nodes of the graph onto the plane spanned by the two principle eigenvectors of the mapping. ¿From the figure, it is clear that both embeddings maintain the original graph structure, and that the two original graphs are well separated. However, compared to the Laplacian embedding, the points in the two original graphs are more densely distributed by the commute time embedding. Another significant advantage of the commute time embedding is that it preserves variance in a maximal way. This is illustrated in 2(b). Here two randomly generated graphs are embedded in two orthogonal planes.

2.2 The Commute Time and the Diffusion Map

Finally, it is interesting to note the relationship with the diffusion map embedding of Lafon *et al* [15]. The method commences from the random walk on a graph which has transition probability matrix $P = T^{-1}A$, where A is the adjacency matrix. Although P is not symmetric, it does have a right eigenvector matrix Ψ, which satisfies the equation

$$P\Psi = \Lambda_P \Psi \tag{11}$$

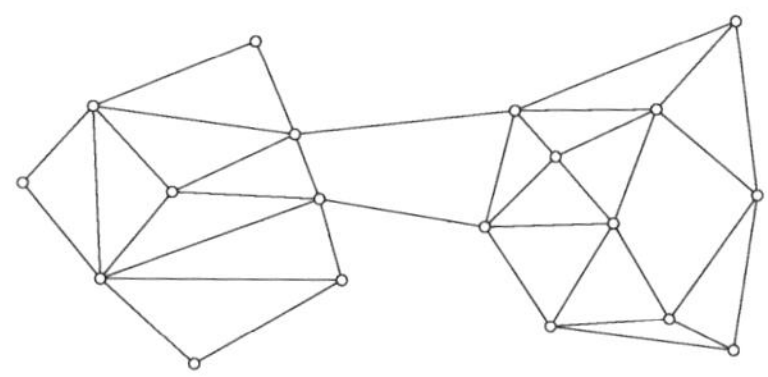

Fig. 1. Original planar graph

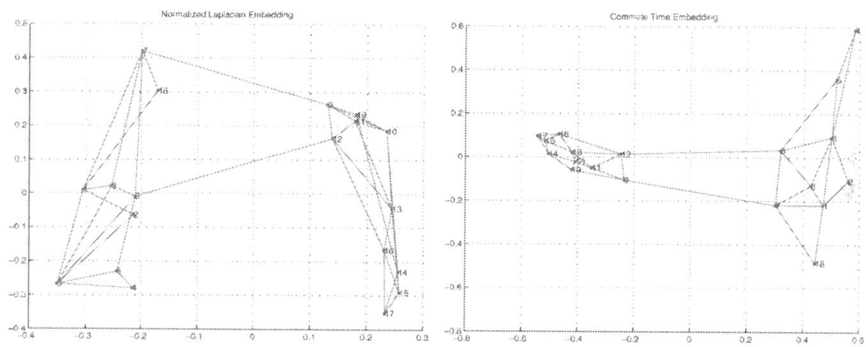

Fig. 2. Graph embedding comparison

Since $P = T^{-1}A = T^{-1}(T - L) = I - T^{-1}L$ and as result

$$\begin{aligned} (I - T^{-1}L)\Psi &= \Lambda_P \Psi \\ T^{-1}L\Psi &= (I - \Lambda_P)\Psi \\ L\Psi &= (I - \Lambda_P)T\Psi \end{aligned} \tag{12}$$

which is identical to Equation 7 if $\mathbf{Z} = \Psi$ and $\Lambda = I - \Lambda_P$ The embedding co-ordinate matrix for the diffusion map is $Y = \Lambda^t \Psi^T$, where t is real. For the embedding the diffusion distance between a pair of nodes is

$$D_t^2(u, v) = \sum_{i=1}^{m} (\lambda_P)_i^{2t} \left(\psi_i(u) - \psi_i(v)\right)^2$$

Clearly if we take $t = -1/2$ the diffusion map is equivalent to the commute time embedding and the diffusion time is equal to the commute time.

The diffusion map is designed to give a distance function that reflects the connectivity of the original graph or point-set. The distance should be small if a pair of points are connected by many short paths, and this is also the behaviour of the commute time. The advantage of the diffusion map or distance is that it has a free parameter t, and this may be varied to alter the properties of the map. The disadvantage is that when t is small, the diffusion distance is ill-posed. The reason for this is that according to the original definition of the diffusion distance for a random walk ($D_t^2(u, v) = \|p_t(u, \cdot) - p_t(v, \cdot)\|^2$), and as a result the distance between a pair of nodes depends on the transition probability between the nodes under consideration and all of the remaining nodes in the graph. As a result if t is small, then the random walk will not have propagated significantly, and the distance will depend only on very local information. There are also problems when t is large. When this is the case the random walk converges to its stationary state with $P^t = T/vol$ (a diagonal matrix), and this gives zero diffusion distance for all pairs of distinct nodes. So it is a critical to control t carefully in order to obtain useful embeddings.

3 Multi-body Motion Tracking Using Commute Time

In this section, we will show how the multi-body motion tracking problem can be posed as one of commute time embedding. Suppose there are N objects moving independently in a scene and the movement is acquired by an affine camera as F frames. In each frame, P feature points are tracked and the coordinate of the ith point in the fth frame is given by (x_i^f, y_i^f). Let X and Y denote two $F \times P$ matrices constructed from the image coordinates of all the points across all of the frames:

$$X = \begin{bmatrix} x_1^1 & x_2^1 & \cdots & x_P^1 \\ x_1^2 & x_2^2 & \cdots & x_P^2 \\ \vdots & \vdots & \ddots & \vdots \\ x_1^F & x_2^F & \cdots & x_P^F \end{bmatrix} \qquad Y = \begin{bmatrix} y_1^1 & y_2^1 & \cdots & y_P^1 \\ y_1^2 & y_2^2 & \cdots & y_P^2 \\ \vdots & \vdots & \ddots & \vdots \\ y_1^F & y_2^F & \cdots & y_P^F \end{bmatrix}$$

Each row in the two matrices above corresponds to a single frame and each column corresponds to a single point. The two coordinate matrices can be stacked to form the matrix

$$W = \begin{bmatrix} X \\ \hline Y \end{bmatrix}_{2F \times P}$$

The W matrix can be factorized into a motion matrix M and a shape matrix S thus, $W_{2F \times P} = M_{2F \times r} \times S_{r \times P}$ where r is the rank of W ($r = 4$ in the case of W without noise and outliers). In order to solve the factorization problem, matrix W can be decomposed by SVD:

$$W = U \Sigma R^T$$

If the features from the same object are grouped together, then U, Σ and R will have a block-diagonal structure.

$$W = [U_1 \cdots U_N] \begin{bmatrix} \Sigma_1 & & \\ & \ddots & \\ & & \Sigma_N \end{bmatrix} \begin{bmatrix} R_1^T & & \\ & \ddots & \\ & & R_N^T \end{bmatrix}$$

and the shape matrix for object k can be approximated by $S_k = B^{-1} \Sigma_k R_k^T$ where B is an invertible matrix that can be found from M.

In a real multi-body tracking problem, the coordinates of the different objects are potentially permuted into a random order. As a result it is impossible to correctly recover the shape matrix S_k without knowledge of the correspondence order. Since the eigenvector matrix V is related to the shape matrix, the shape interaction matrix was introduced by Costeira and Kanade [8,7] to solve the multi-body separation problem. The shape interaction matrix is

$$Q = RR^T = \begin{bmatrix} S_1^T \Sigma_1^{-1} S_1 & 0 & \cdots & 0 \\ 0 & S_2^T \Sigma_2^{-1} S_2 & \cdots & 0 \\ \vdots & \vdots & \ddots & 0 \\ 0 & 0 & \cdots & S_N^T \Sigma_N^{-1} S_N \end{bmatrix} \qquad (13)$$

¿From Equation 13, the shape interaction matrix Q has the convenient properties that $Q(u, v) = 0$, if points u,v belong to different objects and $Q(u, v) \neq 0$, if points u,v belong to the same object. The matrix Q is also invariant to both the object motion and the selection of the object coordinate systems.

Our aim is to use commute time as a shape separation measure. Specifically, we use the commute time to refine the block structure of the Q matrix and group the feature points into objects.

Object Separation Steps

The algorithm we propose for this purpose has the following steps:

1. Use the shape interaction matrix Q as the weighted adjacency matrix Ω and construct the corresponding graph Γ.
2. Compute the Laplacian matrix of graph Γ using $L = T - Q$.
3. Find the eigenvalue matrix Λ and eigenvector matrix Φ of L using $L = \Phi \Lambda \Phi^T$.
4. Compute the commute time matrix CT using Λ and Φ from Equation (3).
5. Embed the commute time into a subspace of R^n using Equation (4).
6. Cluster the data points in the subspace using the k-means algorithm [12].

4 Experiments

We commence in Figure 3 by showing four synthetic examples of point-configurations (left-hand panel) and the resulting commute time embeddings (right-hand panel). Here we have computed the proximity weight matrix Ω by exponentiating the Euclidean distance between points. The main features to note are as follows. First, the embedded points corresponding to the same point-clusters are cohesive, being scattered around approximately straight lines in the subspace. Second, the clusters corresponding to different objects give rise to straight lines that are orthogonal.

Turning our attention to the multi-body tracking example, the top row of Figure 4 shows images from five real-world video sequences. In the second row of the figure, we show the trajectories for the tracked points in each of the video sequences. Here the outliers are successfully removed. The different sequences offer tasks of increasing difficulty. The easiest sequence is the one labelled **A**, where background has a uniform and almost linear relative movement, and the foreground car follows a curved trajectory. There is a similar pattern in the sequence labelled **B**, but here the background movement is more significant. In sequence **C**, there is both camera pan and abrupt object

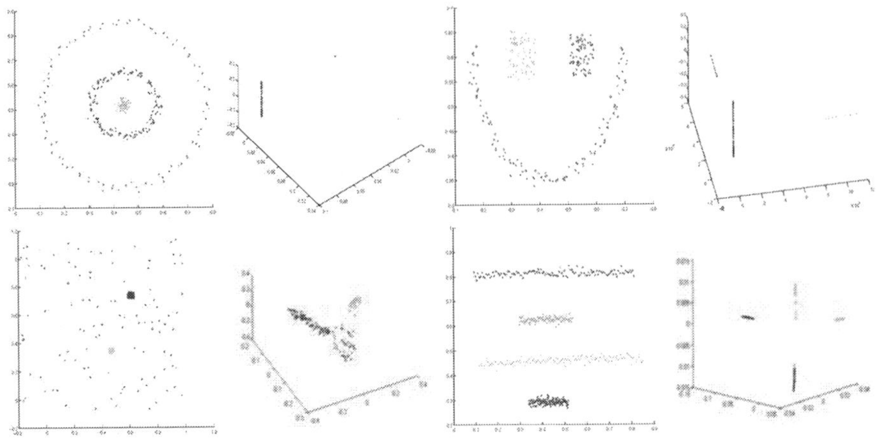

Fig. 3. Commute time embedding examples

movement. Sequence **D** has camera pan and three independently moving objects. In sequence **E** there is background jitter (due to camera shake) and two objects exhibiting independent overall movements together with articulations. Finally, in the third row of the figure, we show the embeddings of the tracked points for the sequences. The feature to note, is that the different moving objects form distinct clusters and are well separated from the background. The colour coding scheme used in the plot is the same as that used in the fifth column of Figure 4.

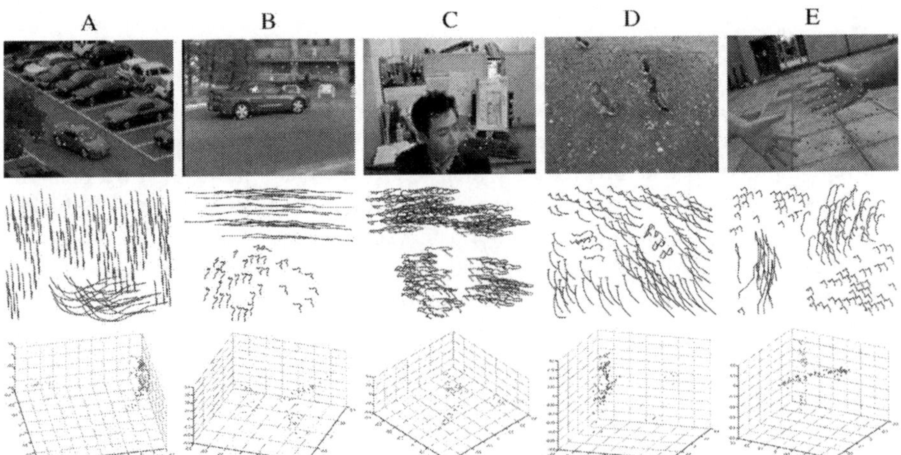

Fig. 4. Real-world video sequences and successfully tracked feature points

5 Conclusion

We have explored the theoretical properties of the commute time embedding, and have established a link with a number of alternative methods in the manifold learning literature. Experiment results show that the embedding maps different point clusters into approximately linear subspaces, that can be easily separated.

References

1. X. Bai, H. Yu, and E.R. Hancock. Graph matching using spectral embedding and alignment. In *ICPR*, pages 398–401, 2004.
2. M. Belkin and P. Niyogi. Laplacian eigenmaps and spectral techniques for embedding and clustering. In *Advances in Neural Information Processing Systems*, pages 585–591, 2001.
3. M. Belkin and P. Niyogi. Laplacian eigenmaps for dimensionality reduction and data representation. *Neural Computation*, 15(6):1373–1396, 2003.
4. T. Caelli and S. Kosinov. An eigenspace projection clustering method for inexact graph matching. *IEEE Trans. Pattern Anal. Mach. Intell.*, 26(4):515–519, 2004.
5. M. Carcassoni and E.R. Hancock. Spectral correspondence for point pattern matching. *Pattern Recognition*, 36(1):193–204, 2003.
6. F.R.K. Chung and S.-T. Yau. Discrete green's functions. In *J. Combin. Theory Ser.*, pages 191–214, 2000.
7. J. Costeira and T. Kanade. A multi-body factorization method for motion analysis. In *ICCV*, pages 1071–1076, 1995.
8. J. Costeira and T. Kanade. A multibody factorization method for independently moving objects. *IJCV*, 29(3):159 – 179, 1997.
9. D. Harel and Y. Koren. Graph drawing by high-dimensional embedding. In *GD '02: Revised Papers from the 10th International Symposium on Graph Drawing*, pages 207–219, 2002.
10. X. He and P. Niyogi. Locality preserving projections. In *NIPS*, pages 585–591, 2003.
11. B. Luo, R.C. Wilson, and E.R. Hancock. Spectral embedding of graphs. 2002 Winter Workshop on Computer Vision, 2002.
12. J. B. MacQueen. Some methods for classification and analysis of multivariate observations. In *Proceedings of the fifth Berkeley symposium on mathematical statistics and probability*, pages 281–297, 1967.
13. H. Qiu and E.R. Hancock. Image segmentation using commute times. In *BMVC*, pages 929–938, 2005.
14. S. Roweis and L. Saul. Nonlinear dimensionality reduction by locally linear embedding. *Science*, 290(5500):2323–2326, 2000.
15. R.R.Coifman, S. Lafon, A.B. Lee, M. Maggioni, B. Nadler, F. Warner, and S.W. Zucker. Geometric diffusions as a tool for harmonic analysis and structure definition of data: Diffusion maps. *National Academy of Sciences*, 102(21):7426–7431, 2005.
16. B. Sch, A. Smola, and K. Muller. Nonlinear component analysis as a kernel eigenvalue problem. *Neural Computation*, 10:1299–1319, 1998.
17. L. Shapiro and J. Brady. Feature-based correspondence: an eigenvector approach. *Image and Vision Computing*, 10(2):283–288, June 1992.
18. J. Shi and J. Malik. Normalized cuts and image segmentation. *IEEE PAMI*, 22(8):888–905, 2000.
19. J. B. Tenenbaum, V. de Silva, and J. C. Langford. A global geometric framework for nonlinear dimensionality reduction. *Science*, 290(5500):2319–2323, 2000.

Graph Based Multi-class Semi-supervised Learning Using Gaussian Process

Yangqiu Song, Changshui Zhang, and Jianguo Lee

State Key Laboratory of Intelligent Technology and Systems,
Department of Automation, Tsinghua University, Beijing, China, 100084
{songyq99, lijg01}@mails.tsinghua.edu.cn, zcs@mail.tsinghua.edu.cn

Abstract. This paper proposes a multi-class semi-supervised learning algorithm of the graph based method. We make use of the Bayesian framework of Gaussian process to solve this problem. We propose the prior based on the normalized graph Laplacian, and introduce a new likelihood based on softmax function model. Both the transductive and inductive problems are regarded as MAP (Maximum A Posterior) problems. Experimental results show that our method is competitive with the existing semi-supervised transductive and inductive methods.

1 Introduction

Graph based semi-supervised learning is an interesting problem in machine learning and pattern recognition fields [15,21]. The idea of using the unlabeled data for training will bring the geometric and manifold information to the algorithm, which may be effective to increase the recognition rate [1,2,8,18,19]. These graph based methods are either transducitve [1,18,19] or inductive [2,8]. For multi-class classification, it is easy to use the one-against-one or the one-against-the-rest method to construct a classifier based on a set of binary classification algorithms. Most of the existing graph based methods solve this as a one-against-the-rest problem. This is reasonable because it is only required to change the form of the labels of the data points, and do not need to modify the algorithm framework.

An alternative method to solve the multi-class problem is using the softmax function [16]. Softmax function is naturally derives from the log-linear model, and is convenient to describe the probabilitiy of each class. Therefore, it is useful to model this conditional probability rather than impose a Gaussian noise in the Bayesian classification framework. Gaussian process is an efficient nonparametric method which uses this softmax function model [16]. Lawrence and Jordan [10] developed a semi-supervised learning algorithm using fast sparse Gaussian Process: Informative Vector Machine (IVM). It is not a graph based algorithm. Zhu and Ghahramani [20] develope a semi-supervised binary classification framework of Gaussian process which is based on Gaussian fields, and they make use of a 1-NN approximation extending to the unseen points.

In this paper, we propose a novel algorithm to solve the multi-class semi-supervised learning problem. Similar to Gaussian process [16], both the training

D.-Y. Yeung et al. (Eds.): SSPR&SPR 2006, LNCS 4109, pp. 450–458, 2006.

and prediction phases can be regarded as MAP (Maximum A Posterior) estimation problems. The difference is that, a prior based on normalized graph Laplacian [5] is used, and a new conditional probability generalized from softmax function is proposed. Using this kind of Gaussian process, we can solve both the transductive and inductive problems.

This paper is organized as follows. Section 2 will present the details of multi-class semi-supervised learning algorithm. Experimental results will be given in section 3. Conclusion is given in section 4.

2 Semi-supervised Gaussian Process

First, we introduce some notations. We denote the input data point as a feature vector $\mathbf{x}_n$ $(n = 1, 2, ..., N)$, and $\mathbf{X}_N \triangleq \{\mathbf{x}_i\}_{i=1}^N$ is the observed data set include both labeled and unlabeled data. The label of the input data point is given by $\mathbf{t}_i$. The label set of all the observed training data is $\mathbf{T}_N = (\mathbf{t}_1, \mathbf{t}_2, ..., \mathbf{t}_N)^T$. For transductive problem, we need to re-estimate the label of the unlabeled data. For inductive problem, we want to estimate the label $\mathbf{t}_{N+1}$ of a new point $\mathbf{x}_{N+1}$. Instead of using the direct process $\mathbf{x} \to \mathbf{t}$, we adopt a latent variable $\mathbf{y}$ to generate a process as $\mathbf{x} \to \mathbf{y} \to \mathbf{t}$. We define the latent variable vector $\mathbf{y}_i = \mathbf{y}(\mathbf{x}_i)$ as functions of $\mathbf{x}_i$, and $\mathbf{Y}_N = (\mathbf{y}_1, \mathbf{y}_2, ..., \mathbf{y}_N)^T$ is the latent variable matrix of the input data. Then, some noisy function on the process $\mathbf{y} \to \mathbf{t}$ could be imposed. $P(\mathbf{Y}_N)$ is the prior, and the conditional probability is $P(\mathbf{T}_N|\mathbf{Y}_N)$.

2.1 The Likelihood

For a C classes problem, the softmax function model [16] sets the label as $\mathbf{t}_i = [0, ..., 0, \mathbf{t}_i^j = 1, 0, ..., 0]$, if $\mathbf{x}_i$ is belonged to class j $(j = 1, 2, ..., C)$. We extend the labels to the unlabeled data, which is set to a zero vector $\mathbf{t}_i = \mathbf{0}$ initially. Thus, the conditional probability $P(\mathbf{t}_i|\mathbf{y}_i)$ $(i = 1, 2, ..., N)$ of a C classes problem is given by:

$$P\left(\mathbf{t}_i|\mathbf{y}_i\right) = \frac{1}{C+1}\prod_{j=1}^{C}\left(\frac{C\exp\mathbf{y}_i^j}{\sum_{j=1}^{C}\exp\mathbf{y}_i^j}\right)^{\mathbf{t}_i^j} \tag{1}$$

We call this model as an extended softmax function model (ESFM), and we have $\sum_{\mathbf{t}_i \in (C+1)\ classes} P(\mathbf{t}_i|\mathbf{y}_i) \equiv 1$. The essence of ESFM is it modifies a C classes problem to be a $C+1$ classes problem. However, there has the difference between a traditional $C+1$ classes problem and our model. When $\mathbf{t}_i = \mathbf{0}$, we have $P(\mathbf{t}_i|\mathbf{y}_i) \equiv 1/(C+1)$. One of the softmax function $\frac{C}{C+1}(\exp\mathbf{y}_i^j/\sum_{j=1}^{C}\exp\mathbf{y}_i^j)$ will be determinately more than $1/(C+1)$. Therefore, this model will never classify a point to be unlabeled. By applying to a semi-supervised learning problem, this model will classify all the unlabeled data to the existent C classes. Then, the

Log-likelihood of the conditional probability is given by:

$$\begin{aligned} L = -\log P(\mathbf{T}_N|\mathbf{Y}_N) &= \sum_{i=1}^{N} L_i = \sum_{i=1}^{N} -\log P(\mathbf{t}_i|\mathbf{y}_i) \\ &= \sum_{i=1}^{N} -\left[\log \tfrac{1}{C+1} + \sum_{j=1}^{C} \mathbf{t}_i^j \left(\log C + \mathbf{y}_i^j - \log \sum_{m=1}^{C} \exp \mathbf{y}_i^m\right)\right] \end{aligned} \tag{2}$$

For computational simplicity, we rewrite the form of matrix $\mathbf{Y}_N$ and $\mathbf{T}_N$ as vectors: $(\mathbf{y}_1^1, \mathbf{y}_2^1, ..., \mathbf{y}_N^1, ..., \mathbf{y}_1^C, \mathbf{y}_2^C, ..., \mathbf{y}_N^C)^T$ and $(\mathbf{t}_1^1, \mathbf{t}_2^1, ..., \mathbf{t}_N^1, ..., \mathbf{t}_1^C, \mathbf{t}_2^C, ..., \mathbf{t}_N^C)^T$. Thus, by differentiating the logarithm of the likelihood probability we have:

$$\begin{aligned} & a_i^j = \sum_{m=1}^{C} \mathbf{t}_i^m \frac{\exp \mathbf{y}_i^j}{\sum_{m=1}^{C} \exp \mathbf{y}_i^m}, \quad b_i^j = \frac{\exp \mathbf{y}_i^j}{\sum_{m=1}^{C} \exp \mathbf{y}_i^m}, \quad c_i^j = a_i^j - \mathbf{t}_i^j \\ & \mathbf{\Pi}_1 = diag\left(a_1^1, a_2^1, ..., a_N^1, ..., a_1^C, a_2^C, ..., a_N^C\right) \\ & \mathbf{\Pi}_2 = \left(diag\left(a_1^1, a_2^1, ..., a_N^1\right), ..., diag\left(a_1^C, a_2^C, ..., a_N^C\right)\right) \\ & \mathbf{\Pi}_3 = \left(diag\left(b_1^1, b_2^1, ..., b_N^1\right), ..., diag\left(b_1^C, b_2^C, ..., b_N^C\right)\right) \\ & \boldsymbol{\alpha}_N = \nabla_{\mathbf{Y}_N}\left(-\log P(\mathbf{T}_N|\mathbf{Y}_N)\right) = \left(c_1^1, c_2^1, ..., c_N^1, ..., c_1^C, c_2^C, ..., c_N^C\right)^T \\ & \mathbf{\Pi}_N = \nabla\nabla_{\mathbf{Y}_N}\left(-\log P(\mathbf{T}_N|\mathbf{Y}_N)\right) = \mathbf{\Pi}_1 - \mathbf{\Pi}_2^T \mathbf{\Pi}_3 \end{aligned} \tag{3}$$

where $\boldsymbol{\alpha}_N$ and $\mathbf{\Pi}_N$ are the gradient vector and the Hessian matrix of $-\log P(\mathbf{T}_N|\mathbf{Y}_N)$ respectively.

2.2 The Prior

In Gaussian process, we take $P(\mathbf{Y}_N)$ as the prior. Many choices of the covariance functions for Gaussian process prior have been reviewed in [11], and covariance will affect the final classification significantly. We adopt the graph or manifold regularization based prior, which is also used for many other graph based methods [2,18]. In our framework, with the defination of $\mathbf{Y}_N$ above, we define:

$$P(\mathbf{Y}_N) = \frac{1}{Z} \exp\{-\frac{\mathbf{Y}_N^T \mathbf{K}_N^{-1} \mathbf{Y}_N}{2}\} \tag{4}$$

$\mathbf{K}_N$ is an $NC \times NC$ block diagonal matrix, which has the Kronecker product form:

$$\mathbf{K}_N^{-1} = \nabla\nabla_{\mathbf{Y}_N}\left(-\log P(\mathbf{Y}_N)\right) = \mathbf{I}_C \otimes \mathbf{\Delta} \tag{5}$$

where $\mathbf{I}_C$ is a $C \times C$ identity matrix, and the matrix:

$$\mathbf{\Delta} = \mathbf{I} - \mathbf{S} \tag{6}$$

is called normalized graph Laplacian [1] in spectral graph theory [5], and the prior $P(\mathbf{Y}_N)$ defines a Gaussian random field (GRF) [19] on the graph. $\mathbf{S} = \mathbf{D}^{-\frac{1}{2}} \mathbf{W} \mathbf{D}^{-\frac{1}{2}}$, $(\mathbf{W})_{ij} = W_{ij}$ and $\mathbf{D} = diag(D_{11}, ..., D_{NN})$. $D_{ii} = \sum_j W_{ij}$, and W_{ij} is called

[1] We also introduce an extra regularization, since the normalized graph Laplacian is a semi definite positive matrix [5]. In particular, we set $\mathbf{\Delta} \leftarrow (\mathbf{\Delta} + \delta\mathbf{I})$, and fix the parameter in the experiments.

the weight function associated with the edges on graph, which satisfies: $W_{ij} > 0$ *and* $W_{ij} = W_{ji}$. It can be viewed as a symmetric similarity measure between $\mathbf{x}_i$ and $\mathbf{x}_j$.

The covariance matrix $\mathbf{K}_N$ is relative to the inverse matrix of the normalized graph Laplacian, so the covariance between two points is depend on all the other training data including both labeled and unlabeled [20]. In contrast, most of the traditional Gaussian processes adopt the Gram matrix based on "local" distance information to construct the covariance [11,20].

2.3 Transduction and Induction

In Gaussian process, both the training phase and the prediction phase can be regarded as MAP (Maximum A Posterior) estimation problems [16] . In the training phase, we estimate the latent variables $\mathbf{Y}_N$ from the training data. By computing the mode of posterior probability $P(\mathbf{Y}_N|\mathbf{T}_N)$ as the estimate of $\mathbf{Y}_N$, which is the negative logarithm of $P(\mathbf{Y}_N|\mathbf{T}_N) = P(\mathbf{Y}_{\mathrm{N}}, \mathbf{T}_N)/P(\mathbf{T}_N)$, we define:

$$\Psi(\mathbf{Y}_N) = -\log P(\mathbf{T}_N|\mathbf{Y}_N) - \log P(\mathbf{Y}_N) \tag{7}$$

$P(\mathbf{T}_N)$ is omitted for it is a constant unrelated to $\mathbf{Y}_N$. $P(\mathbf{Y}_{\mathrm{N}})$ is the prior, which is based on graph regularization. $P(\mathbf{T}_N|\mathbf{Y}_N)$ is the new conditional probability: extended softmax function model (ESFM). Therefore, Laplace approximation method [16] can be used for estimating $\mathbf{Y}_N$ from the posterior. To find the minimum of Ψ in equation (7), the Newton-Raphson iteration [16] is adopted:

$$\mathbf{Y}_N{}^{new} = \mathbf{Y}_N - (\nabla\nabla\Psi)^{-1}\nabla\Psi \tag{8}$$

Where:

$$\nabla\Psi = \boldsymbol{\alpha}_N + \mathbf{K}_N^{-1}\mathbf{Y}_N, \quad \nabla\nabla\Psi = \boldsymbol{\Pi}_N + \mathbf{K}_N^{-1} \tag{9}$$

Since $\nabla\nabla\Psi$ is always positive definite, (7) is a convex problem. When it converges to an optimal $\hat{\mathbf{Y}}_N$, $\nabla\Psi$ will be zero vector. The posterior probability $P(\mathbf{Y}_N|\mathbf{T}_N)$ can be approximated as Gaussian, being centered at the estimated $\hat{\mathbf{Y}}_N$. The covariance of the posterior is $\nabla\nabla\Psi$. After the Laplace approximation, the latent variable vector $\hat{\mathbf{Y}}_N$ will depend on all the training data, and the corresponding $\hat{\mathbf{T}}_N$ could be updated.

In the prediction phase, the objective function is:

$$\Psi(\mathbf{Y}_N, \mathbf{y}_{N+1}) = -\log P(\mathbf{T}_{N+1}|\mathbf{Y}_{N+1}) - \log P(\mathbf{Y}_{N+1}) \tag{10}$$

which is minimized only with respect to $\mathbf{y}_{N+1}$. This leads to:

$$\hat{\mathbf{y}}_{N+1} = \mathbf{K}_{\mathrm{N+1}}{}^T\mathbf{K}_{\mathrm{N}}{}^{-1}\hat{\mathbf{Y}}_N \tag{11}$$

where $\mathbf{K}_{N+1} = \mathbf{I}_C \otimes \mathbf{k}$, and $\mathbf{k}_i = W_{N+1,i} = \exp(-||\mathbf{x}_{N+1} - \mathbf{x}_i||^2/2\sigma^2)$ is the covariance of a new given point and the *ith* training point. Note that, in the training phase, we have: $\nabla\Psi = \hat{\boldsymbol{\alpha}}_N + \mathbf{K}_{\mathrm{N}}{}^{-1}\hat{\mathbf{Y}}_N = 0$. Thus, equation (11) is given by:

$$\hat{\mathbf{y}}_{N+1} = -\mathbf{K}_{\mathrm{N+1}}{}^T\hat{\boldsymbol{\alpha}}_N \tag{12}$$

We substitute the estimated $\hat{\mathbf{Y}}_N$ and $\hat{\mathbf{T}}_N$ to equation (3) to realize the prediction. The form of predictive function (12) has a relationship with the RKHS [14]. Other semi-supervised inductive methods (such as [2]) also show how to find an appropriate RKHS using the Representor Theorem.

The training phase has the same framework as the transductive learning algorithm [18]. [18] imposes a Gaussian noise model and uses the one-against-the-rest way to deal with the multi-class problem. We generalize the softmax function to directly add the unlabeled data to the formulation. Our ESFM will reduce to the one-against-the-rest problem if the off-diagonal blocks of matrix $\mathbf{\Pi}_N$ are zeros. This means $\mathbf{y}_i^j$ and $\mathbf{y}_i^k$ $(k \neq j)$ are decoupled. $\mathbf{y}_i^j$ is larger than the other $\mathbf{y}_i^k$ $(k \neq j)$ if a point $\mathbf{x}_i$ belongs to class j, and $\mathbf{y}_i^j$ is smaller when $\mathbf{x}_i$ is other classes. Although we have modeled the probability of a point belonging to each class, the training phase scales to $O(C^3N^3)$ computational complexity. In [16], the authors point that this could be reduce to $O(CN^3)$ by using Woodbury formula. Further more, the inverse of matrix could be approximated by using Nyström method [17]. Thus, the training phase will reduce to $O(CNM^2)$ computational complexity, where M is the number of a small subset of the training points. On the contrary, the prediction phase has decoupled. We could calculate $\mathbf{y}_i^j$ for each class respectively. The prediction phase is $O(CNN_{test})$ computational complexity.

2.4 Hyper-parameter Estimation

This section mainly follows [16,20]. The hyper-parameter is the standard deviation σ of the RBF kernel. We estimate the hyper-parameter which minimizes the negative logarithmic likelihood:

$$\begin{aligned} J(\sigma) &= \log(-P(\mathbf{T}_N|\sigma)) \\ &\approx \sum_{i=1}^{N}\sum_{j=1}^{C} \mathbf{t}_i^j (\log \sum_{m=1}^{C} \exp \mathbf{y}_i^m - \mathbf{y}_i^j) + \frac{1}{2}\log|\mathbf{K}_{\mathrm{N}}\mathbf{\Pi}_{\mathrm{N}} + \mathbf{I}| + \frac{1}{2}\mathbf{Y}_N^T\mathbf{K}_N^{-1}\mathbf{Y}_N \end{aligned} \tag{13}$$

The derivation of the objective function (13) is:

$$\begin{aligned} \frac{\partial J(\sigma)}{\partial \sigma} &= \boldsymbol{\alpha}_N^T \frac{\partial \mathbf{Y}_N}{\partial \sigma} + \frac{1}{2} tr\left[(\mathbf{I} + \mathbf{K}_{\mathrm{N}}\mathbf{\Pi}_{\mathrm{N}})^{-1}\frac{\partial \mathbf{K}_{\mathrm{N}}\mathbf{\Pi}_{\mathrm{N}}}{\partial \sigma}\right] \\ &+ \frac{1}{2}\left[2\left(\mathbf{K}_N^{-1}\mathbf{Y}_N\right)^T \frac{\partial \mathbf{Y}_N}{\partial \sigma} + \mathbf{y}_N^T \frac{\partial \mathbf{K}_{\mathrm{N}}^{-1}}{\partial \sigma}\mathbf{Y}_N\right] \end{aligned} \tag{14}$$

3 Experiments

3.1 Toy Data

We test the multi-class problem with a toy data. The training data set contains 7 classes. There are three Gaussian, two moon and two round shapes in the figure. As Fig.1 (a) shows, each of the Gaussian distribution has 25 points, and each

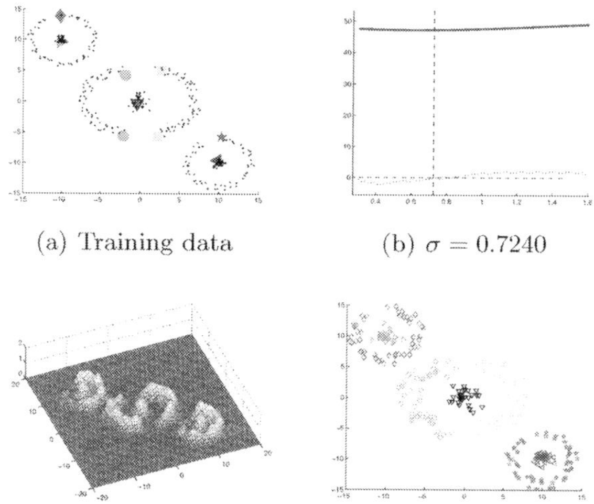

(a) Training data (b) $\sigma = 0.7240$

(c) $\sum_j \exp \mathbf{y}_j$ (d) Test data result

Fig. 1. A multi-class toy problem. The object value is plotted in (b) as blue line, and the derivation is the red dot line.

class of moon and round shape data has 50 points. Only 9 points are labeled. The hyper-parameter estimation result is shown in Fig.1 (b). Fig.1 (c) shows the mash result of the sum of estimated function $\sum_j \exp \mathbf{y}_j$ on the 2-D space. We can see that, the corresponding estimated $\exp \mathbf{y}_j$ be very large if the nearby has training points. The labeled points can also affect the estimated result. See the round shape in figure. The round shape in Fig.1 (a) is not closed, so the estimated function is smaller when the point is far (measured by geodesic distance) from the labeled point on the manifold. Finally, the classification result of the test set are shown in Fig.1 (d).

3.2 Real Data

In this experiment, we test some state-of-the-art algorithms and ours on the real data sets. Some of the images are resized, and all the vectors in these data sets are normalized to the range from 0 to 1. The data sets are (More details of the settings are shown in Table 1):

1. USPS Data [9]: We choose digits "1"-"4", and there are 1269, 929, 824, and 852 examples for each class.
2. 20-Newsgroups Data [18]: There are four topics: "autos", "motorcycles", "baseball" and "hockey". The documents have been normalized in 3970 TFIDF vectors in 8014 dimensional space.
3. COIL-20 Data [12]: We pick the first 5 of the 20 objects to test the algorithms. Each object has 72 images, which were taken at pose intervals of 5 degrees.

4. ORL Face Data [13]: There are 10 different images of each of 40 distinct subjects in the ORL data set. We choose the 15th-30th subjects.
5. Yale Face Data [6]: This data set contains 165 images of 15 individuals.
6. UMIST Face Data [7]: This set consists of 564 images of 20 people. Each covering a range of poses from profile to frontal views. The first 10 subjects are selected.
7. UCI Image Segmentation data [3]: The instances of this set were drawn randomly from a database of 7 outdoor images. We choose 4 classes, and each class has tolally 330 samples including both training and test data.
8. UCI Optical Recognition of Handwritten Digits [3]: We also choose the digits "1"-"4", and there are 571, 557, 572 and 568 examples for each class.

We test several algorithms for comparison: the supervised methods SVM (Support Vector Machine) [4] and RLS (Regularized Least Squares) [2]; transductive method TOG (Transduction On Graphs) [18]; the inductive method LapRLS (Laplacian Regularized Least Squares) [2]; and our method SSGP (Semi-Supervised Gaussian Process). SVM uses the one-against-one scheme, while RLS, LapRLS and TOG use the one-against-the-rest scheme. Each test accuracy of the results is an average of 50 random trials. For each trial, we randomly choose a subset of the data. The selected data are dealt as a splitting seen (include labeled and unlabeled data) and unseen data sets. For supervised methods, we only use the labeled points in the seen data for training. For semi-supervised inductive methods, we use all the seen data to classify the unseen. For TOG, we run two times for each iteration. First, we run it on the seen set, and evaluate the accuracy again on the seen set. Second, we use both the seen and unseen to train another TOG algorithm, and evaluate the accuracy on the unseen set. Moreover, we use the weight $W_{ij} = \exp(-\frac{1}{2\sigma^2}||\mathbf{x}_i - \mathbf{x}_j||^2)$ to construct the graph. Unlike the other data sets, to deal with the 20-Newsgroups data, we construct a 10-NN weighted graph instead of a fully connected graph. The weight on the graph is changed to be $W_{ij} = \exp(-\frac{1}{2\sigma^2}(1 - \frac{\langle \mathbf{x}_i, \mathbf{x}_j \rangle}{||\mathbf{x}_i|| \cdot ||\mathbf{x}_j||}))$. The empirically selected parameters are also shown in Table 1.

Fig.2 shows the results. We can see that, for data distributed on a manifold (Fig.2 (a) (c) (f) (g) (h)), the test accuracy of TOG is the best. This is because it

Table 1. Experimental Settings and Empirically Selected Parameters

Data Set	Class	Dim	N_{subset}	Seen	Unseen	σ_{Graph}	σ_{SVM}	C_{SVM}
USPS	4	256	1000	50%	50%	1.25	5	1
20-Newsgroups	4	8014	1000	50%	50%	0.15	10	1
COIL-20	5	4096	360	80%	20%	1.25	10	1
ORL	15	2576	150	80%	20%	1	10	1
YALE	15	4002	165	70%	30%	2.5	12	3
UMIST	15	2576	265	60%	40%	1	6	5
UCI Image	4	19	800	50%	50%	0.1	1.5	3
UCI Digits	4	64	800	50%	50%	0.15	1.5	1

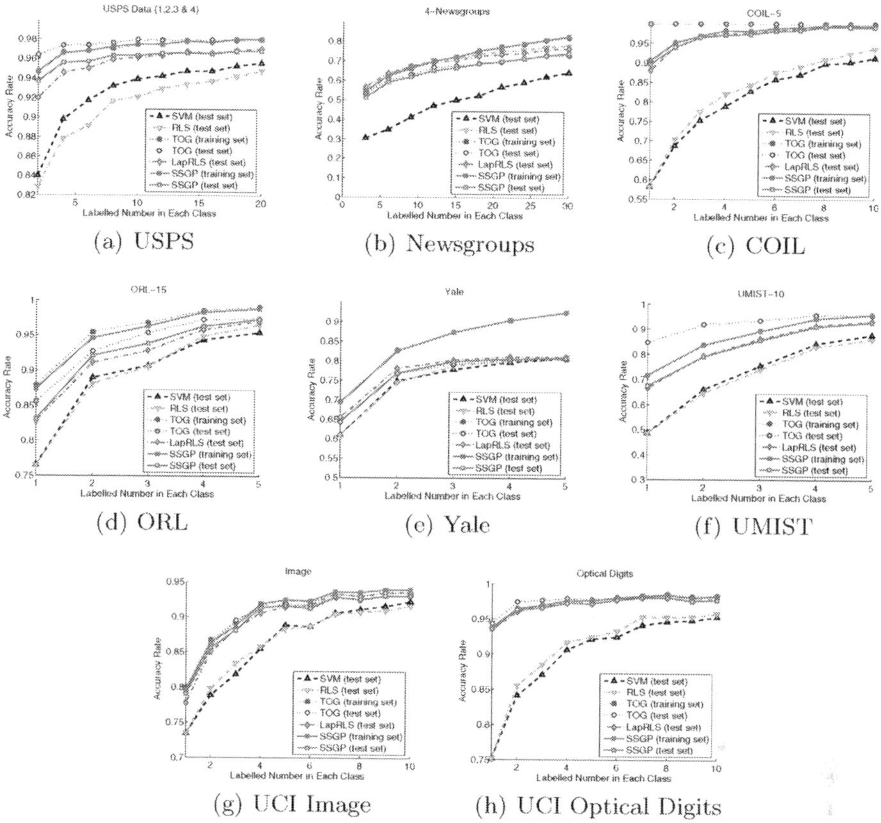

(a) USPS (b) Newsgroups (c) COIL

(d) ORL (e) Yale (f) UMIST

(g) UCI Image (h) UCI Optical Digits

Fig. 2. Real Data

uses both the seen and unseen data, and the unseen data provide more geometric information. For the data do not show explicit manifold information (Fig.2 (b) (d) (e)), the transductive metheds on seen set give the best result. The results of SSGP on the seen set is competetive with TOG on the seen set. Moreover, SSGP is also competetive with the semi-supervised inductive method LapRLS. For most data sets, SSGP and LapRLS do better than the supervised methods SVM and RLS, since semi-supervised methods use the information provided by the unlabeled data. However, see the Yale data set for example, semi-supervised methods only do litter better than the supervised methods, even TOG training based on the whole set could not give much better result.

4 Conclusion

This paper proposes a novel graph based multi-class semi-supervised algorithm. It can work on both seen and unseen data. The accuracy rate is competetive with the existing transducitve and inductive methods. If the data present explicit structure of a manifold, the graph based semi-supervised learning algorithms work efficiently for both transducitve and inductive problems. In the future, we would like to do some research on the methods to speed up the algorithm.

References

1. Belkin, M. and Niyogi, P.: Using manifold structure for partially labeled classification. *NIPS* Vol.15, pp. 929-936, (2003).
2. Belkin, M., Niyogi, P. and Sindhwani, V.: On Manifold Regularization. *AI and Statistics*, pp. 17-24, (2005).
3. Blake, C. L. and Merz, C. J.: UCI Repository of Machine Learning Databases *http://www.ics.uci.edu/ mlearn/MLRepository.html.*
4. Chang C., Lin C.: LIBSVM: A Library for Support Vector Machines. (2001).
5. Chung, F.: *Spectral Graph Theory.* No. 92 in Tegional Conference Series in Mathematics. American Mathematical Society (1997).
6. Georghiades A., Belhumeur P. and Kriegman D.: From Few to Many: Illumination Cone Models for Face Recognition under Variable Lighting and Pose. IEEE Trans on *PAMI*, Vol. 23(6), pp. 643-660, (2001).
7. Graham D. and Allinson N.: Characterizing Virtual Eigensignatures for General Purpose Face Recognition. In *Face Recognition: From Theory to Applications*, Vol. 163, pp. 446-456, (1998).
8. Delalleau, O. Bengio, Y. and Roux, N. L.: Efficient Non-Parametric Function Induction in Semi-Supervised Learning. *AI and Statistics*, pp. 96-103, (2005).
9. Hull, J.: A database for handwritten text recognition research. IEEE Trans. on *PAMI*, Vol. 16(5), pp. 550-554, (1994).
10. Lawrence, N. D. and Jordan, M. I.: Semi-supervised learning via Gaussian processes. *NIPS*, Vol. 17, pp. 753-760, (2005).
11. Mackay D.: Introduction to Gaussian processes. *Technical Report*, (1997).
12. Nene S. A., Nayar S. K. and Murase H.: Columbia Object Image Library (COIL-20), *Technical Report*, (1996).
13. ORL Face Database. *http://www.uk.research.att.com/facedatabase.html.*
14. Seeger M.: Relationships between Gaussian processes, Support Vector machines and Smoothing Splines. *Technical Report*, (1999).
15. Seeger, M.: Learning with Labeled and Unlabeled Data. *Technical Report*, (2000).
16. Williams, C. K. I. and Barber, D.: Bayesian Classification with Gaussian Processes. *IEEE Trans. on PAMI*, Vol. 20(12), pp. 1342-1351, (1998).
17. Williams C. and Seeger M.: Using the Nyström Method to Speed Up Kernel Machines. *NIPS*, Vol. 13, pp. 682-688, (2001).
18. Zhou, D., Bousquet, O., Lal, T. N., Weston, J. and Schölkopf, B.: Learning with Local and Global Consistency. *NIPS*, Vol. 16, pp. 321-328, (2003).
19. Zhu, X., Ghahramani, Z. and LaKerty, J.: Semi-Supervised Learning Using Gaussian Fields and Harmonic Functions. *ICML*, Vol. 20, pp. 912-919, (2003).
20. Zhu, X. and Ghahramani, Z.: Semi-Supervised Learning: From Gaussian Fields to Gaussian Processes. *Technical Report*, (2003).
21. Zhu, X.: Semi-Supervised Learning Literature Survey. *Technical Report*, (2005).

Point Pattern Matching Via Spectral Geometry*

Antonio Robles-Kelly[1] and Edwin R. Hancock[2]

[1] National ICT Australia, RSISE Bldg 115, ANU, ACT 0200, Australia
Antonio.Robles-Kelly@anu.edu.au
[2] Dept. of Comp. Science, The University of York, York YO10 5DD, UK
erh@cs.york.ac.uk

Abstract. In this paper, we describe the use of Riemannian geometry, and in particular the relationship between the Laplace-Beltrami operator and the graph Laplacian, for the purposes of embedding a graph onto a Riemannian manifold. Using the properties of Jacobi fields, we show how to compute an edge-weight matrix in which the elements reflect the sectional curvatures associated with the geodesic paths between nodes on the manifold. We use the resulting edge-weight matrix to embed the nodes of the graph onto a Riemannian manifold of constant sectional curvature. With the set of embedding coordinates at hand, the graph matching problem is cast as that of aligning pairs of manifolds subject to a geometric transformation. We illustrate the utility of the method on image matching using the COIL database.

1 Introduction

The problem of embedding relational structures onto manifolds is an important one in computer science. Furthermore, in the pattern analysis community, there has recently been renewed interest in the use of embedding methods motivated by graph theory. One of the best known of these is ISOMAP [14]. Related algorithms include locally linear embedding which is a variant of PCA that restricts the complexity of the input data using a nearest neighbor graph [11] and the Laplacian eigenmap that constructs an adjacency weight matrix for the data-points and projects the data onto the principal eigenvectors of the associated Laplacian matrix [1]. Lafferty and Lebanon [8] have proposed a number of kernels for statistical learning which are based upon the heat equation on a Riemannian manifold.

Embedding methods can also be used to transform the graph-matching problem into one of point-pattern alignment. The problem is to find matches between pairs of point sets when there is noise, geometric distortion and structural corruption. There is a considerable literature on the problem and many contrasting approaches, including relaxation [4] and optimisation [6], have been attempted. However, the main challenge in graph matching is how to deal with differences in node and edge structure. One of the most elegant recent approaches to the graph matching problem has been to use graph-spectral methods [5], and exploit information conveyed by the eigenvalues and eigenvectors of the adjacency matrix. For instance, Umeyama [16] has developed a method

* National ICT Australia is funded by the Australian Governments Backing Australia's Ability initiative, in part through the Australian Research Council.

D.-Y. Yeung et al. (Eds.): SSPR&SPR 2006, LNCS 4109, pp. 459–467, 2006.

for finding the permutation matrix which best matches pairs of weighted graphs of the same size, by using a singular value decomposition of the adjacency matrices. Scott and Longuet-Higgins [12], on the other hand, align point-sets by performing singular value decomposition on a point association weight matrix. Shapiro and Brady [13] have reported a correspondence method which relies on measuring the similarity of the eigenvectors of a Gaussian point-proximity matrix. Kosinov and Caelli [2] have improved this method by allowing for scaling in the eigenspace.

Our aim in this paper is to seek an embedding of the nodes of a graph which allows matching to be effected using simple point-pattern matching methods. In particular, we aim to draw on the field of mathematics known as spectral geometry, which aims to characterise the properties of operators on Riemannian manifolds using the eigenvalues and eigenvectors of the Laplacian matrix [3]. This approach has a number of advantages. Firstly, our definition of the edge weight is linked directly to the geometry of the underlying manifold. Secondly, the relationship between the Laplace-Beltrami operator and the graph Laplacian provides a clear link between Riemannian geometry and graph-spectral theory [5]. Furthermore, by making use of the Laplace-Beltrami operator to relate the apparatus of graph-spectral theory to Riemannian geometry, the results presented here allow a better understanding of these methods. Finally, the recovery of the embedding coordinates and the geometric transformation via linear algebra yields an analytical solution which is devoid of free parameters.

2 Riemannian Geometry

In this section, we provide the theoretical basis for our graph embedding method. Our aim is to embed the graph nodes as points on a Riemannian manifold. We do this by viewing pairs of adjacent nodes in a graph as points connected by a geodesic on a manifold. In this way, we can make use of Riemannian invariants to recover the embedding of the point pattern on the manifold. With this characterisation at hand, we show how the properties of the Laplace-Beltrami operator can be used to recover a matrix of embedding coordinates. We do this by establishing a link between the Laplace-Beltrami operator and the graph Laplacian. This treatment allows us to relate the graph Laplacian to a Gram matrix of scalar products, whose entries are, in turn, related to the squared distances between pairs of points on the Riemannian manifold.

2.1 Riemannian Manifolds

In this section, we aim to provide a means of characterising the edges of a graph using a geodesic on a Riemannian manifold. The weight of the edge is the cost or energy associated with the geodesic. To commence, let $G = (V, E, W)$ denote a weighted graph with index-set V, edge-set $E = \{(u, v)|(u, v) \in V \times V, u \neq v\}$ and the edge-weight function is $W : E \rightarrow [0, 1]$. If the nodes in the graph are viewed as points on the manifold, then the weight $W_{u,v}$ associated with the edge connecting the pair of nodes u and v can be computed using the the energy $\mathcal{E}_{p_u,p_v}$ over the geodesic connecting the pair of points p_u and p_v on the manifold. To do this, we employ concepts from differential geometry [3, 10]. In this way, we establish a relationship with the curvature tensor, which, in turn, allows us to characterise the sectional curvature of the manifold. The reasons for using the curvature tensor are twofold. Firstly, the curvature tensor is natural, i.e. it is invariant under isometries (that is bijective mappings that preserve distance). Secondly, the curvature tensor can be defined intrinsically through coordinate changes. Hence, the curvature tensor is one of the main invariants in Riemannian geometry.

To commence our development, we require some formalism. Let the vector fields Y, X and Z be the extensions over a neighbourhood of the point $p \in M$ of the vectors $\eta, \xi, \zeta \in M_p$. The curvature tensor, which is quadrilinear in nature [3], is denoted by $R(\xi, \eta)$. To obtain a bilinear form, i.e. the sectional curvature, from the curvature tensor we use two linearly independent vectors $\eta, \xi \in M_p$ and write

$$\mathcal{K}(\xi, \eta) = \frac{\langle R(\xi, \eta)\xi, \eta\rangle}{\mid \xi \mid^2 \mid \eta \mid^2 - \langle \xi, \eta \rangle} \tag{1}$$

Further, consider the parametric geodesic curve $\gamma : t \in [\alpha, \beta] \mapsto M$. We define the Jacobi field along γ as the differentiable vector field $Y \in M_p$, orthogonal to γ, satisfying Jacobi's equation $\nabla_t^2 Y + R(\gamma', Y)\gamma' = 0$, where ∇ is said to be a Levi-Civita connection [3].

With these ingredients, we model the edges in the graph as geodesics in a manifold by substituting the shorthands for the derivative of the parametric geodesic curve $\gamma : t \in [\alpha, \beta]$ with respect to the time parameter t, i.e. γ', and the Jacobi field Y into the expression for the sectional curvature introduced in Equation 1. We get

$$\mathcal{K}(\gamma', Y) = \frac{\langle R(\gamma', Y)\gamma', Y\rangle}{\mid \gamma' \mid^2 \mid Y \mid^2 - \langle \gamma', Y \rangle} \tag{2}$$

To simplify the expression for the sectional curvature further, we make use of the fact that, since Y is a Jacobi field, it must satisfy the condition $\nabla_t^2 Y = -R(\gamma', Y)\gamma'$. Hence, we can write

$$\mathcal{K}(\gamma', Y) = \frac{\langle R(\gamma', Y)\gamma', Y\rangle}{\mid \gamma' \mid^2 \mid Y \mid^2} = \frac{\langle -\nabla_t^2 Y, Y\rangle}{\langle Y, Y\rangle} \tag{3}$$

where we have used the fact that Y is orthogonal to γ', substituted $\mid Y \mid^2$ with $\langle Y, Y\rangle$ and set $\mid \gamma' \mid = 1$. As a result, it follows that $\nabla_t^2 Y = -\mathcal{K}(\gamma', Y)Y$. Hence, the Laplacian operator $\nabla_t^2 Y$ is determined by the sectional curvature of the manifold.

This suggests a way of formulating the energy over the geodesic $\gamma \in M$ connecting the pair of points corresponding to the nodes indexed u and v. Consider the geodesic γ subject to the Jacobi field Y. The energy over the geodesic γ can be expressed making use of the equations above as

$$\mathcal{E}_{p_u, p_v} = \int_\gamma \mid \gamma' + \nabla_t^2 Y \mid^2 dt = \int_\gamma \mid \gamma' - \mathcal{K}(\gamma', Y)Y \mid^2 dt \tag{4}$$

We can provide a physical intepretation of the above result. It can be viewed as the energy associated with the geodesic from the point indexed u to the point indexed v, which is the sum of the kinetic energy and the potential energy contributed by the Jacobi field over γ. Hence, the edge-weight is small if a pair of points are close to one another or the curvature along the geodesic between them is small.

In practice, we will confine our attention to the problem of embedding the nodes on a constant sectional curvature surface. For such a surface, the sectional curvature is constant i.e. $\mathcal{K}(\gamma', Y) \equiv \kappa$. Under this restriction the Jacobi field equation becomes $\nabla_t^2 Y = -\kappa Y$. With the boundary conditions $Y(0) = 0$ and $\mid \nabla_t Y(0) \mid = 1$, the solution is

$$Y(t) = \begin{cases} \frac{\sin(\sqrt{\kappa}t)}{\sqrt{\kappa}}\eta & \text{if } \kappa > 0 \\ t\eta & \text{if } \kappa = 0 \\ -\frac{\sinh(\sqrt{-\kappa}t)}{\sqrt{-\kappa}}\eta & \text{if } \kappa < 0 \end{cases} \tag{5}$$

where the vector η is in the tangent space of M at p_u and is orthogonal to γ' at the point indexed u, i.e. $\eta \in M_{p_u}$ and $\langle \eta, \gamma' \mid_{p_u} \rangle = 0$.

With these ingredients, and by rescaling the parameter t so that $\mid \gamma' \mid = a(u, v)$, we can express the weight of the edge connecting the nodes indexed u and v as follows

$$W(u,v) = \begin{cases} \int_0^1 \left(a(u,v)^2 + \kappa \left(\sin(\sqrt{\kappa}a(u,v)t) \right)^2 \right) dt & \text{if } \kappa > 0 \\ \int_0^1 a(u,v)^2 dt & \text{if } \kappa = 0 \\ \int_0^1 \left(a(u,v)^2 - \kappa \left(\sinh(\sqrt{-\kappa}a(u,v)t) \right)^2 \right) dt & \text{if } \kappa < 0 \end{cases} \tag{6}$$

where $a(u, v)$ is the Euclidean distance between each pair of points in the manifold, i.e. $a(u, v) = \| p_u - p_v \|$.

2.2 Recovery of the Embedding Coordinates

To construct a set of embedding coordinates for the nodes of the graph, we use multidimensional scaling with double centering [15]. We depart from a matrix of embedding coordinates $\mathbf{J}$ obtained from the centred Laplacian using the factorisation $\mathbf{H} = \mathbf{J}\mathbf{J}^T$. The double centering procedure introduces a linear dependency over the columns of the matrix. The double-centered graph Laplacian $\mathbf{H}$ is, in fact, a Gram matrix and, thus, we can recover the node-coordinates making use of a matrix decomposition approach. We construct the centering matrix as follows

$$\mathbf{H} = -\frac{1}{2}\mathbf{B}\mathcal{L}\mathbf{B}^T \tag{7}$$

where $\mathbf{B} = \mathbf{I} - \frac{1}{|V|}\mathbf{e}\mathbf{e}^T$ is the centering matrix, $\mathbf{I}$ is the identity matrix, $\mathbf{e}$ is the all-ones vector and $\mathcal{L} = \boldsymbol{D}^{-\frac{1}{2}}(\boldsymbol{D} - W)\boldsymbol{D}^{-\frac{1}{2}}$ is the normalised graph Laplacian. In the expression above, $\boldsymbol{D}$ is a diagonal matrix such that $\boldsymbol{D} = diag(deg(1), deg(2), \ldots, deg(|V|))$ and $deg(i)$ is the degree of the node indexed i.

It is also worth noting that the double centering operation on the graph Laplacian also has the effect of translating the coordinate system for the embedding to the origin. This allows us to pose the problem of matching as an alignment one that involves only rotation.

To perform this factorisation of the matrix $\mathbf{H}$, we make use of Young-Householder theorem [17]. Let $\Lambda = diag(\lambda_1, \lambda_2, .., \lambda_{|V|})$ be the diagonal matrix with the ordered eigenvalues of $\mathbf{H}$ as elements and $\Phi = (\phi_1|, |\phi_2|, \ldots, |\phi_{|V|})$ be the matrix with the corresponding ordered eigenvectors as columns. Here the ordered eigenvalues and corresponding eigenvectors of the matrix $\mathbf{H}$ satisfy the condition $\mid \lambda_1 \mid\geq\mid \lambda_2 \mid\geq \cdots \geq\mid \lambda_{|V|} \mid> 0$. As a result, we can write $\mathbf{H} = \Phi\Lambda\Phi^T = \mathbf{J}\mathbf{J}^T$, where $\mathbf{J} = \sqrt{\Lambda}\Phi$. The matrix which has the embedding coordinates of the nodes as columns is $\mathbf{D} = \mathbf{J}^T$. Hence, $\mathbf{H} = \mathbf{J}\mathbf{J}^T = \mathbf{D}^T\mathbf{D}$ is a Gram matrix, i.e. its elements are scalar products of the embedding coordinates. Consequently, the embedding of the points is an isometry.

3 Graph Matching by Point Set Alignment

In this section, we show how the graph matching process can be posed as one of manifold alignment. This can be effected by finding the geometric transformation which minimises a quadratic error measure, i.e. least squares distance, between pairs of embedded points. To commence, we require some formalism. Suppose that $\mathbf{H}_D = \Phi_D\Lambda_D\Phi_D^T$ is the centred Laplacian matrix for the set of $\mid V^{\mathcal{D}} \mid$ "data" points whose embedded co-ordinates are given by the matrix $\mathbf{D} = \sqrt{\Lambda_D}\Phi_D^T$. Similarly, $\mathbf{H}_M = \Phi_M\Lambda_M\Phi_M^T$ is the centred Laplacian matrix for the set $\mid V^{\mathcal{M}} \mid$ of "model" points whose embedded co-ordinates are given by the matrix $\mathbf{M} = \sqrt{\Lambda_M}\Phi_M^T$. In practice, the sets of points to be matched may not be of the same size. To accommodate this feature of the data, we assume that the model point set is the larger of the two, i.e. $|V^{\mathcal{M}}| \geq |V^{\mathcal{D}}|$. As a result of the Young-Householder factorisation theorem used in the previous section, the embeddings of the data and model point patterns onto the manifolds $M^{\mathcal{D}} \in \Re^{|V^{\mathcal{D}}|}$ and $M^{\mathcal{M}} \in \Re^{|V^{\mathcal{D}}|}$, respectively, will be assumed to have a dimensionality which is equal to the number of points in the corresponding point-set. Hence, in order to be consistent with our geometric characterisation of the point pattern matching problem, we consider the manifold $M^{\mathcal{D}} \in \Re^{|V^{\mathcal{D}}|}$ to be a covering map, or projection, of the manifold $M^{\mathcal{M}} \in \Re^{|V^{\mathcal{M}}|}$. Here, in order to avoid ambiguities, we are interested in coverings of multiplicity one and, therefore, as an alternative to the matrix $\mathbf{D}$, we work with the matrix of coordinates $\tilde{\mathbf{D}} = [\mathcal{D} \mid \mathbf{n}_{|V^{\mathcal{D}}|+1} \mid \mathbf{n}_{|V^{\mathcal{D}}|+2} \mid \ldots \mid \mathbf{n}_{|V^{\mathcal{M}}|}]$, where $\mathbf{n}_i$ is a vector of length $\mid V^{\mathcal{D}} \mid$ whose entries are null.

With these ingredients, the problem of finding a transformation which can be used to map the data points onto their counterparts in the model point-set can be viewed as that of finding the rotation matrix $\mathbf{R}$ and the point-correspondence matrix $\tilde{\mathbf{P}} = [\mathbf{P} \mid \mathbf{O}]$, where $\mathbf{P}$ is a permutation matrix of order $\mid V^{\mathcal{D}} \mid$ and $\mathbf{O}$ is a null matrix of size $\mid V^{\mathcal{D}} \mid \times \mid V^{\mathcal{M}} - V^{\mathcal{D}} \mid$, which minimise the quadratic error function

$$\epsilon =\| \mathbf{M} - \tilde{\mathbf{P}}\mathbf{R}\tilde{\mathbf{D}} \|^2 \tag{8}$$

To solve the problem, we divide it in to two parts. First, we find the rotation matrix $\mathbf{R}$ by assuming the point-correspondence matrix $\tilde{\mathbf{P}}$ is known. Second, with the optimum rotation matrix at hand, we recover the point-correspondence matrix $\tilde{\mathbf{P}}$.

To recover the rotation matrix $\mathbf{R}$, we make use of the fact that both matrices, $\mathbf{R}$ and $\tilde{\mathbf{P}}$, are orthogonal and write

$$\epsilon = \mathrm{Tr}[\mathbf{M}\mathbf{M}^T] + \mathrm{Tr}[(\tilde{\mathbf{P}}\mathbf{R}\tilde{\mathbf{D}})(\tilde{\mathbf{P}}\mathbf{R}\tilde{\mathbf{D}})^T] - 2\mathrm{Tr}[\mathbf{M}(\mathbf{R}\tilde{\mathbf{D}})^T\tilde{\mathbf{P}}] \tag{9}$$

From the equation above, it is clear that maximising $\mathrm{Tr}[\mathbf{M}(\mathbf{R}\tilde{\mathbf{D}})^T\tilde{\mathbf{P}}]$ is equivalent to minimising ϵ. Further, assuming that the optimum correspondence matrix $\tilde{\mathbf{P}}$ is known, we can view the matrix $\tilde{\mathbf{P}}$ as an augmented permutation matrix, and, hence maximising $\mathrm{Tr}[\mathbf{M}\tilde{\mathbf{D}}^T\mathbf{R}]$ is the same as maximising $\mathrm{Tr}[\mathbf{M}(\mathbf{R}\tilde{\mathbf{D}})^T\tilde{\mathbf{P}}]$. This observation is important, because it implies that the rotation matrix $\mathbf{R}$ is the solution to a Procrustean transformation over the embedding coordinates for the set of data-points. Recall that a Procrustes transformation is of the form $\mathbf{Q} = \mathbf{R}\tilde{\mathbf{D}}$ which minimises $\| \mathbf{M} - \mathbf{Q} \|^2$. It is known that minimising $\| \mathbf{M} - \mathbf{Q} \|^2$ is equivalent to maximising $\mathrm{Tr}[\tilde{\mathbf{D}}\mathbf{M}^T\mathbf{R}]$. This is effected by using Kristof's inequality, which states that, if $\mathbf{S}$ is a diagonal matrix with non-negative entries and $\mathbf{T}$ is orthogonal, we have $\mathrm{Tr}[\mathbf{TS}] \geq \mathrm{Tr}[\mathbf{S}]$.

Let the singular value decomposition (SVD) of $\tilde{\mathbf{D}}\mathbf{M}^T$ be $\mathbf{USV}^T$. Using the invariance of the trace function over cyclic permutation, and drawing on Kristof's inequality, we can write $\mathrm{Tr}[\tilde{\mathbf{D}}\mathbf{M}^T\mathbf{R}] = \mathrm{Tr}[\mathbf{USV}^T\mathbf{R}] = \mathrm{Tr}[\mathbf{V}^T\mathbf{RUS}] \geq \mathrm{Tr}[\mathbf{S}]$. It can be shown that $\mathbf{V}^T\mathbf{RU}$ is orthogonal since $\mathbf{R}$ is orthogonal. Furthermore, the maximum of $\mathrm{Tr}[\mathbf{M}(\mathbf{R}\tilde{\mathbf{D}})^T\tilde{\mathbf{P}}]$ is achieved when $\mathbf{V}^T\mathbf{RU} = \mathbf{I}$. As a result, the optimal rotation matrix $\mathbf{R}$ is given by $\mathbf{R} = \mathbf{VU}^T$.

With the rotation matrix at hand, the correspondence matrix $\tilde{\mathbf{P}}$ can be recovered by noting that the product $\mathbf{M}(\mathbf{R}\tilde{\mathbf{D}})^T = \mathbf{MQ}^T$ is the matrix of pairwise inner products between the embedding coordinates for the data and model point-sets. Since the rotation of $\tilde{\mathbf{D}}$ over $\mathbf{R}$ is optimum, the normalised inner product between pairs of matching points is, in the ideal case, equal to unity, i.e. the angle between normalised coordinate vectors is zero. To take advantage of this, we construct the matrix of normalised pairwise inner products and then use it to recover $\tilde{\mathbf{P}}$. Hence, consider the matrix $\mathbf{Z}$ of order $\mid V^{\mathcal{D}} \mid \times \mid V^{\mathcal{M}} \mid$ whose element indexed i, j is given by the normalised inner product of the respective embedding coordinate vectors, after being aligned by rotation, for the data-point indexed i and j^{th} model-point. The elements of the matrix $\mathbf{Z}$ are hence given by

$$\mathbf{Z}(i,j) = \frac{\sum_{k=1}^{|V^{\mathcal{M}}|} \mathbf{Q}(i,k)\mathbf{M}(j,k)}{\sqrt{\sum_{k=1}^{|V^{\mathcal{M}}|} \mathbf{Q}(i,k)^2}\sqrt{\sum_{k=1}^{|V^{\mathcal{M}}|} \mathbf{M}(j,k)^2}} \tag{10}$$

Since the correspondence matrix $\tilde{\mathbf{P}}$ can be viewed as a matrix which slots over the matrix $\mathbf{Z}$ of normalised pairwise inner products and selects its largest values, we can recover $\tilde{\mathbf{P}}$ from $\mathbf{Z}$ in the following way. We commence by clearing $\tilde{\mathbf{P}}$ and, recursively, do

1.- $\tilde{\mathbf{P}}(i,j) = 1$, where $\{i, j \mid \mathbf{Z}(i,j) = \max_{\mathbf{Z}(i,j)\neq 0}(\mathbf{Z})\}$.
2.- $\mathbf{Z}(i,k) = 0 \; \forall \; k \in \mid V^{\mathcal{M}} \mid$ and $\mathbf{Z}(l,j) = 0 \; \forall \; l \in \mid V^{\mathcal{D}} \mid$.

until $\mathbf{Z} \equiv 0$. The data-point indexed i is then a match to the jth model-point *if and only if* $\tilde{\mathbf{P}}(i,j) = 1$. It is important to note that $\mathbf{Z}$ is the equivalent to the correlation, in a scalar-product sense, between the rows of $\mathbf{M}$ and the columns of $\mathbf{Q}^T$. It can be shown that the matrix $\tilde{\mathbf{P}}$ maximises the trace of $\tilde{\mathbf{P}}^T\mathbf{MQ}^T$ and, hence, minimises the quadratic error function ϵ.

This geometric treatment of the node-correspondence problem and its relationship to the correlation, as captured by the entries of $\mathbf{Z}$, between the rows of $\mathbf{M}$ and the columns of $\mathbf{Q}^T$ lends itself naturally to further refinement via statistical approaches such as EM algorithm [9] or relaxation labelling [4].

4 Experiments

The experimental evaluation of our method is divided into two parts. In Section 4.1, we illustrate the effect of the embedding on a sample point-set. In Section 4.2, we experiment with real world data provided by the COIL data-base.

4.1 Point-Set Deformation

In this section, we illustrate the utility of our method for the purposes of embedding a set of data points in a Riemannian manifold of constant sectional curvature. For this purpose, we have used a set of 25 points sampled regularly from a two-dimensional lattice.

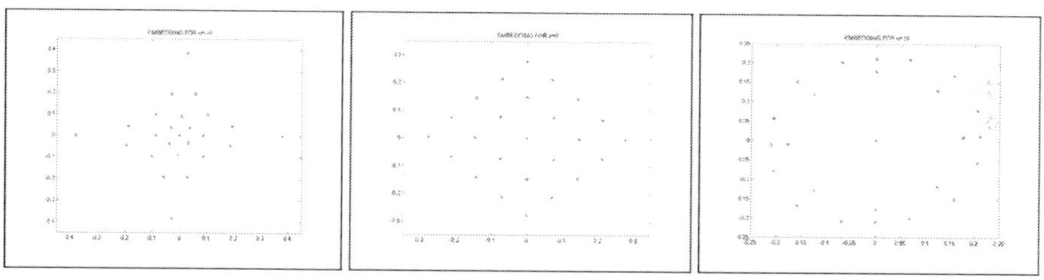

Fig. 1. From left-to-right: embedding results with $\kappa = -10, 0$ and 10 for a point-lattice

In Figure 1 we show, the results obtained by our algorithm for increasing values of κ. From the Figure 1, it is clear that the sectional curvature κ has an important effect in the recovery of the embedding coordinates. For $\kappa = 0$, the embedding is just a rotated version of the original distribution of the points in the plane. When κ is non-zero, then different patterns of behaviour emerge. In the case of negative sectional curvature (i.e. hyperbolic geometry), the embedding "collapses" the distribution of points towards the origin. For positive sectional curvature (i.e. elliptic geometry) the effect is to push the points away from the origin, and the point distribution forms an annulus. This behaviour is consistent with the fact that, for hyperbolic surfaces ($\kappa < 0$) parallel lines diverge. For spherical manifolds ($\kappa > 0$), parallel lines intersect.

4.2 Feature Point Matching

In this section, we aim at assessing the quality of the matching results delivered by our algorithm. As an experimental vehicle, we use the Columbia University COIL-20

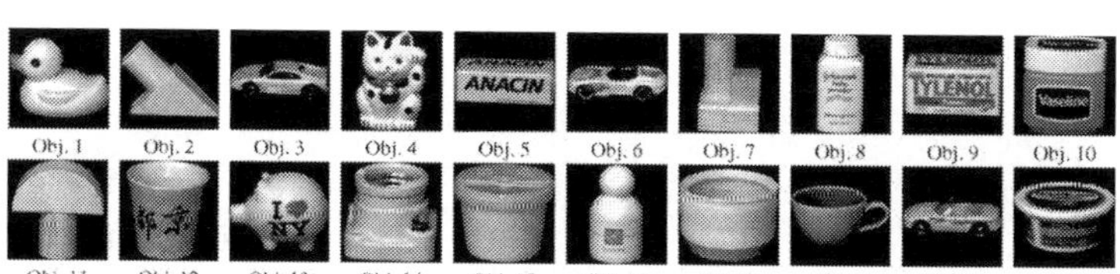

Fig. 2. Sample views for the objects in the Columbia University COIL database

Table 1. Normalised average ratio ε as a function of the sectional curvature κ

Object Index	Normalised average ratio ε of incorrect to correct correspondences			Object Index	Normalised average ratio ε of incorrect to correct correspondences		
	$\kappa=-15$	$\kappa=0$	$\kappa=15$		$\kappa=-15$	$\kappa=0$	$\kappa=15$
1	0.063	0.068	0.071	11	0.064	0.065	0.068
2	0.066	0.071	0.078	12	0.061	0.066	0.069
3	0.064	0.068	0.075	13	0.059	0.067	0.073
4	0.063	0.067	0.071	14	0.062	0.066	0.071
5	0.062	0.066	0.068	15	0.063	0.068	0.072
6	0.064	0.068	0.069	16	0.061	0.067	0.073
7	0.062	0.067	0.071	17	0.062	0.067	0.074
8	0.063	0.068	0.072	18	0.061	0.064	0.072
9	0.065	0.067	0.069	19	0.063	0.069	0.071
10	0.061	0.068	0.071	20	0.063	0.066	0.069

database. The COIL-20 database contains 72 views for 20 objects acquired by rotating the object under study about a vertical axis. In Figure 2, we show sample views for each of the objects in the database. For each of the views, our point patterns are comprised of feature points detected using the Harris corner detector [7].

To evaluate the results of matching pairs of views in the database, we have adopted the following procedure. For each object, we have used the first 15 views, 4 of these are "model" views and the remaining 12 are "data" views. We have then matched, by setting κ to -15, 0 and 15, the feature points for the selected "model" views with those corresponding to the two previous and two subsequent views in the database, i.e. we have matched the feature points for the ith view with those corresponding to the views indexed $i-2, i-1, i+1$ and $i+2$. To provide more qualitative results, we have ground-truthed the correspondences between the "model" and "data" views and computed the normalised average ratio ε of incorrect to correct correspondences $\mu(k,j)$ between the "model" view indexed k and the corresponding "data" view indexed j. The quantity ε is then given by

$$\varepsilon = \frac{1}{4\mid \Pi \mid} \sum_{k\in\Pi} \sum_{j=k-2, j\neq k}^{j=k+2} \frac{\mu(k,j)}{\rho(k,j)} \tag{11}$$

where $\Pi = \{3, 6, 9, 12\}$ is the set of indices for the "model" views and $\rho(k,j)$ is the maximum number of correspondences between the "model" and the "data" view. In Table 1, we show the values of ε as a function of object index for the three values of κ used in our experiments. Note that, in the table, we have used the object indexes in Figure 2.

From the quantitative results shown in Table 1, we conclude that the value of the sectional curvature κ has an important effect in the results delivered by the method. The method performs consistently better for negative values of κ. This is the case in which the alignment, is performed between manifolds that are hyperbolic in nature.

5 Conclusions

In this paper, we have shown how the nodes of a graph can be embedded on a constant sectional curvature manifold. The procedure can be viewed as a transformation to the edge-weights of the graph, which modifies the edge-weights using the sectional curvature. When the sectional curvature is positive, then the effect is to emphasise local or short-distance relationships. When the sectional curvature is negative on the other hand, then the effect is to emphasise long-distance relationships. Using the embedded coordinates corresponding to the nodes of the graph, we show how the problem of graph-matching can be transformed into one of Procrustean point-set alignment.

References

[1] M. Belkin and P. Niyogi. Laplacian eigenmaps and spectral techniques for embedding and clustering. In *Neural Information Processing Systems*, number 14, pages 634–640, 2002.

[2] T.M. Caelli and S. Kosinov. An eigenspace projection clustering method for inexact graph matching. *PAMI*, 26(4):515–519, April 2004.

[3] I. Chavel. *Riemannian Geometry: A Modern Introduction*. Cambridge University Press, 1995.

[4] W. J. Christmas, J. Kittler, and M. Petrou. Structural matching in computer vision using probabilistic relaxation. *IEEE Transactions on Pattern Analysis and Machine Intelligence*, 17(8):749–764, 1995.

[5] Fan R. K. Chung. *Spectral Graph Theory*. American Mathematical Society, 1997.

[6] S. Gold and A. Rangarajan. A graduated assignment algorithm for graph matching. *PAMI*, 18(4):377–388, April 1996.

[7] C. J. Harris and M. Stephens. A combined corner and edge detector. In *Proc. 4th Alvey Vision Conference*, pages 147–151, 1988.

[8] J. Lafferty and G. Lebanon. Diffusion kernels on statistical manifolds. *Journal of Machine Learning Research*, 6:129–163, 2005.

[9] Bin Luo and E. Hancock. Iterative procrustes alignment with the em algorithm. *Image and Vision Computing*, 20:377–396, 2002.

[10] B. O'Neill. *Elementary Differential Geometry*. Academic Press, 1997.

[11] S. T. Roweis and L. K. Saul. Nonlinear dimensionality reduction by locally linear embedding. *Science*, 290:2323–2326, 2000.

[12] G. Scott and H. Longuet-Higgins. An algorithm for associating the features of two images. In *Proceedings of the Royal Society of London*, number 244 in B, pages 21–26, 1991.

[13] L. Shapiro and J. M. Brady. Feature-based correspondance - an eigenvector approach. *Image and Vision Computing*, 10:283–288, 1992.

[14] J. B. Tenenbaum, V. de Silva, and J. C. Langford. A global geometric framework for nonlinear dimensionality reduction. *Science*, 290(5500):2319–2323, 2000.

[15] W. S. Torgerson. Multidimensional scaling I: Theory and method. *Psychometrika*, 17:401–419, 1952.

[16] S. Umeyama. An eigen decomposition approach to weighted graph matching problems. *PAMI*, 10(5):695–703, September 1988.

[17] G. Young and A. S. Householder. Discussion of a set of points in terms of their mutual distances. *Psychometrika*, 3:19–22, 1938.

Graph-Based Fast Image Segmentation*

Dongfeng Han, Wenhui Li, Xiaosuo Lu, Lin Li, and Yi Wang

College of Computer Science and Technology, Key Laboratory of Symbol Computation and Knowledge Engineering of the Ministry of Education, Jilin University, Changchun, 130012, P.R. China
jlu_hdf@126.com

Abstract. In this paper, we describe a fast semi-automatic segmentation algorithm. A nodes aggregation method is proposed for improving the running time and a Graph-Cuts method is used to model the segmentation problem. The whole process is interactive. Once the users specify the interest regions by drawing a few lines, the segmentation process is reliably computed automatically no additional users' efforts are required. It is convenient and efficient in practical applications. Experiments are given and outputs are encouraging.

1 Introduction

Image segmentation is a process of grouping together neighboring pixel whose properties are coherent. It is an integral part of image processing applications like medical images analysis and photo editing. Many researchers are focus on this subject [1], [2], [3], [4], [5], [6], [7], [8], [9] and [10]. However, even to this day, many of the computational issues of perceptual grouping have remained unresolved. Semi-automatic segmentation techniques that allow solving moderate and hard segmentation tasks by modest effort on the part of the user are becoming more and more popular. In the analysis of the objects in images it is essential that we can distinguish between the objects of interest and the rest. This latter group is also referred to as the background. The techniques that are used to find the objects of interest are usually referred to as segmentation techniques: segmenting the foreground from background.

In this paper, we present an efficient method for semi-automated image segmentation for common images. Our method can exactly extract the interest objects such as people, animals and so on. This paper's contribution is twofold.

First, we introduce the nodes aggregation method to the image segmentation. It is a powerful tool for reducing running time.

Second, we propose a novel semi-automation segmentation scheme based on nodes aggregation and Graph-Cuts, which has several favorable properties:

1. Capable of solving interactive segmentation problem.
2. Performs multi-label image segmentation (the computation time does not depend on the number of labels).
3. Fast enough for practical application using nodes aggregation.

* This work has been supported by NSFC Project 60573182, 69883004 and 50338030.

D.-Y. Yeung et al. (Eds.): SSPR&SPR 2006, LNCS 4109, pp. 468 – 474, 2006.

4. It is extensible and allows constructing new families of segmentation algorithms with specific properties.

In section 2 describes our method in detail. Section 3 presents the experiments. Conclusions and future work are given in section 4.

2 Our Method

2.1 Overview

Just as mentioned in [7], the segmentation problems focus on two basic questions:

1. What is the criterion that one wants to segment?
2. Is there an efficient technique to carry out the segmentation?

In this paper, our goal is to give an answer about the two questions. There are two main contributions of our segmentation algorithm. We will introduce each of these ideas briefly below and then describe them in detail in subsequent sections. The first contribution of this paper is a new coarse image representation called nodes aggregation that allows for very fast completing segmentation. The second contribution of this paper is combining nodes aggregation with Graph-Cuts that can produce an elaborate segmentation result.

2.2 Graph-Cuts Based Image Segmentation

Inspired by [3], we use Graph-Cuts as energy minimization technique for the interactive segmentation. Furthermore, at the object marking step, we propose an efficient interactive segmentation algorithm by employing multi-scale nodes aggregation and Graph-Cuts.

We briefly introduce some of the basic terminology used throughout the paper. An image that contains $N = n \times n$ pixels, we construct a graph $G = (V, E, W)$ in which each node $i \in V$ represents a pixel and every two nodes i, j representing neighboring pixels are connected by an edge $e_{i,j} \in E$. Each edge has a weight $w_{i,j} \in W$ reflecting the contrast in the corresponding location in the image. We can connect each node to the four or eight neighbors of the respective pixel, producing a graph. The graph can be partitioned into two disjoint sets, "O", "B" where $O \bigcup B = V$, $O \bigcap B = \phi$. These tasks can be viewed as labeling problems with label 1 representing object and 0 otherwise. Finding the most likely labeling translates to optimizing an energy function. In vision and image processing, these labels often tend to vary smoothly within the image, except at boundaries. Because a pixel always has the similar value with its neighbors, we can model the optimization problem as a MRF. In [3], the authors find the most likely labeling for some given data is equivalent to seeking the MAP (maximum a posteriori) estimate. A graph is constructed and the Potts Energy Model (1) is used as the minimization target.

$$H(G) = \underset{i \in V}{H_{data}}(i) + \underset{\{i, j\} \in N}{H_{smooth}}(i, j) \tag{1}$$

The graph G contains two kinds of vertices: p-vertices (pixels which are the sites in the associated MRF) and l-vertices (which coincide with the labels and will be terminals in the graph cut problem). All the edges present in the neighborhood system N are edges in G. These edges are called n-links. Edges between the p-vertices and the l-vertices called t-links are added to the graph. t-links are assigned weights based on the data term (first term in Equations 1 reflecting individual label-preferences of pixels based on observed intensity and pre-specified likelihood function) while n-links are assigned weights based on the interaction term (second term in Equation 1 encouraging spatial coherence by penalizing discontinuities between neighboring pixels). While n-links are bi-directional, t-links are un-directional, leaving the source and entering the sink. A cut segment the graph.

The object of segmentation is to minimize the energy function (1). The first term reflects individual label-preferences of pixels based on observed intensity and pre-specified likelihood function. The second term encourages spatial coherence by penalizing discontinuities between neighboring pixels. So our goal is minimize the energy function and make it adapt to human vision system. In [3] authors give the construction of the graph in detail. A different way is used to construct the graph in this paper because its construction way is so complex that in practice it will be inconvenient and it will influence the running time.

First a k-Means clustering method is applied on the seed region including "O" and "B". We use the algorithm described in [11] which performs well in practice. (A C code can be found on his homepage) Then, for each node i, the minimum distance from its color $color(i)$ to foreground clusters is computed as:

$$d_i^O = \min_{j \in O} \left\| color(i) - kMeans_j \right\| \tag{2}$$

$$d_i^B = \min_{j \in B} \left\| color(i) - kMeans_j \right\| \tag{3}$$

So the energy of the first term is:

$$H_{data}(i=0) = \begin{cases} CONST & if\ i \in B \\ 0 & if\ i \in O \\ \dfrac{d_i^B}{d_i^B + d_i^O + \varepsilon} & if\ i \neq 1\ or\ 0 \end{cases}$$

$$H_{data}(i=1) = \begin{cases} 0 & if\ i \in B \\ CONST & if\ i \in O \\ \dfrac{d_i^O}{d_i^B + d_i^O + \varepsilon} & if\ i \neq 1\ or\ 0 \end{cases} \tag{4}$$

Where $i = 1\ or\ 0$ means that this node belongs to Object or Background.

The second term $H_{\substack{smooth\\ \{i,j\}\in N}}(i,j)$, which is define as follow:

$$H_{\substack{smooth\\ \{i,j\}\in N}}(i,j) = \exp(\frac{-\|I(i)-I(j)\|^2}{\sigma I}) \times \exp(\frac{-\|L(i)-L(j)\|^2}{\sigma L}) \quad (5)$$

Where *I(i)* is the color of the node *i*, *L(i)* is the location of the *i*. So this term not only reflects the neighbor smoothness but also reflects the influence of two nodes due to distance.

So once a user specifies the seeds of object and background then a graph can be constructed as described above. Then using the Max-Flow algorithm (a more efficient implementation is described in [12]), we can efficient minimize the energy. The output is divided into two parts: one is object the other is background.

2.3 Nodes Aggregation

For a 512×512 pixel image, it will be slow to segment the whole graph straightly. We introduce a multi-scale nodes aggregation method to construct a pyramid structure over the image. Each procedure produces a coarser graph with about half size, and such that Graph-Cuts segmentation in the coarse graph can be used to compute precision segmentation in the fine graph. The coarsening procedure proceeds recursively as follows. The finest graph is denoted by $G^0 = (V^0, W^0, E^0)$ and a state vector $d = \{d_1, d_2, ..., d_N\}$, where $N = \|V\|$ is defined:

$$d = \begin{Bmatrix} 1 & if & i \in O \\ 0 & if & i \in B \end{Bmatrix} \quad (6)$$

$d_i = 1$ means pixel *i* belonging to object, otherwise belonging to background. Supposing $G^s = (V^s, W^s, E^s)$ and d^s is defined at scale *s*. A set of coarse representative nodes $C \subseteq V^s = \{1, 2, 3, ..., n\}$ and $\|V^s\| \leq N$ is chosen, so that every node in $V^{s-1} \setminus C$ is strongly connected to *C*. A node is strongly connected to *C* if the sum of its weights to node in *C* is significant proportion of its weights to nodes outsides *C*. The principle can be found in [9]. Each node in *C* can be considered as representing an aggregation node. Thus we decompose the image into many aggregates. Then a coarse graph $G^{s-1} = (V^{s-1}, W^{s-1}, E^{s-1})$ and a state vector $d^{s-1} = \{d_1^{s-1}, d_2^{s-1}, ..., d_{N^{s-1}}^{s-1}\}$ at scale *s*-1 are defined. The relationship between W^s and W^{s-1}, d^s and d^{s-1} are:

$$w_{k,l}^s = \sum p_{i,k}^{[s-1,s]} w_{i,j}^{s-1} p_{j,l}^{[s-1,s]} \quad (7)$$

It can be written as the matrix formulation:

$$W^s = P^{[s-1,s]T} W^{s-1} P^{[s-1,s]} \quad (8)$$

So relationship between d^s and d^{s-1} are:

$$d^{s-1} = P^{[s-1,s]} d^{s} \tag{9}$$

A more detail description can be found in [9]. So we should find P which is called the inter-scale interpolation matrix in [10]. Once P is found, the pyramid structure can be constructed recursively. The selection of P can be found in [8]. In the end a full pyramid has been constructed. The graph cuts method can be used to segment the coarse graph, then applying a top-down sharpening of the boundaries. In the end each represent node at coarser level is assigned to low level, which produces a nodes reassembling procedure. That means a node i originally belonging to a represent node j can reassembling to a different represent node k due to the weight change. So a node i change its original state value from background to object (supposing it originally belongs to background).

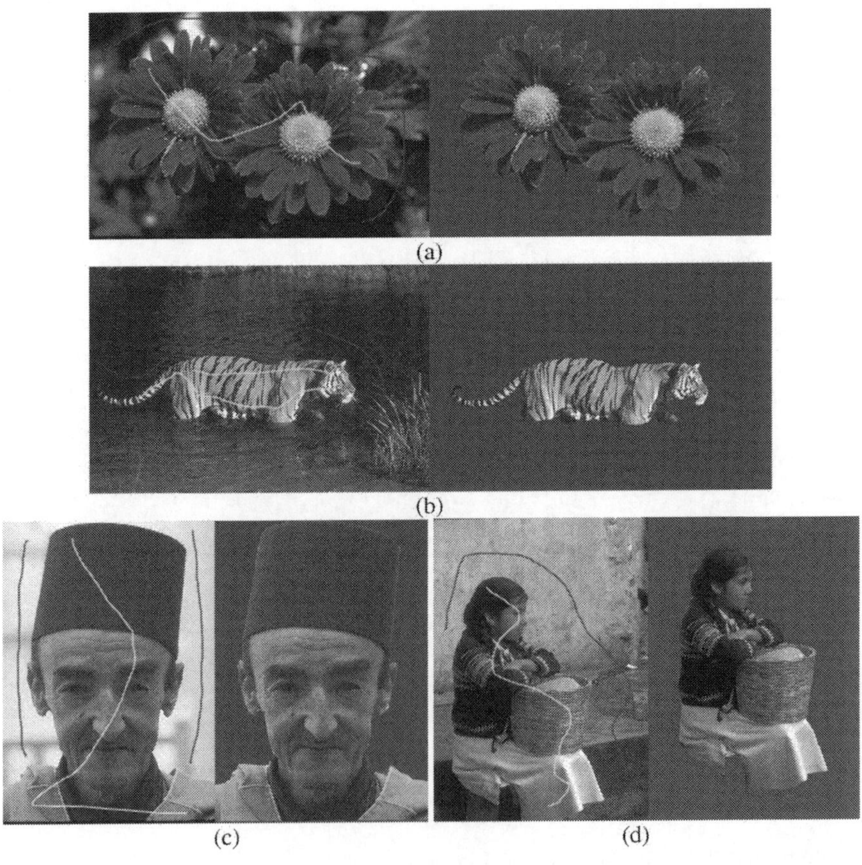

Fig. 1. Some test images and the segmentation results

3 Experiments

We first give the running time for aggregation segmentation and non-aggregation segmentation in Table 1. The test images are shown in Fig. 1 from (a) to (e). The time is the total running time from beginning to end. Because we use nodes aggregation to reduce the number of nodes, the speed can meet the real-time demands. There are some methods for nodes aggregation. We use a simply but efficient method. For more accurate result we can develop new methods for nodes aggregation. Also the ratio of running time is about 20 times faster than non-aggregation. In Fig. 1, some segmentation results are shown. For some hard border images such as Fig. 1 (a) (c), our method can exactly segment the images. For some soft border images such as Fig. 1 (b) (d), our method can also give satisfying results. The outputs are encouraging. We also apply nodes aggregation segmentation method to interest region-based image retrieval system and get a satisfying performance.

Table 1. The comparisons of running time between non-aggregation and aggregation

Images	Non-aggregation	Aggregation
(a)	20s	1.0s
(b)	29s	1.5s
(c)	35s	2.2s
(d)	32s	1.9s

4 Conclusions and Future Work

In this paper an efficient semi-automatic segmentation method is proposed. The main contribution of this paper is introducing nodes aggregation for improving running time which is very important in practical applications. In the future, automatic segmentation algorithm based on nodes aggregation and Graph-Cuts will be developed. We believe it will be a powerful tool in practical applications.

Acknowledgments

This work has been supported by NSFC Project 60573182, 69883004 and 50338030.

References

1. L. Reese.: Intelligent Paint: Region-Based Interactive Image Segmentation, Master's thesis. Department of Computer Science, Brigham Young University, Provo, UT. (1999)
2. E. N. Mortensen and W. A. Barrett.: Interactive segmentation with intelligent scissors. Graphical Models and Image Processing, 60, 5. (1998) 349–384
3. Y. Boykov and M. Jolly.: Interactive graph cuts for optimal boundary & region segmentation of objects in n-d images. Proc. IEEE Conf. Computer Vision. (2001) 105-112
4. C. Rother, A. Blake and V. Kolmogorov.: Grabcut-interactive foreground extraction using iterated graph cuts. In Proceedings of ACM SIGGRAPH. (2004)

5. D. Comanicu, P. Meer.: Mean shift: A robust approach toward feature space analysis. IEEE Trans. Pattern Anal. Machine Intell, May. (2002)
6. J. Shi and J. Malik.: Normalized Cuts and Image Segmentation. Proc. IEEE Conf. Computer Vision and Pattern Recognition. (1997) 731-737
7. J. Shi and J. Malik.: Normalized cuts and image segmentation. PAMI, 22(8) (2000) 888–905
8. E. Sharon, A. Brandt, and R. Basri.: Fast multiscale image segmentation. Proc. IEEE Conf. Computer Vision and Pattern Recognition. (2000) 70–77
9. E. Sharon, A. Brandt, and R. Basri.: Segmentation and boundary detection using multiscale intensity measurements, Proc. IEEE Conf. Computer Vision and Pattern Recognition. (2001) 469–476
10. Meirav Galun, Eitan Sharon, Ronen Basri, Achi Brandt.: Texture Segmentation by Multiscale Aggregation of Filter Responses and Shape Elements, Proc. IEEE Conf. Computer Vision. (2003)
11. Tapas Kanungo, David M. Mount, Nathan S. Netanyahu, Christine D. Piatko, Ruth Silverman, Angela Y. Wu.: A Local Search Approximation Algorithm for k-Means Clustering, In Proceedings of the 18th Annual ACM Symposium on Computational Geometry. (2003)
12. Y. Boykov and V. Kolmogorov.: An Experimental Comparison of Min-Cut/Max-Flow Algorithms for Energy Minimization in Computer Vision, Proc. Int'l Workshop Energy Minimization Methods in Computer Vision and Pattern Recognition, Lecture Notes in Computer Science, Sept. (2001) 359-374

Modeling of Remote Sensing Image Content Using Attributed Relational Graphs*

Selim Aksoy

Department of Computer Engineering, Bilkent University, Ankara, 06800, Turkey
saksoy@cs.bilkent.edu.tr

Abstract. Automatic content modeling and retrieval in remote sensing image databases are important and challenging problems. Statistical pattern recognition and computer vision algorithms concentrate on feature-based analysis and representations in pixel or region levels whereas syntactic and structural techniques focus on modeling symbolic representations for interpreting scenes. We describe a hybrid hierarchical approach for image content modeling and retrieval. First, scenes are decomposed into regions using pixel-based classifiers and an iterative split-and-merge algorithm. Next, spatial relationships of regions are computed using boundary, distance and orientation information based on different region representations. Finally, scenes are modeled using attributed relational graphs that combine region class information and spatial arrangements. We demonstrate the effectiveness of this approach in query scenarios that cannot be expressed by traditional approaches but where the proposed models can capture both feature and spatial characteristics of scenes and can retrieve similar areas according to their high-level semantic content.

1 Introduction

The constant increase in the amount of data received from satellites has made automatic content extraction and retrieval highly desired goals for effective and efficient processing of remotely sensed imagery. Most of the existing systems support building supervised or unsupervised statistical models for pixel level analysis. Even though these models improve the processing time compared to manual digitization, complete interpretation of a scene still requires a remote sensing analyst to manually interpret the pixel-based results to find high-level structures. In other words, there is still a large semantic gap between the outputs of commonly used models and high-level user expectations.

The limitations of pixel-based models and their inability in modeling spatial content motivated the research on developing algorithms for region-based analysis. Conventional region level image analysis algorithms assume that the regions

* This work was supported in part by the TUBITAK CAREER Grant 104E074 and European Commission Sixth Framework Programme Marie Curie International Reintegration Grant MIRG-CT-2005-017504. Initial work for this research was partly supported by the U.S. Army contract W9132V-04-C-0001.

D.-Y. Yeung et al. (Eds.): SSPR&SPR 2006, LNCS 4109, pp. 475–483, 2006.

consist of relatively uniform pixel feature distributions. However, complex image scenes and land structures of interest usually contain many pixels and regions that have different feature characteristics. Furthermore, two scenes with similar regions can have very different interpretations if the regions have different spatial arrangements. Even when pixels and regions can be identified correctly, manual interpretation is often necessary for many applications of remote sensing image analysis like land cover/use classification, urban mapping and monitoring, and ecological analysis in public health studies.

Symbolic representation of scenes and retrieval of images based on these representations are very challenging and popular topics in structural and syntactic pattern recognition. Previous work on symbolic representation attempted to develop languages and data structures to model the attributes and relationships of symbols/icons, and work on symbolic retrieval concentrated on finding exact or partial (inexact) matches between these representations [1,2].

Most applications of syntactic and structural techniques to remote sensing image analysis assumed that object detection and recognition problems were solved. Using structures such as strings, graphs, semantic networks and production rules, they concentrated on the problem of interpreting the scene given the objects. Other related work in the computer vision literature used grid-based representations [3], centroids and minimum bounding rectangles [4]. Centroids and minimum bounding rectangles are useful when regions have circular or rectangular shapes but regions in natural scenes often do not follow these assumptions. Similar work can also be found in the medical imaging literature where rule-based models [5], grid-based layouts [6], and attributed relational graphs [7] were used to represent objects and their relationships given manually constructed rules or delineation of objects by experts. Most of these models are not usable due to the infeasibility of manual annotation in large volumes of images. Different structures in remote sensing images have different sizes so fixed sized grids cannot capture all structures either.

We propose a hybrid hierarchical approach for image content modeling and content-based retrieval. The analysis starts from raw data. First, pixels are labeled using Bayesian classifiers. Then, scenes are decomposed into regions using pixel-based classification results and an iterative split-and-merge algorithm. Next, resulting regions are modeled at multiple levels of complexity, and pairwise spatial relationships are computed using boundary, distance and orientation information. Finally, scenes are modeled using attributed relational graphs that combine region class information and spatial arrangements. Our work differs from other approaches in that recognition of regions and decomposition of scenes are done automatically after the system learns region models with only a small amount of supervision in terms of examples for classes of interest.

The rest of the paper is organized as follows. Decomposition of scenes into regions is described in Section 2. Modeling of regions and their spatial relationships are presented in Section 3. Scene modeling with graphs is discussed in Section 4. Using these graphs in content-based retrieval is described in Section 5 and conclusions are given in Section 6.

2 Scene Decomposition

The first step in scene modeling is to find meaningful and representative regions in the image. An important requirement is the delineation of each individual structure as an individual region. Automatic extraction and recognition of these regions are also required to handle large amounts of data.

In previous work [8], we used an automatic segmentation algorithm based on energy minimization, and used k-means and Gaussian mixture-based clustering algorithms to group and label the resulting regions according to their features. Our newer experiments showed that some popular density-based and graph-theoretic segmentation algorithms were not successful on our data sets because of the large amount of data and the detailed structure in multi-spectral images.

The segmentation approach we have used in this work consists of pixel-based classification and an iterative split-and-merge algorithm [9]. Bayesian classifiers that fuse information from multiple features are used to assign each pixel to one of these classes. Since pixel-based classification ignores spatial correlations, the initial segmentation may contain isolated pixels with labels different from those of their neighbors. We use an iterative algorithm that merges pixels and pixel groups using constraints on probabilities (confidence of pixel classification) and splits existing regions based on constraints on connectivity and compactness.

The algorithms proposed in this paper are evaluated using a LANDSAT scene of southern British Columbia in Canada. The false color representation of this $1,536 \times 1,536$ scene with 6 multi-spectral bands and 30 m/pixel ground resolution is shown in Fig. 1(a), and the region decomposition consisting of 1,946 regions is shown in Fig. 1(b). Spectral, textural and elevation information were used to train the Bayesian classifiers.

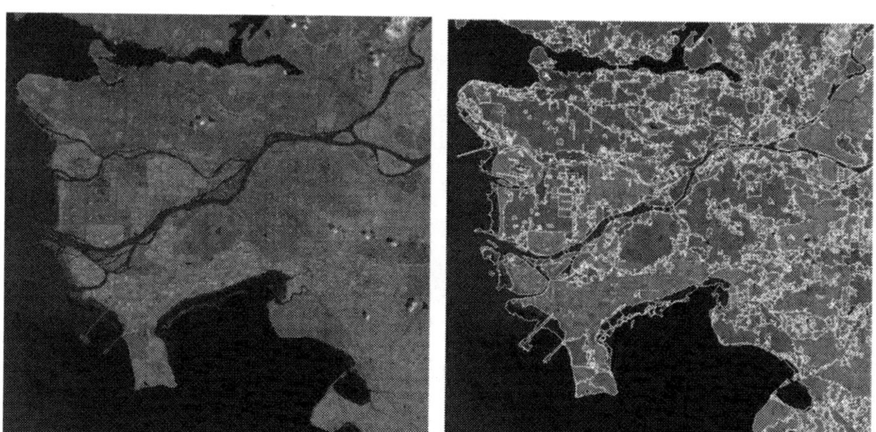

(a) LANDSAT scene (b) Region decomposition

Fig. 1. False color representation of a LANDSAT scene and the region decomposition obtained after applying the split-and-merge algorithm to the results of a pixel-based Bayesian classifier. White pixels in (b) represent region boundaries.

3 Spatial Relationships

3.1 Region Modeling

A straightforward way of representing regions of an image is by using a membership array where each pixel stores the id of the region that it belongs. Hierarchical structures such as quad trees can be used to encode this membership information for faster access. Regions can also be represented using contour-based approaches such as chain codes that exploit the boundary information.

Operations on complex regions with a large number of pixels on the boundary may be computationally infeasible so regions are often modeled using approximations [10,11]. The simplest approximation is the minimum bounding rectangle that can be useful for representing compact regions. Another simple but finer approximation is the grid representation. More detailed approximations such as polygonal representations, B-splines, or scale space representations are often necessary when operations include multiple regions.

In this work, we represent each region using its boundary chain code, polygonal representations at different smoothing levels, grid representation and minimum bounding rectangle. Regions with holes have additional lists for chain codes and polygonal approximations of their inner boundaries. Grid representation, that consists of a low-resolution grid overlaid on the region, stores all grid cells that overlap with the given region and contain at least one more region. In addition, each region has an id (unique within an image) and a label that is propagated from its pixel's class labels as described in the previous section. Example representations are given in Fig. 2. These representations at different levels of complexity are used to simplify the computation of spatial relationships between regions as described in the next section.

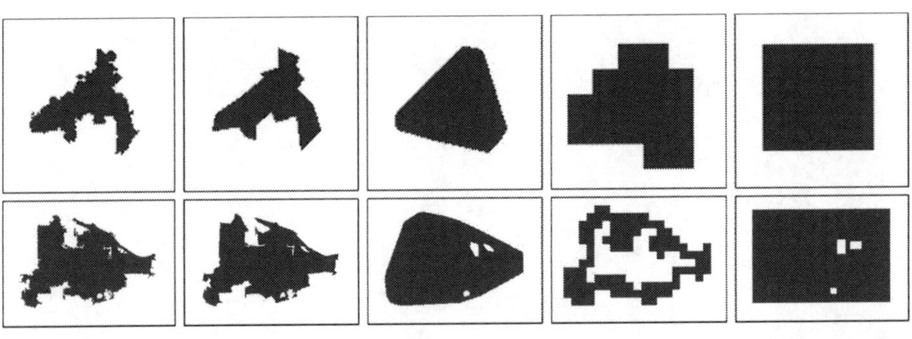

Fig. 2. Region representation examples. Rows show representations for two different regions. Columns represent, from left to right: original boundary, smoothed polygon, convex hull, grid representation, and minimum bounding rectangle.

3.2 Pairwise Relationships

After the images are segmented and the regions are modeled at multiple levels of detail, the next step is the modeling of their spatial relationships. Regions

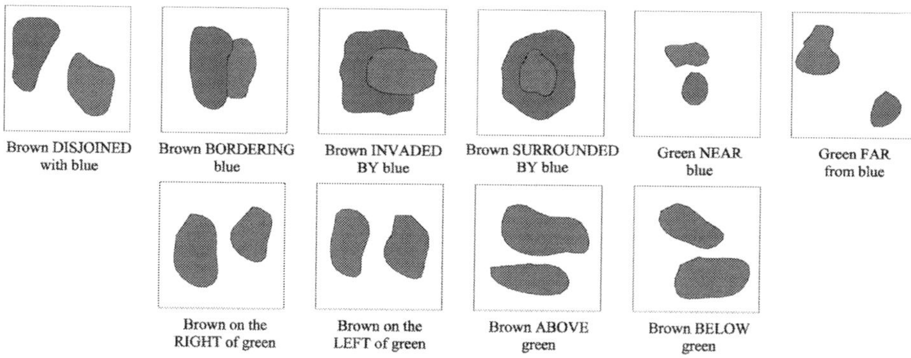

Fig. 3. Spatial relationships of region pairs: *disjoined*, *bordering*, *invaded_by*, *surrounded_by*, *near*, *far*, *right*, *left*, *above* and *below*.

can appear in an image in many possible ways. However, regions of interest are usually the ones that are close to each other. The relationships we compute for each region pair can be grouped as boundary-class relationships (*disjoined*, *bordering*, *invaded_by*, *surrounded_by*), distance-class relationships (*near*, *far*), and orientation-class relationships (*right*, *left*, *above*, *below*) as illustrated in Fig. 3. Boundary-class relationships are based on overlaps between region boundaries. Distance-class relationships are based on distances between region boundaries. Orientation-class relationships are based on centroids of regions.

Since large scenes can easily contain thousands of regions with thousands of boundary pixels, pixel-to-pixel comparison of all possible region pairs to compute their overlaps and distances is not feasible. These computations can be significantly simplified by applying a coarse-to-fine search to find region pairs that have a potential overlap or are very close to each other. In previous work [8,9], we used brute force comparisons of region pairs within smaller tiles obtained by dividing the original scene into manageable sized images. However, regions that occupy multiple tiles may not be handled correctly after that division. The coarse-to-fine search strategy that compares different region approximations in increasing order of complexity enables us to perform exact computations only for very close regions whereas relationships between the remaining ones are approximated using different levels of simpler boundary representations.

Since the relations between two regions can be described with multiple relationships at the same time (e.g., *invaded_by* from *left*, *bordering* from *above*, *near* and *right*), the degree of a region pair having a particular relationship is modeled using fuzzy membership functions. These relationships are based on:

- ratio of the common boundary (overlap) between two regions to the perimeter (total boundary length) of the first region,
- distance between two regions,
- angle between the horizontal (column) axis and the line joining the centroids of the regions.

Details of the membership functions are not included here due to space restrictions but more information can be found in [8].

4 Scene Modeling Using Graphs

At the end of the previous section, each region pair is assigned a degree for each relationship class. In previous work [8], we modeled higher-order relationships (of region groups) by decomposing them into $\binom{k}{2}$ second-order relationships (of region pairs) combined using the fuzzy "min" operator that corresponds to the Boolean "and" operator. In this work, we model higher-order relationships using attributed relational graph (ARG) structures. ARGs are very general and powerful representations of image content. Petrakis *et al.* [7] used ARGs to represent objects and their relationships in medical images. They assumed that the regions were segmented and labeled manually, and concentrated on developing fast matching algorithms for these manually constructed graphs. However, applications of ARGs for representing contents of natural scenes have been quite limited because of inaccurate object recognition and the computational complexity of finding associations between objects in different images. Automatic decomposition of regions in Section 2 and automatic modeling of their spatial relationships in Section 3 gives us an important advantage over the existing methods that require manual segmentation and labeling of the regions.

The ARG can be adapted to model the scenes by representing regions by the graph nodes and their spatial relationships by the edges between such nodes. Nodes are labeled with the class (land cover/use) names and the corresponding confidence values (posterior probabilities) for these class assignments. Edges are labeled with the spatial relationship classes (pairwise relationship names) and the corresponding degrees (fuzzy membership values) for these relationships. The ARG for the LANDSAT scene of Fig. 1 is given in Fig. 4.

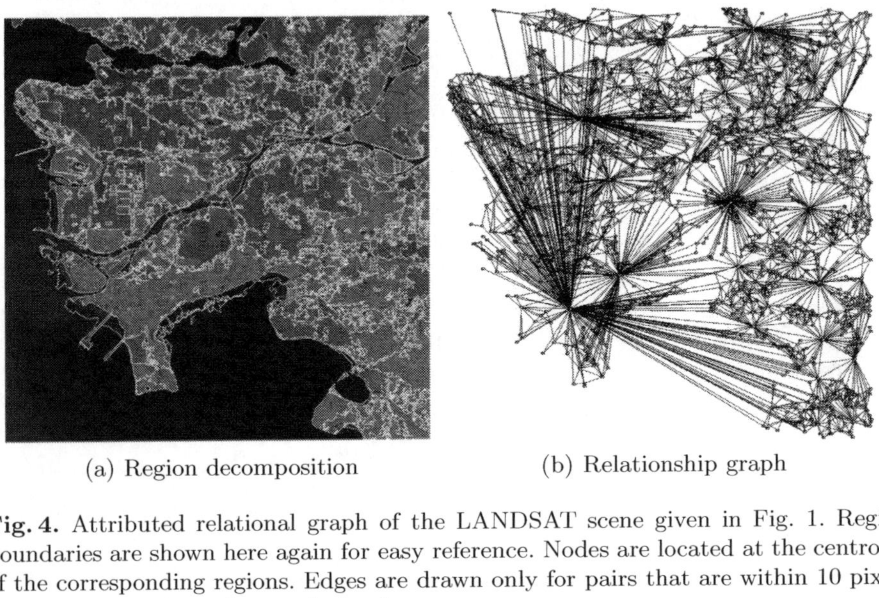

(a) Region decomposition (b) Relationship graph

Fig. 4. Attributed relational graph of the LANDSAT scene given in Fig. 1. Region boundaries are shown here again for easy reference. Nodes are located at the centroids of the corresponding regions. Edges are drawn only for pairs that are within 10 pixels of each other to keep the graph simple.

5 Scene Retrieval

When the scenes are represented using ARGs, image retrieval can be modeled as a relational matching [12] and subgraph isomorphism [13] problem. Relational matching has been extensively studied for structural pattern recognition. We use the "editing distance" [7,14] as the (dis)similarity measure. The editing distance between two ARGs is defined as the minimum cost taken over all sequences of operations (error corrections) that transform one ARG to the other. These operations are defined as substitution, insertion and deletion. The computation of the distance between two ARGs involves not only finding a sequence of error corrections that transforms one ARG to the other, but also finding the one that yields the minimum total cost.

The retrieval scenario starts with the user's selecting of an area of interest (i.e., a set of regions) in an image. The system automatically constructs the graph for that area. Then, this graph is used to query the system to automatically find other areas (i.e., sets of regions) with similar structures in the database. In some cases, some of the relationships (e.g., *above*, *right*) can be too restrictive. Our implementation includes a relationship value named *don't_care* that allows users to constrain the searches where insertion or deletion of graph edges corresponding to relationship classes set as *don't_care* do not contribute any cost in the editing distance. Finally, resulting areas are presented to the user in increasing order of the editing distance between the subgraphs of these areas and the subgraph of the query.

Example queries are given in Figs. 5–7[1]. Traditionally, queries that consist of multiple regions are handled by averaging the features of all regions. However, this averaging causes a significant information loss because it ignores relative spatial organization and distorts the multimodal feature characteristics of the query. On the other hand, our experiments using the scene in Fig. 1 showed that the proposed ARG structure can capture both feature and spatial characteristics of region groups and can retrieve similar areas according to their high-level semantic content.

Experiments also showed that the coarse-to-fine search strategy of Section 3.2 significantly improves the performance. For the example scene with 1,946 regions shown in Fig. 1, computation of all individual region properties (boundary chain code, centroid, perimeter) took 10.56 minutes, and computation of all pairwise spatial relationships took 33.47 minutes using brute force comparisons of regions. On the other hand, computation of all additional region representations (smoothed polygon, grid representation, minimum bounding rectangle) took 2.57 seconds, and computation of all pairwise relationships took 1.7 minutes using coarse-to-fine comparisons. As for the graph search examples, the queries in Figs. 5–7 took 5.52, 7.13 and 15.96 seconds, respectively, using an unoptimized C++-based implementation on a Pentium 4, 3.0 GHz computer running Linux.

[1] Since no ground truth exists for this semantic level of analysis, we provide only qualitative examples in this paper.

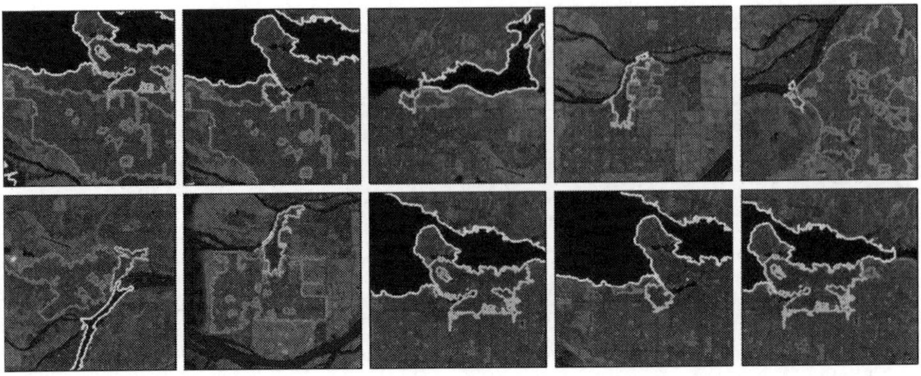

Fig. 5. Searching for a scene where a residential area is *bordering* a city center that is *bordering* water. Orientation-class is set to *don't_care*. Identified regions are marked as cyan, magenta and yellow for city, residential and water, respectively. Scenes are shown in increasing order of their editing distance to the query given on top-left.

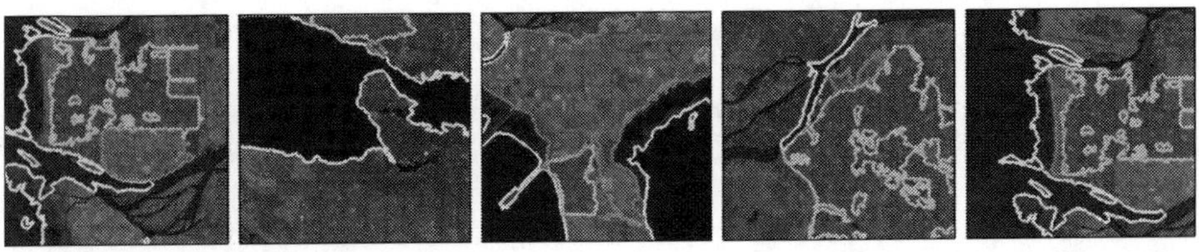

Fig. 6. Searching for a scene where a residential area is *bordering* a field and both are *bordering* water. Identified regions are marked as cyan, magenta and yellow for residential, field and water, respectively.

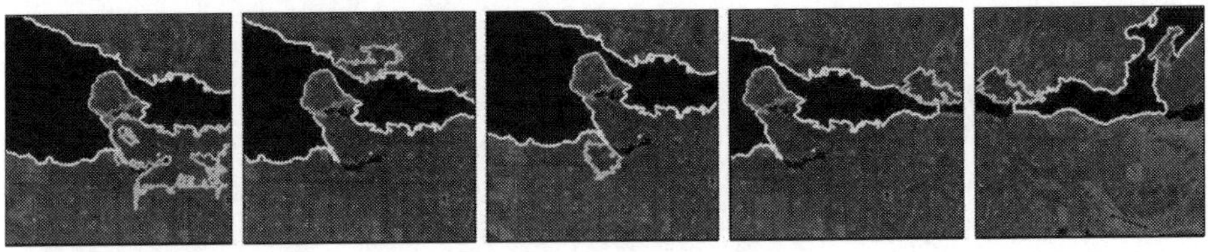

Fig. 7. Searching for a scene where a park is *invaded_by* water and a city center is *bordering* the same water. Identified regions are marked as cyan, magenta and yellow for city, park and water, respectively.

6 Conclusions

We described a hybrid hierarchical approach for image content modeling that involves supervised classification of pixels, automatic grouping of pixels into contiguous regions, representing these regions at different levels of complexity, modeling their spatial relationships using fuzzy membership classes, and encoding scene content using attributed relational graph structures. We demonstrated the effectiveness of this approach for content-based retrieval using queries that provide a challenge where a mixture of spectral and textural features as well as spatial information are required for correct identification of the scenes. The results showed that the proposed models can capture both feature and spatial characteristics of region groups and can retrieve similar areas according to their high-level semantic content. Regarding future work, we believe that improving pairwise relationship models (such as orientation-class relationships where centroids are not always very meaningful for large and non-compact regions) will make the overall representation more powerful and will prove further useful toward bridging the gap between low-level features, representations and semantic interpretation.

References

1. Dance, S., Caelli, T., Liu, Z.Q.: Picture Interpretation: A Symbolic Approach. World Scientific (1995)
2. Conte, D., Foggia, P., Sansone, C., Vento, M.: Thirty years of graph matching in pattern recognition. International Journal of Pattern Recognition and Artificial Intelligence **18** (2004) 265–298
3. Berretti, S., Bimbo, A.D., Vicario, E.: Modelling spatial relationships between colour clusters. Pattern Analysis & Applications **4** (2001) 83–92
4. Smith, J.R., Chang, S.F.: VisualSEEk: A fully automated content-based image query system. In: Proceedings of ACM International Conference on Multimedia, Boston, MA (1996) 87–98
5. Chu, W.W., Hsu, C.C., Cardenas, A.F., Taira, R.K.: Knowledge-based image retrieval with spatial and temporal constructs. IEEE Transactions on Knowledge and Data Engineering **10** (1998) 872–888
6. Tang, H.L., Hanka, R., Ip, H.H.S.: Histological image retrieval based on semantic content analysis. IEEE Transactions on Information Technology in Biomedicine **7** (2003) 26–36
7. Petrakis, E.G.M., Faloutsos, C., Lin, K.I.: Imagemap: An image indexing method based on spatial similarity. IEEE Transactions on Knowledge and Data Engineering **14** (2002) 979–987
8. Aksoy, S., Tusk, C., Koperski, K., Marchisio, G.: Scene modeling and image mining with a visual grammar. In Chen, C.H., ed.: Frontiers of Remote Sensing Information Processing. World Scientific (2003) 35–62
9. Aksoy, S., Koperski, K., Tusk, C., Marchisio, G., Tilton, J.C.: Learning Bayesian classifiers for scene classification with a visual grammar. IEEE Transactions on Geoscience and Remote Sensing **43** (2005) 581–589
10. Ballard, D.H., Brown, C.M.: Computer Vision. Prentice Hall (1982)
11. Zhang, D., Lu, G.: Review of shape representation and description techniques. Pattern Recognition **37** (2004) 1–19
12. Christmas, W.J., Kittler, J., Petrou, M.: Structural matching in computer vision using probabilistic relaxation. IEEE Transactions on Pattern Analysis and Machine Intelligence **17** (1995) 749–764
13. Messmer, B.T., Bunke, H.: Efficient subgraph isomorphism detection: A decomposition approach. IEEE Transactions on Knowledge and Data Engineering **12** (2000) 307–323
14. Myers, R., Wilson, R.C., Hancock, E.R.: Bayesian graph edit distance. IEEE Transactions on Pattern Analysis and Machine Intelligence **22** (2000) 628–635

A Graph-Based Method for Detecting and Classifying Clusters in Mammographic Images

P. Foggia[1], M. Guerriero[2], G. Percannella[2], C. Sansone[1], F. Tufano[2], and M. Vento[2]

[1] Dipartimento di Informatica e Sistemistica, Università di Napoli "Federico II"
Via Claudio, 21 I-80125 Napoli (Italy)
{foggiapa, carlosan}@unina.it
[2] Dipartimento di Ingegneria dell'Informazione e di Ingegneria Elettrica
Università di Salerno - Via P.te Don Melillo, 1 I-84084, Fisciano (SA), Italy
{mguerrie, pergen, ftufano, mvento}@unisa.it

Abstract. In this paper we propose a method based on a graph-theoretical cluster analysis for automatically finding and classifying clusters of microcalcifications in mammographic images, starting from the output of a microcalcification detection phase. This method does not require the user to provide either the expected number of clusters or any threshold values, often with no clear physical meaning, as other algorithms do.

The proposed approach has been tested on a standard database of 40 mammographic images and has demonstrated to be very effective, even when the detection phase gives rise to several false positives.

1 Introduction

Calcifications can be expression of many kinds of pathologies. Microcalcifications found in the breast hold a major diagnostic interest, because they are effective indicators of benign or malignant pathologic modifications [1]. Namely, microcalcifications found in the mammographic images represent the only symptom of neoplasiae in 30-50% of carcinomas that are not recognized in the breast [2].

Physicians agree that the presence of three or more microcalcifications grouped in a cubic centimeter of mammary tissue defines a *cluster*. In general, the higher the number of microcalcifications grouped in a cluster, the higher the probability that a neoplastic pathology is present. Despite the diagnostic relevance, the detection and characterization of single microcalcifications in a mammogram is often a hard task also for expert radiologists. It is not unusual that two radiologists (or even the same radiologist at different times) provide different diagnoses on the same case. This is due to the intrinsic complexity of the problem: the low injection of radiation produces a mammographic image poorly contrasted, making the microcalcifications not easily distinguishable from the mammal tissue in the background.

The availability of an automated tool for detecting clusters could allow the radiologists to concentrate their attention to suspicious areas during the review of a mammogram and to determine more reliably whether further examinations (with additional imaging tests or biopsy) are needed. This approach can improve the radiologists' diagnostic performance, in terms of sensitivity or specificity. These considerations highlight the usefulness of Computer-Aided Detection (CAD)

D.-Y. Yeung et al. (Eds.): SSPR&SPR 2006, LNCS 4109, pp. 484–493, 2006.

techniques [3] for automatically detecting the presence of clusters in mammograms. As a consequence, in the last ten years a large interest with respect to this problem in the field of Pattern Recognition has been registered. Many authors tried to solve this problem (see [4,5] for an in-depth review of the state of the art in this field) so that today some commercial tools are available on the market.

Nevertheless, most research efforts have addressed the problem of the detection of the microcalcifications; however, this aspect is only a part of the whole picture. In fact, the final goal to pursue is the detection of clusters of microcalcifications and the discrimination among benign and malignant ones, but, to the best of our knowledge, just few authors have proposed algorithms which face both these problems.

The typical approach for cluster detection first individuates the microcalcifications within the image and then aggregates them into clusters on the basis of some spatial and morphological features. The main drawback of most algorithms using this approach is that their performance is very often affected by the shape and the size of the cluster, sometimes requiring some a priori knowledge of the presumed number of clusters to be detected. Furthermore, in order to obtain an adequate cluster detection the user is often required to set some thresholds without any clear physical meaning.

In this paper, we propose a method for detecting clusters of microcalcifications which is able to overcome all the above limitations. This result has been obtained through a careful design of the algorithm for grouping the microcalcifications. In the field of Pattern Recognition there is a huge number of clustering approaches. Most algorithms aggregate the points in the feature space on the basis of the average distance of the points within the cluster. Unfortunately, this type of algorithms are not well suited to handle clusters of various shapes and sizes. A particular family of clustering algorithms are those based on graph theory. The algorithms of this family represent the clusters through undirected graphs. Each node is associated to a point in the feature space, while to each edge it is associated the distance of the connected nodes calculated in the feature space. Note that this definition of cluster does not impose any restriction with respect to the size and the shape. Furthermore, when the distance is calculated as the Euclidean distance in the image plane the method resembles the way the radiologists group the microcalcifications. For the above motivations, we decided to adopt the method described in [6], probably the most important graph based clustering method. This method requires to set only a single threshold, i.e. the maximum allowed value associated to an edge. In order to derive automatically the optimal value of this threshold, in this paper we propose an innovative method based on the use of the fuzzy c-means algorithm.

Once a cluster of microcalcifications has been detected, we also provide the information about its malignancy. The classification is carried out by a neural network on the basis of a set of features, which take into account both information computed on the whole cluster and on the single microcalcifications. We also propose and use some new features that are directly computed on the graph-based representation of the cluster.

The proposed cluster detection and classification approach has been tested on a standard database of 40 mammographic images and has shown to be very effective.

The organization of the paper is as follows: in Section 2 and Section 3 the proposed cluster detection and classification approaches are presented respectively. In Section 4 the database used is described together with the tests carried out in order to assess the performance of the proposed method. A comparison with the results obtained by other techniques presented in the literature is also reported. Finally, some conclusions are drawn in Section 5.

2 Cluster Detection

As anticipated in the introduction, a cluster is a group of at least three microcalcifications in a limited area (usually 1 cm^2) of the mammogram. From this definition it derives that the Euclidean distance is the most important feature for clustering microcalcifications. As a consequence, the proposed algorithm first assigns the microcalcifications to candidate clusters on the basis of their relative distances; then, it eliminates clusters composed by less than three microcalcifications. It is clear that the detection of the candidate clusters constitutes the most critical phase of the whole process, especially in presence of falsely detected microcalcifications, and represents also the major innovative contribution provided by this paper.

The proposed clustering method is based on graph theoretical cluster (GTC) analysis. This family of clustering algorithms is capable of detecting clusters of various shapes, at least for the case in which they are well separated. This feature is shared only by few other clustering algorithms. This aspect is very important for the problem at hand since clusters of microcalcifications typically assume various shapes depending on the pathology and are spatially quite well separated.

In order to understand how GTC analysis is used for automatic microcalcifications clustering it is worth to review some basic terminology on the graph theory.

Graph: a graph G can be defined as a set X of *nodes* connected by a set E of *edges*:

$$G = [X, E]$$
$$X = \{x_1, x_2, \ldots, x_n\}$$
$$E = \left\{ e_{ij} = (x_i, x_j) \middle| x_i, x_j \in X \right\}$$

Path: a path P of length L through a graph is a sequence of connected nodes: $P = \langle x_1, x_2, \ldots, x_{L+1} \rangle$, where $\forall i \in (1, L)$, (x_i, x_{i+1}) is in E. A graph is *connected* if for any two nodes there is at least a path connecting them.

Cycle: a graph contains a cycle if there is a path of nonzero length through the graph $P = \langle x_1, x_2, \ldots, x_{l+1} \rangle$, such that $x_1 = x_{l+1}$.

Spanning Tree: a spanning tree of a graph G is a set of n-1 edges that connect all nodes of the graph. A tree is a connected graph $[X,T]$ with no cycles. The graph $[X,T]$ is a tree if and only if exists one and only one path between any pair of vertices.

Minimum Spanning Tree (MST): in general, it is possible to construct multiple spanning trees $[X,T_i]$ with $i > 1$ for a graph G. If a weight $w(e)$ is associated with each edge e, then the minimum spanning tree is the set of edges forming a spanning tree such that

$$w(\text{MST}) = \min_i \left\{ \sum_{e \in T_i} w(e) \right\}$$

The MST of a graph may be derived with Prim's algorithm or Kruskal's algorithm [7]. In this paper we used the Prim's algorithm.

Forest: a graph without cycles and not connected is called a *forest*. Each connected component of the forest is a *tree*.

The proposed method starts by describing with a graph all the microcalcifications detected by an automatic algorithm: graph nodes correspond to microcalcifications, while the edges of the graph encode the spatial relationships between microcalcifications. Each microcalcification is linked by an edge to all the other ones. The weight of each edge is the Euclidean distance in the 2D space between the nodes connected by that edge. After such a graph is obtained, the GTC analysis is employed. It takes the microcalcifications as vertices in the 2D space and constructs the MST on them. By removing all the edges in the tree with weights greater than a threshold λ, we arrive at a forest containing a certain number of subtrees (clusters). In this way, the GTC method automatically groups vertices (microcalcifications) into clusters. Successively, clusters with less than three nodes are eliminated according to the above described rule.

It is worth noting that the optimal value of λ typically depends on the specific mammogram. As a consequence, it is not possible to use a fixed value of λ for every mammogram. Our proposal is then to determine the optimal value of λ by reformulating the problem as the one of partitioning the whole set of edges into two clusters, according to their weights. The cluster of the edges of the MST with small weights will contain edges to be preserved, while the edges belonging to the other cluster will be removed from the MST. In order to solve this problem we employ the Fuzzy C-Means (FCM) clustering algorithm. In particular, FCM is used to separate all the edges of the MST into two clusters. Then, we remove from the MST all the edges belonging to the cluster s whose center exhibits the largest value.

3 Cluster Classification

Cluster classification is aimed at evaluating the benignancy or malignancy of a detected cluster. In order to discriminate between benign and malignant clusters, we have defined a set of features which try to capture their differences.

Malignant clusters are usually characterized by microcalcifications with low brightness and hazy contour so that they can be easily confused with the background, while the microcalcifications of benign clusters show a high contrast with respect to the background; sometime it is possible to find regions affected by noise that are characterized by a high level of brightness. Furthermore, the typical shape of malignant clusters is elliptical, with a noticeable density of microcalcifications. For this reason, we defined a set of features which capture the brightness, density and shape characteristics of the cluster. It is worth noting that since we used a graph structure to represent the cluster the definition of some features exploits some properties of this structure. The defined features are the following:

- **Brightness mean:** the mean of microcalcifications brightness normalized with respect to the brightness of the area covered by the graph

$$Brightness\_mean = \frac{m\_micro}{b\_area}$$

with

$$m\_micro = \frac{\sum_{i=1}^{mpix} bm_i}{nmicro} \text{ and } b\_area = \sum_{i=1}^{apix} ba_i ,$$

where *mpix* is the number of microcalcification pixels in the cluster, bm_i is the brightness of a single pixel of a microcalcification, *nmicro* is the number of microcalcifications in the cluster, *apix* is the number of pixel in the area covered by the graph and ba_i is the brightness of a single pixel of the area.

- **Brightness variance:** variance of microcalcification brightness normalized with respect to the brightness of the area covered by the graph

$$Brightness\_variance = \frac{\sum_{i=1}^{nmicro} (b\_micro_i - m\_micro)^2}{nmicro} \Big/ b\_area$$

with

$$b\_micro_i = \frac{\sum_{j=1}^{mpix_i} bm_{ij}}{mpix_i}$$

where $mpix_i$ is the number of pixels of the microcalcification i and bm_{ij} is the brightness of a single pixel of microcalcification i;

- **Graph density:** ratio between the number of nodes of the graph and the number of pixel of the image covered by the graph

$$Graph\_density = \frac{X^*}{apix}$$

where X^* is the number of nodes of the graph;

- **Diameter/nodes ratio:** ratio between the square of cluster diameter and the number of graph nodes, where the cluster diameter is the maximum distance between two graph nodes.
- **Cluster compactness:** mean distance among graph edges

$$Cluster\_compactness = \frac{\sum_{e \in E} w(e)}{E^*}$$

where E^* is the number of edge of graph;

- **Aspect ratio:** it is the ratio between the sides of the bounding box including the graph, where the bounding box is the smaller rectangle that circumscribe the graph in the image

$$Aspect\_ratio = \frac{w\_box}{h\_box}$$

where *w_box* is the length of horizontal side expressed in pixel and *h_box* is the length of vertical side expressed in pixel.

- **Graph valence:** mean valence of graph nodes, where the valence of a node is the number of edges incident to the node.

Cluster classification is performed by an artificial neural network: in particular, we used a Multi Layer Perceptron (MLP) network with a hidden layer.

4 Experimental Results

In order to evaluate the performance of the proposed method, tests were performed by using a standard database publicly available. It is made of 40 mammographic images, containing in the whole 105 clusters (76 malignant and 29 benign). Images were provided by courtesy of the National Expert and Training Centre for Breast Cancer Screening and the Department of Radiology at the University of Nijmegen, the Netherlands.

The proposed method assumes that microcalcifications have been already detected by a suitable method. To this aim, we chose and implemented the microcalcifications detection algorithm in [9], which uses a hierarchical pyramid neural network (HPNN) that exploits image structure at multiple resolutions for detecting clinically significant features in mammograms. Note that the latter method simply determines if a pixel of the image belongs to a microcalcification, but does not reconstruct the whole microcalcification, as required by our cluster detection and classification method. The contour of the microcalcifications is obtained by using a connected components algorithm.

4.1 Cluster Detection

In order to assess the performance of our method, we referred to the definitions given in [8], where a detected cluster is considered a true positive (*TP*) if it contains two or more microcalcifications within the distance of 1 cm, and is considered a false positive (*FP*) if none of the microcalcifications found in the cluster are inside the ground truth circle; a false negative *(FN)* is counted if a cluster present in the ground truth is not detected.

The performance of the proposed clustering method on the Nijmegen database was measured in terms of *True Positives* and *False Positives per image* rates and is reported in Table 1. Note that the proposed detection method does not need any specific learning procedure as it needs only to set the parameters of the FCM algorithm. In particular, we used typical values for both the fuzziness coefficient $m=2$ and the termination criterion threshold $\varepsilon = 0.05$.

The results reported in Table 1 shows that the proposed method is able to automatically reduce falsely detected clusters, yielding a very low *FP per image* rate.

In order to have a qualitative evaluation of the behavior of the proposed cluster detection method, in Fig. 1 are depicted the outputs of the method on two mammograms of the database. Fig. 1.a is particularly interesting since it includes clusters of different shapes and sizes. Note how the proposed system is able to correctly detect all the clusters within the image. On the other hand, in Fig. 2.b it is possible to appreciate how the system is very effective even when the microcalcification detection gives rise to several false positives.

Table 1. Performance obtained on the Nijmegen database by the proposed cluster detection method

TP rate	FP per image rate
82.83%	0.08%

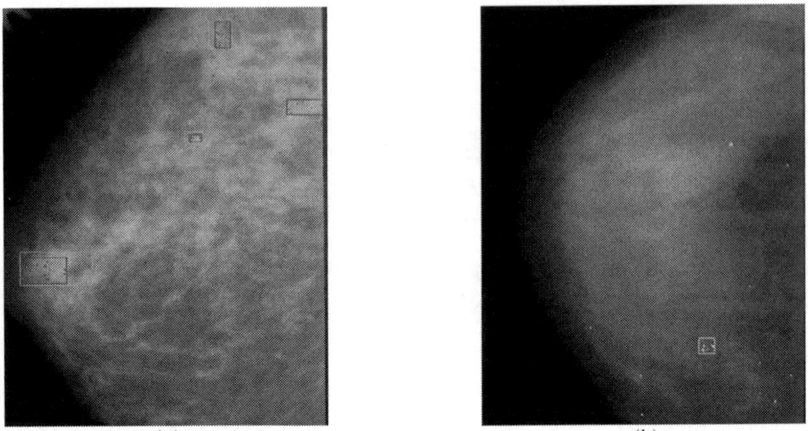

(a) *(b)*

Fig. 1. *(a)* Light gray rectangles represent the cluster detected by the proposed method; dark rectangles account for the ground truth. *(b)* The white points and the white rectangle represent the microcalcification and the cluster detected by the proposed system.

4.2 Cluster Classification

Cluster malignancy is assessed through a neural classifier which requires an adequate training procedure. Unfortunately, the used dataset is quite small; thus, in order to have a more realistic estimate of the recognition capability of the proposed system, a k-fold cross validation was performed. According to this experimental procedure, the whole data set is divided into k disjoint subsets of approximately equal size. Then the classifier is trained k times, each time leaving one of the subsets out of the training and then using the omitted subset to compute the recognition rate. In our case, we performed a ten-fold cross-validation; at each iteration, six parts of the database were used as a training set in each experiment, three parts as a validation set and the remaining one for testing. The validation set was used to avoid to overtrain the neural classifier. Finally, the overall performance has been calculated as the average performance over the ten iterations.

We repeated the above procedure for different number of neurons in the hidden layer of the MLP network. The best performance, reported in Table 2, was obtained when 45 neurons were used.

Data reported in table 2 show that the performance is biased toward the correct classification of the malignant clusters. This is mainly due to the fact that the composition of the dataset is unbalanced; in fact, malignant cases outnumber benign cases.

Table 2. Performance obtained on the Nijmegen database by the proposed cluster classification method; rows represent the true class, while columns the response of the classifier

	BENIGN	MALIGNANT
BENIGN	52.2%	47.8%
MALIGNANT	0%	100%

4.3 Comparison with Other Methods

We have compared our results with those reported in [8, 10-15], since in these papers the same database was used. In particular, in [8, 10-13] cluster detection methods are reported, while in [14] and [15] cluster classification techniques are described.

Among the above cited cluster detection methods, [10] and [8] both employ a Markov Random Field model, but in [10] a Support Vector Machine is used for reducing false positives. Methods presented in [11] and [12] are instead based on a scale-space approach; in [11] a fuzzy-based enhancement of the mammogram is also introduced as a pre-processing step. Finally, in [13] the use of wavelet coefficients together with features extracted from microcalcifications and the co-occurrence matrix is proposed.

Table 1 shows the comparison of the cluster detection results obtained by each method in terms of true positives and false positives. We have denoted our method with *GTC*, while *DEL*, *CHE*, *KAR*, *NET* and *YU* respectively referred to the results obtained in [10], [11], [8], [12] and [13]. From this table, it can be noted that the proposed method is outperformed by four methods in the detection of true clusters, but gives the best results in terms of false positives. The high specificity of our methods makes it particularly appealing for its use as a second-look by radiologists. Almost all the times our system detects a cluster, in fact, it is really present in the mammographic image. Therefore, a radiologist can use the detection of our system for eliminating any doubt about a particular cluster he visually found in the image without a CAD system. More in detail, the comparison with the other MRF-based methods shows that our method provides better or comparable results in terms of true positive rate and performs better in terms of false positives. Slightly worse results are obtained by our method, in terms of true positive rate, with respect to the wavelet-based approach proposed in [13], while significantly better results are reached in terms of false positive. On the other hand, scale-space approaches [11, 12] perform better than our method in terms of true positives, but this is paid with a higher number of false positives. This is especially true for the method presented in [12]. Finally, it must be outlined that in [11] (where the best results in terms of true positive are reported) tests have been performed in selected areas containing all the clusters of the image, while in all our tests the whole mammografic images were used.

Table 3. Comparison of the cluster detection results obtained on the 40 images of the Nijmegen database. Best results are reported in bold.

Method	*TP rate*	*FP per image*
GTC	82.86%	**0.08**
CHE	**90.48%**	0.35
DEL	79.05%	0.30
KAR	83.81%	1.05
NET	88.57%	0.98
YU	85.71%	0.53

As regards cluster classification techniques, it is worth noting that some papers report results obtained only on a limited set of images (in [16], for example, only 18 clusters out of 105 are considered as test set). So, we considered for comparison the methods presented in [14] and [15] whose results refer to all the 40 mammografic images of the database. In particular, in [14] a SVM using a Gaussian kernel is proposed, while MLP classifiers are employed in [15], where a multi-expert approach is also proposed in order to improve the performance of single classifiers. As regards the features for malignancy analysis, both microcalcification features and cluster features are used in [14] and [15]. Table 4 shows the comparison of the cluster classification results obtained by each method in terms of overall accuracy. We have denoted our method with *Graph-based*, while *DES* and *PAP* respectively referred to the results obtained in [15] and [14]. As it is evident, our method exhibits the best performance.

Table 4. Comparison of the cluster classification results obtained on the 40 images of the Nijmegen database. The best result is reported in bold.

Method	***Overall accuracy***
Graph-based	**84.2 %**
DES	75.2 %
PAP	81.0 %

5 Conclusions

Mammography is a powerful tool for early diagnosis of breast cancers. A diagnosis is usually obtained by using human expertise in recognizing the presence of given patterns and types of microcalcifications. So, there are significant motivations for developing computer based support tools able to complement the radiologists work.

In this framework, we proposed a new method based on a graph-theoretical cluster analysis for automatically finding and classifying clusters of microcalcifications in mammograms. The proposed approach was tested on a standard database of mammograms and revealed to be very effective even when the microcalcification detection phase gives rise to several false positives.

References

1. M. Lanyi, Diagnosis and differential diagnosis of breast calcifications, Springer-Verlag, New York, 1986.
2. G. Coopmans De Yoldi, G. Viganotti, S. Bergonzi, C. Gerranti, G.Piragine, E. Cassano, M. Barberini, F. Rilke, U. Veronesi, “Le microcalcificazioni nei carcinomi mammari non palpabili. Analisi di 427 casi”. (in italian), Rad Med, no. 85, pp. 611-614, 1993.
3. A. Lauria, R. Palmiero, M. Imbriaco, G. Selva et al., “Analysis of radiologist performance with and without a CAD system”, European Congress of Radiology, 2002.
4. H.D. Cheng, Xiaopeng Cai, Xiaowei Chen, Liming Hu, Xueling Lou, “Computer-aided detection and classification of microcalcifications in mammograms: a survey”, International Journal on Pattern Recognition, Vol. 36, pp. 2967-2991, 2003.
5. K.Thangavel, M.Karnan, R.Sivakumar, A. Kaja Mohideen, "Automatic Detection of Microcalcification in Mammograms – A Review", International Journal on Graphics, Vision and Image Processing, Vol. 5, pp. 31-61, 2005.

6. C.T. Zahn, "Graph-theoretical methods for detecting and describing gestalt clusters", IEEE Transactions on Computers, Vol. 20(1), pp. 68-86, January 1971.
7. E. Horowitz, S. Sahni, Fundamentals of Computer Algorithms, Computer Science Press, 1978.
8. N. Karssemeijer, "Adaptive Noise Equalization and Recognition of Microcalcification Clusters in Mammograms", Int. Journal of Patt. Rec. and Artificial Intelligence, Vol. 7, no. 6, pp. 1357-1376, 1993.
9. P. Sajda, C. Spence, J. Pearson, "Learning contextual relationships in mammograms using a hierarchical pyramid neural network", IEEE Transactions on Medical Imaging, Vol. 21, No. 3, pp. 239-250, 2002.
10. C. D'Elia, C. Marrocco, M. Molinara, G. Poggi, G. Scarpa, F. Tortorella, "Detection of Microcalcifications Clusters in Mammograms through TS-MRF Segmentation and SVM-based Classification". IEEE International Conference on Pattern Recognition, Vol. 3, pp. 742-745, 2004.
11. T. Netsch and H. Peitgen, "Scale-Space Signatures for the Detection of Clustered Microcalcifications in Digital Mammograms", IEEE Trans. on Medical Imaging, Vol. 18, no. 9, pp. 774-786, 1999.
12. H.D. Cheng, J. Wang and X. Shi, "Microcalcification Detection Using Fuzzy Logic and Scale Space Approach", Pattern Recognition, Vol. 37, pp. 363-375, 2004.
13. S. Yu, L. Guan, "A CAD system for the automatic detection of clustered microcalcifications in digitized mammogram films", IEEE Transactions on Medical Imaging, Vol. 19, no. 2, pp. 115–126, 2000.
14. A. Papadopoulos, D.I. Fotiadis, A. Likas, "Characterization of clustered microcalcifications in digitized mammograms using neural networks and support vector machines", Artificial Intelligence in Medicine, 2006 (in press).
15. M. De Santo, M. Molinara, F. Tortorella, M. Vento, "Automatic classification of clustered microcalcifications by a multiple expert system", Pattern Recognition, Vol. 36, pp. 1467-1477, 2003.
16. B. Verma, J. Zakos, "A computer-aided diagnosis system for digital mammograms based on fuzzy-neural and feature extraction techniques", IEEE Transactions on Inform. Technol. Biomed., vol 5, no. 1, pp. 46–54, 2001.

A Speedup Method for SVM Decision

Yongsheng Zhu[1], Junyan Yang[1], Jian Ye[2], and Youyun Zhang[1]

[1] Key Laboratory of Education Ministry for Modern Design and Rotor-Bearing System, Xi'an Jiaotong University, Xi'an 710049, China
{azhu2005, yjunyan}@gmail.com,
yyzhang1@mail.xjtu.edu.cn
[2] Network & Information Technology Center of Library, Xi'an Jiaotong University, Xi'an 710049, China
yejian@mail.xjtu.edu.cn

Abstract. In this paper, we proposed a method to speed up the test phase of SVM based on Feature Vector Selection method (FVS). In the method, the support vectors (SVs) appeared in the decision function of SVM are replaced with some feature vectors (FVs) which are selected from support vectors by FVS method. Since it is a subset of SVs set, the size of FVs set is normally smaller than that of the SVs set, therefore the decision process of SVM is speeded up. Experiments on 12 datasets of IDA show that the number of SVs can be reduced from 20% to 99% with only a slight increase on the error rate of SVM by the proposed method. The trade-off between the generalization ability of obtained SVM and the speedup ability of the proposed method can be easily controlled by one parameter.

1 Introduction

Support Vector Machines (SVM) is a new type learning machine introduced by V.N. Vapnik and et.al.[1], and was broadly studied and applied in many fields for the comparable classification ability to traditional learning machines and good theoretical bases. However, SVM is slower in test phase than other learning machines such as neural network and decision trees [2-4].

To tackle this problem, several methods were introduced. The decision function of SVM is a weighted linear combination of support vectors (SVs) in feature space[1], therefore the decision speed of SVM is proportional to the number of support vectors and most proposed methods try to accelerate the test phase of SVM by reducing the number of support vectors. The methods can be separated into two types. The first type is pre-processing method, in which some special procedures are adopted before or during training SVM, such as processing the training samples[5], reformulating the training problems of SVM [6,7], as well as adoption of special training strategies during training SVM[8,9]; the second type is post-processing method in which the SVM is firstly trained in the normal way, and some additional process are directly applied to the support vector set after training to reduce the number of SVs. This type method is represented typically by the Reduced Set method [2,4,10], as well as methods reported in [11-13]. Although it can also speeds up the training phase in some degree[8,9], pre-processing methods always deal with the entire training sample

D.-Y. Yeung et al. (Eds.): SSPR&SPR 2006, LNCS 4109, pp. 494–501, 2006.

set in most cases; while post-processing methods just operates on the SVs set, so it is easier in implementation and more practical than pre-processing methods. On the other hand, since post-processing methods directly reduces the size of SVs set, it is more "natural".

In this paper, a new post-processing method is proposed to simplify support vector solutions. In the method, the support vectors(SVs) used by the decision function of SVM are replaced by a subset(named feature vectors, FVs) of SVs set which is selected with Feature Vector Selection Method(FVS)[14]. Experimental results on several standard datasets show that the proposed method can speed up SVM from 20% to 99% with small loss on generalization ability of SVM, and the trade-off between the speedup ability and the performance can be easily controlled by one parameter.

The rest of the present paper is organized as follows. In the second section, the FVS method is briefly described, the decision function of SVM expressed with FVs is also included in this section. Experiments and discussions are described in section 3 and section 4 separately. In section 5, we conclude the paper.

2 Feature Vector Selection Method

2.1 Feature Vector Selection Method

Feature Vector Selection Mehtod(FVS) only involves the operation on the kernel matrix[1] which is defined as:

$$\mathbf{K} = (k_{ij})_{1 \le i, j \le M} \tag{1}$$

where M is the number of samples $\mathbf{x}_i, i = 1, \cdots, M$, $k_{ij} = \mathbf{\Phi}^t(\mathbf{x}_i)\mathbf{\Phi}(\mathbf{x}_j)$, $\mathbf{\Phi}: \mathbf{X} \to \mathbf{F}, \mathbf{x} \to \mathbf{\Phi}(\mathbf{x})$ is a mapping operation from the input space $\mathbf{X}$ into a feature space $\mathbf{F}$. The aim of FVS is to find a basis of $\mathbf{\Phi}(\mathbf{x}_i), i = 1, \cdots, M$ so that each sample can be expressed in $\mathbf{F}$ with the basis.

Let L be the number of selected vectors (named feature vectors in [14], FVs), note $\mathbf{\Phi}(\mathbf{x}_i) = \mathbf{\Phi}_i, i = 1, \cdots, M$ and the FVs by $\mathbf{x}_{s_j}, j = 1, \cdots, L$, note the corresponding images of FVs in $\mathbf{F}$ by $\mathbf{\Phi}(\mathbf{x}_{s_j}), j = 1, \cdots, L$, where $L \le M$. Then for the set of FVs $\mathbf{S} = \{\mathbf{x}_{s_1}, \mathbf{x}_{s_2}, \cdots, \mathbf{x}_{s_L}\}$, we can estimate the mapping of any sample as a linear combination of the images of $\mathbf{S}$ in $\mathbf{F}$. That is, the estimation $\hat{\mathbf{\Phi}}_i$ for $\mathbf{\Phi}_i$ can be formulated as a dot product:

$$\hat{\mathbf{\Phi}}_i = \mathbf{\Phi}_S \boldsymbol{\alpha}_i \tag{2}$$

In which $\mathbf{\Phi}_S = \{\mathbf{\Phi}(\mathbf{x}_{s_1}), \mathbf{\Phi}(\mathbf{x}_{s_2}), \cdots, \mathbf{\Phi}(\mathbf{x}_{s_L})\}$ is a matrix composed with the images of FVs in $\mathbf{F}$; $\boldsymbol{\alpha}_i = \{\alpha_{i1}, \cdots, \alpha_{iL}\}$ is the corresponding coefficient vector to $\mathbf{\Phi}_i$.

As mentioned above, our aim is to find a suitable set $\mathbf{\Phi}_S$ so that all the images of samples in $\mathbf{F}$ can be approximated as accurately as possible by Eq.(2). If the normalized Euclidean distance is used as evaluation criterion of difference between $\hat{\mathbf{\Phi}}_i$ and $\mathbf{\Phi}_i$, the problem of finding $\mathbf{\Phi}_S$ is finally transformed to find a set of samples $\mathbf{S}$ from the entire sample set which maximizes the fitness function J_s [14]:

$$J_s = \max_{\mathbf{S}} \frac{1}{M} \sum_{\mathbf{x}_i \in \mathbf{X}} \left(\frac{\mathbf{K}_{Si}^t \mathbf{K}_{SS}^{-1} \mathbf{K}_{Si}}{k_{ii}} \right) \tag{3}$$

where $\mathbf{K}_{ss}$ is kernel matrix formed by FVs. $\mathbf{K}_{Si} = (k_{s_p i})_{p=1,\cdots,L;i=1,\cdots,M}$ is the kernel matrix calculated with $\mathbf{x}_i, i = 1,\cdots,M$ and FVs. Eq. (3) is an optimization problem, whose solution could be obtained iteratively[14]. In each step, the support vector which is most orthogonal to the already-selected FVs set is chosen as a new feature vector. The iteration continues until an upper limitation value $J_s^{\max}, 0 \le J_s^{\max} \le 1$ of J_s (or an upper limitation of ratio between number of FVs and SVs) is achieved. When $J_s^{\max} = 1$, a complete basis of samples will be found; when $J_s^{\max} < 1$, some unimportant FVs will be ignored and an approximated basis of samples will be found in $\mathbf{F}$. Interestingly, in previous experiments, we found ignoring unimportant FVs leads to the increase of the classification accuracy on the noised data by the obtained SVM[5]. The details about the implementation of FVS algorithm please refers to [14].

2.2 Speedup of SVM Decision

Next, let's turn our attention to the calculation of coefficient $\boldsymbol{\alpha}_i$ appeared in Eq. (2). By left production of $\mathbf{\Phi}_S^t$ to Eq.(2), one easily gets:

$$\mathbf{\Phi}_S^t \hat{\mathbf{\Phi}}_i = \mathbf{\Phi}_S^t \mathbf{\Phi}_S \boldsymbol{\alpha}_i \Leftrightarrow \widehat{\mathbf{K}}_{Si} = \mathbf{K}_{SS} \boldsymbol{\alpha}_i \tag{4}$$

As mentioned above, $\hat{\mathbf{\Phi}}_i$ is a well approximation to $\mathbf{\Phi}_i$, so $\widehat{\mathbf{K}}_{Si}$ can also be considered as a well approximation to $\mathbf{K}_{Si}$ and we denote $\widehat{\mathbf{K}}_{Si}$ as $\mathbf{K}_{Si}$, and $\hat{\mathbf{\Phi}}_i$ as $\mathbf{\Phi}_i$ hereafter. Since $\mathbf{K}_{SS}$ is positively definite as a kernel matrix[1], $\mathbf{K}_{SS}^{-1}$, the inverse matrix of $\mathbf{K}_{SS}$ exists, so $\boldsymbol{\alpha}_i$ is obtained from Eq.(4) as:

$$\boldsymbol{\alpha}_i = \mathbf{K}_{SS}^{-1} \mathbf{K}_{Si} \tag{5}$$

then $\mathbf{\Phi}_i$ is formulated with $\mathbf{\Phi}_S$ as:

$$\mathbf{\Phi}_i = \mathbf{\Phi}_S \mathbf{K}_{SS}^{-1} \mathbf{K}_{Si} \tag{6}$$

Finally, let's try to express the decision function of SVM with FVs. The descision function of SVM is given as follows:

$$y = \operatorname{sgn}\left\{ \sum_{j \in \mathbf{I}_{SV}} \beta_j y_j \left(\boldsymbol{\Phi}_x^T \boldsymbol{\Phi}_j \right) + b \right\} \tag{7}$$

where $\operatorname{sgn}(x) = \begin{cases} +1, x \geq 0 \\ -1, x < 0 \end{cases}$ is the signal function, $\mathbf{I}_{SV}$ is the index set of SVs, β_j is the corresponding Lagrange multiplier of SVs $\mathbf{x}_j$, y_j is the class labels of $\mathbf{x}_j$, b is the corresponding bias, and $\boldsymbol{\Phi}_x = \boldsymbol{\Phi}(\mathbf{x})$ is the image of the test sample $\mathbf{x}$ in $\mathbf{F}$. Substituting Eq. (6) into Eq. (7), one get:

$$\begin{aligned} y &= \operatorname{sgn}\left\{ \sum_{j \in \mathbf{I}_{SV}} \beta_j y_j \left(\boldsymbol{\Phi}_x^T \boldsymbol{\Phi}_S \mathbf{K}_{SS}^{-1} \mathbf{K}_{Sj} \right) + b \right\} \\ &= \operatorname{sgn}\left\{ \sum_{j \in \mathbf{I}_{SV}} \beta_j y_j \left(\mathbf{K}_{Sx}^t \mathbf{K}_{SS}^{-1} \mathbf{K}_{Sj} \right) + b \right\} \\ &= \operatorname{sgn}\left\{ \sum_{j \in \mathbf{I}_{SV}} \beta_j y_j \mathbf{K}_{Sj}^t \mathbf{K}_{SS}^{-1} \mathbf{K}_{Sx} + b \right\} \end{aligned} \tag{8}$$

After applying FVS method to the SVs set, both the SVs set and the FVs set $\boldsymbol{\Phi}_S$ are available, so $\mathbf{K}_{Sj}^t$ and $\mathbf{K}_{SS}^{-1}$ can be calculated prior to the test. Let's denote $\mathbf{A} = \sum_{j \in \mathbf{I}_{SV}} \beta_j y_j \mathbf{K}_{Sj}^t \mathbf{K}_{SS}^{-1}$, then $\mathbf{A}$ can also be calculated prior to the test and the Eq.(8) is simplified as:

$$y = \operatorname{sgn}\left\{ \mathbf{A}\mathbf{K}_{Sx} + b \right\} \tag{9}$$

It is obvious from Eq.(9) that to classify a new sample $\mathbf{x}$, the evaluation of kernel matrix of the new sample and the SVs is replaced by evaluation of kernel matrix of the sample $\mathbf{x}$ and FVs. As a subset of SVs set, the size of FVs set is always smaller than that of SVs, the test phase of the SVM can be speeded up by using Eq.(9).

3 Experiments

To clarify the speedup efficiency of the proposed method, the artificial and real world datasets from the IDA repository [15] are experimented. Every dataset was randomly separated into training and testing samples according to the predefined number ratio, each such a splits form a realization, and every dataset contains several realizations. The experiments are carried out on each realization and the averaged results are reported in this section. Experiments are completed on L2SVM with the commonly used RBF kernel, as more SVs are required by this type of SVM than the regular SVM (L1SVM) in the most situations [9], the reduction of SVs for L2SVM should be more necessary. The penalty coefficient C and the width σ of RBF function were chosen for the first 10 realizations of each dataset with "SVR+UD" method introduced in [16], and then the averaged C and σ over these 10 times choices are applied to each realization.

The method presented in this paper is very similar to that proposed by Downs T., Gates K.E. and Masters A.[12], the only difference is that in their method, a complete basis of SVs in feature space $\mathbf{F}$ is found and used to speed up the test phase of SVM while an approximated basis is used in present method. To clarify the difference of speedup performance between using complete and approximated basis, we first test the method of [12] with four datasets (Thyroid, Banana, Heart, Breast Cancer); the results are given in Table 1. The reduction rate of SVs corresponding to bases is defined as:

$$Red. = \frac{\#SVs - \#Bases}{\#SVs} \times 100 \tag{10}$$

Table 1. Optimized parameters (C, σ), the averaged test error rate (Re), the averaged number of SVs (#SVs), the averaged number of bases (#bases) and the number reduction rate of SVs respect to bases ($Red.$) for four datasets

Dataset	Parameters		Re / %	#SVs	#Bases	$Red.$ / %
	C	σ				
Thyroid	0.89	1.14	3.47 ± 1.87	91.2	91.2	0
Banana	0.35	0.46	10.42 ± 0.44	339.59	302.73	10.85
Heart	0.42	4.39	16.00 ± 3.18	156.54	156.54	0
BreastCancer	7.6	2.1	22.92 ± 4.48	170.75	166.79	2.32

It is shown by table 1 that in most cases, the size of the complete basis of SVs in $\mathbf{F}$ is very close to that of SVs set, so the test phase of SVM just can be speeded up very slightly. This suggests that in present method, we should use $J_s^{\max} < 1$ (or set the upper limitation of the ratio between #FVs and #SVs smaller than 1) to find an incomplete basis so that the decision of SVM can be speeded up. In the next experiment, we tried various $J_s^{\max}$ on the above four datasets to find out its influence on the speedup efficiency and the classification performance; the results are listed in Table 2, where "#FVs" is the averaged number of FVs over all realizations, "$Re\ Incr.$" is the increase of the test error rate with respect to Table 1. The results show that the presented method can reduce the number of SVs obviously (about 40% ~ 80% for different dataset) with only a minor loss in classification performance when $J_s^{\max} = 0.98$; if the loss on classification is acceptable, more 20% reduction of SVs can be achieved by setting $J_s^{\max} = 0.90$ for most datasets. In addition, Appendix A lists the best speedup ability of the proposed method on all 12 datasets of IDA on the premise of no obvious increase on the test error rate (the result of dataset "Titanic" is not presented here since the optimized parameters for this dataset can not be obtained by method of [16]).

Table 2. Experimental results on various J_s^{max}

Dataset	J_s^{max}	*Re* /%	*Re Incr.* /%	#FVs	*Red.* /%
Thyroid	0.98	3.49 ± 1.83	0.02	53.27	41.59
	0.96	3.49 ± 1.83	0.02	53.27	41.59
	0.94	3.52 ± 1.70	0.05	43.19	52.64
	0.92	3.67 ± 1.87	0.20	40.54	55.55
	0.90	3.88 ± 1.90	0.41	38.07	58.26
Banana	0.98	10.41 ± 0.44	-0.01	63.18	81.40
	0.96	10.46 ± 0.46	0.04	53.85	84.14
	0.94	10.52 ± 0.46	0.10	48.46	85.73
	0.92	10.54 ± 0.49	0.12	44.50	86.90
	0.90	10.61 ± 0.49	0.19	41.29	87.84
Heart	0.98	16.03 ± 3.19	0.03	80.23	48.75
	0.96	15.97 ± 3.14	-0.03	61.27	60.86
	0.94	15.81 ± 3.09	-0.19	50.09	68.00
	0.92	16.08 ± 3.15	0.08	42.18	73.05
	0.90	16.23 ± 3.08	0.23	36.31	76.80
Breast Cancer	0.98	24.06 ± 4.00	1.14	103.12	39.61
	0.96	24.19 ± 3.93	1.27	88.26	48.31
	0.94	24.99 ± 4.27	2.07	78.71	53.90
	0.92	25.45 ± 4.24	2.53	71.68	58.02
	0.90	25.38 ± 4.19	2.46	65.93	61.39

4 Discussions

A method which uses FVS method to reduce the number of SVs therefore the computaional complexity of SVM was described in this paper. Compared to the other post-processing speedup methods such as RS method [2,10], our method has several advantages. Firstly, the founded FVs more "meaningful", they are a best approximated basis of SVs set in feature spaces. Secondly, since the FVS method try to find a minimum subset of SVs set to approximate each SVs, theoretically, the proposed method should reduce the complexity of SVM decision furthest among all methods based on the same idea (such as the method proposed in[12]). The third advantage of the proposed method is the potential denoise ability. We have proved in [5] that by ignore the "unimportant" FVs through setting $J_s^{max} < 1$, the classification accuracy on noisy data can be improved. The denosie ability of FVS method is proved again in the present experiments, for Heart and Banana dataset, we can find in table 2 that for some given J_s^{max} , the classification error rate is also decreased.

Since the proposed method is a post-processing method and only operates on the SVs set, it is applicable for other type of kernels such as polynomial and sigmoid, and also applicable for other type of SVMs such as L1SVM, LS-SVM and so on; and also, it is applicable for both support vector classification and support vector regression. For the further speedup and better performance on claiification or regression, the optimized choice method for J_s^{max}, e.g. goden search method, and retraining SVM with the obtained FVs[13], can be applied together with the present method.

Though the proposed methods possess above advantages, the FVS method is still a little expensive in calculation. In future, we plan to improve the present algorithm to make the method more practical in dealing with the large-scale classification problems. Another interesting phenomenon presented by the experiment results is that the dependency of the speedup ability of the present method on the dataset. For instance, the method can reduce number of FVs to 1 for dataset "Flare-solar" without any loss on the generalization ability of SVM while can only reduce 20.02% SVs for dataset "Image" with 0.25% increase on the test error rate. However, the reason for this phenomenon is still under investigation.

5 Conclusions

A new speedup method for SVM decision have been proposed in present paper. The method uses a subset which was chosen from SVs set by FVS algorithm to approximate the decision function of SVM. The trade-off between the speedup ability of the method and the final generaliation abiltiy of SVM can be easily controled by one parameter. As a post-processing method, the proposed method is applicalbe to any kind of SVM with various kernels. Experiments on IDA benchamrk datasets show that the number of SVs can be reduced 20% up to 99% with very slight loss in accuracy, and for some datasets, the reduction on SVs can lead to better classification accuracy.

Acknowledgements. The research is supported by the National Natural Search Foundation of China under Grant Number: 50575179.

References

1. Vapnik, V.N.: The Natural of Statistical Learning Theory. Springer-Verlag, Berlin Heidelberg New York (1995)
2. Burges, C. J. C.: Simplified support vector decision rules. Proc. 13th International Conference on Machine Learning San Mateo, CA. (1996) 71–77
3. Burges, C. J. C.: A tutorial on support vector machines for pattern recognition. Data Miningand Knowledge Discovery. 2 (1998) 121–167
4. Burges, C. J. C., Schoelkopf, B.: Improving the accuracy and speed of support vector learning machines. In: Mozer, M., Jordan, M. and Petsche, T. (eds.): Advances in neural information processing systems. Cambridge, MA: MIT Press. 9 (1997) 375–381
5. Zhu, Y. S., Zhang, Y. Y.: A new type SVM—Projected SVM. Science in China G: Physics, Mechanics & Astronomy Supp. 47 (2004) 21–28
6. Osuna, E., Girosi, F.: Reducing the run-time complexity of Support Vector Machines. Proc.14th International Conference on Pattern Recognition. Btisbane, Austrilia (1998)
7. Tipping, M.E.: Sparse Bayesian Learning and the Relevance Vector Machine. Journal of Machine Learning Research. 1 (2001) 211–244

8. Lee, Y.-J., Mangasarian, O. L.: RSVM: reduced support vector machines. in Proc. 1st SIAM Int. Conf. Data Mining.Chicago (2001)
9. Lin, K.-M.: A Study on Reduced Support Vector Machines. IEEE Trans. Neural Networks. 14 (2003) 1449-1459
10. Schoelkopf, B., Mika, S., Burges, C.J.C., Knirsch, P., Muller, K., Ratsch, G., Smola, A.J.: Input space versus feature space in kernel-based methods. IEEE Trans. Neural Networks. 10 (1999) 1000-1017.
11. Nguyen, D.D., Ho, T.B.: An Efficient Method for Simplifying Support Vector Machines. The 22nd International Conference on Machine Learning (ICML2005). Bonn, Germany (2005)
12. Downs, T., Gates, K.E., Masters, A.: Extract simplification of support vector solutions.Journal of Machine Learning Research. 2 (2001) 293–297
13. Suykens, J.A.K., Lukas, L., Vandewalle, J.: Sparese approximation using least squares support vector machines. In: Proc. of the IEEE International Symposium on Circuits and Systems. Geneva, Switzerland. (2000) II757–II760
14. Baudat, G., Anouar, F.: Feature Vector selection and projection using kernels. Neurocomputing. 55 (2003) 21–38
15. The datasets are from website: http://ida.first.gmd.de/~raetsch/data/benchmarks.htm
16. Zhu, Y.S., Li, C.H., Zhang Y.Y.: A practical parameters selection method for SVM. Lecture Notes in Computer Science (ISNN2004). 3173 (2004) 518–523

Appendix A.

The experimental results on IDA datasets, where column "L2SVM" gives the averaged test error rate of L2SVM without speedup; column "L2SVM+FVs" gives the averaged test error rate of L2SVM speeded up with the proposed method. "#SVs" is the averaged number of SVs and "#FVs" is the averaged number of FVs, " *Red.*" is the number reduction rate of FVs respect to SVs.

Dataset	Test Error Rate (%)		Number of SVs (FVs)		
	L2SVM	L2SVM+FVS	#SVs	#FVs	*Red.* (%)
Banana	10.42 ± 0.44	10.61 ± 0.49	339.59	41.29	87.84
Breast Cancer	22.92 ± 4.48	24.06 ± 4.00	170.75	103.12	39.61
Diabetis	23.56 ± 1.84	23.75 ± 1.84	407.79	40.88	89.98
Flare-solar	34.75 ± 2.74	34.69 ± 2.82	109.23	1.00	99.08
German	23.73 ± 2.21	24.14 ± 2.19	599.95	150.11	74.98
Heart	16.00 ± 3.18	16.23 ± 3.08	156.54	36.31	76.80
Image	2.98 ± 0.63	3.23 ± 0.67	358.60	286.80	20.02
Ringnorm	2.03 ± 0.27	1.91 ± 0.27	141.72	58.91	58.43
Splice	11.58 ± 0.72	12.39 ± 0.74	553.70	387.7	29.98
Thyroid	3.47 ± 1.87	3.88 ± 1.90	91.2	38.07	58.26
Twonorm	2.52 ± 0.15	2.99 ± 0.63	113.32	20.33	82.06
Waveform	10.05 ± 0.42	10.99 ± 0.74	190.76	26.20	86.27

Generalization Error of Multinomial Classifier

Sarunas Raudys

Vilnius Gediminas Technical University
Sauletekio 11, Vilnius LT-10223, Lithuania
raudys@ktl.mii.lt

Abstract. Equation for generalization error of Multinomial classifier is derived and tested. Particular attention is paid to imbalanced training sets. It is shown that artificial growth of training vectors of less probable class could be harmful. Use of predictive Bayes approach to estimate cell probabilities of the classifier reduces both the generalization error and effect of unequal training sample sizes.

Keywords: BKS rule, Complexity, Generalization error, Learning, Imbalanced training sets, Multinomial classifier, Sample size.

1 Introduction

In many pattern recognition tasks, the features are discrete: each measurement can assume one of several possible values, such as, the type of an engine, a sex, profession, a presence of certain disease, etc. Then, asymptotically (as the sample size is increasing) optimal classification rule is the Multinomial one [1, 2]. Theoretical analysis of this method is very important for practice since popular *decision tree classifiers* are in fact pruned and tailored versions of the Multinomial classifier. In multiple classifiers system design, we are faced with Multinomial classifier if local (expert) classifiers produce crisp outputs (class labels) and one uses Behavior knowledge space (BKS) method to fuse expert decisions.

Hughes [3] investigated the two class pattern recognition task with categorical valued features. He converted the analysis to investigation of single categorical feature that can assume one of m possible values characterized by m cell probabilities,

$$P_1^{(i)}, P_2^{(i)}, \dots, P_{m-1}^{(i)}, P_m^{(i)} \quad (\sum_{s=1}^{m} P_s^{(i)} = 1, \; i=1, 2). \tag{1}$$

Hughes used Bayes predictive approach to estimate probabilities (1) assuming that prior distribution of these probabilities is uniform. He derived a mean averaged generalization error, $\overline{P}_N^{\mathrm{M}}$. It is an error averaged over immense variety of potentially possible classification problems defined by uniform prior distribution of probabilities (1). In finite training sample situations, theoretical graphs $\overline{P}_N^{\mathrm{M}} = f(m)$ exhibited clear minima. This results initiated in a number of subsequent small learning sample size investigations (see e.g. [4-8] and references therein).

Characteristic property of many pattern recognition problems is a fact that prior probabilities of pattern classes are different. Available number of training vectors of

D.-Y. Yeung et al. (Eds.): SSPR&SPR 2006, LNCS 4109, pp. 502–511, 2006.

the less probable category is very small often. In order to compensate a shortage of training vectors of this pattern class, sometimes designers make significant efforts in order to collect more training vectors of the minority category. In order to investigate usefulness of this strategy we derive equations for mean generalization error of Multinomial classifier for two category case with unequal prior class probabilities, q_1, q_2, and unequal training sample sizes, N_1, N_2. We show that artificial growth of training vectors of less probable class could be harmful. Use of diverse modifications of predictive Bayes approach to estimate cell probabilities of the Multinomial classifier reduces both the generalization error and effects of imbalanced training sets.

2 Generalization Error of Standard Multinomial Classifier

2.1 The Theory

Let we have E crisp (categorical) valued features, where the e-th feature takes one of m_e states. In two category case, there exist $m = \prod_{e=1}^{E} m_e$ potential combinations (cells, states) of E features, $x_1, x_2, \cdots, x_E$. Denote v_s, u_s, (s =1, 2, … , m) the first and second class cell probabilities. To create the Multinomial classification rule we have to know prior probabilities of the classes, q_1, q_2, and $2(m-1)$ probabilities of the cells. The *Bayes decision rule* allocates vector $\boldsymbol{X} = (x_1, x_2, \cdots, x_E)^T$, falling into the s-th state, according to a maximum of products $q_1 P_s^{(1)}$, $q_2 P_s^{(2)}$. If both products are equal among themselves, we make arbitrary decision. The Bayes probability of error is expressed as

$$P_B = \sum_{s=1}^{m} \min\{q_1 v_s, q_2 u_s\}. \tag{2}$$

In practice, we estimate probabilities (1) from training data. In standard sample based Multinomial and BKS fusion rules, one utilizes maximum likelihood (ML) estimates:

$$\hat{v}_s = n_s^{(1)} / N_1, \quad \hat{u}_s = n_s^{(2)} / N_2, \tag{3}$$

where $n_s^{(i)}$ is a number of training vectors of i-th pattern class in the s-th cell.

Utilization of Eq. (2) results in the plug-in rule. If $q_2 \neq q_1$, $N_2 \neq N_1$, expected generalization error

$$\overline{P}_N^{\text{M}} = \sum_{s=1}^{m} [\, q_1 v_s P\{q_1 \hat{v}_s < q_2 \hat{u}_s\} + q_2 u_s P\{q_1 \hat{v}_s > q_2 \hat{u}_s\} + 0.5(q_1 v_s + q_2 u_s) P\{q_1 \hat{v}_s = q_2 \hat{u}_s\}]. \tag{4}$$

To analyze finite training sample size behavior, we assume that numbers of training vectors in each single cell, $n_s^{(i)}$, are Multinomial random variables. After some algebraic manipulation utilizing properties of multinomial distribution, following expression for the mean of the generalization error is obtained

$$\overline{P}_N^{\mathrm{M}} = \sum_{s=1}^{m} \sum_{j=0}^{N_1} \sum_{t=0}^{N_2} \psi_{jt\beta} \frac{N_1!}{j!(N_1-j)!} v_s^j (1-v_s)^{N_1-j} \frac{N_2!}{v!(N_2-v)!} u_s^t (1-u_s)^{N_2-t}, \qquad (5)$$

where $\psi_{jt\beta} = \begin{cases} q_1 v_s & \text{if } j\beta<t \\ q_2 u_s & \text{if } j\beta>t \\ (q_1 v_s + q_s u_s)/2 & \text{if } j\beta=t \end{cases}$ and $\beta = \frac{q_1}{q_2} \frac{N_2}{N_1}$.

We see that generalization error depends on $2m$ probabilities, t_s and u_s ($s = 1, \dots, m$), and sample sizes, N_1 and N_2. If *both* sample sizes, N_1, N_2, are increasing without bound, expected error, $\overline{P}_N^{\mathrm{M}}$, approaches the Bayes error, i.e.

$$P_B = \lim_{N_1 \to \infty, N_2 \to \infty} EP_N^M \qquad (6)$$

If *only one of the* sample sizes, e.g. N_2, is increasing without bound, expected error, $\overline{P}_{N_2 \to\infty}^{\mathrm{M}}$, becomes

$$EP_{N_2\to\infty}^{\mathrm{M}} = \sum_{s=1}^{m} \sum_{j=0}^{N_1} \psi_{j\beta} \frac{N_1!}{j!(N_1-j)!} v_s^j (1-v_s)^{N_1-j}, \qquad (7)$$

where $\psi_{jt\beta} = \begin{cases} q_1 v_s & \text{if } j\beta<u_s \\ q_2 u_s & \text{if } j\beta>u_s \\ (q_1 v_s + q_s u_s)/2 & \text{if } j\beta=u_s \end{cases}$ and $\beta = \frac{q_1}{q_2 N_1}$.

2.2 Numerical Example: Case $N_2 = N_1$

Calculation of generalization error (Eq. (5)) requires to know the 2(m-1) probabilities, $v_1, v_2, \dots, v_{m-1}, u_1, \dots, u_{m-1}$. If m is very large, it could turn out to be serious difficulty for a practitioner. Therefore, in [8, Section 3.8] for tabulation purposes a simplifying model, **MU**, of distribution of values $v_1, v_2, \dots, v_m, u_1, \dots, u_m$, has been proposed. It was assumed that in two category case, the probabilities of $m/2$ cells of the i-th pattern class are equal among themselves and equal to probabilities of other $m/2$ cells of opposite, the (3-i)-th, class. Thus, in model MU, one deals only with two values of the cells' probabilities, $2P_B/m$, and $2(1-P_B)/m$. We present *a few critical comments* below.

Let the training *sample size be very small.* Then in recognition (test) phase, the class label of the s-th cell having probability $P_s^{(i)} = 2P_B/m$ can be easily "confused" with class label of the cell of opposite category having probability $P_s^{(3-i)} = 2(1-P_B)/m$. In model MU, the probabilities of each half of the cells are artificially equalized. It means that in very small training sample cases, model MU "confuses" class labels more often and overestimates generalization error.

Consider now a situation where *learning sample size is large*. Then in recognition phase, it will be more difficult to confuse the s-th cell of one class having small probability $P_s^{(i)} = 2P_B/m$ with the cell of another class having much higher probability,

Table 1. Expected classification error of Multinomial classifier for "standard" model **MU** (six columns for $m = 26, 38, \ldots, 144$) and real world data based model with 128 cells (two very right columns, theory and experimental evaluation). In both data models, the Bayes error P_B=0.06.

$N\backslash m$	26	36	46	62	106	144	128**RWM**	128**experim**
35	**0.1120**	0.1507	0.1954	0.2309	0.3103	0.3495	**0.1134**	0.1138 (73)
50	0.0837	**0.1100**	0.1374	0.1778	0.2590	0.3037	**0.1080**	0.1087 (72)
70	0.0688	0.0837	**0.1022**	0.1336	0.2082	0.2549	**0.1026**	0.1031 (70)
100	0.0621	0.0603	0.0779	**0.0978**	0.1571	0.2011	**0.0968**	0.0976 (65)
200	0.0600	0.0600	0.0613	0.0650	**0.0868**	0.1121	**0.0863**	0.0873 (51)
300	0.0600	0.0600	0.0601	0.0607	0.0682	**0.0809**	**0.0808**	0.0820 (43)
400	0.0600	0.0600	0.0600	0.0601	0.0627	0.0688	0.0773	0.0887 (37)
500	0.0600	0.0600	0.0600	0.0600	0.0609	0.0638	0.0748	0.0862 (32)
1000	0.0600	0.0600	0.0600	0.0600	0.0600	0.0601	0.0689	0.0705 (21)

$P_s^{(3-i)} = 2(1 - P_B)/m$. In such case, confusions of the class labels will be rare, i.e. model MU begins to underestimate generalization error. In six left columns of Table 1 we tabulated generalization errors for model MU if $q_2 = q_1 = 0.5$, $N_2 = N_1 = N$.

For comparison of model MU with "reality" we have chosen a real-world data obtained in problem of classifying two category eight-dimensional spectral Satellite data by means of multiple classifiers system. Seven multilayer perceptron based classifiers served as seven base experts that produced crisp outputs, 0 (first class) or 1 (the second class). So, E=7, $m = 2^7 = 128$. In Fig. 1 we have a scatter diagrams of 126 bi-variate vectors (v_s, u_s) (s = 2, 3, ... ,127), the cell probabilities. Totally, 15,787 vectors from two pattern classes were used to estimate the probabilities. Probabilities of the first (the expert answers are: 0 0 0 0 0 0 0) and the last cells (the expert answers are: 1 1 1 1 1 1 1) differ from remaining ones basically: $v_1 = 0.0247$, $u_1 = 0.7735$ and $v_{128} = 0.7291$, $u_{128} = 0.0114$. All 128 experimentally evaluated values, $v_1, v_2, \ldots, u_{128}$, have non-zero probabilities. In majority of the cells, however, the probabilities v_s, u_s are very small. The Bayes error $P_B = 0.06$.

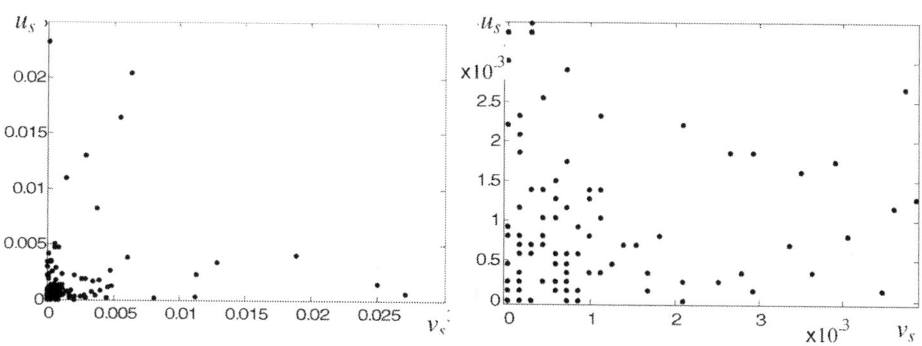

Fig. 1. Scatter diagrams of 126 bi-variate vectors (v_s, u_s) in two scales

We specify this real world data model with diverse 128 cell probabilities as **RWM**. In column "128**RWM**" of Table 1 we present generalization errors calculated according to Equation (5). In the very right column, "128**experim**", we present generalization error evaluated experimentally (*averages* of 5000 independent experiments where N randomly selected 7D binary vectors of each category were used for training of the Multinomial classifier and the test was performed on remaining 15,787 -2N vectors). In brackets we present *standard deviations* multiplied by 10000. The last two columns show very good agreement between theoretical and experimental evaluations. Comparison of column "128**experim**" with six columns calculated for standardized model MU does not show any god agreement: generalization errors calculated for the RWM data model (marked in bold) coincide with that calculated for different MU models characterized by diverse number off cells, m (marked also in bold).

In real world problems usually we have small number of cells with high probabilities v_s, u_s and a large number of ones with small probabilities v_s, u_s. The cells having large probabilities v_s, u_s likely will be classified correctly even in small sample situations. In model MU, however, large probabilities v_s, u_s will be present only in situations where number of cells, m, is small. For that reason, model MU with m comparable with that of real world data overestimates the generalization error in small sample situations (see Table 1).

In large training sample size cases, the cells with moderate differences between probabilities v_s and u_s are not confused any more. For that reason, these cells do not influence an increase in generalization error. There exist, however, large number of cells with small and, therefore, with close probabilities, v_s and u_s. Class labels of these cells can be confused even in large training set situation. For that reason, in large sample case, most important become the cells with small probabilities. When the sample size $N = N_1 = N_2 = 300$, calculation according to Eq. (5) for RWM model gives generalization error $\overline{P}_N^{\mathrm{M}} = 0.0808$. For the standardized data model, MU, this error rate can be obtained if m=144 cells ($\overline{P}_N^{\mathrm{M}}$ =0.0809). It means that almost empty cells are affecting an increase in the expected generalization error. At this time, one can say that "an effective number of cells" is higher: $m_{\mathrm{effective}}$ =144. Thus, in small sample cases (if expected error exceeds the Bayes error 1.5 times and more), standard data model MU overestimates the generalization can, and underestimates it if the sample size is large. Thus, tabulated values for model MU (Table 3.7 in [8] can serve only as a guide of sufficiency of sample size necessary to design BKS fusion rule. In most important pattern recognition problems, one cannot utilize standard data models like MU. One must find a way to evaluate character of distribution of 2(m-1) cells' probabilities and use Equation (5).

2.3 Numerical Example: Case $N_2 \neq N_1$

As a rule, in neural network and pattern recognition literature, however, a sum number of vectors, $n = N_1 + N_2$, has been considered as a single measure of training set size (see e.g. [9 - 11]). Equation (5) points out that for the Multinomial classification rule, the generalization error depends on training set sizes of both pattern classes, N_1 and N_2. The Multinomial classifier and BKS fusion rules are heuristically based plug-in classification rules. Here unknown probabilities of the cells are substituted by their maximum likelihood sample estimates. Consequently, *it is not optimal sample based decision rule.* If the number of training vectors of the less probable category is very small, some researchers compensate a shortage of training vectors of this category by additional collection of more training vectors of the minority pattern class.

In Fig. 2 we present theoretical and experimental graphs "the generalization error, $\overline{P}_N^{\mathrm{M}}$, as function of N_2, the number of training vectors of second pattern class", in experiments with above mentioned 7D binary data. The number of training vectors of the first class was kept constant: $N_1 = 25$. The number of vectors of the second class, N_2, varied between 3 and 1800. Prior probability of the first class, q_1, was set equal to 0.9. Graph 1 in Fig. 1 corresponds to theoretical calculations according to Equation (5). Due to small sample size of the first pattern class, we had large oscillations among different random formations of the training sets. Therefore, 50000 independent experiments with randomly formed learning sets where performed in situations where $N_2 \leq 50$ and 5000 experiments when $N_2 \geq 100$. In Fig. 2 we present mean values.

Both graphs indicate that in situations where prior probabilities of the pattern classes are unequal, artificial increase in training set of smaller category can become harmful: with an increase in N_2, the generalization error increases from its minimum, 0.107, up to 0.173. Important conclusion follows: *an increase in the number of training vectors of one pattern class not always leads to success. To obtain best generalization results one needs to pay attention to correct balance between numbers of vectors in different pattern classes.* Thus, conventional practice where the designer collects extra training vectors of one pattern class not always is a good policy.

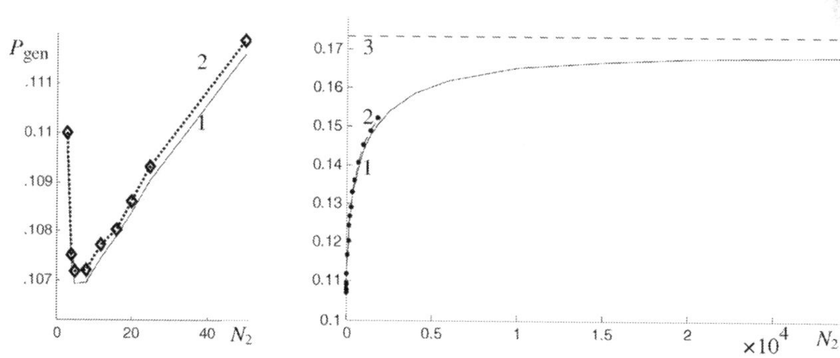

Fig. 2. Generalization error, P_{gen}, as a function of a number of training vectors, N_2: 1 – theoretical values, 2 – experimental evaluations, 3 – asymptotic error $\overline{P}^{\mathrm{M}}_{N_2 \to \infty}$; $N_1 = 25$

Specific application of Eq. (5) arises in analysis of decision tree classifier. Decision tree is a useful strategy to tackle small sample problems. In small sample cases, the number of final leaves must be reduced substantially. Formula (5) can be utilized for provisional evaluation of the generalization error of full decision tree as well as for calculation of partial generalization errors corresponding to distinct branches of the tree. In separate branches, the training sample sizes from opposite categories can differ substantially. It is obvious that formal reduction in a number of the leaves without changing authentic decision rule does not change the classifier's small sample properties. More deep investigation of decision tree classifier suggests that *while simplifying the decision making scheme the number of training errors should increase*.

3 Generalization of the Bayes Predictive Rule

3.1 The Theory: Regularized Multinomial Classifier

To get Bayes predictive rule of Multinomial classifier, one has to know prior distribution of v_s, u_s, $(s = 1, 2, \ldots, m)$. Let prior distribution of the probabilities follow a Dirichlet (polynomial) distribution. To be short, we write formulae for one class only:

$$P_{\text{prior}}(v_1, v_2, \ldots, v_m) = \beta_{mi}(v_1)^{\gamma_1 - 1}(v_2)^{\gamma_2 - 1} \ldots (v_2)^{\gamma_{m-1} - 1}(v_m)^{\gamma_m - 1}, \tag{8}$$

where β_{mi} is a normalizing constant and $\gamma_1, \gamma_2, \ldots, \gamma_m$ are parameters of the prior distribution. Then the Bayes estimate of v_j is

$$\hat{v}_j = (n_j^{(1)} + \gamma_j)/(N_1 + \Sigma_{j=1}^m \gamma_j), \tag{9}$$

If we do not give preference to any particular bin, then $\gamma_1 = \gamma_2 = \ldots \gamma_m = \gamma$. Assume that the prior distribution of probabilities $v_1, v_2, \ldots, v_m$ follow uniform distribution, i.e. γ=1. Then the Bayes predictive estimate of v_j and u_j are

$$\hat{v}_j = (n_j^{(1)} + 1)/(N_1 + m), \quad \hat{u}_j = (n_j^{(2)} + 1)/(N_2 + m). \tag{10}$$

Then, in Equations (5) and (7) one has to use new, index j dependent β value:

$$\beta = \frac{q_1}{q_2}\frac{N_2 + m}{N_1 + m} + \frac{1}{j}\left(\frac{q_1}{q_2}\frac{N_2 + m}{N_1 + m} - 1\right). \tag{11}$$

Calculations show that Bayes approach with uniform prior distribution of parameters v_j and u_j could reduce generalization error substantially if $N_2 \neq N_1$.Uniform prior distribution contains extremely vague information about the parameters of the model. In order to introduce certain prior information a couple of possibilities will be discussed below. Assume at first that *there exist an independent prior data set* of $N_{pi} = \Sigma_{j=1}^m r_{ij}$ vectors to be used to determine prior distribution of $v_1, v_2, \ldots, u_m$, where r_{is} is a number of prior data set vectors of i-th class in the s-th cell. Then

$$P_{\text{apost}}(v_1, v_2, \ldots, v_m) = \beta_{ri}(v_1)^{r_{i1}+1}(v_2)^{r_{i2}+1} \ldots (v_m)^{r_{im}+1}, \tag{12}$$

which could be utilized as new prior distribution. Subsequently, new Bayes estimates

$$\hat{v}_j^* = (n_j^{(1)} + r_{1j})/(N_1 + \textstyle\sum_{j=1}^m r_{1j} + m), \quad \hat{u}_j^* = (n_j^{(2)} + r_{2j})/(N_2 + \textstyle\sum_{j=1}^m r_{2j} + m). \quad (13)$$

In spite of theoretical simplicity utilization of estimate (13) is problematic since researchers do not have *"the independent prior data set"* in typical situations. If such set would be available, it would be merged with training set. At times, one could form such set artificially. One of examples could be a situation where one solves several similar, however, to some extent different pattern recognition tasks. Then specially merged training sets of all tasks could compose *"the independent prior data set"*.

If additional set of vectors is unavailable, one can try using training vectors and some *additional information* to construct "synthetic prior distribution". For example, one can use k - nearest neighbors directed noise injection [12, 13] originally suggested by R. Duin. A noise injection, in fact, introduces *additional non-formal information*: it declares in an inexplicit way that a space between nearest vectors of one pattern class could be filled with vectors of the same category. One can make an assumption (a guess) that L components, $x_1, x_2, \dots, x_E$ of feature vector $\boldsymbol{X}$ are statistically independent. Then parameters $\gamma_1, \gamma_2, \dots, \gamma_m$ in Eq. (8) could be evaluated as a scaled by λ ($0<\lambda<1$) product of E estimates $\hat{P}_{ie} = n_{ie} / N_i$ ($e=1, \dots, E$). Here n_{ie} stands for a number i-th class training vectors where e-th component takes zero value. Similar to the conventional regularized discriminant analysis (RDA) [2, 8], parameter λ plays a smoothing role. Resembling the RDA, optimal value of λ should decrease with an increase in training sample size and with an increase in complexity of distribution of the cell probabilities $v_1, v_2, \dots, u_m$. Thus, it has to be determined in experimental way.

3.2 Experiments

For illustration of usefulness of "shaky prior information" incorporated into Bayes predictive approach in small learning sample situations, we considered two category 7D binary data set already discussed in Sections 2.2 and 2.3. In Fig. 3a, similarly

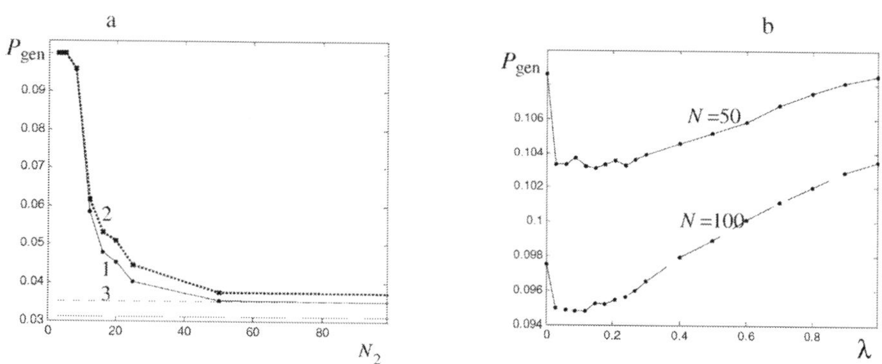

Fig. 3 Generalization error, P_{gen}, as a function of λ, the "regularization" constant (a) and of a number of training vectors, N_2 (b): 1 – theoretical values, Equations (5) and (10), 2 – experimental evaluations, 3 – asymptotic error $\overline{P}^{\text{M}}_{N_2\to\infty}$; $q_1 = 0.9$, $N_1 = 25$

to the graphs in Fig. 2, we have theoretically and experimentally evaluated dependence of generalization error, P_{gen}, on a number of training vectors, N_2 when $N_1 = 25$.

We assumed that q_1=0.9 and used this q_1 value both in theoretical and experimental evaluations of generalization error. Comparison of graphs in Figures 2 and 3a shows that even in case of uniform (almost uninformative) prior distribution Bayes predictive approach allows to reduce generalization error. No increase in generalization error is observed when training set size $N_2 \rightarrow \infty$. Small difference between theoretical experimental graphs observed for large N_2 is caused by the fact that theoretical values are calculated for an entity of situations defined by uniform prior distribution. In the experiment, however, we considered one particular selection of values $v_1, v_2, \ldots, u_m$.

In Fig. 3b we have dependence of generalization error on λ, the "regularization" constant when synthetic prior distribution was obtained from training data, and the feature independence assumption was used. The experiments were performed 5000 times with randomly chosen training sets. A presence of minima indicates that, in principle, the performance of standard Multinomial classifier could be improved by introducing techniques similar to regularized discriminant analysis.

4 Concluding Remarks

We derived equations for generalization error of Multinomial classifier for situations when prior probabilities of the pattern classes are different. It was shown that artificial growth of training vectors of less probable class could be harmful if maximal likelihood estimates were used to design the classifier. Use of predictive Bayes approach to estimate cell probabilities of the Multinomial classifier reduces both the generalization error and negative effect of imbalanced training sets. High accuracy of theoretical formulae was confirmed by experiments with real world data set. A few variants of Bayes predictive approach were considered. It was shown that utilization of certain prior assumptions about the data suggests new ways how to regularize the Multinomial classifier and to reduce the generalization error.

References

[1] Lachenbruch P.A and Goldstein M. Discriminant analysis. *Biometrics* 5(3): 9–85, 1979.

[2] Duda R.O., Hart P.E. and Stork D.G. *Pattern Classification*. 2nd ed. Wiley, NY, 2000.

[3] Hughes G.F. On the mean accuracy of statistical pattern recognizers. *IEEE Trans. on Information Theory* IT-14: 55–63, 1965.

[4] Chandrasekaran B. and Jain A.K. Quantization complexity and independent measurements. *IEEE Trans. Computers*, 23: 102-1-6, 1974.

[5] Duin R.P.W. *On the Accuracy of Satistical Pattern Recognizers*. Ph.D. dissertation. Delft University of Technology, Delft, 1978.

[6] Griskevicius D. and Raudys S. On the expected probability of the classification error of the classifier for discrete variables. In S Raudys (editor), *Statistical Problems of Control*, 38:95–112. Institute of Mathematics and Informatics, Vilnius (in Russian), 1979.

[7] Serych A.P. On use of nonparametric density estimates in pattern recognition. *Proc. of Siberian Physic technical V.D.Kuznetsov Institute*, 63: 13-41, (in Russian), 1973.

[8] Raudys S. *Statistical and Neural Classifiers: An integrated approach to design*. Springer-Verlag, NY, p. 312, 200.

[9] Amari S. A universal theorem on learning curves. *Neural Networks*, 6: 161–66, 1993.

[10] Vidyasagar M. *A Theory of Learning and Generalization*. Springer, London, 1997.

[11] Vapnik V. *Statistical Learning Theory*. John Wiley and Sons, N.Y., 1998.

[12] Skurichina M., Raudys S. and Duin R.P.W. (2000). K-NN directed noise injection in multilayer perceptron training, *IEEE Trans. on Neural Networks*, 11(2): 504–511.

[13] Raudys S. Experts' boasting in trainable fusion rule. *IEEE Trans. on Pattern Analysis and Machine Intelligence*, PAMI 25 (9): 1178-1182, 2003.

Combining Accuracy and Prior Sensitivity for Classifier Design Under Prior Uncertainty

Thomas Landgrebe and Robert P.W. Duin

Elect. Eng., Maths and Comp. Sc., Delft University of Technology, The Netherlands
{t.c.w.landgrebe, r.p.w.duin}@ewi.tudelft.nl

Abstract. Considering the classification problem in which class priors or misallocation costs are not known precisely, receiver operator characteristic (ROC) analysis has become a standard tool in pattern recognition for obtaining integrated performance measures to cope with the uncertainty. Similarly, in situations in which priors may vary in application, the ROC can be used to inspect performance over the expected range of variation. In this paper we argue that even though measures such as the area under the ROC (AUC) are useful in obtaining an integrated performance measure independent of the priors, it may also be important to incorporate the *sensitivity* across the expected prior-range. We show that a classifier may result in a good AUC score, but a poor (large) prior sensitivity, which may be undesirable. A methodology is proposed that combines both accuracy and sensitivity, providing a new model selection criterion that is relevant to certain problems. Experiments show that incorporating sensitivity is very important in some realistic scenarios, leading to better model selection in some cases.

1 Introduction

In pattern recognition, a typical assumption made is that class priors and misallocation costs are known precisely, and hence performance measures such as classification error-rate and classifier loss are typically used in evaluation. A topic that has received a lot of attention recently is the imprecise scenario in which these assumptions do not hold (see for example [9], [2], [1] and [10]), resulting in a number of tools and evaluations suited to this problem. In particular, receiver operator characteristic (ROC) curves [6] have become very popular due to their invariance to both class priors and costs, and are thus used as a basis for performance evaluation and classifier decision threshold optimisation in these imprecise environments. The Area Under the ROC (AUC) measure has thus been proposed, providing a performance evaluation that is independent of priors.

In this paper we argue (and show) that considering the integrated performance (AUC) alone may not be the optimal strategy for model selection in these situations. This is because the AUC measure discounts an important characteristic, namely the performance *sensitivity* across the prior range (we distinguish prior sensitivity from the sensitivity measure often used in medical decision making, which is equivalent to true positive rate). In fact, we show that in some cases, two

D.-Y. Yeung et al. (Eds.): SSPR&SPR 2006, LNCS 4109, pp. 512–521, 2006.

classifiers may compete in terms of AUC, but have significantly different sensitivities over the same prior range i.e. one of the classifiers may have a performance that varies rapidly from low to high values, whereas the other may be more stable. In some problems e.g. medical decision making, the former scenario may be unacceptable, emphasising the fact that this sensitivity should also be considered. A simple criterion is proposed that combines both AUC and sensitivity, called $AccSens$, allowing for a more appropriate criterion for some problems[1].

The paper is organised as follows: Section 2 introduces the notation in the well-defined case, restricted to two-class problems for simplicity, and derives the ROC. In Section 3, the problem of uncertain/varying class priors is considered, discussing the AUC measure, which is invariant of priors. Section 4 discusses the importance of considering prior-dependent sensitivity in conjunction with integrated error, illustrated via a case study, and Section 5 subsequently introduces a new criterion, $AccSens$. A number of real experiments are presented in Section 6 that show some cases in which competing classifiers (using AUC) have significantly different sensitivities (and vice versa). Conclusions are presented in Section 7.

2 Problem Formulation and ROC Analysis

Consider a 2-class classification task between classes ω_1 and ω_2, with prior probabilities $P(\omega_1)$ and $P(\omega_2)$ respectively, and class-conditional probabilities denoted $p(\mathbf{x}|\omega_1)$ and $p(\mathbf{x}|\omega_2)$. Each object is represented by a feature vector $\mathbf{x}$, with dimensionality d. Figure 1 presents an example of a 1-dimensional, two-class example (means at -1.6 and 1.6 respectively, and equal variances of 2), and θ_d represents an equal prior, equal cost operating point.

Two types of of classification errors exist in the two-class case, namely the false positive rate (FP_r), and the false negative rate (FN_r), derived as follows, where θ_w is the classification weight, determining the operating point:

$$
\begin{aligned}
FP_r(\theta_w) &= (1-\theta_w)P(\omega_2)\textstyle\int p(\mathbf{x}|\omega_2)I_1(\mathbf{x}|\theta_w)dx \\
I_1(\mathbf{x}|\theta_w) &= \begin{cases} 1 \text{ if } \theta_w P(\omega_1)p(\mathbf{x}|\omega_1) > (1-\theta_w)P(\omega_2)p(\mathbf{x}|\omega_2) \\ 0 \text{ otherwise} \end{cases} \\
FN_r(\theta_w) &= \theta_w P(\omega_1)\textstyle\int p(\mathbf{x}|\omega_1)I_2(\mathbf{x}|\theta_w)dx \\
I_2(\mathbf{x}|\theta_w) &= \begin{cases} 1 \text{ if } (1-\theta_w)P(\omega_2)p(\mathbf{x}|\omega_2) \geq \theta_w P(\omega_1)p(\mathbf{x}|\omega_1) \\ 0 \text{ otherwise} \end{cases}
\end{aligned} \tag{1}
$$

In the (realistic) case that distributions are not known, but are estimated from data (that is assumed representative), class conditional density estimates are denoted $\hat{p}(\mathbf{x}|\omega_1)$ and $\hat{p}(\mathbf{x}|\omega_2)$, and population prior estimates are denoted π_1 and π_2. These are typically estimated from an independent training set that

[1] Even though we emphasise a varying/uncertain class prior, the theory and analysis in this paper extends also to the related problem of varying misallocation costs [1], since these both have a similar impact from an ROC perspective in that a variation in either prior or cost results in a varying performance, strictly along the ROC [9].

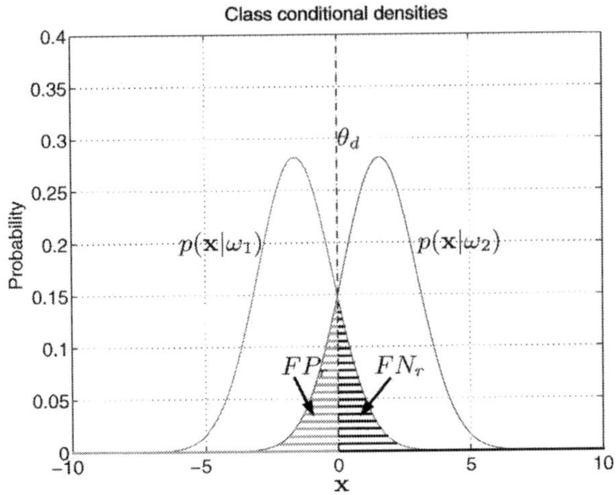

Fig. 1. One-dimensional example illustrating two overlapping Gaussian distributions, and the two error-types associated with an equal error, equal cost operating point θ_d.

is assumed drawn representatively from the true distribution. Equation 1 can then be extended to this case. The classifier weight θ_w allows for FP_r to be traded off against FN_r (and vice-versa) to suit a given application. A particular setting of θ_w results in a single operating point, with a corresponding FN_r and FP_r combination. Varying θ_w (where $0 \leq \theta_w \leq 1$) allows for specification of any desired operating point. An ROC plot [6] consists of a trade-off curve between FN_r and FP_r (as a function of θ_w). As such, the ROC is a useful tool in optimising and evaluating classifiers.

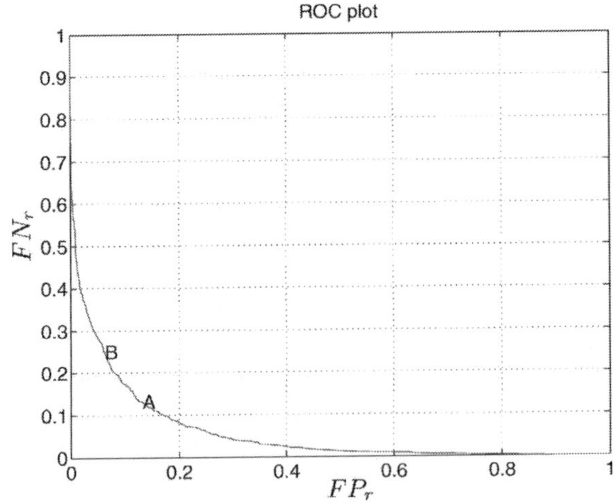

Fig. 2. ROC plot for the example in Figure 1

In the well-defined case that the priors can be estimated sufficiently well, and remain constant (e.g. estimated from training data, and generalising to an application scenario), the classification problem can be optimised (and evaluated) directly using the ROC. Strategies vary, but the most popular ones are as follows (also demonstrated on the ROC plot in Figure 2, which is the ROC plot generated from the example in Figure 1):

- **Equal error optimisation:** In this case, FP_r errors have the same consequences as FN_r errors, and the objective of the optimisation is to select a θ_w such that $FP_r = FN_r$. In Figure 2, point A shows this operating point.
- **Cost-sensitive optimisation:** In some applications e.g. medical decision making, different errors have different misclassification costs (denoted c_1 for FN_r errors, and c_2 for FP_r errors). In this case θ_w should be chosen such that the overall system loss is minimised, where the loss L can be computed as $L = \theta_w c_1 \pi_1 FN_r + (1 - \theta_w) c_2 \pi_2 FP_r$ (profits are ignored here i.e. consequences of correct classifications). In Figure 2, point B illustrates an operating point for the equal prior case, with $c_1 = 0.2$ and $c_2 = 0.8$.

3 Varying Priors, Uncertain Environments

The previous discussion assumed that the priors can be well estimated, and remain fixed in application. However, in many real applications this is not the case (see [9], [2]), confounding the problem of optimising the operating point and model selection (fairly comparing classifiers). In these cases, priors may not be known beforehand, or priors in an independent training set are not representative, or the priors may in fact vary in application. In these cases, even though an immediate optimisation and comparison is not appropriate, several techniques have been proposed for classifier design e.g. [9]. These typically use the ROC plot, since it has the desirable property of being independent of priors/costs (i.e. the same ROC results irrespectively), allowing classifier performance to be inspected for a range of priors (or costs). In particular, the Area Under the ROC (AUC) measure [2] has been derived to give an integrated performance measure, allowing for model comparison independent of the prior. The AUC measure is defined as:

$$AUC = 1 - \int (FN_r) dFP_r \tag{2}$$

This performance measure results in a normalised score between 0 and 1, with 1 corresponding to perfect classification, 0.5 to random classification, and below 0.5 as worse than random (i.e. swap classifier labels). The AUC measure can also be computed over a range of priors/operating points, accounting for knowledge of the degree of uncertainty/variation. Thus, even though priors may be uncertain/varying, the best overall classifier can be chosen based on the most favourable integrated performance[2].

[2] For threshold optimisation, the best strategy may be to use a θ_w corresponding to the centre of the known range, or to apply the minimax criterion [3].

4 The Importance of Incorporating Sensitivity

In this paper we demonstrate that comparing classifiers in uncertain environments on the basis of integrated error (AUC) only may not necessarily be the best strategy to take. This argument arose based on comparison of ROC plots for a number of competing classifiers (the experiments will show some realistic scenarios). It was observed that in some cases, two competing classifiers resulted in a similar AUC score, but inspection of the ROC made it clear that in one case, the performance range was small, but in another, much larger. This implies that for the problem in which priors may vary, the latter classifier may result in very poor performance at one extreme, and very good performance at the other. Depending on the problem, it may be much better to select the former model that is generally more stable over the expected prior range. Next a case study is presented to demonstrate such a scenario.

4.1 Case Study

Figure 3 depicts a demonstration of a model-selection scenario, comparing two different classifiers, denoted A and B respectively. Each classifier is trained on the distribution shown in the left plot, consisting of a two-class problem between ω_1 and ω_2 respectively, where ω_1 objects are drawn from $N(\mu = 3.0, 2; \omega = 1) + \frac{1}{32}N(\mu = -2.0, 5.0; \sigma = 1)$ (N is the normal distribution with mean μ and variance σ), and ω_2 is one class from the banana distribution [4]. In this synthetic problem, 1500 objects are drawn from the true distribution to create a training set, and a further 1500 objects are drawn independently to result in an independent test set[3]. The two classifiers A and B are then trained on the training set, resulting in the decision boundaries at a single operating point as depicted in the left plot. A is a mixture of Gaussians classifier, with two mixtures chosen for ω_1, and one for ω_2. Classifier B is a support vector classifier with a second order polynomial kernel.

In this problem, it is assumed that the priors may vary (in application) such that $0.05 \leq \pi_1 \leq 0.9$, i.e. the abundance of ω_1 varies between 5% and 90%, and the costs are assumed equal (priors at the low and high extremes for ω_1 are denoted π_1^{lo} and π_1^{hi} respectively, computed by analysing where on the ROC the performance drifts to for the new prior, relative to the original operating point). The scatter-plot shows the resultant classifier decision boundaries of the two classifiers at the equal error point (i.e. equal priors). The ROC plot on the right depicts classifier performance for a range of operating points. For the first extreme, i.e. $\pi_1 = 0.05$, A_{lo} and B_{lo} show the respective operating points for the two classifiers. For the second extreme, i.e. at $\pi_1 = 0.9$, A_{hi} and B_{hi} again demonstrate how the operating point shifts. A_e and B_e show the positions of the equal-error points.

It can immediately be observed that the two classifiers have a distinct performance characteristic as a function of the prior values, even though the equal error points are rather similar. Table 1 compares some performance measures

[3] Cross-validation is ignored here as this example is for demonstration purposes only.

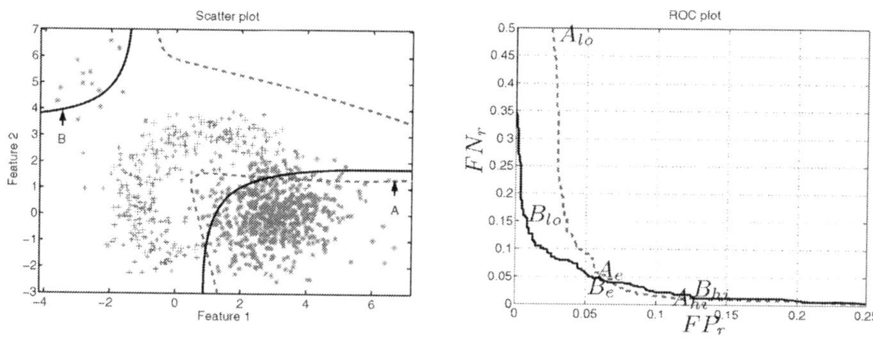

Fig. 3. Case study illustrating performance of two competing classifier models A and B. The left plot shows the data distribution, as well as the respective decision boundaries at a single operating point. The right plot is an ROC-plot for the two models across a range of priors. A_{lo} and B_{lo} are operating points at $\pi_1 = 0.05$, and similarly A_e and B_e are equal-error points, and A_{hi} and B_{hi} correspond to $\pi_1 = 0.9$.

between classifiers A and B. Firstly the error rate shows that both classifiers result in a similar performance for the equal prior case. The AUC measure integrates the classification error over the range of priors (between A_{lo} and A_{hi}), and again this measure shows that both classifiers have similar performance across the prior range as a whole. However, when investigating the sensitivity with respect to the priors, it can be seen that classifier A is much more sensitive than B across the range, with the FN_r varying by up to 47.3%. Prior sensitivity (denoted $Sens$) is computed as the Euclidean distance between the upper and lower prior range, from a π_1^{lo} situation, to π_1^{hi}. This is performed by considering the applicable ranges of FN_r and FP_r:

$$Sens = \frac{1}{\sqrt{2}}\sqrt{((FN_r(\pi_1^{lo}) - FN_r(\pi_1^{hi}))^2 + (FP_r(\pi_1^{hi}) - FP_r(\pi_1^{lo}))^2} \qquad (3)$$

This measure scales between 0 and 1, where a low score indicates the favourable condition of low sensitivity, whereas a high score indicates a large sensitivity to prior variation. Note that $Sens$ is a simple measure in that it subtracts only the extreme values, justified by the fact that an ROC increases monotonically.

In this type of problem, classifier B is clearly more appropriate since it is far less sensitive to a perturbation in prior. It is also clear that the error-rate measure and AUC are not sufficient on their own in this case to choose the best models, and that the prior *sensitivity* across the range of interest should be included to aid in the model selection process.

5 Combining Accuracy and Sensitivity

The case study made it clear that in the uncertain prior situation, classifier sensitivity should be considered in conjunction with integrated error over the prior

Table 1. Performance measures for the synthetic example. Error-rate is denoted ϵ, AUC is the integrated error measure across the prior range, and the sensitivity $Sens$ shows how much the performance varies ($\frac{\%}{100}$) across the prior range.

Model	ϵ	AUC	$Sens$
A	0.057	0.942	0.340
B	0.052	0.945	0.131

range. The next step is to develop a criterion that combines these two performance measures, that is useful for evaluation/model selection in this domain. It is conceivable that some problems may have different consequences for accuracy and sensitivity performances e.g. in some cases a low overall error (i.e. high AUC) may be more important than a low sensitivity, in which case $Sens$ could be weighted lower than AUC. In another case, e.g. medical decision making, a high sensitivity to priors may be more unacceptable than a slightly lower AUC. Thus, for generality, we introduce a weighting corresponding to each term, that can be used to penalise either according to the problem (analogous to misallocation costs). The AUC weight is denoted w_e, and the $Sens$ weight is denoted w_s. We then define the combined measure, called $AccSens$, consisting of the geometric mean of the weighted sum of AUC and $Sens$, as defined in Equation 4. This is appropriate because both measures are scaled between 0 and 1. In the case that w_e and w_s are both set to unity (equal importance), the $AccSens$ error measure also scales between 0 and 1, where a low score is favourable (the $\frac{1}{\sqrt{2}}$ normalises the measure to this range).

$$AccSens = \frac{1}{\sqrt{2}}\sqrt{w_e((1 - AUC)^2) + w_s(Sens^2)} \tag{4}$$

For the case study example (assuming unit weighting), the $AccSens$ errors are 0.244 for model A, and 0.100 for model B, indicating that B is superior.

6 Experiments

A number of experiments on realistic datasets have been undertaken. The objective is to select the most competitive model, considering the problem of varying/uncertain priors, with a known π_1 range: $0.1 \leq \pi_1 \leq 0.9$. Additionally, we assume AUC and $Sens$ are weighted equally. For each model, we investigate an integrated error over the prior range (AUC), the $Sens$ (sensitivity) across the range (Equation 3), the $AccSens$ measure to combine the two, and finally the equal error rate ϵ for comparison purposes. In each experiment, a 10-fold randomised hold-out procedure is performed, effectively resulting in 10 ROC plots upon which the aforementioned statistics are computed. Significance between models is assessed using ANOVA (99.5% significance level). The following datasets are used:

- **Road sign:** A road sign classification dataset [8] consisting of various *sign* and non-*sign* examples represented by images (793 pixels). All *signs* have

Table 2. Results of real experiments, comparing *AUC*, *Sens*, *AccSens*, and ϵ (equal-error point) for a number of models per dataset. Standard deviations are shown.

Model	AUC	Sens	AccSens	ϵ
Road sign				
1) *pca8 mogc 4 4*	0.881(0.026)	0.272(0.039)	0.211(0.029)	0.127(0.022)
2) *pca12 mogc 2 2*	0.886(0.058)	0.180(0.029)	0.154(0.028)	0.093(0.021)
3) *sc svc r 16*	0.951(0.016)	0.149(0.028)	0.111(0.021)	0.052(0.014)
4) *pca17 mogc 2 4*	0.876(0.100)	0.080(0.026)	0.112(0.056)	0.043(0.017)
5) *sc svc r 22*	0.952(0.016)	0.128(0.019)	0.100(0.015)	0.049(0.013)
6) *pca14 mogc 2 4*	0.907(0.061)	0.109(0.021)	0.106(0.033)	0.055(0.016)
Phoneme				
1) *sc knnc3*	0.905(0.013)	0.271(0.049)	0.204(0.028)	0.140(0.011)
2) *sc knnc1*	0.913(0.009)	0.248(0.013)	0.186(0.010)	0.107(0.008)
3) *sc parzenc*	0.891(0.014)	0.294(0.023)	0.222(0.018)	0.128(0.015)
Sonar				
1) *sc knnc3*	0.887(0.027)	0.310(0.107)	0.235(0.073)	0.147(0.039)
2) *sc knnc1*	0.892(0.036)	0.280(0.054)	0.213(0.043)	0.122(0.050)
3) *pca6 parzenc*	0.850(0.050)	0.405(0.069)	0.308(0.046)	0.167(0.054)
4) *sc svc p4*	0.829(0.056)	0.533(0.141)	0.398(0.100)	0.218(0.066)
Ionosphere				
1) *pca0.999 ldc*	0.855(0.039)	0.385(0.118)	0.292(0.084)	0.145(0.043)
2) *fisherm qdc*	0.855(0.037)	0.337(0.053)	0.260(0.041)	0.140(0.036)
3) *fisherm mogc 3 3*	0.834(0.035)	0.365(0.093)	0.285(0.063)	0.160(0.040)
4) *sc svc r 1.0*	0.853(0.171)	0.545(0.231)	0.434(0.095)	0.128(0.044)

been grouped together into a single class (381 objects), to be discriminated from non-*signs* (888 objects).

- **Phoneme:** This dataset is sourced from the ELENA project [5], in which the task is to distinguish between oral and nasal sounds, based on five coefficients (harmonics) of cochlear spectra. In this problem, the "nasal" class (3818 objects) is to be discriminated from the "oral" class (1586 objects).
- **Sonar** and **Ionosphere** are two well-known datasets from the *UCI* machine learning database [7].

Results are presented in Table 2. Various representation and classification algorithms have been used. Preprocessing/representation: *sc* denotes unit variance scaling, *pca* is a principle component mapping followed by the number of components used, or the fraction of variance retained, and *fisherm* is a Fisher mapping. Classifiers: *knnc* denotes the k-nearest neighbour classifier followed by the number of neighbours considered, *parzenc* is a Parzen-window classifier, *ldc* and *qdc* are Bayes linear and quadratic classifiers respectively, *mogc* is a mixture of Gaussians classifier followed by the number of mixtures per class, and *svc* is a support vector classifier, with p denoting a polynomial kernel followed by the order, and r denoting a Gaussian kernel, followed by the variance parameter.

Results show that there are many cases in which incorporation of sensitivity is important for this problem. In the *Road sign* case, an example of this is demonstrated by comparing models 1) and 2). Both show a similar *AUC* score, but 2) is much less sensitive to prior variation. The *AccSens* measure is sensitive to this difference, showing significance (based on an ANOVA hypothesis test). Another interesting comparison is between 3) and 4), in which case model 3) has a significantly higher *AUC*, but 4) has a significantly better *Sens*. Both result in the same *AccSens* score. Models 3), 4), 5), and 6) all compete from an *AccSens* perspective (significantly better than 1) and 2)). In the *Phoneme* dataset, model 3) competes with 1) and 2) in terms of *AUC*, but 2) results in a better *Sens*, and thus results in a superior *AccSens* score (significant). This clearly illustrates the point of the paper once again - without considering sensitivity, model 3) could have been chosen instead of 1) or 2). In the *Sonar* dataset, model 2) appears superior in terms of both *AUC* and *Sens*, and thus there was no benefit of the new measure in this case. Finally, in the *Ionosphere* dataset, models 1), 2) and 4) result in similar *AUC* scores, but 2) appears less sensitive than 4) (not very significant). Using the *AccSens* measure, 1), 2) and 3) are significantly better than 4). As a final general comment on experimental results, it is apparent that there are cases in which a model selection based on *AUC* only is not the optimal procedure. Thus, we argue that in the prior uncertain/unstable environment, prior sensitivity should also be considered, using for example the *AccSens* measure.

7 Conclusions

In this paper the problem of varying/uncertain priors was investigated. ROC analysis has become a standard tool in this domain, with the Area Under the ROC (*AUC*) a popular model selection criterion. We argued that even though this integrated measure can be used to compare classifiers independent of priors, it may also be important to consider how *stable* a model is over the relevant range. A case study and some realistic experiments were presented that demonstrated how classifiers that compete in terms of *AUC* may differ significantly in terms of sensitivity (and vice-versa). It may thus be more sensible for the given problem to consider both. A simple measure, called *AccSens* was proposed, that combines the (weighted) geometric means of *AUC* and sensitivity, allowing for model comparison that considers both integrated accuracy (*AUC*), and prior sensitivity. A few real experiments demonstrated that this methodology is superior in some situations.

Acknowledgements. This research is/was supported by the Technology Foundation STW, applied science division of NWO and the technology programme of the Ministry of Economic Affairs.

References

[1] N.M. Adams and D.J. Hand. Comparing classifiers when misallocation costs are uncertain. *Pattern Recognition*, 32(7):1139–1147, 1999.

[2] A.P. Bradley. The use of the area under the ROC curve in the evaluation of machine learning algorithms. *Pattern Recognition*, 30(7):1145–1159, 1997.

[3] R. Duda, P. Hart, and D. Stork. *Pattern Classification.* Wiley - Interscience, second edition, 2001.
[4] R.P.W. Duin. *PRTools, A Matlab Toolbox for Pattern Recognition.* Pattern Recognition Group, TUDelft, January 2000.
[5] ELENA. European ESPRIT 5516 project. *phoneme dataset*, 2004.
[6] C. Metz. Basic principles of ROC analysis. *Seminars in Nuclear Medicine*, 3(4), 1978.
[7] P.M. Murphy and D.W. Aha. UCI repository of machine learning databases, ftp://ftp.ics.uci.edu/pub/machine-learning-databases. *University of California, Department of Information and Computer Science*, 1992.
[8] P. Paclík. Building road sign classifiers. *PhD thesis, CTU Prague, Czech Republic*, December 2004.
[9] F. Provost and T. Fawcett. Robust classification for imprecise environments. *Machine Learning*, 42:203–231, 2001.
[10] J Yuen. Bayesian approaches to plant disease forecasting. *Plant Health Progress*, November 2003.

Using Co-training and Self-training in Semi-supervised Multiple Classifier Systems

Luca Didaci and Fabio Roli

Dept. of Electrical and Electronic Engineering, University of Cagliari
Piazza d'Armi, 09123 Cagliari, Italy
{luca.didaci, roli}@diee.unica.it

Abstract. Multiple classifier systems have been originally proposed for supervised classification tasks, and few works have dealt with semi-supervised multiple classifiers. However, there are important pattern recognition applications, such as multi-sensor remote sensing and multi-modal biometrics, which demand semi-supervised multiple classifier systems able to exploit both labelled and unlabelled data. In this paper, the use, in multiple classifier systems, of two well known semi-supervised learning methods, namely, co-training and self-training, is investigated by experiments. Reported results on benchmarking data sets show that co-training and self-training allow exploiting unlabelled data in different types of multiple classifiers systems.

1 Introduction

The most of research on semi-supervised learning and classification focused on single classifiers, and few works have dealt with semi-supervised multiple classifier systems (MCS) [1-5]. A survey of semi-supervised learning methods for single classifiers can be found in [6]. The few works on semi-supervised MCS have proposed methods tailored to a specific MCS model which cannot be applied easily to a generic MCS, with the exception of [4] which uses a modified version of the self-training method. On the other hand, there are important pattern recognition applications, such as multi-sensor remote sensing and multi-modal biometrics, which demand semi-supervised multiple classifier systems able to exploit both labelled and unlabelled data. In this paper, the use, in MCS, of two well known semi-supervised learning methods, namely, co-training and self-training, is investigated by experiments. In particular, we extend the single-classifier versions of these algorithms to MCS, and assess the performances achievable for two widely used kinds of MCS. Section 2 first summarizes the co-training and self training methods, then two algorithms for their use in MCS are proposed. In Section 3, experiments with some benchmarking data sets are reported, and results are discussed. Section 4 draws some conclusions.

2 Co-training and Self-training in Semi-supervised Multiple Classifiers

In this section, we briefly describe two techniques for semi-supervised learning, namely, co-training an self-training, and propose their use to design semi-supervised

D.-Y. Yeung et al. (Eds.): SSPR&SPR 2006, LNCS 4109, pp. 522–530, 2006.

MCS. In the following, let us assume to have a set L (usually, small) of labelled data, and a set U (usually, large) of unlabelled data.

2.1 Co-training and Self-training

A co-training approach to semi-supervised learning and classification was proposed by Blum and Mitchell in 1998 [7]. Co-training was proposed for two classifiers and assumes that input features are naturally subdivided into two sets, and each feature subset is sufficient to train an optimal classifier, supposed that enough labelled data are available. Two separate classifiers, one for each feature subset, are trained on the initial, small, labelled data set L. It is assumed that the classifiers will exhibit a low, but better than random, accuracy. Each classifier is then applied to the unlabelled examples in U. For each classifier, the unlabelled examples classified with the highest confidence are added to the labelled data set L, so that the two classifiers can contribute to increase the data set L. Both classifiers are re-trained on this augmented data set, and the process is repeated a given number of times. The rationale behind co-training is that a classifier may assign correct labels to certain examples while it may be difficult for the other classifier to do so. Therefore, each classifier can increase the training set with examples which are very informative for the other classifier.
It is worth pointing out two fundamental assumptions of the co-training method:

1) patterns must be represented with two distinct "views", namely, with two distinct feature sets, and either feature subsets must be sufficient to design an optimal classifier if we have enough labelled data. We need the feature subsets to be conditionally independent so that the examples which are classified with high confidence by one of the two classifiers are *i.i.d.* samples for the other classifier.
2) the classifiers must be "compatible". Compatibility implies that, if we have enough labelled data in the training set, the classifiers C_1 and C_2 provide the same classification labels for all the possible test patterns. A relaxed form of this hypothesis ("partial compatibility") can be also accepted.

In self-training a classifier is initially trained using the labelled data set L. This classifier is then used to assign pseudo-class labels to a subset of the unlabelled examples in U, and such pseudo-labelled data are added to L. Usually, the unlabelled data classified with the highest confidence are selected to increase L. Then the classifier is re-trained using the increased data set L. As the convergence of this simple algorithm can not be guaranteed in general, the last two steps are usually repeated for a given number of times or until some heuristic convergence criterion is satisfied.

2.2 Semi-supervised Multiple Classifiers Using Co-training and Self-training

Co-training of Multiple Classifiers. As pointed out above, the co-training method was introduced under the assumptions of patterns represented with two distinct "views" and "compatible" classifiers. Therefore, co-training can be naturally applied to classifier ensembles made up of compatible classifiers which have distinct feature sets as input. However, such assumptions are not satisfied in many practical cases. On the other hand, Goldman and Zhou showed that co-training also works with two classifiers using the same features [5]. Here we propose to investigate the use of co-training for a generic MCS whose classifiers can be created with different methods (e.g., using different classification algorithms or the bootstrap technique). Figure 1 shows the main step of our extended co-training algorithm applied to multiple classifiers. It should be noted that a similar algorithm is described in [11].

Co-training of Multiple Classifiers

- Given:
  - a set L of labelled training examples
  - a set U of unlabelled examples
  - the P_j prior probabilities of the underlying class distributions
- Create a set U' of examples by choosing *u* examples at random from U
- Loop for *k* iterations:
  - Build N classifiers C_i, $i=1 \ldots N$, using L
  - Run $C_1, \ldots C_N$ on U'
  - For each classifier C_i, for each class *j*,
    select the $n_j \propto P_j$ examples assigned to the class *j* by the classifier C_i
    with the highest classification confidence.
  - Add these $N \times \sum_j n_j$ self-labelled examples to L
  - Randomly choose $N \times \sum_j n_j$ examples from U to replenish U'
- Return a MCS using $\{C_i\}$ as base classifiers

Fig. 1. The extended co-training algorithm applied to multiple classifier systems

Given a set L (usually, small) of labelled data, and a set U (usually, large) of unlabelled data, a set U' is created by choosing *u* examples at random from U. The following steps are executed for a fixed number of iterations. N classifiers are trained on the initial, small, labelled data set L. In order to create different classifiers several methods are possible. For example, different classification algorithms or the bootstrap technique can be used. Each classifier is then applied to the *u* unlabelled examples in U'. For each classifier C_i and for each class *j*, n_j examples assigned to the class *j* by the classifier C_i with the highest classification confidence are selected and they are added to the labelled data set L. The number n_j of selected examples assigned to the class *j* is proportional to the prior probability of the class *j*. Accordingly, if there are no classification errors in the selected examples, the set of the selected examples has the same prior probabilities of the underlying distributions, and the prior probabilities of L remain unchanged. Each classifier selects $\sum_j n_j$ patterns, so that $n = N \times \sum_j n_j$ patterns are moved from U' to L. An equal number *n* of patterns are randomly chosen from U to replenish U'. For the next iterations, all the classifiers are re-trained on the augmented data set L, and the process is repeated a given number of times. At the end of this iterative process, an MCS using $\{C_i\}$ as base classifiers is created. In this work only MCS based on fixed rules [8] (mean, product and majority voting rule) are used.

Ensemble-Driven Self-training. In order to extend the use of the self-training method to MCS, we propose to use the concept of "ensemble-driven" self-training. Each classifier is not self-trained, but it is trained with the examples which are labelled by the MCS. In other words, the MCS is used to assign pseudo-class labels to a subset of the unlabelled examples in U, and such pseudo-labelled data are added to L. Then each classifier of the ensemble is re-trained using the increased data set L. It is worth noting that the self-supervised classifier ensemble proposed by El Gayar exploits the same concept [4], and also the extension of co-training, called "democratic" co-training, proposed by Zhou and Goldman can be regarded as a type of ensemble-driven self-training [5]. Figure 2 shows the main step of our "ensemble-driven" self-training algorithm applied to multiple classifiers.

Ensemble-driven Self -training of Multiple Classifiers

- Given:
  - a set L of labelled training examples
  - a set U of unlabelled examples
  - the P_j prior probabilities of the underlying class distributions
- Create a set U' of examples by choosing *u* examples at random from U
- Loop for k iterations:
  - Build N classifiers C_i, $i=1...N$, using L
  - Create a MCS using $\{C_i\}$ as base classifiers
  - Run the MCS on U'
  - For each class *j*,
    select the $N \times n_j$ examples assigned to the class *j*
    by the MCS with the highest classification confidence ; $n_j \propto P_j$
  - Add these $N \times \sum_j n_j$ self-labelled examples to L
  - Randomly choose $N \times \sum_j n_j$ examples from U to replenish U'
- Return the MCS

Fig. 2. The extended, ensemble-driven, self-training algorithm applied to multiple classifier systems

The MCS is applied to the *u* unlabelled examples in U'. For each class *j*, $N \times n_j$ examples assigned to the class *j* by the MCS with the highest classification confidence are selected and they are added to the labelled data set L. As in the co-training algorithm, the number $N \times n_j$ of selected examples assigned to the class *j* is proportional to the prior probability of the class *j*, in order to do not change the prior probabilities of L. For each step $n = N \times \sum_j n_j$ patterns are moved from U' to L. An equal number *n* of patterns are randomly chosen from U to replenish U'. For the next iterations, all the classifiers are re-trained on the augmented data set L, and the process is repeated a given number of times. At the end of the iterative process the 'self-trained' MCS is returned. It is worth noting that in the co-training algorithm each of the N classifiers selected and labelled n_j patterns per class, so that the number of patterns labelled at each step was $n = N \times \sum_j n_j$. As in the ensemble-driven self-training algorithm patterns are selected and labelled directly by the MCS, $N \times n_j$ pattern per class are selected for each step of the algorithm.

3 Experimental Results

3.1 Data Sets and Experimental Protocol

The performances of the algorithms described in Figures 1 and 2 clearly depend on the classifier ensemble used. The goal of our experiments was to assess such performances using two widely used methods for creating a MCS, and with different data sets. In particular, we assessed performances with classifier ensembles generated by the bootstrap method and using different type of classification algorithms. Table 1 describes the main characteristics of the data sets used, and reports the size of L expressed as percentage of the available training set and as number of patterns. The datasets Letter, BCW and Optdigits come from the UCI Machine Learning repository (http://www.ics.uci.edu/~mlearn/MLRepository.html).

Table 1. Main characteristics of the data sets in terms of the number of classes, features, patterns, and size of the training set L as percentage (%L) of the available training set (the number of patterns in L is given in brackets)

Data set	Classes	Features	Patterns	%L	Reference
Gaussian	2	15	1000	5% (35)	Gaussian data set proposed in [9]
Letter	26	16	20000	15% (2100)	Letter Image Recognition
BCW	2	9	683	5% (23)	Wisconsin Breast Cancer
Optdigits	2	15	3823	5% (190)	Optical Recognition of Handwritten Digits
Feltwell	5	15	10944	2% (100)	Feltwell [10]

Each data set was subdivided randomly into a training set (30% of the patterns) and a test set (70% of the patterns). For the Feltwell data set we maintained the original subdivision in training and test sets, because a random subdivision is known to create an artificial, almost trivial, classification task [10]. Each training set was subdivided randomly into a set L (the labelled data set) and a set U (the unlabelled data set). Each experiment was repeated 10 times, with different random choices of the labelled data set L. For each data set, Table 2 shows the classifier ensembles used. In the case of ensembles generated by bootstrap Table 2 gives the base classifier used. Five different bootstrap replicas of L, that is, ensembles made up of five classifiers, were used. The terms "linearG" and "quadratic" indicate the linear and quadratic Gaussian classifiers. The term "linearLog" indicates the linear classifier that maximize the likelihood criterion using the logistic function. For each step of the algorithms in Figures 1 and 2,

$n = N \times \sum_j n_j$ patterns are selected. The value of n_j is $n_j = \alpha \cdot P_j$ where P_j is the prior probability of the class j and α is an integer value between 3 and 6. The size u of the set U' was chosen as $u = \beta \times n$ where β is 4 or 5. In other words, the set U' is β times greater than the number of patterns selected from U'. The number of iterations k is chosen as following: for large datasets (Feltwell, Letter) k is chosen as 2/3 of the maximum of the possible iterations, that is, as 2/3 of the number of iterations that emptied the set U of unlabelled patterns. For small datasets, the algorithm goes on until there are no more patterns in U.

Table 2. Classifier ensembles and base classifiers used for each data set

Dataset	*Classifier ensemble*	*Base classifier used for bootstrap*
Gaussian	1)[perceptron, linearG, quadratic] 2)[perceptron, perceptron, quadratic]	linearG
Letter	[*k*-nn (k=1); *k*-nn(k=5); parzen]	*k*-nn
BCW	[linearG, linearLog, parzen]	linearG; Parzen
Optdigits	[*k*-nn; MLP; parzen]	*k*-nn; MLP; Parzen
Feltwell	[*k*-nn, MLP, quadratic]	Quadratic; *k*-nn

For each data set, the classifiers were chosen so that the classification error obtained using base classifiers trained on L data set was substantially higher than the error obtained using base classifiers trained on the full training set $L \cup U$. In other words, we selected classifiers for which we expect to obtain a decrease of the error if the semi-supervised mechanism correctly labelled all the patterns of U. For the experiments with co-training, in order to fulfil the compatibility hypothesis, we choose classifiers that agree in their decisions at least for the 90% of the pattern of U when they are trained on the full training set $L \cup U$.

3.2 Results

Tables 3 and 4 report the results of the experiments with the algorithms of Figures 1 and 2. In particular, Table 3 refers to the experiments with classifier ensembles generated by bootstrap, while Table 4 reports the results obtained with ensembles made up of different classifiers (second column of Table 2). The reported values are the test-set error values averaged over ten runs, with different random choices of the initial labelled data set L. Results obtained with different rules for classifier combination are reported.

The 'start' column reports the percentage value of the error before the exploitation of the unlabelled data, that is, using classifiers trained only on the labelled data set L. The Δ column represents the variation of the error dues to the exploitation of the unlabelled data. Positive values of Δ indicate a reduction of the error. As the trend of the error can be non monotonic, we report the value of Δ after a fixed number of iterations of the semi-supervised process (Δ(end)) and the maxim reduction of the error during the semi-supervised process (Δ(best)). Results where Δ(best) >> Δ(end), that is, results for which after a first reduction of the error the error increased, are reported in bold. This non-monotonic trend of the error was observed for the Feltwell dataset (Table 3), and for Gaussian dataset (Table 4). In particular, for the Feltwell dataset, after a first reduction of the error, the co-training and self-training algorithms sometimes provided an error greater than the initial test error (negative values of Δ).

Table 3. Experiments with classifier ensembles generated by bootstrap

		Co-training			**Self-training**		
		start	**Δ (best)**	**Δ(end)**	start	**Δ(best)**	**Δ(end)**
Mean rule	bcw, linearG	12.21	7.65	7.35	11.27	8.24	7.65
	bcw, parzen	11.76	7.30	7.06	5.59	3.09	2.65
	feltwell, quadratic	18.81	**4.43**	**-3.37**	18.86	**4.05**	**1.83**
	feltwell, *k*-nn	18.83	**3.28**	**-1.79**	16.54	2.39	1.10
	letter	14.33	1.49	1.31	15.84	1.77	1.58
	optdigits, *k*-nn	7.05	3.43	3.29	7.14	3.77	3.76
	optdigits, MLP	10.53	4.02	3.92	9.71	3.27	3.27
	optdigit, parzen	7.92	3.66	3.39	8.68	5.12	5.02
	gaussian	18.83	7.67	7.07	18.83	7.10	6.47
Majority voting rule	bcw, linearG	14.51	9.85	9.66	13.38	10.00	9.26
	bcw, parzen	14.17	9.80	9.51	4.90	2.45	1.67
	feltwell, quadratic	18.98	**4.67**	**-3.25**	16.63	**2.53**	**-4.63**
	feltwell, *k*-nn	19.50	**3.34**	**-1.50**	18.30	3.32	2.72
	letter	14.12	0.97	0.68	15.97	1.56	1.31
	optdigits, *k*-nn	7.32	3.38	3.16	8.21	4.99	4.80
	optdigits, MLP	11.27	4.58	4.40	10.61	4.23	4.10
	optdigit, parzen	8.16	3.90	3.64	6.86	2.82	2.31
	gaussian	18.88	7.62	7.08	19.80	9.37	9.00
Product rule	bcw, linearG	12.25	7.75	7.40	9.02	5.78	5.39
	bcw, parzen	12.99	8.53	8.28	3.04	0.69	0.05
	feltwell, quadratic	20.69	**6.37**	**-1.53**	19.70	6.01	5.12
	feltwell, *k*-nn	18.82	**3.31**	**-1.79**	14.07	**1.34**	**-0.17**
	letter	14.36	1.50	1.34	16.59	1.30	1.02
	optdigits, *k*-nn	7.05	3.43	3.29	6.94	3.90	3.90
	optdigits, MLP	13.77	7.18	7.11	10.18	4.16	4.02
	optdigit, parzen	7.15	3.33	3.12	7.67	2.50	2.48
	gaussian	19.55	8.42	7.78	19.67	8.27	7.33

In order to show the typical trend of the test-set error during the semi-supervised process, in Figure 3 we report the error on Letter and Optdigits data sets as a function of the number of iterations of the self-training algorithm applied to classifier ensembles generated using different base classifiers. Similar trends were obtained with the co-training algorithm and for the other data sets.

Table 4. Experiments with ensembles made up of different classifiers (second column of Table 2)

		Co-training			Self-training		
		start	Δ (best)	Δ(end)	start	Δ(best)	Δ(end)
Mean rule	bcw	7.94	4.36	3.97	6.13	3.63	3.14
	feltwell	15.63	2.99	2.26	13.01	2.12	1.42
	letter	14.24	1.47	1.33	14.39	1.21	1.06
	optdigits	7.48	3.52	3.49	7.80	4.62	4.60
	Gaussian 1)	15.83	**5.30**	**2.93**	17.53	**8.57**	**6.60**
	Gaussian 2)	18.83	**6.00**	**4.50**	18.93	**4.43**	**2.00**
Majority voting rule	bcw	8.09	4.56	4.22	7.45	4.31	4.02
	feltwell	14.99	3.59	2.46	13.05	2.83	2.46
	letter	14.29	1.54	1.44	14.30	0.95	0.87
	optdigits	7.51	3.56	3.53	7.63	4.18	4.05
	Gaussian 1)	15.80	**5.33**	**3.33**	16.70	6.57	6.10
	Gaussian 2)	18.87	**5.73**	**3.13**	17.93	2.33	0.30
Product rule	bcw	6.96	3.04	2.30	6.37	3.24	2.55
	feltwell	17.57	4.96	4.38	16.34	4.04	2.64
	letter	14.27	1.51	1.39	14.09	1.46	1.41
	optdigits	7.47	3.54	3.48	7.27	3.85	3.75
	Gaussian 1)	27.73	**16.47**	**14.77**	33.17	**18.23**	**16.57**
	Gaussian 2)	31.80	18.60	17.67	23.90	7.03	5.93

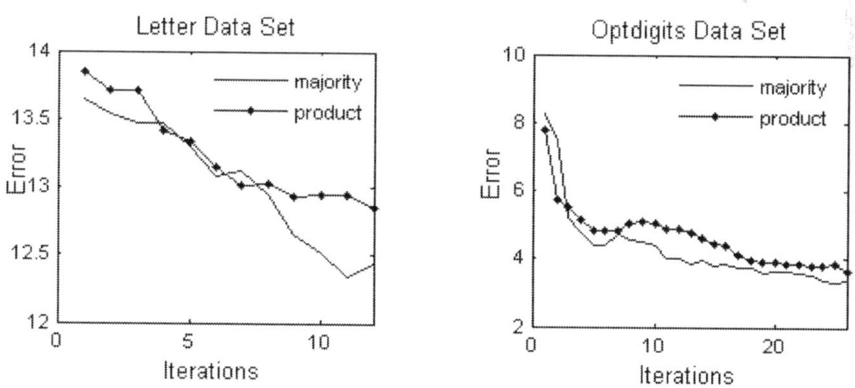

Fig. 3. Examples of the test-set percentage error as function of the number of iterations of the self-training algorithm applied to classifier ensembles generated using different base classifiers

4 Conclusions

This paper's goal was to investigate by experiments the use, in MCS, of two well known semi-supervised learning methods, namely, co-training and self-training. Although final conclusions cannot be drawn on the basis of the limited set of reported experiments, we believe that this work made a first step toward the systematic use of co-training and self-training to design semi-supervised MCS. Reported results show that the extended versions of the co-training and self-training we proposed allow exploiting unlabelled data in two different types of multiple classifiers systems. In addition, our results confirmed a claim of other researchers [5], that is, co-training algorithm can be used even if different feature subsets are not available for the task at hand. As a future work we will continue our experimental investigation and will investigate the trade-off between the "complementarity" of classifiers, useful for MCS, and the request of "compatibility" of the co-training algorithm.

References

1. Roli F.: Semi-supervised Multiple Classifier Systems: Background and Research Directions. 6th Int. Workshop on Multiple Classifier Systems (MCS 2005), Seaside, CA, USA, June 13-15 2005, N.C. Oza, R. Polikar, J. Kittler, F. Roli Eds., Springer-Verlag, LNCS 3541, pp. 1-11
2. D'Alchè-Buc F., Grandvalet Y., Ambroise C.: Semi-supervised marginboost, Neural Information Processing Systems Foundation, NIPS 2002, 2002.
3. Bennet K., Demiriz A., Maclin R.: Exploiting unlabeled data in ensemble methods, Proc. 8th ACM SIGKDD Int. Conf. On Knowledge Discovery and Data Mining, 2002, pp. 289-296
4. N. El Gayar.: An Experimental Study of a Self-Supervised Classifier Ensemble, International Journal of Information Technology, Vol. 1, No. 1, 2004.
5. Y Zhou, S Goldman.: Democratic Co-Learning, Proceedings of the 16th IEEE International Conference on Tools with Artificial Intelligence (ICTAI 2004), pp 594-602
6. X. Zhu, Semi-supervised learning literature survey, Technical report, Computer Sciences TR 1530, Univ. Wisconsis, Madison, USA, Jan. 2006.
7. Blum A., Mitchell T.: Combining labeled and unlabeled data with co-training, Proc. of the Workshop on Computational Learning Theory, 1998, pp. 92-100.
8. Kittler J., Hatef M., Duin R. P. W. and Matas J.: On Combining Classifiers , IEEE Trans. on Patt. Anal. and Machine Intell., Vol. 20, No. 3, pp. 226-239, March 1998
9. Skurichina M. and Duin R. P. W.: Bagging, Boosting and the Random Subspace Method for Linear Classifiers. Pattern Analysis & Applications (2002)5:121–135
10. Giacinto G., Roli F., Bruzzone L.: Combination of neural and statistical algorithms for supervised classification of remote-sensing images, Pattern Recongition Letters, Vol. 21, No. 5, 2000, pp. 385-397.
11. Solyman M., El Gayar N.F.: A Co-training Approach for Semi-supervised Multiple Classifiers, INFO 2006, 4th Int. Conference, Cairo, Egypt, 25-27 March 2006, in press.

MRF Based Spatial Complexity for Hyperspectral Imagery Unmixing

Sen Jia and Yuntao Qian

College of Computer Science, Zhejiang University
Hangzhou 310027, P.R. China
zjujiasen@hotmail.com, ytqian@zju.edu.cn

Abstract. Hyperspectral imagery (HSI) unmixing is a process that decomposes pixel spectra into a collection of constituent spectra (endmembers) and their correspondent abundance fractions. Without knowing any knowledge of HSI data, the unmixing problem is transformed into a blind source separation (BSS) problem. Several methods have been proposed to deal with the problem, like independent component analysis (ICA). In this paper, we introduce spatial complexity that applies Markov random field (MRF) to characterize the spatial correlation information of abundance fractions. Compared to previous BSS techniques for HSI unmixing, the major advantage of our approach is that it totally considers HSI spatial structure. Additionally, a proof is given that spatial complexity is suitable for HSI unmixing. Encouraging results have been obtained in terms of unmixing accuracy, suggesting the effectiveness of our approach.

1 Introduction

Hyperspectral imagery (HSI) records data in hundreds of narrow contiguous spectral bands, which provides the opportunity to identify the ground materials-of-interest. Owing to the spatial resolution of the sensor, disparate materials may contribute to the spectrum measured from a single pixel, causing it a "mixed" pixel and making HSI unmixing a challenging problem in HSI applications. It is a process that decomposes pixel spectra into a collection of constituent spectra (endmembers) and their correspondent abundance fractions [1,2].

At first, to make the problem simple, some methods unmix the HSI data under the circumstances that the knowledge of endmembers, including the spectral signatures and the number of endmembers, is known. And HSI unmixing is converted into a linear problem, which is easy to solve, such as spectral angle mapper [3]. But in most cases, the information of endmembers is not known, and the unmixing problem is transformed into a blind source separation (BSS) problem [4]. Several methods have been proposed to deal with the problem, like independent component analysis (ICA) [5,6]. However, the unmixing results of ICA are not satisfactory. That is, in any case, there are always endmembers incorrectly unmixed [7].

Recently, a temporal complexity based BSS approach was proposed by Stone [8] which has shown success in separating linear mixtures of one-dimensional

D.-Y. Yeung et al. (Eds.): SSPR&SPR 2006, LNCS 4109, pp. 531–540, 2006.

independent signals. It is to seek a weight vector that provides an orthogonal projection of mixtures such that each extracted signal is minimally complex. In this paper, for the sake of utilizing the spatial correlation information of abundance fractions, we extend the algorithm to two-dimensional spatial domain by Markov random field (MRF), named spatial complexity. Different from previous BSS techniques for HSI unmixing [9], our approach totally considers HSI spatial structure. In addition, we prove a theorem, which shows spatial complexity applicable for HSI unmixing.

The paper is organized as follows. First, the MRF based spatial complexity is presented. In Section 3, the corresponding gradient ascent algorithm is described; meanwhile, the pre- and post-processing are discussed. Section 4 presents experimental results of applying our approach to HSI data. Finally, some concluding remarks are given in Section 5.

2 Spatial Complexity Based HSI Unmixing

HSI is a three-dimensional array with the width and length corresponding to spatial dimensions and the spectral bands as the third dimension, which are denoted by I, J and L in sequence. Let $\mathbf{R}$ be the image cube with each spectrum $\mathbf{R}_{ij}$ being an $L \times 1$ pixel vector where the boldface is used for vectors. Let $\mathbf{M}$ be an $L \times P$ spectral signature matrix that each column vector corresponds to an endmember spectrum and P is the number of endmembers in the image. Let $\mathbf{S}$ be the abundance cube (the length of each dimension is I, J and P respectively) and every column $\mathbf{S}_{ij}$ be a $P \times 1$ abundance vector associated with $\mathbf{R}_{ij}$, with each element denoting the abundance fraction of relevant endmember present in $\mathbf{R}_{ij}$. Hence, the linear mixing model can be represented as follows:

$$\mathbf{R}_{ij} = \mathbf{M}\mathbf{S}_{ij} + \mathbf{n} \tag{1}$$

where $\mathbf{n}$ is noise. The unmixing problem is to find a $P \times L$ matrix $\mathbf{W}$ for every pixel vector $\mathbf{R}_{ij}$

$$\mathbf{Y}_{ij} = \mathbf{W}\mathbf{R}_{ij} \tag{2}$$

where $\mathbf{Y}$ is the estimated abundance cube of $\mathbf{S}$ and vector $\mathbf{Y}_{ij}$ expresses the fractional abundances of P endmembers associated with $\mathbf{R}_{ij}$. Figure 1 gives the sketch map of HSI unmixing.

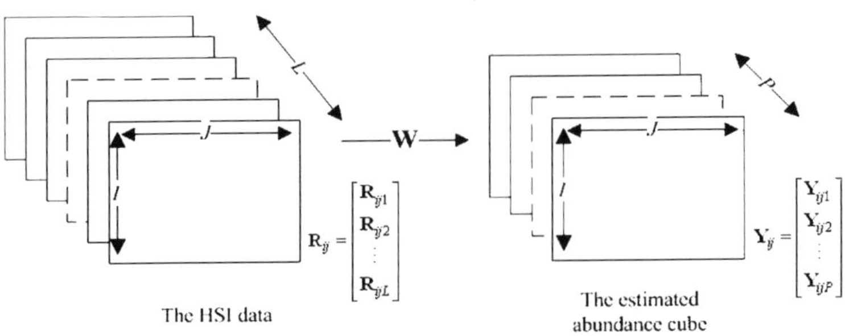

Fig. 1. The HSI unmixing sketch map

5	4	3	4	5
4	2	1	2	4
3	1	$\mathbf{Y}_{ijk}$	1	3
4	2	1	2	4
5	4	3	4	5

(a)

0.05	0.2	0.05
0.2		0.2
0.05	0.2	0.05

(b)

Fig. 2. (a) The five order neighborhood system. (b) The weight matrix for $\lambda_C^{(t)}$.

In order to incorporate the spatial information of HSI data, we extend the method in [8], using spatial predictability to formulate complexity. The maximal predictability corresponds to minimal complexity, and vice versa. For $\mathbf{Y}$, its spatial predictability is defined as

$$F(\mathbf{Y}) = \sum\nolimits_{k=1}^{P} \ln \frac{\sum\nolimits_{i,j=1}^{I,J} (\overline{\mathbf{Y}}_{ijk} - \mathbf{Y}_{ijk})^2}{\sum\nolimits_{i,j=1}^{I,J} (\tilde{\mathbf{Y}}_{ijk} - \mathbf{Y}_{ijk})^2} = \sum\nolimits_{k=1}^{P} \ln \frac{V_k}{U_k} \tag{3}$$

where V_k and U_k reflect the overall and local predictability degree of the kth abundance image, respectively. Taking every pixel $\mathbf{Y}_{ijk}$ into account, $\overline{\mathbf{Y}}_{ijk}$ and $\tilde{\mathbf{Y}}_{ijk}$ reflect its overall and local spatial variability. Their definitions are

$$\begin{aligned} \overline{\mathbf{Y}}_{ijk} &= \frac{1}{I \times J - 1} \sum\nolimits_{t=1} \mathbf{Y}_{ijk}^{(t)} \\ \tilde{\mathbf{Y}}_{ijk} &= \sum\nolimits_{t=1}^{N_C} \lambda_C^{(t)} \mathbf{Y}_{ijk}^{(t)} \end{aligned} \tag{4}$$

where $\mathbf{Y}_{ijk}^{(t)}$,N_C and $\lambda_C^{(t)}$ are taken from the neighborhood system of MRF model.

MRF models are mainly used in feature extraction and image segmentation [10]. For any pixel $\mathbf{Y}_{ijk}$ (for brevity's sake, y is used to express $\mathbf{Y}_{ijk}$), its nth-order neighborhood system is $N_y^n = \{y+r | y+r \in N_y, |r|^2 \leq D[n]\}$, where N_y are the neighbors of $\mathbf{Y}_{ijk}$ (the adjacent pixels except itself; for details, see [10]), $|r|$ denotes the Euclidian distance between sites y and $y+r$, and $D[n]$ is a member of the set of all possible integers defined as $D = \{D[n] | D[n] = p^2 + q^2, p, q \in \mathbb{Z}, D[k] > D[l] \text{ if } k > l\}$. Figure 2(a) displays the five order neighborhood system. It labels first-order spatial neighbors of site $\mathbf{Y}_{ijk}$ as "1", second-order neighbors as "2", and so on. Concerning above three parameters, $\mathbf{Y}_{ijk}^{(t)}$ denotes the tth-order neighbors of $\mathbf{Y}_{ijk}$, N_C is set to 2 for $\tilde{\mathbf{Y}}_{ijk}$, and the corresponding weight matrix of $\lambda_C^{(t)}$ is shown in Figure 2(b). Here, $\overline{\mathbf{Y}}_{ijk}$ and $\tilde{\mathbf{Y}}_{ijk}$ can be regarded as the energy functions of $\mathbf{Y}_{ijk}$ in different range.

However, some conditions must be satisfied for spatial complexity. That is, the number of mixtures is much greater than that of sources, which can be derived from the following theorem.

Theorem 1. *As the number of mixtures increases relative to a fixed number of sources, the difference between the extreme values (both minimum and* *maximum) of mixtures' and sources' spatial complexity decreases, which offers more possibilities to extract all sources.*

The proof of the theorem is in the appendix. It explains why the sources could not be totally separated out when the number of mixtures is close to that of sources, leading to the unsuitability of spatial complexity for multispectral imagery (MSI, only containing several spectral bands). For HSI data, because the number of bands (mixtures number) is much larger than that of endmembers (sources number, $L \gg P$), spatial complexity is a suitable approach for HSI unmixing. In addition, it should be noted that the correlation among neighboring abundance fractions of endmember guarantees its validity.

3 The Gradient Ascent Algorithm and Pre- & Post-processing

To extract the abundance cube in parallel, gradient ascent algorithm is employed. Given $\mathbf{W}_k$ being a $1 \times L$ row vector of $\mathbf{W}$, equation (3) can be rewritten as

$$F(\mathbf{Y}) = F(\mathbf{W}) = \sum\nolimits_{k=1}^{P} \ln \frac{\mathbf{W}_k \overline{\mathbf{C}} \mathbf{W}_k^{\mathrm{T}}}{\mathbf{W}_k \tilde{\mathbf{C}} \mathbf{W}_k^{\mathrm{T}}} \tag{5}$$

where $\overline{\mathbf{C}}$ and $\tilde{\mathbf{C}}$ are both $L \times L$ matrixes of overall and local covariances between mixtures respectively. They can be defined as

$$\begin{aligned} \overline{\mathbf{C}} &= \sum\nolimits_{i,j=1}^{I,J} (\overline{\mathbf{R}}_{ij} - \mathbf{R}_{ij})(\overline{\mathbf{R}}_{ij} - \mathbf{R}_{ij})^{\mathrm{T}} \\ \tilde{\mathbf{C}} &= \sum\nolimits_{i,j=1}^{I,J} (\tilde{\mathbf{R}}_{ij} - \mathbf{R}_{ij})(\tilde{\mathbf{R}}_{ij} - \mathbf{R}_{ij})^{\mathrm{T}} \end{aligned} \tag{6}$$

$\overline{\mathbf{C}}$ and $\tilde{\mathbf{C}}$ only need to be computed once, which greatly alleviate the computational load. $\overline{\mathbf{R}}_{ij} = [\overline{\mathbf{R}}_{ij1}, \overline{\mathbf{R}}_{ij2}, \ldots, \overline{\mathbf{R}}_{ijL}]^{\mathrm{T}}$,$\tilde{\mathbf{R}}_{ij} = [\tilde{\mathbf{R}}_{ij1}, \tilde{\mathbf{R}}_{ij2}, \ldots, \tilde{\mathbf{R}}_{ijL}]^{\mathrm{T}}$, and the definitions of $\overline{\mathbf{R}}_{ijk}$ and $\tilde{\mathbf{R}}_{ijk}$ are similar to equation (4). The derivative of F with respect to $\mathbf{W}$ is

$$\nabla_{\mathbf{W}} F = 2\mathbf{W}\overline{\mathbf{C}}./\mathbf{V} - 2\mathbf{W}\tilde{\mathbf{C}}./\mathbf{U} \tag{7}$$

The sign ./ means "point division", namely, each element of numerator matrix divides the corresponding element of denominator matrix. $\mathbf{V}$ is a $P \times L$ matrix with the kth row vector acquired by replicating V_k. Likewise, the kth row vector of $\mathbf{U}$ is the replication of U_k. The gradient ascent rule is

$$\mathbf{W}_{new} = \mathbf{W}_{old} + \eta \nabla_{\mathbf{W}} F \tag{8}$$

where η is the learning rate. To speed up the convergence of $\mathbf{W}$, a scheme is used on η. In the experiments conducted in Section 4, 0.1 is assigned as its initial value, then 0.9 is multiplied every other twenty steps.

Some methods should be applied to process the HSI data before and after the unmixing algorithm to make the results more accurate. Preprocessing (centering and whitening) through principle component analysis (PCA) [4] is adopted as a means to reduce the relevance of first and second statistics, and to speed up the convergence process. Postprocessing step is to identify the position of the endmembers, and the method in [11,12] is employed. Briefly speaking, the histogram and the skewness of every abundance image are computed first, and the sign of skewness indicates the direction to be searched for the threshold (Positive to the right of the center, otherwise to the left). Then the pixel value where the histogram firstly takes zero is selected as the threshold. Finally, the threshold is used to filter corresponding abundance image. Readers are referred to [11] for details.

At last, the pseudo-code of spatial complexity for HSI unmixing is specified.

PCA($\mathbf{R}$);
compute $\overline{\mathbf{C}}$ and $\tilde{\mathbf{C}}$;
$\mathbf{W}$ is randomized and F_{old} is initialized to zero;
while $F(\mathbf{W}) - F_{old} > \sigma$
  $F_{old} = F(\mathbf{W})$;
  compute $\mathbf{V}$ and $\mathbf{U}$;
  $\nabla_{\mathbf{W}} F = 2\mathbf{W}\overline{\mathbf{C}}./\mathbf{V} - 2\mathbf{W}\tilde{\mathbf{C}}./\mathbf{U}$;
  $\mathbf{W} = \mathbf{W} + \eta \nabla_{\mathbf{W}} F$;
  orthogonalize $\mathbf{W}$ [13] and update η;
end while
every pixel vector $\mathbf{r}$ in $\mathbf{R}$ is multiplied by $\mathbf{W}$ to obtain $\mathbf{Y}$;
threshold every abundance image of $\mathbf{Y}$ using the skewness and histogram;

4 Experimental Results

In this section, two sets of real hyperspectral image data, HYDICE (HYperspectral Digital Imagery Collection Experiment) [14] and PHI (Pushbroom Hyperspectral technique Imager) are applied to evaluate the performance of spatial complexity. Some methods have been proposed to estimate the number of endmembers P, such as virtual dimensionality [15]. In this paper, we assume that it is known in advance to maximize the efficiency of spatial complexity. In addition, the unmixing results of undercomplete ICA (UICA) algorithm [16] are given for comparative analysis.

4.1 HYDICE Data

Figure 3 shows an urban scene of size 307 × 307 extracted from HYDICE data. It is composed of 210 spectral channels with spectral resolution 10nm acquired in the 400nm and 2.5 micron region. After low signal-to-noise ratio (SNR) bands are removed, only 166 bands remain (i.e., L=166). According to the ground truth, this data contain four materials: vegetation, asphalt, soil and roof. So P is set to 4.

Firstly, the UICA algorithm is applied to the dataset. Figure 4 presents the unmixing results. Except that the asphalt and soil are classified in Figure 4(b) and 4(c), the other two images are mixtures, which verify the conclusion of [7]: In any case, there are always endmembers incorrectly unmixed. Then spatial

Fig. 3. The urban scene extracted from HYDICE HSI

(a) mixture (b) asphalt (c) soil (d) mixture

Fig. 4. Unmixing results produced by UICA

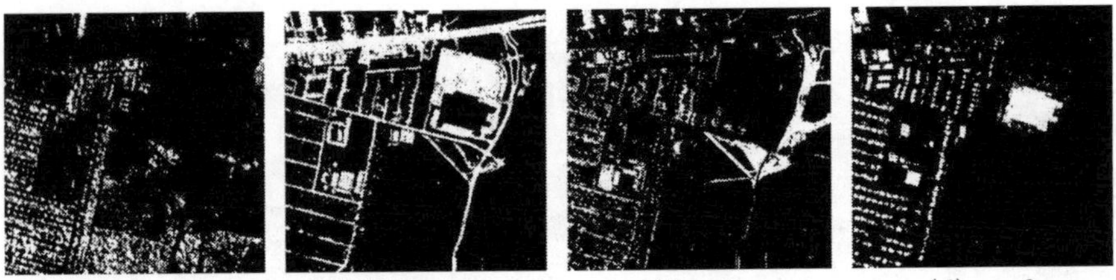

(a) vegetation (b) asphalt (c) soil (d) roof

Fig. 5. Unmixing results produced by spatial complexity

complexity is utilized, and the results are displayed in Figure 5. Different from Figure 4, all the four materials: vegetation, asphalt, soil and roof are successfully extracted. For convenience, we let white stand for the separated materials and black for the background in the unmixing results. Comparing the abundance distribution of asphalt and soil which are both unmixed by the two methods, it is clear that the results of spatial complexity are much better than those of UICA. Concretely speaking, Figure 4(b) misclassifies some vegetation pixels as asphalt, and 4(c) only classifies large pieces of soil but ignores small ones. Contrarily, Figure 5(b) and 5(c) totally extract the two materials. It is worth noting that the unmixing results in Figure 5 have been filtered by the postprocessing method, so they are more distinct. Hence, we can conclude that spatial complexity is more efficient than UICA.

4.2 PHI Data

The PHI data used in the following experiment were directly extracted from the PHI image scene of size 400×331 shown in Figure 6. The image data were ac-

Fig. 6. The subscene of Shanghai World Expo Garden extracted from PHI HSI

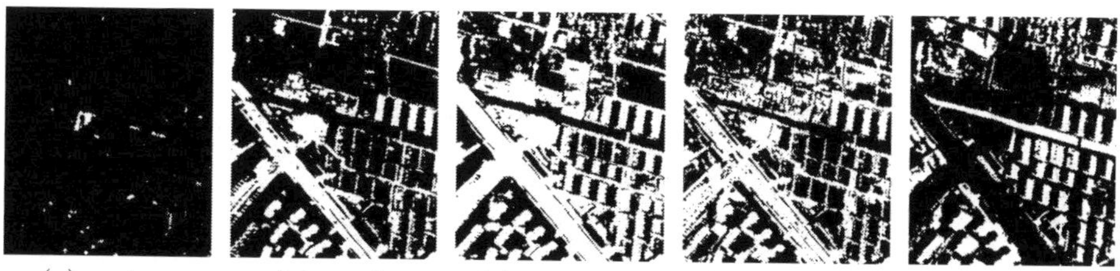

(a) water (b) road (c) mixture (d) mixture (e) mixture

Fig. 7. Unmixing results produced by UICA

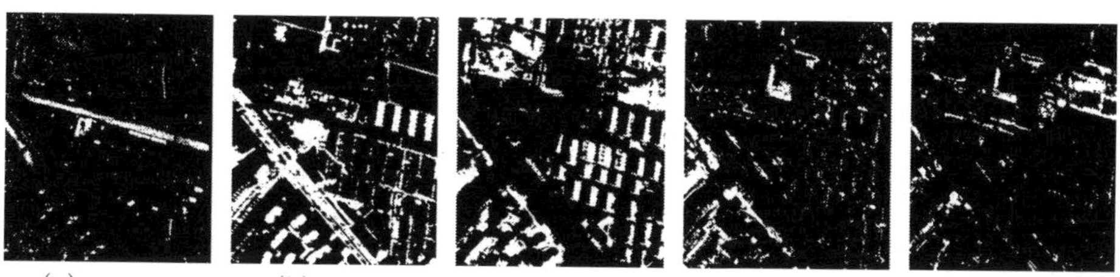

(a) water (b) road (c) roof (d) soil (e) sparse grass

Fig. 8. Unmixing results produced by spatial complexity

quired from the subscene of Shanghai World Expo Garden in October 2003 with the ground sampling distance approximately 1.2m. It has 124 spectral channels ranging from 400nm to 990nm with spectral resolution 5nm. And no spectral bands are removed in the experiments (i.e., L=124). According to the ground truth, this data contain five materials: water, road, roof, soil and sparse grass. So P is assigned to 5.

Same as the above experiment, the UICA algorithm is firstly applied. Figure 7 shows the unmixing results. Except that the water and road are extracted out in Figure 7(a) and 7(b), all the other three images are mixtures, which verify the conclusion of [7] again. The unmixing results generated by spatial complexity are illustrated in Figure 8. All the five materials: water, road, roof, soil and sparse grass, are displayed in sequence. Comparing images (a) and (b) between Figure 7 and 8, it is easy to find that the abundance of water extracted in Figure 7(a) is just a small part of Figure 8(a). So UICA actually separated out only one material: road. And the same conclusion as the above subsection can be drawn.

5 Summary and Conclusions

We have presented a spatial complexity based algorithm for hyperspectral imagery unmixing. The algorithm extends the temporal complexity; the major improvement is the better representation of spatial correlation of abundance image. A proof is presented that spatial complexity is suitable for HSI unmixing. Its effectiveness has been tested by comparison to UICA with data from HYDICE and PHI HSI. The experimental results show that our approach provides a promising method for HSI unmixing.

The reason that ICA could not totally separate out the sources (the abundance cube) ascribes to the dependence among the abundance fractions (i.e., the sum of them associated to each pixel is constant due to physical constraints in the data acquisition process) [7]. Spatial complexity is actually a second order statistics algorithm, which does not use explicitly or implicitly any criterion of statistical independence [4]. Hence, it is reasonable to introduce the method to unmix HSI data, and the experimental results confirm it. However, the conjecture in [8] is proved based on sources independence; so in future work, we will attempt to find out the connection between spatial complexity and sources independence.

Acknowledgment

The authors would like to thank Shanghai Institute of Technical Physics, Chinese Academy of Sciences for providing the PHI data.

References

1. D. Manolakis and G. Shaw. Detection algorithms for hyperspectral imaging applications. *IEEE Signal Processing Magazine*, 19(1):29–43, January 2002.
2. N. Keshava. A survey of spectral unmixing algorithms. *Lincoln Lab Journal*, 14(1):55–73, 2003.
3. R.H. Yuhas, A.F.H. Goetz, and J.W. Boardman. Discrimination among semi-arid landscape endmembers using the spectral angle mapper (sam) algorithm. In *Summaries of the 3rd Annual JPL Airborne Geoscience Workshop*, volume 1, pages 147–149, 1992.
4. A. Cichocki and S. Amari. *Adaptive Blind Signal and Image Processing: Learning Algorithms and Applications.* John Wiley & Sons, 2002.
5. J.D. Bayliss, J.A. Gualtieri, and R.F. Cromp. Analyzing hyperspectral data with independent component analysis. In *Proceeding of SPIE Applied Image and Pattern Recognition Workshop*, volume 3240, pages 133–143, 1997.
6. S.-S. Chiang, C.-I. Chang, J.A. Smith, and I.W. Ginsberg. Linear spectral random mixture analysis for hyperspectral imagery. *IEEE Transaction on Geoscience and Remote Sensing*, 40(2):375–392, February 2002.
7. J.M.P. Nascimento and J.M.B. Dias. Does independent component analysis play a role in unmixing hyperspectral data? *IEEE Transaction on Geoscience and Remote Sensing*, 43(1):175–187, January 2005.
8. J.V. Stone. *Independent Component Analysis: A Tutorial Introduction.* MIT Press, 2004.

9. Q. Du and S. Chakrarvarty. Unsupervised hyperspectral image classification using blind source separation. In *IEEE International Conference on Acoustics, Speech, and Signal Processing 2003*, volume 3, pages 437–440, 2003.
10. H. Deng and D.A. Clausi. Unsupervised image segmentation using a simple mrf model with a new implementation scheme. *Pattern Recognition*, 37(12):2323–2335, 2004.
11. S.-S. Chiang, C.-I. Chang, and I.W. Ginsberg. Unsupervised target detection in hyperspectral images using projection pursuit. *IEEE Transaction on Geoscience and Remote Sensing*, 39(7):1380–1391, July 2001.
12. S.A. Robila and P.K. Varshney. Target detection in hyperspectral images based on independ-ent component analysis. In *Proceeding of SPIE Automatic Target Recognition*, volume 4726, pages 173–182, 2002.
13. A. Hyvärinen. Fast and robust fixed-point algorithms for independent component analysis. *IEEE Transactions on Neural Networks*, 10(3):626–634, 1999.
14. The HYDICE HSI dataset. http:// www.tec.army.mil/Hypercube/.
15. C.-I Chang and Q. Du. Estimation of number of spectrally distinct signal sources in hyperspectral imagery. *IEEE Transaction on Geoscience and Remote Sensing*, 42(3):608–619, March 2004.
16. J.V. Stone and J. Porrill. Undercomplete independent component analysis for signal separation and dimension reduction. Technical report, Psychology Department, Sheffield University, http:// www.shef.ac.uk/ pc1jvs/, 1998.
17. S. Xie, Z. He, and Y. Fu. A note on stone's conjecture of blind signal separation. *Neural Computation*, 17(2):321–330, February 2005.

Appendix: Proof of Theorem 1

The conjecture used in [8] has been modified and proved by Xie et al. [17]. Namely, any mixture has a complexity lying between its least and most complex source signals. Although it aims at one-dimensional signal, it also holds for two-dimensional image (The proof is similar to Xie's, so it is not given), which can be described that the spatial complexity of every mixture lies between its least and most complex sources.

Let $C(\mathrm{X})$ denote the spatial complexity of X, X$\uparrow$ denote the augment of X, X$\rightarrow$Y denote that X approaches to Y, and X$\Rightarrow$Y denote that X implies Y. First, the Xie's conjecture can be formalized as

$$\min(C(\text{sources})) \leq C(\text{mixtures}) \leq \max(C(\text{sources})) \tag{9}$$

Equation (2) indicates that the estimated sources are the "mixing" of mixtures. According to the Xie's conjecture, it is clear that

$$\min(C(\text{mixtures})) \leq C(\text{estimated sources}) \leq \max(C(\text{mixtures})) \tag{10}$$

Second, without loss of generality, due to the randomness of the mixing process [6], the following formulas can be obtained

$$\begin{aligned} &\min_{\text{mixture number}\uparrow}(C(\text{mixtures})) \rightarrow \min(C(\text{sources})) \\ &\max_{\text{mixture number}\uparrow}(C(\text{mixtures})) \rightarrow \max(C(\text{sources})) \end{aligned} \tag{11}$$

which means

$$\underset{\text{mixture number}\uparrow}{R(\text{mixtures})\uparrow \rightarrow R(\text{sources})} \tag{12}$$

where $R(\mathrm{X})$ is defined as

$$R(\mathrm{X}) = \max(C(\mathrm{X})) - \min(C(\mathrm{X})) \tag{13}$$

Correspondingly, from (10),

$$\underset{\text{mixture number}\uparrow}{R(\text{mixtures})\uparrow \Rightarrow R(\text{estimated sources})\uparrow} \tag{14}$$

Considering (12) and (14) simultaneously,

$$\underset{\text{mixture number}\uparrow}{R(\text{estimated sources}) \rightarrow R(\text{sources})} \tag{15}$$

which can derive the following formula

$$\underset{\text{mixture number}\uparrow}{(\text{estimated sources}) \rightarrow \text{sources}} \tag{16}$$

Consequently, the conclusion can be drawn. □

Effectiveness of Spectral Band Selection/Extraction Techniques for Spectral Data

Marina Skurichina, Sergey Verzakov, Pavel Paclik, and Robert P.W. Duin

Information and Communication Theory Group, Faculty of Electrical Engineering, Mathematics and Computer Science, Delft University of Technology, P.O. Box 5031, 2600GA Delft, The Netherlands
m.skurichina@tudelft.nl

Abstract. In the past few years a variety of successful algorithms to select/extract discriminative spectral bands was introduced. By exploiting the connectivity of neighbouring spectral bins, these techniques may be more beneficial than the standard feature selection/extraction methods applied for spectral classification. The goal of this paper is to study the effect of the training sample size on the performance of different strategies to select/extract informative spectral regions. We also consider the success of these methods compared to Principal Component Analysis (PCA) for different numbers of extracted components/groups of spectral bands.

1 Introduction

Densely sampled spectral measurements became a standard tool in many applications such as medical diagnostics or industrial quality control. The amount of data to be dealt with has increased even further due to widespread adoption of hyperspectral imaging sensors capturing spectral readings in spatial raster. The acquired spectral information is, however, largely redundant due to low intrinsic dimensionality of the studied phenomena. Therefore, raw spectral measurements are usually reduced for the sake of data transmission, visualization or data analysis. In this paper, we discuss a specific type of data reduction techniques targeting supervised pattern classification.

The examples of successful methods to find discriminative spectral regions are an Optimal Region Selector (ORS) [1] guided by a genetic algorithm, a top-down and bottom-up multiresolution feature extraction algorithms proposed by Kumar et al. [2], Recursive Band Selection (RBE) [3] etc. The advantage of these techniques is that they make use of the connectivity between neighbouring spectral bins when finding discriminative groups of spectral bands, while the standard feature reduction approaches (such as forward/backward feature selection or PCA [4]) neglect the apriori available information on the ordering of spectral wavelengths. Spectral band selection techniques are also preferred to standard feature reduction techniques, because they allow us to find discriminative regions in spectra instead of single bands or "generalized" features (like in PCA). By this, specialists can make the relation between the informative group of spectral bands found and the physical background of a studied phenomenon. It also implies the possibility to design cheap devices to perform

D.-Y. Yeung et al. (Eds.): SSPR&SPR 2006, LNCS 4109, pp. 541 – 550, 2006.

measurements only for few spectral regions that make sense for discrimination instead of measuring spectra for a wide range of all possible emission wavelengths.

Sometimes it is not possible to find clear discriminative spectral regions especially for spectral data representing mixtures of materials. The information useful for discrimination might be spread over all (or over the majority) spectral features. However, in the case of highly dimensional data with a relatively small amount of available measurements, it is needed to reduce the data dimensionality in order to construct a reliable classification rule [5]. Then PCA may be used, as it insures that all information contained in original features is preserved in extracted principal components. But the first few components (describing the largest variance in the data) do not guarantee the best discrimination between data classes, because PCA is an unsupervised feature extraction technique that does not make use of data class information. For a better classification performance, one may need a larger number of principal components.

Both, the spectral band selection techniques and PCA, have benefits and drawbacks that may depend on the training sample size, the number of desirable components/regions, and on the type of the spectral data they are applied to. What concerns the spectral band selection methods, their success depends on many factors: the exact strategy to find spectral regions, the criterion to select best regions, a merging function to produce a singe value introducing the group of spectral bands and finally the classification rule used to evaluate the success of feature extraction. We can expect that for PCA and the spectral band extraction methods, small training sample sizes may cause problems to find good discriminative components/regions. The PCA may be imperfect when a too small number of principal components is considered. The spectral band extraction methods may tend to select single bands (representing noise in the data) when they are forced to find a large number of spectral regions.

The goal of this paper is to compare the classification performances of different spectral band selection strategies (that extend standard feature selection techniques by using the spectral ordering information) mutually and to PCA for different training sample sizes and different numbers of extracted components/regions. Two real data sets representing two-class problems are used in our study. They are described in section 2. Different feature selection/extraction strategies used to find discriminative spectral regions are introduced in section 3. The results of our experimental study are discussed in section 4. Conclusions are summarized in section 5.

2 Data

Our study is performed on two real-world datasets representing two-class problems.

The first dataset consists of the autofluorescence spectra acquired from healthy and diseased mucosa in the oral cavity. The measurements were performed at the Department of Oral and Maxillofacial Surgery of the University Hospital of Groningen [6]. The spectra were collected from 97 volunteers with no clinically observable lesions of the oral mucosa and 137 patients having lesions in oral cavity. The measurements were taken at 11 anatomical locations with excitation wavelength 365 nm. After preprocessing [6] each spectrum consists of 199 bins. In total, 581 spectra representing healthy tissue and 123 spectra representing diseased tissue were obtained

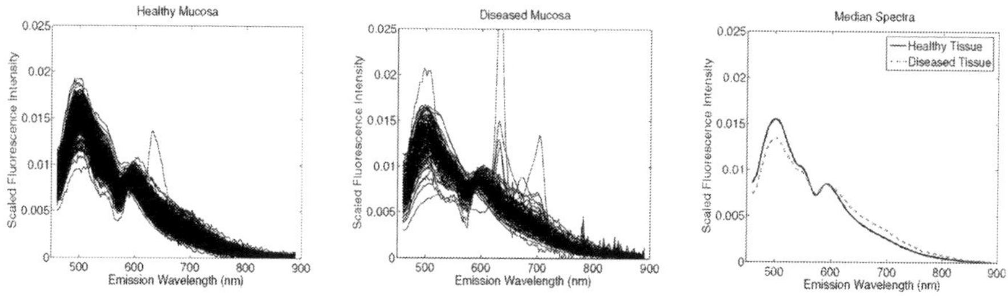

Fig. 1. Normalized autofluorescence spectra for healthy and diseased mucosa in oral cavity

after a thorough inspection of the database and removing all doubtful measurements. In order to reduce a large deviation in a spectral intensity within each data class, spectra were normalized by the unit area. Normalized autofluorescence spectra of healthy and diseased tissues and their median spectra are illustrated in Fig. 1.

The second dataset represents histograms of DNA content of tumour cells in a breast tissue [7]. The data are provided by the Pathology Department of De Wever Hospital of Heerlen. The DNA of all cells in a breast tissue sample is stained with a fluorchrome that emits red light after irradiation with a laser beam. The emitted light photons are collected in a photo multiplier tube in the flowcytometer and converted to electrical pulses that are proportional to the amount of DNA in the cells. After counting 20000 cells, a histogram is made of the DNA content of these cells. Each histogram is described by 256 wavelength channels of flowcytometer. After removing the first two and the last two histogram bins (which contain only noise), each histogram consists of 252 bins. The dataset contains 448 histograms of DNA content representing aneuploid breast tumour cells and 199 histograms describing DNA content of diploid breast tumour cells. The histograms are normalized by the unit area. The examples of these histograms and the median histograms are presented in Fig. 2.

For our experiments, training data sets with 10, 50 and 100 objects per class are considered for both datasets studied. Each time the training objects are chosen randomly from the total set. The remaining data are used for testing. The prior class probabilities are set to be equal as the data are very unbalanced and the real prior class probabilities are unknown. To evaluate the performance of diseased tissue diagnostics when different feature selection/extraction methods are used, we have

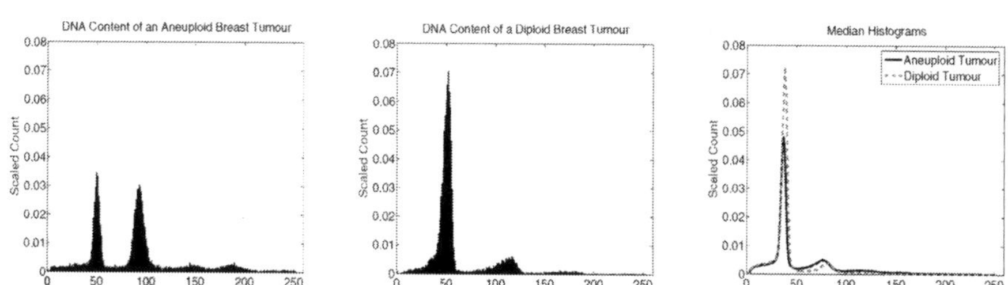

Fig. 2. The median histograms and selected examples of normalized histograms of the DNA content of 20000 cells obtained for aneuploid and diploid breast tumours

chosen the Regularized Linear Classifier (RLC) [8] which constructs a linear discriminant function assuming normal class distributions and using a joint class covariance matrix for both data classes. The value of the regularization parameter used is equal to 10^{-8}. All experiments are repeated 20 times on independent training sets. In all figures the averaged results over 20 trials are presented. The standard deviation of the reported mean generalization errors (the mean per two data classes) is about 0.01 for each considered case.

3 Strategies for Spectral Band Selection

As a rule, each of spectral band selection techniques proposed in the literature uses another criterion to find the discriminative spectral regions and a different function to merge a group of spectral bands into a single value representation. The choice of such a criterion or a merging function can seriously affect the performance of a spectral band selection technique. In order to eliminate the influence of these two factors on the classification performance of the studied spectral band selection strategies, we use the same criterion and the same merging function for all of them.

As a discriminant measure (criterion) to evaluate a discriminative capacity of extracted spectral regions, we use the Mahalanobis Distance (MD) between data classes:

$$MD = (\mu_A - \mu_B)'(p\Sigma_A + (1-p)\Sigma_B)^{-1}(\mu_A - \mu_B), \tag{1}$$

where μ_A, μ_B and Σ_A, Σ_B are the means and the covariance matrices of data classes A and B, respectively; p is the prior probability of the data class A. The larger Mahalanobis distance, the larger discriminative capacity between data classes.

A merging function used by us to reduce the dimensionality of each considered spectral region to a single value representation is the mean function which simply takes the average of spectral intensities in the region.

In our study we consider the following spectral band extraction strategies.

Approach 1. *GLDB-TD.* A top-down multiresolution feature extraction algorithm proposed by Kumar et al. [2], partitions the original p-dimensional spectra into smaller subspaces by using a top-down recursive algorithm. First, the best place to split spectra into two parts is found by computing a discriminant measure between data classes. The criterion value obtained on the parent space is compared with the criterion values calculated on the children subspaces. If the child subspace has a higher discrimination than the parent space, then it is partitioned further. We stop the partitioning, when no child subspace shows an improvement in its discrimination capacity compared to the parent space. The GLDB-TD algorithm is fast, but the final set of spectral regions found is suboptimal, because the optimization is performed only in one-dimensional way: a discrimination capacity is evaluated for each spectral region separately but not for a total set of selected spectral regions.

Approach 2. *GLDB-BU.* A bottom-up generalized local discriminant bases algorithm proposed by Kumar et al. [2], merges p original bands in larger subspaces by using a bottom-up recursive algorithm. First, the best pair to merge among all possible pairs of neighbouring single bands is found by computing a discriminant measure between data classes. The criterion value obtained on the best merged band is compared with the criterion values calculated on its component subspaces. If the merged space has a higher discrimination than the component subspaces, the merge is accepted and we move to the next level. Elsewise, the merge is denied and we consider the second best pair to merge on this level. If no merge is found that gives the better discrimination than component subspaces, then the merging procedure stops. This strategy has the same limitation as GLDB-TD: the optimization is performed only for one spectral group.

Approach 3. *Recursive Band Elimination (RBE).* The RBE technique proposed by Verzakov at al. [3] is a modification of the SVM shaving technique. We apply RBE using the Regularized Linear Classifier (RLC) instead of SVM in order to compare this strategy with other spectral band selection techniques in even conditions. First, the linear classifier is trained on the original p-dimensional spectral data. The absolute values of RLC coefficients $|w_i|$, $i = \overline{1, p}$, are used to split spectra into an initial set of spectral regions. Namely, minima of $|w_i|$ are the splitting points to obtain groups of spectral bands. Then each of S obtained spectral regions is merged into a single feature and the Recursive Feature Elimination (RFE) [9] is applied to perform a backward elimination of spectral regions. At each step of RFE, we train the RLC on the features representing groups of spectral bands and remove the spectral region corresponding to the smallest absolute value of the RLC coefficients.

Both, RBE and GLDB-BU, recursively reduce the number of spectral regions. GLDB-BU starts from p single band regions and merges them iteratively (no spectral band is omitted) until the discrimination cannot be improved anymore by merging the spectral regions. The RBE starts from $S < p$ spectral regions defined by the coefficients of RLC and eliminates them one by one till one last spectral region remains. The RBE performs multivariate optimization for the spectral region selection. The absolute values of RLC coefficients (instead of Mahalanobis Distance) are used as a criterion to select discriminative groups of spectral bands.

Approach 4. *Sequential Partitioning (SP)* [10]. It also performs multidimensional optimization for the spectral region selection. First, spectra are partitioned into two parts by finding the best split (with the optimal criterion value over all possible partitions) in the space of two features extracted from the two spectral regions. When the first split location is anchored, we look for the second optimal split in such a way that the criterion value in a three-dimensional space (on three features extracted from the three spectral regions) is the largest over all possible locations for the second split. We fix the second split location and repeat the procedure until the desired number of spectral regions S is found. In this approach, all spectral bands are used in the partitioning of spectra. However, some of them may be uninformative - introducing only noise. It is good to remove them, as they may deteriorate the classification when they are included in the extracted spectral regions. One way to do this is described below.

Approach 5. *Sequential Partitioning and Elimination of uninformative spectral bands (SPE)* [10]. After a desired number of spectral regions S is found by the previous approach, we shrink the spectral regions removing uninformative bands. We proceed in a sequential way (region by region) moving from the most left region to the most right one. In order to shrink the spectral region, we consider all possible subregions of the reduced size in the region and find the subregion with the largest criterion value in S-dimensional space (where one feature represents a shrunk subregion of the spectral region under consideration and the rest S-1 features are extracted from the other S-1 spectral regions which definitions are fixed for a moment). After shrinking the first spectral region, we anchor its new definition and move to the next spectral region in order to exclude uninformative spectral bands. This method does not guarantee the optimal shrinking for all regions in general, because it is highly dependent on the proceeding order of spectral regions.

Approach 6. *Sequential Selection (SS)* [10]. The discriminative spectral regions are selected sequentially one by one. At each step s ($s = \overline{1, S}$) we consider all possible region definitions (of arbitrary size) in spectra. For each definition we calculate the discriminant measure in s dimensions: one feature represents a current potential pretender for the most discriminative spectral region and other s-1 features are extracted from the previously selected spectral regions. The region (a potential pretender) with the largest criterion value in s-dimensional space is picked as the most discriminative spectral region (in combination with the s-1 previously found optimal regions). In this approach the overlapping and non-overlapping spectral regions may be selected. Some spectral bands might be not selected at all to participate in spectral regions.

Approach 7. *Sequential Selection of Non-overlapping discriminative spectral regions (SSN)* [10]. This approach is identical to SS but the overlapping spectral regions are not allowed to be selected.

Approach 8. *Floating Partition (FP)* [11]. First, spectra are uniformly partitioned to a predefined number of spectral regions S. At each step, we allow the borders between spectral regions to float one spectral bin aside from the current position. Among 3^S possible mutations we select the partition that provides the highest discrimination according to the selected criterion. We repeat the procedure until no improvement in discrimination capacity can be found. This method performs multivariate optimization by simultaneously adjusting all spectral regions. However, it is still a suboptimal procedure because the drifting step for region borders is limited to one spectral bin. The efficacy of this method can be improved by enlarging the drifting step d. But this leads to computational burdens because one has to rank $(2d+1)^S$ cases at each step of the procedure. We could apply this approach upto 10 spectral regions (with d=1) at most.

4 Experiments and Discussion

Before studying the benefits of extracting/selecting the discriminative spectral regions in the comparison with the PCA approach for different training sample sizes N and for different numbers of extracted components/regions B, let us make few remarks on the datasets used. When measuring autofluorescence spectra of healthy and diseased tissues in oral cavity, in reality the autofluorescence spectra of the mixture of materials (skin, tissue under skin, bone and saliva) are obtained. The useful information for lesion diagnostics is hidden in overlapping peculiarities of different materials. The clear-cut discriminative spectral regions do not exist. The useful information for lesion diagnostics is spread over the whole spectrum. What concerns the histograms of the DNA content, the main information is concentrated around the two largest peaks in the left part of the histogram and in the region between them (see Fig. 2). The peaks and the region in between describe different phases of the cell division cycle. The amount of the DNA in these phases characterizes different types of tumour cells. So, all useful discriminative information is concentrated in spectral bands around these peaks of the histogram. Thus, our two datasets represent two extreme cases: when no separate discriminative regions exist (in autofluorescence spectra) and when we have a few well-defined discriminative regions (in histograms). For the first dataset, we can expect that the PCA may be very effective because the principal components aggregate the information represented in all spectral bands. For the second dataset, the spectral band selection techniques, that select regions around the histogram peaks, will be the most beneficial.

When considering results obtained for the autofluorescence spectra (see left plots in Fig.3), we see that for small training sample sizes (N=10+10), all techniques perform similarly with a slight preference for RBE, SSN and FS (the last one only for two extracted spectral regions). The training data are not enough representative to find correctly the discriminative spectral regions. Due to a limited number of training objects, only 19 principal components can be extracted by PCA. When increasing the training sample size, all techniques perform better, but the relative advantages of their generalization errors do not change much. The exceptions are the SP and SPE strategies, which become the best among the studied spectral band selection techniques for large numbers of extracted spectral regions when the training sample size is large (N=100+100). In agreement with our prior knowledge on the dataset, the more regions/components are selected, the better all methods perform. The most successful technique is PCA with a large number of principal components that accumulate useful information spread over the whole spectrum. The PCA is followed by the SP strategy which uses all spectral bands in spectral partitions. Excluding some spectral bands (in SPE) deteriorates the classification performance. However, in order to get a medical insight of the studied phenomenon, for data compression (in remote sensing) or for building filters in the sensor, we are interested in finding few discriminative spectral regions rather than many of them. The PCA is unsupervised feature extraction technique that finds directions in a feature space with the largest variance that are not necessarily discriminative. The spectral band selection methods are supervised techniques that take advantage of data class information. By this, they outperform PCA for small numbers of extracted components/regions. Interestingly, the RBE strategy was the best when 8-15 spectral regions are selected. Usually RBE converges to its best solution for a relatively large number of spectral regions. However, for this dataset, the clear preference for particular discriminative regions does not exist. Probably, RBE outperforms all other strategies due to the superior discriminant measure used (the absolute values of RLC weights are used instead of MD).

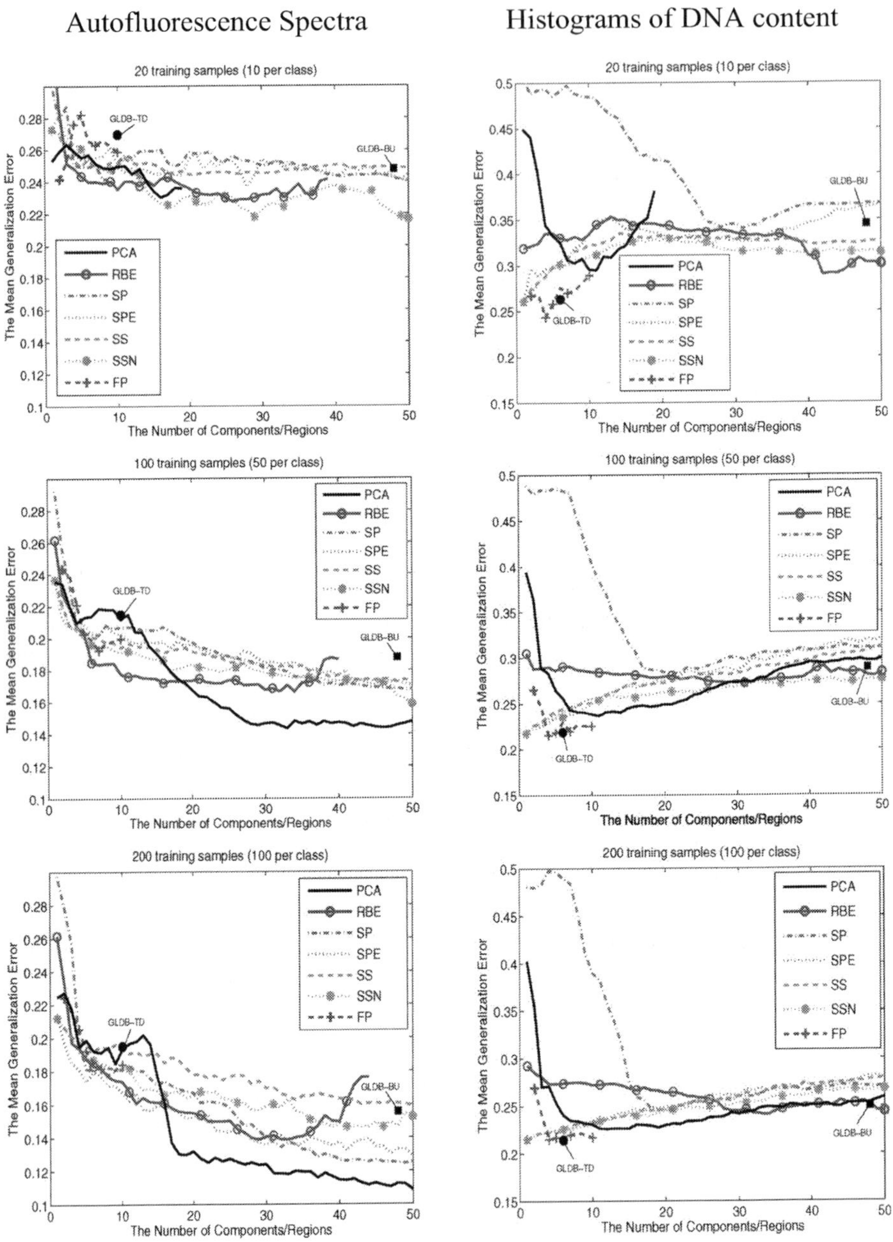

Fig. 3. The mean generalization error (GE) of LDA for training sample sizes 10, 50, and 100 objects per class, when different methods are used to select discriminative spectral regions for autofluorescence spectra measured in oral cavity (*left plots*) and for histograms of DNA content in tumour cells (*right plots*). Because the GLDB-TD and GLDB-BU algorithms terminate automatically using a data-driven criterion, only a single point is given in each plot. The standard deviation of the mean GE is around 0.01.

For histogram data (see right plots in Fig. 3), the performance of all strategies is improved by increasing the training sample size. But the general mutual behaviour of the generalization errors for all methods remains the same. The performance of all techniques worsen when the number of extracted regions/components grows after 8-10 regions. It is logical because mainly the spectral regions around and between the peaks (which are related to the cell division cycle) provide useful information for the classification of tumour cells. Adding more regions is equivalent to adding noise and cannot improve the classification. Indeed, the best results are found by the spectral band selection techniques when less than five spectral regions are retrieved. The spectral band selection strategies SPE, SS, SSN and GLDB-TD perform equally nice and the best among all studied techniques. The first three techniques select the spectral regions around the DNA content peaks, finding the most discriminative parts of spectra. GLDB-TD algorithm can also make a successful split of spectra converging at five-six spectral regions on average. However, the sequential partitioning (SP) completely fails for small numbers of extracted spectral regions. It happens by two reasons. First, all bands (also uninformative) are kept in spectral regions (when uninformative spectral bands are eliminated in SPE, the performance is drastically improved). Second, the partitioning in SP is done in a sequential way. Once the split is found, it cannot be adjusted anymore. The best split found for partitioning into two regions might be very far from the optimal one for partitioning into more regions than two. The FP strategy overcomes this problem by simultaneously adjusting all spectral regions. It competes with other spectral band selection techniques when spectra are split into four-five spectral regions. As the first five principal components extracted by PCA are not discriminative, its performance is poor. Since both, RBE and GLDB-BU, are bottom-up recursive procedures for finding informative spectral regions, they usually converge to the suboptimal solution at a relatively large number of spectral regions (around 45 for RBE and around 95 for GLDB-BU). Hence for this problem (with three-four discriminative spectral regions by definition) the performance of RBE and GLDB-BU is worse than the performance of other feature selection techniques (with exception of SP).

5 Conclusions

The success of spectral band extraction techniques varies over the potential of the spectral data depending on how information useful for classification is introduced: locally (in a few clear-cut discriminative spectral regions) or globally (spread over the majority of spectral wavelengths). The supervised spectral band selection techniques which make use of the connectivity of spectral wavelengths in spectral data (one-dimensional ordering) are more beneficial than unsupervised PCA when one needs to find a small number of discriminative spectral regions/components. However, which spectral band selection technique is preferred seems to be defined by the problem and the criterion used to select the best regions. These issues need more study in the future.

References

1. Nikulin, A., Dolenko, B., Bezabeh, T., Somorjai, R.: Near Optimal Region Selection for Feature Space Reduction: Novel Preprocessing Methods for Classifying MR Spectra. NMR in Biomedicine **11** (1998) 209-216
2. Kumar, S., Ghosh, J., Crawford, M.M.: Best-Bases Feature Extraction Algorithms for Classification of Hyperspectral Data. *IEEE* Transactions on Geoscience and Remote Sensing **39** (2001) 1368 - 1379
3. Verzakov, S., Paclik, P., Duin, R.P.W.: Feature Shaving for Spectroscopic Data. Lecture Notes in Computer Science, Springer-Verlag, Vol. 3138 Berlin Heidelberg New York (2004) 1026-1033
4. Fukunaga, K.: Introduction to Statistical Pattern Recognition. Academic Press (1990)
5. Jain, A.K., Chandrasekaran, B.: Dimensionality and Sample Size Considerations in Pattern Recognition Practice. In: Krishnaiah, P.R., Kanal, L.N. (eds.): Handbook of Statistics, Vol. 2. North-Holland, Amsterdam (1987) 835-855
6. De Veld, D.C.G., Skurichina, M., Witjes, M.J.H., et.al.: Autofluorescence Characteristics of Healthy Oral Mucosa at Different Anatomical Sites. Lasers in Surgery and Medicine **32** (2003) 367-376
7. Verzakov, S., Duin, R.P.W.: The Tangent Kernel SVM for Calibration-Stable Histogram Discrimination. Proceedings of the ASCI conference, ASCI, Delft (2005) 73-80
8. Friedman, J.H.: Regularized Discriminant Analysis. JASA **84** (1989) 165-175
9. Guyon, I., Weston, S., Barnhill, S., Vapnik, V.: Gene Selection for Cancer Classification Using Support Vector Machines. Machine Learning **46**(13) (2002) 389-422
10. Skurichina, M., Paclik P., Duin, R.P.W, et.al.: Selection/Extraction of Spectral Regions for Autofluorescence Spectra Measured in the Oral Cavity. Lecture Notes in Computer Science, Vol. 3138 Springer-Verlag, Berlin Heidelberg New York (2004) 1096-110
11. Meloni, S.: Finding Discriminative Bands in Auto-Fluorescence Spectra for Automatic Cancer Diagnosis. Master Thesis, Cagliari University, Sardinia, Italy (2004)

Edge Detection in Hyperspectral Imaging: Multivariate Statistical Approaches

Sergey Verzakov, Pavel Paclík, and Robert P.W. Duin

Information and Communication Theory Group
Faculty of Electrical Engineering, Mathematics and Computer Science
Delft University of Technology
Mekelweg 4, 2628 CD Delft, The Netherlands
{s.verzakov, p.paclik, r.p.w.duin}@ewi.tudelft.nl

Abstract. Edge detection is well developed area of image analysis. Many various kinds of techniques were designed for one-channel images. Also, a considerable attention was paid to edge detection in color, multispectral, and hyperspectral images. However, there are still many open issues in edge detection in multichannel images. For example, even the definition of multichannel edge is rather empirical and is not well established. In this paper statistical pattern recognition methodology is used to approach the problem of edge detection by considering image pixels as points in a multidimensional feature space. Appropriate multivariate techniques are used to retrieve information which can be useful for edge detection. The proposed approaches were tested on the real-world data.

1 Introduction

The recent development of sensors makes multichannel images usual objects for analysis. One of the important tools for working with multichannel images is edge detection: finding the places where the properties of image undergo considerable changes. First, detected edges allow the visualization of otherwise difficult to represent multichannel image. Second, it allows to localize objects.

The task of edge detection is connected to the segmentation problem which looks for homogeneous image regions (connected or disconnected). Actually, segmentation answers the question whether the pixel belongs to some segment (cluster, class) and with which confidence. In this sense the results of a segmentation can be used for edge detection: e.g. the pixels with ambiguous confidences can be considered as edges. Another way to employ segmentation for edge detection is to mark pixels as edges whenever the order of confidences (or memberships) are changed [1]. So, the solution of the segmentation task can be easily used for edge detection. On the contrary, having solved the edge detection problem, it is not so straightforward to obtain the segmentation. One can state that two close pixels on the the same side (different sides) of edge belong to the same segment (different segments) but it is much more complicated if possible at all to state this for two arbitrary pixels. So, generally speaking, the task of edge detection provides us with less information but this information is more specific.

D.-Y. Yeung et al. (Eds.): SSPR&SPR 2006, LNCS 4109, pp. 551–559, 2006.

For example, in segmentation we need to estimate or get as a prior knowledge the number of segments. This is not needed for the edge detection. One should note that edge detection can be used together with segmentation in order to sharpen edge borders.

The task of detecting edges in gray valued images is very well known. It has a long history and has been thoroughly studied [2,3,4,5]. A review can be found in [6,7]. The same problem for three color, or more general, multichannel images is much less well defined. One of the difficulties in edge detection in multichannel images is the formulation of what is an edge. Indeed, in a gray valued image we can specify the type of intensity profile which we are looking for, i.e. we need to specify a scalar valued function of a scalar argument. (The last statement is not valid for detection of edges in textured images. This task can be converted to edge detection in multichannel images after application of a set of texture detectors.) In a multichannel image we have many more possibilities and it is not always obvious (or it is application dependent) which changes have to be taken into account. The problem of consistent edge definition in multichannel images has not been entirely solved. There are proposals to consider as an overall edge all edges in the separate channels. Hence, possible interaction between channels is neglected. Another approach is to reduce a multichannel image to a gray valued one, e.g., by intensity calculation. It was reported that 90% of the edges detected by this simple approach coincide with edges given by more sophisticated multivariate techniques [8]. However, by this approach we cannot find a change in the color of image which does not involve a change in the intensity level. It implies that channels have to be combined in a non-trivial way: added with different signs or be fused non-linearly [1,9]. But these approaches return again a number of gray valued images. So, the problem of combining is not solved. Very often the question of what is an edge in a multichannel image is not addressed directly but instead gradients of all channels are combined in some way[8,10,11,12].

Other approaches make estimations of the statistical properties of the image in the feature space and learn what can be an edge in this image. Like in [13] where authors proposed to use the "change point" theory for edge detection in gray valued images. The methods employing clustering to extract new channels which are more suitable for the task of edge detection [1,9] can also be considered as such methods, but not completely. They still need to combine the results obtained for different channels. We propose to use the estimation of a joint probability density function (PDF) of two neighbouring pixels. The main idea behind this approach is that pixel combinations typical for edges are rare. So they can be considered as outliers and after learning the joint PDF the edges will be represented by low density regions. One can also think of a modification of this approach which estimates a conditional PDF of the difference between neighbouring pixels. Similar approach was studied in [14], where a complimentary cumulative distribution function was used. However, the authors used a distribution modeling technique (cumulative histograms) relevant only for the small number of channels and small number of possible of gray values. We will use a Parzen density estimation or a mixture of Gaussians. Another important difference between [14] and our work is that we take into account the dependence of distribution on the current pixels location in feature space.

The paper is organized as follows. In the next section we describe some preexisting techniques. Section 3 is devoted to the newly proposed approaches. Then in the section 4 we describe datasets and numerical experiments. The paper is concluded by discussion and conclusions.

2 Preexisting Multichannel Edge Detection Techniques

As we have already mentioned in the introduction, many approaches to edge detection in multichannel images are available. We will review only the most generic ones. All techniques can be split into two large groups. The algorithms of the first type perform image analysis on individual channels and then combine the results (before or after thresholding) without using multivariate statistics in the feature space. However, univariate statistics of gray valued images can still be used for the adaptive selection of the threshold or the size of the filter. We will call this group "Non-statistical or univariate statistical approaches". Another group of algorithms uses multivariate statistics from the beginning and will be referred as "Multivariate statistical approaches". This group can be split into two subgroups. The methods from the first subgroup result again in multichannel images where channels are memberships, confidences, or other types of extracted features. So, combining of the channels is still needed. We will call the methods in this subgroup "Incomplete multivariate statistical approaches". The algorithms from the other subgroup return gray valued images and explicit channel combination is avoided. They are "Complete multivariate statistical approaches".

2.1 Non-statistical or Univariate Statistical Approaches

One of the most popular ways to detect edges in gray valued images is to compute (smoothed) derivatives and then mark as edges all pixels for which the absolute value of the derivative exceeds some threshold and is maximal in some neighbourhood. There are two general ways how to extend this approach to multivariate images. In the first one, spatial partial derivatives are calculated for all channels and combined in some way. For example, $1-$norm, $2-$norm, or $\infty-$norm (max-norm) can be used. In [8] the combination of the gradient magnitudes (instead of its components) is advocated and reported that $\infty-$norm combination gives the best results. More sophisticated approach [10,11] suggests to use the largest eigenvalue of the covariance matrix of the set of partial derivatives as an edge magnitude (LEV combination). The result of this combination is the gray valued images of gradient magnitudes. The standard methods of thresholding and edge thinning can be applied to it. Another type of extension performs edge detection for each channel and then combines binary images by, say, the logical OR operation.

Keeping in mind that we are interested in hyperspectral images mostly, we can state that many channels are highly correlated. Thus a very broad spectral band can obscure more narrow ones during 1−norm, 2−norm, or LEV combinations. The ∞−norm does not suffer from this. However, it also does not make subband averaging which can lead to a better signal to noise ratio. Another problem, which is encountered by all gradient combination techniques, is that derivatives taken at different channels are scaled differently. So, proper scaling and decorrelation have to be applied to hyperspectral images in order to get combinable gradients.

Another popular method of edge detection in gray valued images employs Laplace of Gaussian (LoG) filters in order to compute smoothed second derivatives. Edges now are defined as zero-crossing points. This approach can be extended to multichannel images in two ways. In the first one LoG is applied to all channels, then the results are summed (maybe with some weights) and thresholding takes place on this image. Note, that this approach is equivalent to the conversion of the image in gray valued image and application of the standard univariate LoG edge detection. Also, it is possible to apply edge detection in each channel and combine the binary results by the OR operator.

The hybrid of the two described ways (maximum of the first derivative and the zero-crossings of the second derivatives) is described in [11,12]. At first, edge magnitudes (contrasts) are calculated as LEV combination of the partial derivatives. Then the zeros of directional derivatives of contrasts are taken as edges.

2.2 Incomplete Multivariate Statistical Approaches

In the multivariate statistical approaches to edge detection one typically employs unsupervised pattern recognition techniques (clustering or density estimation) to use feature space information. Having multichannel images, we may consider each pixel as a point in some feature space. This gives us the possibility to look at the data from the statistical point of view. That is, we can base our algorithm on multidimensional distributions. A few approaches, which use channel statistics were proposed in the past.

The first one [1] consists of fuzzy segmentation of the image and considering zero-crossings of memberships differences: $\Delta_{\mathbf{x}}(i,j) = \alpha_{\mathbf{x}}(i) - \alpha_{\mathbf{x}}(j)$. Here, $\mathbf{x}$ is a pixel position, i, j are cluster indices, and α is a membership. It depends on the task which pairs i, j should be considered: only pair with the largest α or pairs for which $\alpha_{\mathbf{x}}(i)$ and $\alpha_{\mathbf{x}}(j)$ are significant.

The second approach [9] suggests to perform channel extraction based on clustering. Namely,

$$J_{\mathbf{x}}(i,j) = \frac{(\boldsymbol{\mu}_i - \boldsymbol{\mu}_j)^T}{||\boldsymbol{\mu}_i - \boldsymbol{\mu}_j||} \boldsymbol{I}_{\mathbf{x}}$$

Here $\boldsymbol{\mu}_i$ is the centroid of i-th cluster. Thus, we get $N(N-1)/2$ new channels. Authors suggest to combine edge magnitudes of these channels by ∞−norm and then apply thresholding. The computational cost can be decreased by taking into account only a few neighbouring clusters at each pixel position.

2.3 Complete Multivariate Statistical Approaches

The technique proposed in [14] involves computation of cumulative multidimensional histogram of pixel differences. In that way, a new distance between pixels is introduced. This approach is appropriate only for images with a small number of channels and not a large number of gray levels. In the next section we propose to use density estimators which are more suitable for high dimensional data.

3 Proposed Statistical Techniques

3.1 Joint Probability Density Functions of Neighbouring Pixels

The main hypothesis which will be used for developing the technique is that edges are pretty rare events. This is a natural assumption for many real-world images. Thus, one can conclude that a pair of neighbouring pixels positioned on different sides of an edge (or pair in which one of the pixels is pure and another is the transitional one, or both are transitional) should be also rare compared to pixel pairs in the interior regions.

Let us define by $\mathbf{I}_\mathbf{x}$ a d-dimensional vector of channel intensities of a multichannel image $\mathbf{I}$ at some pixel position $\mathbf{x} \in \mathbb{R}^2$. Further, suppose that a $\mathcal{N} \subset \mathbb{R}^2$ is a set of local shifts. Then the pixel $\mathbf{y} = \mathbf{x} + \mathbf{r}$ is the neighbouring pixel of the pixel $\mathbf{x}$. The joint PDF $\rho(\mathbf{I}_\mathbf{x}, \mathbf{I}_{\mathbf{x}+\mathbf{r}})$ has to be small if $\mathbf{x}$ is situated at an edge orthogonal (or at least not collinear) to $\mathbf{r}$.

The proposed approach consists of the estimation of $\rho(\mathbf{I}_\mathbf{x}, \mathbf{I}_{\mathbf{x}+\mathbf{r}})$, $\mathbf{r} \in \mathcal{N}$ by an appropriate technique like Parzen or Mixture of Gaussians density estimation. An estimated PDF can be used for edge detection. For the edge direction (i.e. direction along which the edge is locally extended) perpendicular to the shift $\mathbf{r}$ it gives a gray valued image of edge magnitudes $m_{\mathbf{x},\mathbf{r}}$ which is calculated as

$$m_{\mathbf{x},\mathbf{r}} = 1 - \rho(\mathbf{I}_\mathbf{x}, \mathbf{I}_{\mathbf{x}+\mathbf{r}})/R_\mathbf{r}$$
$$R_\mathbf{r} = \max_{\mathbf{x}} \rho(\mathbf{I}_\mathbf{x}, \mathbf{I}_{\mathbf{x}+\mathbf{r}})$$

To detect edges independently of their directions one may combine directional magnitudes as $m_\mathbf{x} = \max_{\mathbf{r}\in\mathcal{N}} m_{\mathbf{x},\mathbf{r}}$.

The binarization of this image can be done by putting threshold at some suitable percentile. To get more thin edges one can consider non-maximum suppression techniques similar to the ones used in gray valued case [7]. One can also think about smoothing obtained edge magnitudes. This is expected to be useful in noisy images.

Another problem is caused by the large dimensionality of the data. We need to estimate a PDF in the doubled feature space ($2d$). A proper dimensionality reduction technique, like PCA, can be used to solve this issue. It seems more reasonable to perform dimensionality reduction in doubled (not original) feature spaces.

3.2 Conditional Probability Density Functions of Neighbouring Pixels Difference

Having defined by $\delta\mathbf{I}_{\mathbf{x},\mathbf{r}} = \mathbf{I}_{\mathbf{x}+\mathbf{r}} - \mathbf{I}_\mathbf{x}$ the difference between neighbouring pixels, the joint distribution can be rewritten as $\rho(\mathbf{I}_\mathbf{x}, \mathbf{I}_{\mathbf{x}+\mathbf{r}}) \equiv \rho(\mathbf{I}_\mathbf{x}, \delta\mathbf{I}_{\mathbf{x},\mathbf{r}})$. So, it is possible to reconsider the above described method as a search for rare combinations of a pixel and a difference. But one can argue that some types of pixels are represented much less often than others. Because of this their pairs are also rare, although they do not represent any edge. To rule out such an unwanted situation, we propose to use conditional PDFs: $\rho(\delta\mathbf{I}_{\mathbf{x},\mathbf{r}}|\mathbf{I}_{\mathbf{x}}) = \rho(\mathbf{I}_{\mathbf{x}}, \mathbf{I}_{\mathbf{x}+\mathbf{r}})/\rho(\mathbf{I}_{\mathbf{x}})$ Then edge magnitudes are defined as

$$\begin{aligned} m^c_{\mathbf{x},\mathbf{r}} &= 1 - \rho(\delta\mathbf{I}_{\mathbf{x},\mathbf{r}}|\mathbf{I}_{\mathbf{x}})/R^c_{\mathbf{r}} \\ R^c_{\mathbf{r}} &= \max_{\mathbf{x}} \rho(\delta\mathbf{I}_{\mathbf{x},\mathbf{r}}|\mathbf{I}_{\mathbf{x}}) \\ m^c_{\mathbf{x}} &= \max_{\mathbf{r}\in\mathcal{N}} m^c_{\mathbf{x},\mathbf{r}} \end{aligned}$$

Consequently, only differences $\delta\mathbf{I}_{\mathbf{x},\mathbf{r}}$ which are rare for the pixel values $\mathbf{I}_{\mathbf{x}}$ will be considered as edges. Note, that to make a consistent conditional density estimation, the dimensionality reduction has to be done in original feature space.

4 Experimental Study

4.1 Datasets

To make a comparison between the discussed techniques a number of datasets have been used. The first one is the hyperspectral image of Washington DC Mall from [15]. This is a 191-channel airborne hyperspectral image of size 1280-by-307. The sensor system used in this case measured a response in 0.4 to 2.4 μm region of the visible and infrared spectrum. The task of edge detection can be formulated as a contour detection of homogeneous areas (roofs, roads, paths, trees, grass, water and shadows). The image itself is too large to be handled at once and we split it into 20 smaller 128-by-153 images. We have used only the upper left one ($DC_{1,1}$). This is a "busy" image with many details and large number of channels. It is expected that the multivariate statistical approach will be more suitable for reliable edge detection than adapted gray valued image analysis techniques.

Another image from [15] is a 12-channel (0.4 to 1 μm) 949-by-220 airborne image. We have split this image into 3 images of sizes 316-by-220 and used the middle one ($FLC1_2$). This image contains much simpler scene and has moderate number of channels. Thus, the usage of adapted image analysis techniques is expected to be enough.

The third collection of images contains 5 microscopic SEM/EDX 8-channel 128-by-128 images of chemical substances [16] from which only CHM_2 has been used . Image is extremely noisy both in the spectral and spatial domains.

4.2 Experiments

We have conducted a set of experiments on the above described images. The results are presented on Fig. 1-3. The first subfigure of each figure shows four typical channel images of multichannel image. The second subfigure represents the edges detected by non-statistical approach. Actually, for all datasets we

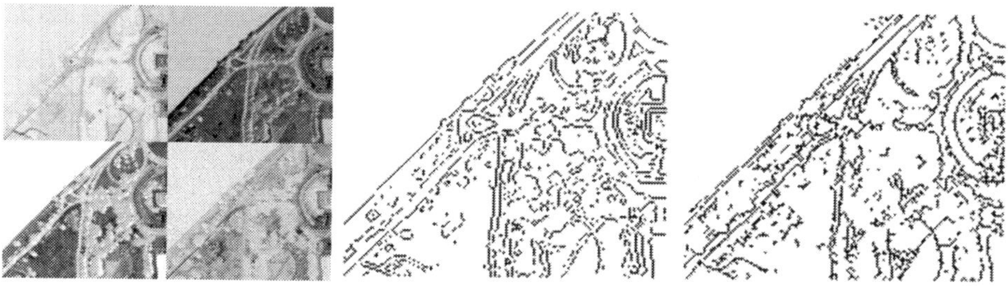

(a) Dataset (b) Sobel method (c) propsed technique

Fig. 1. $DC_{1,1}$

(a) Dataset (b) Sobel method (c) propsed technique

Fig. 2. $FLC1_2$

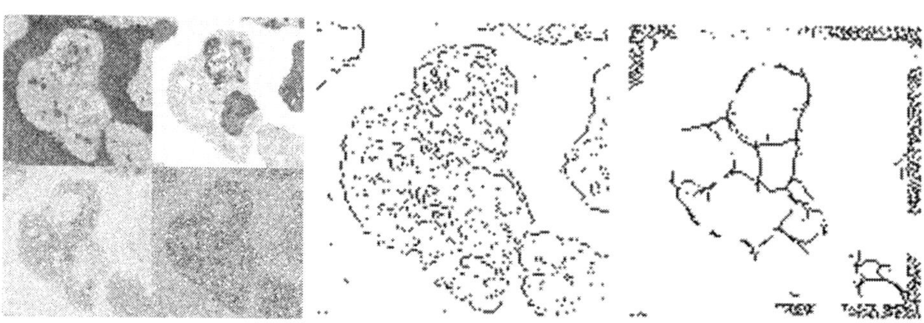

(a) Dataset (b) Sobel method (c) propsed technique

Fig. 3. CHM_2

computed channel gradients with the help of Sobel operator and combined them by ∞−norm. Binarization and non-maxima suppression were performed as it is suggested in [7]. Zero-crossing methods or other channel gradients combination rules give similar or worse results. The rightmost subfigures show the results of the proposed technique. On all images the PCA dimensionality reduction was performed with preserving 95% of total variance. EM-algorithm was used to estimate PDFs as Mixture of Gaussians with 10 components. This number of components proved to be reasonable for all images. The the results obtained by Parzen density estimator are similar to the presented ones. Only the CHM_2 edge magnitudes were smoothed by Gaussian filter with window size 17-by-17 and $\sigma = 2.67$, because other images did not benefit from smoothing.

5 Discussion and Conclusions

In this paper the task of edge detection in multichannel and especially in hyperspectral images was studied. The goal was not to propose a fast real-time algorithm but to try to develop a consistent approach to edge detection for multichannel images. The new approach based on the statistical pattern recognition was proposed. Instead of explicit definition of the edge we try to learn it by looking for improbable pixel combinations. The comparison of the results of conventional methods and proposed one shows that for high-dimensional complicated images detected edges are very similar (Fig. 1). For simpler images the image processing approach gives the better result (Fig. 2). For noisy chemical data our approach allows to obtain closed thin contours. The proposed approach is computationally expensive. So, it is necessary to develop faster density approximation algorithms. Another possible topic for the future research is incorporation spatial relations of pixels into the density estimation.

Acknowledgments

Authors would like to thank J. von Frese, A. Harol, P. Juszczak, M. Loog, and E. Pekalska for the helpful brainstorming. This research was supported by the Technology Foundation STW, applied science division of NWO and the technology program of the Ministry of Economic Affairs.

References

1. T.L. Huntsberger and M.F. Descalzi. Color edge detection. *Pattern Recognition Letters*, 3:205–209, 1985.
2. D. Marr and E. Hildreth. Theory of edge detection. *Proceedings of the Royal Society of London*, B207:187–217, 1980.
3. R. Nevatia and K. R. Babu. Linear feature extraction and description. *Computer Graphics Image Processing*, 13(3):257–269, July 1980.
4. R. M. Haralick. Digital step edges from zero crossing of second directional derivatives. *IEEE Transactions on Pattern Analysis and Machine Intelligence*, 6(1):58–68, 1984.

5. J. Canny. A computational approach to edge detection. *IEEE Transactions on Pattern Analysis and Machine Intelligence*, 8(6):679–698, 1986.
6. T. Y. Young and K.-S. Fu, editors. *Handbook of pattern recognition and image processing.* Academic Press, Inc., 1986.
7. J. S. Lim. *Two-dimensional signal and image processing.* Prentice-Hall, Inc., 1990.
8. T. Kanade and S. Shafer. Image understanding research at cmu. In *Proceedings of an Image Understanding Workshop*, volume I, pages 32–40, 1987.
9. H. Tao and T. S. Huang. Color image edge detection using cluster analysis. In *Proceedings of the 1997 IEEE International Conference on Image Processing (ICIP 1997)*, volume 1, pages 834–837, 1997.
10. S. Di Zenzo. A note on the gradient of a multi-image. *Computer Vision, Graphics, and Image Processing*, 33(1):116–125, 1986.
11. A. Cumani. Edge detection in multispectral images. *CVGIP: Graphical models and Image Processing*, 53(1):40–51, 1991.
12. W. Alshatti and P. Lambert. Using eigenvectors of a vector field for deriving a second directional derivative operator for color images. In *Proceedings of the 5th International Computer Analysis of Images and Patterns Conference (CAIP 93)*, volume 719 of *Lecture Notes in Computer Science*, pages 149–156, 1993.
13. J. S. Huang and D. H. Tseng. Statistical theory of edge detection. *Computer Vision, Graphics, and Image Processing*, 43(3):337–346, 1988.
14. M. Pietikainen and D. Harwood. Edge information in color images based on histograms of differences. In *Proceedings of The 8th International Conference on Pattern Recognition Conference (ICPR 86)*, volume 1, pages 594–596, 1986.
15. D. Landgrebe. *Signal theory methods in multispectral remote sensing.* John Wiley & Sons, 2003.
16. Pavel Paclík, R. P. W. Duin, G. M. P. van Kempen, and R. Kohlus. Segmentation of multi-spectral images using the combined classifier approach. *Image Vision Comput.*, 21(6):473–482, 2003.

Semi-supervised PCA-Based Face Recognition Using Self-training

Fabio Roli and Gian Luca Marcialis

Dept. of Electrical and Electronic Engineering, University of Cagliari
Piazza d'Armi, 09123 Cagliari, Italy
{roli, marcialis}@diee.unica.it

Abstract. Performances of face recognition systems based on principal component analysis can degrade quickly when input images exhibit substantial variations, due for example to changes in illumination or pose, compared to the templates collected during the enrolment stage. On the other hand, a lot of new unlabelled face images, which could be potentially used to update the templates and re-train the system, are made available during the system operation. In this paper a semi-supervised version, based on the self-training method, of the classical PCA-based face recognition algorithm is proposed to exploit unlabelled data for off-line updating of the eigenspace and the templates. Reported results show that the exploitation of unlabelled data by self-training can substantially improve the performances achieved with a small set of labelled training examples.

1 Introduction

Face recognition based on Principal Component Analysis (PCA) operates in two distinct stages: the enrolment stage and the recognition, or authentication, stage [1]. In the enrolment stage, a set of face images is acquired for each user. Then PCA is applied to the enrolled face images to compute the eigenspace associated to the selected eigenvalues. For each user, the enrolled images are projected to the eigenspace and a face template, often computed as the mean of the projected faces, is stored in a gallery. In order to account for variations in the appearance of a user, multiple templates, associated, for example, to different poses, can be stored in the user's gallery. In the recognition stage the input image is projected to the above eigenspace and the system associates the identity of the nearest template to this image. As Uludag et al. pointed out [2], the large intra-class variability of face images, due for example to changes in illumination or pose, can make the templates acquired during the enrolment stage poorly representative of the images to be recognized, so resulting in poor recognition performances. Increasing the size of the galleries of users' templates does not necessarily solve the problem, as the intra-class variability of face images is often due to aging, appearance, expression, and illumination changes which cannot be captured during a single enrolment stage over a short period of time [3]. On the other hand, a lot of new unlabelled face images are made available during the system operation over the time. These new data may be

D.-Y. Yeung et al. (Eds.): SSPR&SPR 2006, LNCS 4109, pp. 560–568, 2006.

exploited to update the templates and re-train the system. It is reasonable to hypothesize that exploiting unlabelled test data to update the eigenspace and the templates may improve the performances on new test data. Liu et al. showed that, in PCA-based face recognition, unlabelled test data can be exploited to update the eigenspace and improve the system's performance [3]. The potential benefits of the automatic, or semi-automatic, updating of the galleries of the users' templates using labelled and unlabelled data have been pointed out in [4-6].

The design of a face recognition system using a small set of labelled faces, collected during the initial enrolment stage, and a large batch of unlabelled face images, collected during the system operation, can be naturally regarded as a problem of semi-supervised learning. In fact, semi-supervised learning deals with the design of recognition systems using both labelled (possibly few) and unlabelled training examples [7]. However, the use of semi-supervised learning methods for face recognition has been poorly investigated so far, the only exception being the work of Balcan et al. [8].

The goal of this paper is to give a contribution to the development of semi-supervised face recognition systems. To this end, the use of a well known semi-supervised learning method, namely, the self-training method, is proposed to develop a semi-supervised version of the standard PCA-based face recognition algorithm. Reported experimental results show that the exploitation of a batch of unlabelled test images by self-training can substantially improve the performance of a PCA-based face recognition system on new test data in comparison with the ones achievable with a small set of labelled training examples.

2 Self-training for Semi-supervised PCA-Based Face Recognition

In this section, first the main steps of the classical supervised PCA-based face recognition method are summarized. Then a semi-supervised version of this method, based on the self-training technique, is proposed.

2.1 Supervised PCA-Based Face Recognition

Enrolment Stage

For each user, a set, usually small, of reference images is acquired. These images constitutes the training set $D_l=(\mathbf{x}_1, \mathbf{x}_2,\ldots, \mathbf{x}_{nl})$ containing n_l labelled data, where each vector $\mathbf{x}_i$, i=1... n_l, represents a face image. An identity label I_k, k=1... K, is associated to each vector. Then PCA is applied to the enrolled face images by the following steps:

Principal components computation

The covariance matrix is computed for the data set D_l. The eigenvalues and eigenvectors of the covariance matrix are computed. A set of eigenvalues with the highest variance is selected. The eigenvectors associated to the selected eigenvalues are named "eigenfaces" and define the matrix W of the "principal components".

Data transformation

The data in D_l are projected to the above eigenspace using the principal components matrix W: $\mathbf{y}_i=W^t(\mathbf{x}_i-\mu)$, i=1… n_l, where $\mu=\frac{1}{n_l}\sum_{i=1}^{n_l}x_i$ is the mean vector of data in D_l.

Template creation

For each user, a face template, often computed as the mean of the projected faces, is stored in a gallery.

Recognition Stage

In the so called "closed set" scenario, the input image is projected to the above eigenspace by the matrix W and the system associates the identity of the nearest template to this image.

2.2 Semi-supervised PCA-Based Face Recognition

Given a set D_l (usually, small) of labeled data, and a set D_u (usually, large) of unlabelled data, semi-supervised methods aim to design recognition systems using both sets. Several semi-supervised methods have been proposed so far, based on expectation-maximization algorithms, self-training, co-training, active learning, transductive learning, and graph-based techniques. We refer the reader to [7] for an overview on semi-supervised learning methods. For the purposes of this work, we summarize here the so called self-training method. In self-training a classifier is initially trained using the labeled data set D_l. This classifier is then used to assign pseudo-class labels to a subset of the unlabelled examples in D_u, and such pseudo-labeled data are added to D_l. Usually, the unlabelled data classified with the highest confidence are selected to increase D_l. Then the classifier is re-trained using the increased data set D_l. As the convergence of this simple algorithm can not be guaranteed in general, the last two steps are usually repeated for a given number of times or until some heuristic convergence criterion is satisfied.

Due to its easy use, we chose self-training as technique to develop a semi-supervised version of the classical PCA-based face recognition algorithm. The main steps of the algorithm developed are summarized in Figure 1. In the enrolment stage a set D_l of labelled images is collected and used to compute the matrix W of the principal components. Data are then projected to the eigenspace defined by W, and, for each user, a face template is computed as mean of the projected faces. During the on-line recognition stage an unlabelled batch of data D_u, to be used in the semi-supervised stage, is collected over a given period of time; to this end, recognition labels obtained as system's outputs are obviously disregarded. It should be noted that the designer should select a period of time that allows collecting a *representative* batch of data. In our experiments, the data set D_u contains the same identity classes, with the same priors, of the set D_l. The semi-supervised stage is then performed off-line. It is assumed that the recognition system carries out this stage either when it is not operating (e.g., during the night) or using a separate processing unit which allows carrying out it in parallel with the recognition stage. The semi-supervised stage starts by projecting the unlabelled data to the eigenspace defined by the matrix W computed in the enrolment stage. Then the semi-supervised cycle goes through *N* iterations. For each iteration, a pseudo-label is assigned to each data in D_u. The pseudo-label coincides with the label of the nearest template. For each identity class, a set P_k is defined which contains all the data pseudo-labelled as belonging to the class I_k. Then, for each identity class, the face image pseudo-labelled with the highest confidence, that is, the one nearest to the class template, is selected from the set P_k, and this image is added to the training set D_l and removed from the set D_u. Therefore, only one pseudo-labelled image per class, the most confident one, is added to the user's gallery during every iteration of the semi-supervision cycle. It is worth noting that the use of a less conservative confidence threshold, which allows adding more than one class example per iteration, could be investigated with the goal to speed up the learning. The increased training set is then used to update the eigenspace by re-computing the principal components matrix W. The labelled and unlabelled data are projected to the updated eigenspace. Finally, the class templates are updated using the augmented training set. In our algorithm the templates are simply the "mean" faces, but more sophisticated methods, based, for example, on clustering, could be used for template update [2]. In our experiments we performed ten iterations of the semi-supervised algorithm. Other stopping criteria could be investigated and used. For example, the iterations could be stopped when no unlabelled data can be added to the users' galleries. To sum up, after the initial enrolment stage, the system goes through on-line recognition phases, with collection of unlabelled data, and off-line semi-supervised phases to update the eigenspace and the templates. Performances on new test data are expected to improve after each semi-supervised stage in comparison with the ones achievable using the previous, non-updated, eigenspace and templates.

3 Experimental Results

3.1 Data Set and Goal of Experiments

The goal of our experiments was to evaluate the capability of the developed semi-supervised algorithm to exploit a batch of unlabelled test images, collected during a given session of the system operation, in order to improve the performance on novel test data in comparison with the ones achievable using the initial training data. To this end, we carried out experiments with the AR data set [9]. This data set contains frontal view faces with different facial expressions, illumination conditions, and occlusions (sun glasses and scarf). Each person participated in two acquisition sessions, separated by two weeks time. Each session is made up of seven images per person. We selected 100 subjects (50 males and 50 females), and manually cropped face images and, after histogram stretching and equalization, resized them at 40x40 pixels. Various subsets of first session images were used for the enrolment stage and the collection of unlabelled data set D_u. Second session images were always used as separate test set.

This experimental setting simulates well the acquisition of unlabelled data during a given session of system's operation, and the performance evaluation, after the semi-supervised stage, on new test data belonging to a separate acquisition session.

Off-line enrolment stage
A set D_l=($\mathbf{x}_1$, $\mathbf{x}_2$,…, $\mathbf{x}_{nl}$) containing n_l labelled images of the users is collected.
Principal components computation
The data set D_l is used to compute the matrix W of the principal components.
Data transformation
The data in D_l are projected to the eigenspace using the principal components matrix W: $\mathbf{y}_i=W^t(\mathbf{x}_i - \mu)$, i=1… n_l, where $\mu = \frac{1}{n_l}\sum_{i=1}^{n_l} x_i$.

Template creation
Set the vector $\mathbf{T}=(\mathbf{t}_1, \mathbf{t}_2,\ldots,\mathbf{t}_K)$ of the mean templates of the K identities.

On-line recognition stage
Input images are projected to the above eigenspace by the matrix W and the system associates the identity of the nearest template to each input image.
Creation of the unlabelled data set
A set D_u=($\mathbf{x}_1$, $\mathbf{x}_2$,…,$\mathbf{x}_{nu}$) containing n_u unlabeled data is collected over the time.

Off-line semi-supervised stage
Initialization – PCA transform of the unlabeled data in D_u: $\mathbf{z}_i=W^t(\mathbf{x}_i - \mu_l)$,
i= $n_{l+1}\ldots n_{l+u}$, where $\mu_l=\frac{1}{n_l}\sum_{i=1}^{n_l} x_i$.

Loop for *N* iterations:

Assign pseudo-labels to the unlabeled data in D_u
For each data in D_u compute the distances from the templates.
Assign the label of the nearest template to each data in D_u.
Assign all the data with pseudo-label I_k to the set P_k (k=1… K).

Increase the set of labelled data with pseudo-labelled data
For each identity class (k=1…K), select from the set P_k the pseudo-labelled data $\mathbf{z}_i$ nearest to the identity template $\mathbf{t}_k$
For each selected $\mathbf{z}_i$, increase the set D_l by setting $D_l = D_l + \mathbf{z}_i$, and remove $\mathbf{z}_i$ from D_u ($D_u = D_u - \mathbf{z}_i$).

Update of PCA transform using the augmented set D_l

PCA transform of data in D_l and D_u using the updated eigenface matrix W

Template updating
Update the vector $\mathbf{T}=(\mathbf{t}_1, \mathbf{t}_2,\ldots,\mathbf{t}_K)$ of the mean templates of the K identities using the augmented set D_l.

Fig. 1. The semi-supervised PCA-based algorithm using self-training

3.2 Results

We assessed performances for different numbers of templates created during the enrolment stage. Reported performances are averaged on five different trials. For each trial, we randomly selected one, two or three images of the first session as templates. The remaining images of the first session were used as unlabelled data set D_u. Second session images were always used as separate test set. Figure 2 shows the percentage accuracy on the test set and the unlabelled data set D_u averaged on five trials. For this experiment, only one face template per person was used. The remaining first-session images (six per person) were used as unlabelled data set D_u. Performances are shown as function of the number of unlabelled data added to the training set during the ten iterations of the semi-supervised stage. Around one-hundred pseudo-labeled data were added during every iteration. Figure 2 shows that the accuracy on second-session images used as test set is very low, around 18%, when unlabelled data are not used. The average accuracy increases substantially with the number of unlabelled data exploited by self-training. The maximum accuracy obtained for test data, around 62%, is anyway low due to the use of a single template per person and the large differences between first and second session images. But the increase of accuracy from 18% to 62% shows the benefits of the exploitation of unlabelled data. The accuracy is much higher for the unlabelled data, as the set D_u contains images of the first session which are more similar to the initial templates. It should be noted the practical interest of results obtained on the unlabelled data set. Unlabelled data are input data collected during a given period of time of the system's operation. Figure 2 shows that the initial accuracy on such batch of data is low, around 50%. After the semi-supervised phase based on self-training the accuracy increases to 89%. This results points out that the semi-supervision process can allow to improve the recognition results previously achieved on a batch of input data. For example, in a video surveillance scenario such a system re-training could allow improving the identification results stored in the data base the day before.

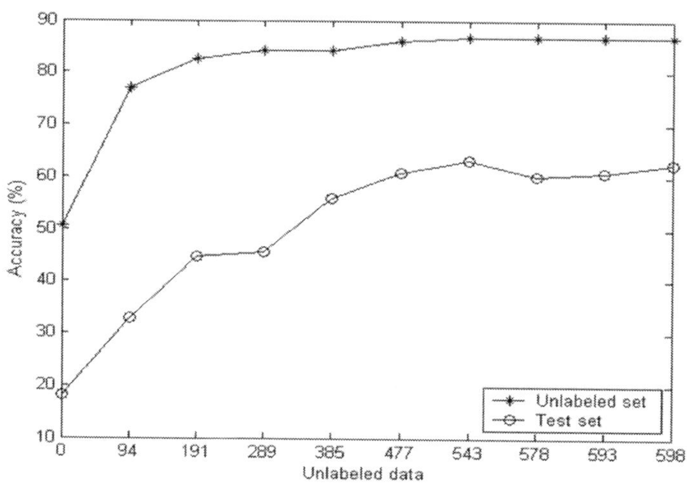

Fig. 2. Average accuracy on the test and unlabelled data sets as function of the number of unlabelled data used in the semi-supervised algorithm of Figure 1

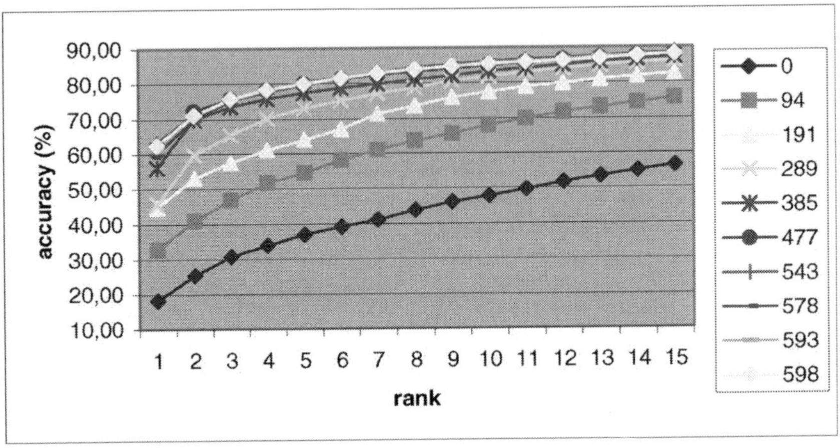

Fig. 3. Rank-order curves for the test data set. Each curve refers to an iteration of the semi-supervised cycle and is labeled with the number of unlabelled data which were pseudo-labeled and added to the training set.

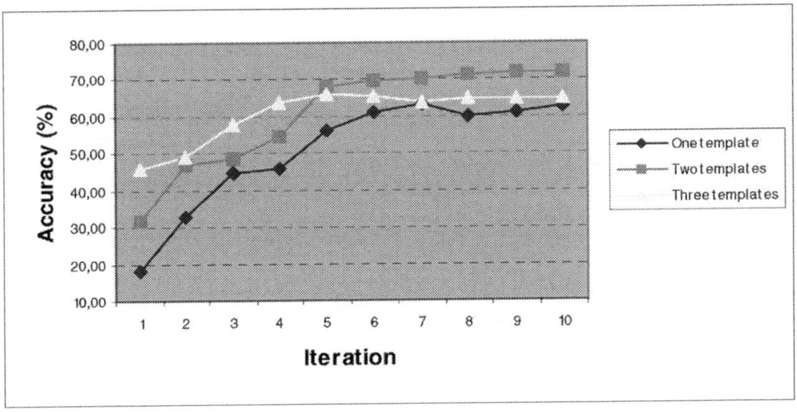

Fig. 4. Average accuracy on the test set as function of the number of iterations of the semi-supervised phase and the number of templates

We also assessed performances in terms of the so called rank-order curves, that is, we assessed the percentage accuracy, averaged on five trials, achieved by considering the fifteen template faces nearest to the input face (Figure 3). Figure 3 clearly shows the improvement of accuracy with the increase of the number of unlabelled data added to the training set during the ten iterations of the semi-supervised cycle.

To investigate how the performance of the semi-supervised algorithm depends on the number of templates collected during the enrolment stage, we performed experiments with one, two and three images of the first session as templates. Figure 4 depicts the three accuracy curves related to the different numbers of templates. Each curve provides the percentage average accuracy as function of the number of iterations of the semi-supervised phase. Around one hundred unlabelled data are added to the training set during every iteration. As one could expect, the greatest benefit of the use of unlabelled data is obtained when only one template per class is used.

Finally, we analyzed how the galleries of users' templates were updated and increased by our semi-supervised algorithm. Figure 5 depicts four examples of the update of users' galleries. For each gallery, the first image on the left is the initial training image used as face template. The remaining images are the unlabelled images which were pseudo-labeled and added to the gallery during the ten iterations of our semi-supervised algorithm. Due to the strict confidence threshold used in our algorithm (i.e., only one pseudo-labelled image per class, the most confident one, is added to the user's gallery during every iteration), images very similar to the template are initially added to the galleries. Then images which exhibit variations of expressions are added, and, in some cases, this causes wrong images to be added to the gallery. It is worth noting that different number of images can be added to different users' galleries.

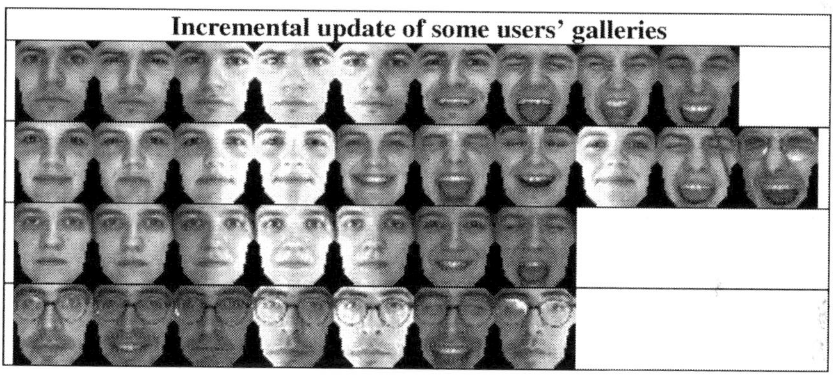

Fig. 5. Examples of the incremental update of users' galleries during the ten iterations of the semi-supervised algorithm. The first image on the left is the initial training image used as face template.

4 Conclusions

Performances of face recognition systems based on principal component analysis strongly depends on how much the face images collected during the enrolment stage are representative of face images to be recognized. Unfortunately, representative face images are difficult to be captured during a single enrolment stage over a short period of time. More representative images might be collected during the system operation over the time. The exploitation of such unlabelled data naturally demands for semi-supervised face recognition systems. Accordingly, we developed and assessed by experiments a semi-supervised version, based on self-training, of the classical PCA-based face recognition algorithm. Reported results show that the exploitation of a batch of unlabelled images by self-training can substantially improve the performance of a PCA-based face recognition system on new test data in comparison with the ones achievable with a small set of labelled training examples. Although final conclusions cannot be drawn on the basis of this work, we believe that a first step towards the development of semi-supervised face recognition systems has been done. As directions for our future work, the use of other semi-supervised learning methods will be investigated. In addition, as the good performances of the proposed system, based on a simple self-training mechanism, are, in a sense, surprising, the conditions under which it can be expected to work well will be analysed further by experiments and, if possible, theoretically.

References

[1] M. Turk and A. Pentland, Eigenfaces for Face Recognition, Journal of Cognitive Neuroscience, 3 (1) 71-86, 1991.

[2] U. Uludag, A. Ross and A. K. Jain, Biometric template selection and update: a case study in fingerprints, Pattern Recognition, Vol. 37, No. 7, pp. 1533-1542, July 2004.

[3] X. Liu, T. Chen, S.M. Thornton, Eigenspace updating for non-stationary process and its application to face recognition, Pattern Recognition, Special issue on Kernel and Subspace Methods for Computer Vision, 2003, 1945–1959.

[4] R. Sukthankar, R. Stockton, Argus: the digital doorman, IEEE Intelligent Systems, March/April 2001, pp. 14-19.

[5] K. Okada, C. von der Malsburg, Automatic video indexing with incremental gallery creation: integration of recognition and knowledge acquisition, Proceedings of ATR Symposium on Face and Object Recognition, pages 153-154, Kyoto, July 19-23, 1999.

[6] K. Okada, L. Kite, C. von der Malsburg, An adaptive person recognition system, Proceedings of the IEEE International Workshop on Robot-Human Interactive Communication, pages 436-441, Bordeaux/Paris, 2001.

[7] X. Zhu, Semi-supervised learning literature survey, Technical report, Computer Sciences TR 1530, Univ. Wisconsis, Madison, USA, Jan. 2006.

[8] M. F. Balcan, A. Blum, P.P. Choi, J. Lafferty, B. Pantano, M.R. Rwebangira, X. Zhu, Person Identification in Webcam Images: An Application of Semi-Supervised Learning, ICML2005 Workshop on Learning with Partially Classified Training Data, Bonn, Germany, 7 August 2005.

[9] A. Martinez, R. Benavente, The AR Face Database. CVC Technical Report #24, June, 1998.

Facial Shadow Removal

William A.P. Smith and Edwin R. Hancock

Department of Computer Science, University of York
{wsmith,erh}@cs.york.ac.uk

Abstract. In this paper we demonstrate how to recover surface shape from single images of faces using shape-from-shading when shadows are present. We make use of a statistical representation of the distribution of surface normal directions based on the equidistant azimuthal projection. This is allows us to develop a statistical model of the variations in facial shape in the surface normal domain. We show how ideas from robust statistics can be used to fit the model to facial images in which there is significant self-shadowing. The method is evaluated on both synthetic and real-world images. It is demonstrated to effectively fill-in the facial surface when more than 30% of the area is subject to self-shadowing.

1 Introduction

The problem of reconstructing the surface height function of a face from a single image is a challenging one that has eluded efforts in shape-from-shading for several decades [1,2]. The goal is to use the image irradiance equation to recover estimates of local surface orientation, and then to recover surface height by integrating the field of surface normals. Unfortunately, there are a number of well documented problems that frustrate this task [1]. The first of these is that the recovered field of surface normals is subject to concave-convex shape ambiguities, and this can lead to the implosion of facial features. This effect can lead to facial features becoming inverted (such as the nose), and the exageration of the relief of others (for instance the cheeks). The second problem is that of self shadowing. This occurs when the light source is at an oblique angle to the face, and the nose casts a shadow and the eye-sockets are also in shadow.

It is for these reasons that the problem of facial shape-from-shading is addressed either using domain specific constraints or using a face-specific shape-model. For instance Zhao and Chellappa [1] have shown how improved shape recovery can be achieved using a symmetric shape-from-shading method. Atick et al [3] have used a model based method constructed from range-data. Smith and Hancock [4] have developed a statistical model of surface orientation distribution trained on gradient data from range images. The model is fitted to intensity data using geometric constraints on the surface normal direction provided by Lambert's law. In related work, Vetter and Blanz have shown how recognition can be performed by fitting a surface model to brightness data, and how the fitted model can be used for realistic view synthesis [2]. Finally, it is

D.-Y. Yeung et al. (Eds.): SSPR&SPR 2006, LNCS 4109, pp. 569–577, 2006.

worth noting that if multiple images from a fixed viewing direction and variable light source direction are available, then photometric stereo can be used for accurate facial surface recovery [5].

Although the methods listed above lead to realistic shape recovery, they do not work well when there is significant self shadowing. Our aim in this paper is therefore to overcome the problem of self-shadowing by fitting a statistical model to image brightness data using robust statistics. The model has been described in our previous work. It captures variations in facial shape using ideas from cartography. Here the distribution of surface normal directions at different locations on the face is captured on a unit sphere. We convert this distribution of surface normal directions into a distribution of points using the azimuthal equidistant projection. The modes of facial variation are captured by applying principal components analysis to the point-distributions. We demonstrate how the statistical model may be fitted to image brightness data using robust statistics, so as to satisfy constraints provided by Lambert's law. Here the robust statistics treat the shadow regions as outliers. We use the difference between measured and fitted brightness values to compute a shadow-weight. The weights are used to exclude shadow regions in the shape-parameter estimation process. The parameter update scheme is based on M-estimators.

We experiment with the resulting shape-recovery method on both synthetic images with known ground truth and real-world images from the Yale-B database. The results indicate that the method works well even when the angle between the light source and the image-normal exceeds 60 degrees. Moreover, the fitting method is able to reliably fill-in shadowed regions of the face.

2 A Statistical Surface Normal Model

Constructing a statistical model that captures the statistical distribution of directional data is not a straightforward task. To overcome the problem, we draw on ideas from cartography. Our starting point is the *azimuthal equidistant* projection [6]. This projection has the important property that it preserves the distances between locations on the sphere. Another useful property of this projection is that straight lines on the projected plane through the centre of projection correspond to great circles on the sphere. We exploit these properties to generate a local representation of the field of surface normals. We commence with a set of needle-maps, i.e. fields of surface normals which in practice are obtained either from range images or shape-from-shading. We begin by computing the mean field of surface normals. The surface normals are represented using elevation and azimuth angles on a unit sphere. At each image location the mean-surface normal defines a reference direction which we use to construct an azimuthal equidistant projection for the distribution of surface normals at this point. The distribution of points on the projection plane preserves the distances of the surfaces normals on the unit sphere with respect to the mean surface normal, or reference direction. We then construct a deformable model over the set of surface normals by applying the Cootes and Taylor [7] point distribution model to the co-ordinates that result from transforming the surface normals from the unit sphere to the tangent plane under azimuthal equidistant projection. More details of this model are given in [4].

3 Fitting the Model to an Image

We may exploit the statistical constraint provided by the model in the process of fitting the model to an intensity image and thus help resolve the ambiguity in the shape-from-shading process. We do this using an iterative approach which is posed as that of recovering the best-fit field of normals from the statistical model, subject to constraints provided by the image irradiance equation.

If $I(i,j)$ is the measured image brightness at location (i,j), then $I(i,j) = \omega(i,j)\,[\mathbf{n}(i,j).\mathbf{s}]$ according to Lambert's law, where $\mathbf{s}$ is the light source direction and ω is the albedo. We begin by assuming constant and unit albedo (the Lambertian remapping process [8] normalises the brightest point to unity) and return to this later. In general, the surface normal $\mathbf{n}$ can not be recovered from a single brightness measurement since it has two degrees of freedom corresponding to the elevation and azimuth angles on the unit sphere. In the Worthington and Hancock [9] iterative shape-from-shading framework, data-closeness is ensured by constraining the recovered surface normal to lie on the reflectance cone whose axis is aligned with the light-source vector $\mathbf{s}$ and whose opening angle is $\alpha = \arccos I$. At each iteration the surface normal is free to move to an off-cone position subject to smoothness or curvature consistency constraints. However, the hard irradiance constraint is re-imposed by rotating each surface normal back to its closest on-cone position. This process ensures that the recovered field of normals satisfies the image irradiance equation after every iteration. The framework is initialised by placing the surface normals on their reflectance cones in the direction opposite to that of the local image gradient.

Our approach to fitting the model to intensity images uses the fields of surface normals estimated using the geometric shape-from-shading method described above. This is an iterative process in which we interleave the process of fitting the statistical model to the current field of estimated surface normals, and then re-enforcing the data-closeness constraint provided by Lambert's law by mapping the surface normals back onto their reflectance cones. If the data is of dimensions $M \times N$, the surface normal model is encapsulated in the $2MN \times K$ matrix $\mathbf{P} = (\mathbf{e}_1|\mathbf{e}_2|\ldots|\mathbf{e}_K)$ formed from the leading K principal eigenvectors of the covariance matrix of the training samples under azimuthal equidistant projection. In other words, they are the modes of variation of the statistical surface normal model. The fitting algorithm can therefore be summarised as follows:

1. Initialise the field of surface normals $\mathbf{n}$.
2. Each normal in the estimated field $\mathbf{n}$ undergoes an azimuthal equidistant projection to give a vector of transformed coordinates $\mathbf{v}$.
3. The vector of best fit model parameters is $\mathbf{b} = \mathbf{P}^T\mathbf{v}$.
4. The corresponding vector of transformed coordinates is $\mathbf{v}' = \mathbf{P}\mathbf{P}^T\mathbf{v}$.

5. Using the inverse azimuthal equidistant projection find $\mathbf{n}'$ from $\mathbf{v}'$.
6. Find $\mathbf{n}''$ by rotating each normal in $\mathbf{n}'$ back to their closest on-cone position.
7. Stop if the difference between $\mathbf{n}$ and $\mathbf{n}''$ indicates convergence.
8. Make $\mathbf{n} = \mathbf{n}''$ and return to step 2.

The method hence combines a strict global constraint (projection onto the statistical model) with a hard local constraint (satisfaction of the image irradiance equation). We find the algorithm converges rapidly and offers stable performance on real world images. However, a number of obstacles are encountered when this simple approach is applied to real images. The simple reflectance model given Lambert's law assumes constant albedo and ignores the effect of cast shadows (regions in which the light source is intercepted by another part of the surface). The result is that the fit to the statistical model is subject to a systematic error and becomes increasingly inaccurate when regions of low albedo dominate (for instance in the presence of facial hair) or when cast shadows become significant (as the light source direction is more extreme). In this case, a significant portion of the face may be in shadow and fitting the statistical model globally results in erroneous shape parameter estimates.

4 Robust Statistics

It is therefore clear that we require a more robust means to fit our statistical model to a potentially noisy observed field of normals, $\mathbf{n}$. The quality of this fit can be measured by calculating the distance between the observed and fitted normals on the projection plane. If $\mathbf{b}$ is the estimated parameter vector, the vector of residuals is given by $\mathbf{R} = \|\mathbf{v} - (\mathbf{Pb})\|$ and the residual at point p is $\eta_p = \sqrt{R(2p-1)^2 + R(2p)^2}$. The standard least squares fit given above, minimises the quantity:

$$\mathbf{b}^* = \arg\min_{\mathbf{b}} \sum_{i=1}^{N} \eta_i^2.$$

This approach is unstable in the presence of outlying data, such as normals erroneously estimated from regions of low albedo or in cast shadow. In particular, the effect of outliers is to severelly distort the estimated facial shape.

In this paper, we turn to the apparatus of robust statistics to help overcome this problem. M-estimators (maximum likelihood type estimators) aim to reduce the effect of outliers by replacing the squared residuals η_i^2 by a kernel function that limits the effects large residuals:

$$\mathbf{b}^* = \arg\min_{\mathbf{b}} \sum_{i=1}^{N} \rho_\sigma(\eta_i) \tag{1}$$

where ρ is a robust kernel with width parameter σ.

The influence of a residual on the parameter estimate under a given M-estimator can be studied by examining its *influence function*, ψ_σ. This is the

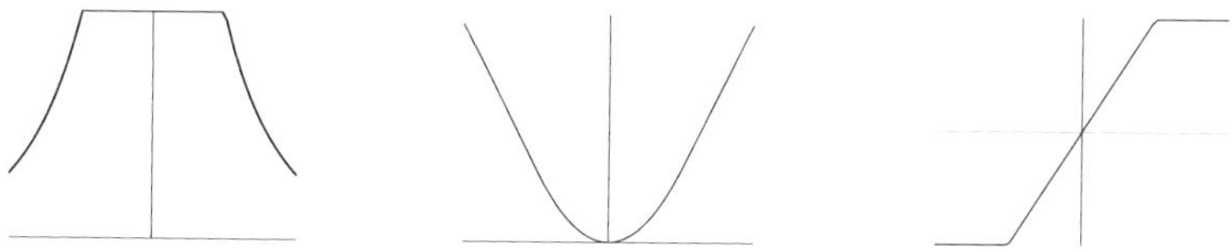

$$w_\sigma(\eta) = \begin{cases} 1 \\ \frac{\sigma}{|\eta|} \end{cases} \qquad \rho_\sigma(\eta) = \begin{cases} \eta^2 \\ 2\sigma|\eta| - \sigma^2 \end{cases} \qquad \psi_\sigma(\eta) = \begin{cases} 2\eta \\ 2\sigma\text{Sign}(\eta) \end{cases}$$

Fig. 1. Huber's M-estimator

derivative of the error kernel: $\psi_\sigma(\eta_p) = \frac{\partial \rho_\sigma(\eta_p)}{\partial \eta_p}$. Hence, in the least squares case where $\rho_\sigma(\eta) = \eta^2$, the influence of a datum is $\psi_\sigma(\eta) = 2\eta$ and therefore increases linearly with the size the error. This is the source of the lack of robustness in least-squares estimation.

We propose a robust solution to (1) using a simple one-step weighted least squares approximation. To do so, we make use of the *weight function*, w_σ, which is related to the influence function by: $w_\sigma(\eta_p) = \frac{\psi_\sigma(\eta_p)}{\eta_p}$. The standard least-squares estimator applies a constant weight to each datum. On the other hand, an error kernel such as Huber's estimator [10] down-weights a datum once its residual exceeds σ:

$$\rho_\sigma(\eta) = \begin{cases} \eta^2 & \text{if } |\eta| < \sigma \\ 2\sigma|\eta| - \sigma^2 & \text{otherwise} \end{cases} \qquad w_\sigma(\eta) = \begin{cases} 1 & \text{if } |\eta| < \sigma \\ \frac{\sigma}{|\eta|} & \text{otherwise} \end{cases} \tag{2}$$

We show the weight function, error kernel and influence function for Huber's M-estimator in Fig 1. This is the M-estimator we use in the remainder of this paper.

We can incorporate the Huber weights into the least squares fit by constructing a diagonal matrix of weights: $\mathbf{W} = \text{diag}(w_\sigma(\eta_1), \ldots, w_\sigma(\eta_N))$. Our one-step weighted least squares approximation of $\mathbf{b}$ is given by:

$$\mathbf{b}^{(t)} = C\mathbf{P}^T\mathbf{W}\mathbf{v}^{(t)} \tag{3}$$

where C is a constant which compensates for the overall scaling effect of $\mathbf{W}$ on $\mathbf{b}$: $C = N\text{Tr}(\mathbf{W}^{-1})$.

Computing (3) requires an initial estimate of $\mathbf{W}$ and hence the residuals. We therefore commence by calculating a least squares fit from which we can calculate the residuals η_p. This is equivalent to calculating (3) with an identity weight matrix, i.e. $\mathbf{W} = \mathbf{I}_N$. For a subsequent iteration t of the algorithm, we can use the weights calculated from the residuals at iteration $(t-1)$.

Implicit in the discussion above is that we have a means to estimate the standard deviation of the residual errors σ, which acts as the width parameter of the function ρ. A robust estimate of σ is required in order to distinguish outliers from inliers. For this reason, we use the *median absolute deviation* (MAD):

$$\text{MAD} = \text{median}\left(|\eta_p - \text{median}(\eta_p)|\right), p = 1 \ldots N$$

which is related to the standard deviation by $\sigma = 1.4826 \times \text{MAD}$.

4.1 Combining and Classifying

Upon convergence, if a pixel p has weight $w_p^{final} \approx 1$, this indicates a high confidence that the on-cone normal $\mathbf{n}_p^{''}$ is reliable. However, as the weight tends to 0, the on-cone normal $\mathbf{n}_p^{''}$ is likely to be erroneous due to violation of the assumptions of Lambert's law, e.g. non-constant albedo or lying in a cast shadow region. In this case, a more accurate estimate of $\mathbf{n}_p^{final}$ is given by the robust fit of the model to the global field of normals.

For this reason, our best estimate of the underlying shape of the face is a weighted combination, in which pixels with a low weight are given a higher proportion of the normal from the model fit $\mathbf{n}_p^{'}$ and a lower proportion of the on-cone normal $\mathbf{n}_p^{''}$, vice-versa for pixels with a high weight. This gives: $\mathbf{n}_p^{combined} = \frac{\overline{\mathbf{n}}_p}{\|\overline{\mathbf{n}}_p\|}$, where $\overline{\mathbf{n}}_p = w_p^{final}\mathbf{n}_p^{''} + \left(1 - w_p^{final}\right)\mathbf{n}_p^{'}$.

With the estimated facial shape to hand, we may now go further and distinguish between pixels of low albedo and those in cast shadow regions. We may recover the surface height z_p by applying a standard surface integration method [11] on the field of normals $\mathbf{n}^{combined}$. Using a simple ray-tracing algorithm, we can assign a binary cast-shadow map:

$$\text{shadow}(z_p, \mathbf{s}) = \begin{cases} 0 & \text{if pixel } p \text{ is in cast shadow} \\ 1 & \text{otherwise} \end{cases} \tag{4}$$

For non-shadow regions, the albedo ω_p can be estimated by rearranging the image irradiance equation: $\omega_p = \frac{I_p}{\mathbf{n}_p^{combined}.\mathbf{s}}$.

5 Experimental Results

In this section we present experiments using the technique described above. We begin by applying the method to known ground truth data allowing us to quantitatively asses the performance of the approach. We then apply the method to real world images, demonstrating the robustness of the approach under real world conditions.

We train our statistical model on a sample of 100 facial needle-maps. The data is acquired from the 3DFS dataset [12] which consists of 100 high resolution scans of subjects in a neutral expression. The scans were collected using a *Cyberware*™ 3030PS laser scanner. The database is pre-aligned, registration being performed using the optical flow correspondence algorithm of Blanz and Vetter [2]. For ground truth, we use a leave-one-out strategy in which we train the model with 99 sets of data, leaving the remaining needle-map as out-of-sample ground truth.

For real world images, we show reconstructions and reilluminations of images from the Yale-B database [5]. These contain albedo variation and cast shadows.

5.1 Ground Truth Data

In Fig. 2 we demonstrate the performance of our method on ground truth data. We apply our algorithm to a selection of images of rendered ground truth

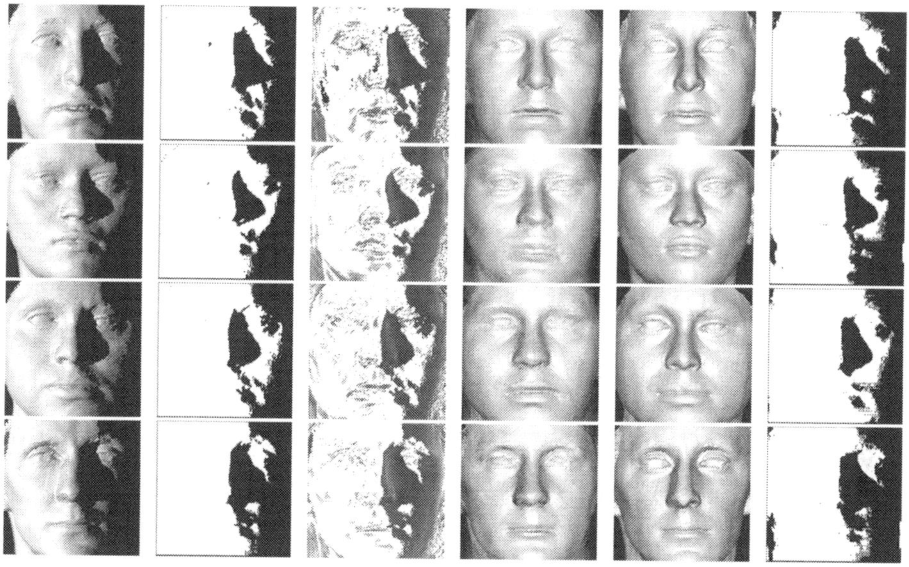

Fig. 2. Fitting to images of ground truth needle-maps rendered with Lambertian reflectance and cast shadows

needle-maps including cast shadows. In column 1 we show the input images. The needle-maps of the out-of-sample subjects are rendered with Lambertian reflectance and a point light source with direction $\mathbf{s} = (-1, 0, 1)$, i.e. 45° from the viewing direction. We also simulate the effect of cast shadows using the shadow map shown in column 2. shadow$(z, \mathbf{s})$ is calculated from ground truth depth data. In column 3 we show the weight function $w_\sigma(\eta_p)$ for each pixel. It is clear that regions in cast shadow have been successfully down-weighted. In column 4 we show the needle-map $\mathbf{n}^{combined}$ calculated from the input image, rendered with frontal illumination. For comparison, in column 5 we show the ground truth needle-map similarly illuminated. There is a good agreement between the two, even in areas in which no information was present in the input image (i.e. those in cast shadows). This suggests that the robust fit of the model has recovered globally accurate shape information, and has filled-in the shadowed areas of the face. The mean surface normal error was typically $< 8°$ across the whole needle-map. Finally in column 6 we show the shadow map shadow$(z^{combined}, \mathbf{s})$, where $z^{combined}$ is the height map integrated from $\mathbf{n}^{combined}$. Again, there is a good agreement between columns 2 and 6, suggesting that this represents a viable means to estimate regions which are in cast shadow.

5.2 Real World Data

In Fig. 3, we demonstrate the quality of the shape information our method can recover from real world images. From the input images in the top row, we use our method to estimate the needle-map $\mathbf{n}^{combined}$ and use the surface recovery

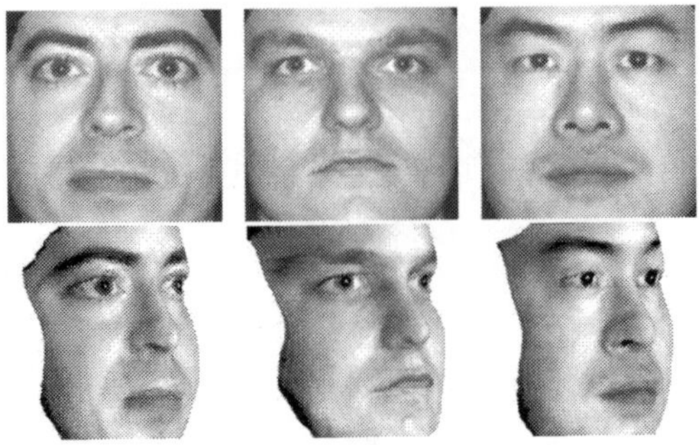

Fig. 3. Novel viewpoint of the surface recovered from the input image in the top row rotated 45° about the vertical axis

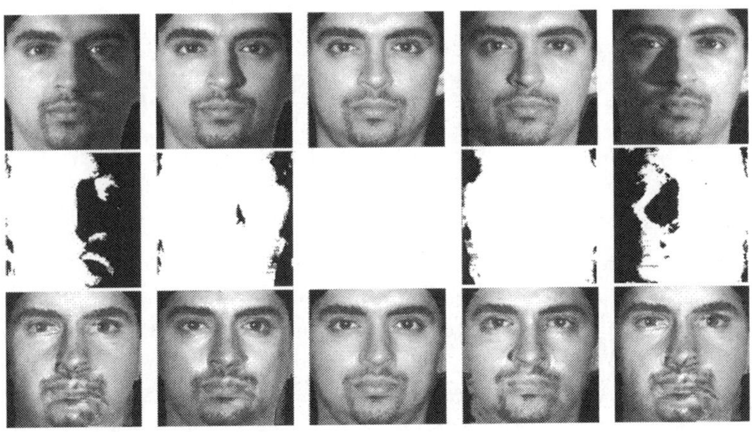

Fig. 4. Novel viewpoint of the surface recovered from the input image in the top row rotated 45° about the vertical axis

method of Frankot and Chellappa [11] to integrate the normals into a surface. We show these surfaces rendered with the estimated albedo and Lambertian reflectance, rotated 45° about the vertical axis. The images show considerable stability under large change in viewpoint.

In Fig. 4, we demonstrate the method on real images which contain cast shadows as well as albedo variations. The first row shows the input images of a single subject under varying illumination. The subject is a challenging choice due to the large albedo variations caused by facial hair. The light source is moved in an arc along the horizontal axis to subtend an angle of −50°, −25°, 0°, 25° and 50° with the viewing direction. We use our method to estimate the normals, albedo and shadow map. We use facial symmetry to fill-in the missing albedo values for the shadow regions. In the second row we show the estimated cast shadow map. Here, the cast shadows caused by the nose seem to correspond well with the input images. Finally in the third row, we show the recovered needle-maps rendered with the estimated albedo and frontal lighting, effectively correcting for variation in input lighting. These synthesised images are of a good quality, even under large changes in illumination and manage to remove much of the effect of the cast shadows.

6 Conclusions

We have shown how a statistical model of facial shape, couched in terms of distributions of surface normal directions, can be fitted to images of shadowed faces using robust statistics. We fit the statistical model globally, but use robust statistics to ensure that regions of low albedo or which fall into a cast shadow have little or no impact on the parameter estimate. The technique is capable of recovering a useful estimate of facial shape, even when significant portions of the face are entirely in shadow.

References

1. Zhao, W.Y., Chellappa, R.: Illumination-insensitive face recognition using symmetric SFS. In: Proc. CVPR. (2000) 286–293
2. Blanz, V., Vetter, T.: Face recognition based on fitting a 3D morphable model. IEEE Trans. PAMI **25** (2003) 1063–1074
3. Atick, J.J., Griffin, P.A., Redlich, A.N.: Statistical approach to SFS: Reconstruction of 3D face surfaces from single 2D images. Neural Comp. **8** (1996) 1321–1340
4. Smith, W., Hancock, E.R.: Recovering facial shape and albedo using a statistical model of surface normal direction. In: Proc. ICCV. (2005) 588–595
5. Georghiades, A., Belhumeur, P., Kriegman, D.: From few to many: Illumination cone models for face recognition under variable lighting and pose. IEEE Trans. PAMI **23** (2001) 643–660
6. Snyder, J.P.: Map Projections–A Working Manual, U.S.G.S. Professional Paper 1395. United States Government Printing Office, Washington D.C. (1987)
7. Cootes, T.F., J.Taylor, C., Cooper, D., Graham, J.: Training models of shape from sets of examples. In: Proc. BMVC. (1992) 9–18
8. Smith, W., Robles-Kelly, A., Hancock, E.R.: Reflectance correction for perspiring faces. In: Proc. ICIP. (2004) 1389–1392
9. Worthington, P.L., Hancock, E.R.: New constraints on data-closeness and needle map consistency for shape-from-shading. IEEE Trans. PAMI **21** (1999) 1250–1267
10. Huber, P.: Robust Statistics. Wiley, Chichester (1981)
11. Frankot, R.T., Chellappa, R.: A method for enforcing integrability in shape from shading algorithms. IEEE Trans. PAMI **10** (1988) 439–451
12. USF HumanID 3D Face Database, Courtesy of Sudeep. Sarkar, University of South Florida, Tampa, FL.

A Sequential Monte Carlo Method for Bayesian Face Recognition

Atsushi Matsui[1,2], Simon Clippingdale[1], and Takashi Matsumoto[2]

[1] Science & Technical Research Laboratories, NHK (Japan Broadcasting Corporation), 1-10-11 Kinuta, Setagaya-ku, Tokyo 157-8510 Japan
[2] Dept. of Electrical Engineering & Bioscience, Waseda University, 3-4-1 Okubo, Shinjuku-ku, Tokyo 169-8555 Japan

Abstract. This paper proposes a Sequential Monte Carlo (SMC) learning algorithm for Bayesian probability distributions that describe model parameters in a video face recognition system based on deformable template matching. The new algorithm achieves significantly improved robustness of recognition against facial expressions and speech movements by comparison with a baseline batch MCMC (Markov Chain Monte Carlo) algorithm, at no additional computational cost. Experimental results demonstrate the effectiveness and computational efficiency of the new algorithm.

1 Introduction

Human faces in broadcast video exhibit substantial variation in position, size, head pose, facial expression and so on, forcing face recognition systems for video indexing to incorporate flexibility in the database and/or matching algorithms used. The authors have introduced a prototype recognition system[1][2] which uses deformable template matching and is based on the Elastic Graph Matching method[3]. Although this sytem can absorb a certain amount of facial deformation due to expressions and speech movements, recognition errors can occur for larger deformations, and additionally there are a number of system hyperparameters which are set in a heuristic fashion.

In this work, we introduce an online learning algorithm for Bayesian posterior probabilities describing faces in video input sequences, which uses a Sequential Monte Carlo (SMC) method [4][5] to perform integrations over a sequence of combined spaces of face model parameters and system hyperparameters. We show that this SMC approach successfully adapts the parameters associated with deformations of each face model, and significantly reduces recognition errors on a video test set showing individuals talking, relative to a baseline batch MCMC (Markov Chain Monte Carlo) algorithm[6][7]. However, it does so at increased computational cost. We then introduce a modification at the resampling stage of the algorithm that restores computational efficiency (to somewhat better than

D.-Y. Yeung et al. (Eds.): SSPR&SPR 2006, LNCS 4109, pp. 578–586, 2006.

that of the baseline MCMC algorithm) without sacrificing the gain in recognition performance achieved by the SMC algorithm.

In section 2 we briefly review the deformable template matching procedure and similarity function used in our original system. In section 3 we introduce the online learning approach where a new likelihood function is proposed, defined in terms of a mixture of von Mises-Fisher distributions, and show how the most probable model can be estimated together with distributions of system parameters. In section 4 we describe the details of the SMC algorithm with experimental results, and introduce the modification of the resampling stage that boosts computational efficiency. The paper concludes with a discussion of the results and possible directions for further work.

2 Deformable Template Matching

The deformable templates used in our original system[1][2] are constructed from face images of target individuals at multiple poses, labeled with feature point positions. Each template consists of the normalized coordinates of $M = 9$ feature points, $\boldsymbol{x}^A = \{\boldsymbol{x}_1^A, ..., \boldsymbol{x}_M^A\}$, together with features $\boldsymbol{c}^A$ computed by convolutions with Gabor wavelets at each of the feature points. The Gabor wavelet at resolution r and orientation n is a complex exponential grating patch with a 2-D Gaussian envelope:

$$g_n^r(\boldsymbol{x}) = \frac{k_r^2}{\sigma^2} e^{-\frac{k_r^2 \|\boldsymbol{x}\|^2}{2\sigma^2}} \times \left[e^{i(\boldsymbol{k}_n^r)^T \boldsymbol{x}} - e^{-\frac{\sigma^2}{2}} \right], \quad \boldsymbol{k}_n^r = k_r \begin{pmatrix} \cos(\frac{n\pi}{N_{orn}}) \\ \sin(\frac{n\pi}{N_{orn}}) \end{pmatrix}, \tag{1}$$

for $N_{orn} = 8$ orientations and $R = 5$ resolutions. This data representation is similar to that used in the Elastic Graph Matching system[3] for face recognition in static images, but the chosen feature points differ, as do the parameters of the Gabor wavelets. The original scheme[1][2] applies templates to input video frames and deforms them by shifting the feature points so as to maximize the similarity to the Gabor features in the template. It then computes an overall match score for each deformed template, incorporating a feature similarity term and a penalty related to the deformation as follows:

$$S_{A,B}^r = 1 - \alpha_f \left(1 - \frac{\langle \boldsymbol{c}_r^A, \boldsymbol{c}_r^B \rangle}{|\boldsymbol{c}_r^A||\boldsymbol{c}_r^B|} \right) - \alpha_s \frac{\sqrt{E_{A,B}}}{\lambda_r}, \tag{2}$$

where A denotes the undeformed template and B the deformed feature points on the image; $\boldsymbol{c}^A$ and $\boldsymbol{c}^B$ are feature vectors of Gabor wavelet coefficients, respectively from the template and measured at the deformed feature point positions $\boldsymbol{x}^B$ on the image; $E_{A,B}$ is the deformation energy between the feature points $\boldsymbol{x}^A$ in the template and the deformed feature points $\boldsymbol{x}^B$ on the image, up to a dilation, rotation and shift; α_f and α_s are weights for the feature similarity and spatial deformation terms respectively; and $\lambda_r = 2\pi/k_r$ is the modulation wavelength of the Gabor wavelet at resolution r. In the sequel we will often omit the A and B superscripts where the meaning is clear.

3 Online Bayesian Learning

Optimizing a target function with penalty terms can be considered as maximizing the Bayesian posterior probabilities of parameters[9]. From the same viewpoint, the most probable model describing faces in input video is defined by the mode of posterior distributions of face models. Though the optimum set of parameters in some case may yield satisfactory performance in other cases, problems can arise if the target probability distribution takes on a more complex form. In general, finding the global maximum of a target function is difficult, and prone to falling into local maxima. Moreover, the optimal parameter set obviously depends on *unknown* input data.

In this paper, instead of searching for the mode, we introduce an online learning algorithm to estimate both the peak and tails of probability distributions of parameters.

3.1 Likelihood Function

Consider the situation where data is given as a video sequence. Let y_n be image data at the nth frame and let $y_{1:n} = \{y_1, y_2 .., y_n\}$ be the image data set up to the current frame. Recall that the feature similarity term in equation (2) depends on an innner product between two normalized feature vectors. Therefore it is natural to consider as a likelihood function for this directional data a mixture of von Mises-Fisher distributions[8]:

$$P(y_n|\boldsymbol{x}_n, \beta_{n,1:R}, \mathcal{H}_j) = \frac{1}{R}\sum_{r=1}^{R} \frac{1}{Z_b(\beta_{n,r})} \exp\left(\beta_{n,r} \frac{\langle \boldsymbol{c}_r^A, \boldsymbol{c}_r^B \rangle}{|\boldsymbol{c}_{n,r}^A||\boldsymbol{c}_{n,r}^B|}\right), \tag{3}$$

$$Z_b(\beta) = \frac{(2\pi)^{\frac{k}{2}} I_{\frac{k}{2}-1}(\beta)}{\beta^{\frac{k}{2}-1}}. \tag{4}$$

where $\boldsymbol{x}_n$ represents a set of feature points at the nth frame, $\beta_{n,r}$ is a hyperparameter for resolution r, and $\mathcal{H}_j$ is a face model (hypothesis or template) with identity number j. $I_p(\beta)$ is the modified Bessel function, and $k = 2M \times N_{orn}$.

3.2 Parameter/Hyperparameter Dynamics

Suppose that we are provided with a set of feature point locations for the jth template, $\boldsymbol{x}_j^A$. We assume a Gaussian predictive distribution for $\boldsymbol{x}_n$:

$$P(\boldsymbol{x}_n|\alpha_n, T_n, \mathcal{H}_j) = \frac{1}{Z_a(\alpha_n)} \exp\left(-\frac{\alpha_n}{2}(T_n^{-1}(\boldsymbol{x}_n) - \boldsymbol{x}_j^A)^T \Lambda_j^{-1}(T_n^{-1}(\boldsymbol{x}_n) - \boldsymbol{x}_j^A)\right), \tag{5}$$

where α_n is a hyperparameter, Λ_j is the covariance matrix of feature point positions for face model $\mathcal{H}_j$, and $Z_a(\alpha_n) = (2\pi)^M \sqrt{\det \Lambda_j/\alpha_n}$ is a normalizing factor.

T_n is a rigid linear transformation of the feature point set, consisting of a dilation by a factor r_n, rotation through an angle θ_n, and translation $(u_n, v_n)^T$.

It expresses the rigid component of the mapping from the template feature point set $\boldsymbol{x}^A$ onto the deformed feature point set $\boldsymbol{x}^B$ on the input image plane, leaving the nonrigid deformation.

$$T_n(\boldsymbol{x}^A) = \begin{bmatrix} r_n \cos\theta_n & -r_n \sin\theta_n & 0 & \dots & & 0 \\ r_n \sin\theta_n & r_n \cos\theta_n & & & & \vdots \\ 0 & & \ddots & & & 0 \\ \vdots & & & & r_n \cos\theta_n & -r_n \sin\theta_n \\ 0 & & \dots & 0 & r_n \sin\theta_n & r_n \cos\theta_n \end{bmatrix} \begin{bmatrix} x_1^A \\ y_1^A \\ \vdots \\ x_M^A \\ y_M^A \end{bmatrix} + \begin{bmatrix} u_n \\ v_n \\ \vdots \\ u_n \\ v_n \end{bmatrix}. \quad (6)$$

For simplicity we will use the symbolic notation T_n to denote the set of mapping parameters $(r_n, \theta_n, u_n, v_n)$, and Θ_n to denote the set of model parameters $(\boldsymbol{x}_n, \alpha_n, \beta_{n,1:R})$. The parameters of T_n determine the size, position, and in-plane rotation angle of a face region. In this paper, we describe a sequential learning algorithm to estimate probability distributions of these parameters given a sequence of input images.

One way of performing online learning is to consider stochastic updates of the parameters in question. Assuming smooth motion of the target face region, we consider as a recursive update $P(T_n|T_{n-1})$ described by

$$\begin{aligned} r_n &= r_{n-1} + \nu_r, & \nu_r &\sim \mathcal{N}(0, \sigma_r^2), \\ \theta_n &= \theta_{n-1} + \nu_\theta, & \nu_\theta &\sim \mathcal{N}(0, \sigma_\theta^2), \\ u_n &= u_{n-1} + \nu_x, & \nu_x &\sim \mathcal{N}(0, \sigma_x^2), \\ v_n &= v_{n-1} + \nu_y, & \nu_y &\sim \mathcal{N}(0, \sigma_y^2). \end{aligned} \quad (7)$$

Similarly, we adopt a transition model for $P(\alpha_n, \beta_{n,1:R}|\alpha_{n-1}, \beta_{n-1,1:R})$:

$$\begin{aligned} \log\alpha_n &= \log\alpha_{n-1} + \nu_\alpha, & \nu_\alpha &\sim \mathcal{N}(0, \sigma_\alpha^2), \\ \log\beta_{n,r} &= \log\beta_{n-1,r} + \nu_\beta, & \nu_\beta &\sim \mathcal{N}(0, \sigma_\beta^2), \end{aligned} \quad (8)$$

where the log-normal distribution guarantees positivity of the hyperparameters.

3.3 Posterior Distribution

Using Bayes' theorem we obtain straightforwardly a recursive formula for $P(\mathcal{H}_j|y_{1:n})$, the posterior distribution of the model $\mathcal{H}_j$ given the data:

$$P(\mathcal{H}_j|y_{1:n}) = \frac{P(y_n|y_{1:n-1}, \mathcal{H}_j)P(\mathcal{H}_j|y_{1:n-1})}{P(y_n|y_{1:n-1})}, \quad (9)$$

where the one-step model marginal likelihood $P(y_n|y_{1:n-1}, \mathcal{H}_j)$ is given by the following integrals:

$$P(y_n|y_{1:n-1}, \mathcal{H}_j) = \int P(y_n|\Theta_n, \mathcal{H}_j)P(\Theta_n, T_n|y_{1:n-1}, \mathcal{H}_j)d(\Theta_n, T_n), \quad (10)$$

$$P(\Theta_n, T_n \mid y_{1:n-1}, \mathcal{H}_j) = \int P(\Theta_n, T_n|\Theta_{n-1}, T_{n-1}, \mathcal{H}_j) \\ P(\Theta_{n-1}, T_{n-1}|y_{1:n-1}, \mathcal{H}_j) d(\Theta_{n-1}, T_{n-1}). \quad (11)$$

At each frame, the system outputs the most probable face model, $\mathcal{H}_{MP}^{(n)}$, which attains the maximum value of the posterior distribution of the model given the sequence of input images: $\mathcal{H}_{MP}^{(n)} = \arg\max_j P(\mathcal{H}_j|y_{1:n})$.

4 Sequential Monte Carlo Algorithm

4.1 Sequential Importance Sampling

We cannot typically compute the recursive posterior distribution analytically, because it requires evaluation of the complex high-dimensional integrals in (10) and (11). Instead we apply a Monte Carlo method to estimate the integral numerically. Sequential Monte Carlo requires proposal distributions from which one can draw samples $\{\Theta_n^{(i)}, T_n^{(i)}\}_{i=1}^{N_j}$ for each model $\mathcal{H}_j$ by standard methods. The proposal distribution in this paper will be given by

$$\pi(\Theta_n, T_n|\mathcal{H}_j) \\ = P(\boldsymbol{x}_n|\alpha_n, T_n, \mathcal{H}_j)P(\alpha_n|\alpha_{n-1}, \sigma_\alpha)P(\beta_{n,1:R}|\beta_{n-1,1:R}, \sigma_\beta)P(T_n|T_{n-1}) \quad (12)$$

from which one obtains the following approximations:

$$P(\Theta_n, T_n|y_{1:n-1}, \mathcal{H}_j) \cong \sum_{i=1}^{N_j} \tilde{w}_{n-1}^{(i)}|_{\mathcal{H}_j} \times \delta\left(||(\Theta_n, T_n) - (\Theta_n^{(i)}, T_n^{(i)})||\right), \quad (13)$$

$$P(y_n|y_{1:n-1}, \mathcal{H}_j) = \int P(y_n|\Theta_n, \mathcal{H}_j)P(\Theta_n, T_n|y_{1:n-1}, \mathcal{H}_j)d(\Theta_n, T_n) \\ \cong \sum_{i=1}^{N_j} P(y_n|\Theta_n^{(i)}, \mathcal{H}_j) \times \tilde{w}_{n-1}^{(i)}|_{\mathcal{H}_j}, \quad (14)$$

where the *normalized importance weights* $\tilde{w}_n^{(i)}$ are equal to:

$$\tilde{w}_n^{(i)}|_{\mathcal{H}_j} = \frac{w_n^{(i)}|_{\mathcal{H}_j}}{\sum_{k=1}^{N_{persons}} \sum_m^{N_k} w_n^{(m)}|_{\mathcal{H}_k}}, \quad (15)$$

$$w_n^{(i)}|_{\mathcal{H}_j} = \frac{P(y_n|\Theta_n^{(i)}, T_n^{(i)}, y_{1:n-1}, \mathcal{H}_j)P(\Theta_n^{(i)}, T_n^{(i)}|\Theta_{n-1}^{(i)}, T_{n-1})}{\pi(\Theta_n^{(i)}, T_n^{(i)}|\mathcal{H}_j)} \times w_n^{(i)}|_{\mathcal{H}_j} \\ = P(y_n|\Theta_n^{(i)}, T_n^{(i)}, \mathcal{H}_j) \times w_{n-1}^{(i)}|_{\mathcal{H}_j}. \quad (16)$$

In the same way, one can evaluate the *sequential model posterior distribution* by

$$P(\mathcal{H}_j|y_{1:n}) = \frac{P(y_n|y_{1:n-1}, \mathcal{H}_j)P(\mathcal{H}_j|y_{1:n-1})}{\sum_{k=1}^{N_{persons}} P(y_n|y_{1:n-1}, \mathcal{H}_k)P(\mathcal{H}_k|y_{1:n-1})} \\ \cong \frac{P(\mathcal{H}_j|y_{1:n-1}) \sum_{i=1}^{N_j} \tilde{w}_n^{(i)}|_{\mathcal{H}_j}}{\sum_{k=1}^{N_{persons}} P(\mathcal{H}_k|y_{1:n-1}) \sum_{i=1}^{N_k} \tilde{w}_n^{(i)}|_{\mathcal{H}_k}}. \quad (17)$$

Unless there are reasons to do otherwise, we set the initial model probabilities uniformly: $P(\mathcal{H}_j|y_0) = 1/N_{persons}$.

4.2 Experimental Results

Table 1 shows recognition results for the new online learning algorithm (Bayesian SMC), compared with a batch learning algorithm (Bayesian MCMC), with which we estimated the predictive distribution of parameters using a Markov Chain Monte Carlo method[7]. Template images and test sequences showed 7 Japanese actors and 3 Japanese actresses in frontal pose against a blue background. For the template images, each individual shows a neutral facial expression, while for the test sequences, subjects were encouraged to show expressions and to talk freely. For the Bayesian SMC system, we used 18 sample images showing six fundamental expressions (happiness, sadness, fear, anger, disgust and surprise) to estimate the covariance matrix Λ_j. For the Bayesian MCMC system, we used the same data to draw N_j samples of parameters. We set the number of Monte Carlo samples for each face model to $N_j = 3600$ for both systems. We assumed that initial face regions were pre-detected so that the center position and radius of the face region were already available. Table 1 shows that the proposed approach reduced the ID error rate from 11.0 % to 2.3 %, but that the total processing time was roughly double that of the original batch algorithm.

Table 1. Face recognition results (ID error rate)

Model	Bayesian MCMC	Bayesian SMC
ID error rate	11.0 %	2.3 %
Processing time	1155 sec.	2329 sec.

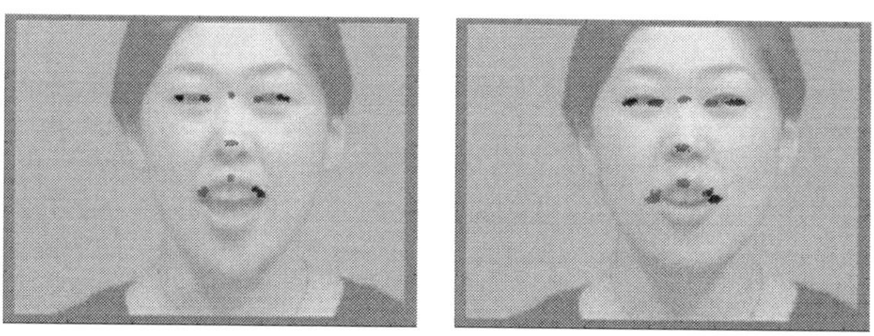

Fig. 1. Examples of SMC sample feature points at frame $n = 0$ (left) and frame $n = 8$ (right). The top 100 samples in order of the importance weights are shown.

Figure 1 shows examples of SMC particles (i.e. sets of feature point parameters) with the top 100 normalized importance weights out of a total of 36,000 particles. It is apparent that the proposed algorithm updates the posterior distribution of feature points more or less successfully despite facial deformations (reference feature points are located at the six eye and mouth corners plus the three locations on the mid-sagittal line).

4.3 Pruned Resampling

Figure 2 shows the evolution of the number of particles N_j drawn by the Bayesian SMC model given a test video sequence of person $j_{true} = 10$. The number of particles for $j = 10$ grows steadily with time (frame number) while the others are either static or gradually decrease. The tendency for a particular model to accumulate ever larger numbers of particles at the expense of the other models reflects the increasing confidence of the system, with increasing volumes of data, that the given model is correct and the others incorrect.

However, much of the associated computation is unnecessary; we do not require so many particles to verify a well-supported hypothesis. Thus we introduce a further normalization into the resampling process such that the number of particles for the most likely model $\widehat{j} = \arg\max_k N_k = \arg\max_k P(y_n|y_{1:n-1}, \mathcal{H}_k)$, rather than the total number $N_{total} = \sum_j N_j$, remains approximately constant. Since the resampling process shares out the total mass of normalized importance weights at the previous step, it is natural to set the new total number of particles N_{total} as follows:

$$N_{total} = N_{\widehat{j}} \times \frac{\sum_{k=1}^{N_{persons}} \sum_{i=1}^{N_k} \tilde{w}_n^{(i)}|_{\mathcal{H}_k}}{\sum_{i=1}^{N_{\widehat{j}}} \tilde{w}_n^{(i)}|_{\mathcal{H}_{\widehat{j}}}} = N_{\widehat{j}} \times \frac{\sum_{k=1}^{N_{persons}} P(y_n|y_{1:n-1}, \mathcal{H}_k)}{P(y_n|y_{1:n-1}, \mathcal{H}_{\widehat{j}})}. \quad (18)$$

This "pruned resampling" maintains nearly constant the number of particles (and hence the volume of computation) at each step associated with the most likely model. In so doing, it reduces the amount of attention paid to increasingly unlikely models faster than does the original resampling scheme.

Table 2 shows recognition results with the pruned resampling scheme. Also shown is the total number of particles in existence $\bar{N}_{total}$, averaged over the 10 input video test sequences and 30 frames per sequence. Figure 3 shows the evolution of the number of particles N_j using the pruned resampling scheme, given the same test data used in Figure 2 ($j_{true} = 10$). Table 2 and Figure 3 show that the new resampling approach successfully prunes redundant particles and computation without introducing any new identification errors. Total processing time is comparable to that of batch MCMC, but the significantly better recognition performance of SMC is not sacrificed.

Table 2. Face recognition results (ID error rate) with and without pruned resampling

Model	Bayesian MCMC	Bayesian SMC	
Pruned resampling?	n/a	no	yes
ID error rate	11.0 %	2.3 %	2.3 %
Processing time	1155 sec.	2329 sec.	1135 sec.
$\bar{N}_{total}$	36000	36000	17488

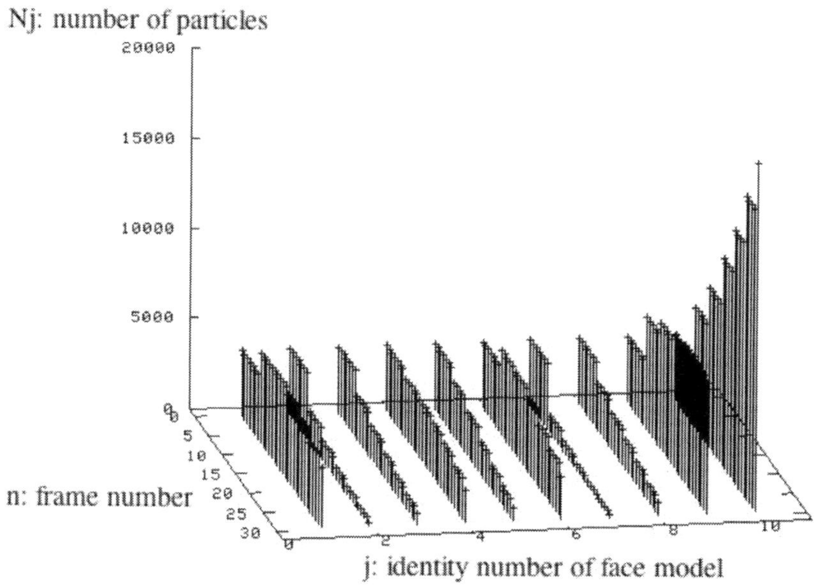

Fig. 2. Evolution of the number of SMC particles N_j with ordinary resampling ($j_{true} = 10$, $N_{total} = N_1 + N_2 + \ldots + N_{10} = 36,000$)

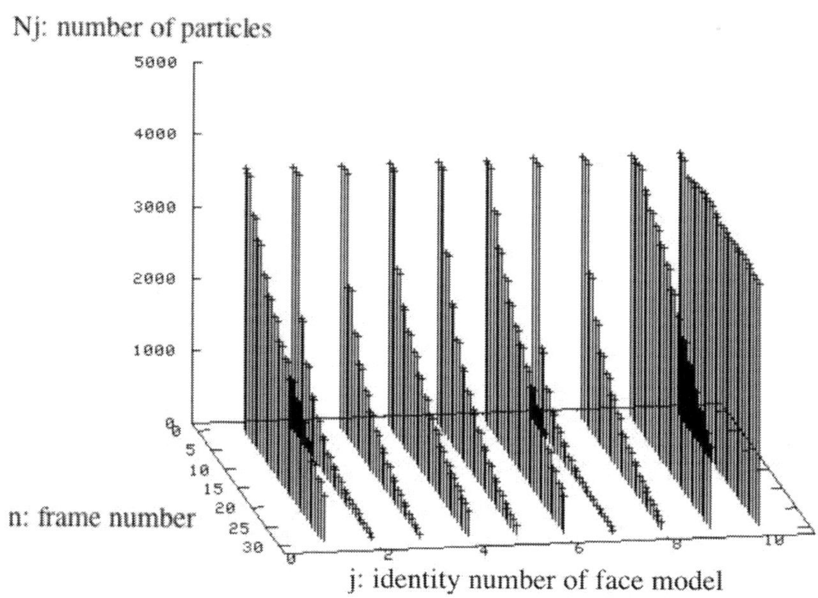

Fig. 3. Evolution of the number of SMC particles N_j with pruned resampling ($j_{true} = 10$, $N_{\hat{j}} = N_{10} \cong 3,600$)

5 Conclusions

We introduced a new Sequential Monte Carlo (SMC) algorithm for online Bayesian learning in the context of a face recognition system based on deformable template matching. The proposed algorithm achieves markedly superior robustness of recognition against facial deformations by comparison to a baseline batch MCMC algorithm. A modification to the resampling stage of the new algorithm restores its computational cost to less than that of the baseline MCMC algorithm without sacrificing any of the gain in recognition performance.

Topics remaining for further work include the automation of the face detection stage and its combination with the SMC algorithm, and extensions to deal with larger image motions and changes in face pose and lighting conditions. The SMC based change detection algorithm described in [10] may be useful in this regard.

References

1. Clippingdale, S., Ito, T.: A Unified Approach to Video Face Detection, Tracking and Recognition. Proc. ICIP'99, Kobe, Japan (1999) 662–666
2. Clippingdale, S., Ito, T.; Partial automation of database acquisition in the FAVRET face tracking and recognition system using a bootstrap approach. Proc. MVA2000, Tokyo, Japan, (2000) 5–8
3. Wiskott, L., Fellous, J. M., Krüger, N., von der Malsburg, C.: Face Recognition by Elastic Bunch Graph Matching. TR96-08, Institut für Neuroinformatik, Ruhr-Universitat Bochum (1996)
4. Doucet, A.: On Sequential Simulation-Based Methods for Bayesian Filtering. Technical report CUED/F-INFENG/TR-310, Cambridge University (1998)
5. Liu, J. S.: Monte Carlo Strategies in Scientific Computing. Springer, New York (2001) 53–77
6. Andrieu, C., Freitas, C. N., Doucet, A., Jordan, M. I.: An Introduction to MCMC for Machine Learning. Machine Learning, **50** (2003) 5–43
7. Matsui, A., Clippingdale, S., Uzawa F., Matsumoto, T.: Bayesian Face Recognition using a Markov Chain Monte Carlo Method. Proc. ICPR2004, **3** (2004) 918–921
8. Mardia, K. V., Jupp, P.: Directional Statistics. John Wiley and Sons Ltd., 2nd edition (2000)
9. Mackay, D. J. C.: Information Theory, Inference, and Learning Algorithms. Cambridge University Press (2003)
10. Matsumoto, T.: Marginal Likelihood Change Detection: Particle Filter Approach. Proc. International Workshop on Bayesian Inference and Maximum Entropy for Science and Engineering, AIP Conf. Proc. **803** (2005) 129–136

Outlier Detection Using Ball Descriptions with Adjustable Metric

David M.J. Tax[1], Piotr Juszczak[1], Elżbieta Pękalska[2], and Robert P.W. Duin[1]

[1] Information and Communication Theory Group
Delft University of Technology
Mekelweg 4, 2628 CD Delft, The Netherlands
[2] School of Computer Science, University of Manchester
Manchester M13 9PL, United Kingdom
D.M.J.Tax@tudelft.nl

Abstract. Sometimes novel or outlier data has to be detected. The outliers may indicate some interesting rare event, or they should be disregarded because they cannot be reliably processed further. In the ideal case that the objects are represented by very good features, the genuine data forms a compact cluster and a good outlier measure is the distance to the cluster center. This paper proposes three new formulations to find a good cluster center together with an optimized ℓ_p-distance measure. Experiments show that for some real world datasets very good classification results are obtained and that, more specifically, the ℓ_1-distance is particularly suited for datasets containing discrete feature values.

Keywords: one-class classification, outlier detection, robustness, ℓ_p-ball.

1 Introduction

In this paper we consider a special classification problem in which one of the classes is sampled well, but in which the other class cannot be sampled reliably [1,2]. An example is machine condition monitoring, where failure of a machine should be detected. It is possible to sample from all normal operation conditions (called the *target* class), but to sample from the failure class (the *outlier* class) is very hard. Furthermore it is also very expensive. Therefore a classifier should be constructed that mainly relies on examples of healthy machines and that can cope with a poorly sampled class of failing machines.

In the most ideal case the target class forms a tight, spherical cluster and all outliers are scattered around this cluster. To identify outliers one has to measure the distance from an object to the cluster center and threshold this distance. Clearly, when the threshold on the distance (or radius of the ball) is increased, the error on the target class decreases but at the cost of the outlier data that is accepted. The optimal ball has a minimum volume while it still encloses a large fraction of the target data.

According to the central limit theorem the target class has a Gaussian distribution when the target objects are noisy instantiations of one prototype disturbed by a large number of small noise contributions. The Mahalanobis distance

D.-Y. Yeung et al. (Eds.): SSPR&SPR 2006, LNCS 4109, pp. 587–595, 2006.

to the cluster center has to be used to detect outliers. But one should take care that a robust estimate of the class shape is used, because outliers in the training set severely deteriorate the maximum likelihood estimates for the Gaussian distribution [3]. The Minimum Determinant Covariance estimator is a practical implementation of a robust mean and covariance estimator [4].

When the assumption of many small noise contributions does not hold, other distance measures can be used. One flexible parameterization of a distance is the ℓ_p-distance. This distance has one free parameter p that rescales distances non-linearly along individual axis before adding the contributions to the final distance. Thresholding this distance defines a ℓ_p-ball as the decision boundary around the target class. The advantage of the ball description is that only few parameters have to be fitted to get a good description of the target class. This is particularly useful when the outlier detector is applied in high dimensional feature spaces and with small training set sizes. A second advantage is that it is possible to compute the volume captured by the ball analytically (see for instance, [5] pg. 11). This allows for an estimate of the error on the outlier class [6] and therefore for model evaluation between outlier detection methods.

In this paper we propose the use of the ℓ_p-distance measure to a center for the description of a class, resulting in a ball-shaped decision boundary. Three models are formulated in section 2. In the first formulation the volume of the ℓ_p distance ball is minimized by weighing the features, while the parameter p and the center of the ball are fixed. In the second we fix the p and the weights of the features, but optimize the center to minimize the volume. In the last formulation we optimize both the center as the p. In section 3 the methods are compared on real world datasets and we end with a conclusion in 4.

2 Theory

We start with a training set $\mathcal{X}^{tr} = \{\mathbf{x}_i, i = 1, ..., l\}$ containing l target objects, represented in an n dimensional feature space: $\mathbf{x} \in \mathbb{R}^n$. This dataset may contain some outliers, but they are not labeled as such. The ℓ_p-distance is defined as:

$$\|\mathbf{x} - \mathbf{z}\|_p = \sqrt[p]{\sum_{j=1}^{n} |x_j - z_j|^p}, \quad p > 0. \tag{1}$$

To detect outliers with respect to the training set $\mathcal{X}^{tr}$, we threshold the distance to some center $\boldsymbol{a}$. This defines the classifier f_p:

$$f_p(\mathbf{x}; \boldsymbol{a}) = \begin{cases} \text{target} & \|\mathbf{x} - \boldsymbol{a}\|_p^p \leq w_0, \\ \text{outlier} & \text{otherwise.} \end{cases} \tag{2}$$

A well performing classifier f_p minimizes both the error on the target class (i.e. the ball encloses almost all the target objects) as the error on the outlier class (i.e. the ball covers a minimum volume in the feature space). By a suitable placing of $\boldsymbol{a}$, by minimizing the threshold (or radius) w_0, weighting of features and optimizing p the two errors are minimized. In the next three sections we propose three formulations to optimize ℓ_p-balls.

2.1 w-Ball: The Weighted-Feature ℓ_p-Ball

In the first formulation, the feature axis are weighted such that the ball has minimum radius w_0. The center $\boldsymbol{a}$ and the parameter p are fixed beforehand. The w_0 is minimized by varying the weight w_j on each individual feature. To avoid the trivial solution of zero weights for all the features, the sum of the weights is fixed to one and all zero-variance directions are removed. [1] To make the solution less sensitive to outliers in the training data, the constraints are weakened by introducing slack variables ξ_i:

$$\min_{\boldsymbol{w},\boldsymbol{\xi}} \; w_0 + C\sum_{i=1}^{l}\xi_i \tag{3a}$$

$$\text{s.t.} \; \sum_j w_j|x_{ij} - a_j|^p \leq w_0 + \xi_i, \quad \xi_i \geq 0 \quad \forall i \tag{3b}$$

$$\sum_j w_j = 1, \quad w_j \geq 0, \quad \forall j \tag{3c}$$

A reweighted ℓ_p-distance is used for the evaluation of a new object. That means that each term in the sum in equation (1) is multiplied by w_j. This formulation is called the ‘weighted-features’ ℓ_p-ball, or w-ball.

In the experiments center $\boldsymbol{a}$ is set to the mean vector of dataset $\mathcal{X}^{tr}$. This formulation is a linear programming problem that can be solved efficiently using standard optimization toolboxes, even for high dimensional feature spaces. Parameter C determines the tradeoff between w_0 and ξ_i. A large C indicates that ξ_i should remain small in comparison to w_0 (see (3a)), resulting in a very large ball. When C is small, the slack ξ_i is allowed to grow and the radius w_0 stays reasonably small. In practice the w-ball is still not robust against outliers [7]. This is caused by the fact that an outlier influences the location $\boldsymbol{a}$ of the ball. Varying C has just a minor effect on the final solution. To get a robust ball description, the center of the ball has to be optimized such that outliers do not have any influence on the solution, even when they are located far away. This is achieved with a formulation given in the next section.

In the left subplot of figure 1 the decision boundaries for the w-ball are shown for $p = 1, 2$ and 6. The optimization reweighs the features such that the balls fit the data best. Depending on p, the shape becomes more diamond-like ($p = 1$) or more box-like ($p = 6$). The two objects on the far right still influence the solution, although they are outside the decision boundary. When the outliers on the right side are moved much further to the right, the weight for this feature is decreased (to satisfy constraint (3b)). When this weight w_1 decreased to zero,

[1] When the k-th feature does not show a variance, the optimal solution is $w_k = 1$ and all other $w_i = 0, i \neq k$.

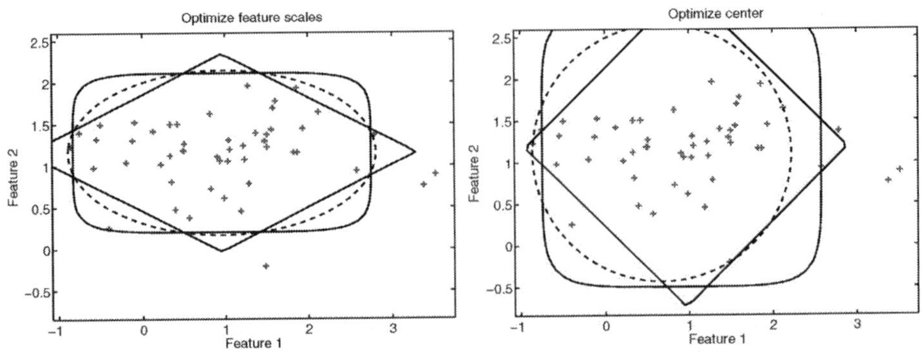

Fig. 1. The decision boundaries of the w-ball (left) and the c-ball (right) on the same dataset, and varying $p, p = 1, 2, 6$. The diamond-shaped boundary is obtained for $p = 1$. For increasing p the boundary becomes more square. For the w-ball $C = 10$ and for the c-ball $f = 0.9$ (see text for explanation of f).

the ball degenerates to a 'strip', effectively performing a feature reduction by removing this feature from the solution.

2.2 c-Ball: The ℓ_p-Ball with Variable Center

For a robust formulation a quantile function is defined. Denote $\tilde{\boldsymbol{y}} = (y_{(1)}, y_{(2)}, ..., y_{(l)})$ the sorted version of $\boldsymbol{y}$, with $y_{(1)} < y_{(2)} < ... < y_{(l)}$. The quantile function is defined as:

$$\mathcal{Q}_f(\boldsymbol{y}) = y_{([fl])}, \tag{4}$$

where $[c]$ returns the nearest integer value of c. Thus, $\mathcal{Q}_0(\boldsymbol{y})$ is the minimum element of $\boldsymbol{y}$, $\mathcal{Q}_1(\boldsymbol{y})$ the maximum and $\mathcal{Q}_{0.5}(\boldsymbol{y})$ the median.

The center $\boldsymbol{a}$ is optimized such that the object furthest away is as near as possible to this center. To be robust against outlier objects in the training set, we only consider a fraction f of the objects. When we define $y_i = \sum_j |x_{ij} - a_j|^p$ the following optimization problem can be formulated:

$$\min_{\boldsymbol{a}} \mathcal{Q}_f((y_1, y_2, ..., y_l)) \tag{5}$$

This formulation is called the the 'centered' ℓ_p-ball, or c-ball. Due to the very non-linear quantile function this optimization cannot be solved very efficiently. In this paper we use a general purpose multivariate non-linear optimizer (based on the Nelder-Mead minimization[8]). It should be noted that this optimization becomes very slow for high dimensional feature spaces (say, $n > 100$). In these cases a standard gradient descent method is applied [2]. On the other hand, the

[2] Note that the gradient and the Hessian of (5) is very simple to compute when the $f\%$ quantile $y_{([fl])}$ has been found. Define $k = [fl]$, then the gradient becomes $\frac{\partial y_k}{\partial a_j} = p \cdot \mathrm{sign}(a_j - x_{kj})|a_j - x_{kj}|^{p-1}$ and the Hessian $\frac{\partial^2 y_k}{\partial^2 a_j} = p(p-1)\mathrm{sign}(a_j - x_{kj})|a_j - x_{kj}|^{p-2}$ and $\frac{\partial^2 y_k}{\partial a_i \partial a_j} = 0$ for $i \neq j$.

solution is insensitive to the most remote $(1-f) \times 100\%$ of the data, making it an estimator with a breakdown value of $\lfloor (1-f)l \rfloor$ [9].

In the right subplot of figure 1 the decision boundaries for the c-ball's are shown for $p = 1, p = 2$ and $p = 6$. The models do not take the difference in variance of the different features into account, resulting in a wider data description than the w-ball. On the other hand, the c-ball is robust against the outlier objects on the right side (the centers are optimized to reject 10% of the data, i.e. $f = 0.9$). Moving these objects even further away will not change the solution as it is shown in the figure. Also notice that the locations of the centers of the balls vary, depending on the p.

2.3 p-Ball: The ℓ_p-Ball with Variable Center and Metric p

In the last formulation also the p in the ℓ_p-distance is optimized, together with the ball center, while fixing the weight per feature. Because p changes, the metric changes and it is not possible to compare solutions in different spaces by just comparing the radii of the balls. To compare balls in different spaces, the volumes of the balls have to be compared. The unit ball of ℓ_p is defined as $B_p^n = \{\mathbf{x} \in \mathbb{R}^n; \|\mathbf{x}\|_p \leq 1\}$. The volume of the unit ball is given by [5]:

$$\mathrm{vol}(B_p^n) = \frac{(2\Gamma(1+1/p))^n}{\Gamma(1+n/p)} \tag{6}$$

The volume of a ball with radius r is $\mathrm{vol}(B_p^n)r^n$. Using this, the following optimization problem can now be formulated:

$$\min_{\boldsymbol{a},p,r} \; \mathrm{vol}(B_p^n)r^n \tag{7a}$$

$$\text{s.t. } \mathcal{Q}_f((\|\mathbf{x}_i - \boldsymbol{a}\|_p^n)) \leq r^n, \quad p > 0 \tag{7b}$$

where r is the ball radius. This is called the p-optimized ℓ_p-ball, or p-ball. Again, the optimization is made more robust by considering the f-fraction quantile.

Notice that in this formulation both the degree p and the center of the ball are optimized, resulting in an even more complex optimization problem. Again a general multivariate nonlinear optimizer has to be used [3]. To avoid problems with the constrained variable p ($p > 0$), a variable substitution is applied and a new unconstrained variable $q = \log(p)$ is introduced. This makes it possible to use an unconstrained optimization procedure.

In figure 2 the decision boundaries for the p-ball are shown for the same data as used in figure 1. The fraction f is set to $f = 0.9, f = 0.8, f = 0.5$. Both the location as the shape of the balls is adapted to capture 90%, 80% or 50% of the data. The resulting optimized values for p are 1.26, 2.21 and 5.39 respectively. Objects outside the decision boundary are completely ignored in the minimization of the ball volume, and can therefore be randomly moved around without affecting the solution.

[3] Here also the gradient and Hessian can be computed, but this is considerably more complicated than in section 2.2.

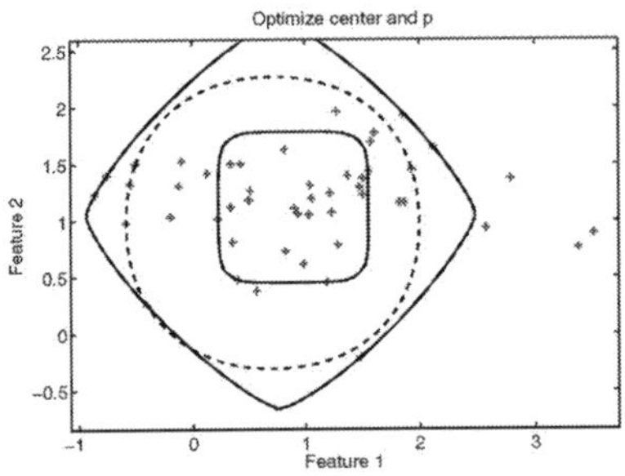

Fig. 2. The decision boundaries of the p-ball for different values of f, $f = 0.9, f = 0.8, f = 0.5$

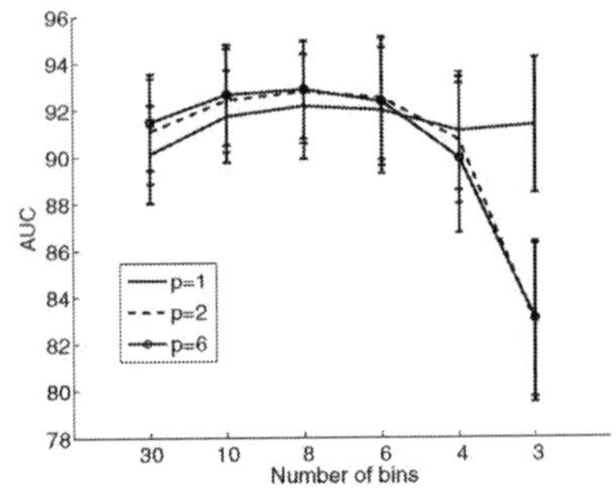

Fig. 3. The AUC performances on dataset 13 for a varying number of bins B in which the features are discretized

3 Experiments

The three methods, the w-ball, c-ball and p-ball, are compared with some standard classifiers on datasets, mainly taken from the UCI repository [10]. These datasets are standard multiclass problems and to convert them into an outlier detection problem, we use one of the classes as target class, and all other classes are considered outlier. Furthermore, we consider datasets for which the target class is reasonably clustered: it does not contain several clusters, and is not distributed in a subspace. The datasets are preprocessed to have unit variance for all features (where the scaling factors are obtained from the training set).

Table 1. Characteristics of the datasets: the number of objects in the target class and the outlier class, and the dimensionality of the data

nr	dataset	tar/out	dim.	nr	dataset	tar/out	dim.
1	Iris virginica	50/100	4	9	Concordia16 digit 3	400/3600	256
2	Breast malignant	458/241	9	10	Colon 2	40/22	1908
3	Breast benign	241/458	9	11	Thyriod normal	93/3679	21
4	Heart diseased	139/164	13	12	Waveform 1	300/600	21
5	Heart healthy	164/139	13	13	Pageblocks text	4913/560	10
6	Biomed diseased	67/127	5	14	Satellite, cotton crop	479/3956	36
7	Arrhythmia normal	237/183	278	15	Satellite, damp grey soil	415/4020	36
8	Ecoli periplasm	52/284	7				

In table 1 the list of datasets with their characteristics is shown. For the two-class Breast and Heart datasets, each of the two classes is used as the target class once. This is to show that for each class separately, different optimal solutions are found. On the datasets some standard classifiers are fitted. First a simple Gaussian distribution is applied, using the maximum likelihood estimates for the mean and covariance matrix. The second method uses the Minimum Covariance Determinant algorithm to estimate a robust covariance matrix [4]. The third method is the Parzen density estimator, that optimizes its width parameter

Table 2. AUC performances of the one-class classifiers on 15 real world datasets. The best performances (and the ones that are not significantly worse according to a t-test, at a 5% confidence level) are indicated in bold. The experiments are done using five times ten-fold stratified cross-validation. The standard deviations are given between brackets.

	datasets				
classifiers	1	2	3	4	5
Gauss	**97.8 (0.5)**	98.5 (0.1)	82.2 (0.2)	63.8 (0.7)	80.0 (0.7)
Min.Cov.Determinant	97.6 (0.2)	NaN (0.0)	73.5 (0.1)	66.7 (1.8)	NaN (0.0)
Parzen	96.8 (0.9)	99.1 (0.1)	68.1 (0.5)	65.6 (0.6)	79.3 (0.4)
k-center	96.0 (0.9)	98.4 (0.2)	72.6 (13.6)	67.3 (2.6)	79.3 (2.3)
Support vector DD	97.3 (0.4)	NaN (0.0)	69.8 (1.0)	64.4 (0.5)	78.4 (0.6)
w-ball $p = 1$	**98.3 (0.4)**	98.0 (0.1)	97.4 (0.2)	**78.9 (0.8)**	73.3 (1.7)
w-ball $p = 2$	**98.0 (0.5)**	97.7 (0.2)	**97.7 (0.1)**	71.3 (3.2)	45.9 (1.3)
w-ball $p = 6$	97.0 (0.4)	97.5 (0.4)	91.1 (0.5)	70.9 (2.7)	40.2 (6.5)
c-ball $p = 1$	96.4 (0.9)	**99.3 (0.1)**	97.5 (0.1)	77.4 (0.4)	**83.9 (0.6)**
c-ball $p = 2$	96.5 (0.5)	99.0 (0.1)	97.3 (0.1)	73.0 (0.3)	82.6 (0.9)
c-ball $p = 6$	96.0 (0.6)	98.5 (0.2)	91.6 (0.2)	63.3 (0.4)	79.5 (0.7)
p-ball	96.0 (0.6)	**99.3 (0.1)**	96.6 (0.3)	72.9 (0.8)	82.6 (0.7)
classifiers	6	7	8	9	10
Gauss	60.8 (0.8)	76.8 (0.4)	92.9 (0.3)	91.3 (0.0)	68.4 (3.6)
Min.Cov.Determinant	53.5 (1.2)	NaN (0.0)	NaN (0.0)	NaN (0.0)	NaN (0.0)
Parzen	48.3 (0.5)	77.3 (0.5)	92.9 (0.5)	92.4 (0.0)	**63.6 (22.4)**
k-center	46.9 (5.2)	77.8 (1.1)	87.0 (2.3)	91.0 (0.6)	68.1 (2.1)
Support vector DD	53.0 (2.1)	52.7 (9.4)	92.2 (1.0)	36.7 (0.5)	**63.6 (22.4)**
w-ball $p = 1$	**71.8 (1.2)**	70.4 (0.8)	91.6 (0.7)	84.4 (0.0)	57.1 (3.6)
w-ball $p = 2$	69.0 (1.1)	**80.9 (0.5)**	91.5 (0.5)	82.9 (0.0)	56.8 (3.0)
w-ball $p = 6$	62.3 (1.1)	70.3 (1.9)	90.1 (0.4)	65.0 (1.1)	56.2 (4.0)
w-ball $p = 1$	**72.7 (0.6)**	78.4 (0.4)	**95.3 (0.4)**	92.6 (0.0)	66.9 (2.1)
w-ball $p = 2$	67.9 (0.4)	78.2 (0.3)	94.6 (0.5)	90.5 (0.0)	71.1 (1.5)
w-ball $p = 6$	61.1 (1.0)	76.2 (0.3)	93.3 (0.4)	85.2 (0.2)	**77.2 (0.9)**
p-ball	66.0 (0.5)	76.5 (0.4)	93.3 (0.4)	**92.6 (0.0)**	70.2 (1.1)
classifiers	11	12	13	14	15
Gauss	84.3 (0.0)	89.9 (0.0)	59.9 (5.9)	88.0 (0.0)	83.0 (0.0)
Min.Cov.Determinant	NaN (0.0)	89.9 (0.0)	93.5 (0.0)	89.6 (0.2)	78.6 (0.1)
Parzen	90.6 (0.0)	90.0 (0.0)	50.6 (5.1)	99.0 (0.0)	39.9 (0.0)
k-center	53.3 (3.0)	87.8 (1.8)	55.9 (3.7)	97.5 (1.5)	85.0 (1.5)
Support vector DD	56.0 (0.0)	41.7 (0.0)	50.1 (5.6)	37.6 (0.0)	21.1 (0.0)
w-ball $p = 1$	96.9 (0.0)	91.2 (0.0)	91.7 (0.1)	**99.1 (0.0)**	91.2 (0.0)
w-ball $p = 2$	99.0 (0.0)	91.6 (0.0)	91.8 (0.1)	98.8 (0.0)	92.3 (0.0)
w-ball $p = 6$	**99.1 (0.0)**	90.5 (0.0)	91.0 (0.1)	98.4 (0.0)	92.4 (0.0)
c-ball $p = 1$	93.1 (0.0)	92.1 (0.0)	92.2 (0.0)	98.7 (0.0)	92.6 (0.0)
c-ball $p = 2$	88.5 (0.0)	93.0 (0.0)	93.0 (0.0)	98.5 (0.0)	**92.7 (0.0)**
c-ball $p = 6$	83.4 (0.0)	91.6 (0.0)	**93.8 (0.0)**	96.9 (0.0)	91.4 (0.0)
p-ball	93.6 (0.0)	**93.0 (0.0)**	93.0 (0.1)	98.5 (0.0)	92.6 (0.0)

using leave-one-out on the training set [11]. The fourth method uses the k-centroid method that places several centers and minimizes the largest distance from any training object to its nearest center. Finally, the support vector data description [12] is used, that is fitting a ball in a Gaussian kernel space. The features are rescaled to unit variance, and therefore the width parameter σ in the RBF kernel was fixed to $\sigma = 1$ which gave acceptable results in most cases.

These standard methods are compared to the w-ball, c-ball and p-ball with varying values for p (when applicable). Five times ten-fold stratified cross-validation is applied, and the average Area Under the ROC curve [13] is reported. The results are shown in table 2. In some cases the classifier could not be trained (for instance, the minimum covariance determinant classifier has a constraint that it cannot be estimated on datasets with more than 50 features). For these cases NaN outputs are shown.

The first observation that can be made is that for many datasets, datasets 1, 2, 4, 5, 6 and 8, the ℓ_1-metric outperforms all the others, even when a different formulation is used (i.e. c-ball instead of w-ball). It appears that all these classes have discrete features, suggesting that the ℓ_1 (city-block) distance is indeed very suited for the description of discrete data. This is tested by discretizing the features of dataset 15 (where the w-ball performs better for higher p) and training an w-ball with $p = 1, 2, 6$. The AUC performances are shown in figure 3. It shows that by reducing the number of bins, the relative performance of the ℓ_1 metric improves while that of the ℓ_2 and ℓ_6 significantly decreases.

Secondly, for high dimensional data, like datasets 7, 10 and 11, the ball-shaped models appear to be simple enough (and therefore stable enough) to outperform more complex models. Often the performance is not that significantly better than that, for instance, of the Gaussian model, but in some cases it can be significant (see dataset 11). The difference in performance between w-ball and c-ball can often be traced to the number of outliers (or the noise) present in the training set. When the ball center can be represented well by the mean of the training set, like in datasets 1, 3, 4, 7, and 14, the w-ball is to be preferred. In other datasets, like 2 and 8, the target class shows a long tail with remote outliers, shifting the mean of the target class out of the main cluster. In these cases the more robust center estimate has to be used.

Finally, the most flexible approach, the p-ball, rarely shows the very best performance, models with a fixed p perform on average slightly better. The p-ball slightly overfits, but fortunately, the optimized value for p is always close to the p of the best performing ball. Clearly, a validation set has to be used to finally judge the best value for p in the w-ball or c-ball. When this validation data is not available, the p-ball is to be preferred.

4 Conclusions

For many outlier detection problems for which the target data is characterized by good features, outliers can be detected well by measuring the distance to a suitable cluster center and thresholding this distance. This paper proposes three new approaches to optimize the cluster center and the distance measure, such that the genuine data is described well by an ℓ_p ball. In the first formulation the feature weights are optimized, by solving a linear programming problem. The second formulation optimizes the cluster center in a robust gradient descent approach. In the last formulation not only the center but also the parameter p is optimized, using a general multivariate nonlinear optimizer.

The results on real world data show that datasets with discretized feature values benefit from the use of the ℓ_1 metric. On the other hand, the optimization of the p in the ℓ_p metric appears to be sensitive to overtraining. When one considers relatively outlier-free data, it is advantageous to fix the center of ℓ_p ball and optimize the scaling of the features. When significant outliers are present, or the target class distribution is significantly asymmetric, the ℓ_p ball has to optimized using a robust procedure.

Obviously, the single ball solution can be extended to a *set* of balls by using the standard k-means clustering algorithm. In k-means clustering often the Euclidean distance to cluster prototypes is used. This can be replaced by the ℓ_p-distance to cluster centers resulting in a generalized Lloyds algorithm [14]. The cluster centers, the feature weights and possibly the p can be optimized using one of the three ball fitting approaches as they are presented in this paper.

Acknowledgments. This research is supported by the Technology Foundation STW, applied science division of NWO and the technology programme of the Dutch Ministry of Economic Affairs.

References

1. Tax, D.: One-class classification. PhD thesis, Delft University of Technology, http://ict.ewi.tudelft.nl/~davidt/thesis.pdf (2001)
2. Koch, M., Moya, M., Hostetler, L., Fogler, R.: Cueing, feature discovery and one-class learning for synthetic aperture radar automatic target recognition. Neural Networks **8**(7/8) (1995) 1081–1102
3. Huber, P.: Robust statistics: a review. Ann. Statist. **43** (1972) 1041
4. Rousseeuw, P., Van Driessen, K.: A fast algorithm for the minimum covariance determinant estimator. Technometrics **41** (1999) 212–223
5. Pisier, G.: The volume of convex bodies and Banach space geometry. Cambridge University Press (1989)
6. Tax, D., Duin, R.: Uniform object generation for optimizing one-class classifiers. Journal for Machine Learning Research (2001) 155–173
7. Barnett, V., Lewis, T.: Outliers in statistical data. 2nd edn. Wiley series in probability and mathematical statistics. John Wiley & Sons Ltd. (1978)
8. Nelder, J., Mead, R.: A simplex method for function minimization. Computer journal **7**(4) (1965) 308–311
9. He, X., Simpson, D., Portnoy, S.: Breakdown robustness of tests. Journal of the American Statistical Association **85**(40) (1990) 446–452
10. Blake, C., Merz, C.: UCI repository of machine learning databases (1998)
11. Duin, R.: On the choice of the smoothing parameters for Parzen estimators of probability density functions. IEEE Transactions on Computers **C-25**(11) (1976) 1175–1179
12. Tax, D., Duin, R.: Support vector data description. Machine Learning **54**(1) (2004) 45–66
13. Bradley, A.: The use of the area under the ROC curve in the evaluation of machine learning algorithms. Pattern Recognition **30**(7) (1997) 1145–1159
14. Lloyd, S.: Least squares quantization in PCM. IEEE Transactions on Information Theory **28**(2) (1982) 129–137

HMM-Based Gait Recognition with Human Profiles

Heung-Il Suk and Bong-Kee Sin

Computer Engineering, Pukyong National University
{daedalos, bkshin}@pknu.ac.kr

Abstract. Recently human gait has been considered as a useful biometric supporting high performance human identification systems. We propose a view-based pedestrian identification method using the dynamic silhouettes of a human body modeled with the hidden Markov model (HMM). Two types of gait models have been developed both with a cyclic architecture: one is a discrete HMM method using a self-organizing map-based VQ codebook and the other is a continuous HMM method using feature vectors transformed into a PCA space. Experimental results showed a consistent performance trend over a range of model's parameters and the recognition rate up to 88.1%. Compared with other methods, the proposed models and techniques are believed to have a sufficient potential for a successful application to gait recognition.

1 Introduction

Recognizing people by their gait, the style of walking of an individual, can be performed without asking them to take any specific actions and even without making them be aware whether they are being watched or not. From Johansson's studies in psychophysics with moving light displays (MLD) attached to body parts, it appeared that humans have the capability of recognizing their acquaintance only through their gait [1].

Recently human motion analysis has been receiving increasing attention from computer vision researchers and it is well explained in the review papers by J. K. Aggarwal et al. [2], C. Cedras et al. [3], and D. M. Gavrila et al. [4]. According to these papers two distinct methods for human motion analysis are distinguished: 'model-based method' using a priori shape models, and the other, called 'appearance-based' or 'view-based', without using explicit shape models. Both methods take a common sequential process of (1) feature extraction, (2) feature correspondence, and (3) high-level processing. The difference between them lies in the way of processing feature correspondence. 'Model-based' methods compare the input features taken from an input image with the parameters of the 2D or 3D body models prepared in advance, and make feature correspondence automatically. On the contrary, the 'appearance-based' methods carry out the feature correspondence by varying the values of position, velocity, shape, color, and so on from consecutive frames.

A brief review is in order. In the works of A. Kale et al. [5, 6], they used the width of the pedestrian's silhouette of the binarized images as the feature vector and developed

D.-Y. Yeung et al. (Eds.): SSPR&SPR 2006, LNCS 4109, pp. 596–603, 2006.

five basic stances from k-means clustering. Then a new feature vector composed of the Euclidean distance between an input feature vector and each of the 5 basic stances was the final feature vector for training HMM and recognizing people. On the other hand, J. J. Little et al. [7] extracted the frequency and phase of the gait derived from optical flow as the feature vector. In [8], C. BenAbdelkader et al. encoded the planar dynamics of a walking person in a 2D plot consisting of the pairwise image similarities of the sequence of images of the person and proposed a recognition method using k-nearest neighbor after reducing the dimension of the feature vector. Whereas R. Collins et al. [9] introduced a view-dependent method using template matching of body silhouettes. The key frames from a test sequence were extracted by performing cyclic gait analysis and those frames were compared to training frames using normalized correlation. The recognition was performed by nearest neighbor matching among correlation scores.

In this paper, we use people's silhouette as a profile vector and represent the characteristics of the gait with the sequences of the profile vectors. Since body shape alone is not enough for gait recognition, we need to take account of the gait dynamics which can be modeled by a 'Markov chain'. A hidden Markov model or HMM is a Markov chain variant that is very powerful for modeling highly variable and noisy patterns. We have chosen this model for gait recognition. In this work, the profile vector used for the recognition is composed of a fixed number of horizontal distance measurements of the left, right boundary of a person from the center and then normalized with respect to the height of the person. This feature can be directly compared with the features of A. Kale et al. [5, 6] and it has turned out that the features used in our work showed better performance. We have designed two separate recognizers, each using discrete hidden Markov models (DHMM) and continuous hidden Markov models (CHMM) respectively. For the training of DHMMs, each feature vector is quantized to a codeword by a self-organization map (SOM). For the training of CHMMs, feature vector is mapped to a point in the PCA space to reduce the dimension. In this way we kept the models from being overfitted due to the lack of training data.

The contribution of this paper is found in the way of representing the structural characteristics and the modeling dynamic characteristics of the human gait. The structural characteristics are represented by the left and right boundaries of the person and the dynamical characteristics are well models by an HMM. The HMM has one-way directed circular ring topology fitted to recognizing a person in a video sequence without external segmentation into a single gait cycle. This type of an HMM does not require a specific starting point in a gait cycle. Another benefit of this model we believe is that the longer the input sequences are, the better the recognition rate is.

2 Profile Extraction

One of the simplest and most direct ways to represent the shape of a pedestrian is the silhouette against the background. The silhouette can be described as an ordered sequence of boundary points and we define it as a profile vector that represents the shape characteristics of the person. A profile vector is composed of the horizontal distances from the horizontal center to the left and right boundary of the human blob B (Fig.1). The horizontal center of the person is calculated as

$$X_c = \frac{\sum_{(x,y)\in B} x}{|B|} \tag{1}$$

where, B is a set of foreground image pixels, X_c is the horizontal center of B, $|B|$ is the number of pixels and (x, y) is the coordinate of the pixel in the image. If we consider all the boundary pixels of the silhouette in composing a profile vector, its dimension of the vector will be too high and variable from one frame to another. So, we choose only 40 pixels of the silhouette of the human blob, 20 pixels from the left and 20 pixels from the right sampled at a fixed vertical interval from the bottom. The elements of the vector are then normalized with respect to the height of the human blob.

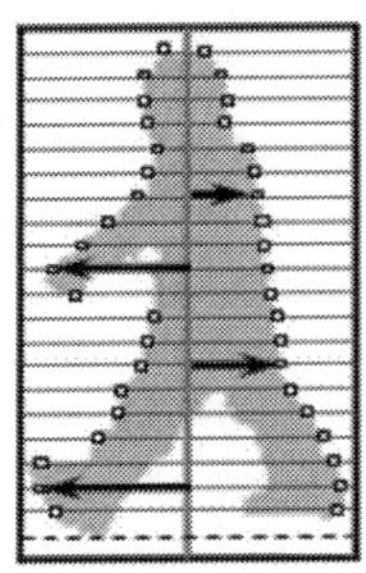

Fig. 1. Composition of a profile vector

3 Model of Human Gait

In this paper, we create two kinds of gait models using the discrete hidden Markov model (DHMM), and the continuous hidden Markov model (CHMM), to represent the dynamic characteristics of a human gait. The input data for the DHMM are the codeword sequences quantized from the profile vector sequences by a self-organization map (SOM) and those for the CHMM are the sequence vector transformed into a subspace of PCA. We first explain the vector quantization for the DHMM and then the PCA for the CHMM.

3.1 SOM

SOM is a topology preserving feature map. It being simple, we chose this for a vector quantizer of the feature vectors. Taking account of the fact that a human gait is composed of a cyclic sequence of stereotypic stances, we designed the output layer of the SOM to have a ring topology with each state corresponding to a typical stance in the sequence. The SOM is trained with the profile vectors extracted from the input image sequences. The set of trained weight vectors are considered as a codebook for the gait pattern space. With this SOM we quantize each input profile vector into a codeword.

$$codeword = \arg\min_{k} \|\mathbf{x} - \mathbf{w}_k\|. \tag{2}$$

Here $\|\cdot\|$ is the Euclidean norm of the difference between the input profile vector and the code vector, $\mathbf{x}$ is an input vector and $\mathbf{w}_k$ is the codeword vector. We choose the codeword whose code vector has the minimum Euclidean distance from the input profile vector.

3.2 Principal Component Analysis

The principal component analysis is the method that can be used to reduce the dimension of any vectors by considering the variance of and the relationship among variables, while minimizing the loss of the information. It reduces the dimension of a vector by transforming it into the direction that has a large variance. This is originated from the fact that there is more information in the direction having a large variance compared to the direction having the small variance. We use this method to reduce the dimension of the profile vectors for the training of the CHMM.

It is known that the sum of the eigenvalues of the covariance matrix is equal to the total variance of the original variables. We can consider a mapping to a reduced dimension by specifying that the new components must account for at least a fraction of d of the total variance. Choosing a value of d, which is a fraction of the total variance, between 70% and 90% preserves most of the information in $\mathbf{x}$ [10]. We then choose k so that

$$\sum_{i=1}^{k} \lambda_i \geq d \sum_{i=1}^{p} \lambda_i \geq \sum_{i=1}^{k-1} \lambda_i \tag{3}$$

where, λ_i denotes the eigenvalues and p is the rank of the covariance matrix.

3.3 Gait Model

Recognizing people using structural features of body silhouette is not easy since simple structural characteristic can be shared by many individuals. We can exploit the observation that people have their own dynamic characteristics or temporal features, even though they have similar structural characteristics. These dynamic characteristics can be represented by the Markov chain in an HMM.

The transition of the states in an HMM is modeled by the 'Markov process'. The sequence of the transition of the states is not observable, i.e. hidden, and it is possible to estimate it through an observable data.

In this paper, we design the topology of an HMM as a circular ring (Fig. 2) considering that human gait is periodic. The parameters of the DHMM are trained using

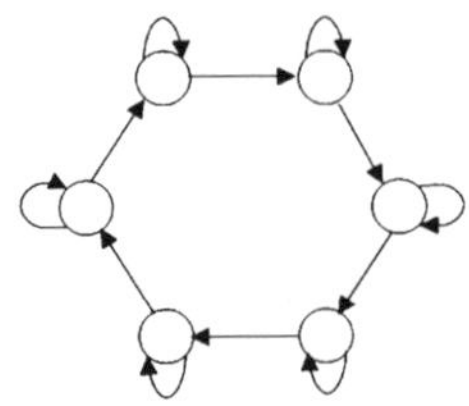

Fig. 2. Topology of a gait HMM

the codeword sequences quantized by the SOM and those of the CHMM are trained using the reduced vectors transformed into a PCA space.

Since the HMMs are trained for each person in the data set, it can be used as a way to represent the characteristics of the specific person's gait. In other words, when there is an input vector sequence taken from the video sequences of a person's gait, the HMM which best represents the characteristics of the person's gait will produce the highest likelihood for the input vector sequence. The likelihood of each HMM is computed by the forward algorithm.

$$\hat{i} = \arg\max_{i} P\left(\mathbf{X} \mid \lambda_i\right). \quad (3)$$

Here, $\mathbf{X}$ is the input vector sequence, λ_i is the HMM model for the ith person.

4 Experiment and Analysis

The proposed method was tested on a video database consisting of seven sequences for each of the six subjects, taken from the web site of the University of Calgary in Canada [11]. Those video sequences were captured at 15 frame rate, 24 bit colors and an image size of 320×160. Each sequence includes 85 frames and 6 gait cycles on average.

4.1 Data Coding

Fig. 3 shows the example of the vector quantization for a half cycle of a person's gait using an SOM with seven output nodes. The quantized vector sequences are used for the DHMM training.

A profile vector contains the 40 boundary points as marked in Fig. 4(a). Each of the elements is the horizontal distance from the vertical center to the left and right boundary of a human blob and is numbered in clockwise starting from the left bottom (Fig. 4(a)). The covariance matrix of all the profile vectors used for training motivates us to reduce the dimension. As shown in Fig. 4(b), only a small fraction of elements (covariances) are significant implying great variation and correlation. Those numbers in each axis indicate the nth element of a profile vector. It is conjectured that these bright colored parts correspond to key features to be used to recognize people. Those components are feet/legs (1-5, 35-40) and hands/arms (11-15) regions.

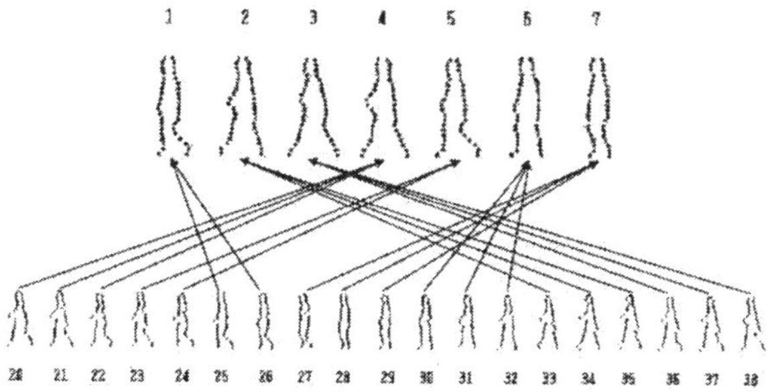

Fig. 3. Vector quantization example for half cycle

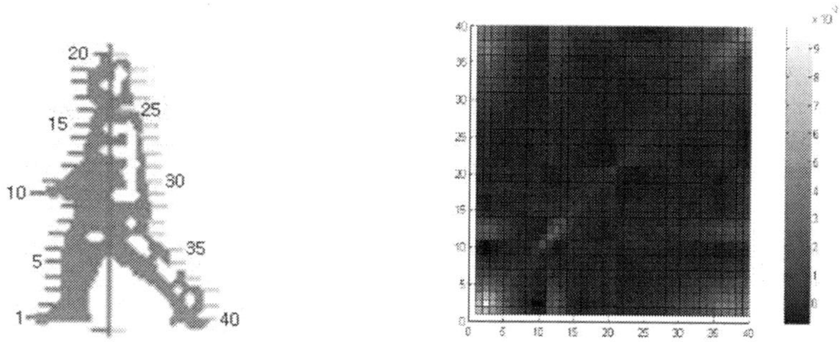

(a) (b)

Fig. 4. (a) The elements of the profile vector (b) Visualization of the covariance matrix of the profile vectors

4.2 Recognition

Due to the lack of the test data, cross validation was used. In the DHMM test by varying the number of codeword and states, it showed the highest performance, 88.10%, with seven states and seven code vectors. The detailed results over a number of states with fixed codebook size are shown in Table 1.

Table 1. Results of DHMM (codebook size = 7)

DHMM test					
# States	5	6	7	8	9
Hits(%)	40.48	47.30	88.10	69.05	0.0

For the test of the CHMM, we varied the reduced-dimension of each profile vectors. The highest performance was obtained with seven states and eight dimensions. The result with seven states is shown over a number of dimensions in Table 2.

Table 2. Results of CHMM (# of states = 7)

CHMM test					
Dimension	5	6	7	8	9
Hits(%)	76.19	73.81	69.05	88.10	76.19

4.3 Performance

In the third experiment, we compared the performance with those of the two other methods by J. J. Little et al. [7] and A. Kale et al. [5] using the same data. In the J. J. Little et al. research, they achieved a recognition rate of 90.5% by composing feature vectors with the frequency and phase taken from optical flow using the video sequences of twice as big as the frame size and the frame rate of our data. Direct comparison shows that J. J. Little et al reported higher performance than ours, but it was on different (or more detailed) data set. On the other hand, A. Kale et al's method achieved 64.3% on our data set with five HMM states. The comparison between our method and that of A. Kale et al. is shown in Fig. 5. Although we can not strongly argue for the statistical significance of the result, our system's performance was higher in all cases with a single candidate.

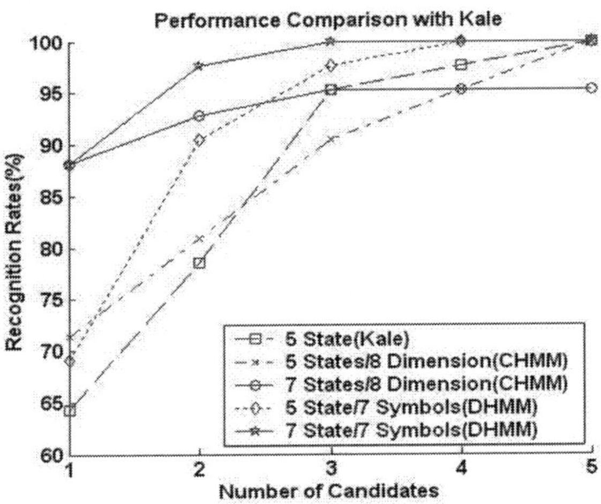

Fig. 5. Performance comparison with Kale's

5 Conclusion

We proposed an improved gait recognition method using HMMs. It added new features including SOM-based codebook and a profile vector representation for each frame, and the cyclic HMM topology for the gait dynamics. When tested on the video data from the web site of the University of Calgary in Canada by varying the number of states, codebook sizes and the number of dimensions, we achieved the highest recognition rates with seven states in both the DHMM and the CHMM.

In this work, we used a background subtraction technique to extract a silhouette without considering clothes, illumination, camera angle and walking angle. And we used a small data set for test. These are the direction of immediate future work. One advantage of the model-based approach is that we can estimate the inner structure of the pedestrian from the SOM and HMM as shown in Fig. 6. This result can be used to confirm the silhouette shape and/or to enable further understanding of human motion of activity.

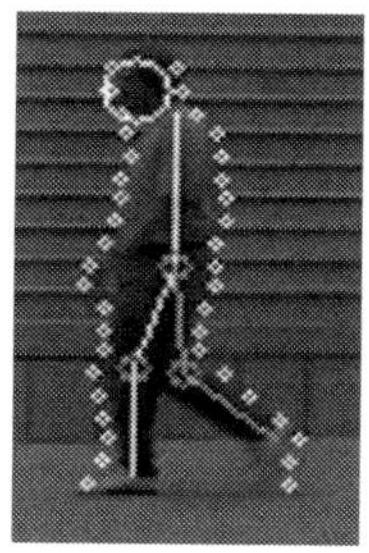

Fig. 6. Pre-estimated stick figure

References

1. G. Johansson.: Visual perception of biological motion and a model for its analysis, *Perception and Psychophysics*, vol. 14. no. 2 (1973) 201-211
2. J. Aggarwal and Q. Cai.: Human motion analysis - a review, *Computer Vision and Image Understanding*, vol. 73, no. 3 (1999) 428-440
3. C. Cedras and M. Shah.: Motion-based recognition - a survey, *Image and Vision Computing*, vol. 13, no. 2. (1995) 129-155
4. D.M. Gavrila.: The visual analysis of human movement - a survey, *Computer Vision and Image Understanding*, vol. 73. (1999) 82-98
5. A. Kale, A. Rajagopalan, N. Cuntoor, and V. Kruger.: Gait-based recognition of humans using continuous HMMs, *Proceedings of the IEEE Conference on Automatic Face and Gesture Recognition*, (2002) 321-326
6. A. Kale, A. N. Rajagopalan, A. Sundaresan, N. Cuntoor, A. RoyChowdhury, V. Krueger, R. Chellappa.: Identification of humans using gait, *IEEE Transactions on Image Processing*, September (2004)
7. J.J. Little and J.E. Boyd.: Recognizing people by their gait: the shape of motion, *Videre*, vol. 1, no. 2. (1998) 1-32
8. C. BenAbdelkader, R.Cutler, and L.Davis.: Motion-based recognition of people in eigengait space. *IEEE Conf Automatic Face and Gesture Recognition*, (2002) 254-259
9. R. Collins, R. Gross, and J. Shi.: Silhouette-based human identification from body shape and gait, *IEEE Conf Automatic Face and Gesture Recognition*, (2002) 351-356
10. Andrew R. Webb, *Statistical Pattern Recognition Second Edition,* John Wiley and Sons, 2002.
11. http://pages.cpsc.ucalgary.ca/~boyd/gait/experiment.html

Maxwell Normal Distribution in a Manifold and Mahalanobis Metric

Yukihiko Yamashita, Mariko Numakami, and Naoya Inoue

Graduate School of Science and Engineering, Tokyo Institute of Technology,
2-12-1-S6-19, O-okayama, Meguro-ku, Tokyo 152-8553, Japan
{yamasita, numakami, n708i}@ide.titech.ac.jp

Abstract. The normal distribution in Euclidean space is used widely for statistical models. However, for pattern recognition, since pattern vectors are often normalized by their norm, they are on a hyper-spherical surface. Therefore, we have to study a normal distribution in a non-Euclidean space. Here, we provide the new concept of geometrically local isotropic independence and define the Maxwell normal distribution in a manifold. We also define the Mahalanobis metric, which is an extension of the Mahalanobis distance in Euclidean space. We provide the Mahalanobis metric equation, which is covariant with coordinate transformation. Furthermore, we show its experimental results.

1 Introduction

In many fields, such as pattern recognition and data analysis, a normal distribution in Euclidean space is used for a statistical model. However, for pattern recognition, pattern vectors are often normalized by their norm and they are not in Euclidean space but on a hyper-spherical surface. Therefore, we have to study a normal distribution in a non-Euclidean space.

The normal distribution is characterized by the equality between sample mean and maximum likelihood, isotropic independence, maximum entropy or maximum number of cases, and limits such as those provided by the central limit theorem. In this paper, we extend the second characterization. We propose the concept of geometrically local isotropic independence (GLII) and define the Maxwell normal distribution in a manifold (MNDM). From the definition, we give MNDM on the n-dimensional hyper-spherical surface S^n and show that in the case of S^2, it coincides with the Fisher distribution [1].

The Mahalanobis distance [2], which is a distance normalized by a variance-covariance matrix, is also used for pattern recognition and data analysis. By using it, we can know that a distance between two vectors that is not large in terms of Euclidean distance may be large from the viewpoint of probability (Fig. 1(a)).

In the n-dimensional Euclidean space E^n, the normal distribution with average μ and variance-covariance matrix Σ is expressed by the following probability

D.-Y. Yeung et al. (Eds.): SSPR&SPR 2006, LNCS 4109, pp. 604–612, 2006.

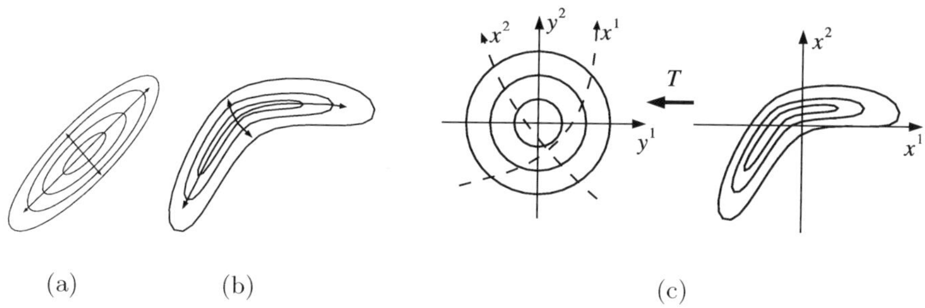

Fig. 1. Mahalanobis metric

distribution function (p.d.f.):

$$p(x) = \frac{1}{\sqrt{(2\pi)^n \det(\Sigma)}} \exp(-\langle \Sigma^{-1}(x-\mu), x-\mu \rangle /2), \qquad (1)$$

where $\langle \cdot , \cdot \rangle$ denotes the inner product and Σ is assumed to be nonsingular. When x is mapped to y by a linear transformation $y = \Sigma^{-1/2}x$, p is transformed to the p.d.f. of the standard normal distribution of y. The inner product that expresses the Mahalanobis distance is given as the pull back of the inner product as $\langle y, y' \rangle = \langle \Sigma^{-1}x, x' \rangle$.

Suppose that the counter of a p.d.f. is given as in Fig. 1(b). It is natural that the distance is evaluated not by a straight edge but by a curve. By extending the linear transform to a diffeomorphism, we obtain the Mahalanobis metric. It is very interesting that its differential equation does not depend on the coordinate system or the original metric. Furthermore, the diffeomorphism will disappear in it.

Information geometry [3] uses a manifold for probability distributions. However, since it uses a manifold as the set of distributions, there is no direct relationship to our theorem.

In Section 2, we describe the characterization of the normal distribution in Euclidean space. In Section 3, we define GLII and provide its differential equation. In Section 4, we define MNDM as a solution of the equation and provide a solution on S^n. In Section 5, we define the Mahalanobis distance and provide its differential equation. In Section 6, we present experimental results of a simulation.

In this paper, we restrict manifolds to the Riemannian manifold and use the Levi-Chivita connection. We assume that every probability distribution function (p.d.f.) is infinitely continuously differentiable and does not become 0.

2 Characterization of Normal Distribution

We explain the characterization of a normal distribution in Euclidean space. In some of the following characterizations we assume that the ensemble average is zero without loss of generality.

The normal distribution is often called the Gaussian distribution. The following characterization in E^1 was given by C.F. Gauss [4]. Let μ and x be the true and observed values, respectively. Assume that the p.d.f. depends only on the absolute error $|x - \mu|$ and the samples x_i $(i = 1, 2, \cdots, m)$ are extracted independently. If the maximum likelihood estimator of μ is always given by the sample mean $(\sum_{i=1}^{m} x_i)/n$, the distribution should be a normal distribution whose average is μ.

The velocity distribution of the ideal gas is given as a 3-dimensional normal distribution (Maxwell distribution) [5]. Below, we briefly describe its characterization in E^2. Let (x^1, x^2) and (y^1, y^2) be two 2-dimensional stochastic variables such that x^1 and x^2 are independent of each other and so are y^1 and y^2. We have

$$\begin{cases} y^1 = x^1 \cos\omega + x^2 \sin\omega, \\ y^2 = -x^1 \sin\omega + x^2 \cos\omega \end{cases} \tag{2}$$

for some $\omega \neq n\pi/2$, where $n = 0, \pm 1, \pm 2, \cdots$. Then, the distribution of (x^1, x^2) is a 2-dimensional normal distribution whose variance-covariance matrix is given by $\sigma^2 I$, where σ is a positive number and I is a unit matrix.

The characterization by the entropy in E^1 is given as follows. Let $p(x)$ be a p.d.f. defined in $(-\infty, \infty)$. We assume that $p(x)$ maximizes the entropy

$$-\int_{-\infty}^{\infty} p(x) \log p(x) dx \tag{3}$$

under the condition that

$$\int_{-\infty}^{\infty} x^2 p(x) dx = \sigma^2. \tag{4}$$

Thus, $p(x)$ is given by the normal distribution whose average and variance are 0 and σ^2, respectively. Similarly we can derive the normal distribution from the Maxwell-Boltzmann distribution [6], [7].

The central limit theorem is given as follows [5]. For simplicity, we neglect several conditions for convergence. Let x_i $(i = 1, \ldots, m)$ be independently extracted samples of the same arbitrary distribution. Then, $\sum_{i=1}^{m} x_i/\sqrt{n}$ converges to a normal distribution whose average and variance coincide with those of the original distribution, respectively. From the viewpoint of convergence, the normal distribution is also characterized as the limit of the Poisson distribution.

In a manifold, it is very difficult to define average, variance, and dilation uniquely. The facts show the difficulty of defining a normal distribution in a manifold by extending the characterization by Gauss or entropy or the central limit theorem. Furthermore, in physics, since the variance, which is given by a square term, means the energy, the condition of its preservation is justified. However, in general, the reason why we fix the variance is not clear.

Therefore, in this paper, we extend the characterization by Maxwell. It is based on the independence. The assumption of independence of a distribution with respect to orthogonal coordinates is reasonable in many cases. In the following discussion, we assume that the dimension of manifolds is not less than 2.

3 Geometrically Local Isotropic Independence (GLII)

In E^2, we can describe the independence as $p(x, y) = p(x)p(y)$. However, in a manifold we cannot construct a global orthogonal coordinate system based on geodesics that correspond to the orthogonal coordinate system in E^n. Thus, we cannot use the characterization by Maxwell in E^n directly. We propose the concept of *geometrically local independence* (GLI). Since we cannot define the global direction in a manifold either and only the isotropic independence is necessary to extend the characterization by Maxwell, we define the *geometrically local isotropic independence* (GLII).

Let $\{x^\mu\}$ be a local coordinate system in a manifold. Let $g_{\mu\nu}$ be a metric tensor on the coordinate system $\{x^\mu\}$. Let $g = \det(g_{\mu\nu})$. Let p be a p.d.f. on the coordinate system $\{x^\mu\}$. Let q be its p.d.f. normalized by $\sqrt{g}$ (We call it the normalized p.d.f. for short), which is written as

$$q = \frac{p}{\sqrt{g}}. \tag{5}$$

This q is a scaler and it is invariant for the coordinate transformation. We define $\delta_{\mu\nu}$ as

$$\delta_{\mu\nu} = \begin{cases} 1 \ (\mu = \nu) \\ 0 \ (\text{else}) \end{cases} . \tag{6}$$

Here, let $\{x^\mu\}$ be a normal coordinate system [8]. There exists a normal coordinate system at every point in a Riemannian manifold. At the origin ($x^\alpha = 0,\ \alpha = 1, 2, \cdots, n$) of the normal coordinate system, we have $g_{\mu\nu} = \delta_{\mu\nu}$ and all Levi-Chivita connections $\Gamma^\mu_{\nu\gamma}$ vanish. Now, we define GLI.

Definition 1. *(Geometrically local independence, GLI) Let $\{x^\mu\}_{\mu=1}^n$ be a normal coordinate system. We define a normalized p.d.f. q as being geometrically locally independent (GLI) with respect to x^μ and x^ν $(\mu \neq \nu)$ at the origin $(x^\alpha = 0,\ \alpha = 1, 2, \cdots, n)$ if and only if $\frac{1}{q}\frac{\partial q}{\partial x^\mu}$ (the changing rate of q normalized by q with respect to x^μ) does not depend on x^ν at the origin with the approximation of the first order of coordinates.*

Theorem 1. *Let q be a normalized p.d.f. At the origin of the normal coordinate system, the probability distribution is GLI with respect to x^μ and x^ν $(\mu \neq \nu)$ if and only if at the origin we have*

$$\frac{\partial^2}{\partial x^\nu \partial x^\mu} \log q = 0. \tag{7}$$

The proof of this theorem is clear. Theorem 1 yields that x^μ and x^ν are commutative in Definition 1. Thus, we can say 'GLI with respect to x^μ and x^ν'.

GLI is an extension of the independence since in E^2 GLI for any parallel translation of a coordinate system is equivalent to the independence.

Theorem 2. *Let $\{x^1, x^2\}$ be the orthogonal coordinate system of E^2. Let $\{y^1, y^2\}$ be an orthogonal coordinate system, which is given by its parallel translation. A p.d.f. is GLI with $\{y^1, y^2\}$ for every parallel translation if and only if x^1 and x^2 are independent.*

The proof of this theorem is also clear since every coordinate system in E^2 is a normal coordinate system. Now, we define GLII.

Definition 2. *(Geometrically local isotropic independence, GLII) Let $\{x^\mu\}$ be a normal coordinate system. Let $\{x'^\mu\}$ be a normal coordinate system given by a rotation at the origin. A p.d.f. is said to be geometrical isotropically locally independent (GLII) at the origin if and only if the normalized p.d.f. is GLI with respect to any pair of coordinates y^μ and y^ν ($\mu \neq \nu$) of a system which is given by an arbitrary rotation of $\{x^\mu\}$ whose center is the origin.*

Theorem 3. *Let q be a normalized p.d.f. At the origin, q is GLII if and only if*

$$\frac{\partial^2}{\partial x^\mu \partial x^\nu} \log q = 0, \qquad \frac{\partial^2}{\partial (x^\mu)^2} \log q = \frac{\partial^2}{\partial (x^\nu)^2} \log q \tag{8}$$

for any pair of x^μ and x^ν ($\mu \neq \nu$).

This theorem can be proved by the transformation of partial derivatives. We extend Theorem 3 from a normal coordinate system to a general coordinate system in the following Theorem 4. We denote the covariant differential by ∇_μ.

Theorem 4. *A normalised p.d.f. q is GLII if and only if*

$$\nabla_\mu \nabla_\nu \log q = f g_{\mu\nu} \tag{9}$$

with a scalar f.

By reducing eq. (9) with $g^{\mu\nu}$, we can get f. Then, eq. (9) is equivalent to

$$\left(\nabla_\mu \nabla_\nu - \frac{1}{n} g_{\mu\nu} \Delta \right) \log q = 0, \tag{10}$$

where $\Delta = g^{\alpha\beta} \nabla_\alpha \nabla_\beta$ and n is the dimension of the manifold.

Since $\nabla_\nu \log q$ is a covariant differential of a scalar, it is equivalent to a partial differential, that is, eq. (9) is equivalent to $\nabla_\mu \partial_\nu \log q = f g_{\mu\nu}$. However, we describe it by a covariant differential in order to see the symmetry easily.

Proof. First, let $\{x^\mu\}$ be a normal coordinate system. From equations in (8), the distribution is GLII if and only if

$$\frac{\partial^2}{\partial x^\mu \partial x^\nu} \log q = f \delta_{\mu\nu}. \tag{11}$$

Since x^μ is a normal coordinate and $\log q$ is a scalar, $\frac{\partial^2}{\partial x^\mu x^\nu} \log q$ and $\delta_{\mu\nu}$ are equal to $\nabla_\mu \nabla_\nu \log q$ and $g_{\mu\nu}$, respectively. Then, eq. (11) can be described as

$$\nabla_\mu \nabla_\nu \log q = f g_{\mu\nu}. \tag{12}$$

We can let f be a scalar since f is the same for any normal coordinate system at the point. Then, both sides of eq. (12) are transformed as 2nd order covariant tensors so that eq. (12) holds for any coordinate system.

Conversely, if eq. (12) holds for a scalar f in a coordinate system, then eq. (11) holds in the normal coordinate system. This completes the proof. □

4 Maxwell Normal Distribution in a Manifold (MNDM)

For the GLII probability distribution in E^n, we have the following theorem.

Theorem 5. *If a probability distribution is GLII at every point in E^n, it is an uncorrelated normal distribution whose variables have the same variance.*

This theorem can be proved since the variables of the partial differential equation can be separated. From Theorem 5, we propose the following definition.

Definition 3. *(Maxwell normal distribution in a manifold (MNDM)) A probability distribution is MNDM if and only if it is GLII at every point.*

As an example of MNDM in a manifold, we provide it in S^n.

Theorem 6. *The normalized p.d.f. of MNDM on S^n depends only on the angle x^1 from some point on S^n, and for a constant κ it is given by and for a constant κ it is given by*

$$q = C \exp(\kappa \cos x^1), \tag{13}$$

where C is a constant given by

$$C = \frac{\kappa^{(n-1)/2}}{(2\pi)^{(n+1)/2} I_{(n-1)/2}(\kappa)}, \tag{14}$$

where $I_p(\kappa)$ is the deformed Bessel function.

The proof of this theorem is so long we provide it in another opportunity.

Remark. MNDM in S^2 coincides with the Fisher distribution, which was originally given as a Maxwell-Boltzmann distribution of a magnetic dipole in a magnetic field, since its energy is proportional to $\cos x^1$, where x_1 is the angle between the magnetic dipole and the magnetic field. When x^1 is small, we have $\cos x^1 \simeq 1 - (x^1)^2/2$, so it is approximated by a normal distribution in Euclidean space. Although the Fisher distribution has similar characterizations by Gauss and by entropy, the distance used for the definition of the average or variance is not the ordinary distance on S^n but the distance in E^{n+1} embedding S^n. One of the advantages of the characterization by GLII is that such embedding is not needed. Another advantage is that it can provide the Mahalanobis metric, which we discuss in the next section.

5 Mahalanobis Metric

First, we describe how a p.d.f is transformed with a diffeomorphism (Fig. 1 (c)). Let M and M' be manifolds. Let $\{x^\mu\}$ and $\{y^\mu\}$ be the coordinate systems, and let $g_{\mu\nu}$ and ${g'}_{\mu\nu}$ be the metric tensors in M and M', respectively. Let $g = \det(g_{\mu\nu})$ and $g' = \det(g'_{\mu\nu})$. Let $T : M \to M'$ be a diffeomorphism.

Let q' be a normalized p.d.f. in M'. Since we can unify elements in M' and M, we can consider a normalized p.d.f. q in M as corresponding to q'. We define $\det T$ as

$$\det T = \sqrt{\frac{g'}{g}} \det \left(\frac{\partial y^\mu}{\partial x^\nu} \right). \tag{15}$$

Note that $\det T$ is invariant with coordinate transformations in both M and M'. Then, we have

$$q(x) = \det T \, q'(T(x)). \tag{16}$$

We propose the following definition of the Mahalanobis metric.

Definition 4. *(Mahalanobis metric) We assume that there exists a transformation $T : M \to M'$ and the normalized p.d.f. q' of MNDM in M' such that a p.d.f p is given by transforming q' by T. The Mahalanobis metric $\tilde{g}_{\mu\nu}$ with respect to p in M is given as the pull back of $g'_{\mu\nu}$.*

From the assumption, we have

$$\nabla'_\mu \nabla'_\nu \log q' = f g'_{\mu\nu} \tag{17}$$

where ∇'_μ is the covariant differential defined by the metric tensor $g'_{\mu\nu}$ in M'. Let $\tilde{g} = \det(\tilde{g}_{\mu\nu})$. Since we have

$$\det \left(\frac{\partial y^\mu}{\partial x^\nu} \right) = \sqrt{\frac{\tilde{g}}{g'}}, \tag{18}$$

eq. (15) and $p(x) = \sqrt{g}\, q(x)$ yield that

$$q'(T(x)) = p(x)/\sqrt{\tilde{g}}. \tag{19}$$

Furthermore, the transformation from $(x, \tilde{g}_{\mu\nu})$ to $(y, g'_{\mu\nu})$ by T is equivalent to the coordinate transformation. Then, we get the following theorem by rewriting $\tilde{g}_{\mu\nu}$ as $g_{\mu\nu}$ and calculating $\nabla_\mu \nabla_\nu \log g$.

Theorem 7. *A metric $g_{\mu\nu}$ is a Mahalanobis metric with respect to p if and only if*

$$\nabla_\mu \nabla_\nu \log p - \frac{\partial}{\partial x^\nu} \Gamma^\alpha_{\alpha\mu} + \Gamma^\eta_{\nu\mu} \Gamma^\alpha_{\alpha\eta} = f g_{\mu\nu}. \tag{20}$$

with a scalar f.

Remarks. Eq.(20) is covariant with the coordinate transformation. Since eq. (20) does not include a term that depends on the original metric of M, the Mahalanobis metric does not depend on the original metric either. That is an extension of Mahalanobis distance not depending on the original inner product [9]. Although we introduced a diffeomorphism T for its definition, terms with respect to T and M' disappear in eq. (20).

6 Experiment on Mahalanobis Metric

It may be difficult to solve eq. (20) directly since it includes an arbitrary scalar f. When we assume that $M' = E^n$ and the variance of the normal distribution in M' is one, we have $f = -1$. (For another example, if $M' = S^n$, we have $f = -\log q$.) Then, the equation can be solved. Since this Mahalanobis metric $g_{\mu\nu}$ is given as a metric transformed from E^n, we have $R^{\mu}_{\nu\alpha\beta} = 0$. We add this equation to the criterion.

For the experiment, we let $M' = E^2$ and a diffeomorphism be

$$y^1 = \alpha x^1, \tag{21}$$

$$y^2 = \alpha(x^2 + \beta h(x^1)h(x^2)), \tag{22}$$

where $\alpha = 3$, $\beta = 0.3$, and

$$h(x) = \begin{cases} e^{4\left(1-\frac{1}{1-x^2}\right)} & (-1 < x < 1) \\ 0 & (\text{else}) \end{cases} . \tag{23}$$

From this diffeomorphism we can calculate p as shown in Fig. 2(a). We transform the differential equation (20) and $R^{\mu}_{\nu\alpha\beta}$ to difference equations on a 51×51 mesh. We let the differentials of $g_{\mu\nu}$ on the boundary be zero. Let F be the squared sum of errors of the difference equations. We calculate the Mahalanobis metric by the steepest descent method.

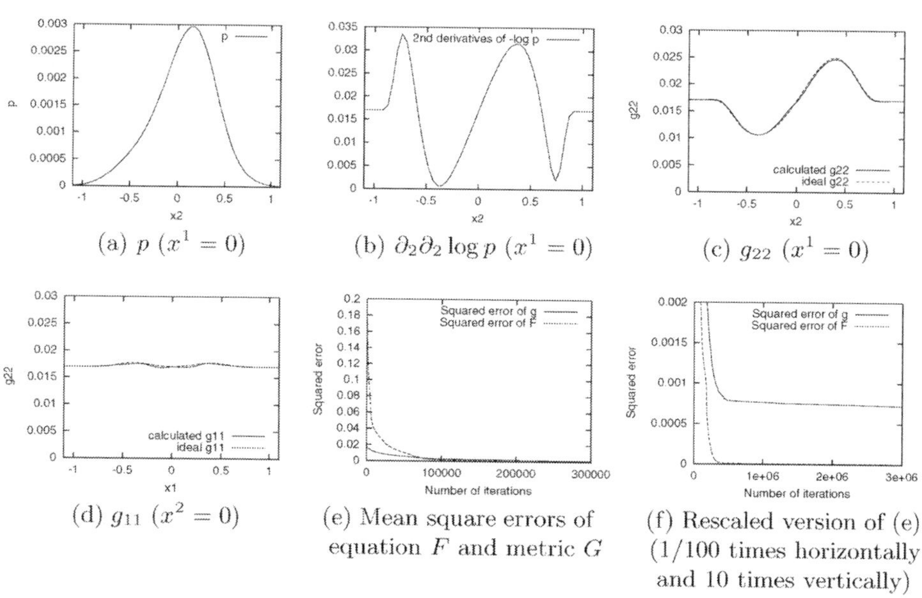

(a) p ($x^1 = 0$) (b) $\partial_2\partial_2 \log p$ ($x^1 = 0$) (c) g_{22} ($x^1 = 0$)

(d) g_{11} ($x^2 = 0$) (e) Mean square errors of equation F and metric G (f) Rescaled version of (e) (1/100 times horizontally and 10 times vertically)

Fig. 2. Experimental results

Figures 2(c) and (d) illustrate the calculated Mahalanobis metric tensor components g_{22} and g_{11}. Since we know the diffeomorphism here, we know the ideal Mahalanobis metric shown by dashed lines in the figures. From these figures, we can see that the calculated metric can approximate the ideal Mahalanobis metric. Figure 2(b) illustrates $\partial_2\partial_2 \log p$. From $g_{\mu\nu} = -\nabla_\mu\nabla_\nu \log p$, the connection terms correspond to the difference between Figs. 2(b) and (c), which are very large. Figure 2(e) and its rescaled version (f) illustrate the convergence of F and the square error of the Mahalanobis metric denoted by G. From this figure, we can see it takes a long time to calculate it. We have to develop a more efficient method.

7 Conclusions

In this paper, we defined GLI, GLII, and MNDM. We clarified that they are extensions of independence, isotropic independence, and Maxwell distribution in Euclidean space, respectively. We provided MNDM on a hyper-spherical surface. We gave a definition of the Mahalanobis metric, which is an extension of the Mahalanobis distance. In future work, we will obtain MNDM in other manifolds, develop a method of obtaining the Mahalanobis metric from samples, and apply it to nonlinear dimensional reduction.

References

1. Mardia, K.: Statistics of directional data. Academic Press, London and New York (1972)
2. Mahalanobis, P.: Normalization of statistical variates and the use of rectangular co-ordinates in the theory of sampling distributions. Sankhay **3** (1937) 1–40
3. Amari, S.: Differential geometrical method in statistics. Springer-Verg, Berlin Heidelberg (1985)
4. Maistrov, L.: Probability theory, a historical sketch. Academic Press, New York and London (1974) (Translated and Edited by Kotz, S.).
5. Feller, W.: An introduction to probability theory and its application. third edition edn. Volume 1. John Wiley & Sons, New York (1968)
6. T.C., F.: Probability and its engineering uses. second edition edn. D. Van Nostrand Company, Princeton, New Jersey (1965)
7. Tien, C., Lienhard, J.: Statistical Thermodynamics. Revised printing edn. Hemisphere Publishing Corporation, Washington (1985)
8. Spivak, M.: A comprehensive introduction to differential geometry. Volume 1–5. Publish or Perish, Inc., Houston, Texas (1979)
9. Yamashita, Y., Ogawa, H.: Optimum image restoration and topological invariance. System and Computers in Japan **24** (1993) 53–63

Augmented Embedding of Dissimilarity Data into (Pseudo-)Euclidean Spaces

Artsiom Harol[1], Elżbieta Pękalska[2], Sergey Verzakov[1], and Robert P.W. Duin[1]

[1] Information and Communication Theory group
Faculty of Electrical Engineering, Mathematics and Computer Science
Delft University of Technology, The Netherlands
{a.harol, e.pekalska, s.verzakov, r.p.w.duin}@ewi.tudelft.nl
[2] School of Computer Science
University of Manchester, United Kingdom
pekalska@cs.man.ac.uk

Abstract. Pairwise proximities describe the properties of objects in terms of their similarities. By using different distance-based functions one may encode different characteristics of a given problem. However, to use the framework of statistical pattern recognition some vector representation should be constructed. One of the simplest ways to do that is to define an isometric embedding to some vector space. In this work, we will focus on a linear embedding into a (pseudo-)Euclidean space.

This is usually well defined for training data. Some inadequacy, however, appears when projecting new or test objects due to the resulting projection errors. In this paper we propose an augmented embedding algorithm that enlarges the dimensionality of the space such that the resulting projection error vanishes. Our preliminary results show that it may lead to a better classification accuracy, especially for data with high intrinsic dimensionality.

1 Introduction

Pattern recognition relies on the description of regularities in observations of classes of objects. How this knowledge is extracted and represented is of importance for learning. Representations which are alternative to feature-based descriptions should be studied as they may capture different characteristics of a problem we want to analyze [1,4].

An example of such a representation is a proximity representation, where every object is described by some continuous nonnegative symmetric function of two variables. Learning from such representations relies on embedding of the proximity data into some vector space. It is usually desirable to find a mapping such that the initial topology is preserved as much as possible. The simplest way to do that is to construct an isometric mapping, which preserves all given distances.

However the broad range of proximity functions, satisfying only the conditions described above, may not allow one to construct an isometric embedding into

D.-Y. Yeung et al. (Eds.): SSPR&SPR 2006, LNCS 4109, pp. 613–621, 2006.

Euclidean space. In that case one needs to look for a more general space, with smaller number of restrictions. The solutions might be to design a mapping into pseudo-Euclidean space.

Embedding algorithms are usually defined on the basis of some representation objects, called prototypes. The projection accuracy for new data is proportional to the number of dominated intrinsic dimensions, described by them. If one has sufficient amount of prototypes, the projection error is of a little significance. But, if the cost to get more data for a space representation is very high, augmented embedding might be a good solution. It reconstructs given proximity information by means of one (Euclidean) or two (pseudo-Euclidean) extra dimensions. Nevertheless, it does not help much in cases when data has large intrinsic nonlinearities, since it is based on a global linear projection.

The paper is organized as follows. In section 2, a linear embedding of distance data into a pseudo-Euclidean space is presented. In section 3 augmented embedding for proximity data is presented. Data sets with experiments are described in Section 4. Conclusions are presented in Section 5.

2 Linear Embedding in (pseudo-) Euclidean Spaces

In this section we focus on linear isometric embedding of distance-based information into pseudo-Euclidean spaces. The results also hold for Euclidean cases, i.e. when the Gram operator derived from distances is positive definite, and coincide with the classical scaling [7,8]. The technique described in this chapter is standard and can be found in [1,4].

The formalism is as follows. Suppose we have a pair $(\mathbb{X}, d)$, where $\mathbb{X}$ is a finite set of n elements equipped with a pairwise continuous non-negative symmetric distance functions d_{ij}. These distance functions define a matrix $\mathbf{D}$ of size $n \times n$.

Having these properties of proximity functions, the whole finite representation $\mathbf{D}$ can be embedded into pseudo-Euclidean space.

By definition, a *pseudo-Euclidean* space $\mathbb{R}^{(p+q)}$ [4] of signature (p, q) is a pair $(V, \boldsymbol{\Phi})$, where V is a vector space under the field of real numbers of dimension $(p+q)$ and $\boldsymbol{\Phi}$ is a non-degenerate symmetric bilinear form, which represents the generalized inner product in V. Given an orthonormal (w.r.t $\boldsymbol{\Phi}$) basis $e = (e_1, e_2, \cdots, e_n)$, the generalized inner product between two vectors in $\mathbf{x}, \mathbf{y} \in V$ is expressed as

$$\langle \mathbf{x}, \mathbf{y} \rangle_{pq} = \sum_{i=1}^{p} x^{(i)} y^{(i)} - \sum_{j=p+1}^{p+q} x^{(j)} y^{(j)}. \tag{1}$$

Any *pseudo-Euclidean* space admits a decomposition into a direct orthogonal sum of two non-commensurate Euclidean subspaces of dimensions p and q respectively, i.e. $\mathbb{R}^{(p+q)} = \mathbb{R}^p \dot{+} \mathbb{R}^q$. The inner product is positive definite in $\mathbb{R}^p$ and negative definite in $\mathbb{R}^q$. The *pseudo-Euclidean* space corresponds to a Euclidean space in case of $q = 0$.

From the definition it is clear that the notion of inner product in *pseudo-Euclidean spaces* is relative, since its is not necessary positive definite and the *square-distance*, defined as $\|\mathbf{x}-\mathbf{y}\|^2 = \langle \mathbf{x}-\mathbf{y}, \mathbf{x}-\mathbf{y}\rangle = (\mathbf{x}-\mathbf{y})^T \mathcal{J}_{pq}(\mathbf{x}-\mathbf{y})$ can be negative. Here, $\mathcal{J}_{pq} = \begin{pmatrix} \mathbf{I}_{p\times p} & 0 \\ 0 & -\mathbf{I}_{q\times q} \end{pmatrix}$ is the canonical matrix of the symmetric bilinear form, corresponding to the orthogonal (w.r.t $\mathbf{\Phi}$) basis $e = (e_1, e_2, \cdots, e_n)$ of V and $\mathbf{I}$ represents an identity matrix.

Based on linear relations between square pseudo-Euclidean distances $\mathbf{D}^2 = (d_{ij}^2)$ and inner products in $\mathbb{R}^{(p+q)}$ space [4], one can write:

$$\mathbf{D}^2 = \mathrm{diag}(\mathbf{G})\mathbf{1}^T + \mathbf{1}\mathrm{diag}(\mathbf{G})^T - 2\mathbf{G}, \tag{2}$$

where $\mathbf{1}$ is a column vector of ones and $\mathbf{G}$ is a Gram operator, defined as:

$$\mathbf{G} = \mathbf{X}\mathcal{J}_{pq}\mathbf{X}^T. \tag{3}$$

Here, $\mathbf{X}$ is a matrix of object coordinates in that space.

Assuming that only distances between a set of objects are given, the sought coordinates can be determined based on the relations between distances and inner products, as presented above. Note that having found one set of coordinates, another one can be created by a rotation and(or) a translation.

The mapping is constructed such that the origin coincides with the mean of $\mathbf{X}$. It is done by using a centering matrix $\mathbf{J} = \mathbf{I} - \frac{1}{n}\mathbf{1}\mathbf{1}^T$. So, $\mathbf{G} = -\frac{1}{2}\mathbf{J}\mathbf{D}^2\mathbf{J}$. The underlying configuration $\mathbf{X}$ can be found as an eigendecomposition:

$$\mathbf{G} = \mathbf{Q}|\Lambda|^{1/2} \begin{pmatrix} \mathcal{J}_{pq} & \\ & 0 \end{pmatrix} |\Lambda|^{1/2}\mathbf{Q}^T, \tag{4}$$

where Λ is a diagonal matrix of the first decreasing p positive and q negative eigenvalues ($k = p + q$), followed by zero(s). $\mathbf{Q}$ is a matrix of the corresponding eigenvectors. Consequently,

$$\mathbf{X} = \mathbf{Q}_k \Lambda_k^{\frac{1}{2}} \mathbf{P}^T, \quad k \le n, \tag{5}$$

where only k eigenvectors are taken into account. Here, $\mathbf{P}$ is some matrix, which brings the unique solution by fixing the rotation and satisfying the constraint:

$$\mathbf{P}\mathcal{J}_{pq}\mathbf{P}^T = \mathcal{J}_{pq}. \tag{6}$$

$\mathbf{X}$ in a k-dimensional space is determined from the matrix $\mathbf{D}$. If $k \ll n$ then a smaller $\mathbf{D}$ could be used to determine this k-dimensional space. The projections of new objects, represented by the distances to objects from $\mathbb{X}$ can be done by linear operations.

3 Augmented Embedding

Suppose we have selected k prototype patterns $\mathbf{x}_i \in \mathbb{X}$. We may construct $(k-1)$ dimensional pseudo-Euclidean space based on them, where each object from this set of selected objects has coordinates $\mathbf{x}_i = (x_i^{(1)}, x_i^{(2)}, \ldots, x_i^{(p)}, x_i^{(p+1)}, \ldots,$

$x_i^{(p+q)})^T$, $p+q=(k-1)$ and $i=1,\ldots,k$. Let us now assume that our configuration lives in a (k+1)-dimensional space such that we add one dimension to represent the positive subspace p and one dimension to represent the negative subspace q. The configuration (5) stays the same, except that the coordinates for these two extra dimensions are zeros. Objects from some new set $\tilde{\mathbb{X}}$ may be projected on this space by given their distances to $\mathbf{x}_i$. For every $\mathbf{x}_s \in \tilde{\mathbb{X}}$ $(s=1,\ldots,m)$ it can be done as follows:

$$d_{si}^2 = \sum_{l=1}^{p}\left(x_s^{(l)} - x_i^{(l)}\right)^2 - \sum_{l=p+1}^{k-1}\left(x_s^{(l)} - x_i^{(l)}\right)^2 + \varepsilon^2, \tag{7}$$

where $\varepsilon^2 = \varepsilon_p^2 - \varepsilon_q^2$ stands for the projection error and might be negative. We also assume that the center of mass lies in the origin: $\sum_{i=1}^{k+1}\mathbf{x}_i = 0$, remembering that the last coordinates for each prototype in our space are $x_i^{(k)} = x_i^{(k+1)} = 0$. Summing up among all k prototypes we receive the following equation:

$$\sum_{i=1}^{k} d_{si}^2 = \sum_{i=1}^{k}\sum_{l=1}^{p}\left(x_s^{(l)} - x_i^{(l)}\right)^2 - \sum_{i=1}^{k}\sum_{l=p+1}^{k-1}\left(x_s^{(l)} - x_i^{(l)}\right)^2 + k\varepsilon^2 \tag{8}$$

Opening brackets and recalling that the norm of any vector $\mathbf{x}_s$ can be expressed as:

$$\|\mathbf{x}_s\|^2 = \sum_{l=1}^{p}\left(x_s^{(l)}\right)^2 - \sum_{l=p+1}^{k-1}\left(x_s^{(l)}\right)^2 + \varepsilon^2 \tag{9}$$

we receive:

$$\|\mathbf{x}_s\|^2 = \frac{1}{k}\sum_{i=1}^{k-1}\left(d_{si}^2 - \|\mathbf{x}_i\|^2\right) \tag{10}$$

Substituting this result into equations (7) and after some computations we receive the following solution for the projected object $\mathbf{x}_s$ into $(k-1)$-dimensional space as $\mathbf{x}_s{}'$:

$$\mathbf{x}_s' = \frac{1}{2}|\Lambda|^{-1}\mathcal{J}_{pq}\mathbf{X}_i'^T\left(\mathrm{diag}(\mathbf{G}_i) - \mathbf{d}_s^2\right), \tag{11}$$

Here $\mathbf{X}_i'$ is a matrix of prototype coordinates, $\mathbf{G}_i$ is a Gram matrix for objects from $\mathbf{X}_i'$ and $\mathbf{d}_s^2$ is a vector of distances from an object $\mathbf{x}_s$ to all prototypes $\mathbf{x}_i$.

One should remember that the solution for $\mathbf{x}_s'$ is unique within the fixed rotation: $\mathbf{x}_s' = \mathbf{Q}|\Lambda|^{1/2}\mathbf{P}^T$, that satisfies the constraint (6). Finally, the sought vector of coordinates for the projected object can be derived as follows:

$$\|\mathbf{x}_s\|^2 = \|\mathbf{x}_s'\|^2 + \varepsilon^2 \tag{12}$$

On the other hand, recalling (10) and rewriting it in the matrix form:

$$\|\mathbf{x}_s\|^2 = -\frac{\mathbf{1}^T}{k}\left(\mathrm{diag}(\mathbf{G}_i) - \mathbf{d}_s^2\right), \tag{13}$$

we can derive ε^2.

However,

$$\varepsilon^2 = \varepsilon_p^2 - \varepsilon_q^2. \tag{14}$$

It means, that the all possible solutions for ε_p and ε_q, lie on a hyperbola (14) in the augmented subspace.

Our task is to optimize both positive and negative parts simultaneously to get a unique solution. It can be done in different ways. First, in a non-regularized version, one may just check the sign of ε^2 and depending on that assume the existence of only one ε_p or ε_q of the variable, calculating it as $sign(\varepsilon^2)\sqrt{|\varepsilon^2|}$. It means that the only one ε_p or ε_q encodes the projection error while the other is zero. As a result the objects will be projected directly on the axes of the augmented 2D subspace.

More advanced techniques, taking some assumptions about possible solutions, could also be constructed, assuming the simultaneous existence of both ε_p and ε_q variables. We will focus on looking for the so-called regularized normal solutions (solutions near the origin) that take the history into account, i.e. values close to the positive and negative class means, averaged among all axis in the space of dimension $(p+q)$. For this we will minimize the following functional:

$$F(\varepsilon_p, \varepsilon_q) = (\varepsilon_p - \hat{\mu_p})^2 + (\varepsilon_q - \hat{\mu_q})^2 \mapsto \min, \tag{15}$$

where

$$\begin{aligned} \hat{\mu_p} &= \frac{\mid \mu_p \mid}{p} \\ \hat{\mu_q} &= \frac{\mid \mu_q \mid}{q} \end{aligned} \tag{16}$$

expresses the averaged absolute values of positive and negative distribution means for each class in the $(k-1)$-dimensional space. This functional by the construction is convex and has a unique solution. It should be noted that the overall positive and negative means of the representation set is at the origin due to the centering procedure we have done. But, once the projection is made for the training set, the mean values μ_p and μ_q shift.

Moreover, in our regularization algorithm we choose to optimize the position of the test objects taking into account the class means from the training set. For each test object, the closest class mean is determined based on the pseudo-Euclidean distances. It means that $\hat{\mu}_p$ and $\hat{\mu}_q$ constitute now the mean vector of that class.

So, for every new object to project, the task (15) can be solved by the standard method of Lagrangian multipliers, taking into account the restriction (14).

$$L = F(\varepsilon_p, \varepsilon_q) + \lambda\left(\varepsilon^2 - \varepsilon_p^2 + \varepsilon_q^2\right), \tag{17}$$

where λ is some constant. Constructing Euler equations we receive:

$$\begin{aligned} \frac{\partial L}{\partial \varepsilon_p} &: \quad \varepsilon_p^2 = \frac{\hat{\mu_p}^2}{(1-\lambda)^2} \\ \frac{\partial L}{\partial \varepsilon_q} &: \quad \varepsilon_q^2 = \frac{\hat{\mu_q}^2}{(1+\lambda)^2} \end{aligned} \tag{18}$$

Substituting ε_p^2 and ε_q^2 we receive the following equations with respect to λ:

$$\varepsilon^2 = \frac{\hat{\mu_p}^2}{(1-\lambda)^2} - \frac{\hat{\mu_q}^2}{(1+\lambda)^2} \tag{19}$$

Solving this fourth-order equation, we get four solutions. Two of them we reject since they are imaginary. Among remaining two we select the one that brings the minimum to our functional (15).

4 Experimental Setup

Ionosphere data set. The data set describes radar returns from the ionosphere and is obtained from the UCI repository [5]. The targets are free electrons in the ionosphere. "Good" radar returns are those showing evidence of some type of structure in the ionosphere. "Bad" returns are those that do not; their signals pass through the ionosphere.

Received signals were processed using an autocorrelation function whose arguments are the time of a pulse and the pulse number. There were 17 pulse numbers for the used system. Instances in this database are described by 2 attributes per pulse number, corresponding to the complex values returned by the function resulting from the complex electromagnetic signal.

The number of instances is 351, the number of attributes is 34 plus the class attribute. All 34 predictor attributes are continuous; the 35th attribute is either "good" or "bad". This is a binary classification task with no missing values.

The dissimilarity matrix computed on the Ionosphere data set and used in our experiments is Euclidean. Moreover, distances in the matrix are scaled to be in $[0, 1]$.

Chicken pieces data set. This data set consists of 446 images of chicken pieces [2]. Each piece belongs to one of five categories, which represents specific parts of the chicken: wing (117 samples), back (76), drumstick (96), thigh and back (61), and breast (96). Each image is in binary format containing the silhouette of a particular piece. Pieces were placed in a natural way without considering orientation.

To extract string representations, some preprocessing had been done and provided to us by the group of prof. Bunke [6]. First, edge detection was performed. Secondly, the edges were approximated by straight line segments of fixed length. The sequence of angles between the segments were chosen as the string representation. Such string representations are then compared by edited distances. The cost of substitution is the absolute difference between the angles, while the costs of insertion and deletions are fixed. In our experiments we have used the segments of length 25 and the insertion and deletion costs 60. Final dissimilarity matrix computed on a data set appears to be non-Euclidean. Again, distances are rescaled to be in $[0, 1]$.

In all our experiments we use the classification error of the 1-NN classifier averaged over 20 repetitions as a performance criterion for our embedding techniques. For both data sets we set uniform prior probabilities for each of the

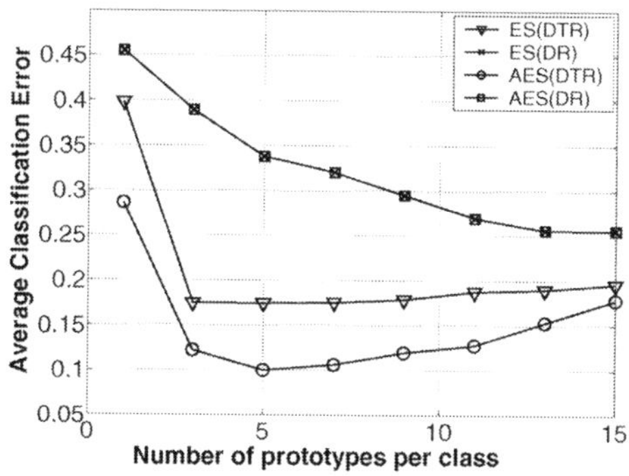

(a) Ionosphere data set

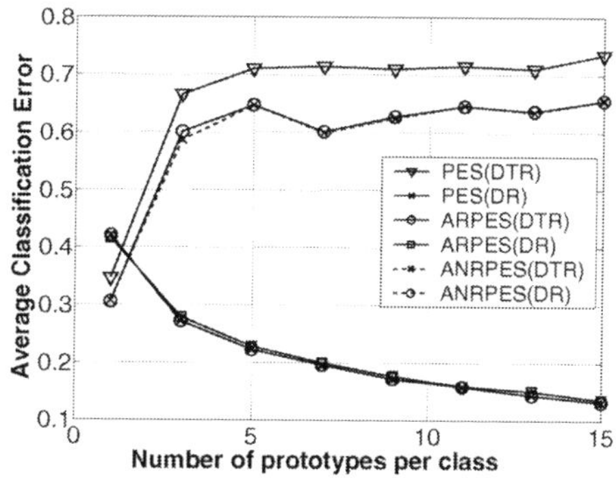

(b) Chicken pieces data set

Fig. 1. Averaged classification error (over 20 repetitions) for 1-NN classifier. Euclidean distance matrix is computed on Ionosphere data set, while pseudo-Euclidean distances are computed for Chicken pieces data set.

classes. Training data is divided into two parts. The first part is used for the representation, randomly chosen from the training set, and defining the pseudo-Euclidean embedding described in section 2. The remaining part is projected in this space. In such an augmented space the complete training data is used for the performance evaluation of the 1-NN rule. The test data are also projected to this augmented space and the distances to the training objects are recomputed according to the pseudo-Euclidean distance of that space. The obtained results are averaged out.

The choice of the 1-NN is justified since all high level classifiers require the construction of probabilistic models in pseudo-Euclidean spaces, which are not defined yet in pattern recognition literature, while the 1-NN rule operates directly with distances obtained via an embedding algorithm. However, the whole idea described in this paper should be seen as a first step towards the construction of advanced classification methods which are left for our future research.

In figure 1 we use the following notation. "ES" and "PES" denote the usage of Euclidean or pseudo-Euclidean spaces. The entire training data is denoted as "DTR", while for the selected representation set as "DR". The regularized or not regularized versions of the augmented embedding are denoted either as "AR" or "ANR".

In figure 1 for both different data sets the idea of augmentation helps, especially when one wants to operate with sufficiently small-dimensional spaces. However, in pseudo-Euclidean spaces the projection of the training set does not lead to better classification accuracy, like traditionally in Euclidean spaces. Moreover, it decreases drastically. Our opinion is that the data is linearly projected on a very nonlinear space, possibly equipped with curvature and torsion. In cases the representation set is small to describe all nonlinearities present in data, the classification possibilities are weak.

Standard deviations for the Ionosphere data set are less than 0.0251, while for the Chicken data set are less 0.0182.

Figures 2(a) and 2(b) illustrate the regularized vs. non-regularized versions of the augmented embedding in pseudo-Euclidean spaces, and bring an intuition behind them for future high-level pseudo-Euclidean classifiers, despite the fact that for the 1-NN rule the difference is of little significance. Here, the axes represent two augmented dimensions, the positive and the negative ones. These plots visualize how objects from the chicken pieces data are projected into this 2D augmented subspace in both cases: on the axes themselves (non-regularized version) or when their positions are optimized (regularized version).

Of course, other regularization of the augmented embedding may be constructed within this framework. For example, the position on the augmented subspace may be found in such a way, that some trained distortion function for projected training objects is minimal and applied to test data. However, this is only feasible when one has small number of prototypes (to make use of the whole idea of augmentation) but sufficiently large number of projected training objects (to train distortion parameters).

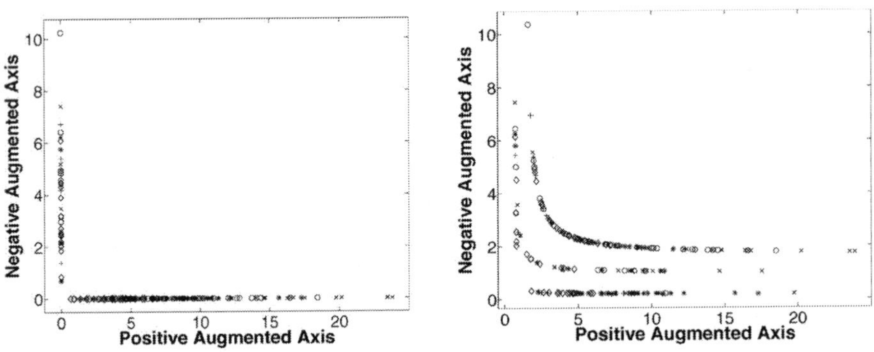

(a) No regularization (b) With regularization

Fig. 2. Chicken pieces data set. Projection of test objects in a space, spanned by 10 prototypes. Two pictures represent augmented subspaces, constructed either with and without regularization. The use of regularization helps to prevent object overlap.

5 Conclusion

In this paper we have presented an idea of an augmented embedding which can be seen as a first step towards statistical learning in pseudo-Euclidean spaces. The method helps to reconstruct projection errors made by existing linear embedding algorithms. It may bring higher level of topology preservation than the standard methods, especially in cases of small amount of prototypes to construct a proper space. We have showed that by adding one (in a Euclidean case) or two (in a pseudo-Euclidean case) extra dimensions it becomes possible to retrieve projection errors back made by existing linear embedding methods, leading to better classification. Our experiments support this statement. However, we should accept that the projection distortion may take high values, especially in spaces with large initial non-linearities between objects.

Acknowledgments

This work is supported by the Dutch Organization for Scientific Research (NWO). The authors would like to thank Prof. Dr. Horst Bunke for providing the *Chicken pieces data set.*

References

1. Pekalska E and Duin RPW. The Dissimilarity Representation for Pattern Recognition. Foundations and Applications. World Scientific, Singapore, 2005.
2. Andreu G, Crespo A, and Valiente JM. Selecting the toroidal self-organizing feature maps (TSOFM) best organized to object recognition. Proceedings of ICNN'97, 2:1341-1346, June 1997. Houston, Texas (USA). IEEE.
3. Pekalska E, Paclik P, and Duin, RPW. A generalized kernel approach to dissimilarity based classification. Journal of Machine Learning Research, 2:175-211, 2002.
4. Goldfarb L. A New Approach to Pattern Recognition. Progress in Pattern Recognition, Elsevier Science Publishers BV, 2:241-402, 1985.
5. Newman DJ, Hettich S, Blake CL and Merz CJ. UCI Repository of machine learning databases. University of California, Irvine, Dept. of Information and Computer Sciences, 1998, http://www.ics.uci.edu/~mlearn/MLRepository.html
6. Spillmann B. Description of the Distance Matrices, University of Bern, Institute of Computer Science and Applied Mathematics, Computer Vision and Artificial Intelligence (FKI), 2004.
7. Borg I. and Groenen P. Modern Multidimensional Scaling, Springer-Verlag, 1997.
8. Cox T.F. and Cox M.A.A. Multidimensional Scaling, Chapman & Hall, 1994.

Feature Over-Selection

Sarunas Raudys

Vilnius Gediminas Technical University
Sauletekio 11, Vilnius, LT-10223, Lithuania
raudys@ktl.mii.lt

Abstract. We propose probabilistic framework for analysis of inaccuracies due to feature selection (FS) when flawed estimates of performance of feature subsets are utilized. The approach is based on analysis of random search FS procedure and postulation that joint distribution of true and estimated classification errors is known *a priori*. We derive expected values for the FS bias, a difference between actual classification error after FS and classification error if ideal FS is performed according to exact estimates. The increase in true classification error due to inaccurate FS is comparable or even exceeds a training bias, a difference between generalization and Bayes errors. We have shown that there exists overfitting phenomenon in feature selection, entitled in this paper as feature over-selection. The effects of feature over-selection could be reduced if FS would be performed on basis of positional statistics. Theoretical results are supported by experiments carried out on simulated Gaussian data, as well as on high dimensional microarray gene expression data.

1 Introduction

Well known peaking (over-fitting) phenomenon relates generalization error of pattern recognition algorithm and a number of features in finite learning sample situations: the generalization error decreases at first with an increase in feature dimensionality. Then it saturates and starts increasing afterwards. After discovery [1], this phenomenon was transferred to proper selection of the complexity of a classifier: in small training-set cases, often it is preferable to use simple structured classification rules than the complex ones, and, vice versa, in large training-set cases, complex classifiers can be used more efficiently (the scissors' effect, [2, 3], see also [4], Section 1.5]). In neural network training, this effect is known under a name of overtraining (overfitting) [5]: with an increase in the number of training iterations the generalization error decreases at first, saturates and starts increasing afterwards. Like in the problem with input feature dimensionality, here we face an increase in complexity of the classifier with a progress of learning procedure. If before training the single layer perceptron based classifier, a data mean is shifted to a centre of coordinates, one starts training from initial weight vector with zero components and training sample sizes in two pattern classes $N_2= N_1=N/2$, then after the first iteration performed in a batch mode, one obtains simple Euclidean distance classifier. Next, iterative training process gradually moves the perceptron to six more complex classifiers [4] (for an introduction into statistical pattern recognition, see e.g. [6]).

D.-Y. Yeung et al. (Eds.): SSPR&SPR 2006, LNCS 4109, pp. 622–631, 2006.

Peaking phenomenon requires adjusting the dimensionality of input features to training sample size and the complexity of the classification algorithm. To reduce the number of features, FS procedures are utilized usually. There are four examples: a) evaluate the quality of p original features independently and select r best ones, b) forward selection, c) backward selections and d) random search where from p original features one generates *a group of m random feature subsets* composed of r features ($r < p$). Then one evaluates the quality of all m subsets and selects the best.

From point of view of a complexity, the algorithm "a" is the simplest. An answer which algorithm is more complex, "b" or "c", depends on p, r and the data. The complexity of random search feature selection algorithm is determined by number m. In spite of algorithmic simplicity, often random search is comparable in performance with more sophisticated FS algorithms. Therefore, algorithm "d" could be utilized as an undemanding model to study the complexity of FS problem.

If the features are selected incorrectly, generalization error of the classification system increases. Main factors that are affecting FS success in finite sample size situations are: 1) correctness of determination of the number of final features, r, in dependence on complexity of the classifier and training set size, 2) the accuracy of the criterion and validations sample size utilized to evaluate the quality of feature subset and 3) an excellence of the feature selection algorithm.

Determination of optimal dimensionality was considered in [1, 4, 6, 7]. Accuracy of the criteria (a bias, a variance) was considered while comparing methods to estimate the classification error [4, 6]. Comparative complexities of various FS schema have been studied in [8, 9] and references therein. Very often inaccuracies of feature quality determination were ignored. Exceptions are few, papers [10-16].

In order to separate effects of FS from that of training, *we do not study training sample size and complexity relations*. We assume there that a variety of already trained classifiers exist. Each of them is based on individual feature subset of the same dimensionality. On a basis of independent validation set one needs to select the best feature subset (classifier). We investigate both the accuracy of performance estimate (variance) and the complexity of the FS schema. We use probabilistic framework suggested in [10, 11], improve computer simulation tools, derive equations for an increase in expected classification error due inexact FS and show that with an increase in complexity of the feature subset selection schema, classification error rate exhibits peaking behavior. Theoretical and experimental analysis show that while applying random search FS schema, in order to obtain better result, one needs consider smaller amount of feature subsets and do not select apparently the best subset of features. Ng [13] gave reasons for not selecting the hypothesis with the lowest validation error. He demonstrated this by analyzing very artificial schema. Presently, we demonstrate such effect both analytically and experimentally for realistic feature selection tasks.

2 Statement of the Problem

In this section we will elucidate the factors influencing FS accuracy by considering as simple pattern recognition problem as possible. Consider two class problem with multivariate p-dimensional Gaussian classes with different means, μ_1, μ_2, and sharing common covariance matrix Σ. In this demonstrative example, p=150; only several features were "really good": $\mu_1 - \mu_2 =$ [1.45 1.15 0.95 0.80 0.70 0.60 0.55 0.50 0.45

0.42 0.40 0.375 0.370 0.3679 0.3657 0.3635 .... 0.0776 0.0755]^T. All variances were equal to 1.0 and correlations between all pairs of features, ρ=0.667. The designer needs to create standard linear Fisher classifier based on a best r-dimensional feature subset (r=8<< p). Note, that in this *example with equal correlations*, a subset of eight individually best features is not the best: this subset results Bayes error P_B=0.1830, while one of randomly formed subset composed of 1, 2, 3, 54, 95, 113, 127, and 113th features gives much better, P_B=0.0983.

In our analysis we assume that there exists a variety of already trained classifiers. The classes are Gaussian. Therefore, the Bayes errors, $\Phi(-½\delta)$ are known to me (δ stands for Mahalanobis distance). The designer does not know performances of feature subsets, however, he/she has independent validation set. On a basis of his information the designer needs to select a best subset (classifier) from $C_p^r \approx 5.26\times10^{12}$ potentially possible ones. We consider in this section that the designer uses the sample based Mahalanobis distance, $\hat{\delta}$, as a measure of feature subset's quality (the classification error, $\hat{P}_{\text{error}} = \Phi(-½\hat{\delta})$).

In Figure 1a we present a histogram of Bayes errors obtained in M=50,000 random generations of 8-dimensional feature subsets. Such multimodal density is rather typical for many real-world pattern recognition problems where one has a small number of relatively good features. In this example we used a single 150D validation set composed of N = 10+10 vectors. Very small validation set size (N=20 vectors), is specially tailored to to-day's microarray gene expression data experiments to be discussed later. In Figure 1b we present a scatter diagram of m=5,000 2-dimensional (2D) vectors (P_{B^i}, $\hat{P}_{\text{error}^i}$), i = 1, 2, ... , m selected uniformly out of M subsets (having M=50,000 subsets, it is possible to do this in $C_{50000}^{5000} \approx 9.68\times10^{32}$ different ways).

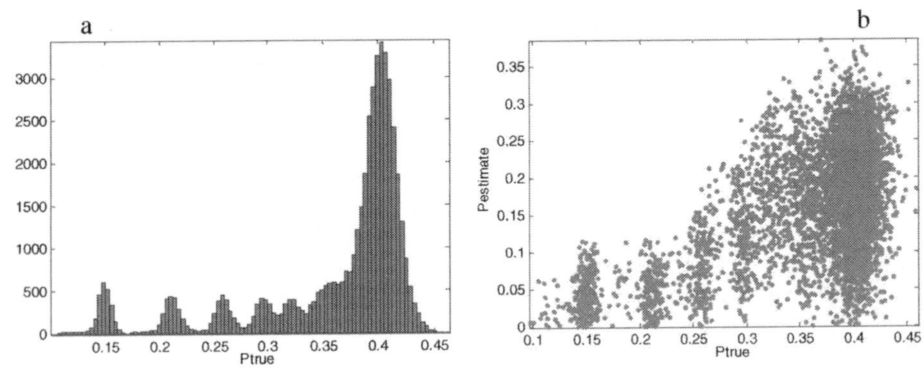

Fig. 1. a – a histogram of 50,000 values of P_B, b – a scatter diagram of vectors (P_B, $\hat{P}_{\text{error}}$)

Scatter diagram 1b shows that a great number of subsets with practically zero estimate $\hat{P}_{\text{error}}$ of classification error exist. True classification error for these subsets varies between 0.12 and 0.40. In mimicking random search FS strategy performed by classifier designer, we formed C_{50000}^{m} virtual groups composed of m feature subsets (m = 1, 3, 6, 10, …, 10000). In each group we found a subset (say s-th subset) with smallest estimate $\hat{P}_{\text{error}^s}$ and this subset's true error, P_{B^s}. An average of C_{50000}^{m} values of P_{B^s} we call "*a mean of the true classification error after feature selection*". The average was calculated by specially combinatory algorithm developed by Pikelis (see Appendix A.4 in [4]). In Figure 2a we have a graph, True1, of dependence of the mean of the true error in feature selection on m, the group size, the number of feature subsets in the group. At the same time we calculated average of C_{50000}^{m} values of $\hat{P}_{\text{error}^s}$ which is called "*a mean of an apparent classification error after feature selection*", graph Apparent in Figure 2a. In similar way, we can find average of "*ideal classification error after feature selection*" where we seek for subset with smallest P_{B^g} in the group (graph Ideal in Figure 2a). We see, a bias between True1 and Ideal is rather wide. The bias increases with an increase in m, the size of the group.

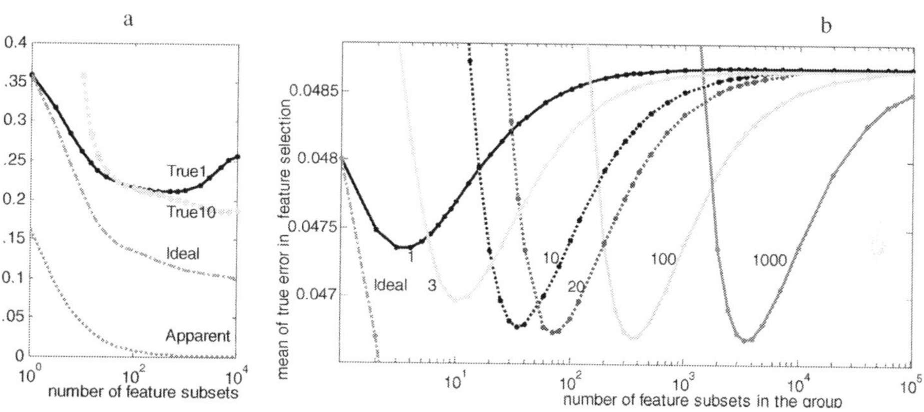

Fig. 2. Dynamics of true, ideal and apparent errors in feature selection: a - experiments with artificial Gaussian data; b – theoretical graphs

Graph True1 in Fig. 2a demonstrates peaking behavior: with an increase in the group size, true classification error after FS decreases at first, saturate and then starts increasing - we fit to validation set too much. We name this effect *feature over-selection. In order to obtain better result*, one need to consider smaller amount of feature subsets and/or do not select apparently the best subset of features. While inspecting 2D vectors in each group, we select the j-th positional statistics, i.e. the j-th feature subset according to estimates $\hat{P}_{\text{error}^s}$. In order to find averages from virtually formed $C_{50000-j+1}^{m-j+1}$ groups of feature subsets, a new combinatory algorithm was developed. The FS performed according to positional statistics helped in many pattern recognition tasks. Graph True10 in Figure 2a shows a mean value of true error after

FS if the j-th positional statistics (j=10) was used to select best feature subset. We see "*not choosing the best*" strategy allowed to reduce classification error substantially. This selection strategy firstly was analyzed by Ng in [13], for rather simplified artificial model of the hypotheses selection. Now we demonstrate its usefulness for standard pattern recognition task with dependent variables. In next section, we will suggest probabilistic framework to analyze feature over-selection theoretically.

3 Probabilistic Framework to Analyze Feature Selection Bias

A main declaration utilized in our analysis, is consideration of random search feature selection procedure. Then we may assume that m 2D vectors $(P_{B^i}, \hat{P}_{\text{error}^i})$ utilized to select certain subset of features are random vectors extracted from 2D population. Following a standard probability theory, probability density function of random vector $(P_B, \hat{P}_{\text{error}})$ may be expressed as a product of conditional and unconditional densities

$$f_1(P_B, \hat{P}_{\text{error}}) = f_2(\hat{P}_{\text{error}} \mid P_B) f_3(P_B) = f_4(P_B \mid \hat{P}_{\text{error}}) f_5(\hat{P}_{\text{error}}). \quad (1)$$

In derivation of expected value of true classification error after FS we postulate that conditional density, $f_2(\hat{P}_{\text{error}} \mid P_B)$*, and unconditional one,* $f_3(P_B)$*, are known a priori.* As a result, final conclusions are conditioned by $f_1(P_B, \hat{P}_{\text{error}})$. Standard theory gives

$$f_5(\hat{P}_{\text{error}}) = \int f_2(\hat{P}_{\text{error}} | P_B) f_3(P_B) dP_B \text{ and} \quad (2)$$

$$f_4(P_B \mid \hat{P}_{\text{error}}) = f_2(\hat{P}_{\text{error}} \mid P_B) f_3(P_B)/f_5(\hat{P}_{\text{error}}), \quad (3)$$

where integration is performed over all interval of varying P_B. If one defines the distribution of P_B over a set of discrete values, integration is substituted by summation.

Equation (3) could be used in order to evaluate a mean of true classification error if certain estimate, $\hat{P}_{\text{error}^k}$, is already picked out of a pool of values, $\hat{P}_{\text{error}^1}, \hat{P}_{\text{error}^2}, \ldots, \hat{P}_{\text{error}^m}$. In original paper [11], a situation with extreme (minimal) value of the error estimate $\hat{P}_{\text{error}}$ was considered. Inspired by the author's multiple experimental observations that utilization of positional statistics sometimes outperforms usage of minimal values (one of them is presented in previous section) and Ng [13] considerations, in this paper we will move from analysis of minimal value to the k-th positional statistics, $\hat{P}_{\text{error}^k}$, the k-th value in a ranged sequence $\hat{P}_{\text{error}^{k_1}} \leq \hat{P}_{\text{error}^{k_2}} \leq \ldots, \hat{P}_{\text{error}^{k_{1m}}}$. Statistical theory of extreme value distributions gives

$$f_6(\hat{P}_{\text{error}^k}) = \frac{\Gamma(m+1)}{\Gamma(k)\Gamma(m-k+1)} [F_5(\hat{P}_{\text{error}})]^{k-1} [1 - F_5(\hat{P}_{\text{error}})]^{m-k} f_5(\hat{P}_{\text{error}}), \quad (4)$$

where $F_5(\hat{P}_{\text{error}})$ is cumulative distribution of random variable $\hat{P}_{\text{error}}$.

Use of distribution density of the k-th positional statistics (4) results expected value of true classification after feature selection and that of average of apparent error

$$EP_{\text{true}k} = \iint f_4(P_B|\hat{P}_{\text{error}^k}) f_6(\hat{P}_{\text{error}^k}) d\hat{P}_{\text{error}^k} dP_B \text{ and} \quad (5)$$

$$E\hat{P}_{\text{apparent}k} = \int \hat{P}_{\text{error}^k} f_6(\hat{P}_{\text{error}^k}) d\hat{P}_{\text{error}^k} . \quad (6)$$

The integrations (or summations) in Eq. (5) are performed along intervals of variations of error estimate, $\hat{P}_{\text{error}}$, and true error, P_B. Both expected values are conditioned by serial number of positional statistics. If k=1, we deal with extreme (minimal) value. Hypothetical characteristics, an expectation of ideal classification error, we have in case where from m randomly formed subsets of features we select the best one according to true classification error values, P_{B1}, P_{B1}, ... , P_{Bm}. Probability density function of extreme value is given by equation

$$F_7(P_B) = m\,[1 - F_3(P_B)]^{m-1} f_3(P_B), \quad (7)$$

where $F_3(P_B)$ is cumulative distribution function of random variable P_B.

Then the expectation of ideal classification error could be found as

$$EP_{\text{ideal}} = \int P_B f_7(P_B) dP_B . \quad (8)$$

Above equations allow to investigate behavior of expected values of true, apparent and ideal classification errors after feature selection, We remind that the conditional density, $f_2(\hat{P}_{\text{error}} \,|\, P_B)$, and unconditional density, $f_3(P_B)$, should be known *a priori*. Due to complexity of the problem with extreme values and positional statistics we do not have explicit expressions. So, further analysis should be performed by numerical integration. In Fig. 2b we depict dependence of expected values of true classification errors when feature selections were performed according to extreme value (graph 1) and the 3th, 10th, 20th, 100th and 2000th positional statistics. Moreover, it was postulated that distribution density of 2D vectors (P_B, $\hat{P}_{\text{error}}$) was a mixture of two Gaussian densities

$$f_3(P_B, \hat{P}_{\text{error}}) = q_1 \times f_N((P_B, \hat{P}_{\text{error}}), \mathbf{M}_1, \mathbf{S}_1) + (1 - q_2) \times f_N((P_B, \hat{P}_{\text{error}}), \mathbf{M}_2, \mathbf{S}_2),$$

where $f_N((x_1, x_1), \mathbf{M}, \mathbf{S})$ denotes Gaussian probability density function of 2D vector, (x_1, x_1), having mean $\mathbf{M}$ and covariance matrix $\mathbf{S}$. Note that density $f_3(P_B, \hat{P}_{\text{error}})$ depends on random validation set critically. After analysis of two dozens of very left parts of distributions similar to that depicted in Fig. 1a,b, in a variety of situations with single validation sets of size (N=20), we selected for this illustration: q_1=0.1,

$$\mathbf{M}_1 = \begin{bmatrix} 0.03 \\ 0.01 \end{bmatrix}, \mathbf{S}_1 = \begin{bmatrix} 0.0002^2 & 0.0 \\ 0.0 & 0.005^2 \end{bmatrix}, \mathbf{M}_2 = \begin{bmatrix} 0.05 \\ 0.013 \end{bmatrix}, \mathbf{S}_2 = \begin{bmatrix} 0.00025^2 & 0.0 \\ 0.0 & 0.009^2 \end{bmatrix}.$$

We pay readers attention to a fact that for single validation set, the variances of hold out error counting error estimates (right bottom elements of matrices $\mathbf{S}_1$, $\mathbf{S}_2$) are much smaller as variance, s^2, predicted by theory for diverse independent validation sets, $s^2 = P_B(1 - P_B)/N$. In Figure 2b we also depict the very left part of graph "Ideal", the ideal error after feature selection, which decreased rapidly until 0.03 (for $m \approx 30$) and saturated at this level. Apparent classification error after feature selection started to decrease from 0.013 level, and for $m \approx 200$ it became practically zero. Experimentation with artificially generated feature selection problems revealed that a character of distribution $f_1(P_B, \hat{P}_{error})$ greatly depends on a way how dependencies between original variables of the data are generated, individual qualities of the features and, most important, on particular randomly chosen p-dimensional validation set.

4 Analysis of Over-Selection Phenomenon in Real World Task

We performed experiments with 7129-dimensional two class leukaemia data set [16, 17]. In order to select $r = 8$ features we employed standard linear Fisher classifier. From 72 examples we used 35 samples for training. Remaining 37 vectors constituted the validation set. Such large dimensionality/sample size ratio is frequent in many biomedical investigations, especially in to-day's analysis of microarray gene expression data. *Three millions* random feature subsets were generated. We examined the accuracy problem in a situation where training set (re-substitution) error estimates *with certain correction of "training bias"* were used as evaluations of "true" error. The validation set estimates were used to pick up "the best" feature subsets.

Re-substitution error estimates are optimistically biased. For Fisher classifier, asymptotically as training sample size, N, and dimensionality, r, are large, expected value of classification error can be found from simple, however, exact asymptotic formula $EP_N \approx \Phi(-\frac{1}{2}\delta / T_{bias})$, where $T_{bias} = \sqrt{(N-p)/N/(1+4p/N/\delta^2)}$ [4, 7].

We are also interested in the bias of re-substitution error, which can be expressed as $E\hat{P}_R \approx \Phi(-\frac{1}{2}\delta \times T_{bias})$. Double use of one-dimensional interpolation for $\hat{P}_R$ value allows to obtain, approximate (restored) value of generalization error, $\hat{P}_g$.

The histograms of "*restored*" generalization, $\hat{P}_g$, and validation, $\hat{P}_V$, error estimates are presented in Figure 3ab. We see that in spite of simplicity of asymptotic formulae, training bias elimination was performed quite correctly: left and middle parts of both histograms are almost identical. Fig. 3c shows a scatter diagram of distribution of 2000 vectors ($\hat{P}_R$, $\hat{P}_V$). Due to very small sample size, many subsets have the same $\hat{P}_R$ and $\hat{P}_V$ values. For better data visualization, a small uniform noise was added to each component. Two dimensional distribution of 10,474 vectors with smallest re-substitution and validation errors is detailed by 2D scatter diagram in Fig. 3d. We see that FS strategies based on the validation set and the modified training set estimates pick different feature subsets. This conclusion follows also from conditional mean of $\hat{P}_V$ presented in Fig. 3e: the smallest generalization error values could be obtained if subsets with *two* validation errors would be selected! We plot mean values of *restored generalization error* rates conditioned on 17 distinct validation error values. In Figure 3f we present dynamics of true error after feature selection performed according to the smallest validation error (True1) and that performed according to 30th and 300th positional statistics. We also present "Ideal" and "Apparent" classification errors in the same way as it was done in the experiments with artificial Gaussian data (Section 2, Figure 2a).

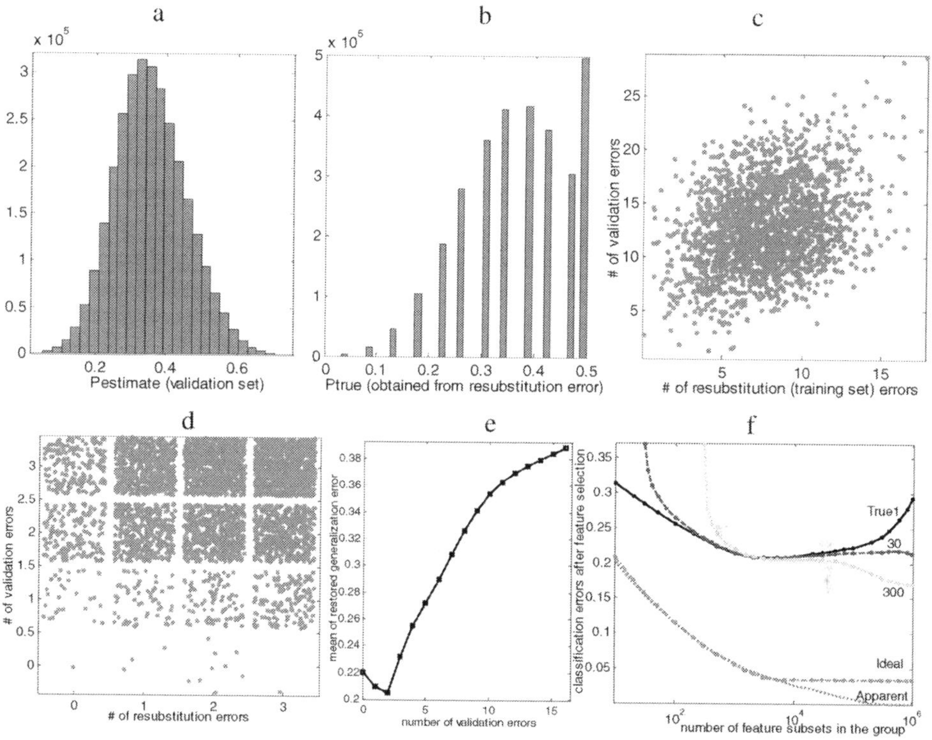

Fig. 3. The histograms of 3,0000,000 values of validation error estimates (a) and restored generalization error (b); c, d - scatter diagrams of distribution of the number of misclassification errors in training and validation sets; e – average "restored" generalization error as a function of estimated error, $\hat{P}_{\text{error}}$, f – dynamics of true, ideal and apparent errors

5 Concluding Remarks

While designing the classifiers from training set we obtain training bias, a difference between generalization and Bayes errors, EP_N - P_B, In present paper we consider feature selection bias (the difference between True1 and Ideal)) which arise when size of validation set is finite. We present further development of probabilistic framework started in [10, 11]. This approach is based on analysis of random search FS procedure and postulation that joint distribution of true and estimated classification errors is known *a priori*. Theoretical and experimental results advocate that feature selection bias can be very large if the size of feature subset group is very large. Calculation of expected generalization error of Fisher classifier according to $EP_N \approx \Phi(-½\delta / T_{bias})$ for N=20 and p=8 gives that for P_B=0.2 (for $m\approx$5, inspect Fig. 2a), $EP_N \approx$0.3 and for P_B=0.1 ($m\approx$10,000), $EP_N \approx$0.19. Similar estimates we obtained for gene expression data. It means that FS bias is comparable with training bias provided the training set size is equal to that of validation set and linear Fisher discriminant is utilized as the classifier.

We also showed that there exists overfitting phenomenon in feature selection, named in this paper as feature over-selection. This effect is validation set dependent. It was observed when validation set size was very small. The feature over-selection phenomenon could be diminished if FS would be performed on basis of positional statistics. Development of practical recommendations is a problem of future research.

Acknowledgement. A part of the experiments with gene expression data were performed when the author visited the Institute for Biodiagnostics, NRCC, Winnipeg, Canada (NATO Expert Visit Grant SST.EAP.EV 980950). The author is thankful to Richard Baumgartner and Ray Somorjai for useful and challenging discussions.

References

1. G. F. Hughes. On the mean accuracy of statistical pattern recognizers. *IEEE Transactions on Information Theory* IT-14:55–63, 1965.
2. S. Raudys. On the problems of sample size in pattern recognition. In V.S. Pugatchiov (editor) *Detection, Pattern Recognition and Experiment Design*, 2:64–76. Proceedings of the 2nd All-Union Conference Statistical Methods in Control Theory. Nauka, Moscow, 1970 (in Russian).
3. L. Kanal, B. Chandrasekaran. On dimensionality and sample size in statistical pattern classification. *Pattern Recognition* 3:238–55, 1971.
4. S. Raudys. *Statistical and Neural Classifiers - An integrated approach to design.* Springer-Verlag London, 2001.
5. S. Haykin. *Neural Networks: A comprehensive foundation.* 2nd edition. Prentice-Hall, Englewood Cliffs, NJ, 1999.
6. K. Fukunaga. *Introduction to Statistical Pattern Recognition.* 2nd Ed. Acad. Press, 1990.
7. S. J. Raudys, A. K. Jain. Small sample size effects in statistical pattern recognition: Recommendation for practitioners. *IEEE Trans. on Pattern Analysis and Machine Intelligence*, 13 (3) 242-254, 1991.
8. P. Pudil, J. Novovicova and J. Kittler. Floating search methods in feature selection. *Pattern Recognition Letters*, 15(11):1119-1125, 1994.
9. I. Guyon and A. Elisseeff, An introduction to variable and feature selection, *J. of Machine Learning Research,* 3:1157-1182, 2003.
10. S. Raudys. Classification errors when features are selected. In S. Raudys (editor), *Statistical Problems of Control*, 38: 9-26, 1979. Institute of Mathematics and Informatics, Vilnius, 1979 (in Russian).
11. S. Raudys. Influence of sample size on the accuracy of model selection in pattern recognition. In *Statistical Problems of Control* (S. Raudys, ed., Institute of Mathematics and Informatics, Vilnius) 50 9-30, 1981 (in Russian).
12. G. D. Murray. A cautionary note on selection of variables in discriminant analysis. *Appl. Statist.* 26 (3) 246-250, 1997.

13. A. Ng. Preventing "overfitting" of cross-validation data, *Proc. of the Fourteenth International Conference on Machine Learning*, Morgan Kaufman, 245-253, 1997.
14. J. Ye. On measuring and correcting the effects of data mining and model selection. *J. of American Statistical Association*, 93 (441) 120-131, 1998.
15. P. Domingos. Process-oriented estimation of generalization error. In Proceedings of the Sixteenth International, *Joint Conf on Art. Intell.*, Morgan Kaufman, 714-722, 1999.
16. C. Ambroise and G. J. McLachlan. Selection bias in gene extraction on the basis of microarray gene-expression data. *Proc. Natl. Acad. Sci. USA* 99 (10) 6562-6566, 2002.
17. T.R. Golub et al. Molecular classification of cancer: Class discovery and class prediction by gene expression monitoring. *Science* 286 531-537, 1999.

Flexible-Hybrid Sequential Floating Search in Statistical Feature Selection

Petr Somol[1,2], Jana Novovičová[1,2], and Pavel Pudil[1,2]

[1] Dept. of Pattern Recognition, Institute of Information Theory and Automation, Academy of Sciences of the Czech Republic, 182 08 Prague, Czech Republic
{somol,novovic}@utia.cas.cz
http://www.utia.cas.cz/user_data/PR_dept

[2] Faculty of Management, Prague University of Economics, Czech Republic
pudil@fm.vse.cz
http://www.fm.vse.cz

Abstract. Among recent topics studied in context of feature selection the hybrid algorithms seem to receive particular attention. In this paper we propose a new hybrid algorithm, the flexible hybrid floating sequential search algorithm, that combines both the filter and wrapper search principles. The main benefit of the proposed algorithm is its ability to deal flexibly with the quality-of-result versus computational time trade-off and to enable wrapper based feature selection in problems of higher dimensionality than before. We show that it is possible to trade significant reduction of search time for negligible decrease of the classification accuracy. Experimental results are reported on two data sets, WAVEFORM data from the UCI repository and SPEECH data from British Telecom.

1 Introduction

Feature selection, as a pre-processing step to machine learning and pattern recognition applications, has been effective in reducing dimensionality. It is sometimes the case that such tasks as classification or approximation of the data represented by so called feature vectors, can be carried out in the reduced space more accurately than in the original space. Liu and Yu [1] provide a comprehensive overview of various aspects of feature selection. Their paper surveys existing feature selection algorithms for classification and clustering, evaluation criteria and data mining tasks and outlines some trends in research and development of feature selection.

Many existing feature selection algorithms designed with different evaluation criteria can be categorized into *Filter* [2], [3] *Wrapper* [4] and *Hybrid* [5], [6]. Filter methods rely on general characteristics of the training data to select some features independently of the subsequent learning algorithm. Therefore they do not inherit any bias of a learning algorithm. The wrapper methods require one predetermined learning algorithm in feature selection and use its performance to

D.-Y. Yeung et al. (Eds.): SSPR&SPR 2006, LNCS 4109, pp. 632–639, 2006.

evaluate and determine which features are selected. These methods tend to give superior performance as they find features better suited to the predetermined learning algorithm, but they also tend to be more computationally expensive. When the number of features becomes very large, the filter methods are usually to be chosen due to computational efficiency. To combine the advantages of both methods, algorithms in a hybrid approach have recently been proposed to deal with high dimensional data.

In this paper we introduce a flexible hybrid version of the floating search, the *hybrid sequential forward floating selection* (hSFFS) as well as its backward counterpart (hSBFS) that cross the boundary between filters and wrappers. We show that it is possible to trade significant reduction of search time for negligible decrease of the classification accuracy.

2 Motivation for Hybrid Algorithms

Filter methods for feature selection are general preprocessing algorithms that do not rely on any knowledge of the learning algorithm to be used. They are distinguished by specific evaluation criteria including distance, information, dependency. Since the filter methods apply independent evaluation criteria without involving any learning algorithm they are computationally efficient. Wrapper methods require a predetermined learning algorithm instead of an independent criterion for subset evaluation. They search through the space of feature subsets using a learning algorithm, calculate the estimated accuracy of the learning algorithm for each feature before it can be added to or removed from the feature subset. It means, that learning algorithms are used to control the selection of feature subsets which are consequently better suited to the predetermined learning algorithm. Due to the necessity to train and evaluate the learning algorithm within the feature selection process, the wrapper methods are more computationally expensive than the filter methods.

The main advantage of filter methods is their speed and ability to scale to large data sets. A good argument for wrapper methods is that they tend to give superior performance. Because of the success of the *sequential floating search* methods of filter type introduced by Pudil et al. [7] on many datasets and our focus on real-world datasets with potentially large number of features and small training sets, we have developed a *hybrid floating selection* algorithm that crosses the boundary between filter and wrapper methods and emphasizes some of the advantages of wrapper methods.

3 Hybrid Floating Sequential Search

Floating search methods [7], [8], sequential forward floating selection (SFFS) and sequential backward floating selection (SBFS), are now considered to be standard feature selection tools, providing good performance and close-to-optimum or optimum results in most tasks [9], [10]. In the following we will focus on the sequential *forward* floating selection because it has proven appropriate for most real-world datasets. The definition of the backward algorithm is analogous.

Starting from empty feature set, the SFFS procedure consists of applying after each forward (feature adding) step a number of backward (feature removing) steps as long as the resulting subsets are better than previously evaluated ones at that level. Consequently, there are no backward steps at all if the performance cannot be improved. The algorithm allows a 'self-controlled backtracking' so it can eventually find good solutions by adjusting the trade-off between forward and backward steps dynamically. It is possible to say that, in a certain way, it computes only what it needs without any parameter setting. In this way it overcomes effectively the so-called *nesting problem* inherent to older methods [11].

The same scheme can be used both in filter and wrapper context, as the floating algorithms put no restrictions on the behavior of criterion functions (unlike, e.g., Branch & Bound, which requires monotonic criteria). Here we introduce a flexible hybrid version of the floating search, hybrid sequential forward floating selection (hSFFS) that crosses the boundary between filters and wrappers. We show, that only a fraction of full wrapper computational time is sufficient to obtain results not too different from the full wrapper ones. This is accomplished by taking use of filter criteria to avoid less promising subsets in wrapper computation. The proportion of subsets to be passed to wrapper-based evaluation can be specified by the user. In this way one can decide the trade-off between the length of computation and criterion maximization effectiveness.

3.1 Formal Description of hSFFS

For the purpose of formal hSFFS description we use the following notion and abbreviations: Let the number of all features be D and the full feature set be $X_D = \{x_i, i = 1, \ldots, D\}$. Due to the hybrid nature of the algorithm to be defined we will distinguish two criterion functions. $J_F(.)$ denotes the faster but possibly less appropriate *filter* criterion, $J_W(.)$ denotes the slower *wrapper* criterion. The *hybridization coefficient*, defining the proportion of feature subset evaluations to be verified by wrapper means, is denoted by $\lambda \in \langle 0, 1\rangle$. Here $\lfloor \cdot \rfloor$ denotes value rounding. Let SFS, SBS denote sequential forward selection and sequential backward selection [11], respectively.

It is required that at each stage k all the so-far best subsets X_i and corresponding criterion values $J_i = J(X_i)$ are known for $i = 1, \ldots, \tilde{k}$ with $\tilde{k}$ denoting the largest subset size tested so-far ($k < \tilde{k}$ while backtracking).

Hybrid SFFS Algorithm

Initialization: The algorithm is initialized by setting $k = 0$ and $X_0 = \emptyset$. Then, Step 1 is called twice to obtain feature sets X_1 and X_2; to conclude the initialization let $J_1 = J_W(X_1), J_2 = J_W(X_2)$ and $k = 2$.

STEP 1: (*Adding*) By analogy to the SFS method, select from the set of available features, $X_D \setminus X_k$ the best feature with respect to the set X_k, say x^+, and add it to the current set X_k to form new feature set X_{k+1}; to achieve this, first pre-select c_k^+ most promising candidate features by maximizing $J_F(\cdot)$, then decide according to the best $J_W(\cdot)$ value, i.e.:

$$c_k^+ = \max\{1, \lfloor\lambda(D-k)\rfloor\} \tag{1}$$

$$C_k^+ = \{x_{i_t}, t = 1, \ldots, c_k^+ : J_F(X_k \cup \{x_{i_t}\}) \geq J_F(X_k \cup \{x_j\}) \, \forall j \neq i_t\} \tag{2}$$

$$x^+ = \arg\max_{x \in C_k^+} J_W(X_k \cup \{x\}), \quad X_{k+1} = X_k \cup \{x^+\}. \tag{3}$$

STEP 2: (*Inferior search path cancellation*) If J_{k+1} is known from before and $J(X_{k+1}) < J_{k+1}$, set $k = k + 1$ and go to Step 1.
STEP 3: (*Conditional removal*) By analogy to the SBS method find the worst feature in the set X_{k+1}, say x^-; to achieve this, first pre-select c_k^- most promising candidate features by maximizing $J_F(\cdot)$, then decide according to the best $J_W(\cdot)$ value, i.e.:

$$c_k^- = \max\{1, \lfloor\lambda k\rfloor\} \tag{4}$$

$$C_k^- = \{x_{i_t}, t = 1, \ldots, c_k^- : J_F(X_k \setminus \{x_{i_t}\}) \geq J_F(X_k \setminus \{x_j\}) \, \forall j \neq i_t\} \tag{5}$$

$$x^- = \arg\max_{x \in C_k^-} J_W(X_{k+1} \setminus \{x\}). \tag{6}$$

If $J_W(X_{k+1} \setminus \{x^-\}) = J_k$, i.e., no better solution has been found, set $J_{k+1} = J(X_{k+1})$, $k = k + 1$ and go to Step 1; otherwise remove this feature from the set X_{k+1} to form a new feature set $X_k^{'}$, i.e.

$$X_k^{'} = X_{k+1} \setminus \{x^-\}. \tag{7}$$

Note that now $J(X_k^{'}) > J(X_k) = J_k$. If $k = 2$, then set $X_k = X_k^{'}$ and $J_k = J(X_k^{'})$ and go to Step 1, otherwise set $k = k - 1$ and repeat Step 3.

Remark 1: Definitions (1) and (4) ensure that for any $\lambda \in \langle 0, 1\rangle$ at least one evaluation of $J_W(\cdot)$ is done in each algorithm step for each tested subset size.

Remark 2: Algorithm Step 2 can be considered optional. It is defined to prevent possible criterion decrease that may occur when the algorithm returns to higher dimensionality after backtracking. Keeping intermediate criterion values as high as possible is certainly desirable, yet as such cannot guarantee a better result.

3.2 Simplified Flowchart of the hSFFS

A simplified flowchart of the hSFFS algorithm is given in Fig. 1. The alternative terminating condition $k = d + \delta$ in the flowchart allows premature termination of the search process, should there be no reason to evaluate cardinalities greater than d. In such a case it is good to let the algorithm reach a little higher dimensionality ($d + \delta$) to allow possible find of a better solution for d by means of backtracking. The value of δ can be selected arbitrarily, or estimated heuristically, e.g., as the longest backtracking sequence performed so-far. Nevertheless, letting the algorithm finish (reach dimensionality D) is to be recommended. The fact that floating search finds solutions for all cardinalities in one run is one of its key advantages.

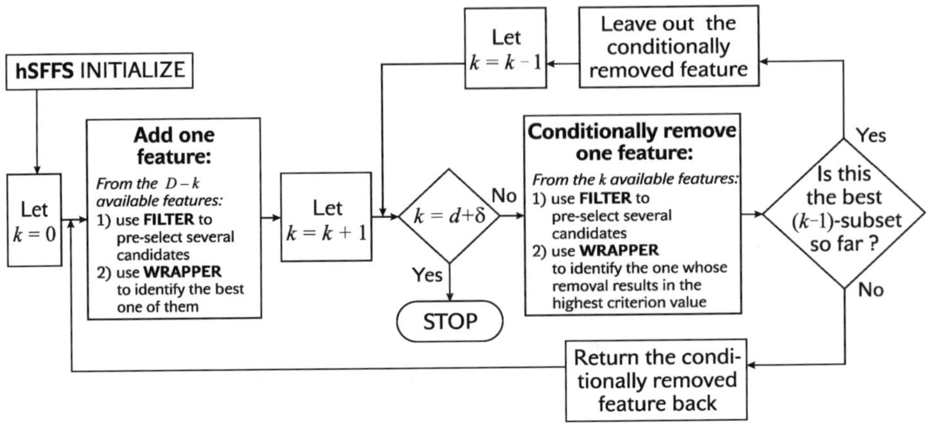

Fig. 1. Simplified diagram of the *hybrid* SFFS

4 Experiments

4.1 Datasets

The performance of the proposed algorithm is illustrated on two datasets. We used WAVEFORM data (40 features, 1692 samples from class 1 and 1653 samples from class 2) obtained via the UCI repository [12] and SPEECH data originating from British Telecom (15 features, 682 word "yes" and 736 word "no" samples), obtained from the Centre for Vision, Speech, and Signal Processing of the University of Surrey, UK.

4.2 Feature Subset Selection Criteria

We suppose, that the class-conditional densities are multivariate Gaussian, but the parameters of the densities (i.e. mean vectors and covariance matrices) are unknown and are replaced by their maximum likelihood estimates.

In the case of the filter model we used estimation of Bhattacharyya distance as the independent criterion $J_F(\cdot)$. A dependent criterion $J_W(\cdot)$ used in the wrapper model is the classification rate of the Bayes Gaussian plug-in classifier. All classification rates have been verified by a 25-fold cross-validation.

4.3 Experimental Results

For each dataset the results are presented in two graphs. The first graph (Figures 2 and 3) shows the Gaussian classifier correct classification rate on best feature subsets selected by the *hybrid* SFFS for different values of the hybridization coefficient λ as well as results of the filter SFFS and the wrapper SFFS. The second graph (Figure 4) shows the times of complete hSFFS(λ) runs for each λ. Note that floating search yields all subsets in one run, thus the graph of time contains just a single line.

It can be observed that especially for lower subset sizes the increase of λ quickly improves the classification rate. The improvement of the classification rate does not depend linearly on increased computational time. For the values of λ less than roughly 0.5 the classification rate tends to increase considerably faster than the time (an exception being, e.g., the 11 features case in Fig. 2). This is quite important. It suggests that investing some additional time into hybrid search with $\lambda \leq 0.5$ brings relatively more benefit than investing all the time needed for full wrapper based feature selection. The results for $\lambda \approx 0.5$ tend to be closer to those of wrappers than those of filters. This positive effect can be understood as an illustration of the ability of Bhattacharyya distance to pre-select reasonable feature candidates for further evaluation in the Gaussian wrapper. However, it also shows the limits of this Bhattacharyya ability. A hypothetically perfect filter criterion would cause the hSFFS yield for each λ the same best solution. The lack of such perfect criteria is the reason for using wrapper based search.

Remark: This is not to say that the time complexity of the proposed hybrid search is negligible. Obviously, it is to be expected considerably slower than the time complexity of filter search, yet only a fraction of the time complexity of wrapper search.

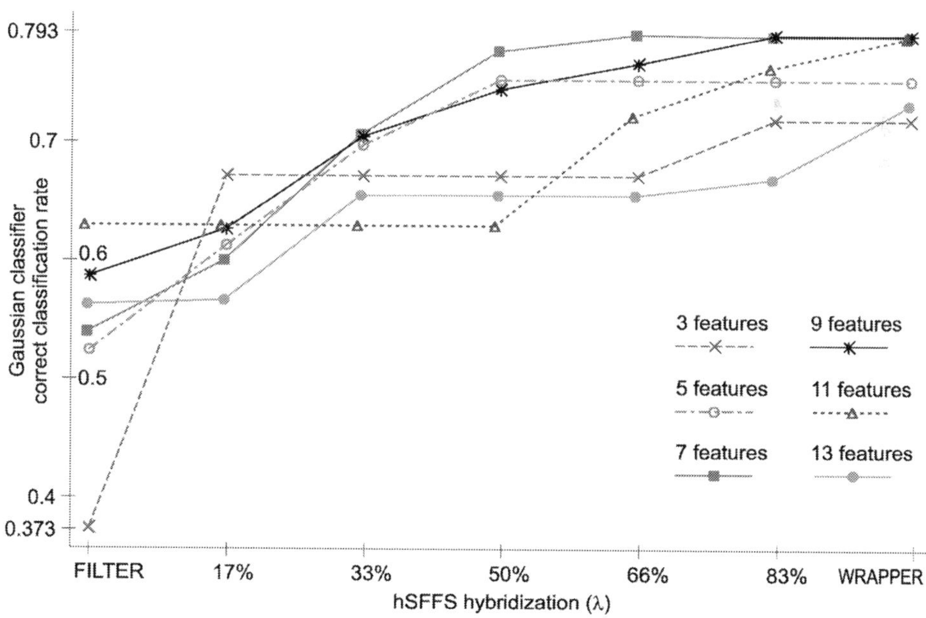

Fig. 2. SPEECH dataset: Comparison of classifier performance on feature subsets selected by the hSFFS for different λ, the *filter* SFFS and the *wrapper* SFFS

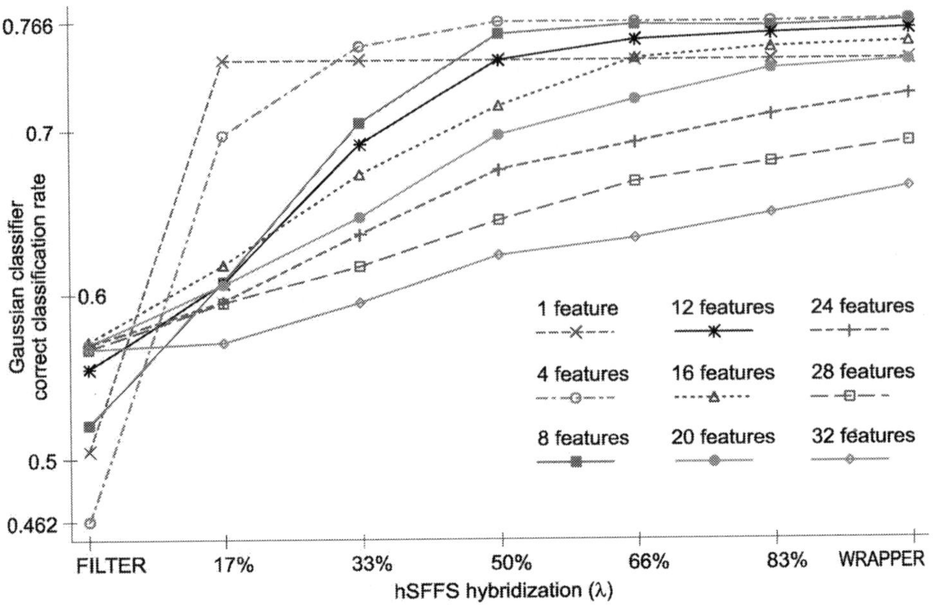

Fig. 3. WAVEFORM dataset: Comparison of classifier performance on feature subsets selected by the hSFFS for different λ, the *filter* SFFS and the *wrapper* SFFS

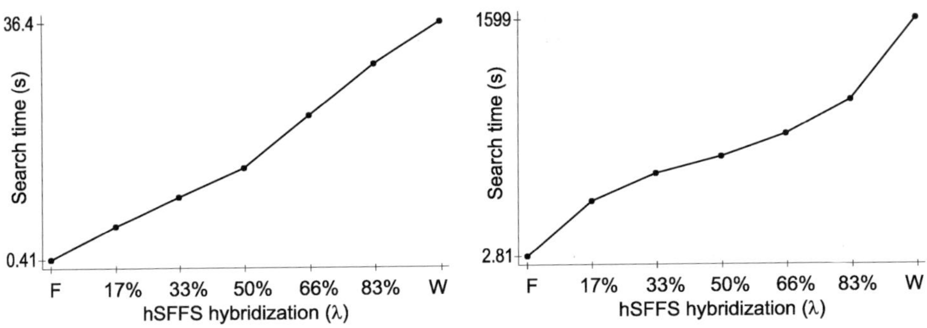

Fig. 4. SPEECH and WAVEFORM datasets: Time complexity of the *filter* SFFS, the hSFFS as a function of λ and the *wrapper* SFFS

5 Conclusions and Future Work

We have defined a flexible hybrid version of floating search methods for feature selection. The main benefit of the proposed floating search hybridization is the possibility to deal flexibly with the quality-of-result vs. computational time trade-off and to enable wrapper based feature selection in problems of higher dimensionality than before. We have shown that it is possible to trade significant reduction of search time for often negligible decrease of the classification accuracy.

In the future we intend to "hybridize" other search methods in a similar way as presented here and to investigate in detail the hybrid behavior of different combinations of various probabilistic measures and learning methods.

Acknowledgement. The work has been supported by the following grants: CR MŠMT grant 1M0572 DAR, EC project FP6-507752 MUSCLE, Grant Agency of the Academy of Sciences of the Czech Republic (CR) No. A2075302, and CR Grant Agency grant No. 402/03/1310.

References

1. Liu, H., Yu, L.: Toward Integrating Feature Selection Algorithms for Classification and Clustering. IEEE Trans. on Knowledge and Data Engineering **17** (2005) 491-502
2. Yu, L., Liu, H.: Feature Selection for High-Dimensional Data: A Fast Correlation-Based Filter Solution. In: Proc. 20th Intl Conf. Machine Learning (2003) 856-863
3. Dash, M., Choi, K., Scheuermann, P., Liu, H.: Feature Selection for Clustering - a Filter Solution. In: Proc. Second Int. Conf. Data Mining (2002) 15-122
4. Kohavi, R., John, G.H.: Wrappers for Feature Subset Selection. Artificial Intelligence **97** (1997) 273–324
5. Das, S.: Filters, Wrappers and a Boosting-Based Hybrid for Feature Selection. In: Proc. 18th Intl Conf. Machine Learning (2001) 74-81
6. Sebban, M., Nock, R.: A Hybrid Filter/Wrapper Approach of Feature Selection using Information Theory. Pattern Recognition **35** (2002) 835–846
7. Pudil, P., Novovicova, J., Kittler, J.: Floating Search Methods in Feature Selection. Pattern Recognition Letters **15** (1994) 1119-1125
8. Pudil, P., Novovicova, J., Somol, P.: Recent Feature Selection Methods in Statistical Pattern Recognition. In: Pattern Recognition and String Matching, Springer-Verlag, Berlin Heidelberg New York (2003)
9. Jain, A.K., Zongker, D.: Feature selection: evaluation, application and small sample performance. IEEE Trans. PAMI **19** (1997) 153–158
10. Kudo, M., Sklansky, J.: Comparison of algorithms that select features for pattern classifiers. Pattern Recognition **33** (2000) 25–41
11. Devijver, P.A. Kittler, J Pattern Recognition: A Statistical Approach. Prentice-Hall (1982)
12. Murphy, P.M., Aha, D.W.: UCI Repository of Machine Learning Databases [Machine-readable data repository]. University of California, Department of Information and Computer Science Irvine CA (1994)

EM Cluster Analysis for Categorical Data*

Jiří Grim

Institute of Information Theory and Automation
of the Czech Academy of Sciences,
P.O. Box 18, 18208 Prague 8, Czech Republic
grim@utia.cas.cz
http://ro.utia.cas.cz/mem.html

Abstract. Distribution mixtures with product components have been applied repeatedly to determine clusters in multivariate data. Unfortunately, for categorical variables the mixture parameters are not uniquely identifiable and therefore the result of cluster analysis may become questionable. We give a simple proof that any non-degenerate discrete product mixture can be equivalently described by infinitely many different parameter sets. Nevertheless a unique result of cluster analysis can be guaranteed by additional constraints. We propose a heuristic method of sequential estimation of components to guarantee a unique identification of clusters by means of EM algorithm. The application of the method is illustrated by a numerical example.

1 Introduction

The cluster analysis of categorical data is well known to be a difficult problem. Let us recall that arithmetical operations and therefore means and variances are undefined for categorical variables. Generally, the values of categorical variables are neither ordered nor there is any reasonable and commonly acceptable way to define a distance or similarity measure. Binary variables as a special case may appear to be naturally ordered but often there is no reliable argument to prefer one of the two possibilities to assign the values 0 and 1. For these and other reasons the standard clustering algorithms are not directly applicable to multivariate categorical data.

One of the first statistical methods of cluster analysis of categorical data is due to Lazarsfeld [14]. Motivated by sociological research he proposed fitting of multivariate Bernoulli mixtures to binary data to identify possible latent classes of respondents by means of mixture components. Wide application of the latent class (latent structure) analysis was enabled by the computationally efficient EM algorithm [4]. Discussion of latent class analysis from a statistical point of view can be found in Fielding [5]. Other approaches to clustering and latent variable models are discussed e.g. by Vermunt et al. [18], (see also [1], [6], [15]).

* This research was supported by the EC project no. FP6-507752 MUSCLE, by the grant No. 1ET400750407 of the Grant Agency of the Academy of Sciences CR and partially by the project MŠMT 1M0572 DAR and GAČR 402/03/1310.

D.-Y. Yeung et al. (Eds.): SSPR&SPR 2006, LNCS 4109, pp. 640–648, 2006.

Multivariate Bernoulli mixture is only a special case of the general conditional independence model which can be defined for general discrete variables (categorical, qualitative or nominal) as a finite mixture of product components. The application of discrete product mixtures to cluster analysis corresponds with the original approach of Lazarsfeld, however, in both cases there is a problem to justify the obtained solutions. The conditional independence model is not uniquely identifiable in case of categorical variables and therefore the result of cluster analysis becomes questionable.

In the present paper we first introduce the conditional independence model for unordered categorical variables and briefly describe the corresponding version of EM algorithm for estimation of mixtures (Sec. 2). Then we discuss the problem of identifiability of distribution mixtures with product components (Sec. 3) in connection with the recently considered concept of "practical identifiability" of multivariate Bernoulli mixtures [3]. In Sec. 4 we propose the method of sequential identification of mixture components as a tool to obtain unique clusters. The application of the method is illustrated by a numerical example. Finally we summarize the main results in the Conclusion.

2 Conditional Independence Models

Let $\xi_1, \ldots, \xi_N$ be a finite number of general discrete random variables. In particular, we assume that each variable $\xi_n \in \mathcal{X}_n$ takes on some categorical (nominal, qualitative) values from a finite set $\mathcal{X}_n$ without any type of ordering. Simultaneously, let μ be a discrete random variable taking on values from a finite set of integers $\mathcal{M}$

$$P\{\mu = m\} = w_m, \quad m \in \mathcal{M}, \quad \sum_{m\in\mathcal{M}} w_m = 1, \quad \mathcal{M} = \{1, \ldots, M\}. \qquad (1)$$

We suppose that the random variables ξ_n are conditionally independent given the value of μ. In other words we assume that the conditional probability distributions $F(\boldsymbol{x}|m), m \in \mathcal{M}$ of the random vector

$$\boldsymbol{\xi} = (\xi_1, \ldots, \xi_N) \in \mathcal{X}, \quad \mathcal{X} = \mathcal{X}_1 \times \ldots \times \mathcal{X}_N, \quad \mathcal{N} = \{1, \ldots, N\}$$

can be expressed as a product of univariate conditional distributions $f_n(x_n|m)$:

$$P\{\boldsymbol{\xi} = \boldsymbol{x}|\mu = m\} = F(\boldsymbol{x}|m) = \prod_{n\in\mathcal{N}} f_n(x_n|m), \quad \boldsymbol{x} \in \mathcal{X}, \quad m \in \mathcal{M}. \qquad (2)$$

In view of Eqs. (1), (2) the unconditional joint probability distribution of the random vector $\boldsymbol{\xi}$ can be expressed in the form of a finite distribution mixture of product components:

$$P(\boldsymbol{x}) = \sum_{m\in\mathcal{M}} w_m F(\boldsymbol{x}|m) = \sum_{m\in\mathcal{M}} w_m \prod_{n\in\mathcal{N}} f_n(x_n|m), \quad \boldsymbol{x} \in \mathcal{X}, \quad (w_m > 0). \quad (3)$$

Here the probability w_m is usually called the weight of m-th component and $f_n(x_n|m)$ are the conditional (component specific) univariate distributions of the variables ξ_n respectively. In this sense the distribution mixture (3) is defined by the parameter set $\Theta = \{M, w_m, f_n(\cdot|m), m \in \mathcal{M}\}$. [1]

In case of dichotomous variables $\xi_n \in \{0,1\}$ the probability distribution (3) becomes the well known multivariate Bernoulli mixture

$$P(\boldsymbol{x}) = \sum_{m \in \mathcal{M}} w_m \prod_{n \in \mathcal{N}} \vartheta_{nm}^{x_n}(1-\vartheta_{nm})^{1-x_n}, \quad \boldsymbol{x} \in \{0,1\}^N, \quad 0 < \vartheta_{nm} < 1 \tag{4}$$

which is a special case of the general conditional independence model (3).

The standard way to estimate the conditional independence models from data is to compute maximum-likelihood estimates of mixture parameters by means of the iterative EM algorithm [4],[13]. In particular, let $\mathcal{S}$ be a set of independent observations of the random vector $\boldsymbol{\xi}$:

$$\mathcal{S} = \{\boldsymbol{x}^{(1)}, \boldsymbol{x}^{(2)}, \ldots, \boldsymbol{x}^{(K)}\}, \quad \boldsymbol{x}^{(k)} \in \mathcal{X} \tag{5}$$

which are identically distributed with some unknown distribution mixture of the form (3). To compute the m.-l. estimates of the unknown parameters $w_m, f_n(\cdot|m)$ we maximize the likelihood function

$$L = \frac{1}{|\mathcal{S}|}\sum_{x \in \mathcal{S}} \log\, P(\boldsymbol{x}) = \frac{1}{|\mathcal{S}|}\sum_{x \in \mathcal{S}} \log\left[\sum_{m \in \mathcal{M}} w_m F(\boldsymbol{x}|m)\right] \tag{6}$$

by means of the basic EM iteration equations

$$q(m|\boldsymbol{x}) = \frac{w_m F(\boldsymbol{x}|m)}{\sum_{j \in \mathcal{M}} w_j F(\boldsymbol{x}|j)}, \quad w_m^{'} = \frac{1}{|\mathcal{S}|}\sum_{x \in \mathcal{S}} q(m|\boldsymbol{x}), \quad m \in \mathcal{M}, \tag{7}$$

$$f_n^{'}(\xi|m) = \frac{1}{\sum_{x \in \mathcal{S}} q(m|\boldsymbol{x})}\sum_{x \in \mathcal{S}} \delta(\xi, x_n) q(m|\boldsymbol{x}), \quad \xi \in \mathcal{X}_n, \quad n \in \mathcal{N}, \tag{8}$$

where $w_m^{'}, f_n^{'}(\cdot|m)$ are the new parameter values and $\delta(\xi, x_n)$ denotes the usual delta-function, i.e. $\delta(\xi, x_n) = 1$ for $\xi = x_n$ and otherwise $\delta(\xi, x_n) = 0$.

The EM algorithm generates a nondecreasing sequence $\{L^{(t)}\}_0^\infty$. As the criterion (6) is bounded above ($L < 0$) the monotonic property implies convergence of the sequence $\{L^{(t)}\}_0^\infty$ to a possibly local maximum of the function (6) in the parameter space (for more details cf. e.g. [13]). Obviously, a local maximum may be starting-point dependent.

Given the estimated distribution mixture (3) we can characterize any data vector $\boldsymbol{x} \in \mathcal{X}$ by its affinity with the mixture components in terms of the conditional probabilities. The conditional posterior weights $q(m|\boldsymbol{x})$ are particularly useful if there is some interpretation of the mixture components, e.g. if the components correspond to some "latent classes" [14], "hidden causes" [15] or

[1] Here and in the following we assume that the parameter sets differing only by the order of components are identical.

"clusters" having a specific meaning. This idea is closely related to the original latent structure analysis of Lazarsfeld.

There are also some other theoretical arguments justifying the finite distribution mixture (3) as a "latent class" model. It should be emphasized that the statistical relations among the random variables $\xi_1, \ldots, \xi_N$ are wholly explained by their dependence on the variable μ which is sometimes called the latent variable. Given the value of the latent variable μ, the random variables ξ_n are statistically independent, i.e. their interdependence is removed. In this sense the values of the variable μ can be viewed as "hidden causes" which cannot be observed directly but remove the statistical interaction between the observed variables $\xi_1, \ldots, \xi_N$. Once specified, the hidden cause μ would permit us to treat the visible variables ξ_n in a simple way as if they were mutually independent [15]. In view of these arguments the conditional independence model (3) is assumed to be "the most universal and distinctive characteristics featured by the notion of causality" (cf. [15], [16]).

Remark 1. It is easily verified that the conditional independence model (3) is not restrictive in the sense that any discrete probability distribution $P(\boldsymbol{x})$ on $\mathcal{X}$ can be expressed as a mixture (3) provided that the number of components may be chosen sufficiently large. In particular, let $P(\boldsymbol{x})$ be a general discrete probability distribution on $\mathcal{X}$ defined by a table of probabilities. Considering a numbering of the points of $\mathcal{X}$, we can write

$$\mathcal{X} = \cup_{k=1}^{K}\{\boldsymbol{x}^{(k)}\}, \quad P\{\boldsymbol{\xi} = \boldsymbol{x}^{(k)}\} = P(\boldsymbol{x}^{(k)}) = p^{(k)}, \quad \boldsymbol{x}^{(k)} \in \mathcal{X}, \quad k = 1, \ldots, K$$

where $p^{(k)}$ is the table probability attached to $\boldsymbol{x}^{(k)} \in \mathcal{X}$ and $K = |\mathcal{X}|$. Then the distribution mixture of the form (3) equivalent to the given table of values is obtained by setting

$$w_m = p^{(m)}, \quad F(\boldsymbol{x}|m) = \prod_{n \in \mathcal{N}} \delta(x_n, x_n^{(m)}), \quad m = 1, \ldots, K. \tag{9}$$

In other words, the components $F(\boldsymbol{x}|m)$ in (9) are reduced to Dirac distributions $\delta(\boldsymbol{x}, \boldsymbol{x}^{(m)})$ positioned at the points $\boldsymbol{x}^{(m)} \in \mathcal{X}$ while the component weights w_m are equal to the respective table probabilities $p^{(m)}$.

3 Problem of Identifiability

The conditional independence model has been used by many authors in different areas as a tool of cluster analysis [18]. One of the most popular application fields appears to be the bacterial taxonomy. Gyllenberg et al. [12] recall about thirty references relating to a widely used method of classification of bacteria known as probabilistic numerical identification. The method is based on estimating parameters of the multivariate Bernoulli mixtures (4) from the observed data. The resulting components of the Bernoulli mixture are then used to identify the individual classes of bacteria (so called taxons). The posterior probability $q(m|\boldsymbol{x})$

(cf. (7)) is known in bacterial identification as the Willcox probability that the observed bacteria strain $\boldsymbol{x}$ belongs to the m-th class (taxon). In this sense the estimated Bernoulli mixture (4) defines the taxonomic structure of bacteria.

It is obvious that, before estimating the mixture (4), we should verify that it can be estimated uniquely, since otherwise we could obtain several different taxonomic structures for a given set of bacterial data. If the distribution mixture (3) or (4) is not defined uniquely then the corresponding interpretation of data in terms of clusters or latent classes becomes questionable (cf. [14], [12]). Unfortunately, multivariate Bernoulli mixtures are not identifiable, i.e. different parameter sets $\Theta = \{M, w_m, f_n(\cdot|m), m \in \mathcal{M}\}$ can correspond to exactly the same Bernoulli mixture.

Essentially, the proof of this assertion follows from the early papers of Teicher [17] and Blischke [2]. More recently Gyllenberg et al. [12] alternatively repeated the proof of Teicher for the specific case of discrete distributions by showing that the conditional independence model (3) is identifiable if and only if the mixtures of univariate discrete distributions $f_n(x_n|m)$ are identifiable. Then, by using a theorem of Blischke (cf. [2]), they show the mixtures of univariate Bernoulli distributions to be non-identifiable as a special case of the non-identifiable binomial distributions and therefore multivariate Bernoulli mixtures are non-identifiable, too. In the following we give a simple and intuitive proof of this property for a more general class of discrete mixtures with product components (cf. [9]).

Lemma 1. Any discrete distribution mixture of the form

$$P(\boldsymbol{x}) = \sum_{m\in\mathcal{M}} w_m F(\boldsymbol{x}|m), \quad F(\boldsymbol{x}|m) = \prod_{n\in\mathcal{N}} f_n(x_n|m), \quad \boldsymbol{x} \in \mathcal{X}, \ (w_m > 0). \tag{10}$$

can be equivalently described by infinitely many non-trivially different parameter sets $\Theta^{'} = \{M^{'}, w_m^{'}, f_n^{'}(\cdot|m), m \in \mathcal{M}^{'}\}$ if at least one of the univariate conditional distributions $f_n(\cdot|m)$ is non-singular in the sense that

$$f_n(x_n|m) < 1, \quad \text{for all } \ x_n \in \mathcal{X}_n. \tag{11}$$

Proof. One can easily verify that any univariate discrete distribution $f_n(\cdot|m)$ which is non-degenerate in the sense of the inequality (11) can be expressed as a convex combination of two different distributions in infinitely many ways, e.g. $(0 < \alpha < 1, \ \ \beta = 1 - \alpha)$:

$$f_n(\cdot|m) = \alpha f_n^{(\alpha)}(\cdot|m) + \beta f_n^{(\beta)}(\cdot|m), \quad f_n^{(\alpha)}(\cdot|m) \neq f_n^{(\beta)}(\cdot|m). \tag{12}$$

Now, by means of the substitution (12), we can express the component $w_m F(\boldsymbol{x}|m)$ as a weighted sum of two different components $F^{(\alpha)}(\boldsymbol{x}|m)$, $F^{(\beta)}(\boldsymbol{x}|m)$

$$w_m F(\boldsymbol{x}|m) = w_m^{(\alpha)} F^{(\alpha)}(\boldsymbol{x}|m) + w_m^{(\beta)} F^{(\beta)}(\boldsymbol{x}|m), \quad \boldsymbol{x} \in \mathcal{X} \tag{13}$$

where

$$w_m^{(\alpha)} = \alpha w_m, \quad F^{(\alpha)}(\boldsymbol{x}|m) = f_n^{(\alpha)}(x_n|m) \prod_{i\in\mathcal{N}, i\neq n} f_i(x_i|m),$$

Table 1. Example of the 16-dimensional Bernoulli mixture from the paper [3]

w_m	ϑ_1	ϑ_2	ϑ_3	ϑ_4	ϑ_5	ϑ_6	ϑ_7	ϑ_8	ϑ_9	ϑ_{10}	ϑ_{11}	ϑ_{12}	ϑ_{13}	ϑ_{14}	ϑ_{15}	ϑ_{16}
$0.222\dot{2}$	0.80	0.80	0.80	0.80	0.20	0.20	0.20	0.20	0.20	0.20	0.20	0.20	0.20	0.20	0.20	0.20
$0.194\dot{4}$	0.20	0.20	0.20	0.20	0.80	0.80	0.80	0.80	0.20	0.20	0.20	0.20	0.20	0.20	0.20	0.20
$0.166\dot{6}$	0.20	0.20	0.20	0.20	0.20	0.20	0.20	0.20	0.80	0.80	0.80	0.80	0.20	0.20	0.20	0.20
$0.138\dot{8}$	0.20	0.20	0.20	0.20	0.20	0.20	0.20	0.20	0.20	0.20	0.20	0.20	0.80	0.80	0.80	0.80
$0.111\dot{1}$	0.80	0.20	0.20	0.20	0.80	0.20	0.20	0.20	0.80	0.20	0.20	0.20	0.80	0.20	0.20	0.20
$0.083\dot{3}$	0.20	0.80	0.20	0.20	0.20	0.80	0.20	0.20	0.20	0.80	0.20	0.20	0.20	0.80	0.20	0.20
$0.055\dot{5}$	0.20	0.20	0.80	0.20	0.20	0.20	0.80	0.20	0.20	0.20	0.80	0.20	0.20	0.20	0.80	0.20
$0.027\dot{7}$	0.20	0.20	0.20	0.80	0.20	0.20	0.20	0.80	0.20	0.20	0.20	0.80	0.20	0.20	0.20	0.80

$$w_m^{(\beta)} = \beta w_m, \quad F^{(\beta)}(\boldsymbol{x}|m) = f_n^{(\beta)}(x_n|m) \prod_{i \in \mathcal{N}, i \neq n} f_i(x_i|m)$$

and therefore, after substitution (13) in (10), we obtain two formally different mixtures defined by different parameter sets $\Theta \neq \Theta'$ which describe exactly the same probability distribution $P(\boldsymbol{x})$. •

It can be seen that the non-identifiability of any non-degenerate Bernoulli mixture (4) directly follows from Lemma 1. It appears that the question of uniqueness has been neglected in the literature on numerical taxonomy (cf. discussion in [12]) but, surprisingly, this circumstance does not seem to have any serious practical consequences. Moreover, it has been observed that in numerical experiments the mixture parameters can often be uniquely identified from sufficiently large randomly generated samples of data vectors.

Thus e.g. Carreira-Perpinan et al. [3] in a re-identification experiment generated randomly a set of 10000 of 16 dimensional binary vectors from a specific Bernoulli mixture of $M = 8$ components (cf. Tab. 1). Using this data they estimated repeatedly Bernoulli mixtures of different number of components by means of EM algorithm. For $M = 8$ the original parameters were re-identified 9 out of 10 times. When using fewer components ($M = 4$) the EM algorithm reproduced some of the original components and linear combinations of the remaining ones. A more complex mixture model ($M = 10$) always reproduced the eight original components with the last two being their slight modifications or linear combinations. We have observed similar results in our early paper [7]. It appears that "well separated" components in high-dimensional spaces are "practically identifiable" (cf. [3]) if the data set $\mathcal{S}$ is large enough.

4 Sequential Identification of Components

For obvious reasons the theoretically possible ambiguity in estimating discrete models of conditional independence is a serious disadvantage from the point of view of practical applications. Nevertheless, we can achieve a unique result of cluster analysis e.g. by introducing additional constraints. One intuitively

Table 2. Parameters of the 16-dimensional Bernoulli mixture obtained by re-estimating the mixture parameters from Tab.1 by using sequential adding of components

w_m	ϑ_1	ϑ_2	ϑ_3	ϑ_4	ϑ_5	ϑ_6	ϑ_7	ϑ_8	ϑ_9	ϑ_{10}	ϑ_{11}	ϑ_{12}	ϑ_{13}	ϑ_{14}	ϑ_{15}	ϑ_{16}
.2220	.800	.800	.800	.800	.200	.200	.200	.200	.200	.200	.200	.200	.200	.200	.200	.200
.1943	.200	.200	.200	.200	.800	.800	.800	.800	.200	.200	.200	.200	.200	.200	.200	.200
.1666	.200	.200	.200	.200	.200	.200	.200	.200	.800	.800	.800	.800	.200	.200	.200	.200
.1388	.200	.200	.200	.200	.200	.200	.200	.200	.200	.200	.200	.200	.800	.800	.800	.800
.1109	.800	.200	.200	.200	.800	.200	.200	.200	.800	.200	.200	.200	.800	.200	.200	.200
.0832	.200	.800	.200	.200	.200	.800	.200	.200	.200	.800	.200	.200	.200	.800	.200	.200
.0555	.200	.200	.800	.200	.200	.200	.800	.200	.200	.200	.800	.200	.200	.200	.800	.200
.0277	.200	.200	.200	.800	.200	.200	.200	.800	.200	.200	.200	.800	.200	.200	.200	.800
.0008	.442	.392	.371	.355	.395	.348	.327	.312	.373	.327	.307	.292	.359	.314	.294	.280

acceptable and easy to apply method is a sequential adding of new components to the estimated mixture.

In particular, applying EM algorithm, we start with a mixture having a single component and arbitrary (e.g. randomly chosen) initial parameters. In this case the EM algorithm converges in one iteration to a component defined as a product of univariate marginal distributions. In the next phase a new component is added, initialized as a product of univariate uniform distributions with equal initial weight, i.e. $w_1 = w_2 = 0.5$. Then the EM iterations are continued until sufficient convergence. When the relative increase of the likelihood function is less then some small positive threshold ϵ, a new uniform component is added again and the component weights are normed to obtain $w_3 = w_2$ and $w_1 + w_2 + w_3 = 1$. The EM iterations are then started again with the new initial parameters. In this way the new component defined as a product of uniform marginals is added repeatedly as long as it is "accepted" by the previous mixture model. The computation is stopped when the weight of the new added component is less than a suitably chosen low threshold after the convergence is achieved.

There is no theoretical support of the proposed method to guarantee some qualitative properties of the resulting mixture. Nevertheless, it can be heuristically justified by some computational properties. Let us note first that, from the computational point of view, the resulting mixture model is defined uniquely. In practice, the only source of uncertainty may be the limited parameter accuracy at the end of each convergence phase. Moreover, the method avoids random influences of initial values and represents a reasonable mechanism to choose a proper number of components. Note that the newly added uniform component tends to "fit" to data points insufficiently "covered" by the previous model and for this reason the weight of the added component is usually decreasing in final stages of computation when the number of components is sufficiently large.

It should be emphasized that by including a new (uniform) component to the previously adjusted model we may violate the monotonic property of EM algorithm in the following iteration. Moreover the new component interferes with the previously estimated parameters and partly devaluates the preceding convergence phase.

We illustrate the proposed sequential identification method by considering the re-identification problem from the paper [3]. In order to avoid random influences of a given finite sample $\mathcal{S}$ we re-estimated the original multivariate Bernoulli mixture (cf. Tab.1) in a way which is equivalent to an infinite sample size. Note that for the sample size $|\mathcal{S}|$ approaching infinity we can write

$$L^* = \lim_{|\mathcal{S}|\to\infty} \frac{1}{|\mathcal{S}|} \sum_{x\in\mathcal{S}} \log\left[\sum_{m\in\mathcal{M}} w_m F(\boldsymbol{x}|m)\right] = \sum_{x\in\mathcal{X}} P^*(\boldsymbol{x}) \log\left[\sum_{m\in\mathcal{M}} w_m F(\boldsymbol{x}|m)\right], \tag{14}$$

i.e. the sum over the infinite sample $\mathcal{S}$ can be equivalently replaced by summing over all $\boldsymbol{x} \in \mathcal{X}$ whereby $P^*(\boldsymbol{x})$ denotes the asymptotic relative frequency of $\boldsymbol{x}$.

In order to maximize the asymptotic likelihood function L^* we have modified the basic EM iteration Eqs. (7), (8) in analogy with Eq. (14):

$$w_m^{'} = \sum_{x\in\mathcal{X}} P^*(\boldsymbol{x})q(m|\boldsymbol{x}), \quad m \in \mathcal{M}, \tag{15}$$

$$\vartheta_{nm}^{'} = \frac{1}{\sum_{x\in\mathcal{X}} P^*(\boldsymbol{x})q(m|\boldsymbol{x})} \sum_{x\in\mathcal{X}} x_n P^*(\boldsymbol{x})q(m|\boldsymbol{x}), \quad n \in \mathcal{N}. \tag{16}$$

Instead of generating a given number of pseudo-random binary vectors we have computed and stored the values $P^*(\boldsymbol{x})$ for all the 65536 binary vectors $\boldsymbol{x}$ from the 16-dimensional binary cube. By using the "asymptotic" likelihood function L^* and the corresponding version of EM algorithm we have the possibility to avoid any random small sample fluctuations. In other words we can verify the properties of the proposed method in the extreme case of infinite sample size.

The method of sequential adding of components based on the asymptotically modified EM iteration equations (15), (16) has been applied repeatedly to re-estimate the mixture parameters from Tab.1. A new component has been added whenever the relative increase of the maximized criterion L^* was less than a chosen threshold $\epsilon = 10^{-12}$. The estimated parameters from Tab.2 have been obtained after 3000 iterations. The threshold ϵ has been varied between 10^{-9} and 10^{-12} with very similar results. In all computational experiments we have observed a clear tendency to suppress the weight of superfluous components. The 10th component was not added and the weight of the last added 9th component was by two or three orders less than w_8, i.e. $w_9 \approx 10^{-4} - 10^{-5}$ (cf. Tab. 2).

5 Conclusion

The models of conditional independence have been proposed repeatedly as a tool of cluster analysis of multivariate categorical data since the standard approaches are usually not directly applicable. A serious drawback of the conditional independence models follows from the fact that they are not uniquely identifiable. We give a simple and intuitive proof that any non-degenerate discrete mixture with product components can be equivalently described by infinitely many different parameter sets and therefore it is non-identifiable. We propose to guarantee a unique result of cluster analysis by introducing additional constraints, in particular by sequential adding of components in EM algorithm.

Let us recall finally that there are numerous application possibilities of the conditional independence models based on approximating unknown probability distributions (cf. e.g. [7], [8], [10], [11]). In application to practical problems of pattern recognition and statistical modelling the approximation accuracy is of primary importance. The non-identifiability of estimated mixtures is less relevant and may be even useful in view of increased flexibility of mixture models.

References

1. Bartholomew D.J.: Factor analysis for categorical data. *J. Roy. Statist. Soc. B*, 3 **42** (1980) 293-321
2. Blischke W.R. : Estimating the parameters of mixtures of binomial distributions. *Journal Amer. Statist. Assoc.*, **59** (1964) 510-528
3. Carreira-Perpignan M.A., Renals S.: Practical identifiability of finite mixtures of multivariate Bernoulli distributions. *Neural Computation*, **12** (2000) 141-152
4. Dempster, A.P., Laird, N.M. and Rubin, D.B.: Maximum likelihood from incomplete data via the EM algorithm. *J. Roy. Statist. Soc. B*, **39** (1977) l-38
5. Fielding A.: Latent structure models. In: *The Analysis of survey data*, (Eds. O'Muircheaxtaigh C.A, Payne C.), London: Wiley, (1977) 125-157
6. Gibson W.A.: Three multivariate models: Factor analysis, latent structure analysis and latent profile analysis. *Psychometrika*, **24** (1969) 229-252
7. Grim J.: Multivariate statistical pattern recognition with nonreduced dimensionality, *Kybernetika*, **22** (1986) 142-157
8. Grim J., Boček P., Pudil P. (2001): Safe dissemination of census results by means of interactive probabilistic models. In: *Proceedings of the ETK-NTTS 2001 Conference*, (Hersonissos (Crete), European Communities 2001, **2** (2001) 849-856
9. Grim J.: Latent Structure Analysis for Categorical Data. Research Report UTIA, No. 2019, 13 pp., Academy of Sciences, Czech Republic, Prague (2001)
10. Grim J., Haindl M.: Texture Modelling by Discrete Distribution Mixtures. Computational Statistics and Data Analysis, 3-4 **41** (2003) 603-615
11. Grim J., Kittler J., Pudil P., Somol P.: Multiple classifier fusion in probabilistic neural networks. *Pattern Analysis & Applications*, 7 **5** (2002) 221-233
12. Gyllenberg M., Koski T., Reilink E., Verlaan M.: Non-uniqueness in probabilistic numerical identification of bacteria. *Journal of Applied Prob.*, **31** (1994) 542-548
13. McLachlan G.J. and Peel D.: *Finite Mixture Models*, John Wiley & Sons, New York, Toronto, (2000)
14. Lazarsfeld P.F., Henry N.: *Latent structure analysis.* Houghton Mifl.: Boston (1968)
15. Pearl J.: *Probabilistic reasoning in intelligence systems: networks of plausible inference.* Morgan-Kaufman, San Mateo, CA (1988)
16. Suppes, P.A.: *Probabilistic theory of causality.* North-Holland: Amsterdam, (1970)
17. Teicher, H. : Identifiability of mixtures of product measures. *Ann. Math. Statist.*, **39** (1968) 1300-1302
18. Vermunt J.K., Magidson J.: Latent Class Cluster Analysis. In: *Advances in Latent Class Analysis*, (Eds. Hagenaars J.A. et al.), Cambridge Univ. Press (2002)

Two Entropy-Based Methods for Learning Unsupervised Gaussian Mixture Models

Antonio Peñalver, Francisco Escolano, and Juan M. Sáez

Robot Vision Group
Alicante University, Spain
a.penalver@umh.es, {sco, jmsaez}@dccia.ua.es

Abstract. In this paper we address the problem of estimating the parameters of a Gaussian mixture model. Although the EM (Expectation-Maximization) algorithm yields the maximum-likelihood solution it requires a careful initialization of the parameters and the optimal number of kernels in the mixture may be unknown beforehand. We propose a criterion based on the entropy of the pdf (probability density function) associated to each kernel to measure the quality of a given mixture model. Two different methods for estimating Shannon entropy are proposed and a modification of the classical EM algorithm to find the optimal number of kernels in the mixture is presented. We test our algorithm in probability density estimation, pattern recognition and color image segmentation.

1 Introduction

Gaussian Mixture models have been widely used for density estimation, pattern recognition and function approximation. One of the most common methods for fitting mixtures to data is the EM algorithm [6]. However, this algorithm is prone to initialization errors and it may converge to local maxima of the log-likelihood function. In addition, the algorithm requires that the number of elements (kernels) in the mixture is known beforehand (model-selection).

A d-dimensional random variable $\mathbf{y}$ follows a finite-mixture distribution when its pdf $p(\mathbf{y}|\Theta)$ can be described by a weighted sum of known pdf's named kernels. When all these kernels are Gaussian, the mixture is named in the same way:

$$p(\mathbf{y}|\Theta) = \sum_{i=1}^{K} \pi_i p(\mathbf{y}|\Theta_i) \tag{1}$$

where $0 \leq \pi_i \leq 1, i = 1, ..., K$, and $\sum_{i=1}^{K} \pi_i = 1$, being K the number of kernels, $\pi_1, ..., \pi_k$ the *a priori* probabilities of each kernel, and Θ_i the parameters describing the kernel. In Gaussian mixtures, $\Theta_i = \{\mu_i, \Sigma_i\}$, that is, the average vector and the covariance matrix. The set of parameters of a given mixture is $\Theta \equiv \{\Theta_1, ..., \Theta_k, \pi_1, ..., \pi_k\}$. Obtaining the optimal set of parameters Θ^* is usually posed in terms of maximizing the log-likelihood of the pdf to be estimated:

D.-Y. Yeung et al. (Eds.): SSPR&SPR 2006, LNCS 4109, pp. 649–657, 2006.

$$\ell(Y|\Theta) = \log p(Y|\Theta) = \log \prod_{n=1}^{N} p(y_n|\Theta) = \sum_{n=1}^{N} \log \sum_{k=1}^{K} \pi_k p(y_k|\Theta_k). \quad (2)$$

With $\Theta^* = \arg\max_\Theta \ell(\Theta)$ and $Y = \{y_1, ...y_N\}$ is a set of N i.i.d. samples of the variable Y. The EM (Expectation-Maximization) algorithm [6][12] generates a sequence of estimations of the set of parameters $\{\Theta^*(t), t = 1, 2, ...\}$ by alternating an expectation step and the maximization one until convergence. The equations are:

$$p(k|\mathbf{y}_n) = \frac{\pi_k p(\mathbf{y}^{(n)}|k)}{\sum_{j=1}^{K} \pi_j p(\mathbf{y}^{(n)}|k)} \quad (3)$$

$$\begin{aligned} \pi_k &= \tfrac{1}{N} \textstyle\sum_{n=1}^{N} p(k|\mathbf{y}_n), \;\; \mu_k = \frac{\sum_{n=1}^{N} p(k|\mathbf{y}_n)\mathbf{y}_n}{\sum_{n=1}^{N} p(k|\mathbf{y}_n)}, \\ \Sigma_k &= \frac{\sum_{n=1}^{N} p(k|\mathbf{y}_n)(\mathbf{y}_n-\mu_k)(\mathbf{y}_n-\mu_k)^T}{\sum_{n=1}^{N} p(k|\mathbf{y}_n)}, \end{aligned} \quad (4)$$

A detailed description of this classic algorithm is given in [12]. Here we focus on the fact that if K is unknown beforehand it cannot be estimated through maximizing the log-likelihood because $\ell(\Theta)$ grows with K.

In a classical EM algorithm with a fixed number of kernels density can be underestimated giving a poor description of the data. The so called model-selection problem has been addressed in many ways [16][17][8][7][14]. In this paper we propose a method that starting with only one kernel, finds the maximum-likelihood solution. In order to do so, it tests whether the underlying pdf of each kernel is Gaussian and otherwise it replaces that kernel with two kernels adequately separated from each other. In order to detect non-Gaussianity we compare the entropy of the underlying pdf with the theoretical entropy of a Gaussian. After the kernel with worse degree of Gaussianity has been splited in two, new EM steps are performed in order to obtain a new maximum-likelihood solution. In the next sections we describe two different entropy estimation techniques to test whether a given kernel describes properly the underlying data.

2 Entropy Estimation

Entropy is a basic concept in information theory [4]. For a discrete variable Y with $y_1, ..., y_N$ a the set of values, we have:

$$H(Y) = -E_y[\log(P(Y))] = -\sum_{i=1}^{N} P(Y = y_i) \log P(Y = y_i). \quad (5)$$

A fundamental result of information theory is that Gaussian variables have the maximum entropy among all the variables with equal variance. Consequently the entropy of the underlying distribution of a kernel should reach a maximum when such a distribution is Gaussian. This theoretical maximum entropy is given by:

$$H_{max}(Y) = \frac{1}{2} \log[(2\pi e)^d |\Sigma|]. \quad (6)$$

Then, in order to decide whether a given kernel is truly Gaussian or must be replaced by two other kernels, we compare the estimated entropy of the underlying data with the entropy of a Gaussian.

The estimation of the Shannon entropy of a probability density given a set of samples has been studied widely in the past [1]. In this paper we present results with two different methods: "plug-in" and "non plug-in".

2.1 Entropy Estimation with Parzen's Windows

The Parzen's windows approach [11] is a non-parametric method for estimating pdf's for a finite set of patterns. The general form of these pdf's using a Gaussian kernel and assuming diagonal covariance matrix $\psi = Diag(\sigma_1^2, ... \sigma_{N_a}^2)$ is:

$$P^*(Y, a) \equiv \frac{1}{N_a} \sum_{y_a \in a} K_\psi(y - y_a), \tag{7}$$

where $K_\psi(y - y_a)$ is a gaussian kernel centered y y_a, a is a sample of the variable Y and N_a is the size of the sample. In [15] a method for adjusting the widths of the kernels using maximum likelihood is proposed. Given the definition of entropy in Equation 5, we have:

$$H_b(Y) \equiv -E_b[\log(P(Y))] = -\frac{1}{N_b} \sum_{y_b \in b} \log(P(y_b)) \tag{8}$$

where b is a sample of the variable Y and N_b is the size of the sample. If expression in Equation 7 is plugged into Equation 8 then the entropy is estimated by:

$$H^*(Y) = \frac{1}{N_b} \sum_{y_b \in b} \log \left(\frac{1}{N_a} \sum_{y_a \in a} K_\psi(y_b - y_a) \right) \tag{9}$$

2.2 Renyi's Entropy and Entropic Spanning Graphs

Entropic Spanning Graphs obtained from data to estimate Renyi's α-entropy[10] belong to the "non plug-in" methods of entropy estimation. Renyi's α-entropy of a probability density function f is defined as:

$$H_\alpha(f) = \frac{1}{1-\alpha} \ln \int_z f^\alpha(z) dz \tag{10}$$

for $\alpha \in (0, 1)$. The α entropy converges to the Shannon entropy $-\int f(z) \ln f(z) dz$ as $\alpha \rightarrow 1$, so it is possible to obtain the second one from the first one.

A graph G consists of a set of vertices $X_n = \{x_1, ..., x_n\}$, with $x_n \in R^d$ and edges $\{e\}$ that connect vertices in graph: $e_{ij} = (x_i, x_j)$. If we denote by $M(X_n)$ the possible sets of edges in the class of acyclic graphs spanning X_n (spanning trees), the total edge length functional of the Euclidean power weighted Minimal Spanning Tree is:

$$L_\gamma^{MST}(X_n) = \min_{M(X_n)} \sum_{e \in M(X_n)} | e |^\gamma \tag{11}$$

with $\gamma \in (0, d)$ y $| e |$ the euclidean distance between graph vertices.

The MST has been used as a way to test for randomness of a set of points. In [9] it was showed that in d-dimensional feature space, with $d \geq 2$:

$$H_\alpha(X_n) = \frac{d}{\gamma}\left[\ln \frac{L_\gamma(X_n)}{n^\alpha} - \ln \beta_{L_\gamma,d}\right] \tag{12}$$

is an asymptotically unbiased, and almost surely consistent, estimator of the α-entropy of f where $\alpha = (d-\gamma)$ and $\beta_{L_\gamma,d}$ is a constant bias correction depending on the graph minimization criterion, but independent of f. Closed form expressions are not available for $\beta_{L_\gamma,d}$, only known approximations and bounds: (i) Monte Carlo simulation of uniform random samples on unit cube $[0,1]^d$; (ii) Large d approximation: $(\gamma/2)\ln(d/(2\pi e))$ in [2].

We can estimate $H_\alpha(f)$ for different values of $\alpha = (d-\gamma)/d$ by changing the edge weight exponent γ. As γ modifies the edge weights monotonically, the graph is the same for different values of γ, and only the total length in expression 12 needs to be recomputed.

Entropic spanning graphs are suitable for estimating α-entropy with $\alpha \in [0,1[$, so Shannon entropy can not be directly estimated with this method. Figure 1 on the left hand shows that the shape of the function does not depend neither on the nature of data nor on their size.

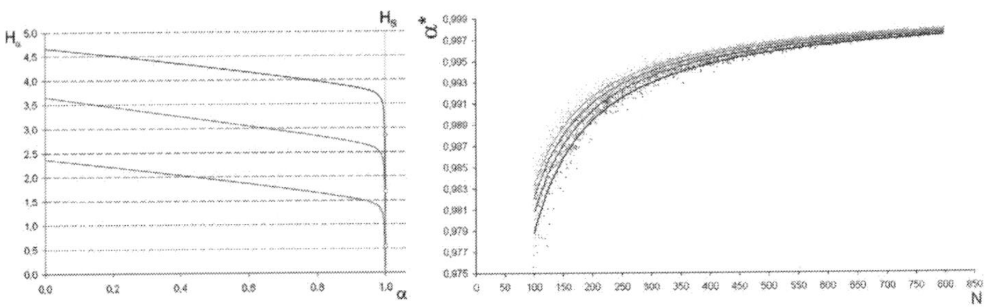

Fig. 1. Left: H_α for gaussian distributions with different covariance matrices. Right: α^* for dimensions between 2 and 5 and different number of samples.

We will approximate the value of H_α for $\alpha = 1$ by means of a continuous function that captures the tendency of H_α in the environment of 1. From a value of $\alpha \in [0,1[$, we can calculate the tangent line $y = mx + b$ to H_α in this point, using $m = H_\alpha^{'}$, $x = \alpha$ and $y = H_\alpha$. In any case, this line will be continuous and we will be able to calculate its value for $x = 1$.

From now on, we will call α^* to the α value that generates the correct entropy value in $\alpha = 1$, following the described procedure.

As H_α is a monotonous decreasing function, we can estimate α^* value in the Gaussian case by means of a dichotomic search between two well separated α values for a constant number of samples, problem dimension and different covariance matrices. Experimentally, we have verified that α^* is almost constant for diagonal covariance matrices with variance value greater than 0.5.

In order to appreciate the effects of the dimension and the number of samples on the problem, we calculated α^* for a set of 1000 distributions with random $2 \leq d \leq 5$ and number of samples. Experimentally we have verified that the shape of the underlying curve adjusts suitably to a function of the type: $\alpha^* = 1 - \frac{a + b \exp^{cD}}{N}$, where N is the number of samples, D is the problem dimension and a, b, c are three constants to estimate. In order to estimate these values, we used Monte Carlo Simulation, minimizing the mean square error between expression and data. We obtained $a = 1.271, b = 1.3912$ and $c = -0.2488$. Figure 1 on the right hand shows α_* for different dimension an number of samples.

3 Entropy-Based EM Algorithm

Comparing the estimations given for Equations 6 with 9 and 12, we have a way of quantifying the degree of Gaussianity of a given kernel. Given a set of kernels for the mixture (initially one kernel) we evaluate the real global entropy $H(y)$ and the theoretical maximum entropy $H_{max}(y)$ of the mixture by considering the individual pairs of entropies for each kernel, and their prior probabilities:

$$H(Y) = \sum_{k=1}^{K} \pi_k H_k(Y) \quad \text{and} \quad H_{max}(Y) = \sum_{k=1}^{K} \pi_k H_{max_k}(Y) \,. \tag{13}$$

If the ratio $H(y)/H_{max}(y)$ is above a given threshold we consider that all kernels are well fitted. Otherwise, we select the kernel with the lowest individual ratio and it is replaced by two other kernels that are conveniently placed and initialized. Then, a new EM with $K+1$ kernels starts.

A low $H(y)/H_{max}(y)$ local ratio indicates that multi-modality arises and thus the kernel must be replaced by two other kernels. In the split step the original covariance matrix needs to generate two new matrices with two restrictions: overall dispersion must remain almost constant and the new matrices must be positive definite. This is an ill-posed problem because the number of equations is less than the number of unknowns [13][18].

From definition of mixture in equation 1, considering that the $K*$ component is the one with lowest Gaussianity threshold, it must be decomposed into the K_1 and K_2 components with parameters $\Theta_{k_1} = (\mu_{k_1}, \Sigma_{k_1})$ and $\Theta_{k_2} = (\mu_{k_2}, \Sigma_{k_2})$. The corresponding priors, the mean vectors and the covariance matrices should satisfy the following split equations:

$$\begin{aligned} \pi_* &= \pi_1 + \pi_2 \\ \pi_* \mu_* &= \pi_1 \mu_1 + \pi_2 \mu_2 \\ \pi_*(\Sigma_* + \mu_* \mu_*^T) &= \pi_1(\Sigma_1 + \mu_1 \mu_1^T) + \pi_2(\Sigma_2 + \mu_2 \mu_2^T) \end{aligned} \tag{14}$$

Recently, in [5] a spectral decomposition of the actual covariance matrix is performed and the original problem is replaced by estimating the new eigenvalues and eigenvectors of new covariance matrices.

Let $\sum_* = V_* \Lambda_* V_*^T$ be the spectral decomposition of the covariance matrix $\sum_*$, with $\Lambda_* = diag(\lambda j*^1, ..., \lambda j*^d)$ a diagonal matrix containing the eigenvalues of $\sum_*$ with increasing order, $*$ the component with the lowest entropy ratio,

π_*, π_1, π_2 the priors of both original and new components, μ_*, μ_1, μ_2 the means and $\sum_*, \sum_1, \sum_2$ the covariance matrices. Let also be D a dxd rotation matrix with columns orthonormal unit vectors. D is constructed by generating its lower triangular matrix independently from $d(d-1)/2$ different uniform $U(0,1)$ densities. The proposed split operation is given by:

$$\begin{gathered} \pi_1 = u_1\pi_*, \;\; \pi_2 = (1-u_1)\pi_* \\ \mu_1 = \mu_* - (\textstyle\sum_{i=1}^d u_2^i \sqrt{\lambda_*^i} V_*^i)\sqrt{\frac{\pi_2}{\pi_1}}, \;\; \mu_2 = \mu_* - (\textstyle\sum_{i=1}^d u_2^i \sqrt{\lambda_*^i} V_*^i)\sqrt{\frac{\pi_1}{\pi_2}} \\ \Lambda_1 = diag(u_3)diag(\iota - u_2)diag(\iota + u_2)\Lambda_* \frac{\pi_*}{\pi_1} \\ \Lambda_2 = diag(\iota - u_3)diag(\iota - u_2)diag(\iota + u_2)\Lambda_* \frac{\pi_*}{\pi_2} \\ V_1 = DV_*, \;\; V_2 = D^T V_* \end{gathered} \tag{15}$$

where, ι is a d x 1 vector of ones, $u_1, u_2 = (u_2^1, u_2^2, ..., u_2^d)^T$ and $u_3 = (u_3^1, u_3^2, ..., u_3^d)^T$ are $2d+1$ random variables needed to construct priors, means and eigenvalues for the new component in the mixture. They are calculated as

$$\begin{gathered} u_1 \sim be(2,2), u_2^1 \sim be(1,2d), \\ u_2^j \sim U(-1,1), u_3^1 \sim be(1,d), u_3^j \sim U(0,1) \;\; \text{and} \;\; j = 2, ..., d \end{gathered} \tag{16}$$

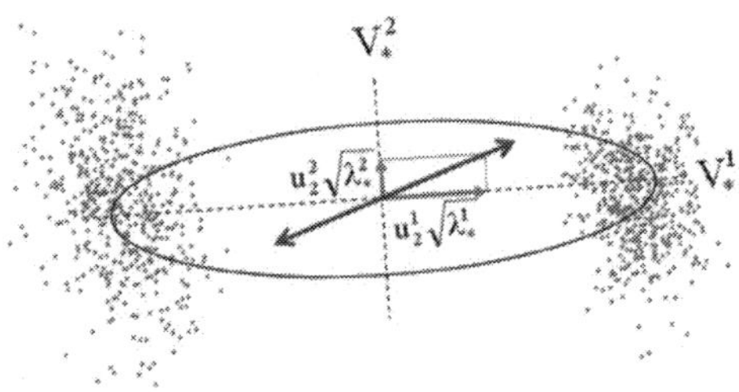

Fig. 2. 2-D Example of splitting one kernel into two new kernels

A graphical description of the splitting process in the 2-D case is showed in Fig.2. Directions and magnitudes of variability are defined by eigenvectors and eigenvalues of the covariance matrix. Otherwise, a completed algorithmic description of the process is showed in Fig. 3.

4 Experiments and Discussion

In order to test our approach we have performed several experiments with synthetic, real and image data. In the first one we have generated 2500 samples from 5 bi-dimensional Gaussians with different prior probabilities, averages and covariance matrices. We have used a Gaussianity threshold of 0.95, and a convergence threshold of 0.001 for the EM algorithm. In both, "plug-in" and "non plug-in" entropy estimation approaches our algorithm converges after 30 iterations finding correctly $k = 5$. In Figure 4 we show the evolution of the algorithm.

ENTROPY BASED EM ALGORITHM
Initialization: Start with a unique kernel.
$K \leftarrow 1$. $\Theta_1 \leftarrow \{\mu_1, \Sigma_1\}$ with μ_1 = data average and Σ_1 = data covariance.
repeat: *//Main loop*
  repeat: *//E, M Steps*
    Estimate log-likelihood in iteration i: ℓ_i
  until: $|\ell_i - \ell_{i-1}| <$ CONVERGENCE_TH
  Evaluate $H(Y)$ and $H_{max}(Y)$ globally
  if $(H(Y)/H_{max} <$ ENTROPY_TH)
    Select kernel K_* with the lowest ratio and decompose into K_1 and K_2
    Initialize parameters Θ_1 and Θ_2(Eq.15)
      Initialize new averages: μ_1 and μ_2
      Initialize new eigenvalues and eigenvector matrices: Λ_1, Λ_2, V_1 and V_2
      Set new priors: π_1 and π_2
  else Final $\leftarrow$ True
until: Final = True

Fig. 3. Entropy Based EM algorithm

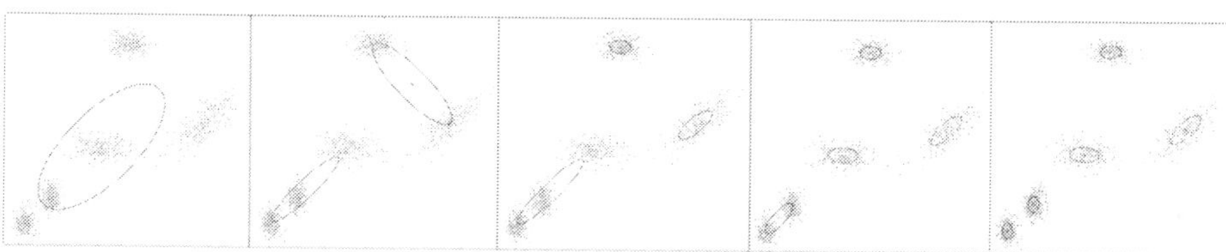

Fig. 4. Evolution of our algorithm from 1 to 5 final kernels

We have also tested our algorithm in unsupervised color image segmentation. At each pixel i in the image we compute a 3-dimensional feature vector x_i with the components in the RGB color space. We obtain the number of components (classes) M and $y_i \in [1, 2, ..., M]$ to indicate from which class the pixel i_{th} came. Therefore our image model sets that each pixel is generated by one of the Gaussian densities in the Gaussian mixture model. We have used different entropy thresholds and a convergence threshold of 0.1 for the EM algorithm. In Fig. 5 we show some results obtained from three different images. The greater it is the demanded threshold the higher is the number of kernels (colors) generated. In the "non plug-in" approach, a random selection of 1000 points has been made to estimate the MST due to memory problems. The results obtained with both methods are identical.

Finally, we have applied the proposed method to the well known *Iris* [3] data set, that contains 3 classes of 50 (4-dimensional) instances referred to a type of iris plant: *Versicolor*, *Virginica* and *Setosa*. 50 samples are insufficient to construct the pdf using Parzen. In order to test our method, we have generated 300 training samples from the averages and covariances of the original classes

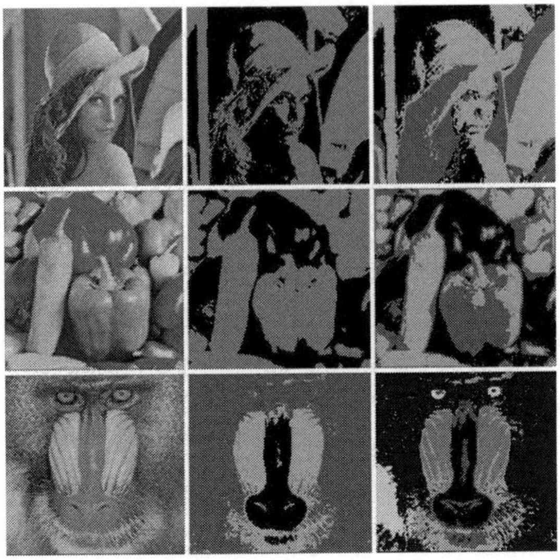

Fig. 5. Color image segmentation with increasing gaussianity thresholds

and we have checked the performance in a classification problem with the original 150 samples. Starting with $K = 1$, the method correctly selected $K = 3$. Then, a maximum a posteriori classifier was built, with classification performance of 98%. With the MST approach, with no pdf estimation required, the algorithm can be executed with the original data set with the same classification performance.

5 Conclusions and Future Work

In this paper we propose a method for finding the optimal number of kernels in a Gaussian mixture based on maximum entropy. The algorithm starts with only one kernel overcoming the local convergence of the usual EM algorithm. The "plug-in" entropy estimation approach is suitable for low-dimensional problems with large data, while the "non plug-in" approach is appropriate for high-dimensional settings with a reduced data set. The algorithm is efficient for density estimation, pattern recognition and unsupervised color image segmentation. We are currently exploring methods to remove noisy features from data.

References

1. E. Beirlant, E. Dudewicz, L. Gyorfi, and E. Van der Meulen. Nonparametric entropy estimation. *International Journal on Mathematical and Statistical Sciences*, 6(1):17–39, 1996.
2. D.J. Bertsimas and G. Van Ryzin. An asymptotic determination of the minimum spanning tree and minimum matching constants in geometrical probability. *Operations Research Letters*, 9(1):223–231, 1990.

3. C.L Blake and C.J. Merz. Uci repository of machine learning databases. *University of California, Irvine, Dept. of Information and Computer Sciences*, 1998.
4. T. Cover and J. Thomas. *Elements of Information Theory*. J. Wiley and Sons, 1991.
5. P. Dellaportas and I. Papageorgiou. Multivariate mixtures of normals with unknown number of components. *Statistics and Computing*, To appear.
6. A. Dempster, N. Laird, and D. Rubin. Maximum likelihood estimation from incomplete data via the em algorithm. *Journal of The Royal Statistical Society*, 39(1):1–38, 1977.
7. M.A.T Figueiredo and A.K. Jain. Unsupervised selection and estimation of finite mixture models. In *International Conference on Pattern Recognition. ICPR2000*, Barcelona, Spain, 2000. IEEE.
8. M.A.T Figueiredo, J.M.N Leitao, and A.K. Jain. On fitting mixture models. *Energy Minimization Methods in Computer Vision and Pattern Recognition. Lecture Notes in Computer Science*, 1654(1):54–69, 1999.
9. A.O. Hero and O. Michel. Asymptotic theory of greedy aproximations to minnimal k-point random graphs. *IEEE Trans. on Infor. Theory*, 45(6):1921–1939, 1999.
10. A.O. Hero and O. Michel. Applications of entropic graphs. *IEEE Signal Processing Magazine*, 19(5):85–95, 2002.
11. E. Parzen. On estimation of a probability density function and mode. *Annals of Mathematical Statistics*, 33(1):1065–1076, 1962.
12. R.A. Redner and H.F. Walker. Mixture densities, maximum likelihood, and the em algorithm. *SIAM Review*, 26(2):195–239, 1984.
13. S. Richardson and P.J. Green. On bayesian analysis of mixtures with unknown number of components (with discussion). *Journal of the Royal Statistical Society B*, (1), 1997.
14. N. Ueda, R. Nakano, Z. Ghahramani, and G. E. Hinton. Smem algorithm for mixture models. *Neural Computation*, 12(1):2109–2128, 2000.
15. P. Viola, N. N. Schraudolph, and T. J. Sejnowski. Empirical entropy manipulation for real-world problems. *Adv. in Neural Infor. Proces. Systems*, 8(1), 1996.
16. N. Vlassis and A. Likas. A kurtosis-based dynamic approach to gaussian mixture modeling. *IEEE Trans. Systems, Man, and Cybernetics*, 29(4):393–399, 1999.
17. N. Vlassis and A. Likas. A greedy em algorithm for gaussian mixture learning. *Neural Processing Letters*, 15(1):77–87, 2000.
18. Z. Zhang, K.L. Chan, Y. Wu, and C. Chen. Learning a multivariate gaussian mixture models with the reversible jump mcmc algorithm. *Statistics and Computing*, (1), 2004.

Maximum Likelihood Estimates for Object Detection Using Multiple Detectors

Magnus Oskarsson and Kalle Åström

Centre For Mathematical Sciences, Lund University, Lund, Sweden
{magnuso, kalle}@maths.lth.se
http://www.maths.lth.se/index.html

Abstract. Object detection in real images has attracted much attention during the last decade. Using machine learning and large databases it is possible to develop detectors for visual categories that have a very high hit-rate, with low false positive rates. In this paper we investigate a general probabilistic framework for context based scene interpretation using multiple detectors. Methods for finding maximum likelihood estimates of scenes given detection results are presented. Although we have investigated how the method works for a specific case, namely for face detection, it is a general method. We show how to combine the results of a number of detectors i.e. face, eye, nose and mouth detectors. The methods have been tested using detectors trained on real images, with promising results.

1 Introduction

During the last decade much interest in the computer vision community has been put into the areas of object detection, recognition and classification. Using machine learning and large databases it is possible to develop detectors for visual categories such as faces, eyes, cars, bicycles, animals etc.

Many of these detectors work in the following semantic way. A binary classifier is obtained for objects at a predefined scale and position using a large database of positive and negative examples, and machine learning. This classifier is then applied to different parts of the image, i.e. at different positions and different scales. A typical result is shown in Figure 1. Some kind of clustering algorithm is then applied in order to reduce the number of detections that originate from the same object.

In this paper, we discuss (i) how this clustering method can be improved using scene models and (ii) how multiple detectors can be used in conjunction in order to improve the results.

Geometric constraints have been used to improve detection in a number of publications. The constraints can be more or less explicit. In [3] the concept of body plans was used to detect humans and horses. The geometric relationship between parts was explicitly given in the model. In [6] a hierarchical model of components of humans was used in a SVM-framework to detect humans in still images. In [10] probabilities for walking humans were learned with good results. In [5] they use a number of detectors to improve face-detection. Here they optimize over a weighted sum of individual likelihoods from each detector. Context in the form of appearance around object was

D.-Y. Yeung et al. (Eds.): SSPR&SPR 2006, LNCS 4109, pp. 658–666, 2006.

exploited in [4]. In [8] statistics of spatial relationships were learned to improve object detection. For surveys on context and face detection see e.g. [2,13].

In our work we are interested in how one can use scene models and context to, on the one hand, get better detection rates. But another goal is to get higher level information about the detected objects, how they are related to each other geometrically and their properties such as appearance and pose. As will be seen in the experiments on face detection in section 3 one of the biggest gains in combining detectors in a contextual manner is in the precision of detection. We achieve this not by starting with the detections and combine them in some manner, but instead formulate a hypothetical scene and calculate how likely this scene is given the detection result. The problem is then to search the scene space for the most probable scene. By the most probable scene we mean the scene with the highest likelihood given the detections.

Fig. 1. Typical detection result. Each circle represents a face detected at the center of the circle and the radius corresponds to at which scale the detection was made. At most positions and scales there are no detections.

2 Scene Modeling and Model Estimation

2.1 Scene Modeling

We assume that the scene contains a number of objects of different types, Ω_i, and that for each such object there is a corresponding pose parameter p. In the examples below pose includes position in the image and scale. However, depending on the structure of the detector, the pose parameter could include more or fewer parameters. One might, for instance, have a detector that is trained on faces with a given position, distance and orientation relative to the camera. Or the pose parameter could contain shape variational parameters of deforming objects.

For simplicity we assume that the object types form a hierarchy (or a directed acyclic graph). In the simplest case there is only one type of object, e.g. eyes. In the slightly more advanced case there is an hierarchy, e.g. faces are modeled in the scene and then on the next level mouth, nose and eyes pose are conditioned on the pose of the face.

Using images with ground truth position we estimate the probability $P(\#\Omega_i = n_i)$, of a scene containing n_i objects of type Ω_i at the top level. We then estimate the joint probability density function $f(p_{i,1},\ldots,p_{i,n_i})$ of the pose parameters $(p_{i,1},\ldots,p_{i,n_i})$ for these n_i objects of type Ω_i at the top level.

We then go on estimating the probability $P(\#\Omega_{ij}=n_{ij})$ for object type Ω_{ij} at the next level of hierarchy and the corresponding pose parameter joint probability density function conditioned on the pose parameter of the ascendent p_i.

The resulting scene model x is a collection of objects, $\{\Omega\}$, of different types, where each object has a different pose p.

Simulation of such models x can be made by sampling the number of objects and their pose parameters at the top level, followed by sampling of objects and poses of objects on lower level in the hierarchy.

Furthermore, for each scene model x, the likelihood can be determined as

$$L(x) = \Pi_{i=1}^{n}(P(\#\Omega_i = n_i)f(p_{i,1},\ldots,p_{i,n_i})\cdot\ldots$$

$$\ldots\cdot\Pi_{j=1}^{m_i}(P(\#\Omega_{ij}=n_{ij})f(p_{ij,1},\ldots,p_{ij,n_{ij}}))). \qquad (1)$$

2.2 Detection Modeling

As was described in the previous section, we have developed a number of detectors for different types of objects. Each such detector is evaluated at a discrete number of positions, $Y = p_1, p_2, \ldots, p_N$, in the corresponding pose space. We assume that the probability of obtaining a detection at pose p, given a scene with an object at true pose $\bar{p}$ close to p, is only dependent on this true object. This implies that we believe that each object has limited reach, i.e. a detection of an object is plausible only at close distances from the object. At larger distances there is still a possibility of 'false' detections, which we assume to be constant in the pose space.

In the experiments we have furthermore assumed a kind of stationarity in this respect. We have used pose spaces such as (position+scale) and have assumed that the probability depends on relative position and relative scale between the detected pose p and the actual pose $\bar{p}$. This probability

$$P_{detect}(\text{detection at pose } p|\text{closest pose is } \bar{p})$$

is estimated from databases with objects of known pose.

Each studied image is analyzed by running all of the detectors of type Ω_{ij} at all sampled positions Y_{ij}. The result y is typically a small number of positive detections of different types and a large number of non-detections of different types. Such a result y is shown in Figure 1, where each circle represents a face detected at the center of the circle and the radius corresponds to at which scale the detection was made. One may view y as a large boolean vector of length $N=\sum_{ij} N_{ij}$.

For each such detection result it is possible to determine the likelihood of the detection given a scene model x,

$$P(y|x) = \Pi_k^N P_{detect}(y_i|x). \qquad (2)$$

2.3 Scene Model Estimation from Ensemble Detections

Assume that we have an image, on which we have run multiple detectors and have obtained a detection result y. The scene model estimation is then given by maximum aposteriori estimation

$$\tilde{x} = \arg\max_{x} P(y|x) f(x). \tag{3}$$

One problem with this approach is that the prior $f(x)$ for different scene models vary in magnitude with different number of degrees of freedom for the scene model observation space, i.e. scene models containing many objects have probability density function values several magnitudes lower than those of few objects.

For an example of scene modeling see section 3.2.

Finding the optimum over all possible scene models x is a large optimization problem. The idea here is not that the scene model estimation should be performed by exhaustive search of this space. The main point is that the likelihood serves as a basis for comparing and choosing between different results obtained by heuristics, e.g. clustering of detection results. Using this model we hope to obtain better results than ad hoc methods which are used today, e.g. [3,5].

3 Experiments on Face Detection

3.1 Detection Probabilities

In these experiments we have used four detectors for faces, eyes, noses and mouths. Each detector has been trained on different positions and scales, i.e. the pose space is three dimensional (x, y, s). The detectors are based on boosting as described in [12] with cascades.

Typically the detection pose set, as illustrated in Figure 2, has coarser sampling in space (x, y) for coarser (larger) scales.

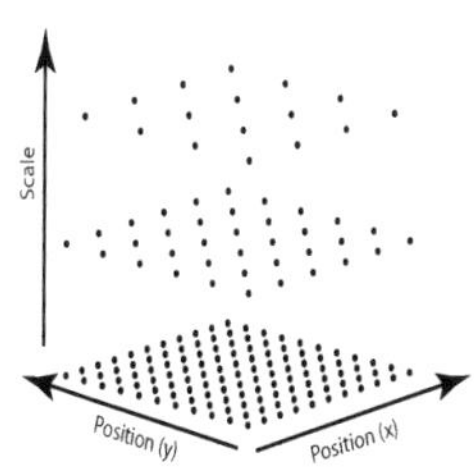

Fig. 2. Object detection is performed at a discrete set of pose points as illustrated in the figure

A typical detection (before clustering) is shown in Figure 1.

As can be expected when detecting an object at pose p_m, the detector will return true for many tested poses close to that position. By studying 697 images containing

995 faces with known poses, we estimate the probability of detecting an object at pose p_d when the true pose is p_m. Here only the relative pose is used

$$p_{rel} = \begin{pmatrix} (p_{d,x} - p_{m,x})/p_{m,s} \\ (p_{d,y} - p_{m,y})/p_{m,s} \\ \log(p_{d,s}/p_{m,s}) \end{pmatrix}. \tag{4}$$

Notice that the relative pose p_{rel} is invariant under a common translation and scale. Notice also that $p_d = p_m$ implies that $p_{rel} = \mathbf{0}$. For each type of object, (face, eyes, nose and mouth), the probability of detection

$$P_{det}(p_{rel}) = P(\text{detect at relative pose } p_{rel})$$

is estimated using a three-dimensional histogram. The result is shown in Figure 3.

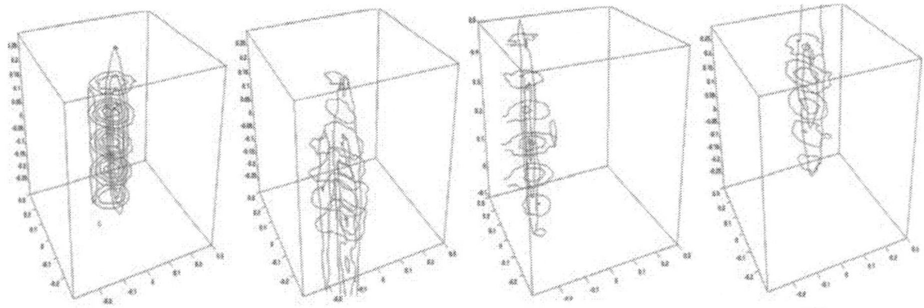

(a) (b) (c) (d)

Fig. 3. Estimated detection probability densities as a function of relative pose for (a) faces, (b) eyes, (c) noses and (d) mouths. Here the position are on the x- and y-axes and the scale is on the z-axis. Notice the bias in nose position. This is due to a difference in nose center definition between the training of the detector and the ground truth modeling made here.

We approximate these detection probabilities as scaled Gaussian functions, i.e.

$$P_{det} = a\frac{1}{(2\pi)^{3/2}\sqrt{|\Sigma|}}e^{(p_{rel}-m)^T\Sigma^{-1}(p_{rel}-m)/2}, \tag{5}$$

with different parameters (a, m, Σ) for each detector. Notice that there might be a trade-off here between having a very specific detector with high detection probability only in a small region, or a very unspecific detector with a broad detector response and perhaps not so high detection probability. As can be seen in the examples the detectors are quite specific in position, but not so specific in terms of relative scale.

3.2 Scene Modeling

In this experiments we modeled the scene as having a number of faces, each of which contained two eyes, one nose and one mouth. We estimated model probabilities from

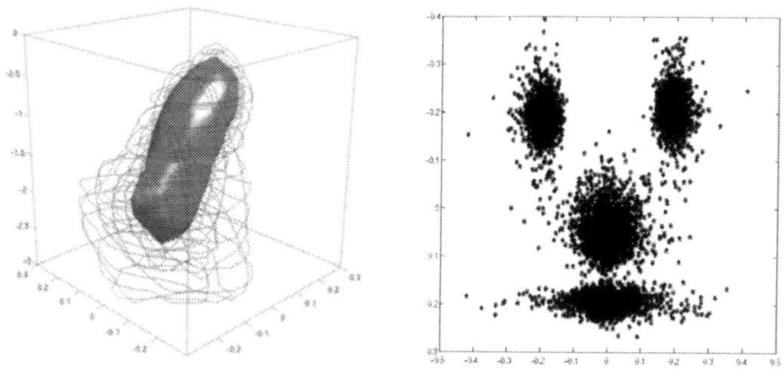

(a) (b)

Fig. 4. (a) Estimated probability density function for faces in pose space. Notice that the y-coordinate is typically higher for smaller faces (lower scales) and that there is larger variations in position (both x and y) for smaller faces. Larger faces tend to be more centrally positioned and with smaller variance in position. (b) The distribution of eyes, noses and mouths relative to the face poses in the database.

2268 images. These contained different number of faces. In total there were 2551 faces in these images. Images come in different sizes and shapes. We adopted a scene coordinate model with origin in the middle of the image and base scale equal to the square root of the image area in pixels. We represented image pose in this space with scale coordinate equal to the logarithm of the face width divided by the base scale. Figure 4 illustrates the estimated probability density function of a random face in this pose space.

In the experiment the following approximations were made. The positions of faces in a random image were assumed to be independently distributed according to Figure 4a with the addition that no two faces were allowed to spatially overlap more than 30 percent. This is not entirely realistic, since an image with many faces will probably be biased towards smaller scales.

Furthermore we assume that the relative pose of the two eyes, the nose and the mouth can be approximated by Gaussian distributions. Figure 4b shows the distribution of such facial positions relative to the face in the 2551 faces in the database.

3.3 Optimizing Likelihood

A two step maximization of the Likelihood was made:

$$\max_{scene} \log P(detection|scene). \tag{6}$$

Initial guesses for scenes were found from detected faces, noses, and mouths. From each detected feature a face with two eyes, a nose and a mouth was estimated using estimated relative mean positions. The procedure works in the following way,

1. Randomly select a feature from all detected features.
2. From this feature estimate a whole face. Add this to the current scene.
3. Calculate the likelihood for the new test-scene, and if this is greater than the current maximum, this is the new optimum.

The procedure starts with an empty scene and is repeated for a number of iterations. The best scene is then kept and the whole procedure is done all over again but starting with the current best scene instead of an empty scene. This procedure is repeated for a number of iterations, i.e. as many times as the maximum number of faces to be expected. This means that the number of faces is chosen automatically, the only restriction being that the maximum number of faces that can be found is limited by the number of iterations.

Given an initial scene we can then optimize the scene by moving each individual feature in the scene, i.e. the faces, eyes, noses and mouths, in position and scale. This is done by a simple neighborhood search.

3.4 Results

Using the estimated image and detector probabilities, we have tested the detection system on a number of images. For these images we obtained ground truth by manually marking positions in the images. A typical detection result can be seen in Figure 5. The database used for testing consisted of color images of groups of people collected from the Internet, with one to fifteen fronto-parallel faces. In total we used 300 images with a total of 1437 faces. This database is completely different from the database used for estimating all the detection and model probabilities, and can be acquired from the authors upon request. The results from the four detectors were collected and the maximum likelihood estimate of the scene was calculated according to section 3.3. The recall and precision were calculated for the face, nose and mouth detectors as well as for the combined result. The recall or hit-rate is defined as the total number of correct detections divided by the total number of faces in the images. The precision is defined as the number of correct detections divided by the total number of detections. The result can be seen in Table 1. Notice that only a single precision-recall (P/R) value is given. As our method is based on finding a maximum likelihood estimate, the result is based on the detection results from the different detectors, and the estimated detector statistics. This means that it is not straight-forward to to get a P/R curve by tuning a detection rate. By tuning the different detectors in different ways one would probably get different P/R curves depending on which detector one changed. One way to get a P/R curve might be to change the detector statistics, but as there is no detection threshold there are many parameters which one can vary. Also since these parameters are estimated and hopefully close to the true parameters, changing them would change the model, and the final detection would then not be the maximum likelihood estimate of the scene. As can be seen, the precision of the mouth and nose detectors aren't very high. The eye detector is not included in the table, but its precision is even lower. The precision of the face detector is quite high, but the recall is not very high. The missed faces are in most cases

Table 1. Recall and precision for the different detectors as well as for the maximum likelihood estimate

	Nose	Mouth	Face	Combined
Precision	0.45	0.50	0.97	1.00
Recall	0.81	0.79	0.75	0.84

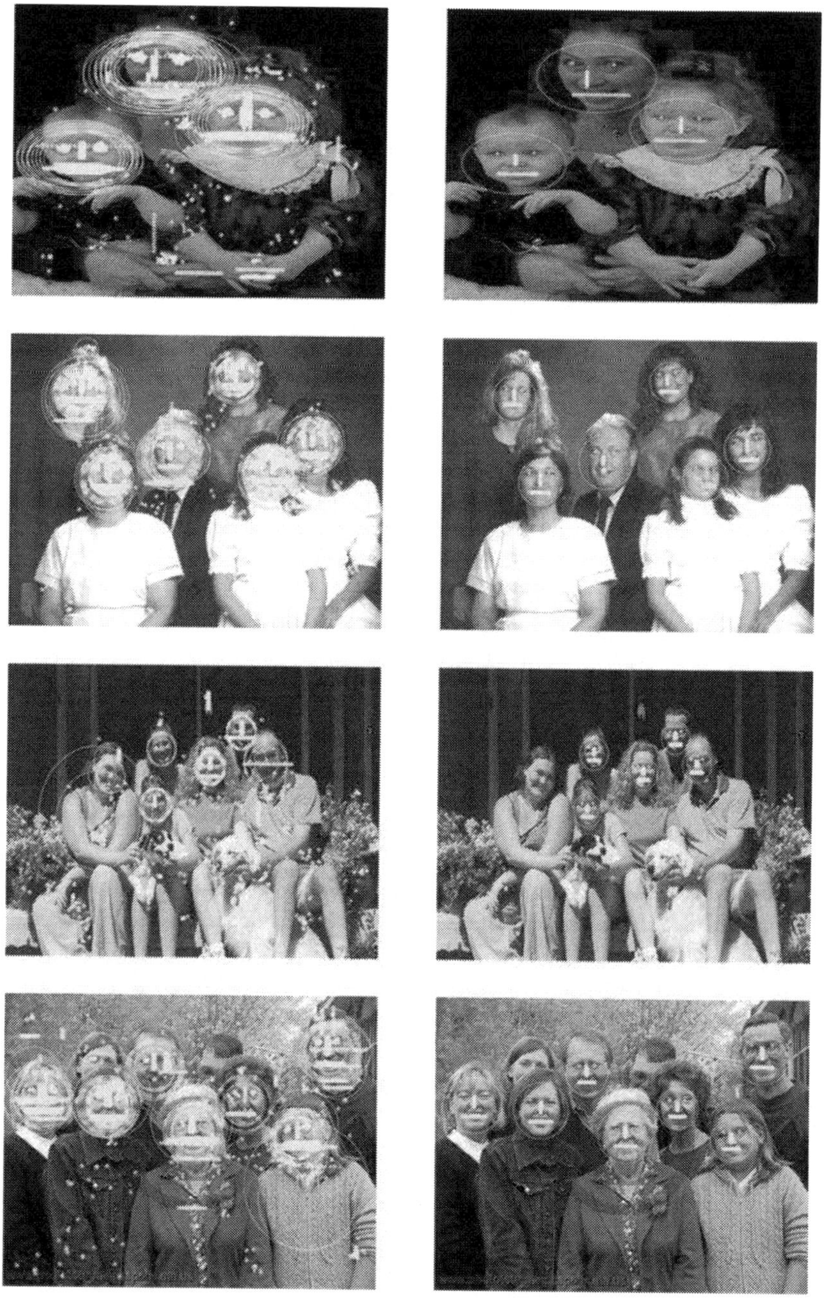

Fig. 5. On the left hand side images the initial unclustered detection results on a number of images is shown. The face detections are indicated by blue circles, the eyes by magenta stars, the mouths by red horizontal lines and the noses by yellow vertical lines. The sizes indicate the scale of the detections. On the right hand side the maximum likelihood estimates of the scenes are shown.

children or older people which were scarce in the training database for the detectors. One can see that the precision of the combined detector is very high; there were only three false positive faces in all the tested images.

4 Conclusions

In this paper we have investigated a probabilistic framework for context based scene interpretation using multiple detectors. Methods for finding maximum likelihood estimates of scenes given detection results were presented. The benefits of optimizing a scene model given a detection result is manifold. Firstly we get a clustering method that gives valid results in the sense that they adhere to a given real world model. Secondly we get higher recall rates than just using the individual detectors. But the most important gain is that we get a better understanding of the detected scene and the objects in it. This leads to both higher precision detections and the possibility to infer properties of the imaged object such as e.g. pose. One draw-back is of course that we need to use more computational effort compared to just using one specific mid-level detector.

References

1. D. Crandall, P. Felzenszwalb, and D. Huttenlocher. Spatial priors for part-based recognition using statistical models. In *Proc. Conf. Computer Vision and Pattern Recognition, San Diego, USA*, pages I: 10–17, 2005.
2. I. L. Dryden, K. V. Mardia, and A. N. Walder. Review of the use of context in statistical image analysis. *J. of Applied Statistics*, 24(5):513–538, 1997.
3. D. Forsyth and M. Fleck. Body plans. In *CVPR, Puerto Rico, USA*, pages 678–683, 1997.
4. H. Kruppa and B. Schiele. Using local context to improve face detection. In *BMVC, Norwich, England*, pages 3–12, 2003.
5. C. Mikolajczyk, K.and Schmid and A. Zisserman. Human detection based on a probabilistic assembly of robust part detectors. In *ECCV, Prague, Czech Republic*, 2004.
6. A. Mohan, C. Papageorgiou, and T. Poggio. Example-based object detection in images by components. *IEEE Trans. Pattern Analysis and Machine Intelligence*, 23(4):349–361, 2001.
7. S.J.D. Prince, J.H. Elder, Y. Hou, M. Sizintsev, and Y. Olevskiy. Statistical cue integration for foveated wide-field surveillance. In *Proc. Conf. Computer Vision and Pattern Recognition, San Diego, USA*, pages II: 603–610, 2005.
8. H. Schneiderman. Learning statistical structure for object detection. In *CAIP, Groningen, Netherlands*, pages 434–441, 2003.
9. H. Schneiderman and T. Kanade. Object detection using the statistics of parts. *Int. Journal of Computer Vision*, 56(3):151–177, February 2004.
10. H. Sidenbladh and M. Black. Learning image statistics for bayesian tracking. In *ICCV, vancouver, canada*, pages 709–716, 2001.
11. A. Torralba. Contextual priming for object detection. *Int. Journal of Computer Vision*, 53(2):169–191, July 2003.
12. P. Viola and M. Jones. Rapid object detection using a boosted cascade of simple features. In *Proc. Conf. Computer Vision and Pattern Recognition*. IEEE Computer Society Press, 2001.
13. M.H. Yang, D.J. Kriegman, and N. Ahuja. Detecting faces in images: A survey. *IEEE Trans. Pattern Analysis and Machine Intelligence*, 24(1):34–58, January 2002.

Confidence Based Gating of Multiple Face Authentication Experts

Mohammad T. Sadeghi[1,2] and Josef Kittler[2]

[1] Signal Processing Research Lab., Department of Electronics
University of Yazd, Yazd, Iran
[2] Centre for Vision, Speech and Signal Processing
School of Electronics and Physical Sciences
University of Surrey, Guildford GU2 7XH, UK
{M.Sadeghi, J.Kittler}@surrey.ac.uk

Abstract. We address the problem of fusing experts employing diverse similarity measures in LDA face space. The gradient direction measure is reviewed and experimentally compared with the normalised correlation in two different conditions, when the face images are well registered and when the registration process is performed automatically. We show that by combining the gradient direction measure and normalised correlation using a confidence based gating, the resulting decision making scheme consistently outperforms the best method. The gating is based on a novel decision confidence measure proposed in the paper.

1 Introduction

In certain pattern recognition applications the training sets are notoriously small. A typical example is biometric person recognition where only a few training data points are available for each individual. An extreme case of the small sample set situation arises in image and video database retrieval, where only a single exemplar is available to define the class of objects of interest.

The usual approach to such problems is to base the decision making on some form of similarity measure, or scoring function, which relates unknown patterns to the query object template. If the degree of similarity exceeds a prespecified threshold, the unknown pattern is accepted to be the same as the query object. Otherwise it is rejected. The similarity concept can also be used in recognition scenarios where the unknown pattern would be associated with that class, the template of which is the most similar to the observed data.

The similarity score is computed in a suitable feature space. Commonly, similarity would be quantised in terms of a distance function, on the grounds that similar patterns will lie physically close to each other. Thus smaller the distance, the greater the similarity of two entities. The role of the feature space in similarity measurement is multifold. First of all the feature space is selected so as to maximise the discriminatory information content of the data projected into the feature space and to remove any redundancy. However, additional benefits sought after from mapping the original pattern data into a feature space is to

D.-Y. Yeung et al. (Eds.): SSPR&SPR 2006, LNCS 4109, pp. 667–676, 2006.

simplify the similarity measure deployed for decision making. A classical example of this is the use of the Euclidean distance metric in Linear Discriminant Analysis (LDA) feature spaces as the within class covariance matrix in the LDA space becomes an identity matrix and such metric becomes theoretically optimal. LDA was introduced to the face verification area by Belhumeur in 1996 [1]. Despite the theoretical optimality of Euclidean metric in the LDA space, in [3] , it has been demonstrated that it is outperformed by the Normalised Correlation (NC).

However, in [3] it has been further demonstrated that the Gradient Direction (GD) scoring function is even more effective. In this method the distance between a probe image and a model is measured in the gradient direction of the aposteriori probability of the hypothesised client identity. A mixture of Gaussian distributions with Identity covariance matrix has been assumed as the density function of the possible impostors. In [5] the GD metric was further generalised by considering a general covariance matrix for the components of the Gaussian mixture model(GGD metric). The main problem with the Gradient Direction metric is its computational complexity. In [4], an approximation to the Gradient Direction metric (AGD) was developed. The AGD metric is defined as the difference between the mean (template) of the claimed identity and the local mean of other identities representing the anti-class (impostors). Although not as powerful as the Gradient Direction method, we showed that this approximate Gradient Direction metric gives good performance, in comparison with normalised correlation and is significantly simpler to implement than the Gradient Direction metric method.

The previous studies were performed on the BANCA database [1] using an internationally agreed experimental protocols by applying a geometric face registration method based on manually annotated eyes positions. One of the main issues for assessing the performance of a similarity measure in an automatic face authentication system is how robust the approach is to miss-registration errors. In this paper, the performance of the NC scoring function is compared with the GD metric and its extensions in experiments involving automatically registered faces. Our experimental studies show that overall the NC function is less sensitive to miss-registration error but in certain conditions GD metric performs better. In order to gain maximum benefit from the complementary merits of these scoring functions , we propose a combined strategy which fuses the scores using a confidence function based weighting. In the proposed method, in the training stage the statistical distribution of the miss-classified scores is estimated for both NC and GD metrics. Then in the test stage, the confidence of the resulting scores are measured. The score value with the higher confidence level is finally adopted for decision making.

The paper is organised as follows. In the next section the Normalised Correlation and Gradient Direction metrics are reviewed. The proposed method of score fusion is then introduced in Section 3. A description of the experimental design including the face database used in the study, the experimental protocols

[1] http://www.ee.surrey.ac.uk/banca/

and the experimental setup is given in Section 4. The experimental results using different scoring functions and the fusion results are presented and discussed in Section 5. Finally a summary of the main findings and conclusions can be found in Section 6.

2 Similarity Score Functions

In a face verification system, a matching scheme measures the similarity or distance of the test sample, $\mathbf{x}$ to the template of the claimed identity, $\boldsymbol{\mu}_i$. Note that $\mathbf{x}$ and $\boldsymbol{\mu}_i$ are the projections of the test sample and class mean into the feature space respectively. The simplest similarity measure, s, for matching the probe and the ith client mean is the *Euclidean Distance* between the vectors $\mathbf{x}$ and $\boldsymbol{\mu}_i$, i.e.

$$s_E = \sqrt{(\mathbf{x} - \boldsymbol{\mu}_i)^T(\mathbf{x} - \boldsymbol{\mu}_i)} \tag{1}$$

In [3], it has been demonstrated that a matching score based on *Normalised Correlation* (NC) scoring function, defined by Equation 2, is more efficient.

$$s_N = \frac{||\mathbf{x}^T\boldsymbol{\mu}_i||}{\sqrt{\mathbf{x}^T\mathbf{x}\boldsymbol{\mu}_i^T\boldsymbol{\mu}_i}} \tag{2}$$

In [3] an innovate metric called the *Gradient Direction* (GD) metric has been proposed. In this method the distance between a probe image and a model is measured in the gradient direction of the aposteriori probability of the hypothesised client identity. A mixture of Gaussian distributions with Identity covariance matrix has been assumed as the density function of the possible classes of identity. In [5], we revisited the theory of the Gradient Direction metric and extended it to a Generalised Gradient Direction metric. We demonstrated that applying GD metric using either a general covariance matrix derived from the training data or an isotropic covariance matrix with a variance of the order of the variation of the image data in the feature space is even more efficient than the NC function. The proposed optimal matching score is defined as

$$s_O = \frac{||(\mathbf{x} - \boldsymbol{\mu}_i)^T\nabla_O P(i|\mathbf{x})||}{||\nabla_O P(i|\mathbf{x})||} \tag{3}$$

where $\nabla_O P(i|\mathbf{x})$ refers to the gradient direction. In the generalised form of the GD metric, the optimal direction would be

$$\nabla_G P(i|\mathbf{x}) = \boldsymbol{\Sigma}^{-1}\sum_{\substack{j=1 \\ j \neq i}}^{m} p(\mathbf{x}|j)(\boldsymbol{\mu}_j - \boldsymbol{\mu}_i) \tag{4}$$

where $p(\mathbf{x}|j)$ is the j-th client measurement distribution. Considering an isotropic structure for the covariance matrix, i.e. $\mathbf{\Sigma} = \sigma\mathbf{I}$, equation 4 could be simplified as:

$$\nabla_I P(i|\mathbf{x}) = \sum_{\substack{j=1 \\ j \neq i}}^{m} p(\mathbf{x}|j)(\boldsymbol{\mu}_j - \boldsymbol{\mu}_i) \tag{5}$$

Note that the magnitude of the σ will affect the direction through the values of $p(\mathbf{x}|j)$.

3 Selecting Similarity Functions

One of the most exciting research directions in the field of pattern recognition and computer vision is classifier fusion. It has been recognised that the classical approach to designing a pattern recognition system which focuses on finding the best classifier has a serious drawback. Any complementary discriminatory information that other classifiers may capture is not tapped. Multiple expert fusion aims to make use of many different designs to improve the classification performance. In the case considered here, as different metrics span the feature space in different ways, it seems reasonable to expect that a better performance could be obtained by combining the resulting classifiers. Our experimental study (reported in section 5) demonstrates that in most of the cases just one of the NC and GD metrics fails to make the correct decision. Therefore, we expect that by dynamically selecting the experts using the respective metrics the performance of the verification system can be improved. In this study, a simple method for combining the NC based classifier and the GD metric one is proposed.

Suppose that s_N and s_O refer to the NC and GD scores for a test sample, x. Let $p_e(s)$ denote probability of error. Then, if $p_e(s_N) < p_e(s_O)$, the NC metric should be used, otherwise the GD metric will give a better result. Now,

$$p_e(s) = p_e(s|C)p_e(C) + p_e(s|I)p_e(I) \tag{6}$$

where $p_e(s|C)/p_e(s|I)$ refers to the probability of error if x is classified as client/impostor and $p_e(C)$ and $p_e(I)$ refer to the probability of client and impostor errors respectively.

In the evaluation step, in addition to the threshold(s), the probability density functions of the distances corresponding to the miss-classified samples ,$P_e(s|C)$ and $P_e(s|I)$, can be estimated. These functions are determined for both NC and GD metrics. In this study a simple unimodal Gaussian function was used for modelling the density functions. $p_e(C)$ and $p_e(I)$ are in fact the False Rejection and False Acceptance error in the evaluation step. Then, in the test step, the error probabilities of the measured distances are calculated using Equation 6 for both metrics. These values are considered as the confidence levels of the measurements made. The value with the higher confidence level (lower error probability) is used finally to make the decision.

4 Experimental Design

In this section the face verification experiments carried out on images of the BANCA database are described. The BANCA database is briefly introduced first. The main specifications of the experimental setup are then presented.

4.1 BANCA Database

The BANCA database has been designed in order to test multi-modal identity verification systems deploying different cameras in different scenarios (Controlled, Degraded and Adverse). The database has been recorded in several languages in different countries. Our experiments were performed on the English section of the database. Each section contains 52 subjects (26 males and 26 females). Experiments can be performed on each group separately.

Each subject participated to 12 recording sessions in different conditions and with different cameras. Sessions 1-4 contain data under *Controlled* conditions whereas sessions 5-8 and 9-12 contain *Degraded* and *Adverse* scenarios respectively. Each session contains two recordings per subject, a true client access and an informed impostor attack. For the face image database, 5 frontal face images have been extracted from each video recording, which are supposed to be used as client images and 5 impostor ones. In order to create more independent experiments, images in each session have been divided into two groups of 26 subjects (13 males and 13 females). Thus, considering the subjects' gender, each session can be divided into 4 groups. The decision function can be trained using only 5 client images per person from the same group and all client images from the other groups.

In the BANCA protocol, 7 different distinct experimental configurations have been specified, namely, Matched Controlled (MC), Matched Degraded (MD), Matched Adverse (MA), Unmatched Degraded (UD), Unmatched Adverse (UA), Pooled test (P) and Grand test (G).

4.2 Experimental Setup

The performance of different decision making methods based on the Normalised Correlation (s_N) and the Gradient Direction (s_I) metrics are experimentally evaluated on the BANCA database using the configurations discussed in the previous section. The evaluation is performed in the LDA space. The original resolution of the image data is 720×576. The experiments were performed with a relatively low resolution face images, namely 64×49. The results reported in this article have been obtained by applying a geometric face normalisation based on the eyes positions. The eyes positions were localised either manually or automatically. A fast method of face detection and eyes localisation was used for the automatic localisation of eyes centre [2]. The XM2VTS database [2] was used for calculating the LDA projection matrix.

[2] http://www.ee.surrey.ac.uk/Research/VSSP/xm2vtsdb/

The thresholds in the decision making system have been determined based on the Equal Error Rate criterion, i.e. by the operating point where the false rejection rate (FRR) is equal to the false acceptance rate (FAR). The thresholds are set either globally (*GT*) or using the client specific thresholding (*CST*) technique [5]. As we mentioned earlier, in the training sessions of the BANCA database 5 client images per person are available. In the case of global thresholding method, all these images are used for training the clients template. The other group data is then used to set the threshold. In the case of the client specific thresholding strategy, only two images are used for the template training and the other three along with the other group data are used to determine the thresholds. Moreover, in order to increase the number of data used for training and to take the errors of the geometric normalisation into account, 24 additional face images per each image were generated by perturbing the location of the eyes position around the annotated positions.

In the previous studies [5] [4], it was demonstrated that the Client Specific Thresholding (CST) technique was superior in the matched scenario (Mc, Md, Ma and G) whereas the Global Thresholding (GT) method gives a better performance on the unmatched protocols. The results reported in the next section were acquired using this criterion.

5 Experimental Results and Discussion

Tables 1 contains a summary of the results obtained on the test set when manually annotated eyes position were used for the face geometric normalisation. The values in the table indicate the FAR, FRR and Total Error Rates (TER), i.e. the sum of false rejection and false acceptance rates. In the GD metric the impostor distributions have been approximated by isotropic Gaussian functions with a standard deviation, σ, of the order of 10^4 . The order of σ is related to the order of the standard deviation of the input data (gray level values). This order is the consequence of normalising the length of the LDA axes to unity.

Table 1. ID verification results using the Normalised Correlation and Gradient Direction methods with Global and Client Specific Thresholding techniques for unmatched and matched protocols respectively. FAR: False Acceptance Rate, FRR: False Rejection Rate and TER: Total Error Rate.

	NC			GD		
	FAR	FRR	TER	FAR	FRR	TER
MC	2.98	5.77	8.75	1.25	3.97	**5.22**
MD	4.14	8.20	12.34	1.25	7.05	**8.30**
MA	5.96	10.00	15.96	1.35	6.53	**7.88**
UD	13.65	13.72	**27.37**	13.94	15.26	29.2
UA	20.19	21.92	42.12	16.06	16.15	**32.21**
P	14.01	14.23	28.24	11.57	10.64	**22.21**
G	8.20	3.33	11.54	2.02	1.58	**3.60**

Table 2. ID verification results using the Normalised Correlation and Gradient Direction methods with Global and Client Specific Thresholding techniques for unmatched and matched protocols respectively. Eyes were localised automatically for the face registration purpose.

	NC			GD		
	FAR	FRR	TER	FAR	FRR	TER
MC	5.096	10.9	**16.00**	8.558	8.333	16.89
MD	10.38	14.1	**24.49**	12.98	14.1	27.08
MA	10.67	10.9	21.57	11.92	9.103	**21.03**
UD	18.85	24.62	**43.46**	23.65	23.59	47.24
UA	24.13	24.49	48.62	23.08	21.41	**44.49**
P	19.36	20.09	39.44	19.33	19.1	**38.43**
G	16.47	10.3	26.77	14.87	9.316	**24.19**

A comparison of the NC results against the results using the GD metric in Table 1 clearly demonstrates that the GD metric outperforms the NC metric.

As we mentioned earlier one of the most important criteria for adopting a similarity measure is the robustness of the method against miss-registration errors. In spite of the significant advances in face detection and localisation algorithms, the success of the methods in systems operating in realistic, dynamic scenarios is still very limited. The face pose variation, illumination changes and the size of the face images can degrade the performance of the face localisation algorithms. Table 2 contains the results of similar experiments when the face registration step was performed based on automatically localised eyes position[2].

These results demonstrate that unlike the results using manually localised eyes position, the GD method is not the outright winner in the automatic face verification system. In most of the cases, the NC metric gives a better or comparable verification rate i.e. overall the NC metric seems slightly less sensitive to errors in face registration. In the next section, it is demonstrated that by combining the NC and GD based classifiers, the performance of the face verification system can be improved.

Figure 1 shows a summary of a statistical study of the False Acceptance and False Rejection errors using the NC and GD metrics. In these plots *BF* and *OF*, respectively, stand for *B*oth metrics *F*ail and *O*ne (and only one) metric *F*ail. These results demonstrate that, in most of the cases just one of the metrics fails to make the correct decision. Therefore, we expect that by combining the NC-based and GD-based classifiers the performance of the verification system can be improved.

We adopted the decision level fusion strategy proposed in section 3 in order to combine the NC and GD metrics. Tables 3 and 4 contain the combined verification results using manually and automatically registered data respectively. These results demonstrate that, overall, a better performance is achieved using the combined method especially on the unmatched scenarios. The main reason that not much better results are obtained for the matched protocols is that as

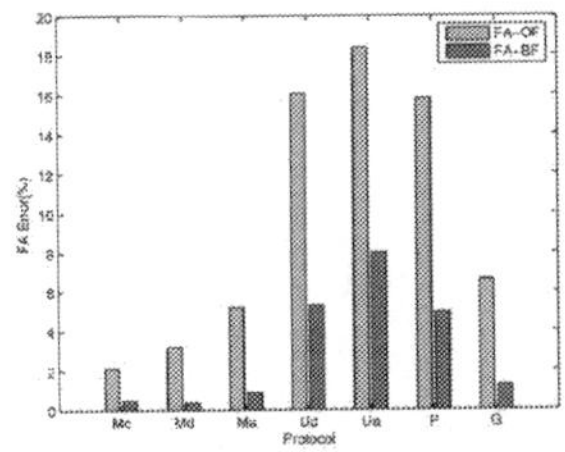

(a) FA error (Manual registration)

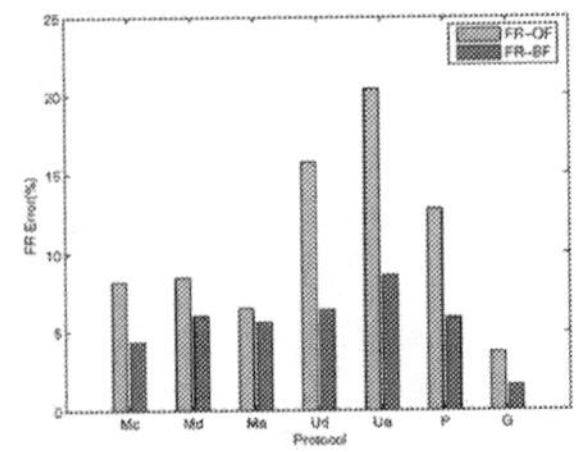

(b) FR error (Manual registration)

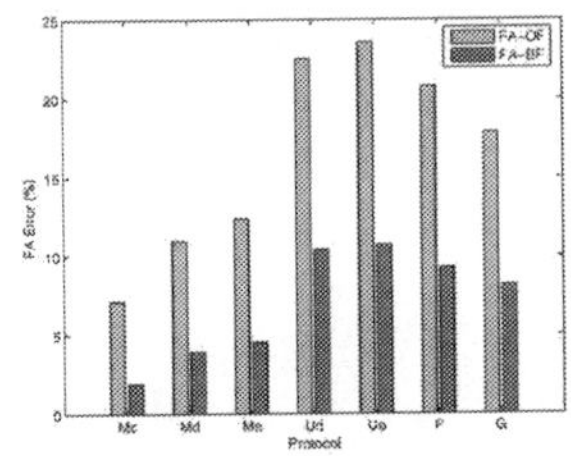

(c) FA error (Automatic registration)

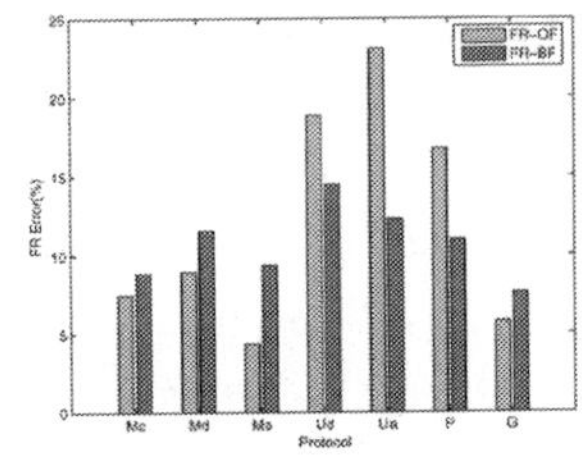

(d) FA error (Automatic registration)

Fig. 1. Percentage of the False Acceptance and False Rejection errors when only One (OF) or Both (BF) metrics fail to make the correct decision

Table 3. ID verification results on BANCA protocols using CST method for matched and GT method for unmatched protocols, manual registration, combining NC and GD scores

	Evaluation			Test		
	FAR	FRR	TER	FAR	FRR	TER
Mc	0.2811	0.2821	0.5631	1.538	4.231	5.769
Md	1.739	1.128	2.867	1.827	7.436	9.263
Ma	1.317	0.5641	1.881	2.5	6.667	9.167
Ud	12.69	12.56	25.26	9.423	14.1	23.53
Ua	19.23	19.74	38.97	14.81	17.44	32.24
P	14.01	13.89	27.9	8.365	12.22	20.59
G	2.387	1.111	3.498	3.59	1.41	5

we mentioned earlier we adopted the CST technique for these protocols. We only have a few clients per subject and in some cases in the evaluation stage all the clients for each subject are successfully classified. So, there is not enough data available for estimating the distribution of the miss-classified client distance values, $P_e(s|C)$. Therefore, in the CST method, although the thresholds are client specific, only two global models were estimated as $P_e(s|C)$ and $P_e(s|I)$ using the miss-classified samples in the evaluation stage. These models are not as representative as required.

Table 4. ID verification results on BANCA protocols using CST method for matched and GT method for unmatched protocols, automatic registration, combining NC and GD metrics

	Evaluation			Test		
	FAR	FRR	TER	FAR	FRR	TER
Mc	0.391	0.7436	1.135	4.808	10.38	15.19
Md	1.154	2.077	3.231	7.885	14.62	22.5
Ma	1.513	1.846	3.359	10.87	10.64	21.51
Ud	19.04	19.62	38.65	18.37	23.97	42.34
Ua	25	25	50	20.87	23.21	44.07
P	20.13	19.74	39.87	17.44	19.4	36.84
G	2.302	3.077	5.379	14.23	11.03	25.26

6 Conclusions

The problem of measuring similarity in LDA face space has been considered. First, recently proposed gradient direction measures were reviewed and experimentally compared with normalised correlation. The experiments were conducted on the Banca database, using the standard Banca face verification protocols. Although the gradient direction measures were shown to be significantly more discriminative than normalised correlation when probe face images were well registered, in poor registration conditions normalised correlation performed better. As the extent of misregistration was largely a function of the imaging conditions in which the probe data was acquired, the best performing method was effectively scenario dependent.

We showed that by combining the gradient direction measure and normalised correlation using a confidence based gating, the resulting decision making scheme consistently outperformed the best method. The gating is based on a novel decision confidence measure proposed in the paper. The measure, developed in the Bayesian framework, involves estimating the probability distribution of errors for each similarity measure on the evaluation set. The proposed scheme has the advantage that it renders the verification process fully scenario independent.

References

1. P.N. Belhumeur, J.P. Hespanha, and D.J. Kriegman. Eigenfaces vs. Fisherfaces: Recognition using class specific linear projection. *IEEE Trans. on Pattern Recognition and Machine Intelligence*, 19(7):711–720, 1997.
2. M. Hamouz, J. Kittler, J.-K Kamarainen, P. Paalanen, H. Kalviainen, and J. Matas. Feature-based affine-invariant localization of faces. *IEEE Transactions on Pattern Analysis and Machine Intelligence*, 27(9):1490–1495, September 2005.
3. J. Kittler, Y. P. Li, and J. Matas. On matching scores for LDA-based face verification. In M Mirmehdi and B Thomas, editors, *Proceedings of British Machine Vision Conference 2000*, pages 42–51, 2000.

4. J. Kittler and M. Sadeghi. Approximate gradient direction metric for face authentication. In *Joint IAPR International Workshops on Syntactical and Structural Pattern Recognition and Statistical Pattern Recognition, S+SSPR 2004*, pages 797–805, Lisbon, Portugal, 18-20 August 2004.
5. M. Sadeghi and J. Kittler. Decision making in the LDA space: Generalised gradient direction metric. In *the 6th International Conference on Automatic Face and Gesture Recognition*, pages 248–253, Seoul, Korea, May 2004.

Diversity Analysis for Ensembles of Word Sequence Recognisers

Roman Bertolami and Horst Bunke

Institute of Computer Science and Applied Mathematics
University of Bern, Neubrückstrasse 10, CH-3012 Bern, Switzerland
{bertolam, bunke}@iam.unibe.ch

Abstract. In this paper we propose a general framework for analysing the diversity of ensembles of word sequence recognition systems. The goal of the framework is to enable the application of any diversity measure developed for standard multi-class classification problems to ensembles of word sequence recognisers. Experiments with several diversity measures are conducted on artificial as well as on real world data and show the effectiveness of the proposed approach.

1 Introduction

Ensemble methods have been applied to many different fields of pattern recognition [1]. In handwriting recognition, improvements have been reported in isolated character [2] as well as in single word recognition [3,4]. Only recently work has been published on ensemble methods for word sequence recognition [5,6,7], which is still a field with many challenges.

The goal of ensembles methods is to correct the errors of one ensemble member with the output of the other ensemble members. To achieve this goal we need a certain diversity among the ensemble members. Intuitively speaking, the members should make no coincident errors, i.e. the errors of one classifier should be independent of the errors of the other classifiers. A high diversity of a classifier ensemble is considered to be a strong hint to good performance. Hence, measuring diversity allows one to predict the performance of an ensemble without the need of conducting computationally expensive experiments. Several diversity measures have been proposed in the literature for multi-class problems. Surveys can be found in [8,9,10].

Two different groups of diversity measures for ensembles of classifiers can be distinguished: pairwise measures and nonpairwise measures. Pairwise measures derive the final diversity of an ensemble of N classifiers from the $N(N-1)$ pairwise diversity values. Usually the mean of these values is used as ensemble diversity. Popular members of this group of diversity measures are correlation, disagreement, and double fault. On the other hand, nonpairwise measures consider all the classifier of an ensemble together and calculate one diversity value directly. Popular nonpairwise diversity measures are entropy and Sharkey-Level based measures.

D.-Y. Yeung et al. (Eds.): SSPR&SPR 2006, LNCS 4109, pp. 677–686, 2006.

All diversity measures currently known have been developed for conventional multi-class classification problems. To the knowledge of the authors, no diversity measures for word sequence recognition have been proposed until know.

The aim of the current paper is to provide a generic framework to apply diversity measures, developed for conventional multi-class classification problems, to the task of word sequence recognition. Word sequence recognition is a difficult problem because not only a single class, but a sequence of word classes of unknown length has to be returned by the recogniser. Even though appropriate diversity measures for word sequence recognition are potentially useful in the ensemble generation process, no such measures have been published yet. The contribution of this paper is to make classical diversity measures available for word sequence recognisers.

The remaining part of the paper is organised as follows. Section 2 introduces the diversity analysis framework and provides an example in detail. Experiments conducted on artificial as well as on real world data are described in Sect. 3. Finally, conclusions are drawn in the last section of the paper.

2 Methodology

Assume we have an ensemble where each of the n recognisers outputs a word sequence $W_i = (w_{i_1}, \ldots, w_{i_{m_i}})$; $i = 1, \ldots, n$. The number of words m_i in these sequences may differ and therefore a synchronisation process is required first. The synchronised results of the ensemble members are stored in a Word Transition Network (WTN) [11]. Because the class labels are used to calculate the diversity measures, the segments of the WTN have to be labelled with the ground truth. Any diversity measure available for multi-class problems can then be applied to the segments and based on these measures the final ensemble diversity is derived. Next, we describe various aspects of the proposed diversity analysis framework in greater detail. An extensive example of the entire process is provided in Section 2.3.

2.1 Synchronisation of the Word Sequences

Let the n ensemble members have recognised their individual word sequences $(W_1, \ldots, W_n)$. Each of these sequences might contain a different amount of words and therefore an alignment procedure is necessary to synchronise the n word sequences. Any possible string alignment procedure can be used for this purpose. However, because the optimal alignment of multiple strings is an *NP*-complete problem [12], we suggest to use a heuristic approach (e.g. incremental) for the alignment. For details see [11]. The result of the alignment is a WTN which consists of m segments. Each arc in the WTN is labelled with a word out of $(W_1, \ldots, W_n)$. Null transition arcs ε occur when the number of words in $(W_1, \ldots, W_n)$ is not equal for each W_i $(i = 1, \ldots, n)$.

Next, the ensemble result is derived applying some decision rule to each segment of the WTN. Any kind of decision strategy can be used, but for the sake

Table 1. Probabilities of coincident errors between classifier C_i and C_j

	C_i correct	C_i wrong
C_j correct	a	b
C_j wrong	c	d

of simplicity we just use a voting procedure. The resulting sequence of decision results constitute the combination result $\hat{W}$. Note that if the decision result of a segment is a null transition, this segment does not contribute any word to $\hat{W}$.

Once we have the combination result $\hat{W}$ and the ground truth, we can label the segments of the WTN. Therefore, we first align $\hat{W}$ with the ground truth. Using the alignment we can map the words in the ground truth to the WTN segments.

2.2 Application of Diversity Measures

Existing diversity measures can then be applied to the labelled segments of the WTN, similarly to conventional multi-class classification problems. The average of these measures gives the final ensemble diversity.

In the present work we apply pairwise as well as Sharkey-Level based diversity measures to word sequence recognition. Pairwise diversity measures consider only a pair of recognisers at a time. Any possible pair of ensemble members produces a diversity value. The average across all pairs gives the final diversity. Based on the probabilities of coincident errors between two recognisers C_i and C_j (Tab. 1) several measures have been proposed.

Correlation. Because the output of two recognisers can be considered as numerical values (1 for correct and 0 for wrong), we can calculate the correlation coefficients.

$$\rho_{i,j} = \frac{ad - bc}{\sqrt{(a+b)(c+d)(a+c)(b+d)}} \tag{1}$$

Disagreement. The disagreement measure is a very intuitive measure of diversity. It is the probability that the two considered recognisers will disagree on their decision.

$$D_{i,j} = b + c \tag{2}$$

Double Fault. Another quite intuitive measure is the double fault measure which is the probability that both recognisers make a wrong decision.

$$DF_{i,j} = d \tag{3}$$

The second kind of diversity measures are derived from the Sharkey-Levels introduced in [13]. Each segment of the WTN can be assigned to one of the following levels:

Level 1. No coincident errors. Each ensemble member produces the correct result.

Level 2. Some coincident errors, but the majority of the ensemble members provide the correct result.
Level 3. Majority is not correct but some of the members produce the correct result.
Level 4. The correct result is not output by any of ensemble members.

The frequencies of the different levels are then used as diversity measures. Thus, L_i is the frequency that a segment of the WTN belongs to Level i, where $i = 1, \ldots, 4$.

2.3 Example

Next, we will provide an example of the entire process of calculating diversity measures for ensembles of handwritten text line recognisers. The input is the handwritten text line *They will be asked to comment* shown in Fig 1. Features are extracted and each ensemble member performs the recognition step. The recognised word sequences $(W_1, \ldots, W_9)$ are shown in Fig. 2.

Next, the word sequences $(W_1, \ldots, W_9)$ are synchronised. An iterative alignment is used for this purpose [11]. The result of this alignment step is shown in Fig. 3. Note that null transition arcs have to be inserted at the beginning of some text lines to align the additional word in W_1 and W_7.

Once the alignment is performed, we can calculate the combination result for each alignment segment. We apply a majority voting and get the word sequence *they will be asked to council.* To label the alignment segments we first have to align the combination result $(\hat{W})$ with the ground truth (T) as shown in Fig. 4.

Based on this information we label the segments of the aligned word sequences of Fig. 3. The result of the labelling process is shown in Fig. 5.

Now we are able to apply any of the described diversity measures, originally developed for conventional multi-class classification problems, to our sequence recognition problem. E.g. the Level 1 diversity measure yields 4/7, whereas the Level 3 diversity measure is equal to 1/7.

3 Experiments and Results

One of the main motivations for computing ensemble diversity is to predict the performance of an ensemble of recognisers without the need to to run computationally expensive experiments. Hence, the aim of the experiments is to show relationships between the recognition accuracy and the different diversity measures. Experimental evaluation of the proposed framework is conducted on a synthetic and a real world data set.

3.1 Synthetic Word Sequences

To be able to test our framework, we generate artificial ensemble results. First the ground truth is created and then the ensemble results are generated.

Fig. 1. Handwritten input text

W_1: if they will be asked to council
W_2: they will be asked to comment
W_3: it will be asked to comment
W_4: they will be asked to council
W_5: they will be asked to council
W_6: they will be asked to comment
W_7: if it will be asked to comment
W_8: they will be asked to council
W_9: they will be asked to council

Fig. 2. Results from the different ensemble members

W_1:	if	they	will	be	asked	to	council
W_2:	ε	they	will	be	asked	to	comment
W_3:	ε	it	will	be	asked	to	comment
W_4:	ε	they	will	be	asked	to	council
W_5:	ε	they	will	be	asked	to	council
W_6:	ε	they	will	be	asked	to	comment
W_7:	if	it	will	be	asked	to	comment
W_8:	ε	they	will	be	asked	to	council
W_9:	ε	they	will	be	asked	to	council

Fig. 3. Alignment of the ensemble results in a WTN

$\hat{W}$:	they	will	be	asked	to	council
T:	They	will	be	asked	to	comment

Fig. 4. Alignment of the combination result and the ground truth

W_1:	if	they	will	be	asked	to	council
W_2:	ε	they	will	be	asked	to	comment
W_3:	ε	it	will	be	asked	to	comment
W_4:	ε	they	will	be	asked	to	council
W_5:	ε	they	will	be	asked	to	council
W_6:	ε	they	will	be	asked	to	comment
W_7:	if	it	will	be	asked	to	comment
W_8:	ε	they	will	be	asked	to	council
W_9:	ε	they	will	be	asked	to	council
T:	ε	They	will	be	asked	to	comment

Fig. 5. Labelling the segments

To build the ground truth for a word sequence, the number of words n is first defined by a random number in a given range. Next, n words are randomly chosen from the underlying lexicon and then used as the ground truth word sequence.

Next, the ensemble results are generated iteratively. Given the ground truth word sequence, the first ensemble member's result sequence is created randomly with a given accuracy. The next member's result sequence is then generated with respect to a defined correlation to the previously generated word sequence. This process is continued until the desired number of results has been obtained.

We generated ensembles with five, ten, fifteen, and twenty members with a lexicon of 10,000 word classes. In our experiments we also varied the correlation between two successive members to obtain ensembles with different diversities.

The advantage of the artificial data is that we can control the correlation between the ensemble members. Also the number of classes can be chosen. Therefore, we are able to simulate different application domains, e.g. character sequences ($\sim$ 80 classes) or word sequences ($\sim$ 10,000 classes). Furthermore, an arbitrarily large amount of data can be produced relatively fast. On the other hand, it may happen that the generated ensemble member results do not sufficiently well model the results produced by real word ensemble members.

3.2 Ensembles of Handwritten Text Line Recognisers

In the second part of the experimental evaluation we use real word data from offline handwritten text line recognition. The data originate from [14] where three different ensemble member selection strategies have been validated. For this purpose, many different ensembles of various sizes have been built and tested which we now use to evaluate the proposed diversity framework.

The ensemble members were derived from specific integration of a statistical language model in the hidden Markov model based recognition system. The handwritten text lines that were used originate from the IAM[1] database [15]. For further details about the experimental setup of the recognition and ensemble generation step we refer to [14].

3.3 Diversity Analysis Results

In the experiments described in this section we investigate which of the proposed diversity measures, applied in our framework, is a better indicator of the accuracy of the combined ensemble results. Furthermore, we analyse the differences between the artificial and the real world data results.

The results for the pairwise measures Correlation, Disagreement, and Double Fault are shown in Fig. 6. While clear tendencies can be seen in the outcomes of the artificial ensemble, only the Double Fault measure seems to be a useful indicator for the accuracy of the ensemble performance on the real world data. The four different lines that can be observed in the artificial data plots originate from the four different sizes of the analysed ensembles.

Figure 7 shows the results of the level based diversity measures. On the real world data set, the best indicator of a good performance is Level 4. This means

[1] The IAM database is publicly available for download at http://www.iam.unibe.ch/~fki/iamDB

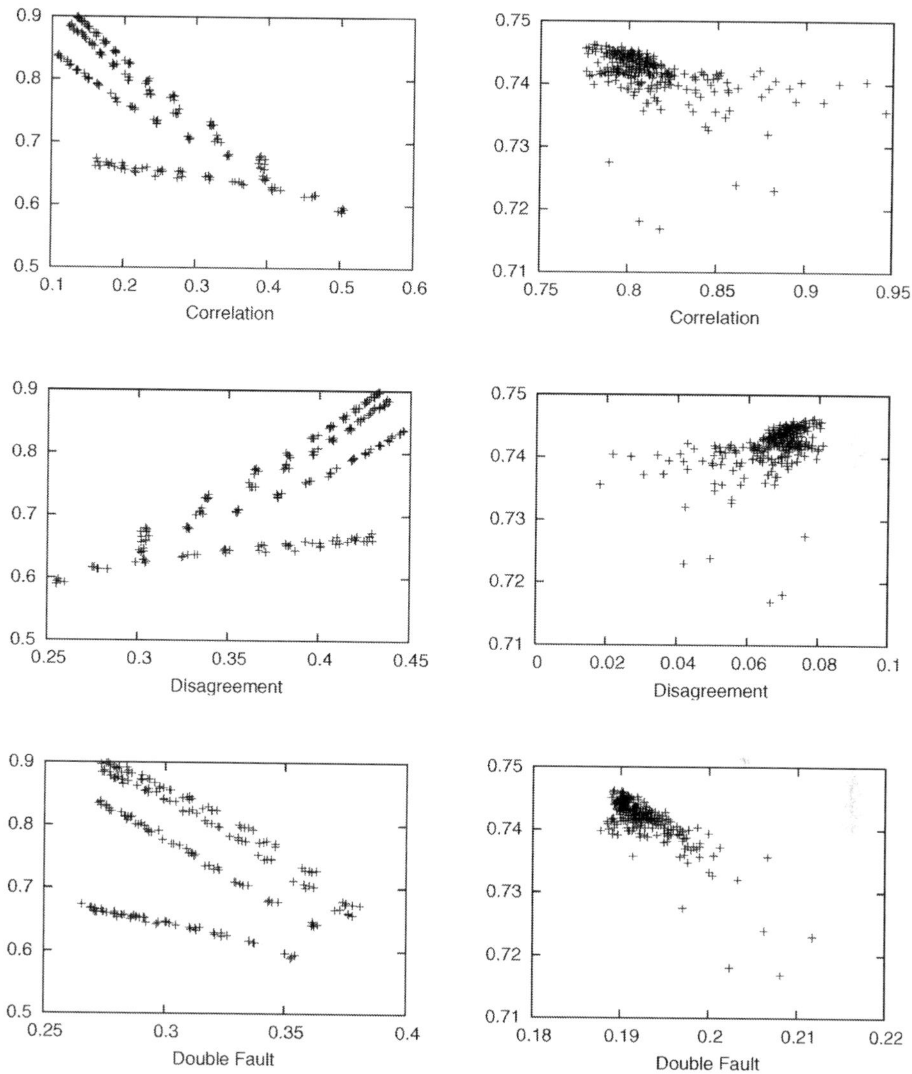

Fig. 6. Evaluation of pairwise diversity measures. The x-axis shows the values of the diversity measure whereas the ensemble accuracy is displayed on the y-axis. The results with artificial data are shown in left column, whereas the outcome on the real world data is shown on the right.

that if we are able to decrease the number of segments where the target word does not even occur a single time, we can expect that the overall performance of the ensemble increases.

The correlation coefficients between the different diversity measures and the recognition accuracy are listed in Tab 2. These values support the optical impression of Fig. 6 and Fig. 7.

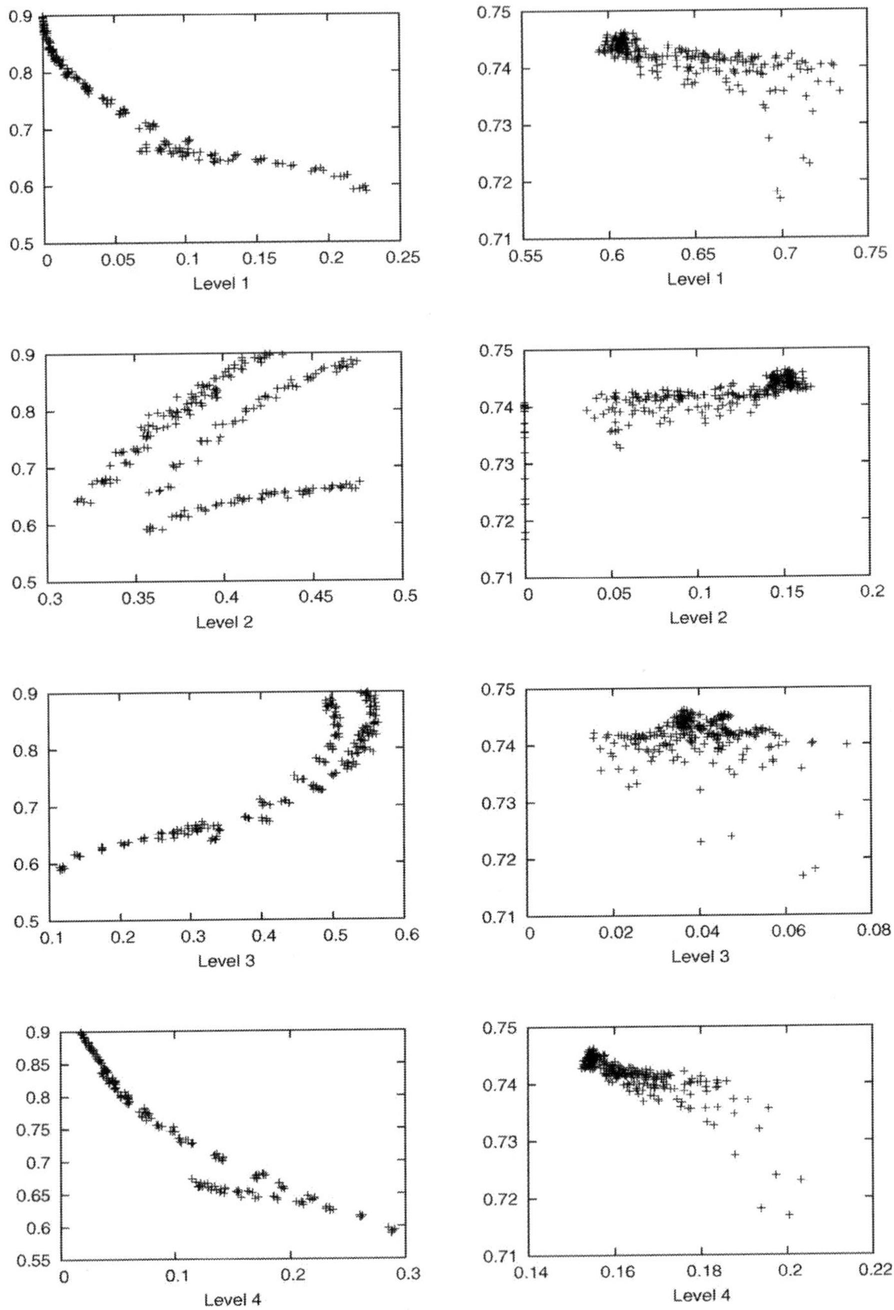

Fig. 7. Evaluation of level based diversity measures. The x-axis shows the values of the diversity measure whereas the ensemble accuracy is displayed on the y-axis. The results with artificial data are shown in left column, whereas the outcome on the real world data is shown on the right.

Table 2. Correlation coefficients of the different diversity measures

	Artificial Data	Real Data
Correlation	-0.8	-0.45
Disagreement	0.77	0.42
Double Fault	-0.45	-0.81
Level 1	-0.92	-0.66
Level 2	0.3	0.7
Level 3	0.9	-0.15
Level 4	-0.95	-0.83

4 Conclusions

We have proposed a framework for diversity analysis of ensembles of word sequence recognition systems. The framework allows one to apply any diversity measure available for conventional multi-class classification problems to ensembles of word sequence recognisers.

In the proposed framework, the word sequences of the individual ensemble members are synchronised in a sequence of segments first. Next, these segments are labelled with the ground truth. Once each segment is labelled, diversity measures for multi-class problems can be applied to the segments. The average of these values gives then the final diversity of the ensemble.

Experiments have been conducted with artificial as well as with real world data from offline handwritten text line recognition. Several pairwise and nonpairwise diversity measures have been applied to both tasks to show the effectiveness of the different measures within the proposed framework. Some of these diversity measures seem to be useful indicators of good ensemble performance.

In the current paper we have exclusively considered the task of word sequence recognition. However, the proposed framework is applicable to continuous speech recognition and other domains as well, where an individual classifier outputs a sequence of classes, rather than just a single class.

Acknowledgement

This research was supported by the Swiss National Science Foundation (Nr. 200020-19124/1). Additional funding was provided by the Swiss National Science Foundation NCCR program "Interactive Multimodal Information Management (IM)2" in the Individual Project "Visual/video processing".

References

1. Oza, N., Polikar, R., Kittler, J., Roli, F., eds.: Multiple Classifier Systems, 6th International Workshop, Springer LNCS 3541 (2005)
2. Huang, T., Suen, C.: Combination of multiple experts for the recognition of unconstrained handwritten numerals. IEEE Transactions on Pattern Analysis and Machine Intelligence **17** (1995) 90–94

3. Gader, P., Mohamed, M., Keller, J.: Fusion of handwritten word classifiers. Pattern Recognition Letters **17** (1996) 577–584
4. Günter, S., Bunke, H.: Ensembles of classifiers for handwritten word recognition. International Journal on Document Analysis and Recognition **5** (2003) 224 – 232
5. Nakano, Y., Hananoi, T., Miyao, H., Maruyama, M., Maruyama, K.: A document analysis system based on text line matching of multiple ocr outputs. In: Document Analysis Systems VI, 6th International Workshop, Florence, Italy. (2004)
6. Wilczok, E., Lellmann, W.: Adaptive combination of commercial OCR systems. In Andreas Dengel, Markus Junker, A.W., ed.: Reading and Learning: Adaptive Content Recognition, Springer-Verlag Heidelberg (2004) 124–136
7. Bertolami, R., Bunke, H.: Multiple handwritten text recognition systems derived from specific integration of a language model. In: 8th International Conference on Document Analysis and Recognition, Seoul, Korea. Volume 1. (2005) 521–524
8. Windeatt, T.: Diversity measures for multiple classifier system analysis and design. Information Fusion **6** (2004) 21–36
9. Brown, G., Wyatt, J., Harris, R., Yao, X.: Diversity creation methods: A survey and categorisation. Information Fusion **6** (2005) 5–20
10. Kuncheva, L.I.: Combining Pattern Classifiers: Methods and Algorithms. John Wiley & Sons Inc (2004)
11. Fiscus, J.: A post-processing system to yield reduced word error rates: Recognizer output voting error reduction. In: IEEE Workshop on Automatic Speech Recognition and Understanding (ASRU), Santa Barbara. (1997) 347–352
12. Wang, L., Jiang, T.: On the complexity of multiple sequence alignment. Journal of Computational Biology **1** (1994) 337–348
13. Sharkey, A., Sharkey, N.: Combining diverse neural networks. The Knowledge Engineering Review **12** (1997) 231–247
14. Bertolami, R., Bunke, H.: Ensemble methods for handwritten text line recognition systems. In: International Conference on Systems, Man and Cybernetics, Hawaii, USA. (2005) 2334–2339
15. Marti, U.V., Bunke, H.: The IAM-database: an English sentence database for offline handwriting recognition. International Journal on Document Analysis and Recognition **5** (2002) 39 – 46

Adaptive Classifier Selection Based on Two Level Hypothesis Tests for Incremental Learning

Haixia Chen[1], Senmiao Yuan[1], and Kai Jiang[2]

[1] College of Computer Science and Technology, Jilin University, Changchun 130025, China
hxchen2004@sohu.com
[2] China Electronics Technology Group Corporation No. 45 Research Institute, Beijing 101601, China
kjiang2004@sohu.com

Abstract. Recently, the importance of incremental learning in changing environments has been acknowledged. This paper proposes a new ensemble learning method based on two level hypothesis tests for incremental learning in concept changing environments. We analyze the classification error as a stochastic variable, and introduce hypothesis test as mechanism for adaptively selecting classifiers. Hypothesis tests are used to distinguish between useful and useless individual classifiers and to identify classifier to be updated. Classifiers deemed as useful by the hypothesis test are integrated to form the final prediction. Experiments with simulated concept changing scenarios show that the proposed method could adaptively choose proper classifiers and adapt quickly to different concept changes to maintain its performance level.

1 Introduction

Most supervised learning algorithms assume stationary target concept over time, require a larger number of training examples beforehand and learn the target concept in batch mode. In real-world applications, however, data are always collected over an extended period of time, and the target concept underlying the data is changing. Thus incremental learning of a classification system must have specially designed mechanism for dealing with drifting concepts.

Methods dealing with concept drift can be classified into two categories: the instance based approaches and the ensemble based approaches. The former one tries to select instances most relevant to the current concept to keep up with the drifting concepts. The FLORA family algorithms [1] and the AQ series algorithms [2] are typical examples of this approach. Problem with those approaches is that information learned from historical data is discarded in order to adapt to the current concept. This is called catastrophic forgetting in literature [3]. The ensemble approaches settle down this problem by storing knowledge learned from historical data. To adapt to the conflicting concepts, they should dynamically delete, reactivate or create new ensemble members based on the base classifiers' consistency with the current data. The first concept drift handling system STAGGER falls into this category [4].

The ensemble based incremental learning approaches can be further divided into two categories: the boosting style method and the bagging style method. Examples of the former one include the Learn++ series algorithms [5] and Adaptive Boosting [6].

D.-Y. Yeung et al. (Eds.): SSPR&SPR 2006, LNCS 4109, pp. 687 – 695, 2006.

In those methods, a new classifier is generated from a skewed distribution adjusted according to the joint performance of former classifiers. There is a strong dependent relationship between those classifiers, and it is problematic to select only partial of those classifiers to reflect drifting concepts. So, it is more beneficial to use bagging style method in concept drifting environments.

Street et al. [7] uses a simple majority vote of classifiers on sequential chunks and responds to concept drift by replacing an unnecessary classifier with a new classifier. Wang et al. [8] uses weight inversely proportional to the error of the classifier on the current chunk to combine their outputs. A few best classifiers are selected to reflect concept drifting. However, in those methods the classifier selection process is managed by some predefined parameters and can't adapt well to the concept change rate. When the concept change rate is low, we hope to keep more classifiers in the system to reduce the classification variance and to make full exploitation of former knowledge. However, when the concept change rate is high, or a conflicting concept appears, we hope to discard those unnecessary former classifiers as soon as possible. So a more flexible classifier evaluation mechanism is needed which can adaptively select base classifiers for integration according to the concept change rate and degree. Chu et al. [9] views the setting of weights for base classifiers as an optimization problem. They use EM algorithm to find outliers and used logistic regression to calculate weights that fit best to the data set excluding those deemed outliers. Computational cost for EM algorithm is high and the validity of model assumption is hardly hold in concept changing environments. Furthermore, for classification systems, fine tuning of the parameters is always a main cause of overfitting and should be avoided.

In this paper, we present a method to incremental Learning in concept drifting environments by explicitly detecting and adapting to drifting concepts. The method is based on two level hypothesis tests. At the first level, the hypothesis test is used to identify the classifiers to be updated on-line. New classifier is built only when no classifiers are updated. So the increase of the ensemble size is controlled. At the second level, the hypothesis test is used to distinguish between useful and useless individual classifiers so as to avoid the interference of conflicting classifiers and make full use of the useful ones. The method belongs to the bagging style ensemble learning method. Compared to the former mentioned approaches, it explicitly monitors the concept changes and can select classifiers accordingly. There have been many work related to explicit detection of concept drift. Methods based on comparison between current performance and a confidence interval adaptively derived from former data batches have been recently proposed [10] [11] [12] [13]. However, since the performance is a random variable, it is difficult to distinguish between a concrete concept drift and a random fluctuation, especially when the environments are noisy and the concept change rate varies too. Chu and Zaniolo [6] viewed the drift detection as a choice between two candidate distributions. Performance probability conditioned on those two distributions are calculated and compared. However, the comparison is based on only one performance observation. So it is highly biased. Furthermore, the comparison threshold is difficult to set. Our method is statistically well-grounded and no complex parameter tuning is necessary.

The rest of this paper is organized as follows. We give a detailed explanation of the principle of the proposed method in Section 2. Experimental results are given in Section 3, followed by conclusion in Section 4.

2 Incremental Learning Based on Two Level Hypothesis Tests for Changing Concepts

In the concept changing environments, the problem of incremental learning boils down to decide how to detect the change of the target concept and how to adapt to these changes.

2.1 Detection of Concept Change

Take the prediction of the classifier h as a stochastic event which has two possible outcomes δ for an example: $\delta = 0$ for correct prediction and $\delta = 1$ for wrong. Let R be the proportion of $\delta = 1$ in experiments with n test examples. Then R follows a Binomial distribution with parameters n and p:

$$\Pr(R = r) = \binom{n}{nr} p^{nr} (1-p)^{n-nr}, nr = 0, 1, \ldots, n. \quad (1)$$

Where p is the expected error rate of classifier h with respect to target concept f and distribution $\mathcal{D}$: $p = error_{\mathcal{D}}(h) \equiv \Pr_{x \in \mathcal{D}}[f(x) \neq h(x)]$. Moreover if the sample size are large ($n \geq 30$), the distribution of R is approximately Normal (a well-known result derived by the Central Limit Theorem). For a sequence $r_1, \ldots, r_k$ with k observations of R, statistic $t = \sqrt{k}(\overline{r} - p)/s$ can be used for a t test to check if its expectation p has changed (alternative hypothesis) or not (null hypothesis). Where $\overline{r}$ is the mean value of the sequence, s is the standard deviation of the sequence.

The expected error rate p of a classifier is composed of three parts: Bayes error determined by the distribution underlying the target concept, bias imposed by the learning algorithm and variance caused by finite training examples[14]. For a classifier, bias imposed by the learning algorithm is fixed. For fixed batch size, variance caused by the finite training examples is stable. Thus, the change of the expected error rate is mainly caused by the change of the distribution. Therefore, the change of the expected error rate of a classifier on different data batches with the same size is a good indicator of the change of the target concept.

However, the expected error rate of a classifier on its target concept is unavailable. We can only estimate it from the training data set used to induce the classifier. In our implementation, we sample with stratified sampling method n ($n \geq 30$ for justification of the normal distribution approximation) examples from the training dataset for a performance test. The process repeated k times. And the expected error rate is estimated as the average of those tests. Given a classifier h and a new data set D, the pseudo code for hypotheses test is shown in Fig. 1. First, k test data sets are generated using the sampling method mentioned before, and base classifiers are tested on those data set. L-1 sequences, each with k observations and for each classifier, are generated. Then, the t statistic for each sequence is calculated and is compared to $t_{\alpha/2}(k-1)$, where α is the significant level. If $|t_i| \geq t_{\alpha/2}(k-1)$, i=1,..., L-1, then concept change is suspected. Otherwise, the concept is deemed as stationary.

```
function HypothesisTest(h,D,k,n,α){
  [SD(1),...,SD(k)]=SampleDataset(D,k,n);
  for(i=1;i<=k;i++){
    err(i)=Error(h, SD(i));
  }
  t=Statistic(err,h);
  if( abs(t)>=t_{α/2}(k−1) ){
    return FALSE;
  }
  return TRUE;
}
```

Fig. 1. Pseudo code for hypothesis test of concept change

2.2 Update of the Classification System

We have developed a general algorithm for handling changing concepts based on two level hypothesis tests. The pseudo code is presented in Fig. 2. Let H be the classifiers ensemble to be updated, $H=\{h_l\}$. $h_l : X \rightarrow Y$ is a base classifier in the classification system, $l=1,\ldots,L$. L is the number of base classifiers in the ensemble. Let $D(t)$ be the data batch used to update the classification system at time t. In each time step, the algorithm begins by running through the current classification system to find the type of concept change for each base classifier. If the concept is stationary, then the classifier is updated online using $D(t)$ and is denoted as useful. If the concept is suspected to drift(change gradually), then the classifier is denoted as useful. If the concept is suspected to shift(change abruptly), the classifier is denoted as useless. After a pass through the ensemble, if no classifier is updated, then a new classifier will be induced from scratch using $D(t)$ as training data. To give the final predication of a new example, all the classifiers denoted as useful are integrated by weights inversely proportional to the error of the classifier on the current chunk for the final prediction.

Fig.3 gives an illustration of the effect of α_1,α_2 in determining the types of concept changes. These two parameters divide the t distribution into three regions: D1 is denoted as the blank region under the probability curve, D2 is denoted as the bright gray region under the probability curve, and D3 is denoted as the dark gray region under the probability curve. If the tested statistic value falls in D3, then the hypothesis that the underlying concept is stationary is hold. In this case, we can update the base classifier by online learning algorithm. It should be noted that, in addition to update the classification model, the sample error rate p should also be updated. If the tested statistic value falls in D2, this is unlikely under the null hypothesis, and the target concept should be deemed as changed. But from a stricter perspective, the change is not so serious, and the tested classifier can also provide useful information for classification of the current data set. So the classifier is tagged as useful but not updated to avoid interference of different concepts. If the tested statistic value falls in D3, we suppose that a completely different concept or even a conflicting one appears, and the classifier for the historical data is eliminated from integration, because it may be conflictive with the current one.

```
function SystemUpdate(D(t)){
  S=0; // usefulness flag vector for each classifier
  bUpdate=FALSE;
  for(i=0;i<L;i++){
if(HypothesisTest(H(i),D(t),k,n,α1)==TRUE){//stationary
  OnLineLearning(H(i),D(t));
  S(i)=1; // the classifier is useful
  bUpdate=TRUE;
}elseif(HypothesisTest(H(i),D(t),k,n,α2)==TRUE){//drift
  S(i)=1;
}else{ // shift
  S(i)=0;
}
  }
  if(bUpdate == FLASE){
    BatchLearning(h,D(t)); // train a new classifier
    AddNewClassifier(H,h); // Add it to the ensemble
    AddElement(S,1); // turn on its useful flag
  }
}
```

Fig. 2. Pseudo code for updating the classification system with new data batch

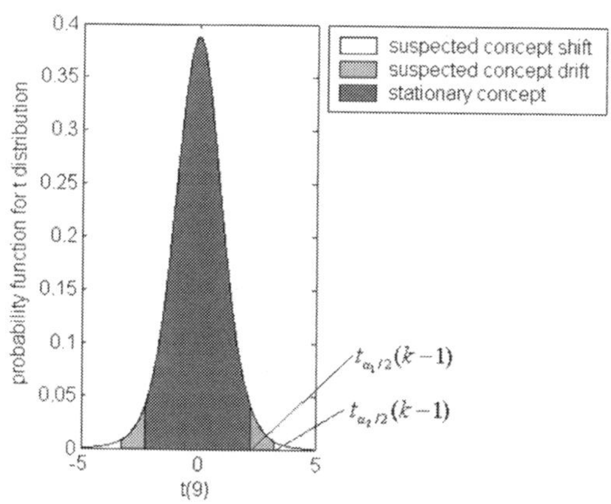

Fig. 3. Illustration of the three kinds of concept changes

3 Experiments

In the experiments, we compare the performance of the proposed algorithm, abbreviated as NBE, to other three algorithms: batch learning using all available training data (abbreviated as BAT), sliding window trained with data only in the current batch (abbreviated as WIN), and ensemble learning with weights proportional to their accuracy on the most recent data block as suggested in Ref. [8] (abbreviated as ENS). At most 10 base classifiers are kept in ENS. Parameters for NBE are: $n=30$, $k=$

$10, \alpha_1 = 0.05, \alpha_2 = 0.01$. We set n for reason addressed above, set k based on pilot experimental study, and set α_1, α_2 following a general suggestion from statistics. Naïve Bayes [15] are used as base classifiers because it is a stable and robust classification algorithm with little bias and we could focus our attention on detecting and adapting mechanisms of the ensemble but not on fine tuning of the base classifiers.

We use two groups of concepts generally used in literature to simulate different types of concept drifts. The use of artificial datasets allow us to control the points where the concept drifts and recurs, and the degree to which the concept drifts. The first data set is to simulate sudden changes in target concepts and is generated from a group of concepts that is first introduced by Schlimmer and Granger in STAGGER [4]. The concepts are defined on three attributes: $color \in$ {green,blue,red} , $shape \in$ {triangle,circle,rectangle}, $size \in$ {small,medium,large}. We define the target concepts as: concept1 $\Leftrightarrow$ $color$ = red $\wedge$ $size$ = small, concept2 $\Leftrightarrow$ $color$ = green $\vee$ $shape$ = circle, concept3 $\Leftrightarrow$ $size$ = medium $\vee$ $size$ = large, concept4 $\Leftrightarrow$ concept1. Concept2 and concept3 are conflicting with concept1, so behaviors of approaches dealing with conflictive concepts could be observed. Concept4 is identical with concept1. This is designed to investigate the behavior of those approaches when they encounter with recurring concept. For every target concept, ten data chunks, each with 300 examples, are generated sequentially. In each chunk 100 examples are used for training and the other 200 examples for testing.

The second data set is proposed to simulate gradual changes in target concepts with a hyperplane in a d-dimensional space:

$$\sum_{i=1}^{d} a_i x_i = a_0 . \tag{2}$$

Where x_i is the coordinate of the ith dimension/feature, $x_i \in \{0,1\}$, and a_i is the weight for the feature, $a_i \in [0,1]$. If $\sum_{i=1}^{d} a_i x_i \geq a_0$, the example is classified as positive. Otherwise it is labeled as negative. By adjusting the weight of each feature, the target concept could change smoothly [16]. We set d=30, and initialized a_i randomly. To simulate a smoothing concept drift, a concept is formed by randomly choosing a feature and increasing its weight by 0.1. The weight is subtracted by 1 if it surpassed 1. In this way, a sudden concept change is inserted in. 40 data batches are generated, each for a new concept. 300 examples are generated in all data batches, among which 100 examples are used for training and the other 200 examples for testing. For each new concept, a data chunk with 300 examples are generated randomly, among which 100 examples are used for training and the other 200 examples for testing.

Fig. 4 compares the results of BAT, WIN, ENS and NBE on the first data set representing four concepts averaged over 30 runs. In general, NBE performs better than the other three compared approaches, especially when concept drifts. BAT performs well for stationary concept (time from 1 to 10) as it learns from all cumulated data. However, when concept drifts (time from 11 to 40), historical data becomes a burden for learning new concepts, and its performance degenerates seriously with more different concepts appear. WIN retains its performance for all time steps, even when the target concept drifts. However, it couldn't use all available examples when the target concept is stationary or when the same concept reappear. Both of the ensemble learning method are better than the above two methods, because they could retain former knowledge in different base classifiers and adapt to new concept by discarding unnecessary ones. But NBE outperforms ENS especially when conflicting concepts appears (time from 11 to 30) or the same concept recurs (time from 31 to 40). We think this is because that number of base classifiers in ENS is a predefined parameter, so the algorithm could not adapt quickly to the sudden change of target concept. When concept changes suddenly, though those unnecessary former classifiers are assigned with relatively low weights, they still work for a period. And they are dominating in quantity. However, NBE find a better compromise between adaptation to drift and waste of knowledge. As it could detect concept drift explicitly, classifiers build for conflicting targets couldn't interfere with the prediction of the current one, and classifiers generated from the same concept could be used jointly with the current one for a better performance.

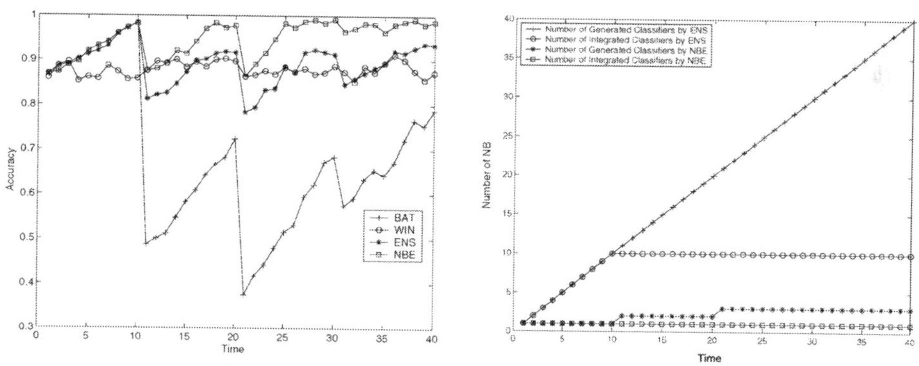

(a) (b)

Fig. 4. Performance comparisons between BAT, WIN, ENS and NBE for the first data set. (a) accuracy; (b) number of base classifiers.

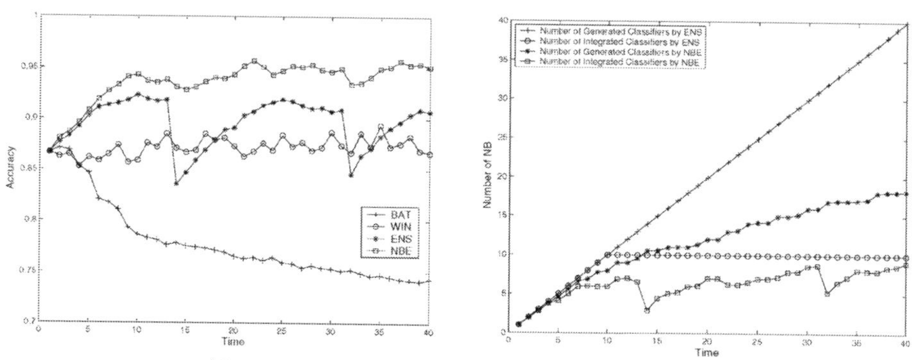

(a) (b)

Fig. 5. Performance comparisons between BAT, WIN, ENS and NBE for the second data set. (a) accuracy; (b) number of base classifiers.

Fig. 5 compares the results of BAT, WIN, ENS and NBE on the second data set representing forty concepts averaged over 30 runs. It is noted that NBE performs the best and maintain its performance at a relatively stable level. Although ENS performs better than the other two algorithms in most cases, its performance degenerated seriously sometimes. Fig. 5(b) clearly shows that NBE don't create NB for each data batch and don't use all classifiers in the ensemble for integration. The number of classifiers in ensemble increases sublinearly with the time steps.

4 Conclusion

Nothing remains static. The world around us is evolving all the time. As a mirror of the real world, the classification system should also adapt to the changing target concepts. This paper studies the incremental learning of classification system in concept changing environments. A new ensemble learning method based on two level hypothesis tests is proposed. Different kinds of concept changes are detected by the two level hypothesis tests and treated accordingly. Experiments show that the proposed algorithm could adapt quickly to different kinds of concept changes and achieve better performance by adaptively selecting and integrating individual classifiers. In addition, the number of individual classifiers produced in the learning process is fewer than the other ensemble learning method.

References

1. Widmer, G., Kubat, M.: Learning in the presence of concept drift and hidden contexts. Machine Learning, 23, 1996, 69-101
2. Maloof, M.A., Michalski, R.S.: Incremental learning with partial instance memory. Artificial Intelligence, 154, 2004, 95-126
3. McCloskey, M., Cohen, N.: Catastrophic interference in connectionist networks: the sequential learning problem. The Psychology of Learning and Motivation. Vol. 24, 1989, 109-164
4. Schlimmer, J.C., Granger, R.H. Jr.: Incremental learning from noisy data. Machine Learning, 1, 1986, 317-354
5. Polikar, R., Udpa, L., Udpa, S.S., Honavar, V.: Learn++: an incremental learning algorithm for supervised neural networks. IEEE Transactions on Systems, man, and Cybernetics-Part C: Applications and Reviews, Vol.31, No.4, 2001, 497-508
6. Chu, F., Zaniolo, C.: Fast and light boosting for adaptive mining of data streams. H. Dai, R. Srikant, and C. Zhang (Eds.) PAKDD 2004, LNAI 3056, 2004, 282-292
7. Street, W., Kim, Y.: A streaming ensemble algorithm (SEA) for large-scale classification, Proc. 7th ACM SIGKDD, ACM Press, 2001, 377-382
8. Wang, H., Fan, W., Yu, P.S., Han, J.: Mining concept-drifting data streams using ensemble classifiers. SIGKDD'03, August 24-27, 2003, Washington, DC, USA
9. Chu, F., Wang, Y., Zaniolo, C.: Mining noisy data streams via a discriminative model. E. Suzuki and S. Arikawa (Eds.) DS 2004, LNAI 3245, 2004, 47-59
10. Klinkenberg, R., Renz, I.: Adaptive information filtering: Learning in the presence of concept drifts. Learning for Text Categorization, Menlo Park, CA, USA, AAAI Press, 1998, 33-40

11. Fung, G.P.C., Yu, J.X., Lu, H.: Classifying text streams in the presence of concept drifts. H. Dai, R. Srikant, and C. Zhang (Eds.) PAKDD 2004, LNAI 3056, 2004, 373-383
12. Natwichai, J., Li, X.: Knowledge maintenance on data streams with concept drifting. J. Zhang, J.H. He, and Y. Fu (Eds.) CIS 2004, LNCS 3314, 2004, 705-710
13. Gama, J., Medas, P., Castillo, G., Rodrigues, P.: Learning with drift detection. A.L.C. Bazzan and S. Labidi (Eds.) SBIA 2004, LNAI 3171, 2004, 286-295
14. Duda, R.O., Hart, P.E.: Pattern classification and scene analysis. Second Edition, New York, Willey and Sons, 2001
15. Domingos, P., Pazzani, M.: On the optimality of the simple Bayesian classifier under zero-one loss. Machine Learning, 29, 1997, 103-129
16. Hulten, G., Spencer, L., Domingos, P.: Mining time-changing data streams. In: ACM SIGKDD International Conference on Knowledge Discovery and Data Mining, San Francisco, CA, USA, 2001, 97-106

Combining SVM and Graph Matching in a Bayesian Multiple Classifier System for Image Content Recognition

Bertrand Le Saux and Horst Bunke

Institut für Informatik und Angewandte Mathematik,
University of Bern, Neubrückstrasse, 10, CH-3012, Bern, Switzerland
{lesaux, bunke}@iam.unibern.ch

Abstract. In this paper, we propose an approach to image content recognition that exploits the benefits of different image representations to associate meaning with images. We choose classifiers based on global appearance, scene structure and region type occurrence, and define confidence measures on their output. The resulting posterior probabilities of the classifiers are combined in a Bayesian framework. We show that this method leads to a robust and efficient system that contributes to reducing the semantic gap between low level image features and higher level image descriptions.

1 Introduction

In recent years, many approaches to image classification and scene recognition have been proposed. They usually utilize binary classifiers that learn from a set of indexed images to recognize a given concept (such as an image with people, an outdoor image, etc.) and strongly depend on the image representation they use. Such programs satisfy needs in image retrieval and computer vision, and could possibly be applied to a wide range of areas including robotics, digital libraries, and web searching.

Image classification has been performed by using support vector machines (SVM) on image histograms [1] or more recently hidden Markov models on multi-resolution features that capture more information [2]. These methods do not take into account that human description of an image content is rarely global but often specific to an image part. Several approaches use flexible models to localize objects in the image and perform object class recognition [3]. For scene recognition, both background and foreground objects are important, and thus image segmentation has usually been the basis to include local information in the image representation. The resulting indexes were more and more meaning-oriented: from image blocks [4] and attributed relational graphs [5] to presence vectors that check the occurrence of given region types [6].

Simple models for image representation are efficient when they fit salient features of the class of scenes to recognize. But they can be totally useless in other cases. Thus more and more complex image representations were proposed. These methods rely on the capacity of powerful classifiers to separate different image categories. Unfortunately, these classifiers are prone to overfitting if not enough training data are available, as is often the case in practice. We show in this paper that a system of classifiers that use

D.-Y. Yeung et al. (Eds.): SSPR&SPR 2006, LNCS 4109, pp. 696–704, 2006.

simple image representations related to different semantic levels, from global features to region type, and finally to region relationship information, is more suitable to learn and recognize a concept. These classifiers can be combined efficiently using a Bayesian framework.

This paper is organized as follows. Section 2 presents the various classifiers we use and shows their respective advantages. Section 3 introduces the Bayesian framework that allows us to combine these classifiers. In Section 4, posterior probabilities are defined for the classifiers. We present experiments and discuss their results in Section 5. Finally, conclusions are drawn in Section 6.

2 Image Classifiers

In this section we present the different classifiers and the scene features they characterize. For every class we build a binary classifier on each of the three representations. In the recognition task, images are tested with classifiers corresponding to different concepts. As a result, every image can be assigned to no, one or more than one concept.

SVM on Color Histograms. In [1], images are represented by simple 16-bin histograms of the color distribution in the Luv color space. A SVM with a Gaussian kernel is used to learn a decision rule in this feature space. This classifier, however, is inappropriate as soon as the distinction between the classes depends on local image regions. In this case it tends to overfit on the zones of the images that have no discriminative power, like the background. However, when a type of scene has a strong density peak on some colors or is visually consistent over all images of a class, the simplicity of this model insures that it is particularly efficient.

k-NN on Image Graphs. While the image representation described in Section 2 conveys only low-level and global information about the image, our second representation gives insight about the structure of the image and thus provides semantic information at the medium level. In [7], images are represented as graphs of regions; nodes of the graph are region histograms and an edge exists between two nodes if the corresponding regions are adjacent. A graph-edit distance based on the A^* algorithm [8] is used to match image representations. In this approach, one needs to define a set of possible graph-edit operations and assign a cost to each of them. We use the distance between the color histograms that represent the regions as the vertex substitution cost. To make vertex deletion easier on large graphs, we define its cost as the inverse of the number of vertices. To have comparable costs, this is extended to all deletion and insertion costs. Based on the graph-edit distance, a k-Nearest-Neighbor (k-NN) procedure classifies the images.

The semantic information provided by this classifier is twofold: first, it encodes the proximity of the regions in the image and second, the use of a graph-edit distance allows one to implicitly build a model of the class to learn by selecting the more similar regions between two images.

SVM on Presence Vectors. In this approach, images are indexed by boolean vectors that indicate the presence of particular region types, such as greenery, sky, or skin, chosen from a 50-region lexicon [6]. This lexicon is built by unsupervised clustering of a set of regions based on color. The same training set is used to build the lexicon and train the classifiers and thus no manual intervention is needed to index regions nor images. A decision rule based on the resulting presence vectors is then evaluated by means of a SVM with a polynomial kernel. In the linear case, this approach checks the co-occurrence of region types. This image classifier is efficient when some region types are particularly salient, eg. skin regions to recognize images with people. But it depends on the quality of the region lexicon and might suffer from region types that have been badly estimated.

This kind of classifier offers even more semantic information than the previous ones since the learned decision rules make the composition of image classes explicit. For example, a "street" picture usually contains "building" and "road" region types, but no "field" or "snow" ones.

3 Bayesian Combination

Multiple Classifier Systems (MCS) have successfully been used to solve difficult classification problems [9]. In this paper we present a MCS based on the three classifiers described in Section 2. This section presents various combination rules based on a Bayesian framework [10,11].

Let $\mathcal{I}$ denote the set of images, and X a random variable on $\mathcal{I}$ standing for the distribution of images. We denote by Y a boolean random variable for the class to predict, i.e. the type of scene to associate with the image. Let us assume that we have R image representations denoted as $x_1, \ldots, x_R$, that are modeled as functions $f : \mathcal{I} \rightarrow \mathcal{F}_k$ where $\mathcal{F}_k$ is the feature space associated with x_i. These representations correspond to the input of each classifier. We denote by $X_1 = x_1(X), \ldots, X_R = x_R(X)$ the random variables associated with these representations.

From a Bayesian point of view, an image x should be assigned to the class with the maximum probability according to all classifiers, i.e x belongs to the class to predict if:

$$P(Y = 1|x_1, \ldots, x_R) \geq P(Y = 0|x_1, \ldots, x_R) \quad (1)$$

Practically, we can have only some estimates of the individual posterior probabilities $P(Y = 1|x_i)$ of each classifier i. Combination rules combine these estimates to approximate $P(Y = 1|x_1, \ldots, x_R)$ in the best possible way.

3.1 Mean Rule

If we use individual classifiers that are good enough, we can assume that they are wrong on a few data samples only and that their answer differs from the overall answer by a small classifier error with no bias. All estimates can be used to obtain a better overall estimate by computing their average. From (1), we get the mean rule:

$$\frac{1}{R}\sum_{i=1}^{R} P(Y = 1|x_i) \geq \frac{1}{R}\sum_{i=1}^{R} P(Y = 0|x_i) \quad (2)$$

If there are outliers among the posterior probabilities, a robust estimate of the average like the median can be used to prevent them from affecting the final decision. The median rule is:

$$\text{med}_{i=1}^{R} P(Y=1|x_i) \geq \text{med}_{i=1}^{R} P(Y=0|x_i) \tag{3}$$

3.2 Product Rule

Using Bayes rule, (1) is equivalent to:

$$\begin{aligned} P(x_1,\ldots,x_R|Y=1)P(Y=1) \geq \\ P(x_1,\ldots,x_R|Y=0)P(Y=0) \end{aligned} \tag{4}$$

If the features are different, we can assume that the feature spaces $\mathcal{F}_1,\ldots,\mathcal{F}_R$ are conditionally statistically independent: $P(x_1,\ldots,x_R|Y=y) = \prod_{i=1}^{R} P(x_i|Y=y)$. We obtain that:

$$P(Y=1)\prod_{i=1}^{R} P(x_i|Y=1) \geq P(Y=0)\prod_{i=1}^{R} P(x_i|Y=0) \tag{5}$$

or, by using once more Bayes rule, in terms of posterior probabilities:

$$\begin{aligned} (P(Y=1))^{-(R-1)}\prod_{i=1}^{R} P(Y=1|x_i) \geq \\ (P(Y=0))^{-(R-1)}\prod_{i=1}^{R} P(Y=0|x_i) \end{aligned} \tag{6}$$

Eq. (6) is known as the product rule. It is critical in the sense that a single classifier with an output close to 0 can heavily influence the result of the whole ensemble.

3.3 Sum Rule

If we assume that $P(Y=1|x_i) = P(Y=1)(1+\delta_i)$, i.e that the posterior probabilities do not differ too much from the prior probabilities, then, by substituting it in the left term of (6) and keeping only the first-order terms, we obtain:

$$(P(Y=1))\prod_{i=1}^{R}(1+\delta_i) = P(Y=1) + \sum_{i=1}^{R}\delta_i \tag{7}$$

And finally by expressing δ_i according to the probabilities, we find the sum rule, that outputs the class label if:

$$\begin{aligned} (1-R)P(Y=1) + \sum_{i=1}^{R} P(Y=1|x_i) \geq \\ (1-R)P(Y=0) + \sum_{i=1}^{R} P(Y=0|x_i) \end{aligned} \tag{8}$$

3.4 Other Combination Strategies

For comparison purpose, we present here other rules that are commonly used to combine classifiers. The first two ones also use the posterior probabilities. The max rule checks if:

$$(1-R)P(Y=1) + R\max_{i=1}^{R} P(Y=1|x_i) \geq$$
$$(1-R)P(Y=0) + R\max_{i=1}^{R} P(Y=0|x_i) \quad (9)$$

The min rule is:

$$(P(Y=1))^{-(R-1)} \min_{i=1}^{R} P(Y=1|x_i) \geq$$
$$(P(Y=0))^{-(R-1)} \min_{i=1}^{R} P(Y=0|x_i) \quad (10)$$

By using the majority voting rule, the MCS simply takes the same decision as the majority of classifiers. Only their binary outputs $\Delta_i^y,\ y \in [0,1]$ are considered:

$$\sum_{i=1}^{R} \Delta_i^1 \geq \sum_{i=1}^{R} \Delta_i^0 \quad (11)$$

4 Classifier Outputs

Since the three classifiers we use have a simple binary output, we need to define how to estimate their posterior probabilities to assign a class in order to combine them in a system more accurately than the simple voting rule.

4.1 SVM Case

SVM methods used for classification [12] project input data to a space of large (or infinite) dimension and use a linear decision boundary to separate different classes. Data x_i are classified by using the signed normalized distance d_{x_i} to the separating hyperplane.

In our application, only data with positive value are classified as belonging to the considered category. Data x_i with the signed distance d_{x_i} in the $]-1;1[$ interval fall within the margin and are considered ambiguous. We define an estimate of the posterior probability of the SVM-based classifiers from the output value d_{x_i} by the following linear mapping, which approximates the real unknown distribution:

$$P(Y=1|x_i) = \begin{cases} 1 & \text{if } d_{x_i} \geq 1 \\ (d_{x_i}+1)/2 & \text{if } -1 < d_{x_i} < 1 \\ 0 & \text{if } d_{x_i} \leq -1 \end{cases}$$

4.2 k-NN Case

For classifiers based on the k-NN procedure, the posterior probabilities can be estimated simply by counting the number of positive neighbors k^+ among all the considered neighbors k:

$$P(Y = 1|x_i) = \frac{k^+}{k} \tag{12}$$

We use $k = 15$ in the graph-based classifier. This ensures that the probability estimates have enough different values.

5 Experiments

5.1 Data Set

The data set is composed of 200 images collected from the web. Four classes, consisting of 40 images each, contain instances of a particular scene type: *snow landscape, countryside, streets* and *people*. A fifth one consists of various generic images aimed

Fig. 1. Typical results of the histogram classifier (1), image graph classifier (2) and the presence vector classifier (3). Left-hand side shows images that are correctly retrieved by the considered classifier and only by this one. Right-hand side shows images that are typically misclassified by using this image representation.

to cover the whole spectrum of possible real scenes and thus used as negative samples for the classifiers. In the experiments, a training set of 150 instances (30 per category) is extracted randomly from the data set. The remaining instances are used as a test set. Error rates are averaged on 25 runs of holdout cross-validation. Some examples of each class are shown in Figure 1.

5.2 Strong and Weak Points of the Classifiers

Figure 1 shows images that illustrate the particular behavior of each classifier. The left-hand side of this figure shows the true positives that are recognized only by this kind of classifier and not by the other two, while the right-hand side shows images that are not recognized, i.e. either false positives of false negatives.

The histogram classifier recognizes the image [1-a] on the basis of the color only, while the other classifiers do not find the structure of the mountain landscapes they have learned on other images of the category. Due to the global and continuous image representation, this classifier also has a better tolerance to luminosity shifts than individual region matching (cf. image [1-b]). On the contrary, it hardly learns concepts characterized by images that have flat histograms, so there are many false negatives for concepts like *streets* (cf. image [1-c]) or *people*.

The image graph classifier allows to retrieve the highly structured image [2-a] although colors are almost lost completely in shadow. This can turn into a drawback when the image segmentation into regions is not reliable: in the close-up portrait of image [2-b], the face was segmented into several artificial regions, resulting in a meaningless graph representation for which the classifier is not able to find a good match.

Previous classifiers fail on images [3-a] and [3-b], which are partly overexposed (and so have density peaks on white colors), contain some heterogeneous elements (like the house in the country image) and have unusual structures. The presence vector classifier that tests the occurrence of some meaningful region types proves to be better suited to recognize these images. But it can also fail to find a good balance between two opposite clues, like the grass area and the face regions of image [3-c].

5.3 Comparison of Different Combination Schemes

The first round of experiments intends to test the complementarity of the various classifiers. The use of a MCS is promising if for one classifier that misclassifies some data, there is another one that can correctly recognize it. The oracle rule checks if for each test data there is at least one classifier that has the right answer. Table 1 shows the test errors of the individual classifiers and the oracle answer. We conclude that, for all concepts, all images can be recognized by at least one classifier, depending of the features

Table 1. Individual classifier errors and oracle prediction

	presence-vectors	region-graphs	histograms	oracle
snow landscape	15.15%	11.15%	3.52%	0.0%
country	6.30%	13.45%	7.64%	0.0%
people	10.30%	12.24%	11.88%	0.0%
streets	9.70%	17.94%	13.45%	0.0%

Table 2. Various combinations for the multi classifier system

	product	sum	max	min	mean	median	voting
snow landscape	7.15%	7.15%	8.24%	7.63%	**5.58%**	5.82%	6.67%
country	4.73%	6.55%	6.67%	6.79%	**4.61%**	4.61%	5.09%
people	8.48%	11.03%	12.73%	9.45%	**6.18%**	7.03%	6.91%
streets	7.88%	10.06%	11.15%	8.97%	**6.42%**	6.75%	7.39%

Table 3. Overall performances

	presence-vectors	region-graphs	histograms	mean-rule MCS
mean error	10.36%	13.70%	9.13%	5.70%
standard deviation	6.31%	5.16%	7.74%	1.40%

the image representation focuses on. This is an indication that the individual classifiers have a high degree of complementarity.

In Table 2, we test various schemes for classifier combination. The sum, max and min rules give the worst results and do not perform better than the individual classifiers. The sum rule relies on the assumption that the posterior probabilities do not differ too much from the prior. This is true when the classifier outputs are ambiguous, due to noisy image representations for example, but all the classifiers we use are not weak learners and give good results, different from the prior. The max and min rules might be considered as rough approximations of the sum rule that emphasizes one of the classifiers. Performance of the max rule is below the sum rule, while the min rule performs slightly better on classes *streets* and *people*.

Among the other combining schemes, both the median and voting rules can be considered as variations of the mean rule. The median is another estimate of the mean, but here it gives slightly less good results due to the small number of classifiers. The voting procedure is equivalent to a mean rule applied to the binary outputs of the classifiers: the better results of the mean rule validate the estimates of the posterior probabilities that we have defined.

Both the product rule and the mean rule (and its derivates) can be considered as successful since they globally improve the classification rates. The product rule corresponds to the true Bayes formula under conditional independence assumption and thus should lead to more precise results. However by multiplying the posterior probabilities, it gives too much importance to the classifier that assigns low probabilities to a class. By contrast, the role of the mean rule is to reduce the effect of the errors of the individual classifiers. In this case it acts on both the classifier uncertainties and the error from the probability estimation, and thus achieves the best results.

5.4 Comparison of the MCS vs. the Individual Classifiers

The comparison of error rates shows that the MCS with the mean rule outperforms all the individual classifiers (cf. Tables 1 and 2). The histogram-based classifier does better for one class only, the *snow landscape* one. This is due to the fact that images from this category do not have much variance in their color distribution (it corresponds mainly to ice and snow regions).

Table 3 shows the error rates of the individual classifiers and of the MCS on average over different image categories. The MCS allows to reduce error rates by more than 3 percent. It does not suffer from the inadequacy of a particular image representation and thus is more robust. Moreover, the standard deviation is highly reduced, which means that the MCS is more general-purpose and able to deal with different kinds of visual concepts that can meet the needs of various users.

6 Conclusion

We have presented in this article a novel approach to scene recognition based on a multiple classifier system. It uses image classifiers that characterize different semantic features: global appearance, image structure and presence of meaningful region types. We define posterior probability estimates of the classifier outputs and combine them in order to maximize the posterior probability of the whole system.

This approach achieves very good performance without using a classification method that requires a fine tuning or complex image representation like multi-resolution features. On the contrary, we use simple individual classifiers and rely on their combination to arrive at the best decision. This makes it really robust and suitable for recognizing various kinds of scenes. Thus the combination of multiple classifiers, as proposed in this paper, can bring great benefits to image indexing and search in image databases.

References

1. Chapelle, O., Haffner, P., Vapnik, V.: Svms for histogram-based image classification. IEEE Trans. on Neural Networks **10** (1999) 1055–1065
2. Li, J., Wang, J.Z.: Automatic linguistic indexing of pictures by a statistic modeling approach. IEEE Trans. on Pattern Analysis and Machine Intelligence **25** (2003) 1075–1088
3. Fergus, R., Perona, P., Zisserman, A.: Object class recognition by unsupervised scale-invariant learning. In: IEEE Conference on Computer Vision and Pattern Recognition. (2003) 264–271
4. Minka, T., Picard, R.: Interactive learning using a society of models. Pattern Recognition **30** (1997) 565–581
5. Beretti, S., Del Bimbo, A., Vicario, E.: Efficient matching and indexing of graph models in content-based retrieval. IEEE Trans. on Pattern Analysis and Machine Intelligence **23** (2001) 1089–1105
6. Le Saux, B., Amato, G.: Image recognition for digital libraries. In: ACM Multimedia / International Workshop on Multimedia Information Retrieval. (2004) 91–98
7. Le Saux, B., Bunke, H.: Feature selection for graph-based image classifiers. In: IAPR Iberian Conference on Pattern Recognition and Image Analysis. (2005) 147–154
8. Nilsson, N.J.: Principles of Artificial Intelligence. Tioga, Palo Alto, CA (1980)
9. Kittler, J., Roli, F., eds.: Multiple Classifier Systems, First International Workshop, MCS 2000, Cagliari, Italy, Springer Verlag (2000)
10. Kittler, J., Hatef, M., Duin, R.P., Matas, J.: On combining classifiers. IEEE Trans. on Pattern Analysis and Machine Intelligence **20** (1998) 226–239
11. Tax, D.M., van Breukelen, M., Duin, R.P., Kittler, J.: Combining multiple classifiers by averaging or by multiplying? Pattern Recognition **33** (2000) 1475–1485
12. Vapnik, V.: The Nature of Statistical Learning Theory. Springer Verlag, New-York (1995)

Comparison of Classifier Fusion Methods for Classification in Pattern Recognition Tasks

Francisco Moreno-Seco, José M. Iñesta,
Pedro J. Ponce de León, and Luisa Micó

Department of Software and Computing Systems
University of Alicante
P.O. box 99, E-03080 Alicante, Spain
{paco, inesta, pierre, mico}@dlsi.ua.es
http://grfia.dlsi.ua.es

Abstract. This work presents a comparison of current research in the use of voting ensembles of classifiers in order to improve the accuracy of single classifiers and make the performance more robust against the difficulties that each individual classifier may have. Also, a number of combination rules are proposed. Different voting schemes are discussed and compared in order to study the performance of the ensemble in each task. The ensembles have been trained on real data available for benchmarking and also applied to a case study related to statistical description models of melodies for music genre recognition.

1 Introduction

Combining classifiers is one of the most widely explored methods in pattern recognition in the recent years. These techniques have been shown to reduce the error rate in classification tasks in opossite to single classifiers. Also, the combination of different techniques to make a final decision makes the performance of the system more robust against the difficulties that each individual classifier may have on each particular data set. Different reasons have been argued for this behaviour, amongst others, statistical, computational or representational reasons [1].

Several different approaches have been used to obtain classifier ensembles. As stated in a recent work by Duin [2], base classifiers should be different, but they should be comparable as well. Also, works on this subject point out the importance of the concept of *diversity* in classifier ensembles, with respect to both classifier outputs and structure [3,4,5,6]. This points out that a trade-off between comparability and diversity is desirable when combining different classifiers.

Classifiers for an ensemble can be generated using different initializations (like in neural networks), different parameter choices (like the number of neighbors in the k-NN rule), different classification schemes or, for example, different training sets from the same target problem. A set of classifiers generated in one of these ways is called to be consistent.

D.-Y. Yeung et al. (Eds.): SSPR&SPR 2006, LNCS 4109, pp. 705–713, 2006.

In this work, the base classifiers used to combine are comparable in terms that they are applied to the same data sets and using the same partitioning, and are diverse since they come from different pattern recognition paradigms: a *k-nearest-neighbor* (k-NN), a *multi-layer perceptron* (MLP), a *support vector machine* (SVM), a *decision tree* (DT), and a *naïve Bayes classifier* (NB). All the base classifiers have been trained in the same feature spaces and with the same training set.

Current research and new proposals on the decision combination of the base classifiers is presented in this paper. First, the classification techniques based on them are described, along with the different ensemble schemes for combining classifier decisions. Following this, the results for the ensembles are presented and compared with single classifier results for data sets from the UCI/Statlog project [7], and for a data set based on the classification of music styles using MIDI files. Finally, the conclusions drawn from the results are discussed, pointing the research to further work lines.

2 Base Classifiers

Five conceptually different classification techniques have been used in this work: the k-nearest-neighbour classifier (k-NN), the naive Bayesian classifier (NB), a support vector machine (SVM), a multi-layer perceptron (MLP), and a decision tree (DT). For the first case, given a sample $\mathbf{x}_i$, the distances to the prototypes in the training set are computed, and the class labels of the closest k are taken into account to classify the sample into the most frequent class among them. After some initial testing on the performance of this particular classifier on some of the utilized datasets, a single value $k = 3$ was established for this classifier in all the experiments for simplicity. The rest of the classifiers have been applied using the default parameters established for them in the open source software project WEKA, using the Explorer interface [8]. The decision tree is the J48.

Each base classifier has been trained using the same training set, and its accuracy has been estimated using the same test set. Two methods have been used to train the classifiers, and the ensembles: first, for the UCI/Statlog project data sets, a total of 50 pairs of train/test sets were generated, using 10 random seeds for generating 5 cross-validation pairs (with approximately an 80% of the data for training, and the rest for testing). The base classifiers have been run 50 times with different train and test sets from the same data (each data sample has been classified 10 times). The error rate of the classifier has been estimated by counting the total number of errors over the 50 experiments, divided by the total number of samples classified (that is 10 times the size of the data set).

Once the ensembles have been trained with the UCI/Statlog project data sets, a validation experiment has been run, using a new random seed for generating another 5 pairs of train/test sets. The base classifiers have also been run with the validation data, in order to obtain a reference. Obviously, the validation data is not unseen data for the classifiers, as it should be, but the results can be a reference for future experiments on completely unseen data.

The training of the base classifiers in the music genre classification task was made under a more realistic approach: each data set has been divided into 5 subsets with approximately the same size. The division has been made at the level of MIDI files. Given the 5 subsets, 3 of them have been used to train the classifier, 1 for test (and for training the ensembles), and the last one for validation. The partitions have been rotated 5 times, in order to obtain more significant results.

3 Ensemble Design: Voting Schemes

Designing a suitable method of decision combinations is a key point for the ensemble's performance. In this paper, different possibilities have been explored and compared. In particular, several weighted voting methods, along with the unweighted plurality vote (the most frequent class is the winner class). In the discussion that follows, N stands for the number of samples, contained in the training set $\mathcal{X} = \{\mathbf{x}\}_{i=1}^{N}$, M is the number of classes in a set $\mathcal{C} = \{c_j\}_{j=1}^{M}$, and K classifiers, C_k, are utilized.

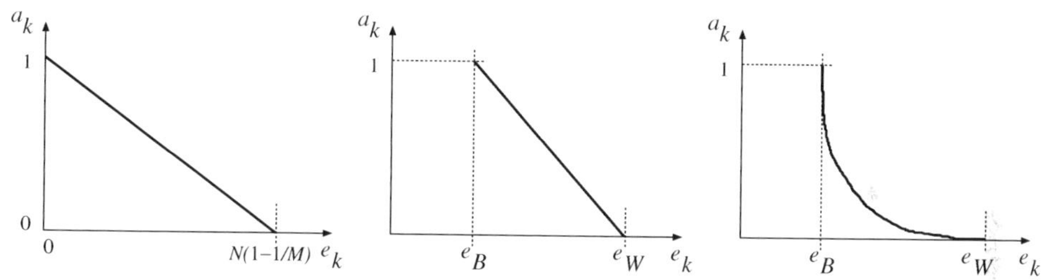

Fig. 1. Different models for giving the authority (a_k) to each classifier in the ensemble as a function of the number of errors (e_k) made on the training set

3.1 Unweighted Methods

1. Plurality vote (PV). Is the simplest method. Just count the number of decisions for each class and assign the sample $\mathbf{x}_i$ to the class c_j that obtained the highest number of votes. The problem here is that all the classifiers have the same 'authority' regardless of their respective abilities to classify properly. In terms of weights it can be considered that $w_k = 1/K \; \forall k$.

3.2 Weighted Methods

2. Simple weighted vote (SWV). The decision of each classifier, C_k, is weighted according to its estimated accuracy (the proportion of successful classifications, α_k) on the training set [9]. This way, the authority for C_k is just $a_k = \alpha_k$. Then, its weight w_k is:

$$w_k = \frac{a_k}{\sum_l a_l} \tag{1}$$

Also for the rest of weighting schemes presented here (except the last one), the weights are the normalized values for a_k, as shown in this equation.

The weak point of this scheme is that an accuracy of 0.5 in a two-class problem still has a fair weight although the classifier is actually unable to predict anything useful. This scheme has been used in other works [10] where the number of classes is rather high. In those conditions this drawback may not be evident.

3. Re-scaled weighted vote (RSWV). The idea is to assign a zero weight to classifiers that only give N/M or less correct decisions on the training set, and scale the weight values proportionally. As a consequence, classifiers with an estimated accuracy $\alpha_k \leq 1/M$ are actually removed from the ensemble. The values for the authority are computed according to the line displayed in figure 1-left. Thus, if e_k is the number of errors made by C_k, then

$$a_k = \max\{0, 1 - \frac{M \cdot e_k}{N \cdot (M-1)}\}$$

4. Best-worst weighted vote (BWWV). In this ensemble, the best and the worst classifiers in the ensemble are identified using their estimated accuracy. A maximum authority, $a_k = 1$, is assigned to the former and a null one, $a_k = 0$, to the latter, being equivalent to remove this classifier from the ensemble. The rest of classifiers are rated linearly between these extremes (see figure 1-center). The values for a_k are calculated as follows:

$$a_k = 1 - \frac{e_k - e_B}{e_W - e_B} \quad ,$$

where

$$e_B = \min_k\{e_k\} \quad \text{and} \quad e_W = \max_k\{e_k\}$$

5. Quadratic best-worst weighted vote (QBWWV). In order to give more authority to the opinions given by the most accurate classifiers, the values obtained by the former approach are squared (see figure 1-right). This way,

$$a_k = (\frac{e_W - e_k}{e_W - e_B})^2 \quad .$$

6. Weighted majority vote (WMV). The theorem 4.1 of Kuncheva's book [11, p. 124] states that accuracy of the ensemble is maximized by assigning weights

$$w_k \propto log\frac{\alpha_k}{1-\alpha_k}$$

where α_k is the individual accuracy of the classifier. In order to use a voting method of this type as a reference for the previously proposed methods (numbers 3 to 5), in this case the weight of each classifier is computed as:

$$w_k = log\frac{\alpha_k}{1-\alpha_k}$$

Classification by the Weighted Methods. Once the weights for each classifier decision have been computed, the class receiving the highest score in the voting is the final class prediction. If $\hat{c}_k(\mathbf{x}_i)$ is the prediction of C_k for the sample $\mathbf{x}_i$, then the prediction of the ensemble can be computed as

$$\hat{c}(\mathbf{x}) = \arg\max_{c_j \in \mathcal{C}} \sum_k w_k \delta(\hat{c}_k(\mathbf{x}_i), c_j) \quad , \tag{2}$$

being $\delta(a, b) = 1$ if $a = b$ and 0 otherwise.

Since the weights represent the normalized authority of each classifier, it follows that $\sum_{k=1}^{M} w_k = 1$. This makes it possible to interpret the sum in Eq. 2 as $P(\mathbf{x}_i|c_j)$, the probability that $\mathbf{x}_i$ is classified into c_j.

4 Experiments

Two different experiments have been carried out in order to compare the voting schemes proposed (numbers 3 to 5) with those of reference (1, 2 and 6). The first experiment tries to study the performance of the voting schemes when used with benchmarking data. For that, 19 data sets from the public available UCI/Statlog projects have been utilized. Each data set has been partitioned as explained in section 2. In total, 50 pairs of train/test sets were generated, so a total number of 50 experiments for each data set have been run in order to train the weights of the ensembles. The error rates of each base classifier were computed as the total number of errors made (on the 50 experiments) divided by the total number of samples classified. Finally, in order to test the ensembles, another 5 pairs of train/test sets were generated for validation. Recall from section 2 that the validation data are not unseen data.

The table 1 presents the error rates of the validation experiments for the datasets, with the best results for each data set emphasized in boldface. Note that the result for the best single classifier classifier is showed as a reference.

To summarize the results, the ensembles outperform the best classifier in 8 out of 19 data sets, the best classifier wins in 5 data sets, and in the remaining 6 data sets they obtain the same error rate. Specially significant is the result for the glass database, where the ensembles obtain an error rate which is almost 4% below the error rate of the best classifier. Note that the quadratic best-worst has performed the best, being 8 times one of the winner schemes. Note that the best single classifier was not always the same (1 NB, 1 SVM, and 3 MLP) and there are not analytic methods to decide which is the best classifier to be used according to the data. Thus, the ensembles seem a better option for designing a classification system.

A Case Study. In order to test on a real new problem the experiences we have learned from the first study, the same approach is now applied to a real problem related to music information retrieval. The goal is to classify a digital music score into a set of genres. In this case, jazz and classical music have been consider due to a general agreement among the experts about their definitions

Table 1. Error rates (in %) of the different ensembles with the UCI/Statlog data sets, together with the result of the best individual classifier (BEST) column. The winning classifications schemes in terms of accuracy for each data set have been highlighted.

DATA SET	PV	SWV	RSWV	BWWV	QBWWV	WMV	BEST
australian	**13.04**	**13.04**	**13.04**	13.62	14.64	**13.04**	14.64 (SVM)
balance	12.64	11.36	11.36	10.56	10.56	11.20	**8.80** (MLP)
cancer	3.37	3.37	3.37	3.37	3.37	3.37	**3.22** (SVM)
diabetes	23.18	23.18	23.18	23.44	**22.66**	23.18	**22.66** (SVM)
german	24.30	24.30	24.30	**23.5**	23.70	24.30	23.70 (SVM)
glass	32.71	30.84	30.84	**28.51**	29.91	**28.51**	32.24 (MLP)
heart	15.93	15.93	15.93	15.19	15.19	15.93	**14.07** (NB)
ionosphere	**9.12**	**9.12**	**9.12**	11.11	11.11	**9.12**	9.40 (MLP)
liver	36.81	36.81	35.36	33.62	**31.88**	35.36	**31.88** (MLP)
monkey1	3.60	3.60	**0**	**0**	**0**	**0**	**0** (MLP)
phoneme	16.78	16.78	16.78	13.53	**12.31**	16.78	**12.31** (3-NN)
segmen	3.51	3.07	3.07	**2.55**	**2.55**	3.07	3.77 (DT)
sonar	24.04	24.04	24.04	23.08	23.08	24.04	**22.12** (MLP)
vehicle	21.75	21.04	21.04	20.33	20.33	20.33	**18.91** (MLP)
vote	4.37	4.37	4.37	**3.69**	4.14	4.37	4.14 (DT)
vowel	14.02	11.74	11.74	5.87	**4.92**	5.68	**4.92** (3-NN)
waveform21	14.74	14.70	14.70	13.36	**13.3**	14.70	**13.30** (SVM)
waveform40	14.50	14.50	14.50	13.96	**13.74**	14.50	**13.74** (SVM)
wine	**1.69**	**1.69**	**1.69**	2.25	2.25	**1.69**	**1.69** (NB)

and taxonomy. The JvC (Jazz vs. Classical) corpus is made up of samples extracted from standard MIDI files[1] files from jazz and classical music and it has been already utilized in former works[2] [12,13].

MIDI files contain music in symbolic format (a sort of digital score). The files used here contain a melody track from which descriptors are extracted. All melodies are monophonic sequences of notes (at most one note is playing at any time). The corpus is composed of a total of 150 MIDI files, 65 of them being classical music and 85 being jazz. This dataset represents more than 8 hours of music.

Each sample is a vector of musical descriptors for a number of feature categories that assess melodic, harmonic and rhythmic properties of a melody. These descriptors are mainly descriptive statistics like, for example, average note pitch, standard deviation of note durations, pitch interval range, etc. A total of 28 descriptors are available.

From the set of MIDI files two datasets have been built. The first one composed of 150 samples, one sample per melody track. The second one is made up of 7125 samples. For this second dataset, each sample corresponds to a fragment of a melody, extracted applying a 50-bar wide sliding window on each melody track.

[1] http://www.midi.org

[2] This dataset is available for research purposes on request to the authors.

The window is shifted one bar at each time along the track, until the end of the track is reached. Each time the window is shifted, a new sample is extracted. Being ω the size of the window, the first dataset corresponds to a value $\omega = \infty$, and the second dataset for $\omega = 50$.

The experiments with the JvC data sets have been carried out using a train, test, and validation scheme. Random partitions are not advisable since for $\omega = 50$ attention has to be paid to samples belonging to the same melody do not appear in both training and test or validation. This fact would underestimate the error estimation. Each data set has been splitted into 5 partitions (keeping in the same partition those samples belonging to the same MIDI file). 3 of them have been used for training, 1 for test, and the remaining one for validation. The experiment has been repeated 5 times, rotating the partitions. The results of the validation presented in table 2 are average error rates from the 5 experiments.

Table 2. Average error rates (in %) of the different ensembles with the JvC data sets, together with the results of the base classifiers

Ensemble/classifier	Data set	
	JvC, $\omega = \infty$	JvC, $\omega = 50$
Plurality	7.33	9.28
SWV	7.33	9.28
RSWV	7.33	9.16
BWWV	**6.00**	**6.31**
QBWWV	**6.00**	8.29
WMV	**6.00**	9.46
3-NN	**6.00**	11.80
DT	13.33	15.66
MLP	8.00	13.30
NB	16.00	15.56
SVM	10.67	11.08

The results for $\omega = \infty$ show that even when the best single classifier (the 3-NN classifier) is much better than all the other single classifiers, the ensembles still perform adequately. For the $\omega = 50$ data set, the ensembles perform much better than any base classifier, specially the BWWV, which obtains an error rate 4.5% below the rate of the best classifier (SVM). The results shown in table 2 confirm that the ensembles performance is better in the general case (although in some cases may be slightly worse than a single particular classifier).

5 Conclusions

We have proposed three weighted voting methods (RSWV, BWWV, and QB-WWV) for classifier ensembles, and we have tested their performance with the UCI/Statlog project data sets (a widely known repository of real data sets), and also with a case study of music genre classification. In both cases the proposed ensembles have shown a more robust performance in general than individual classifiers, and with some data sets the results of the best ensemble is much better than that of a classifier.

Among all the voting schemes tested, the approaches based on scaling the weights to a range established by the best and the worst classifiers have shown the best classification accuracy in most of the data sets.

Future work includes a more adequate validation scheme for the UCI/Statlog project data sets, and using more base classifiers for testing the ensembles. Also, we plan to study more carefully the results of each ensemble on the data sets to find out the reasons of the (good or bad) performance of the ensemble, and develop new voting methods to improve these results.

Acknowledgments

The authors wish to thank Roberto Paredes from the Politechnical University of Valencia for providing us with the UCI/Statlog data sets, and the program for generating the partitions. The authors are also in debt with Miguel Sánchez Molina from University of Alicante, for his valuable help with the WEKA tool. This work was supported by the Spanish project CICyT TIC2003–08496–C04, the project GV06/166 from Generalitat Valenciana, and the IST Programme of the European Community, undes the Pascal Network of Excellence, IST-2002-506778.

References

1. Dietterich, T.G.: Ensemble methods in machine learning. Lecture Notes in Computer Science **1857** (2000) 1–15
2. Duin, R.: The combining classifier: to train or not to train? In: Proceedings of the International Conference on Pattern Recognition ICPR'2002. Volume II., Quebec (Canada) (2002) 765–770
3. Dietterich, T.: Ensemble methods in machine learning. In: First Internacional Workshop on Multiple Classifier Systems. (2000) 1–15
4. Kuncheva, L.I.: That elusive diversity in classifier ensembles. In: Proc. 1st Iberian Conf. on Pattern Recognition and Image Analysis (IbPRIA'03). Volume 2652 of Lecture Notes in Computer Science. (2003) 1126–1138
5. Kuncheva, L.I., Whitaker, C.J.: Measures of diversity in classifier ensembles. Machine Learning **51** (2003) 181–207
6. Partridge, D., Griffith, N.: Multiple classifier systems: Software engineered, automatically modular leading to a taxonomic overview. Pattern Analysis and Applications **5** (2002) 180–188
7. Blake, C., Merz, C.: UCI repository of machine learning databases (1998)
8. Witten, I., Frank, E.: Data Mining: Practical machine learning tools and techniques. 2nd edn. Morgan Kaufmann, San Francisco (USA) (2005)
9. Opitz, D., Shavlik, J.: Generating accurate and diverse members of a neural-network ensemble. In Touretzky, D., Mozer, M., Hasselmo, M., eds.: Advances in Neural Information Processing Systems. Volume 8. (1996) 535–541

10. Stamatatos, E., Widmer, G.: Music performer recognition using an ensemble of simple classifiers. In: Proceedings of the European Conference on Artificial Intelligence (ECAI). (2002) 335–339
11. Kuncheva, L.: Combining Pattern Classifiers: methods and algorithms. Wiley (2004)
12. Ponce de León, P.J., Iñesta, J.M.: Statistical description models for melody analysis and characterization. In: Proceedings of the 2004 International Computer Music Conference, International Computer Music Association (2004) 149–156
13. Pérez-Sancho, C., Iñesta, J.M., Calera-Rubio, J.: Style recognition through statistical event models. In: Proceedings of the Sound and Music Computing Conference, SMC'04. (2004)

AUC-Based Linear Combination of Dichotomizers

Claudio Marrocco, Mario Molinara, and Francesco Tortorella

Dipartimento di Automazione, Elettromagnetismo,
Ingegneria dell'Informazione e Matematica Industriale
Università degli Studi di Cassino
03043 Cassino (FR), Italy
{c.marrocco, m.molinara, tortorella}@unicas.it

Abstract. The combination of classifiers is an established technique to improve the classification performance. The combination rules proposed up to now generally try to decrease the classification error rate, which is a performance measure not suitable in many real situations and particularly when dealing with two class problems. In this case, a good alternative is given by the Area under the Receiver Operating Characteristic curve (AUC). This paper presents a method for the linear combination of two-class classifiers aimed at maximizing the AUC. The effectiveness of the approach has been confirmed by the tests performed on standard datasets.

1 Introduction

Linear combination is a widely diffused technique for combining multiple classifiers. The main reasons are both its simplicity and effectiveness [1]. This can be particularly useful in two-class problems that require highly discriminating classifiers. To this aim, linear combiners are generally built with the aim of minimizing the classification error. However, the considered applications frequently involve cost matrices and class distributions both strongly asymmetric and dynamic and in such cases the overall error rate, usually employed as a reference performance measure in classification problems, is not a suitable metric to evaluate the quality of the classifier [2]. An important tool to correctly analyse the performance of the classifier under different class and cost distributions is given by the Receiver Operating Characteristic (ROC) curve that provides a description of the performance of the dichotomizer at different operating points independently of the prior probabilities of the two classes. Moreover, the geometrical properties of the ROC curve can be profitably used to optimise the performance of a dichotomizer with reference to various metrics and classification requirements. In this case, to compare different dichotomizers it could be useful to employ a single value measure that describes the quality of the classifier. To this aim, the most widely used single measure is the *Area Under the ROC Curve* (AUC) that represents a more effective and discriminating performance measure than the accuracy to evaluate the quality of dichotomizers [3].

In this paper, focusing on the simplest and most widely used implementation of linear combiners, which consists of assigning a nonnegative weight to each individual classifier, we propose a method to directly optimise the AUC. In fact, we consider the linear combination of several classifiers and propose a method to achieve the optimal weight vector of the combination. To this aim, an analysis of the dependence of the

D.-Y. Yeung et al. (Eds.): SSPR&SPR 2006, LNCS 4109, pp. 714–722, 2006.

AUC of the linear combiner on the weighting is presented for two classifiers and a greedy approach to extend the combination rule to several classifiers is proposed together with the results of experiments performed on standard datasets that confirmed the effectiveness of the approach.

The rest of the paper is organized as follows: in the next section we present a short description of the AUC measure while section 3 shows the proposed method for two and N>2 classifiers. Then the obtained experimental results are reported and in the last section some conclusions and possible future developments are drawn.

2 AUC of a Dichotomizer

In two-class classification problems, the goal is to build a dichotomizer *f*(*z*) (i.e. a two-class classifier) which assigns a pattern *z* coming from an instance space *S* to one of two mutually exclusive classes that can be generically called *Positive* (*P*) class and *Negative* (*N*) class. Without loss of generality, let us assume that *f*(*x*) is in the range $(-\infty,+\infty)$ and provides a confidence degree that the sample belongs to one of the two classes, e.g. the class *P*. The sample should be consequently assigned to the class *N* if $f(x)\rightarrow-\infty$ and to the class *P* if $f(x)\rightarrow+\infty$; actually the output of the dichotomizer is compared with a threshold *t* to decide which is the class the sample should be assigned to. In order to evaluate performance of *f*(.), we can consider the outputs provided by the trained classifier on a set *S* containing n_+ positive samples and n_- negative samples $S=\{\mathrm{p}_i\in P, i=1\dots n_+\}\cup\{\mathrm{n}_j\in N, j=1\dots n_-\}$ and build the empirical ROC curve [4], whose points $(FPR(t),TPR(t))$ are given by:

$$TPR(t)=\frac{1}{n_+}\sum_{i=1}^{n_+}\Pi(f(\mathrm{p}_i)\geq t) \qquad FPR(t)=\frac{1}{n_-}\sum_{j=1}^{n_-}\Pi(f(\mathrm{n}_j)\geq t) \tag{1}$$

where *t* is in the range $(-\infty,+\infty)$ and $\Pi(.)$ is a predicate which is 1 when the argument is true and 0 otherwise. Based on the ROC curve, a widely used single measure is the *Area Under the ROC Curve* (AUC), which intuitively provides an estimate of the quality of the dichotomizer (AUC=0.5 for a non discriminating dichotomizer, AUC=1 for a perfectly discriminating dichotomizer). The AUC of the dichotomizer *f*(.) could be easily evaluated by numerically integrating the corresponding ROC. However, there is a useful equivalence [5] between the AUC of a dichotomizer and the Wilcoxon-Mann-Whitney (WMW) statistic which is defined as:

$$\frac{\sum_{i=1}^{n_+}\sum_{j=1}^{n_-}I(f(\mathrm{p}_i),f(\mathrm{n}_j))}{n_+\cdot n_-} \tag{2}$$

where *I*(*x*,*y*) returns 1 if *x*>*y*, 0.5 if *x*=*y* and 0 if *x*<*y*. In this way, it is possible to evaluate the AUC of *f* directly through (1) without explicitly plotting the ROC curve and estimating the area with a numerical integration.

Thanks to this equivalence, the AUC has a useful physical meaning: if we consider the outputs $f(x_P)$ and $f(x_N)$ provided by the dichotomizer on two samples z_P and z_N randomly extracted from *P* and *N*, it can be demonstrated [6] that the WMW statistic in (1), and thus the AUC, provides an unbiased estimate of the probability

$P(f(z_P) > f(z_N))$. In other words, the AUC of a dichotomizer measures the probability of correct pair-wise ranking [7] and thus evaluates the performance of the classifier considered as a *ranker*.

3 AUC-Based Linear Combination of Dichotomizers

The purpose of the method we are going to introduce is to construct a linear combination of dichotomizers aimed at maximizing the AUC of the resulting classification system. We focus first on the combination of two dichotomizers and then extend to N>2 dichotomizers.

3.1 Linear Combination of Two Dichotomizers

Let S be a set of samples defined as above. Let us consider two dichotomizers f_0 and f_1 whose outputs on positive and negative samples are:

$$x_i^0 = f_0(\mathrm{p}_i) \quad y_j^0 = f_0(\mathrm{n}_j) \qquad\qquad x_i^1 = f_1(\mathrm{p}_i) \quad y_j^1 = f_1(\mathrm{n}_j) \tag{3}$$

The AUC's for the two dichotomizers evaluated according to the WMW statistic are:

$$\mathrm{AUC}_0 = \frac{\sum_{i=1}^{n_+}\sum_{j=1}^{n_-} I(x_i^0, y_j^0)}{n_+ \cdot n_-} \qquad \mathrm{AUC}_1 = \frac{\sum_{i=1}^{n_+}\sum_{j=1}^{n_-} I(x_i^1, y_j^1)}{n_+ \cdot n_-} \tag{4}$$

Let us now consider a linear combination of f_0 and f_1. Without any loss of generality, the resulting classifier can be represented by:

$$f_{lc} = f_0 + \alpha \cdot f_1 \tag{5}$$

where α is the relative weight of f_1 with respect to f_0. The outputs of f_{lc} to p_i and n_j will be consequently:

$$\xi_i = f_{lc}(\mathrm{p}_i) = x_i^0 + \alpha \cdot x_i^1 \qquad \eta_j = f_{lc}(\mathrm{n}_j) = y_j^0 + \alpha \cdot y_j^1 \tag{6}$$

According to the WMW statistic, the AUC of f_{lc} is given by:

$$\mathrm{AUC}_{lc} = \frac{\sum_{i=1}^{n_+}\sum_{j=1}^{n_-} I(\xi_i, \eta_j)}{n_+ \cdot n_-} \tag{7}$$

and depends on the value of α. Therefore the optimal choice for the weight is the value maximizing $\mathrm{AUC}_{\mathrm{lc}}$. To this aim, let us analyze the term $I(\xi_i, \eta_j)$ and study how it depends on the values of $I(x_i^0, y_j^0)$ and $I(x_i^1, y_j^1)$; for the following analysis we consider a tie as an error and thus we group together the cases $I(x,y) = 0.5$ and $I(x,y) = 0$. With this assumption, the set of all the pairs on which AUC_{lc} is evaluated can be split in four subsets S_{00}, S_{11}, S_{01}, S_{10}, where S_{uv} is defined as:

$$S_{uv} = \left\{ (i,j) \middle| I(x_i^0, y_j^0) = u \text{ and } I(x_i^1, y_j^1) = v \right\} \tag{8}$$

Moreover, let C_{uv} denote the number of pairs container in the set S_{uv}: it can be simply verified that the expression for AUC_{lc} can be written as:

$$AUC_{lc} = \frac{1}{n_+ \cdot n_-}\left[\sum_{(i,j)\in S_{00}} I(\xi_i,\eta_j) + \sum_{(i,j)\in S_{11}} I(\xi_i,\eta_j) + \sum_{(i,j)\in S_{10}\cup S_{01}} I(\xi_i,\eta_j)\right] = \frac{1}{n_+ \cdot n_-}\left[0 + C_{11} + \nu(\alpha)\right] \quad (9)$$

In other words, while the pairs on which both dichotomizers are wrong do not contribute to AUC_{lc} and the pairs correctly ranked by both the dichotomizers give a contribution independent of the value of α, the dependence of AUC_{lc} on α is limited to the set of pairs on which the dichotomizers disagree. Therefore, the larger the set $S_{10}\cup S_{01}$ (i.e., the higher the disagreement between f_0 and f_1), the higher the value of AUC_{lc} which, in principle, can be obtained. Taking into account eq. (9), the optimal value for α can be defined as $\alpha_{opt} = \arg\max \nu(\alpha)$. In order to find such value let us make explicit the dependence of $I(\xi_i,\eta_j)$ on α; to this aim, recall that the indicator function is not null only if $\xi_i > \eta_j$, i.e. if:

$$\left(x_i^0 - y_j^0\right) + \alpha \cdot \left(x_i^1 - y_j^1\right) > 0 \quad (10)$$

To simplify the following calculations, let us call *Score Difference Ratio* (SDR) the quantity $-\frac{x_i^0 - y_j^0}{x_i^1 - y_j^1}$ and denote it with $\Delta_1^0(i,j)$; for pairs (i,j) belonging to S_{01} or S_{10} this value is positive because in both cases the differences have opposite signs. The condition (10) leads to different constraints on α depending on which of the two sets S_{01}, S_{10} we consider. In particular we obtain:

$$\alpha < \Delta_1^0(i,j) \text{ if } (i,j) \in S_{10} \qquad \alpha > \Delta_1^0(i,j) \text{ if } (i,j) \in S_{01} \quad (11)$$

If such conditions were verified for each pair $(i,j) \in S_{10}\cup S_{01}$, we would obtain the max value allowable for $\nu(\alpha)$, i.e. $C_{10}+C_{01}$, but in general this cannot be obtained since the distributions of the SDRs coming from the sets S_{10} and S_{01} are not separated. In any case, α_{opt} has to be found by maximizing the number of the pairs satisfying eq. (10). To this aim, let us evaluate how many pairs of each set are correctly ranked for a given value α for the weight of the linear combination. In order to simplify the notation in the following analysis, let us disregard the dependence on the particular samples in the SDRs and denote with δ_h^{01} with $h = 1,\ldots,C_{01}$ the SDR value of the h-th pair contained in S_{01}; in a similar way, let δ_k^{10} with $k = 1,\ldots,C_{10}$ indicate the SDR value of the k-th pair contained in S_{10}. With this notation, we define the *Correctly Ranked Rate* on S_{01}, $CRR_{01}(\alpha)$, and the *Wrongly Ranked Rate* on S_{01}, $WRR_{01}(\alpha)$, as:

$$CRR_{01}(\alpha) = \frac{\#\left\{\delta_h^{01} < \alpha,\, h = 1\ldots C_{01}\right\}}{C_{01}} \qquad WRR_{01}(\alpha) = \frac{\#\left\{\delta_h^{01} \geq \alpha,\, h = 1\ldots C_{01}\right\}}{C_{01}} \quad (12)$$

Both indices are in the range [0,1] and are not independent since they sum up to 1. In a similar way, it is possible to evaluate the same indices on the set S_{10}:

$$CRR_{10}(\alpha) = \frac{\#\left\{\delta_k^{10} > \alpha,\, k = 1\ldots C_{10}\right\}}{C_{10}} \qquad WRR_{10}(\alpha) = \frac{\#\left\{\delta_k^{10} \geq \alpha,\, k = 1\ldots C_{10}\right\}}{C_{10}} \quad (13)$$

Since for each set the indices are dependent on each other, it is sufficient to know only one index for each set in order to have the corresponding value for $\nu(\alpha)$. A possible choice could be to consider only WRR_{01} and CRR_{10} and to represent them as coordinates in a plane: in this way, the values produced by a particular α individuate a point in the unit square whose corners are the points (0,0), (1,0), (0,1) and (1,1). When the value of the weight α varies between 0 and ∞ the quantities WRR_{01} and CRR_{10} vary accordingly, thus drawing a curve running from (1,1) to (0,0). We call it *Difference Ratio Operating Characteristic* curve (*DROC* curve).

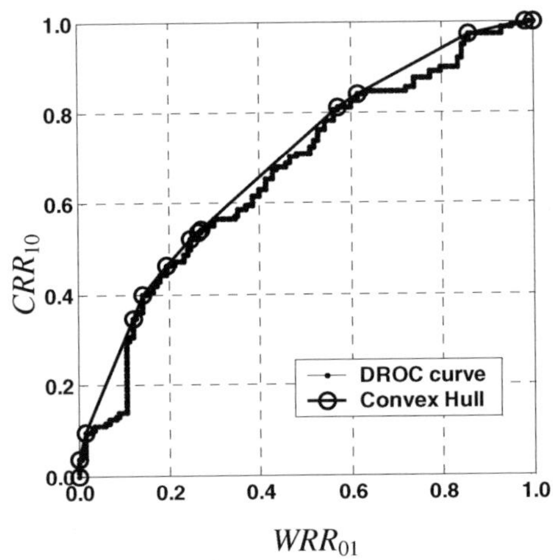

Fig. 1. A *DROC* curve together with the related convex hull

The DROC can be profitably used also for selecting the optimal value of the weight α_{opt}. To this aim, let us point out that the quantity to be maximized in eq. (9) can be written as:

$$\nu(\alpha) = C_{10} \cdot CRR_{10}(\alpha) + C_{01} \cdot (1 - WRR_{01}(\alpha)) \tag{14}$$

In the case the DROC curve is defined by means of a finite number of experimental points connected with straight lines (similar to the curve shown in fig. 1), it can be demonstrated that α_{opt} can be determined by locating the point where a line with slope $m = \frac{C_{01}}{C_{10}}$, moving down from above, touches the DROC curve and taking the corresponding value of α. In particular, such point lies on the *DROC Convex Hull*, i.e. the smallest convex set containing the points of the DROC curve.

3.2 Linear Combination of N>2 Dichotomizers

Let us now consider the linear combination of several (say N) classifiers:

$$f_{lc} = \alpha_0 f_0 + \alpha_1 f_1 + \ldots + \alpha_{N-1} f_{N-1} \tag{15}$$

In order to find the optimal weight vector $\alpha_{opt} = (\alpha_0\ \alpha_1\ \cdots\ \alpha_{N-1})$ that maximizes the AUC associated with f_{lc}, we should extend the algorithm proposed in the previous section. Unfortunately, eq. (11) cannot be generalized to N>2 dichotomizers in such a way that the maximization of the resulting AUC is computationally feasible. For this reason, we adopted a suboptimal algorithm that approximates the solution using a greedy approach. Hence, rather than considering every possible combination in its entirety, we iteratively find the optimal weight of the linear combination of two dichotomizers so as to evaluate all the combination weights in N-1 steps. In this context an important role is played by the order of combination, i.e. which pair of classifiers should be combined first. Since we know from previous sections that the greater is the diversity among the classifiers to be combined the greater is the improvement to the performance of the base classifiers which could be gained, we choose the pair that exhibits the maximum disagreement coefficient. Once the weight has been computed, the two dichotomizers are replaced by their combination and thus the dichotomizers to be combined decrease from N to N-1. At this point, we have to evaluate the disagreement between the new classifier and the other classifiers. It is worth noting that, for this step, it is not necessary to compute the output of the new classifier, since its score differences can be directly evaluated as the weighted sum (with the same weight estimated for the combination) of the score differences of the combined classifiers. These steps are repeated until all the dichotomizers have been combined: in each iteration it is chosen the pair of dichotomizers with the highest disagreement coefficient. At the end, we obtain the weight vector α_{opt}. It is worth noting that also in this case one of the weights of the vector will be equal to 1, but this will not imply any loss of generality.

4 Experimental Results

In order to evaluate the performance of the proposed method, it has been tested on three datasets publicly available at the UCI Machine Learning Repository [8]. All of them have two classes and a variable number of numerical input features. The features were previously rescaled so as to have zero mean and unit standard deviation. To avoid any bias in the comparison, 10 runs of a multiple hold out procedure have been performed on all data sets. In each run, the dataset has been parted into three sets: a training set to train the classifier, a validation set to estimate the optimal weight vector and a test set to assess the reliability of the proposed method. More details for each dataset are given in table 1.

Table 1.

Datasets	# Features	# Samples	% Pos.	% Neg.	Train. Set	Valid. Set	Test Set
Australian Credit Approval	14	690	44.5	55.5	483	103	104
Contraceptive Method Choice	9	1473	57.3	42.7	1031	221	221
Pima Indian Diabetes	8	768	34.9	65.1	538	115	115

The employed dichotomizers are Support Vector Machines (SVM) and Multi Layer Perceptrons (MLP). The SV-based classifiers have been implemented by means of SVMlight tool [9] while for the MLPs we have used the NODElib library [10]. Six different kernels have been used for the SVMs (linear, polynomial of degree 2 and 3, gaussian with σ^2=5,2,1) while for the MLPs we have considered three classifiers with different numbers of units (2,4,6) in the hidden layer, all trained for 10000 epochs using the back propagation algorithm with a learning rate of 0.01.

For the sake of comparison, we have also considered another method which evaluates the weight vector maximizing the AUC by employing a bound constrained global optimization algorithm, called *Multilevel Coordinate Search* (MCS) [11]. It is based on a multilevel coordinate search that balances global and local search; the local search is done via sequential quadratic programming and it is not exhaustive. Of course, the aim here is not to provide another algorithm to construct the optimal combination, but to obtain a reliable estimate of the weight vector maximizing the AUC on the validation set (even though with a computationally expensive algorithm) with which to compare the results provided by our method.

In order to choose a set of combinations of dichotomizers significant for the comparison, we have preliminarily ordered the single classifiers according to their mean AUC evaluated on the validation set and then we have built 7 different groups by putting together the *K* best classifiers (with *K* ranging from 2 to 8) and other 7 groups collecting the *K* worst classifiers. For each of these groups we have built two linear combinations, the first obtained with our DROC-based method while the second employs the weight vector estimated by means of the MCS algorithm.

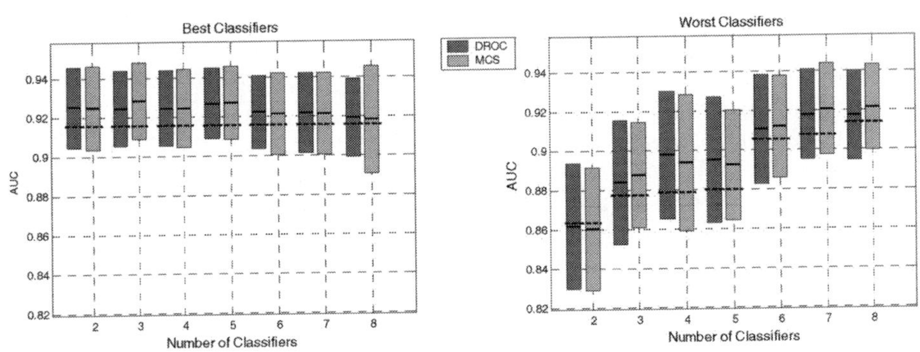

Fig. 2. The results on the Australian data set

In figures 1-3 we report the results in terms of AUC obtained on the test set for each data set. In particular, we show the mean value and the error bars relative to each method together with the performances of the best single classifier of the group (dashed line). From these results we can see that the DROC based method is actually able to determine the best weight vector: in fact the two methods provide quite the same performance (both in terms of average AUC and standard deviation) in all the examined cases. Obviously, they are both more proficient when working with the weaker classifiers.

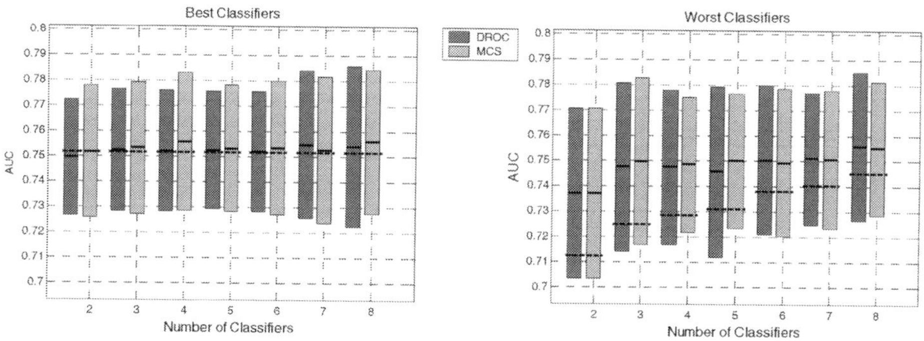

Fig. 3. The results on the Contraceptive data set

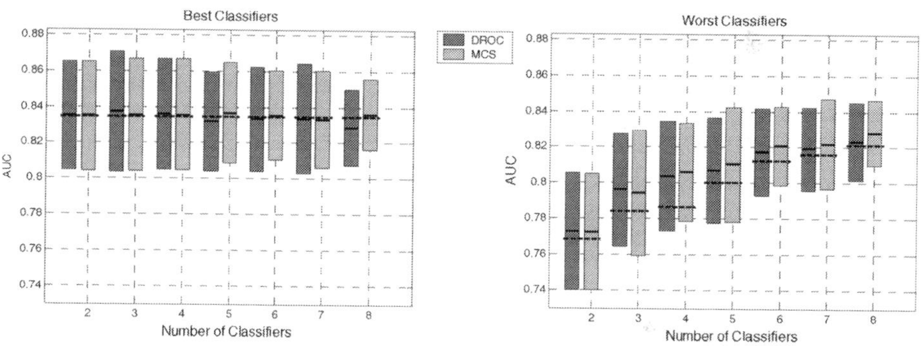

Fig. 4. The results on the Pima data set

5 Conclusions

In this paper we have introduced a method for the linear combination of dichotomizers with the aim of maximizing the AUC of the resulting classification system. The method is based on an analysis of the dependence of the AUC of the linear combiner on the weight α in the case of two dichotomizers and is extended to the general case through a greedy approach. The experiments have demonstrated that the algorithm actually allows the optimal value of the weight vector to be found. The future researches on this topic will consider the comparison with other linear combination methods (simple and weighted average) as well as a more thorough analysis of the direct relation found between the performance improvement attainable with the combination and the measure of the disagreement existing between the dichotomizers.

References

1. Fumera G., Roli F., 2005. A Theoretical and Experimental Analysis of Linear Combiners for Multiple Classifier Systems. IEEE Trans. on Patt. Anal. and Mach. Intell., vol. 27, no. 6, 942-956.
2. Provost, F., Fawcett, T., Kohavi, R., 1998. The Case against Accuracy Estimation for Comparing Induction Algorithms. Proc. ICML-1998, Morgan Kaufmann, 445-453.

3. Huang, J., Ling, C.X., 2005. Using AUC and Accuracy in Evaluating Learning Algorithms. IEEE Transactions on Knowledge and Data Engineering 17, 299-310.
4. A. Webb, 2002. Statistical pattern Recognition, 2nd ed., Wiley, USA.
5. Pepe, M., 2003. The Statistical Evaluation of Medical Tests for Classification and Prediction. Oxford University Press, Oxford, UK.
6. Lehmann, E.L., 1975, Nonparametrics. Statistical Methods Based on Ranks, Holden-Day, S. Francisco, USA.
7. Hanley J.A., McNeil B.J., 1982. The Meaning and the Use of the Area under a Receiver Operating Characteristic Curve. Radiology 143, 29-36.
8. Blake, C., Keogh, E., Merz, C.J.: UCI Repository of Machine Learning Databases. (1998) [www.ics.uci.edu/~mlearn/MLRepository.html]
9. Joachims, T.: Making Large-Scale SVM Learning Practical, in Schölkopf, B., Burges, C.J.C., Smola, A.J., eds., Advances in Kernel Methods, MIT Press (1999) 169-184.
10. Flake, G.W., Pearlmuter, B.A., 2000. Differentiating Functions of the Jacobian with Respect to the Weights. In S. A. Solla, T. K. Leen, and K. Müller, eds., Advances in Neural Information Processing Systems 12, The MIT Press.
11. Huyer, W., A. Neumaier, A., 1999. Global optimization by multilevel coordinate search, J. Global Optimization 14, 331-355.

Confidence Score Based Unsupervised Incremental Adaptation for OOV Words Detection

Wei Chu, Xi Xiao, and Jia Liu

Department of Electronic Engineering,
Tsinghua University,
Beijing 100084, China
chuwei@stu.ee.tsinghua.edu.cn

Abstract. This paper presents a novel approach of distinguishing in-vocabulary (IV) words and out-of-vocabulary (OOV) words by using confidence score-based unsupervised incremental adaptation. The unsupervised adaptation uses Viterbi decode results which have high confidence scores to adjust new acoustic models. The adjusted acoustic models can award IV words and punish OOV words in confidence score, thus obtain the goal of separating IV and OOV words. Our Automatic Speech Recognition Laboratory has developed a Speech Recognition Developer Kit (SRDK) which serves as a baseline system for different speech recognition tasks. Experiments conducted on the SRDK system have proved that this method can achieve a rise over 41% in OOV words detection rate (from 68% to 96%) at the same cost of a false alarm (taken IV words as OOV words) rate of 10%. This method also obtains a rise over 11% in correct acceptance rate (from 88% to 98%) at the same cost of a false acceptance rate of 20%.

1 Introduction

Nowadays speech recognition system can perform quite well on isolated words recognition if only providing IV word utterances as input and a vocabulary which is not very large. But the situation gets worse as the appearance of OOV words. In real world, OOV words input problem should not be overlooked, because the recognizer is faced with the OOV words spoken by users all the time.

Confidence score is utilized to evaluate the reliability of recognition results by S. Cox [1]. Later on, many approaches of calculating confidence score are introduced. T. J. Hazen has done prominent work in summarizing and devising confidence scores in word-level and utterance-level [2]. But for our practical short isolated words recognition, it is hard to distinguish IV words from OOV words in confidence score domain.

One major reason is that the acoustic models used in SRDK can not generate confidence scores which are separable for IV and OOV words. In this paper, a confidence score-based unsupervised incremental adaptation method is used to adjust the acoustic models used in SRDK system.

During the adaptation, we first send adaptation data including IV and OOV words into SRDK system, then use the Viterbi decode results of the recognizer which have high confidence scores to guide the model adaptation. A Threshold for confidence

D.-Y. Yeung et al. (Eds.): SSPR&SPR 2006, LNCS 4109, pp. 723–731, 2006.

score is set in order to ensure that almost all the words used for adaptation are correctly recognized IV words.

Because the adaptation data are limited, we adopt maximum likelihood linear regression (MLLR) + maximum a posteriori (MAP) adaptation method. Our experiments have proved this unsupervised adaptation procedure can greatly improve the later performance of OOV words detection.

2 Word Confidence Scoring

In this OOV words detection task, two classical but proved to be efficient confidence scores are employed. For computational reasons, we adopt a two-pass search strategy in which a semi-syllable based confidence score [3] is calculated in the first-pass, and a filler model based confidence score [4] is calculated in the second-pass. Finally, we combine the two confidence scores into a single dimensional confidence score through a simple linear discrimination method.

2.1 Semi-syllable Based Likelihood Ratio

$C_{ss}(X,W_0)$ is the semi-syllable based confidence score of the best hypothesis W_0 when the observed vector sequence is X which focuses on likelihood ratio:

$$C_{ss}(X,W_0)=\frac{P(X|W_0)}{P(X)}. \tag{1}$$

Consider the states alignment of the observed vectors, we can express $P(X|W_0)$ as:

$$P(X|W_0)=P(X_1,X_2,\cdots,X_m|h_1,h_2,\cdots,h_m), \tag{2}$$

where h_i is the semi-syllable alignment of the observed vector sequence X_i, The corresponding relationship between h_i and X_i is determined during Viterbi match. Assuming the observed vectors X_i are independent of each other, we have:

$$P(X)=\prod_{i=1}^{m}P(X_i). \tag{3}$$

Furthermore, we assume that semi-syllables h_i are independent of each other. We represent the conditional probability $P(X|W_0)$ as:

$$P(X|W_0)=P(X_1,X_2,\cdots,X_m|h_1,h_2,\cdots,h_m)=\prod_{i=1}^{m}P(X_i|h_i). \tag{4}$$

So, we get

$$CS_{ss}(X,W_0)=\prod_{i=1}^{m}\frac{P(X_i|h_i)}{P(X_i)}. \tag{5}$$

For each segmented observed vector X_i,

$$P(X_i) = \sum_{j=1}^{M} P(X_i | h_j). \tag{6}$$

In our system based on semi-syllable, if h_j is matched as a consonant (or vowel), M represents the amount of all the consonants (or vowels). Consequently, $CS_{log\_ss}(X, W_0)$ in the log domain can be described as:

$$CS_{log\_ss}(X, W_0) = \log P(W_0 | X) = \sum_{i=1}^{m} [\log P(X_i | h_i) - \log \sum_{j=1}^{M} P(X_i | h_j)]. \tag{7}$$

2.2 Filler Model Based Likelihood Ratio

$CS_{fl}(X, W_0)$ is the filler model based confidence score of the best hypothesis W_0 when the observed vector sequence is X.

$$CS_{fl}(X, W_0) = \frac{P(X | W_0)}{P(X | H_{Filler})}. \tag{8}$$

In our system, online garbage model $H_{Online\_Garbage}$ is considered to work as filler model H_{Filler}.

In the back-end, N-best hypotheses are listed out. Besides the best hypothesis W_0, the left $N-1$ hypotheses are called online garbage. $P(X | H_{Online\_Garbage})$ is as follows:

$$P(X | H_{Online\_Garbage}) = \frac{1}{N-1} \sum_{i=1}^{N-1} P(X | W_i). \tag{9}$$

So the normalized confidence score $CS_{log\_fl}(X, W_0)$ in the log domain is expressed as follows:

$$CS_{log\_fl}(X, W_0) = \frac{1}{n_X} \left\{ \log P(X | W_0) - \log \left[\frac{1}{N-1} \sum_{i=1}^{N} P(X | W_i) \right] \right\}, \tag{10}$$

where n_X represents the frame numbers of word W_0, and makes words with different frame numbers comparable in confidence score domain.

2.3 Confidence Score Combination

Since the two confidence scores are of different information, better performance will be achieved while combining them together.

For computational reasons, Fisher linear discrimination is used to find the projection vector $\boldsymbol{p}^{\mathrm{T}}$, and generate a linear discriminating plane between the IV and OOV words. Now we obtain a single dimensional confidence score $CS_{single}(X, W_0)$:

$$CS_{single}(X, W_0) = \boldsymbol{p}^{\mathrm{T}} \boldsymbol{CS}_{multi}(\boldsymbol{X}, \boldsymbol{W_0}) = [\alpha \quad \beta] \left[CS_{log\_ss}(W_0) \quad CS_{log\_fl}(W_0) \right]^{\mathrm{T}}. \tag{11}$$

3 Unsupervised Incremental Adaptation

After finishing calculating confidence score, we find that IV and OOV words can not be separated easily in confidence score domain. It is hard for us to detect OOV words by confidence score. The main reason for this phenomenon is that the initial acoustic models are trained with generic speech data. This initial acoustic models can perform speaker-independent speech recognition tasks quite well when providing only IV word utterances as input. But when adding OOV word utterances into the input sequence, these acoustic models can not generate separable confidence scores for IV and OOV words. For this reason, we have an idea that we can use specific IV word utterances to adjust suitable acoustic models. The acoustic model is a specific context-dependent phoneme HMM to our OOV word detect task. D. Wang has used this confidence score-based unsupervised adaptation method to improve the performance of speech recognition [5]. Our experiments have proved that these acoustic models after adaptation can also award IV words and punish OOV words in confidence score domain.

Because we use IV word utterances to perform unsupervised incremental adaptation, it is possible that wrongly recognized results will degrade the model parameters accuracy. Our strategy is to select only correctly recognized IV word utterances with high confidence scores for the adaptation.

3.1 MAP Adaptation

In MAP adaptation, the following formulas are used in each step of re-estimation, for each Gaussian pdf [6]:

$$\mu = \frac{\frac{N_{prior}}{N_{Init}} \sum_{i \in Init} x_i + \sum_{j \in Adapt} w(x_j) x_j}{N_{prior} + \sum_{j \in Adapt} w(x_j)}, \tag{12}$$

$$\sigma^2 = \frac{\frac{N_{prior}}{N_{Init}} \sum_{i \in Init} x_i^2 + \sum_{j \in Adapt} w(x_j) x_j^2}{N_{prior} + \sum_{j \in Adapt} w(x_j)} - \mu^2, \tag{13}$$

where N_{prior} is a control parameter of the adaptation process. The fewer N_{prior} is, the more adaptation utterances are taken into account with respect to prior data. $w(x_j)$ is a weighting factor to determine in what way the utterances should be used in the adaptation process. In our system, We adopt a strict strategy for $w(x_j)$:

$$\begin{cases} w(x_j) = 1 \text{ if } CS_{single}(x_j, W_0) > Th \\ w(x_j) = 0 \text{ if } CS_{single}(x_j, W_0) \le Th \end{cases}. \tag{14}$$

In our experiment we find that when the confidence scores of recognition results exceed a certain threshold, all the Viterbi decoder output is right. Only utterances with confidence score above Th can be used for adaptation in order to ensure that wrong Viterbi decode results will not perform negative effect on model parameters.

3.2 MLLR Adaptation

MLLR adaptation [7] is suitable when the amount of adaptation data is small or limited. MLLR adaptation performs faster than MAP adaptation when given the same amount of adaptation data. For each Gaussian pdf, μ_{ik} is transformed by using the following formula:

$$\tilde{\mu}_{ik} = A_c \mu_{ik} + b_c, \tag{15}$$

where A_c is a regression matrix and b_c is an additive bias vector associated with some broad class c, which can be either a broad phone class or a set of tied Markov states. We also only utilize those utterances with confidence scores over Th in MLLR adaptation, just as in our MAP adaptation.

4 Experiment Results

To show the effectiveness of the proposed method, we conduct experiments on our SRDK system. The initial acoustic models of the SRDK system are trained from approximately 100 hour speech data.

The vocabulary size of IV words is 200. Because our OOV words detection is expected to perform well in adverse situation, it is assumed that OOV word utterances occupy 50% of the total utterances. 3000 IV words and 3000 OOV word utterances are prepared as input utterances.

We use one third of the total input utterances for MLLR+MAP unsupervised incremental adaptation. Left 4000 utterances including IV and OOV word utterances are taken as OOV words detection test set.

In MLLR+MAP adaptation, to find an optimum value for Th, we compare OOV words detection performance under different Th. The results are depicted in Figure 1. The work point refers to the OOV words detection rate at the point where OOV words detection rate + false alarm rate = 1. Given a Th, each work point represents the best work condition under this Th. We want to mention that all our following experiments are conducted under this optimum Th (4×10^5).

The original OOV words detection point is 82.5% before our unsupervised adaptation. When the Th is higher than the optimum Th (4×10^5), the amount of utterances which used in adaptation decreases, and the work point falls, but always over the initial work point. When the Th is lower than the optimum Th, the performance of OOV words detection falls greatly. It is mainly because the incorrectly recognized words have performed negative effects on the unsupervised adaptation. We observed that when the input data used in adaptation contain a few OOV word utterances, the work point after adaptation is still higher than the initial work point.

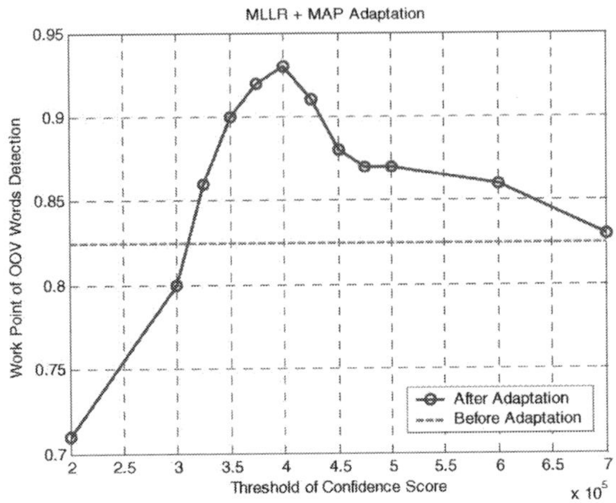

Fig. 1. Change in work points of OOV words detection depending on the changes in threshold of confidence score after MLLR+MAP adaptation

Figure 2 and Figure 3 below illustrate that IV and OOV words become easier to separate in confidence score domain after adaptation. During the incremental adaptation procedure, the acoustic model parameters are gradually adjusted according formulas (12), (13) and (14), thus can generate higher confidence scores for the later coming utterances used in adaptation. So it is also feasible to gradually lower *Th* as the adaptation procedure goes gradually. But we want to perform a robust unsupervised incremental adaptation, thus a fixed *Th* is used during the adaptation procedure to prevent the possible underlying instabilities which may perform negative effects on OOV word detection.

In Figure 4, false alarm rate in Figure 4 refers to the rate of false acceptance of IV word as OOV words. Figure 4 shows that the proposed method has achieved a rise

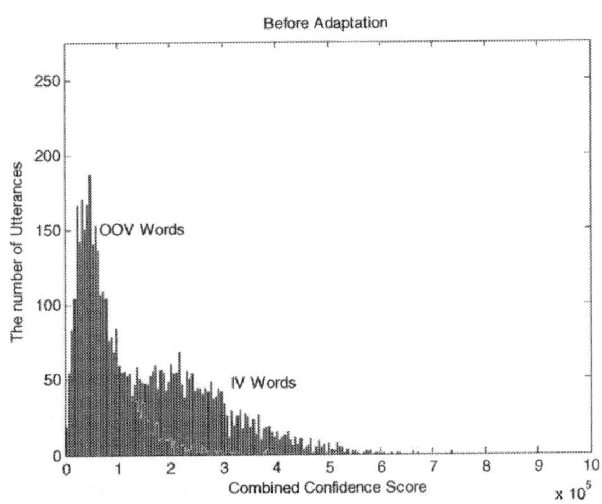

Fig. 2. Confidence score distribution of IV and OOV words before adaptation

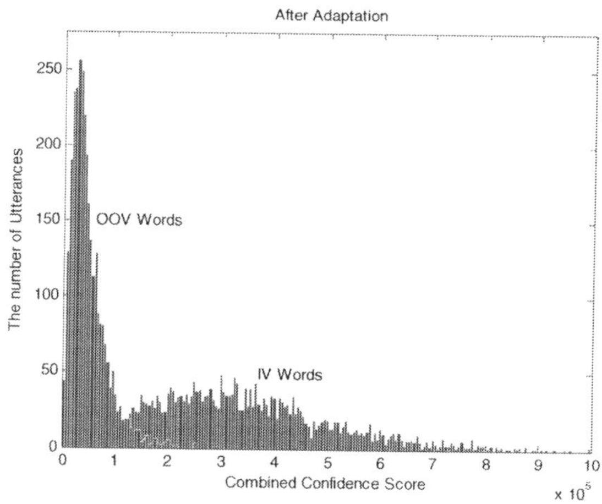

Fig. 3. Confidence score distribution of IV and OOV words after adaptation. Under Th = 4×10^5

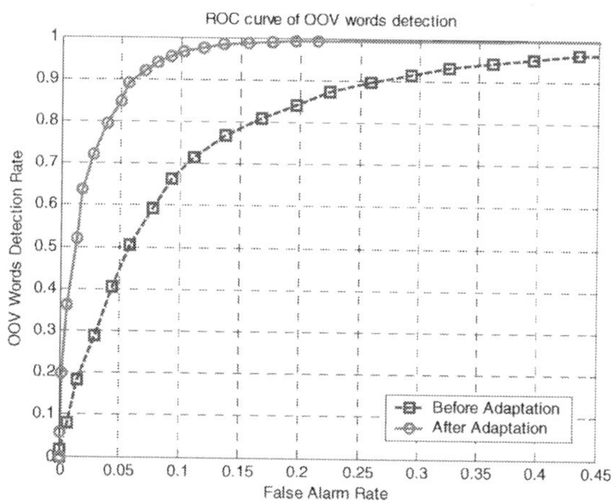

Fig. 4. OOV words detection performance before adaptation and after adaptation. Under Th = 4×10^5.

over 41% in OOV words detection rate (from 68% to 96%) at the same cost of a false alarm rate of 10%.

In Figure 5, false acceptance rate refers to the percentage of wrongly recognized words which are accepted. The correct acceptance rate refers to the percentage of correctly recognized words which are accepted. It is essential for us to examine the recognition performance of the adapted models. Figure 5 depicted that we can achieve a rise in correct acceptance rate (from 88% to 98%) at a false acceptance rate of 20%, when the input data are composed of 50% IV word utterances and 50% OOV word utterances. But when the input data are all IV word utterances, we observe degradation in correct acceptance rate (from 88% to 68%) at a false acceptance rate of

20%. The main reason is that the adjusted acoustic models is task-oriented (best fit for 50% IV + 50% OOV), and its performance relies greatly on the proportions of IV and OOV word utterances in the input data.

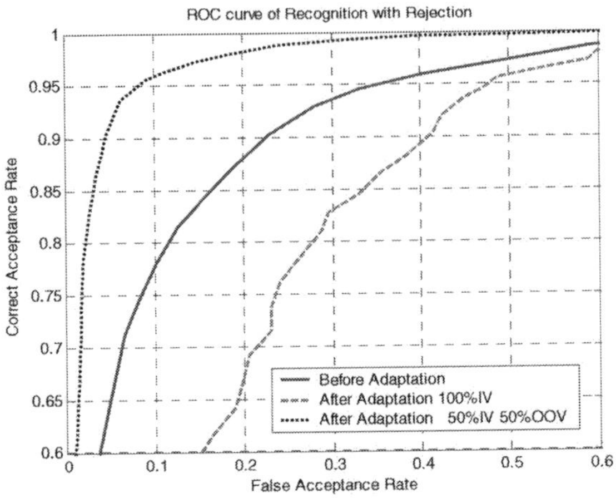

Fig. 5. ROC curve of recognition with rejection before adaptation and after adaptation. Under Th$=4\times10^5$.

5 Conclusions

This paper presented a new method for improving the OOV word detection rate by using unsupervised incremental adaptation based on confidence score. The effectiveness of this method has been proved by experiments on our SRDK system. Our future work will include applying this idea not only to the acoustic models, but also to the language model of a real-time human-machine interactive system in which input utterances are composed of isolated words and sentences. It is important to find a balance point between OOV words detection rate and recognition rate in practical usage according to the real feelings of the users.

Acknowledgements

The research is supported by National Natural Science Funds of China (Item No. 60572083). The authors would like to thank everyone who contributed to building and improving the SRDK system.

References

1. Cox, S. and Rose, R.: Confidence measures for the SWITCHBOARD database. In Proceedings of ICASSP 1996. Atlanta. (1996) 511-514
2. Hazen, T. J., Seneff, S. and Polifroni, J.: Recognition confidence scoring and its use in speech understanding systems. Computer Speech and Language. 16(1). (2002) 49-67

3. Sankar, A. and Wu, S.-L.: Utterance verification based on statistics of phone-level confidence scores. In Proceedings of ICASSP 2003. Menlo Park. (2003) 584–587
4. Boite, J., Bourlard, H., D'hoore, B. and Haesen, M.: A new approach towards keyword spotting. In Proceedings of Eurospeech 93. Berlin. (1993) 1273-1276
5. Wang, D., and Narayanan, S. S.: A confidence-score based unsupervised map adaptation for speech recognition. In Proceedings of 36th Conference on Signal, Systems and Computers. Pacific Grove. (2002) 222–226
6. Charlet, D.: Confidence-measure-driven unsupervised incremental adaptation for HMM-based speech recognition. In Proceedings of ICASSP 2001. Salt Lake City. (2001) 357–360
7. Leggetter, C. J. and Woodland, P. C.: Maximum likelihood linear regression for speaker adaptation of continuous density hidden Markov models. Computer Speech and Language. Vol. 9, No. 2. (1995) 171-185

Polynomial Network Classifier with Discriminative Feature Extraction

Cheng-Lin Liu

National Laboratory of Pattern Recognition (NLPR)
Institute of Automation, Chinese Academy of Sciences
PO Box 2728, Beijing 100080, P.R. China
liucl@nlpr.ia.ac.cn

Abstract. The polynomial neural network, or called polynomial network classifier (PNC), is a powerful nonlinear classifier that can separate classes of complicated distributions. A method that expands polynomial terms on principal subspace has yielded superior performance. In this paper, we aim to further improve the performance of the subspace-feature-based PNC. In the framework of discriminative feature extraction (DFE), we adjust the subspace parameters together with the network weights in supervised learning. Under the objective of minimum squared error, the parameters can be efficiently updated by stochastic gradient descent. In experiments on 13 datasets from the UCI Machine Learning Repository, we show that DFE can either improve the classification accuracy or reduce the network complexity. On seven datasets, the accuracy of PNC is competitive with support vector classifiers.

1 Introduction

Artificial neural networks (ANNs) with supervised learning have shown superior classification performance in many experiments [1]. Frequently used neural classifiers include the multi-layer Perceptron (MLP), radial basis function (RBF) network, polynomial network, etc. The polynomial network is also known as higher-order neural network (HONN), functional link network, polynomial regression [2], or generalized linear discriminant function [3]. In this paper, we call this classifier structure as polynomial network classifier (PNC). Since the outputs of PNC are the weighted combinations of higher-order nonlinear functions of input features, it is powerful to separate pattern classes of complicated distributions.

The PNC can be viewed as a single-layer neural network with the input features and their polynomial terms as the network inputs. For d features, the total number of polynomial terms up to r-th order is [4]

$$D = \sum_{i=0}^{r} \binom{d+i-1}{i} = \binom{d+r}{r}. \quad (1)$$

With large d, the polynomial network will suffer from high computation complexity and will give degraded generalization performance. The complexity can

D.-Y. Yeung et al. (Eds.): SSPR&SPR 2006, LNCS 4109, pp. 732–740, 2006.

be reduced by either reducing the number of input features or selecting expanded polynomial terms [2]. The former way is more computationally efficient and performs fairly well in practice. A PNC with dimensionality reduction by principal component analysis (PCA) has shown superior performance to multilayer neural networks in previous experiments [5,6].

On the other hand, constrained polynomial structures with moderate complexity have been proposed, like the pi-sigma network (PSN) [7], the ridge polynomial network (RPN) [4], and the reduced multivariate polynomial model (RMPM) [8]. The general HONN is a sigma-pi network in that it combines the products of features. Rather, the output of a PSN is the product of weighted combinations of features. Its number of weights is thus linear with the number of summation units (the order of polynomials). The output of RPN is the summation of pi-sigma units of different orders, and the order can be increased incrementally. The RMPM combines the univariate polynomials, the polynomial of sum of features and its product with the weighted sum of features. These networks actually involve all the polynomial terms of input features up to certain order, but the weights of polynomials are highly constrained. They hence need polynomials of fairly high order (say, 5 or 6) to approximate complicated functions, and cannot guarantee the precision of approximation in difficult cases.

The PNC with full polynomials on reduced features still have higher complexity than the above constrained networks, but usually, a low order (say, 2 or 3) can achieve a reasonable precision of function approximation. The behavior of a lower-order network on feature subspace is easy to explain and to control. Nevertheless, its performance largely depends on the technique of feature selection or dimensionality reduction. Supervised subspace learning methods, like the Fisher linear discriminant analysis (LDA) [3] and heteroscedastic discriminant analysis [9,10], may lead to better separability than the unsupervised PCA. These methods, nevertheless, are based on parametric density assumptions and the learning criterion is only loosely connected to classification error.

In this paper, we propose a subspace-feature-based PNC with discriminative feature extraction (DFE). With any classifier structure, DFE optimizes the subspace parameters together with the classifier parameters under a classification-related objective on a training sample set [11]. The subspace thus learned is totally classification-oriented and the subspace learning and classifier learning are best fitted. Overfitting can be overcome by adjusting the dimensionality of subspace and the order of classifier. DFE is mostly based on the minimum classification error (MCE) criterion of Juang and Katagiri [12], and has been successfully applied to many pattern recognition problems [13,14]. It has not been combined with polynomial networks, however. Despite that the MCE criterion is applicable to any classifier structures, for neural networks with sigmoid outputs, the minimum squared error (MSE) criterion works well and is easy to optimize by stochastic gradient descent [15].

We have evaluated the classification performance of PNC on 13 datasets from the UCI Machine Learning Repository [16]. The results show that compared with the PNC with PCA, DFE either improves the classification accuracy or reduces the network complexity. The complexity of PNC is much lower than support vector classifiers (SVCs) [17], and on seven of the 13 datasets, the PNC with DFE competes with SVCs in accuracy.

2 Subspace-Feature-Based PNC

We consider second-order (binomial) and third-order polynomial networks, and to save space, we only give the details of binomial networks. The structure and the learning algorithm of third-order networks are similar to binomial ones.

For M-class classification, the PNC has M output units. On a d-dimensional feature vector $\mathbf{x} = [x_1, \ldots, x_d]^T$, the output of binomial network for class ω_k is computed by

$$y_k(\mathbf{x}) = \sigma\Big[\sum_{i=1}^{d}\sum_{j=i}^{d} w_{kij}^{(2)} x_i x_j + \sum_{i=1}^{d} w_{ki}^{(1)} x_i + w_{k0}\Big], \tag{2}$$

where $\sigma(a)$ is the sigmoid activation function:

$$\sigma(a) = \frac{1}{1+e^{-a}}.$$

In classification, the input pattern (feature vector) is classified to the class of maximum output. The sigmoid function is used in training, and is not necessary in classification. Without the sigmoid function, the network weights can also be estimated by (non-iterative) pseudo inverse [2]. Since the sigmoid function makes the network outputs approximate posterior class probabilities, the trained weights with it are more suitable for classification than for regression.

By principal component analysis (PCA), the feature vector is projected onto an m-dimensional principal subspace ($m < d$):

$$\mathbf{z} = \Phi^T \mathbf{x} = [\phi_1^T \mathbf{x}, \ldots, \phi_m^T \mathbf{x}]^T = [z_1, \ldots, z_m]^T, \tag{3}$$

where $\Phi = [\phi_1, \ldots, \phi_m]$ is the transformation matrix (subspace basis) composed of the eigenvectors of covariance matrix $E[\mathbf{x}\mathbf{x}^T]$ corresponding to the m largest eigenvalues. We assume that the origin of the feature space has been shifted to the mean of samples. On the subspace features, the network outputs are computed by

$$y_k(\mathbf{x}) = \sigma\Big[\sum_{i=1}^{m}\sum_{j=i}^{m} w_{kij}^{(2)} z_i z_j + \sum_{i=1}^{m} w_{ki}^{(1)} z_i + w_{k0}\Big]. \tag{4}$$

On a training set of N samples $(\mathbf{x}^n, c^n)$, $n = 1, \ldots, N$ (c^n is the class label of $\mathbf{x}^n$), the connecting weights of PNC are adjusted to minimize the regularized squared error:

$$E = \frac{1}{2N}\Big\{\sum_{n=1}^{N}\sum_{k=1}^{M} [y_k(\mathbf{x}^n) - t_k^n]^2 + \beta \sum_{w\in W} w^2\Big\}, \tag{5}$$

where β is a coefficient of weight decay (excluding the biases); t_k^n is the target value of class k, with value 1 for the genuine class and 0 otherwise. The weights and biases are initialized to small random values, and by stochastic gradient descent, they are iteratively updated on the training samples until the squared error approaches the minimum. In training, the subspace basis Φ remains unchanged, and the polynomials of projected features can be viewed as the inputs of a single-layer network, for which the training process converges fast.

3 PNC with Discriminative Feature Extraction

A problem with the subspace-feature-based PNC is that the subspace does not necessarily lead to optimal classification because it is learned independently of the network weights. The subspace learned by PCA does not even consider the class information of training samples. Supervised subspace learning techniques, like LDA and heteroscedastic discriminant analysis, are expected to give better separability than PCA, but do not guarantee the optimality. We aim to learn a better subspace for PNC using discriminative feature extraction (DFE) [11].

By DFE, we adjust not only the network weights in supervised learning, but also the subspace basis simultaneously. Consider that $z_i = \phi_i^T \mathbf{x}$, $i = 1, \ldots, m$, let us re-write the network outputs of (4) as

$$y_k(\mathbf{x}) = \sigma\Big[\sum_{i=1}^{m}\sum_{j=i}^{m} w_{kij}^{(2)} \phi_i^T \mathbf{x} \phi_j^T \mathbf{x} + \sum_{i=1}^{m} w_{ki}^{(1)} \phi_i^T \mathbf{x} + w_{k0}\Big] = \sigma(s_k(\mathbf{x})), \qquad (6)$$

where $s_k(\mathbf{x})$ denotes the weighted sum of output unit k.

In the PNC with DFE, since the projected feature $z_i = \phi_i^T \mathbf{x} = \sum_{j=1}^{d} \phi_{ij} x_j$ is a weighted combination of original features and the weights (subspace parameters ϕ_{ij}, $j = 1, \ldots, d$) are now adjustable, an m-th order polynomial as $\prod_{i=1}^{m}(\phi_i^T \mathbf{x})$ is actually a pi-sigma unit of the ridge polynomial network (RPN). However, our network has more polynomial terms and needs a lower order than the RPN. Interpreting ϕ_i, $i = 1, \ldots, m$, as subspace basis vectors or feature extractors, a lower-order polynomial network on this feature subspace has decision boundaries of moderate complexity.

The network weights and the subspace basis parameters are adjusted to minimize the regularized square error (5) on a training sample set. The subspace parameters can be initialized to small random values as the network weights. Alternatively, the subspace learned by PCA or LDA is a good start of parameter search. The weights and subspace parameters are then adjusted by stochastic gradient descent on training samples. At time t, the parameters are adjusted on a training sample $\mathbf{x}$ by

$$\begin{aligned} w_{kij}^{(2)}(t+1) &= w_{kij}^{(2)}(t) - \epsilon(t)\big[(y_k - t_k)y_k(1-y_k)z_i z_j + \tfrac{\beta}{N} w_{kij}^{(2)}(t)\big], \\ w_{ki}^{(1)}(t+1) &= w_{ki}^{(1)}(t) - \epsilon(t)\big[(y_k - t_k)y_k(1-y_k)z_i + \tfrac{\beta}{N} w_{ki}^{(1)}(t)\big], \\ w_{k0}(t+1) &= w_{k0}(t) - \epsilon(t)(y_k - t_k)y_k(1-y_k), \\ \phi_i(t+1) &= \phi_i(t) - \epsilon(t)\textstyle\sum_{k=1}^{M}(y_k - t_k)y_k(1-y_k)\frac{\partial s_k}{\partial \phi_i}, \\ & k = 1, \ldots, M,\ i = 1, \ldots, m,\ j = i, \ldots, m, \end{aligned} \qquad (7)$$

where $\epsilon(t)$ is the learning step, which is set to a small value initially and decreases gradually in the training process. The partial derivative of ϕ_i is specified as

$$\frac{\partial s_k}{\partial \phi_i} = \left(2w_{kii}^{(2)} z_i + \sum_{j<i} w_{kji}^{(2)} z_j + \sum_{j>i} w_{kij}^{(2)} z_j + w_{ki}^{(1)}\right)\mathbf{x}. \quad (8)$$

In discriminative learning, we keep the unit norm of basis vectors but not the orthogonality. On adjusting the basis vectors on a training sample, each vector is normalized to unit norm ($\|\phi_i\| = 1$).

By stochastic gradient descent, the training samples are fed to the PNC for a number (40 or more in our experiments) of cycles. The learning step decreases linearly until it vanishes at the end of training. On every input sample, the network weights and subspace parameters are updated according to (7). The network weights and the subspace vectors have remarkably different magnitudes of derivatives. To accelerate the convergence of training, they are set two different learning steps, ϵ_1 for weights and ϵ_2 for subspace vectors, and $\epsilon_1 \gg \epsilon_2$ holds.

Another factor affecting training convergence and classification performance is the scale of projected features. We normalize the scale with the square root of the largest eigenvalue λ_1 of $E[\mathbf{x}\mathbf{x}^T]$ (estimated on training samples and fixed):

$$z_i = \frac{\phi_i^T \mathbf{x}}{\sqrt{\lambda_1}}. \quad (9)$$

All the feature vectors are subtracted from the mean of the training samples. For datasets that have significantly different scales among feature dimensions, it is helpful to uniform the standard deviation of all dimensions of training data (and test data accordingly). This is done before subspace projection.

4 Experiments

We evaluated the classification performance of subspace-feature-based PNC on 13 datasets from the UCI Machine Learning Repository [16], as summarized in Table 1. We selected the multi-class datasets that have at least 10 features. Some data sets have been partitioned into standard training and test subsets. For the others, we arrange the samples in random order and evaluate in 5-fold cross-validation.

Some datasets have appreciable variability of scale among different dimensions. We normalized them by dividing each dimension with $(0.9\sigma_i^2 + 0.1\sigma_0^2)^{1/2}$, where σ_i^2 is the dimension-wise variance and σ_0^2 is the average variance, both estimated on training data.

We compare the PNC-DFE (PNC combined with DFE) with PNC-PCA, one-versus-all support vector classifiers with polynomial and RBF kernels (SVC-poly and SVC-rbf), and the k-nearest neighbor (k-NN) classifier. For the SVC-poly, the feature vectors are uniformly scaled such that the average self-inner product of training vectors is one, and so, the kernel $k(\mathbf{x}_1, \mathbf{x}_2) = (1 + \kappa \mathbf{x}_1 \cdot \mathbf{x}_2)^r$ with

Table 1. Summary of 12 datasets from UCI Repository. The right two columns shows the selected polynomial order and subspace dimensionality (multiple of m_1).

Name	#class	#feature	#train	#test	Normal.	Order	m_1
Waveform	3	21	50,000	5-fold	No	2	1
Wine	3	13	178	5-fold	Yes	2	2
Soybean-small	4	35	47	5-fold	Yes	2	1
Vehicle	4	18	846	5-fold	Yes	2	2
Dermatology	6	34	358	5-fold	Yes	2	2
Segment	7	19	2,310	5-fold	Yes	3	3
Thyroid	3	21	3,772	3,428	Yes	2	4
Satimage	6	36	4,435	2,000	No	2	5
Optdigit	10	64	3,823	1,797	No	2	10
Pendigit	10	16	7,494	3,498	No	3	3
Vowel	11	10	528	462	No	2	2
Isolet	26	617	6,238	1,559	No	2	25
Letter	26	16	16,000	4,000	No	3	3

$\kappa = 2^i$ performs fairly well. For the SVC-rbf, the average within-class variance is scaled to one, such that in the kernel function $k(\mathbf{x}_1, \mathbf{x}_2) = \exp(-\frac{\|\mathbf{x}_1-\mathbf{x}_2\|^2}{2\sigma^2})$, a parameter value of $\sigma^2 = 0.5 \times 2^i$ performs fairly well. For both, i is an integer selected from -4 to 4.

For the k-NN classifier, SVC-poly, and SVC-rbf, we tried several values of k, polynomial order and κ, or σ^2 such that the classification accuracy on each test set is maximized.

For PNC-PCA and PNC-DFE, we set the number of subspace features $m = l \cdot m_1$, $l = 1, \ldots, 5$. m_1 is dependent on the dataset. The selected values of polynomial order and m_1 are listed in the right columns of Table 1. As seen, three datasets (Segment, Pendigit and Letter) are used 2nd-order and the others are used 3rd-order.

The test accuracies (%) of PNC (with full polynomials and dimensionality reduction by PCA and DFE) on the 13 datasets are shown in Table 2. For the "Vowel" dataset, there is no dimensionality reduction when $m = 10$. For each dataset, the accuracy of full PNC is shown below the title of dataset, and the accuracies of PNC-PCA and PNC-DFE with variable subspace dimensionality are listed in two rows. For the "Isolet" dataset, we do not give the accuracy of full PNC because the number of features is too large.

We can see that on four datasets (Vehicle, Segment, Satimage, and Letter), the full PNC gives the highest accuracy. This can be explained that the four datasets have small number of features and are difficult to classify, so dimensionality reduction by either PCA or DFE cannot improve the classification accuracy. For the other datasets, except for "Soybean-small" and "Isolet", subspace-feature-based PNC performs significantly better than the full PNC.

Comparing the accuracies of PNC-PCA and PNC-DFE, it is evident that except for two datasets (Waveform and Satimage), PNC-DFE mostly give higher accuracies than PNC-PCA, especially on subspaces of lower dimensionality. On

Table 2. Test accuracies (%) of PNC (full and PCA) and PNC-DFE on 12 datasets

Dataset	PCA=m_1	PCA=$2m_1$	PCA=$3m_1$	PCA=$4m_1$	PCA=$5m_1$
Full PNC	DFE=m_1	DFE=$2m_1$	DFE=$3m_1$	DFE=$4m_1$	DFE=$5m_1$
Waveform	63.78	87.02	**87.22**	87.12	86.98
84.92	60.64	86.90	86.96	86.72	86.64
Wine	77.53	90.45	92.13	92.70	92.70
92.13	79.79	92.13	**93.82**	93.82	93.82
Soybean-small	74.47	100	100	100	100
100	91.49	100	100	100	100
Vehicle	53.19	67.38	71.75	77.30	78.37
84.16	71.51	77.42	78.72	79.67	80.02
Dermatology	77.65	89.94	96.37	96.37	96.37
96.37	93.30	**96.93**	96.93	96.65	96.37
Segment	61.08	84.16	92.21	92.38	92.25
96.41	92.90	94.33	94.89	95.50	95.80
Thyroid	92.65	93.49	93.17	93.49	95.51
94.78	96.06	97.32	97.67	97.72	**97.87**
Satimage	86.75	87.10	87.75	88.00	87.95
88.65	86.45	87.45	87.90	88.15	88.05
Optdigit	95.72	98.05	98.61	98.61	98.55
98.50	97.16	98.50	**98.72**	98.50	98.66
Pendigit	85.11	95.77	97.68	98.03	**98.37**
98.23	89.57	97.17	97.91	98.17	**98.37**
Vowel	43.72	50.09	58.01	60.39	
59.52	57.14	60.17	**64.72**	61.47	
Isolet	93.33	95.19	95.51	**96.28**	96.28
	95.57	95.96	95.96	**96.28**	96.09
Letter	32.35	73.17	85.38	91.47	94.03
94.70	57.00	80.88	89.47	92.70	94.40

some datasets (Waveform, Soybean-small, Dermatology, Optdigit, Isolet), the PNC-DFE achieves the best or nearly best accuracy on a very low-dimensional subspace as $m = 2m_1$.

The highest accuracies of PNC (full and PNC-PCA), PNC-DFE, SVC-poly, SVC-rbf, and k-NN classifier on the 13 datasets are compared in Table 3. On the "Soybean-small" dataset, all these classifiers achieves perfect classification. Among the other datasets, SVC-poly or SVC-rbf gives the highest accuracies on seven datasets, and PNC or PNC-DFE performs best on five datasets. Expect for four datasets (Soybean-small, Segment, Satimage, and Letter), PNC or PNC performs significantly better than the k-NN classifier. The accuracy of PNC or PNC-DFE is comparable or higher than SVC on seven datasets (Waveform, Wine, Soybean-small, Vehicle, Thyroid, Optdigit, Vowel).

We did not implement the reduced multivariate polynomial model (RMPM) [8], but results on 10 of our 13 datasets were reported in the literature. Though the datasets were partitioned in different ways, nine of the 10 best accuracies of RMPM (Waveform 83.3%, Soybean-small 95.0%, Vehicle 82.3%, Segment 94.1%,

Table 3. Highest accuracies of PNC (full and PCA), PNC-DFE, SVCs and k-NN classifier

	PNC	PNC-DFE	SVC-poly	SVC-rbf	k-NN
Waveform	**87.22**	86.96	87.14	87.08	85.24
Wine	92.70	**93.82**	92.13	93.26	87.08
Soybean-small	100	100	100	100	100
Vehicle	**84.16**	80.02	81.56	81.21	71.99
Dermatology	96.37	96.93	**97.77**	97.21	96.09
Segment	96.41	95.80	96.62	**96.88**	96.71
Thyroid	95.51	**97.87**	96.70	95.36	94.28
Satimage	88.65	88.15	90.70	**91.40**	90.35
Optdigit	98.61	98.72	98.66	**98.89**	98.00
Pendigit	98.37	98.37	**98.77**	98.74	97.80
Vowel	60.39	**64.72**	56.06	64.50	59.52
Isolet	96.28	96.28	**96.92**	96.86	92.69
Letter	94.70	94.40	96.78	**97.65**	95.83

Thyroid 94.0%, Satimage 88.2%, Optdigit 95.3%, Pendigit 95.7%, Letter 74.1%) are lower than our best accuracies of PNC or PNC-DFE. The complexity of PNC mainly depends on the number of features, and is much lower than SVC and k-NN classifier. The k-NN classifier stores all training samples and compares them with each test pattern. The SVC has a large number of support vectors, ranging from 10% to 70% of all training samples. Due to the limited space, we do not discuss the computational complexity in details.

5 Conclusion

We proposed to improve the performance of subspace-feature-based polynomial network classifier (PNC) using discriminative feature extraction (DFE), which optimizes the subspace parameters together with the network weights on training samples. Under a regularized squared error criterion, the parameters are efficiently adjusted by stochastic gradient descent. In our experiments on 13 datasets of UCI Machine Learning Repository, DFE mostly improves the accuracy of subspace-feature-based PNC. At moderate complexity, the PNC (full or subspace-feature-based) outperforms the k-NN classifier on nine datasets and competes with support vector classifiers on seven datasets.

Acknowledgements

This work is supported by the Hundred Talents Program of Chinese Academy of Sciences. The author thanks the anonymous reviewers for valuable comments.

References

1. L. Holmström, P. Koistinen, J. Laaksonen, E. Oja, Neural and statistical classifiers—taxonomy and two case studies, *IEEE Trans. Neural Networks*, 8(1): 5-17, 1997.
2. J. Shürmann, *Pattern Classification: A Unified View of Statistical and Neural Approaches*, Wiley Interscience, 1996.
3. K. Fukunaga, *Introduction to Statistical Pattern Recognition*, 2nd edition, Academic Press, 1990.
4. Y. Shin, J. Ghosh, Ridge polynomial networks, *IEEE Trans. Neural Networks*, 6(3): 610-622, 1995.
5. U. Kreßel, J. Schürmann, Pattern classification techniques based on function approximation, *Handbook of Character Recognition and Document Image Analysis*, H. Bunke and P.S.P. Wang (Eds.), World Scientific, 1997, pp.49-78.
6. C.-L. Liu, K. Nakashima, H. Sako, H. Fujisawa, Handwritten digit recognition: benchmarking of state-of-the-art techniques, *Pattern Recognition*, 36(10): 2271-2285, 2003.
7. Y. Shin, J. Ghosh, The Pi-sigma network: an efficient higher-order neural network for pattern classification and function approximation, *Proc. 1991 IJCNN*, Seattle, Vol.1, pp.13-18.
8. K.-A. Toh, Q.-L. Tran, D. Srinivasan, Benchmarking a reduced multivariate polynomial pattern classifier, *IEEE Trans. Pattern Anal. Mach. Intell.*, 26(6): 740-755, 2004.
9. H. Brunzell, J. Eriksson, Feature reduction for classification of multidimensional data, *Pattern Recognition*, 33(10): 1741-1748, 2000.
10. M. Loog, R.P.W. Duin, Linear dimensionality reduction via a heteroscedastic extension of LDA: the Chernoff criterion, *IEEE Trans. Pattern Anal. Mach. Intell.*, 26(6): 732-739, 2004.
11. A. Biem, S. Katagiri, B.-H. Juang, Pattern recognition using discriminative feature extraction, *IEEE Trans. Signal Processing*, 45(2): 500-504, 1997.
12. B.-H. Juang, S. Katagiri, Discriminative learning for minimum error classification, *IEEE Trans. Signal Processing*, 40(12): 3043-3054, 1992.
13. X. Wang, K.K. Paliwal, Feature extraction and dimensionality reduction algorithms and their applications in vowel recognition, *Pattern Recognition*, 36(10): 2429-2439, 2003.
14. X. Yang, G. Pang, N. Yung, Discriminative training approaches to fabric defect classification based on wavelet transform, *Pattern Recognition*, 37(5): 889-899, 2004.
15. H. Robbins, S. Monro, A stochastic approximation method, *Ann. Math. Stat.*, 22: 400-407, 1951.
16. UCI Machine Learning Repository, http://www.ics.uci.edu/ mlearn/ ML Repository.html.
17. C.J.C. Burges, A tutorial on support vector machines for pattern recognition, *Knowledge Discovery and Data Mining*, 2(2): 1-43, 1998.

Semi-supervised Classification with Active Query Selection

Jiao Wang and Siwei Luo

School of Computer and Information Technology, Beijing Jiaotong University, Beijing 100044, China
Wangjiao0828@163.com

Abstract. Labeled samples are crucial in semi-supervised classification, but which samples should we choose to be the labeled samples? In other words, which samples, if labeled, would provide the most information? We propose a method to solve this problem. First, we give each unlabeled examples an initial class label using unsupervised learning. Then, by maximizing the mutual information, we choose the samples with most information to be user-specified labeled samples. After that, we run semi-supervised algorithm with the user-specified labeled samples to get the final classification. Experimental results on synthetic data show that our algorithm can get a satisfying classification results with active query selection.

1 Introduction

Recently, there has been great interest in Semi-supervised classification. The goal of semi-supervised learning is to use unlabeled data to improve the performance of standard supervised learning algorithms. Since in many fields, obtaining labeled data is hard or expensive, semi-supervised learning methods with small labeled sample size is of great use.

In case the unsupervised learning methods can separate the points well (see e.g. Fig.1a), there is no need for semi-supervised methods. However, in case of noise (see e.g. Fig.1b), or in case of two modes which belong to two different classes overlap (see e.g. Fig.1c), semi-supervised learning with a few labeled points in each class can improve the performance significantly.

A number of algorithms have been proposed for semi-supervised learning, including EM [8], Co-training [1, 14], Tri-training [15], random field models [9, 12], graph based approaches [2, 6, 13]. Different methods have different assumptions, and can be used in different situation. Especially, when data resides on a low-dimensional manifold within a high-dimensional representation space, semi-supervised learning methods should be adjusted to work on manifold. Belkin gives a solution to this problem with manifold Regularization methods in [4].

Query selection is extensively studied in the supervised framework. In [10], the queries are selected to minimize the version space size for support vector machine. In [7], a committee of classifiers is employed, and a point is queried whenever the committee members disagree. Many other methods are proposed to actively choose the samples in supervised learning, but few are done to choose samples in semi-supervised learning.

D.-Y. Yeung et al. (Eds.): SSPR&SPR 2006, LNCS 4109, pp. 741 – 746, 2006.

The labeled samples play an important role in semi-supervised learning. Then a question rises: which samples should be the labeled samples? Among the existing semi-supervised learning methods, some choose the labeled samples manually[6], to do this, one has to have some domain knowledge of which samples need most to be labeled; some choose the labeled samples randomly, which may not contain the "right" samples; and in [11], Zhu et al. choose the samples actively by greedily selecting queries from the unlabeled data to minimize the estimated expected classification error, but Zhu's active learning method can only be used together with his semi-supervised learning method.

In this paper, we give a more general and automatic query selection method in the semi-supervised framework. Our method can be applied to most of the existing semi-supervised learning methods. It is the pre-process of the existing semi-supervised methods. The main idea is to consider which samples, if labeled, would give more information. Following this idea, we use the mutual information $I(Y; y_*)$ (y_* represents one sample's class label and Y represents the whole sample's class labels) as a measure of active query selection. By maximizing the mutual information, we get the sample which needs most to be labeled. Using this method, we can choose the samples to be labeled actively and automatically, and it does not need any domain knowledge.

In this paper, in order to explain our method, we work with the Laplacian Eigenmaps [3] and manifold Regularization [4] of Belkin to show the entire process. We can see how the active query selection method works on manifold. We do not claim that this method can only use on manifold, and indeed we aim to illuminate that applying our method, to any semi-supervised method, would always yield satisfying results.

This paper is organized as follows: In section 2, we introduce our algorithm in a brief way. Section 3 gives details of every part of our algorithm. Experimental results on synthetic data are shown in section 4, followed by conclusions in section 5.

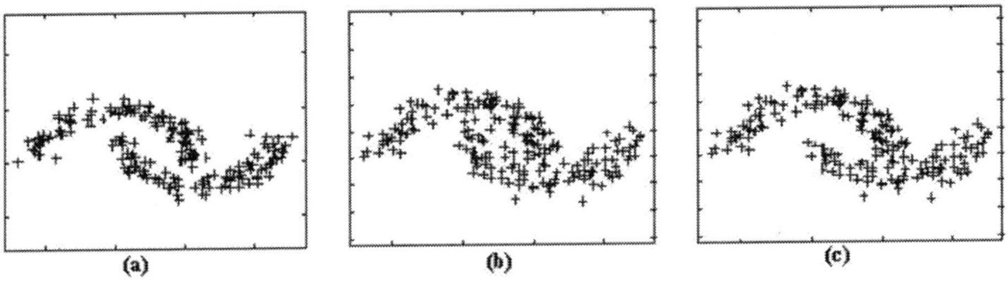

Fig. 1. (a) Example of situation in which unsupervised learning methods (here we use Laplacian Eigenmaps) can work well. (b)(c) Examples of situations in which unsupervised learning can not give satisfying results, and they need some labeled samples to help.

2 Our Algorithm

To explain the entire process of our active query selection method, we work with the Laplacian Eigenmaps [3] and manifold Regularization [4] of Belkin. The steps are as follow:

Step 1. Give each sample (unlabeled) an initial class label using unsupervised learning. Here, we use Laplacian Eigenmaps to map the sample to a real value function f.

Step 2. By maximizing the mutual information $I(Y; y_*)$ (y_* represents one sample's class label and Y represents the whole sample's class labels), we actively choose the samples with most uncertain class labels to be user-specified labeled samples.

Step 3. Give the chosen samples their class label.

Step 4. Run the semi-supervised algorithm (here, we use the manifold Regularization algorithm) with the user-specified labeled samples to get the final classification.

3 Details of Our Method

3.1 Using Laplacian Eigenmaps to Get Initial Class Label

Given a sample set $x_1,\ldots,x_n \in R^m$. Construct its neighborhood graph $G=(V,E)$, whose vertices are sample points $V=\{x_1,\cdots,x_n\}$, and whose edge weights $\{w_{ij}\}_{i,j=1}^n$ represent appropriate pairwise similarity relationships between samples. For example, w_{ij} can be the radial basis function:

$$w_{ij} = \exp(-\frac{1}{\sigma^2}\sum_{d=1}^{m}(x_{id}-x_{jd})^2) \tag{1}$$

where σ is a scale parameter. The radial basis function of w_{ij} ensure that nearby points are assigned large edge weights.

We first consider two-class situation. Assume that f is a real value function whose value is bounded from 0 to 1 (0 and 1 each represents a class label). $y_i = f(x_i), Y=(y_1,y_2,\ldots,y_n)^T$. Laplacian Eigenmaps try to minimize the following objective function

$$\sum_{ij}(y_i - y_j)^2 W_{ij} \tag{2}$$

By minimizing this objective function, we get $y_1, y_2,\ldots, y_n$, the initial class label of each sample, with $y_i \in [0,1]$.

3.2 Using Mutual Information to Choose the Samples with Most Uncertain Class Labels

This is our active query selection step. And we use the mutual information $I(Y; y_*)$ (y_* represents one sample's class label and Y represents the whole sample's class labels) as a measure of query selection. By maximizing the mutual information, we get the sample which would give most information, that is, which needs most to be labeled.

In order to calculate $I(Y; y_*)$, inspired by the work of [5], we define a Gaussian random field on the vertices of V

$$p(y) \propto \exp\{-\lambda y^T \Delta y / 2\} \tag{3}$$

where $\Delta = D - W$, and D is a diagonal matrix given by $D_{ii} = \sum_{j=1}^{n} W_{ij}$.

The mutual information between Y and y_* is the expected decrease in entropy of Y when y_* is observed:

$$\begin{aligned} I(Y; y_*) &= H(Y) - E\{H(Y \mid y_*)\} \\ &= (1/2)\log(1 + p(y_*)(1 - p(y_*))x_*^T H^{-1} x_*) \end{aligned} \tag{4}$$

where $H = \nabla^2(-\log p(Y))$, and ∇^2 is the Hessian matix.

The best sample to label is the one that maximizes $I(Y; y_*)$. And the mutual information is largest when $p(y_*) \approx 0.5$, *i.e.*, for samples with most information.

3.3 Using the User-Specified Labeled Samples to Get the Final Classification

After we actively choose the sample to give label, we can run the semi-supervised classification methods to get the final result. In Manifold Regularization methods of Belkin, the author minimizes the following cost function

$$\min_{f \in H} H[f] = \frac{1}{l} \sum_{i=1}^{l} (f(x_i) - y_i)^2 + \gamma_A \|f\|_K^2 + \gamma_I \sum_{i,j=1}^{n} (f(x_i) - f(x_j))^2 W_{ij} \tag{5}$$

Where l is the number of labeled samples. γ_A, γ_I are regularization parameters. $\|f\|_K^2$ is some form of constraint to ensure the smoothness of the learned manifold.

Here, the l samples in the above cost function are not chosen randomly or manually as in the original work of Belkin. But rather, they are chosen with the active query selection methods discussed in 3.2.

4 Experimental Results

As we point out at the beginning of this paper, unsupervised learning can not work well in case of noise, and in case of two modes which belong to two different classes overlapped. In these situations, semi-supervised learning with a few labeled samples can help.

Using some synthetic data, we show that our active query selection method can choose the most informative samples to give labels. Fig.2 (a) is a noise case of fig.1 (a), and without labeled samples, the Laplacian Eigenmaps can not find a satisfying classification (the yellow curve). Using our active query selection method, the algorithm chooses some samples to be labeled, these samples are shown in (b) with purple color. After that, user give the class label of these chosen samples (the red and blue samples in (c), each color represents a class), then, with these user-specified labeled samples, manifold regularization method find the more satisfying classification as shown in (c) (the yellow curve).

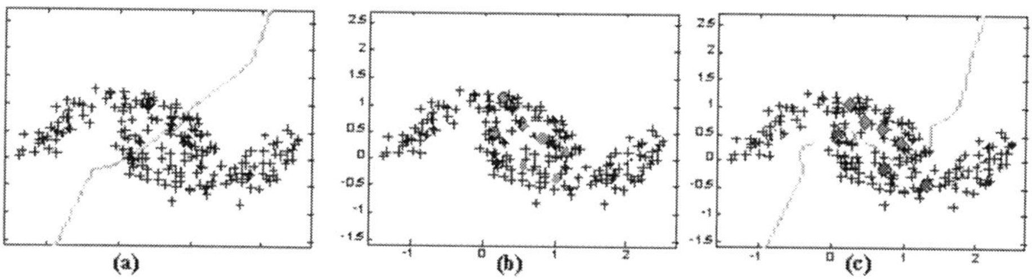

Fig. 2. (a) Laplacian Eigenmaps can not find a satisfying classification without labeled samples. (b) The samples automatically chosen to be labeled. (c) The manifold regularization results with the labeled samples.

5 Conclusions

A key problem of semi-supervised learning is to choose the most informative samples to be labeled at the very beginning of semi-supervised algorithms. Using mutual information, we give a solution to this problem. Our method of samples chosen can apply to most of the existing semi-supervised learning methods, and in this paper, we combine it with manifold regularization to show how it works.

We also do experiments on some synthetic data, and yield satisfying results. In future works, we will try this method on some real world experiments. Another problem of semi-supervised learning is how many labeled sample are suitable, for example, should we choose five samples to give label, or, should we choose ten? In future work, we will consider this problem in the framework of the active query selection of this paper.

Acknowledgements

The research is supported by national natural science foundations of china (60373029), the Research Fund for the Doctoral Program of Higher Education of China (20050004001) and Co-Construction Project of Key Subject of Beijing.

References

1. A. Blum and T. Mitchell: Combining labeled and unlabeled data with co-training. In Proceedings of the 11th Annual Conference on Computational Learning Theory. Madison, WI, pp.92–100, (1998).
2. A. Blum, S. Chawla: Learning from Labeled and Unlabeled Data using Graph Mincuts. ICML (2001).

3. Belkin M., Niyogi P: Laplacian Eigenmaps for Dimensionality Reduction and Data Representation. Neural Computation, June (2003)
4. Belkin M., Niyogi P., Sindhwani V: On Manifold Regularization. Department of Computer Science, University of Chicago, TR-2004-05.
5. B. Krishnapuram, D. Williams, Ya Xue, A. Hartemink, L. Carin, and M. A. T. Figueiredo: On Semi-Supervised Classification. NIPS (2004).
6. D. Zhou, O. Bousquet, T.N. Lal, J. Weston and B. Schoelkopf: Learning with Local and Global Consistency. NIPS (2003).
7. Freund, Y., Seung, H. S., Shamir, E., & Tishby, N.: Selective sampling using the query by committee algorithm. Machine Learning, 28, 133-168. (1997).
8. K. Nigam: Using Unlabeled Data to Improve Text Classification. PhD thesis, Carnegie Mellon University Computer Science Dept, (2001).
9. M. Szummer and T. Jaakkola: Partially labeled classification with markov random walks. NIPS (2001).
10. Tong, S., and Koller, D.: Support vector machine active learning with applications to text classification. ICML (2000).
11. Xiaojin Zhu, J. Lafferty, and Z. Ghahramani: Combining active learning and semi-supervised learning using Gaussian fields and harmonic functions. ICML (2003).
12. Xiaojin Zhu, Z. Ghahramani, and J. Lafferty: Semi-supervised learning using Gaussian fields and harmonic functions. ICML (2003).
13. Xiaojin Zhu: Semi-Supervised Learning with Graphs. PhD thesis, Carnegie Mellon University Computer Science Dept, (2005).
14. Zhou, Z.-H., & Li, M.: Semi-supervised regression with co-training. International Joint Conference on Artificial Intelligence, (2005).
15. Zhou, Z.-H., & Li, M.: Tri-training: exploiting unlabeled data using three classifiers. IEEE Trans. Knowledge and Data Engineering, 17, 1529–1541, (2005).

On the Use of Different Classification Rules in an Editing Task

Luisa Micó[1], Francisco Moreno-Seco[1], José Salvador Sánchez[2], José Martinez Sotoca[2], and Ramón Alberto Mollineda[2]

[1] Dept. Llenguatges i Sistemes Informàtics, Universitat d'Alacant
E-03071 Alacant (Spain)
[2] Dept. Llenguatges i Sistemes Informàtics, Universitat Jaume I
Av. Sos Baynat s/n, E-12071 Castelló de la Plana (Spain)

Abstract. Editing allows the selection of a representative subset of prototypes among the training sample to improve the performance of a classification task. The Wilson's editing algorithm was the first proposal and then a great variety of new editing techniques have been proposed based on it. This algorithm consists on the elimination of prototypes in the training set that are misclassified using the k-NN rule. From such editing scheme, a general editing procedure can be straightforward derived, where any classifier beyond k-NN can be used. In this paper, we analyze the behavior of this general editing procedure combined with 3 different neighborhood-based classification rules, including k-NN. The results reveal better performances of the 2 other techniques with respect to k-NN in most of cases.

Keywords: Pattern recognition, classification, nearest neighbor, prototype selection, editing.

1 Introduction

The k-Nearest Neighbor (k-NN) rule is a well known non-parametric classification approach. This rule classifies an unknown sample into the class most represented among its k nearest neighbors according to some metric [5]. Although it is mainly used for classification, the k-NN rule is widely used also for *edition.*

Given a set $\mathcal{T}$ of prototypes, an editing technique consists on the selection of a subset, $\mathcal{S} \subseteq \mathcal{T}$ where the overlapping among different classes has been reduced. The removed prototypes can be either those which belongs to overlapping regions, those erroneously labeled or atypical prototypes (*outliers*). The use of this technique improves the performance of the 1-NN classifier.

The Wilson's editing algorithm [2] was the first proposal related with the elimination of misleading prototypes from the training set $\mathcal{T}$. This technique retains in $\mathcal{T}$ only the correctly classified samples by a *leaving one out* strategy with a k-NN classifier. However, a more general editing scheme can be derived from Wilson's, by considering the error estimation strategy and the classification rule as editing scheme parameters.

D.-Y. Yeung et al. (Eds.): SSPR&SPR 2006, LNCS 4109, pp. 747–754, 2006.

In this work, an exhaustive evaluation of such general editing scheme based on three different neighborhood-based classification rules has been done. The main purpose is to compare the performances of these three classifiers in an editing task over a wide variety of known datasets. The three neighborhood-based rules are the well-known k-NN rule [3,7], the k-NCN rule [4] and the new k-NSN rule [9].

The k-NN rule classifies a sample into the majority class among its k nearest neighbors in $\mathcal{T}$. The k-NCN rule classifies a sample in the class most represented among the k neighbors whose centroid is the closest to the sample. These k neighbors are not usually the k nearest neighbors. The results achieved by the k-NCN rule are very interesting, outperforming the k-NN rule in many cases, specially with small training sets (which is what usually happens in practice). Finally, the k-NSN rule considers the k best neighbors selected by fast NN search algorithms when looking for the NN.

The structure of the paper is as follows. Section 2 presents the general editing scheme. In section 3 we shall briefly describe the distance-based rules that have been considered, and some details of their uses for edition. Section 4 consists of exhaustive experiments with 12 datasets and a discussion of results. Finally, we will conclude and outline some future work in section 5.

2 A General Editing Scheme

The classification accuracy of the NN rule can be improve by eliminating outliers and cleaning possible overlapping among classes in the original training set. This is the main goal of any editing technique.

A general editing procedure can be straightforward derived from Wilson's scheme. Given a training set $\mathcal{T}$, an error estimation strategy ξ, and a classification rule δ, let $\mathcal{R} \subseteq \mathcal{T}$ be the subset of samples incorrectly classified by δ using ξ. The edited subset $\mathcal{S}$ is obtained by removing from $\mathcal{T}$ those samples in $\mathcal{R}$. This process can be repeated until a certain condition η is satisfied. Figure 1 illustrates an schematic description of such procedure. Once the training set has been edited, the 1-NN rule is used to classify new samples.

In the experiments, this editing procedure is combined with 3 neighborhood-based classification rules, that is, rules which take into account the distances to a number of close neighbors and their classes to decide the class of a new sample. These classifiers are the plain k-NN rule [3], and two other related decision rules named the k-NCN rule [4] and the new k-NSN rule [9], respectively. Next section describes in details these techniques.

3 Neighborhood-Based Classification Rules

3.1 The k-NN Rule

One of the most widely studied non-parametric classification approaches corresponds to the k-NN rule. Given a set of n previously labeled prototypes or

Input: a training set of labeled samples $\mathcal{T} = \{(x_i, c_i)\}$
a classification rule δ
an error estimation strategy ξ
an ending condition η
Output: a final edited subset $\mathcal{S}$
Method:

repeat
let $\mathcal{R} \subseteq \mathcal{T}$ be the subset of samples incorrectly classified by ξ, δ
$\mathcal{S} \longleftarrow \mathcal{T} - \mathcal{R}$
$\mathcal{T} \longleftarrow \mathcal{S}$
until η is true

Fig. 1. A general editing scheme

training set (TS), the k-NN classifier [7] consists of assigning an input sample to the class most frequently represented among the k closest prototypes in the TS, according to a certain dissimilarity measure. A particular case of this rule is when $k = 1$, in which each input sample is assigned to the class indicated by its closest neighbor.

The asymptotic classification error of the k-NN rule (that is, when n grows to infinity) tends to the optimal Bayes error rate as $k \to \infty$ and $k/n \to 0$. Moreover, if $k = 1$, the error is bounded by approximately twice the Bayes error [8]. This behavior in asymptotic classification performance combines with a conceptual and implementational simplicity, which makes it a powerful classification technique capable of dealing with arbitrarily complex problems, provided there is a large enough TS available.

However, in many practical settings, this theoretical behavior can hardly be achieved because of certain inherent weaknesses that significantly reduce the applicability of k-NN classifiers in real-world tasks. For example, the performance of these rules, as with any non-parametric approach, is extremely sensitive to incorrectness or imperfections in the TS.

That is the reason why a considerable amount of works have been devoted to improve the NN classification accuracy by eliminating outliers from the original TS and also cleaning possible overlapping among classes. This strategy has generally been referred as to *editing* [8], whereas the corresponding classifier has been called *edited NN rule.*

3.2 The k-NCN Rule

The nearest centroid neighborhood [6] refers to a concept in which neighborhood is defined taking into account not only the proximity of prototypes to a given input sample but also their *symmetrical distribution* around it. From this general idea, the corresponding classification rule, the k-nearest centroid neighbors (k-NCN) [4], has been proven to overcome the traditional k-NN classifier in many practical situations.

Now the editing approach presented here corresponds to a slight modification of the original work of Wilson and basically consists of using the *leaving one out* error estimate with the k-NCN classification rule.

3.3 The *k*-NSN Rule

Recently, a new distance-based classification rule, the *k nearest selected neighbor* rule (k-NSN) has been proposed. The k-NSN rule is based on a class of fast NN search algorithms, those who search iteratively for the nearest neighbor: in each step, these algorithms select a candidate to nearest neighbor, then compute its distance to the sample, update the current nearest neighbor, prune the training set, and look for a new candidate. This process is repeated until no new candidates may be found. The k-NSN rule classifies the sample using the k nearest candidates selected while looking for the nearest neighbor, the so called k nearest selected neighbors. Of course, the performance of the k-NSN rule depends highly on the underlying fast NN search algorithms, and usually the fastest algorithm is the one with which the k-NSN results are the poorest, and vice versa. When the training set is large and/or the dimensionality of the data is high, the k-NSN rule obtains results that are similar to those of the k-NN rule (in fact, the k-NSN rule uses in these cases almost all of the k-NN for classification), but without the extra computational effort of finding exactly the k-NN.

Wilson's Editing with *k*-NSN. The fast NN search algorithms in which is based the k-NSN rule all require a certain data structure (usually a tree) to be built during the training phase, prior to classification. When using a *leaving one out* scheme for error estimation or for Wilson editing, these algorithms may need to rebuild its data structures many times, and this is usually a very time consuming step. In this work the k-NSN rule has been used only with one fast NN algorithm, the LAESA [10] algorithm, which is one of the simplest algorithms with which the k-NSN rule has been tested.

The LAESA algorithm uses a reduced matrix of distances between a subset of *base prototypes* and the rest of the prototypes in the training set. As the number of base prototypes required depends on the dimensionality of the data and not on the size of the training set, the spatial complexity is lineal on that size. The base prototypes are selected in the training phase as those that are maximally separated, and then the reduced matrix is computed.

In a *leaving one out* procedure, the algorithm should recompute the base prototypes and the reduced matrix each time the training set changes (i.e. each time a prototype is left out). However, in many cases the base prototypes would be the same as for the whole training set, so the matrix would be (almost) the same. Only in a few cases the result would differ. The set of base prototypes affects the number of distances computed, and thus the number of selected neighbors, so the performance of the k-NSN rule may be slightly different, but not too much. The simplest way to avoid recomputing each time the base prototypes set and the reduced matrix is to compute the set and the matrix for the whole training set, and then, if the prototype left out is one of the base prototypes, simply ignore the row corresponding to that prototype in the matrix for further computations.[1]

4 Experiments and Discussions

Experiments involved 12 datasets from the UCI Machine Learning Repository (`http://www.ics.uci.edu/$\sim$mlearn`). Table 1 summarizes the main characteristics of each data set: number of classes, attributes, and prototypes.

Table 1. A brief summary of the UCI databases

Data set	No. Classes	No. Features	Size
Cancer	2	9	685
Clouds	2	2	5002
Concentric	2	2	2501
Diabetes	2	8	770
Gauss	2	2	5002
German	2	24	1002
Glass	6	9	216
Heart	2	13	272
Liver	2	6	347
Phoneme	2	5	5406
Sonar	2	60	210
Waveform21	3	21	5001

To guarantee the statistical significance of results, all classification tasks were designed following a 5-fold cross validation. The 5 training partitions derived from each dataset were edited with the 3 editing techniques resulting from the combination of the general procedure of section 2 with the 3 distance-based classification rules (k-NN, k-NCN, k-NSN) described in section 3. In the case of k-NSN, the LAESA algorithm [10,9] was used as the fast NN search algorithm. Only the parameter $k = 3$ was used for all editing.

Then, resulting edited partitions were used to classify with the 1-NN rule their corresponding test partitions for each dataset. An additional baseline 1-NN clasification task was performed with the original training partitions (without any edition) and their corresponding test partitions (called **nedit**).

For each pair of dataset and editing technique, the average size of edited partitions and the average 1-NN classification accuracy on test partitions were collected. Table 2 provides a summary of these results.

[1] The matrix is used to compute a lower bound of the distance of each prototype to the sample, so the candidate to nearest neighbor is selected as that whose lower bound is the lowest.

Table 2. Average size of the edited sets and average 1-NN classification accuracies on test partitions (standard deviations are in brackets). Values in bold type indicate the highest accuracy for each database.

	Cancer		**Clouds**	
scheme	*edit*	*accuracy*	*edit*	*accuracy*
NOEDIT	547	95.17(2.38)	4000	84.66(0.96)
k-NSN	530(2.94)	96.19(2.08)	3391(18.43)	87.66(0.66)
k-NN	528(3.83)	**96.34**(1.90)	3498(12.09)	**88.26**(0.55)
k-NCN	528(2.64)	95.61(2.44)	3504(13.00)	**88.26**(0.44)
	Concentric		**Diabetes**	
scheme	*edit*	*accuracy*	*edit*	*accuracy*
NOEDIT	1999	81.59(1.26)	614	67.32(4.15)
k-NSN	1978(4.71)	81.59(1.26)	428(3.31)	70.83(3.62)
k-NN	1976(2.79)	81.59(1.26)	425(4.17)	71.75(2.22)
k-NCN	1984(3.49)	81.59(1.26)	436(8.95)	**72.01**(2.12)
	Gauss		**German**	
scheme	*edit*	*accuracy*	*edit*	*accuracy*
NOEDIT	4000	64.94(0.90)	800	65.61(2.22)
k-NSN	2592(24.58)	**68.32**(0.90)	544(8.47)	68.41(1.79)
k-NN	2688(17.98)	64.72(0.48)	543(10.09)	69.30(1.02)
k-NCN	2708(16.13)	64.72(0.48)	563(7.47)	**70.61**(1.69)
	Glass		**Heart**	
scheme	*edit*	*accuracy*	*edit*	*accuracy*
NOEDIT	171	65.21(14.95)	216	58.17(5.31)
k-NSN	110(7.86)	60.61(12.59)	131(2.53)	65.91(1.25)
k-NN	109(7.24)	59.66(10.08)	139(1.33)	63.68(1.27)
k-NCN	111(8.26)	**67.11**(11.41)	139(2.58)	**66.25**(3.78)
	Liver		**Phoneme**	
scheme	*edit*	*accuracy*	*edit*	*accuracy*
NOEDIT	216	65.21(7.36)	4323	69.72(7.28)
k-NSN	115(5.38)	63.21(6.32)	3936(44.91)	72.24(6.44)
k-NN	116(7.73)	66.95(6.68)	3898(51.08)	**72.83**(6.29)
k-NCN	120(4.17)	**70.54**(6.26)	3937(52.62)	72.33(6.25)
	Sonar		**Waveform21**	
scheme	*edit*	*accuracy*	*edit*	*accuracy*
NOEDIT	166	52.11(10.75)	3999	77.96(2.58)
k-NSN	136(4.83)	56.49(12.73)	3249(19.81)	80.70(2.05)
k-NN	137(4.82)	**56.97**(13.03)	3250(19.63)	80.70(2.05)
k-NCN	140(4.49)	55.55(13.58)	3245(22.52)	**80.74**(2.00)

In all datasets, edited partitions improve the 1-NN classification results of the corresponding original partitions (**NOEDIT**). Note that in all those cases the number of prototypes of edited partitions is lower than the size of original training partitions. These relations between classification accuracies and sizes are really common, but do not necessarily occur for all datasets. Their presence denote that there is some overlapping that can be removed by edition. Therefore, these datasets can better illustrate the behavior of editing techniques.

With respect to the comparison among the three different editions, it can be observed that they produce similar results both in the number of prototypes removed and in 1-NN classification accuracies. But, in most of cases, the editing scheme derived from k-NCN leads to better accuracies than the other 2 rules and, specifically, than the k-NN. These differences are more notable in the datasets with small number of prototypes (Glass, Heart, Liver). The importance of this observation is that small size datasets are very frequent in real world problems and they are usually a challenge for researchers. In addition, the use of the k-NSN rule also produces good edited partitions, with the lowest number of prototypes in many cases and with a similar accuracy in most of situations. These results confirm the applicability of this new rule for editing tasks.

5 Conclusions, Discussions, and Future Work

An editing process consists basically of removing from a training set those samples which may disorient a classifier training (samples in overlapping regions and outliers). This paper focuses on the comparison of 3 editing methods, which are particular instances of a general editing procedure directly derived from Wilson's scheme. Given that this general procedure allows a classifier as a parameter, each specific editing method is defined by a neighborhood-based classification rule. The 3 classifiers considered are the plain k-NN [3,7] (the original classifier of the Wilson's scheme), the k-NCN [4] and a more recent k-NSN [9]. The k-NCN searches those k neighbors whose centroid is closest to a given sample, while k-NSN uses a fast NN search algorithm to find the k reference neighbors.

Exhaustive experiments were conducted over 12 datasets to compare the performances of these rules when used in editing tasks. A 5-fold cross validation strategy was defined for classifier evaluation. Editing methods were applied on training partitions and resulting edited partitions were used for 1-NN classification of their corresponding test partitions. Although average results were similar among editing methods, the k-NCN was better than k-NN in most of cases, considering the 1-NN classification accuracy. The differences were more significant in datasets with a small number of samples, which is a very frequent situation. Finally, and in spite of its approximated strategy, the k-NSN produces good results both in sizes of edited subsets and in classification accuracies on test partitions.

The main conclusion of this paper is the appropriateness of k-NCN in classification tasks on small size problems with respect to k-NN. An interesting question arises from this fact. How related is this conclusion with samples density? This feature is probably the most important condition in the behavior of k-NN techniques, but depends not only on the number of samples but also on the volume where samples are distributed. So, small size datasets are not necessarily those with low density. Future analysis should involve some measure to evaluate density, and a methodology for relating density, size, and dimensionality with the use of each neighborhood-based classification rule.

Acknowledgments

This work has been supported in part by grant TIC2003–08496 from the Spanish CICYT (Ministerio de Ciencia y Tecnología), GV06/166 from Generalitat Valenciana, and the IST Programme of the European Community, undes the Pascal Network of Excellence, IST-2002-506778.

References

1. Bernardo, E., Ho, T.-K.: On classifier domain of competence, In: Proc. 17th. Int. Conf. on Pattern Recognition, Cambridge, UK (2004) 136–139.
2. Wilson, D.L.: Asymptotic properties of nearest neighbour rules using edited data, IEEE Trans. on Systems, Man and Cybernetics **2** (1972) 408–421.
3. Cover, T.M., Hart, P.E.: Nearest Neighbor Pattern Classification, IEEE Trans. on Information Theory **IT-13(1)** (1967) 21–27.
4. Sánchez, J.S. *et. al.*: Analysis of new techniques to obtain quality training sets, Pattern Recognition Letters **24** (2003) 1015–1022.
5. Duda, R., Hart, P.: Pattern Classification and Scene Analysis. Wiley (1973).
6. Chaudhuri, B.B.: A new definition of neighborhood of a point in multi-dimensional space, Pattern Recognition Letters **17** (1996) 11-17.
7. Dasarathy, B.V.: Nearest Neighbor Norms: NN Pattern Classification techniques. IEEE Computer Society Press (1991), Los Alamos, CA.
8. Devijver, P.A., Kittler, J.: Pattern Recognition: A Statistical Approach. Prentice Hall (1982), Englewood Cliffs, NJ.
9. Moreno-Seco, F., Micó, L., Oncina, J.: Extending fast nearest neighbour search algorithms for approximate k-NN classification, In: Lecture Notes in Artificial Intelligence **2652** (2003) 589–597.
10. Micó, L., Oncina, J., Vidal, E.: A new version of the nearest neighbour approximating and eliminating search algorithm (AESA) with linear preprocessing-time and memory requirements, Pattern Recognition Letters **15** (1994) 9–17.

Mono-font Cursive Arabic Text Recognition Using Speech Recognition System

M.S. Khorsheed

Computer & Electronics Research Institute,
King AbdulAziz City for Science and Technology (KACST)
PO Box 6086, Riyadh 11442, Saudi Arabia
mkhorshd@kacst.edu.sa

Abstract. This paper presents a system to recognise cursive Arabic typewritten text. The system is built using the Hidden Markov Model Toolkit (*HTK*) which is a portable toolkit for speech recognition system. The proposed system decomposes the page into its text lines and then extracts a set of simple statistical features from small overlapped windows running through each text line. The feature vector sequence is injected to the global model for training and recognition purposes. A data corpus which includes Arabic text of more than 100 *A4*−size sheets typewritten in Tahoma font is used to assess the performance of the proposed system.

1 Introduction

Among the branches of pattern recognition is the automatic reading of a text, namely, text recognition. The objective is to imitate the human ability to read printed text with human accuracy, but at a higher speed.

Most optical character recognition methods assume that individual characters can be isolated, and such techniques, although successful when presented with Latin typewritten or typeset text, cannot be applied reliably to cursive script, such as Arabic. Previous research on Arabic script recognition has confirmed the difficulties in attempting to segment Arabic words into individual characters [1].

Hidden Markov Models (HMMs) [2] are among other classification systems that are used to recognise character, word or text. They are statistical models which have been found extremely efficient for a wide spectrum of applications, especially speech processing. This success has motivated recent attempts to implement HMMs in character recognition whether on-line [3] or off-line [4]. The HMM provides an explicit representation for time-varying patterns and probabilistic interpretations that can tolerate variations in these patterns. In off-line recognition systems, the general idea is to transform the word image into a sequence of observations. The observations produced by the training samples are used to tune the model parameters whereas those produced by the testing samples are used to investigate the system performance.

HMMs have been also used to Arabic word recognition. Following are the approaches introduced in twofold: the global approach and the analytical approach. The global approach treats the word as a whole. Features are extracted from the

D.-Y. Yeung et al. (Eds.): SSPR&SPR 2006, LNCS 4109, pp. 755–763, 2006.

unsegmented word and compared to a model [5]. The analytical approach decomposes the word into smaller units, which may correspond to a character or part of a character [6]. Another research [7] proposed a system which depends on the estimation of character models, a lexicon, and grammar from training samples. The training phase takes scanned lines of text coupled with the ground truth, the text equivalent of the text image, as input. Then, each line is divided into narrow overlapping vertical windows from which feature vectors are extracted. The character modelling component takes the feature vectors and the corresponding ground truth and estimates the character models. The recognition phase follows the same step to extract the feature vectors which are used with different knowledge sources estimated in the training phase to find the character sequence with the highest likelihood $P(O|\lambda)$.

This paper presents a HMM–based system to recognise cursive Arabic script offline. Statistical features are extracted from the text line image and fed to the recogniser. The system is built on the Hidden Markov Models Toolkit (HTK) [8]. This is primarily designed for building HMM-based speech processing tools in particular recognisers. The proposed system is lexicon free and it depends on the technique of character models and grammar from training samples.

2 System Overview

Fig 1 shows the block diagram of the proposed system. The global model is a network of interconnected character models. Each character-model represents a letter in the alphabet. The system may be divided into stages. The first stage is performed prior to HTK, and includes: image acquisition, preprocessing and feature extraction. The objective is to acquire the document image, preprocess it and then decompose it into text line images. Each line image is transfered into a sequence of feature vectors. Those features are extracted from overlapping vertical windows along the line image, then clustered into discrete symbols.

Stage two is performed within HTK. It couples the feature vectors with the corresponding ground truth to estimate the character model parameters. The

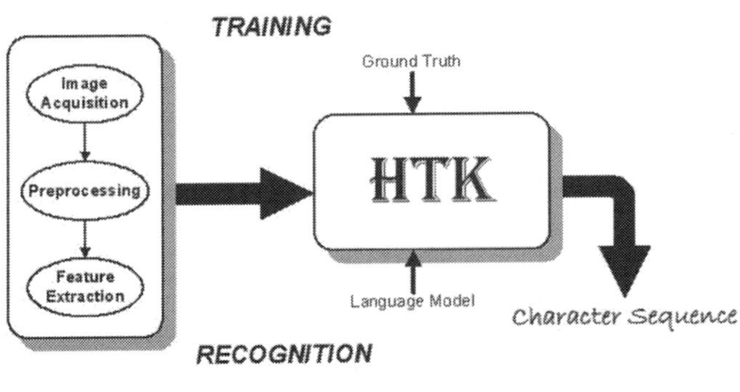

Fig. 1. A block diagram of the HTK-based Arabic recognition system

final output of this stage is a lexicon–free system to recognize cursive Arabic script. During recognition, an input pattern of discrete symbols representing the line image is injected to the global model which outputs a stream of characters matching the text line.

2.1 Feature Extraction

This research implements HMMs to recognise the input pattern. This implies that the feature vector, extracted from the text, is computed as a function of independent variable. In speech, the cepstral features are extracted from the speech signal with respect to time. Similarly, in on–line handwritten recognition a feature vector is computed as a function of time also. In off–line recognition system the case is different; there is no independent variable. Moreover, the whole page image needs to be recognised. In this research, the text line has been chosen as the unit for training and recognition purposes.

Now, assuming that the horizontal position along the text line is the independent variable, a sliding window is scanning the line from right to left. A set of simple features is extracted from pixels falling within that window. The resulting feature vector is mapped against predefined codebook vectors, and replaced with the symbol representing the nearest codebook vector. This step transfers the text line image into a sequence of discrete symbols.

Each line image is divided into overlapped narrow windows, see Fig 2. These windows are then vertically divided into cells where each cell includes a predefined number of pixels. Those cells are used for feature extraction.

Features extracted from the text could be structural [9], spectral [10] or as per here statistical. Statistical features are easy to compute and script independent. They avoid any segmentation at word or character level. These features are:

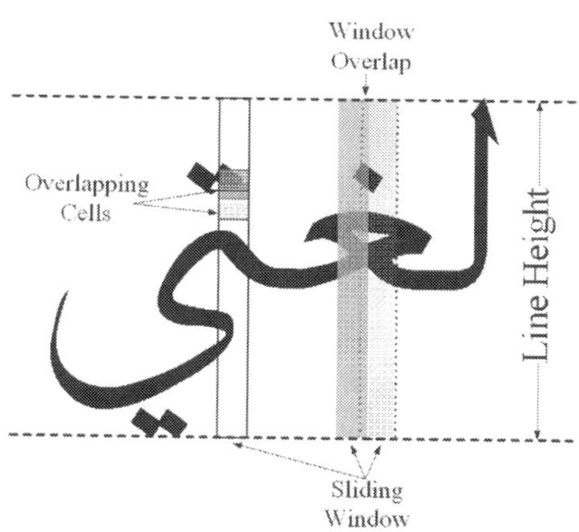

Fig. 2. Dividing the line into windows and cells

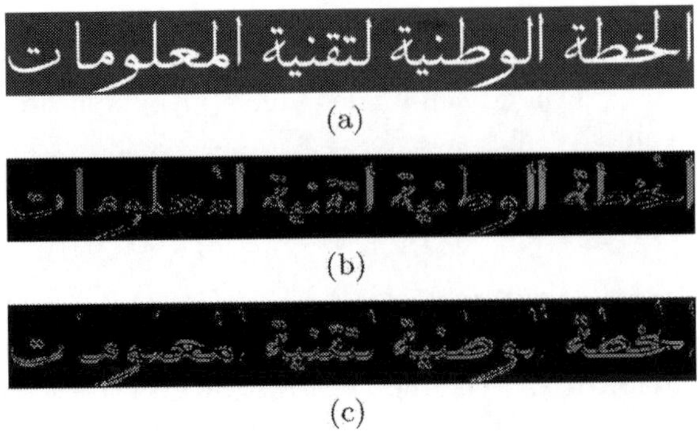

Fig. 3. Feature extraction: a)original line image. b)vertical derivative of (a). c)horizontal derivative of (a).

intensity, *intensity of horizontal derivative* and *intensity of vertical derivative*, see Fig 3.

The intensity feature represents the number of ones in each cell. The intensity of horizontal derivative detects the edge through X–axis, then computes the number of ones in the resulting cell. The intensity of vertical derivative detects the edge through Y–axis, then computes the number of ones in the resulting cell. The feature vector of one narrow window is built by stacking features extracted from each cell in that window.

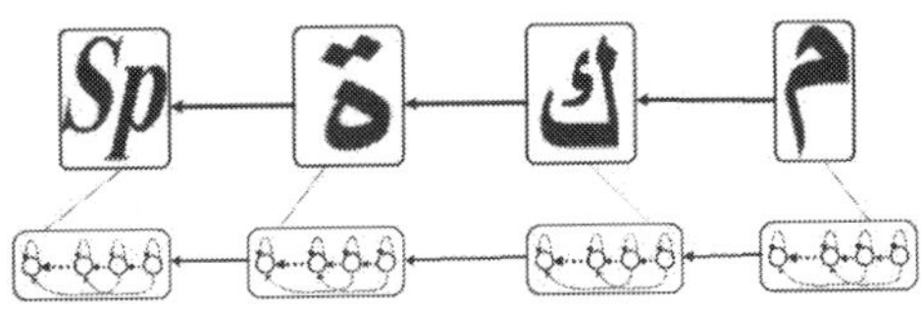

Fig. 4. HMM structure of the word *Makkah* "¡mkT¿", *sp* denotes space character

2.2 HTK Inference Engine

The hidden Markov model toolkit (HTK) [8] is a portable toolkit for building and manipulating hidden Markov models. It is primarily designed for building HMM–based speech recognition systems. HTK was originally developed at the Speech Vision and Robitics Group of the Cambridge University Engineering Department (CUED).

Much of the functionality of HTK is built into the library modules available in C source code. These modules are designed to run with the traditional command line style interface, so it is simple to write scripts to control HTK tools execution. The HTK tools are categorized into four phases: data preparation, training, testing and result analysis tools.

The data preparation tools are designed to obtain the speech data from data bases, CD ROM or record the speech manually. These tools parametrize the speech data and generate the associated speech labels. For the current work, the task of those tools is performed prior to the HTK, as previously explained, then the result is converted to HTK data format.

HTK allows HMMs to be built with any desired topology using simple text files. The training tools adjusts HMM parameters using the prepared training data, representing text lines, coupled with the data transcription. These tools apply the Baum–Welch re–estimation procedure [2] to maximise the likelihood probabilities of the training data given the model.

HTK provides a recognition tool to decode the sequence of observations and output the associated state sequence. The recognition tool requires a network to describe the transition probabilities from one model to another. The dictionary and language model can be input to the tool to help the recogniser to output the correct state sequence.

The result analysis tool evaluates the performance of the recognition system by matching the recogniser output data with the original reference transcription. This comparison is performed using dynamic programming to align the two transcriptions, the output and the ground truth, and then count the number of: *substitution* (S) and *deletion* (D).

The optimal string match calculates a score for matching the output sample with respect to the reference line. The procedure works such that identical labels match with score 0, a substitution carries a score of 10 and a deletion carries a score of 7. The optimal string match is the label alignment which has the lowest possible score. Once the optimal alignment has been found, the *correction rate* (CR) is then:

$$CR = \frac{N - D - S}{N} \times 100\% \qquad (1)$$

where N is the total number of labels in the recogniser output sequence.

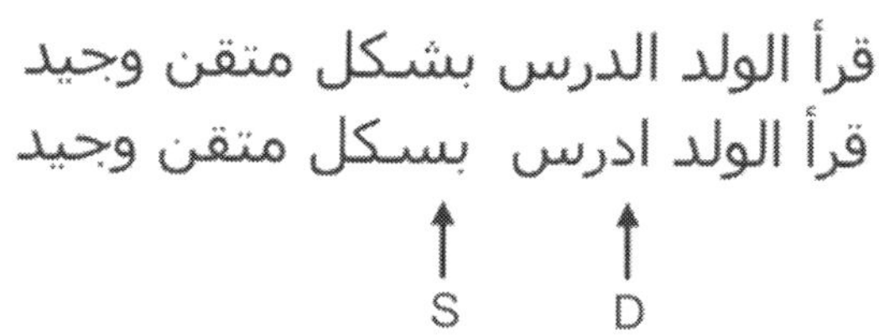

Fig. 5. Optimal string matching. The upper line is the reference line.

Fig 5 illustrates an Arabic reference line and a possible system output. The sentence includes 30 lables. The system output includes one subtitution, one deletion and two insertions. The correction rate equals:

$$CR = \frac{30 - 1 - 1}{30} \times 100 = 93.33\%$$

The example showes how the HTK analysis tool measures the performance of the recognition system.

3 Recognition Results

The performance of the proposed system was assessed using a corpus which includes more than 100 pages of Arabic text in Tahoma font, see Fig 5. Tahoma is a simple font with no overlap or ligature. The data corpus includes 18413 words and 100724 letters, not including spaces.

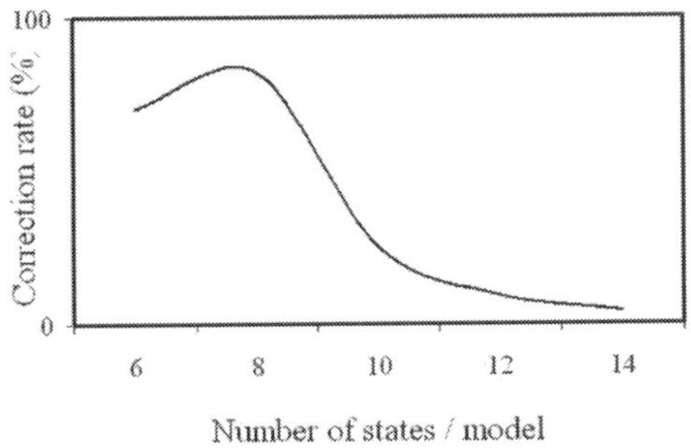

Fig. 6. The impact of the number of states per model on system performance

After noise elimination and deskewing, the 100 pages were decomposed into more than 2500 line images. A set of experiments were performed using 1500 line images for training and 1000 line images for testing. All character models had the same number of states; 8 states. There is no mathematical method to calculate the optimal number of states per model. Alternatively, various values were examined to select the best number of states per model, see Fig 6.

Text line height is proportional to the font size. The line image is measured in pixels. For the corpus under consideration the line image height varies from 35 pixels to 95 pixels depending on the font type and size. To eleminate this dependency, all line images were resized to a single height value; 60 pixels. This

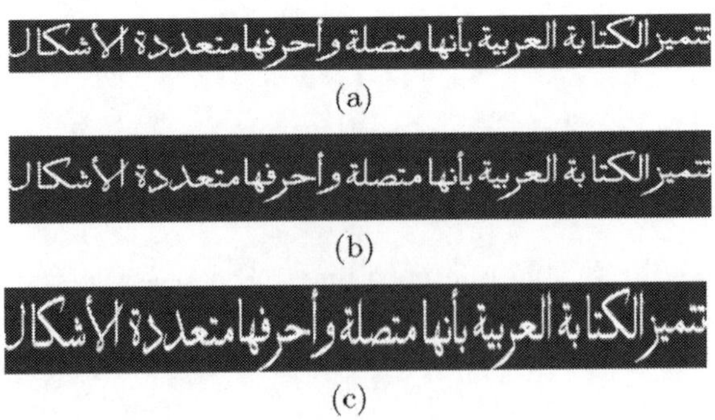

Fig. 7. Text lines with different line heights: (a) 43 pixels, (b) 60 pixels and (c) line image in (a) resized to 60 pixel size

value equals the mean of image heights of all text lines in the data corpus, see Fig 7.

3.1 Cell Size

At any horizontal position, the sliding window is divided into a number of cells, as shown in Fig 2. These cells may or may not overlap. The overlap can be vertical or horizontal and it increases the amount of features generated from a single line and hence increases the processing time.

Table 1. Cell size categories

Category	Cell Size	Horizontal Overlap	Vertical Overlap
CS_1	3×3	-	-
CS_2	3×3	1 pixel	-
CS_3	3×3	2 pixels	-
CS_4	5×5	2 pixels	-
CS_5	5×5	2 pixels	2 pixels

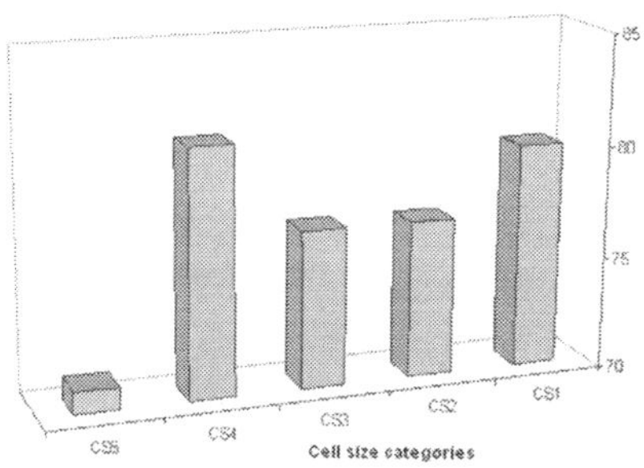

(a) All categories

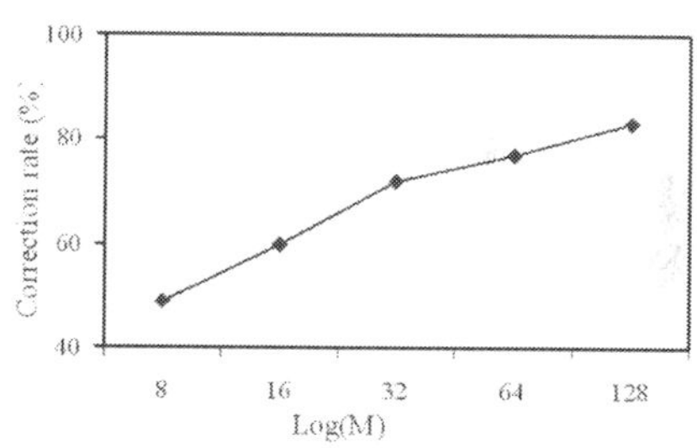

(b) $CS2$

Fig. 8. System performance for the Tahoma font

The first set of experiments studied the cell size parameter. Variuos combinations of cell sizes and overlaps were tested, see Table 1.

Fig 8$-a$ shows the correction rate, CR, of the five categories for the Tahoma font. The codebook here includes 64 clusters. The results illustrates that HTK was more efficiently tuned for Tahoma font rather than for Thuluth font. This is sensible due to decorative curves found at the end of some letters in Thuluth font which overlap with succeeding letters.

3.2 Codebook Size

After extracting features from a line image, the feature vector dimensionality is reduced from 2D to 1D using K-means clustering algorithm [2]. Mapping the 2D feature vector to the nearest cluster is based on the minimum Euclidean distance measure.

This set of expriments studied the codebook size parameter. Five different codebook size values were examined for each font for each category: $8, 16, 32, 64$ and 128. A total number of 50 codebooks were created for this purpose. Fig $8-b$ shows the correction rates of CS_2. It illustrates the improvement in the system performance as the codebook size increases. This conclusion is also applicable to other cell size categories.

3.3 Tri–HMMs

Character models implemented so far are independent; isolated from preceding and succeeding character models. This type is referred to as a mono–model. It has two main advantages: (1) it is easy to train since the total number of models is small; 60 models, (2) the labelling procedure is simple and straightforward. A more efficient type, though more complex, is referred to as a tri–model. Here, each HMM represents a character, its predecessor and its successor. The total number of HMMs jumps to 9393 models. Training and recognition procedures are the same for both types. However, the labelling procedure is more complex with tri-models. The context-dependant tri–HMMs increases the system performance from 84% to 92%.

4 Conclusion

A new system to recognise cursive Arabic text has been presented. The proposed system is based on HMM Toolkit basically designed for speech recognition purpose. Various model parameters have been studied using a corpus that includes data typewritten in a computer–generated font; Tahoma. The system was capable to learn complex ligatures and ovelaps. The system performance has been improved when implementing the tri–model scheme. Future work will concentrate on enlarging the data corpus to include more fonts.

References

1. M. S. Khorsheed, "Off-line Arabic character recognition – A review," *Pattern Analysis and Applications*, vol. 5, no. 1, pp. 31–45, 2002.
2. L. Rabiner and B. Juang, *Fundamentals Of Speech Recognition.* Prentice Hall, 1993.
3. H. Kim, K. Kim, S. Kim, and J. Lee, "On-line recognition of handwritten chinese characters based on hmms," *Pattern Recognition*, vol. 30, no. 9, pp. 1489–1500, 1997.
4. W. Kim and R. Park, "Off-line recognition of handwritten korean and alphanumeric characters using hmms," *Pattern Recognition*, vol. 29, no. 5, pp. 845–858, 1996.
5. M. Pechwitz and V. Maergner, "Hmm based approach for handwritten arabic word recognition using ifn/enit database," in *The 7th International Conference on Document Analysis and Pattern Recognition*, pp. 890–894, 2003.

6. T. Sari, L. Souici, and M. Sellami, "Offline handwritten arabic character segmentation algorithm: Acsa," in *The 8th International Workshop on Frontiers in Handwriting Recognition*, pp. 452–457, 2002.
7. I. Bazzi, R. Schwartz, and J. Makhoul, "An omnifont open-vocabulary OCR system for English and Arabic," *IEEE Trans. on Pattern Analysis and Machine Intelligence*, vol. 21, no. 6, pp. 495–504, 1999.
8. S. Young, G. Evermann, D. Kershaw, G. Moore, J. Odell, D. Ollason, V. Valtchev, and P. Woodland, *The HTK Book*. Cambridge University Engineering Dept., 2001.
9. M. S. Khorsheed, "Recognising handwritten Arabic manuscripts using a single hidden markov model," *Pattern Recognition Letters*, vol. 24, no. 14, pp. 2235–2242, 2003.
10. M. S. Khorsheed, "A lexicon based system with multiple hmms to recognise typewritten and handwritten Arabic words," in *The 17th National Computer Conference*, (Madinah, Saudi Arabia), pp. 613–621, 5-8 April 2004.

From Indefinite to Positive Semi-Definite Matrices

Alberto Muñoz[1] and Isaac Martín de Diego[2]

[1] University Carlos III de Madrid, c/ Madrid 126, 28903 Getafe, Spain
alberto.munoz@uc3m.es
[2] University Rey Juan Carlos, c/ Tulipán s/n, 28933 Móstoles, Spain
isaac.martin@urjc.es

Abstract. Similarity based classification methods use positive semi-definite (PSD) similarity matrices. When several data representations (or metrics) are available, they should be combined to build a single similarity matrix. Often the resulting combination is an indefinite matrix and can not be used to train the classifier. In this paper we introduce new methods to build a PSD matrix from an indefinite matrix. The obtained matrices are used as input kernels to train Support Vector Machines (SVMs) for classification tasks. Experimental results on artificial and real data sets are reported.

1 Introduction

Classification methods generally rely on the use of a (symmetric) similarity matrix. In many situations it is convenient to consider more than one similarity measure. For instance, in Web Mining problems we have an asymmetric link matrix among Web pages, A. A_{ij} is 1 when there is a link between page i and page j and it is 0 when there is not a link. Two different matrices are defined from A: the co-citations ($A^T A$) and co-references (AA^T) matrices. Another matrix is defined from the terms by documents (or web pages) matrix, D. $D_{ij} = 1$ if term i appears in web page j and it is 0 when it does not appear. The 'document by document' matrix is defined by $D^T D$. The co-citations, co-references and 'document by document' matrices correspond to different similarity representations focusing on different data aspects. Several methods have been proposed to combine similarity matrices [11,8,10] in order to create a new single representation for which a classifier is trained. If the similarity representations are not equivalent, a better classification performance should be expected if we are able to combine them. Often, the resulting combination matrix is not positive semi-definite (PSD), that is, it has one or more positive eigenvalues and one or more negative eigenvalues. Then, it is not possible to embed the data into a Euclidean space. The combination matrix is not appropriate to train most used classifiers, and thus it must be modified.

In this paper we afford a deep review of the existing techniques to obtain a PSD matrix from an indefinite one, and propose new methods specially useful for classification tasks. The process of obtaining a PSD matrix from an indefinite

D.-Y. Yeung et al. (Eds.): SSPR&SPR 2006, LNCS 4109, pp. 764–772, 2006.

matrix will be called *Euclideanization* in the following. We will use the resulting matrix as kernel to train a Support Vector Machine (SVM) classifier.

The rest of the paper is organized as follows. In Section 2, we review the existing Euclideanization methods. In Section 3 we propose several Euclideanization methods, adapting them to the classification context. The experimental setup and results on artificial and real classification problems are described in Section 4. Section 5 concludes.

2 Classical Methods

Let K be a real $n \times n$ symmetric indefinite matrix. By the spectral decomposition theorem K can be written as $K = U_n \Lambda_n U_n^T = \sum_{i=1}^n \lambda_i u_i u_i^T$, where Λ_n is a diagonal matrix of eigenvalues of K (first, p positive eigenvalues with decreasing values, next q negative ones with decreasing magnitude, and finally, zero values), and U_n is an orthogonal matrix whose columns u_i are standarized eigenvectors.

2.1 Multidimensional Scaling

The first Euclideanization method considers the matrix $Z = U_r \Lambda_r^{\frac{1}{2}}$, where $r \leq p$ [2]. The new matrix is defined as follows:

$$K^*_{MDS} = ZZ^T = U_r \Lambda_r U_r^T \,. \tag{1}$$

This is equivalent to consider only those eigenvalues larger than a positive constant ϵ, (if $\epsilon = 0$, then $r = p$). In the case of indefinite matrices, the magnitudes of negative eigenvalues suggest the deviation from Euclideaness [13]:

$$r_{mm} = 100 \frac{|\lambda_{min}|}{\lambda_{max}} \,, \quad r_{neg} = 100 \frac{\sum_{\lambda_i < 0} |\lambda_i|}{\sum_{i=1}^n |\lambda_i|} \,. \tag{2}$$

Now, consider a classification problem involving a sample $x_1, \ldots, x_n$ and an indefinite kernel matrix K, where $K_{ij} = K(x_i, x_j)$. In order to use the kernel matrix K^*_{MDS} with an SVM classifier, we should be able to calculate $K^*(x, x_i)$ for a new point x, given that the SVM classifier takes the form $f(x) = b + \sum_i \alpha_i K^*_{MDS}(x, x_i)$. Let $K(x, \cdot) \in \mathbb{R}^{1 \times n}$ be the vector of original kernel values for the new point, then:

$$K^*_{MDS}(x, \cdot) = K(x, \cdot) U_r \Lambda_r^{-\frac{1}{2}} \Lambda_r^{\frac{1}{2}} U_r^T = K(x, \cdot) U_r U_r^T \,. \tag{3}$$

2.2 Pseudo-euclidean Space

An alternative solution to MDS is to use both positive and negative eigenvalues of K to represent the data set in a Pseudo-Euclidean space [3], $Z = U_k |\Lambda_k|^{\frac{1}{2}}$, where $k = p + q$. The new matrix is defined as follows:

$$K^*_{Pseudo} = ZZ^T = U_k |\Lambda_k| U_k^T \,. \tag{4}$$

In this case, the kernel expression for new points is given by [12]:

$$K^*_{Pseudo}(x,\cdot) = K(x,\cdot)U_k|\Lambda_k|^{-\frac{1}{2}}M|\Lambda_k|^{\frac{1}{2}}U_k^T = K(x,\cdot)U_kMU_k^T\,, \quad (5)$$

where $M = \begin{bmatrix} I_p & 0 \\ 0 & -I_q \end{bmatrix}$. An alternative method is to consider a positive constant ϵ and only those eigenvalues such that $|\lambda_i| \geq \epsilon$.

2.3 Adding a Quantity to the Diagonal of the Matrix

In this method, a positive constant λ is added to the diagonal of the original matrix, large enough to make positive all the eigenvalues of the kernel matrix ($\lambda > |\lambda_{min}|$ will do):

$$K^*_{Add} = K + \lambda I = U(\Lambda + \lambda I)U^T\,. \quad (6)$$

3 Alternative Methods

3.1 Square Transformation

First, we propose a very intuitive, computationally cheap and free parameter method to build a kernel matrix from a symmetric indefinite matrix K as follows:

$$K^*_{ST} = K^2 = KK = U\Lambda U^T U\Lambda U^T = U\Lambda^2 U^T\,. \quad (7)$$

For the new points the kernel values can be calculated by: $K^*_{ST}(x,\cdot) = K(x,\cdot)K$.

3.2 Bending

Hayes and Hill propose in [4] a method termed 'bending' for the modification of estimates of covariance matrices in the construction of genetic selection indices. Bending is an iterative process of updating a matrix when a weighting matrix is given to control the relative importance of the elements of the original matrix [6]. Let K be a symmetric indefinite matrix and let W be a weighting matrix for the elements of K. The Bending process is resumed in Algorithm 1.

Let $n = 0$, $K_0 = K$, ε a positive constant and $\odot$ denotes the Hadamard product.
while K_n is indefinite **do**
  Calculate the decomposition: $K_n = U_n\Lambda_n U_n^T$.
  Replace Λ_n with Λ_n^*, where $\lambda_i^* = g(\lambda_i, \varepsilon)$.
  Calculate a new matrix: $K_{n+1} = K_n - \left[K_n - U_n\Lambda_n^* U_n^T\right] \odot W$.
  $n = n + 1$.
end while

Algorithm 1. Bending

In the original Bending algorithm, $g(\lambda_i, \varepsilon) = max(\lambda_i, \varepsilon)$. Alternatively, we propose to use $g(\lambda_i, \varepsilon) = max(|\lambda_i|, \varepsilon)$ as in Pseudo method, $g(\lambda_i, \varepsilon) = \lambda_i + \lambda$ as in the method of adding a constant to the eigenvalues, or $g(\lambda_i, \varepsilon) = \lambda_i^2$ as in the Square Transformation (ST) method. If the weighting matrix W is such that all its elements are equal, then the Bending method is equivalent to the MDS method. If $W_{ij} = 0$ then the value in K_{ij} does not change at this step. We propose to calculate $K^*(x, \cdot)$ for a new point x in a similar way to the MDS method. Then, to modify $K^*(x, \cdot)$, a weighting matrix for the new points should be known in advance:

$$\begin{aligned} K_{n+1}(x,\cdot) &= K_n(x,\cdot) - (K_n(x,\cdot) - K^*_{MDS}(x,\cdot)) \odot W(x,\cdot) \\ &= K_n(x,\cdot) - \left(K_n(x,\cdot) - K_n(x,\cdot)UU^T\right) \odot W(x,\cdot) \\ &= K_n(x,\cdot) - K_n(x,\cdot)\left(I - UU^T\right) \odot W(x,\cdot)\,. \end{aligned} \tag{8}$$

The Bending method can be used when we are dealing with a distance matrix but the matrix under consideration is an indefinite matrix, to guarantee that the diagonal elements of the PSD final matrix become 0.

3.3 Alternating Projections

Alternating Projections [14] is a theoretically powerful method for computing best approximations from a closed convex set K that is the intersection of a finite number of closed convex sets, $K = \cap_{m=1}^{M} K_m$. We will use this method to find the nearest matrix at the intersection of the sets of PSD matrices and the matrices which diagonal elements are fixed to a given value. This method works as an iterative algorithm that reduces the problem to find best approximations from the individual sets.

Consider the following problem:

$$\begin{array}{ll} min_A & \|K - A\|_W \\ s.t. & A = A^T\,, \\ & A \succeq 0\,, \\ & diag(A) = c\,, \end{array} \tag{9}$$

where $\|X\|_W = \|W^{1/2}XW^{1/2}\|_F$, $\|\cdot\|_F$ denotes the Frobenius norm, W is a symmetric PSD matrix, and c is a vector in $\mathbb{R}^n$. This problem appears in the finance industry when given a symmetric matrix K (for example correlations between stocks), the nearest symmetric PSD matrix K^* with unit diagonal (the nearest correlation matrix) is required.

The solution to (9) is a matrix in the intersection of the set of symmetric PSD matrices (S) and the set of symmetric matrices with diagonal equals to the vector c (U), that is closest to K using a weighted Frobenius norm. Since S and U are both closed convex sets, it can be shown that the minimun in (9) is achieved and the solution is unique [7].

Let P_S and P_U be the projections onto S and U respectively. To find the nearest matrix in the intersection of the sets S and U we can iteratively project by repetating the operation:

$$A \leftarrow P_U(P_S(A)) \,. \tag{10}$$

It can be shown [5] that:

$$P_S(A) = W^{-1/2}\left((W^{1/2}AW^{1/2})^*_{MDS}\right)W^{-1/2} \,. \tag{11}$$

In practice, we suggest to use a diagonal matrix W. Then, it it easy to show that:

$$P_U(A) = A - (diag(A) - c) \,. \tag{12}$$

For a new point x, a matrix of weights $W(x,\cdot)$ is needed. We propose to calculate $P_S(K(x,\cdot))$ as in the MDS method in (11), and to obtain $P_U(P_S(K(x,\cdot)))$, only the values of c for the new point are needed.

Next, we present two new Euclideanization methods. In the first one, a kernel matrix is modified to be as similar as possible to the indefinite matrix, without losing the PSD property. In the second method, a linear combination of kernels as similar as possible to the original indefinite matrix is built.

3.4 Conformal Transformation

Let A be a given PSD matrix similar to an indefinite matrix K. In our context A could be the average of several kernel matrices (see Section 4 for details). Let W be a diagonal matrix in $\mathbb{R}^{n\times n}$. Consider the problem:

$$min_W \; \|K - WAW\|^2_F \,. \tag{13}$$

Note that if A is PSD, so is WAW. Given a PSD matrix A and an indefinite matrix K, we look for a Conformal Transformation (CT) of A such that the resulting matrix K^* is the closest to the input matrix K.

We propose an iterative method to solve problem (13). W is initialized as the identity matrix of order n. The elements of W are modified iteratively (adding or subtracting a fix constant), while a better approximation between matrices WAW and K (a lower Frobenius norm value) is being obtained. The w value for a new point x is:

$$w_x = \frac{\sum_{i=1}^n w_i K(i,x)A(i,x)}{\sum_{i=1}^n (w_i A(i,x))^2} \,, \tag{14}$$

where $w = diag(W) \in \mathbb{R}^{n\times 1}$.

Instead of a diagonal matrix, a more complicated expression for W could be used in (13). Note that $WAW = A \odot w * w^T$. Instead of using a matrix defined by a single column w, we extend our method by considering a matrix $V \in \mathbb{R}^{n\times r}$ of r columns of weights. The expression for the new matrix is $A \odot V * V^T$.

3.5 Conformal Linear Combination

Let K an indefinite matrix and let $K_1, \ldots, K_M$ a set of M PSD matrices (kernels). Consider the problem of finding the PSD linear combination of those matrices, closest to K:

$$\begin{array}{ll} min_{\lambda_m} & \|K - \sum_{m=1}^{M} \lambda_m K_m\|_F^2 \\ s.t. & \sum_{m=1}^{M} \lambda_m = 1\,, \\ & \lambda_m \geq 0 \quad \forall m = 1, \ldots, M\,. \end{array} \tag{15}$$

It is easy to show that this problem is equivalent to a simple quadratic programming problem. We will label this method as 'conformal linear combination' (CLC). In the particular case of $M = 2$ kernels, the solution is:

$$\lambda_1 = \frac{\langle K_1 - K_2, K - K_2 \rangle}{\|K_1 - K_2\|_F^2}, \quad \lambda_2 = 1 - \lambda_1\,. \tag{16}$$

4 Experiments

To test the performance of the proposed methods, SVMs have been trained on artificial and real data sets using the kernel matrices K^* previously constructed. To evaluate the accuracy of the classifiers, the classification error, the sensitivity: (True '+' recovered/Total true '+') and the specificity: (True '-' recovered)/(Total true '-') measures are used. In all cases, the results have been averaged over 10 runs.

4.1 Artificial Data Set

This data set consists of 400 two-dimensional points (200 per class). Each group corresponds to a normal cloud with mean vector μ_i and diagonal covariance matrix $\sigma_i^2 I$. Here $\mu_1 = (3,3)$, $\mu_2 = (5,5)$, $\sigma_1 = 0.7$ and $\sigma_2 = -0.9$. We have defined two kernels from the projections of the data set onto the coordinate axes. We have used 75% of the data for training and 25% for testing. The interest of this example lies in the fact that, separately, both kernels achieve a poor result (a test error higher than 15%).

We have used the *pick-out* method [11] to combine the two kernels involved. For a pair of elements in the sample, the pick-out method chooses the maximun of the kernels involved if the two elements belong to the same class and the minimun of the kernels under consideration if the two elements belong to different classes. The output matrix obtained is not necessarily PSD. The eigenvalues of the output matrix for the artificial data set are represented on Figure 1. Although the first three eigenvalues are clearly higher than the rest, half of the eigenvalues are negative. The deviation from Euclideaness can be measured using (2): $r_{mm} = 1.6 \pm 0.3$ and $r_{neg} = 2.8 \pm 0.6$ ($mean \pm s.d.$), which suggests a moderate deviation from Euclideaness.

Table 1 shows the classification results. The MDS and Pseudo subscripts represent the value of the positive constant ϵ. The MDS, Bending and AP methods achieve the lowest test error. The support vectors obtained with the MDS method were used to define the weighting matrices needed in the Bending and AP methods. The MDS and Pseudo classification results strongly depend on the value of the parameter ϵ. The best results were achieved using $\epsilon = 5$, which implies the

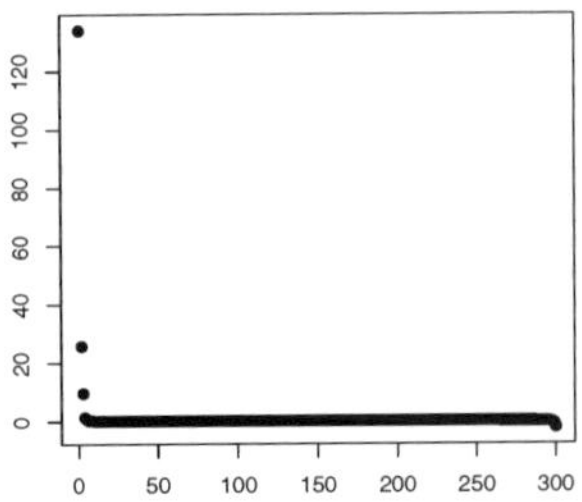

(a) *Eigenvalues*

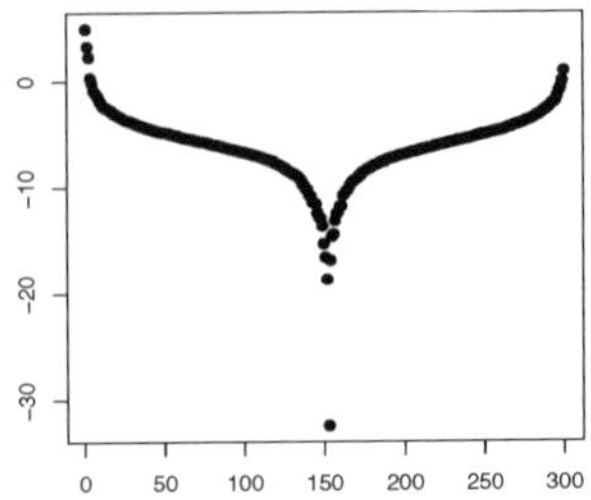

(b) $Log|Eigenvalues|$

Fig. 1. Eigenvalues of the pick-out output matrix for the artificial data set

Table 1. Percentage of misclassified data, sensitivity (Sens.), specificity (Spec.) and percentage of support vectors (S.V.) for the kernels with complementary information

Method	Train	Test	Sens.	Spec.	S.V.
$\mathbf{MDS}_0$	3.0	10.0	0.956	0.845	19.4
$\mathbf{MDS}_1$	3.2	7.1	0.952	0.911	19.7
$\mathbf{MDS}_5$	4.4	4.3	0.935	0.978	21.5
$\mathbf{Pseudo}_0$	5.3	10.8	0.966	0.821	19.4
$\mathbf{Pseudo}_1$	3.6	7.5	0.939	0.913	18.8
Adding λI	5.2	5.2	0.899	0.996	60.1
ST	5.2	7.0	0.951	0.911	3.9
Bending	4.4	4.3	0.935	0.978	21.5
AP	4.4	4.3	0.935	0.978	21.5
AKM	6.4	6.5	0.868	1.000	35.4
$\mathbf{CT}_{AKM}$	6.3	6.1	0.876	1.000	35.1
$\mathbf{CLC}_{AKM}$	6.4	6.7	0.864	1.000	35.9

selection of the two highest eigenvalues. The ST method involves significantly less support vectors than the other methods. On the other hand, adding a quantity to the diagonal of the eigenvalue matrix increases the percentage of support vectors. The conformal transformation method outperforms the average of the kernels method (AKM [10]) when the AKM method was used to initialize the transformation. The starting matrices of the CLC method are the two original kernels. The classification results were similar to that obtained from the AKM. Both kernels, individually, achieve poor classification results, and thus, given the definition of the kernel, it is not possible to define a linear combination of the kernels able to significantly improve the AKM results.

4.2 A Real Data Set Classification Problem

In this section we have dealt with a database from the UCI Machine Learning Repository: the Johns Hopkins University Ionosphere database [1]. The data

Table 2. Percentage of misclassified data, sensitivity (Sens.), specificity (Spec.) and percentage of support vectors (S.V.) for the ionosphere data set

Method	Train	Test	Sens.	Spec.	S.V.
MDS$_0$	1.9	6.4	0.973	0.874	43.0
MDS$_1$	3.3	6.5	0.969	0.878	34.4
MDS$_5$	5.4	7.7	0.964	0.851	34.1
Pseudo$_0$	1.9	6.7	0.952	0.901	44.7
Pseudo$_1$	3.3	6.5	0.969	0.878	34.4
Adding λI	1.6	6.4	0.983	0.855	65.9
ST	2.1	5.9	0.965	0.901	21.5
Bending	2.0	6.0	0.966	0.895	44.5
AP	1.7	5.9	0.977	0.881	43.5
CT$_{RBF}$	4.0	6.0	0.987	0.859	53.0
AKM	2.3	6.7	0.982	0.851	45.0
CLC$_{AKM}$	2.2	5.9	0.980	0.875	45.5

set consists of 351 observations with 34 continous predictor attributes variables each. We have used 60% of the data for training and 40% for testing.

For this data set we have combined several RBF kernels $K_m(x,z)=e^{-\|x-z\|^2/c_m}$ with $c_m = 10 + 5 * (m-1)$ and $m = 1, \ldots, 10$. We have used a linear kernel $K(x,z) = x^T z$, and a polynomial kernel $K(x,z) = (1 + x^T z)^2$ as well. We have considered the following transformation: $K(x,z) = \frac{K(x,z)}{\sqrt{K(x,x)}\sqrt{K(z,z)}}$ to make comparable the different kernels values. The KWS method [9] (Kernel Weighting Scheme) has been used to combine these kernels. In this method we use the kernel value, the neighbourhood of the elements and the label information to assign different weights to each element into the kernel matrix. The output matrix is not necessarily PSD. The percentage of negative eigenvalues is 19.0%, and their relative importance is significant: $r_{mm} = 3.2 \pm 0.3$ and $r_{neg} = 4.6 \pm 0.4$.

The classification results are shown on Table 2. The ST, AP and CLC methods achieve the best results. Similar results were obtained with the CT method which improves the a priori kernel matrix used (RBF with parameter $c = 55$: 7.0% in test error). The CLC method clearly outperforms the AKM method. The performance of MDS and Pseudo methods is related to the choice of the parameter ϵ. The support vectors obtained with the MDS method with $\epsilon = 0$ were used to define the weighting matrices needed in the Bending method. When the diagonal elements of the output matrix were fix to be 1, the AP method outperforms the MDS method.

5 Conclusions

In this paper, we propose new techniques to build a PSD matrix from an indefinite one. The obtained PSD matrix is used as input kernel to train a SVM classifier. The classification results strongly depend on the method used to build the kernel. The Square Transformation method implies the lowest number of support vectors. The Alternating Projections and Bending methods have been shown to be good alternatives to the classical techniques. The Conformal Transformation method clearly improves the results obtained from an a priori kernel. The Conformal Linear Combination method has been shown to be an alternative to the average of the kernels method.

Acknowledgments

This work was partially supported by Spanish grant SEJ2004-03303.

References

1. C.L. Blake and C.J. Merz, C.J. UCI *repository of Machine Learning databases. University of Carolina, Irvine, Department of Information and Computer Sciences. http://www.ics.uci.edu/~mlearn/MLRepository.html,* 1998.
2. T.F. Cox and M.A.A. Cox. *Multidimensional Scaling.* Chapman-Hall, 1994.
3. L. Goldfarb. *A new approach to Pattern Recognition.* Progress in Pattern Recognition, 2, 241-402, 1985.
4. J.F. Hayes and W.G. Hill. *Modification of estimates of parameters in the construction of genetic selection indices ("Bending").* Biometrics, 37 (1981) 483-493.
5. N. Highman. *Computing the nearest correlation matrix- a problem from finance.* IMA Journal of Numerical Analysis, 22 (2002) 329-343.
6. H. Jorjani, L. Klei and U. Emanuelson. *A simple method fot weighted bending of genetic (co)variance matrices.* J. Dairy Sci, 86 (2003) 677-679.
7. D.G. Luenberger. *Optimization by Vector Space Methods.* New York: Wiley, 1969.
8. I. Martín de Diego, J.M. Moguerza and A. Muñoz. *Combining Kernel Information for Support Vector Classificacion.* Proc. MCS (2004), LNCS 3077, Springer, 102-111.
9. I. Martín de Diego, A. Muñoz, and J.M. Moguerza, *Methods for the Combination of Kernel Matrices within a Support Vector Framework.* Submitted.
10. J.M. Moguerza, A. Muñoz and I. Martín de Diego. *Improving Support Vector Classificacion via the Combination of Multiple Sources of Information.* Proc. SSPR and SPR (2004), LNCS 3138, Springer, 592-600.
11. A. Muñoz, I. Martín de Diego and J.M. Moguerza. *Support Vector Machine Classifiers for Assymetric Proximities.* Proc. ICANN (2003), LNCS 2714, Springer, 217-224.
12. E. Pekalska, P. Paclík and R.P.W. Duin. *A Generalized Kernel Approach to Dissimilarity-based Classification.* JMLR, Special Issue on Kernel Methods 2 (2) (2002) 175-211.
13. E. Pekalska, R.P.W. Duin, S. Günter and H. Bunke. *On Not Making Dissimilarities Euclidean.* Proc. SSPR and SPR (2004), LNCS 3138, Springer, 1145-1154.
14. J. von Neumann. *The Geometry of Orthogonal Spaces.* Functional operators-vol.II. Annals of Math. Studies, no. 22. Princeton University Press. 1950.

A Multi-stage Approach for Anchor Shot Detection

L. D'Anna[1], G. Marrazzo[1], G. Percannella[1], C. Sansone[2], and M. Vento[1]

[1] Dip. di Ingegneria dell'Informazione ed Ingegneria Elettrica, Università degli Studi di Salerno
Via Ponte Don Melillo, I, I-84084, Fisciano (SA), Italy
{ldanna, gmarrazzo, pergen, mvento}@unisa.it
[2] Dipartimento di Informatica e Sistemistica, Università degli Studi di Napoli "Federico II"
Via Claudio 21, I-80125 Napoli, Italy
carlo.sansone@unina.it

Abstract. In this paper we present a novel algorithm for anchor shot detection (ASD). ASD is a fundamental step for segmenting news video into stories that is among key issues for achieving efficient treatment of news-based digital libraries.

The proposed algorithm creates a set of audio/video templates of anchorperson shots in an unsupervised way, then classifies shots by comparing them to the templates. Audio similarity is evaluated by means of a new index and helps to achieve better performance than a pure video approach. The method has been tested on a wide database and compared with other state-of-the-art algorithms, demonstrating its effectiveness with respect to them.

1 Introduction

Story segmentation is a basic step towards effective news video indexing. All the solutions to this problem proposed in the literature may be ascribed to one of the two following approaches. According to the first, segmentation is accomplished by directly finding the story boundaries. Such boundaries are typically obtained by looking for the occurrences of some specific event (a sequence of black frames, the co-occurrence of a silence in the audio track and a shot boundary in the video track, etc.), or an abrupt change of some features at a high semantic level, as a topic switch. The main limitation of this approach relies on the fact that the overall performance depends in the first case to the validity of the hypothesis that a story boundary is associated to a specific event in the audio or the video stream, while in the second case to the possibility of reliably deriving high semantic level features.

The other approach performs story segmentation according to the following news program model assumption: given that each shot of the news video can be classified as an anchor shot or a news report shot, then a story is obtained by linking each anchor shot with all successive shots until another anchor shot, or the end of the news video, occurs. Using this model for the stories, news boundaries correspond to a transition from a news report shot to an anchor shot, or from an anchor shot to another. According to the above news story model, automatic anchor shot detection (ASD) becomes the most challenging problem to partition a news video into stories. It has to be noted that the main limitation of this approach relies on the validity of the above described news story model. However, some papers [1,2] have shown that such

D.-Y. Yeung et al. (Eds.): SSPR&SPR 2006, LNCS 4109, pp. 773 – 782, 2006.

a model is valid for most TV networks. Consequently, we preferred to follow this approach, directing our efforts to provide a solution to the ASD problem.

In the scientific literature there are many papers that propose ASD algorithms: the majority exploits only video information. They can be roughly grouped into two categories. First approaches rely on the definition of a set of models of anchor shots, so that ASD is done by matching news video shots with the models [3,4,5]. Typically, a distinctive frame, called *key-frame*, is extracted for each shot and used for detection purposes. In [3] color classification and template matching are used. In [4] a unique anchor shot model is defined, while in [5] an anchor shot is modeled as a sequence of frame models. All these approaches strongly depend on the specific video program model. This is a severe limitation, since it is difficult to construct a general model able to represent all the different kind of news and since the style of a particular news program can change over the time.

In order to overcome this limitation, some authors [6] proposed to build a key-frame model in an unsupervised way. However, since a single model is chosen for each news video, the authors implicitly assume that different anchorperson models share the same background. This is not true for most news stations: because of different camera angles, different models can have different backgrounds. Other authors [2,7] propose unsupervised methods that look for shots with similar visual contents that repeatedly occur during the whole news video. In particular, in [2] a graph-theoretical cluster analysis method is employed to classify video shots. As pointed out by the authors, this approach fails when identical news report shots appear in different stories of the same news program, or when an anchor shot model is present once in a program (for example, the case in which the anchorperson appears for most times in the right or left part of the screen and only once at the center). In [7] shot classification is also performed on the basis of the motion features of the anchor shot. For the authors, it is reasonable to assume that in an anchorperson shot both the camera and the anchorperson are almost motionless. In some TV news, however, zooming effects can be used also in an anchor shot; if motion is contained also in the background, the anchor shot can be missed.

In the last years the use of audio as a good additional source of information for video segmentation has been rapidly raised up. There is, in fact, a number of systems that integrate audio and video features in the context of news segmentation by performing multi-modal analysis [8]. However, the majority of the presented proposals use audio features for directly individuating news boundaries, by means of a silence or a speaker change detector, in order to strengthen or to weaken the boundaries provided by the analysis based on video techniques.

Among the approaches that use the audio track for ASD, in [9] the authors propose a Hidden Markov Model (HMM) to classify frames on the basis of the statistics of the frames present in a news video. The features used are the difference image between frames, the average frame color and also the audio signal. The HMM parameters are evaluated during a training phase by using the ground truth of a given news program, and then are dependent on the style of the specific news video edition. In [10] it is proposed a technique that performs segmentation and clustering of portions of video with similar audio and video contents and tries to find temporal synchronization between pairs of clusters. In particular, ASD is performed by separately clustering and then comparing audio clips and video key-frames. When sufficient overlap is found between an audio and a video cluster then an anchor shot is detected. Unfortunately, parallel audio and video clustering often lead to dissimilar grouping solutions, e. g. when a news report is commented by the anchorperson, or when there is a speaker change within a shot.

All summarizing, the major drawbacks of former approaches are the following: *i*) supervised model-based techniques are not general enough, as they require *a priori* definition and construction of an anchor shot model; *ii*) the definition of a unique anchor shot model for a news edition is restrictive and gives rise to missed detections when different backgrounds are present in a news video edition; finally, *iii*) indiscriminate use of audio information is not effective due to its incoherence with video and then yields to a misleading shot classification.

In this paper, we propose a two stage audio/video ASD method that is able to overcome all the above limitations. In the first stage the method builds in an unsupervised way a set of templates, each one representing a different anchor shot model within a video. The second stage uses a video similarity metric to retrieve a set of candidate anchor shots, which might have been missed by the first stage, and classify them by evaluating the audio similarity with respect to the templates. Note that, differently from the formerly described approaches, we do not use audio information as-it-is, but we perform audio-based classification only on the set of candidate shots and by employing a suitably defined similarity metric.

We tested our method on a significant database made up of several video news editions from the two main Italian broadcasters (RAI 1 and CANALE 5), obtaining a very good performance. Moreover, we also compared our algorithm with other three state-of-the-art unsupervised ASD algorithms achieving significant performance improvements.

The organization of the paper is as follows: in Section 2, the proposed algorithm is described; in Section 3, the database used is reported together with the tests carried out in order to assess the performance of the proposed algorithm; finally, in Section 4, some conclusions are drawn.

2 The Proposed Approach

As stated in the introduction, we propose a two stage analysis which is able to achieve ASD in an unsupervised way. The first stage extracts a set of audio/video templates from the news video under analysis, so avoiding any training procedure or manual definition of the templates. The second stage selects a set of candidate anchor shots by means of a video similarity metric with respect to the templates, then validates them by exploiting both audio similarity and the presence of faces. Details about template nature and similarity metrics used for shot classification will be given in next subsections.

2.1 First Stage: Anchor Shot Template Extraction

The preliminary task is to define and build a set of templates as audio/video touchstones for all the shots in the TV news program. The main difference with the previous papers in the literature is the definition of several anchor shot templates: in fact, during a news program there are typically several kinds of anchor shot camera settings, due to different angle, background and distance with respect to anchor person. A single anchor shot template would not match all these situations and would introduce many missed items in the anchor shot detection task. In our approach each template practically corresponds, from the video point of view, to a single frame of the considered shot (i.e., its key-frame), while, from the audio point of view, to the audio shot characterization in a given feature space.

After shot segmentation, all video shots are processed in order to build the templates. As anchor shots are mainly identifiable from their high visual similarity, we preliminarily need a clustering technique in order to group similar shots and find candidate anchor shot clusters. Then, several heuristics can be used to discard false detected clusters: anchor shots occur at least two times with the same angle, have a large temporal spanning along the news program and are characterized by the presence of a face. All these features are taken into account in the extraction of the anchor shot templates.

A first clustering task groups together shots with same visual appearance. We used a graph-theoretical clustering (GTC) analysis, which considers shot key-frames as nodes of a complete graph in a given feature space. Each edge of the graph is assigned a weight corresponding to a distance between pairs of nodes, then the *minimum spanning tree* (MST) is built on the graph. The distance between nodes is defined in this way: each key-frame is divided into 16 rectangular regions of the same size, then color histograms of corresponding regions of the two nodes are compared; the eight regions with the most similar histograms are individuated and finally the sum of their histograms differences is considered as the distance between nodes. After constructing the MST and removing all the edges in the tree with weights greater than a threshold λ, a *forest* containing a certain number of subtrees (*clusters*) is obtained. Each cluster correspond to a group of visually homogeneous shots. As anchor shots occur repeatedly during a news edition, clusters with less than 2 nodes are discarded. For determining the optimal value of λ, we used a *Fuzzy C-Means* clustering algorithm [11]. This algorithm builds a list of all edges, sorted by their weights, and finds the best separation threshold which partitions the list in two parts. Edges belonging to the sub-list with the highest weights are pruned.

The second step discards the clusters with low *lifetime*, i.e. the time interval that includes all the shots of the cluster. Anchor shot occurrences are typically temporally sparse, so clusters with lifetime lower than a threshold δ are removed. Threshold setting can be performed in a straightforward way, on the basis of the length of the specific news program: some details about this point are provided in the next section. In addition, it is worth noting that this setting is not critical, since anchor shots missed because of a non optimal setting of the threshold, can be recovered by the successive stage. Lifetime analysis is very effective in managing interviews during news reports, where recurrent camera shots on the interviewed person may generate large clusters; checking the lifetime allows to classify the shots within such clusters as news reports since their lifetime is typically very small.

The last step discards the shots without faces. We introduced a robust face detection method which requires presence of a face all along a shot. To this aim we extract three frames from each shot (the first, the middle and the last frames) and apply the face detection algorithm [12] on them. If there are more than one frame without faces then the shot is removed.

After this step, we can again eliminate clusters with less than 2 shots, since they violate the assumption that an anchorperson occurs at least twice in the video. Moreover, some clusters may have changed their lifetime, so we apply again the lifetime control. This procedure gives rise to the final set of anchor shot clusters.

Finally, for each remained cluster we build a set N_{As} containing all the shots belonging to that cluster and extract a unique key-frame from each N_{As}: these key frames represent our anchor shot templates from a video point of view. Moreover, from an audio point of view, we assume that the visual presence of the anchorperson corresponds to his/her speech. Consequently, we assume as audio template the union of the audio portions relating to all the shots belonging to the set N_{As}, represented in an adequate feature space. Figure 1 summarizes the complete template extraction algorithm.

1) Construct a complete graph G such that:
   a) its nodes Kf_i correspond to the key-frames of each shot;
   b) each of its edges $e_{ij}=(Kf_i, Kf_j)$ is characterized by a weight $w_{ij}=d(Kf_i, Kf_j)$, where $d(Kf_i, Kf_j)$ is calculated on the basis of the color histogram differences between Kf_i and Kf_j;
2) determine the minimum spanning tree (MST) of G;
3) remove from the MST those edges with *large* weights by using the *Fuzzy C-Means* algorithm, in order to create shot clusters;
4) remove clusters with only one node;
5) remove all clusters with lifetime lower than a threshold δ;
6) extract 3 frames from each shot S_j belonging to the set of remaining clusters;
7) remove from clusters those shots which have less than 2 frames which contain a face;
8) apply steps 4) and 5) to remaining clusters;
9) for each remained cluster build the set N_{As} by extracting all anchor shots from that cluster.
10) extract a unique key-frame from each cluster and the whole audio track from the set N_{As}, giving rise to the set of audio/video anchor shot templates.

Fig. 1. The proposed template extraction algorithm

2.2 Second Stage: Shot Classification

Classification must be performed on all the shots outside the final set of clusters, as the first stage might have missed those anchor shots occurring only once along the news program or which have low lifetime. These shots can be seen as missed templates of anchor shots which correspond to special camera settings, due to lighting, angle or special background with respect to the anchorperson. However, we observed that during a news edition there are only few of such special kinds of anchor shot, so we expect that only few candidate shots need to be recovered. Consequently, we perform a preselection of these candidates on the basis of the video similarity with respect to the already found templates. Finally, the candidates are classified by both audio and face detection. This framework prevents us from performing audio classification on the whole set of shots, as audio processing is highly time-consuming. Furthermore, audio classification would also bring to false detected anchor shots, due to cases where the anchorperson directly comments a news report.

During the candidate selection step, we consider one template at a time and compare it to the key-frame of each of the discarded shots: then, we select only the three candidate shots with the highest similarity with respect to any of the templates. To define similarity we use the metric presented in [13], which is computationally efficient and sensible to global features such as the general studio setting. The need for a more global metric lies in the fact that if we used the same technique as in the clustering step, we would discard the same shots as in the first phase; moreover, a more global similarity metric is more permissive and let us take into account, for the classification task, also those shots which are not so strictly related to the templates.

Audio classification requires shot characterization in an adequate feature space. By using the results reported in [14], a set of 48 features (20 MFCC, 14 LPCC, 14 PF, see [14] for further details) is extracted from each frame in the audio track of the shots. In this case, a frame is a segment of 1024 audio samples.

After feature extraction, each shot is filtered in order to remove non-voiced and silence portions by means of an appropriate masking module. This is a useful step because voiced segments correspond to vocal cords movement, so they characterize a speaker from an acoustic point of view. Moreover, filtering out audio portions helps in speeding-up the shot classification process. Our masking module is based on the analysis of Energy and ZCR (Zero Crossing Rate) and segments audio into three classes of clips: voiced, non-voiced and pause. Our method is an improvement of the method presented in [15], obtained by discarding frequent audio class transitions and by merging those voiced clips which are separated by a single pause segment.

Audio shot classification is carried out by computing for each shot to be classified the value of an adequate similarity index, namely *D-index*, which expresses similarity between a shot and a template from an audio point of view.

Each shot S_i is considered as a cluster of audio feature vectors, so it can be represented by its centroid, namely C_i. For a generic pair of shots (S_m, S_k) belonging to N_{As} we assume:

$$D_{m,k} = 1 - \frac{d(C_m, C_k)}{d_{\max}} \tag{1}$$

where $d(C_m, C_k)$ is the Euclidean distance between C_m and C_k, while d_{max} is the diameter of the cluster of centroids of the shots belonging to N_{As}.

```
Candidate selection
1) Select a template;
2) compare all the discarded shots with the template;
3) build the candidates list Lc of the three most similar shots (accord-
   ing to a suitably defined similarity metric), sorted in descending
   order;
4) repeat steps 1), 2) and 3) for every template and refresh Lc;
Classification
5) Extract audio feature vectors for all shots in NAs and Lc;
6) remove unvoiced and silent segments by means of a masking module;
7) calculate Di for each candidate shot Si, by comparison with the audio
   templates and make audio classification decision for Si;
8) apply face detection to candidates;
9) apply AND rule between face detection and audio classification to
   reach the final decision.
```

Fig. 2. The shot classification algorithm

Consequently, we can consider d_{max} as the diameter of the set of templates. Given a generic shot S_i to be classified, we calculate its *D-index*, namely D_i, as the average of all the $D_{i,k}$ obtained by considering all the shots S_k in the set N_{As}. If $D_i > 0$ then S_i is classified as anchor shot. The case $D_i > 0$ implies that the average distance between the shot S_i and the set of templates is lower than the diameter d_{max} of the set of templates. Consequently, we classify S_i as an anchor shot. On the contrary, if $D_i < 0$ then the shot S_i is sufficiently far away from the cluster of templates, so it is discarded as a news report shot.

Finally, a further face detection module helps to validate the classification, so that only those candidates which are classified as anchor shots by both audio similarity and face detection are stated as anchor shots.

A scheme of the classification stage is provided in Figure 2.

3 Experimental Results

In order to assess the performance of our approach, we have collected a database composed by several videos from the two main Italian broadcasters, namely, RAI 1 and CANALE 5. Our database includes the main news editions from each broadcaster. A similar dataset, developed by the Linguistic Data Consortium for the TREC Video Retrieval Evaluation contest and composed of 70 hours of news video from ABC and CNN, was unfortunately not publicly available. Table 1 shows some details of the considered database.

Table 1. The video database used in this paper

Broadcaster	No. editions	Total length	No. anchor shots	No. news reports
RAI 1	26	7:32:19	377	4602
CANALE 5	16	9:21:12	250	4269
TOTAL	**42**	**16:53:31**	**627**	**8871**

The performance of our system is expressed in terms of *Precision* and *Recall* [16]. The *F*-measure has also been used, since it combines the former indexes in a single figure of merit according to the following formula:

$F = (2 * Precision * Recall) / (Precision + Recall)$.

In Table 2 the performance of the first stage alone, considered as a preliminary classification stage, are reported.

Table 2. Performance after the first stage of the proposed method

	Precision	*Recall*	*F*
RAI 1	0.994	0.924	0.957
CANALE 5	0.967	0.891	0.927

The second stage classifies the set of three candidates, so recovery is effective only if missed (or "recoverable") anchor shots are selected as candidates. The performance of the second stage is shown in Table 3. In order to evaluate the effectiveness of the audio similarity method proposed in this paper, in Table 3 we have reported also the highest achievable performance, calculated as that obtained if the second stage introduced no further false item and recovered all missed anchor shots. Note that almost all anchor shot are recovered, while only few new falsely detected anchor shots were introduced.

Table 3. Performance of the second stage of the proposed method

	Recovered / Recoverable anchor shots	*Falses / Introducible falses*	*Precision / Max Precision*	*Recall / Max Recall*	*F / Max F*
RAI 1	19/19	5/29	0.968/0.994	0.979/0.979	0.974/0.987
CANALE 5	14/15	7/63	0.940/0.969	0.952/0.956	0.946/0.963

During experimental phase we had to tune the threshold δ on the lifetime value. We verified that δ can change in a wide range with almost no effect on performance, and that its optimal value depends only on the length of the news edition. We fixed δ=2' for news editions shorter than fifteen minutes and δ=4' otherwise. In this sense, our algorithm can be really considered as unsupervised.

The proposed method has also been compared with three state-of-the-art unsupervised ASD algorithms [2, 6, 7]. Each of these algorithms is characterized by several thresholds, so different operating points can be obtained in a *Precision-Recall* plane [16]. We decided to choose the values of the thresholds that maximize *F* over the whole set of videos. This has been done separately for each of the two TV-networks. Obviously, this is an overestimation of the real performance of the algorithms, since such maximization should be done on a different set of news videos. It is also worth noting that, as experimentally demonstrated in [17], the choice of the operating point is crucial for the algorithms, as their performance dramatically depends on the choice of the thresholds.

Table 4. Performance comparison of our algorithm with respect to the algorithms in [2, 6, 7]

	RAI 1			CANALE 5		
	Precision	*Recall*	*F*	*Precision*	*Recall*	*F*
Our algorithm	0.968	0.979	0.974	0.940	0.952	0.946
Gao and Tang [2]	0.827	0.928	0.875	0.810	0.889	0.848
Hanjalic et *al.* [6]	0.681	0.617	0.647	0.889	0.533	0.667
Bertini et *al.* [7]	0.987	0.822	0.897	0.867	0.796	0.830

In Table 4 the performance of our algorithm is compared with those obtained by the methods in [2, 6, 7] on our dataset, so demonstrating its effectiveness with respect to them.

4 Conclusions

In this paper a novel algorithm for anchor shot detection is presented. An effective audio/video anchor shot template matching algorithm is introduced, in order to gain effectiveness against unavoidably missed items left out by pure video analysis. Moreover, a new audio similarity index is discussed, which allows the definition of a synthetic and unique value to quantify resemblance of a generic shot to the template from an audio point of view. The method has been tested on a news video database consisting of about 20 hours, providing significant improvements with respect to other state-of-the-art algorithms.

Future work will include a refinement in the anchor shot audio recognition module, which will be able to detect whether there is one or two anchor persons, so to consequently define a single or a pair of values for the similarity index for each shot.

References

1. M. De Santo, G. Percannella, C. Sansone, M. Vento, "An Unsupervised Shot Classification System for News Video Story Detection", in A.F Abate, M. Nappi, M. Sebillo (eds.) Multimedia Database and Image Communication, World Scientific Publ., pp. 93-104, 2005.
2. X. Gao, X. Tang, "Unsupervised Video-Shot Segmentation and Model-Free Anchorperson Detection for News Video Story Parsing", IEEE Transactions on Circuits and Systems for Video Technology, vol. 12, No. 9, pp. 765 776, 2002.
3. B. Gunsel, A. M. Ferman, A. M. Tekalp, "Video Indexing Through Integration of Syntactic and Semantic Features", Proc. of Workshop Applications of Computer Vision, Sarasota, FL, pp. 90-95, 1996.
4. D. Swanberg, C.F. Shu, R. Jain, "Knowledge Guided Parsing in Video Databases", Proc. of SPIE Symposium on Electronic Imaging: Science and Technology, San Jose, CA, pp. 13-24, 1993.
5. S. W. Smoliar, H. J. Zhang, S. Y. Tao, Y. Gong, "Automatic Parsing and Indexing of News Video", Multimedia Systems, vol. 2, no. 6, pp. 256-265, 1995.
6. A. Hanjalic, R. L. Lagendijk, J. Biemond, "Semi-Automatic News Analysis, Indexing, and Classification System Based on Topics Preselection", Proc. of SPIE, Electronic Imaging: Storage and Retrieval of Image and Video Databases, San Jose (CA), 1999.
7. M. Bertini, A. Del Bimbo, P. Pala, "Content-Based Indexing and Retrieval of TV News", Pattern Recognition Letters, vol. 22, pp. 503-516, 2001.
8. C.G.M. Snoek, M. Worring, "Multimodal Video Indexing: A Review of the State-of-the-art", Multimedia Tools and Applications, vol. 25, pp. 5-35, 2005.
9. S. Eickeler, S. Muller, "Content-based video indexing of TV broadcast news using Hidden Markov Models", ICASSP '99, pp. 2997-3000, 1999.
10. W. Qi, L. Gu, H. Jiang, X. R. Chen, H. J. Zhang, "Integrating Visual, Audio and Text Analysis for News Video", 7th IEEE International Conference on Image Processing, Vancouver, British Columbia, Canada, 2000.
11. J. C. Bezdek, "Pattern Recognition with Fuzzy Objective Function Algorithms", Plenum Press, New York, 1981.
12. P. Viola, M. Jones, "Rapid Object Detection Using a Boosted Cascade of Simple Features", Proc. of the IEEE CVPR Conference, vol. 1, pp. 511-518, 2001.
13. H. Y. Lee, H. K. Lee, Y. H. Ha, "Spatial Color Descriptor for Image Retrieval and Video Segmentation", IEEE Transactions on Multimedia, vol. 5, no. 3, pp. 358-367, 2003.

14. L.P. Cordella, P. Foggia, C. Sansone, M. Vento. "A Real-Time Text-Independent Speaker Identification System". 12th International Conference on Image Analysis and Processing, IEEE Computer Society Press, Mantova, Italy, pp. 632 - 637, 17 - 19 September, 2003.
15. D. Wang, L. Lu, H-J Zhang, "Speech Segmentation Without Speech Recognition", ICASSP 2003 vol. I, pp.468-471, 2003.
16. U. Gargi, R. Kasturi, S.H. Strayer, "Performance Characterization of Video-Shot-Change Detection Methods", IEEE Trans. on Circuits and Systems for Video Technology, vol. 10, no. 1, pp. 1-13, 2000.
17. M. De Santo, G. Percannella, C. Sansone, M. Vento, "A Comparison of Unsupervised Shot Classification Algorithms for News Video Segmentation", Lecture Notes in Computer Science vol. 3138, Springer, Berlin, pp. 233-241, 2004.

An Improved Possibilistic C-Means Algorithm Based on Kernel Methods

Xiao-Hong Wu[1,2] and Jian-Jiang Zhou[1]

[1] College of Information Science & Technology, Nanjing University of Aeronautics and Astronautics, Nanjing, 210016, China
wxhong@nuaa.edu.cn
[2] College of Electrical & Information Engineering, Jiangsu University, Zhenjiang, 212013, China

Abstract. A novel fuzzy clustering algorithm, called kernel improved possibilistic c-means (KIPCM) algorithm, is presented based on kernel methods. KIPCM is an extension of the improved possibilistic c-means (IPCM) algorithm. Different from IPCM which is applied in Euclidean space, KIPCM can make data clustering in kernel feature space. With kernel methods the input data can be implicitly mapped into a high-dimensional feature space where the nonlinear pattern now appears linear. It is unnecessary to calculate in this high-dimensional feature space because we directly calculate inner products from the input data by kernel function. KIPCM can identify clusters of complex shapes and solve nonlinear separable problems better than IPCM and FCM (fuzzy c-means). Our experiments show that the proposed algorithm compares favorably with FCM and IPCM.

1 Introduction

Fuzzy clustering is one of the important unsupervised learning algorithms and fuzzy clustering always has significant advantages over traditional clustering. The well-known fuzzy clustering is the fuzzy c-means (FCM) algorithm [1]. FCM algorithm makes the memberships of a data point across classes sum to 1 by the probabilistic constraint. And FCM is appropriate to interpret memberships as probabilities of sharing. However, the memberships of FCM do not always correspond to the intuitive concept of degree of belong or compatibility. Furthermore, the FCM is sensitive to noises or outliers [2]. To overcome these disadvantages Krishnapuram and Keller have presented the possibilistic c-means (PCM) algorithm [2] by abandoning the constraint of FCM and constructing a novel objective function. The PCM can cluster noisy data and noisy data have low degrees of compatibility in all clusters, so their effects on the clustering can be neglected. But PCM is very sensitive to good initialization and it has an undesirable tendency to produce coincident clusters [3] that because the columns and rows of the typicality matrix are independent of each other. PCM attaches importance to the notion of typicality that alleviates the undesirable effect of noises but neglects the membership that makes the class centroid close to data points. To overcome the shortcoming of PCM, Zhang and Leung have proposed improved possibilistic c-means (IPCM) algorithm [4]. IPCM solves the noise sensitivity defect of FCM, and also overcomes the coincident clusters problem of PCM.

D.-Y. Yeung et al. (Eds.): SSPR&SPR 2006, LNCS 4109, pp. 783 – 791, 2006.

However, FCM, PCM and IPCM have the same drawback that they use point prototypes and a norm-induced distance, as a consequence, they obtain good clustering results only when the data set contains clusters of roughly the same size and shape. To identify clusters of various shapes which are complex topological structures in the same data set, kernel methods [5] has been introduced into fuzzy c-means clustering [6]. In this paper we propose kernel improved possibilistic c-means (KIPCM) algorithm based on kernel methods. With kernel methods the input data samples can be mapped implicitly into a high-dimensional feature space where the nonlinear pattern now appears linear and IPCM algorithm is carried out. We need not calculate in high-dimensional feature space because the kernel function can do it just in input space.

The rest of this paper is organized as follows: in section 2 the IPCM algorithm is introduced, and in section 3 the KIPCM algorithm is presented. Some tests and conclusions are given in later section.

2 Improved Possibilistic C-Means Algorithm

Given an unlabeled data set $\mathbf{X}=\{\mathbf{x}_1,\mathbf{x}_2,\ldots,\mathbf{x}_n\} \subset \Re^p$, find the partition of **X** into 1<c<n fuzzy subsets by minimizing the following objective function

$$J_m(\mathbf{U},\mathbf{T},\mathbf{V}) = \sum_{i=1}^{c}\sum_{k=1}^{n} u_{ik}^m t_{ik} D_{ik}^2 + \sum_{i=1}^{c}\eta_i \sum_{k=1}^{n} u_{ik}^m (t_{ik}\log t_{ik} - t_{ik} + 1) \tag{1}$$

subject to the constraints: $0 \le u_{ik}, t_{ik} \le 1, D_{ik} = \|\mathrm{x}_k - \mathrm{v}_i\|$, and $\sum_{i=1}^{c} u_{ik} = 1, \forall k$.

Where c is the number of clusters and n is the number of data points, u_{ik} is the membership of x_k in class i, t_{ik} is the possibilistic (typicality) value x_k in class i, and m is a weighting exponent, m $\in [1, \infty)$.

Then $\min_{(\mathbf{U},\mathbf{T},\mathbf{V})} J_m(\mathbf{U},\mathbf{T},\mathbf{V})$ is optimized under constraints and the following equations are obtained

$$u_{ik} = \left[\sum_{j=1}^{c} \left(\frac{\eta_i (1-\exp(-\frac{D_{ik}^2}{\eta_i}))}{\eta_j (1-\exp(-\frac{D_{jk}^2}{\eta_j}))} \right)^{\frac{2}{m-1}} \right]^{-1}, \forall i,k \tag{2a}$$

$$t_{ik} = \exp(-\frac{D_{ik}^2}{\eta_i}), \forall i,k \tag{2b}$$

$$\mathrm{v}_i = \frac{\sum_{k=1}^{n} u_{ik}^m t_{ik} \mathbf{x}_k}{\sum_{k=1}^{n} u_{ik}^m t_{ik}}, \forall i \tag{2c}$$

Where v_i is the cluster center or prototype of u_i. η_i in PCM is described as

$$\eta_i = K \frac{\sum_{k=1}^{n} u_{ik}^m D_{ik}^2}{\sum_{k=1}^{n} u_{ik}^m}, K > 0 \tag{3}$$

Here, K is always chosen to be 1 in equation (3) and (4).

If D_{ik} >0 for all i and k, m>1, and **X** contains c<n distinct data points, then the algorithm described below is called IPCM-AO algorithm:

Initialization

1) Fix c,m and w, 1<c<n, 1<m, w <+∞; Set iteration counter r=1 and maximum iteration r_{max};

2) Run FCM until termination to get initial membership $\mathbf{U}^{(0)}$ and initial cluster centers $V^{(0)}$. Then use equation (3) to get η_i.

Repeat

Step 1 Update typicality matrix $\mathbf{T}^{r}$ by equation (2b) ;
Step 2 Update membership matrix $\mathbf{U}^{r}$ by equation (2a);
Step 3 Update $\boldsymbol{V}^{\mathbf{r}}$ by equation (2c);
Step 4 Increment r;

Until ($\left\| \mathrm{U}^{r} - \mathrm{U}^{r-1} \right\| < \varepsilon$) or r> r_{max}

3 Kernel Improved Possibilistic C-Means Algorithm

With the theory of Mercer kernel [7], the input space X is mapped into a novel high dimensional feature space F:

$$\mathrm{X} = (x_1, \ldots, x_M) \rightarrow \Phi(\mathrm{X}) = (\phi(\mathrm{X}_1), \ldots, \phi(\mathrm{X}_N)) \tag{4}$$

Kernel function K satisfies:

$$K(x_i, x_j) = < \phi(x_i) \cdot \phi(x_j) > \tag{5}$$

Scalar product calculation in input space is transformed into kernel function calculation by nonlinear mapping:

$$< x_i \cdot x_j > \rightarrow < \phi(x_i) \cdot \phi(x_j) >= K(x_i, x_j) \tag{6}$$

Then the objective function (1) is transformed as follows

$$\begin{aligned} J_{m,w}(\mathbf{U},\mathbf{T},\mathbf{V}) &= \sum_{i=1}^{c}\sum_{k=1}^{n} u_{ik}^m t_{ik} \left\|\phi(\mathrm{x}_k) - \phi(\mathrm{v}_i)\right\|^2 \\ &+ \sum_{i=1}^{c} \eta_i \sum_{k=1}^{n} u_{ik}^m (t_{ik}\log t_{ik} - t_{ik} + 1) \end{aligned} \tag{7}$$

$$\begin{aligned} \left\|\phi(\mathrm{x}_k) - \phi(\mathrm{v}_i)\right\|^2 &=< [\phi(\mathrm{x}_k) - \phi(\mathrm{v}_i)] \cdot [\phi(\mathrm{x}_k) - \phi(\mathrm{v}_i)] > \\ &= K(\mathrm{x}_k, \mathrm{x}_k) + K(\mathrm{v}_i, \mathrm{v}_i) - 2K(\mathrm{x}_k, \mathrm{v}_i) \end{aligned} \tag{8}$$

In this paper, we use Gaussian kernel function:

$$K(x, y) = \exp(-\frac{\|x - y\|^2}{2\sigma^2}) \tag{9}$$

So the equation (8) can be written as

$$\left\|\phi(\mathrm{x}_k) - \phi(\mathrm{v}_i)\right\|^2 = 2 - 2K(\mathrm{x}_k, \mathrm{v}_i) \tag{10}$$

To minimize equation (7), subject to the constraints m>1, $0 \le u_{ik}, t_{ik} \le 1$, and $\sum_{i=1}^{c} u_{ik} = 1, \forall k$, we obtain the following equations

$$u_{ik} = \left[\sum_{j=1}^{c} \left(\frac{\eta_i (1 - \exp(-\frac{2 - 2K(\mathrm{x}_k, \mathrm{v}_i)}{\eta_i}))}{\eta_j (1 - \exp(-\frac{2 - 2K(\mathrm{x}_k, \mathrm{v}_j)}{\eta_j}))} \right)^{\frac{2}{m-1}} \right]^{-1}, \forall i, k \tag{11a}$$

$$t_{ik} = \exp(-\frac{2 - 2K(\mathrm{x}_k, \mathrm{v}_i)}{\eta_i}), \forall i, k \tag{11b}$$

$$\phi(\mathrm{v}_i) = \frac{\sum_{k=1}^{n} u_{ik}^m t_{ik} \phi(\mathbf{x}_k)}{\sum_{k=1}^{n} u_{ik}^m t_{ik}}, \forall i, j \tag{11c}$$

Here, equation (11c) can not be calculated directly, and by multiplying $\phi(x_j)^T$ on the left sides of equation (11c), the following equation is obtained

$$K(\mathbf{x}_j, v_i) = \frac{\sum_{k=1}^{n} u_{ik}^m t_{ik} K(\mathbf{x}_k, \mathbf{x}_j)}{\sum_{k=1}^{n} u_{ik}^m t_{ik}}, \forall i, j \tag{11d}$$

In kernel fuzzy c-means (KFCM) algorithm [6], $K(\mathbf{x}_j, v_i)$ is calculated as

$$K(\mathbf{x}_j, v_i) = \frac{\sum_{k=1}^{n} u_{ik}^m K(\mathbf{x}_k, \mathbf{x}_j)}{\sum_{k=1}^{n} u_{ik}^m}, \forall i, j \tag{12}$$

Using kernel methods to equation (3), they are transformed into equation (13)

$$\eta_i = K \frac{\sum_{k=1}^{n} u_{ik}^m (2 - 2K(x_k, v_i))}{\sum_{k=1}^{n} u_{ik}^m}, K > 0 \tag{13}$$

If $D_{ik} = \|\phi(x_k) - \phi(v_i)\| > 0$ for all i and k≥1, and **X** contains c<n distinct data points, then the algorithm described below is called KIPCM-AO algorithm:

1. Initialization

1) Fix c, m and η, 1<c<n, 1<m, w <+∞. Set iteration counter r=1 and maximum iteration r_{max}.

2) Run FCM until termination to get initial membership $\mathbf{U}^{(0)}$ and initial cluster centers $V^{(0)}$. Then use Eq. (13) to get η_i.

3) Calculate the kernel matrix $K_{vx}^{(0)}$ with $\mathbf{V}^{(0)}$ by equation (12);

2. Repeat

Step 1 Update $\mathbf{T}^{(r)}$ with equation (11b), $K_{vx}^{(r-1)}$ and η_i;

Step 2 Update membership $\mathbf{U}^{(r)}$ with equation (11a), $K_{vx}^{(r-1)}$ and $\mathbf{T}^{(r)}$;

Step 3 Update $K_{vx}^{(r)}$ with equation (11d), $\mathbf{U}^{(r)}$ and $\mathbf{T}^{(r)}$;

Step 5 Increment r;

Until ($\|U^{(r)} - U^{(r-1)}\| < \varepsilon$) or r> r_{max}

4 Experiments

We first do experiment with $\mathbf{X}_{12}$ which is a two dimensional data set with 11 points whose coordinates are given in Table 1. The data set $\mathbf{X}_{12}$ comes from N. R. Pal [8] and Figure 1 shows its distribution in coordinates. There are ten points (except $\mathbf{x}_6$ and $\mathbf{x}_{12}$) form two diamond shaped clusters with five points each on the left and right sides of the y axis. We can see $\mathbf{x}_6$ and $\mathbf{x}_{12}$ as noisy points and each has the same distance from the two clusters. The initialization of cluster centers

$$V_0 = \begin{bmatrix} 0.08 & 0.36 \\ 0.41 & 0.99 \end{bmatrix} \tag{14}$$

Computational condition: ε =0.00001, maximum number of iterations=100. m=2.0, w =2.0, the width σ of Gaussian kernel function is 10.

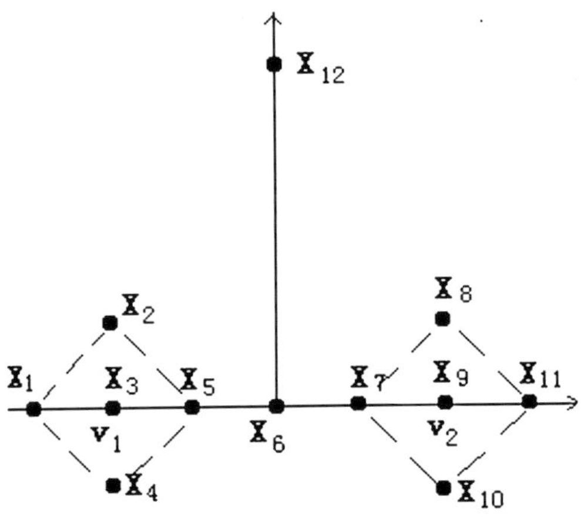

Fig. 1. Data set $\mathbf{X}_{12}$

Table 1 shows the terminal membership values of FCM by applying FCM-AO and Table 2 shows the terminal membership values and typicality values of IPCM by applying IPCM-AO to $\mathbf{X}_{12}$. The memberships of $\mathbf{x}_6$ and $\mathbf{x}_{12}$ in both FCM and IPCM are 0.500 in each cluster. From Table 2 IPCM provides both membership and typicality information but FCM provides only membership information. For example, the typicality values of $\mathbf{x}_6$ and $\mathbf{x}_{12}$ are 0.27 and 0.00, that is to say, $\mathbf{x}_{12}$ is more atypical than $\mathbf{x}_6$ for either cluster. So IPCM can distinguish between $\mathbf{x}_6$ and $\mathbf{x}_{12}$ and distinguish them from other data. FCM considers $\mathbf{x}_6$ identical with $\mathbf{x}_{12}$ but it is not the fact. So FCM can not distinguish noises from input data. In conclusion FCM is more sensitive to noises than IPCM.

Table 3 shows the terminal membership values and typicality values of KIPCM by applying KIPCM-AO to $\mathbf{X}_{12}$. Both KIPCM and IPCM can avoid the influence of

Table 1. Data set $\mathbf{X}_{12}$ and terminal **U** from FCM for $\mathbf{X}_{12}$

	Data set $\mathbf{x}_{12}$		FCM	
Pt.	x	y	U_1^T	U_2^T
1	-5.00	0.00	0.94	0.06
2	-3.34	1.67	0.97	0.03
3	-3.34	0.00	0.99	0.01
4	-3.34	-1.67	0.90	0.10
5	-1.67	0.00	0.92	0.08
6	0.00	0.00	0.50	0.50
7	1.67	0.00	0.08	0.92
8	3.34	1.67	0.03	0.97
9	3.34	0.00	0.01	0.99
10	3.34	-1.67	0.10	0.90
11	5.00	0.00	0.06	0.94
12	0.00	10.00	0.50	0.50

Table 2. Terminal **U** and **T** from IPCM for $\mathbf{X}_{12}$

Pt.	U_1^T	U_2^T	T_1^T	T_2^T
1	0.90	0.10	0.67	0.00
2	0.92	0.08	0.70	0.00
3	1.00	0.00	1.00	0.00
4	0.92	0.08	0.70	0.00
5	0.93	0.07	0.73	0.05
6	0.50	0.5	0.27	0.27
7	0.07	0.93	0.05	0.73
8	0.08	0.92	0.00	0.70
9	0.00	1.00	0.00	1.00
10	0.08	0.92	0.00	0.70
11	0.10	0.90	0.00	0.67
12	0.50	0.50	0.00	0.00

noises or outlier .In Table 2 the typicality values of $\mathbf{x}_6$ and $\mathbf{x}_{12}$ from KIPCM are smaller than that from IPCM. So KIPCM is more insensitive to noises than IPCM.

The other example is that we perform experiments on IRIS data set [9]. The computational condition is ε =0.00001, maximum number of iterations=100, m=2.0, w =2.0, the width σ of Gaussian kernel function is 3. The clustering accuracy from FCM, IPCM and KIPCM on IRIS data set is illustrated in Table 4. The KIPCM algorithm has better clustering accuracy than the other two algorithms evidently.

Table 3. Terminal U and T from KIPCM for $\mathbf{X}_{12}$

Pt.	U_1^T	U_2^T	T_1^T	T_2^T
1	0.78	0.22	0.47	0.00
2	0.80	0.20	0.50	0.00
3	0.94	0.06	0.74	0.00
4	0.80	0.20	0.50	0.00
5	0.81	0.19	0.53	0.03
6	0.50	0.50	0.17	0.17
7	0.19	0.81	0.03	0.53
8	0.20	0.80	0.00	0.50
9	0.06	0.94	0.00	0.74
10	0.20	0.80	0.00	0.50
11	0.22	0.78	0.00	0.47
12	0.50	0.50	0.00	0.00

Table 4. Clustering accuracy from FCM, IPCM and KIPCM for IRIS data set

Data set	FCM	IPCM	KIPCM
IRIS	89.3%	92.0%	93.3%

5 Conclusions

Based on kernel methods we propose kernel improved possibilistic c-means (KIPCM) algorithm as an extension of improved possibilistic c-means (IPCM) algorithm. The KIPCM computes both membership and typicality values the same as IPCM. However, KIPCM can map input data points to a high-dimensional feature space where clustering unlabeled data is carried out. By using kernel method the KIPCM can deal with noises or outliers better than IPCM. Furthermore KIPCM can deal with linear non-separable problem better than FCM and IPCM. Experiments show the better performance of KIPCM.

References

[1] Bezdek, J. C.: Pattern Recognition with Fuzzy Objective Function Algorithms, Plenum Press, New York (1981)

[2] Krishnapuram, R., and Keller, J.: A Possibilistic Approach to Clustering, IEEE Trans. Fuzzy Systems,1 (2) (1993) 98-110

[3] Barni, M., Cappellini, V., and Mecocci, A.: Comments on "A Possibilistic Approach to Clustering", IEEE Trans. Fuzzy Systems, 4(3) (1996) 393-396

[4] Zhang, J.-S., Leung, Y.-W.: Improved possibilistic C-means clustering algorithms, IEEE Trans. Fuzzy Systems, 12(2) (2004) 209-217

[5] Vapnik, V.: Statistical Learning Theory. Wiley (1998)

[6] Girolami, M.,: Mercer kernel based clustering in feature space, IEEE Trans. on Neural Networks, Vol.13, No.13 (2002) pp. 780-784
[7] Aizerman, M., Braverman, E., and Rozonoer, L.: Theoretical foundations of the potential function method in pattern recognition learning. Automation and Remote Control 25: (1964) 821-837
[8] Pal, N. R., Pal, K., and Bezdek, J. C.: A mixed c-means clustering model, Processings of the IEEE Trans. Fuzzy Systems, Spain (1997) pp.11-21
[9] Anderson, E.: The Iris of Gasp Peninsula, Bulletin of American Iris Society, 59 (1935) 2-5

Identifiability and Estimation of Probabilities from Multiple Databases with Incomplete Data and Sampling Selection

Jinzhu Jia, Zhi Geng, and Mingfeng Wang

School of Mathematical Sciences, LMAM, Peking University, Beijing 100871, China

Abstract. For an application problem, there may be multiple databases, and each database may not contain complete variables or attributes, that is, some variables are observed but some others are missing. Further, data of a database may be collected conditionally on some designed variables. In this paper, we discuss problems related to data mining from such multiple databases. We propose an approach for detecting identifiability of a joint distribution from multiple databases. For an identifiable joint distribution, we further present the expectation-maximization (EM) algorithm for calculating the maximum likelihood estimates (MLEs) of the joint distribution.

1 Introduction

With development and popularity of computers, various databases have been built, which contain different variables or attributes, and whose data are collected in different conditions. For example, in medical research, some researchers collect data of these variables, but other researchers may collect data of other variables; On the other hand, some data are from follow-up studies, but other data may be from case-control studies. Distributions of diseases and associations among variables should be evaluated by combining all databases from different researches. There are several statistical approaches for combining multiple databases, such as file-matching for large databases and split questionnaire survey sampling [5, 6]. Multiple databases are depicted as a hypergraph in [1, 4]. In these approaches, a database involving a subset of variables is treated as a sample from a marginal distribution of these variables. In this paper, we consider that a database involving a subset of variables may be a sample drawn conditionally on some other variables, called designed variables.

Identifiability and estimation of a joint distribution of variables are two important problems for data mining from multiple databases. In this paper, we show conditions for identifiability of a joint distribution from marginal and conditional distributions of observed variables, and we propose an approach for detecting identifiability. For an identifiable joint distribution, we present the expectation-maximization (EM) algorithm for calculating the maximum likelihood estimates (MLEs) of the joint distribution [3, 4]. In the partial imputation EM algorithm [4], there must be a database containing all variables and other databases are

D.-Y. Yeung et al. (Eds.): SSPR&SPR 2006, LNCS 4109, pp. 792–798, 2006.

drawn from marginal distributions. In this paper, we show that the database containing all variables may be unnecessary for identifying the joint distribution of all variables, and our algorithm for finding MLEs also deals with databases drawn from conditional distributions.

Section 2 gives notation and definitions. In Section 3, we show conditions for identifiability of a joint distribution from multiple databases. In Section 4, we apply the EM algorithm to calculate MLEs of the joint distribution. Section 5 gives a numerical example to illustrate our approach. Finally, Section 6 presents a simulation to evaluate the estimates.

2 Notation and Definitions

Let $V = \{x_1, \ldots, x_p\}$ denote the set of all variables included in an interested system, and assume that all variables in V are discrete. Let $p(V = v)$ denote the distribution of variables in V. Suppose that there are K databases, $D_1, \ldots, D_K$. For the database D_k, let A_k denote the set of designed variables which is used to stratify the population into subpopulations with different values of A_k, let B_k denote the set of observed variables, and let $V_k = A_k \cup B_k$ represent the set of variables involved in the database D_k. Denote the database D_k as $[B_k|A_k]$, which means that variables in B_k are observed conditionally on variables in A_k. For the database D_k, let $n_k(b_k|a_k)$ denote the frequency of observed individuals with value $B_k = b_k$ in the subpopulation of $A_k = a_k$. Below we first give a general definition of identifiability [2].

Definition 1. *Consider a vector Y of random variables having a distribution $F(y;\theta)$ that depends on an unknown parameter vector θ. θ is identifiable by observation of Y if distinct values for θ yield distinct distributions of Y, that is, $\theta_1 \neq \theta_2 \Rightarrow F(y;\theta_1) \neq F(y;\theta_2)$.*

Informally, identifiability means that these databases contain sufficient information such that the joint distribution can be uniquely determined. In the case of multiple databases, we say that the joint distribution $P(V)$ is identifiable by databases $[B_1|A_1] \ldots, [B_K|A_K]$ if distinct values for $P(V)$ yield distinct values for vector $(P(B_1|A_1), P(B_2|A_1), \cdots, P(B_K|A_K))$.

3 Conditions for Identifiability from Multiple Databases

In this section, we show a condition for identifiability of the joint distribution of V from multiple databases $[B_1|A_1], \ldots, [B_K|A_K]$. We first show several lemmas which are used to prove the condition for identifiability.

Lemma 1. *Suppose that there are only two databases $D_1 = [B|A]$ and $D_2 = [A|B]$ where $A \cup B = V$. Then the joint distribution $P(V)$ of variables in V is identifiable.*

Proof. Since $P(B|A)/P(A|B) = P(B)/P(A)$, we have $\sum_B [P(B|A)/P(A|B)] = \sum_B [P(B)/P(A)] = 1/P(A)$. Thus we can find $P(A, B) = P(B|A)P(A) =$

$P(B|A)/[\sum_B P(B|A)/P(A|B)]$. Since $P(A,B)$ is a function of $P(A|B)$ and $P(B|A)$ and since they are identifiable by the databases $[A|B]$ and $[B|A]$ respectively. So we get that $P(A,B)$ is identifiable by the databases $[B|A]$ and $[A|B]$.

Lemma 2. *Suppose that there are only two databases $D_1 = [B_1|A_1]$ and $D_2 = [B_2|A_2]$ and that $A_i \cup B_i = V$ for $i = 1$ and 2. Define $A = A_1 \cap A_2$ and $B = V \setminus A$. Then the conditional probability $P(B|A)$ is identifiable by the databases, where the conditional probability $P(B|\emptyset)$ is defined as the marginal probability $P(B)$.*

Proof. If $A = \emptyset$, then the result can be obtained immediately from Lemma 1. Below we consider the case of $A \neq \emptyset$. Define $C_1 = A_1 \setminus A$ and $C_2 = A_2 \setminus A$. Then $A_1 = C_1 \cup A$, $A_2 = C_2 \cup A$, $C_1 \cap C_2 = \emptyset$ and $C_i \cap A = \emptyset$ for $i = 1, 2$. We have

$$\frac{P(B_1|A_1)}{P(B_2|A_2)} = \frac{P(A_2)}{P(A_1)} = \frac{P(C_2, A)}{P(C_1, A)}.$$

Thus,

$$\sum_{C_2} \frac{P(B_1|A_1)}{P(B_2|A_2)} = \sum_{C_2} \frac{P(C_2, A)}{P(C_1, A)} = \frac{P(A)}{P(C_1, A)} = \frac{1}{P(C_1|A)}.$$

Further, we have

$$P(B_1|A_1)P(C_1|A) = P(B_1|C_1, A)P(C_1|A) = P(B_1, C_1|A) = P(B|A).$$

We get that $P(B|A)$ is a function of $P(B_1|A_1)$ and $P(A_1|B_1)$ and thus $P(B|A)$ is identifiable by $[B_1|A_1]$ and $[B_2|A_2]$.

Lemma 3. *Suppose that there are K databases, $[B_1|A_1], \ldots, [B_K|A_K]$, and $A_k \cup B_k = V$ for all k. The joint distribution $P(V)$ is identifiable if $\bigcap_{i=1}^n A_i = \emptyset$.*

Proof. This result can be obtained immediately by applying Lemma 2 repeatedly to each pair of databases.

Now we propose an algorithm for detecting identifiability of the joint distribution $P(V)$ from K databases, $[B_1|A_1], \ldots, [B_K|A_K]$, where $A_k \cup B_k \subseteq V$.

Algorithm: Detect identifiability of $P(V)$ from the K databases.

1. Initialize $t = 0$, $[B_k^{(0)}|A_k^{(0)}] = [B_k|A_k]$ for all k, and $V^{(0)} = V$. Below we repeatedly check whether $P(V^{(t)})$ is identifiable by databases $[B_k^{(t)}|A_k^{(t)}]$ for all k.
2. Find a database $[B_i^{(t)}|A_i^{(t)}]$ such that $B_i^{(t)} \cup A_i^{(t)} = V^{(t)}$. If there is no such a database, then $P(V)$ is not identifiable and stop the algorithm.
3. Let $V^{(t+1)} = A_i^{(t)}$, $[B_k^{(t+1)}|A_k^{(t+1)}] = [B_k^{(t)} \setminus B_i^{(t)}|A_k^{(t)} \setminus B_i^{(t)}]$ for all k, and $t = t + 1$.
4. Repeat Steps 2 and 3 until $V^{(t)} = \emptyset$.

If the algorithm returns an empty set $V^{(t)}$, then the probability $P(V)$ is identifiable. Below we first give an example to illustrate the algorithm, and then we show the correctness of the algorithm.

Example 1. Consider that there are seven variables $V = \{1, 2, 3, 4, 5, 6, 7\}$ and five databases [456|1237], [147|23], [45|6], [2|1347], [3|45]. We apply the above algorithm to the databases to detect identifiability of $P(V)$.

For the first iteration, at Step 2, we find $[B_1|A_1] = [456|1237]$ with $A_1 \cup B_1 = V$. At Step 3, we reset the set to be identified as $V^{(1)} = A_1 = [1237]$, remove variables 4, 5 and 6 in B_1 from all databases and obtain databases for the next iteration as [17|23], [2|137] and [3]. For the second iteration, at Step 2, we find $A_1^{(1)} \cup B_1^{(1)} = V^{(1)} = [1237]$. At Step 3, set $V^{(2)} = A_1^{(1)} = [23]$, remove variables 1 and 7 in $B_1^{(1)}$ and obtain databases [2|3], [3]. For the third iteration, we get $V^{(3)} = [3]$ and database [3]. For the fourth iteration, we get $V^{(4)} = \emptyset$ and thus we say that P(V) is identifiable.

Lemma 4. *Suppose that there are two databases $[B_1|A_1]$ and $[B_2|A_2]$. Define $V_i = A_i \cup B_i$ for $i = 1$ and 2, $V_0 = V_1 \cap V_2$ and $A = A_1 \cap A_2$. Then $P(V_0 \setminus A|A)$ is identifiable by the two databases if $V_1 \setminus V_2 \subseteq B_1$ and $V_2 \setminus V_1 \subseteq B_2$.*

Proof. From $[B_1|A_1]$ we can get $P(B_1|A_1)$, and in turn we can obtain $P(B_1 \setminus (V_1 \setminus V2)|A_1)$. Similarly we can obtain $P(B_2 \setminus (V_2 \setminus V1)|A_2)$. From the conditions $V_1 \setminus V2 \subseteq B_1$ and $V_2 \setminus V_1 \subseteq B_2$, we have $A_1 \subseteq V_2$ and $A_2 \subseteq V_1$ respectively. Thus we get $B_1 \setminus (V_1 \setminus V_2) \cup A_1 = V_1 \cap V_2$ and $B_2 \setminus (V_2 \setminus V1) \cup A_2 = V_1 \cap V_2$. By Lemma 2, we know that $P(V_0 \setminus A|A)$ can be obtained.

Corollary 1. *Suppose that $V_2 \subseteq V_1$ and A_1 can be partitioned into two sets A and B (that is, $A_1 = A \cup B$ and $A \cap B = \emptyset$) such that $A \subseteq A_2$ and $B \subseteq B_2$. Then $P(B|A)$ is identifiable and thus $P(V_1 \setminus A|A)$ is identified.*

Proof. Since $A_1 = A \cup B$, we get $V_2 = A_2 \cup B_2 \supseteq A \cup B = A_1$. Thus we have $V_1 \setminus V_2 \subseteq B_1$. From supposition we have $V_2 \setminus V_1 = \emptyset \subseteq B_2$. From Lemma 4, we obtain that $P(B|A)$ is identifiable. We have

$$P(B_1|A_1) \times P(B|A) = \frac{P(V_1)}{P(A_1)} \times \frac{P(A_1)}{P(A)} = P(V_1 \setminus A|A).$$

Thus we showed that $P(V_1 \setminus A|A)$ is identifiable.

Example 2. Suppose that $V = \{1, 2, 3, 4, 5, 6, 7\} = V_1$ and $V_2 = \{1, 2, 4, 5, 7\}$ and that we have two databases $[B_1|A_1] = [1236|457]$ and $[B_2|A_2] = [24|157]$. Partition A_1 as $A = \{5, 7\}$ and $B = \{4\}$. Since $A \subseteq A_2$ and $B \subseteq B_2$, we get from Corollary 1 that $P(4|57)$ and $P(1, 2, 3, 4, 6|57)$ are identifiable.

Theorem 1. *The probability $P(V)$ is identifiable by databases $[B_1|A_1]$, ..., $[B_K|A_K]$ if the algorithm returns an empty set $V^{(t)}$.*

Proof. We first show the result following the iterations of the algorithm. At the first iteration, we have immediately that $P(B_i^{(0)}|A_i^{(0)})$ is identifiable by the database $[B_i^{(0)}|A_i^{(0)}]$ where $A_i^{(0)} \cup B_i^{(0)} = V$. At the second iteration, we partition $V^{(1)} = A_i^{(0)}$ into two sets $A_i^{(1)}$ and $B_i^{(1)}$, which must be contained by A_k and B_k of some original database $[B_k|A_k]$ respectively. By Corollary 1, we have that $P(V \setminus A_i^{(1)}|A_i^{(1)})$ is identifiable. Following the iterations, we can obtain that $P(V \setminus A_i^{(t)}|A_i^{(t)})$ is identifiable. If $A_i^{(t)} = \emptyset$ is obtained at the t-th iteration, then $P(V)$ is identifiable.

4 Maximization Likelihood Estimation of Probabilities

For a joint probability $P(V)$ which is identifiable, we propose the EM algorithm for calculating MLEs from databases $[B_1|A_1], \ldots, [B_K|A_K]$. Data in each database, say $D_k = [B_k|A_k]$, may be incomplete (i.e. A_k is a pure subset of V), and they may be obtained conditionally on designed variables B_k. For simplicity, we show only a simple case of two variables $V = \{X, Y\}$ and two databases $[X|Y]$ and $[Y|X]$. The algorithm can be easily extended to a general case.

Let $n_1(x|y)$ denote the observed frequency of individuals with $X = x$ conditional on $Y = y$ in the first database for $x = 1, \ldots, I$; and let $n_2(y|x)$ denote that with $Y = y$ conditional on $X = x$ in the second database for $y = 1, \ldots, J$. Below we present the EM algorithm for calculating MLEs $\{\hat{p}_{xy}\}$ of the joint probabilities $P(X = x, Y = y)$ for all x and y. We treat the frequencies $\{n_1(x|y), \forall x\}$ as the y-th column of an $I \times J$ contingency table $\{n_1^{(y)}(i,j)\forall i,j\}$. Similarly, $\{n_2(y|x), \forall y\}$ as the x-th row of an $I \times J$ contingency table $\{n_2^{(x)}(i,j), \forall i,j\}$. Assume that the total frequency $n_k^{(m)} = \sum_{i,j} n_k^{(m)}(i,j)$ follows a Poisson distribution with parameter $\lambda_k(m)$ and that $\{n_k^{(m)}(i,j), \forall i,j\}$ follows a multinomial distribution with parameters $\{p_{ij}\}$ given a total frequency $n_k^{(m)}$.

At the E-step of the EM algorithm, we find the expected frequencies of $\{n_1^{(y)}(i,j), \forall ij\}$ and $\{n_2^{(x)}(i,j), \forall ij\}$. At the M-step, we find the MLEs $\{\hat{p}_{ij}, \forall i,j\}$ and $\{\lambda_k(m), \forall k, m\}$. The EM algorithm for calculating MLEs is given below.

1. E-step:
$$\begin{aligned}\hat{n}_k^{(m)}(i,j) &= E[n_k^{(m)}(i,j)|\{n_1(x|y)\}, \{n_2(y|x)\}, \{\lambda_k^{(t)}(m)\}, \{p_{ij}^{(t)}\}] \\ &= \lambda_k^{(t)}(m) \times p_{ij}^{(t)},\end{aligned}$$
for i, j, k and m.
2. M-step:
$$\hat{\lambda}_k^{(t+1)}(m) = \sum_{i,j} \hat{n}_k^{(m)}(i,j),$$
$$\hat{p}_{ij}^{(t+1)} = \frac{1}{N} \sum_{k,m} \hat{n}_k^{(m)}(i,j),$$

where the total frequency N equals $\sum_{k,m,i,j} \hat{n}_k^{(m)}(i,j)$.

3. Repeat the E-step and M-step until some convergence criterion is achieved.

5 A Numerical Example

In this section, we use an artificial data to illustrate our approach. Suppose that we have a case-control study and a follow-up study on association between smoking and lung cancer, as shown in Tab. 1. In the case-control study with databases $[X|Y]$, we have 709 cases and 709 controls, and then ask their smoking history. In the follow-up study with databases $[Y|X]$, we have a smoker group of 2000 individuals and a non-smoker group of 2000 individuals, and follow their states for years. Combining these databases, we can identify the joint distribution of X and Y and the marginal distributions of X and Y (see Tab. 1), any of which is not identifiable by using only one of $[X|Y]$ and $[Y|X]$. Especially, the MLE of the relative risk, $\hat{P}(Y = 1|X = 1)/\hat{P}(Y = 1|X = 0) = 3.1160$, can be found by combining all data in these databases. It is more efficient than the MLE of the relative risk $(8/2000)/(2/2000) = 4$ obtained by using data only from the follow-up study since there are a few of cancer cases in the follow-up study.

Table 1. A case-control study and a follow-up study on smoking and lung cancer

	Case-control study			Follow-up study		MLEs of probabilities	
	Smoker	Non-smoker	Total	Smoker	Non-smoker	Smoker	Non-smoker
	$X = 1$	$X = 0$		$X = 1$	$X = 0$	$X = 1$	$X = 0$
Cancer ($Y = 1$)	688	21	709	8	2	0.0067	0.0002
Control ($Y = 0$)	650	59	709	1992	1998	0.9082	0.0849
Total				2000	2000		

6 Simulation

Consider two binary variables $V = \{X_1, X_2\}$ with values 0 and 1. Suppose that there are two databases $[X_1|X_2]$ and $[X_2|X_1]$. From the results in Section 3, we know that the joint probabilities of X_1 and X_2 are identifiable from these databases. Below we give a numerical simulation to illustrate estimation of the joint probability $P(x_1, x_2)$. From the true distribution given in Tab. 2,

Table 2. True probabilities and their estimates

		Estimates	
	True value	Mean	Standard error
p_{00}	0.1000	0.0993	0.0230
p_{01}	0.2000	0.2004	0.0372
p_{10}	0.3000	0.3007	0.0499
p_{11}	0.4000	0.3994	0.0507

we generate each of databases $[X_1 = i|X_2 = j]$ and $[X_2 = j|X_1 = i]$ for all i and j from a binomial distribution with sample size 50. Then we apply the EM algorithm to calculate MLEs $\hat{p}_{ij}$ with the initial values $(p_{00}, p_{01}, p_{10}, p_{11}) = (0.25, 0.25, 0.25, 0.25)$ and $\lambda = 100$. The desired convergence accuracy is 10^{-6}. We repeat the simulation 1000 times. The results of means and standard errors are given in Tab. 2. It can be seen that estimates are very close to their true probabilities.

Acknowledgements

We would like to thank the two referees for their valuable comments and suggestions that improved the presentation of this paper. This research was supported by NSFC 10431010, NBRP 2005CB523301 and MSAR.

References

[1] C. Beeri, R. Fagin, D. Maier, M, Yannakakis, On the desirability of acyclic database schemes, J. Association for Computing Machinery 30 (1983) 479-513.

[2] P. J. Bickel, K. A. Doksum, Mathemetical Statistics. Holden-Day, Oakland, 1977.

[3] A. P. Dempster, N. M. Larid, D. B. Rubin, Maximum likelihood estimation from incomplete data via the EM algorithm(with disscussion), J. R. Stat. Soc. Ser. B. **39** (1977) 1-38.

[4] Z. Geng, K. Wan, F. Tao, Mixed graphical models with missing data and the partial imputation EM algorithm, Scan. J. of Stat. **27** (2000) 433-444.

[5] R. J. A. Little and D. B. Rubin, Statistical Analysis with Missing Data, 2nd Ed. Wiley, New York, 2002.

[6] S. Rassler, Statistical Matching. Lecture Notes in Statistics, 168, Springer, New York, 2002.

Unsupervised Image Segmentation Using a Hierarchical Clustering Selection Process*

Adolfo Martínez-Usó, Filiberto Pla, and Pedro García-Sevilla

Dept. Lenguajes y Sistemas Informáticos, Jaume I Univerisity
{auso, pla, pgarcia}@uji.es
http://www.vision.uji.es

Abstract. In this paper we present an unsupervised algorithm to select the most adequate grouping of regions in an image using a hierarchical clustering scheme. Then, we introduce an optimisation approach for the whole process. The grouping method presented is based on the maximisation of a measure that represents the perceptual decision. The whole strategy takes profit from a hierarchical clustering to find a maximum of the proposed criterion. The algorithm has been used to segment real images as well as multispectral images achieving very accurate results on this task.

1 Introduction

Image segmentation has been a very focused topic in the literature. Looking for the main regions that perceptually compose an image has been the target for many researchers [10,3,5,8]. This task has been used as an aim in itself or as a preprocessing step. In both cases, the difficulties involved in this process are very well-known [4].

The main motivation of this work has been to obtain a robust and completely unsupervised segmentation criterion based on perceptual similarities among the image regions and the contour information. Thus, on one hand, the method will be guided by a merging process using an agglomerative hierarchical clustering and, on the other hand, the method will satisfactorily select the optimal, or sub-optimal, partition in this hierarchical structure combining region and boundary information.

Thus, the method presented is based on two basic steps which define our segmentation strategy:

- A **hierarchical clustering**, in order to create a structure of non-overlapping partitions from an oversegmented representation until the stage with just one region (cluster). The hierarchical structure will serve as the guide to an optimisation strategy. It is supposed that the hierarchical structure will contain the optimal (or close to optimal) segmentation.
- A **global similarity measure** as a criterion function to be optimised, which is applied through the above named structure looking for a partition that optimises the criterion function. This partition will represent the final segmentation result.

* This work has been partly supported by projects ESP2005-07724-C05-05 from Spanish CICYT and P1-1B2004-36 from Fundació Caixa-Castelló.

D.-Y. Yeung et al. (Eds.): SSPR&SPR 2006, LNCS 4109, pp. 799–807, 2006.

We apply our algorithm not only to classical real images from the literature, which is useful to test the performance of the method, but especially to multispectral images with more than three bands as well. For instance, in some applications, like fruit quality inspection tasks, there exist some types of defects that can only be detected in certain bands of the spectrum, and most of the defects have a specific range of bands where they can be better discriminated. Therefore, for some inspection tools, multispectral images are becoming useful to achieve better recognition and classification results. These results regard to an application in which we are working on as a part of a quality fruit estimation project on oranges.

The whole segmentation process is summarised in the following algorithm. It is particularly interesting to emphasise the two final steps of the algorithm where the most relevant contribution of this work is.

1. Preliminary steps on cluster initialisation (Sect. 2).
2. The method compares clusters and iteratively merges neighbouring regions to create a structure that constitutes a hierarchical family of derived clusters (Sect. 3).
3. One of those previous partitions will maximise a criterion function used to choose the right clusters, being the chosen partition the final segmentation result (Sect. 4).

Finally, many segmentation techniques have high computational requirements, especially the iterative ones. These requirements become higher when the image size increases. Thus, a study of the evolution of the distances among the image regions is also presented (see Sect. 5). This study will allow us to introduce an important optimisation in the process without making worse the final segmentation results.

2 Preliminary Steps for Clustering Initialisation

This phase of the algorithm identifies the main homogeneous regions in the image. This initialisation uses a quadtree structure (QT) to represent the image, which is a multiresolution representation commonly used to split or decompose a given image into similar square regions. The QT also serves to represent the spatial relationships among regions.

First merging, while the QT is growing up: The hierarchical structure developed by a QT has several levels of resolution made by nodes and leaves. Every time a level is developed, the algorithm checks if any two of the current leaves can be merged. Because of the spatial information contained in the QT representation, neighbourhood operations are completely defined from the QT structure [12].

Second merging, look for high gradient magnitudes: Every region formed in the previous step starts growing up while the gradient magnitude increases at the region border. Thus, each region border grows towards the high gradient magnitude values. The gradient magnitude considered for each pixel is the maximum gradient value found in all bands.

This behaviour is usually formulated as an active contour functional [9], whose internal energy has the discrete form $E_i = -\sum_{x \in \beta_i} |\nabla g(x)|$, where, β_i represents the set of boundary pixels in region i and the value $|\nabla g(x)|$ returns the gradient magnitude at pixel x. Regions previously formed are used as the active contours initialisation. Of course, on this functional, the energy E in each contour region has to be minimised. It is important to note that if two regions compete for a third region, which is neighbour of both of them, the algorithm computes the average of the gradient magnitudes of each region border, and merges the regions with the highest difference.

Smoothing regions: After the previous steps, regions very often present isolated small holes inside of them, that is, many times there is a gradient magnitude peak due to impulsive or other type of noise. Thus, this step looks for these kind of isolated regions and merges them with the larger regions which contain the isolated ones.

3 Hierarchical Clustering in the Image Domain

Some relevant recent works for image segmentation are based on clustering or grouping processes [6,7]. These methods are designed to discover and extract hidden structures in data sets [2]. Regarding to the procedure we use, it starts with as many clusters as the initialisation step provides. From these initial clusters, the method forces a merging operation in each iteration, that is, the algorithm always eliminates a cluster in each step. This means that each iteration compares every pair of spatially connected clusters and calculates the distance (D) between them. The two clusters, that is, image regions, with more similar values are forced to merge. Distance D is defined as follows:

$$D_{ij} = d_{ij}(1 + \delta_{ij}) \tag{1}$$

where,

$$d_{ij} = (\mu_i - \mu_j)^T \left(\Sigma_i^{-1} + \Sigma_j^{-1}\right)(\mu_i - \mu_j) \qquad \delta_{ij} = \frac{\sum_{x \in \beta_{ij}} |\nabla g(x)|^2}{\|\beta_{ij}\|} \tag{2}$$

Mean values on the intensity of clusters i and j are represented by μ_i, μ_j whereas their covariance matrices are represented by $\sum_i$, $\sum_j$. In δ_{ij} function, β_{ij} is the set of pixels that belong to the boundary between regions i and j. Function $\|\cdot\|$ returns the cardinality of a set.

Note that the proposed measure D_{ij} can be evaluated analytically as a mixture of two expressions. Under the assumption of considering normally distributed clusters, d_{ij} value is a Mahalanobis distance between two distributions, defined in the pixel domain (gray level, colour, multispectral, ...). This term in the distance expression D accounts for the similarity in pixel values of the distribution of pixel values in both regions considered to be merged. On the other hand, δ_{ij} averages the gradient magnitude values on the boundary between these two connected clusters. Thus, this term accounts for the strength of the discontinuity between the distributions of pixel values between the considered clusters, including a spatial measure of discontinuity in the distance function.

4 Clustering Assessment

While the algorithm completes the process described in the previous section, a non-parametric estimation of the goodness of the data partition is performed. That is, the final result is selected without any a priori information about the final number of clusters or the shape of the resulting regions. An example of this type of approach can be found in [11] which is very similar to ours. However, their method suffers from problems on images where the boundary information has special relevance.

The criteria to select an optimal clustering from the previous hierarchical structure is completely independent of the way in which the hierarchical tree structure is constructed using the distance measure introduced in the previous section.

The algorithm has been formulated as the maximisation of a criterion function S that represents, given an image partition, how well the pixels fit to the corresponding regions. In such a way, it maximises the following expression:

$$S = S_i \cdot S_e \tag{3}$$

being S_i a inner measure and S_e a external measure of each partition in the hierarchical tree structure. S_i can be considered as an average measure that a pixel in the image belongs to the region it has been assigned to, and S_e an average measure that a pixel does not belong to its neighbouring regions. Therefore, the criterion function S takes into account that the pixels in the image belong to the assigned regions in the partition and, simultaneously, that the pixels do not belong to neighbouring regions in the image domain. This can be considered as a perceptual measure of the grouping, in such a way that it estimates how "well" the pixels are grouped and they are consistent internally and, at the same time, different enough from spatially nearby clusters.

Given a particular partition, the inner and external average pixel measures, S_i and S_e respectively, are defined as follows:

$$S_i = \frac{1}{N} \sum_R \sum_{x \in R} S(x, R) = \frac{1}{N} \sum_R \sum_{x \in R} e^{\frac{-(x-\mu_R)^2}{2\sigma^2}} \tag{4}$$

$$S_e = \frac{1}{N} \sum_R \sum_{x \in R} S(x, NR(R)) = \frac{1}{N} \sum_R \sum_{x \in R} \prod_{R'}^{NR(R)} S(x, R') =$$
$$= \frac{1}{N} \sum_R \sum_{x \in R} \prod_{R'}^{NR(R)} (1 - S(x, R')) = \frac{1}{N} \sum_R \sum_{x \in R} \prod_{R'}^{NR(R)} \left(1 - e^{\frac{-(x-\mu_{R'})^2}{2\sigma^2}}\right) \tag{5}$$

In these equations, N represents the total number of pixels in the image. R, R' are regions and $NR(R)$ is the set of neighbouring regions of region R. The pixel value is represented by x whereas μ_R is the average value of the pixels in the region R. $S(x, R)$ is a similarity measure between pixel x and region R.

Equation (4) represents the sum of the inner values for each cluster. That is, it expresses a measure that image pixels belong to its current assigned cluster. This is estimated assuming a Gaussian distribution of region pixel values, characterised by a mean and an expected variance. The expected variance is fixed and it is a smoothing parameter of the expected segmentation.

On the other hand, equation (5) provides the external values. It expresses a measure that image pixels do not belong to its current neighbouring clusters. This is also estimated assuming a Gaussian distribution of the pixel values of the neighbouring clusters.

The way this value is expressed is the complementary of the measure that a pixel belongs to the neighbouring regions $S(x, NR(R))$, which is defined as the measure that a pixel does not belong to any of the neighbouring regions, that is, $S(x, NR(R)) = \prod_{R'}^{NR(R)} (1 - S(x, R'))$.

In these equations σ is always a fixed value representing the variance threshold we permit on the density-estimation. This variance represents a smoothing parameter of the expected segmentation. The lower σ, the more clusters will have the chosen partition using function (3). Nevertheless, in the method here described, once the σ was selected experimentally, it was fixed to the same for all the experiments.

5 Optimising the Merging Process

It is a well-known fact that the main drawback of iterative methods is how much time/resources the algorithm needs to perform these iterations. The problem becomes worse when the algorithm has to deal with complex structures and the process advances bit by bit due to the need to check the clustering robustness. In this sense, we can try to optimise the process of merging regions.

Graphs in Fig. 1 show the distance values between any pair of neighbouring regions arranged by this value for two different images at three different stages of the process. That is, distances calculated as described in section 3 are arranged from lower to higher value. Each time two regions are merged, the algorithm has to recalculate the affected distances. Fig. 1 shows the number of regions NR to merge under each graph. The x-axis corresponds to the initial number of distances computed for the corresponding image and the y-axis represents the distance value.

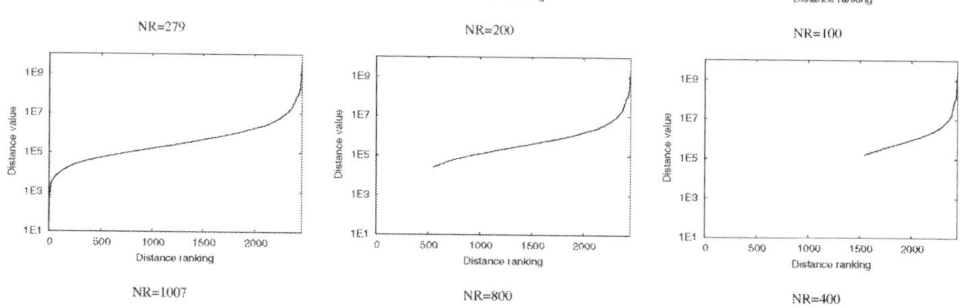

Fig. 1. Distance evolution examples. First row, toys RGB image. Second row, orange multispectral image. Coordinate axes: number of distances (x) and distance values (y). NR gives the initial number of regions considered.

As we can see from the graphical results shown on this figure, the distance evolution presents a very regular behaviour and, therefore, a very predictable evolution. In such cases, it is clearly desirable to reduce as much as possible this phase in order to take profit of the particular nature of this evolution. In this sense, instead of merging only two clusters at each iteration, we start from the pair of clusters with the minimum value of D and move to the pair of clusters which provided the next smallest distance value if both clusters has not been previously merged at this iteration. We keep merging pairs of clusters according to their distance value until the distance value reaches a threshold which is established depending on the maximum distance value.

6 Results

Maximising the inner and external average measures product for all partitions considered gives us the final segmentation result. Fig. 2 shows two graphical results[1] drawing the behaviour of criterion function (3). The graph axes represent the number of clusters (x-axis) and the corresponding S value of the partition (y-axis). First graph in Fig. 2 has been obtained for the *toys* image, where the maximum value is reached at 25 clusters. Next graph, obtained from the multispectral image of an *orange* (the one on the first row in Fig. 4), has its maximum value at 8 clusters. The segmentation results for these images can be seen in figures 3 and 4. In figure 3, other three results are shown for classical images as *peppers*, *tree* and *beans*. As we can see, the results in all the images are consistent with a perceptual interpretation, with well defined contours and all important regions detected. Note that the results are presented using an edge image where the darker the edge, the greater difference between mean values of neighbouring regions.

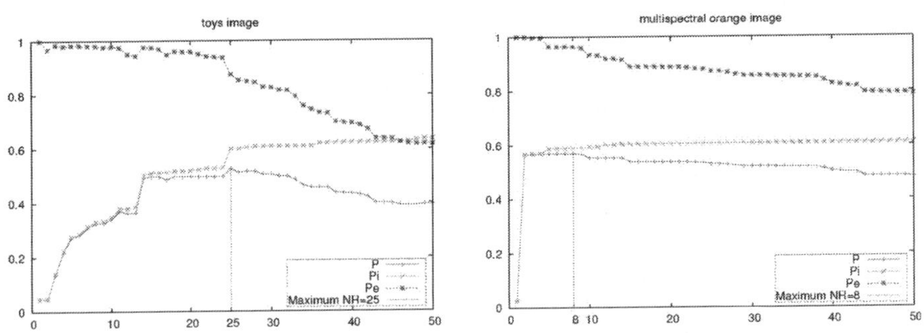

Fig. 2. Maximum likelihood estimation. On the left a real image (*toys*), on the right a multispectral image of an orange.

Although this algorithm has achieved satisfactory results on gray level and RGB colour images, it is expected that with more complex images, like multispectral images, it may also provide consistent results. In this sense, a collection of multispectral images

[1] In all the graphical results, we have marked on the x-axis the position which provided the maximum value of S.

Fig. 3. Examples of results on classical images. From left to right, toys, peppers, tree and beans.

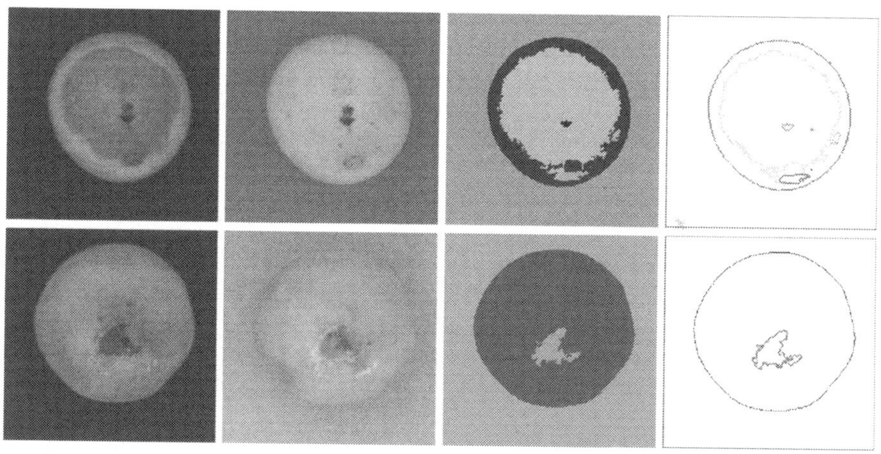

Fig. 4. Multispectral images of oranges. Two first columns show two example bands from each orange (420/640 nm and 680/1000 nm). The following columns show the segmentation results with random colours and the edges for the same results.

of oranges obtained by an imaging spectrograph (RetigaEx, Opto-knowledged Systems Inc., Canada) has been used. The spectral range is extended from 400 to 1050 nm with a spectral resolution of 10 nm for each band. The database used includes several kinds of orange defects. Fig.4 shows the result of applying the algorithm on two of these images. The orange on the top presents overripe and scratch defects, whereas the orange on the bottom suffers from a rotten defect. Note that it has correctly segmented the orange parts, labelling defect skin and healthy skin as different regions. We can also see the graph (Fig.2) obtained for the first orange image where, according with the results shown in Fig.4, the maximum value is at 8 regions. The average values S_i and S_e can also indicate the accuracy of the segmentation results. That is, low values in

Table 1. The "/" character separates values for original segmentation / values for optimised segmentation. NR column presents the final number of regions. S is the final value considered. Finally, last column gives the Borsotti value for each segmentation.

image	NR	S	Borsotti value
beans	38/38	0.2662/0.2661	11886.23/11884.31
toys	25/25	0.5258/0.5258	719.18/721.27
peppers	88/90	0.2931/0.2674	3155.00/4510.90
tree	100/95	0.2861/0.2633	7797.52/7425.97
orange (top)	8/6	0.5698/0.5802	1795.00/1578.76
orange (bottom)	3/3	0.4122/0.4125	2689.28/2689.28

these parameters may result on poor segmentation images, probably, because the input images are corrupted or affected by noise. Empirically, results where $S_i \cdot S_e \leq 0.2$ are discarded. As we can see on graphs from Fig. 2, the S value is noticeable bigger than this value, indicating the reliability of the resulting segmentations.

In addition, we also present the improved results derived from the use of the optimisation proposed in section 5. For our experimental evaluation purposes, and due to the fact that differences between the original segmentation results and the optimised ones are not clearly visible (they seem identical), quantitative results for the images used on this paper are given in table 1, demonstrating the performance of the proposed optimisation. In [1], Borsotti et al. proposed a measure to estimate the quality of the segmentation results. This value measures the intra-region uniformity and the inter-region contrast. It also has a penalisation factor inversely proportional to the number of regions in segmented images [13]. Thus, we also present this well-known value on table 1 as an indicator of how close the two segmentation results are[2].

Finally, it is important to stress that the final number of regions (based on the number of clusters) is determined by the algorithm with no prior knowledge about the image. The σ value used is the same in all the experiments carried out for this work. In this case, $\sigma = \sqrt{10}$.

7 Conclusions

An unsupervised image segmentation algorithm has been presented. It performs the clustering in the image domain, using spatial information to build a hierarchical clustering structure. The hierarchical clustering is performed using a proposed distance function that tries to integrate the similarity in the distribution of pixel values, and the edge information, in order to adapt region borders to image edges. In our experiments, this process has been improved by a significant optimisation in terms of speed and without affecting the solution quality. Moreover, a criterion function based on the maximisation of a similarity measure has been introduced. This function estimates the right number of clusters in the hierarchical structure, in such a way that it is in accordance with a perceptual decision about the grouping.

[2] Note that we had to develop an extension of Borsotti quality estimation equation to multispectral values.

The proposed method has been tested in gray level, colour and multispectral images, providing satisfactory results in the chosen groupings and the segmentation results, where regions are well defined by contour edges.

Although the criterion function used behaves as expected, further work is directed to achieve a smoother criterion function that can avoid local oscillations and, therefore, trying not to get stuck in local maxima around the global maximum, in order to achieve a more accurate decision about the final grouping representing the segmentation.

References

1. M. Borsotti, P. Campadelli, and R. Schettini. Quantitative evaluation of color image segmentation results. *Pattern Recognition Letters*, 19:741–747, 1998.
2. Joachim Buhmann. Data clustering and learning. The Handbook of Brain Theory and Neural Networks, 2nd Edition:308–312, 2002.
3. Heng-Da. Cheng, X.H. Jiang, Ying Sun, and Jingli Wang. Color image segmentation: Advances and prospects. *Pattern Recognition*, 34(12):2259–2281, 2001.
4. E.M. Gurari and H. Wechsler. On the difficulties involved in the segmentation of pictures. *PAMI(4)*, (3):304–306, 1982.
5. R.H. Haralick and L.G. Shapiro. Image segmentation techniques. *Computer Vision, Graphics, and Image Processing*, (29):100–132, 1985.
6. J. Keuchel, M. Heiler, and C. Schnörr. Hierarchical image segmentation based on semidefinite programming. 3175:120–128, 2004.
7. S. Makrogiannis, G. Economou, S. Fotopoulos, and G.B. Nikolaos. Segmentation of color images using multiscale clustering and graph theoretic region synthesis. *IEEE Transactions on Systems, Man, and Cybernetics*, 35(2):224–238, 2005.
8. X. Muñoz, J. Freixenet, X. Cufí, and J. Martí. Strategies for image segmentation combining region and boundary information. *PRL*, 24(Issue 1-3):375–392, 2003.
9. Mark Nixon and Alberto S. Aguado. *Feature Extraction in Computer Vision and Image Processing*. 2002.
10. N.R. Pal and S.K. Pal. A review on image segmentation techniques. *Pattern Recognition*, 26(9):1277–1294, 1993.
11. E.J. Pauwels and G. Frederix. Finding salient regions in images: Non-parametric clustering for image segmentation and grouping. *Computer Vision and Image Understanding*, 75(1/2):73–85, 1999.
12. H. Samet. *Applications of Spatial Data Structures: Computer Graphics, Image Processing and GIS*. 1990.
13. Yu Jin Zhang. A review of recent evaluation methods for image segmentation. *Proceedings of the Sixth ISSPA*, 1:148–151, 2001.

Application of a Two-Level Self Organizing Map for Korean Online Game Market Segmentation

Sang-Chul Lee[1], Jae-Young Moon[2], Jae-Kyeong Kim[2], and Yung-Ho Suh[2]

[1] Department of Management Information Systems, Korea Christian University, San 204, Hwagok 6-Dong, Kangseo-Ku, Seoul 157- 722, South Korea
leecho@kcu.ac.kr

[2] School of Business Administration, Kyung Hee University, 1 Hoegi-Dong, Dongdaemoon-Gu, Seoul 130-701, South Korea
{moonlight, jaek, suhy}@khu.ac.kr

Abstract. Online game industry has encountered higher competition in global market. To survive successfully in today's competitive online game markets, they need to determine who the target customers are and what motivates them. The purpose of our research is to identify the critical variables and to implement a new methodology for online game market segmentation using self organizing map. Our research tested the model with Korean online game users.

1 Introduction

A new revolutionary period of e-commerce has begun since the early 2000's [21, 22]. Recently, the global online game industry has been grown rapidly and has been developed into the core of the world cultural industries. With the rapid growth, many online game companies hoped that the first mover would be successful and recklessly entered into online game markets without understanding the core needs of those audiences. However, the lack of consideration has forced many online game companies to fail to survive in game market [16]. To survive in today's competitive markets, online game companies need to determine who the target customers are and what motivates them. This process is called market segmentation, by which companies are able to understand their loyal customers and concentrate their limited resources into them [20].

However, previous research had problems of both methodologies and variables. First, the traditional methodology for market segmentation was based mainly on statistical clustering techniques; hierarchical and partitive approaches. However, hierarchical method can not provide a unique clustering because a partitioning to cut the dendrogram at certain level is not precise. This method ignores the fact that the within-cluster distance may be different for different clusters [6, 29]. Partitive method predefines the number of clusters, before performing it. It can be part of the error function and can not identity the precise number of clusters [7, 25, 29]. Additionally, these algorithms are known to be sensitive to noise and outliers [4, 5, 29].

To settle these problems, we segment Korean online game market using a two-level Self-Organizing Map (SOM): SOM training and clustering [29]. Instead of clustering the data directly, a large set of prototypes is formed using the SOM. The prototypes can be interpreted as proto-cluster, which are combined in the next phase from the actual clusters. The benefit of using this method is to effectively reduce the

D.-Y. Yeung et al. (Eds.): SSPR&SPR 2006, LNCS 4109, pp. 808 – 816, 2006.

complexity of the reconstruction task and to reduce the noise. Our research implements this method into marketing research field.

Secondly, variables of previous studies could not be accepted since they have been conducted mainly from the technological and psychological perspective. The main concern of technological research on online game is to design and develop a more attractive and effective online game environment [1, 26, 31]. However, no matter how sophisticated the technologies applied, users would not revisit the game site if it failed to reflect their needs. Our research identifies the primary factors for online game from a business perspective.

The purpose of our research is to identify the critical variables and to implement a new methodology for online game market segmentation. Our research approach is categorized into two phases. The first phase is using a statistical approach (Structural Equation Model: SEM) to find the critical segmentation factors. The second phase is conducted by a two-level Self Organizing Map (SOM) to indicate the actual clusters. To implement our methodology, Korean online game data was analyzed because Korean online game market was located in the center of those trends. Therefore, research about Korean online game markets will be helpful for other countries to understand the change of global game markets.

2 Theoretical Background

2.1 Determinant Variables

In online game, more emphasis is being placed on the impact of a flow using a business perspective [2, 9, 10, 15, 18, 23, 28]. Through the review of the relevant literature, we identify the primary factors for online game from a business perspective as follows: the convenience of operator, the suitability of feedback, the reality of design, the precision of information and the involvement of virtual community. Our research hypothesizes that these determinants have a positive effect on flow.

The convenience of the operator was defined as the manipulatability of operators to play games [27]. Operator is an important determinant of influencing interaction between users and games [2, 12, 30]. Feedback is the reaction from online games [3, 10]. For example, when players kill a monster within NCsoft's Lineage, they receive feedback upgrading their level. The reality of design is defined as the design of interface making gamers feel online games as part of the real world [1, 26, 32]. Information is the contents from online game to achieve the stated goals. Gamers who received more precise information about how to play the games tended to achieve online game goals and experience flow easier [10, 24]. Virtual community is defined as computer-mediated spaces with potential for integration of member-generated content and communication [14]. Online game users should solve problems together interacting with other users in virtual communities [10].

2.2 A Two-Level SOM

Vesanto and Alhoniemi proposed a two-level SOM: SOM training and clustering. A two-level SOM was combined SOM, K-means and DB Index [29]. In the first level (SOM training), the data were clustered directly in original SOM to form a large set of prototypes. SOM was developed by Kohonen, which was very suitable for clustering in that it implemented an ordered dimensionality-reducing mapping of the training data and has prominent visualization properties [19].

In the second level (SOM clustering), the prototypes of SOM are clustered using k-means and the validity of clusters is evaluated using DB index. K-means clustering is to partition a data set into a set of group, minimizing distances within and maximizing distances between clusters. To select the best one among different partitioning, each of these can be evaluate using some kind of validity index. Generally, there are four validity indices; DB (Davies-Bouldin) index [11], Dunn's index [13], CH (Calinski-Harabasz) index [8] and index I [25]. DB index is suitable for evaluation of k-means partitioning because it gives low values, indicating good clustering results for spherical cluster [29]. Our research used DB index.

DB index is a function of the ratio of the sum of within-cluster scatter to between-cluster separation. The DB index is defined as equation (1).

$$DB(U) = \frac{1}{c}\sum_{i=1}^{c} \max_{i \neq j}\left\{\frac{\Delta(X_i) + \Delta(X_j)}{\delta(X_i, X_j)}\right\} \tag{1}$$

The scatter within the ith cluster, ΔX_i, is computed as equation (2), the distance between cluster X_i, and X_j, is denoted as equation (3). Here, Z_i represents the ith cluster centre.

$$\Delta X_i = \frac{1}{|C_i|}\sum_{x \in C_i} \{\| x - z_i \|\} \tag{2}$$

$$\delta(X_i, X_j) = \| z_i - z_j \| \tag{3}$$

Conclusively, the proper clustering is achieved by minimizing the DB index.

3 Research Methods

3.1 Research Framework

To segment the online game market and develop marketing strategies, our research approach is categorized into two phases. Firstly, the confirmatory factor analysis (CFA) and structural equation model (SEM) are used to identify the critical segmentation variables for clustering. Secondly, a two-level SOM is used to segment online game market. The first level develops the prototypes from large data set and the actual clusters are developed from the prototypes in the second level.

After segmentation of the markets, we use ANOVA to recognize the characteristics of sub-divided clusters. Finally, we target a segment market with the highest customer loyalty, and used those results as the starting point for the marketing strategies.

3.2 Data and Measurement

To test the model, a Web-based survey was employed. We developed the web-questionnaire page using a common gateway interface (CGI). We sent a mail to customer within OZ intermedia in Korea, which explained the objectives of the research and contained the link to the Web-Survey. Conclusively, the 1704 complete data is available for analysis, after elimination of missing data.

Our research developed multi-item measures for each construct. Twenty-one items for five determinants are selected. We asked respondents to indicate on a five point Likert scale to what extent the determinants influence on flow in online game. We used CFA to evaluate convergent validity for five constructs and 15 items remained within our model. All the fit statistics of the measurement model were acceptable.

4 Results

4.1 Identification of Critical Factors

To find the critical factors for segmentation, we used AMOS 4.0 in structural equation modeling (SEM). The structural model was well converged. The results indicated that the chi-square of the model was 295.82 with d.f. of 104, the ratio of chi-square to d.f. was 2.844, GFI was 0.982, AGFI was 0.973, RMSR was 0.031 and NFI was 0.975; all the fit statistics were acceptable. Additionally, the squared multiple correlations (R^2) indicated that the present model explains 51 % of the variance in flow.

Table 1. The results of Stuctural Equeation Model

Path			Estimate	S.E.	t	p
O	-->		0.037	0.024	1.456	0.145
FB	-->		0.108**	0.032	3.482	0.000
IF	-->	F	0.081*	0.035	2.499	0.012
D	-->		0.274**	0.036	8.798	0.000
C	-->		0.437**	0.031	14.794	0.000

* $p<0.05$, ** $p<0.01$

O: The convenience of operator, FB:The suitability of feedback
IF: The precision of information, D: Reality of Design
C: The involvement of virtual community, F: Flow,

Four of the five paths were statistically significant and the path from the convenience of operator to flow was insignificant, as shown in Table 1. The critical variables for marketing segmentation are the suitability of feedback, the reality of design, the precision of information and the involvement of virtual community.

4.2 Market Segmentation

To segment the Korean online game market, our research was conducted using a two-level SOM. In the experiments, the first level was SOM training. 1704 data samples of the Korean were collected using the test variables: the suitability of feedback, the precision of information, the reality of design and virtual community except the convenience of operator. A SOM was trained using the sequential training algorithm for

Korean data samples. A neighborhood width decreased linearly 5 to 1 using the Gaussian function. A map was used by 19*11 matrix and 209 prototypes were developed.

The second level was SOM clustering. The partitive clustering of 209 SOM's prototypes was carried out using batch K-means algorithm. The K-means ran multiple times for each k. The DB index was used to select the best clustering in Fig. 2. The analysis of the DB index resulted in the development of ten market segments.

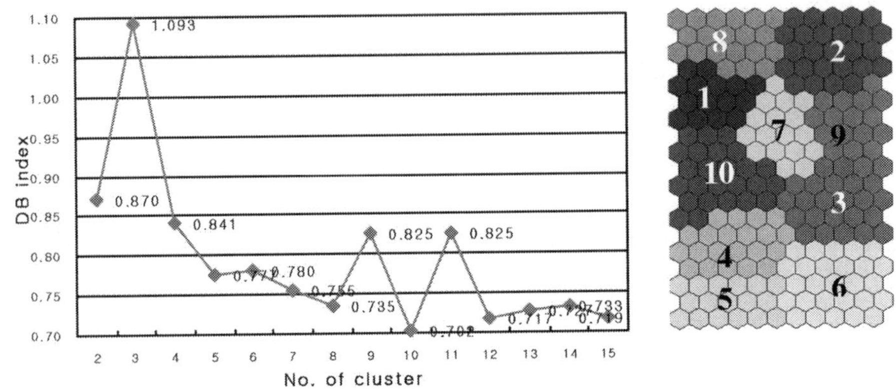

Fig. 2. DB Index and visualization of clusters

4.3 Determination of Target Market

After segmenting the markets, we used ANOVA to recognize the variable characteristics of each cluster. According to results of ANOVA, all variables (components) were significant; F=154.34 to 588.19 and p=0.00. To precisely recognize the variable characteristics of clusters, we categorized the effectiveness of the variables into 3 levels; high, middle and low. The middle level ranged between 3 ± 2.5 because our research measurement was used on a five point Likert scale. The high score suggested that the cluster was influenced by the variables positively, the middle score was normal, the low score was negative.

Additionally, to identify the structure of the clusters, we conducted on the analysis of the demographic and behavioral variables: gender, age, income level, i_year (how long did gamers use the Internet), i_day (how many hours did gamer use the Internet per day), and g_day (how many hours did gamer play online games per day). The characteristics and structure of clusters are summarized in Table 2.

The analysis of customer loyalty indicated that the ranks of clusters are as follows: cluster 6 (4.02 for average) > cluster 5 (3.94) > cluster 3 (3.76) > cluster 4 (3.69) > cluster 9 (3.54) > cluster 7 (3.48) > cluster 10 (3.28) > cluster 1 (3.24) > cluster 2 (3.07) > cluster 8 (2.86) in Table 2. The other analysis of the intention of revisit and WOM (Word of Mouth) indicated the same results. As a result, cluster 6 was indicated as the primary target market.

Table 2. Profiles of clusters

	C 1 (n=116)	C 2 (n=204)	C 3 (n=130)	C 4 (n=130)	C 5 (n=176)	C 6 (n=224)	C 7 (n=103)	C 8 (n=196)	C9 (n=161)	C 10 (n=264)
FB	2.72 (L*)	1.52 (L)	2.26 (L)	3.10 (M)	3.72 (H)	2.72 (L)	2.44 (L)	1.91 (L)	1.80 (L)	3.03 (M)
IF	2.69 (L)	2.50 (L)	3.17 (M)	3.22 (M)	3.76 (H)	3.63 (H)	2.72 (L)	2.18 (L)	3.01 (M)	2.92 (M)
D	2.77 (M)	2.47 (L)	3.56 (H)	3.25 (M)	3.60 (H)	3.80 (H)	2.97 (M)	2.55 (L)	3.22 (M)	3.08 (M)
C	2.68 (L)	3.45 (H)	3.59 (H)	3.49 (H)	3.67 (H)	4.00 (H)	3.35 (H)	2.38 (L)	3.56 (H)	3.05 (M)
Gender	male	male	female	female	male	female	both	male	female	female
Age	26-30	21-25	21-25	26-30	21-25	26-30	21-30	26-30	26-30	26-30
Income ($)	1,001-2,000	- 500	1,001-2,000	1,001-2,000	501-1,000	501-1,000	- 500	- 500	1,001-2,000	501-1,000
i_year	3	3	2-4	3	3	2-4	3	4	2-5	2-3
i_day	2	5	3-5	3-5	5	5, 10	5	3	5	3
G_day	1	2	3	1-2	1	2	1	1	1	1
Revisit	3.29	3.15	3.77	3.72	3.94	4.02	3.51	2.96	3.61	3.33
WOM	3.18	3.00	3.75	3.67	3.93	4.02	3.45	2.75	3.47	3.22
Loyalty**	3.24	3.07	3.76	3.69	3.94	4.02	3.48	2.86	3.54	3.28
Rank	8	9	3	4	2	1	6	10	5	7

* L=Low, M=Middle, H=High

** Loyalty is estimated by average of revisit and WOM

O: The convenience of operator, FB:The suitability of feedback

IF: The precision of information, D: Reality of Design

C: The involvement of virtual community, F: Flow,

5 Implication

The results of our research have the following implications for Korean online game companies. To attract the primary target audiences, companies should develop strategies depending on the effectiveness of the variables and the demographic and behavioral characteristics of cluster 6.

The characteristics of target audiences indicate that the members are positively influenced by the suitability of feedback, the reality of design and the involvement of virtual community. The strategies of the reality of the suitability of feedback proposed that companies should provide gamers with a higher level faster, items and more cybermoney, when gamers completed their mission. The strategies of the reality of design proposed that companies should make an interface where the game site looks real. For example, the interface of recent games changed 2D such 'Lineage' into 3D such as 'MU', 'Lagnarok' and 'Laghaim'. For virtual community, companies need to provide a Role Playing Game (RPG) where the gamer cooperates with each other rather than shooting games where the gamer compete with each other. Furthermore, the different villages and guilds which were harmonized with customer needs were provided. For example, 'Lineage' provided 15 villages to satisfy the different gamers' needs.

As to the demographic information, there are more female members in the target group than male members are. Popular ages range from 26 to 30 in the group. Monthly income level of the members of the group ranges from $500 to $ 1,000. They have used the Internet for over 2-4 years, use the Internet for 5 or 10 hours per day, and play online games for more than 2 hours per day.

The result indicated that online game companies should develop diverse types of online games considering the extension of the age of online game users. The number of female users is growing fast and the needs of online game users become diversified [21, 33]. To better satisfy their needs, online game companies should cluster similar customers into specific market segments with different demands and then develop marketing strategy based on their properties. Especially, our research shows that the middle-aged and female users are classified as target customers as well as adolescents. This finding is consistent with the statistics in the Korean Game White Paper, which indicates that female users increased from 31% of the game population in 2001 to 47% in 2003 and the middle-aged users increased from 2% in 2001 to 21% in 2003.

These implications were proven to be true through NCsoft's example, which is the primary Korean online game company. They recognized that online game customers' needs have been changed and encountered higher competition with foreign online game competitors. To survive in this changing environment, they developed the games for male and female separately. For instances, the background of the recent game 'Lineage' was medieval, the type was combatable, and their target audiences were adolescents and younger male, while 'Shining Lore' is developed to target female customers who might be more interested in sweet and exciting stories [16].

6 Conclusion and Limitation

The results of our study have several contributions to academia and business world. Our research identifies the new primary factors for online game markets which may not be found in the previous researches from the technological perspectives. Additionally, our research proposes a new methodology for market segmentation using a two-level SOM and marketing strategies for the survival in competitive online game market.

Even though our research is conducted on Korean online game market, these implications are able to be applied into those of other countries because Korean online game market is the frontiers of global online game market. And the research about Korean online game market is thought to be helpful for other countries to understand the change of their own online game markets. However, other countries are able to develop their own marketing strategies more exactly using our methods with considering and adjusting their market environment, instead of accepting the results of our research.

For further study, more demographic and behavioral variables might be necessary to segment the markets more precisely. Secondly, a cross-national analysis can be added to our research in order to better understand the loyal customers in different countries.

References

1. Ackley, J.: Roundtable Reports: Better Sound Design. Gamasutra. (1998). http://www.Gamasutra.com /features/gdc_reports/cgdc_98/ ackley.htm
2. Agarwal, R., Karahanna, E.: Time Files When You're Having Fun: Cognitive Absorption and Beliefs About Information Technology Usage. MIS Quarterly, Vol.24, No.4. (2000) 665-694

3. Baron, J.: Glory and Shame: Powerful Psychology in Multiplayer Online Games. Gamasutra. (1999). http: //www.gamasutra.com/features/19991110/Baron_01.htm
4. Bezdek, J. C.: Some New Indexes of Cluster Validity. IEEE Transactions on Systems, Man, and Cybernetics, Part B, Vol.28, No.3. (1998) 301-315
5. Blatt, M., Wiseman, S., Domany, E.: Super-paramagnetic Clustering of Data. Physical Review Letters, Vol.76, No.8. (1996) 3251-3254
6. Boudaillier, E., Hebrail, G.: Interactive Interpretation of Hierarchical Clustering. Intelligent Data Analysis, Vol.2, No.3. (1998) 41-
7. Buhmann, J., KÜhnel, H.: Complexity Optimized Data Clustering by Competitive Neural Networks. Neural Computation, Vol.5, No.3. (1993) 75-88
8. Calinski, R. B., Harabasz, J.: A Dendrite Method for Cluster Analysis. Communication in Statistics, Vol.3. (1974) 1-27
9. Cho, N. J., Back, S. I., Ryu, K. M.: An Exploratory Investigation of Player Loyalty to Online Games. Journal of the Korean Operations Research and Management Science Society, Vol.26, No.2. (2001) 85-97
10. Choi, D. S., Park, S. J., Kim, J. W.: A Structured Analysis Model of Customer Loyalty in Online Games. Journal of MIS Research, Vol.11, No.3. (2001) 1-20
11. Davies, D. L., Bouldin, D. W.: A Cluster Separation Measure. IEEE Transactions Pattern Analysis and Machine Intelligence, Vol.1. (1979) 224-227
12. Davis, F. D., Bagozzi, R. P., Warshaw, P. R.: Extrinsic and Intrinsic Motivation to Use Computers in the Workplace. Journal of Application Society Psychology, Vol.22, No.14. (1992) 1111-1132
13. Dunn, J. C.: A Fuzz Relative of the ISODATA Process and Its Use in Detecting Compact Well-Separated Clusters. Journal of Cybernetics, Vol. 3. (1973).32-57
14. Hagel, J., Armstrong, A.: Net Gain: Expanding Markets through Virtual Communities. Harvard Business School Press, Boston. (1997)
15. Hoffman, D. L., Novak, T. P.: Marketing in Hypermedia Computer Mediated Environments: Conceptual Foundations. Journal of Marketing, Vol.60, No.3. (1996) 50-68
16. ICA: 2003 Cases of Information and Telecommunication Export, ICA, Seoul. (2003)
17. KGDI: 2003 Korean White Game Paper. KGDI, Seoul. (2003)
18. Kim, N.H., Lee, S.C., Suh, Y.H.: Strategy of Market Penetration in Japanese Internet Market: Comparing Online Game Loyalty between Korea and Japan with MSEM. Journal of the Korea Society for Quality Management, Vol.31, No.1. (2003) 21-41
19. Kohonen, T.: Self-organizing Map. Proceedings of the IEEEE, Vol.78, No.9. (1990) 1469-1480
20. Kotler, P.: Marketing Management: Analysis, Planning, Implementation and Control, 9th edn. A Simon and Schuster Co, New Jersey. (1997)
21. Laudon, K. C., Traver, C. G.: E-Commerce: Business, Technology, Society. Addison Wesley, Boston. (2002)
22. Lee, H. K.: Overview and Problems within the Korean Online Game Industry. Information and Telecommunication Policy, Vol.13. (2000) 20-37
23. Lee, S. C., Kim, N. H., Suh, Y. H.: The Effect of Flow and Addiction upon Satisfaction and Customer Loyalty in Online Games. Korean Management Review, Vol.32, No.3. (2003) 1479-1501
24. Lewinski, J. S.: Developer's Guide to Computer Game Design. Wordware Publishing Inc, Texas. (2000)
25. Maulik, U., Bandyopadhyay, S.: Performance Evaluation of Some Clustering Algorithms and Validity Indices. IEEE Transactions of Pattern Analysis and Machine Intelligence, Vol.24, No.12. (2002) 1650-1654

26. Sanchez-Crespo, D.: 99 from a Game Development Perspective. Gamasutra. (1999). http://gamasutra.com/ features/19990802/siggraph_01.html
27. Spector, W.: Remodeling RPGs for the New Millennium. Gamasutra. (1999). http://www.gamasutra.com/ features/game_desing/19990115/remodeling_01.htm
28. Sujan, H., Weitz, B. A., Kumar, N.: Learning, Orientation, Working Smart, and Effective Selling. Journal of Marketing, Vol.58, No.3. (1994) 39-52
29. Vesanto, J., Alhoniemi, E.: Clustering of the Self-organizing Map. IEEE Transactions on Neural Networks, Vol.11, No.3. (2000) 586-600
30. Webster, J., Martocchio, J. J.: Micro Computer Playfulness: Development of a Measure with Workplace Implications. MIS Quarterly, Vol.16, No.2. (1992) 201- 226
31. Wells, J. D., Fuerst, W. L., Choobinceh, J.: Managing Information Technology for One to One Customer Interaction. Information and Management, Vol.35, No.1. (1999) 53-62
32. Woodcock, W.: Game AI: the State of the Industry. Gamasutra. (1999). http:// www.gamasutra.com /features/19990820/game_ai_01.html
33. Yu, S. S.: Overview of Information and Telecommunication Industries: Software and Internet Contents. Information and Telecommunication Policy (2002) 131-144

Clustering Based on Compressed Data for Categorical and Mixed Attributes

Erendira Rendón[1] and José Salvador Sánchez[2]

[1] Lab. Reconocimiento de Patrones, Instituto Tecnológico de Toluca
Av. Tecnológico s/n, 52140 Metepec, Mexico
erendon@ittoluca.edu.mx

[2] Dept. Llenguatges i Sistemes Informàtics, Universitat Jaume I
Av. Sos Baynat s/n, E-12071 Castelló de la Plana, Spain
sanchez@uji.es

Abstract. Clustering in data mining is a discovery process that groups a set of data so as to maximize the intra-cluster similarity and to minimize the inter-cluster similarity. Clustering becomes more challenging when data are categorical and the amount of available memory is less than the size of the data set. In this paper, we introduce CBC (*Clustering Based on Compressed Data*), an extension of the Birch algorithm whose main characteristics refer to the fact that it can be especially suitable for very large databases and it can work both with categorical attributes and mixed features. Effectiveness and performance of the CBC procedure were compared with those of the well-known K-modes clustering algorithm, demonstrating that the CBC summary process does not affect the final clustering, while execution times can be drastically lessened.

1 Introduction

Clustering constitutes a very effective technique for exploratory data analysis and has been widely studied for several years. It has found applications in a broad variety of areas such as pattern recognition, statistical data analysis and modelling, data and web mining, image analysis, marketing and many other business applications. The basic clustering problem consists of grouping a data set into subsets (typically called clusters) such that items in the same subset are similar to each other, whereas items in different subsets are as dissimilar as possible. The general idea is to discover a structure that is already present in the data. Most of the existing clustering algorithms can be classified into two main categories, namely hierarchical (agglomerative or divisive) and partitioning algorithms [8].

Pattern recognition and data mining practical applications frequently require dealing with high volumes of data (thousands or millions of records with tens or hundreds of attributes). This characteristic excludes the possibility of using many of the traditional clustering algorithms. Furthermore, this type of application is often done with data containing categorical attributes, thus becoming more difficult.

It has to be noted that much of the data in the databases are categorical, that is, fields in tables whose attributes cannot naturally be ordered as numerical values. The problem of clustering categorical data involves complexity not encountered in the corresponding

D.-Y. Yeung et al. (Eds.): SSPR&SPR 2006, LNCS 4109, pp. 817–825, 2006.

problem for numerical data. While considerable research has been done on clustering numerical data, there has been much less work on the important problem of clustering categorical data.

The present paper focuses on clustering databases with categorical and/or mixed (both categorical and numerical) attributes. To this end, a new clustering algorithm is here introduced, which is based on summarizing the data by using a balanced tree structure. The resulting tree is then utilized to group the data into clusters, which will be finally labelled by means of a nearest neighbor rule. Moreover, it is worth noting that this algorithm allows its application to very large databases (data sets of size greater than the size of available memory).

2 Related Algorithms

Clustering numerical data has been the focus of substantial research in various domains for decades, but there has been much less work on clustering categorical data. Recently, the important problem of clustering categorical data started receiving interest. In this section, we briefly review a number of algorithms belonging to the area of clustering categorical records.

The Rock algorithm [5] is an adaptation of an agglomerative hierarchical algorithm. The procedure attempts to maximize a goodness measure that favors merging pairs with a large number of "links". Two objects are called neighbors if their similarity exceeds a certain threshold given by the user. The number of links between two objects is the number of common neighbors. The Rock algorithm selects a random sample from the databases after a clustering algorithm that employs links is applied to the sample. Finally, the obtained clusters are used to assign the remaining objects on the disk to the appropriate clusters. Huang proposes the K-modes algorithm [6], which is an extension of the K-means procedure for categorical data. The way to compute the centroid is substituted by a vector of modes. It also proposes two measures of similarity for categorical data. The final clustering of the K-modes algorithm depends on the initial selection of the vector of modes, very much the same as with its predecessor K-means.

The Coolcat algorithm proposed by Barbará et al. [2] is an incremental algorithm that aims to minimize the expected entropy of the clusters. Given a set of clusters, the algorithm will place the next point in the cluster where it minimizes the overall expected entropy. Similar to this proposal is the Limbo technique [1], which constitutes a scalable hierarchical categorical clustering algorithm that builds on the Information Bottleneck (IB) framework for quantifying the relevant information preserved when clustering. As a hierarchical algorithm, Limbo has the advantage that it can produce clusters of different sizes in a single execution. Moreover, Limbo handles large data sets by producing a memory bounded summary model for the data.

Cactus [3] represents an agglomerative hierarchical algorithm that employs data summarization to achieve linear scaling in the number of rows. It requires only two scans of the data. This scheme is based on the idea of co-occurrence for pairs of attributes-values. The Birch algorithm [10] constructs a balanced tree structure (the CF-tree), which is designed for a multi-phase clustering method. First, the database is scanned to build an initial in-memory CF-tree which can be seen as a multi-level compression of the data that tries to preserve the inherent clustering structure of the data.

Second, an arbitrary clustering algorithm can be used to cluster the leaf nodes of the CF-tree.

3 The CBC Algorithm

The new CBC clustering algorithm here introduced consists of three main stages: (1) summary, (2) clustering, and (3) labelling. In the first stage, the data are summarized in a balanced tree structure. The summary obtained in the first step is then grouped by means of a clustering procedure. Finally, in the the third stage we perform a scan over the database and assign each object to the representative (that is, the composite object) closest to the clusters obtained in the previous phase.

Next, we provide the definition of several concepts that will be importantly used by the CBC algorithm.

Definition 1. *An* **event** *is a pair relating features and values. It can be denoted by* $[X_i = E_i]$*, indicating that the feature* X_i *takes the values of* E_i *and* $E_i \subset U_i$*.* E_i *is the subset of values that the feature* X_i *takes,* U_i *is the representation domain of* X_i*. An example of event is* $e_1 = [color = green, blue, red]$*.*

Definition 2. *A* **categorical object** *is a logical join of events, relating values and features, where features may take one or more values [4]. It is denoted by* $X = [X_1 = E_1] \wedge \ldots \wedge [X_d = E_d]$*. A categorical object can be represented by the Cartesian product set* $E = E_1 \times \ldots \times E_d$*. The domain of categorical object* X *is represented by* $U^{(d)} = U_1 \times \ldots \times U_d$*, that is, the d-dimensional feature space. For example, the categorical object represented by* $X = [HairColor = black, brown] \wedge [BloodType = B+, A+]$ *has the following features: HairColor is black or brown; BloodType is B+ or A+.*

The intersection of two categorical objects $E_i = E_{i1} \times \ldots \times E_{id}$ *and* $E_j = E_{j1} \times \ldots \times E_{jd}$ *is defined as* $E_i \bigotimes E_j = (E_{i1} \bigotimes E_{j1}) \times \ldots \times (E_{id} \bigotimes E_{jd})$*. Analogously, the union of* E_i *and* E_j *is* $E_i \bigoplus E_j = (E_{i1} \bigoplus E_{j1}) \times \ldots \times (E_{id} \bigoplus E_{jd})$*.*

Definition 3. *Let* $E_i = E_{i1} \times \ldots \times E_{id}$ *and* $E_j = E_{j1} \times \ldots \times E_{jd}$ *be two objects defined in* $U^{(d)}$*, then a* **composite object** *is the result of combining* E_i *and* E_j *as* $E_i \bigoplus E_j = (E_{i1} \bigoplus E_{j1}) \times \ldots \times (E_{id} \bigoplus E_{jd})$*. For example, consider two objects* $A = \{green, B+, high\}$ *and* $B = \{black, O+, high\}$*, then the composite object formed from these is* $A \bigoplus B = \{\{green, black\}, \{B+, O+\}, high\}$*.*

3.1 A Similarity Metric for Categorical Objects

In the present work, we have used a distance metric similar to that proposed by Ichino and Yaguchi [7]. The distance between objects $X_i = [X_{i1} = E_{i1}] \wedge \ldots \wedge [X_{id} = E_{id}]$ and $X_j = [X_{j1} = E_{j1}] \wedge \ldots \wedge [X_{jd} = E_{jd}]$ in $U^{(d)}$ is computed by:

$$d_p(X_i, X_j) = [\sum_{k=1}^{d} C_k \psi(E_{ik}, E_{jk})^p]^{1/p} \qquad p \geq 1 \tag{1}$$

where $C_k > 0$ $(k = 1, \ldots, d)$ is a weighting factor to control the relative importance of the event E_k or $C_k = 1/d$ when all the events have the same relevance, and

$$\psi(E_{ik}, E_{jk}) = \frac{\phi(E_{ik}, E_{jk})}{|U_k|} \tag{2}$$

being $|U_k|$ the number of possible values in the domain U_k, and $\phi(E_{ik}, E_{jk}) = |E_{ik} \cup E_{jk}| - |E_{ik} \cap E_{jk}|$

Now we can transform the distance measure in Eq. 1 into a similarity metric [8], such as $S(X_i, X_j) = 1 - d_p(X_i, X_j)$.

3.2 Summary of the Database

The central task of the first stage of CBC is the construction of a tree structure to compress the data. It uses a weight-balanced tree, called *CO-tree*, which needs three parameters: the number of entries for non-leaf nodes B, the number of entries for leaf nodes L, and an absorbtion threshold T. Each non-leaf node contains at most B entries, each one with a pointer to its child node, and the composite object represented by this child. A leaf node contains at most L entries with composite objects. Moreover, all entries in a leaf node must satisfy a condition with respect to the absorbtion threshold T. Fig. 1 illustrates an example of the CO-tree structure.

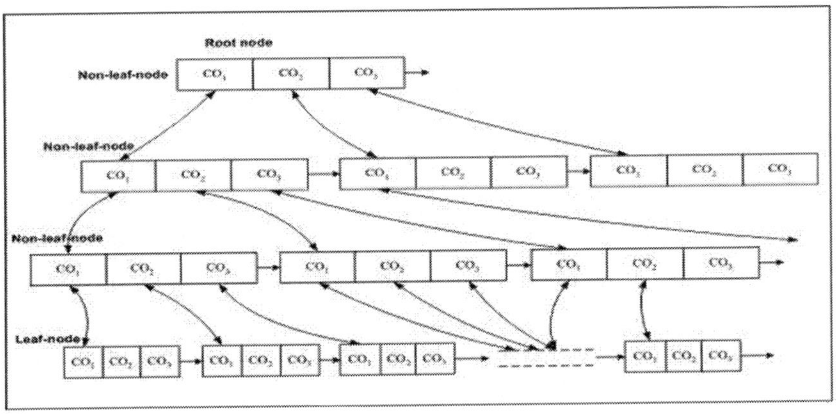

Fig. 1. An example of the CO-tree structure ($B = 3$, $L = 3$)

Insertion into a CO-Tree. Here we present the algorithm for inserting a new entry, say ent, into a given CO-tree.

1. Identifying the appropriate leaf: starting from the root, it recursively descends the CO-tree by choosing the closest child node according to the similarity metric given in Sect. 3.1.
2. Updating the leaf: when reached a leaf node, it finds the closest leaf entry, say L_i, and then tests whether L_i can absorb ent without violating the threshold condition. If so, the new entry is absorbed by L_i. If not, it tests whether there is space for this new entry on the leaf. If so, a new entry for ent is added to the leaf. Otherwise, we must split the leaf node. Splitting is done by choosing the farthest pair of entries as seeds, and redistributing the remaining entries based on the closest criteria.

3. Updating the path to the leaf: after inserting ent into a leaf, we must update the information for each non-leaf entry on the path from the root to the leaf. If no splitting was done, this simply involves the leaf entry that absorbed ent. Conversely, a leaf split means that a new non-leaf entry has to be inserted into the parent. If the parent has space for this new entry, at all higher levels, we only need to update the corresponding entries to reflect the addition of ent. However, we may have to split the parent as well, and so on up to the root. If the soot is split, the tree height increases by one.

The first stage of CBC starts with an initial threshold value $T = 0$, scans the data, and inserts entries into the CO-tree. In this phase, it is possible to carry out an additional step of the summary by using a reconstruction algorithm. This reconstruction algorithm is applied when a more compact CO-tree is required or when the memory is depleted and some objects in the database have not been inserted yet. When rebuilding the CO-tree (smaller than the initial one), the same insertion algorithm is used, but increasing the threshold value $T = T + 0.2$. It reinserts the leaf entries of the old tree and then, the scanning of the data is resumed from the point at which it was interrupted. This process continues until all objects in the database have been scanned. The leafs of the resulting CO-tree contain a summary of the database in the manner of composite objects.

3.3 Clustering the Leaf Nodes

The second stage of the algorithm basically consists of going through the leaf nodes of the resulting CO-tree in order to obtain composite objects that will be representatives of the clusters.

After the construction of the CO-tree, that is, when the entire database has been summarized, we cluster the composite objects present in the leaf nodes, thus producing groups of composite objects. In this phase, any clustering algorithm could be applied. However, we propose a new clustering algorithm called EA. The EA clustering algorithm is described in detail below.

Definition 4. *Let $B = \{X_1, X_2, \ldots, X_n\}$ be a set of composite objects in $U^{(d)}$ obtained from the final CO-tree. Then $C \subseteq B$, $C \neq \emptyset$ will be a* **cluster** *if and only if the following conditions are satisfied:*

a) $\forall X_j \in B\ \ [X_i \in C \wedge \max_{X_t \in B, X_t \neq X_i} \{S(X_i, X_t)\} = S(X_i, X_j) \geq \beta] \Rightarrow X_j \in C$
b) $[X_t \in C \wedge \max_{X_p \in B, X_p \neq X_i} \{S(X_i, X_p)\} = S(X_p, X_t) \cdot \beta] \Rightarrow X_p \in C$
c) $|C|$ *is minimum*

where $0 \leq \beta \leq 1$ is a similarity threshold.

EA Clustering

1. Scan all leaf entries (composite objects) present in the final CO-tree.
2. Compute the similarity matrix using Eq. 1.
3. Compute β as an average of the values in the similarity matrix. It can also be given by a human expert.
4. Compute the clusters C.
5. Compute the composite objects of each cluster formed in Step 4.

4 Experimental Analysis

In this section, we evaluate the performance of CBC and compare its effectiveness for clustering with that of the classical K-modes algorithm. We performed two groups of experiments: in the first one, we ran the CBC algorithm with respect to parameters L, B and memory size. In the second one, we compared CBC with the K-modes algorithm in terms of misclassified objects and running times. We consider an object as misclassified when the original tag was different from that assigned by CBC.

4.1 Description of the Data Sets

The present experimental study was carried out by using three well-known benchmark databases taken from the UCI Machine Learning Database Repository (`http://www.ics.uci.edu/ Mlear/MLRepository.html`).

- Mushroom: each data record contains information that describes the physical characteristics (e.g., color, odor, size, shape) of a single mushroom. The mushroom database has a total of 8124 records belonging to two classes: 4208 edible mushrooms and 3916 poisonous mushrooms.
- Connect-4: it consists of 67557 records, each one described by 42 characteristics. The three classes are *win* with 44473 records, *loss* with 16635, and *draw* with 6449 records.
- Kr-vs-kp: Chess and -Game-King+Rook versus King+Pawn contains 3196 records, each one described by 36 attributes. The two classes are *white-can-win* with 1669 and *white-cannot-win* with 1527 records.

4.2 Results

The first experiment pursues to analyze the effect of memory size on running times and percentage of misclassified objects. To this end, we have tested the CBC algorithm when varying the memory size and keeping constant the values of parameters L and B. For the Mushroom database, $L = 5$ and $B = 3$. For the Kr-vs-kp data set, $L = 6$ and $B = 3$. For the Connect-4 database, $L = 4$ and $B = 3$. Table 1 reports the results corresponding to these experiments. The third column provides the running times for the summary stage, that is, the construction of the CO-tree to summarize the data. Columns 4, 5, and 6 correspond to the percentage of misclassified objects for different values of the parameter β in the second stage of the algorithm. Finally, the seventh column shows the average running times (the time required for the summary along with the time for the clustering stage).

From the results in Table 1, one can see that the time for the summary stage is close to linear with respect to the memory size. On the other hand, the parameter β used in the clustering stage of CBC significantly affects to the quality of the clusters obtained. In general, it seems that the lower the values of β, the lower the percentage of misclassified objects. Finally, the size of the memory does not affect to the result of the CBC clustering algorithm in terms of percentage of misclassified objects.

The second experiment tries to study the effect of the parameter L. Correspondingly, Table 2 provides the results when varying the value of parameter L, while keeping

Table 1. Effect of memory size (Kbytes) on running times (seconds) and percentage of misclassified objects

	Memory size	Summary time	$\beta = 0.2$	$\beta = 0.1$	$\beta = 0.05$	Average time
Mushroom	5	1.34	12.81	7.03	6.08	2.48
	10	1.33	12.81	7.03	6.08	
	20	2.50	12.81	7.03	6.08	
	25	3.00	12.81	7.03	6.08	
Kr-vs-kp	2	1.69	40.47	14.29	10.43	1.42
	3	1.88	46.38	16.05	10.66	
	4	2.19	47.88	20.19	12.58	
	5	2.18	47.88	19.19	12.31	
Connect-4	40	51.21	34.16	35.82	34.16	45.00
	60	50.11	34.16	34.16	34.16	
	80	68.62	34.16	36.14	34.38	
	100	69.17	34.16	36.14	29.58	

Table 2. Effect of parameter L

	L	Summary time	$\beta = 0.2$	$\beta = 0.1$	$\beta = 0.05$	Average time
Mushroom	3	2.50	15.12	6.94	4.55	2.50
	5	2.50	12.81	7.03	6.08	
	7	2.80	18.52	9.56	11.38	
	9	3.38	16.37	13.90	12.83	
	11	3.54	18.63	10.56	10.56	
Kr-vs-kp	4	1.49	38.27	42.25	26.48	1.64
	5	1.83	26.55	16.59	11.53	
	6	1.69	40.57	14.29	10.43	
	7	2.00	47.20	14.03	11.33	
Connect-4	2	50.29	33.56	31.53	31.53	48.09
	3	57.74	30.98	30.98	30.98	
	4	64.94	34.22	34.22	34.22	
	5	60.86	34.16	29.99	30.98	

constant the memory size and the value of B. The memory size is 20, 2, and 20 for Mushroom, Kr-vs-kp, and Connect-4 databases, respectively. For the three data sets we have used $B = 3$. Most comments drawn for the first experiments become valid for the present analysis. In this sense, one can observe that the time for the summary stage is close to linear with respect to the value of L. Also, it seems that the percentage of misclassified objects does not depend on the parameter L.

Analogously, the third group of experiments are devoted to study the effect of the parameter B. Table 3 reports the running times and percentage of misclassified objects when varying the value of the parameter B. The size of available memory is 20, 2, and 20 for Mushroom, Kr-vs-kp, and Connect-4 databases, respectively. For the three data sets we have used $L = 3$. Like in the previous experiments, the value of the parameter B does not affect to the quality of the clusters given by the CBC algorithm.

Table 3. Effect of parameter B

	B	Summary time	$\beta = 0.2$	$\beta = 0.1$	$\beta = 0.05$	Average time
Mushroom	3	2.50	15.12	6.94	4.55	2.52
	5	2.10	18.17	7.88	5.93	
	7	2.56	22.44	12.51	8.34	
	9	2.51	18.66	10.06	6.35	
	11	2.58	16.84	7.33	7.11	
Kr-vs-kp	4	1.53	38.27	35.42	30.03	1.63
	5	1.52	38.27	28.16	22.51	
	6	1.62	40.06	19.90	11.85	
	7	1.51	39.55	17.29	12.26	
Connect-4	3	57.74	30.98	30.98	30.98	49.98
	4	61.21	32.36	32.36	32.36	
	5	58.07	34.16	40.23	40.23	
	6	63.61	34.15	34.15	34.15	

Table 4. Comparison of CBC and K-modes in percentage of misclassified objects

	CBC		K-modes
	Worst	Best	
Mushroom	22.44	4.55	7.42
Kr-vs-kp	47.88	10.43	34.01
Connect-4	40.23	29.99	45.03

Finally, Table 4 allows to compare the percentage of misclassified objects by means of the CBC procedure with that of the well-known K-modes clustering algorithm. As can be seen, in all domains the CBC approach has shown a better behavior than the K-modes algorithm. The most important differences are with the Kr-vs-kp database, in which the CBC algorithm obtains a 10.43% of error rate, while that of the K-modes is 34.01%. On the other hand, it has to be noted that in the case of Connect-4, even the worst cluster given by CBC (40.23%) outperforms the result of K-modes (45.03%).

5 Concluding Remarks

This paper introduces the CBC algorithm, *Clustering Algorithm Based on Compressed Data*, which builds a summary of the database in the main memory using a balanced tree structure called CO-tree. The CBC algorithm works with categorical features, and also with mixed data. Another important characteristic of the new algorithm refers to the fact that it can handle large data sets.

The CBC clustering algorithm consists of tree main stages: (1) summary, (2) clustering, and (3) labelling. Although the databases have to be summarized, the results of our experimental study with three benchmark databases are very encouraging because it clearly outperforms the K-modes in terms of percentage of misclassified objects.

Possible extensions to this work are in the direction of testing the CBC algorithm with other similarity measures. Also, the possibility of using a different clustering approach in the second stage becomes especially important in order for improving the quality of the resulting clusters. Finally, a more exhaustive empirical analysis is necessary to corroborate the conclusions given in the present paper.

Acknowledgments

This work has been partially supported by research grants TIC2003-08496 from the Spanish CICYT (Ministry of Science and Technology).

References

1. Andritsos, P., Tsaparas, P., Miller, R.J., Sevcik, K.C.: LIMBO: scalable clustering of categorical data, In: Proc. 9th Intl. Conf. on Extending Database Technology (2004) 123–146.
2. Barbará, D., Li, Y., Couto, J.: COOLCAT: an entropy-based algorithm for categorical clustering, In: Proc. 11th Intl. Conf. on Information and Knowledge Management (2002) 582–589.
3. Ganti, V., Gehrkeand, J., Ramakrishanan, R.: CACTUS — Clustering categorical data using summaries, In: Proc. 5th ACM Sigmod Intl. Conf. on Knowledge Discovery in Databases (1999) 73–83.
4. Gowda, K., Diday, E.: Symbolic clustering using a new dissimilarity measure, Pattern Recognition **24** (1991) 567–578.
5. Guha, S., Rastogi, R., Shim, K.: ROCK: A robust clustering algorithm for categorical attributes, In: Proc. of the IEEE Intl. Conf. on Data Engineering (1999) 512–521.
6. Huang, Z.: A fast clustering algorithm to cluster very large categorical data sets in data mining, In: Proc. SIGMOD Workshop on Research Issues on Data Mining and Knowledge Discovery, Tech. Report 97–07, UBC, Dept. of Computer Science (1997).
7. Ichino, M., Yaguchi, H.: Generalized Minkowski metrics for mixed feature-type data analysis, IEEE Trans. on Systems, Man and Cybernetics **24** (1994) 698–708.
8. Kaufman, L., Rousseeuw, P.J.: *Finding Groups in Data: An Introduction to Cluster Analysis*. John Wiley & Sons, New York (1990).
9. Milenova, B.L., Campos, M.M.: Clustering large databases with numeric and nominal values using orthogonal projection, In: Proc. 29th Intl. Conf. on Very Large Databases (2003).
10. Zhang, T., Ramakrishnan, R., Livny, M.: BIRCH: an efficient data clustering method for very large databases, In: Proc. ACM-SIGMOD Intl. Conf. on Management of Data (1996) 103–114.

On Optimizing Kernel-Based Fisher Discriminant Analysis Using Prototype Reduction Schemes*

Sang-Woon Kim[1] and B. John Oommen[2]

[1] Dept. of Computer Science and Engineering, Myongji University, Yongin, 449-728 Korea
kimsw@mju.ac.kr
[2] School of Computer Science, Carleton University, Ottawa, ON, K1S 5B6, Canada
oommen@scs.carleton.ca

Abstract. Fisher's Linear Discriminant Analysis (LDA) is a traditional dimensionality reduction method that has been proven to be successful for decades. Numerous variants, such as the Kernel-based Fisher Discriminant Analysis (KFDA) have been proposed to enhance the LDA's power for nonlinear discriminants. Though effective, the KFDA is computationally expensive, since the complexity increases with the size of the data set. In this paper, we suggest a novel strategy to enhance the computation for an *entire family* of KFDA's. Rather than invoke the KFDA for the entire data set, we advocate that the data be first reduced into a smaller representative subset using a Prototype Reduction Scheme (PRS), and that dimensionality reduction be achieved by invoking a KFDA on *this* reduced data set. In this way data points which are ineffective in the dimension reduction and classification can be eliminated to obtain a significantly reduced kernel matrix, K, without degrading the performance. Our experimental results demonstrate that the proposed mechanism *dramatically* reduces the computation time without sacrificing the classification accuracy for artificial and real-life data sets.

1 Introduction

The "Curse of Dimensionality": Even from the infancy of the field of statistical Pattern Recognition (PR), researchers and practitioners have had to wrestle with the so-called "curse of dimensionality". The situation is actually quite ironic : If the patterns to be recognized are represented in a feature space of small dimensions, it is likely that many crucial discriminating characteristics of the classes are ignored. However, if on the other hand, the dimensions of the feature space are large, we encounter this "curse", which brings along the excess

* The second author was partially supported by NSERC, the Natural Sciences and Engineering Research Council of Canada. This work was generously supported by the Korea Research Foundation Grant funded by the Korea Government (MOEHRD-KRF-2005-042-D00265).

D.-Y. Yeung et al. (Eds.): SSPR&SPR 2006, LNCS 4109, pp. 826–834, 2006.

baggage of all the related problems associated with learning, training, representation, computation and classification [1], [2]. The "dimensionality reduction" problem involves reducing the dimension of the input patterns and yields the advantages clearly explained in [1] and [2].

The literature reports numerous strategies that have been used to tackle this problem. The most well-known of these is the Principal Components Analysis (PCA) (the details of which are omitted here) to compute the basis (eigen) vectors by which the class subspaces are spanned, thus retaining the most significant aspects of the structure in the data [1]. While the PCA finds components that are efficient for *representation*, the class of Linear Discriminant Analysis (LDA) strategies seek features that are efficient for *discrimination* [1]. LDA methods effectively use the concept of a within-class scatter matrix, S_w, and a between-class scatter matrix, S_b, to maximize a separation criterion, such as $J = tr(S_w^{-1} S_b)$. The advantage of an LDA is that it is non-recursive. Being essentially linear algorithms, neither the PCA nor LDA can effectively classify data which is inherently nonlinear. Consequently, a vast body of research has gone into resolving this limitation, and a detailed review of this is found in [2]. This is exactly the focus of this paper. In this paper, we suggest a novel strategy to enhance the computation for an *entire family* of KFDA's. Rather than invoke the KFDA for the entire data set, we advocate that the data be first reduced into a smaller representative subset using a Prototype Reduction Scheme (PRS) (explained and briefly surveyed presently), and that dimensionality reduction be achieved by invoking a KFDA on *this* reduced data set.

The state-of-the-art in dealing with nonlinear methods include an adaptive method utilizing a rigorous Gaussian distribution assumption [3], a complete PCA plus LDA algorithm [4], two variations on Fisher's linear discriminant [5], Kernel-based PCA (KPCA) [6], Kernel-based FDA (KFDA) (for two classes by Mika *et al.* [7] and for multi-classes by Baudat and Anouar in [8]), Kernel-based PCA plus Fisher LDA (KPCA+LDA) [9], and LDA extensions which use the Weighted Pairwise Fisher Criteria and the Chernoff Criterion [10].

Methods for Handling Nonlinearity: The KPCA (or KFDA) provides an elegant way of dealing with nonlinear problems in an input space $\mathcal{R}^d$ by mapping them to linear ones in a feature space, F. That is, a dot product in space $\mathcal{R}^d$ corresponds to mapping the data into a possibly high-dimensional dot product space F by a nonlinear map $\Phi : \mathcal{R}^d \rightarrow F$, and taking the dot product in the latter space [6]. All of them utilize the *kernel trick* to obtain the kernel PCA components by solving a linear eigenvalue problem similar to that done for the linear PCA. The only difference is that the size of the problem is decided by the *number* of data points, and not by the dimension. In both the KPCA and KFDA, to map the data set $\{\boldsymbol{x}_1, \boldsymbol{x}_2, \cdots, \boldsymbol{x}_n\}$, (where each $\boldsymbol{x}_i \in \mathcal{R}^d$) into a feature space F, we have to define an $n \times n$ matrix, K, the so-called kernel matrix, (of dimension n) which is analogous to the $d \times d$ covariance matrix of the linear PCA (or LDA).

To solve the KPCA-associated computational problem, a number of methods, such as the techniques proposed by Achlioptas and his co-authors [11], [12], the power method with deflation [6], the method of estimating K with a subset of the data [6], the Sparse Greedy matrix Approximation (SGA) [13], the Nystom method [14], and the sparse kernel PCA method based on the probabilistic feature-space PCA concept [15], have been proposed. In [6], a method of estimating the matrix K from a subset of $n'(< n)$ data points, while still extracting principal components from all the n data points, was considered. Also, in [13], an approximation technique to construct a compressed kernel matrix K' such that the norm of the residual matrix $K - K'$ is minimized, was proposed. Indeed, pioneering to the area of reducing the complexity of kernel-based PCA methods are the works of Achlioptas [12] and his co-authors. Their first category includes the strategy of artificially introducing sparseness into the kernel matrix, which, in turn, is achieved by *physically* setting some randomly-chosen values to *zero*. The other alternative, as suggested in [11], proposes the elimination of the underlying data points themselves. This is the spirit of the strategy we advocate.

Optimizing KFDA: To solve the computational problem for KFDA methods, a number of schemes, such as the efficient leave-one-out cross-validation method [16], the techniques proposed by Xu and his co-authors [17], [18], and the method of using a minimum squared-error cost function and the orthogonal least squares algorithm [19], have been proposed. They are *briefly* (for space limitations) described below, but the details can be found in [26].

Cawley and Talbot [16] showed that the leave-one-out cross-validation of kernel Fisher discriminant classifiers, namely, $f(\boldsymbol{x}_i) = \boldsymbol{w} \cdot \Phi(\boldsymbol{x}_i) + b$, (where $\boldsymbol{w} = \sum_{i=1}^{n} \alpha_i \Phi(\boldsymbol{x}_i)$), can be implemented with a computational complexity of only $O(n^3)$ operations rather than the $O(n^4)$ complexity of a naive implementation, where n is the number of training samples. Xu and his co-authors [17], [18] proposed a reformative kernel Fisher discriminant method (for two and multiple classes respectively) which only computes the kernel matrix, K, between the test pattern and a *part* of the training samples, called the "significant nodes" (assuming that the eigenvectors for larger nonzero eigenvalues led to superior discriminant vectors) which are a few training samples selected from the entire data set. In [19], Billings and Lee suggested that after selecting n_s important terms, $\{\boldsymbol{x}'_i\}_{i=1}^{n_s}$, from all the training patterns, $\{\boldsymbol{x}_j\}_{i=1}^{n}$, using the orthogonal least squared (OLS) algorithm when one has to test the sample $\boldsymbol{z}$, the authors classified it as class ω_1 if $\sum_{i=1}^{n_s} \alpha'_i k(\boldsymbol{z}, \boldsymbol{x}'_i) > c$, where α'_i are the estimated coefficients; Otherwise it is classified as belonging to class ω_2.

Unlike the results mentioned above in [16], [17], [18] and [19], we propose an alternate strategy, akin to the one suggested in [11] for the KPCA family of algorithms – which is a fairly straightforward concept, yielding a *significant* computational advantage. Quite simply put, we propose to solve the computational problem in KFDA by reducing the *size* of the design set without sacrificing the performance, where the latter is achieved by using a Prototype Reduction Scheme (PRS). Thus, the contribution of this paper is that we show that the computational burden of a KFDA can be reduced significantly by not considering the "original" kernel matrix *at all*. Instead, we rather define a reduced-kernel matrix by first preprocessing the training points with a PRS scheme. Further, the PRS scheme does not necessarily have to *select* a reduced set of data points.

Indeed, it can rather *create* a reduced set of prototypes from which, in turn, the reduced-kernel matrix is determined. All of these concepts are novel to the field of designing Kernel-based FDA methods and have been rigorously tested for artificial and real-life data sets.

Prototype Reduction Schemes: Various PRSs[1], which are useful in nearest-neighbour-like classification, have been reported in the literature - two excellent surveys are found in [20], [21]. Bezdek *et al* [20], who composed the second and more recent survey of the field, reported that there are "zillions!" of methods for finding prototypes (see page 1459 of [20]). One of the first of its kind, was a method that led to a smaller prototype set, the Condensed Nearest Neighbor (CNN) rule [22]. Since the development of the CNN, other methods [23] - [25] have been proposed successively, such as the Prototypes for Nearest Neighbor (PNN) classifiers [23] (including a modified Chang's method proposed by Bezdek), Vector Quantization etc. Apart from the above methods, we mention the following: Support Vector Machines (SVM) [24] can also be used as a means of selecting prototype vectors. Observe that these new vectors can be subsequently adjusted by means of an LVQ3-type algorithm. Based on this idea, a new PRS (referred to here as HYB) of hybridizing the SVM and the LVQ3 was introduced in [25]. Based on the philosophy that points near the separating boundary between the classes play more important roles than those which are more interior in the feature space, and that of *selecting* and *adjusting* the reduced prototypes, a new hybrid approach that involved two distinct phases was proposed in [25]. Due to space limitations, the details of other schemes are omitted, but can be found in [26].

2 Optimizing the KFDA with PRSs

The fundamental problem that we encounter when optimizing any KFDA is that of reducing the dimensionality of the training samples. This, in turn, involves four essential phases, namely that of computing the kernel matrix, computing *its* eigenvalues and eigenvectors, extracting the principal components of the kernel matrix from among these eigenvectors, and finally, projecting the samples to be processed onto the reduced basis vectors. We observe, first of all, that all of these phases depend on the size of the data set. In particular, the most time consuming phase involves computing the eigenvalues and eigenvectors of the kernel matrix.

There are a few ways by which the computational burden of the kernel method can be reduced. Most of the reported schemes [11], [12], [13], [14], [15] resort to using the specific properties of the underlying kernel matrix, for example, its sparseness. Our technique is different. The method we propose is by reducing the *size* of the training set. However, we do this, by not significantly reducing the accuracy of the resultant training samples. This is achieved by using a PRS.

The rationale for the proposed method can be conceptually explained using Fig. 1. Fig. 1(a) represents the original training samples presented to the clas-

[1] Our overview is necessarily brief, but additional details can be found in [26].

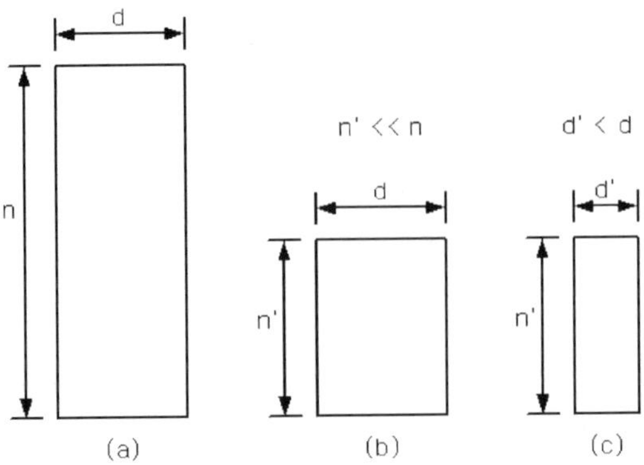

Fig. 1. The rationale for the proposed method. (a) The training samples, where n and d are the number of samples and the dimensionality, respectively. (b) The condensed prototypes extracted from the training samples using a PRS, where $n' \ll n$. (c) The prototype vectors whose dimensionality has been reduced with a KFDA, where $d' \ll d$.

sifier system, where n and d are the number of samples and the dimensionality, respectively. Fig. 1(b) represents the condensed prototypes which are extracted from the training samples using a PRS, where $n' \ll n$. Using the latter data, Fig. 1(c) represents the resultant prototype vectors in which the dimensionality has been reduced by invoking a KFDA, where $d' \ll d$. Observe that since the fundamental problem with *any* kernel-based scheme is that it increases the time complexity from $O(d^3)$ to $O(n^3)$, we can see that the time complexity for the dimensionality reduction from d to d' is sharply decreased from $O(n^3)$ to $O(n'^3)$.

The question now is essentially one of determining which of the training points we should retain. Rather than deciding to discard or retain the training points, we permit the user the choice of either *selecting* some of the training samples using methods such as the CNN, or *creating* a smaller set of samples using the methods such as those advocated in the PNN, VQ, and HYB. This reduced set effectively represents the new "training" set. Additionally, we also permit the user to migrate the resultant set by an LVQ3-type method to further enhance the quality of the reduced samples.

The PRS serves as a preprocessor to the n d-dimensional training samples to yield a subset of n' potentially new points, where $n' << n$. The "kernel" is now computed using this reduced set of points to yield the so-called *reduced-kernel* matrix. The eigenvalues and eigenvectors of *this* matrix are now computed, and the principal components of the kernel matrix are extracted from among *these* eigenvectors of smaller dimension. Notice now that the samples to be tested are projected onto the reduced basis directions represented by *these* vectors.

To investigate the computational advantage gained by resorting to such a PRS preprocessing phase, we observe, first of all, that the time used in determining the reduced prototypes is *fractional* compared to the time required for the expensive matrix-related operations. Once the reduced prototypes are obtained, the eigenvalue/eigenvector computations are significantly smaller since these computations are now done for a much smaller set, and thus for an $n' \times n'$ matrix. The net result of these two reductions is reflected in the time savings we report in a later section in which we discuss the experimental results obtained for artificial and real-life data sets.

3 Experimental Results: Artificial/Real-Life Data Sets

Experimental Data: The proposed method has been rigorously tested and compared with many conventional ones. This was done by performing experiments on both "artificial" and "real-life" data sets.

The data set described as "Random"is generated randomly with a uniform distribution but with irregular decision boundaries. In this case, the points are generated uniformly, and the assignment of the points to the respective classes is achieved by *artificially* assigning them to the region they fall into, as per the manually created "irregular decision boundary". The data set named "Non_normal 2", which has also been employed as a benchmark experimental data set [1], and [25] for numerous experimental set-ups was generated from a mixture of four 8-dimensional Gaussian distributions. The data sets "Iris2", "Ionosphere" (in short, "Iono"), "Sonar", "Arrhythmia" (in short, "Arrhy") and "Adult4", which are real benchmark data sets, are cited from the UCI Machine Learning Repository[2]. Their details can be found in the latter site, and also in [26] and omitted here in the interest of compactness. In the above data sets, the data set for class ω_j was randomly split into two subsets, $T_{j,T}$ and $T_{j,V}$, of equal size. One of them was used for choosing the initial prototypes and training the classifiers, and the other one was used in their validation (or testing). Later, the role of these sets were interchanged.

As in all learning algorithms, choosing the parameters of the PRS and KFDA play an important role in determining the quality of the solution. The parameters for the PRS, the KPCA and the KFDA, are summarized as follows:

1. The kernel function employed is the polynomial $k(x_i, x_j) = (1 + x_i' x_j)^2$.
2. The number of features to be selected is 2 for all the KFDAs.
3. The constant μ is chosen as $\mu = 0.001$ for regularization in KFD, and the fusion coefficient θ in CKFDA is chosen as $\theta = 1.4$.

Selecting Prototype Vectors: In order to evaluate the proposed dimensionality reduction mechanisms, we first selected the prototype vectors from the experimental data sets using the CNN, the PNN and the HYB algorithms. In the HYB, we selected initial prototypes using a SVM algorithm. After this selection, we invoked a phase in which the optimal positions (i.e., with regard to classification) were learned with an LVQ3-type scheme [25]. For the SVM and LVQ3 programs, we utilized two publicly-available software packages[3].

---

[2] http://www.ics.uci.edu/mlearn/MLRepository.html

[3] These packages can be available from http://www-ai.cs.uni-dortmund.de/SOFTWARE/SVM_LIGHT/svm_light.eng.html and http://cochlea.hut.fi/research/som_lvq_pak.shtml, respectively.

Table 1. The classification accuracies of the proposed computational mechanisms for the artificial and real-life data sets. The details of the entries and how the values were obtained are explained in the text.

Type	Dataset	PRSs	WHL	KPCA	KFD	GDA	KPCA+LDA	CKFDA
Artificial Data	Rand	WHL	96.50	79.50	88.50	88.50	88.50	91.75
		CNN	96.25	59.75	80.50	80.50	80.50	85.25
		PNN	95.75	61.25	83.00	83.00	83.00	89.25
		HYB	89.50	70.50	85.75	85.75	85.75	85.75
	Non_n2	WHL	92.50	92.50	92.50	92.50	92.50	92.50
		CNN	91.90	91.90	91.90	91.90	91.90	91.90
		PNN	92.10	92.10	92.10	92.10	92.10	92.20
		HYB	94.00	94.00	94.00	94.00	94.00	94.10
Real-life Data	Iris2	WHL	92.00	71.00	94.00	94.00	94.00	92.00
		CNN	89.00	63.00	93.00	93.00	93.00	95.00
		PNN	94.00	56.00	89.00	91.00	91.00	89.00
		HYB	94.00	72.00	95.00	92.00	92.00	93.00
	Ionos	WHL	78.65	75.85	76.14	88.64	88.64	83.52
		CNN	81.82	45.17	69.89	88.07	88.07	75.85
		PNN	82.68	43.19	80.11	84.09	84.09	84.94
		HYB	83.24	48.01	80.68	84.94	84.94	83.24
	Sonar	WHL	82.22	52.89	84.14	83.18	83.18	82.21
		CNN	79.81	53.37	77.89	79.81	79.81	77.41
		PNN	82.69	48.56	79.33	81.73	81.73	82.69
		HYB	80.77	50.97	82.21	79.81	79.81	81.73
	Arrhy	WHL	97.57	79.87	99.78	99.78	99.78	99.78
		CNN	96.47	49.78	99.12	99.78	99.99	99.78
		PNN	99.12	53.54	97.57	99.78	99.78	99.33
		HYB	99.12	84.07	99.78	99.33	99.33	99.11
	Adult4	WHL	93.40	91.85	92.97	92.07	92.07	92.73
		CNN	91.58	81.85	83.67	84.40	84.40	85.87
		PNN	89.36	79.35	80.82	81.04	81.04	83.58
		HYB	92.78	59.41	86.41	82.39	82.39	87.32

From the experiments, we can see that the kernel matrix dimensions to be processed in the KFDA computations can be reduced significantly by first employing a PRS. Thus, for the artificial data set "Non_n2" data set, the dimensionality reduced from 500×500 to 63×63 when the HYB method was used as the PRS, and for the real data set "Arrhy", the dimensionality reduced from 226×226 to 8×8 when the PNN method was used as the PRS. Both of these[4] are truly *significant* by any metric of measurement. It should also be mentioned that the reduction rate increased *dramatically* when the size of the data sets was increased. The reduction in the resultant KFDA processing time follows as a natural consequence!

Experimental Results: Table 1 shows the classification accuracies of the proposed computational mechanisms for the data sets. In WHL, the test data sets

[4] The results of the other data sets are omitted here, but can be found in [26].

were classified with the NN rule by utilizing the entire training sets as the code-book vectors. On the other hand, for KPCA, KFD, GDA, KPCA+LDA, and CKFDA classifications, we first chose prototype samples from the training data sets with the CNN, PNN and HYB methods respectively. After selecting the prototype vectors, we reduced their dimensionality using the KPCA, KFD, GDA, KPCA+LDA, and CKFDA methods. Finally, the test data sets were classified with the respective decision rules, where the prototype vectors of reduced dimensionality were utilized as the code-book vectors. The experiments were repeated by exchanging the roles of the data sets, and the two results were then averaged.

Consider the processing times for the "Non_n2" data set. If the entire set of size 500 was processed, the times taken for the KPCA, KFD, GDA, KPCA+LDA and CKFDA are 32.02, 78.44, 138.21, 30.20 and 30.36 seconds, respectively. However, if the same sets were first preprocessed by the CNN, to yield the CNN-KPCA, CNN-KFD, CNN-GDA, CNN-KPCA+LDA and CNN-CKFDA procedures[5], the processing times are 2.54, 3.28, 7.51, 2.53 and 2.55 seconds respectively - which represent a 10-fold to 20-fold reduction ! Notice that the classification accuracies of the method are almost the same as shown in Table 1. Identical comments can also be made about the PNN and HYB schemes[6]. The results of the other data sets are omitted here in the interest of brevity, but is in [26].

4 Conclusions

In this paper, we suggest a computationally superior mechanism to solve the computational problem for KFDA methods. Rather than define the kernel matrix and compute the principal components using the entire data set, we propose that the size of the data be reduced into a smaller prototype subset using a PRS Since the PRS yields a smaller subset of data points that effectively samples the entire space to yield subsets of prototypes, this alleviates the computational burden significantly. The experimental results demonstrate that the proposed schemes can improve the extracting speed of the proposed methods by an order of magnitude, while yielding almost the same classification accuracy.

References

1. K. Fukunaga.: *Introduction to Statistical Pattern Recognition, Second Edition.* Academic Press, San Diego, 1990.
2. F. Camastra.: Data Dimensionality Estimation Methods: A Survey. *Pattern Recognition*, **36** 2945–2954, 2003.
3. R. Lotlikar and R. Kothari.: Adaptive Linear Dimensionality Reduction for Classification. *Pattern Recognition*, **33** 185–194, 2000.
4. J. Yang and J. -Y. Yang.: Why can LDA be performed in PCA transformed Space?. *Pattern Recognition*, **36** 563–566, 2003.
5. T. Cooke.: Two Variations on Fisher's Linear Discriminant for Pattern Recognition. *IEEE Trans. Pattern Anal. and Machine Intell.*, **PAMI-24(2)** 268–273, Feb. 2002.

[5] Here, the notation of CNN-KPCA means that the dimensionality reduction is done with KPCA *after* extracting prototypes with the CNN method.

[6] It should be mentioned that such an increase/decrease in accuracy is insignificant.

6. B. Schölkopf, A. J. Smola, and K. -R. Müller.: Nonlinear component analysis as a kernel eigenvalue problem. *Neural Comput.*, **10** 1299–1319, 1998.
7. S. Mika, G. Ratsch, B. Schölkopf, A. Smola, J. Weston, and K. R. Müller.: Fisher Discriminant Analysis with Kernels. *Proc. of IEEE International Workshop Neural Networks for Signal Processing IX*, 41–48, Aug. 1999.
8. G. Baudat and F. Anouar.: Generalized Discriminant Analysis Using a Kernel Approach. *Neural Comput.*, **12** 2385–2404, 2000.
9. J. Yang, A. F. Frangi, J. -Y, Yang and D. Zhang.: KPCA plus LDA: A Complete Kernel Fisher Discriminant Framework for Feature Extraction and Recognition. *IEEE Trans. Pattern Anal. and Machine Intell.*, **PAMI-27(2)** 230–244, Feb. 2005.
10. M. Loog and R. P. W. Duin.: Linear dimensionality reduction via a Heteroscedastic extension of LDA: The Chernoff criterion. *IEEE Trans. Pattern Anal. and Machine Intell.*, **PAMI-26(6)** 732–739, June 2004.
11. D. Achlioptas and F. McSherry.: Fast computation of low-rank approximations. *Proc. of the Thirty-Third Annual ACM Symposium on the Theory of Computing*, Hersonissos, Greece, ACM Press, 611–618, 2001.
12. D. Achlioptas, F. McSherry and B. Schölkopf.: Sampling techniques for kernel methods. *Advances in Neural Information Processing Systems*, **14**, MIT Press, Cambridge, MA, 335–342, 2002.
13. A. J. Smola and B. Schölkopf.: Sparse greedy matrix approximation for machine learning. *Proc. of ICML'00*, Bochum, Germany, Morgan Kaufmann, 911–918, 2000.
14. C. Williams and M. Seeger.: Using the Nystrom method to speed up kernel machines. *Advances in Neural Information Processing Systems*, **13**, MIT Press, Cambridge, MA, 2001.
15. M. Tipping.: Sparse kernel principal component analysis. *Advances in Neural Information Processing Systems*, **13**, MIT Press, Cambridge, MA, 633–639, 2001.
16. G. C. Cawley and N. L. C. Talbot.: Efficient leave-one-out cross-validation of kernel Fisher discriminant classifiers. *Pattern Recognition*, **36** 2585–2592, 2003.
17. Y. Xu, J. -Y. Yang and J. Yang.: A reformative kernel Fisher discriminant analysis. *Pattern Recognition*, **37** 1299–1302, 2004.
18. Y. Xu, J. -Y. Yang, J. Lu and D.-J. Yu.: An efficient renovation on kernel Fisher discriminant analysis and face recognition experiments. *Pattern Recognition*, **37** 2091–2094, 2004.
19. S. A. Billings and K. L. Lee.: Nonlinear Fisher discriminant analysis using a minimum squared error cost function and the orthogonal least squares algorithm. *Neural Networks*, **15(2)** 263–270, 2002.
20. J. C. Bezdek and L. I. Kuncheva.: Nearest prototype classifier designs: An experimental study. *Int'l. Journal of Intelligent Systems*, **16(12)** 1445–11473, 2001.
21. B. V. Dasarathy.: *Nearest Neighbor (NN) Norms: NN Pattern Classification Techniques.* IEEE Computer Society Press, Los Alamitos, 1991.
22. P. E. Hart.: The condensed nearest neighbor rule. *IEEE Trans. Inform. Theory*, **IT-14** 515–516, May 1968.
23. C. L. Chang.: Finding prototypes for nearest neighbor classifiers. *IEEE Trans. Computers*, **C-23(11)** 1179–1184, Nov. 1974.
24. C. J. C. Burges.: A tutorial on support vector machines for pattern recognition. *Data Mining and Knowledge Discovery*, **2(2)** 121–167, 1998.
25. S. -W. Kim and B. J. Oommen.: Enhancing prototype reduction schemes with LVQ3-type algorithms. *Pattern Recognition*, **36(5)** 1083–1093, 2003.
26. S.-W. Kim and B. J. Oommen, "On using prototype reduction schemes to optimize kernel-based Fisher discriminant analysis". *Unabridged version of this paper.*

Sparse Covariance Estimates for High Dimensional Classification Using the Cholesky Decomposition

Asbjørn Berge and Anne Schistad Solberg

Department of Informatics
University of Oslo, Norway

Abstract. Results in time series analysis literature state that through the Cholesky decomposition, covariance estimates can be stated as a sequence of regressions. Furthermore, these results imply that the inverse of the covariance matrix can be estimated directly. This leads to a novel approach for approximating covariance matrices in high dimensional classification problems based on the Cholesky decomposition. By assuming that some of the targets in these regressions can be set to zero, simpler estimates for class-wise covariance matrices can be found. By reducing the number of parameters to estimate in the classifier, good generalization performance is obtained. Experiments on three different feature sets from a dataset of images of handwritten numerals show that simplified covariance estimates from the proposed method is competitive with results from conventional classifiers such as support vector machines.

1 Introduction

Many modern pattern recognition problems face the researchers with the problem of feature spaces of high dimensionality coupled with a sparsity of available labeled samples to be used for training. Further compounding the problem in many cases is that features are highly correlated, this adding a redundancy to the data that in some cases may obscure the information important for classification. When using parametric methods, such as the Gaussian Maximum Likelihood (GML) classifier, the parameter estimates, most importantly the covariance matrix estimate, will become increasingly unstable when the number of labeled samples is low compared to the dimensionality of the feature space. A wealth of approaches for dealing with the curse of dimensionality have been proposed in the literature, ranging from dimensionality reduction of the feature space to regularization of parameter estimates by biasing them toward simpler and more stable estimates. Still, many of these methods have slight weaknesses which would be gainful to try to resolve. Among these is the need for inversion of covariance matrices when evaluating the classifier, and estimation of redundant parameters in the full dimensional feature space. Especially when matrices are near-singular, which they tend to be if the ratio between labeled samples

D.-Y. Yeung et al. (Eds.): SSPR&SPR 2006, LNCS 4109, pp. 835–843, 2006.

and dimensionality is ill posed, inversion is plagued with numerical instabilities. Therefore, a direct estimation of the inverse covariance matrix would be useful.

Direct estimation of the inverse covariance matrix was suggested mainly for computational convenience in [1]. In that paper it was furthermore noted that for many statistical problems the inverse covariance matrix has many zero or near-zero values, and a direct feature selection approach was applied to choose which elements could be set to zero. Obviously this approach is computationally infeasible for high dimensional data with covariance matrices with thousands or tens of thousands of elements. We propose an approach that relies on the fact that a modified Cholesky decomposition of an inverse covariance matrix defines coefficients in a regression. By choosing targets in this regression to be zero, we can find simpler models for the covariance matrix with fewer parameters to estimate. A heuristic is suggested for searching for these parameters, guided by measuring classification performance on a ten-fold cross-validation, with the goal of finding sparse inverse covariance matrices where only the elements useful for classification are estimated. By reducing the number of parameters to estimate, variability in these covariance estimates is reduced. The results suggest that classifiers based on these sparse covariance matrices have improved generalization performance.

The main contribution of this paper is a novel approach for expressing and estimating sparse covariance approximations for high dimensional classification problems. We propose a heuristic for only estimating the necessary parts of the class-wise covariance matrices based on a simple search algorithm. The reduction in the number of parameters to estimate reduces the variability in the remaining parameters, while we still are using the full dimensional feature space for classification, gaining increased class separability.

2 Sparse Class Conditional Covariance Matrices

Consider a classification problem with k classes, assuming class conditional distributions to be Gaussian with mean μ_k and class-wise covariance matrices Σ_k. It is well known that this reduces to comparing the k quadratic discriminant functions $g_k(x) = -\frac{1}{2}\log|\Sigma_k| - \frac{1}{2}(x-\mu_k)'\Sigma_k^{-1}(x-\mu_k) + \log\pi_k$, where π_k is the a priori probability for class k. Noting that $-\log|\Sigma_k| = \log|\Sigma_k^{-1}|$, it is clear that there is no need for matrix inversion when classifying data, if we have a method for estimating the inverse covariance matrices directly.

2.1 Parametrization of the Inverse Covariance by the Modified Cholesky Decomposition

We decompose the inverse covariance matrix as a modified Cholesky decomposition [2]

$$\Sigma^{-1} = LDL^T,$$

where L_i is a lower triangular matrix with ones on the diagonal

$$L = \begin{bmatrix} 1 & & & & \\ -\alpha_{2,1} & 1 & & & \\ -\alpha_{3,1} & -\alpha_{3,2} & 1 & & \\ \vdots & & & \ddots & \\ -\alpha_{p,1} & & -\alpha_{p,2} & -\alpha_{p,p-1} & 1 \end{bmatrix}$$

and D a diagonal matrix. If we were to consider the features of each sample as a time-series, the elements in L can be seen row-wise as parameters in autoregressive processes of the same order as the row r. Several authors in the time series literature have noted this [3], [4], [5]. We will use this fact to transform the task of approximating covariance matrices into a sequence of regressions. For each row, r, one could then "predict" the next feature based on the $r-1$ preceding features. Keeping with the earlier notation, and assuming zero mean for readability, this can be expressed as:

$$x_r = \sum_{j=1}^{r-1} \alpha_{r,j} x_j + \varepsilon_r \tag{1}$$

where the rth diagonal entry of $D_{r,r} = \text{var}(\varepsilon_r)$ This parametrization has the effect is that the resulting covariance matrix will still be positive definite, as long as the diagonal elements of D are positive.

2.2 Search for a Sparse Representation of the Class-Wise Covariance Matrices

As pointed out earlier, [1] proposed to choose the sparsity of the inverse covariance matrices using a sequential forward feature selection. Clearly this is infeasible for high dimensional data where the number of unique elements in the covariance matrix is in the thousands or tens of thousands, thus we have to resort to a heuristic. The general idea of the proposed method is to find a sufficiently complex covariance matrix to solve our classification problem, by evaluating a search space that is small enough to handle.

Search Algorithm. From the regression formulation in equation 1 we can argue that a zero $\alpha_{r,j}$ indicates that when predicting x_r, x_j does not carry much interesting information. For all rows, if we were to zero the coefficient of a preceding feature, we could, using time-series terminology, argue that we ignore a specific *lag* when predicting the next feature. If we ignore a specific lag for all rows in our sequence of regressions, all elements in an off-diagonal in L can be set to zero. Consider the illustration of a sparse L in figure 1, where the sparse L matrix has only two off-diagonals where we estimate parameters.

We can also observe that zeros in the inverse covariance matrix means conditional independence. It is well known that when there is a zero in position (i, j) of the inverse covariance matrix, x_i and x_j are independent, conditioned on the rest of the features [6]. Informally this means that given all other features, x_i does not carry information regarding x_j and vice versa. For data that has some specific order, for example discretized curves, the intuition that the correlation between neighboring features does not carry information useful for discrimination between classes, and in some cases is even detrimental to classifier results, was established in [7]. Our proposed heuristic for reducing the search space for a sparse representation of the inverse covariance matrix is based on the same intuition, that some of the correlations between features do not help in separating the classes, and thus can be dropped.

The general idea is to start by approximating the covariance matrices with the simplest possible models, i.e. diagonal matrices, and add parameters to the approximation until the classification performance of the model does no longer improve. With regard to our proposed heuristic, we search for which off-diagonals in the covariance matrices that needs to be estimated in order to improve classification performance on the training data. The search, guided by ten-fold cross-validation (10-CV) as a performance measure, can be described by the following steps:

1. *Initialization* - Estimate diagonal inverse covariances for all classes k, Σ_k^{-1} and calculate 10-CV performance. The parameters to estimate is the variances in D_k.
   No parameters in L_k are estimated.
2. *Search* - Select off-diagonals in L_k to be nonzero in a sequential forward manner
   (a) Find the 10-CV performance gain when adding each individual off-diagonal to the pool of parameters to estimate
   (b) Add the one off-diagonal that gives the largest improvement in 10-CV
   (c) Loop from a) until 10-CV performance does not improve further.

3 Maximum Likelihood Inverse Covariance Estimates

In regard to our motivation in the previous sections, we can develop Maximum Likelihood estimates for the inverse covariance matrix. By the modified Cholesky decomposition $\Sigma_k^{-1} = L_k D_k L_k^T$ [2], the log-likelihood function for the class-wise inverse covariance for class k can be expressed

$$l(\Sigma_k^{-1}) = \sum_{l=1}^{N_k} [-\frac{1}{2}\log(|\Sigma_k|) - \frac{1}{2}(x_l - \mu_k)^T L_k D_k L_k^T (x_l - \mu_k)]$$

where N_k is the number of samples in class k. Express $L_k = I - B_k$, where B_k is a lower triangular matrix with zeros on the diagonal. It is clear that the parameters we need to estimate are the diagonal elements of D_k and the lower triangular elements of B_k. We adopt the following notation: Let $x_{l,r}$ be the r'th feature of the l'th sample $x_l = x_{l,1:p}$, where p is the dimensionality of the feature space. Let $B_{k,r,1:(r-1)}$ be the nonzero elements of row r of B_k, i.e. lower triangular elements of the matrix in the given row. See the illustration in figure 1 for an illustration of which matrix elements in B_k that are estimated for row r. Likewise, $x_{l,1:(r-1)}$

is the $r-1$ first features of sample l in the dataset. To simplify the expression, we write $v_{l,k} = x_l - \mu_k$, which gives the further expressions $v_{l,k,r}$ and $v_{l,k,1:(r-1)}$ using the same notation as before. We can rewrite the likelihood using these definitions, letting r index diagonal elements $\sigma^2_{k,r}$ of D_k and observing that the log-determinant of Σ_k can be written as the sum of the diagonal elements of D, since the determinant of a matrix product can be written as a product of determinants, and further that $|L_k| = 1$ by definition. The likelihood becomes

$$
\begin{aligned}
l(\cdot) &= \frac{1}{2}\sum_{l=1}^{N_k} \log(|D_k|) - ((I-B_k)^T v_{l,k})^T D_k((I-B_k)^T v_{l,k}) \\
&= \frac{1}{2}\sum_{l=1}^{N_k}\sum_{r=1}^{p} \log\sigma^2_{k,r} - \sum_{r=1}^{p}[(I - B^T_{k,r,1:(r-1)})v_{l,k}]^2\sigma^2_{k,r} \\
&= \frac{1}{2}\sum_{l=1}^{N_k}\sum_{r=1}^{p} \log\sigma^2_{k,r} - \sum_{r=1}^{p}[v_{l,k,r} - B^T_{k,r,1:(r-1)}v_{l,k,1:(r-1)}]^2\sigma^2_{k,r}
\end{aligned}
$$

To estimate the elements of the diagonal matrix D_k, differentiate by $\sigma^2_{r,k}$, and set to zero

$$
\sigma^2_{r,k} = \frac{N_k}{\sum_{l=1}^{N_k}[v_{l,k,r} - B^T_{k,r,1:(r-1)}v_{l,k,1:(r-1)}]^2}
$$

Furthermore, we find the estimate of B_k row-wise by differentiating the log-likelihood by $B_{k,r,1:(r-1)}$ and set to zero. This gives

$$
\sum_{l=1}^{N_k}[\sigma^2_{r,k}(v_{l,k,r} - B^T_{k,r,1:(r-1)}v_{l,k,1:(r-1)})v^T_{l,k,1:(r-1)}] = 0,
$$

which after some rearranging leads to

$$
B_{k,r,1:(r-1)} = [\sum_{l=1}^{N_k} v_{l,k,1:(r-1)}v^T_{l,k,1:(r-1)}]^{-1}[\sum_{l=1}^{N_k} v_{l,k,r}v^T_{l,k,1:(r-1)}],
$$

which is the result of regression of $v_{l,k,r}$ onto all previous elements in $v_{l,k}$, i.e. $v_{l,k,1:(r-1)}$.

Sparse Regressions of $B_{k,r,1:(r-1)}$. The sequence of regressions can be simplified if we assume that some elements of $B_{k,r,1:(r-1)}$ are always zero. This way we can simply remove the corresponding predictors, $v_{l,k,1:(r-1)}$, and thus only estimate the nonzero parameters. Consider figure 1 where it can be seen that for row r of B_k has only two targets in the regression not defined to be nonzero. The implicit sparsity in the representation of the inverse covariance matrix can be considered a feature selection in each regression. This reduces the size of the matrix to be inverted in the regressions, which might give a classifier that is more resilient to low sample counts, since the number of samples needed to make this matrix inversion ill-conditioned might be much lower than in the conventional case.

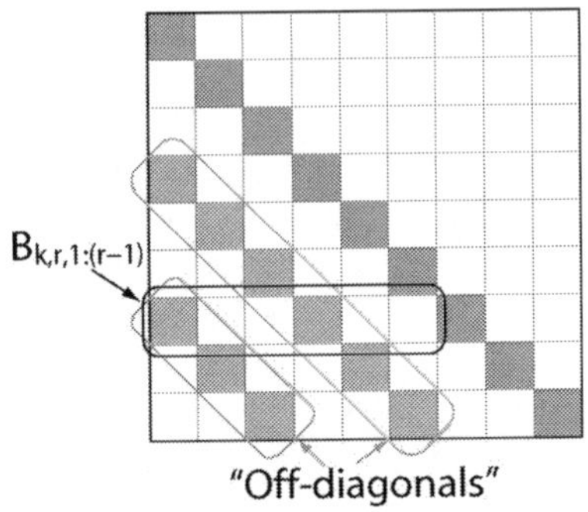

Fig. 1. Illustration of a matrix of correlations, L, for the inverse covariance matrix . The matrix is lower triangular, with 1 on the diagonal, the elements to estimate is below-diagonal and is represented with B in the text. The sparsity in the covariance estimate is obtained by only estimating the matrix elements in *some* off-diagonals. The matrix is estimated by a sequence of regressions, one for each row in the matrix B. Thus for row r, we estimate the elements $B_{k,r,1:(r-1)}$. These regressions can be simplified if we define that all elements not in the chosen off-diagonals are zero.

4 Experiments

In our experiments we used the *mfeat* dataset [8], which is a set of images of handwritten numerals. The data consists of 10 classes, each having 200 samples, and the dataset was split randomly in half to generate training and test data. Performance when training was measured using ten-fold cross-validation. From the images, three different feature sets were considered, 47 Zernike moments, 64 Karhunen-Loève coefficients, and 76 Fourier coefficients. The classifier used in our proposed method is a Gaussian Maximum Likelihood classifier, assuming class-wise covariance matrices (GML-quadratic). All covariance matrices are approximated with the same off-diagonals according to the results from the search strategy. In table 1, results of some comparable classifiers on these data are given. The results from the proposed method is included for reference. The classifiers are Gaussian ML classifiers assuming common covariance (GML-linear) and class-wise covariances (GML-quadratic), support vector machines with linear (SVM-linear) and quadratic (SVM-quadratic) kernels, and Parzen density estimation. Note that Zernike moments and Fourier are rotation invariant features, so much of the error in the classification is actually confusion between 6 and 9.

In figures 2(a), 2(b) and 2(c), the error rate by cross-validation and on test data is given as a function of the fraction of covariance elements compared to a full model. A full model estimates the entire covariance matrix for each class, just as a GML-quadratic classifier, but still avoids inverting the covariance matrix. Interestingly, avoiding matrix inversion in the classifier seems to make the classifier slightly more stable than the conventional GML classifier. Figure 2(a) considers the Zernike moment feature set. The classification performance by

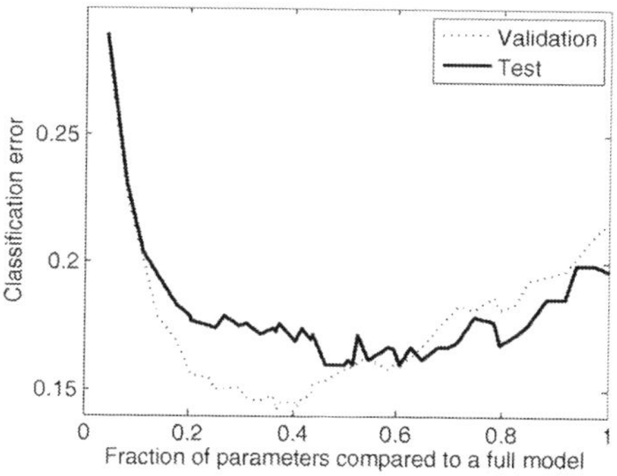

(a) Zernike moments

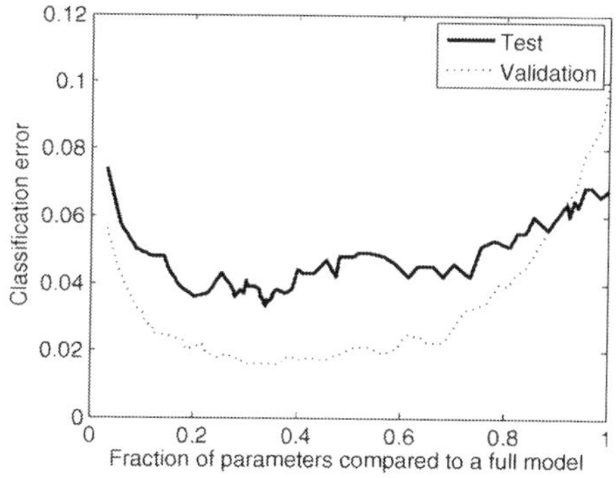

(b) Karhunen-Loève coefficients

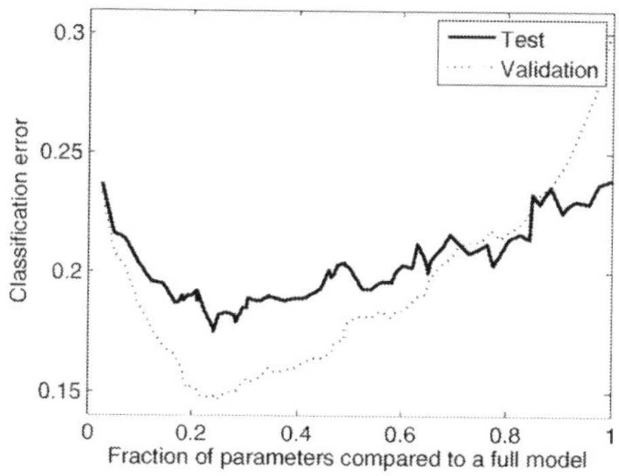

(c) Fourier coefficients

Fig. 2. Error rates for the proposed method by cross-validation and on test data compared to the fraction of covariance parameters of a full model for (a) 47 dimensional Zernike moments feature set, (b) 64 dimensional Karhunen-Loève feature set and (c) 76 dimensional Fourier feature set. For the Zernike and PCA feature sets, around 30% of the parameters of a full model seems sufficient for good generalization performance. For the Fourier feature set the number of parameters sufficient for a good classifier is around 25%. These choices is clearly suggested by the cross-validation classification error.

cross-validation has a minimum at 36.8% of the covariance parameters, and the mean result on the test data for that fraction of parameters is 17.4%. The same results for the Karhunen-Loève feature set are given in figure 2(b). The minimum by crosss-validation is here at 29.6% of the parameters, and the mean classification result for this experiment is 3.7%. For the Fourier coefficient feature set the results are shown in figure 2(c). This feature set had a minimum classification error by cross-validation at 25.1% of the parameters, and the classification result on the test set was 18.2%. Note the far right on the figures, since the number of samples available for training is nearing the dimensionality of the dataset, a full covariance model will be near singular and even the proposed model collapses. However, the decline is very graceful, and does not start until 80% of the

Table 1. Error rates (in percent) for classifiers on test data (100 samples per class for training), and on simplified models found by the proposed method on the three feature sets, Zernike moments, Fourier coefficients and Karhunen-Loève coefficients.

Classifier	Zernike	Fourier	K-L
GML-quadratic	19.8	23.9	6.8
GML-linear	18.2	18.5	4.5
SVM-linear	18.3	19.6	6.4
SVM-quadratic	15.7	15.9	2.1
Parzen	18.5	17.0	3.0
Proposed method	17.4	18.2	3.7

parameters of the model is used. These results are summarized in table 1, and compared with results for other classifiers.

All experiments indicated a gradual decline in performance as the number of features estimated increased, however, at the same time the performance curves in figures 2(a), 2(a) and 2(c) indicate that for three different feature sets, there is a fairly wide area where the classification performance on the test set is good. In all the experiments, the minimum classification error by cross-validation occurred in this area.

Considering the results presented in table 1, we observe that the proposed method is certainly competitive with conventional methods.

5 Conclusion and Future Work

Using results from time series analysis, we have proposed a novel approach for estimating sparse covariance matrices in full dimensional feature spaces for Gaussian ML classifiers. Experiments on different feature sets of a handwritten numeral classification problem indicates that it performs equally, or better than conventional classifiers. The results from these initial experiments are encouraging, and suggests the usefulness of further research in this direction.

We envision that this approach will be useful for reducing the number of parameters to estimate in a mixture of Gaussians classifier, where the motivation for using the sparsest possible model for component distributions is even stronger.

Future work with this method will focus on improving the selection heuristic. The present heuristic of choosing off-diagonals in L to estimate is intuitive when there is some clear correlation structure in the data that is present in the entire feature set. An example of this is the strong correlation between neighboring features when using discretized curves as input. Another improvement would be to consider selecting different off-diagonals to estimate for each class.

References

1. Dempster, A.: Covariance selection. Biometrics (1) (1972) 157–175
2. Golub, G.H., Van Loan, C.F.: Matrix computations, 3rd ed. John Hopkins University Press (1996)

3. Smith, M., Kohn, R.: Parsimonius covariance matrix estimation for longitudinal data. Journal of the American Statistical Association **97**(460) (2002) 1141–1153
4. Bilmes, J.A.: Factored sparse inverse covariance matrices. In: Proc. IEEE International Conference on Acoustics, Speech, and Signal Processing(ICASSP'00). Volume 2. (2000) 1009–1012
5. Pouhramadi, M.: Foundations of Time Series Analysis and Prediction Theory. Wiley (2001)
6. Whittaker, J.: Graphical models in applied multivariate statistics. Wiley (1990)
7. Hastie, T., Tibshirani, R., Buja, A.: Flexible discriminant analysis and mixture models. Journal of the American Statistical Association **89**(428) (1994) 1255–1270
8. Jain, A.K., Duin, R.P., Mao, J.: Statistical pattern recognition: A review. IEEE Trans. Pattern Anal. Machine Intell. **22**(1) (2000) 4–37

Generic Blind Source Separation Using Second-Order Local Statistics

Marco Loog

The Image Group IT University of Copenhagen
Copenhagen Denmark
marco@itu.dk
www.itu.dk/~marco/

Abstract. Exploiting the fact that one is dealing with time signals, it is possible to formulate certain blind source (or signal) separation tasks in terms of a simple generalized eigenvalue decomposition based on two matrices. Many of the techniques determine these two matrices using second-order statistics, e.g., variance, covariance, autocorrelation, etc.

In this work, we present a second-order, covariance-based method to determine the independent components of a linear mixture of sources. This is accomplished without the use of a possible temporal variable on which the data may depend, i.e., we explicitly avoid the use of autocorrelations, time delay, etc. in our formulation. The latter makes it possible to apply the simple eigenvalue decomposition-based technique to general pattern recognition methods and as such to find possible independent components of generic point clouds.

1 Introduction

Blind source or signal separation (BSS) and independent component analysis (ICA) are—often linear; the case we consider—unsupervised learning methods that attempt to decompose a (set of) random vector, e.g. feature vectors, into components which are independent, or at least as independent as possible in a certain sense. As such, these techniques improve upon well-known principal component analysis which does not go beyond the decorrelation of the respective input signals.

Although many of the approaches to and variations on BSS and ICA are generally applicable [7,17], most of them are specifically designed to solve source separation problems for time-series. Under certain assumptions (see below), exploiting characteristics of time-dependent signals, some of the source separation problem can be solved particularly easy. These approaches merely involve the determination of two matrices and a generalized eigenvalue decomposition using these matrices, i.e., performing a simultaneous diagonalization of them. Typical assumptions are related for example to the presence of autocorrelation within the sources or stronger assumptions such as periodicity of the individual components. Often, the first one of the two matrices to be determined is simply the covariance matrix of the total data, i.e., it uses second-order information. The entries of the second matrix may be determined using time-delayed observations, higher-order statistics, frequency components, etc.

In this paper, we formulate a BSS method that only exploits second-order moments, i.e. variances and covariances and is eigenvalue decomposition-based, but which does

D.-Y. Yeung et al. (Eds.): SSPR&SPR 2006, LNCS 4109, pp. 844–852, 2006.

not need any kind of temporal assumptions. This implies that the method is also easily and generally applicable to common pattern recognition data: A collection of feature vectors in n-dimensions that are independently and identically distributed.

The idea for our approach comes in part from the work of Holland and Wang [14,15] in which regional or local dependencies of a probability density function are defined. These local dependencies can be directly related to localized versions of Pearson's correlation coefficient [19] or, more generally, to local moments of a probability density function [21]. In Section 3, it is explained in what way specifically local covariances are of use to our approach. This is done right after, in Section 2, we introduced the BSS problem we consider and the general joint diagonalization approach pursued by several authors. First however, we give a, necessarily, brief overview of related approaches.

Related Methods. [23] provides a nice introduction to BSS using the technique of joint diagonalization. Some further remarks and references to BSS based on eigendecompositions can be found in [25] and, more generally, in [4]. [8] discussed more broadly the current state of BSS and ICA.

A classic work, based on second-order moments, uses a time-delayed approach to signal separation into independent components [20]. The main assumption here is that the signal is non-stationary, or better, that sources do not have a constant power profile over time. Another well-known work is [26], which exploits the cross-correlation present in non-white signals also using time-delays for this. A third paper [5] uses yet another possible characteristic and assumes non-Gaussianity of the sources. Like the first two, its solution is based on a generalized eigendecomposition, however here both second- and fourth-order moments are used to determine the two matrices involved. Non-stationarity also plays a role in [24] and [2], while [16] exploits the non-stationarity of the variance of the signal, employing a fixed-point algorithm to optimize the separation criterion. An approach in which the sources may be temporally white but spatially colored is presented in [10].

A different approach utilizes a cyclostationarity assumption [11,12]. A similar approach is pursued in [1], but here a more involved iterative algorithm is needed. [18] makes the very strong assumption of periodic sources.

[22] using temporal dependence between the sources and presents an approach which works in the Fourier domain like [3]. Both methods again use that one is dealing with time signal. [9] uses time-lags and higher-order statistics. Similar work is presented in [6], in which non-stationarity, the temporal structure of sources, and time-delayed correlation matrices are exploited.

Contrary to all previously mentioned works, in this work, we solve the BSS task based on second-order moments only without the need to make assumptions about, e.g., temporal dependencies in the sources.

2 Eigenvalue Decomposition-Based BSS

In the BSS problem, we are given T n-dimensional vectors $x(t)$. Typically, t is the time instance at which the n (random) observations, stored in $x(t)$, took place. (However, as already indicated, a main point in this work, is that t can be considered merely an index to the several xs.) Now, in addition, we assume that the observations $x(t)$ are obtained by linearly mixing n statistically independent source signals[1]. Denoting the n-dimensional vector of sources as $s(t)$, we more precisely assume that there exists an invertible $n \times n$ mixing matrix A for which $x(t) = As(t)$ for all t. (We assume, throughout the paper and without loss of generality, that all signals have zero expectation.) The actual BSS problem is to determine a good estimate for an inverse transformation A^{-1}, the so-called unmixing matrix.

Now, to start with, we note that various cross-statistics of the observations $x(t)$ are obtained by the matrix A and the diagonal cross-statistics of the sources $s(t)$ [23]. As an example, for the t-averaged covariance matrices, we have the following equivalence:

$$C_x := \sum_t \mathsf{E}[x(t)x'(t)] = \sum_t A\mathsf{E}[s(t)s'(t)]A' =: AC_sA' .$$

Assuming independent sources, C_s is diagonal. Similar relations can be deduced based on more specific assumption, like the ones mentioned in the introduction. As an example: For non-white sources, one can use time-lagged second-order statistics [23,26]:

$$C_x(\tau) := \sum_t \mathsf{E}[x(t)x'(t+\tau)] = \sum_t A\mathsf{E}[s(t)s'(t+\tau)]A' =: AC_s(\tau)A' ,$$

in which τ is the time lag and where again the source matrix is diagonal if the sources are independent.

Such two conditions, or similar ones, are sufficient for source separation and the inverse transformation matrix A^{-1} can be recovered, up to permutations and scaling, as the matrix V fulfilling the generalized eigenvalue equation $C_xV = C_x(\tau)V\Lambda$ in which Λ is a diagonal matrix with all eigenvalues on its diagonal (in many articles cited in the introduction, one can find the derivation for this). Another way to solve it, which will later on be our actual method of choice, is as follows (see e.g. [13]). Firstly, whiten the data $x(t)$ by transforming it with $C_x^{-1/2}$. Secondly, determine $C_x(\tau)$ for the *transformed* data and the eigenmatrix E of this matrix, i.e., the matrix with all eigenvectors of $C_x(\tau)$ as columns. Finally, set $A^{-1} = E'C_x^{-1/2}$.

Now, in order to make such simple technique applicable to data not necessarily subject to temporal or any other kind of ordering, we need to come up with a second matrix that does not assume such ordering. Note that for the first matrix we can simply take the data's global covariance matrix.

3 Local and Global Covariance

Consider the covariance localized by a kernel K in the n dimensional signal space

$$C_K = \mathsf{E}_K[(x(t) - \mathsf{E}_K(x(t))][x(t) - \mathsf{E}_K(x(t))]' .$$

[1] We do not consider the overdetermined and underdetermined cases here. We only consider the exactly determined case, i.e., the number of sources is equal to the number of observed signals.

E_K indicates that the kernel weighted expectation is taken.

It is not hard to demonstrate that if K is separable, i.e.,

$$K(x(t)) = \prod_{i=1}^{n} K_i(x_i(t)),$$

and all n variables are independent of each other, that C_K becomes diagonal; for independent, variables the covariance splits out in two terms that both integrate to zero—just as in the calculation for a global covariance matrix with independent variables—and the off-diagonals all become zero.

Shrinking the kernel size to zero, one can define point-local (co)variances [21]. These local second-order moments are directly related to the localized version of Pearson's correlation coefficient [19], which in turn can be simply derived from the so-called local dependence function proposed by Holland and Wang [14,15]. An interesting property of these measures is that they are zero everywhere if and only if all n variables are mutually independent.

In the light of these results, we can state that C_K can be non-diagonal for certain K if and only if the variables, i.e., the observed signals, are dependent. This means that if we have such kernel, we can use the resulting local covariance matrix C_K as the second matrix in the generalized eigenvalue decomposition to solve the BSS problem. Of course, this definition of the second matrix leaves quite some room for choices. For the experiments, we stick to one particular choice, which is given in the next paragraph together with a brief recapitulation of the whole approach.

Specific Approach used in the Experiments. The specific approach, which we use in the next section to demonstrate the methodology, is based on a very simple block kernel

$$K(x(t)) = \prod_{i=1}^{n} 1_{[-r,r]}(x_i(t)),$$

i.e., the product of n indicator functions on the interval $[-r, r]$ around the origin, i.e., a (hyper)cube centered at the origin and with sides of length $2r$.

In more detail the specific source separation approach used in the next section to demonstrate the possible performance is as follows (recall the end of Section 2).

1. Centralize the data and calculate the data's global (or total) covariance matrix C_x.
2. Whiten the centralized data: $x(t) \mapsto C_x^{-1/2} x(t)$.
3. Determine on the transformed data the localized covariance matrix C_K, where r is set to two[2]. Equivalently, determine all data points that are within the (hyper)cube induced by K and determine the sample covariance matrix of these points.
4. Eigendecompose C_K: $C_K = E \Lambda E'$.

[2] $r = 2$ is a bit of an arbitrary choice. Generally, with too large an r, all data will lay in the (hyper)cube and therefore C_K will be equivalent to C_x and so the eigenvalue problem is still underdetermined. For r to small, estimating C_K might be based on too few data points, possibly leading to instable results. The current choice ensures that about 80% of the observations are used in the estimation process.

5. Let the estimate $\widehat{A^{-1}}$ of the inverse of the mixing matrix A equal $E'C_x^{-1/2}$.
6. Transform the original centralized data to obtain the estimated source outputs: $x(t) \mapsto \widehat{A^{-1}}x(t)$.

4 Some Illustrative Experiments

To demonstrate the performance of our method, we generated two rather different 3-dimensional data sets consisting of 10,000 points on which we tested it. Figures 1 and 2 give a pictorial overview of both data sets, their mixed versions, and the source separation obtained with the new approach. The mixing matrix A is given by

$$A = \begin{pmatrix} .86 & .31 & .41 \\ .88 & .67 & .47 \\ .37 & .97 & .94 \end{pmatrix},$$

which introduces a lot of dependency between the observed signals.

In the four rows of the figures, we see, from top to bottom: The histograms of the three sources; the three pairwise 2D projections of the three sources; the three projections of the mixed data along the axes; the three pairwise 2D projections of the unmixed data, i.e., the source estimates. We note again that sources can only be reconstructed up to permutations of the sources and scaling of them. This is also visible in the resulting unmixed signals.

As is clear from the figures, the method is capable of finding the independent components with a reasonable precision although some seems not to have been recovered entirely satisfactory, which also can be seen from the slight tilting of the distribution in the 2D projections.

An additional check of how good the performance of the method is, is to look at the matrix $\widehat{A^{-1}}A$ which should be a permutation matrix of a perfectly diagonal matrix—in our case this diagonal should only consist of -1s and $+1$s—if the BSS would work flawlessly. For the data displayed in Figure 1, we have

$$\widehat{A^{-1}}A = \begin{pmatrix} 1.000 & -0.016 & 0.014 \\ -0.028 & -0.014 & 1.000 \\ 0.004 & 1.000 & 0.028 \end{pmatrix}.$$

For the data displayed in Figure 2, we have

$$\widehat{A^{-1}}A = \begin{pmatrix} 0.013 & 0.021 & -1.000 \\ 0.013 & 1.000 & 0.032 \\ 1.000 & -0.004 & 0.005 \end{pmatrix}.$$

Both these matrices are quite close to a permuted $(-1, +1)$-diagonal matrix, however several "off-diagonal" entries are possibly not entirely negligible.

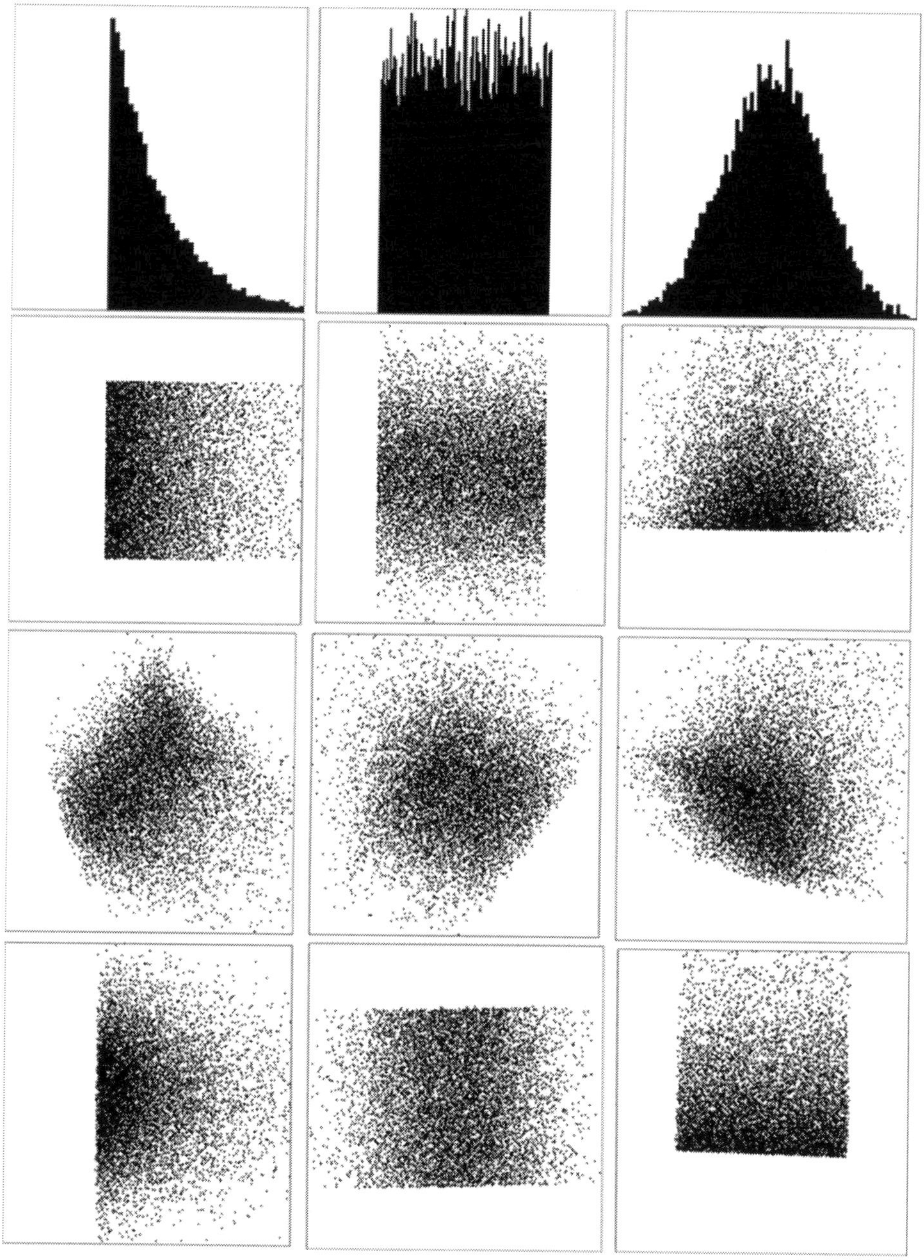

Fig. 1. Results on first artificial data set. From top to bottom row: Histograms of three independent sources (from left to right: exponential, uniform, and normal distribution), 2D pairwise scatter plots of three sources, 2D projections along axes of mixed data, 2D pairwise scatter plots of three estimated, unmixed sources.

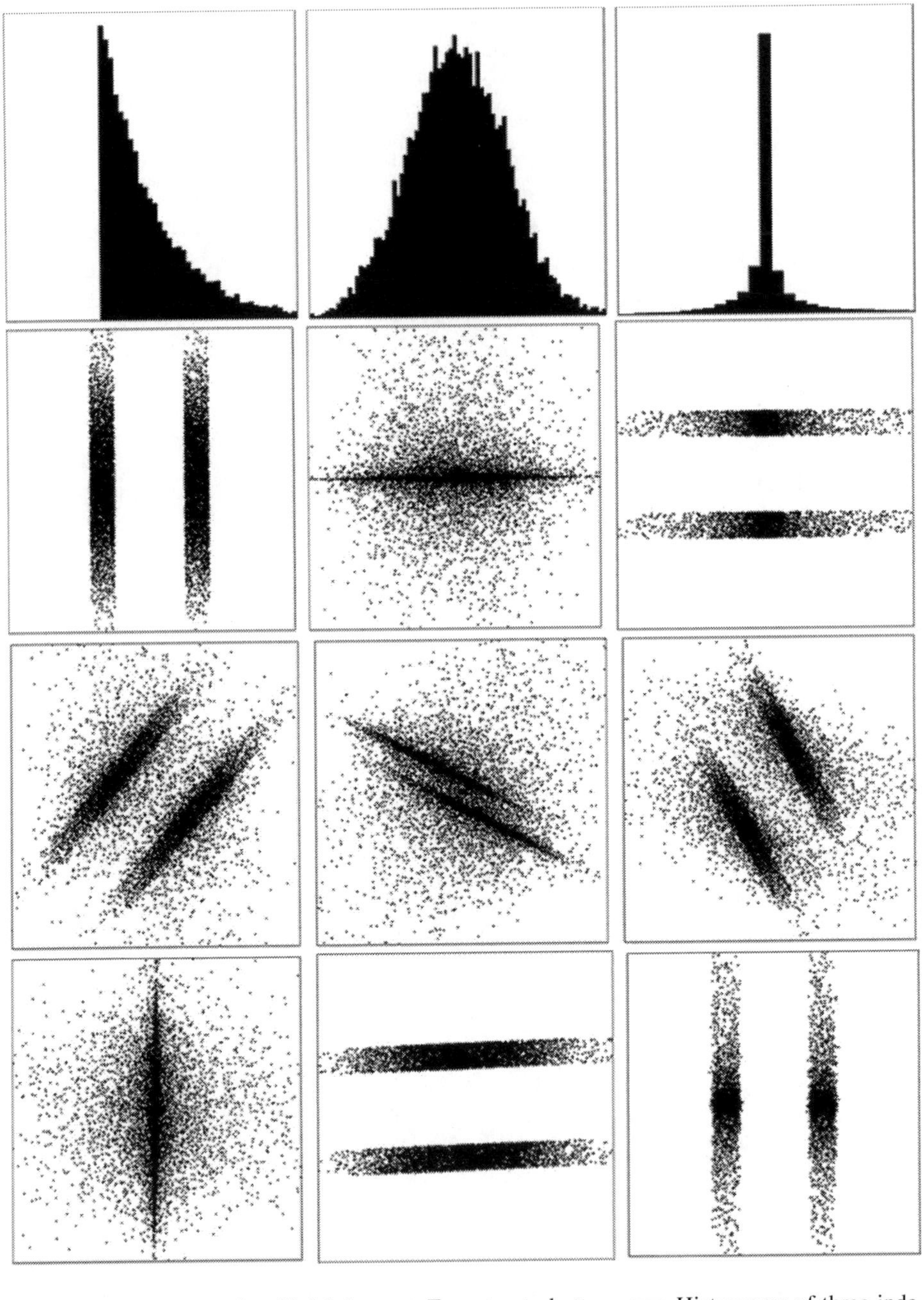

Fig. 2. Results on second artificial data set. From top to bottom row: Histograms of three independent sources (from left to right: 'double' uniform, normal, and normal 'to the power of three' distribution), 2D pairwise scatter plots of three sources, 2D projections along axes of mixed data, 2D pairwise scatter plots of three estimated, unmixed sources.

5 Discussion and Conclusion

A novel generalized eigendecomposition-based approach to the blind source separation problem is presented, which only exploits second-order covariance structures and does not build on any assumptions coming from the presence of any kind of intrinsic ordering of the data, e.g. no time dependence is assumed. An important consequence is that this makes it possible to use these type of simple source separation techniques within the general pattern recognition setting.

The approach presented here is built upon determining the global covariance matrix of the data and a notion of local covariance matrix. Having these two matrices leads to a fully determined generalized eigenvalue problem based on which the BSS problem can in principle be solved. The notion of local (co)variances is inspired by the work on local dependence functions and moments [14,15,19,21].

The experiments on two artificial data sets showed that reasonable performance can be obtained based on a very simple choice of local covariance.

One way to possibly improve this, is by designing the local covariance matrix (i.e., the kernel K) more carefully. A better approach, is to estimate several local covariance matrices and perform a joint (approximate) diagonalization of these matrices. This is the approach often pursued in the eigendecomposition-based BSS techniques discussed in the Introduction (see [23] and, for example, [24]). Often these techniques also have to base their analysis on more than two matrices because of their sensitivity to outliers, noise, small samples, etc. Carrying out such additional experiments is part of future research.

Possibly more interesting is the question how the technique relates to supervised dimensionality reduction approaches. The generalized eigendecomposition approach to BSS is, for example, very reminiscent of the calculations performed in Fishers's linear discriminant analysis (LDA) [13], where the second local covariance matrix is replaced by an estimate of the average within-class scatter. Relating such techniques, one may gain deeper insight in their behavior.

In conclusion we presented a promising framework, based on which interesting further research may be conducted. The method proposed is generally applicable, easy to use, and able to perform BSS on generic n-dimensional data vectors.

References

1. Karim Abed-Meraim, Yong Xiang, Jonathan H. Manton, and Yingbo Hua. Blind source separation using second-order cyclostationary statistics. *IEEE Transactions on Signal Processing*, 49:694–701, 2001.
2. A. Belouchrani, K. A. Meraim, J.-F. Cardoso, and E. Moulines. A blind source separation technique using second order statistics. *IEEE Transactions on Signal Processing*, 45:434–444, 1997.
3. Herbert Buchner, Robert Aichner, and Walter Kellermann. A generalization of blind source separation algorithms for convolutive mixtures based on second-order statistics. *IEEE Transactions on Speech and Audio Processing*, 13:120–134, 2005.
4. Jean-François Cardoso. Blind signal separation: Statistical principles. *Proceedings of the IEEE*, 86:2009–2025, 1998.

5. Jean-François Cardoso and A. Souloumiac. Blind beamforming for non-Gaussian signals. *IEE Proceedings-F*, 140:362–370, 1993.
6. Seungjin Choi, Andrzej Cichocki, and Adel Beloucharni. Second order nonstationary source separation. *Journal of VLSI Signal Processing*, 32:93–104, 2002.
7. Andrzej Cichocki and Shun-ichi Amari. *Adaptive Blind Signal and Image Processing*. John Wiley & Sons, first edition, 2002.
8. Andrzej Cichocki and Jacek M. Zurada. Blind signal separation and extraction: Recent trends, future perspectives, and applications. In *Proceedings of the 7th International Conference on Artificial Intelligence and Soft Computing (ICAISC 2004)*, pages 30–37, Zakopane, Poland, June 7–11 2004.
9. Yannick Deville, Mohammed Benali, and Frédéric Abrard. Differential source separation for underdetermined instantaneous or convolutive mixtures: concept and algorithms. *Signal Processing*, 84:1759–1776, 2004.
10. Da-Zheng Feng, Xian-Da Zhang, and Zheng Bao. An efficient multistage decomposition approach for independent components. *Signal Processing*, 83:181–197, 2003.
11. Anne Ferréol and Pascal Chevalier. On the behavior of current second and higher order blind source separation methods for cyclostationary sources. *IEEE Transactions on Signal Processing*, 48:1712–1725, 2000.
12. Anne Ferréol, Pascal Chevalier, and Laurent Albera. Second-order blind separation of first- and second-order cyclostationary sources—application to am, fsk, cpfsk, and deterministic sources. *IEEE Transactions on Signal Processing*, 52:845–861, 2004.
13. K. Fukunaga. *Introduction to Statistical Pattern Recognition*. Academic Press, New York, 1990.
14. P. W. Holland and Y. J. Wang. Dependence function for continuous bivariate densities. *Communications in Statistics – Theory and Methods A*, 16:863–876, 1987.
15. P. W. Holland and Y. J. Wang. Regional dependence for continuous bivariate densities. *Communications in Statistics – Theory and Methods A*, 16:193–206, 1987.
16. Aapo Hyvärinen. Blind source separation by nonstationarity of variance: A cumulant-based approach. *IEEE Transactions on Neural Networks*, 12:1471–1474, 2001.
17. Aapo Hyvärinen, Juha Karhunen, and Erkki Oja. *Independent Component Analysis*. Wiley-Interscience, first edition, 2001.
18. Maria G. Jafari and J. A. Chambers. A novel adaptive algorithm for the blind separation of periodic sources. In *Proceedings of the 2004 IEEE International Joint Conference on Neural Networks*, volume 1, pages 25–29, Budapest, Hungary, July 2004. IEEE.
19. M. C. Jones. The local dependence function. *Biometrika*, 83:899–904, 1996.
20. L. Molgedey and H. G. Schuster. Separation of a mixture of independent signals using time delayed correlations. *Physical Review Letters*, 72:3634–3636, 1994.
21. Hans-Georg Müller and Xin Yan. On local moments. *Journal of Multivariate Analysis*, 76:90–109, 2001.
22. Danielle Nuzillard and Jean-Marc Nuzillard. Second-order blind source separation in the fourier space of data. *Signal Processing*, 83:627–631, 2003.
23. Lucas Parra and Paul Sajda. Blind source separation via generalized eigenvalue decomposition. *Journal of Machine Learning Research*, 4:1261–1269, 2003.
24. Dinh-Tuan Pham and Jean-François Cardoso. Blind separation of instantaneous mixtures of nonstationary sources. *IEEE Transactions on Signal Processing*, 49:1837–1848, 2001.
25. K. J. Pope and R. E. Bogner. Blind signal separation: I. linear, instantaneous combinations. *Digital Signal Processing*, 6:5–16, 1996.
26. Ehud Weinstein, Meir Feder, and Alan V. Oppenheim. Multi-channel signal separation by decorrelation. *IEEE Transactions on Speech and Audio Processing*, 1:405–413, 1993.

Hyperspectral Data Selection from Mutual Information Between Image Bands

José Martínez Sotoca and Filiberto Pla

Dept. Llenguatges i Sistemes Informàtics, Universitat Jaume I,
E-12071, Castelló, Spain
{sotoca, pla}@lsi.uji.es

Abstract. This work presents a band selection method for multi and hyperspectral images using correlation among bands based on mutual information measures. The relationship among bands are represented by means of the *transinformation matrix*. A process based on a Deterministic Annealing optimization is applied to the *transinformation matrix* in order to obtain a reduction of this matrix looking for the image bands as less uncorrelated as possible between them. Some experiments are presented to show the effectiveness of the bands selected from the point of view of pixel classification.

Keywords: Multispectral images, mutual information, deterministic annealing, unsupervised feature selection.

1 Introduction

Hyperspectral sensors acquire information in large quantities of spectral bands, which generate hyperspectral data in high dimensional spaces. These systems use multispectral image representations in order to estimate and analyze the presence of vegetation pathologies, substances or chemical compounds, pathologies, and so on, providing a qualitative and quantitative evaluation of those features.

When having available hyperspectral data, a common question to be solved is how to select the right spectral bands to characterize the problem. The main objective of band selection in multispectral imaging is to avoid redundant information and reduce the amount of data to be processed. Therefore, from the point of view of remote sensing, we would be interested in feature selection [3] rather than in feature extraction [7]. For instance, obtaining a new set of reduced image representations from a linear combination of the whole set of original image bands is not desirable, since we would need the total amount of information to obtain the new features. On the other hand, selecting a subset of relevant bands from the original set, allows the process of image acquisition to be reduced to a certain number of bands instead of dealing with the whole amount of data, making simpler the image acquisition and analysis.

In the framework of multispectral imaging, another possible answer to the problem of feature selection would be using an unsupervised approach [4][2]. In this work, a Deterministic Annealing (DA) approach is used to analyze the

D.-Y. Yeung et al. (Eds.): SSPR&SPR 2006, LNCS 4109, pp. 853–861, 2006.

amount of information contained in the *mutual information matrix*, which represents the relations of information between pairs of spectral bands. The proposed algorithm uses a Deterministic Annealing (DA) approach to look for groups of bands as less correlated as possible, representing correlation between image bands by means of mutual information. Selected bands are further used in pixel classification tasks to assess the performance of proposed technique.

2 Deterministic Annealing for Rank Reduction

Let us consider a pair of random variables A_i and A_j, representing the image bands i and j. The amount of information contained in both images can be expressed as the joint entropy $H(A_i, A_j) = \sum p(a_i, a_j) \log_2 \frac{1}{p(a_i,a_j)}$, where $p(a_i, a_j)$ represents a joint probability distribution.

For two images i and j, the joint probability distribution $p(a_i, a_j)$ of both images can be estimated as $p(a_i, a_j) = \frac{h(a_i,a_j)}{MN}$, where $h(a_i, a_j)$ is the joint gray level histogram, and the normalizing factor, MN (M columns and N rows) is the image size.

Mutual information $H(A_i{:}A_j)$ is a basic concept in information theory [1]. It measures the interdependence between random variables. In the case of two images, the mutual information is defined as:

$$H(A_i : A_j) = H(A_i) + H(A_j) - H(A_i, A_j) \tag{1}$$

where $H(A_i)$, $H(A_j)$ are the entropy of images i and j, and $H(A_i, A_j)$ is the joint entropy.

One way to establish the interdependence between a set of features is defining the *transinformation matrix*. In our framework, this is a square matrix representing the mutual information between pairs of image bands. The diagonal terms represent the entropy of single bands, and contiguous bands in the spectrum tend to be highly correlated (brighter values in Fig 1).

The technique here proposed is aimed at reducing the rank of the *transinformation matrix* by selecting a given number of features that minimize the correlation among them. Therefore, we look for a global minimum without carrying out a search of subsets of features in the feature space. The process must be capable of picking up a subset of bands, in order to obtain as better performance as possible from the classification point of view reducing the feature space.

Discretizing the mutual information and representing the *transinformation matrix* as an "image" with gray levels (see Fig 1), defines a spreading measure of the information in the gray level distribution of the matrix. This measure will estimate the information contained about the appearance of the different regions of the spectrum in the matrix. Thus, we can analyze the probability that the event (value associated with each position of the matrix) takes place. This probability n_{ij} can be calculated as $n_{ij} = h_{ij}/D^2$, where h_{ij} is the value in the

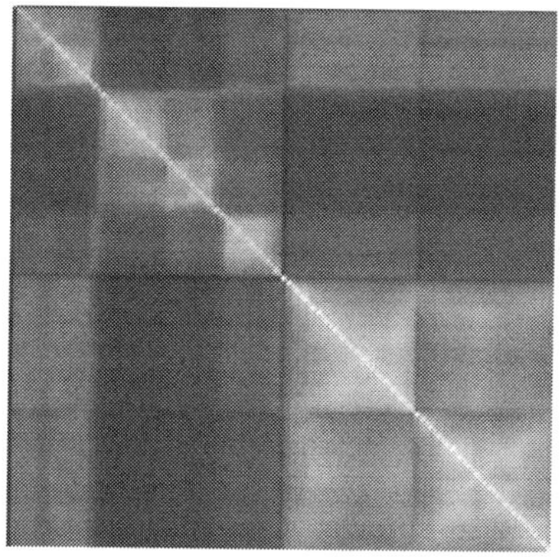

(a)

Fig. 1. *Transinformation matrix* for a multispectral image with 128 wavebands. Darker values represent less correlation.

histogram for the gray level at i and j. Including the probability n_{ij} in each position in the matrix, we define the following function of information as:

$$I_{ij} = n_{ij} H(A_i : A_j). \tag{2}$$

From the function I_{ij} , we are interested in associating a probability of significance $p(I_{ij}|ij)$ for each position i and j in the matrix. This probability will mean how relevant is the interaction of band i and j for the problem. Therefore, a probabilistic model is applied over each position of the matrix $p(I_{ij}|ij)$. It is, thus, possible to utilize DA to obtain the image bands that contain higher values of significance in the matrix. To apply DA in such a framework, the following requirements must be fullfiled:

- The entropy S of the distribution of probabilities $p(I_{ij}|ij)$ associated to this representation of "level of uncertainty" must be maximum.
- The sum of probabilities are normalized to one.
- The product of $p(I_{ij}|ij)$ per the value of I_{ij} between pairs of bands, provides a value about the amount of information I associated to the ensemble.

Therefore, we can establish the the following relation:

$$S = -\sum_{i=1}^{D}\sum_{j=1}^{D} p(I_{ij}|ij) \log \frac{p(I_{ij}|ij)}{p_{ij}} \tag{3}$$

subject to

$$\sum_{i=1}^{D}\sum_{j=1}^{D} p(I_{ij}|ij) = 1 \quad \text{and} \quad \sum_{i=1}^{D}\sum_{j=1}^{D} p(I_{ij}|ij) I_{ij} = I \tag{4}$$

where p_{ij} is proportional to the prior contribution of each relation between pairs of bands. Thus, S is the entropy relative to some "measures" p_{ij} that has to be maximized [5]. To maximize S subject to the constraint Eq 4, we can introduce Lagrangian multipliers α and β,

$$S - \alpha \left(\sum_{i=1}^{D} \sum_{j=1}^{D} p(I_{ij}|ij) - 1 \right) - \beta \left(\sum_{i=1}^{D} \sum_{j=1}^{D} p(I_{ij}|ij) I_{ij} - I \right) \tag{5}$$

Setting the partial derivative of Eq 5 with respect $p(I_{ij}|ij)$ to zero, we obtain the following expression,

$$-\log p(I_{ij}|ij) - 1 + \log p_{ij} - \alpha - \beta I_{ij} = 0 \tag{6}$$

where

$$p(I_{ij}|ij) = p_{ij} e^{-\alpha - 1 - \beta I_{ij}} \tag{7}$$

Taking into account that the sum of probabilities are normalized to one, then

$$\sum_{i=1}^{D} \sum_{j=1}^{D} p_{ij} e^{-\beta I_{ij}} = e^{1+\alpha} = Z \tag{8}$$

where Z is the so-called *partition function* and

$$p(I_{ij}|ij) = \frac{p_{ij} e^{-\beta I_{ij}}}{Z} \tag{9}$$

Taking $\beta = \frac{1}{T}$, our probability function is expressed as

$$p(I_{ij}|ij) = \frac{p_{ij} e^{-I_{ij}/T}}{\sum_{i=1}^{D} \sum_{j=1}^{D} p_{ij} e^{-I_{ij}/T}}$$

and

$$p_{ij} = I_{ij} p(I_{ij}|ij)$$

The result is the Bayes' Theorem, where we can obtain the posterior probability distribution for each position through the exponential function of the values observed in the matrix multiplied by the prior probability p_{ij}.

The initialization of DA starts with large enough values of T, and a uniform distribution of probabilities $p(I_{ij}|ij) = 1/D^2$. The initial set of features X to choose is empty. As $T \rightarrow 0$ a reduction of the amount of information I is carried out. In practice, the system is annealed to a low temperature, such the amount of information I ("level of dependence" of the matrix) is sufficiently small.

On the other hand, we express the probability contributions of each band A_i accumulating for each row or column i (the matrix is symmetrical) as:

$$B_i = \sum_{j=1}^{D} p(I_{ij}|ij) \tag{10}$$

While T decreases, the difference between the values of $p(I_{ij}|ij)$ grow up. As T goes down, the probability contributions of some bands $B_i \rightarrow 0$, but it is possible that further in the annealing, with lower T, previous low values of B_i grow up for the new circumstances. Only if $B_i \cong 0$, we can almost assure that the corresponding band will not contribute in the probability distribution in the next iterations.

Summarizing, a brief sketch of the algorithm is as follows:

1. *Initialize:* $T = T_0$, $p(I_{ij}|ij) = 1/D^2$ and $|X| = 0$
2. *Minimize:* $F = I - TS$
3. *Calculate:* $B_i = \sum_{j=1}^{D} p(I_{ij}|ij)$
4. *If* $B_i \cong 0$ *then:* $X \leftarrow (X \cup A_i)$
5. *Count the number of image bands* R *such:* $B_i > 1/D$
6. *Lower Temperature:* $T \leftarrow q(T)$
7. *Go to step 2 while* $R \geq 2$

In our experiments, we used an exponential schedule to reduce T, $q(T) = \alpha T$, where $\alpha < 1$, but other annealing schedules are possible. At the end of the algorithm, the probability contributions B_i are concentrated in the two best bands with values about $\simeq 0.5$.

3 Empirical Results

To test the proposed approach, different databases of multispectral images are used in the experiments:

1. Multispectral images of oranges obtained by an imaging spectrograph (Retiga-Ex, Opto-knowledge System Inc. Canada). This database was captured in to spectrum range, VIS collection (400-720 nm in the visible) and NIR collection (650-1050 nm in the near infrared). In both cases, the camera has a spectral resolution of 10 nanometers. The database includes several kinds of defects. It has eight classes, obtaining 1463346 labelled pixels from VIS and 1491888 labelled pixels from NIR.
2. Spectral image (700 X 670 pixels) acquired with the 128-bands HyMap spectrometer during the DAISEX-99 campaign (http:/io.uv.es/projects/daisex/), and six different classes were considered in the area (see Fig 2 (b))
3. Spectral image (145 X 145 pixels) acquired with the AVIRIS data set with 220 bands collected in June 1992 over the Indian Pine Test site in Northwestern Indiana (see Fig 2 (c)). The data set is designated as *92AV3C*, and it has seventeen classes.(http:/dynamo.ecn.purdue.edu /∼biehl/MultiSpec)

In order to assess the performance of the method, a Nearest Neighbor (NN) classifier was used to classify pixels into the different classes. The performance of the NN classifier was considered as the validation criterion to compare the significance of the subsets of selected image bands obtained by the proposed approach and other methods (two supervised and one unsupervised approaches) in

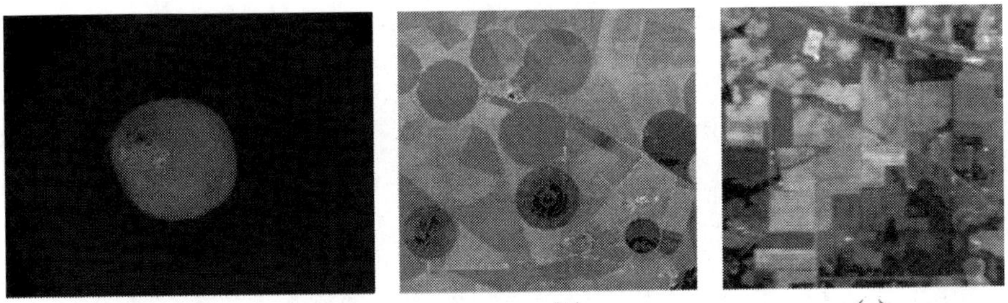

(a) (b) (c)

Fig. 2. (a) Example of RGB composition for an orange image in the Visible spectrum. (b) HyMap RGB composition, Barrax, Spain. (c) RGB composition of AVIRIS (92AV3C: NW Indiana's Indian Pine test site).

the experiment carried out. To increase the statistical significance of the results, the average values over five random partitions were estimated.

In the case of supervised approaches, the main motivation is that the labelled data contains information about the distribution of classes exiting in the hyperspectral data, and they allow the search for relevant feature subsets. Comparing the performance with those approaches, we can measure the capability to obtain subsets of relevant features (image bands) by the introduced DA approach without a prior knowledge of the class distributions in the multispectral image.

The first method is the well-known *ReliefF* algorithm [6] based on pattern distances. The second technique is related to divergence measures between classes. One of the best-known distance measures utilized for feature selection in multiclass problems is the average Jeffries-Matusita (JM) distance. To obtain suboptimal subsets of features, we have applied a search strategy based on a Sequential Forward Selection applying this distance ((SFS) JM distance) [3].

Moreover, we evaluated an unsupervised method presented in a previous work based in information measures between image bands [8]. This approach called "Minimization of the Dependent Information" (MDI) measures the region of dependence given a number bands for a multispectral image, and obtains a minimum interdependence.

3.1 Performance Evaluation

During the image labelling process, there are always pixels in an image that are not assigned to any class of interest, mainly because they are pixels that either do not clearly belong to some of the predefined classes or they are assigned to a complementary class. The pixels that have not been assigned to any class are labelled as "unknown" class.

The experimental results shown in this section about the classification rates correspond to the average classification accuracy obtained by the NN classifier over the five random partitions described previously. The samples in each partition were randomly assigned to the training and test set with equal sizes as follows: VIS = 43902 pixels, NIR = 44758 pixels, HyMap = 37520 pixels, 92AV3C = 2102 pixels.

On the other hand, given the huge size of the data sets and the trouble in computational cost to apply the supervised approaches, particularly in the case of VIS, NIR and HyMap, the following independent partitions with respect to the data sets were randomly extracted maintaining the prior probability of the classes: VIS = 87805 pixels, NIR = 89516 pixels, HyMap = 93804 pixels and 92AV3C = 10512 pixels. Using these databases, the proposed DA and the others methods were applied in order to obtain a ranking of relevance of the features, that is, of bands.

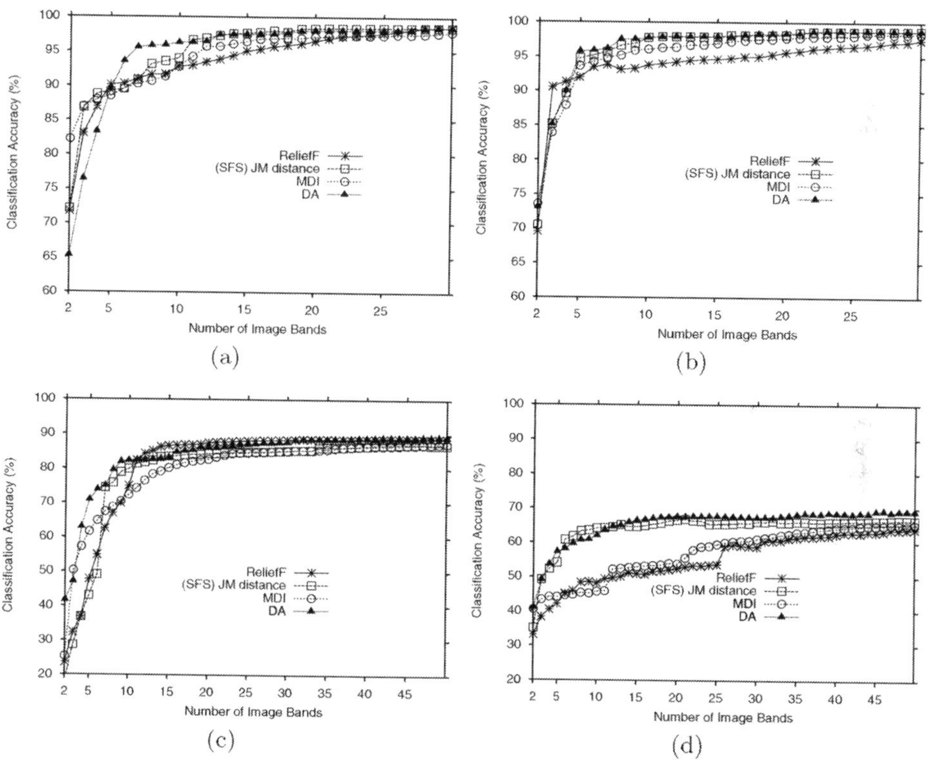

Fig. 3. (a) Results over oranges in VIS. (b) Results over oranges in NIR. (c) Results over spectral image with HyMap spectrometer. (d) Results over 92AV3C spectral image. In all cases, it is shown the performance of the *NN* classifier with respect to the number of features obtained by DA, (SFS) JM distance and *ReliefF*.

Fig 3 represents the classification rate with respect to the subset of N bands selected by each method. Note that the proposed DA method obtained better performance with respect to the rest of methods in the case of database of VIS, and similar accuracy for the best of the other approaches for NIR, HyMap and 92AV3C. It is worthwhile mentioning that the DA approach has a good behavior

Table 1. Computational cost in minutes (m) when selecting all features except for (SFS) JM distance, where it is shown for 30 features (VIS and NIR) and 50 features (HyMap and 92AV3C)

Criteria	Time (m)			
	VIS	NIR	HyMap	92AV3C
ReliefF	198 m	237 m	423 m	20 m
(SFS)JM distance	17 m	49 m	152 m	151 m
MDI	349 m	407 m	2337 m	2446m
DA	4 m	8 m	130 m	102 m

in all cases when choosing the smaller sets of bands (first one to ten), where the decision is more critical.

ReliefF performs poorer with respect to the other approaches except with HyMap image, where the performance of (SFS) JM distance is worse. *ReliefF* obtains a ranking of relevance for each single feature and the computational cost grows exponentially with respect to the number of samples in the data set.

(SFS) JM distance provides a high classification accuracy, but the computational cost grows exponentially with respect to the number of dimensions. Table 1 shows the computational time in minutes for the tested methods.

MDI provides similar classification accuracy respect to *ReliefF* but its nature is completely unsupervised. Moreover, it is not efficient from the computational point of view to obtain subsets in spaces with high dimensionality. This is mainly due to the cost of computing the joint probability distributions for each combination of bands.

In the case of DA, the principal problem arises when the *transinformation matrix* is built. Thus, the different co-occurrences of pixels in each pair of image bands are calculated [8], which represents an quadratic cost in time. On the other hand, when the matrix is built, the proposed DA method obtain the selected features very quickly.

Therefore, for the band selection problem, where there exists high correlation among different features (image bands), the principle of looking for non correlated bands from the different regions of the spectrum, by reducing the mutual information in the ensemble of image bands, has proven to be an effective approach to obtain subsets of selected image bands that also provide satisfactory results from the classification accuracy point of view.

4 Concluding Remarks

In this work, correlation among image bands in multispectral images has been established in terms of mutual information. The relationships between bands can be represented by the *transinformation matrix*. Using this representation, an approach to rank reduction of the *transinformation matrix* using Deterministic Annealing has been proposed to look for a given number of bands as less correlated as possible among them.

Although the proposed method has not been established in terms of class separability for supervised training sets, it has been shown in the experimental results that the image bands selected by DA provide very satisfactory results with respect to classification accuracy when using the selected bands. This effect is more noticeable when choosing small sets of features, when the decision is more critical. These two advantages, its unsupervised nature and the ability to choose relevant bands in the case of small sets, represent the more relevant characteristics of the proposed approach.

Acknowledgements

This work has been supported in part by grants P1-1B2004-08 from Fundació Caixa Castelló-Bancaixa and ESP2005-07724-C05-05 from the Spanish CICYT.

References

1. J. Aczel, J., Daroczy, Z.: On measures of information and their characterization. New York: Academic Press, 1975.
2. Agrawal, R., Gehrke, J., Gunopulos, D., Raghavan, P.: Automatic subspace clustering of high dimensional for data mining applications. In Proceedings of the ACM SIGMOD International Conference on Management of Data, Seattle, WA, June (1998), 94–105
3. Bruzzonne, L., Roli, F., Serpico S.B.: An extension to multiclass cases of the Jeffreys-Matusita distance. IEEE Transactions on Geoscience and Remote Sensing, **33** (1995) 1318–1321
4. Groves, P., Bajcsy, P.: Methodology for hyperspectral band and classification model selection. IEEE Workshop on Advances in Techniques for Analysis of Remotely Sensed Data. An Honorary Workshop for Prof. David A. Landgrebe, Washington D.C., 2003.
5. Jaynes, E.T.: Prior Probatilities. IEEE Transations on System Science and Cybernetic, SSC-4, (1968) pp. 227–241. Reprinted in Concepts and Applications of Modern Decision Models, V.M. Rao Tummala and R. C. Henshaw, eds., (Michigan State University Business Studies Series, 1976).
6. Kononenko, I.: Estimating attributes: analysis and extensions of RELIEF. In Proceedings of 7th European Conference on Machine Learning, Catania, Italy,(1994) 171–182
7. Kumar, S., Ghosh, J., Crawford, M.M.: Best basis feature extraction algorithms for classification of hyperspectral data. IEEE Transactions on Geoscience and Remote Sensing, **39**, no. 7, (2001) 1368–1379
8. Sotoca, J.M., Pla F., Klaren A.C.: Unsupervised band selection for multispectral images using information theory. In 17th. International Conference on Pattern Recognition, Cambridge (UK),**3**, (2004) 510–513

Model Selection Using a Class of Kernels with an Invariant Metric

Akira Tanaka[1], Masashi Sugiyama[2], Hideyuki Imai[1], Mineichi Kudo[1], and Masaaki Miyakoshi[1]

[1] Division of Computer Science,
Graduate School of Information Science and Technology, Hokkaido University, Sapporo, 060-0814, Japan
{takira, imai, mine, miyakosi}@main.eng.hokudai.ac.jp
[2] Department of Computer Science, Tokyo Institute of Technology, Meguro-ku, Tokyo, 152-8552, Japan
sugi@cs.titech.ac.jp

Abstract. Learning based on kernel machines is widely known as a powerful tool for various fields of information science such as pattern recognition and regression estimation. The efficacy of the model in kernel machines depends on the distance between the unknown true function and the linear subspace, specified by the training data set, of the reproducing kernel Hilbert space corresponding to an adopted kernel. In this paper, we propose a framework for the model selection of kernel-based learning machines, incorporating a class of kernels with an invariant metric.

1 Introduction

Learning based on kernel machines[1] is widely known as a powerful tool for various fields of information science such as pattern recognition and regression estimation. Many kernel machines, represented by the support vector machines[2] and the kernel ridge regression[3,4], are proposed. In these methods, kernels are recognized as useful tools to calculate the inner product in high-dimensional feature spaces[3,4].

On the other hand, according to the theory of reproducing kernel Hilbert spaces[5,6], the essence of using kernels in learning problems is that the unknown target (classifiers in pattern recognition problems, unknown true functions in regression estimation problems, and so on) belongs to the reproducing kernel Hilbert space corresponding to the adopted kernel. On the basis of this essence, Ogawa formulated a learning problem as an inversion problem of a linear operator from the reproducing kernel Hilbert space corresponding to the adopted kernel onto a certain vector space concerned with the given training data set and constructed a series of learning machines, named "(parametric) projection learning", that gives a good approximation of the orthogonal projector of the unknown true function onto the linear subspace, specified by the given training data set, of the reproducing kernel Hilbert space corresponding to the adopted kernel[7].

D.-Y. Yeung et al. (Eds.): SSPR&SPR 2006, LNCS 4109, pp. 862–870, 2006.

In the field of machine learning based on kernel machines, the model selection, that is, the selection of a kernel (or its parameters) is one of the most important problems. In this paper, we construct a framework of the kernel selection on the basis of the projection-learning-based interpretation of learning problems, incorporating a class of kernels with an invariant metric.

2 Mathematical Preliminaries for the Theory of Reproducing Kernel Hilbert Spaces

In this section, we prepare some mathematical tools concerned with the theory of reproducing kernel Hilbert spaces.

Definition 1. *[5] Let $\mathbf{R}^n$ be an n-dimensional real vector space and let $\mathcal{H}$ be a class of functions defined on $\mathcal{D} \subset \mathbf{R}^n$, forming a Hilbert space of real-valued functions. The function $K(\boldsymbol{x}, \tilde{\boldsymbol{x}})$, $(\boldsymbol{x}, \tilde{\boldsymbol{x}} \in \mathcal{D})$ is called a reproducing kernel of $\mathcal{H}$, if*

1. *For every $\tilde{\boldsymbol{x}} \in \mathcal{D}$, $K(\boldsymbol{x}, \tilde{\boldsymbol{x}})$ is a function of $\boldsymbol{x}$ belonging to $\mathcal{H}$.*
2. *For every $\tilde{\boldsymbol{x}} \in \mathcal{D}$ and every $f \in \mathcal{H}$,*

$$f(\tilde{\boldsymbol{x}}) = \langle f(\boldsymbol{x}), K(\boldsymbol{x}, \tilde{\boldsymbol{x}}) \rangle_{\mathcal{H}}, \tag{1}$$

where $\langle \cdot, \cdot \rangle_{\mathcal{H}}$ denotes the inner product of the Hilbert space $\mathcal{H}$.

The Hilbert space $\mathcal{H}$ that has a reproducing kernel is called a reproducing kernel Hilbert space (RKHS). The reproducing property Eq.(1) enables us to treat a value of a function at a point in $\mathcal{D}$. Note that reproducing kernels are positive definite [5]:

$$\sum_{i,j=1}^{N} c_i c_j K(\boldsymbol{x}_i, \boldsymbol{x}_j) \geq 0, \tag{2}$$

for any $N, c_1, \ldots, c_N \in \mathbf{R}$, and $\boldsymbol{x}_1, \ldots, \boldsymbol{x}_N \in \mathcal{D}$. In addition, $K(\boldsymbol{x}, \tilde{\boldsymbol{x}}) = K(\tilde{\boldsymbol{x}}, \boldsymbol{x})$ for any $\boldsymbol{x}, \tilde{\boldsymbol{x}} \in \mathcal{D}$ is followed[5]. If a reproducing kernel $K(\boldsymbol{x}, \tilde{\boldsymbol{x}})$ exists, it is unique[5]. Conversely, every positive definite function $K(\boldsymbol{x}, \tilde{\boldsymbol{x}})$ has the unique corresponding RKHS [5].

Next, we introduce the Schatten product [8] that is a convenient tool to reveal the reproducing property of kernels.

Definition 2. *[8] Let $\mathcal{H}_1$ and $\mathcal{H}_2$ be Hilbert spaces. The Schatten product of $g \in \mathcal{H}_2$ and $h \in \mathcal{H}_1$ is defined by*

$$(g \otimes h)f = \langle f, h \rangle_{\mathcal{H}_1} g, \quad f \in \mathcal{H}_1. \tag{3}$$

Note that $(g \otimes h)$ is a linear operator from $\mathcal{H}_1$ onto $\mathcal{H}_2$. It is easy to show that the following relations hold for $h, v \in \mathcal{H}_1$, $g, u \in \mathcal{H}_2$.

$$(h \otimes g)^* = (g \otimes h), \quad (h \otimes g)(u \otimes v) = \langle u, g \rangle_{\mathcal{H}_2} (h \otimes v), \tag{4}$$

where the super script * denotes the adjoint operator.

3 Formulation of Learning as Linear Inverse Problems

Let $\{(y_i, \boldsymbol{x}_i) | i = 1, \ldots, \ell\}$ be a given training data set with $y_i \in \mathbf{R}$, $\boldsymbol{x}_i \in \mathbf{R}^n$, satisfying

$$y_i = f(\boldsymbol{x}_i) + n_i, \tag{5}$$

where f denotes the unknown true function and n_i denotes a zero-mean additive noise. The aim of machine learning is to estimate the unknown function f by using the given training data set and statistical properties of noise.

In this paper, we assume that the unknown function f belongs to the RKHS $\mathcal{H}_K$ corresponding to a certain kernel function K. If $f \in \mathcal{H}_K$, then Eq.(5) is rewritten by

$$y_i = \langle f(\boldsymbol{x}), K(\boldsymbol{x}, \boldsymbol{x}_i) \rangle_{\mathcal{H}_K} + n_i, \tag{6}$$

on the basis of the reproducing property of kernels. Let $\boldsymbol{y} = [y_1, \ldots, y_\ell]'$ and $\boldsymbol{n} = [n_1, \ldots, n_\ell]'$ with the super script $'$ denoting the transposed matrix (or vector), then applying the Schatten product to Eq.(6) yields

$$\boldsymbol{y} = \left(\sum_{k=1}^{\ell} [\boldsymbol{e}_k^{(\ell)} \otimes K(\boldsymbol{x}, \boldsymbol{x}_k)] \right) f(\boldsymbol{x}) + \boldsymbol{n}, \tag{7}$$

where $\boldsymbol{e}_k^{(\ell)}$ denotes the k-th vector of the canonical basis of $\mathbf{R}^\ell$. For a convenience of description, we write

$$A_K = \left(\sum_{k=1}^{\ell} [\boldsymbol{e}_k^{(\ell)} \otimes K(\boldsymbol{x}, \boldsymbol{x}_{\boldsymbol{k}})] \right). \tag{8}$$

The operator A_K is linear one that maps an element of $\mathcal{H}_K$ onto $\mathbf{R}^\ell$ and Eq.(7) can be written by

$$\boldsymbol{y} = A_K f + \boldsymbol{n}, \tag{9}$$

which represents the relation between the unknown true function f and an output vector $\boldsymbol{y}$. The information of input vectors is integrated in the operator A_K. Therefore, a machine learning problem can be interpreted as an inversion problem of Eq.(9) [7].

Based on the model Eq.(9), a novel learning framework named "(parametric) projection learning" was proposed[7,9,10,11]. The projection learning gives the minimum variance unbiased estimator of the orthogonal projection of the unknown true function f onto $\mathcal{R}(A_K^*)$ (the range of A_K^*), and the parametric projection learning gives its improvement, incorporating a relaxation of the unbiasedness of the projection learning. The parametric projection learning includes the projection learning as a special case. The parametric projection learning is defined as follows:

Definition 3. *[10,11] The parametric projection learning B_{PPL} is defined by*

$$B_{PPL}(\gamma) = \mathrm{argmin}_B [\mathrm{tr}[(BA_K - P_{\mathcal{R}(A_K^*)})(BA_K - P_{\mathcal{R}(A_K^*)})^*] \\ + \gamma E_{\boldsymbol{n}} ||B\boldsymbol{n}||^2], \tag{10}$$

where $P_{\mathcal{R}(A_K^*)}$ *and* γ *denote the orthogonal projector onto* $\mathcal{R}(A_K^*)$ *and a real positive parameter that controls the trade-off of the two terms, which works as a relaxation of the unbiasedness, respectively.*

One of the solutions of the parametric projection learning is given by

$$B_{PPL}(\gamma) = A_K^*(A_K A_K^* + \gamma Q)^+ \tag{11}$$

as shown in [10,11], where the super script $^+$ denotes the Moore-Penrose generalized inverse [12] and Q denotes the noise correlation matrix defined by

$$Q = E_{\boldsymbol{n}}[\boldsymbol{n}\boldsymbol{n}'].$$

Finally, the solution of the parametric projection learning is given by

$$\hat{f}(\boldsymbol{x}) = B_{PPL}\boldsymbol{y},$$

and the concrete form of it is written by

$$\begin{aligned} \hat{f}(\boldsymbol{x}) &= \left(\sum_{i=1}^{\ell}\left[K(\boldsymbol{x},\boldsymbol{x_i}) \otimes \boldsymbol{e}_i^{(\ell)}\right]\right)(G+\gamma Q)^+\boldsymbol{y} \\ &= \sum_{i=1}^{\ell} \boldsymbol{y}'(G+\gamma Q)^+\boldsymbol{e}_i^{(\ell)} K(\boldsymbol{x},\boldsymbol{x_i}), \end{aligned} \tag{12}$$

where $G = A_K A_K^*$ is the Gram's matrix of K written by $G = (g_{ij})$, $g_{ij} = K(\boldsymbol{x}_i, \boldsymbol{x}_j)$, which is easily confirmed by using the properties Eq.(4) of the Schatten product. Note that the assumption $Q = O$ yields the solution based on the Moore-Penrose generalized inverse of A_K.

4 Model Selection Using a Class of Kernels with an Invariant Metric

In general, the solution of kernel-based learning machines is given by a linear combination of $K(\boldsymbol{x}, \boldsymbol{x}_i)$ that spans $\mathcal{R}(A_K^*)$. Thus, the validity of the model depends on $||f - P_{\mathcal{R}(A_K^*)}f||^2_{\mathcal{H}_K}$. However, we can not directly evaluate it, since f in unknown. In this section, we construct a framework of selection of a good kernel that minimizes $||f - P_{\mathcal{R}(A_K^*)}f||^2_{\mathcal{H}_K}$ by incorporating a class of kernels with an invariant metric.

Let K_0 be a specific kernel and let $\mathcal{K}$ be a class of kernels satisfying

$$\mathcal{H}_K \subset \mathcal{H}_{K_0} \tag{13}$$

and

$$\langle f, g\rangle_{\mathcal{H}_K} = \langle f, g\rangle_{\mathcal{H}_{K_0}}, \tag{14}$$

for any $K \in \mathcal{K}$ and any functions $f, g \in \mathcal{H}_K$. Let

$$\mathcal{S}_{\mathcal{K}} = \{f | f \in \mathcal{H}_K \textit{ for all } K \in \mathcal{K}\}. \tag{15}$$

We assume that $\mathcal{S}_\mathcal{K} \neq \phi$. Thus, $\langle f, g\rangle_{\mathcal{H}_K}$ is invariant for any $K \in \mathcal{K}$ and any $f, g \in \mathcal{S}_\mathcal{K}$, which means that $K \in \mathcal{K}$ has the invariant metric that is the same with that of $\mathcal{H}_{K_0}$ for any $f \in \mathcal{S}_\mathcal{K}$. Note that $||f||^2_{\mathcal{H}_K}$ is also invariant for any $K \in \mathcal{K}$ and any $f \in \mathcal{S}_\mathcal{K}$.

We assume that $f \in \mathcal{S}_\mathcal{K}$ and let

$$f = P_{\mathcal{R}(A_K^*)}f + (I - P_{\mathcal{R}(A_K^*)})f \tag{16}$$

be a decomposition of f with $K \in \mathcal{K}$, then

$$\begin{aligned} ||f||^2_{\mathcal{H}_K} &= ||P_{\mathcal{R}(A_K^*)}f||^2_{\mathcal{H}_K} + ||(I - P_{\mathcal{R}(A_K^*)})f||^2_{\mathcal{H}_K} \\ &= ||P_{\mathcal{R}(A_K^*)}f||^2_{\mathcal{H}_{K_0}} + ||(I - P_{\mathcal{R}(A_K^*)})f||^2_{\mathcal{H}_{K_0}} \end{aligned} \tag{17}$$

holds and it immediately follows that

$$||f||^2_{\mathcal{H}_K} \geq ||P_{\mathcal{R}(A_K^*)}f||^2_{\mathcal{H}_{K_0}}. \tag{18}$$

Thus, it is guaranteed that $||(I - P_{\mathcal{R}(A_K^*)})f||^2_{\mathcal{H}_K} (= ||(I - P_{\mathcal{R}(A_K^*)})f||^2_{\mathcal{H}_{K_0}})$ is minimized by

$$K_{opt} = \text{argmax}_{K\in\mathcal{K}} ||P_{\mathcal{R}(A_K^*)}f||^2_{\mathcal{H}_{K_0}}, \tag{19}$$

which means that the selection of the best kernel from $\mathcal{K}$ is achieved.

As is mentioned in the previous section, a minimum variance unbiased estimator of $P_{\mathcal{R}(A_K^*)}f$ is given by the projection learning. However, its variance may be too large to use the solution as an approximation of $P_{\mathcal{R}(A_K^*)}f$. Thus, we may have to use an another solution, such as that based on a regularization scheme, as an approximation of $P_{\mathcal{R}(A_K^*)}f$, for instance.

5 Numerical Examples

In this section, we show a numerical example of a regression estimation of a one-dimensional function in order to investigate the properties of the proposed framework of a kernel selection.

We adopt L^2 as $\mathcal{H}_{K_0}$ and the sinc kernel defined by

$$K_S^\alpha(x, \tilde{x}) = \frac{\sin\alpha(x - \tilde{x})}{\pi(x - \tilde{x})},\ \alpha \in [\alpha_s, \alpha_e],\ 0 < \alpha_s < \alpha_e. \tag{20}$$

as a class of kernels with an invariant metric. In fact, the sinc kernel has the same metric with L^2 as shown in [13]. Moreover,

$$\mathcal{H}_{K_S^{\alpha_1}} \subset \mathcal{H}_{K_S^{\alpha_2}} \tag{21}$$

holds for any $\alpha_1 \leq \alpha_2$, since the RKHS corresponding to K_S^α is the space of band-limited functions in $[-\alpha, \alpha]$ in the Fourier domain. According to the monotonicity of the RKHSs corresponding to the sinc kernels,

$$\mathcal{S}_\mathcal{K} = \{f | f \in \mathcal{H}_{K_S^\alpha}\ \textit{for all}\ \alpha \in [\alpha_s, \alpha_e]\} = \{f | f \in \mathcal{H}_{K_S^{\alpha_s}}\}. \tag{22}$$

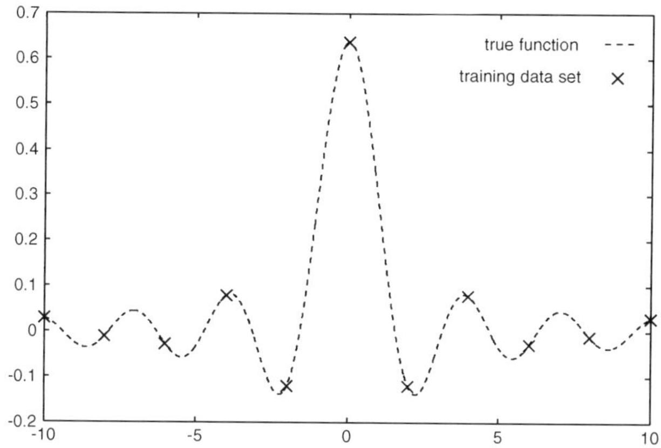

Fig. 1. The relation of the training data set and the unknown true function

Thus, the unknown true function f must belong to $\mathcal{H}_{K_S^{\alpha_s}}$ to make our framework to be consistent for any $\alpha \in [\alpha_s, \alpha_e]$.

We use

$$f(x) = \frac{\sin 2x}{\pi x} \tag{23}$$

as the unknown true function f and

$$\{(f(x_i), x_i) | x_i \in \{-10, -8, \ldots, -2, 0, 2, \ldots, 8, 10\}\} \tag{24}$$

as the given training data set. Figure 1 shows the relation of the training data set and the unknown true function. We adopt A_K^+ as a learning machine, since $Q = O$ in this case.

We dare to adopt $[1.5, 2.5]$ for the interval of the parameter searching. Note that when $\alpha \in [1.5, 2)$, the condition $f \in \mathcal{H}_{K_S^\alpha}$ is broken, that is, the estimated function obtained by A_K^+ is no longer the orthogonal projection of f. The result with the condition $\alpha \in [1.5, 2)$ could reveal the importance of the condition $f \in \mathcal{H}_K$ in machine learning problem. On the other hand, when $\alpha \in [2, 2.5]$, the consistency of our framework is guaranteed and the result based on it could reveal the validity of our framework. Figure 2 shows the transitions of $||\hat{f}||_{L^2}^2$, $||f - \hat{f}||_{L^2}^2$, and the sum of them with respect to α. Figures 3 $\sim$ 5 show the learning results with the parameters $\alpha = 1.5,\ 2.0,\ 2.5$, respectively.

According to the result shown in Fig.2 with $\alpha \in [1.5, 2)$, it is confirmed that $f \neq \mathcal{H}_{K_S^\alpha}$ causes the fail of estimation of the orthogonal projection of f. In fact, the norm of $\hat{f}$ is larger than that of f. Thus, it is concluded that adopting the kernel whose RKHS does not include f makes no sense for learning.

On the other hand, when $\alpha \in [2, 2.5]$ is satisfied, that is, $f \in \mathcal{H}_{K_S^\alpha}$ holds, it is confirmed that $\hat{f}$ is the orthogonal projection of f, since the sum of $||\hat{f}||_{L^2}^2$ and $||f - \hat{f}||_{L^2}^2$ is nearly equal to $||f||_{L^2}^2$. Moreover, it is confirmed that the maximizer of $||\hat{f}||_{L^2}^2$, satisfying $f \in \mathcal{H}_{K_S^\alpha}$, actually catches the best parameter $\alpha = 2$, which supports the validity of our framework.

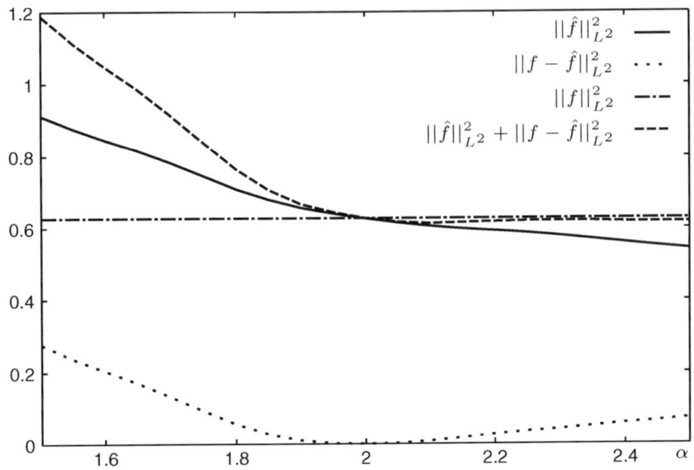

Fig. 2. Transitions of the squared norm of the estimated function, that of the error, and the sum of them with respect to α

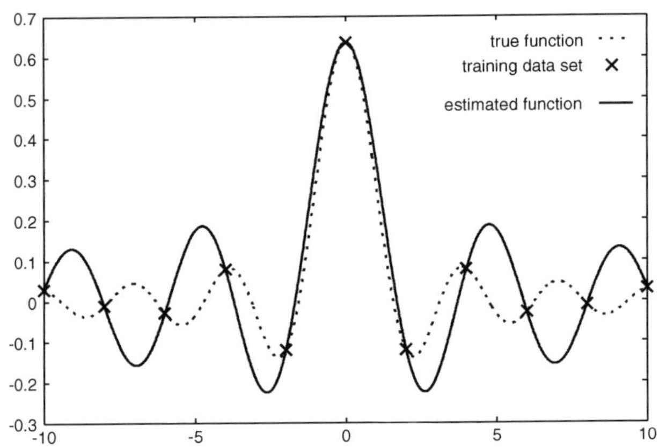

Fig. 3. The learning result with $\alpha = 1.5$

Remarks

We used a noise-free case in the numerical example. However, it is inevitable to consider the noise in practical cases.

As mentioned in the previous section, when the noise exists, the solution based on the projection learning is not robust in general. Thus, we may have to use a regularization scheme such as parametric projection learning with the optimal parameter chosen by a parameter selection criterion such as the SIC[14].

Although we adopted the sinc kernel as a class of kernels with an invariant metric in the numerical example, the sinc kernel is not so useful, since the intersection of the corresponding RKHSs is reduced to the RKHS corresponding to the minimum parameter of the interval for the parameter searching due to the

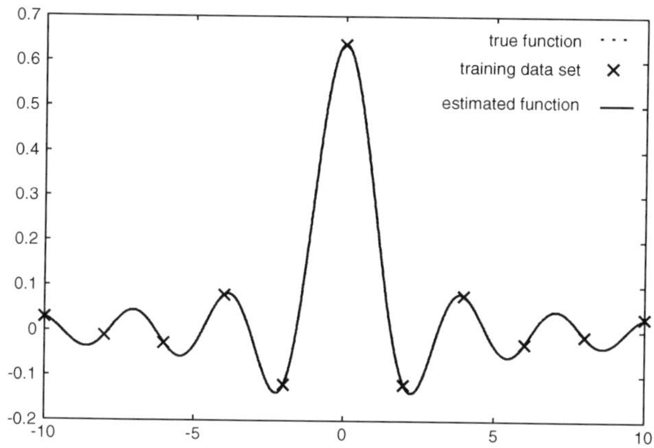

Fig. 4. The learning result with $\alpha = 2.0$

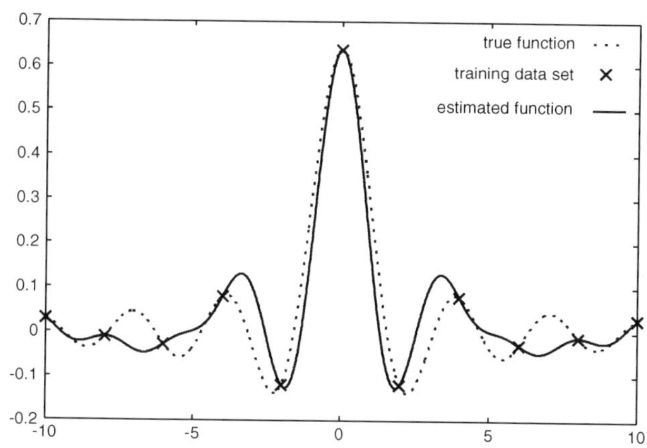

Fig. 5. The learning result with $\alpha = 2.5$

monotonicity of the corresponding RKHSs, which means that we can not adopt the interval that includes the unknown true parameter. Therefore, it is one of very important problems to construct a wide class of kernels with an invariant metric whose intersection includes a wide class of functions.

6 Conclusion

In this paper, we constructed a framework of a kernel selection on the basis of the projection-learning-based interpretation of learning problems, incorporating a class of kernels with an invariant metric. Coping with the noise and construction of a class of kernels with an invariant metric that is suitable for practical problems are future works.

References

1. Muller, K., Mika, S., Ratsch, G., Tsuda, K., Scholkopf, B.: An introduction to kernel-based learning algorithms. IEEE Transactions on Neural Networks **12** (2001) 181–201
2. Vapnik, V.N.: The Nature of Statistical Learning Theory. Springer, New York (1999)
3. Shawe-Taylor, J., Cristianini, N.: Kernel Methods for Pattern Recognition. Cambridge University Press, Cambridge (2004)
4. Cristianini, N., Shawe-Taylor, J.: An Introduction to Support Vector Machines and other kernel-based learning methods. Cambridge University Press, Cambridge (2000)
5. Aronszajn, N.: Theory of Reproducing Kernels. Transactions of the American Mathematical Society **68** (1950) 337–404
6. Mercer, J.: Functions of Positive and Negative Type and Their Connection with The Theory of Integral Equations. Transactions of the London Philosophical Society **A** (1909) 415–446
7. Ogawa, H.: Neural Networks and Generalization Ability. IEICE Technical Report **NC95-8** (1995) 57–64
8. Schatten, R.: Norm Ideals of Completely Continuous Operators. Springer-Verlag, Berlin (1960)
9. Sugiyama, M., Ogawa, H.: Incremental Projection Learning for Optimal Generalization. Neural Networks **14** (2001) 53–66
10. Imai, H., Tanaka, A., Miyakoshi, M.: The family of parametric projection filters and its properties for perturbation. The IEICE Transactions on Information and Systems **E80–D** (1997) 788–794
11. Oja, E., Ogawa, H.: Parametric Projection Filter for Image and Signal Restoration. IEEE Transactions on Acoustics, Speech and Signal Processing **ASSP–34** (1986) 1643–1653
12. Rao, C.R., Mitra, S.K.: Generalized Inverse of Matrices and its Applications. John Wiley & Sons (1971)
13. Saitoh, S.: Integral Transforms, Reproducing Kernels and Their Applications. Addison Wesley Longman Ltd, UK (1997)
14. Sugiyama, M., Ogawa, H.: Subspace Information Criterion for Model Selection. Neural Computation **13** (2001) 1863–1889

Non-Euclidean or Non-metric Measures Can Be Informative

Elżbieta Pękalska[1,2], Artsiom Harol[1], Robert P.W. Duin[1], Barbara Spillmann[3], and Horst Bunke[3]

[1] Faculty of Electrical Engineering, Mathematics and Computer Sciences, Delft University of Technology, The Netherlands
[2] School of Computer Science, University of Manchester, United Kingdom
[3] Institute of Computer Science and Applied Mathematics, University of Bern, Switzerland
{e.m.pekalska, a.harol}@tudelft.nl, r.duin@ieee.org, {spillman, bunke}@iam.unibe.ch

Abstract. Statistical learning algorithms often rely on the Euclidean distance. In practice, non-Euclidean or non-metric dissimilarity measures may arise when contours, spectra or shapes are compared by edit distances or as a consequence of robust object matching [1,2]. It is an open issue whether such measures are advantageous for statistical learning or whether they should be constrained to obey the metric axioms.

The k-nearest neighbor (NN) rule is widely applied to general dissimilarity data as the most natural approach. Alternative methods exist that embed such data into suitable representation spaces in which statistical classifiers are constructed [3]. In this paper, we investigate the relation between non-Euclidean aspects of dissimilarity data and the classification performance of the direct NN rule and some classifiers trained in representation spaces. This is evaluated on a parameterized family of edit distances, in which parameter values control the strength of non-Euclidean behavior. Our finding is that the discriminative power of this measure increases with increasing non-Euclidean and non-metric aspects until a certain optimum is reached. The conclusion is that statistical classifiers perform well and the optimal values of the parameters characterize a non-Euclidean and somewhat non-metric measure.

1 Introduction

Many currently available data are non-vectorial by origin. Although some ways exists to represent particular information in a vectorial form, these may be unnatural, of poor quality for the final prediction or very difficult to obtain. Vectorial representations are convenient since there exists a plethora of powerful learning techniques [4]. These are developed in inner product spaces or normed spaces, in which the inner product or norm defines the corresponding metric. On the other hand, if objects contain an inherent, identifiable structure or organization such as contours, shapes, spectra, images or texts, then structural descriptions are advisable. Objects can then be compared by suitable (min-max or edit) distances.

D.-Y. Yeung et al. (Eds.): SSPR&SPR 2006, LNCS 4109, pp. 871–880, 2006.

In other words, a collection of objects can be represented in a relative way, by a vector of dissimilarities (proximities) to a given set of representative examples. This is the so-called dissimilarity (proximity) representation [5,3]. Since proximity can be defined in both quantitative and qualitative contexts, it becomes a natural bridge between structural and statistical pattern recognition.

Kernel methods offer also an alternative to vectorial representations [6]. A kernel K is a (conditionally) positive definite (cpd) function of two variables, interpreted as a generalized inner product, hence similarity, in a Hilbert space $\mathcal{H}$ induced by K. Thanks to the reproducing property of K, the support vector machine (SVM) is built in $\mathcal{H}$ as a linear combination of kernel values to the so-called support vectors. The class of admissible kernels is, however, very limited due to the strong requirement of their being cpd. This is equivalent to stating that the corresponding distance $d(x,y) = \sqrt{K(x,x)+K(y,y)-2K(x,y)}$ is Euclidean for finite kernels [3]. In our terminology, kernels are an example of general proximity representations for which other learning strategies can successfully be applied.

Although proximity measures are widely used for matching and object comparison [1,2,7], classification often relies on assigning a new object to the class of its nearest neighbor. Alternative generalization frameworks exist that handle general proximity measures. They represent dissimilarity information in suitable representation vector spaces [5,8,9,3] or deal with indefinite kernels [10,3]. In case of non-Euclidean or non-metric dissimilarity data, researches usually either rely on the nearest neighbor distances, or choose to constrain/correct the measure to make it obey the metric axioms, e.g. by adding an appropriate constant or using a suitable transformation. In kernel methods, this is equivalent to regularizing the kernel by adding a proper constant to the diagonal.

If a highly non-metric/non-Euclidean measure describes the problem well (as judged by experts), corrections will likely lead to a significant loss of information [11,10,8]. If such deviations are small, they may be neglected as noise. Understanding is, therefore, necessary to identify under which circumstances and to what extent non-metric or non-Euclidean measures are advantageous in statistical learning. We contribute to this issue by presenting an empirical study in which the performance of dissimilarity-based statistical classifiers is related to indices measuring their departure from the Euclidean or metric behavior.

2 Representation Spaces and Classifiers

Assume a training set $T = \{t_1, \ldots, t_N\}$ of N objects and a representation set $R = \{p_1, p_2, \ldots, p_n\} \subseteq T$ of n prototypes. Given a dissimilarity measure d, a dissimilarity representation is an $N \times n$ matrix $D(T,R)$ with the elements $d(t_i, p_j)$. An object $t_i \in T$ is represented by an n-element vector of dissimilarities $D(t_i, R)$. The k-NN rule can directly be applied to such data. While it has good asymptotic properties for metric data, its performance deteriorates for small training (representation) sets. In such cases, alternative learning strategies can be more advantageous. They determine a suitable vector space equipped with the algebraic structure of either an inner product or norm in which the proximity information is represented. In such a vector space, the traditional learning algorithms can appropriately be adapted. Two simplest approaches are a linear isometric embedding into a pseudo-Euclidean space or the use of the so-called dissimilarity spaces [12,5,3].

In this paper, $D(\cdot, R)$ is interpreted as a data-dependent mapping $D(\cdot, R): X \rightarrow \mathbb{R}^n$ from some initial representation X (such as a vector space, images, strings or graphs) to a vector space defined by R. This is the *dissimilarity space*, in which each dimension $D(\cdot, p_i)$ corresponds to a dissimilarity to a prototype $p_i \in R$. The property that dissimilarities should be small for similar objects (belonging to the same class) and large for distinct objects, gives them a discriminative power. Hence, $D(\cdot, p_i)$ can be interpreted as 'features' and traditional classifiers built in vector spaces can be adapted [9,3]. The simplest are linear and quadratic classifiers, which are weighted combinations of the dissimilarities $d(x, p_i)$ between an object x and the prototypes p_i. The classifiers are optimized on $D(T, R)$, hence on the complete set T, even if only R is used for their representation. They can outperform the k-NN rule since they become more global in their decisions (suppressing the influence of individual noisy examples).

Classifiers. Normal density based (Bayesian) classifiers [4] tend to perform well in dissimilarity spaces [3,5,9]. This especially holds for summation-based dissimilarity measures, summing over a number of components with similar variances. The reason is that such dissimilarities will be approximately normally distributed thanks to the central limit theorem (if one or few variances are dominant, then they will approximate the χ^2 distribution) [3].

For a two-class problem, a quadratic normal density based classifier (NQC), is given by $f(D(x,R)) = \sum_{i=1}^{2} \frac{(-1)^i}{2} (D(x,R) - \mathbf{m}_i)^\top S_i^{-1} (D(x,R) - \mathbf{m}_i) + \log \frac{p_1}{p_2} + \frac{1}{2} \log \frac{|S_1|}{|S_2|}$, where $\mathbf{m}_i$ are the mean vectors and S_i are the class covariance matrices, all estimated in the dissimilarity space $D(\cdot, R)$. p_1 and p_2 are the class prior probabilities. If S_1 and SS_2 are replaced by the average covariance matrix, then a linear classifier is obtained. If the covariance matrices become singular, they need to be regularized. Here, we choose the following regularization $S_i^\kappa = (1-\kappa) S_i + \kappa p_i \operatorname{diag}(S_i)$, $\kappa \in [0,1]$, which leads to the RNQC, i.e. the regularized NQC. In our implementation, the normal-density functions are estimated per class and the final decision relies on the maximum a posteriori probability.

Another useful strategy for dissimilarity data is offered by sparse linear programming machines (LPM), which construct hyperplanes in the corresponding dissimilarity spaces. They are able to automatically determine a prototype set R (or if trained on $D(T, R)$, they may reduce the set R further on) which defines the final classifier. Two variants are here considered: the μ-LPM and the auc-LPM. The μ-LPM is a form of the ℓ_1-SVM with $\mu \in [0, 1)$, where μ is related to the expected classification error [13,9]. The auc-LPM is defined to maximize the area under the ROC curve, as recently proposed in [14].

We also define a new linear classifier, which is a nonnegative least square classifier (NLSQC). Let D denote $D(T, R)$, $R \subseteq T$, $|T| = N$ and $|R| = n$. Consider a two-class problem with the corresponding labels $y_i = +1/-1$. Let $Y_T = \operatorname{diag}(\mathbf{y}^T)$ and $Y_R = \operatorname{diag}(\mathbf{y}^R)$, where $\mathbf{y}^T$ and $\mathbf{y}^R$ are the label vectors for the sets

T and R, respectively. We define our classifier as $f(D(x,R) = sign(h(D(x,R)))$, where $h(D(x,R)) = -\mathbf{w}^{\mathsf{T}} Y_R D(x,R) + w_0$, $w_i \geq 0$, $i = 0,1,\ldots,n$. (Since w_i are multiplied by y_i^R, so $w_i y_i^R$ can be of any sign.) The classifier will assign 1 to x if $h(D(x,R)) = a > 0$ and –1 if $h(D(x,R)) = a < 0$. By fixing $a = 1$, it yields $y_i^T h(D(t_i,R)) > 1$ for the training objects t_i. The weights are now sought to minimize the sum of square differences $(y_i^T h(D(t_i,R)) - 1)^2$ for all t_i. We formulate the following problem:

$$\mathrm{Min}_{\mathbf{w}} \;\; ||D_{YY} \left[\begin{smallmatrix} \mathbf{w} \\ w_0 \end{smallmatrix} \right] - \mathbf{1}||_2^2, \text{ subject to } \;\; w_i \geq 0, \;\; i = 0,1,\ldots,n \tag{1}$$

where $D_{YY} = [Y_T(-D)Y_R \;\; -\mathbf{y}^T]$ and $\mathbf{1}$ is a vector of all ones. This can be solved by a standard nonnegative least square method that gives a sparse solution in terms of R. The non-zero weights correspond to the selected subset R' of R. In our case, the sparsity will not be large because of the choice of $-D$ in D_{YY} (or, equivalently, because of non-positive weights $-\mathbf{w}$ in the function h). In this quadratic criterion, $-D$ acts as a similarity (large values in $-D$, hence small distances, indicate large similarity) and requires many objects of R to support the decision boundary. On the contrary, if we choose D instead $-D$ in $h(D(x,R))$, i.e. $D_{YY} = [Y_T D Y_R \;\; \mathbf{y}^T]$, this will lead to a *very sparse* solution determined by dominating, possibly outlier distances only, hence to a poor discrimination. Non-positive weights $-w_i$ diminish the influence of large distances and shift the 'focus' towards the objects with small distances to the other class.

Equation (1) can be extended to $\mathbf{w}^{\mathsf{T}}(Y_R D^{\mathsf{T}} D Y_R)\mathbf{w}^{\mathsf{T}} + 2\mathbf{w}^{\mathsf{T}} Y_T D Y_R \mathbf{1} + 2\mathbf{1}^{\mathsf{T}}\mathbf{y}^T w_0 + N(w_0^2+1)+1$, in which the first term is the same as in the formulation of a linear SVM defined in a 'feature space' $X = D$. (There, in the dual problem, one would minimize $\frac{1}{2}\mathbf{w}^{\mathsf{T}}(Y_R D^{\mathsf{T}} D Y_R)\mathbf{w}^{\mathsf{T}} - \mathbf{w}^{\mathsf{T}}\mathbf{1}$ given that $\mathbf{w}^{\mathsf{T}}\mathbf{y}^R = 0$ and $w_i \geq 0$ [6].) Such an SVM would work in a *entire* dissimilarity space as it selects the support vectors in the form of $D(t_j,R)$ from T (and not from R)! Hence, a linear SVM in a dissimilarity space is not sparse. The advantage of the NLSQC is that it is a linear function with no additional parameters, which optimizes a square error and is, thereby, competitive to a quadratic classifier. Although it cannot outperform the SVM, it may compete with other LPMs applied to dissimilarity data. These LPMs are usually trained on a complete representation $D(T,T)$ and determine both R and the weights of the classifiers. These representation sets R may be used to train the NLSQC on $D(T,R)$ to enhance the sparsity.

3 Indices Characterizing Data

Assume K classes, $\omega_1, \omega_2, \ldots, \omega_K$ such that $|\omega_i| = n_i$ and $N = \sum_i n_i$. Two indices are defined to reflect the class separability. The first one is $J^1_{\text{sep}} = \frac{\sum_{i=1}^K n_i A_{ii}}{\sum_{i=1}^K n_i/(N-n_i)\sum_{j\neq i} n_j A_{ij}} \in (0,1)$, where A_{ij} is the average dissimilarity between the i-th and j-th classes (hence A_{ii} is the between-class average dissimilarity). The second index focusses on the nearest neighbor distances. $J^2_{\text{sep}} = \frac{1}{K}\sum_{i=1}^K B_i$, where $B_i = \frac{1}{n_i}\sum_{k=1}^{n_i} \frac{\min_{x\in\omega_i} d_{NN}(t_k,x)}{\min_{z\notin\omega_i} d_{NN}(t_k,z)}$ is the average ratio of the nearest neighbor within-class to between-class distances. The smaller the values, the better separability. Note that if $J^2_{\text{sep}} \approx 1$ or larger, than (on average) the nearest neighbor distances to objects from other classes are similar or smaller than the nearest neighbor distances within the class, hence the 1-NN rule cannot perform well.

Concerning the departure from the Euclidean behavior, it is known that a symmetric $N \times N$ dissimilarity matrix $D = D(T,T)$ has a Euclidean behavior iff the corresponding Gram matrix $G = -\frac{1}{2} J D^{\star 2} J$, where $D^{\star 2} = (d^2_{ij})$ and $J = I - \frac{1}{N}\mathbf{1}\mathbf{1}^{\mathsf{T}}$, is positive semidefinite [12,5,3]. It means that all eigenvalues λ_i of G are nonnegative. Hence, the magnitudes of negative eigenvalues manifest the amount of deviation from the Euclidean behavior. An indication of such a deviation is given by $J_{\text{eigM}} = |\lambda_{min}|/\lambda_{max}$, i.e. the ratio of the absolute value of the smallest negative eigenvalue to the largest positive one. The overall contribution of negative eigenvalues is estimated by $J_{\text{eigS}} = \sum_{\lambda_i<0} |\lambda_i| / \sum_{j=1}^{N} |\lambda_j|$.

Concerning non-metric aspects, any symmetric D can be made metric by adding a suitable constant γ to all off-diagonal elements of D. In a first attempt, such a constant can be found as $\gamma_0 = \max_{p,q,t} |d_{pq} + d_{pt} - d_{qt}|$ [3]. This estimation is however largely overpessimistic. Starting from this initial γ_0, we find a better estimation of $\gamma \in (0, \gamma_0)$ by an iterative bisection method. Our index is therefore $J_\gamma = \gamma \geq 0$ and it should be judged wrt the actual dissimilarity values. Another way to characterize the deviation from the metric behavior is by J_{ineq} equal to the total number of disobeyed triangle inequalities.

4 Data, Experiments and Results

In our study we use the Chicken Pieces Silhouettes data [15], available from `http://algoval.essex.ac.uk/data/sequence/chicken`. This set consists of 446 binary images from chicken pieces, labeled to one of the five classes,which represent specific parts of the chicken: wing (117 examples), back (76), drumstick (96), thigh and back (61), and breast (96). After edge detection applied to these silhouettes, the edges were approximated by straight line segments of a fixed length L, taking values between 5 and 40 pixels. Since chicken pieces are placed in arbitrary position in an image and mirror symmetry occurs, the line segments may not be the most appropriate. Instead, the sequence of angles between the neighboring segments was chosen as the initial string representation. Additionally, the approximate algorithm of Bunke and Bühler [16] was applied to handle the rotation invariance and axis symmetry. Given such string representations a family of edit distances [17] is considered with fixed insertion and deletion costs equal to some C and a substitution cost of the absolute difference between the angles. Consequently, we deal with an (L,C)-family of edit distance measures parameterized by L and C. In our case, we set $L = 5, 7, 10, 15, 20, 25, 29, 30, 31, 35, 40$ pixels and $C = 45, 60, 90, 120$ (angle degrees), which give rise to 44 different dissimilarity data, in total. The distances were originally asymmetric and are made symmetric by averaging, $d_{ij} = \frac{d_{ij}+d_{ji}}{2}$.

In our classification experiments we perform 50 runs of 2-fold cross-validation. In each run, all objects are first randomly split into the training set T and test set S. Then, classifiers are trained on $D_{L,C}(T,T)$ and tested on $D_{L,C}(S,T)$.

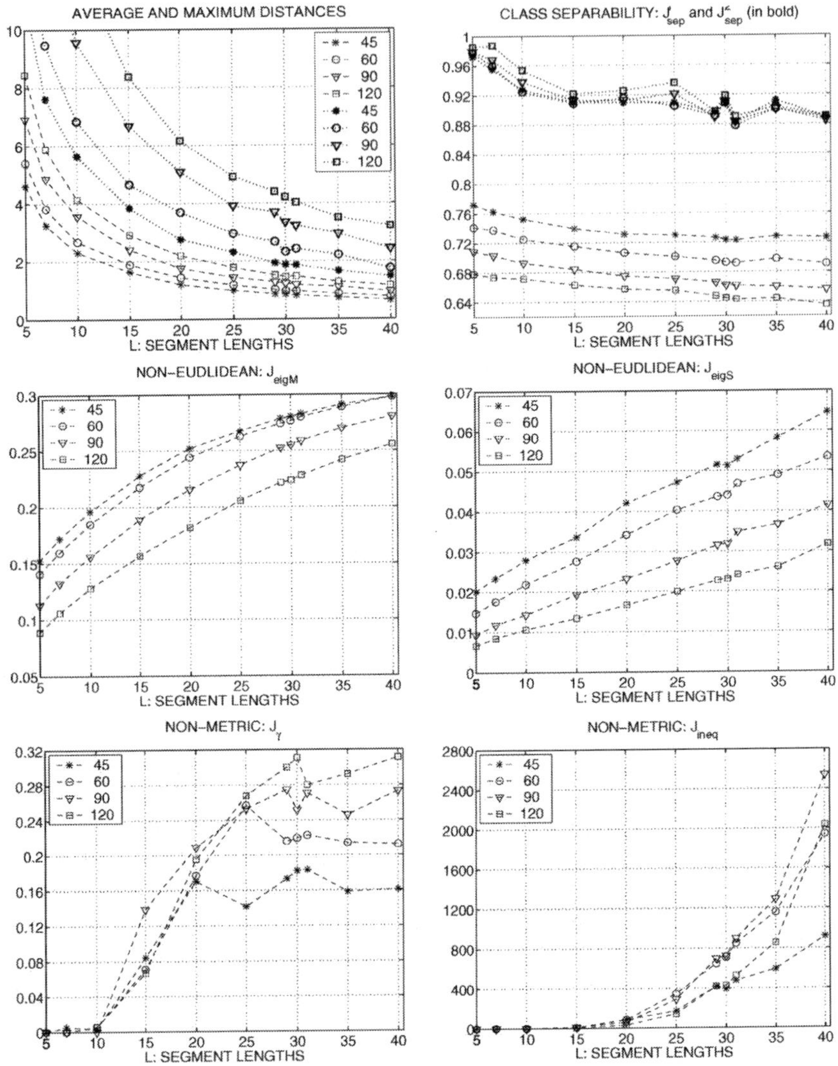

Fig. 1. Indices characterizing dissimilarity data. Legend values refer to C. Markers describing the same value of C are connected by lines to enhance the visibility.

Next, in the second fold, the classifiers are trained on $D_{L,C}(S,S)$ and tested on $D_{L,C}(T,S)$. Finally, the errors, weighted by prior probabilities, are determined. This is repeated 50 times and the results are averaged out. To avoid too large distance values, all dissimilarities are scaled by $\frac{1}{\sqrt{N}}$, where $N = |T|$. The following classifiers are used: the 1-NN and k-NN rules directly applied to the dissimilarity representation $D(T,T)$ (k is optimized in a LOO approach), edited-and-condensed nearest neighbor (CNN) [18], μ-LPM, with $\mu = \max\{0.01, 1.3 \cdot \text{LOO-NN-error}\}$, the auc-LPM (with the trade-off parameter set to 20) [14], the NSQLC and the RNQC with $\kappa = 0.05$. Additionally, the NSQLC is trained on the representation sets determined by the μ-LPM and auc-LPM, and denoted as the NSQLC(μ) and the NSQLC(auc), respectively. Remember that the LPMs and the NSQLC determine $R \subset T$ and that all (multi-class) linear classifiers are derived in an one-against-all strategy.

The properties of dissimilarity data are characterized by the indices introduced in Sect. 2. These will reflect the character of the dissimilarities, the class separability and the deviation from both Euclidean and metric behaviors. The indices are derived in the same setup as above. Their values are first averaged over two folds in a cross-validation scheme, and then over 50 runs.

Results. The indices defined in Sec. 2 were evaluated on 44 dissimilarity data with varying parameters L and C of the (L, C)-edit distances. By observing the results in Fig. 1, the following conclusions can be drawn:

- The average dissimilarities decrease with growing L. The smaller L, the larger maximal distances. The average and maximal distances grow with increasing C.
- The classification task is difficult since J^2_{sep} takes values close to 0.9 or 1. This means that the NN distances within a class are not much smaller than the NN distances to the objects outside the class. $C = 31$ seems to be optimal. Concerning the J^1_{sep}, the smaller C, the better the separability.
- None of the dissimilarity data set has a Euclidean behavior. The deviation becomes larger with the increasing L and decreasing C, as judged by J_{eigS} and J_{eigM}.
- The (L, C)-edit distances are practically metric up to $L \leq 10$; they become non-metric for larger values of L. The deviation from the metric behavior becomes larger with increasing C and is the smallest for $C = 45$. For $L \geq 30$, the additive constant γ that makes the dissimilarity measure metric roughly equals to $16 - 30\%$ of the average distance.

The classification results are shown in Fig. 2. In general, we observe that the performance of all classifiers improves with the increasing value of L up to a certain optimum and then starts to decrease. Most classifiers perform the best or nearly the best for $L = 30$. Concerning C, the classifiers reach the highest accuracy for $C = 45$ and gradually decrease their performance for larger values of C.

We will now provide the average *total* number of prototypes, i.e. $|R|$, determined by sparse linear classifiers. These numbers, presented as '$\cdot - \cdot - \cdot$', are averaged over C as only minor differences occur. The numbers taking the places of the first, second and third dot refer to $L = 5$, $L = 30$ and $L = 40$, correspondingly. We have: the μ-LPM: 223-123-112, the auc-LPM: 120-86-85, the NLSQC: 217-191-188, the NLSQC(μ): 217-116-106 and the NLSQC(auc): 119-84-83. For the CNN, the condensed sets vary over C and vary from 38 to 45.

The CNN relies on the smaller condensed set R but it performs the worst of all. The auc-LPM needs a relatively small R, but it also does not work well; it cannot compete with the 1-NN and k-NN rules unless $L \leq 15$. Other linear

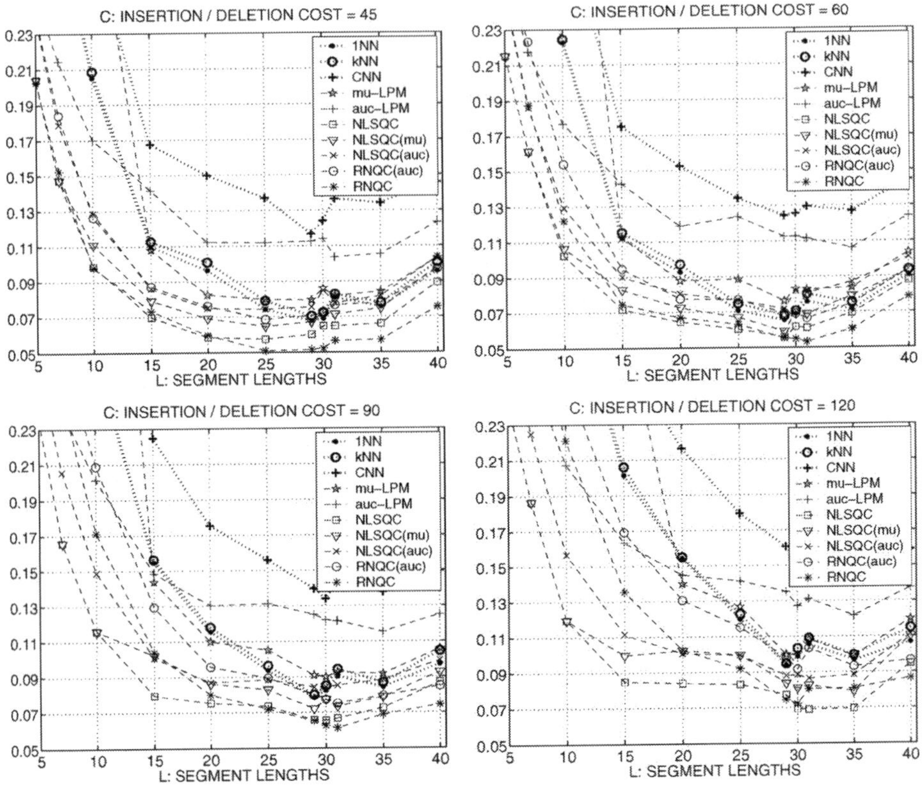

Fig. 2. Average 2-fold cross-validation errors (over 50 runs) for four values of C. Markers describing the same classifier are connected by lines to enhance the visibility. The standard deviations of the average errors are $0.0017 - 0.002$ on average (depending on C) for all classifiers. Their maximal values bounded by 0.0027, except for the μ-LPM, for which the maximal values are 0.011 for $C <= 10$ (where μ-LPM fails). The differences (between the average errors) larger than 0.01 are statistically significant.

classifiers outperform the auc-LPM, expect for the μ-LPM and $L \leq 15$. In general, the μ-LPM does not perform better than the NN rules (with little exceptions) and it deteriorates for $L \leq 15$, which is caused by the fact that the hyperplane cannot be determined (μ-LPM fails) and in our set-up the pseudo-Fisher classifier is automatically trained instead. However, if the representation objects determined by the auc-LPM or the μ-LPM are used to train the NLSQC, the performance drastically increases. The NLSQC(μ) is the third best performing classifier, which provides a good trade-off between the total cardinality of R and the classification accuracy. The representation objects preselected by the auc-LPM seem to make a n over-optimized set for the NLSQC(auc). Interestingly, the performance of the NLSQC(auc) is similar or much better than that of the RQNC(auc). For all C, our NLSQC performs the best or second best, after the RNQC if $L \geq 30$. Nearly all training objects are, however, needed for the representation. For the RNQC, always holds that $R = T$.

5 Conclusions

In this paper, examples of a parameterized family of (L, C)-edit distances are evaluated for the classification task on chicken pieces silhouettes. The deviation from non-Euclidean behavior grows with increasing L and decreasing C, while the deviation from non-metric behavior grows with both increasing L and C. Linear or quadratic classifiers built in dissimilarity spaces can outperform the direct k-NN rule and reach the optimal (or nearly optimal) results for $L = 30$. Our new linear classifier, the NLSQC, reaches the highest accuracy for most values of L and C. The best overall performance is reached for $L = 30$ and $C = 45$ which gives rise to a highly non-Euclidean and somewhat non-metric dissimilarity data. This is very interesting, since many researchers try to avoid non-metric data and define edit distances as metric measures. Our results suggests that non-Euclidean and/or non-metric distances can be informative and useful in statistical learning. We hope to explore these issues in the future research.

Acknowledgments. This work is supported by the Dutch Organization for Scientific Research (NWO).

References

1. Dubuisson, M., Jain, A.: Modified Hausdorff distance for object matching. In: ICPR. Volume 1. (1994) 566–568
2. Jacobs, D., Weinshall, D., Gdalyahu, Y.: Classification with Non-Metric Dist.: Image Retrieval and Class Representation. TPAMI **22** (2000) 583–600
3. Pękalska, E., Duin, R.: The dissimilarity representation for pattern recognition. Foundations and applications. World Scientific (2005)
4. Hastie, T., Tibshirani, R., Friedman, J.: The Elements of Statistical Learning. Springer Verlag (2001)
5. Pękalska, E., Paclík, P., Duin, R.: A Generalized Kernel Approach to Dissimilarity Based Classification. JMLR **2** (2002) 175–211
6. Schölkopf, B., Smola, A.: Learning with Kernels. MIT Press (2002)
7. Veltkamp, R., Hagedoorn, M.: State-of-the-art in shape matching. Technical Report UU-CS-1999-27, Utrecht University, The Netherlands (1999)
8. Pękalska, E., Duin, R., Günter, S., Bunke, H.: On not making dissimilarities Euclidean. In: S+SSPR. (2004) 1145–1154
9. Pękalska, E., Duin, R., Paclík, P.: Prototype selection for dissimilarity-based classifiers. Pattern Recognition **39** (2005) 189–208
10. Haasdonk, B.: Feature space interpretation of SVMs with indefinite kernels. TPAMI **25** (2005) 482–492
11. Laub, J., Müller, K.R.: Feature discovery in non-metric pairwise data. JMLR (2004) 801–818
12. Goldfarb, L.: A unified approach to pattern recognition. Pattern Recognition **17** (1984) 575–582

13. Graepel, T., Herbrich, R., Schölkopf, B., Smola, A., Bartlett, P., Müller, K.R., Obermayer, K., Williamson, R.: Classification on proximity data with LP-machines. In: ICANN. (1999) 304–309
14. Tax, D., Veenman, C.: Tuning the hyperparameter of an auc-optimized classifier. In: BNAIC. (2005) 224–231
15. Andreu, G., Crespo, A., Valiente, J.M.: Selecting the toroidal self-organizing feature maps (TSOFM) best organized to object recogn. In: ICNN. (1997) 1341–1346
16. Bunke, H., Bühler, U.: Applications of approximate string matching to 2D shape recognition. Pattern recognition **26** (1993) 1797–1812
17. Bunke, H., Sanfeliu, A., eds.: Syntactic and Structural Pattern Recognition Theory and Applications. World Scientific (1990)
18. Devijver, P., Kittler, J.: Pattern recognition: A statistical approach. Prentice/Hall (1982)

Merging and Arbitration Strategy Applied Bayesian Classification for Eye Location

Eun Jin Koh and Phill Kyu Rhee

Dept. of Computer Science & Engineering Inha University
Biometrics Engineering Research Center
Yong-Hyun Dong, Incheon, Korea
supaguri@im.inha.ac.kr, pkrhee@inha.ac.kr

Abstract. Based on template facial features and image segmentation, this paper demonstrates a novel method for automatic detection of eyes in grayscale still images. A decision model of eye location is instituted by the priori knowledge of template facial features. After roughly detection of face, we apply three steps for system to locate eyes. Firstly, the Bayesian eye detector is used to find eye patterns in the upper region of the face image. This vector based Bayesian classifier adopts Haar transform as vectorize because we know that is robust at illumination variation. Secondly, merging and arbitration strategy are applied. It can manage variations of around eye regions due to spectacle rims or eye brows. Finally, Gaussian-projection function can locate robust precision eye position. The experimental results show that the proposed method can achieve higher performance at any test data.

1 Introduction

Constructing automatic face recognition system has been a promising field of computer vision and pattern recognition. The face recognition task is achieved in three steps, i.e. 1) face detection, 2) marking facial feature points and 3) face recognition.

The face detection determines whether or not there are any faces in the image and, if exist; notify the face location and range of each face. The face recognition identifies or authenticates one or more persons in the detected image using a stored database of faces. In general, face recognition system needs those remarkable facial landmarks such as eyes must be located before any other processing is performed. Since recognition algorithm is based on template matching, face must be accurately aligned before other recognition processing, which is usually achieved based on the location of eyes. Because marking facial feature is necessary step to the template matching, it is an important step to face recognition and eyes are crucial points of facial features. Therefore, automatically locating eyes is very significant stage.

There is a series of techniques that can effectively detect eyes in frontal upright face images. But they suffer from bad light conditions and a rim of glasses. As a matter of fact, eyebrows or thick glasses frame enough to be confused with eye that the classifier often makes a incorrect decision.

To detect eyes in rough face image which is detected by face detector, the size window is applied to every pixel position in the image. To detect eyes which are larger than the window size, size of input image is repeatedly reduced by method of

D.-Y. Yeung et al. (Eds.): SSPR & SPR 2006, LNCS 4109, pp. 881 – 888, 2006.

super sampling or b-spline with factor 1.2 for each step, and the window is applied at each size. We define this method as multi-resolution. The multi-resolution method has some invariance at position and scale. The amount of invariance determines the number of windows which is must be applied at one image and it directly influences time of computation.

This paper introduces methodology for developed eye location procedure and depicts on both learning and estimating components. The experimental results are gained using the standard BioID, FERET and INHADB facial image database.

Detailed explanation of training images collection and methods are given in Section 2. The classifier architecture, postprocessing and arbitration strategy are given in Section 3. In Section 4, the experimentation of the system is described. Conclusions and directions for future research are presented in Section 5.

2 Bayesian Discriminant Training Method

In General, classification is the issue of predicting the class of a sample with probabilistic theories. The standard Bayesian algorithm produces a posterior probability - the probability of the class given that feature value has been measured.

Let ω_1 and ω_2 be the finite set of classes, and let x be the feature vector of eye. Assume that x is a d-dimension component vector and let be the conditional probability density function for x with the probability density function for x conditioned on being the true class. And let be the prior probability that class is, then the posterior probability can be computed from by Bayes formula. The extraction of features and training are learned supervised method.

To train the classifier a large number of eye and non-eye images are needed. 1,970 eye images were collected from face databases at FERET probe. The images include eyes of similar sizes, orientations, positions, and various intensities. These samples were used to normalize each eye to the same scale at pixels. These samples are converted to vectors and measured Mahalanobis distance using Bayesian discriminant method and PCA. In paper [5], to make vector of face they use 1D Haar wavelet representation, but we use 2D Haar wavelet representation. Because an eye image is 2D data, 2D Haar more suitable for representing features of eye than 1D Haar. To classify eye and non-eye using Bayesian method, at least two models [5] are required. One is eye class model and the other is non-eye class model.

In order to classify eye image from non-eye images, we need notion of distance. We refer to Mahalanobis distance instead of general Euclidean distance. It is useful measure of determining similarity of a test image set to a known one. It is different from Euclidean distance in that it takes into compute the correlations of the data set. Formally, the Mahalanobis distance from a group of values with mean $m = (m_1, m_2, m_3, ..., m_n)$ and covariance matrix Σ for a multivariate vector is defined:

$$d(x) = \sqrt{(x-m)^t \, \Sigma^{-1} (x-m)} \tag{1}$$

3 The Eye Location Method Under Bayesian Discriminant Framework

The block diagram of the proposed method is shown in Fig. 1. When a rough face region is presented to the system, apply preprocessing to face image. Then creating vector and getting Mahalanobis distance are applied to each window. The detected positions which are classified as eye according to their distance are stored. The merging and arbitration strategy applied to outputs which are gathered from advance operation. In particular case, merging applied regions can be arbitrated. Because we know that one face image contains exactly two eyes, finally accepted regions are restricted in two.

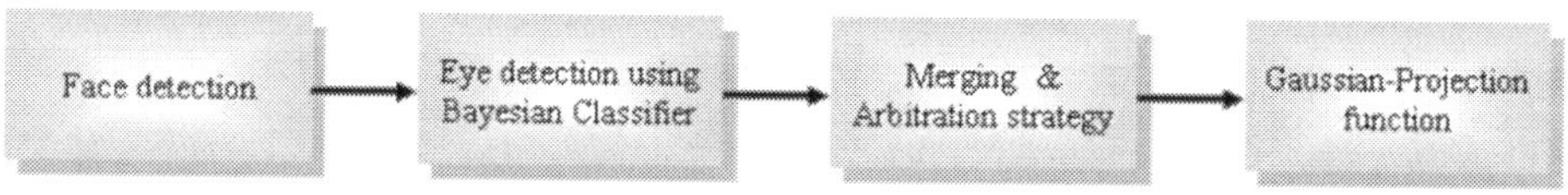

Fig. 1. Block diagram of the proposed eye location method.

3.1 Eye location Using Bayesian Discriminant Method with Mahalanobis Distance

We get Mahalanobis distance for each window. Suppose $d(x)_e$ is Mahalanobis distance for eye class, $d(x)_n$ is Mahalanobis distance for non-eye class. $d(x)_e$ and $d(x)_n$ can be calculated from the input pattern x, the eye class parameters (the mean eye, the covariance matrix), and the non-eye class parameters (the mean non-eye, the covariance matrix). We use two thresholds, θ and τ as follows:

$$\begin{aligned} &\theta = \max(d(\alpha)_e) && \textit{for sample } \alpha \textit{ that make-up of eye class} \\ &\tau = \max(d(\beta)_e - d(x)_n) && \textit{for sample } \beta \textit{ make-up of non-eye class} \end{aligned} \tag{2}$$

These are constant values, which are calculated when training time. Bayesian classifier offer classifying rule to the eye detection system, such that,

$$x \in \begin{cases} \omega_e & \textit{if } (d(x)_e < \theta) \textit{ and } (d(x)_e - d(x)_n < \tau) \\ \omega_n & \textit{otherwise} \end{cases} \tag{3}$$

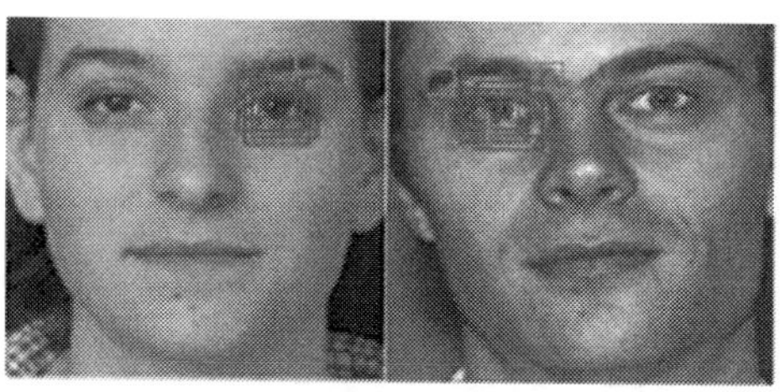

Fig. 2. Examples of detection

Detected some examples are shown in Fig. 2. In the figure, each box represents the location and size of a window to which the Bayesian discriminant method gives a positive response. The blue box is correct location and the red box is false location. The Bayesian classifier has some invariance to position and scale, which results in multiple boxes around some faces. Note also that there are some false locations; they will be eliminated by methods presented in Section 3.3 and 3.4.

3.2 Merging Strategy

The examples in Fig. 2 showed that the raw output from Bayesian classifier may contain some false detection. In this section, we present one strategy to improve the credibility of location: merging overlapped detections [4] from Bayesian classifier.

There are many detected rectangles at multiple nearby eyes, while false detections usually arise with less frequency. This discovery gives us a heuristic that can eliminate much false detection. For each position and scale, the number of detected window within bounds of a specified neighborhood of that position can be counted. If the number is greater than specific number, the position is classified as an eye. The result position is indicated by the centroid of the neighbor detections, therefore duplicated detections can be eliminated. This method is operated as following manner:

a) The detections are recorded.
b) The centers of windows are calculated.
c) The centers are "spread out" and a threshold is applied.
d) The centroid of windows which satisfy threshold in scale and position are computed, and the centers of windows are collapsed to single point.
e) If another window overlaps the rectangle, the window is eliminated because it regarded as false detection.

This method is good at not only improving accept rate but decreasing false detection. If a specific position is correctly classified as an eye, then all other detected positions which overlap it are regarded as errors, therefore these can be eliminated. The position with the higher number of detections is conserved, and the position with the lower detections is eliminated. This method is affected by two variances that are threshold and size of spread out. Only accept a detection if there are at least threshold detections within a region (spread out along x, y, and scale) in the detections. The size of the region is determined by size, which is the number of pixels from the center of the region to its edge. We need to decide reasonable threshold and size for improve performance of system. Performance of eye location at 42 randomly extracted images

Table 1. Performance of eye location according to threshold and size

Data Set	merging (2, 2)	merging (2, 4)	merging (4, 2)	merging (4, 4)	merging (6, 2)	merging (6, 4)	merging (8, 2)	merging (8, 4)
FERET	64.29%	69.05%	95.24%	97.62%	76.19%	78.57%	76.19%	76.19%
BioID	69.05%	64.29%	95.24%	95.24%	76.19%	76.19%	73.81%	78.57%

merging *(size, threshold): Only accept a eye candidate window if there are at least* threshold *detections within a square (spread out along x, y). The area of the square is determined by* size, *which is the distance from the center of the square to its edge.*

among each dataset according to threshold and size is shown in Table 1. And examples about this are shown in Fig. 3.

From the Table, we can see that the system generates highest accept rate when size is 4 and threshold is 4. Therefore, we apply this value to the system.

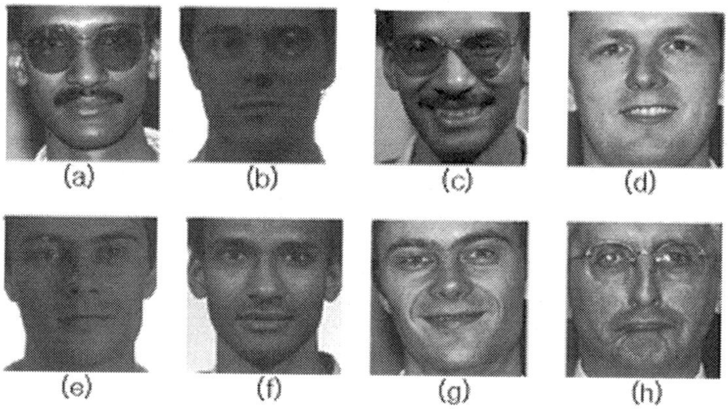

Fig. 3. Detection examples according to variable values

3.3 Arbitration Strategy

Sometimes even though merging is applied, two or more detected windows which gather around one eye are remained. As a matter of fact, eyebrows or thick rims of spectacle often look so similar to closed eyes that the classifier often makes a wrong decision. So both the eyes and neighborhood of eyes should be considered. Merging operates to remain region which has high detection density and to eliminate overlap windows. But if two or more high density regions exist independently around one eye, merging may product wrong result. Some example of errors is shown in Fig. 3. Red marks of (a), (b) and (h) are representative incorrect outputs via merging. These false outcomes are occurred by reason of peculiarity of their density distribution of windows which has produced by Bayesian discriminant classifier. Bayesian discriminant classifier often misclassifies eyebrows as eyes. Because the feature of eyebrow is similar to feature of eye, many detected rectangles are occurred around eyebrow. Since the positions with the higher number of detections are conserved at merging step, if density of detections at eyebrow is higher than that of eye, detections around eyebrow are reserved, but on the other hand detections around eye-center are eliminated by merging. In Fig. 4, (a) is detection around eyebrow which has highest density, (b) is detection around eye-center which has lower density than (a). Because the detection (b) overlap the detection (a) and density of (a) is higher than (b), detection (b) is eliminated by merging. Therefore, detection (d) and (e) are conserved only. In Fig. 5 we can see that detection (d) is upper region of eye-center and detection (f) is lower region of eye-center. A point (g) is centroid of detection (d) and (f). The point (g) is very close to center of eye. This heuristic method is called "Arbitration strategy". We showed case of arbitration among two error detections only, but Arbitration strategy is work under the situation that three or more error detections exist. Because we know that only one eye is in existence at left or right region of image, if there are two or more detections at left or right region, the system can derive correct location of eye by using Arbitration strategy.

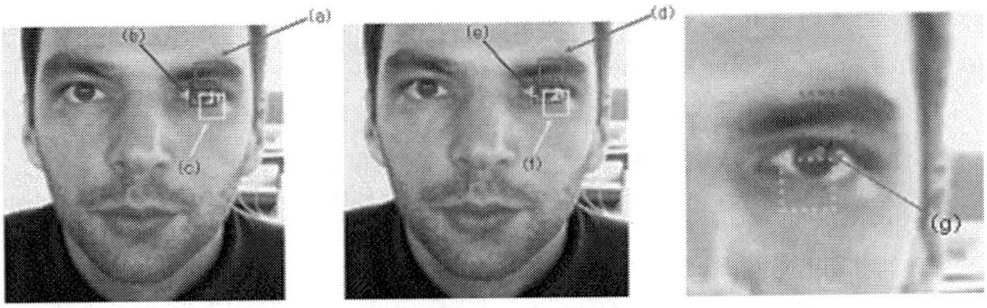

Fig. 4. Example of error result of merging. (a) is detection around eyebrow which has highest density, (b) is detection around eye-center, (c) is detection around eye which has lower density than (b). Because density of (a) is higher than density of (b), detection (b) is eliminated as (e). Therefore, finally detection (d) and (f) can be conserved only. A point (g) is centroid of detection (d) and (f). The point (g) is very close to center of eye.

4 Experimental Results

4.1 Data Sets

The training set is obtained from FERET database, and totally 1970 eyes of 985 faces are extracted and normalized for training. Experimental test set consists of BioID (1521 images), FERET (3816 images), INHADB (1200 images), and totally 6537 faces are concerned in the evaluation of localization performance. The three test sets are from diverse sources to cover diverse eye variations in view angles, sizes, illumination, and glasses. Experiments based on such various sets should be able to test the generalization performance of our eye location algorithm.

4.2 Evaluation Protocol

To estimate the accuracy of eye location, a scale independent localization measure [3] is used. This relative error measure compares the eye automatic detected positions which are results of our system with the manually marked positions of each eye. C_l and C_r are the manually assigned left and right positions, $C_l^{'}$ and $C_r^{'}$ are the automatic detected position, d_l is the Euclidean distance between $C_l^{'}$ and C_l, d_r is the Euclidean distance between $C_r^{'}$ and C_r, d_{lr} is the Euclidean distance between d_l and d_r. The relative error of detection is defined as follows:

$$err = \frac{\max(d_l, d_r)}{d_{lr}} \tag{4}$$

4.3 Comparison with Other Eye Location Methods

Two different eye location methods are implemented and evaluated on the test set. Method 1: Method of detecting eyes using only Bayesian Classifier without arbitration strategy. Method 2: Algorithm of adding arbitration strategy step to Method1.

Table 2. Testing performance of Method 1 (err < 0.14)

data	source	images	accepted faces	false detects	accept rate
SET1	FERET	3816	3422	394	89.68%
SET2	BioID	1521	1395	126	91.72%
SET3	INHADB	1200	1090	110	90.83%
Total	—	6537	5907	630	90.36%

Table 3. Testing performance of Method 2 (err < 0.14)

data	source	images	accepted faces	false detects	accept rate
SET1	FERET	3816	3646	170	95.55%
SET2	BioID	1521	1428	93	93.89%
SET3	INHADB	1200	1132	68	94.33%
Total	—	6537	6206	331	94.94%

Fig. 5. Some eye location results from test sets

Performance of tow methods under err < 0.14 are shown in Table 1 and Table 2. From the tables, we can see that Method 2 is better than Method 1. So we can conclude that arbitration strategy is effective for eye location. In addition, the average processing time per face of method 2 on an AMD Barton 2500+ PC system is 50 ms without code optimization. We show some outputs for visual inspection at Fig. 5.

Method 2 is compared with other systems. In paper [3], a detection rate is 99.1% under err < 0.2. On the other hand, our detection rate is 99.31% when err < 0.2 at all test sets. In paper [2], a detection is considered to be correct if err < 0.25. Their detection rate on BioID dataset is 94.81%. We evaluate method 2 on BioID under the same evaluation protocol. The detection rate of our system is 95.20% under err < 0.14, and the detection rate is 97.63 % under err < 0.25. And their system achieve on 97.18% of JAFFE data set. But backgrounds and illumination conditions of JAFFE are not as complex and diverse as these of BioID.

5 Conclusion

This paper describes a novel arbitration strategy applied Bayesian classifier for eye location. The system, which is trained on images from only a portion of one database, yet works on test images from diverse sources, displays robust generalization performance. The novelty of this paper comes from the combination of the 2D Haar based Bayesian classifier, the statistical modeling of eye and non-eye classes, and the arbitration strategy for growing performance. The arbitration strategy applied Bayesian classifier is trained with 1511 eye images and 3100 random natural (non-eye) images. Experimental results using 6537 images (containing a total of 13074 eyes) from various image sources. The novel system achieves 94.94 percent eye detection accuracy under err < 0.14.

In addition, because the arbitration strategy and Bayesian discriminant method don't localized for eye, it can be totally applied for location of other face organs such as nose or mouth.

References

[1] O. Jesorsky, K. Kirchberg, R. Frischholz, "Robust face detection using the Hausdorff distance," In: J. Bigun, F. Smeraldi Eds. Lecture Notes in Computer Science 2091, Berlin: Springer, 2001, pp.90-95.

[2] H. Zhou, X. Geng, "Projection functions for eye detection," Pattern Recognition, 2004, in press.

[3] Y. Ma, X. Ding, Z. Wang, N. Wang, "Robust precise eye location under probabilistic framework", IEEE International Conference on Automatic Face and Gesture Recognition, 2004

[4] Henry A. Rowley, Shumeet Baluja, Takeo Kanade, "Neural Network-Based Face Detection," IEEE Trans. Pattern Analysis and Machine Intelligence, vol. 20, NO. 1, 1998

[5] C. Liu, "A Bayesian Discriminating Features Method for Face Detection" IEEE Trans. Pattern Analysis and Machine Intelligence, vol. 25, no. 6, pp. 725-740, 2003

[6] C. Liu and H. Wechsler, "Gabor Feature Based Classification Using the Enhanced Fisher Linear Discriminant Model for Face Recognition," IEEE Trans. Image Processing, vol. 11, no. 4, pp. 467- 476, 2002.

[7] S.Lucey, S.Sridharan, V.Chandran, "Improved facial feature detection for AVSP via unsupervised clustering and dicriminant analysis", EURASIP Journal on Applied Signal Processing, vol 3, pp.264-275, 2003.

[8] T.Kawaguchi, D.Hikada, and M.Rizon, "Detection of the eyes from human faces by hough transform and separability filter," Proc. of ICIP, pp.49-52, 2000.

[9] P.Viola, M. Jones, "Rapid object detection using a Boosted cascade of simple features," Proc. of IEEE Conf. on CVPR, pp. 511 –518, 2001.

[10] Yongsheng Gao, Leung, "Face recognition using line edge map", Pattern Analysis and Machine Intelligence, IEEE Transactions on Volume 24, Issue 6, June 2002 Page(s):764 - 779

[11] J. Huang, X.H. Shao, H. Wechsler. "Pose Discrimination and Eye Detection Using Support Vector Machines", Proceeding of NATO-ASI on Face Recognition: From Theory to Applications, 1998.

[12] H.Schneiderman, T.Kanade. "A statistical model for 3D object detection applied to faces and cars," Proc. of IEEE Conf. on CVPR, 2000.

Recognizing Face or Object from a Single Image: Linear vs. Kernel Methods on 2D Patterns

Daoqiang Zhang[1,2], Songcan Chen[1], and Zhi-Hua Zhou[2]

[1] Department of Computer Science and Engineering
Nanjing University of Aeronautics and Astronautics,
Nanjing 210016, China
{dqzhang, s.chen}@nuaa.edu.cn
[2] National Laboratory for Novel Software Technology
Nanjing University, Nanjing 210093, China
zhouzh@nju.edu.cn

Abstract. We consider the problem of recognizing face or object when only single training image per class is available, which is typically encountered in law enforcement, passport or identification card verification, etc. In such cases, many discriminant subspace methods such as Linear Discriminant Analysis (LDA) fail because of the non-existence of intra-class variation. In this paper, we propose a novel framework called 2-Dimensional Kernel PCA (2D-KPCA) for face or object recognition from a single image. In contrast to conventional KPCA, 2D-KPCA is based on 2D image matrices and hence can effectively utilize the intrinsic spatial structure information of the images. On the other hand, in contrast to 2D-PCA, 2D-KPCA is capable of capturing part of the higher-order statistics information. Moreover, this paper reveals that the current 2D-PCA algorithm and its many variants consider only the row information or column information, which has not fully exploited the information contained in the image matrices. So, besides proposing the unilateral 2D-KPCA, this paper also proposes the bilateral 2D-KPCA which could exploit more information concealed in the image matrices Furthermore, some approximation techniques are developed for improving the computational efficiency. Experimental results on the FERET face database and the COIL-20 object database show that: 1) the performance of KPCA is not necessarily better than that of PCA; 2) 2D-KPCA almost always outperforms 2D-PCA significantly; 3) the kernel methods are more appropriate on 2D pattern than on 1D patterns.

1 Introduction

Face and object recognition have been an active research area of computer vision and pattern recognition for decades, and many powerful recognition algorithms have been proposed [16]. Among them, subspace methods such as principal component analysis (PCA) [8], linear discriminant analysis (LDA) [11][2] and Bayesian algorithm [4] have been extensively studied and many variants of them have been proposed [9][16]. Recently, the popular 'kernel trick' [7][11] and matrix-based (or more generally, tensor-based) representation of faces or objects without image-to-vector transformation [12][13][15][2] have been introduced into subspace based face recognition, and

D.-Y. Yeung et al. (Eds.): SSPR&SPR 2006, LNCS 4109, pp. 889–897, 2006.

accordingly, the so-called kernel PCA (KPCA) [7], kernel LDA (KLDA) [11], 2DPCA [12] and 2DLDA [2] have been proposed independently.

In some specific scenarios such as law enforcement, passport or identification card verification, etc, there may be only single image per class can be used for training the face recognition system. This brings great trouble to many existing algorithms such as LDA and Bayesian algorithm, which require at least two training samples per class to obtain the so-needed intra-class variation. we only consider PCA and its variant in this paper for face or object recognition from a single image. In [10], a method called $(PC)^2A$ was proposed as an extension of the standard PCA, which combines the original face image with its first-order projected image and then performs PCA on the enriched version of the image. In [1], the enhanced $(PC)^2A$ was proposed to use higher order projected images. In [14], the singular value decomposition (SVD) was adopted to generate virtual samples and then perform PCA on the combined images. In [3], a probabilistic approach was described, in which the model parameters were estimated by using a set of images generated around a so-called representative sample image.

As mentioned above, KPCA and 2DPCA are two important variants of PCA based on the kernel trick and matrix-based image representation respectively. The basic idea of kernel trick is to perform the linear analysis by nonlinearly transforming the original input space into a higher or even infinite dimensional feature space and expect that the nonlinear problems in original space can be converted into a linear one in the transformed space. On the other hand, the key idea of 2DPCA is to represent images as matrices without image-to-vector transformation and expect to utilize the underlying spatial structure information for efficient feature extraction and recognition. Although KPCA and 2DPCA have been successfully used for face and object recognition with multiple images per class, their performance evaluation on single image per class remains unknown.

In this paper, a novel framework called 2D Kernel PCA (2D-KPCA) is first proposed, which integrates both advantages of KPCA and 2D-PCA. In contrast to KPCA, 2D-KPCA is based on 2D image matrices and hence can effectively utilize the intrinsic spatial structure information of the images, which is ignored in traditional KPCA after the image-to-vector transformation. On the other hand, in contrast to 2D-PCA, 2D-KPCA is capable of capturing part of the higher-order statistics information, while the linear 2D-PCA can address at most the second order statistics. Moreover, this paper reveals that the current 2D-PCA algorithm and its many variants consider only the row information or column information, which has not fully exploited the information contained in the image matrices. So, besides proposing the unilateral 2D-KPCA, this paper also proposes the bilateral 2D-KPCA which could exploit more information concealed in the image matrices Furthermore, some approximation techniques are developed for improving the computational efficiency. Then a comparative study is made on performances of the above four methods on recognizing the face and object from a single image. Experiments are carried out on two well-known databases: the partial FERET face database [6] and the COIL-20 object database [5]. The results show that when recognizing the face and object from only a single image: 1) the performance of KPCA is not necessarily better than that of PCA (in fact if without kernel parameters optimization, KPCA is inferior to PCA in most cases in our experiments); 2) 2D-KPCA nearly always outperforms 2D-PCA significantly; 3) the kernel methods are more appropriate on 2D pattern than on 1D patterns.

The rest of this paper is organized as follows: In Section 2, we present the 2D-KPCA framework. Section 3 gives the experimental results on partial FERET face database and COIL-20 object database. Finally, we conclude in Section 4.

2 Two-Dimensional Kernel PCA

2.1 Unilateral 2D-KPCA

Given *M* training face or object images, denoted by *m* by *n* matrices $A_k\ (k=1,2,...M)$. In traditional KPCA, a kernel-induced mapping function maps the data vector from original input space to a higher or even infinite dimensional feature space. Define the kernel mapping on matrices as

$$\Phi(A)=\left[\phi(A^1)^T,\phi(A^2)^T,...,\phi(A^m)^T\right]^T \tag{1}$$

where A^i is the *i*-th row vector (1 by *n*) of the matrix A and ϕ is conventional kernel mapping on vectors. Let $S_t=\sum_{K=1}^{M}(\Phi(A_k)-\bar{\Phi})^T(\Phi(A_k)-\bar{\Phi})$, here $\bar{\Phi}=1/M\sum_{k=1}^{M}\Phi(A_k)$. Without loss of generality, assume that $\bar{\Phi}=0$, then

$$S_t=\sum_{K=1}^{M}\Phi(A_k)^T\Phi(A_k)=\sum_{K=1}^{M}\sum_{i=1}^{m}\phi(A_k^i)^T\phi(A_k^i)\triangleq\Phi^T\Phi \tag{2}$$

here $\Phi=\left[\phi(A_1^1)^T,...,\phi(A_1^m)^T,...,\phi(A_M^1)^T,...,\phi(A_M^m)^T\right]^T$.

In Unilateral 2D-KPCA (denoted as U2D-KPCA), the following criterion is adopted to compute the optimal projective vector v

$$J(v)=trace\left(v^TS_tv\right)=v^T\Phi^T\Phi v \tag{3}$$

which is equivalent to solve the eigenvalue problem: find $\lambda\geq 0$ and eigenvectors $v\in span\{\phi(A_k^i)^T,i=1,...,m;k=1,...,M\}$, satisfying $\lambda v=\Phi^T\Phi v$.

If we follow the conventional kernel analysis as in KPCA, there exist *mM* samples to span the kernel feature space $\{\phi(A_k^i)^T,i=1,...,m;k=1,...,M\}$, which will result in heavy computational cost for subsequent optimization procedure. To alleviate the computational cost, in this paper, we use *M* samples to approximate the kernel feature space: $\tilde{\Phi}=\left[\phi(\bar{A}_1)^T,...,\phi(\bar{A}_M)^T\right]^T$, here $\bar{A}_k$ is the mean of the *m* row vectors of A_k. So $v=\tilde{\Phi}^Tq$, and we have the following equivalent problem

$$\lambda K_m q=K^TKq \tag{4}$$

where $K_m=\tilde{\Phi}\tilde{\Phi}^T$ is the *M* by *M* kernel matrix and $K=\Phi\tilde{\Phi}^T$ is the *Mm* by *M* kernel matrix.

Suppose $R=[q_1,q_2,...,q_d]\in \Re^{M\times d}$ are the solutions of Eq. (4) corresponding to the largest d eigenvalues, then $v_i=\tilde{\Phi}^T q_i, i=1,...,d$ is the solutions of Eq. (3). For extracting features for a new pattern $A\in \Re^{m\times n}$ with unilateral 2D-KPCA, one simply projects the mapped pattern $\Phi(A)$ onto $v_1,...,v_d$

$$Y=\Phi(A)[v_1,...,v_d]=\Phi(A)\tilde{\Phi}^T R=K_{row}R \tag{5}$$

Here K_{row} is the m by M kernel matrix, and Y is the extracted m by d feature matrix .

The essence of aforementioned U2D-KPCA can be seen as performing conventional KPCA on the rows of the image matrices when each row is treated as an individual element. Similarly, we can construct the alternative U2D-KPCA if treating each column of images as an individual element.

Denote $\Phi(A)=\left[\phi(A^1),\phi(A^2),...,\phi(A^n)\right]$, where A^i is the i-th column vector (m by 1) of the matrix A, then

$$S_t=\sum_{K=1}^{M}\Phi(A_k)\Phi(A_k)^T=\sum_{K=1}^{M}\sum_{i=1}^{n}\phi(A_k^i)\phi(A_k^i)^T \triangleq \Phi\Phi^T \tag{6}$$

Here $\Phi=\left[\phi(A_1^1),...,\phi(A_1^n),...,\phi(A_M^1),...,\phi(A_M^n)\right]$.

The objective function for alternative U2D-KPCA is

$$J(v)=trace\left(v^T S_t v\right)=v^T\Phi\Phi^T v \tag{7}$$

Let $\tilde{\Phi}=\left[\phi(\overline{A}_1),...,\phi(\overline{A}_M)\right]$, here $\overline{A}_k$ is the mean of the n column vectors of A_k, so $v=\tilde{\Phi}\alpha$, and we have the following equivalent problem

$$\lambda K_m\alpha=KK^T\alpha \tag{8}$$

where $K_m=\tilde{\Phi}^T\tilde{\Phi}$ is the M by M kernel matrix and $K=\tilde{\Phi}^T\Phi$ is the M by Mn kernel matrix.

Suppose $L=[\alpha_1,\alpha_2,...,\alpha_d]\in \Re^{M\times d}$ are the solutions of Eq. (8) corresponding to the largest d eigenvalues, then $v_i=\tilde{\Phi}\alpha_i, i=1,...,d$ is the solutions of Eq. (7). For extracting features for a new pattern $A\in \Re^{m\times n}$ with alternative U2D-KPCA, one simply projects the mapped pattern $\Phi(A)$ onto $v_1,...,v_d$

$$Z=[v_1,...,v_d]^T\Phi(A)=L^T\tilde{\Phi}^T\Phi(A)=L^T K_{col} \tag{9}$$

Here K_{col} is the M by n kernel matrix, and Z is the extracted d by n feature matrix.

2.2 Bilateral 2D-KPCA

As analyzed above, U2D-KPCA and alternative U2D-KPCA are essentially KPCA on rows and columns of images respectively. However, both U2D-KPCA and alternative U2D-KPCA only consider the dependency (correlation) among the row or column vectors of the image matrix and neglects the other one. Therefore, some useful information for recognition may be lost in them. Considering this, the bilateral 2D-KPCA is proposed by integrating U2D-KPCA (Eq. (5)) and alternative U2D-KPCA (Eq. (9)) together, which could exploit more information concealed in the image matrices.

After performing U2D-KPCA (Eq. (5)) and alternative U2D-KPCA (Eq. (9)), m by d feature matrix Y and d by n feature matrix Z are obtained for each image. They are combined together for recognition. In this paper, we propose two ways for combining feature matrices Y and Z. In the first bilateral 2D-KPCA method (denoted as B2D-KPCA-1), Y and Z are firstly transformed into 1D vectors independently for each images, and then PCA is applied onto these vectors (Ys and Zs) respectively. Finally, two shorter vectors are further combined into one vector for classification. In the second bilateral 2D-KPCA method (denoted as B2D-KPCA-2), Y and Z are firstly transformed into 1D vectors and then combined into one 1D vectors for each images, and then perform PCA on the combined vectors.

It is worthy noting that for comparison, we also implemented the bilateral 2D-PCA algorithms (denoted as B2D-PCA-1 and B2D-PCA-2 respectively) according to a similar procedure as 2D-KPCA. And accordingly, the unilateral 2D-PCA introduced in Section 2 is denoted as U2D-PCA.

3 Experimental Results

In this section, a series of experiments are presented to evaluate the performances of the proposed 2D-KPCA including U2D-KPCA, B2D-KPCA-1 and B2D-KPCA-2, compared with existing PCA, KPCA and 2D-PCA methods on single training image per class recognition. These algorithms are tested on two well-known datasets, FERET face database [6] and COIL-20 object database [5]. In our experiments, we adopted the Gaussian kernel function: $k(x, y) = \exp\left(\frac{\| x - y \|^2}{2\sigma^2}\right)$ for KPCA and 2D-KPCA, and kernel width σ are chosen as the standard variation of training data. It is worthy noting that for fair comparison, we don't perform any kernel or parameters optimization for both KPCA and 2D-KPCA. And in all the experiments, the nearest neighbor classifier is employed for classification.

3.1 FERET Face Database

In this experiment, a partial FERET face database containing 400 gray-level frontal view face images from 200 persons are used, each of which is cropped with the size of 60×60. There are 71 females and 129 males; each person has two images (**fa** and **fb**) with different facial expressions. The **fa** images are used as gallery for training while the **fb** images as probes for test.

Figure 1 gives the comparisons of accuracies of linear and kernel methods under different feature dimensions on FERET face database. Here total 4 pairs of methods are compared: (a) PCA vs. KPCA; (b) U2D-PCA vs. U2D-KPCA; (c) B2D-PCA-1 vs. B2D-KPCA-1; (d) B2D-PCA-2 vs. B2D-KPCA-2. It can be seen from Fig. 1 that except KPCA, the other three kernel methods outperform the corresponding linear methods greatly. Table 1 gives comparison of accuracies of linear and kernel methods on FERET database, including results of three recent methods for single image face recognition on the same database. And Table1 also shows that except KPCA, the other three kernel methods proposed in this paper outperforms much better than the corresponding linear methods.

Then why the kernel methods perform better on 2D patterns than on 1D patterns? We guess one reason maybe that the 2D representations in some sense enlarge the size of samples through treating each rows or columns of images as individual samples, and hence the image covariance matrix in kernel-induce feature space is more accurately evaluated than in 1D representation where each class has only single sample.

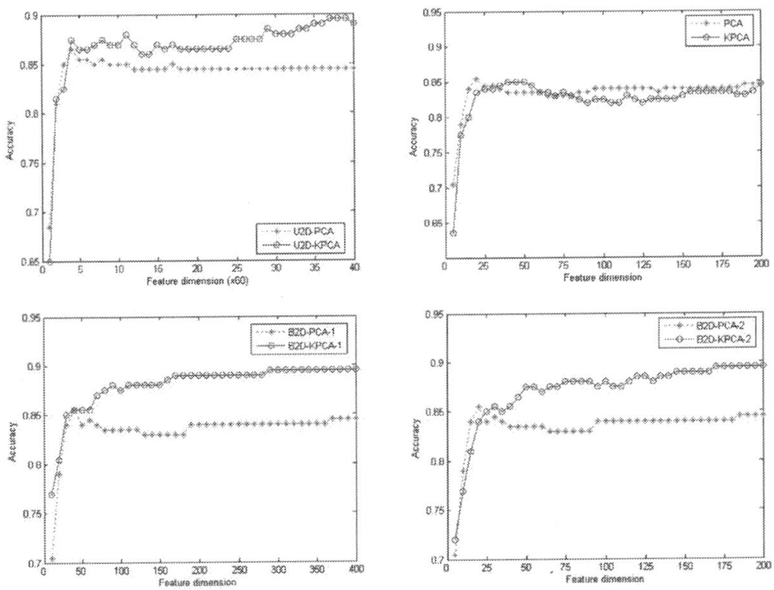

Fig. 1. Comparisons of accuracies of linear and kernel methods under different feature dimensions on FERET face database

3.2 COIL-20 Object Database

COIL-20 is a database of gray-scale images of 20 objects. The objects were placed on a motorized turntable against a black background. The turntable was rotated through 360 degrees to vary object pose with respect to a fixed camera. Images of the objects were taken at pose intervals of 5 degrees, which corresponds to 72 images per object. In our experiments, each of the 1440 images were cropped with the size of 64x64. For each of the 20 objects, we only use the first image per object as the training image, and the rest 71 images for testing.

Figure 2 gives comparisons of accuracies of linear and kernel methods under different feature dimensions on COIL-20 object database. Figure 2 shows that the performance of KPCA is not necessarily better than that of PCA. In fact, KPCA is inferior to PCA in most cases in this experiment. On the other hand, it can be also seen from Fig. 2 that the proposed three kernel methods nearly always outperform the corresponding linear methods on this database.

Table 2 gives the detailed comparisons of the relative recognition ability between linear and kernel methods. For each of the 20 objects, the first image is used for training, and the rest 71 images for testing. On each of the 71 images, if the accuracy of the kernel method is higher than that of corresponding linear method, then the count of 'win' plus 1, and vice versa. Then the counts are averaged on the 20 objects and different dimensions. Table 2 indicates that except KPCA, the performances of the other three kernel methods are better than corresponding linear ones.

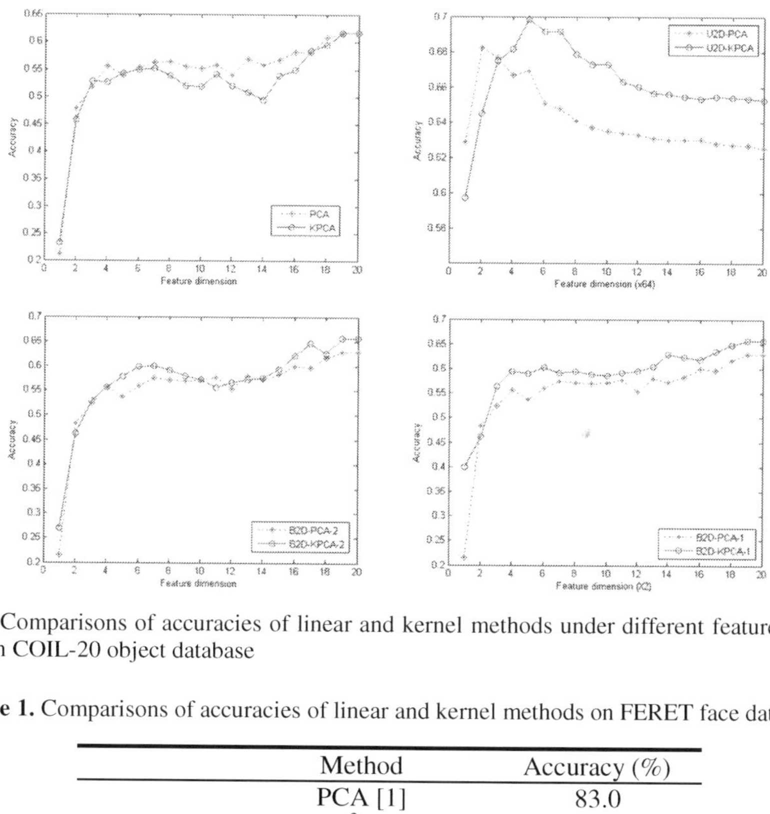

Fig. 2. Comparisons of accuracies of linear and kernel methods under different feature dimensions on COIL-20 object database

Table 1. Comparisons of accuracies of linear and kernel methods on FERET face database

	Method	Accuracy (%)
Linear	PCA [1]	83.0
	$(PC)^2A$ [10]	83.5
	$E(PC)^2A$ [1]	85.5
	$(2D)^2PCA$ [15]	85.0
	U2D-PCA	84.5
	B2D-PCA-1	84.5
	B2D-PCA-2	84.5
Kernel	KPCA	83.5
	U2D-KPCA	89.0
	B2D-KPCA-1	89.5
	B2D-KPCA-2	89.5

Table 2. Comparisons of relative recognition ability between linear and kernel methods on COIL-20 object database

Match	win	stand-off	lose
KPCA vs. PCA	11.3	32.6	27.1
U2D-KPCA vs. U2D-PCA	33.5	16.6	20.9
B2D-KPCA-1 vs. B2D-PCA-1	34.1	17.7	19.2
B2D-KPCA-2 vs. B2D-PCA-2	28.3	19.2	23.5

4 Conclusions

In this paper, we propose a novel framework called 2D-KPCA (including U2D-KPCA, B2D-KPCA-1 and B2D-KPCA-2) for face and object recognition from a single image. Then we make a comparative study on performances of the linear and kernel methods on recognizing the face and object from a single image on two well-known databases: the partial FERET face database and the COIL-20 object database. The experimental results suggest that, when recognizing the face and object from only a single image: 1) the performance of KPCA is not necessarily better than that of PCA; 2) 2D-KPCA nearly always outperforms 2D-PCA significantly; 3) the kernel methods are more appropriate on 2D pattern than on 1D patterns.

Acknowledgements

This work was supported by NSFC (60505004, 60473035), JiangsuSF (BK2004001, BK2005122), the Jiangsu Postdoctoral Research Foundation, and funds from Shanghai Key Laboratory of Intelligent Information Processing. Portions of the research in this paper use the FERET database of facial images collected under the FERET program.

References

1. Chen, S.C, Zhang, D.Q., Zhou, Z.-H.: Enhanced (PC)2A for face recognition with one training image per person. Pattern Recognition Letters, 25 (2004) 1173-1181
2. Kong, H., Wang, L., Teoh, E.K., Wang, J.G., Venkateswarlu, R.: A framework of 2D Fisher discriminant analysis: applications to face recognition with small number of training samples. In: IEEE Conf. CVPR, 2005
3. Martinez, A.M.: Recognition imprecisely localized, partially occluded, and expression variant faces from a single sample per class. IEEE Trans. on Pattern Analysis and Machine Intelligence, 25(6) (2002) 748-763

4. Moghaddam, B., Jebara, T., Pentland, A.: Bayesian face recognition. Pattern Recognition 33(11) (2000) 1771-1782
5. Nene, S.A., Nayar, S.K., Murase, H.: Columbia Object Image Library (COIL-20). Technical Report CUCS-005-96, February 1996
6. Phillips, P.J., Wechsler, H., Huang, J., Rauss, P.J.: The FERET database and evaluation procedure for face-recognition algorithms. Image and Vision Computing, 16 (5) (1998) 295-306
7. Scholkopf, B., Smola, A., Muller K.: Nonlinear Component Analysis as a Kernel Eigenvalue Problem. Neural Computation 10(5) (1998) 1299-1319
8. Turk, M., Pentland, A.: Eigenfaces for recognition. J. Cognitive Neuroscience, 3 (1) (1991) 71-86
9. Wang, X., Tang, X.: A unified framework for subspace face recognition. IEEE Trans. on Pattern Analysis and Machine Intelligence, 26(9) (2004) 1222-1228
10. Wu, J., Zhou, Z.-H: Face Recognition with one training image per person. Pattern Recognition Letters 23(14) (2002) 1711-1719
11. Yan, S.C, Xu, D., Zhang, L., Zhang, B.Y., Zhang, H.J.: Coupled kernel-based subspace learning. In: IEEE Conf. CVPR, 2005
12. Yang, J., Zhang, D., Frangi, Yang, J.Y.: Two-dimensional PCA: a new approach to appearance-based face representation and recognition. IEEE Trans. on Pattern Analysis and Machine Intelligence, 26 (1) (2004) 131-137
13. Zhang, D.Q., Chen, S.C., Liu, J.: Representing image matrices: Eigenimages vs. Eigenvectors. In: Proceedings of the 2st International Symposium on Neural Networks (ISNN'05), 659-664, Chongqing, China, 2005
14. Zhang, D.Q., Chen, S.C., Zhou, Z.-H.: A new face recognition method based on SVD perturbation for single example image per person. Applied Mathematics and Computation, 163(2) (2005): 895-907
15. Zhang, D.Q., Zhou, Z.-H.: (2D)2PCA: 2-directional 2-dimensional PCA for efficient face representation and recognition. Neurocomputing, 69(1-3) (2005): 224-231.
16. Zhao, W., Chellappa, R., Rosenfeld, R., Phillips, P.J.: Face recognition: a literature survey. [http://citeseer.n j.nec.com/374297.html], 2000

A Coupled Statistical Model for Face Shape Recovery

Mario Castelán*, William A.P. Smith, and Edwin R. Hancock

Department of Computer Science, University of York, York Y01 5DD, UK

Abstract. We focus on the problem of developing a coupled statistical model that can be used to recover surface height from brightness images of faces. The idea is to couple intensity and height by jointly modeling their combined variations. The models are constructed by performing Principal Component Analysis (PCA) on the shape coefficients for both intensity and height training data. By fitting the model to intensity data, the height information is implicitly recovered from the coupled shape parameters. Experiments show that the methods generate accurate surfaces from out-of training intensity images.

1 Introduction

One of the simplest approaches to facial shape recovery using shape-from-shading is to extract a field of surface normals and then recover the surface height function by integrating the surface normals [4,8,14]. Unfortunately, there are a number of obstacles that are encountered when this simple strategy is applied to real-world data. The most important of these is that when integrated, the concave/convex ambiguities in the needle-map can lead to the distortion of the topography of the reconstructed face. One of the most serious instances of this problem is that the nose can become imploded.

In general, shape-from-shading is an under-constrained problem since a surface normal has two degrees of freedom corresponding to the elevation and azimuth angles on the unit sphere which can not be recovered from a single brightness measurement. Domain specific constraints have been used to overcome this problem. Several authors [15,11,5,10] have shown that, at the expense of generality, the accuracy of recovered shape information can be greatly enhanced by restricting a shape-from-shading algorithm to a particular class of objects. For instance, both Prados and Faugeras [10] and Castelán and Hancock [5] use the location of singular points to enforce convexity on the recovered surface. Zhao and Chellappa [15] have introduced a geometric constraint which exploited the approximate bilateral symmetry of faces.

On the other hand, Atick et al. [1] proposed a statistical shape-from-shading framework based on a low dimensional parametrization of facial surfaces. Principal components analysis was used to derive a set of 'eigenheads' which compactly captures 3D facial shape. Unfortunately, it is surface orientation and not height which is conveyed by image intensity. Therefore, fitting the model to an image equates to a computationally expensive parameter search which attempts to minimise the error between the rendered surface and the observed intensity. Dovgard and Basri [7] combined the statistical constraint of Atick et al. and the geometric constraint of Zhao and Chellappa into a single

* Supported by National Council of Science and Technology (CONACYT), Mexico, under grant No. 141485.

D.-Y. Yeung et al. (Eds.): SSPR&SPR 2006, LNCS 4109, pp. 898–906, 2006.

shape-from-shading framework. However, asymmetry in real face images results in errors in the recovered surfaces. Blanz and Vetter [3] decoupled surface texture from shape and performed PCA on the two components separately. Their framework could be used regardless of pose and illumination changes, but linear combinations of shape and texture had to be formed separately for the eyes, nose, mouth and the surrounding area. In addition, expensive alignment and parameter fitting procedures had to be carried out. The results delivered by fitting this morphable model proved to be accurate enough to generate photo-realistic views from an input image, though sacrificing efficiency and simplicity.

The aim in this paper is to explore how coupled statistical models can be used to overcome these difficulties. We couple height surface with intensity, developing a coupled statistical model that jointly describes variations in image brightness and height data over the surface of a face. The coupled model is inspired by the active appearance model developed by Cootes, Edwards and Taylor [6], which simultaneously models 2D shape and texture.

2 Principal Component Analysis

In this section we describe how the intensity and 3D data are represented, and how eigenspace models are constructed for these data. Here we follow the approach adopted by Turk and Pentland who were among the first to explore the use of principal components analysis for performing face recognition [13]. Further, we make use of the technique described by Sirovich et al. [12] to render the method efficient.

2.1 Generating an Intensity Model

The image data is vectorized by stacking the image columns to form long column vectors $\mathbf{p}$. If the K training images contain M columns and N rows, then the pixel with column index j_c and row index j_r corresponds to the element indexed $j = (N-1)j_c+j_r$ of the long column vector. The long column vectors are centered by computing the mean $\mathbf{m}_p = \frac{1}{K}\sum_{k=1}^{K} \mathbf{p}_k$.

From the centered vectors an $MN \times K$ data matrix $\mathbf{P} = (\mathbf{p}_1-\mathbf{m}_p|\mathbf{p}_2-\mathbf{m}_p|...|\mathbf{p}_K-\mathbf{m}_p)$ is constructed, whose covariance matrix is $\mathbf{\Sigma}_p = \frac{1}{K}\mathbf{P}\mathbf{P}^T$. Unfortunately, since it is of size $MN \times MN$ the computation of the eigenvalues and eigenvectors of $\mathbf{\Sigma}_p$ becomes computationally impossible for large sets of data. However, the numerically efficient method proposed in [12] can be used to overcome these difficulties. This involves computing the eigen-decomposition of the $K \times K$ matrix $\frac{1}{K}\mathbf{P}^T\mathbf{P} = \mathbf{U}_p\mathbf{\Lambda}_p\mathbf{U}_p^T$, where the ordered eigenvalue matrix $\mathbf{\Lambda}_p$ and *temporal* eigenvector matrix $\mathbf{U}_p$ are both real. The *spatial* eigenvectors (or eigenfaces) of the covariance matrix $\mathbf{\Sigma}_p = \frac{1}{K}\mathbf{P}\mathbf{P}^T$ are given in terms of the eigenvectors of $\mathbf{P}^T\mathbf{P}$ by $\widetilde{\mathbf{P}} = \mathbf{P}\mathbf{U}_p$.

We deform the mean long-vector of image intensities in the directions defined by the eigenvalue matrix $\widetilde{\mathbf{P}}$. If we truncate $\widetilde{\mathbf{P}}$ after the L leading eigenvectors then the deformed long vector is $\mathbf{p}^* = \mathbf{m}_p + \widetilde{\mathbf{P}}\mathbf{b}_p$, where $\mathbf{b}_p = [b_{p_1}, b_{p_2}, \cdots, b_{p_L}]^T$ is a column vector of real valued parameters of length L. Suppose that $\mathbf{p}^o$ is a centered long-vector of measurements to which we wish to fit the statistical model. We seek the parameter vector $\mathbf{b}_p^*$ that minimizes the squared error. The solution to this least-squares estimation problem is

$$\mathbf{b}_p^* = \widetilde{\mathbf{P}}^T \mathbf{p}^o. \tag{1}$$

In order to be valid examples of the class represented by the training set, the values of the coefficients $\mathbf{b}_p^*$ should be constrained to fall in the interval $\mathbf{b}_p \in [-3\sqrt{\mathbf{\Lambda}_p e}, +3\sqrt{\mathbf{\Lambda}_p e}]$, where $e = [1, 1, \cdots, 1]^T$ is the all-ones vector.

2.2 Generating a Height Model

We aim to train a surface height model corresponding to the image intensity data using range images. However, for range data there are alternative representations, One of the most commonly used alternatives is a representation that uses cylindrical coordinates [1,2]. Using cylindrical coordinates, the surface of a human face (or head) can be parameterized by the function $R(\theta, \ell)$, where R is the radius, and θ and ℓ are the height and angular coordinates respectively. This representation has been adopted since it captures the linear relations between basis heads. Unfortunately, it can lead to ambiguity since different data can be fitted to the same head-model.

An alternative, which overcomes this problem, is to use a Cartesian representation [7], in which each surface point is specified by its (x, y, z) co-ordinates, where the z-axis is in the direction of the viewer. The Cartesian coordinates are related to the cylindrical coordinates through $(x, y, z) = (x_0 + R(\theta, \ell)\sin\theta, y_0 + \ell, z_0 + R(\theta, \ell)\cos\theta)$, where (x_0, y_0, z_0) is a reference shift. In this paper we will use the Cartesian form.

Each of the K range images (which are registered with the intensity images) in the training set may be represented by long vectors of height values $\mathbf{h}$ in the same way as the intensity data. The mean height vector $\mathbf{m}_h$ is given by $\mathbf{m}_h = \frac{1}{K}\sum_{k=1}^{K} \mathbf{h}_k$. We form the $MN \times K$ matrix of centered long vectors $\mathbf{H} = (\mathbf{h}_1 - \mathbf{m}_h | \mathbf{h}_2 - \mathbf{m}_h | ... | \mathbf{h}_K - \mathbf{m}_h)$. We can perform PCA to extract the set of spatial modes of variations of $\mathbf{H}$, $\widetilde{\mathbf{H}} = \mathbf{H}\mathbf{U}_h$. In the same manner, a centered long vector of height values $\mathbf{h}^o$ can be projected onto the eigenheads and represented using the vector of model coefficients $\mathbf{b}_h^* = \widetilde{\mathbf{H}}^T \mathbf{h}^o$.

3 Coupling Surface Height with Intensity

We now show how the intensity and the height models described above can be combined into a single coupled model. Each training sample can be summarized by the parameter vectors $\mathbf{b}_p$ and $\mathbf{b}_h$, representing the intensity and height of the sample respectively. In both models, we may consider small scale variation as noise. Hence, if the i_{th} eigenvalue for the intensity model is λ_p^i (where $\boldsymbol{\lambda}_p$ is the diagonal vector of the eigenvalue matrix $\mathbf{\Lambda}_p$), we need only S eigenmodes to retain p percent of the model variance. We choose S using $\sum_{i=1}^{S} \lambda_p^i \geq \frac{p}{100}\sum_{i=1}^{K} \lambda_p^i$. Similarly, for the height model we keep T eigenmodes to capture p percent of the variance.

For the k_{th} training sample we can generate the concatenated vector of length $S+T$:

$$\mathbf{b}_c^k = \begin{pmatrix} \mathbf{W}\mathbf{b}_p^k \\ \mathbf{b}_h^k \end{pmatrix} = \begin{pmatrix} \mathbf{W}\widetilde{\mathbf{P}}^T(\mathbf{p}_k - \mathbf{m}_p) \\ \widetilde{\mathbf{H}}^T(\mathbf{h}_k - \mathbf{m}_h) \end{pmatrix}, \tag{2}$$

where $\mathbf{W}$ is a diagonal matrix of weights for each intensity model parameter, allowing for the different relative weighting of the intensity and height models. As the elements of $\mathbf{b}_p$ and $\mathbf{b}_h$ represent different classes of data (grayscale and height), they can not be compared directly. We follow Cootes and Taylor [6] and set $\mathbf{W} = r\mathbf{I}$, where r^2 is the ratio of the total height variance to the total intensity variance. The coupled model data matrix is $\mathbf{C} = (\mathbf{b}_c^1|\mathbf{b}_c^2|...|\mathbf{b}_c^K)$. We apply a final PCA to this data to obtain the coupled model:

$$\mathbf{b}_c = \widetilde{\mathbf{C}}\mathbf{c} = \begin{pmatrix} \widetilde{\mathbf{C}}_p \\ \widetilde{\mathbf{C}}_h \end{pmatrix} \mathbf{c}, \tag{3}$$

where $\widetilde{\mathbf{C}}$ are the eigenvectors and $\mathbf{c}$ is a vector of coupled parameters controlling the intensity and height models simultaneously. The matrix $\widetilde{\mathbf{C}}_p$ has S rows, and represents the fist S eigenvectors, corresponding to the intensity subspace of the model. The matrix $\widetilde{\mathbf{C}}_h$ has T rows, and represents the final T eigenvectors, corresponding to the height subspace of the model.

We may express the vectors of projected intensity and height values directly in terms of the parameter vector $\mathbf{c}$:

$$\mathbf{p} = \mathbf{m}_p + \widetilde{\mathbf{P}}\mathbf{W}^{-1}\widetilde{\mathbf{C}}_p\mathbf{c}. \tag{4}$$

$$\mathbf{h} = \mathbf{m}_h + \widetilde{\mathbf{H}}\widetilde{\mathbf{C}}_h\mathbf{c}. \tag{5}$$

For compactness we write: $\mathbf{Q}_p = \mathbf{W}^{-1}\widetilde{\mathbf{C}}_p$.

A plot of cumulative variance versus number of eigenmodes is shown in Figure 1. The height, intensity and coupled models are represented by the dashed, solid and dotted lines respectively. It is evident that fewer eigenmodes are required to capture variance in facial height than in facial intensity. This is because the intensity model has to deal with

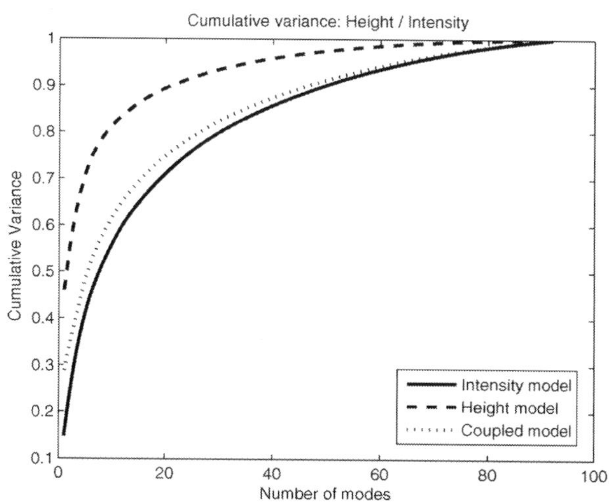

Fig. 1. Plot of cumulative variance versus number of eigenmodes used for intensity model (solid line), height model (dashed line) and coupled model (dotted line)

changes in shape and illumination, while the height model only deals with changes in shape. We retained 65 dimensions of the height model and 85 dimensions of the intensity model (each accounting for 95% of the variance). For the coupled model we retained 80 modes.

3.1 Fitting the Model to Intensity Data

Fitting the model to intensity data involves estimating the parameter vector $\mathbf{c}$ from input images of faces. To do this we seek the coupled model parameters which minimize the error between the best fit parameters $\mathbf{b}_p^*$ and the recovered parameters $\mathbf{Q}_p\mathbf{c}$. In doing so, we implicitly recover the surface which is also represented by the coupled model parameters.

Suppose that $\mathbf{p}^o$ is a centered vector of length MN that represents an intensity image of a face. Its best fit parameter vector, $\mathbf{b}_p^*$, is calculated through Equation 1. We fit the model to data seeking the vector $\mathbf{c}^*$ of length $S+T$ that satisfies the condition

$$\mathbf{c}^* = \arg\min_{\mathbf{c}} \{(\mathbf{b}_p^* - \mathbf{Q}_p\mathbf{c})^T(\mathbf{b}_p^* - \mathbf{Q}_p\mathbf{c})\} \tag{6}$$

The corresponding best fit vector of height values is given by

$$\mathbf{h} = \mathbf{m}_h + \widetilde{\mathbf{H}}\widetilde{\mathbf{C}}_h\mathbf{c}^* \tag{7}$$

We used a Matlab implementation of a quasi-Newton minimization procedure to solve Equation 7, constrained such that each coupled parameter lies within ± 3 standard deviations from the mean. One input image took around 5 seconds to converge to the best solution.

4 Experiments

In this section we report experiments focused on out-of-training characterization for the coupled model. The face database used for building the models was provided by the Max-Planck Institute for Biological Cybernetics in Tuebingen, Germany [2]. This database was constructed using Laser scans ($Cyberware^{TM}$) of heads of young adults, and provides head structure data in a cylindrical representation. For constructing the height based model, we converted the cylindrical coordinates to Cartesian coordinates and solved for height values. We were also provided with the intensity maps for each 3D face.

We used 93 out-of-training cases. We calculated the fractional height difference error $\|Ground\_truth - Recovered\_surface\|/Ground\_truth$ as an average over the 93 surfaces and over all points. For the purposes of analysis, we ordered the out-of-training cases so that the first examples were those close in appearance to the mean shape $\mathbf{m}_p$. We used the sum of the first ten values of $\mathbf{b}_p$ (to account for at least 50% of the variability), i.e., $\sum_{i=1}^{10} b_{p_i}$ as a similarity measure.

In Figure 2 we show surface recovery results for three cases. The figure is divided into five columns. The different rows are for the three different subjects. In the first column, the three panels show the input image together with its frontal and profile views. The second panel contains the recovered best-fit intensity image. The third and

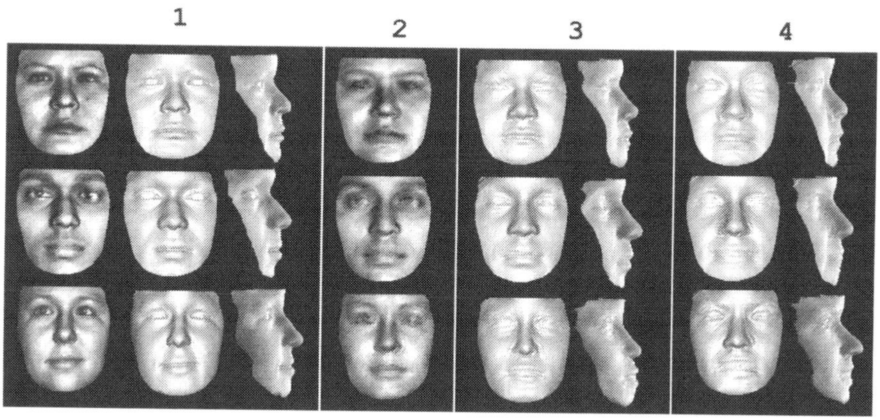

Fig. 2. Surface recovery results for three cases. The figure is divided in four columns. The first column shows the input image together with its frontal and profile views. The second column presents the best-fit intensity recovery. The next two panels present frontal and profile views from the intensity-height coupled model (third column) and the single height model (fourth column).

fourth columns contain panels which show frontal and profile views. The results of applying the full model are shown in the third column and those of applying the single height model are shown in the fourth column (i.e. the height data for the surface in panel 1 was used as an input for the single model $\tilde{\mathbf{H}}$). The first two rows present cases where a percentage of height error around 1% while the last row shows cases with percentage of error bigger then 2%. As expected, the results shown in columns 3 seem to match

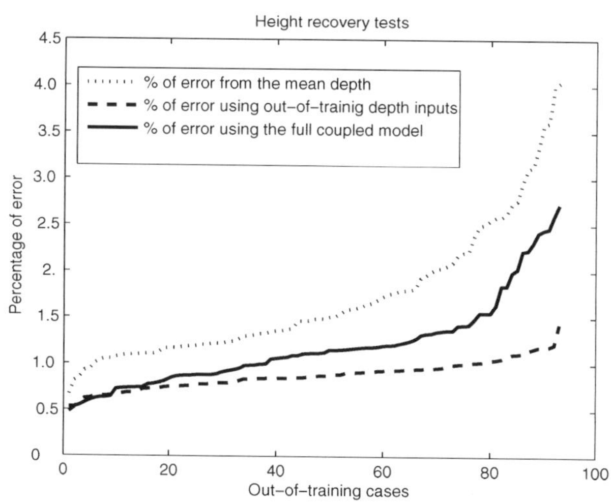

Fig. 3. Percentage of height difference for the 93 cases, ordered in an ascending way. The solid lines shows the full coupled model performance. The dashed line presents the error using a single height model (using height information as input), while the dot-dashed line shows the error from the mean height shape.

the best-fit intensity in column 2 rather than the original data in panel 1. However, even in those cases that differ significantly from the mean (last row), there is a good resemblance to the original data. This may be a consequence of basing surface recovery on the best-fit parameters directly from an intensity image. From the recovered faces, one can infer that in column 3, the surface recovery was led by the appearance of an input image. On the other hand, a visual analysis of the profiles in column 4 suggests that surface recovery was determined by an input height map.

In Figure 3 we plot the fractional height difference for the 93 cases. The solid line shows the coupled model performance. The dashed line shows the error obtained using a single height model (using height information as input), while the dot-dashed line shows the error from the mean height shape. The results were ordered in an ascending way for the purposes of comparison. The average error for the simple height model, coupled model and from the mean height are respectively 0.08%, 1.19% and 1.71%. Observe that many out-of-training examples whose intensities cannot be accurately recovered will generate less accurate height maps. However, considering that we are comparing two kinds of inputs (height and intensity), we can say that the coupled model delivers encouraging results.

Finally, we illustrate the utility of the coupled model with real world face images. These are drawn from the Yale B database [9] and are disjoint from the data used to train the statistical model. In the images the faces are in the frontal pose and were illuminated by a point light source situated approximately in the viewer direction. We aligned each

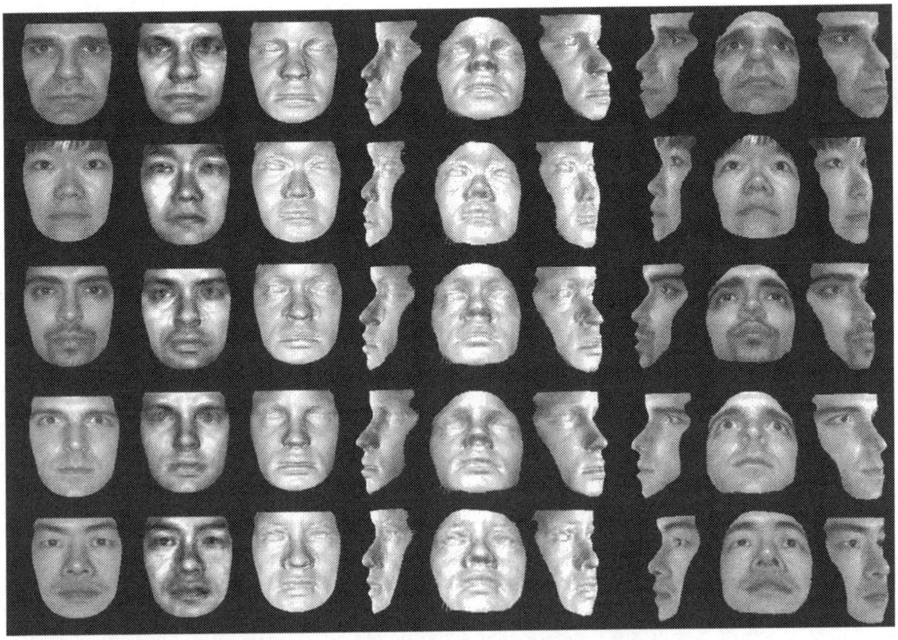

Fig. 4. Height recovery results using five examples from the Yale B database. From left to right: input image, intensity best-fit recovery, frontal illumination of the recovered height and profile and close-to-profile views with warped albedo-free and input images.

image with the mean intensity shape so that the eyes, nose tip and mouth center were in the same position. The surface recovery results are shown in Figure 4, where we present, from left to right: input image, intensity best-fit recovery, frontal illumination of the recovered height and profile and close-to-profile views with warped albedo-free and input image. Notice that even when the best-fit recovered intensity image is of lower quality than those in Figure 2(2), the surface reconstructions from the best-fit intensity parameters are sufficiently good to render novel views.

5 Conclusions

We have explored a way for coupling intensity and height information to construct statistical models of facial shape. Our coupled model strongly links the best-fit coefficients for intensity and height into a single statistical model. To recover the parameters of the coupled model, and hence reconstruct height, requires an optimization method. In this way best-fit intensity parameters can be calculated directly from an input image, and then used to recover height through the optimization search. The process only take some few seconds to converge to a minimum. The coupled model proved to be good enough to generate accurate surfaces from real world intensity imagery in an efficient way. Future research will focus on exploring the effect of these approaches to alternative representations including surface gradient, azimuth and zenith angles, and surface normals.

References

1. G. P. Atick, J. and N. Redlich. Statistical approach to shape from shading: Reconstruction of three-dimensional face surfaces from single two-dimensional images. *Neural Computation*, 8:1321–1340, 1996.
2. V. Blanz and T. Vetter. A morphable model for the synthesis of 3d faces. In *SIGGRAPH '99: Proceedings of the 26th annual conference on Computer graphics and interactive techniques*, pages 187–194, New York, NY, USA, 1999. ACM Press/Addison-Wesley Publishing Co.
3. V. Blanz and T. Vetter. Face recognition based on fitting a 3d morphable model. *IEEE Trans. Pattern Anal. Mach. Intell.*, 25(9):1063–1074, 2003.
4. H. E. Bors, A.G. and R. Wilson. Terrain analysis using radar shape-from-shading. *IEEE Trans, on Pattern Analysis and Machine Intelligence*, 25(5), 2003.
5. M. Castelán and E. Hancock. Acquiring height maps of faces from a single image. In *Proc. IEEE 3DPVT*, pages 183–190, 2004.
6. E. G. Cootes, T.F. and C. Taylor. Active appearance models. In *Proc. European Conference in Computer Vision*, pages 484–498, 1998.
7. R. Dovgard and R. Basri. Statistical symmetric shape from shading for 3d structure recovery of faces. In *Proc. European Conference on Computer Vision*, pages 99–113, May 2004.
8. R. Frankot and R. Chellapa. A method for enforcing integrability in shape from shading algorithms. *IEEE Transactions on Pattern Analysis and Machine Intelligence*, 10:438–451, 1988.
9. B. D. Georghiades, A. and D. Kriegman. From few to many: Illumination cone models fro face recognition under variable lighting and pose. *IEEE Transactions on Pattern Analysis and Machine Intelligence*, pages 634–660, 2001.

10. E. Prados and O. Faugeras. Unifying approaches and removing unrealistic assumptions in shape from shading: Mathematics can help. In *Proc. European Conference on Computer Vision*, pages 141–154, May 2004.
11. D. Samaras and D. Metaxas. Illumination constraints in deformable models for shape and light direction estimation. *IEEE Trans. PAMI*, 25(2):247–264, 2003.
12. L. Sirovich and R. Everson. Management and analysis of large scientific datasets. *The International Journal of Supercomputer Applications*, 6(1):50–68, 1992.
13. M. Turk and A. Pentland. Face recognition using eigenfaces. In *Proc. IEEE Conference on Computer Vision and Pattern Recognition*, pages 586–591, 1991.
14. Z. Wu and L. Li. A line integration based method for depth recovery from surface normals. *CVGIP*, 43(1):53–66, 1988.
15. W. Zhao and R. Chellapa. Illumination-insensitive face recognition using symmetric shape-from-shading. In *Proc. Conference on Computer Vision and Pattern Recognition*, pages 286–293, 2000.

Analysis and Selection of Features for the Fingerprint Vitality Detection

Pietro Coli, Gian Luca Marcialis, and Fabio Roli

Department of Electrical and Electronic Engineering – University of Cagliari
Piazza d'Armi – I-09123 Cagliari – Italy
{pietro.coli, marcialis, roli}@diee.unica.it

Abstract. Although fingerprint verification systems have attained a good performance, researchers recently pointed out their weakness under fraudulent attacks by fake fingers. In fact, the acquisition sensor can be deceived by fake fingerprints created with liquid silicon rubber. Among the solutions to this problem, the software-based ones are the cheapest and less intrusive. They use feature vectors made up of measures extracted from one or multiple impressions (static measures) or multiple frames (dynamic measures) of the same finger in order to distinguish live and fake fingers. In this paper, we jointly use both static and dynamic features and report an experimental investigation aimed to compare them and select the most effective ones.

1 Introduction

Fingerprint matching algorithms are widely used for automatic personal verification [1]. Although the fingerprint verification systems have shown a good degree of accuracy, their weak point is the acquisition sensor, which can be of optical or solid-state type [1]. It has been shown by Matsumoto et al. that commonly used sensors can be deceived by submitting a "gummy" finger, made up of liquid silicon rubber and similar materials [2]. The image produced by this kind of fingers is processed as well as a "live" image.

Although reproducing fingerprint is not simple, the academic and commercial interest on spoof attempts is increasing. In order to prevent the fraudulent attempts by fake fingers, several solutions have been proposed. Most of them are based on the use of additional hardware, embedded in the sensor, which can detect the "vitality" of the finger, e.g., through the heartbeat detection.

A novel approach to the vitality detection of fingerprints has been proposed in [3-5]. It is based on the extraction of features which can discriminate between live and fake fingerprints by using the images acquired by the sensor. These software-based solutions are obviously less intrusive and cheaper than the hardware-based ones. So far, the state-of-the-art consists of two main approaches: the first one is based on the dynamic measure of some "intrinsic" features [3-4], while the second one uses relative measures between an unknown fingerprint image and another one which is known to be "live" [5]. The search of some physiologic or physic characteristics of the fingerprint, in order to derive some vitality measures, follows the observation that

D.-Y. Yeung et al. (Eds.): SSPR&SPR 2006, LNCS 4109, pp. 907–915, 2006.

the "vitality" is a property of the fingertip and not of the fingerprint. Therefore, the main features considered so far are based on the perspiration of the skin through the pores and the "deformation" properties of the skin. To measure the first kind of features a temporal analysis of the fingerprint was adopted [3-4], whilst a static analysis was adopted for the second kind [5].

So far, static and dynamic features have been used separately. However, it is reasonable to argue that both features provide discriminant information about live and fake fingers. Accordingly, their joint contribution should be investigated. This is the purpose of this paper, which reports an experimental comparison aimed to select the most discriminant subsets of features for optimising the performance of automatic fingerprint vitality detection systems.

Section 2 presents the features we used for our experiments. Section 3 describes the used data set and the experiments performed for the feature selection process. Section 4 draws some preliminary conclusions.

2 Static and Dynamic Features for the Fingerprint Vitality Detection

The vitality detection by software-based solutions starts from the measure of some features extracted from the fingerprint image acquired by standard acquisition sensors (e.g., optical or capacitive [1]).

At present, the main approaches proposed in literature to perform vitality detection are based on the study of the skin perspiration through the pores and the elastic properties of the skin.

For the first kind of measures, the user hold his finger on the scanner surface, and the biometric system acquires different frames in the time. Features are extracted according to the differences among these frames.

For the second kind of measures, the vitality information is given by the different elastic response which is hypotesized to coming from true and fake fingers. In order to derive this type of information the user has to repeat the acquisition process on the scanner surface.

In the following, we refer to the features derived from these approaches by the terms "dynamic" and "static". In fact, to measure the first kind of features the "dynamic" acquisition of multiple frames of the same finger is required, whilst the "static" acquisition of multiple impressions of the same finger is required for the second kind.

The following sections describe the features we adopted for our experiments.

2.1 Static Features

The use of elastic measures has been adopted in previous works in order to correct the non-linear deformations introduced during the acquisition stage, and so improving the alignment between different impressions of the same fingerprint [6].

Recently, an elastic model has been adopted into a vitality detection system [5]. In this work Chen et al. compute an elastic model to study the different deformation of a fake fingerprint from a live one. They have observed that the contact of the fingertip with a plane surface involves an elastic deformation: the flow of papillary ridges changes from a 3D to a 2D-pattern. Through the mathematical model proposed in [6], Chen et al. showed that it is possible to study the elastic behaviour of a live and a fake finger.

The first step for the computation of this static measure is the extraction of a fixed and ordered set of minutiae from the fingerprints. For the purpose of this work we manually extracted the minutiae in order to eliminate all possible error sources. From a fingerprint k we manually extracted 20 minutiae $M^k=(m^k{}_1,m^k{}_2,\ldots m^k{}_{20})$.

Then, starting from a couple of set of minutiae extracted from two fingerprints, namely, a "template" fingerprint and an input impression of the same subject, $M^t=\{m^t{}_1,m^t{}_2,\ldots m^t{}_{20}\}$ and $M^c=\{m^c{}_1,m^c{}_2,\ldots m^c{}_{20}\}$, we computed the TPS (Thin Plate Splines) model to obtain the complete correspondence of the 20 minutiae:

$$F(M^c) = M^t \tag{1}$$

Each minutia is characterised by a triplet $\bar{u} = \{x, y, \vartheta\}$ where $\{x, y\}$ are its Cartesian coordinates in the image and ϑ is its orientation. The correspondence is defined according to:

$$F(\bar{u}) = c + A\bar{u} + Ws(\bar{u}) \tag{2}$$

Where the c and A parameters are referred to the rigid transformation and W indicates the non-linear transformation that includes the elastic deformation, and $s(u)=u^2 \log u$ is the basis function.

The amount of the deformation is given by a function known as "bending energy": this value can be computed by the elastic parameters W and s. Further details for the bending energy computation can be found in [5]. We refer to this feature with the name "SF2".

Moreover, we added some "morphological" features which give a general description of the fingerprint pattern based on its geometrical properties. The first one of these static measures is the mean of the intra-distances among the set of the extracted minutiae (named SF1). For each of these it has been computed the sum of the distances with the other ones. Finally we have considered the mean of this values. The second one is the mean of the ridge width (named SF3). By following the skeleton extracted for each fingerprint images, we computed the value of the width of the correspondent ridge. In fact, the creation of the fake finger (e.g., by the consensual method) includes a sequence of steps which can involve a modification of the width of the ridges and furrows with respect to the correspondent live finger. We give details about the fingerprint reproduction method we followed in Section 3.1.

2.2 Dynamic Features

The perspiration is a unique feature of the skin: the perspiration of the pores, which are in every part of the skin, allows to conserve the body temperature into a constant value (body temperature homeostasis). During the contact of a finger on the sensor surface, the perspiration produces a modification of the skin wetness and so of the corresponding image. With a synthetic finger used to spoof a biometric system, the perspiration phenomenon is not present.

In [3-4] this physiological feature of the skin is used in order to extract some vitality measures from fingerprint images. By using some standard optical and capacitive sensors a time sequence of fingerprint images is captured. The user keeps his fingertip in contact with the surface of the sensor for about 5 seconds, and the sensor captures a certain number of images (frames). The grey-levels variation of two sequential fingerprint frames is a dynamic measure of the perspiration process and, therefore, of the vitality of the finger. These dynamic measures are used as feature vectors submitted to a machine learning algorithm for classification into the "live" and the "fake" fingerprint classes. In order to obtain a good trade-off between the reliability of dynamic features and the usability of the biometric system, an acquisition time of five seconds has been considered [3].

The algorithm for the measure of the perspiration is based on the grey-level values along the ridges path according to the steps described in [4] which we followed:

1) acquisition of two frames of the same fingerprint temporally separated from 5 seconds;
2) separation of the two fingerprint images from the background;
3) binarization and thinning of the last frame;
4) computation of two mono-dimensional signal C_1 and C_2, containing the grey-level profile along the skeleton extracted at step 3.

The grey-level variations from C_1 to C_2 represent the dynamic variations of the fingerprint moisture during the acquisition process. Accordingly, a set of dynamic features can be extracted. In [3-4] six measures were computed. However some of these features have not been used in the classification stage. As it is remarked in [3-4], the dynamic measures strongly depend on the acquisition characteristics of the sensor, in particular on the device's dynamic. Therefore, in this work we have selected the following dynamic features from those described in [4]: the time difference of mean grey-level on the skeleton (DF1), the dry saturation percentage change (DF2), and the wet saturation percentage change (DF3). These measures have shown to be appropriate for the capture device we used in our experiments.

Moreover, we added other two dynamic measures: the time variation of the grey-level mean value of the whole image (DF4) and the L1-distance of its grey-levels histogram (DF5).

3 Experimental Results

3.1 The Data Set

The data-set we used is made up of twenty-eight different fingerprint images from a male population aged between 20 and 40. Images were taken by the Biometrika FX2000 optical sensor.

The data-set is made up of:

- twenty-eight couples of frames (temporally separated from 5 seconds) from live fingerprints;
- twenty-eight couples of frames (temporally separated from 5 seconds) from fake fingerprints of the same subjects;
- twenty-eight additional live impressions of the same subjects.

The fake data have been created from twenty-eight reproductions of each live fingerprint with the cooperation of the subject, which put his finger on a plasticine-like material. These moulds have been then filled with liquid silicon rubber to create wafer-thin silicon replicas. About one day has been necessary for the solidification of the rubber. Fig. 1 illustrates the main steps of our fingerprint reproduction method.

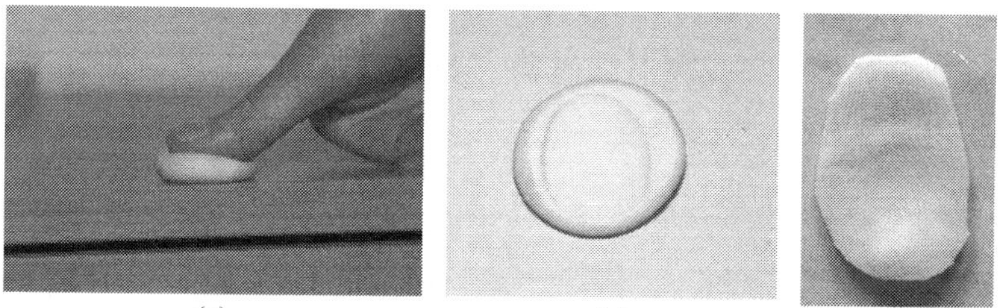

(a) (b) (c)

Fig. 1. Basic steps of our fingerprint reproduction by "consensual" method. (a) The user put his finger on the plasticine-like material, thus creating the mould. (b) The liquid silicon rubber is dripped over the mould. (c) After one day, the solidification of the rubber is completed and it can be removed from the mould. It can be used as a fingerprint stamp.

Our data set has size and characteristics similar to the ones of other data sets used for the vitality detection [3-4]. For example, the data set used in [3] is made of eighteen live finger images, eighteen fake finger images and eighteen finger images from cadavers.

3.2 Experimental Protocol

In order to extract the above features from fingerprint images, we adopted the following protocol:

- the second impression has been considered as the template of the fingerprint stored in the system database. The minutiae-points were manually detected in order to avoid errors due to the minutiae detection algorithm;
- the first and the second frame of the first impression have been considered as the images provided by the system during an access attempt. Only attempts related to fingerprints of the same subject were considered ("genuine" attempts by live and fake fingers). Even for these images the minutiae-points were manually detected.

In particular:

- to compute the morphological features (SF1 and SF3), we used the second frame of the first impression;
- to compute the bending energy (SF2), we used twenty-eight images of the second impression as the template of the fingerprint verification system. We computed the bending energy on the basis of the comparison between the template and the second frame of the first impression;
- to compute the dynamic features (from DF1 to DF5) we used the first and the second frame of the first impression;
- each feature was normalised according to:

$$f_i^{(n)} = \frac{f_i - \mu_i}{\sigma_i} \quad (1)$$

Where $f_i^{(n)}$ is the i-th normalised feature (i = 1,…, 8), μ_i and σ_i are the mean and the standard deviation of f_i over all available patterns.

Thus, we obtained fifty-six feature vectors made up of three static features from live and fake fingers, and fifty-six feature vector made up of five dynamic features from live and fake fingers.

We used the k-Nearest Neighbour classifier to discriminate between the live and fake fingerprint images characterized by such feature vectors. We performed trials with values of the parameter k between 1 and 27. For each trial, the accuracy of the k-Nearest Neighbour classifier was assessed by the leave-one-out method, namely, the accuracy values reported are averaged on fifty-six trials and the k value corresponding to the highest accuracy is selected.

3.3 Results

Table 1 shows the correlation coefficient between features. Some subsets of features (dynamic features especially) exhibit a significant degree of correlation. For example, DF4 and DF5 are negatively correlated with DF1. On the other hand, these features are less correlated with static ones than DF1. Therefore, these results suggest that an appropriate feature selection step is necessary in order to exploit static and dynamic features at best.

Table 2 reports the best performance achieved by using only one feature, in particular the maximum classification accuracy obtained using values of k ranging in the interval {1,…,27}. It is easy to see that static features perform generally worse than dynamic ones, except for SF3.

Table 1. Correlation coefficient of static and dynamic features

	SF2	**SF3**	**DF1**	**DF2**	**DF3**	**DF4**	**DF5**
SF1	0,08	0,29	0,36	0,15	0,31	-0,27	-0,27
SF2		0,09	0,19	-0,04	0,04	-0,16	-0,13
SF3			0,66	0,68	0,04	-0,43	-0,43
DF1				0,65	0,30	-0,85	-0,89
DF2					0,08	-0,47	-0,54
DF3						-0,50	-0,24
DF4							0,78

Table 2. Best classification percentage accuracy achieved by the individual features over the investigated values of k (k=1,…,27)

	SF1	**SF2**	**SF3**	**DF1**	**DF2**	**DF3**	**DF4**	**DF5**
Accuracy (%)	53,6	57,1	62,5	85,7	60,7	82,1	75,0	71,4

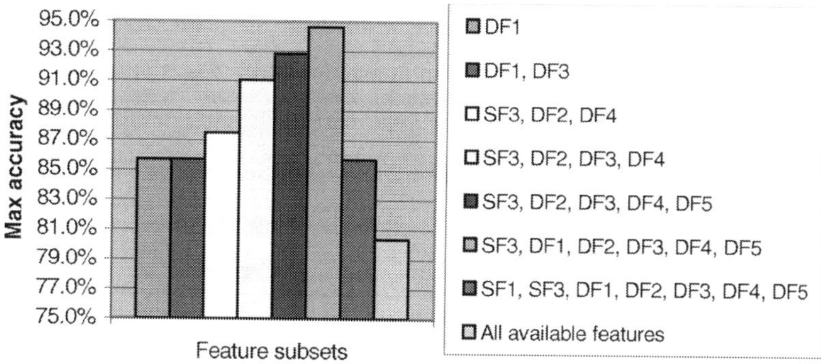

Fig. 2. Overall percentage accuracy over the investigated values of k (k=1,…,27) for the best subset of each group of n features (n = 1,…,8)

In order to detect the best features subsets, we performed classification of fake and live fingers by using the k-NN classifier with all possible subsets of the available features. In other words, we computed the accuracy on the subsets obtained by grouping n features (n = 2,…,8). Then we selected the best group of each subset by using the best accuracy over k ranging from 1 to 27. Fig. 2 shows the overall accuracy of each best group of n features.

Reported results point out that the performance achieved with the best subset is higher than that of the best individual feature (about 10% more than DF1's accuracy) and that six features allow to obtain the best performance. This subset is made up of SF3 and all the dynamic features (Fig. 2). It is worth noting that the best classification accuracy obtained by using all static features only is 62,5% and that obtained by using all dynamic features only is 82,1%.

With regard to the best feature subset, it can be noticed that adding the remaining static features negatively affects the performance. However this does not necessarily mean that SF1 and SF2 are not useful for discriminating between live and fake fingers in general, due to the small sample size of the used data set.

Nevertheless, the good performance achieved by combining the dynamic features with the SF3 static feature only can be explained by the physical interpretation that can be associated to these features (especially to dynamic ones [3-4]), which gives more evidence that these results can be expected to hold also on larger data sets. To investigate this hypothesis, let us consider as an example the DF4 and SF3 features.

The first one corresponds to the time difference of the grey-levels mean on the whole images. The expressive power of this feature is evident by looking at Figs. 3(a-d), which shows two frames of the same live finger taken at 0 and 5 seconds (Figs. 3(a-b)), and two sequential frames of the correspondent fake finger (Figs. 3(c-d)). It is evident that the second "live" frame exhibits a higher number of "dark" pixels than that of the first "live" frame, due to the perspiration effect. This does not happen for the two "fake" frames due to the absence of the perspiration phenomenon (see also [3-4]). Therefore, by computing the average difference of the pixels grey-levels between the first and the second frame, we can expect that its value for fake fingers is lower than for live fingers. Similar observations can be made for other dynamic measures.

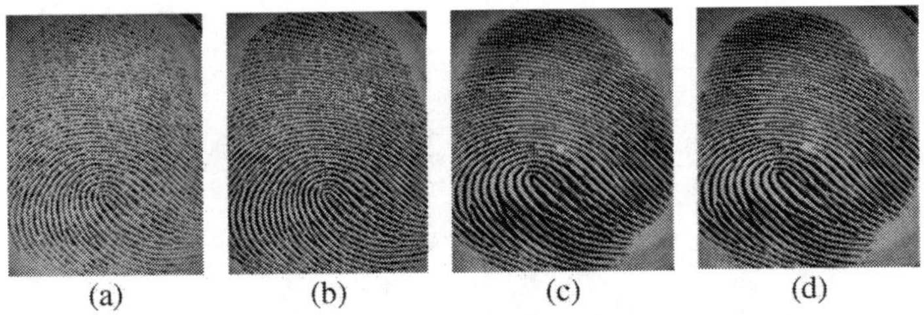

Fig. 3. (a-b) Two sequential frames of the same live fingerprint acquired by an optical sensor at 0 seconds (a) and 5 seconds (b). (c-d) Two sequential frames of the fake finger correspondent to that of (a-b).

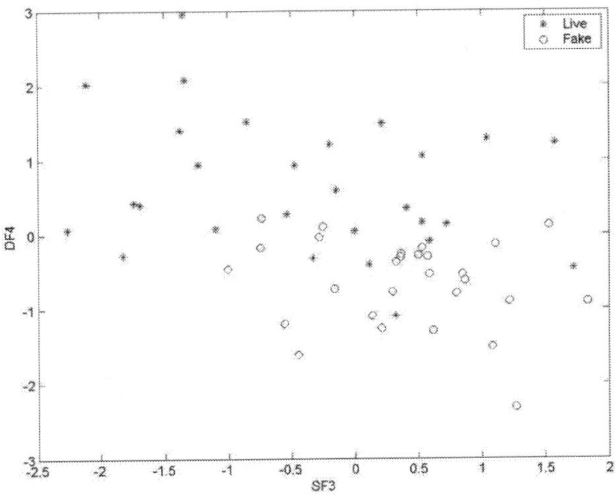

Fig. 4. The patterns of the used data set projected to the SF3-DF4 feature subspace

The second one is the average width of the ridges in a fingerprint image. In fact, the fake fingers creation process involves steps which lead to a different ridges width of the stamp with respect to that of the correspondent live finger (Fig. 1). We observed that the ridges of fake fingerprints images were wider than those of live fingerprints images. Figs. 3(b, d), which is related to a live finger impression and to the correspondent fake finger, clearly show that the ridges of the fake sample are wider than those of the live sample. Therefore, by computing the average width of the fingerprint ridges, we can expect a higher value for fake fingers than for live fingers.

Accordingly, the plot of the pattern projected on the SF3-DF4 feature subspaces should be characterised by fake patterns around low values of the DF4 feature and high values of the SF3 feature and by live patterns spread over the range of each feature, due to the variability of the perspiration phenomenon and the skin characteristics of live fingers each others. Fig. 4 shows the Live and Fake patterns projected to the SF3-DF4 feature subspaces. Distributions of Live and Fake patterns follow the given physical interpretation, thus supporting our hypothesis about the reduction of the impact of the small sample size issue on reported results.

4 Conclusions

In this paper, we investigated static and dynamic features for the vitality detection of fingers. The adopted static features were based on the elastic and morphological properties of the skin, whilst the dynamic ones were based on the perspiration phenomenon as observed in previous works.

So far, no studies on joint static and dynamic features are present in the literature. However, it can be argued that both static and dynamic features could jointly help in distinguishing between live and fake fingers. This preliminary study was aimed to support such hypothesis and also to indicate the most promising of the investigated features. Reported results showed that it was possible to find a subset of static and dynamic features which performed much better than those made up of only one feature or features of the same type. Despite the small size of the used data set, the possibility of associating a physical interpretation to some features suggested that our results could be confirmed even on larger data sets.

References

1. D. Maltoni, D. Maio, A.K. Jain, S. Prabhakar, *Handbook of fingerprint recognition*, Springer, 2003.
2. T. Matsumoto, H. Matsumoto, K. Yamada, S. Hoshino, Impact of artificial "gummy" fingers on fingerprint systems, Proc. of SPIE Vol. 4677, Optical Security and Counterfeit Deterrence Techniques IV, pp.24-25, 2002.
3. R. Derakhshani, S. Schuckers, L. Hornak, L. O'Gorman, Determination of vitality from a non-invasive biomedical measurement for use in fingerprint sensors, *Pattern Recognition*, 36 (2) 383-396, 2003.
4. S. Parthasaradhi, R. Derakhshani, L. Hornak, S. Schuckers, Time-series detection of perspiration as a vitality test in fingerprint devices, *IEEE Transactions on Systems, Man and Cybernetics*, Part C, 35 (3) 335-343, 2005.
5. Y. Chen, A.K. Jain, S. Dass, Fingerprint deformation for spoof detection, Biometric Symposium, Cristal City, VA, 2005.
6. A. Ross, S. Dass, A.K. Jain, A deformable model for fingerprint matching, *Pattern Recognition*, 38 95-103, 2005.

Recognizing Facial Expressions with PCA and ICA onto Dimension of the Emotion

Young-suk Shin

Department of Information and telecommunication Engineering, Chosun University, #375 Seosuk-dong, Dong-gu, Gwangju, 501-759, Korea
ysshin@chosun.ac.kr

Abstract. This paper addresses the problem of facial expressions recognition using principal component analysis and independent component analysis onto dimension of the emotion. To reflect well the changes in facial expressions, a representation based on principal component analysis (PCA) excluded the first 2 principal components is presented, ICA representation from this PCA representation is developed. Facial expression performance in two dimensional structure was significant 90.9% in pleasure/displeasure dimension and 66.6% in the arousal/sleep dimension. The findings indicate that the two dimensional structure of emotion may reflect various emotion states as a stabled structure for the facial expression recognition.

1 Introduction

In the field of facial expression recognition, most research has been made in trying to recognize expressions of discrete emotions suggested by Ekman[1]. Such studies provide a convenient framework [2, 3, 4]. But these studies have limitations for recognition of natural facial expressions which consist of several other emotions and many combinations of emotions. Thus, when developing methods for analyzing facial expressions in human-computer interaction, dimension approach is needed.

The dimensions of emotion can be overcome this limitation. The two most common dimensions are "arousal" (calm/excited), and "valence" (negative/positive) [5, 6]. To recognize facial expressions in various emotion states, we worked with dimensions of emotion instead of with basic emotions or discrete emotion categories. The dimensions of emotion proposed are pleasure/displeasure dimension and arousal/sleep dimension.

Several methods for representing facial expression images have been proposed such as PCA (Principal Component Analysis), ICA (Independent Component Analysis), Optic flow and Geometric tracking method, and Gabor representation [7, 8,9]. ICA filters as features on facial expression recognition were demonstrated the successful classifying twelve facial actions of the upper and lower face [9]. At recently study, PCA representation excluded the first 1 principal component in full face was applied to input features of neural network classifier in work for facial expression recognition [10]. PCA representation excluded the first 1 principal component can remove neutral expressions. ICA is a generalization of PCA which learns the high-order moments of the data in addition to the second-order moments [11]. Therefore we thought that ICA

D.-Y. Yeung et al. (Eds.): SSPR&SPR 2006, LNCS 4109, pp. 916–922, 2006.

representation using PCA images excluded neutral expression components in full face could be used effectively in the facial expression recognition as well.

This paper develops a method to recognize facial expressions on dimension of emotion using a combination of PCA and ICA. Section 2 indicates a representation of facial expression images based on PCA and ICA for feature extraction of facial expressions. Section 3 describes the classification of facial expressions on two dimensional structure of emotion. Section 4 concludes with discussion.

2 Feature Extraction

This section provides a database based on dimension structure of emotion and the representation of facial expression images for feature extraction of facial expressions. The representation of facial expression images is developed as two steps. In the first step, we present a representation based on PCA excluded the first 2 principal components. Second step, ICA representation from this PCA representation was developed.

2.1 Database of Dimension Structure

The database [12] with two dimension structure of emotion contained 498 images, 3 females and 3 males, each image using 640 by 480 pixels. Expressions were divided into two dimensions(Pleasure/Displeasure and Arousal/Sleep dimension) according to the study of internal emotion states through the semantic analysis of words related with emotion by Younga et al. [13] using 83 expressive words.

Each expressor of females and males posed 83 internal emotional state expressions when 83 words of emotion are presented. 51 experimental subjects rated pictures on the degrees of expression in each of the two dimensions on a nine-point scale. The images were labeled with a rating averaged over all subjects. The result of the dimension analysis of emotion words related to internal emotion states is shown in figure. 1.

2.2 PCA Representation Excluded Neutral Expressions

Facial expression images were centered with coordinates for eye and mouth locations, and then cropped and scaled to 20x20 pixels to minimize reconstruction error. The luminance was normalized in two steps. The rows of the images were concatenated to produce 1×400 dimensional vectors. The row means are subtracted from the data set, X. Then X is passed through the zero-phase whitening filter, V, which is the inverse square root of the covariance matrix:

$$V = E\{XX^T\}^{-\frac{1}{2}}, \quad Z = XV \tag{1}$$

From this process, Z removes much of the variability due to lightening. Atick and Redlich [14] have argued for such compact, decorrelated representations as a general coding strategy for the visual system. Redundancy reduction has been discussed in relation to the visual system at several levels. A first-order redundancy is mean luminance. The variance, a second order statistic, is the luminance contrast. PCA is a way of encoding second order dependencies in the input by rotating the axes to corresponding to directions of maximum covariance.

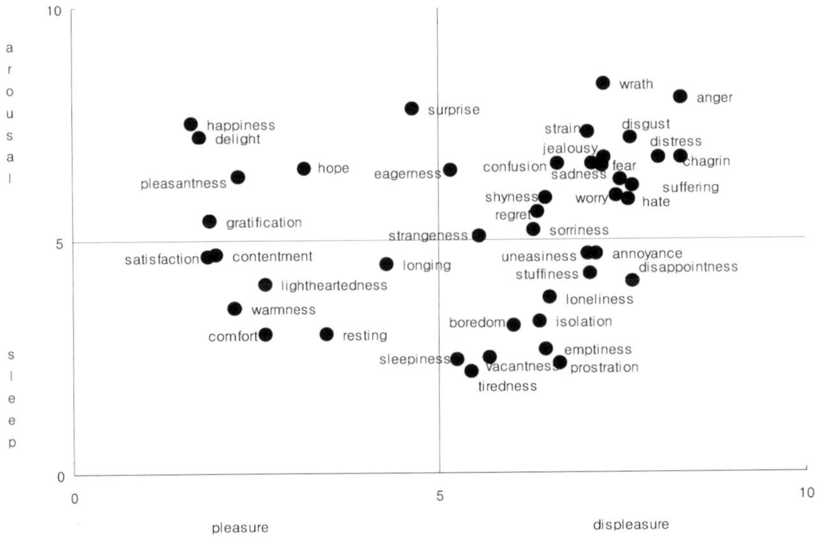

Fig. 1. The dimension analysis of emotion words related to internal emotion states

The first 1 or 2 principal components of PCA do not address the changes of facial expressions. It just displays the neutral face. That is to say, the neutral face means redundant codes in facial expressions. Figure 2(a) shows PCA representation that included the first 2 principal components. But selecting intermediate ranges of components that excluded the first 2 principal components of PCA do address well the changes in facial expression (Figure 2(b)).

To extract information of facial expression excluded redundant codes such as neutral expressions in facial expressions, we employed the 200 PCA coefficients, P_n,

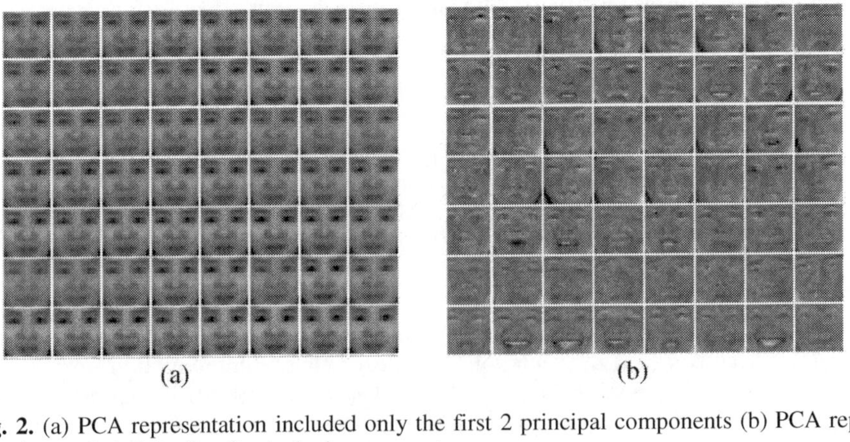

(a) (b)

Fig. 2. (a) PCA representation included only the first 2 principal components (b) PCA representation excluded the first 2 principal components excluded the first 2 principal components of PCA of the face images. 200 principal components excluded the first 2 principal components provided best performance on facial expression recognition.

The principal component representation of the set of images in Z in Equation(1) based on P_n is defined as $Y_n = Z * P_n$. The approximation of Z is obtained as $\overline{Z} = Y_n * P_n^T$. The columns of Y_n consist of input data for ICA representation.

2.3 ICA Representation

Independent component analysis (ICA) is a generalization of principal component analysis, which decorrelates the high-order moments of the input. Much of the important information is contained in the high-order statistics of the images. In a task such as facial expression recognition, a representational basis in which the high-order statistics are decorrelated should consider changes in facial expressions. Therefore, we applied images after excluding the high-order statistics such as neutral expressions for feature extraction of facial expressions to ICA representation.

The images were converted to vectors and comprised the rows of a 252x200 data matrix, Y. We assume the facial images in Y to be a linear mixture of an known set of statistically independent source images U, where $A = W^{-1}$ is an unknown mixing matrix. The sources, U are gained by a matrix of learned filters, W. ICA representation is generated according to the following linear model [15, 16]

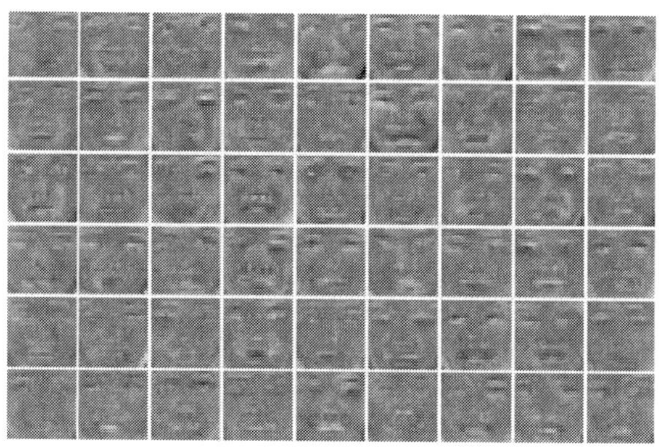

Fig. 3. Basis images for the ICA factorial representation $(A = W^{-1})$

Fig. 4. ICA factorial representation=($u_{1,}\ u_{2,} \ldots\ldots, u_n$)

$$U = WY \ . \tag{2}$$

The weight matrix, *W*, was obtained by using the FastICA algorithm [17]. The FastICA algorithm computes the independent components that become uncorrelated by a whitening process and then maximizes non-Gaussianity of data distribution by using kurtosis maximization. The columns of the ICA output matrix, $WY = U$ provided a factorial code for the training images in Y . Each column of U contained the coefficients of the basis images in A for reconstructing each images in Y . The columns of $A = W^{-1}$ consist of basis images for the ICA factorial representation (Fig. 3). Figure 4 shows the factorial code representation in facial expression image. The representational code for the test images was found by $WY_{test} = U_{test}$. The matrix excluded the first 2 principal components of test images is Y_{test} and W is the weight matrix gained by performed ICA on the training images.

3 Recognizing Facial Expressions

252 images for training and 66 images excluded from the training set for testing are used. The 66 images for test include 11 expression images of each six people. Facial expression recognition in various emotion states was evaluated by the nearest neighbor classifier in two dimensional structure of emotion on pleasure/displeasure dimension and arousal/sleep dimension. The coefficient vectors U in each of the two dimensions are given as vectors of U_{train} and U_{test} . Coefficient vectors in each test set were assigned to the class label of the coefficient vector in the training set that was most similar as evaluated by *S*:

$$S = \frac{U_{train} \cdot U_{test}}{\|U_{train}\| \|U_{test}\|} \min\left(\frac{\|U_{train}\|}{\|U_{test}\|}, \frac{\|U_{test}\|}{\|U_{train}\|}\right) \tag{3}$$

The class label consists of four section on two dimensional structure of emotion. C1 class is described with pleasure/displeasure dimension ranging from 5 up to 9 and arousal/sleep dimension ranging from 1 up to 4. C2 class is described with pleasure/displeasure dimension ranging from 5 up to 9 and arousal/sleep dimension ranging from 5 up to 9. C3 class is described with pleasure/displeasure dimension ranging from 1 up to 4 and arousal/sleep dimension ranging from 5 up to 9. C4 class is described with pleasure/displeasure dimension ranging from 1 up to 4 and arousal/sleep dimension ranging from 1 up to 4.

The first test verified with 252 facial images trained already. The recognition result that was produced by 252 images trained previously showed 100% recognition rates. The recognition result of test set showed 90.9% in the pleasure/displeasure dimension and 66.6% in the arousal/sleep dimension. Table 1 describes a part of facial expression recognition derived from three people in all six people on two dimensions of emotion.

Table 1. The result of facial expression recognition derived from three people. (Abbreviation: P-D, pleasure/displeasure; A-S, arousal/sleep;).

Named emotional word of Pictures(person)	Class label	Test Set		Recognized Class label
		P – D	A – S	
depression(a)	1	6.23	4.43	3
crying(a)	1	6.47	4.10	1
gloomy(a)	2	7.37	5.53	1
strange(a)	1	6.17	5.17	1
proud(a)	4	3.07	4.47	4
confident(a)	3	3.47	4.57	1
despair(a)	2	6.23	5.97	2
sleepiness(a)	4	5.00	1.80	1
likable(a)	3	1.97	4.23	3
delight(a)	3	1.17	4.20	3
boredom(a)	1	6.77	5.50	2
pleasantness (b)	3	1.40	5.47	3
depression (b)	1	6.00	4.23	1
crying(b)	2	7.13	6.17	2
gloomy(b)	1	5.90	3.67	1
strangeness(b)	2	6.13	6.47	1
proud(b)	3	2.97	5.17	3
confident(b)	4	2.90	4.07	2
despair(b)	1	7.80	5.67	2
sleepiness(b)	4	6.00	1.93	3
likable(b)	4	2.07	4.27	2
delight(b)	3	1.70	5.70	2
gloomy(c)	1	6.60	3.83	1
strangeness(c)	2	6.03	5.67	2
proud(c)	4	2.00	4.53	4
confident(c)	4	2.47	5.27	4
despair (c)	1	6.47	5.03	2
sleepiness(c)	1	6.50	3.80	1
likable(c)	4	1.83	4.97	4
delight(c)	3	2.10	5.63	4
boredom(c)	2	6.47	5.73	1
tedious(c)	1	6.73	4.77	1
jealousy(c)	2	6.87	6.80	2

4 Conclusion

A new approach method to recognize facial expressions in various emotion states with ICA and PCA has been discussed in this paper. Facial expression performance in two dimensional structure was evaluated by the nearest neighbor classifier. The result of facial expression recognition with ICA and PCA onto dimension structure of emotion shows significant conclusions as follow.

First, the two dimensional structure of emotion provided a stabled structure for the facial expression recognition. Second, pleasure-displeasure dimension was analyzed as a more stable dimension than arousal-sleep dimension. Pleasure/Displeasure dimension was significant 90.9%, while Arousal-Sleep dimension was significant 66.6%. We suggest that the two dimensional structure of emotion may provide a structure for the facial expression recognition as close to real life as possible.

References

1. Ekman, P.: An argument for basic emotions. Cognition and emotion 6(3) (1993) 169-200
2. Cohen, I., Sebe, N., Garg, A., Chen, L. S., Huang, T. S.: Facial expression recognition from video sequences:temporal and static modeling. Computer Vision and Image Understanding 91(1-2) (2003) 160-187
3. Dailey, M.N., Cottrell, G.W.,Padgett, C., Adolphs, R.:EMPATH:a neutral network that categorizes facial expressions. Journal of Cognitive Neuroscience 14 (2002) 160-187
4. Smith, E., Bartlett, M.S., Movellan, J.: Computer Recognition of Facial Actions: A Study of Co-articulation Effects, Proceedings of the Eight Annual Joint Symposium on Neural Computation (2001)
5. Peter J. L.: The emotion probe: Studies of motivation and attention. American Psychologist, 50(5) (1995) 372-385
6. Russell, J. A.: Evidence of convergent validity on the dimension of affect. Journal of Personality and Social Psychology, **30**, (1978) 1152-1168
7. Donato, G., Bartlett, M., Hager, J., Ekman, P. and Sejnowski, T.: Classifying facial actions, IEEE PAMI, 21(10) (1999) 974-989
8. Pantic, M., Rothkrantz, L.J.M.: Towards an Affect-Sensitive Multimodal Human Computer Interaction, Proc. Of IEEE. 91 1370-1390
9. Bartlett, M.: Face Image analysis by unsupervised learning, Kluwer Academic Publishers (2001)
10. Youngsuk S., Youngjoon A.: Facial expression recognition based on two dimensions without neutral expressions, LNCS(3711) (2005) 215-222
11. Comon, P.: Independent component analysis - a new concept? Signal processing 36 (1994) 287-314
12. Saebum, B., Jaehyun, H., Chansub, C.: Facial expression database for mapping facial expression onto internal state. '97 Emotion Conference of Korea, (1997) 215-219
13. Younga, K., Jinkwan, K., Sukyung, P., Kyungja, O., Chansub, C.: The study of dimension of internal states through word analysis about emotion. Korean Journal of the Science of Emotion and Sensibility, **1** (1998) 145-152
14. Atick, J., Redlich, A.: What does the retina know about natural scenes?, Neural Computation (4) (1992) 196-210
15. Olshasen, B. Field, D.: Natural image statistics and efficient coding, Network:computation in neural systems, 7(2) (1996) 333-340
16. Bell,A. Sejnowski, T.: The independent components of natural scenes areedge filters, Vision Research, 37(23) (1997) 3327-3338
17. Hyvarinen, A., Karhunen, J., Oja, E.: Independent component analysis, John Wiley & Sons, Inc. (2001)

An Audio Copyright Protection Schemes Based on SMM in Cepstrum Domain*

Shenghong Li[1], Lili Cui[1], Jonguk Choi[2], and Xuenan Cui[2]

[1] School of Information Security Engineering,
Shanghai Jiaotong University
Shanghai, 200030, China
shli@sjtu.edu.cn,
cuilili@sjtu.edu.cn

[2] Copyright Protection Research Institute,
Sangmyung University, 5096, Korea
juchoi@sangmyung.ac.kr,
cuixuenan00@163.com

Abstract. In this paper, we present an audio scheme protective of copyright protection using information hiding. We propose visually recognizable binary image and text information as watermark (copyright) information embedded in audio signal. Cepstrum representation of audio can be shown to be very robust to a wide range of attacks. We apply SMM(statistical mean manipulation) theory in embedding image watermarking, and address attacks against lossy audio compression like MP3, white Gaussian noise and so on. A blind detection watermarking can be realized with the proposed scheme.

Keywords: copyright protection, information hiding, watermark, statistical mean manipulation, cepstrum domain.

1 Introduction

The digital watermark technique is a technique to solve the copyright problem. The media owner can use this technique to insert some information into the media. There has been a fair amount of research on diverse applied techniques of audio watermark, i.e. Spread Spectrum method [1-4], echo hiding [5-7], a method Replica Signal [9] etc.

However in most audio watermarking methods, the embedding algorithms embed a chaos sequence or pseudo-random array to be watermarking in the content, insert mean information is very peculiar. In this paper we will insert a still binary image being audio watermarking into the ceptrum domain. Extensive experimental results prove that the embedded watermark is inaudible and robust.[1]

* The work is fully supported by the international co-operation project of the ministry Science and Technology of Korea: Co-Development of Broadcasting Sync. Equipment and DRM Watermark Chipset for Digital Broadcasting Content based on Original Watermarking Technology of Korea. (Project No: M60401000150-05A0100-15010).

D.-Y. Yeung et al. (Eds.): SSPR&SPR 2006, LNCS 4109, pp. 923 – 927, 2006.

2 Details of the Proposed Algorithm

The cepstrum domain analysis is used commonly in speech application, such as recognition area. In speech recognition, the cepstral coefficients are regarded as the main features of a voice. The cepstral coefficients vary less after general signal processing then samples in time domain .Due to the advantage of cepstral coefficients, Li and Yu [10] proposed a robust audio data hiding technique in cepstrum domain.

Cepstral analysis utilizes a form of homomorphic system which converts the convolution operation an addition operation. It consists of three consecutive steps: Fourier transform, take logarithm and inverse Fourier transform. It is easy to see that those three operations are all linear It should be noted that the logarithm we take at the second step is complex logarithm and $X(n)$ is formally called "*complex cepstrum*". But in practice, people often define the real part of complex cepstrum to be the "real" cepstrum for convenience.

$$X(n) = IFFT\left(\log\left(REAL\left(FFT\left(x(n)\right)\right)\right)\right) \tag{1}$$

And we can exactly recover the original signal in time domain from its cepstrum domain representation by taking correspondent inverse operations

$$x(n) = IFFT\left(\exp\left(REAL\left(FFT\left(X(n)\right)\right)\right)\right) \tag{2}$$

Cepstrum coefficients are around zero except the last, therefore we shall modify small cepstrum coefficients except the last. Experimental studies have shown that most common signal processing could change individual cepstrum coefficients dramatically, but their statistical mean often experiences much less disturbance, offering an appropriate candidate for information carrying.

3 Scheme on Binary Image Watermark Embedding

In embedding process, we adopt the concept of the cepstrum, and embed the data based on statistical mean theory which is much more robust, especially for attacks destroying synchronization structure of audio signal. We shall focus on the statistical mean of cepstrum coefficients to be a real number for embedding '1',and another number for embedding '0', then we can detect the watermarking by adjudging the threshold derived from the two number. The detail watermarking embedding works as following:

1. Transform time domain signal to cepstrum domain.
2. Divide audio cepstrum into frames, which is depend on the size of binary image.
3. Calculate the mean of each frame of cepstral coefficients. Modify the mean of cepstral coefficients to zero. Then the embedding algorithm is following:

To embedding '1':

$$X(n)' = X(n) + \alpha * W_m(n) \tag{3}$$

To embedding '0':

$$X(n)' = X(n) \tag{4}$$

Where α is the factor controlling the allowable distortion for individual cepstrum component $X(n)$. $W_m(n)$ is watermarking information, m denotes the number of frame.

4. Create the final watermarked audio.

4 Binary Image Watermark Detecting

The watermark should be extractable even if common signal processing (including data compression and some kinds of noise attacks) operations are applied to the host audio. In detection procession, don't need original audio signal, is total blind detection process. The detection method is base on statistical mean manipulation, calculate the sum of every frame cepstrum coefficients, and set the threshold Td to identify the watermark information.

5 Experiment Results

In the experiment, Matlab6.1 is used as emulation software, the music used as the watermarked media is 102.06 seconds music, 11025Hz of sampling rate and 16 bit recorded for each sampling. The embedding capacity is 62kbps.The watermark is 64x64 binary image , given in Fig1 (a). A blind listening test was used to confirm the transparency of the watermarked signal and most listeners couldn't distinguish the difference of the watermarked signals.

The following are the test results, where Fig1 (a) is original watermark (binary image), and Fig1 (b) is picked up without attacked by our detection method.

(a) (Sim=1) (b) (Sim=1) (c) (Sim=0.805) (d) (Sim=0.624)

Fig. 1. (a) Original watermark, (b) Detected watermark, (c) MP3 compression at 64kbps, (d) MP3 compression at 32kbps

To test the robust of our scheme, we evaluate the performance of the watermark against lossy attacks by diving the test results into four subtest, the performance can refer Fig1 and Fig4:

Subtest1 (MP3 Attack): We compare the effect marked the audio and the decoded audio given by MP3 compression at different bit rate. Fig1 (c) is under attack of MP3 compression at the rate of 64kbps, and provides transparent audio quality. Fig1 (d) is at 32kbps. Each similarity value corresponding to the compression rate is shown in Fig2. In Fig2 with the rate of MP3 increase, accordingly the similarity value increase, here we list four kinds of conditions, the lowest rate is 32kbps, and under 32kbps, the image can't be extracted. So we can see that embedded image can be extracted for MP3 compression at the rate of above 32kbps.

Subtest2 (White Gaussian Noise Attack): Fig3. (a), (b) are under attack of white Gaussian noise with mean zero, covariance 1 and 0.1. We can see our proposed scheme

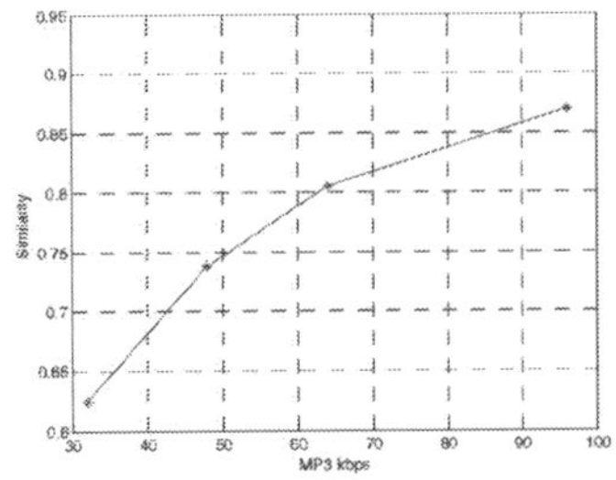

Fig. 2. Similarity comparison with various Mp3 compression rate

(a) (Sim=0.89) (b) (Sim=0.81) (c) (Sim=0.88)

Fig. 3. (a)White Gaussian noise (0,1); (b) White Gaussian noise (0,0.1); (c) Median filter

demonstrates good survivability with (0,1) white Gaussian noise. But if mean is nonzero, or covariance is above 0.1, we can't detect the watermark.

Subtest3 (Filter Attack): Fig3 (c) shows the detection performance after median filter, which is nonlinear filter. It can be seen from Fig3 that our scheme demonstrates good robustness.

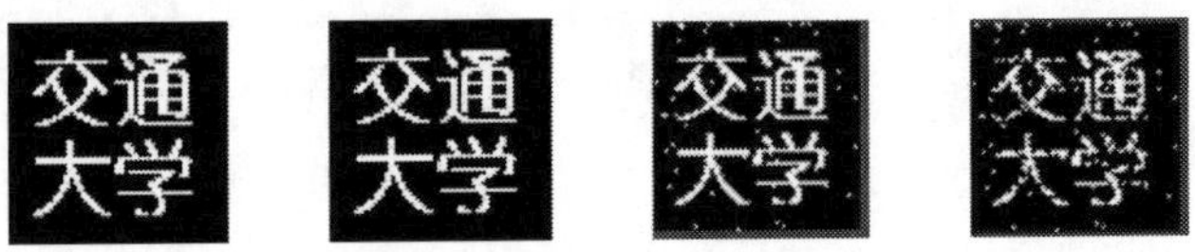

(a) (Sim=0.95) (b)(Sim=0.86) (c)(Sim=0.98) (d)(Sim=0.58)

Fig. 4. (a) 22050Hz (b) 44100Hz (c) 8000Hz (d)8 bit

Subtest4 (Repeat Sampling and Repeat Quantification): for repeat sampling test we subsampling watermarked signal at 22050Hz and 44100Hz and 8000Hz ,then revert original sampling frequency, Fig4 (a-c) show the performance under these condition.for repeat quantification test we quantify watermarked signal from 16 bit at first to 8 bit, then restore signal, the performance just as (d) show.

$$W_m' = \begin{cases} 1 & \sum_{i=1}^{N} x_m(i) > T_d \\ 0 & \sum_{i=1}^{N} x_m(i) < T_d \end{cases} \tag{5}$$

where the $x_m(i)$ denote the mth frame of the audio signal, W_m' is detected watermarking information, the value is '1' or '0'. We embed 1 bit each frame. To show the performance of our test, we compare the extracted watermark with original watermark. In this comparison, we use the similarity measure given in (6).

$$Sim(W,W') = W \cdot W' / \sqrt{W \cdot W} \tag{6}$$

6 Conclusion

In this paper, an audio scheme protective of copyright protection using information hiding is proposed. The binary image watermark scheme based on SMM(statistical mean manipulation) theory, and divide frames to embedding watermark. The audio scheme is robustness against the data compression and some kinds of attacks such as MP3, Audio Stirmark, white Gaussian noise and repeat sampling and repeat quantification.

The following is our future work:

(1) Study on the performance of SMM further.
(2) Research for the robust performance of the other audio attack.
(3) Research for the robust performance of the other embedding domain.
(4) Multi-watermark embedding.
(5) Study on the performance of Text as watermarks.

References

1. P.Bassia,I.Pitas, and N. Nikolaidis: Robust audio watermarking in the time domain. IEEE Transactions on Multimedia, vol. 3, June (2001), pp. 232-241.
2. D.Kirovski and H.Malvar: Robust spread–spectrum audio watermarking. IEEE International Conference on Acoustics, Speech, and Signal processing, vol. 3, (2001). pp. 1345-1348
3. L.Boney, A.H.Tewfik, and K.N. Hamdy: Digital watermark for audio signals. In International Conference on Multimedia Computing and Systems, IEEE, Hiroshima, Japan, June, (1996), pp.473-480.
4. H.Malik, S.Khokhar, and A.Rashid: Robust audio watermarking using frequency selective spread spectrum theory, In International Conference on Accoustic, Speech and Signal Processing, IEEE, Montreal, Canada, May, (2004). pp. 385-388.
5. D.Gruhl, A.Lu, W.Bender: Echo hiding. in Proc. Information Hiding Workshop , University of Cambridge, U.K., (1996), pp. 295-315.
6. S. W. Foo, T. H. Yeo, and D. Y. Huang: An Adaptive Audio Watermarking System. Electrical and Electronic Technology, TENCON. Proceedings of IEEE Region 10 International Conference on, Vol2, (2001), pp.509-513.
7. H. O. Oh, J. W. Seok, J.W. Huang and D. H. Youn: New echo embedding technique for robust and imperceptible audio watermarking. Acoustics, Speech, and Signal Processing, Proceedings. 2001 IEEE International Conference on, Vol3, (2001), pp.1341-1344.
8. S. Shin, J. W. Kim, J. Choi: Audio watermarking using Digital Filter. Korea Information Security, Conference, vol. 11. No.1 (2001). pp.464-468.
9. R. Petrovic: Audio signal watermarking based on replica modulation. Telecomfmunications in Modern Satellite, Cable and Broadcasting Service, TELSIKS 2001. 5th International Conference on vol 1.(2001). pp.227-234.
10. X. Li, H. H. Yu: Transparent and Robust Audio Data hiding in cepstrum Domain. ICME2000, vol. 1, (2000). pp.397-400.

Combining Features to Improve Oil Spill Classification in SAR Images

Darby F. de A. Lopes, Geraldo L.B. Ramalho, Fátima N.S. de Medeiros, Rodrigo C.S. Costa, and Regia T.S. Araújo

Image Processing Research Group, Universidade Federal do Ceara
60455-760 - Fortaleza, CE, Brazil
{darby, fsombra, rodcosta, regia}@deti.ufc.br, glbramalho@gmail.com
http://www.gpi.deti.ufc.br/

Abstract. As radar backscatter values for oil slicks are very similar to backscatter values for very calm sea areas and other ocean phenomena, dark areas in Synthetic Aperture Radar (SAR) imagery tend to be misinterpreted. In this paper three feature sets are used to identify the oil slicks in SAR images. These images are submitted to different MLP architectures to verify the separability performance over each feature set. This analysis is very suitable for remote sensing of environment applications concerning marine oil pollution. The estimated resulting performance points out which feature set is the best suitable for the suggested application.

1 Introduction

Since the last decade Synthetic Aperture Radar (SAR) systems have played an important role in remote sensing of environmental disasters. These systems provide oil spills detection and monitoring, that seriously affect the marine ecosystem, providing a more rigorous and effective environment monitoring. Furthermore, SAR images have considerably contributed to understand atmospheric phenomena, land use mapping and monitoring, deforestation assessment, geographic evolution, urban growing rates assessment, agricultural crops monitoring and so on. The potential damage for the environment and economy of the area at stake requires that agencies be prepared to rapidly detect, monitor, and clean up any large spill [1]. Remote sensing of dark spots in the sea is a complex process, due to the simultaneous movement of radar and spots. The presence of an oil film on the sea surface damps out the small waves and reduces the rough surface due to the increased viscosity of the top layer and drastically reduces the measured backscattering energy, resulting in darker areas in SAR imagery [2]. The interest in appraising texture features in this work becomes from the different rough degrees presented in SAR images. Oil spill images are characterized by being less rough when compared to the similar slicks. Moreover, the procedures to extract texture features are independent of segmentation methods. The diffusion of the electromagnetic waves in the surface of the sea depends, mainly, on the rough surface which is influenced by the presence of winds, currents, waves

D.-Y. Yeung et al. (Eds.): SSPR&SPR 2006, LNCS 4109, pp. 928–936, 2006.

and parameters of the radar, such as incidence angle, frequency, polarization and resolution. The sea behaves as a specular surface when there are not waves and winds. However, dark areas might not be oil slicks but merely local wind effects or natural oil films due to low winds [3].

Automatic identification of oil spills in SAR images is a very complex task because similar images of oil spills frequently occur, particularly in low-wind conditions [4] requiring a careful interpretation. In general, the human interpreter determines if a dark object is an oil spill or a look-alike one. The contrast between oil spectral and water radiance around the oil determines which might be oil slicks. Studies have been carried out to improve methods to detect oil spills in satellite images. Liu *et al.*[5] proposed algorithms to detect and track mesoscale oceanic features employing multiscale wavelet analysis using the 2-D Gaussian wavelet transform to track oil slicks, eddies, fronts, whirlwinds and icebergs. The authors concluded that the wavelet analysis can provide a more cost-effective monitoring program that would keep track of changes in important elements of the coastal watch system. In [4] it was proposed a semi-automatic algorithm for spots detection which identifies objects in the scene with larger probability of being oil spills. A neural network approach for oil spills detection in European Remote Sensing Satellite-Synthetic Aperture Radar (ERS-SAR) imagery has been explored as an alternative tool in [2]. Del Frate *et al.* [2] proposed an algorithm to classify spots based on a set of geometric features extracted from real oil spots and look-alike ones. The input of the network consisted of a set of features regarding an oil spill candidate and the output concerns the probability for the candidate to be a real oil spill. The authors reported that the introduction of physical characteristics related to atmospheric conditions such as wind speed and water temperature could improve the algorithm results.

Concerning evaluation of feature selection issue, Jain and Zongker [6] applied feature selection algorithms to SAR images in order to classify land use combining features of four different texture models. The researchers also evaluated the potential difficulties of performing feature selection in small sample size situations due to the curse of dimensionality.

This paper proposes an analysis of the discrimination power of three different feature sets, comparing the performance of a classifier based on neural networks applied to each set: the physical-geometrical feature set generated by statistical measures on geometric characteristics [7], the texture feature set obtained as described in Bevk *et al* [8], and the third one as a composition of the previous sets. To minimize the computational effort of the classifier, principal component analysis (PCA) is used to reduce the dimensionality of each feature set and the reduced sets are compared with the original ones. The overall performance of the classifier is evaluated for different feature sets based on geometrics and texture attributes aiming at optimizing oil spills detection in SAR images. The proposed method can be used to support environmental remote monitoring.

This paper is organized as follows. The next section describes the methodology, the feature extraction process and the approach used to detect oil spills. Section 3 presents the simulation results and the last section concludes the paper.

2 Methodology

The feature data sets are generated using SAR images collected from different sources. After extracting the features from the spots, the data is divided into three feature sets and two different analysis are made: a) a classifier processes the original feature sets and b) a classifier processes the reduced sets using principal component analysis. The classifier estimated performance states the discrimination power of each set. Fig. 1 exhibits the block diagram of the previously described proposed methodology.

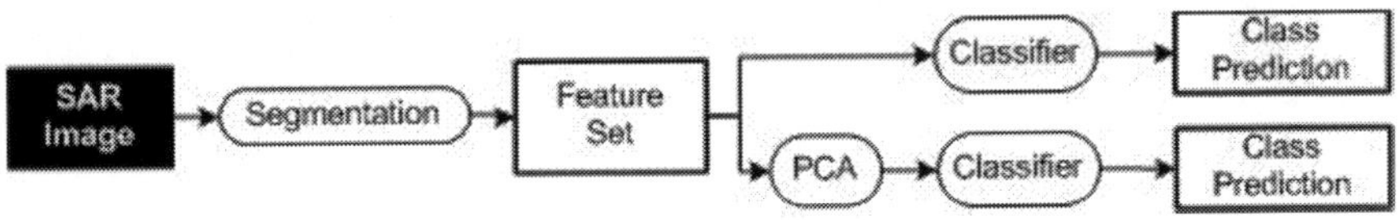

Fig. 1. General steps for feature sets evaluation of dark spots in SAR images

2.1 Feature Extraction

Texture analysis is able to provide an automatic classification of features presented in SAR images [1]. In general, texture characteristics are important for surface or object identification from aerial, satellites or biomedical images and for other applications such as industrial monitoring or product quality, remote sensing of natural resources, and medical diagnosis with tomography [9]. Despite its importance and ubiquity in image data, a formal approach or precise definition of texture does not exist [10]. The term is used to point to intrinsic properties of surfaces, especially those that do not vary smoothly in intensity. Texture includes intuitive properties like roughness, granulation and regularity. More formally, it can be defined as the set of local neighborhood properties of image grey levels [11].

Statistical information of texture characteristics is based on the representation of texture using properties governing the distribution and relationships of grey level values in the image [12]. The spatial grey level dependence matrix proposed in [13] is used to extract features, i.e., energy, contrast or entropy. In this paper the first and second order statistics of the segmented images are extracted to provide the textural features of oil spills.

The first-order probability distribution of the amplitude of a quantized image may be defined as:

$$H(g) = \frac{n_g}{N}; \quad g = 0, 1, \ldots, G-1 \tag{1}$$

where N represents the total number of pixels in the image, G denotes the number of grey levels and n_i denotes the number of pixels of grey value i in a given image. The histogram is a probability function of pixel values, therefore we can characterize its properties with a set of statistical parameters (also called first-order statistics). Many parameters may be derived from the histogram such as its mean, variance and percentiles. The following parameters are also computed:

mean (S_M), standard deviation or image contrast (S_D), skewness (S_S), kurtosis (S_K), entropy (S_{Ent}) and energy (S_E) [8].

Second-order statistics operate on the probability function ($P(i,j|d,\theta)$), that measures the probability of observing a pair of pixel values that are some vector $\overrightarrow{d}$ apart in the image [8].

The grey level cooccurrence can be specified in a matrix of relative frequencies $P_{i,j}$ with which two neighboring pixels separated by distance d in a given direction, occur on the image, one with grey level i and the other with grey level j. Generally, the cooccurrence matrix is computed for a finite number of pixel orientations, formally for angles in intervals of 45°. The cooccurrence matrices are symmetric.

The results of the grey level coocurrence are averaged for each angle with its transposed matrix as follows:

$$S(i,j) = \sum_{\theta=0,45,90,135^\circ} \frac{P(i,j|\theta,d) + P(i,j|\theta,d)^t}{8} \tag{2}$$

The second order statistics are extracted from the matrix shown in equation 2. Based on this matrix the following texture measures are computed: autocorrelation (A), cluster proeminence (CP), cluster shade (CS), contrast (C), correlation ($Corr$), covariance (Cov), energy (E), entropy (Ent), local homogeneity (H) and maximum probability (MAX). More detailed definitions of these features can be found in [13].

Another set of features used to describe a dark spot is extracted after the segmentation step. These measures are the physical-geometrical characteristics. Del Frate *et al* [7] state that some of these characteristics take into account the geometry and the shape of the dark spot, other part contains information about the backscattering intensity (calculated in dB) gradient along the border of the analyzed dark spot and others focus on the backscattering in the dark spot and/or in the background. The following measures, corresponding to the physical-geometrical set, are computed: area (Ar), average backscattering inside the area ($ABIA$), standard deviation of the backscattering inside the area ($SDBIA$), average backscattering outside the area ($ABOA$) and standard deviation of the backscattering outside the area ($SDBOA$). From the previous ones the following parameters are calculated: ratio between area and perimeter (AP), ratio between average backscattering inside and outside the area ($RBIO$), ratio between average backscattering and its standard deviation inside the area ($RBSDI$), ratio between average backscattering and its standard deviation outside the area ($RBSDO$), ratio between backscattering standard deviation inside and outside the area ($RSDIO$) and ratio between $SBSDI$ and $RBSDO$ ($RBSDIO$).

2.2 Principal Component Analysis

The use of more features extracted from patterns may lead to a better characterization and thus a better classification with a lower error rate, but in practice, the opposite is observed. For a given problem the error rate initially drops with an increasing number of features, but at a certain point the error rate saturates or rises if additional features are included. This phenomenon is called curse of dimensionality. The origin of this phenomenon is the fact that classifier design relies on the inference of statistical properties from the data such as the estimation of the likelihoods or the estimation of the parameters of a distribution [14].

The problem of feature selection is defined as follows: given a set of candidate features, select a subset that performs the best under some classification system. This procedure can reduce not only the cost of recognition by reducing the number of features that need to be collected, but in some cases it can also provide a better classification accuracy due to finite sample size effects [5]. The term feature selection is taken to refer to algorithms that output a subset of the input feature set [6]. Principal components analysis (PCA) is a multivariate procedure which rotates the data such that maximum variabilities are projected onto the axes, mapping the image data into a new, uncorrelated co-ordinated system or vector space [15]. It produces a space in which the data has the most variance along its first axis, the next largest variance along a second mutually orthogonal axis, and so on. The later principal components would be expected, in general, to show little variance. These could be considered therefore to contribute little to separability and could be ignored, thereby reducing the essential dimensionality of the classification space and thus improving the classification speed. It is useful to know that due the nonlinearity of some data sets, the PCA space transformation not always leads to an optimal feature subspace. In this case further analysis using another space transformation methods are necessary to achieve better results.

3 Simulation Results

The experiments were obtained by using a set of 20 real dark spot images, where half of them are oil spill images and the other half consist of look-alike images. Figure 2a and Figure 2b are SAR image examples of a typical oil slick and a natural film, respectively. The first two sets are physical-geometrical features ($S1$) and texture features ($S2$). The third one is formed by the union ($S3 = S1 \bigcup S2$) of the both cited. The sets $S1$, $S2$ and $S3$ are respectively 8, 15 and 23-dimensional.

The classifiers performance assessment is shown in Figures 3 and 4. The results were obtained by running the classifier algorithm 100 times using a hold-out method varying the training size from 10% to 90% of the whole sample set. As the performance for the compound set degrades due to its higher dimensionality, we also tested different MLP architectures. Using the same N inputs and M outputs, where N is the size of the input vector and M is the number of different classes, we changed the number of neurons in the hidden layer from 2 until 20. Indeed, we experimented individual higher classification rates as the classifier fitted more the data and noise. This can be observed in Fig. 3(a) for the feature set $S3$. We decided to use the 5 hidden neurons MLP architecture, beside its

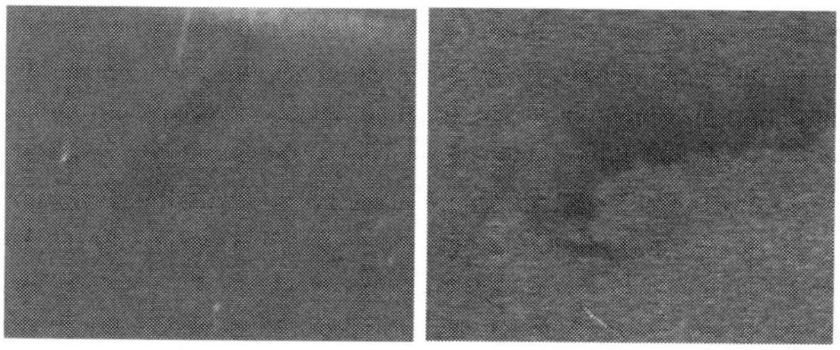

(a) (b)

Fig. 2. SAR image examples of (a) an oil slick and (b) a natural film

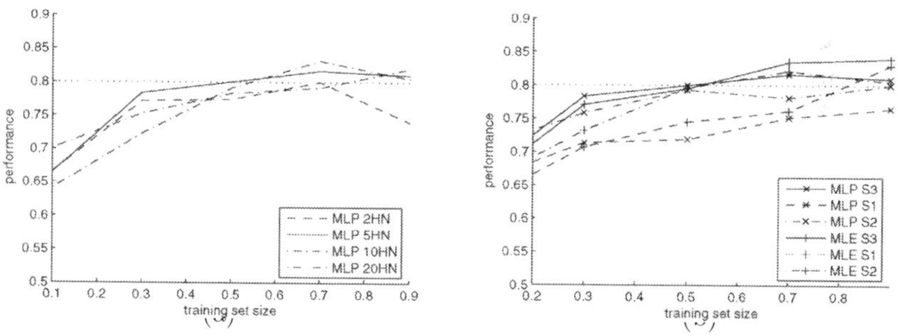

Fig. 3. MLP performance comparison (a) using different number of hidden neurons over the $S3$ feature space and (b) with a reference model MLE over all three feature sets

higher computational training cost, because of the generalization loss caused by *overfitting* when using the MLPs with more hidden neurons.

In Fig. 3(b) we provide a comparison between a three layer MLP with 5 hidden neurons and a *maximum-likelihood* estimator (MLE) [16] used as a reference model. The *maximum-likelihood* estimator tries to fit one gaussian probability function to each class centered on their means using unitary covariances and based on assumption of data independence. The maximum class probability is taken to assign a class label to the sample. The error probability is computed according the bayesian decision rule: $P_e = p_1 P(e|C_1) + p_2 P(e|C_2)$, where $P(e|C_n)$ is the conditional error probability for the input vector classified as belonging to class C_n and p_n is the *a priori* probability for the classes.

The use of PCA to reduce the dimensionality has achieved a better classification performance. Fig. 4 shows the PCA transformed data set presents a slightly better separability. Unfortunately this varies as the linearity changes from one data set to another. Thus, for $S3$ it is a good solution, but to the rest of the data sets the classifier performance is worse than working on the original space or quite the same.

Table 1 shows a rounded average *confusion matrix* computed from 100 classification rounds. The feature set $S3$ was applied to a MLP classifier with 5

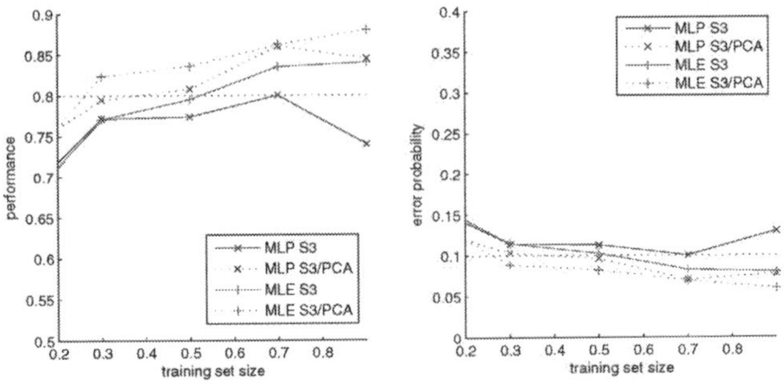

Fig. 4. MLP x Naïve Bayes performance and error probability comparison between the original space and the PCA transformed space

Table 1. Confusion matrix for $S3$ feature set applied to MLP classifier with 5 hidden neurons

Predicted Class	True Class C_1	C_2
C_1	9	2
C_2	1	8

Table 2. Variance comparison between original and PCA transformed feature spaces

	Original space			PCA space		
Classifier	S1	S2	S3	S1	S2	S3
MLP	0.0201	0.0255	0.0135	0.0215	0.0208	0.0172
BAYES	0.1010	0.0570	0.0712	0.0928	0.0339	0.0681

hidden neurons using 70% training size. Oil Spill samples are represented by class C_1 and the look-alike ones by the class C_2. It is worthy of notice that this low false-alarm rate was achieved using only 20 image samples.

The classifier variances obtained in the experiments are shown in Table 2. The variances for the original feature space and for the PCA transformed ones are very similar. The result obtained by adding the texture features (feature set $S3$) has shown that a better classification performance can be reached without loss of generality. Although the Naïve Bayes classifier has achieved higher correct classification rates, as expected, the MLP has provided better generalization.

4 Conclusions

This paper presented a methodology to improve oil spill classification in SAR images. In this approach, a small set of images is described by a large number of features. Thus, for this purpose a non-parametrical classifier like MLP is more suitable than the statistical parameters based ones, like Fisher Discriminant

Analysis (FLDA) for example. This occurs because the higher-order moments, necessary to establish the discriminant, are poorly estimated which leads to errors. The maximum-likelihood estimator, used in this paper, can give only a good point of observation, which we use to compare the performances of the classifiers. The overall misclassification achieved with a MLP classifier is low enough but we have a lot of work to do in order to reduce false alarms to permit the use of this methodology in reliable marine surveillance applications. Further investigation is required to choose a more robust classifier in order to achieve a higher rate of correct classification and improve its reliability for environment surveillance applications.

The error probability is smaller as the number of training samples grows up. We believe that with a larger data set it is possible to develop a MLP architecture that can reach even higher performances. Finally, the feature sets tested on these experiments have shown that textural features provide important effect in the performance improvement for oil spill detection application. The results reported in this paper point out that the use of texture features can add significantly discrimination power for oil spill detection applications without loss of generality. This improvement is reached when using that set combined with physical-geometrical features. As the use of PCA transformation also accomplished a less complex classifier, the overall computational cost was maintained low. It is noteworthy that a very small data set was used, furthermore we concluded that any performance improvement can be a very hard task to perform with this set. Although we consider these results an advance for automatic oil spill detection systems, the misclassification rate is not lower enough. In future works, we will investigate improvements on this approach by using methods for automatic feature selection using classifier combination.

Acknowledgement

The authors would like to thank CNPq (#476177/2004-9) and FUNCAP for their financial support.

References

1. Marghany, M.: Radarsat automatic algorithms for detecting coastal oil spill pollution. Asian Journal of Geoinformatics **3** (2001) 191–196
2. Frate, F.D., Salvatori, L.: Oil spill detection by means of neural networks algorithms: a sensitivity analysis. IEEE International Geoscience and Remote Sensing Symposium **2** (2004) 1370–1373
3. Calabresi, G., Frate, F.D., Lichtenegger, J., Petrocchi, A.: Neural networks for the oil spill detection. IEEE International Geoscience and Remote Sensing Symposium **1** (1999) 215–217
4. Solberg, A.H.S., Dokken, S.T., Solberg, R.: Automatic detection of oil spills in envisat, radarsat and ers sar images. IEEE International Geoscience and Remote Sensing Symposium **4** (2003) 2747–2749

5. Liu, A., Peng, C., Chang, S.S.: Wavelet analysis of satellite image for coastal watch. IEEE International Journal of Oceanic Engineering **22** (1997) 9–17
6. Jain, A., Zongker, D.: Feature selection: Evaluation, application, and small sample performance. IEEE Transactions on Pattern Analysis and Machine Intelligence **19** (1997) 153–158
7. Nirchio, F., Sorgente, M., Giancaspro, A., Biaminos, W., Parisato, E., Ravera, R., Trivero, P.: Automatic detection of oil spills from sar images. International Journal of Remote Sensing **26** (2005) 1157–1174
8. Bevk, M., Kononenko, I.: A statistical approach to texture description of medical images: a preliminary study. Proceedings of the 15th IEEE Symposium on Computer-Based Medical Systems (2002) 239–244
9. Chang, T., Kuo, C.C.J.: A wavelet transform approach to texture analysis. IEEE Transactions on Image Processing **4** (1992) 429–441
10. Haralick, R.M., Shapiro, L.G.: Computer and robot vision. Addison-Wesley, New York (1992)
11. Livens, S.: Image Analysis for Material Characterization. PhD thesis, Universiteit Antwerpen (1998)
12. Castellano, G., Bonilha, L., Li, L.M., Cendes, F.: Texture analysis of medical images. Clinical Radiology **59** (2004) 1061–1069
13. Haralick, R.M., Shanmugam, K., Dinstein, I.: Textural features for image classification. IEEE Transactions on Systems, Man, and Cybernetics **3** (1973) 610–621
14. de Wouwer, G.V.: Wavelets for Multiscale Texture Analysis. PhD thesis, Universiteit Antwerpen (1998)
15. Richards, J., Jia, X.: Remote Sensing Digital Image Analysis - An Introduction. Springer (1999)
16. Webb, A.R.: Statistical Pattern Recognition. 2 edn. Wiley, England (2002)

Author Index

Lecture Notes in Computer Science

For information about Vols. 1–4016

please contact your bookseller or Springer

Vol. 4144: T. Ball, R.B. Jones (Eds.), Computer Aided Verification. XV, 564 pages. 2006.

Vol. 4127: E. Damiani, P. Liu (Eds.), Data and Applications Security XX. X, 319 pages. 2006.

Vol. 4121: A. Biere, C.P. Gomes (Eds.), Theory and Applications of Satisfiability Testing - SAT 2006. XII, 438 pages. 2006.

Vol. 4115: D.-S. Huang, K. Li, G.W. Irwin (Eds.), Computational Intelligence and Bioinformatics, Part III. XXI, 803 pages. 2006. (Sublibrary LNBI).

Vol. 4114: D.-S. Huang, K. Li, G.W. Irwin (Eds.), Computational Intelligence, Part II. XXVII, 1337 pages. 2006. (Sublibrary LNAI).

Vol. 4113: D.-S. Huang, K. Li, G.W. Irwin (Eds.), Intelligent Computing, Part I. XXVII, 1331 pages. 2006.

Vol. 4112: D.Z. Chen, D. T. Lee (Eds.), Computing and Combinatorics. XIV, 528 pages. 2006.

Vol. 4111: F.S. de Boer, M.M. Bonsangue, S. Graf, W.-P. de Roever (Eds.), Formal Methods for Components and Objects. VIII, 447 pages. 2006.

Vol. 4109: D.-Y. Yeung, J.T. Kwok, A. Fred, F. Roli, D. de Ridder (Eds.), Structural, Syntactic, and Statistical Pattern Recognition. XXI, 939 pages. 2006.

Vol. 4108: J.M. Borwein, W.M. Farmer (Eds.), Mathematical Knowledge Management. VIII, 295 pages. 2006. (Sublibrary LNAI).

Vol. 4106: T.R. Roth-Berghofer, M.H. Göker, H. A. Güvenir (Eds.), Advances in Case-Based Reasoning. XIV, 566 pages. 2006. (Sublibrary LNAI).

Vol. 4104: T. Kunz, S.S. Ravi (Eds.), Ad-Hoc, Mobile, and Wireless Networks. XII, 474 pages. 2006.

Vol. 4099: Q. Yang, G. Webb (Eds.), PRICAI 2006: Trends in Artificial Intelligence. XXVIII, 1263 pages. 2006. (Sublibrary LNAI).

Vol. 4098: F. Pfenning (Ed.), Term Rewriting and Applications. XIII, 415 pages. 2006.

Vol. 4097: X. Zhou, O. Sokolsky, L. Yan, E.-S. Jung, Z. Shao, Y. Mu, D.C. Lee, D. Kim, Y.-S. Jeong, C.-Z. Xu (Eds.), Emerging Directions in Embedded and Ubiquitous Computing. XXVII, 1034 pages. 2006.

Vol. 4096: E. Sha, S.-K. Han, C.-Z. Xu, M.H. Kim, L.T. Yang, B. Xiao (Eds.), Embedded and Ubiquitous Computing. XXIV, 1170 pages. 2006.

Vol. 4094: O. H. Ibarra, H.-C. Yen (Eds.), Implementation and Application of Automata. XIII, 291 pages. 2006.

Vol. 4093: X. Li, O.R. Zaiane, Z. Li (Eds.), Advanced Data Mining and Applications. XXI, 1110 pages. 2006. (Sublibrary LNAI).

Vol. 4092: J. Lang, F. Lin, J. Wang (Eds.), Knowledge Science, Engineering and Management. XV, 664 pages. 2006. (Sublibrary LNAI).

Vol. 4090: S. Spaccapietra, K. Aberer, P. Cudré-Mauroux (Eds.), Journal on Data Semantics VI. XI, 211 pages. 2006.

Vol. 4088: Z.-Z. Shi, R. Sadananda (Eds.), Agent Computing and Multi-Agent Systems. XVII, 827 pages. 2006. (Sublibrary LNAI).

Vol. 4079: S. Etalle, M. Truszczyński (Eds.), Logic Programming. XIV, 474 pages. 2006.

Vol. 4077: M.-S. Kim, K. Shimada (Eds.), Advances in Geometric Modeling and Processing. XVI, 696 pages. 2006.

Vol. 4076: F. Hess, S. Pauli, M. Pohst (Eds.), Algorithmic Number Theory. X, 599 pages. 2006.

Vol. 4075: U. Leser, F. Naumann, B. Eckman (Eds.), Data Integration in the Life Sciences. XI, 298 pages. 2006. (Sublibrary LNBI).

Vol. 4074: M. Burmester, A. Yasinsac (Eds.), Secure Mobile Ad-hoc Networks and Sensors. X, 193 pages. 2006.

Vol. 4073: A. Butz, B. Fisher, A. Krüger, P. Olivier (Eds.), Smart Graphics. XI, 263 pages. 2006.

Vol. 4072: M. Harders, G. Székely (Eds.), Biomedical Simulation. XI, 216 pages. 2006.

Vol. 4071: H. Sundaram, M. Naphade, J.R. Smith, Y. Rui (Eds.), Image and Video Retrieval. XII, 547 pages. 2006.

Vol. 4070: C. Priami, X. Hu, Y. Pan, T.Y. Lin (Eds.), Transactions on Computational Systems Biology V. IX, 129 pages. 2006. (Sublibrary LNBI).

Vol. 4069: F.J. Perales, R.B. Fisher (Eds.), Articulated Motion and Deformable Objects. XV, 526 pages. 2006.

Vol. 4068: H. Schärfe, P. Hitzler, P. Øhrstrøm (Eds.), Conceptual Structures: Inspiration and Application. XI, 455 pages. 2006. (Sublibrary LNAI).

Vol. 4067: D. Thomas (Ed.), ECOOP 2006 – Object-Oriented Programming. XIV, 527 pages. 2006.

Vol. 4066: A. Rensink, J. Warmer (Eds.), Model Driven Architecture – Foundations and Applications. XII, 392 pages. 2006.

Vol. 4065: P. Perner (Ed.), Advances in Data Mining. XI, 592 pages. 2006. (Sublibrary LNAI).

Vol. 4064: R. Büschkes, P. Laskov (Eds.), Detection of Intrusions and Malware & Vulnerability Assessment. X, 195 pages. 2006.

Vol. 4063: I. Gorton, G.T. Heineman, I. Crnkovic, H.W. Schmidt, J.A. Stafford, C.A. Szyperski, K. Wallnau (Eds.), Component-Based Software Engineering. XI, 394 pages. 2006.

Vol. 4062: G. Wang, J.F. Peters, A. Skowron, Y. Yao (Eds.), Rough Sets and Knowledge Technology. XX, 810 pages. 2006. (Sublibrary LNAI).

Vol. 4061: K. Miesenberger, J. Klaus, W. Zagler, A. Karshmer (Eds.), Computers Helping People with Special Needs. XXIX, 1356 pages. 2006.

Vol. 4060: K. Futatsugi, J.-P. Jouannaud, J. Meseguer (Eds.), Algebra, Meaning, and Computation. XXXVIII, 643 pages. 2006.

Vol. 4059: L. Arge, R. Freivalds (Eds.), Algorithm Theory – SWAT 2006. XII, 436 pages. 2006.

Vol. 4058: L.M. Batten, R. Safavi-Naini (Eds.), Information Security and Privacy. XII, 446 pages. 2006.

Vol. 4057: J.P.W. Pluim, B. Likar, F.A. Gerritsen (Eds.), Biomedical Image Registration. XII, 324 pages. 2006.

Vol. 4056: P. Flocchini, L. Gąsieniec (Eds.), Structural Information and Communication Complexity. X, 357 pages. 2006.

Vol. 4055: J. Lee, J. Shim, S.-g. Lee, C. Bussler, S. Shim (Eds.), Data Engineering Issues in E-Commerce and Services. IX, 290 pages. 2006.

Vol. 4054: A. Horváth, M. Telek (Eds.), Formal Methods and Stochastic Models for Performance Evaluation. VIII, 239 pages. 2006.

Vol. 4053: M. Ikeda, K.D. Ashley, T.-W. Chan (Eds.), Intelligent Tutoring Systems. XXVI, 821 pages. 2006.

Vol. 4052: M. Bugliesi, B. Preneel, V. Sassone, I. Wegener (Eds.), Automata, Languages and Programming, Part II. XXIV, 603 pages. 2006.

Vol. 4051: M. Bugliesi, B. Preneel, V. Sassone, I. Wegener (Eds.), Automata, Languages and Programming, Part I. XXIII, 729 pages. 2006.

Vol. 4049: S. Parsons, N. Maudet, P. Moraitis, I. Rahwan (Eds.), Argumentation in Multi-Agent Systems. XIV, 313 pages. 2006. (Sublibrary LNAI).

Vol. 4048: L. Goble, J.-J.C.. Meyer (Eds.), Deontic Logic and Artificial Normative Systems. X, 273 pages. 2006. (Sublibrary LNAI).

Vol. 4047: M. Robshaw (Ed.), Fast Software Encryption. XI, 434 pages. 2006.

Vol. 4046: S.M. Astley, M. Brady, C. Rose, R. Zwiggelaar (Eds.), Digital Mammography. XVI, 654 pages. 2006.

Vol. 4045: D. Barker-Plummer, R. Cox, N. Swoboda (Eds.), Diagrammatic Representation and Inference. XII, 301 pages. 2006. (Sublibrary LNAI).

Vol. 4044: P. Abrahamsson, M. Marchesi, G. Succi (Eds.), Extreme Programming and Agile Processes in Software Engineering. XII, 230 pages. 2006.

Vol. 4043: A.S. Atzeni, A. Lioy (Eds.), Public Key Infrastructure. XI, 261 pages. 2006.

Vol. 4042: D. Bell, J. Hong (Eds.), Flexible and Efficient Information Handling. XVI, 296 pages. 2006.

Vol. 4041: S.-W. Cheng, C.K. Poon (Eds.), Algorithmic Aspects in Information and Management. XI, 395 pages. 2006.

Vol. 4040: R. Reulke, U. Eckardt, B. Flach, U. Knauer, K. Polthier (Eds.), Combinatorial Image Analysis. XII, 482 pages. 2006.

Vol. 4039: M. Morisio (Ed.), Reuse of Off-the-Shelf Components. XIII, 444 pages. 2006.

Vol. 4038: P. Ciancarini, H. Wiklicky (Eds.), Coordination Models and Languages. VIII, 299 pages. 2006.

Vol. 4037: R. Gorrieri, H. Wehrheim (Eds.), Formal Methods for Open Object-Based Distributed Systems. XVII, 474 pages. 2006.

Vol. 4036: O. H. Ibarra, Z. Dang (Eds.), Developments in Language Theory. XII, 456 pages. 2006.

Vol. 4035: T. Nishita, Q. Peng, H.-P. Seidel (Eds.), Advances in Computer Graphics. XX, 771 pages. 2006.

Vol. 4034: J. Münch, M. Vierimaa (Eds.), Product-Focused Software Process Improvement. XVII, 474 pages. 2006.

Vol. 4033: B. Stiller, P. Reichl, B. Tuffin (Eds.), Performability Has its Price. X, 103 pages. 2006.

Vol. 4032: O. Etzion, T. Kuflik, A. Motro (Eds.), Next Generation Information Technologies and Systems. XIII, 365 pages. 2006.

Vol. 4031: M. Ali, R. Dapoigny (Eds.), Advances in Applied Artificial Intelligence. XXIII, 1353 pages. 2006. (Sublibrary LNAI).

Vol. 4029: L. Rutkowski, R. Tadeusiewicz, L.A. Zadeh, J.M. Zurada (Eds.), Artificial Intelligence and Soft Computing – ICAISC 2006. XXI, 1235 pages. 2006. (Sublibrary LNAI).

Vol. 4028: J. Kohlas, B. Meyer, A. Schiper (Eds.), Dependable Systems: Software, Computing, Networks. XII, 295 pages. 2006.

Vol. 4027: H.L. Larsen, G. Pasi, D. Ortiz-Arroyo, T. Andreasen, H. Christiansen (Eds.), Flexible Query Answering Systems. XVIII, 714 pages. 2006. (Sublibrary LNAI).

Vol. 4026: P.B. Gibbons, T. Abdelzaher, J. Aspnes, R. Rao (Eds.), Distributed Computing in Sensor Systems. XIV, 566 pages. 2006.

Vol. 4025: F. Eliassen, A. Montresor (Eds.), Distributed Applications and Interoperable Systems. XI, 355 pages. 2006.

Vol. 4024: S. Donatelli, P.S. Thiagarajan (Eds.), Petri Nets and Other Models of Concurrency - ICATPN 2006. XI, 441 pages. 2006.

Vol. 4021: E. André, L. Dybkjær, W. Minker, H. Neumann, M. Weber (Eds.), Perception and Interactive Technologies. XI, 217 pages. 2006. (Sublibrary LNAI).

Vol. 4020: A. Bredenfeld, A. Jacoff, I. Noda, Y. Takahashi (Eds.), RoboCup 2005: Robot Soccer World Cup IX. XVII, 727 pages. 2006. (Sublibrary LNAI).

Vol. 4019: M. Johnson, V. Vene (Eds.), Algebraic Methodology and Software Technology. XI, 389 pages. 2006.

Vol. 4018: V. Wade, H. Ashman, B. Smyth (Eds.), Adaptive Hypermedia and Adaptive Web-Based Systems. XVI, 474 pages. 2006.

Vol. 4017: S. Vassiliadis, S. Wong, T.D. Hämäläinen (Eds.), Embedded Computer Systems: Architectures, Modeling, and Simulation. XV, 492 pages. 2006.

FSC
www.fsc.org
MIX
Papier | Fördert gute Waldnutzung
FSC® C083411

Kolibri 360 GmbH
Ferdinand-Jühlke-Straße 7
99095 Erfurt, Deutschland
produktsicherheit@kolibri360.de